I0344024

# Alphabetical Guide *for* Beekeepers

*Ken Stevens*

**Alphabetical Guide for Beekeepers**


The publishers wish to thank Mrs Sue Taylor of East Sussex without whose help this new edition of The Alphabetical Guide for Beekeepers would not have been possible, Bob Curtis of Curtis Photography, Brighton for the cover image and of course Ken Stevens for his huge amount of work over decades in compiling this Guide.

Readers may wish to advise Northern Bee Books of factual areas of improvement and additions to the entries which can be incorporated in future editions

ISBN 978-1-908904-21-8

Published by Northern Bee Books, 2012
Scout Bottom Farm
Mytholmroyd
Hebden Bridge
HX7 5JS (UK)

Design and artwork
D&P Design and Print
Worcestershire



Printed by Lightning Source, UK

# Alphabetical Guide *for* Beekeepers

*Ken Stevens*

*I dedicate this book to Jan Barnard who has given me every encouragement to bring this work to a conclusion.*

Northern Bee Books

# List of Subject Indexes

# 1 - Abbreviations

| | |
|---|---|
| ABD | American Bee Disease (removes foul) |
| ABF | American Beekeeping Federation |
| ABJ | American Bee Journal |
| ABPV | Acute Bee Paralysis Virus |
| ACAS | Agricultural Chemicals Approved Scheme |
| AEA | Apicultural Education Association |
| AFB | American Foul Brood |
| A.I. | Artificial insemination |
| AIV | Apis Iridescent Virus |
| A.L. | Advisory Leaflet |
| APC | The APC Bee |
| ATP | Adenosine Phosphate |
| BHP | Bald-headed Brood |
| BBJ | British Bee Journal |
| BBKA | British Beekeeper's Association |
| BBNO | Bee Books New & Old |
| BDI | Bee Disease Insurance |
| BFA | Bee Farmers' Association |
| BHB | bald-headed brood |
| BHC | hydrochlorohexane |
| BIBBA | British Isles Bee Breeders' Association |
| BKA | Beekeepers' Association |
| B.L. | Bacillus larvae |
| BOD | Biological oxygen demand |
| BSI | British Standards Institute |
| CAB | Central Association of Beekeepers |
| CBL | County Beekeeping Lecturer (extant) |
| c.c. | cubic centimetre |
| CNS | Central nervous system |
| cps | Cycles per second |
| DARG | Devon Agricultural Research Group |
| DCA | Drone congregation area |
| DD | Disappearing disease |
| DDT | Di-chloro-diphenyl-tri-chlorethane |
| DIY | Do it Yourself |
| D.M.P. | Di-methyl phalate |
| DMSO | Di-methyl sulfoxide |
| DNA | Deoxyribonucleic acid |
| DVAV | Dorsal/ventral abdominal vibration |
| EDB | Ethyldibromide |

| | |
|---|---|
| ECC | European Economic Community |
| E.F.B. | European Foul Brood |
| EPP | Effective pollination period |
| FERA | Food & Environment Research Agency |
| GRP | Glass resin plastic |
| Gk | Ancient Greek |
| GPO | General Post Office |
| GSL | Geographical stress lines |
| HMF | Hydroxymethylfurfuraldehyde |
| HPA | Honey Procucers I Association |
| HPLC | High performance liquid chromatography |
| HWP | Honey working party |
| IAA | Indole acetic acid |
| IBAP | Intervention Board for Agric. protection |
| IBRA | International Bee Research Association |
| ICBB | International Convention for Bee Botany |
| IgG | Immunoglobulin |
| JAR | Journal for Apicultural Research |
| KTBH | Kenyan top bar hive |
| LD | Lethal dose |
| L/D | laevulose/dextrose ratio |
| N & P | nectar and pollen |
| OSR | oil-seed rape |
| ppm | parts per million |
| RIE | rocket immunoelectrophoresis |
| RNA | ribonucleic acid |
| SEM | scanning electron microscope |
| S.I. | Systeme International d'unites |
| sp. | species |
| STP | standard temperature and pressure |
| TEM | transmission electron microscope |
| VMD | Vetinary Medicines Directorate |
| WBE | whole bee extract |

# 2 - Anatomy & Physiology

abdomen
accommodation
abdominal contents
accustom
abdominal glands
acetycholine
abdominal muscles
absorption
activation hormone
acid gland
adrenal
acinus
acorn cup
Aedeagus
adipose tissue
adrenaline
adult worker
air sacs
alimentary canal
alitrunk
alkaline gland
amnion
anatomy
anguiculi
ankle
antenna
antenna cleaner
antenna comb
antennal vesicle
anterior intestine
anti-body
antigen
anus
aorta
Apodeme
appendage
appendicular glands
appreciation
arcus
appreciation – sound
aerolium
arthromere
association fibres
atrium
auricle
axillary sclerite
barb
basalare
basal
basal joint
basement membrane
basilar
basisternite
basitarsus
bee
bee milk
bee stings
beeswax
blastocoel
blastoderm
blind gut
blood
bee stings
blood corpuscle
blood vessel
body of a bee
bouton
brain
breathing
brood food glands
bulb of endophallus
bulb of sting shaft
bursa
bursal (cornua)
central nerve cord
central nervous system
cervix
champagne cork organ
chemoreceptors
chiasma

chordotal organ
chromatid
chromosome
chyle stomach
cibarium
circulation
circulatory system
circum-oesophageal connectives
claspers
clavate hairs
claw
clypeus
CNS
colon
commissure
compound eye
condyle
cone
connectives
conoid hairs
convoluted
copulatory pouch
corbicula
cornea
cornua
corpora allata
corpora cardiacum
corpora pedunculata
costal
coxa
crop
crystalline cone
cubital index
cutical
dart
dermis
detoxificqtion
deutocerebum
diaphragm
diffusion
digestion
digestive tract
direct wing muscles
diverticulum
dorsal diaphragm
dorsal light reaction
dorsal plate
dorsal vessel
duct
ecdysis
ecdysone
ectoderm
effector organs
ejaculatory duct
embodium
embryology
emunctory
endocrine glands
endocuticle
endoderm
endophallus
endoskeleton
endosternite
enteric canal
epicuticle
epiopticon
epipharynx
epithelium
exohormone
exoskeleton
exteroceptor
eye compound
facet
fat body
feeding
femur
fibular
first skeletal ring
flabellum
flagellum
flexor muscle
flight
follical cells
fore gut
fore legs
fore wings
fossa
frons
furca
furcula
galea
ganglion
gaster
generative organs
genhormones
geranic acid
germ band
gestation
gland
glabrous

glossa
glossal canal
glossometre
gnathsoma
gonopophyses
groove
gustatory sense
gut
haemocyte
haemolymph
hair
haltere
hamulus
head
head glands
heat production
hearing sense
heat receptor
hind gut
hind legs
hind wings
homologous
honey crop (sac)
honey mouth
honey sac
honey stomach
honey stopper
hamuli
hygroreceptor
hypodermis
hypopharyngeal gland
hypostoma
ileum
indirect flight muscle
ingurgitate
innervate
interesting facts
intersegmental membrane
intestine
invagination
invertase
jaw
Johnson's organ
joint
kidney
Kozcheznikov gland
labellum
labial glands
labial palp
labium
labrum
lacinia
lamina externa
lancets
large intestine
laying queen
laying rate
laying worker
leg
length of tongue
longitudinal flight muscles
lorum
lymphocytes
macrotrichia
malgihian tubules
mandible
mandibular gland
mating sign
maxilla
manubrium
masque of Reamur
mass crawling
mast cell
median segment
Mellisa
membrane
mentum
mesenteron
mesoblast
mesoderm
mesodermic originations
mesothorax
metascutellum
metathorax
middle legs
mid-gut
mid-rib
monstrosity
morphology
mosaic vision
motor nerve
motor neurons
mouth
mouthparts
muscle
mushroom bodies
myogenic rhythm
neck
nectar load
nerolic acid

nerve cells
nerve fibres
nerve trunk
nervure
neurohormones
neuromere
neuropile
neurosecretary cells
notum
nutrition
nymph
oblique bands
oblong plate – sting
occipital foramen
occiput
ocellus
oenocyte
oesophagus
old bees
olfactory receptors
ommatidium
operculum
optic
opticon
optic rod
oral plate
ostium
ovariole
oviduct
ovipositor
ovoid appendage
ovule
palma
palp
palp labial
palp maxillary
palp sting
parameral plates
pectin
pedicel
pellicle
periopticon
peripheral nervous system
peritrophic membrane
phallotreme
phallus
pharyngeal gland
pharynx
poison gland
poison sac
polar filament
pollen basket
pollen brush
pollen collection
pollen load
pollen press
pollen rake
pollen tube (anat.)
pore (anat.)
post cerebral gland
postmentum
prementum
proboscis
process
proctiger
proctodeum
pronotum
propodeal spiracles
propodeum
prothoracic collar
prothoracic glands
prothorax
protocerebrum
proventricular valve
pulvillus
pupation
pylorus
quadrate plate
queen anatomy
queen mandibular gland
queen ovaries
queen's sting
queen stubby
rastellum
receptaculum seminis
receptor
rectal ampulla
rectum
reflex
reproductive organs
reproductive system
respiratory system
respiratory movement
respiratory system
reticulation
retina
retinula
rhabdom
rhabdomeric microvilli
ribosome

rudimentary
sac
saliva
salivarium
salivary canal
salivary gland
salivary syringe
sarothrum
scape
scent gland
scenting
sclerite
sclerotin
scolophore
scopa
scutellum
scutum
secondary wing
segment
segmentation
semen
seminal duct
senses of bees
sense of food presence
sense of colour
sense of danger
sense of form
sense hair
sense of hearing
sense of location
sense organ
sense peg
sense plate
sense pore
sense of sight
sense of smell
sense of taste
sense of touch
sensillum
sensillum campaniform
sensillum chordotal
sensory hair
serous
seta
sex organs
silk glands
simple eye
sinus
skeletal system
small intestine
somatic cells
somite
sperm
spermatheca
spermathecal area
spermathecal duct
spermathecal gland
spermathecal pump
spermathecal valve
spermatogonium
spermatophore
spermatogonium
spermatophore
sphincter
spicule
spine
spinneret
spiracle
spiracle abdominal
spiracular plate
spiracle propodeal
spiracles thoracic
spiral thickening
spur
sternite
stigma
sting
sting at rest
sting cavity
sting components
sting glands
sting palps
stipe
stomach
stomodeum
stomogastric nerve system
streamlined
strigilis
stylet
subalare
submentum
suboesophageal ganglion
supraoesophageal ganglion
suture
swarming strain (DNA)
sympathetic nerve system
synapse
tactile
taenidium
tarsomere

tarsus
tegmentum
tegula
tenth segment
tentorium
terebra
tergite
tergum
terminal filament
testicular tubule
testis
thoracic salivary
thorax
tibia
tibio-tarsal joint
tomentum
tongue
touch
trachea
trichogen
tritocerebrum
trochanter
trophy
trophocyte
turpenoid (scent gland)
umbrella valves
under wing
unquiculi
urate cells
vacuole
vagina
valve
valve fold
valve heart
valve-like plate
valve oesophagus
valve ventricular
vas deferens
vein
velum
venation
venom
venom gland
ventral diaphragm
ventral nerve trunk
ventral plate
ventrical glands
ventricular valve
ventriculus
vertex
vesicle
vesicular seminalis
vestibule
vestigial organ
viscous
vitelline membrane
waist
wax glands
wax nippers
wax pockets
wax scales
wax shears
weapon
wing beats
wing power
wings
wing venation
winter cluster.
wrist

# 3 - Apiaries

bee garden
bee garth
bee house
bee shelter
beekeeping in schools
beekeeping shed
Buckfast Abbey
damp
direction of hive entrance
district
drainage
drone zone
electric fence
enclosure
establish
evacuate
flight path
flood
flora
footing
garden
garth
glade
green belt
ground
hive clay
hive density
hive protection
hive site
hive spacing
hives per hectare
hive stand
hives under trees
home yard
honey flow
honey forest
honey house
horse
hut
increasing forage
indoors
level
livestock
market garden
marsh
meadow
mobile site
mountain
New Forest
opening
out-apiary
out-doors
outyard
overturn
panel pin
portable bee house
prevailing wind
protective clothing
punch
queen mating station
ramada
roof garden
ruderal
sea
seaside
secluded
settling
shade
shed
shelter
shore
shut
sites
straight line
stream
surroundings
swamp
swarm's new home
terrain
undergrowth

undesirable
uneven
wasteland
water course
waterlogged
weed
wilderness
wind breaks
windy site
winter chambers
wood ant

# 4 - Apparatus

Adam tensioner
aerosol bomb
alighting board
anti-spark tube
aperture
apidictor
apparatus
apparel
appliance
Arnaba foundation
Ashforth feeder
Beddoes frame cleaner
bee blower
bee brush
bee cap
bee dress
bee escape
bee garb
bee gloves
bee hat
Beekeepers rule
bee sounds (apidictor)
bee space
bee suit
bee tent
bee trap
bee veil
bellows
bent nosed smoker
black netting
blowlamp
boiler suit
bottle feeder
bottling
box
Brand's capper
branding
brass cone
brood frame
bucket feeder
bung
burlap
cartridge
cell protector
cell punch
centrifugal extractor
Chantry queen cage
clearer
clearer board
cleaver
Cobana
comb cutter
comb foundation
compartment separator
compass
compound microscope
computer
cone
cone escape
contact feeder
containers
converter clip
cork
corrugated packing
cover cloth
crate
crimper
crown board
Crowther
cut comb container
densitaster
device
devil's snuff (smoker
disc entrance
dissecting pin
divider
division board
DN frames
double screen
drain

draw quilt
dress
drink fountain
drinking trough
drip tray
driving bees
drone trap
dryer
dummy board
effect of propolis
elasticated cuff
elasticated veil
electric blower
electric embedder
electric extractor
electric fence
electricity
electron microscope
embedder
entrance adjustable
entrance block
entrance slide
equipment
essentials
excluder (queen)
expanded metal
extractor
feeder
feed hole
fence
fermentation valve
flanged cork
flash heater
flashlight
flight board
floor board
flowed-in
foam rubber
follower board
forceps
fork lift truck
frame gripper
frame holder
frame jig
frame lifter
frame nailer
frame sizes
freezer
front slides
fume board
gable roof
gadget
gap
garb
gauntlet
gimp pin
glass quilt
glossometer
glove
goose feather
grafting tool
Granton (tray)
grate
gum boots
hair curler
H & J
handhold
Hanneman excluder
hanging section holder
heather press
Hertzog
hive carrier
hive ceramic
hive clips
hive cloth
hive escape
hive fastening
hive fillet
hive furniture
hive legs
hive lift
hive maintenance
hive materials
hive roof
hive runner
hive scale
hive scraper
hive stand
hive strapping
hive to;
hive twin
Hoffman
holding down cage
honey bear
honey bottle
honey carton
honeycomb section
honey dispenser
honey extracting house
honey extractor

honey filter
honey gate
hone y grading glasses
honey house
honey hydrometer
honey loosener
honey plant (equipment)
honey press
honey pump
honey ripener
honey settling tank
honey sign
honey slinger
honey spoon
honey strainer
honey twin spin
honey valve
hoop
hydrometer
implement
inner bodies
inner cover
interchanging
in-the-flat
introductory cage
jar
joints
knap sack
label
ladder
laminated
Langsroth frame
leaking feeders
length of frame
lens
lid
lift
liner
lingel
liquefaction
lock slide
lock spring
lug
machine
mailing cage
maintenance
mandril
manipulating cloth
mat
match
material
mating hive
mechanical smoker
mesh
metal divider
metal ends
metal feeder
metal queen excluder
metal roof
netals
metal work
method of filling bottles
M.G. wax extractor
microscope
microtome
microwave oven
Miller feeder
minicosy
mininucleus
irror
mixed equipment
model hive
modified Snelgrove board
monocular
mould
mouse guard
movable frame
nails
nailing
nailing tool
netting
newspaper
nylon
OAC strainer
objective
overall
organza
overall feeder
pad
pallet
parallel radial extractor
Penrose uncapper
perforated partition
perforated screen
perforextractor
Perspex
polarimeter
pollen trap
polypropylene
polystyrene

polythene
polyurethane
polyvinylchloride PVC
popular smoker
porch
porter bee escape
post
postal cage
pot
Pratley uncapping
projector
protective clothing
psychrometer
puff ball
puffer
pump
punch
punk
PVC
pyknometer
Quebec board
queen cage
queen catcher
queen cell protector
queen cup
queen excluder
queen marking outfit
queen marking paint
queen mating station
quilt
rabbet
radial honey extractor
radial symmetry
rails
rampin
rebate
repairs
reversible extractor
riddle
ripener
roof rack
rope hive lift
runner
sacking
saw-cut top bar
scalpel
scanning electron microscope
scraper
screen
screen travelling
screw-eye spacer
seat
secateurs
second-hand equipment
section
section crate
seeker
SEM
separator
settling tank
shed
Shepherd tube
shim
shipping cage
showcase
side bar
side rail
sieve
simple microscope
siphon
skewer
skid board
slide
sliding entrance blocks
slinger
slotted floor
slotted frame spacers
slotted separators
slotted queen excluder
slow feeder
smoke bomb
smoker cartridge
smoker fuels
smoker structure
Snelgrove board
SN frames
software
soft wood
solar wax extractor
spaced dummy
spatular
specification
spectacles
spiral cell protector
spirit level
split board
sponge rubber
sprayguard entrance
spring balance
spring block

spur embedder
stainless steel
stand
staple
stapling gun
stocking
stoneware honey crocks
stopper
straight nosed smokers
strainer
straining cloth
straw skep
substandard
super clearer
super glue
swarfega
swarm box
swarm catcher
syringe
tailboard
tailgate loader
tangential extractor
tap
tapes
tap strainer
Taranov board
Teeswain
TEM
thermometer
thermostat
timber
tin
tin snips
tissue
tool
top bar
top feeder
touchwood
trailer
transferring tool
travelling box
travelling cage
travelling screen
trestle
truck
tube
tweezers
twin dissecting pin
tyre
unassembled
uncapping fork
uncapping knife
uncapping plane
uncapping tray
valve
valve – honey
variable entrance size
varnish
Vaseline
ventilation cone
ventilation of honey house
ventilation screen
ventilator
vibrating knife
vision (veils)
votator
wad
Waldron
wall
Wardecker waterer
washer
water fountain
water butt
wax foundation mill
wax press
wax smelter
wearing gloves
wedge
weld
well made
West cell protector
whirl
wick
winter packing
wire
wired foundation
wire gauze
wire netting
wire queen excluders
wire tamer
wiring board
with – withy
woiblet
wood
wooden
word processor
Worth cage
zinc queen excluder
zip

# 5 - Beekeeping

advantages of two broods
after effect
after swarm
ageing
age of brood
aggressive
agricultural changes in UK
agriculture
airspeed
alarm odour
alien
align
alive
anaesthetic
anaphylactic shock
anchoring effect
angry bees
ankle
antagonism
antidote for bee stings
antipathy
aperture
apiarist
apiary
apiary health
apiary hygiene
apiary inspector
apiary site
apicultural reverses
apifactory
apifuge
artificial feeding
artificial heating
artificial insemination
artificial queen cups
assessing weight
atrophied
attack
attendants
autumn
autumn feeding
avoidance of stings
baby bee
baby drones
baby queens
baby workers
Babylonians
backyard beekeeping
backlot beekeeping
bacteriostatic
bait combs
balance
bald-headed brood
balling
Barbeau
BBKA comprehensive insurance
bee blower
bee farmer
bee food
bee forage
bee garden
bee garth
bee glue
bee house
beekeeper
beekeeper's rule
beekeeping
beekeeping in towns
beekeeping instructor
beekeeping shed
beekeeping standards
bee man
bee milk
bee pasture
bee proof
bee widow
bees attract bees
bee's choice of site
bees killed
bee sounds

bee space
bee stings
bee tent
bee venom activity
beggar bees
blowing bees
bottling
bottom bee space
bottom entrance
brace comb
branding
bronchial spasm
brood
brood age of
brood and a half
brood cells
brood cycle
brood food
brood rhythm
brood spreading
brood trial
build up
cast
caste
catch
cell
cell base
cell protector
cell rearing colony
cells of honeycomb
cessation of flow
challenge
check list
chill coma
chilled brood
chimneying
clean floorboardcleaing
clearing supers
close driving
close-ended side bars
closing a hive
closing down for winter
cluster
cohesion
cohort
cold
cold way
coloured brood
comb
comb of brood
comb building
comb collapse
comb renewal
combs
comb spacing
comb sterilization
commercial beekeeper
competent
consanguinity
constancy
contraction of brood nest
convex cappings
cottagers
court
cover cloth
cross breed
cross wind
crush
cup
cuprinol
damage
dampness
dead brood
deadman's floorboard
demise
demonstration
density of winter cluster
departure
de-queen
desire
device
dialect
dimensions
disposition
disqualification
distaste
distension
dividing colonies
division
division of labour
DIY
docile
domed cappings
domesticate
dormant
double brood chamber
double screen
doubling
drainage
draw

drawback
drawn comb
draw quilt
drifting
drink fountain
drip
driven bees
drone breeder
drone slaughter
drone trap
drum
dry
dry swarm
dwarf bees
dwindling
eating bees
entrance
entrance block
equalizing
essential
establish
examine
expansion
experiment
expert
expose
external cluster
fatal
February
feed hole
feeding
food attractants
finding queen
finishing colony
fixed conbs
flank combs
flat
fliers
flight board
floor
floorboard
flow
flowers rarely visited
flowers per tree
foreign proteins
forest beekeeping
freedom
fluid
follow the sheep
food chamber
forest honey
fowls
froth
fuel
Galbanum
Galeria
gap
geographical origin
gone through
Good candy
grafting
granulated sugar
granulation
grass
gravid
gravity
Greek beekeeping
greenhouse crops
gregarious
grip
groundspeed
half depth super
handling queencells
ham fisted
harbinger
hardiness
harvest
harvesting
hatchability
heap of dead drones
hearing (sense)
hefting
hiss
hive – baited
hive closing
hive colour
hive decoy
hive density
hive double
hive entrance
hive fastening
hive feet
hive insulation
hive large
hive maintenance
hive manufacture
hive mates
hive mind
hive – movable frame
hive number

hive protection
hive records
hive spacing
hives under tree
hive twin
hive wooden
hiving
hiving package bees
hiving a swarm
hiving 'sough'
hobbyist beekeeper
Hoffman
homing instinct
homoiothermic
honey bar
honeybee pollinator
honey canopy
honey cappings
honey cells
honey characteristics
honey chemistry
honeycomb section
honey crop (harvest)
honey dew
honey fermentation
honey flavour
honey flow
honey guide bird
honey harvest
honey impairment
honey keeping qualities
honey potential
honey processing
honeyproduction
honey pump
honey ripe
honey ripening
honey storage
honey sucker
honey surface layer
honey take
honey thermal activity
hum
humidity
hunger swarm
hybrid vigour
identifying
identity odour
immunize
immunity
immunoglobin
immuno therapy
inanition
inanition
increase
infection
infestation
inquisitive bee
instrumental insemination
interchanging
interior
interstitial cells
intruder
invade
inventory
inversion of queen cells
irritation
July
June gap
knock out
labour
large scale
larva
larval colour
larval food
larval growth
late
lay flat
laying
laying queen
laying rate
laying worker
leaking feeders
learner
learning by bees
length of life
length of active season
length of time to mate and lay
length of winter
let-alone beekeeping
lethargic
liaison bees
light compass reaction
listen
listless
little bee
little blacks
living
loading
local

location
longevity
maceration
magnetite
maintenance
making increase
making wax foundation
mate
mammal
management
manipulation
marked queen
market garden
mass
massacre of drones
master beekeeper
mated queen
match
mating swarm
maturation
May
meliponi culture
meliponae
meliponins
memory
method
method of destruction
microvibration
mixed equipment
mixed pollen loads
mobile site
model hive
morale
moribund
moth eaten
mouldy comb
moused
movable comb
movable frame
movement
movement of entrance
moving bees
mow
mud
multiple matings
multiple race crosses
multiple queen casts
multi-queen colonies
nadiring
narrow spacing
nectar
nectar load
nectary deep
neglected brood
nest cavity
nest – mouse
new comb
newspaper uniting
night
notes
notice
November
novice
N& P
nucleus
nucleus mini
number on one site
nursery cage
nutrition
NZ beekeeping
occupy
odd
odour
odour – alarm
odour hive
odour wax
oenocyte
oil immersion microscope
old bees
old comb
old queens
open
open brood
opening
opening hives
optimum
optimum temperature
orchard
order
organizer
orientation
outfit
overcrowded
over heating
over stocking
over wintering
oviparous
ovoid
ovulate
ownership

package bees
packing bees
packing hives for winter
Pagden
palynology
pap
paper
papering
paralysis
parent colony
partial acceptance
pastime
peak brood cycle
peak colony strength
pellicle
Penicillin waksmanii
pepper pot
perambulate
perforated partition
perforated screen
pheromone alarm
pheromone aggregation
pheromone clustering
pheromone defence
pheromone foot-print
pheromone forage marking
pheromone mutual recognition
pheromone trail marking
pheromone sex
photography
police
pollen barrier
pollen collector
pollen cells
pollen colour
pollen dispenser
pollen intake
polymorphism
pop hole
post
postage stamps
post code
post nuptial
pour
preparation
preparation for swarming
pressure
prevailing wind
prevention of disease
prevention of drifting
prevention of fighting
preventing of robbing
prevention of swarming
preventative
prick
prime swarm
processing honey
processing wax
produce
production of beekeepers
production of bees
production of lecturers
production of pollen
production of propolis
production of queens
propagate
properties of honey
propolis collection
propolis dermatitis
propolized
prostaglandins
protection against
public relations
pulled virgin
pump
pupa
pupal skin
pupate
pupation
pursuer
QC
qualities of good queen
queen accepted
queen attendants
queen bank
queen bee
queen cell supersedure
queen colour of the year
queen rejection
queen replacement
quell
quiescent
ragged cappings
rain water
recognition odour
recognition
records recruits
recycle
regicidal knot
reinforce

rejection
release
reluctance
rent
repairs
repellent
reproductive isolation
re-queen
re-queening
research
residual effect
reserve
reservoir bees
resonance
restive
restless
retinue
return
reverse side
re-worker
rheumatism
robbed
robber
rogue bees
rotation
rotten
rustling
sacrificial combs
safety
sag
sale of bees, honey etc.
salts in honey
saturated
scattered brood
scent trail
school beekeeping
scorch
scout bees
seams of bees
searcher bees
second-hand equipment
security
self spacing
semen
sending samples of bees
sending samples spray damage
senility
sense of food presence
sense of colour
sense of danger
September
sepulchre
servicing
sex of eggs
shade
shed
shelter
shield entrance (robbing etc.)
shirty
shook swarm
shutting down for winter
siblings
side bar
sideliner
signs in and out of hive
simulate
sisters
sites
situation
size of brood chamber
size of entrance
skeppist's terminology
skills
skip
slaughter of drones
smoke
smoking bees
smothering
sneeze
Snelgrove method swarm control
snowglare
soft bees
solidify
solution
specialization
sperm storage
split
spotting
spray mortality
spread brood
spread diseases
spring development factors
spring management
squat
staggered spacing
syrup
staining of washing
stand
starter or primer
starter colony

starting beekeeping
starvation
static electricity
steal
step comb
stick
stimulation
sting reaction
sting treatment
stir
stock
stone
storage
straighten
strength of colony
subdue
submerge
submissive
substances bees collect
substances harmful to bees
substandard
successful mating
suffocation
sugar syrup
sunken cappings
sunlight
supered
surge
swarm
swarm management
switched
symptom
system
Take care,
talk
television
temper
temperatures
tense
terror
test
threshold
thrift
throwing out
tie
tiering
timetable
top entrance
top feeder
top packing
top heavy
topple over
top spacing
top ventilation
torn
toxic subjugants
transfer
transporting bees
tropical apiculture
ubiquitous
unblock
uncapping
uncover
under or over
undress
unfecundated
unite
unmarked queen
unmated queen
unopened queen cell
unsealed upward ventilation
vandalism
vacated queen cell
venue
vicious colonies
warmth
warm way
water requirements
water soluble
wearing gloves
weekend beekeeping
weighing bees/hives
well found
well kept
wet
wheeze
white eyed brood
wide open entrances
wide spacing
wild colonies
windbreaks
wind speed
windy site
wing power
winter feeding
winter kill
winter passages
women
woodwork
worker cells etc.

worn out
wrinkle
wrist
yield
zest

# 6 - Behaviour

abandon
abdominal glands
abscond
absconding swarm
acceptance
accident
accommodation
activity
adult
adult workers
ageing
agemates
aggressive
alight
allergen
allergic reaction
allergy
anchoring
anger
apparent
apple fertilization
Avetisyan
avoidance of stings
axillary sclerite
back ache
backyard beekeepers
back lot beekeeping
bearing
bee's choice of site
bee's duties
bee-tight
bee venom
bee way
beggar bees
boost
breathing
brood nest
burden
burrowings
buzz
cast
catalepsy
cattle
chemotaxis
cibarium
circulation
citric acid
cleansing dance
cleansing flight
cloud of bees
cold blooded
collide
colony behaviour
colony morale
colony rhythm
colony
colour
comb building
corbicula
corpora allata
corpora cardiacum
cranberry
crazy dance
crescent dance
crippled bees
crop fidelity
cross bees
cross wind
crowding
dance
dark comb (travel stain)
darkness
day
DD
dead bees
dead queen
dearth
death feigning
Death's head hawkmoth
deposition nectar

direction finding
dorsal light reaction
drink trough
drone behaviour
drone brood
drone mating
drone slaughter
droppings
Duchet Francois (wax)
duties
DVAV dance
eat
eating bees
ectohormone
effect of smell
effects of stings
effects vibration
effects of water
electricity
epipharynx
escort
essentials
etiquette
evacuate
excitable
exhibiting
eye simple
faeces
fanners
feed hole
feeding larvae
feign death
festoon
first flight
flabellum
flagellum
flight
flight path
float
floccule
flood
fly
flying
flying speed
follower bee
food transmission
foot bath
fuel consumption
function of a honeybee colony
general characters
glands
glossa
glossal canal
grass
gravity (von Frisch dance)
greater wax moth
grip
grooming
guard bee
guttation
hamulus
hatching
heat production
heat reception
hind legs
hind wings
histamine
hitch-hiking
hive bee
hive closure
hive – to
homing
honey arch
honeycomb
honey crystallization
hamuli
hormone
humble bee
hysterisis
indirect flight muscles
individual
inflorescence
infuriate
ingurgitate
inquisitive bee
instinct
insulation
intercommunication
interesting facts
irritability
jaw
jittery flight
job
Johnson's organ
joy dance
judging honey
judging wax
jumpiness
labellum
labial glands

labial palp
labour
lancets
larval faeces
larval skin
light compass reaction
light intensity
load
longitudianal flight muscles
making comb
making honey
mandible
mandibular gland
mandibular pheromone
mating
mating sign
mating swarm
maxilla
manipulation
mannose
manubrium
marketing
massage dance
masticate
micro vibration
middle 'c'
mid-gut
migratory beekeeping
mirror
mood
moult
mouth parts
mutual recognition
myogenic rhythm
Nasanov gland
navigation
nectar concentration
nectariferous plants
nest bees
nest brood
nest cavity
nest mates
nursing duties
oblong plate
odour colony
olfactory
olfactory reception
ommatidium
organophosphorous
orientation flights
ovariole
oviposit
oviposition
ovulate
pallbearer bee
paraglossa
pattern of behaviour
payload
peach
pecten
peeping
penis
perforated cappings
perfume
pheromone foot print
pheromone (forage marking)
pheromone (mutual recognition)
pheromone (trail marking)
phoresy
pollen basket
pollen cellulose coat
pollen collection
pollen collector
pollen consumption
pollen digestion
pollen gatherer
pollen going in
pollen grain
pollen load
pollen mould
pollen pack
pollen pellet
pollen pickled
pollen storage
pollen transfer
pollen viability
pollination
pollination hand
pollination self
pose
posture
preparation micro slides
prepare
preparing show wax
proboscis
progressive feeding
properties of propolis
propolis collection
propolized
protogynous

proventricular valve
provisioning – mass
provisioning progressive
Psithyrus (bumbles)
pupal changes
pupal skin
pylorus
quadrate plate
queen activity
queen anatomy
queen cage
queen catcher
queen failure
queen piping
queen mandibular glands
queen ovaries
queen cramps
rate
raw honey
rectum
red clover
reflex
regurgitation
reservoir bees
respiration
respiratory movement
response
rest
restriction of the brood
retina
return
reverse side
robbing
rocking
rope hive lift
rotation
round dance
run
salivary syringe
scent
scent gland secretions
scenting
scent pheromone
scent trail
Schwirrlauf
scolophore
scramble
scraper
scrubbing dance
scutellum
searcher bees
secateurs
secondary wing
selfless
sense of form
sense of location
sense organ
sense plate
sense of queenlessness
sense of queen's presence
sense of colour
sense of danger
sense of sight
sense of smell
sense of taste
sense of touch
sensillum
sensitivity to movement
sensory transduction
servicing
sex characteristics
sex pheromones
sex attraction
shake
shaking bees
shutting down for winter
side effect
sight
signalling behaviour
silent robbing
simple eye
simple microscope
sinus
slaughter of drones
small intestine
smell
smoker fuction
smoker structure
sneeze
sniffing
soliciting food
speed of bee
spermatheca
spermathecal duct
spermathecal gland
spermathecal pump
spermathecal valve
spermatogenesis
spermatozoon
spin

spinneret
spinning
spiracle
spiracle propodeal
stand-still
static electricity
steal
stealing nectar
sting
sting – first aid
sting function
sting palps
stings – effects of
sting reaction
straight line
strain
stress
strop
subjugation
subjugation by smoke
submerged antipathy
submissive
substrate borne sound
successful mating
sugar spectrum
sun – solar orb
supersedure characteristics
swaling
Swammerdam (Biblia naturae)
swarm
swarm cluster
swarm control
swarmed
swarming habit
swarming impulse
swarming strain
swarm prevention
swelling
swimming
swivel
symmetrical
synergist
systole
take off
taking a swarm
tall
tamping
taste
temperature control
thanetosis
touch
trehalose
trophalaxis
trophy
uncap
uncapped
uncapped honey
unstable
upright
urate cells
UV
valve – heart
valve – oesophagous
vein
venom
venom sac
ventilation control of
ventilation in summer
ventilation in winter
ventilation of honeybee colonies
ventral diaphragm
ventral nerve trunks
ventriculus
vesicle
viscosity
vision
void faeces
wag-tail dance
walk
woible
washboard dance
washing
water balsam
water carrier
water gatherer
water sources
wax formation
wax glands
wax producing bees
wax secretion
wax shears
wean
wing beats
wing movements
wings
winter cluster
winter confinement
winter defence
winter dwindling
wipe

wiring frames and foundation
worker brood
worker comb
worker jelly
worker's duties
working distance
worn out

# 7- Biology

air sacs
adenosine triphosphate
adipose tissue
adrenaline
adult
adult worker
aerobic
aerotaxis
Aetology
age of brood
Aggregation pheromone
A.I.
akinsesis
alarm odour
albino
allelomorphs
allergen
allogamy
allomone
Amnion
Amoeba
anabolism
anaerobic
analagous
analysis
anaphase
ancestry
Androgenetic
anemorphilous
Angiosperm
Annelida
anterior
anti-biotics
antibody
antidote
antigen
antitoxin
Apis
Apitherapy
Apocrita
Apoidea
Apomict
apple fertilization
appreciation
Apterygota
Arachnida
arenaria
Arthropoda
arvenis
-aze
asexual reproduction
assay
assimilation
association
atopy
atrophied
autogamy
autophilous
auto pollination
azobacter
Bacillus
bacteriophage
bacterium
bear fruit
bee venom activity
bee venom components
bilateral symmetry
biosphere
blastokinesis
bleaching
bleaching slides
blue
botanical name
brood pheromones
burden
caste
caste differentiation
catabolism
catalaze
caterpillar

catkin
cell
central nerve cord
centrifugal force
cephalic
chalones
characteristics
Chelifer
chemotaxis
chiasma
chiropterophilous
chitin
chitinaze
chlorophyll
chorion
chromatid
chromosomal mutation
chromatography
chromosome
chrysalis
circadian rhythm
clearing
cleavage
clone
cocoon
co-enzyme
coition
colony odour
colony rhythm
commensalism
compatibility
condensation
cover slip
c.p.s.
cross pollination
crossing over
cross section
cubital index
cytoplasm
cytology
dimorphism
diploblastic
diploid
dissecting microscope
dissecting pin
dissection
dissection acarine
dissolve
distal
diurnal rhythm
division of labour
DNA
dominant
Dzierzon
ecdysis
ecdysone
eclosion
ecomorph
ectoderm
egestion
egg
egg-embryology
egg laying
egg sterile
egg unfertilized
ejection of worker
embodium
embryo
embryology
emerging adult
emerging brood
emunctory
endocrine glands
endoderm
energy
energy profit
Entomology
enzyme
apicuticle
epidermis
epilepsies
E.P.P.
eusocial
eye colour
father
fertile worker
fertilization
fixation
floral nectary
florescence
floret
flower seed
flowers per tree
focus
footprint pheromone
forceps
frequency
Fuchsin
function
fundatrix

fungi
Galeria
gamete
gametogenesis
ganglion
gas liquid chromatography
gastrulation
genetics
genotype
genus
germ
germ band
germ cell
germinate
Gerstung
gestation
glycogen
gravid
gregarine
gregarious
guttation
haemocyte
haemolymph
hair
half brother
hatching
hatching eggs
hemizygous
heredity
hermaphrodite
hive mind
homeostasis
homoiotheermic
homologous
honeydew symbiosis
honey stickiness
HPLC
hydathode
hysteresis
ICBB
igG
inflorescence
inheritance
inhibine
inervate
inoculum
inquiline
integument
intersegmental
invagination
invertebrate
in vitro
ion
juvenile hormone
kataphase
kinesis
Lachnids
laevulorotatory
learning by bees
lens
leucocytes
life cycle
locus
lumen
m
magnification
mammal
man
meiosis
melting point
meront
mesoblast
mesoderm
mesodermal originations
metabolism
metamorphosis
meta phase
microbe
micropyle
microscope
microscopical analysis
microsome
microsporidia
muscle
mushroom bodies
mutant
mutation
mitochondrion
mitosis
monocotyledon
monoecious
monostrain
motile
motor nerve
moult
moulting hormone
mounting slides
mutagen
mutant
natural selection

nectar guide
nerve cells
nerve fibres
neuromere
neuropile
neurosecretory cells
nexine
nigrosin
nucleic acid
nucleus
oecotrophobiosis
oenocyte
offspring
oil immersion
oligotrophic
oocyte
oogenesis
oogonia
organism
osmosis
panicle
pantoporate
pantothenic acid
pap
paper chromatography
parasite
parasitoid
parthenocarpy
parthenogenesis
parthenogenic females
patrilinage
peduncle
perianth
pericarp
pericystis
permeability
phage
phagocyte
pharae
phloem sap
polarimeter
pollen release
pollen signature
pollen slide
pollen terminology
pollinating agents
pollination
pollination close
pollinium
polyethism
polymorphism
polyploidy
pore
pose
preparation microscope slides
prepare
prepupa
preserving fluid
processes – micro slides
pronucleus
prophase
prophylactic
protandrous
protogynous
protoplasm
provisioning – mass
provisioning –progressive
pubescence
pupa
pupal changes
pupal skin
pupate
pupation
queen activity
raceme
recessive
rectal ampulla
red
reduction division
reflex
refractive index
refugium
relative humidity
research
repiration
resting cell
Rhisobium
R.I.
riboflavin
ribonucleic acid
ribosome
R.I..E.
RNA
royal jelly
R.Q. respiration
sap
scent pheromone
scent trail
sclerotin
secretion

secretion of nectar
sectioning
sediment
seed
seed germination
seeker
self pollinating
SEM
seminal vesicle
sense organ
sense peg
sensillum
sensory transduction
sex characteristics
sex differences
sexlinkage
sex of eggs
sexual cell
sexual maturity
s.g.
sight
silk glands
sinus
sisters
sitology
smear
sniffing
snow
solute
solution
somatic cells
somite
sorbitol
sp.
specie
specific gravity
specific heat
specific rotation
specimen
spectrum nectar sugars
spectrophotometric method
spectrospermatosoon
spermatogenesis
spore
sport
stadium
staining (microcopy)
sting effects
sublimation
sucrase
sugar chemistry
sugar spectrum
supersaturated
surface tension
survival
suspension
simbiont
simbiosis
synapsis
syndrome
synergist
system
systemics
telophase
temperatures
test for beeswax/honey
tetraploid
thanetosis
theory
thermal activity
thiamine
thin layer chromatography
toxic
trait
trial
triploid
tropism
turbid
turgid
unisexual
vaporize
vapour pressue
variation (genetics)
vasculum
velocity light/sound
ventral
virus
viscosity
viviparous
VPD
warmth
warmth brood nest
watch glass
water vapour
wing – beats
working day
Xeromoorphic
yolk
zygote

# 8 - Books

ABC & XYZ of Bee Culture
A bee is born
About bees and honey
About honey
About pollen
Advanced bee culture
Advantages of the House Apiary
African bees
African Bee Journal
A.L.
All about mead
Allen A.L. Sandeman
Amateur beekeeping
Anatomy and Dissection of Honeybee
Anatomy and Physiology of honeybee
The Ancient Mariner's Farewell
Ants. Bees and Wasps
Ants Bees and Dragon flies
The Apiarians Guide
Apiarians manual
The Apiary
The Apiary laid open
Apiculture for Schools
The Archaeology of Beekeeping
Armitt
The Art and Adventure
The Art of Beekeeping
Atkins How to succeed with bees
Avebury Lord
Backyard beekeeping
Background to beekeeping
Bagster S.
Basic beekeeping
Bath (Ants bees wasps)
Bazin Natural History Bees
BBNO
Beck B.F. Honey for Health
Beckley Peter Keeping Bees
Bee Books new and old
The Bee Craftsman
The Bee-hive
Beekeeping book
Beekeeping at Buckfast Abbey
Beekeeping Do's and Dont's
Beekeeping for all
Beekeeping in Antiquity
Beekeeping in the Tropics
Beekeeping New and Old
Beekeeping Q's and A's (Dadant)
Beekeeping techniques
Beekeeping – gentle craft
Beekeeping techniques
Bee Master of Warrilow
beemasters (past)
Bee matters and masters
Bees and beekeeping
Bees & honey
Bees & Mankind
Bees and people
Bees and wasps
Bees flowers and fruit
Bees Sense and Language
Bees wasps and allied insects
Bees wasps and ants
Beeswax Ron Brown
Bible
Butler (Feminine Monarchy)
catalogue
children's bee books
clippings
Complete Guide Beekeeping
Cowan beekeeper's guide etc.
Crane (Beekeeping Honey Archaeology)
Dadant (Hive and the Honeybee)
Dewey
Edwards
The Female Monarchy
Galton 1000 yrs. Beekeeping in Russia
Gedde John
Geoponica

Golding (the shilling book)
Gruzinian bee (Hive & Honeybee)
Harwood British Bee Plants
Hive & the Honeybee (classic)
Hive Skyscraper
Hodges Dorothy
Hogs at the honeypot(Teach Yourself)
Homer (Odyssey)
Hooper
House Apiary (Spiller)
Howes F.N.
Huish (treatise)
Insecticide toxicity
J.A.R. – IBRA
journals
Keys (Ancient Bee Master)
Koran
Laboratory diagnosis of honeybee
Langstroth Hive & Honeybee
Language (International bee terms)
leaflets
library
Life of the bee
literature
Loudon (cottager's manual)
Mace (bee farming etc.)
Maeterlinck (life of Bee)
magazine
Manley (Honey Farming etc.)
microfiche
Miller (years among the bees)
More D.
mustard (poem)
newsletter
Northern Bee Books
Oldfield Honey for Health
Pamphlets
Pchelovodstvo
periodicals
Pollen loads of the honeybee
press
publications
reprints
R.H.B. 1000 years beekeeping Devon
rhymes
The behaviour and social life of honeybee
The ABC & XYZ Root A.I.
Snelgrove method
Spiller (House apiary)
subscription
summary
Survey 1000 years Russia
Honey for Health Tonsley
Trees and Shrubs
Varro (De rustica)
Vernon (Hogs @ honeypot)
volume
Wedmore A manual of beekeeping
The ventilation of beehives
Wildman Thos. (complete guide)
Winnie the Pooh
Xenophon (Persian Expedition)
Year book.

# List of fifty five notable bee books by His Honour, David Smith QC

| | | |
|---|---|---|
| 1568 | Thomas Hill | *A pleasaunt instruction of the parfit orderinge of bees annexed to The proffitable arte of gardening* |
| 1593 | Edmund Southerne | *A treatise concerning the right use and ordering of bees* |
| 1606 – 1704 | Charles Butler | *The feminine monarchie* |
| 1634 | John Levett | *The ordering of bees* |
| 1637 | Richard Remnant | *A discourse or historie of bees* |
| 1655 | Samuel Hartlib | *The reformed commonwealth of bees* |
| 1675 | John Gedde | *A new discovery of an excellent method of bee-houses and colonies* |
| 1676 | John Worlidge | *Apiarium* |
| 1679 - 1687 | Moses Rusden | *A further discovery of bees* |
| 1712 – 1765 | Joseph Warder | *The true amazons* |
| 1721 | John Gedde | *The English apiary* |
| 1744 | Gilles Bazin | *The natural history of bees* |
| 1744 – 1774 | John Thorley | *Melisselogia or the female monarchy* |
| 1768 – 1778 | Thomas Wildman | *A treatise on the management of bees* |
| 1773 – 1819 | Daniel Wildman | *A complete guide for the management of bees* |
| 1796 – 1814 | John Keys | *The antient bee-master's farewell* |
| 1806 – 1841 | Francis Huber | *New observations on the natural history of bees* |
| 1815 – 1844 | Robert Huish | *A treatise on the nature, economy, and practical management of bees* |

| | | |
|---|---|---|
| 1827 - 1870 | Edward Bevan | *The honey bee* |
| 1832 - 1848 | Thomas Nutt | *Humanity to honeybees* |
| 1838 – 1880 | Henry Taylor | *The bee-keeper's manual* |
| 1842 | William Cotton | *My bee book* |
| 1849 – 1860 | T.B. Miner | *The American beekeeper's manual* |
| 1853 – 1927 | Lorenzo Langstroth | *The hive and the honey-bee* |
| 1853 – 1918 | Moses Quinby | *Mysteries of bee-keeping explained* |
| 1865 – 1878 | Alfred Neighbour | *The apiary* |
| 1870 – 1889 | A. Pettigrew | *The handy book of bees* |
| 1875 – 1884 | John Hunter | *A manual of bee-keeping* |
| 1880 – 1904 | B.B.K.A. | *Modern beekeeping* |
| 1881 – 1924 | Thomas Cowan | *British bee-keepers' guide book* |
| 1886 – 1921 | Frank Cheshire | *Bees and bee-keeping* |
| 1887 – 1928 | Samuel Simmins | *A modern bee farm* |
| 1888 – 1947 | W.B. Webster | *The book of bee-keeping* |
| 1901 – 1929 | Maurice Maeterlinck | *The life of the bee* |
| 1904 – 1950 | J.G. Digges | *The practical bee guide*<br>First edition entitled *The Irish bee guide* |
| 1908 – 1949 | Tickner Edwardes | *The lore of the honey-bee* |
| 1917 – 1960 | Charles Dadant | *First lessons in bee-keeping* |

| | | |
|---|---|---|
| 1925 – 1952 | A.B. Flower | *Beekeeping up to date* |
| 1930 – 1937 | William Herrod-Hempsall | *Bee-keeping new and old* |
| 1932 – 1945 | E.B. Wedmore | *A manual of bee-keeping* |
| 1934 – 1963 | L.E. Snelgrove | *Swarming: its control and prevention* |
| 1942 – 1964 | Edwin Way Teale | *The golden throng* |
| 1948 | R.O.B. Manley | *Bee-keeping in Britain* |
| 1952 | Dorothy Hodges | *The pollen loads of the honeybee* |
| 1953 | C.R. Ribbands | *The behaviour and social life of honeybees* |
| 1954 – 1974 | Colin Butler | *The world of the honey bee* |
| 1954 | Karl von Frisch | *The dancing bees* |
| 1956 | R.E. Snodgrass | *Anatomy of the honey bee* |
| 1962 | H.A. Dade | *Anatomy and dissection of the honey-bee* |
| 1975 | Eva Crane | *Honey: a comprehensive survey* |
| 1983 | Eva Crane | *The archaeology of beekeeping* |
| 1990 | Eva Crane | *Bees and beekeeping: science, practice and world resources* |
| 1999 | Eve Crane | *The world history of beekeeping and honey hunting* |
| 2003 | Lesley Goodman | *Form and function in the honey bee* |
| 2004 – 2009 | Kesseler & Harley | *Pollen: the hidden sexuality of flowers* |

# 9 - Breeding

acquired characteristics
activation hormone
adult
albino
allelomorphs
Alley/Miller
anaphase
ancestry
artificial insemination
asexual reproduction
backcross
Barret
bee food
breed
breeder colony
breeder drones
breeding
cell starting colony
characteristics
chiasma
chromosomal mutation
consanguinity
cross
crossbreed
cull
displacement crossing
double grafting
double hybrid
drone behaviour
drone breeder
drone trap
ecdysone
eclosion
ecomorph
embryo
emergency feeding
emerging adult
emerging brood
family tree
father
fatherless drone
fertilization
finishing colony
follower bee
gamete
gametogenesis
gastrulation
gene
generation
genotype
germ band
gestation
grafting
grafting tool
grub
gynandromorphy
half brother
handling queen cells
haploid
harbinger
hatchability
hatching
hatching eggs
homozygous
hybrid
hybrid vigour
hybridize
hygroscopic
immunity
incompatibility
inbreeding
ingestion
inheritance
larva
larval colour
larval food
line
line breeding
linkage
male

male gamete
malformation
mating hive
mating sign
Mendelism
metamorphosis
mid-gut
mitosis
mongrel
monoecious
monostrain
mini nucleus
multiple matings
multiple race crosses
mutant
nucleic acid
nucleus
old queens
oogenesis
oogonia
P1
parent
pedigree
progeny
pronucleus
pure line
queen bank
recessive
reciprocal hybrids
reduction division
relative humidity
reproductive isolation
sexual reproduction
siblings
sperm
sperm storage
spiracle
split board
sport
stadium
stamina
starter colony
strain
strain conformity
stud
survival
telophase
top crossing
transferring tool
variation (genetics)
valley
zygote

# 10 - Chemistry & Physics

abietic acid
absolute alcohol
Acaricide
acetic acid
acetone
acetylcholine
acetyldromedol
acidity
acids
acid value
adenosine triphosphate ATP
adrenaline
aggregation pheromone
alarm odour
alcohol
alcohols
aldehyde
aliphatic ketone
alkali
akkergen
amino acids
ammonium nitrate
amyl acetate
anaesthetic
analysis
andromedotoxin
anemonal
Anisol
anti-biotics
anti-freeze
anti-histamine
antiseptic
Aphox
apifuge
api milbien
aromatic compounds
arsenicals
ATP
attractant
azobacter
bearing
bee-go
beeswax
beeswax composition
benzene
BHC
Bio-assay
Biochemistry
bioflavonoid
biological control
biotin
bleaching
brimstone
brood pheromones
butyric acid
catalaze
caustic potash
caustic soda
CDA
cedar wood oil
cellusolve
certan
chalk
chemicals
chitinaze
chlorides
chlorinated hydrocarbons
chloroform
chlorophyll
chrysin
citral
citric acid
clove oil
cold cream
contact insecticide
cosmetics
cough mixture
coumarin
creosote
cuprinol

cyanides
cyanogas
danger from insecticides
DDT
de-natured sugar
desiccate
detox
devil's snuff
dew (pesticide)
dew point
dextrin
dextrorotatory
dextrose
dextrose/laevulose
diastaze
diatomaceous
dichlorvos
diffusion
dilatant
dimensions
direct contact
Diphenhydramine
disaccharide
disinfectant
DMP
DMSO
ecdysone
ectohormone
effect insecticides
effect of propolis
effect of repellents
effect of smell
effect of smoke
effervescence
electricity
electronic
electrophoresis
elements
enzyme
Epinephrine
equilibrium
erucic acid
ester
ether
ethylene dibromide
ethylene oxide
evaporate
excipient
face lotion
face pack
fat
ferment
flame test
flash heater
flavone
floor polish
fluid
fluid ounce
fluorescence
focus
Folbex
folic acid
follow the sheep
formaldehyde
formalin
formic acid
formulation
fructose
fruit sugar
Fumagillin
Fumidil B
galangine
Gamma ACH
gamma radiation
gas liquid chromatography
geranic acid
glucose
glucose oxidase
glue
glycogen
granulated sugar
grape sugar
grease
green sulphur
Gustathion
half-life
harmful substances
heating honey
herbicide
heterosaccharide
hexose
histamine
hive paint
hive preservative
HMF
honey
honey acids
honey adulteration
honey analysis
honey anti-biotic

honey aroma
honey chemistry
honey composition
honey constituents
honey crystallisation
honey crystals
honeydew
honey electrical conductivity
honey elements\
honey enzymes
honey fermentation
honey hygroscopic
honey mineral content
honey's optical properties
honey's physical properties
honey pigments
honey ripening
honey spectrum (colours)
honey stability
honey substitute
honey sugar content
honey surface layer
honey tannins
honey thermal activity
honey viscosity
honey vitamins
hormone
hormone inhibitor
Hostathion
HPLC
humectant
humidity
hyaluromidase
hydrogen peroxide
hydrolize
hydrolysis
hydrometer
hydroliphic
hydroxymethylfurfuraldehyde
icing sugar
ideal wood preservative
inhibine
inorganic
insecticidal fumigants
insecticide
insecticide formulations
insecticide toxicity
inversion
invertase
invert sugar
in vitro
in vivo
ion
iron
irradiate
iso-amyacetate
isomer
isomerose syrup
isopentyl acetate
isoprenaline
juvenile hormone
kairomones
ketones
lacquer
lactic acid
laevulorotatory
larval food
LD value
L/D value
leaching
lesser wax moth
laevulorotatory
levulose
ligroin
linseed oil
liquid
liquid paraffin
liquefaction
litmus
lye
magnesium
magnetite
malic acid
maltose
mandibular pheromone
material
Maury yeast
mannitol
manuring
mass
melezitose
melting point
metal
metals
methylated spirits
methyl salicylate
mineral
miscible
mixture
moisture

moisture content of honey
moisture equilibrium of honey
molecular attraction
monel metal
monosaccharide
mother of vinegar
moulting hornone
mounting slides
mucus
nectar quality
neoepinine
nerolic acid
niacine
Nicotinic acid
nigrosin
nine (9-)
nitrate
nitric acid
nitrobacteria
nitrobenzene
nitrogen
nitrogen fertilizer
nitrogen fixing bacteria
nitrous oxide
non reducing sugar
nucleic acid
odour – alarm
oil
oil of almonds
oilseed comparisons
oligosaccharides
opaque honey
optical density
optical rotation
organic
organochlorine
-ose
osmosis
oxalic acid
oxidization
oxygen
paint for hives
pantothenic acid
paper chromatography
para
paraffin oil
paraffin wax
pasteurize
pectolase
penicillin
peptide 101
pH value
Pharmacia (G.B.)
phenol
pheromone-alarm
aggregation pheromone
phloem sap
phosphate
phospholipase
phosphorous
phostoxin
polarized light
poisonous honey
poisonous sugars
pollen compatibility
pollen – composition
polysaccharide
potash
potassium
potassium nitrate
precipitation
preparation micro slides
preserving fluid
production of royal jelly
prolene
properties –honey
properties – wax
prophylactic
proprionic anhydride
propolis – constituents
prostaglandins
protein
PVC
pyrethrin
pyrethroids
pyridoxine
queen marking paint
queen paint
queen substance
races
raffinose
reducing sugar
reduction
refractive index
relative humidity
releasing agent
residual effect
resin
rhododendron honey
riboflavin

ribonucleic acid
ribosome
ribo nucleic acid
RNA
royal jelly
R.Q. respiration
saccharine
safrol oil
salt
saltpetre
salts in honey
sandarac
saponin
saturated
scent gland secretions
scent pheromone
scrubbing solution
sending sample – spray damage
sex pheromone
silicone
slow release units
soda
sodium
sodium metabisulphite
soft soap
soft water
soil properties
solubility
solute
solution
solvent
sorbitol
spatular
spice
sporopollenin
spray
syrup
starch
stearin
sterilant
stimulation - nectar secretion
sting components
stopper
streptomycin
sublimation
substances bees collect
substrate
succinic acid
sucrase
sucrose
sucrose acetate
sugar
sugar candy
sugar chemistry
sugar content of nectar
sugar spectrum
sugar tolerant yeasts
sulphates
sulphathiozole
sulphur
super saturated
sweet
synergist
synthesis
syrup
systemic
systemic compound
systemics
RASFP
tallow
tannin
tartaric acid
terpenes
terramycin
test for insecticide
thermal vaporizer
thiamine
thixotrophy
threshold
thymol
Torulopsis
toxic sugars
translocated herbicide
transpiration
trehalose
terpenoid
turpentine
unstable
urea
UV
vapona
vaporize
vapour pressure
varnish
varostan
venom
vinyl chloride
viscin
vitamin
volatile

votarol
watch glass
water
waterglass
water of crystallization
water vapour
wax bleaching
wax bloom
wax composition
wax – properties
wax solubility
wax tenacity
wax texture
web worm (acetic)
weedicide
wetting agent
white spirit
whole bee venom
wood preservatives
xylene
yeast
yeast nutriments
zymase

# 11 - Classification

-acea
Aculeata
aculeate
Angiosperm
anthophilous
Arthropoda
biotype
botanical name
class
classification
classification honeybee
Dewey
Endopterygota
evolution
family
genotype
genus
Hexapoda
hierachical system
Holometabola
Holotype
Homoptera
honeybee (Apoidea)
honeybee classification
Hymenoptera
insect
Kingdom
Linnaeus
Lepidoptera
maritime
muralis
nomenclature
odonata
odontobombus
origin – bees
pollen classification
pollen identification
pratensis
protozoan
pterygota
race
sp.
specie
specific name
spermatophyte
spp.
sub family
subgenus
suborder
sub species
super family
super organism
sylvatica
Symphyta
taxon
taxonomic
taxonomic endings
tribe
type
variety

# 12 - Countries

Anastomo kofini
Anatolia
Ancient Greece
Apis mellifera adami (Crete)
A.m.jemenitica (Jemen)
A.m.unicolor Madagascar)
Argentina
Australia
bee eater Iran
beekeeping outfit (America)
Belize
bell bird NZ
Brazil
Bulgaria
Canada
China
Crete
Cyprus
Czechoslovakia
Denmark
dragon fly Russia
driven bees Australia
Duchet Fraancois (Switzerland)
Dutch bees (Holland)
Dutch clover
Dutch elm
Dzierzon (Polland)
earthenware Malta
eat Australia
Echium Australia
Egypt
English honey
Essenos (Holy Land)
European honey
evergreen NZ
France
Germany
Giant honeybee India Sri Lanka
goat
G.P.O.
Great Britain
Hanneman excluder (Bulgaria)
Hawii
hive ceramic Malta
hive density (Czech,Greece,Swiss, Russia, China, U.S.A.)
hive manufacture
Holland
humeli Crete
Hungary
Hybla (Sicilly)
hymetus (Greece)
India
Israel
Italy
Janscha (Slovenia)
Japan
Jerusalem pine
Kahmir virus
Kenya
Khalifman Russia
krupnik (Polland)
Libya
Luneburg Heide
Macedonia (Jugoslavia)
Malta
Mauritzio (Swiss)
mellarius (Rome)
Mendelism (Austria)
mespilus (Portugal)
Mexico
Nile
North America
Norway
Nuka
NZ forage
package bees
Palestinian bees
parthegenetic females (S.Africa)
Pchelovodstvo (Russia)

Peace River Valley Canada
persimmon (Japan,Australia)
Philippines
Poland
pollen identification
Portugal
Prokopovich (Russia)
Quebec board Canada
pseudoscorpion (Sri Lanka)
Rig Veda
Romania
rosebay (Siberia)
rubber India
Russia
Shira (Germany)
Scotland
South Africa
Spain
Sri Lanka
stingless bees (Brazil)
suspend Africa
Switzerland
Syrian bees
tajonal (Yucatan)
Tanzania
Tasmania
Trebizond
Tunisia
tupelo (Mississippi)
Turkey
United States America
USA Beekeeping control
varostan (Japan)
Vocano Island
Wales
Woodbury Australia
wood swallow Australia
Yucatan
Yugoslavia

# 13 - Diseases

ABD
ABPV
Acarapis
Acarapis woodii (life cycle)
Acari
addled brood
addled egg
adult bee disease
adult disease diagnosis
Aetology
AFB
AFB life cycle
AFB symptoms
AIV
American Foul Brood
Amoeba
apiary hygiene
apiary inspector
api – milbien
Apimyiasis
Apis iridescent virus
Arkansas bee virus
Ascosphaera apis
Ascosphaera alvei
Asiatic mite
Asperfillus flavus
Asperfillus fumigatus
Bacillus
Bacillus alvei
Bacillus antisepticus
Bacillus larvae
Bacillus laterosphoros
Bacillus thuringengis
B. eurydice (EFB)
Bailey Dr.
bald-headed brood
bee health
bee louse
bee paralysis
bee virus x,y
benefits from Nosema free colonies
black queen cell virus
black robber disease
brood disease
CBPV
chalk brood
chilled brood
Chinese slipper
cloudy wing virus
comb sterilization
concave cappings
crawlers
crippled bees
cyst
DD
dead bees
dead brood
dead colony
demise
destruction
deutonymph
diagnosis
disappearing disease
discoloured brood
disease
dissecting for Acarine
drifting
droppings
Drosophila
drugs
dust
Dutch bees
Dutch elm
dwindling
dysentery
ecto parasite
EDB
AFB
EFB cycle
effect of drugs

# 14 - Drones

accessory glands
aedeagus
air-sacs
antenna
artificial insemination
bee trap
chromosomes
comet of drones
cornua
Dadant
DCA
diploid males
drifting
drone
drone anatomy
drone assembly
drone behaviour
drone brood
drone congregation
drone excluder
drone genitalia
drone life cycle
drone mating
drone slaughter
drone zone
egg – drone
ejaculatory duct
endophallus
eversion
expulsion of drones
eye compound
gene
genitalia
head
heap of drones
male
mating
maturation
massacre of drones
meeting place
mucus glands
oblique bands
old virgin
olfactory
olfactory receptors
parthenogenesis
penis
phallotreme
phallus
post nuptial
precursor
proctiger
propodeal spiracles
pupa
quarry
rejection
release
reproductive organs
sea
semen
seminal duct
seminal vesicle
sex
sex differences
sexual maturity
sexual reproduction
simbles
skywards
slaughter of drones
sperm
spermatozoon
stud
successful mating
supersedure characteristics
tenth segment
territory
undersized
valve-like plate
vas deferens
vestibule
volatile
wedding certificate

# 15 - English/Other languages

a
academic
-acae
adaptation
adversary
agent
agriculture
airspeed
albino
alien
align
alive
allergy
alveole
American terms
analogous
analysis
anaphylactic shock
ancestry
anchoring
anecbalic
anemophilous
anger
Angiosperm
anguiculi
animal
Annelida
annual
antagonism
anterior
anther
anthesis
Anthophila
anthropogenic
antibody
antidote
automasia
Apiact
Apian
Apiarian
apiarist
apiary
apiculturalist
apidictor
apifactory
Apiology
apistical
apparent
appreciation
aroma
article
association
atrophied
Aurelia
autophilous
azobacter
Bacillus
back ache
back cross
back end
backlot beekeeping
backyard beekeeping
bacteriophage
bacterium
balance
balling (regicidal knot)
basal
batik
beard
bear fruit
bee
bee bee tree (Evodia)
bee bole
bee culture
beekeeper
beekeeping
beekeeping (subject)
bee master
bee pest
bee proof

bee protein
bees
beestings
beeswax
beeswax definition
bee tight
bee tree
bee veil
bee way
bee widow
bee yard
beginner
biennial
boost
brood
brooding
burrowings
cat
chamber
colony
cleptolecty
climacterial number
colt
comb
competition
contribution
craft
cross combing
cut-off
cycle
darkness
dart
DCA
death feigning (thanetosis)
demise
demonstration
dense
dentistry
departure
dermis (hypodermis)
Deseret
desire
destruction
device
diagnose
diagnosis
diastole
digit
dilatant
discoverer
distal
distinguish
draw
draw back
drawn comb
drift
dross
duct
ductile
eat
eclectic
eclosion
Ecology
Ecosphere
Ecosystem
Ecotype
edaphic
edible
eduction
effervescence
efflorescence
egestion
eidetic
ejection
eke
elusive
embryo
emergence
empty
enclosure
empty
enclolsure
endemic
endo
entomophilous
enzootic
ephemeral
equipment
escape
escort
establish
etiquette
etiology
-etum
eusocial
evacuate
evaporate
eversion
evolution
excitable
excretion
farina

fatherless drone
fecund
feign death
fellowship
female
fence
feral
ferment
festoon
fibre glass
flat
flavour
flower
flower constancy
flower seed
fluorescence
focus
foraging
force
freedom
gadget
garner
gelatinous (ling)
genetics
geographical line (Era)
germ
gimmick
glabrous
glean
glue
gone through
grading glasses
grate
gravid
gregarious
group
growth
grub
gustatory
habit
habitat ('ophilous)
ham fisted
harbinger
harvest
hefting
heredity
hints
histology
hive box
hive mind
hiving sough
HMF
hobbyist beekeeper
homing
homologous
honeydew
honey extracting house
honey flavour
honeymoon
honey organic
honey sugary
horizontal
horticulture
host
hum
humectant
husbandry
hyaline
hybrid
hybridize
hydrophilic
hyper
hypo
image
imago
inanition
incline
increase
index
industry
infuriate
ingestion
ingurgitate
ingot
inhibitor
inquiline
inseminate
integument
intercommunication
interference
interior
in-the-flat
inventory
irritation
-iso
jargon
killer bees
kinesis
king bee
knock out

large scale
lateral
leaching
learner
lecture
lee (lees)
Lee James
leg honey
level
lid
lore
loss of
lot
lug
macro
making comb
making increase
malformation
mandible
manna
marking
match
mature
mead making
mel
melliphagus
mellipherous
mellifluous
melivorous
member
memory
metabolism
method
micro
micromillmeter (micron)
multi
must
miscible
misdescription
mixture
mnemonic
modify
molest
mongrel
monstrosity
mow
native
nest
nidificate
nucleus
number
off course
oligotrophic
oophilous
oral
organ
organoleptic
outstanding
overkill
ovoid
pabulum
para
parameter
partheno
parthenogenesis
perambulate
peri
peristalsis
phage
phenology
phoresy
poison
pollen pellet
pollen terminology
pollen transfer
pollinate
pollination value
pollinator
pollinium
pollinizer
polyandry
pome
porous
portal
pose
posterior
posture
powder
powerful
pp.
precipitation
predators
prepare
prevention robbing
preventative
probability
process
produce
programming
prolific

protein
pubescence
pummel
pupate
qualification
queenless
quell
race
ramify
range
rate
rear
record
regicidal knot
regulation
reserve
resistance
resonance
riparian
rusca
rowl
ruche
sag
sample
saprophyte
sayings
scale
scap
sclerotin
seams
security
segment
senility
sensory receptors
setbacks
shirty
shut
sign
sincere
sitology
skeppist
skills
skip
slinger
snippets
solidify
solute
solution
sororicide
specialization
specific name
speck
specimen
spectrum
spelling
squatters
sterile
stigma
stir
sub
submission
sinstrata
subtend
sulcus
surplus
swaling
swarm names
sweet
syllabus
symbiosis
synonymous
synopsis
system
tactile
tamping
taxis
technique
terminology
throw
terment
toxic
transfer
tunnel
type
ubiquitous
unassembled
uncap
undigested
unsatisfactory
Utah
viable
virile
viscosity
vulnerable
weak
wild
wipe
withstand
work
yield

# 16 - Hints

admix
advantages of 2 brood chambers
after 8
anaesthetic
back ache
bait combs
balling
bee brush
bee gloves
bee proof
black
black netting
blow
blow lamp
bottling
bottom bee space
branding
brightness
brood
brood and a half
brood spreading
brood trial
brush
buttercup
cat
Celsius
check list
chicken
chicken feathers
choice of
cleaning
comb collaps
communal feeding
cooling
corrugated packing paper
cover cloth
crafts
cream of tartar
cross wind
crush
crystallization
cup
cut (cutting)
cut comb container
darkness
dirty
disadvantageous factors
disinfectant
dislodge
dismantle
disrobe
eat
effects of movement
expose
eyelet
fire
foreign matter
forewarning
frame holder
free for all
Fuller's earth
funnel
Good candy
grafting
grip
handhold
honey aroma
June
knock out
late
leaking feeder
level
lighting a smoker
lime trees
Limnanthes
loading
lubrication
mixed equipment
monocular
nailing
netting
newspaper

newspaper cutting
notes
nucleus
observation hives
obstruct
odd
odour of wax
one hour queen introduction
orchard
over pollination
overturn
paraffin wax
paraglossa
perspiration
phostoxin
pollen compatibility
pollination
pollination – cross
potassium nitrate
pour
precursor
preparation of micro slides
preparing show wax
preservation
preserve
prevention of disease
prevention of drifting
prevention of fighting
prevention of robbing
prevention of swarming
prick
problems
production of propolis
protective clothing
queen excluder slots
queen glabrous
queen imported
queen introduction
queens long-lived
queen marking
queen mediocre
queen not found
queen off colour
queen old
queen paint
queen clipping
queen replacement]
queen right
rampin
rapid granulation
recipes
removal of propolis (wax)
rendering wax
re-queen
seasonal hints
sense of sight
shake
shaking bees
shallow
shampoo
shed
shield entrance from ...
shim
shock
showing vinegar
showing wax
sign
signs in and out of hive
sitology
size of entrance
skim
slug
smaller wax moth
smoker cartridge
smoker fuels
smoker function
smoking bees
smothering
sneeze
snow
sodium metabisulphite
solubility
solvent
spark
spiral cell protector
spirit level
sponge rubber
spread honey
spring
spring balance
spring management
squat
staple
starter colony
starvation
starved bees
sterilization
stick (verb)
stings effects of
sting treatment

stone
stopping
storage
stormy fermentation
straighten
strain
strainer
straining cloth
straw
strength of beeswax
stress
strop
submerge
substances bees collect
sugar syrup
sun (solar orb)
sunflower
sunlight
super clearer
surface tension
surge
swarmed
swarming impulse
swarm lure
swelling
swimming
swipe
switched
swivel
symmetry
taking samples
taking a swarm
taking exams
tap
tape recordings
tap strainer
threshold
throwing out
tips
topple over
top ventilation
torch
torment
transferring tool
transporting bees
trigger off
'T' shaped
turn
tweezers
twin dissecting pin
undergrowth
under or over
uniting aids
unscrew
unsound
upper entrance]
upright
upward ventilation
urgent
useful dimensions
uses for honey
uses of propolis
uses of royal jelly
uses of wax
vibration
wasps continued
watch strap
water proofing – canvass,leather. shoes
water soluble
wax
wax glands
wax moth
wax plastic
wax producing bees
wax reclamation
weak
wearing gloves
web spider
wedge
weld
well made
wet
wind speed
winter
winter passages
wire
wire gauze
wire netting
wiring frames and foundation
worried
wound dressing
zip.

# 17 - Hives

advantages of two broods
alighting board
Alveary hive
Anastomo kofini
artificial heating
bar frame
bar hive
bark hive
basket hive
Beaton hive
bee bole
bee gum
beehive
bell glass
belt
board
body
Bolton hive
bottom bar
bottom entrance
box
box hives
brood chamber
Buckeye hive
butt
bye hive
Carr Wm. Broughton
catenary
CDB hive
chamber
claustral
cog
coiled straw
colour
commercial hive
compartment separater
conqueror hive
cork hive
cottager hive
crown board
damage
deadman's floorboard
double brood chamber
double screen
double walled
dry
Dugat
earthen ware
eclipse
education
egress
eke
entrance
entrance adjustable
entrance block
entrance – bottom
entrance slide
fatal
Father of beekeeping
fixed comb
flax
floor
frame sizes
gable roofed
Gedde John
glass quilt
Glen
grain
Grange
Greek bar hive
gums
hackle
half depth super
handhold
Harbison
hard board
heather board
hive
hive assembly
hive –bar

hive carrier
hive ceramic
hive clay
hive clips
hive collateral
hive ceramic
hive cloth
hive collateral
hive colour
hive Dadant
hive double (skyscraper)
hive escape
hive fastening
hive feet
hive fillet
hive fixed comb
hive floor
hive furniture
hive insulation
hive large
hive leaf
hive legs
hive lift (body)
hive -log
hive – long idea
hive maintenance (types)
hive materials
hive model
hive movable frame
hive multi storey
hive national
hive observation
hive paint
hive plans
hive plastic
hive protection
hive records
hive roof
hive runner
hive scale
hive site
hivc sky scraper
hive Smith
hive stand
hive standard
hive Stewarton
hive strain
hive straw
hive top
hive twin
hive WBC
hive – wide idea
hive wooden
honey home
honey house
hoop
ideal wood preservative
identifying
improved national
inner bodies
inner cover
in-the-flat
jar hives
joint
jumbo hives
KTBH (Kenya)
Kenya hive
KerrLangstroth hive
leaf hive
lift
lock spring
log hive
long idea hive
lug
maintenance
material
mating hive
modified commercial
modified Dadant
modified national
modify
nadir
nadiring
nailing tool
narrow spacing
national major hive
national minor hive
needle – skep
Neighbour (bar hive)
nucleus
nucleus – mini
number of cells
Nutt's collateral
oblong hives
observation hives
octagonal hive
open topped hive
osier
paint for hives
paraffin wax

# 18 - Honey & Nectar

Acceton
adulteration of honey
Alergoba
alfalfa
almond
alsike clover
amino
analysis
andromedotoxin
Anemonal
Anise hissop
anti freeze
antiseptic
apple
Athole Brose
Australian honey
Australian h. colours
Baker's honey
beech honey dew
beeswax
bell heath
benefits from honey
biotin
blended honey
blossom honey
blue weed
botulism
Brassica
brightness
bubbles in honey
bubble size
bubble test
buckthorn
buckwheat
bulk comb honey
burn ointment
ceral honey
cessation of flow
charlock
cherry
chickweed
Chinese tallow tree
chunk honey
clarified honey
clear honey
clematis
Clethra alnifolia
clover
Coccoidea
colloid
comb honey
corne honey
cotton
coumarine
creamed honey
cruciferae
crystal
crystallised honey
cucumber
currant
cut comb
dandelion
dark honey
dearth
dense
density
density of honey
deposition of nectar
dextrin
dextrorotatory
dextrose
dextrose/laevulose
diastase
diatomaceous
Didymus
digestive tract
dilatant
drain
drained honey
dried honey
dripped honey
drum

leaf honey
length of tongue
levulose
liquid honey
liquefaction
loading
locust
Lucerne
maiden
maiden honey
making honey
malodorous
maltose
mead
manna (honeydew)
manuka
manuring
maple
Marchalina
majorum
marking
marking on containers
medium honey
mel
melezitose
Melilotus spp.
meliponae
Melissodes
melissopalynology
melliferous
mellifluous
melliorous
mesquite
Michaelmas daisy
microscopical analysis of honey
microwave
mignonette
mulberry
mustard
mineral content of honey
mistletoe
moisture content of honey
moisture equilibrium of honey
moor
mother-wort
Narbonne
naturally crystallized
nectar
nectar concentration
nectar flow
nectar quality
Nicotaniana
non floral nectar
nucleus
NZ beech honey
NZ forage
NZ honey
oak
odour of nectar
odour – honey
ointments
old man's beard
onion
opalescent honey
opaque honey
optical density
optical propereties of honey
optical rotation
optimum temperature
organic
organza
overheating
packing tins
paper chromatography
parsnip
pasteurize
Paterson's curse
pea
peach
pear
pearly white honey/wax
phacelia
Penrose uncapper
peppermint
perforextractor
Pfund grader
pH value
phloem sap
polarized light
Polaroid
poisonous honey
poisonous plant
polarimeter
pollen clogged
pollen identification
pollen signature
polyflora
potassium
pound
pour

preserve
pressed honey
prick
primer]privet
processing honey
production of honey
production of judges
prolene
properties – honey
prosopis
pump
purple loosestrife
purslane
pyknomiter
quantities of nectar
quickthorne
radius
ragwort
rape
rapid granulation
raw honey
receptacle
recrystalize
red clover
reducing sugar
refractive index
regurgitation
relative humidity
rhododendron
rhododendron honey
Rhynchota
Ribes grossulaaria
riboflavin
Rig Veda
ripe honey
ripener
Robinia
rock cress
Romania
rosebay
rosemary
rowan
rubber
ruby honey
running honey
safflower
sage
salts in honey
sanfoin
salvia
scabious
scum
sea - green honey
seeding honey
self-heal
service tree
set honey
settling tank
shampoo
shim
silver lime
single species honey
siphuncle
snowberry
soft set
soil fertility (heather)
Solidago
solidify
sooty fungus
sourwood
soya bean
spatular
specific gravity
specific heat
specific rotation
speck
speed of bee
spin
spinnbarkeit
spread honey
spruce
spun honey
starter or primer
stoneware honey crocks
storing
strainer
strawberry tree
succinic acid
sucrose
sugar content or nectar
sugar spectrum
sugary honey
sunflower
super
supered
supering
surface layer of honey
surplus
suspension
sweet

# 19 - Honeycomb

accommodation cells
alveole
angles of cell walls
Bar frame
black comb
brace comb
burr comb
cell
cell base
cells of honeycomb
cell walls
dark comb
drawn comb
exchange comb
fixed comb
Florea
follower board
gravity
hanging section holder
Harbison
hexagonal
hive clips
hive fixed comb
honey characteristics
honey chunk
honey class
honey comb
honey comb section
lime trees
metal divider
old comb
packing sections
rhomb
section
separator
septum
showing comb honey
slotted separator
spring block
stamp
starter or primer
step comb
storage
suspend
transition cells
travel stain
triangular
'T' shaped
uncap
uncapped
uncapping fork
uncapping knife
uncapping plane
uneven
vertical
vertical mode
vibrating knife
wax cappings
wax nippers
wax particles
wax plastic
weeping
well filled
white cappings

# 20 - Insects

African bee
African Brazillian bee
Africanized bee
allied insects
Andrena
Andrena amata
Anodontobombus
ant
APC bee
Apidae
Apimyiasis
Apinae
Apis cerana
Apis dorsata
Apis florea
Apotrigona
Asilidae
Avebury Lord
bambara
bee
bee fly
bee killer
bee moth (wax)
bee moth
bee parasites
bee pollinators
bee wolf
beneficial insects
big fly
black bee
blue orchard bee
Bombidae
Brazillian bee
British black bee
Buckfast queens
bumble bee
bumble parasites
bumble varieties
Caucasian bee
Chaleicodoma
Coccoidea
Coleoptera
cordovan
cuckoo bee
Cyprian bee
darkness (wasps)
Death's head hawkmoth
Dewey (books)
dorsata
dragon fly
Drosophila
Dutch bees
dwarf bees
earwig
Eastern honeybee
Egyptian bee
encourage bees
Endopterygota
enemies
energy profit
Entomology
Ephestia
Euvarroa
Fasciata
Florea
fly
fly catcher
forest honey
French bees
Galleria
German bees
Giant honeybee
gnat
greater wax moth
Greek bee
green fly
Gruzinian bee
hairless bee
hairy willowherb (hawkmoth)
Halictinae

tree wasp
trophophallaxis
Tunisia
tutu (honeydew)
Ultracoelostoma
Vespa crabro
wasp
wasps cont'd
wax moth
web spider
webworm
weevil
white ant
wide spacing (fiant bee)
wood ant
worker
yellow bee pirate

# 21 - Journals

ABC - ABC & XYZ of Bee Culture
ABJ American Bee Journal
Advisory Leaflets - Defra etc.
African Bee Journal
Apiact
The Apiarist
Apicultural Abstracts
Apicutlture in Tropical Countries
article
Australian Beekeeper
Australian Bee Journal
BBJ
BBKA News
Bee Craft
Beekeepers Calendar
Beekeepers' Quarterly
Beekeeping Newsletters
Bee Press
Bee World (I.B.R.A.)
Gleanings
Herrod Hempsall

# 22 - Legalities

Act
Acts & Regulations
Agenda
agent
Apiary Health
Approved Scheme
avoidance of stings
BBKA Comprehensive Insurance
Beekeeping Instructor
Bees Act
brains trust
branch
British beekeepers
British standad
British standard frame
B.S.I.
bubble test
buildings
bulletin
CBL
chairman
codex alimentarius
committee
communication
competition
comprehensive insurance
Congress
Convention
County Beekeeping Lecturer
Crystal Palace
display
disqualification
Domesday book
E.E.C.
flight path
G.P.O.
grading glasses
honey analysis
honey definition
honey enzymes
honey extracting house
honey grading glasses
honey house
honey judge
honey judging
honey label
honey marketing
honey regulations
honey show
honey show schedule
(HMF) Hydroxymethylfurfuraldehyde
I.B.A.P.
The Importance of Bees Order
industrial honey
industry
insurance
judge
judge's steward
judging candles
Junior exam
label
labeling
lecturer
legality
misdescription
moderator
moisture content of honey
Meetings
National Diploma
notice
novice
official
oral
order
organizer
ownership
Parliament
points
pollen identification
pollination agreements
pollination fee
post code

Practical examination
Preliminary examination
preparation for exams
President
Press
prize
production of beekeepers
production of beekeepers
production of bees
production of judges
production of lecturers
proposition
public relations
qualification
queen posting
queen mailed
questionaire
quiz
referee
reference
registration of beekeepers
regulation
reprints
research
rules
rustling
Sale of honey
Sale of honeybees U.K.
School beekeeping
scum
secretary
senior course
senior exam
settling
show
showcase
showing candles
showing comb honey
showing mead
showing vinegar
sideliner
small claims
Speaker
specialization
specialize
spectrophotometric
spray poisoning
stamp
stand-still
statistics
Statutory instrument
steal
Stoneleigh
student
sub committee
subjugation
subscription
sucrose-octo-acetate
sugar beet
sugar de-natured
swaling
syllabus
symbolic – bees/hives
Symposium
taking exams
Talmud
teaching aid
terror
test
tests
theory
thesis
third party
thrift
time table
treasurer
trial
trophy
U..K. beekeeping
unpolished
unpopular flowers
USA beekeeping control
vandalism
venue
VHC
viva voce
washing
wax modelling
wax purity

# 23 - Management

After 8
after effect
anchoring effect
antiseptic
avoidance of stings
azobaacter
bee blower
bee escape
bee –proof
equalizing
excitable
excluder
hive bar
hive closing
hive closure
hive cloth
homing instinct
management
manipulation
queen clipping
queen rejection
queen replacement
U.K. beekeeping

# Manufacturers

Abbott Laboratories
Dadant & Sons Inc
Lees
Taylors
Thomas
Thornes

# 24 - Mead (Honey wine)

acetobacter
activate
aenomel
alcohol
All about mead
anaerobic
beer
honey beer
beeswing
body
botchet
bouquet
bracket
braggot
bragwort
clarre
conditum
corma
cyser
ginger beer
hydromel
mead
melomel
metheglin
morat
mulsum
must
odour of mead
oenomel
omphacomel
paraffin
pectolase
precipitation
pyment
re–inforce
rhodomel
Saccharomyces
sack mead
sediment
settling
showing mead
showing vinegar
siphon
spice
starter or primer
stormy fermentation
strainer
sugar tolerant yeasts
tannin
thermostat
Torulopsis
turbid
uncapping
usquebaugh
varii
vinegar
wassail
wet
wild yeast
wine
yeast
yeast nutriment
zymase
zythus

# 25 - Measurements (Math)

acre
angles of cell walls
Ångström
aperture
armour
Artemis
bee space
bee tight
bloom
blooming times
bottom bar
brood cells
brood cycle
brood food
brooding temperature
brood nest
brood rearing
c.c.
cell walls
Celsius
centimeter
chill coma
crop (anat)
dead bees
density honey/wax
density winter cluster
dew point
dimensions
facts
floor polish
floret
fluid ounce
flying speed
foraging radius
fuel consumption
gallon
gap
geographical (Era)
glossometre
grain
gram
hectare
hefting
hive scale
honey hydrometer
hydrometre
hygrometre
inch fractions
-iso
joule
Kelvin
kilogram
kilometre
l – (litre)
label
labelling
larva
lateral
L.D. value
L/D. value
length
length of day
length frame
length of life
length of active season
length of time to mate
length of tongue
length of winter
lettering
litre
lumen
mb (millibar)
marking containers
mathematics
measurements
mega
melting point
metre
metric abreviations
metric prefixes

# 26 - Mythology

Aesculapus
Aelian
aenomel
ambrosia
Anabasis
Anacron
Ancient feeding techniques
Antiope
apiarius
apicure
Aqua mulsa
Aristaeus
Aristomachas
Aristotle
A swarm of bees in May
Azalea pontica
Babylonians
bar
bee bole
The Bee Boy's Song
bee byke
bee fold
bee garden
bee garth
bee herd
bee hunting
Bee lippen
bee scep
bee stall
beeswax candle
bee walk
beo ceorl
bestiary
bier
bike
botchet
braggot
bull swarm
bum
butt
byke
cashered stocks
Cassius Dionysis
castling
caul
cecrops
Cephenes
Cerinthus
ceromancy
chaste honey
church candles
cierge
clicket
cloom
close driving
coiled straw
coul staff
cowcloom
cross sticks
custos apium
Cyclops
death (human)
Deseret
Diodorus
Dioscorides
Didymus
dog star
Egypt (ancient)
embalming
epomphalia
Eros
Essenos
Evelyn John
flora
follow the sheep
Galbaum
Geoponica
giant puff ball
Ancient Greece
Greek bee hives

# 27 - People

Abbottt Charles Nash
Advantages of the House Apiary (Spiller)
Ambushady
Adam Bro.
Aelian
Allan Harry
Allen The Rev Montague Yate
Aristaeus
Aristomachus
Aristotle
Armbruster
Armitt
Atkinson John
Bailey Dr.
Baldensperger
Barret
Bath W.F.
Bee farmer
Bee garden (Howes F.N.
The Beehive (Galton Dorothy)
Beekeeping Do's and Don't's (Tickner Edwards)
Beekeeping in Antiquity (H.M. Fraser)
Beekeeping in Britain
Beekeeping in Schools
Beekeeping New & Old (H. Hempsall)
Beekeeping Q's & Ans. (Dadant)
Beekeeping Techniques (A.S Deans)
Bee Mason
Bees' sense and language (von Frisch)
Bevan
Bielby
Bingham
Blow
Butler Rev.
Butler Dr. Colin
Carr Wm. Broughton
Cassius Dionysis
Cato
Cheshire
Cleaters
Clergyman beekeepers
cobana (Zbikowsky)
Collin Abbe
Columella
Cooke Samuel
Cooper B.A.
Cornelius
Cotton Rev.
Cowan
Crane Eva
Creighton R. & C.)
Cumming Rev.
Dade Major
Dines A.
Diodorus
Digges
Doubling (H.H. Herrod Hempsall)
Doolittle
Duchet Francois
Dugat
Dyce
embalming (Frazer)
energetic reward (Dr. Corbet)
Evelyn John
evolution (Darwin)
examiner
Fabre
face pack (Cleopatra)
famous sayings (Doolittle)
Father of beekeeping
fennel (Varro)
The Feminine Monarchy (Butler)
Fibonacci
finger print (Dr. Louveaux)
fixed comb (Langstroth)
Flashman
Frisch
Frow
Galangine (Villaneuva)

# 28 - Pests

accompany
Acherontia
Achroia
Aeroglyphus robustus
Aechna grandis
Aethina tumida
alpine swift
Anagasta kuchniella
Annelida
ant
anti-pollinator
ape
aperture
Aphis
apivorous
Araachnida
Asilidae
Australian bee eater
B. thuringengis
badger
bear
bee eater
bee enemies
bee killer
bee pest
Bees flowers and fruit (Mace)
bird
black bear
blackcurrant
blind louse
blue titmouse
Bombylidae
Braula coeca
Bufo
bumble bee parasites
caterpillar
Ceuthorhynchus
dog
droppings
dwindling
earwig
ecto parasite
Edwards
endoparasite
enemies
Ephestia
flycatcher
forficular
frog
Galleria
gnat
gnatosoma (Varroa)
Greater wax moth
great spotted woodpecker
green woodpecker
hatching
hatching eggs
heather beetle
hedgehog
hive escape
hive feet
hive legs
hive maintenance
hive protection
honey badger
honeybee parasites
honey buzzard
honey characteristics
honeycomb moth
honey eater (bird)
honey moth
honey possum
incubation
inquiline
insectivore
intruder
invade
Kenya hive (badger)
Lanius
lesser wax moth

# 29 - Plants

Aaron's rod
Acacia
Acer
Aconite
Aesculus hippocastanum
Agricultural changes in UK
Ailanthus altissima
Alder
alfalfa
almond
alsike clover
Althea rosea
Alyssum maritimum
Anchusa capensis
Anenome
Anglica archangelica
Anise Hyssop
annual
anther
Aphis
Apiaceous
apple
apple varieties
apricot
Arabian thistle
Arabis
Armenia maritima
aroma
arvensis
asparagus
Aster
Astralagus
attack
Aubretia
Australian honey
autophilous
autumn crocus
Azalea ponticum
balsam
Banksia
barberry
basil
bean
bee balm
bee bee tree (Chinese)
beech
bee flowers
bee garden
bee glue
bee gum
bee pasture
bee plant
bee tree
beet – sugar
bell heath
biennial
bilberry
bindweed
birch
bird cherry
bird's foot trefoil
blackberry
blackcurrant
bloom
bluebell
blueberry
blue weed
borage
Boraginaceae
botanical name
botany
bottle brush
box
bracken
bract
branch
Brassica
Brassica arvensis
Brassica napus
broad bean

broom
buckthorn
buckwheat
bud
Buddleia
bugloss
bullace
bunt
bush
butter cup
button wood
Buxus
Castanea
catkin
catmint
celandine
chamomile
Chapman honey plant
charlock
cherry
chestnut
chicory
chickweed
Chinese bee tree
Chinese tallow tree
Chionodoxa
chives
Christmas rose
Christmas tree
chrysanthemum
clarkia
Clematis
clover
clover varieties
coconut
coffee
colt's foot
comfrey
common bird's foot
compositae
conifer
coreopsis
corncockle
cornflower
corolla
Cotoneaster
cotton
crab apple
cranberry
cranesbill
crimson clover
crocus
cross pollination
Cruciferae
cuckoo flower
cucumber
culinary herbs
cultivar
current
cynoglossum
Daboecia
Dahlia
damson
dandelion
dioecious
dichogamy
Dicotyledon
dogwood
Dutch clover
Dutch elm disease
Echium lycopsis
Ecotype
efflorescence
elm
enclosure
entomophilous
Erica
Ericaceae
Escallonia
escape
-etum
Eucalyptus
evening primrose
evergreen
everlasting pea
extra-floral nectary
Fagaceae
family
fennel
fickle plants
fir
firethorn
fireweed
fixation
flax
floccule
flora
floral nectary
florescence
floret

flower
flowering currant
flowering plant
flowers rarely visited
flowers per tree
fluorescence
forest honey
forgt-me-not
fox glove
French honeysuckle
fruit
Fuschia
furze
garden campanula
garden escape
gean
geranium
glands
glass house crops
globe thistle
Glory of the snow
goat's rue
Golden rain tree
golden rod
gooseberry
gorse
grape
grape fruit
grass
greenhouse crops
guajillo
gums
hairy willowherb
hardheads
harebell
Harwood
hawkweed
hawthorn
hazel
heartsease
heath
heather
hedge
Hellenium
Heliotrope
Helliborus
herb
Hesperis matronalis
Himalayan balsam
hobbyist beekeepers
hogweed
holly
hollyhock
holmoak
honesty
honey aroma
honey brassica
honey Erica
honey Eucalyptus
honey forest
honey fungus
honey leaf
honey suckle
honey tree
hornbeam
horse chestnut
horsemint
hound's tongue
huckleberry
humulus
hybrid
hydathode
Hypericum
hyssop
Iceland poppy
iceplant
ideal weather conditions
Ilex
impatiens
Indian bean tree
inflorescence
intine
ivy
Japanese candle
Japanese knotweed
Japanese quince
Japonica
jasmine
Jerusalem pine
keel
kex
king bloom
knapweed
knotgrass
Labiatae
laburnum
laurel
lavender
Legume
Leguminosae flower

# 30 - Pollen

Alder
alfalfa
almond
Angiosperm
anemone
auricle
autumn crocus
bear fruit
bee bread
berseem
Biopol
bird's foot trefoil
blackberry
blackcurrant
blackthorn
Bokhara clover
box
Brassica
Brassica arvensis
broof
broom
buckwheat
Buddleia
catkin
celandine
charlock
cherry
chestnut
chickweed
clematis
clover
comfrey
cotoneaster
cotton
crab apple
cranesbill
crocus
cross pollination
dandelion
endexine
evening primrose
extine
facts
farina
fat body
flax
floccule
flower
forget-me-not
fungistat
geographical honey
glean
goat's rue
golden rod
gooseberry
gorse
grain
hardheads
harebell
harvest
hawkweed
hawthorn
Haydak's formula
hay fever
hazel
heath
heather
Hodges
holly
hollyhock
holm oak
honesty
honey analysis
humulus
horse chestnut
Hypericum
I.A.A.
Impatiens
incompatibility
intine

Japonica
knapweed
lesser celandine
loganberry
lucerne
lupin
mallow
Malus floribunda
manuka
maple
majorum
Meligethes
Melilotus
melissopalynology
melliferous
Michaelmas daisy
microscopial analysis of honey
microspore
mignonette
mulberry
mullein
mustard
minerals in pollen
mistletoe
mixed pollen loads
monocolporate
mother-wort
nexine
nutritional value of pollen
onion
orange
palynology
parsnip
Paterson's curse
peach
pear
pearly white (Morning glory)
pecten
phacelia
pollen
pollen analysis
pollen barrier
pollen basket
pollen beetle
pollen brush
pollen colour
pollen compatibility
pollen composition
pollen consumption
pollen digestion
pollen dispenser
pollen gatherer
pollen going in
pollen grain
pollen identification
pollen intake
pollen load
pollen loads of the bee
pollen mite
pollen mould
pollen nutritition
pollen pack
pollen pellet
pollen pickled
pollen ploidy
pollen potentential
pollen press
pollen production
pollen rake
pollen release
pollen reserve
pollen signature
pollen slide
pollen sources
pollen storage
pollen substitute
pollen supplement
pollen terminology
pollen transfer
pollen trap
pollen tube
pollen viability
pollinate
pollinating agents
pollination
pollination of crops
pollination – cross
pollination fee
pollination hand
pollination self
pollination value
pollinator
pollinium
pollinizer
poplar
poppy
privet
production of pollen
prosopis
protein

# 31 - Pollination

abrade
allogamy
anemophilous
anther
anti-pollinator
apple varieties
auricle
autogomy
autophilous
autopollination
bear fruit
bee pollinators
colony density
colony morale
effect of rain
emergency feding
energy profit
Entomophily
E.P.P.
flowers per tree
glasshouse crops
herbicide
heterosis
hive density
hives per hectare
honeybee pollinators
hystersis
king bloom
large scale
Long Ashton
malformation
Mellissodes
orchard
osmia
rent
uses of honeybee colonies
wind pollinated

# Propolis

abietic acid
bee glue
bee gum
flavone
galangine
propolis
propolis collection
propolis constituents
propolis dermatitis
propolis gargle
propolising small holes
propolized
solubility
solute

# 32 - Queens

acceptance
accompany
acetone
aedeagus
akinesis
alight
Alley/Miller
antagonism
antipathy
aperture
Apiary health
appendicula glands
appreciation
artificial insemination
artificial queen cups
attendants
baby queens
balling
Barbeau
The Beekeepers' rule
breeder queen
catalepsy
catch
cell
challenge
Chantry queen cage
chorion
chromosome
chrysalis
clipped
clustering pheromone
colour
colour of the year
copulation
copulatory pouch
dead queen
departure
de-queen
desire
dimpled
Dichlorvos
discovery of queen substance
double grafting
drifting
drone assembly
drone behaviour
drone breeder
drone congregation area (meeting)
drone eggs
drone laying queen
Dugat
DVAV
dwarf bee
Dzierzon
egg
egg – drone breeder
egg laying
egg laying worker
egg –queen
egg - worker
elusive
embedding
emergency queen cells
epilepsis
escort
eversion
expulsion
factors
failing queen
fatal
fecund
female
finding queen
fliers
flying
follicle cells
foot print pheromone
freeze
glabrous
Good candy

G.P.O.
gravid
grip
hair curler
hairless bee
handling queen cells
Hanneman excluder
health certificate
heterosis
hind legs
hive colour
hive mates
hive odour
hive –to
holding down cage
honey bar
icing sugar
identifying
identity odour
imago
imperfect queen
incompatibility
increase
ingestion
inseminate
instrumental insemination
interchanging
interesting facts
introductory cage
inversion of queen cells
Janscha
juvenile hormone
Kozhevnikov gland
laying
laying queen
laying rate
length of time to mate and lay
maiden
mandril
marked queen
mated queen
mating
mating hive
mating sign
mating swarm
maturation
melanosella
melanosis
metal queen excluder
meta scutellum
middle 'c'
multiple matings
mnemonic
mother
motor neurons
nervure (wing cutting)
noise
nucleus
nucleus – mating
nucleus mini
nuptial flight
nurse bees
nurse colonies
nursery cage
nursing duties
offset
old bees
old queens
old virgin
one hour queen introduction
open brood
oocyte
oogenesis
oogonia
ovariole
ovary
oviduct
ovulate
ovum
parent
parent colony
partial acceptance
peak brood cycle
peak egg laying
pedigree
perambulate
pheromone (clustering)
pheromone sex
polyandry
post nuptial
prevention of drifting
prevention of robbing
prevention of swarming
prime swarm
princess
production of bees
production of queens
progeny
prolific
pulled virgin

scent trail
screening
scrub queen
scutum
selected tested queens
selection
selective breeding
seminal pump
seminal vesicle
sense of queenlessness
sense of queen's presence
sex
sexual maturity
sexual reproduction
shipping cage
side effect
signalling behaviour
skywards
slotted queen excluder
Snelgrove method
Sororicide
sperm
spermatheca
spermathecal area
spermathecal duct
spermathecal pump
spermathecal valve
spermatozoon
Sphaerularia bombi
spinning
spiral cell protector
split
splits
sponge rubber
spotting
spotty brood
stamina
stand still
sterile egg
sting at rest
storage
strain conformity
stud
submerge
submerged antipathy
submissive
substrate borne sound
successful mating
sugar for bees
summer
supersedure cell
supersedure characteristics
supersedure swarm
susceptible
swallow
swift
swimming
tenth segment
terebra
tergite
terminal filament
territory
test comb
test guage
tethered queens
thanetosis
thorax
tracheae
trail odour
transferring tool
travelling cage
trigger off
trophocyte
twelve apostles
uncapped queen cell
under sized
unfecundated
unmarked queen
unopened queen cell
untested queen
upside down
useful dimensions
vacated queen cell
vagina
vas deferens
virgin queen
void faeces
vulva
walk
warmth of broodnest
water method
wedding certificate
weights
well proportioned
West cell protector
wind speed
winter cluster
wire queen excluder
worker egg
worker jelly

yearly queen colour
young adult
young queen

# 33 - Races of bees

Adansonil
African bees
African.Brazillian
Africanized bees
alkali bee
Anatolian bee
Apis
A.adonsonii
A.capensis
A.carnica
A.carpatica
A.caucasia
A.cerana
A.dorsata
A.florea
A.m.fasciata
A.m.intermisssa
A.m.jemenitica
A.m. lamarckii
A.m.lehzeni
A.m.lugistica
A.m.major litorea
A.m.meda
A.m.mellifera
A.m.monticola
A.m. nigritum
A.m.rubica
A.m.scutellata
A.m.syriaca
A.m.unicolor
A.m.varieties
Banat bees

# Royal jelly

acetylcholine
brood food
brood food glands

# 34 - Recipes - Remedies

autumn feeding
bee food
beer
borax
broof
burn ointment
car polish
cerate
cleansing cream
Cleopatra's facial balm
cold cream
cough mixture
cream of tartar
floor polish
fudge
furniture cream
furniture polish
Good candy
grate
hair cream
hand cream
Haydak's formula
hay fever
honey aroma
honey baked apples
honey beer
honey bread
honey butter
honey cakes
honey cappings
honey cream
honey creamed
honey drinks
honey fermentation
honey fudge
honey hand cream
honey ice cream
honey mask
honey mint sauce
honey recipes
honey seeding
honey skin treatment
honey toffee
honey valve
honey vinegar
honey whipped
hydromel
icing sugar
krupnik
leather polish
lee (lees)
lipsalve
making candy
mead
mead making
marmalade
meer
meloja
melomel
metheglin
morat
mother of vinegar
mulberry (morat)
mulsum
must
nourishing cream
ointments
oxymel
paraffin wax
Pharmacia
Pharmalgen
propolis gargle
queen cage candy
racking
Rgus candy
recipes
saltpeter
Scholtz
scrubbing solution
shampoo

shaving cream
shoe polish
silicone
skin lotion
soda
soft soap
sore throat medicine
stearin
sting antidote
strawberry jam
Syriac Book of Medicines
TAFSP
tallow
tartaric acid
terebra
thymol
anguentum
vinegar
water-proofing canvass
water-proofing shoes

# 35 - Skies Environment & Calendar

Agricultural changes
anti-cyclone
appreciation of colour
Artemis
autumn
bearing
benty
biome
biosphere
biotic
blooming times
BOD
bog
break crop
brickearth
bright
brightness
British black bee
Bunter sandstone
bush
chalk
chalk and oolite
clay
climacterial number
climate
coal bearing rocks
coast
cold
Columella (Arcturus)
common
condition
coppice
cretaceous
cross wind\cutting
cyclone
dampness
darkness
day
dew
dextrorotatory
Didymus
Downs
drainage
drift
dust
Ecosphere
ecosystem
ecotone
edaphic
elements
environment
Epochs
Erica
essentials
excrement
fauna
Feb
feral
fir
fire
fixation
flashlight (red light)
flicker
flight path
float
flood
flora
flow
flowering plants
flowers per tree
fluid
flying
forest
frost
funeral
gales
garden
garth
gault
geology

glade
grain (wood)
grassland
grass verge
gravity
Green belt
greensand
ground
ground speed
GRP
habitat
half-life
hard water
heath
heathen
heather stance
hedge
hedge trimming
hippocras
Hippocrates
hive density
hive feet
hive site
hives under trees
homeostasis
homing instinct
honey forest
honey harvest
horizontal
humidity
hygrometer
ideal weather conditions
incline
increasing forage
Indian summer
industry
inversion
ion
iron pan
isohyet
isophene
isopleth
January
July
June
Keuper marl
keuper sandstone
lake
late
lawn
leaching
lee
length of day
length of life
length of active season
length of winter
level
lias
light
light compass reaction
limestone
lithosphere
loam
local
location
loess
lore
lower lias
luminescence
magnetite
March
maritime
market garden
marl
marsh
mass
massacre of drone
May
mb
meadow
meeting place
meteorology
micro climate
middle 'c'
mid-summer
milk and honey
Millstone grit
mineral
moisture
molecular attraction
mono cropping
monsoon
mood
moon
moor
motor ways
mountain
mud
muralis
museum

# 36 - Stings

adrenaline
allergen
allergy
anaphylactic shock
anti body
antidote
antidote for bee stings
antigen
anti-histamine
apimine
Apitoxin
Apium virus
atopy
asthma
avoidance of
barb
bee gloves
bee stings
bee venom
bee venom activity
bulb of sting shaft
central nervous system
components
dart
desentiziation
Dimethyl chloride
Diphen hydramine (Benadryl)
Drenamist
effects of stings
excrutiating
furcular
gonapophyses
histamine
hyalbromidaze
immunize
immunity
immunoglobin
immunotherapy
infuriate
iso-amyl acetate
isoprenaline
lancets
leucocytes
mandibular gland
mast cell
melittin
neo-epinine
nerve trunk
nip
oblong plate
ointments
onion
pain
parameral plates
peptide 101
Pharmacia (Ltd)
Pharmalgen
phospholipase
Pleiades
poison gland
poison sac
pratensis
prick
proctiger
provoke
persuer
queen – sting of
quadrate plate
ramus
releaser
rheumatism
safety
sense of danger
sex differences
sneeze
spiteful
sting
sting wound
stylet
subjugation by smoke

swelling
tablet
temper
touch
triangular
trigger off
turgid
tweezers
umbella valves
unscre
urticaria
venom
venom gland
watch strap
WBE
weal
weapon
whole bee venom

# 37 - Swarming

abscond
absconding swarm
acorn cup
after swarm
anchoring effect
anebalic
Apidictor
artificial swarm
bait combs
beard
bee bob
bees' choice of site
bee trap
cast
cluster frames
dead queen
departure
dislodge
dry swarm
elusive
emergence
emergency swarming
fliers
foundation
Gerstung foundation press
hive baited
hive –to
hiving
hiving a swarm
impulse swarming
maiden swarm
mating swarm
memory
middle 'c'
multi queen casts
natural swarm queen cells
nest cavity
old queens
Pagden
pheromone (clustering)
preparation for swarming
prime
prime swarm
Schwirrlauf
secateurs
shook swarm
silver swarm
Snelgrove board
spotting
squib
starvation swarm
supersedure swarm
swarm
swarm box
swarm catcher
swarm cluster
swarm control
swarm names
Swarm of bees in May
swarmed
swarming fever
swarming habit
swarming impulse
swarming preparations
swarming season
swarm lure
swarm management
swarm prevention
take
take off
taking a swarm
throw
truant swarm
upside down
vitality
weekly inspection

# 38 - Wax & Oddments

acetobacter
bee escape
beekeeping in towns
beer
bee veil
beeswax candle
dark honey (honey shows)
dubbin
encaustic
face lotion
face pack
flame test
floor polish
Gayre (mead)
propolis (See various entries)
Sale of honey in UK
Salicaceae
Various waxes (under 'wax')

# A

**a**
Used as a prefix to mean without or not acellular (non cellular), amorphous etc. Before many words this prefix is used to imply 'the opposite for example atypical (not typical). Derived from Greek 'without' – 'not'

**abdomen** ***
The third and hindmost of the three parts of a honeybee. A short 'petiole' connects its blunt forward end to the thorax. The plates of its nine segments overlap allowing for expansion and contraction as when breathing or for the enlargement of honey crop, rectum or ovaries. Ventrally the worker has four pairs of wax glands, pockets and mirrors and a dorsal scent gland on the penultimate segment. Its shape and alignment with the head and thorax lends itself to an aerodynamic flying shape. The queen's abdomen accommodates the enlarged ovarioles and extends to nearly twice its original size. The abdomen of drone and queen enclose their respective genitalia.
See: gaster and abdominal contents

**abdominal contents**
The abdomen is more flexible and variable in size than other parts of the body. Two extendable portions are the honey sac and the rectum. The circulation of haemolymph and of air and gases are powered from the abdomen. In females the sting apparatus is 'in situ' within the abdominal tip unless extended, likewise the male reproductive organs in the drones. The queen's abdomen as well as permitting fecundation can become greatly distended to accommodate enormous numbers of eggs. Twin nerve cords lie beneath the ventral diaphragm while the tubular heart is in a corresponding position over the dorsal diaphragm. Also floating in the abdominal cavity are the emunctory malpighian tubules while the digestive system extends from the stomach valve where it joins the honey sac to the anus in the proctiger. There are also four pairs of ventrally placed wax glands with associated mirrors and the dorsal scent gland.

**Abdominal glands**
The ventricular epithelium is composed of proliferating digestive cells which are of a glandular nature becoming detached, mixing within the food mass and yielding digestive enzymes. The dorsal 'scent gland' or Nasonov gland produces a 'pheromone' which is an attractant.
The four pairs of ventral wax glands secrete wax which oozes out onto the wax mirrors forming platelets.
See: acid, alkaline, Nasonov, scent, sting and wax glands.

**abdominal muscles**
Retractor muscles pull abdominal segments together. Extension is produced by short protractor muscles that arise on projecting lobes on the front margins of the terga and sterna and posteriorly on the overlapping margins of preceding plates. By contraction they shorten the overlap of plates pushing segments apart. Compressors are crossed muscles that draw tergum and sternum together vertically by pulling the edges of the overlapping plates together. The pulsations of the abdomen are transmitted by blood pressure to the air sacs and so assist respiration.
See: emunctory and proctiger

**Aaron's rod** ***Verbascum thrapsus*** **(Scrophulariaceae)** ***

Giant mullein. A hardy biennial that grows to 1.5m and flowers in late summer. Not much of a nectar plant but bees are attracted to it for pollen. A striking plant with silver, downy leaves that stands up stark and conspicuous when it attains its full height. *V. nigrum* (black mullein) and *V. thrapsus.*

**abandon**

To desert or forsake utterly. Usually bees have so much wealth as a result of their labours: waxen honeycomb, a home, stores of pollen and honey and for much of the year, brood that they seldom abandon their nest unless driven away and this especially applies when they have brood. However bees do sometimes abscond, for instance queenless bees may join a queenright colony, or starving bees may leave home en masse. Could we include CCD – colony disappearance disorder?

**Abbott**

Responsible for self-spacing top bars (Victorian)

**Abbott, Charles Nash**

Founded BBJ in May 1873 the first British bee periodical. Grandfather of C. P. Abbott – film distributor, Southall, Middx, author of '*Queen Breeding for Amateurs*' 1947. BBKA expert and Examiner.
See: Abbott Laboratories

**Abbott Laboratories Ltd**

Queenborough, KENT ME11 SEL
One of their products is *Fumidil B* As this us only required for the treatment of honeybee colonies (it inhibits the vegetative development of Nosema within the bee) and few beekeepers buy it, the cost in this country has steadily risen. It has been available at lower prices abroad.

***ABC and XYZ of Bee Culture***

Author A.I.Root, Publisher A.I.Root Co, Medina, Ohio, U.S.A. An encyclopaedia pertaining to scientific and practical culture of bees. Written in 1877 this splendid book has been revised and new editions published many times and must be one of the most widely read bee publications.

**ABD**

An attempt to remove the word 'foul' from brood disease otherwise called American Foul Brood which disease occurs all over the world and has been known since Roman times

***A bee is born***

Book by H. Doering

**abeille**

French for 'bee' (f) *Les Abeilles* – a technical film for beekeepers 1957 18 min sound – colour 16mm Cr. Jean Painleve. A study of different kinds of bees
Distributors: French Embassy Film Service

**ABF**

American Beekeeping Federation

**abietic acid**

$C_{19}H_{29}CO_0H$ and $C_{20}H_{30}O_2$ A yellow crystalline acid and an important ingredient of propolis that may be associated with its anti-biotic properties. It is a carboxylic acid and can be obtained from resin and is a derivative from phenanthrene and contains two unsaturated groups. In 1965 Prof. Reny Chauvin discovered 53% in propolis. It is used in driers, soaps varnishes and is derived from a species of pine.

**A.B.J.**

American Bee Journal. It claims to be the oldest bee magazine printed in the English language founded 1861
See: American Bee Journal

***About Bees and Honey***
Author: Althea Braithwaite Pub. National Trust by Dinosaur Publications Ltd
32pp 8 ¾ x 5 ¾ numerous colour illustrations – *Bumbles, Solitary & Honeybees*

**About Honey**
Author P.E. Norris Nature's elixir for Health & Energy-(Bee Books New Old)

**About Pollen**
Author J.Binding from BBNO

**ABPV**
Acute bee paralysis virus. Dr. Bailey has found that this virus has 'x' and 'y' strains, the former having a hexagonal crystalline structure while the 'y' strain has a larger hexagonal structure. Infection seems to reach a summer peak.
See: 'Infectious Diseases of the Honeybee', 'Bee World' and recent discoveries relating to apparent linkages between virus, other diseases and also paralysis.

**abrade**
To wear or scrape off. Dr. E. Margaret Adey refers to the significance of insect pollination even in many self-fertile species when their visitations for pollen abrade the stigma, thus rupturing its membrane and enabling pollen tubes to penetrate in the same species this is more likely to lead to fertilisation than the actions of more gentle agencies, such as gravity, water or wind.
See: hysteresis

**abscond**
To make a sudden departure. Western honeybees replace their queen by swarming or supersedure but a colony with a nest and brood rarely absconds. A cast or swarm having left the parent colony may abscond from a newly offered home if it is not found to be acceptable. Hunger, vibration, smells, interference and other abnormal influences will occasionally cause a colony to vacate its nest (or hive) When the colony leaves en masse sometimes leaving brood behind, it is referred to as an absconding swarm.

**absconding swarm**
Some races such as florea, dorsata and cerana seasonally depart from their nest before the onset of poor weather (monsoons). However the same characteristic is found in A. mellifera when all the adult bees and the queen if one were present, clean up the hive and go. Some circumstances may compel a colony o leave food behind, as for instance when interfered with by an evil-smelling creature like a shrew. Environmental conditions such as pollution by ash (Mt St. Helena U.S.A.1981) caused colonies to exhibit a strong tendency to swarm. Usually any food present is cleaned out of the combs a for example when a queenless colony joins one that is queenright or a fickle cast leaves on a mating flight. Hunger swarms or desperation swarms may occur due to lack of food, robbing, smell, vibration or other environmental impositions. Such swarms differ from reproductive swarms in that few workers and no adult or viable immature queens are left behind in the original colony.
See: hunger and mating swarms

**absolute alcohol**
See: alcohol

**absorption**
The digestive process which involves the passage of nutritive material through living cells into the protoplasm.
The sucking up of minerals and water by the root hairs of plants.
The drawing in of liquid by a porous substance, e.g. the use of a wick for feeding or used as a pad for any volatile prophylactic fluid.
See: assimilation

**Abushady A.Z.**
An Egyptian student an Oxfordshire who started an international organisation the Apis Club with Bee World as its magazine.

**Acacia spp.** ***(*mimosa*) A spectacular

early flowering shrub in the south west but although the pollen may be useful, little is heard of its nectar or honeydew. It is sometimes confused with *Robinia pseudoacacia* (false acacia). Mimosa is a common plant in many warmer regions such as: Mediterranean and Australia.
See: mimosa, wattle

**academic**

Relating to higher education and the establishments of learning associated therewith.

**Acapulco Miel**

Founded in Mexico in 1960 with the object of establishing 25,000 colonies. A visit during the 1981 Apimondia Congress, showed a business with 11,000 hives scattered over almost 100 miles of countryside.

**Acarapis spp.** Schneider, Morganthaler *et al*

Acari that live externally on the honeybee. There is some possibility of confusion between A.dorsalis externus (longer hind-legs) and vagans. The latter are however thought to live near the hind wing roots but they do not seem to be common. Dorsalis spend their lives in the v-shaped groove between the mesoscutum and the mesoscutellum (Morgenthaler). Externus are found on the veins of the bee's hind wings and on the surface of the forewing, propodeum and first abdominal segment of the worker. On drones they may be found all over the body.

***Acarapis woodi* (Rennie)** ***

A small, parasitic mite that lodges and breeds within the first pair of thoracic tracheae. A migrant adult climbs to the end of a hair and awaits its chance to transfer to another bee. To survive it must find a young bee, less than five days old, and it may well be guided to the appropriate spiracles by the breathing of a young bee. Once the fringe of hair on the 'lobe' that covers the first pair of spiracles harden, the mite can no longer gain entrance. They pierce the tracheal lining and in feeding on the haemolymph can cause septicaemia but in any case block the air-flow through these tracheae which are all-important for flight. Such non-flyers are called 'crawlers'. Dr. Bailey refer to this pathogen as enzootic and opines that they are not likely to cause serious disease symptoms unless associated with a virus such as CBPV.
See: *Acarapis woodi* (life cycle)

***Acarapis woodi*** (life cycle)

An adult seeks a younger bee (more easily found in the spring and early summer) under twelve days old and more likely under 8 days. 2-5 eggs, almost as large as the adult, are laid. The eggs become larvae which feed for 3-6 days and become nymphs for 1-4 days, the females taking longer than the males. Two cycles are possible before bees become foragers – if they are fit enough to do this. Swarms can carry infested bees. The brood and stores, therefore in the combs, are not affected, only live or very recently dead bees can transmit the mite.

**Acari Arthropoda Arachnida**

Sub-class or order containing Mites and Ticks Gk. akar(i) – mite
See: *Acarapis woodi, A.dorsalis, A.externus* and *A.vagans*

**Acaricide**

Preparations formulated to exterminate the mite *Acarapis woodii*
See: Frow, Folbex, Phenothiazine and Italian workers have found tetrachlorodiphenyl sulphone reasonably effective – Frediani D Pisa.

**Acarine disease**

Colony weakness can develop as a result of the disease and associated disorders that are caused by the mite Acarapis woodi entering and breeding inside the first pair of thoracic tracheae*. Bees infested become crawlers, climbing grass stems and any other aid in vain attempts to become airborne. This is one of the first diseases to be suspected

when large numbers of bees look poorly in an otherwise clean and well-found colony (occasionally pesticides may produce similar symptoms. The disease is of course found in conjunction with other diseases. (the mite acting as a vector?). A colony can survive while being affected by several diseases at the same time. Virus diseases, nosema, acarine, and foul brood have all been found in a single colony that still clung to life. AL's 330 AL 362. *** The tracheae (sometimes unilaterally) become stained ranging from a 'bubbly' appearance, copper-coloured and jet-black.
See: '*Acarapis woodi*', 'A.w (life cycle)', 'Frow', 'Folbex', 'acaricide'

**ACAS Agricultural chemicals approved scheme**
See: PSPS

**acceptance**
The acceptance by a colony of a new queen, that is to say a new queen has been allowed to take over the role of mother of the hive and becomes the only laying queen in the colony. Various methods of queen introduction are used to encourage complete acceptance. Partial acceptance is common and often induces supersedure.
See: partial acceptance

**accessory glands**
These are large club-shaped sacs in the drone's genitalia which fill with mucous and increase in size as a drone matures. They are U-shaped and joined at their bases where the seminal vesicles are connected.

**accident**
A mishap, undesirable or unfortunate happening. These do occur in beekeeping and hints of subsequent behaviour would fill a large tome. Food spilled or discovered by the bees may be substituted by a saucer of syrup which, when cleaned up, results in a 'stop visiting' signal (pheromone) to the colonies) involved. The escape of bees from a nucleus, observation hive etc., during transport may necessitate leaving the hive where it is until nightfall. A leak (via an unsuspected hole or crack) could be plugged with foam rubber, gunge or whatever' at once.– he that fears the bee stings is not worthy of the honeycomb

**acclimatise**
To become habituated to a new climate. Workers live for too short a time to appreciate climate. One might emerge, fulfil its role in life and never experience rain (or sunshine). Any change such as a new colony rhythm of brooding and peak development is likely to come from the longer-lived queen. Little is known about the ability of queens to acclimatize but it is assumed that this must take a generation or two at least.
See: adaption

**accommodation**
1. Focusing the eye by changes in shape of the crystalline lens to suit various distances so that a sharp image falls onto the retina. Bees cannot do this as the lenses of their eyes are of fixed shape.
2. The trend towards a negative response as a result of continuous stimulation.
3. The place provided for a particular activity.

**accommodation cells**
Also known as interstitial cells these are cells in the comb which do not conform to the usual hexagonal pattern (and are therefore not used for brood) but are either intermediate between drone and worker cells or have three, four or five sides. The cells adjoining a flat surface will usually be in this category. They can be used for storage hence the name 'accommodation cells'.
See: interstitial

**accompany**
1. A queen sent by mail is normally accompanied by about a dozen workers. These do more than just accompany her as she is dependent on them for glandular secretions which form her normal food

and without which her condition would decline. With healthy nurse bees she is kept in a reasonably fit condition. However if the accompanying workers (attendant bees) die, she will not out-live them.
2. Parasites such as certain mites may also be said to 'accompany' their host. Bees do not accompany one another to locate forage but will travel together as a swarm.

See: dance

**accustom**

To get bees used to a certain familiar behaviour such as the use of a watering place or the acceptance of a new queen. Approach to their entrances can be changed within limits (e.g. long tubes leading into an observation hive) but one is limited to a few feet when it comes to moving a hive.
See: moving bees

**acea (-aceae)**

A suffix used after the Latin names of classes and orders of animals and plants - Crustacea, Rosaceae etc.

**Acer spp. (Aceraceae)**

These include trees and shrubs such as Field Maple *A.campestre*, Norway Maple *A.platanoides*, *Sycamore pseudoplatanus*. They are monoecious – fruit a pair of keys with elongated wings.
See: sycamore

**acetic acid $CH_3COOH$**

A colourless liquid which when at its maximum concentration is known as 'glacial' acetic acid which solidifies at 10C and is the acid constituent of vinegar. AL 473 gives details of its use for dealing with combs (even when stores are present) that have been infected by *Nosema apis*. A Russian suggestion was that O.3 ml per kg of syrup fed to the bees at the end of August helps overwintering and accelerates spring build-up. Sya Boldyrev USSR Pchelovodstvo 1977. Its fumes are heavier than air and concentrated forms should not be allowed to touch the skin and being corrosive it attacks metal ends, runners (and embedded wires) and even cement floors.

**acetobacter** (*A.acetu*)

This is the agent that transforms ethyl alcohol into acetic acid and water, working most efficiently in aerobic conditions
See: aerobic, honey vinegar and mother of vinegar

**acetone** $(CH_3)_2CO$

A colourless, volatile, flammable liquid, useful as a solvent. Also used in smokeless powders and varnishes. Harmful to bees though it soon evaporates. Always allow queen marking paints to dry and keep her away from bees until any smell has worn off as the smell of acetates infuriates bees.

**acetylcholine** $C_7H_{17}O_3N$

An alkaline organic compound which is a nerve 'transmitter' activated when required by an enzyme. It is present in the nerve synapses enabling nerve transmissions to occur. It is a vitamin of the B group which together with pantothenic acid is present in royal jelly – Dade. Its administration can cause parasympathetic stimulation. Vines &Rees
See: adrenalin, botulism and organophosphorus.

**acetylandromedol**

See: andromedotoxin

**Acherontia atropos**

See: Death's head hawk moth

***Achroia grisella*** (Fabricus)

A drab, light-fawn coloured moth with a 1.5 cm wing span (158mm). It is very common in the U.K. and will cause catastrophic damage to combs that are not protected by bees or fumigants. Eggs laid late in the season may not mature until the spring and the female will only lay in the dark. Eggs will not withstand a hard frost.

The grub feigns death when disturbed. Webbed tunnels filled with droppings are a clear sign of the larva's ravages. Adult female moths enter hives after dark – they have little trouble passing guards as they can run faster than bees.
See: *Galleria mellonella.*

**acid gland**
See: venom gland.

**acidity**
The degree to which a substance is acid. pH less than 7.0. Although the bee has an alkaline gland to form a constituent of 'venom' a bee's gut, honey, mead, royal jelly, sting venom are all acidic as of course is honey vinegar'
See: beeswax acid value

**acids**
Information concerning aids of interest to beekeepers include; abietic, acetic, amino, carbolic, citric, erucic, folic, formic, geranic, gluconic, honey acids, lactic, linolenic, malic, nerolic, nicotinic, nucleic, oxalic, linolenic, malic, nerolic, nicotinic, nucleic, oxalic, pyrogallic, ribonucleic, succinic, tartaric.

**acid value**
The number of milligrams of N/10 potassium hydroxide required to neutralize the free fatty acids in one gram of a fat, oil or resin. Also: 'acid number' Wedmore gives 16.5 to 22 for beeswax. To check acidity dissolve 24 grains of wax in 1oz warm alcohol, filter, add equal quantity of water: blue litmus should redden in 15 minutes.

**Acinus** ***pl. acini***
Sac-like bulges at the end of gland tubules in particular the translucent, post cerebral glands where they are arranged in small bunches on a branching system of tubules.

**Acceton (ancient)**
See: virgin honey.

**Aconite – winter** ***Eranthis hyemalis***
**Ranunculaceae** ***
A perennial with yellow flowers between January and April. 15-20cm tall. Bees work them for the yellow pollen and the honey, when obtained it is of a pale shade.

**acorn cup** ***
This is a descriptive term for the small inverted cups which are built as the swarming season approaches. They become significant and indicative of a possible swarm seven days after an egg appears in them.
See: play cell and queencup.

**Advisory Committee on Pesticides**
See: ACAS, PSPS and BAA.

**acquired characteristics**
These are features which become apparent during a living organisms development and it is not considered possible for them to be passed from parent to offspring. However there is evidence to cast doubt on that opinion. Honeybee colonies provide, due to their longevity, opportunities to see this.
See: genotype and phenotype

**acre**
A unit of land measure common before decimalisation. Equal to 0.40469ha.
A square with sides ten 'chains' long or 4840 square yards. Its wide use in the realm of agriculture may well cause it to last into posterity. Acreage, God's half-acre and a dozen derivations spring to mind.
See: foraging area and hectare

**Act**
A law produced as a statute by Parliament. In the 80's the threat of Varroa has led to the passing of a new Act regarding the importation of bees.
See: Bees Act

**Acts & Regulations**
Regarding the laws applying to the sale and supply of honey produced in the U.K. Food & Drugs Act 1955. The labelling of Food

Regulations. 1975 amended The Honey Regulations 1976 The Materials and Articles in contact with Food Regulations 1978. Weights and Measures Acts 1963 -1979 The Weights & Measures (Marking of Goods and Abbreviations of units regulations 1975 as amended. The Weights & Measures (Marking of Goods & Abbreviations of Units Regulations 1975 as amended. The Weights and Measures Act 1963 (Honey) Order 1976. Trade Descriptions Acts 1968 &1972 The Trade Description (Indications of Origin) (Exemption No.1) Directions 1972 Glazed Ceramic Ware (Safety) Regulations 1975. Consumer Safety Act 1978, Trading Standards Dept., Devon County Council March 1980.

**activate**

To initiate or hasten a reaction or to make a thing more active.
In starting mead off this implies arranging conditions of sweetness and temperature to the degree at which the yeast is activated. Also called 'proving' the yeast. A prophylactic may need to be activated as for example the lighting of a Folbex strip.
See: catalyst.

**activation hormone**

Produced by the neurosecretory cells of the bee's brain. Influences moulting and pupation of larvae but some contend that it performs the overall regulation of the endocrine system.

**activity**

A vigorous quality of action which may be applied to bees or beekeeper. At the hive entrance, various conditions of light and temperature coupled with the presence or absence of a honey flow, or other cause of activity, may be described as fitful, considerable or even furious. These signs give us much information as to what is likely to be happening inside the hives, whether robbing is in progress, whether the sky has darkened and rain is imminent and so on.

**Aculeata**

This sub-order of Hymenoptera includes bee, wasps and ants, many of which are notorious for their ability to sting. The larva of aculeate Hymenoptera are quite helpless and are supplied with food either by their female parent or in the case of some social insects, by their adult sisters
They include 531 spp. bees, Apoidea, ants Formicoidea and Vespoidea and Sphecoidea. Numerous families of wasps are characterized by their stinging adults.

**aculeate**

Having sharp points, prickles or a sting.

**Acute bee paralysis virus**

See: ABPV

**Adam Bro. OSB, OBE**

Wrote: *"In Search of the Best Strains of Bee"* 1968, and *"Beekeeping at Buckfast Abbey"*, 1975 At Buckfast since 1910 and took over from Bro Columban in 1915. An apiarist of world renown, a notable bee breeder and investigator of honeybee races. His research led to the production of the 'Buckfast strain' which was in demand all over the world.
It was propagated in Israel and at Weaver apiaries in U.S.A. Buckfast Abbey was the home of this great man who has scoured Europe and beyond in search of bees with the best characteristics. A very good mead was made at the Abbey which had access to several good out-apiary sites on the heather.
See: Books mentioned and *'Breeding the Honeybee'* A Central Assn. book and '1000 years of Devon Beekeeping' by Ron Brown.

**Adams tensioner**

A device invented by Mr. Adam of E.H.Thornes for tightening straps for transportation of hives.

**Adansonii** *–Apis mellifera adansonii*

A sub-species of the honeybee, indigenous to S. Africa. Considered to be more aggressive than other honey bees but suited to a tropical habitat. A hard working bee. It is

inclined to be swarmy and at times of nectar shortage, absconding swarms are likely.
A worker takes only 19 days to develop as opposed to 21 days with other honeybees.
It is likely that this was the type of bee that escaped and caused such a commotion in the U.S.A (e.g. 'killer bee').
See: *Apis mellifera scutellata*, African/ Brazilian and Africanised bee.

**adaptation**
The process which an organism passes through to fit it to its environment. This may be short or long term and natural selection coupled with mutation can result in phenotypes which resemble one another in certain respects while having different basic 'genes'.
See: acclimatise and accommodation

**ADAS**
Department of what was MAFF – Agricultural development and advisory service.
It has responsibility for advice, development and investigation. 'Cut-backs' in government spending in 70's-80's did not help though there has been increased streamlining and efficiency to make the best uses of resources available. ADAS is frequently referred to instead of MAFF in documents and literature.
See: Luddington.

**ADAS Publications**
Advisory leaflets, short term leaflets, Bulletins, films and other visual aids
At the date of writing this the unit concerned was MAFF Publications Unit, Lion House. Willowburn Trading Estate, Alnwick, Northumberland.

**addled brood**
Brood of any age that fails to develop normally due in most cases to a genetic fault (as with much inbred queens). Not a particularly serious condition in itself but likely to cause all but the expert to consider that more serious diseases might be present. Neglected brood – as when the balance of a colony is upset because of prolonged queenlessness or for other reasons -. Treatment consists of requeening.

**addled egg**
An unsound egg, an egg that fails to hatch or does not develop normally.
This could be caused by a genetic fault due to the queen (such as the wrong ploidy) or to environmental circumstances such as the presence of bacteria, chilling or other pathogens may also be responsible. Such eggs are normally removed, possibly they are eaten by the bees. Their absence gives rise either to empty cells that are surrounded by brood or a young occupant that is surrounded by older brood. Patchy brood usually calls for the replacement of the queen.

***adenosine triphosphate* ATP**
It is present in all living cells and presents a store from which energy is released in the bee as it oxidizes the sugar in honey in its muscle and other tissues producing warmth and energy. Glucose in the muscle undergoes anaerobic glycolysis forming the energy-rich ATP.
See: catabolism

**adhesive**
A suitable type for hives under full exterior conditions is Aerodux599 + hardener501 a resorcinol-phenol-formaldehyde formulation which should last upwards of 25 years.
Best results obtained from thoroughly dry timber. The adhesive can be mixed with sawdust to fill cracks and splits.
(poisonous). (Check latest)
See: glue

**adipose tissue**
Animal tissue where fat is located.
See: fat body

**admix**

Mixing that implies mingling as when bees from different colonies are intermingled. A dusting with flour or icing sugar may help.

**adrenal**

Situated near kidneys
See: malpighian.

**adrenaline $C_9H_{13}O_3N$**

A hormone produced by the kidneys Pre-charged syringes of adrenalin for self-administration are available from International Medication Systems (UK) Ltd. Vital in the treatment of a person who has been rendered unconscious through allergy to bee stings. Its administration causes stimulation of the sympathetic nervous system speeding heart action and contracting blood vessels. Approved Prescriptions Services Ltd, Whitecliff Road, Cleckheaton, West Yorkshire, BD19 3BZ. Several proprietary forms now available.

**adult**

A fully developed bee, one that has gone through all stages of metamorphosis and has emerged from its cell. An imago. Adults are not fully functional when they emerge. Queens require a day or two to fully mature, drones almost a fortnight, while the workers, not being capable of sexual maturity, simply need a day or two to harden their wings and stings for the efficient subsequent use of these organs.

**adult bee disease**

Diseases that affect the adults of a honeybee colony. These include Nosema, Acarine and Amoeba while the more recent Varroasis affects both adults and brood. For convenience honeybee diseases are put into two groups: Brood and Adult bee diseases. In Britain these are dealt with by (now FERA) or the ADAS National Beekeeping Unit when the numbers are large or when there is no CBL (County Beekeeping Lecturer – a post long since brought to a close), help can be might be obtained from FERA at National Bee Unit Room 02F19 Sand Hutton, York. YO41 1XF or perhaps Agricultural Colleges
(Bring this up-to-date)
See: Adult bee disease diagnosis and ADAS

**adult bee disease diagnosis**

Since so many cuts have been made in these services the latest situation is likely that these notes may well be totally out-of-date. Such diagnosis however entails the microscopic examination of adult bees (the larger the sample the more accurate the diagnosis though 30 bees in a match box is enough. Information sent could include: were they collected dead or alive. How many seams of bees in the colony? Combs of brood, age of queen, fighting or robbing, amount of food available, recently given combs/foundation/equipment, were they treated or straight from store, newly purchased? Location of site, recent weather conditions. Photographs can be useful too.
See: adult bee disease, CBL (FERA?)

**adulteration of honey**

Cheaper sugars, other additives or the contamination or overheating of honey can lead to its adulteration. Standard tests for HMF and diastase, are principally aimed at preventing overheated honey from being imported. Cane, beet, invert, corn (maize) and isomerose syrups will all blend into honey and often require laboratory tests to determine their presence. Natural honey, though mainly a glucose/fructose solution, can have almost two hundred identifiable ingredients. Pollen identification alone can pin its origin down to a small area and other chemical, physical and electrical conductivity tests enable us to check a honey's purity.
See: finger print, bio-assay, honey adulteration and honey analysis

**adulteration of wax**

Beeswax is a unique substance but it will readily absorb aromas, many toxins and other waxes including propolis. Once combined there is no easy way of separating

and isolating the pure wax. In the form of foundation many formulations and arrangements have been made, ostensibly to strengthen or reduce the price. As pure wax is most acceptable to the bees and has a high market value, any move that might impair its quality is to be deprecated.
See: tests for purity

**adult worker**

After emerging for 2-3 days they clean themselves, take food from older bees, remain on brood combs and clean cells. Then from 4th to 6th day they begin to function as nurses feeding older larvae with pollen-honey mixture which in turn stimulates their own brood food glands. From 6 to 9 days they proceed to feed younger larvae and perhaps the queen and during this period will begin to take orientation (play flights) at about 7 days. From day 12 -18 as house bees they become involved in cleaning, ventilating and comb building, then 18-20 days before becoming foragers they move nearer the entrance to unload returning foragers or to act as guards and become recruited to out-of-hive activities It is interesting to note that a worker will have done much useful work before facing the risk of losing life as a guard bee.
See: division of labour

**Advanced bee culture (book)**

Author: Hutchinson W.Z 1911 NBB.

**Advantages of two brood boxes**

1. Reduces swarming by allowing plenty of room.
2. Queen cell check by splitting a lifting upper chamber.
3. Maximum population for honey flow.
4. Demaree and other swarm preventative measures possible.
5. Queen rearing under supersedure impulse
6. Crowd for sections by removing one box.
7. Easy to spare a comb of brood for weaker colony.
8. Increase can be made during flow without reducing surplus.
9. Adequate stores for successful wintering.
10. Stores well positioned for wintering.
11. Enables beekeeper to provide a 'food' chamber.

***The Advantages of the 'House-Apiary'***

Booklet by J. Spiller published Taunton 1952.

**adversary**

Opponent – not man against bee though the bees may see things differently.
See: enemies

**Advice to Intending Beekeepers**

AL 283 and DARG'S Bee-way code

**Advisory leaflet A.L.**

A pamphlet produced and distributed free by MAFF (dated organization) who utilized the services of a small group of experienced beekeepers. They are updated, improved and re-printed from time to time and give helpful advice to beekeepers though some rather unhelpful comments have been included in the past as to the best use of bees for pollination Each leaflet has a title and number as for example 'acarine' A.L.330.
See: other numbers under A.L.

**A.E.A.**

See: Apicultural Education Association

**Aebi, Ormond and Harry**

Wrote '*The art and adventure of beekeeping*'

**Aedeagus (aedoeagus)**

The male intromittent organ in which the semen and mucus are transiently stored during copulation. In *A.mellifera* drones the bulbous portion of the aedeagus has a thick lining bearing large scelerites. This lining becomes detached and often lodges in the queen's sting chamber and gives rise to what beekeepers have observed and called the 'mating sign' (nick-named ' the wedding

certificate'
See: mating sign

**Aelian Claudius Celianus AD200 – 235**
A priest and Encyclopaedist referred to by Dr. F. Malcolm Fraser as a writer about bees. He wrote in Greek with a flavour of Herodotus* & Xenophon and he tended to humanize bees, using them to point out morals. He reported that honey from Pontus cured epilepsy.
*some of whose ideas he refuted.

**aenomel**
Mulsum or honeyed wine used by the Romans.

**aerobic**
Requiring free oxygen in order to live.

***Aeroglyphus robustus***
A pest of stored grain in Canada and northern U.S. reported in all stages in a queen cell containing a dead queen. Dec 1981 A.B.J.
See: grain itch mite, Mediterranean flour moth.

**aerosol**
A container or system holding minute colloidal particles in suspension, usually less than one micron is size dispersed in a gas. By the 70's types had become ubiquitous. The main danger being the release of halogenated fluorocarbons which are thought to rise into the upper atmosphere and upset the earth's receipt of radiation from the sun and favouring more u/v which is harmful. Used for paint, in smoke bombs for the beekeeper and the fine application of other types of chemical mist. Wax release agents such as silicones may also be obtained from aerosols.
See: smoke bomb

**Aerosol bomb**
An aerosol of insecticide as used for agricultural purposes.

**aerotaxis**
The movement of certain bacteria and micro-organisms towards or away from oxygen.

***Aeshna grandis L.***
A dragonfly that is reported to have taken a large toll of honey bees in Germany and to have been so numerous as to have given trees a 'glassy' appearance. Bees have been reported as having learned to remain in their hives when Epiaeschna heros were flying so strongly as to darken the skies in Florida. Other dragon flies said to eat bees include: *Anax junious* Drury – the American dragon fly: *Coryphaeschna ingens Rambur* the wood flying adder. *Cordulegaster diastatops Selys* and the damsel fly *Aeschna cyanea* Muller from R.A. Morse

**Aesculapius**
The god of medicine and healing who held the snake as sacred and as an emblem of health and recovery. The snakes were fed on honey.

***Aesculus hippocastanum***
See: horse chestnut

***Aethina tumida* (Small hive beetle)**
Occasionally found in beehives and no remedies can be quoted. Reported as having become more common (2011).

**Aetiology**
This is the study of causation, as in the realm of pathology. The science of causes.

**AFB (ABD)**
See: American Foul Brood.

**AFB symptoms**
In the earliest stages only a cell containing the ropey, larval remains gives away the fact that the disease is present. Our highly trained appointed officers have checked many an outbreak by spotting the disease at this stage. As it develops though, a 'pepper-pot' brood pattern i.e. holes or dead larvae

amongst sealed brood, will be seen, the colony itself becoming weak and unable to reach its potential. It will become short of adult bees and later when the remains of dead larvae have dried into hard, black scales, these may be seen by looking over the top of a vertical comb and tilting it so that the lower sides of cells can be viewed. These scales fluoresce under u/v.
See: American Foul Brood

**Aflatoxin B1**
Toxins produce by fungi in mycotic diseases such as 'stone brood' by varieties of Aspergillus spp. Bees fed on aflatoxin contaminated diets die early. It is a metabolite in some
See: *A.flavus*, parasiticus strains.

**African bee**
In the north of that country a reddish-brown bee resembling the Egyptian bee is found. They work well in their own environment but are poor winterers in the UK.
See: *A.m.capensis* and *adansonii*, also Africanized bee.

**African bees**
A book by T.D.A. Cockerell.

**African Bee-Keeping Journal (part English)**
P.O. Box 4, Bergvliet C.P., S. Africa.

African/Brazilian bee
In 1956 a geneticist took 26 colonies of *Apis mellifera* adansonii (now called '*A.m. scutellata*) to near Rio Claro, Brazil. By 1977 they had spread to Guyana, Venezucla, Bolivia, Peru, Paraguay, Uruguay and N. Argentina. They are more inclined to sting than other races of *mellifera*, swarm excessively and abscond. By 1981 their progress was reported as far as the Isthmus of Panama. Sensationally described by some members of the press as 'killer-bees' they do at worst seem a little more disruptive than the honeybees that we are used to. There is little to distinguish them visually from Italian bees. They are said to be hybridizing and are not thought to spread as fast nor winter well in cooler climates. Their honey-getting capabilities are reported as being good.

**Africanized bee**
African honey bees were introduced into Brazil by Dr. W.E.Kerr in 1956. Queen restrainers were inadvertently removed and swarms escaped. Subsequently the variety spread and being somewhat more savage than many other varieties a bad reputation accompanied them. The African bee *Apis mellifera adansonii* (now *scutellata*) is the culprit but it is reported that a southerly climatic limit around the 34th parallel has been reached whereat this bee fails to overwinter as it does not form a winter cluster. It is hoped that a similar northerly climatic barrier will hinder its advance into the U.S.A.
See: 'African/Brazilian bee'

**after 8**
Manipulations with bees after eight o'clock (20.00) in the evening. The majority of bee-work at the hives is conducted during the day and it is possible to overlook certain advantages or differences that occur after bees have finished flying and when any possible marauding should have ceased. Queen introduction and uniting are said to be more easily carried out when flying has ceased. A red light has been used to work A.scutellata by night to minimize their interference with the neighbourhood. In the dark bees will cling to you and crawl on you giving an intermittent buzz until they find a tender spot.

**after-effect**
The result or consequence or delayed effect resulting from a given course of action/ This has to be considered relative to the manipulation of a honeybee colony. Sometimes when a colony has been taken to pieces by a novice it can take several days for the colony to return to normal and

more harm might have been done than if the colony had been left to its own devices. Nevertheless skilful and knowledgeable handling can increase honey crops and help colonies to remain vigorous and healthy. See: let-alone beekeeping.

**after swarm (cast)**
"They settled in two swarms which soon coalesced into one" Gilbert White. By hiving into a box of combs with an excluder (to prevent the queen from running through) it is possible to separate and select a queen when more than one is present. See: casts and bull swarm.

**ageing**
Although walking is more energy consuming for a bee than flying, the queen who walks by far the most, lives longest. There is little doubt that a queen's life will be governed by her total egg-laying which in turn is governed by the workers who respond to the environmental conditions encountered. The full utilization of brood food glands shortens bees' lives and those who make little or no contribution to feeding brood or queen either live longer in the winter cluster or are capable of extended foraging. Foraging is also a life-shortening exercise as is apparent from the elongated life of winter cluster bees. The traces of juvenile hormone present also have in influence on ageing.
See: brood food, drone and foraging

**age-mates**
The various kinds of work required by a colony are shared out largely according to ability based on physical suitability (glands etc.), colony requirements and competition for the work by other colony members. Those of any age group, which have been referred to as 'age-mates' may well specialise and carry out very particular duties. Work is being done to discover whether any kind of cast system exists within a cast.
See: pallbearer bee and phoresy.

**Agenda**
Matters to be brought before a committee. The intelligent planning of an agenda by the officers and members of a committee can vastly enhance the benefits to be derived from their meeting.

**Agent**
Something that is used for or able to cause a specific result. The honeybee and the wind are pollinating agents.

**age of brood**
Days: 1 - 3 egg
3 – 8 open larvae
9 – 13 grubs change to bee-like shape
14 eyes pink
15 eyes red
16 queens emerge
17 eyes purple, thorax yellow, abdomen white
18 little change abdomen yellow
19 antennae darken
21 wings extend – workers emerge
24 drones emerge.
See: drone life-cycle, eye colour, queen life-cycle, worker bee

**aggregation pheromone**
Dr. Butler's discovery of 'queen substance' was a gateway to the investigation of pheromones produced and utilized by the honeybee in pursuance of her varied behavioural patterns. Aggregation pheromone is that component of queen substance which causes bees to form a cluster and to repeatedly investigate any material that the queen has touched.
See: 902 and Apis cerana

**aggressive**
Being inclined to attack. This may be a genetic trait in some honeybees, like *A.m.scutellata*, or a colony may become aggressive due to adverse circumstances. P. Metcalf suggests the abbreviation (Ag) for aggressive worker behaviour.
See: temper

**Agricultural changes in the U.K.**
Eastern England in particular has become increasingly mechanized from the agricultural point of view and in the 80's had fewer hedges, was more windswept and used more herbicides than ever before. Larger and larger areas planted with oilseed rape. Heather in the south has steadily given way to bracken or been converted into farmland. The Downs have been repeatedly doused with fertilizers to give maximum barley yields. Meadows and animal forage has included less clover. In 1984 a serious fire ruined acres of heather on Woodbury Common, bracken was one of the first plants to re-establish itself.
See: agricultural reverses

**agricultural chemicals**
Chemical substances that have been formulated to help counteract the harmful effects of certain pathogens or organisms by acting as pesticides. There is an agreement between their manufacturers and the authorities to control their distribution and use in the U.K. A yearly book of 'approved products was published by MAFF.. Field trials have to be backed by actual evidence. Aphox has been suspected of causing bees to be rejected at the entrance. All chemicals should be 'cleared' by MAFF (?) before widespread use can be made of them.
See: spray poisoning, approved scheme (ACAS).

**agriculture**
Farming with plants. Horticulture is a branch that specializes in fruit and glasshouse culture. For clover pollination honeybees are the most usable insects and all varieties of white, alsike and red clover, Lucerne (alfalfa) and trefoil will only produce a quarter to one tenth as much seed if bees are not effectively used. ADAS Labs, Kierton Boston, Lincs.

**A.I.**
Abbreviation for artificial or instrumental insemination.
See: artificial and instrumental insemination

***Ailanthus altissima* (Simarubaceae)**
See: Tree of heaven

**air-sacs**
Thin-walled extensions (diverticula) of the tracheae. They are not spirally thickened and can be expanded and collapsed in breathing. Their inflation improves the aerodynamic aspects of the bee and in the case of the drone must necessarily be inflated prior to coition. The sacs are bilaterally symmetrical and to a certain extent interconnected. They are very large in the abdomen and taper rearwards while in the thorax and the head they are positioned posteriorly. Abdominal pulsations cause a pumping action which circulates air and waste gases so that air is drawn into or puffed out of the tracheae at the spiracles.
See: abdominal muscles

**airspeed**
Speed through the air as opposed to ground speed. A tethered object (experimental queen) will possibly have airspeed but no ground speed.
See: organ of Johnston, flying speed and speed of a bee

**AIV**
Apis iridescent virus. When injected into adult *Apis mellifera*, cyctoplasmic crystalline aggregates are formed especially in the fat body and hypopharyngeal glands, also in the gut wall and proximal ends of malpighian tubules. Thought to cause 'clustering disease' in Kashmir bees but later found to be present in samples affected by acarine. Work at Rothamstead showed virus isolated from *A.cerana* multiplied when introduced to *A.mellifera*. Rothamstead – Bailey, Ball and Woods

**akinesis**
Rigid immobility as when a worker assumes a rigid position with wings spread and abdomen contracted and lowered. When

an intruder is interrogated by guard bees it may offer food but will assume an attitude of akinesis if further investigated or threatened. If it does not then it is likely to be attacked and thrown out. Queens feign death. Wax worms can remain quite still once uncovered.
See: catalepsy, thanatosis, epilepsy and freeze

**A.L.**

Advisory leaflet. These cover a wide range of subjects and were published by MAFF and issued free. They are revised, up-dated, reprinted and new editions appeared from time to time. They are also supported by other temporary or additional notes. For details contact FERA Publications Unit, the previous MAFF address was Lion House, Willowburn Trading Estate, Alnwick Northumberland, NE66 2PF
this may be totally out-of-date Those concerning beekeeping include:
A.L.283,,306,328,330,344,347,362,367,377,411,412,445,451,46,473,485,486,561, 574,549,566,834 ?????
See: Bulletin

**alarm odour**

The pheromones responsible for this have been identified iso-pentyl acetate (Bock, Shearer, Stone and Johnson) It was formerly called iso-amyl acetate and secondary alarm odour produced by the mandibular glands 2-heptanone is released by guard bees to alert those close to hand and aggressiveness can spread rapidly throughout a colony. The latter is released when a bee stings. Alarm pheromones fade rapidly so that a colony is not kept in an undue state of alarm after the danger has passed. Some of these chemicals may not be correct isobutyl acetate ?? The speed, intensity and duration of responses are said to increase with temperature, and higher humidity increases intensity.
See: alarm odour, 2-decenoic...2-heptanone...

**albino**

Most animals and many flowers seem to produce white 'sports' from time to time. As this is a congenital pigment deficiency continued selective breeding can stabilise this absence of colour. In honey bees one sometimes comes across individual bees with white eyes and the absence of pigment would cause them to be blind

**alcohol (Ethyl alcohol) $C_2H_5OH$**

A colourless, flammable liquid – the intoxicating principle of fermented drinks. In the making of mead, because alcohol can dissolve both propolis and wax these should always be filtered out of the 'must' or flavour will be impaired. Stores that have been allowed to become moist, will ferment and these are bad for the bees. Methylated spirit is a denatured (poisonous) form. Absolute alcohol (99%) is used for microscopy (its freedom from water can be checked with anhydrous copper sulphate). Such alcohol is dutiable and requires Custom's clearance. Royal jelly may be preserved in it. It has a great affinity for water. Alcohols form a large group of organic compounds.
See: alcohols

**alcohols:**

methyl alcohol $CH_3OH$ – methanol or wood spirit.
ethyl alcohol $C_2H_5OH$ - ethanol or ordinary alcohol
propyl alcohol $C_3H_4OH$ - propanol
butyl alcohol $C_4H_9OH$ - butanol
amyl alcohol $C_5H_{11}OH$ - pentanol
Plus succeeding members to $CnH_2n+_1OH$
Poisonous methanol is used to denature ordinary alcohol.
See: alcohol

**aldehyde**

Organic chemicals with - CHO group attached to hydrocarbon - alcohol deprived of its hydrogen element (as for instance by oxidization). They are volatile fluids producing flavour and aroma and may be possible sources of irritants and repellents as for example benzaldehyde, adjectival form: 'aldehydic'.
See: benzaldehyde, HMF and formaldehyde

**Alder (Alnus spp, *glutinosa*) Betulaceae*****
A catkin bearing tree that flowers Feb.to Mar. or earlier. It can be a useful source of fresh, early pollen greenish/yellow when the weather permits bees to work it.

**Alergoba**
A rapidly granulating Hawaiian honey.

**alfalfa *Medicago sativa* Leguminosae** ***
Lucerne, grown on limey soils in southern counties as cattle forage and it frequently forms meadows that are cut before flowering. A visit by a honeybee trips the sexual column of the flower causing it to separate quite violently thus striking the standard petal and often hitting the bee on the ventral side of the head. This can cause the bees to tire of working the crop and fresh colonies need to be brought in for optimum pollination. In the U.S.A. alkali bees are also used. The honey is light and spicier than typical clovers but the yellow pollen is inadequate in protein content. Safflower (*Carthamus tinctorius*) is sometimes grown adjacent to alfalfa fields so that brood rearing can be sustained.
See: lucerne

**alien**
Foreign or strange. A bee from another colony. Drifting can result in alien bees being accepted by a strange colony especially if they are laden and adopt a submissive attitude. Drones are unable to be aggressive and are usually accepted should they enter a colony other than their own, provided the colony needs drones. A queen, desired and protected by her own bees, is still regarded as alien when transferred to another colony and care must be taken to make her acceptable.
See: akinesis, hive mates and queen introduction.

**alight**
To land or settle. A worker must make contact with its feet before it is able to use its sting. Foragers use flower structure – lines, shapes, colours and access as means of determining where best to alight to scrabble for pollen or delve for nectar. A swarm is encouraged to alight by those bees nearest the queen who expose their scent glands and fan. Should they fail to find her they will accumulate around the largest knot of bees before returning whence they came.
See: bee-bob and settle.

**alighting board** ***
A flat, usually sloping board, that projects out in front of the entrance to the hive. It is an integral part of a WBC hive which also includes a porch that helps to keep the alighting board dry. Bees' wings often stick helplessly to flat, wet surface. Most bee farmers who require to transport their hives from one place to another do without them, though as famous a beekeeper as Bro. Adam insists that they help heavily laden returning bees to make it to the entrance. For transport his boards were detachable. They seem to suit guard bees and fanners and make the hiving of swarms easier. Sandpaper them if smooth or sprinkle sand onto a freshly painted alighting board, as bees require a good 'grip'.

**align**
To arrange in a uniform line. Such an arrangement of hives may be more helpful and satisfying to the beekeeper rather than the bees. Grass cutting and orchard culture may suggest such an alignment but colonies are best able to go about their business when well-spaced out and where drifting is discouraged by breaking the uniformity of a straight line.
See: direction of hive entrance and drifting

**alimentary canal (enteric)**
The gut or nutritive tract concerned with the digestion and absorption of food. It reaches from the mouth in the base of the head to the rectum which tapers to the posterior vent or anus. Its parts include the cibarium (food chamber and sucking pump) oesophagus, honey sac, proventriculus, ventriculus (with associated malpighian

tubes which open into the posterior end) anterior intestine and posterior intestine or rectum. In the larva it does not become continuous until the feeding stages are completed thus preventing contamination of its food.
See: under part names - digestive tract and gaster

**alimentary glands**
See: abdominal glands

**aliphatic ketone**
General formula R-CO-R where the carbon atoms are linked in chains and not rings. It is secreted by workers' mandibular glands and probably plays its part in identification and may well be involved in the mastication of wax.
See: ketone

**alitrunk**
The part of the thorax of an insect where it is fused to the first segment of the abdomen. It is the wing-bearing division of the thorax also called the propodeum.

**alive**
Functioning as a living unit. When bees are not flying as by night or in winter it may not be obvious whether or not a colony is still alive. A tap on the side of the hive wall should elicit a response in the form of a sharp, short hiss or buzz or there may be detectable warmth over the quilt or crownboard. If each of these tests is negative then the hive may need to be 'opened-up'. The queen is always kept alive until one of the very last.

**alkali**
Soluble hydroxides of the alkali metals and less soluble hydroxides of the alkaline earth metals. Industrially the term includes soluble substances that perform like bases and have a pH value of between 7 and 14. E.g. Sodium carbonate, caustic soda, water glass, ammonia. They neutralize acids, saponify wax and loosen and dissolve propolis. Solutions used hot to scrub wood or in which metal ends and Porter escapes can be boiled are often recommended.
See: caustic, lye, pH value and scrubbing solution

**alkali bee (*Nomia melanderi*, Aculeata)**
Also called the APC bee and a native to N.W. America but confined to localities where the soil is sub-irrigated over hard-pan layer. It nests in bare alkali spots and has been developed for the pollination of alfalfa (lucerne) hence the alfalfa pollen collecting bee. Pits lined with polythene and stocked with a core of soil containing nests from natural sites. Capable of setting greater yields than leaf-cutter bees (*Megachile rotunda*) though erratic because of diseases.
See: predators and pesticides

**alkaline gland**
A gland associated with the sting apparatus though almost certainly having no influence on the constituents of the venom. Its secretion enters the sting chamber. It has been speculated that it is used as a lubricant for the moving parts of the sting, an adhesive for gluing eggs to cell bases and that it is pheromonic.
See: sting glands

**All about Mead**
Author: S.W. Andres 1975?

**allelomorphs (alleles)**
Genes are described as alleles of one another if they occupy the same locus (relative position) on homologous (similar) chromosomes i.e. the corresponding chromosomes derived from each parent and control the development of a particular characteristic.
See: homozygous and heterozygous

**Allen Harry**
Well known and liked as BBJ writer and CBL an expert of apicultural matters. His tape recording 'The influence of soil in beekeeping' was available from BBKA and

has contributed much on the subject in this work.

**Allen The Rev. Montague Yate, M.A. (Oxon) 1880**

He began beekeeping in 1891 and favoured English blacks. He wrote: 'European Bee Plants and their Pollen' 1937, which included a microscopic study of pollen. Lecturer, Judge and Secretary of YBKA.
See: Pollen identification for Beekeepers

**Allen A.L. Sandeman**

Wrote 'Beekeeping with twenty hives' 1943

**allergen (atopen)**

A substance that produces an allergy like phospholipase and which is found in bee venom, bee protein and pollen and in some cases even honey, wax and propolis.
See: allergy and bee venom

**allergic reaction**

A condition produced when a foreign substance such as pollen or venom enters the body and combines with protein to form a specific antibody which in turn produces distressing side-effects and reactions. It is estimated that it occurs in about four out of every thousand people in the general population. Death is rare but moderate allergies show in the form of: 'hives' 'diarrhoea', 'stomach cramps', 'wheezing' 'difficult breathing' 'tightness of the throat', 'swollen joints' and sometimes 'faintness.
See: allergy and allergen

**allergy**

The word has been used for mental and physical conditions and means altered capacity to react. It is used in various ways to describe the proneness of certain people to a more severe reaction than normal when allergies can apply to a large number of external influences which specifically play upon internal faults. It often implies a changed reactivity on second or subsequent infections or poisoning. (Dr. A Frankland)' A specific reaction depending upon previous sensitisation – immunity gone wrong'. Allergic reaction results from IgE attaching itself to mast cells and other receptor sites. Recent advances in the treatment of Bee Venom hypersensitivity – HRC Riches MD FRCP Harefield Hosp. Middx.
See: atopy, allergic, anaphylactic, immunity, propolis dermatitis and IgG Bee sting allergy.

**Alley/Miller**

A method of queen cell raising that involves cutting 'V' shapes along the bottom of comb newly filled with eggs, for transfer to a cell-raising stock to optimise the number of queen cells that can be cut free when completed.

**allied insects**

Creatures that come close to the honeybee in the evolutionary scale – Apidae; though some may take this further to include other members of Hymenoptera. Under this heading are included: alkali, bumble, blue orchard, carder, carpenter, cuckoo, hive, homeless, leaf-cutting, long-tongued, mason, mining, short-tongued, solitary, stingless. Tree, wild etc.
See: Bees

**allogamy**

Cross-fertilisation, the opposite to autogamy.
See: cross-fertilisation

**allomone**

Interspecific semiochemical substance which favours the producer but is active between different species.
See: pheromone

**almond *Prunus amygdalus* spp. Rosaceae *****

An ornamental nut tree whose early flowering (mid February to early April) makes it unsuitable for commercial use in G.B. It can produce a good honey flow and is very useful in the U.S.A. (crops have suffered due to CDD – 2OO8). Its

delicate pink blossoms bespangle the trees before its leaves appear and are a welcome sign of spring. It also provides extra-floral nectaries. The honey is light and has a compelling flavour, the pollen described as ochreous or brown. Varieties include: dulcis, communis, migdalarbo.
See: extra-floral nectary

**alpine swift** ***
A bird that is known as the 'bee-eater'.

**Alsike clover** ***Trifolium hybridium***
Leguminosae
Alsike Nr Upsala – Sweden White, pink-tipped flowers, many branches. Can cope with a more acid soil than most clovers. It produces pale honey typical of clovers and flowers May – October.

***Althea rosea*** **Malvaceae**
See: Hollyhock

**Alveary hive**
Alvearia hive for bees - Herod Hempshall. A small wicker hive in early times. Latin 'alveus' a hollow vessel. 'adverium' a beehive.
See: alveole and cloom

**Alveole**
Cell in comb, hence alveolate - pitted like honeycomb.

***Alyssum maritimum*** **(madwort) Cruciferae**
A small perennial plant, seldom exceeding 10cm in height with white honey-scented flowers. It blooms in late spring and summer, also A.saxatile – popular yellow flowers known as 'Golden Dust' Pollen yellow.

**Amateur Beekeeping**
Edward Lloyd Sechrist 1958 G.B.

**Ambrose**
See: St. Ambrose

**ambrosia**
Pythagorean milk and honey the food and drink of the Gods which gave immortal youth and beauty to those who ate it. Described as nectar, bee-bread, fragrant, delicious, heavenly, sweet and pleasing a finely-flavoured beverage. Delightful to taste and smell. That which imparts a sense of divinity – poetic inspiration, heavenly music, worthy of the Gods. Ambrosian – relating to St. Ambrose, Bishop of Milan in the 4th century.
See: St. Ambrose

**America**
See: U.S.

**American Bee Breeders Assn.**
Treas. Louis C. Harbin, P.O.Box 218 Theodore, Ala.36582
(Perhaps out-of-date)

**American Bee Journal**
Hamilton, Illinois, U.S.A. Available through Thornes (still ?)

**AFB Life cycle**
The causative pathogen is Paenibacillus larvae a motile rod-shaped bacterium. Spores ingested germinate on reaching larval gut where pH is less acid. Vegetative rods penetrate lumen and grow in abundance in haemolymph. Death of larvae occurs after nine days and millions of spores are formed. House bees vary in their ability to remove the glutinous remains which may dry into a black scale that adheres firmly to the base of the cell and remains infective (being highly resistant to desiccation, heat and disinfectants). Old combs from colonies long since dead can therefore continue to spread the disease.
See: AFB symptoms and American Foul Brood

**American foulbrood (ABD)** ***
Causative agent: Paenibacillus larvae transmitted to the larvae via its food. It multiplies in the larval gut and unless

ejected by the nurse bees death usually occurs on about the twelfth day (after the cell has been capped). Such cells have an abnormal look losing their clean domed appearance and becoming flat, sunken, cracked and greasy-looking while ragged holes are frequently nibbled in them. At this stage an inserted match stick may be given a twist to make the putrid content adhere when a ropey, brown glue will follow it a cm or more from the cell. Once suspected every precaution should be taken to prevent its spread and FERA advised at once.
See: 'AFB life cycle, symptoms, and hot-dip

**American terms and abbreviations**
Cleat, yard, parade, fall, side-liner, ramada, out-yard, singer, lumber, bubbler, buttonwood, basswood, outfit.

**amino acids**
A group of organic compounds (which act as chemical building blocks) some twenty of which form the fundamental constituents of living matter. Each contains amino ($NH_2$) and carboxyl (COOH) groups. Proteins are derived from these acids which in turn can be hydrolysed from protein when the 'H' molecule of a fatty acid is replaced by an amino group. We have automatic amino acid analysers that give values of amino acids per gram of honey for six of the most common ones: proline, histidine, glutamic acid, phenylalanine, aspartic acid and alanine.

**ammonium nitrate ($NH_4NO_3$)**
White crystals that can be burnt to produce nitrous oxide an anaesthetic gas that will knock out wasps. It is not as dangerous as cyanide but is still a hazard if misused as the compound can explode if heated under certain conditions.
See: anaesthetic and saltpetre.

**amnion**
The inner embryonic membrane of insects, foetal membrane of mammals, viscous envelope of some ovules.

**Amoeba**
The name given to a disease of honeybees associated with the presence of a parasite known as *Malpighamoeba mellificae* Pell. This interferes with the functioning of the Malpighian tubules (the bees' equivalent of kidneys) and produces cysts that pass from the bee with its faeces. It is more harmful when present in combination with other diseases or conditions such as dysentery. Requeening coupled with comb sterilization usually brings it under control.

**amyl acetate $CH_3COOC_{55}H_{11}$**
A colourless liquid that has a smell of pear drops. It is useful for 'despatching a sacrificial sample of bees that are required for dissection as, unlike chloroform, it relaxes the insect's body making it easier to handle. It is used in perfumes, as a solvent and is a good solvent.
See: knock-out

**Anabasis**
A military expedition such as a march from the coast to the interior as described by Xenophon. This included poisoning of soldiers who ate local honey.
See: Xenophon

**anabolism**
Constructive metabolism. Just as katabolism refers to the breaking up of complicated organic molecules so anabolism is the reverse process – the synthesis of complicated molecules from simple ones. The constructive processes in living organisms.
Opp. 'katabolism'

**Anacron**
A Greek who died at the age of 115 and attributed his long life to the daily use of honey.

**anaerobic**
An anaerobic organism is one capable of living in the absence of free oxygen. Yeasts that decompose the sugars of honey to

form ethyl alcohol and carbon dioxide live in the absence of gaseous or dissolved oxygen. They are therefore called anaerobic yeasts and this is why fermentation valves are required to exclude any oxygen from the liquid to be fermented into wine (mead) which requires yeast that works anaerobically
See: aerobic, mead and vinegar

**Anaesthetic**
Hence anaesthetisation. This is an agent that produces insensibility in a living organism. Nitrous oxide (laughing gas) has been used to render a difficult colony insensible and carbon dioxide has been used to keep queens quite still during artificial insemination. Others include carbon tetrachloride (though poisonous), chloroform, ether, burning puff-ball. Amyl acetate kills but relaxes small insects whereas chloroform stiffens them.
See under chemical headings and subjugation

***Anagasta kuchniella***
The Mediterranean flour moth.
See: predators

**analogous**
With similar function though not structure or development. One might talk about analogous organs, for example mammalian kidneys and honeybee malpighian tubules. The term is used in genetics when referring to the functions of genes, zygotes etc. It can be used to describe the similarity of function of organs that are not homologous.

**analysis**
The determination of the components of a substance quantitatively or qualitatively. The substance may have to be decomposed or reacted with another material to enable separation identification and quantification to be made. Techniques, some of which are used to examine the sugar spectrum of nectar or honey or the contamination of bees or honey by insecticides etc. include: 'high performance chromatography', 'paper chromatography' 'gas liquid chromatography on silica gel', 'thin layer chromatography (on silica gel) 'ELISA 13C nuclear magnetic resonance' 'enzyme linked immunoabsorbent assay' etc.
See: assay and bio-assay

**anaphase**
In mitosis this constitutes the phase during the divergence of daughter chromosomes up to the division of chromatin into chromosomes. In meiosis where haploid cells are produced from diploid cells, anaphase constitutes the fourth stage.
See: meiosis and mitosis

**anaphylactic shock**
In the case of acute reactions from a bee sting allergens can cause a sudden vascular collapse leading to shock, and death may occur in a matter of minutes. In such a case immediate physical help and or an adrenaline injection is needed. Four minutes with a stationary heart leads to irreversible brain damage. An individual who knows they are sensitive should recognise the early symptoms and administer adrenaline by inhalation (i.e. she should carry an inhaler) or have injections.
See: anaphylaxis, bee venom and sting

**Anastomo kofini**
A hive made from woven wicker coated with cow dung and clay and found in Attica and in the eastern regions of the Peloponnese's. Another variant the amphisteno kofini – a woven wicker cylinder with sloping sides, used upright and open at each end.
See: Greek bar hive, Vraski

**Anatolian bee**
Extensively tested by Bro. Adam. This bee is small, gentle and dark (queens not easy to find) though its looks belie its ability to function on a lower honey consumption than many races while at the same time it is capable of above average honey yields. Anatolia is the plateau that spreads between

the Black Sea and the Mediterranean.

**anatomy**
The structure of an animal or plant. Under subject headings the names of anatomical parts of the bee and associated things like flowers are given in alphabetical order.
See: bee anatomy and botany

**Anatomy and dissection of the Honeybee**
Author H.A.Dade – Bee Research Assn. 1962, '78 reprint IBRA - International Bee Research Association

**Anatomy and Physiology of the HB**
R.E.Snodgrass 1925 McGraw Hill

**ancestry**
The earlier lineage or previous generations of gene donors.
See: origin of the honeybee

**anchoring effect**
In the early part of the breeding season a colony may well become anchored by a small brood nest which will when colder conditions supervene prevent the bees from utilising food stored on combs that are out of reach because they cannot cross the cold gap. Also a swarm or cast that might otherwise abscond, even though carefully hived, may be anchored by giving them some unsealed brood to look after.
See: isolation starvation

***Anchusa capensis*** **Evergreen Alkanet** ***
Boraginaceae A hardy annual flowering May/June. It has sky-blue flowers and produces nectar and pollen. Other species include: 'italica', 'officinalis' and 'sempervirens'

**The ancient bee-master's farewell**
Written by J.Keys 1796

**ancient feeding techniques**
Skeps might have the hollow stem of a reed pushed into the flight hole, having first been primed with honey; or a plateful of honey would be given below the combs. Honey of course was cheaper than sugar in those times. An Elder stem, split in half with the pith removed would be pushed in. Ale and brown sugar was also mentioned – perhaps less damaging in the summer when the dysenteric effect might be lost on the wing.
See: kex

***Andrena armata*** **Andrenidae**
A little red solitary bee formerly known as *A.fulva*, the tawny mining bee. It prefers to nest in sandy soils under short grass – lawns, golf courses. They make a burrow with cells leading off and the female lays a single egg on a ball of pollen/nectar (mass provisioning) the cell is sealed with a plug of loose soil moistened with saliva. There is a small mound of excavated soil with a small hole in the middle which is the entrance to the burrow. Andrenidae are short-tongued bees. A. alicans, the early mining bee and A.cineraria the grey-haired bee.

**androgenetic**
Having arisen from supernumerary germ with only paternal chromosomes. Biological word element 'andro' meaning 'man' or 'male'.

**andromedotoxin**
One of the world's most deadly poisons, Acetylandromedol, also known as andromedotoxin, has been extracted from the nectar of some of the most scented varieties of rhododendrons and their hybrids such as mountain laurel honey.
See: anemonal and toxic nectar

**anecbalic**
A word coined by Dr. Wallon a Brussels physician to describe bees of a non-swarming strain Greek A, negating prefix, N, transition consonant, Ek – outside of, ballo, to throw or cast out, i.e. to display non-swarming behaviour. The English equivalent 'anecbalic' might be used for this characteristic.

**Anemonal**

Bee poison found in some buttercup species. See: toxic nectar

**Anemone spp.** ***

The perennial wood anemone, A.nemorosa is common in many woods and yields useful pollen in the early spring. Garden varieties e.g. A.pulsatilla grown in window boxes and such places have been visited by bees in the dead of winter. White pollen.

**Anemophilous**

Flowering plants that are pollinated by means of the wind. Trees with catkins such as elm and hazel are examples. Literally 'wind-loving'. Usually small flowers with perianths insignificant or lacking, pollen grains shed singly in great quantities, pollen carried by air currents. Gk. 'anemo – wind' See: entomophilous

***Angelica archangelica*** ***

Five to six feet tall. Green/white flowers in globular umbels July/August. A hardy perennial yielding nectar and pollen. Sow in damp places or riverbanks. L.angelic – herb

**anger**

The reaction that bees usually display when they are interfered with without subjugation. Many are the words used to describe this condition such as: aggression, defence characteristic, intractability, irritability, snottiness, temper, tetchiness etc. *A.adonsonii* (*scutellata*) acquired a bad reputation as excitable, hot, testy, touchy, unmanageable and so on. With normal races requeening with a more docile queen will usually put things right. This is important, especially where bees are kept in built-up areas, not only because bad-tempered bees are anti-social but because the queens of nearby beekeepers may well mate with drones that carry genes of the bad-tempered strains.
See: alarm odour, angry bees, subjugation and temper.

**Angiosperm Angiospermae**

This term has become more specific since its earliest usage in 1690. It now includes all the flowering plants other than gymnosperms (naked ovules as in the Yew). It is the Class (within the Division Spermatophyta) whose seeds are enclosed within specialised carpels which form an ovary that ripens into a true fruit.

**Angiosperm (pollen grain)** ***

In section this has a central zone which is the living cell, two nuclei and cytoplasm. On a suitable stigma it germinates to form a pollen tube along which the nuclei pass and if successful fertilise the ovule (female gametophyte). Then there is the intine, a middle layer of cellulose membrane that forms a complete surround to the cytoplasm. Finally the tough external layer (exine) that is made of a remarkably stable material 'sporopollenin' which has enabled us to identify flora and the eating habits of fauna back through history. When bees store pollen they will have treated it so as to prevent pollen tube growth.
See: pollen grain

**angles of cell walls**

The angle subtended between two adjacent walls of a hexagonal cell is 60 degrees. The base however is formed by three rhombs and has angles that are not quite as easy to measure. It requires the calculus to determine the most economical use of wax to provide cells with rounded bases on both sides of the comb, which favour the occupant and are suitable for storage purposes, though the remarkable thing is that bees worked it out long before man was capable of doing so.
See: honeycomb

**angry bees**

Certain races such as *A.m.adansonnii* (*scutellata*) have a predisposition to bad temper and many mellifera strains are easily provoked. Weather, actions by the beekeeper, or interference from unknown sources can cause a colony to seem spiteful, high temperature, abnormal states of static electricity a sudden cessation of a flow, are all blamed at various times for making bees difficult to handle. Precautions to avoid upsetting a colony's temper should always be taken and those with inbuilt vicious traits should be requeened.

**Ångström**

An Angstrom is a measure of length, a unit which is one ten millionth part of a millimetre or ×0001 micron. An 'A' unit equals 10-10m but has now been superseded by the S.I. system the preferred unit being a nanometer (nm) which equals 10-09m.
See: nanometre

**anguiculi**

Claws

**animal**

Living things other than plants, capable of motion. Many animals benefit by sharing this planet with honeybees. Horses seem rather sensitive to bee stings whereas cows seem relatively immune. The relationship of several domestic animals to bees and of course their many predators are dealt with under respective headings: badger, bear, mouse, moth etc.

**Anise Hyssop *Agastache anethiodora* - Labiatae *****

A herb with aromatic leaves and lavender coloured flowers blooming from June through to October. It is a special favourite in America where the light, minty-flavoured honey is said to carry the fragrance of the plant. Slow to granulate the honey is light, delicate flavoured, has a delightful aroma and heavy body. Twenty five colonies per acre gave a surplus of 100/125 lbs. each in July 1982 (ABJ). Giant Hyssop – A.urticifo lia
See: Hyssop

**Anisol**

A few drops placed at the hive entrance controls robbing. It is strongly scented mixture of aniseed oils and it can be used when uniting colonies.
See: uniting aids

**ankle**

The flexible portion above the foot that connects it to the leg. Certain bees have been referred to as 'ankle stingers'. This is likely to happen where only socks are relied upon as protection against bees that have been alerted to the defence of their colony. Gum boots are the wisest precaution (though a plastic bag may be drawn over the feet and ankles in an emergency). Quick movements, rough textured clothing and even when departing with arms held against the body, the quick movement of the feet is a give-away to the over-vigilant guards.

**Annelida**

Ringed or segmented worms. These are often parasitic and have been found in the alimentary canal of certain wasps.
See: worm

**annual (yearly)**

A plant that grows from seed, produces flowers and then seeds itself and dies within the year. Subscriptions to bee journals, associations etc. are usually due at yearly intervals.
See: biennial, ephemeral and perennial

**Anodontobombus**

This group of pollen storing bumble bees within which we find at least nine species.
See: Odontobombus

**ant *** Formicoidea spp.** ***Aculeata***

A small, social insect having a king and queen which are temporarily winged for August mating-swarms. Helpless larvae reared in an underground nest by relatives. Wood ants help bees indirectly by shepherding sucking insects on pine trees to produce honey-dew. Ants can be serious predators working by day and more especially by night sometimes removing

large amounts of honey and aggravating the bees. Remedies include: orange peel, a sprig of mint, catnip, green black walnut leaves, alcohol, sodium fluoride, borax powder salt, sulphur, leaves of tansy, crushed chrysanthemum leaves, turpentine or creosote round hive legs and the base of hives and stands. Nests frequently occur under paving slabs when these are used underneath hives also under plastic sheeting.
See: wood ant

**antagonism**

1. Counteraction or active opposition against something. This could be an anti-pollinator or an anti-hormone.
2. A colony's antagonism towards a new queen or its being united to other bees may be increased by robbing or by disturbance of the bees but reduced by a flow or when the bees are confused (as by driving, shaking or the use of masking odours.
See: antipathy and acceptance

**antenna pl. antennae** ***

A pair of sensitive appendages sometimes called 'feelers' that protrude from the head just above the clypeus. Each consists of a rigid 'scape' long enough to hold the flexible 'flagellum' well clear of the eyes. The scape fits into a rotatable joint at the head and is followed distally first by the 'pedicel' (containing the organ of Johnston) and then by ten other sections in females, but eleven in drones whose antennae are also thicker. The drones have a million sensilla, workers over 10,000 and queens around 5000. Communication, identification, hive work, mating, foraging and navigation all depend on the mobile antenna which can taste (organ of Johnston) feel and measure air-speed.
See: antenna cleaner, flagellum and pedicel

**antenna cleaner (strigilis)** ***

There are antennae cleaners on each of the fore legs. The basitarsus is provided with an antenna-sized, semi-circular notch. This is lined with a fringe of stiff hairs that form a comb. The antenna on the appropriate side is locked into this notch by a short protrusion on the adjacent tibia, known as the 'fibula (velum), and wiped clean. This is done by stropping the antenna to clear away occluding material such as pollen and is a common action amongst bees just prior to their becoming airborne. A vital action before take-off…..pre-flight checks one might say.

**antenna comb**

The strigilis
See: antenna cleaner.

**antennal vesicle**

A small globular bladder under the base of each antenna. It is continuous with a small vessel which passes along to the distal tip and pulsates transporting nourishment to the multitudinous nerve endings that are found in this highly sensitive area.

**anterior**

Before, to the front or at the forward end. For example the anterior intestine is that part which lies in front of the rectum.
Opp. 'posterior'

**anterior intestine**

The intestines of a bee lie between the ventriculus and the anus. The forward part is a small tube sometimes called the small intestine. This is the anterior intestine. It is looped or coiled according to the position of the stomach and the Malpighian tubules duct their waste products into the anterior end where it joins the ventriculus. In the larvae the foregut does not connect with the anterior intestine until feeding is complete thus avoiding contamination with its food.

**anther** ***

The terminal portion of a flower's stamen. It contains the pollen sacs. The male sexual part of a flower attached to the sterile filament. Cross-pollination is encouraged in many flowers by the anthers and therefore the pollen, not becoming functional at the same time as the stigma.

**anthesis**

Flowering, period of flowering, 'efflorescence', 'inflorescence'. Full bloom or the period or act of flower expansion when the flower bud opens and the stamens mature.
See: efflorescence and inflorescence

**Anthophila**

(flower loving)
A division of Hymenoptera which includes bees. Gk. 'anthos' flower, 'philein' to love. Anthophilous: attracted to or feeding on flowers.

**Anthropogenic**

Produced or caused by man.

**antibiotics**

Substances that inhibit or destroy life – particularly parasitic organisms. The continuous use of antibiotics is a self-defeating measure as it selects out resistant strains of the organisms, or, alternatively other organisms will occupy the vacated niche. This is a widely accepted view expressed by T.S.K. and M.P. Johnsson in Bee World Vol. 58 1977.

**antibody**

An animal protein produced to combine with and to neutralise a specific antigen or which prevents the harmful effects of a toxin. Dr. Riches tells us that it is in the gamma fraction of the globulin component of the human's blood protein that antibodies are found. The immunoglobulins formed include five major structural classes. It is an immune response caused by an antigen that produces antibodies. A series of carefully controlled injections can help to achieve the right response. This desensitisation (say for bee stings) takes several months and is best carried out in the winter.
See: antigen

**anti-cyclone**

A weather pressure pattern where an extensive area is affected by high atmospheric pressure . Although fog, gloom, thunderstorms and extreme cold can accompany these conditions according to the time of year. It is usually in anticyclonic conditions that we get light winds and lots of sunshine. As summer honey flows are more likely to occur when a stable 'high' has established itself, a barometer and knowledge of flowering times can help us to anticipate when more supers will be needed and swarm control practiced.
See: cyclone, pressure and weather

**antidote**

A remedy that counteracts or prevents injurious effects. Many are those that have been advocated for bee stings in the past.
See: antidote for stings and sting antidote

**antidotes for bee stings**

Calcium gluconate may help to counteract toxic reactions. The use of epinephrine (or similar drug) at the onset of 'shock' reaction, may prove life-saving. Avil (Pheniramine) 50mg tablets, and Benadryl (Diphenhydramine)…
See: anti-histamines, sting relief and Wasp-eze

**anti-freeze**

Honey lowers the freezing point of water. At 68% the solution freezes at minus 12 degrees Celsius.

**antigen**

A foreign substance which causes changes in the blood resulting in the formation of antibodies. IgE antibody is a dangerous response to an antigen which may cause a generalised rash, drop in blood pressure, state of shock. difficulty with breathing or at worst, death may occur within a few minutes.
See: anaphylactic, antibody, desensitisation and stings

**anti-histamine**

Chemicals which when introduced into the body aurally can help to reduce the

histamine liberating effects of bee venom. There are many proprietary brands such as Benadryl, Anthisan, Banistyl, Dimotane, Fabahistin, Histryl, Phenergan, Piriton, Tavegil or the newer Beconase and Syntaris. Anti-asthma drug Intal as Rynacom and Lomusol
These were drugs in use years ago. Piriton (chlorpheniramine) 4mg one hour before opening a hive though this can cause excessive sedation in some people.

**Antiope**

Name of the prince in after-swarms (Butler) The 'prince being the virgin queen.
A second swarm or one that wanders – derived from Greek mythology.
See: Marpesia and Oreithya

**antipathy**

Repugnance or dislike. Workers of a colony will show this to a newly introduced queen even when their salvation depends upon her acceptance. In such cases a longer period of introduction is essential. When the bees cling tenaciously to the queen cage, without fanning and making as if to eat their way through the mesh, suspect that the colony has not yet been conditioned to receive her possibly having a queen of their own, a virgin or laying workers
See: acceptance, antagonism, balling and laying workers

**anti-pollinator**

Agents that hinder or prevent pollination. Spiders, well-camouflaged and hidden may wait near a flower and catch an insect pollinator. Similarly snakes found hiding near flowers that attract humming birds. Insecticides which may prove repellent or deadly to bees. The spraying of flowers in bloom may kill or repel bees or upset the viability of the pollen.

**antiseptic**

1 A substance that can be used to kill or inhibit the growth of bacteria without causing harm to the host. 2 A substance that prevents putrefaction or the establishment of bacteria by destroying harmful micro-organisms.
The bactericidal (antiseptic) property of honey is thought to be due to its being so hydroscopic and to the formation of hydrogen peroxide.

**anti-spark tube**

The part of a smoker which directs air from the bellows to the underside of the grate at the base of the fire chamber. It is arranged with a gap between the bellows and the tube and cut away underneath as an anti-spark precaution. In drawing air into the bellows it would be highly inadvisable to allow a spark to enter.
See: smoker

**antitoxin**

A substance that the body develops to neutralise, bind or isolate a toxin
See: antibody

**antonomasia**

The identification of a person by an epithet or appellative instead of using his actual name. A beekeeper is frequently referred to by casual acquaintances as ' the bee-man'
Ants, bees and wasps A book by J. Lubbock
Ants, bees and wasps Book by Lord Avebury Baron
Ants, bees, dragon-flies, earwigs, crickets and flies Book by W. Bath

**anus pl. anni**

The bee's rectum tapers to a terminal opening the 'anus' which is an aperture in the proctiger (the reduced 10th segment). Both solid and liquid waste in the form of faeces pass out of the body via the anus. It is over the actual sting and within the sting chamber of females.
See: faeces, proctiger and tenth segment

**aorta**

A narrow tube in the bee that ducts blood from the heart to the head. It has a series of loops in the posterior part of the thorax and

delivers blood pulses to the brain. These are the only blood vessels in the otherwise open circulatory system of the bee.
See: heart

**Apamine**
Bee venom fraction.
See: DMSO and venom components

**APC bee**
Alfalfa pollen collecting bee.
See: alkali bee

**ape**
See mention under 'African Honey Bird'

**aperture**
A worker bee can squeeze through a slot if it exceeds 2×9mm. A square hole needs to be 3×2mm before it can pass and a circular hole even larger at 3×6mm. For queens slots (as in excluders) should not exceed 4×2mm. Mouse guards should have 9mm holes though a mouse can squeeze through slots of that depth.
See: slots and useful dimensions

**aphid. pl. aphides Aphidoidea**
A genus of hemipteran insects such as green-fly. They can cover the upper surfaces of the leaves of poplar, oak, lime and sycamore with a sweet, shiny layer of honey-dew. This is exuded from their bodies as a fine spray, they themselves adhering to the under surfaces for protection. Action taken to restrict or eliminate them can also harm the bee and advice as to the best means of dealing with them should reflect sympathy for the bee.
See: sucking insects.

**Aphox**
Pirimicarb in the dispersible grain formulation. Used to control aphids on field beans. A carbamate insecticide of short persistence. Suspected of causing bees to be rejected when they returned to their own entrance.

**Apiaceous**
Celery (Apium graveolens) and parsley are apiaceous plants belonging to the umbelliferous genus Apium.

**Apiact**
The international technical magazine of apicultural & economic information, published in English, French, \Russian, German and Spanish. 50pp quarterly review. Available from Thornes (still?)
See: Apimondia

**Apian (Latin *apianus*)**
Adjective appertaining to bees.

**Apiarian**
Relating to bees or apiculture. A term not used frequently now.
The apiarians guide. Book written by J.H. Payne 1838

**Apiarians Manual.**
Book written by T.M. Howatson 1827

**apiarist**
An apiculturist or beekeeper. A person who owns or is able to handle colonies in an apiary – stocks of bees.
See: beekeeper

**The Apiarist**
New Zealand beekeeping journal published every second month by Alliance Bee Supplies Co, P.O. Box 5055 Christchurch, N.Z.
?? Still available ??

**apiarius**
Pl. apiarii – Roman apiarist.
See: mellarius

**apiary** ***
A place where hives are put. Best chosen carefully away from other apiaries so that the bees have forage capability and protection from wind exposure, interference, habitations, roads and paths. The health of the bee population in the area

is also important. Easy access by transport , security of tenure and storage facilities will be added advantages. The word 'bee-yard' is often used by Americans.
See: hive site, site and out-apiary

**The Apiary.**
Written by A.Neighbour 1865.

**Apiary health**
There was an apiary health certification scheme for bee breeders which at one time encouraged bee/queen breeders to submit their apiaries for inspection and if possible to gain a certificate of health. Very few were able to achieve the standards required. A few counties were willing to provide education in such matters and together with the efforts of Bee Diseases Ltd and the MAFF (Defra?) raised the standard of hygiene in the apiary.
See: apiary hygiene and appointed officers

**apiary hygiene**
The efficient management of bees so that bee diseases are kept out of the apiary as far as possible and methods of cleaning and sterilisation are adopted so as to minimise any chance of exceeding disease thresholds or passing pathogens from one colony to another. General rules are to avoid exposing any bee material (e.g. do not leave honey-wet equipment for bees to lick clean) replace brood comb regularly, sterilise all material when bee-less by the use of boiling, blow torch or fumigation as appropriate.
See: spring clean and sterilisation

**Apiary inspector**
(this is included for interest though well out-of-date)
While this official title does not apply in the U.K. it is used for a senior official, a qualified beekeeper who visits and inspects apiaries and within the limits of time available gives advice or help. Inspections under the Foul Brood order have been undertaken with a view to examining bees for disease hopefully at intervals not exceeding three years. In the absence of regulations there are in the main only part-time appointments of apiary inspectors in G.B. and they are under hard pressure to fulfil the title as applicable in countries where apiculture forms a more profitable part of the nation's business.
Up-date as necessary.

**The Apiary laid open**
Written by W. Dyer in 1781

**apiary site**
A place for beehives. Even in the most densely populated areas sites can be found for bees and in London Victoria Station and Fleet Street are instances of unlikely sounding places where bees have been kept. The 'Beeway Code' by DARG should be read and followed. Errant swarms and accidental interruptions of flight paths can lead to trouble. Lofts, roof tops, garden sheds and other less likely places are used though reasonably sized gardens, quiet plots and farms are usually more suitable. All too often when someone first starts with bees they have the erroneous idea that one's own home is the only place to keep a beehive.
See: Beeway Code, hive site and out-apiary

**apiator**
apiarist (Keys)

**Apicultural abstracts**
Journal by IBRA
See: IBRA

**Apicultural Education Association**
Formerly called Beekeeping Education Association which was originally the grouping together of County Beekeeping Lecturers but later open to members in full time beekeeping education who were proposed and seconded by its members.
Its function was to act as a vehicle for conveying ideas between those with a professional interest in beekeeping education, to Government Departments and other bodies.

**apicultural reverses**

There was a huge drop in the demand for wax following the Reformation. More foreign wines were imported reducing the amount of mead that was used. Sugar began its inexorable replacement of honey as the prime sweetening agent. Agricultural practices took little or no account of the bee and the 'concrete jungle' made inroads into pastures. In 1982 a world threat from the Varroa mite drastically reduced imports of bees (especially queens) into the U.K. By 2009 so many factors seem to have combined against the bee: pesticides, herbicides, the loss of hedges and the introduction GM crops and monoculture to suit anticipated needs would seem to be snags enough but over all a massive spread of electrical waves passing through every one of us day and night has in its turn been blamed for causing what has been termed 'colony disappearing disease' (CDD), though with so many indications of the Government's lack of concern for bees in the past, the present state of the honeybee is to say the least 'parlous'.

**Apiculture for Schools**

Edited by Ken Stevens 1975 Devon.

**Apiculture in tropical climates**

Full reports of conferences London1976 and Delhi 1980 Edited by Dr. Eva Crane IBRA

**apiculturist**

A person engaged in apiculture. Words like agriculture, horticulture and apiculture take the suffix 'ist' and not 'alist'. Perhaps there has become a tendency to use this word in preference to 'beekeeper'

**apicure**

One of the so-called remedies for bee 'ills', especially foul brood, that were so ardently recommended just after the turn of the century ( time of the I.O.W. disease).

**Apidae**

The family of related bees with long tongues and hind tibiae equipped with a 'corbicula' and that feed their larvae on pollen and honey. It has two sub-families Apis, together with bumble, carpenter, mason, leaf-cutter and stingless bees. A part of the super-family Apoidea. Sub-family Socials – *A.mellifera*, *Bombus terrestris*, agorum. Andrenidae: Adrena, Dasygastris, Megachile, Osmia and Chalcidoma. (Whew!)

**apidictor (1956)**

An electronic device developed by E .F.Woods, a B.B.C. engineer. It enabled the operator to filter out the volume of 'warbling' noise made by nurse bees that have a surplus of royal jelly. Once the volume reached a certain level swarming was possible and as it increased swarming became likely and then - imminent. Because the sound received had to be interpreted by the operator and some skill was needed for this, a visual display of green, amber and red signals was later incorporated. In skilled hands the 'apidictor' is accurate and saves time and labour as the presence of queen cells could be identified without having to open the hive and disturbing the stock.

**apifactory**

An apiary kept for economic purposes.

**apifuge (Grimshaw's)**

One time repellent for bees. Appis – bee, fugio – I flee

**Api milbien**

An Acarine fumigant once marketed by Woodland Apiaries. Remove stopper, insert a rag or wick into the bottle and place it on the floor at the back of the hive. See: oil of wintergreen

**Apimondia**

The World Beekeeping Organisation
The International Federation of beekeepers' associations. A Congress is held every other year with a chosen theme.
E.g. in 1979 Athens, 1981 Acapulco (Theme The Economic and Social Significance

of Beekeeping) - 1983 in Budapest, 1985 Nagoya Japan and presumably these International Congresses continue.
See: Apiacta and International beekeeping congress

**Apimyiasis (Villeneuve 1916)**
Wedmore refers to Apimyiase angellozi a condition that bumbles and less often honeybees suffer from due to parasitisation by the larva of a fly Senotainia tricuspis. The flies deposit a tiny larva with sharp mandibles which enable it to penetrate the softer intersegmental membrane and enter the blood. Here it grows but does not complete its development by eating solid tissue until the bee dies. Pupation takes place in the soil.
See: Tachinidae

**Apinae**
Sub-family comprising meliponines and Apis. Though it has been proposed that the stingless bees should be assigned to a separate sub-family Meliponinae. It is one of the four tribes which include honeybees and compose Apoidea. The genus Apis is now the sole representative extant as earlier primitive Apini (such as Electropis found in Baltic amber) are extinct.

**Apiology**
The Literature of beekeeping – the study of Apis.

**Apiotherapy Society**
The North American Apitherapy Society was formed in November 1978. Its elected officers were Charles Mraz executive director, Dr.Jurgen von Bredow president, Harry Proelick sec, and Dr. Helene Guttman treasurer. Annual membership $10.00 open to any interested person

**Apis**
The generic name for the four honeybee species in Aculeata. The cult of Apis originated in Egypt in the days of Hyksos shortly after Exodus. Apis – the sacred bull, was very much venerated in Egypt – Velikovsky "Worlds in Collision" quoting "The Apis inscription of Necho – Wahibre in Breasted, Records of Egypt IV 1976.
See: Apini and stingless bees

***Apis cerana* (*cerana cerana*, China and *cerana indica*, Sri Lanka)**
The Eastern honeybee, formerly referred to as *'Apis indica'* (*Apis indica japonica*). Slightly smaller than the Western honeybee, it is found in the warmer regions of the East. It is farmed for honey but has the habit of absconding, spaces its combs more closely and queen substance is interestingly different from A. mellifera in that it only inhibits laying workers' ovaries and stabilises those parts of a swarm cluster that are 'above' the queen. The drone attractant affects mellifera and could interfere with mating though crosses are impossible. Times of nubile queen's flight have been suggested as varying from those of A. mellifera.

**Apis club**
Forerunner of IBRA. In 1934 a booklet on Bee diseases was published in 1949 and cost ten shillings and six pence per annum. This was a week's wages for many people then.
See: Abushady

***Apis dorsata* *****
The giant tropical honeybee, largest of the four species, also known as the 'rock' bee They are gregarious colonisers and live on the rocky mountain side in places as in India and Sri Lanka. Like its small tropical cousin florae it builds a single comb in the open-air, usually under a rock ledge or similar projection. As it is such a large bee (about the size of a hornet) it gets a reputation for being savage but a colony will leave the nest 'en masse' if heavily smoked. The comb is very thick at the top for honey storage and the larger cells are used for queens and drones; the lower part of the comb tapers down to worker brood size. Both honey and wax are of superior quality, the latter being particularly useful for dentistry. In the

Andaman Isles, sap from *Amomum aculeatum* is used to tranquilise dorsata.

*Apis florea* ***

The little honeybee and the smallest of the four honeybees. It builds nests in caves or rock niches or under twigs. It can be farmed in the open-air provided the nests are shaded from direct sunlight. In Iran colonies migrate between sunny and shaded positions according to the time of year. It builds a simple comb, about 30cm wide ´ 20cm deep. When working with these bees in Sri Lanka commensals (Chelifae?) almost half the size of the bees, were found to be hanging onto quite a few of the bees in a swarm. Florea lengthen cells in the upper part of the comb, once brood has emerged, for honey storage. The comb is protected by a 3-deep curtain of workers which seems to pulsate as a whole to put predators off. They migrate seasonally, absconding from their former nests. Drones and queens are reared in the autumn. Crops vary from 200g to 3kg with a higher dextrin content than other honeys and it does not granulate.

*Apis iridescent virus*

Only found in A.cerana in Kashmir and Northern India especially in colonies infested with a mite that resembles A.woodi – Bailey. Close bodily contact between overcrowded bees during the active season links this symbiotic situation with that of CPBV and Acarine in bees in the U.K.

*Apis mellifera adami*

Name given to a race of bees in Crete and the Aegean islands in honour of Bro. Adam Their body is of medium size, broad abdomen, broad teguments, uniform pigmentation, - with yellow bands on first abdominal tergites prevailing in workers while drones and queen uniformly dark coloured. Low cubital index – tendency to swarm, mild and use little propolis.
See: Greek bees

*Apis adansonii*

Lowland bees of central and southern Africa.
See: *A.m.scutellata*

*Apis capensis*

A black bee found only in the south-western Cape of S. Africa, not aggressive like adansonii and adapted to the local, cool, wet conditions. It forms small colonies and is fairly easy to handle but not very productive. In queenless colonies laying workers are said to have 10 to 30 egg tubes (compared with 1-10 in other races) and have a small but non-functional spermatheca. It is asserted that these laying workers are able to produce small quantities of queen substance and their eggs can be diploid and give rise to parthenogenetic virgins!
See: capensis, parthenogenetic females and South Africa

*Apis m. carnica* **(Pollman)**

The Carniolan bee, somewhat smaller than mellifera but with large cubital index and much longer legs and proboscis. Said to be hardier than Italian and good honey-getting hybrids have resulted from Italian drones ´ Carniolan queens. They originated from the Eastern Danube Valley – Macedonia, S. Austria, Yugoslavia, Hungary, Romania and Bulgaria. Valuable non-swarmers bred in Austria though Mace says of them generally 'marked swarming tendency second only to the Dutch'. Slender, greyish-black with short, dense brownish-grey hairs. The chitin is dark and the abdomen marked with spots or bands of whitish hairs. The queen is distinct with red legs, the drones are black. They are quiet, gentle, not inclined to rob or use much propolis and make white cappings. They excel in collecting power and overwinter well.

*Apis m. carpatica*

The Carpathian bee. Romanian type of Carnica bee akin to Macedonian found in Russia and neighbouring countries.

### *Apis caucasica* (Gorb)

A large bee with lead-grey hairs from the mountainous regions of the central Caucasus in southernmost USSR. It has a wide metatarsus, large discoidal wing segment and a long proboscis. Said to work well in bad weather though not considered much good for the heather. Temper is good but their vices include the need for much food to successfully overwinter and a tendency to rob and to use too much propolis. The Russians say about the originals that they have high productivity, low swarming, gentleness and exceptional pollination qualities. There are also dark and yellow northern varieties.

### *A.m.fasciat*a

The Egyptian bee. There is much evidence of its importance in ancient Egypt. Described as of medium size with slender punted abdomen, first and third segments light yellow to reddish-brown and coloured with greyish hairs The queen's abdomen reddish-brown on first segment. Bad tempered, swarmy and do not like cold. Now queens still have reddish-bronze tone and they are described as similar to Cyprian and Italian bees though slightly smaller and less predictable under smoke. Prolific bees of the Nile Valley, north of Aswan are synonymous with A.m.lamarkii resembling Cyprians though swarmier and more difficult to handle. They have yellow bands and white tomenta.

### *A.m.intermissa*

Punic, Tellian or Tunisian bees found in N.W. Africa including Algeria, Malta, Morocco and Tunisia. Dark brown or black these small bees are good fighters, hasty tempered, bad propolisers, make watery cappings and winter badly. Their small dark queens are not easily found at first sight look much like dark workers in the U.K. Their comb has cells visibly smaller than A.m.mellifera.

### *A.m.jemenitica*

A bee limited to Yemen, Dhofar and the Rostaq area. It is assumed that its ancestors came to Arabia from the African mainland in the straights of Bab al Mandeb. Colonies tend to be found in caves and on cliff faces and consist of several combs - Dr. Paul Boyles.

### *A.lamarckii*

A bee found on the West Red Sea coast – Syria, Egypt etc. It tends to build smaller colonies than European bees and to defend them more vigorously.
See: A.m. fasciata

### *Apis mellifera lehzeni*

The dark brownish-black bee known as the British black which was widespread throughout the British Isles in the days of skeps but which was decimated (some say 'wiped-out') by the I.O.W.disease just after the turn of the century. C.1905 – 1910.

### *Apis ligustica* Spin.

Italian bees. They have been bred to such an extent outside Italy that there are more Italian bees in other parts of the world than there are in Italy. Some of the best known varieties are yellow in colour, usually calm and gentle, prolific and able to do well if their colony peak coincides with a really good honey-flow. They do not winter economically in G.B. as do some hardier races. Of the leather-coloured and darker varieties the descriptions 'long-working' and 'good comb making' have been used but also robbing, darker sealed combs, inclination to drift and variable swarming are other traits mentioned.

### *Apis m. litorea*

A small yellow banded bee found on the Tanzanian coast. It spreads along the forage rich lowlands up to 500m where a coastal strip receives rainfall all the year round.

### *Apis m.major*

A race of honeybees found in the Rift

mountains of Morocco and described by Ruttner

***A.m.major rubica***

Saharan bee which Br.Adam says suffers from two serious defects: susceptibility to paralysis and inability to ripen heather honey properly

.

***A.m.meda* var. cecropia**

Name given to early Greek bee indicating that it was of near east origin. Kiesenwetter 1860

***A.m.mellifera***

The Western honeybee. From Europe and west of the Alps and Central Russia. The German or dark bee. The Dutch central and northern France bees are probably the same variety. A large dark bee, somewhat nervous and not very swarmy. It has long hair, short tongue and low cubital index, makes a small brood nest, is good at wintering using little stores. It uses little propolis, is comparatively tolerant of wax moth but is not as docile or manipulative as some races. Does well on a heather flow.

***A.m.monticola***

A gentle race of honeybee found between 1500 an1850 ft. up in the Tanzanian mountains. K.I.Kigatiira tells us they occupy the cool tropical mountains in Kenya between 2400 - 3000m, where there is thick forest cover with sparse but continuous flowering and temperatures that often go below zero.

***A.m.nigritum***

A honeybee described in 1936 by Lepletier. It is found in Central West Africa.

***A.m.rubica***

A race of honeybees found in the Sudan and described by Ruttner.

***A.m.scutellata***

Formerly called adansonii a sub-species of the honeybee indigenous to S. Africa. More aggressive than other honeybees and suited to a tropical habitat. A hard working bee but inclined to be swarmy and when nectar is scarce absconding swarms are likely. In Africa a worker takes only 18/19 days to develop as opposed to our bee 21. Found naturally between the Sahara and Kalahari deserts. Imported into Brazil in 1956 and spread to Venezuela by 1973. Tend to ball queens when moved (so find and cage). Does not form a proper winter cluster so does not overwinter in cool climates. Susceptible to EFB. Bees are smaller (20 more worker cells per unit of comb) Forage at lower temperatures and light intensities than European bees. Recorded as working by moonlight (!) and in light rain.
See: African/Brazilian bee

***Apis m. syriaca***

Docile, prolific Israeli bee. An Italian bee of Californian origin improved and developed in Israel.

***A.m.unicolor***

A black honeybee found in Madagascar, Reunion and Mauritius. Probably introduced to Reunion about 1666 and Mauritius soon after 1722. Not easy to handle and bees of European origin introduced around 1900 are more successful and produce about twice as much honey as unicolor. Many unicolor bees were demolished by Acarine brought in from Europe in the early 1950's.

***Apis mellifera* varieties**

These are Western honeybees (as opposed to Eastern) and include: European, adonsonii, capensis, carnica, capatica, caucasica, cecropia, cypria, fasiata, intermissa, lamaarkii, Lehzeni ligustica, meda, mellifera, monticola, punica, scutellata syriaca, and unicolor.See under A.m.headings, Banat, Buckfast and other breeders names e.g. 'midnite'

**apistical**

Of or belonging to beekeeping.

**Apitherapy**

The study of the healing properties of all bee products.
See: apitherapy.

**Apitoxin**

A protein found in honeybee venom.
See: hyaluronidase and phospholipase A

**Apium virus**

Apis or honeybee poison, used by homeopathic physicians since Rev Brauns a German clergyman, who found in 1835 that it was useful in the cure of humans and animals suffering from obstinate diseases.

**apivorous**

A creature that feeds on bees; birds for example. Fortunately no creature seems to feed entirely on bees.

**Aprocrita Insecta: Hymenoptera**

This is a sub-species of Hymenoptera and it is divided into two sections: Parasitica – ichneumons and other parasitic wasps like gall wasps and Aculeata which includes bee, wasps and ants.
See: Aculeat

**Apodeme pl. apodema**

A chitinous inner projection within the bee's exoskeleton which serves as a reinforcement and as an anchor for the attachment of muscles.
See: phragma

**Apoidea**
**Super-family within sub-order Apocrita**

Feathery (plumose) body hairs. Use nectar and pollen as food. Includes social bees, bumbles and stingless bees and sub-social and solitary bees (leaf cutting and carpenter). Apoidea with Formicoidea and Vespoidea form the sub-order Apocrita.

**Apomic**

A biotype produced asexually e.g. by parthenogenesis. Meiosis does not take place yet the gamete is diploid. Hence apomictic, apomixes.

**Apotrigona**

African stingless bee.
See: stingless bees

**apparatus**

An assemblage of articles for use in a particular way. The apidictor was an early electronic apparatus used to detect swarming potential. Device, appliance, gadget, implement, trappings, impedimenta – might all be applied to items used in the pursuit of apiculture. The manufacturers of beekeeping equipment show a preference for the word 'appliance'
See: machine

**apparel** ***

Clothes, dress, attire, garb. The clothes worn by a beekeeper will reflect his knowledge and familiarity with bees and possibly the weather conditions in which he is forced to work. It does not follow that the more clothes one wears the safer one will be from stings. Basically bees will tend to climb upwards if they are able to get into one's clothes. So gum boots, a zipped boiler suit, hat and veil (which may be integral with the upper part of the whole of the suit) and gloves with gauntlets overlapping sleeves to keep bees out. Many experienced beekeepers manage in shirt sleeves with a veil to hand and yet get fewer stings than the much protected tyro.
See: protective clothing

**apparent**

Obvious or seeming to be. Guard against misreading beekeeping situations such as colonies that seem quiet or feeders that remain full. Investigation is often needed.

**appearance of wax**

Pure beeswax varies in colour from white through yellows to dark brown. A favourite shade for show wax is primrose. Some plants may impart a distinct pigmental influence to the wax, sainfoin associated with saffron yellow and heather with a pinkish tinge. Thin sheets of wax have a semi-translucent quality at room temperatures. Any surface 'bloom' may be removed to produce a high degree of shine making wax objects exceedingly beautiful. Wax has a soapy feel and appearance.

**appendage**

A subordinate attachment of body projection such as an antennae, wing or leg. It can be a complete organ itself or a part attached to a trunk, branch or limb. A bee has six perambulatory appendage and four wings.

**appendicular glands Appendicular: hanging or suspended.**

There are tube-like glands found on a fertile queen's spermatheca and they produce the nourishment to keep the contents (at times millions of spermatozoa) in a virile condition.

**apple Malus pumila Rosaceae** ***

The most widely cultivated British fruit tree having many cultivars. There are early, mid-season and late varieties and inflorescence can vary from one year to another. They do best on the non-acid soil (pH5). Lime every 3 years plus a small amount of boron (borax) for calcium movement in the blossom nectar. Bees are vitally useful for pollination but careful preparation and arrangements should be made by the beekeeper for the best results. Nectar is produced at 55° F though bees cannot cope with windy conditions or wet at these low temperatures, nor does pollen function well. Honey ranges from light to dark amber with a superb flavour and rapid, even granulation. Pollen is pale yellow on anthers and pale, greenish-yellow in corbinculae. The grain is 23-28μ Crab apples make useful pollinizers.
See: apple fertilisation, apple types and flowers per tree

**apple bud** ***

Annual forecasting of the likely date of flowering is revised as the season progresses. When the flower bud approaches inflorescence and shows colour (pink bud) this is the last moment for the application of insecticides until petal-fall. Accurate assessments of the moment when the king blossoms burst forth is necessary so that colonies of honeybees can be positioned at the very moment when they can 'tune-in' to the crop and be most effective as pollinators.

**apple fertilisation**

Fertilisation only takes place when the pollen grain has germinated and sent its tube right down through the stigma (style) into the ovary and the nucleus of the male germ cell has fused with the female germ cell of the ovum. The growth of the tube can take almost the length of the flower's life, especially at marginal temperatures. The arrival of hundreds (as opposed to a mere dozen) of grains of suitable pollen vastly enhances the chances of this happening in time to lead to the formation of a perfect apple.
See: apple bud, cross-pollination and E.P.P.

**apple varieties**

Cross-pollination requires a donor plant (pollinizer) with compatible pollen.
Numbers below give matching types:
0 Early Victoria 1 9 Quarrenden 5,11
1 Stirling 0 10 Sudeley 3 2 Derby
3,4,5 11 Worcester 0,1,15,16 3
Lane 1,11 12 Grieve 0,1,11 4
Newton 2,3,5 13 Blenheim 4,5,9
5 Bramley 1,3,9 14 Rose 0,1,12
6 Wellington 5,15 15 Cox
0,1,11,12 7 Bath 8,16
16 Allington 0,1,2,11
8 Gladstone 7,11,15
The single flowering crab is also a tough and useful pollinizer.
See: Malus floribunda.
Numerous newer varieties should be included

**appliance**

A tool, or implement or device. Hence beekeeping appliances and appliance manufacturers. There are certain essential appliances associated with present day apiculture including attire, feeders, hives, honey and wax extracting equipment. There are also 'gimmicks' gadgets and new inventions which may or may not prove useful in the fullness of time. Appliances range from metal ends to fork-lift trucks, queen-marking discs to honey handling plants (appliances).

**appliance manufacturers As of 1980**
Burgess J.T. and Son, Busy-Bee Bkpg. Supplies, Bray Richard, Chiltern Honey Farm, Dartmoor Bee Farm, Four-acres hives, Harris & Son, Hillside Apiaries, Honey Farmer Ltd, Honeyfields Apiary, Kemlea Bee Supplies, Lee Robert, Maisemore Apiaries, M.K. Hardware, Sherrif B.J., South Down Honey Farms, Steele & Brodie, Taylor E.H., Thorne E.H., Wills-Bees Apiaries, Yorkshire Apiaries
These should of course be up-dated.
See: CONBA, catalogue and suppliers

**Appointed officers**
Situation as of 1980 now check with DEFRA etc.
An officer employed by ADAS to carry out the provisions of the Foul Brood Disease of Bees Order and such Orders that have followed.
See: bee-inspector and authorised officers.

**appreciation**
Of colour, polarized light, sound, u/v, vibration
See under these headings.

**appreciation of colour**
Von Frisch has shown how bees are able to distinguish between different colours. All colours are dependent on light and bees are not sensitive to anything that only reflects red light. As regards flower colours other factors would normally be more influential than colour in attracting bees. Bees will fly towards artificial light and do not fly in the dark while poor light discourages them and the greatest foraging activity occurs in bright daylight. Colour should be used to help bees identify their own entrance especially returning mated queens, and small coloured shapes are useful for that purpose. Light coloured clothing is less likely to provoke attack from guard bees than darker colours, black, brown and dark blue being particularly unsuitable.

**appreciation of sound**
Bees shown little reaction to sound though they are responsible for several characteristic sounds. One example is a 'piping' queen or similar 'squeak' which causes them to 'freeze'. Sound is of course a form of vibration and bees are able to utilise this as a means of communication. Dances include a vibratory 'waggle', stinging is accompanied by a high-pitched buzz. A knock on a hive or heavy footstep nearby brings an immediate response. Queenrightness may be checked by a knock producing a longer 'moan' instead of the brief 'buzz' of a queenright lot.
See: vibration

**approved scheme**
For agricultural chemicals administered by MAFF include Departments of Agriculture and Health and Trade Organisations. The Pesticide Advisory Committee consider all aspects of any chemical submitted and then recommend to Government Depts. whether it can be used safely for the intended purpose and what precautions should be taken.
See: ACAS and PSFS

**apricot Prunus armeniaca – Rosaceae**
A stone fruit that likes a warm climate and is therefore only grown under the most favourable conditions here. It flowers in early spring with the plums and is well-liked by the bees as it provides both nectar and yellow pollen for them while benefiting from their pollination services. It yields well in warm countries.

**April**

This month heralds the start of the active beekeeping season. Feeders give way to supers. Water which might have been a nuisance in winter is now required in large quantities. Days are longer now and bulbs, dandelions and willow are useful. The intake of pollen and warmth from the brood nest should help to satisfy the beekeeper that all is well. Clean floor boards may be given but postpone inspections involving the lifting out of brood combs. Provided colonies are healthy then the development of the brood nest will reach its peak making swarms in late April or early May a distinct possibility in the South. Give extra space, keep an eye on the food supply and start queen cell checks once drones are on the wing. a sudden change over from 'rags to riches' often occurs.

**Apterygota Sub-class of Insecta**

The truly wingless insects.

**aqua mulsa**

Pliny wrote: Add one third of honey to rainwater and keep in the sun for forty days from the rising of the Dog Star (Sirius).
See: star

**Arabian thistle Onopordum arabicum**

A hardy biennial yielding nectar and pollen, seven to eight feet (2-2.5m) high with large, mauve flower heads from July to September. Leaves and stems covered with white hair.

**Arabis Cruciferae**

A hardy perennial 4" (10cm) high that is often seen in rock gardens. It flowers early in the year about March and in the shelter of a rock garden can be a useful source of nectar as well as green pollen.

**Arachnida Arthropoda**

A class in the phylum Arthropoda including spiders, scorpions, crabs, ticks and mites. Their bodies are divided into two regions: prosoma and opisthosome. Respiration by tracheae. Four pairs of perambulatory appendages, no antennae.
See: spider

**arch**

See: honey canopy

**The archaeology of beekeeping Publisher IBRA**

Author Dr. Eva Crane. The first book to explore the rich heritage of beekeeping archaeology with over250 carefully chosen photographs and drawings.

**arcus**

Described by Dade as part of the arolium. A chitinous arc forming the shape of the arolium and attached to the manubrium of the foot. It spreads out when the arolium is brought into play enabling an adhesive pad to grasp on to a slippery surface, the claws having folded back.

**arenaria**

As a specific name this implies that a plant or creature is usually found on sand. Several of the solitary but gregarious bees use sandy nesting sites.

**Argentina**

One of the world's largest producers and exporters of honey. Over 50% of the crop is exported. 1980 crop down to 24,0000 metric tonnes, due to drought during the flow. 28,000 1983.
See: number of colonies

**Aristaeus**

In 'Archaeology of beekeeping' Eva Crane tells us that according to Greek legend beekeeping was invented by Aristaeus.

**Aristomachus**

58 years devoted to the bees – said Pliny – (Dr.H.Malcolm Fraser)

**Aristotle**

Greek philosopher and writer 384-322BC Probably wrote Historia Animalum during a 2 year stay at Mytilene 344-342bc in various books he gave insights into the workings of a hive. His was followed by pseudo-Aristotlean writings by beekeepers even more skilled. He wrote about bees and honey observing t e division of labour, bees flower fidelity, regurgitation and ripening of nectar, that bee bread was carried on the legs and that the queen was female while drones male. He noted there were water carriers but thought that pollen was beeswax and was uncertain about the generation of bees. He was a pupil of Plato.

**Arkansas bee virus**

A virus discovered by Dr. Bailey.
See: virus

**Armbruster Prof. L. 1920**

Bee Breeding Theory which influenced Bro. Adam

**Armeria maritima Plumbagineae**

See: thrift

**Armitt J.H. FRES**

Author of 'Beekeeping for Recreation and profit' This able beekeeper (BBKA Hons. Certificate & Expert, describes a double brood chamber technique for controlling swarming while fully utilising the queen's potential and for maximum honey production. A book relating craftsmanship; to the bee colony's natural development and instinctive behaviour.

**armour**

See: Safeguard (Keys) Dorothy Galton

**Arnaba foundation**

Moulded plastic foundation by Arnaba Kaneohe, Hawaii proved successful in Kaneohe 1980.
See: Stapla, Pierco.

**arolium**

Found on the pretarsus (foot of the honeybee) A median pad (or lobe) reinforced by the chitinous arcus. When the claws are extended the arolium is closed and folded in the raised position between the claws. When the claws fail to grip, on a smooth surface, the continued action of the muscles extend and flatten this arolium thus making maximum contact with the flat smooth surface.
See: arcus and empodium

**aroma**

A sweet, perfume-like smell or fragrance often with a subtle pervasive influence. A noticeable characteristic of many flowers, nectar and honey made from these e.g. mead and vinegar. The particular aroma may help to identify the floral sources and according to personal taste, help to determine the popularity of the item. Wax, honey, propolis and a colony itself each have distinctive aromas. Our taste may not coincide with that of the bees and the use of certain shampoos for example can lead to aggression from the bees. Unwanted aromas associated with fresh honey, often disappear with maturity.
See: odour, smell, scent, aroma of wax and bouquet

**aroma of wax**

Beeswax has a pleasing, characteristic aroma when fresh. Many associate the subtle smell with that of honey. Perfumes are sometimes incorporated into wax bases as these do not detract from the pleasure of the aroma. A sweet, subtle and characteristic smell. Show 'cakes of wax' are sometimes kept immersed in honey to lock in or enhance natural aroma. Certain processes of heating and filtration can destroy the aroma.

**aromatic compounds**

A substance that has an aroma – scent, sweet or fragrant containing an unsaturated ring of carbon atoms – benzene, naphthalene, anthracene and their derivatives.
See: benzene

**arsenicals**

Arsenic is a highly destructive poison, harmful to bees and to humans fortunately it has given way to pesticides just as effective but with a shorter residual effect.

**Artemis Greek Diana**
The moon-goddess, often figured in the shape of a bee.

**arthromere**
A segment or part of an arthropod's body e.g. tarsomere. Gk 'meros' meaning part.

**Arthropoda Metazoa Animalia**
A group of creatures belonging to the largest phylum in the animal kingdom. They are segmental, have an exo-skeleton of chitin and bilateral symmetry. They include the class Insecta under which the honeybee is classified.

**article**
A column or so of writing under a chosen heading with possible sub-headings the nature of beekeeping articles might cover a wide range but tend to be other than 'reports' 'stories or 'instructions'. A composition on a certain topic as used by newspapers, magazines/ reviews and other periodicals.
See: feature articles and cutting

**artificial feeding**
The provision of honey or pollen or substitutes as food. Unlike some animals bees will always regard a human provider of food in the same light as one who steals their food. It is generally agreed that the bees benefit best from being allowed to use their own natural food. However, up to 50% of their carbohydrate requirements for overwintering can be replaced by a 2:1 sugar syrup fed by mid-September and in special cases such as stimulation for early pollination, the feeding of pollen substitutes when natural pollen is scarce and the application of certain prophylactics are occasions when artificial feeding will also be undertaken.
See: candy, emergency feeding, feeder, pollen substitutes, sugar syrup and AL412

**artificial heating of hives**
A method of heating hives using a one-candle-power oil lamp was produced in the late 1940's by T. Alton of West Hartlepool. A metal sheet at the base of the hive was thermostatically controlled from the crown-board. It was claimed that the bees did not cluster, queens were kept in full-lay, supers were seen drawn out and honey yields improved.

**artificial insemination A.I.**
The introduction of the male gamete to the female ovum under artificial (usually instrumental) conditions. This has been carried out successfully for years as when using prize bulls to sire large numbers of cows. Sperm can be frozen in liquid nitrogen and later used as required. It has been possible to send drone sperm to the USA whereas live-bee imports were not permitted. There have been reports of damaged sperm showing up in the granddaughter queens. As a means of genetic control however, this technique provides a much more positive choice of parentage than natural methods. The sperm is taken from drones eleven days after emergence.
See: instrumental insemination.

**The art and adventure of beekeeping**
**Author: Aebi Ormond and Harry 1975**
**184pp NBB**

**artificial queen cup**
A small round cup (akin to an acorn cup) made out of beeswax and more recently from coloured plastics. When liquid beeswax is used a mandril is dipped in to a depth of about 10mm and allowed to cool. Repeated dippings to a lesser extent, allow a cup to be built up with sides of tapering thickness, yet with a solid base Cool water is used to wet mandril first and to help wax to solidify. A row of such cups is set up on a bar across a normal frame and eggs or very young larvae grafted (transferred onto a little 'royal jelly' that is put into each cup The frame with its prepared cells then goes into a cell-starting

colony.
See: grafting

**artificial swarm**
The removal of the queen together with a number of bees usually from the same hive. As a basis of many systems of swarm control or prevention, the queen is put into a new brood chamber while all the combs containing brood, together with adhering bees are moved away to another site. The flying bees remain with their queen and this acts as a substitute for a natural swarm. It is a useful way of rearing new queens and reducing congestion in the brood nest.
See: shook swarm

**The art of beekeeping Author W. Hamilton Herald 2nd impression 1971**

**arvensis**
Specific name used to imply that a plant or creature is usually founding arable fields, e.g. such as Sonchus arvensis or corn/sow thistle.
Ascosphaera apis The pathogen responsible for Chalk brood. The other genus in Ascosphaeraceae is the pollen mould- Bettsia alvei. Mycocidin licenced in US for the control of *A. apis*.
See: Chalk brood

***Ascisohaera alvei* spp.**
Akin to the fungus *A.apis* that causes chalk brood. *A.alvei* attacks neglected pollen turning it into a hard lump that is difficult for the bees to remove.
See: Bettsia and 'pollen mould

**-ase (aze)**
This suffix is used to indicate that it is an enzyme that has control of the substance that forms the root of the word, e.g. invertase as produced by workers.
See: chitinase

**asexual reproduction**
Reproduction without the fusion of gametes.
Ash Fraxinus excelsior Oleaceae
A deciduous native tree that bears inconspicuous flowers in April possibly yielding some pollen for the bees though many other sources are usually available at this time, The purplish-green male and female flowers appear well before the leaves and are mainly wind-pollinated.
"Oak before ash – we're in for a splash Ash before oak – we're in for a soak!"

**Ashforth feeder**
Like the Miller this is a box about 10cm deep that covers the whole hive, fitting on like a shallow super. It differs from the Miller in that the access slot is at one end of the box so that when this is at the front of the hive a slight forward incline enables the bees to lick up every scrap of syrup. Should the cluster be near the centre when it is required to feed, the Ashforth may be more difficult for the bees to utilise than the Miller with its central slots. It has to be used with a bee-proof cover such as a crown board.
See: overall feeder and Miller feeder

**Asiatic mite**
*Varroa jacobsoni*

**Asilidae**
Robber flies – predaceous insects that are widely distributed and having certain species that pay particular attention to honeybees in some areas. This leads to local names such as the 'black bee-killer'. Although many honeybees are captured and eaten no species singles out honeybees entirely but they include other insects amongst their prey.
See: fly

***Asparagus officinalis* Liliaceae**
Grown in the vegetable garden for its tips. It needs pollination by bees to set seed well. It grows wild in some coastal areas . The amber honey is not particularly exciting but pollen is also obtained. A perennial flowering June-July.

***Aspergillus flavus***
Causative agent of stone brood. Other species include A.fumigatus from which Fumidil is derived.
See: Fum B and stone brood

*Aspergillus fumigatus*
See: *Fumidil B*

**assay**
To examine by trial and thus make an assessment of the constituents of a substance. For instance the assaying of minute quantities of sugar in solution (nectars) has been rendered far easier by the spectrophotometric method. The processes become more numerous as science advances. Various forms of chromatography are also used for the analysis of various substances.
See: analysis, bio-assay, chromatography, fluorescence

**assessing weight**
Three supers of honey will, when processed by the bees, only fill one super with honey. Often, while this is going on large amounts of food are being fed to the bees. Beware therefore of 'counting your chickens before they are hatched' in other words take care not to over estimate the success of colonies in the spring. When hefting, it is necessary to know the 'feel' of an empty hive – wet wood, combs of pollen, perhaps some heavy item like a slate over the feed-hole, can easily give a false indication of weight.
A spring balance or visual reference to the hive contents can help to improve your ability to assess weight.

**assimilation**
The absorption of foodstuffs when the simple products of digestion are built into complex substances required by the organism. The conversion of the ingested and digested nutrient material unto protoplasm
(anabolism).
See: anabolism'

**association**
Biologically this can refer to associated units of vegetation or neurons adjoining other neurons etc.
Also used in the sense of its being an organized body of persons with a common aim. It is widely used for various groups of people within the beekeeping fraternity. They include BKA, BBKA, IBRA, CAB, BFA, AEA, BIBBA, CONBA, IBRA. Most beekeeping associations run on a county basis and are affiliated to the main British Beekeepers Association known as BBKA. County associations are composed of district oriented local branches which may in large counties be grouped into Divisions.

**Association of Beekeeping Appliance Manufacturers of G.B.**
A voluntary grouping of firms producing beekeeping equipment, the advantages of their co-operation could be the standardisation of equipment and the equitable distribution of stocks for the benefit of the consumers.
See: CONBA

**association fibres**
Sensory nerves are joined to the central nerve system while motor nerves function outwards from the centre to body muscles, glands etc. A third intermediary set are known as the association fibres and connect incoming to outgoing impulses which allows feed-back and reaction so that the bee can decide when to operate a gland or alter course to avoid an obstacle.
See: fibres and nerve cells

**Aster spp.**
The genus includes annuals and perennials like Michaelmas daisy and golden rod. In all there are over 400 species all good bee plants tending to autumn flowering Pollen yellow, honey amber.
See: Michaelmas daisy

**Athole Brose**
Equal part Honey and whisky the former

preferably from Scottish heather and the latter from the pot still. There was also athol porridge, a mixture of whisky and honey made in the Highlands of Scotland.
See: ambrosia

**Atkins E.W. & K.**
Wrote 'How to succeed with bees.

**Atkinson John NDB**
In his capacity as MAFF Beekeeping Adviser based at Trawcoed, Wales he specialised in disease control and was responsible for the technical side of foul brood control. Some fifty issues of technical notes to Foul Brood officers have been embodied in 'Beekeeping – the technical series' He is a fluent Esperantist and produced an excellent sound and colour film 'Nimfopesto' or Know your AFB

**atopy**
A genetically transmitted tendency to develop excessive quantities of IgE (immunoglobulin E type) antibodies in response to a wide variety of environmental factors.
See: allergy and immunoglobulin

**ATP**
See: adenosine triphosphate

**Astalagus spp**
See: vetch

**A swarm of bees in May**
The well-known rhyme handed down from generation to generation implies that an early swarm could be put to good use and would have the best chance of producing a sizeable crop. The silver spoon, linked with June suggests a reduced but never-the-less useful swarm. July swarms were not looked upon with favour and likened to a 'pig's eye' or a 'fly' August - disgust and you'll always remember the one in September.

**atrophied**
An organ that is atrophied is said to have wasted away through lack of nourishment. Its development can also have been inhibited by hormones and these might similarly have been absent so that the organ is not stimulated into activity. The ovaries of the honeybee worker are said to be 'atrophied' and remain vestigial as long as the pheromone 'queen substance' is received. In certain circumstances when the supply of this pheromone is cut off the ovaries develop and become functional.
See: laying worker and 'rudimentary

**asthma**
Laboured breathing due to a paroxysmal disorder. This can be brought on by bees, bee protein or stings.
See: bronchial spasm

**atrium**
The small vestibule or porch in the exo-skeleton at the entry to the propodial and abdominal spiracles. A court before a house – Dade

**attack**
To make an aggressive or offensive move. Worker bees can and will attack using their stings without hesitation then having scent-marked the victim, will cling and attempt to bite with their mandibles. Their action will excite other bees to follow suit. As man's association with bees extends back over so long a period of time most people are aware of this possibility but know that when not interfered, with bees normally go assiduously about their business of seeking food. Unfortunately at or in the proximity of their hives bees can be provoked into an attack.

**attendants**
A steadily changing group of worker bees. In the case of a normal colony during the active season they will be young nurse bees with well-developed brood food glands. Queens need to be adequately supplied with secretions from 'compatible' nurse bees and these receive 'queen substance' which is then

passed on to the other bees of the colony. It is these bees which control her egg-laying and at times the likelihood of swarming. Wherever the queen walks, nearby workers tend to turn towards her making her easier to spot in an undisturbed colony. Those bees who touch, feed or caress her will include 'attendants' Court, courtiers ,retinue and other names are used and those that escort a queen in a mailing cage come into this category.
See: escort

**attractant**
A Canadian Patent covers the composition of the attractant octodeca-trans-2,cis-9, trans-12-trienoic acid which has been extracted from pollen and makes feed more attractive.

***Aubretia A.deltoidea* Cruciferae**
A small, hardy perennial that hangs in masses over walls and banks its flowers yielding nectar and pollen being especially useful in sheltered, sunny spots. Its flowers linger until after the first frost of winter.

**audio-visual aids AVA**
Aids to teaching which incorporate sight and sound. Charts, diagrams and films, film loops or strips, illustrations overhead projector sheets, slides, transparencies and computers are examples.
See: under appropriate headings.

**August**
Honey is traditionally harvested this month. Be especially careful to prevent any possibility of robbing. Late flows from red clover, mustard, blackberry, rosebay or water balsam can occur but usually apart from ling and ivy it is all over. Decide which colonies with new queens can be strengthened for winter. This is the only month during the active season when bees can be left to their own devices (schools make note). Wait until next month to 'winter feed' but never let colony reserves fall below 10kg. Colonies that were prepared for the heather should be on their new sites as early in the month as possible, certainly by the 10th in the south.

**aurelia**
'Improper' use for pupa (Bevan) The chrysalis of an insect, from its golden colour

**auricle**
The ear-shaped upper end of the metatarsus. It is part of the hind leg's pollen press and takes the form of a toothed, sloping shelf that is fringed with hairs. Above it the distal end of the adjacent tibia has a corresponding surface which enables the auricle to slide against it thus compressing the pollen and pushing it into the corbicula. The tibial extremity is lined with a row of stiff spines 'the rastellum' which combs pollen from the opposite leg's pollen brushes and deposits it onto the auricle.

**Australasian Beekeeper**
Published by Pender Bros, Pty Ltd P.O. Box 20, Maitland, N.S.W. Australia. It claims to be the leading bee journal in the Southern Hemisphere.

**Australia**
One source states that the first honeybees reached Australia by sailing ship in 1812. An 1810 attempt had failed as the bees died. In 1822 Captain Wallace of the Isabella successfully delivered some black, German-type bees. In 1884 Kangaroo Island introduced pure Italians and this became the sanctuary for this pure race of bees. In 1983, 21.5 thousand metric tonnes of honey were produced and over half was exported and there were 2,500 beekeepers with 100,000 hives. Langstroth type hives are almost universal and migratory beekeeping, largely for tree honey, was the main beekeeping activity in (I980). In 2009 as they were one of the few areas free of Varroa they were able to supply great numbers of bees to U.S.A.

**Australian bee-eater *Merops ornatus* Meropidae**

A beautiful 10" bird (inc. its 6" tail) which includes honeybees it its food. It is golden green with azure blue throat feathers and rich yellow bill.
See: bee-eater and wood swallow

**Australian Bee Journal**

P.O. Box 313, Wangaratta 3677, Victoria. The official journal of the Victorian Apiarist Assn. See: 'Australian Bee-keeper'

**Australian honey**

The Federal Government sponsors The Australian Honey Board 647 George St, Sydney N.S.W. The Western states derive honey from; coral gum, mallee, karri, marrri, red bell, jarrah, parrot bush etc. while best known flavour in N.S.W. is yellow box. The Eastern states deriving honey from a multitude of trees such as: Caley's iron bark, spotted gum, stringy bark, and mallee while plentiful yields can be had from clover, lucerne and Salvation Jane.

**Australian Honey Board Films**

Were available from The Film Librarian, c/o Agent-General for Western Australia, 115 Strand, London, WC2 R OAJ.

**Australian honey colours**

Dark from messmate, blackbutt, blood wood, river red gum, jarrah and some mallee. Light from pink gum, dusky-leaved iron bark, green mallee, yellow box, napunjah, yarrell, Paterson's curse and orange blossom. Napunjah. and yarrell – quick to granulate.

**Austria**

Queen rearing in isolated valleys. Use of Bee houses. Journal 'Bienenvater' founded 1869.

**autogamy**

Self-fertilisation, the opposite to 'allogamy'
See: self-fertilisation

**autophilous**

Self-pollinated
See: self-fertile

**autopollination**

This sometimes occurs in plants where the flower parts are arranged so that the pollen can fall from the anthers onto the flower's own stigma so that pollination occurs automatically, without the intervention of an external agent. Usually autopollination is prevented or has to be supplemented by the action of an external agent.

**autumn**

The season between summer and winter. As little can be done with the bees during the inactive winter period autumn is the time when colonies are 'prepared' for the next active season. With honey crop taken a colony should still have adequate stores for overwintering. Strong healthy, well-protected colonies with young, vigorous queens make for maximum potential next year.
See: Sep, Oct, Nov and autumn feeding

**autumn crocus Colchicum autumnale Iridaceae**

Howes tells us that it exists in the wild as well as cultivated. Harwood refers to the saffron crocus Crocus sativa as a purple-lilac autumn flowering variety that is cultivated for its yellow stigma from which saffron is obtained for colouring purposes.

**autumn feeding**

This is likely to be carbohydrate in the form of a 2:1 sucrose solution. 2 lb. to the pint or 2.0 kg to 1.0 litre of water. Fed rapidly until in G.B. a total of 40 lb. of stores exists in the hive. No additives are recommended (save for Fumidil B in certain circumstances) and feeding should be completed before the end of September. The arrangements of the combs should not be changed when feeding has been completed.

**AVA**
See: audio visual guides
**Avebury Lord**
Wrote Ants, bees and wasps

**Avetisyan's hypothesis Prof. G.A.Avetisyan USSR**
Colonies that are moved from one region to another continue the rhythm of foraging behaviour customary to their home locale.

**avoidance of stings**
To minimize the likelihood of stings, only approach beehives when wearing protective clothing. As bees are naturally inclined to be aggressive when dark, rough, hairy, smelly, jerky creatures move near the hives, it behoves one to look to personal hygiene, move slowly and carefully and to wear white or light coloured clothing. Also make sure that it is properly adjusted before getting into the vicinity of the bees. Sensible methods of subjugation and the use of a docile strain of bee help. Should weather or other conditions cause a colony to adopt an abnormal reaction, postpone opening them up until conditions have improved.
See: aggression, docility, strain and temper

**awareness**
See: sense of

**axillary sclerite**
There is a lst, 2nd and 3rd axillary sclerite associated with the articulation of the root of the forewing. These are known as articular sclerites and they are small sclerotized plates tilted by direct muscles (basilar and subalar) inside the thorax. They pull alternately on the leading and trailing edges of the forewing or the basilar and subalar so changing the 'pitch' of the wing and helping the bee to perform its characteristic 'figure of eight' wing movement and controlling yaw, roll and elevation.

***Azalea pontica* Rhododendron spp. Ericaceae**
Said to have been responsible for the poisonous honey referred to by Xenophon in 'Anabasis'. Bearing in mind the translations through various languages and the distance away in time and space, positive identification of plant bee and honey are not likely now.

**Azotobacter**
Azo means nitrogen. It is a nitrogen fixing bacterium that is found in the root nodules of many leguminous plants. It works best in chalky soils and takes nitrogen from the atmosphere, using it together with carbon dioxide and water from the soil to produce amino acids. It obtains its energy by the breakdown of carbohydrates, setting free carbon dioxide. The amount of nitrogen fixed is proportional to the amount of organic material decomposed.
See: rhizobium

# B

**baby bee**
The honeybee babies are the brood which ranges from eggs to the emerging imagines. Although a bee that has just emerged may be referred to as a baby bee, looking downy (pubescent), hesitant and enquiring, brood is the baby form of the bees.
See: brood, larvae and young adults.

**baby drones**
Drones not yet adult, range from the egg (which is visibly indistinguishable from that of an egg destined to become either a queen or a worker) to the emerging adult. The normal drone is reared in a drone cell approximately 6mm across, and the plump larva is obviously larger than the equivalent worker. They are downy-looking on emergence and are not properly mature for another two weeks.
See: drone brood and dwarf drone

**baby queens**
The young of the honeybee are referred to as 'brood' A baby queen would be the 'princess' starting life as an egg identical to any other female egg. Its life as a larva however is spent in a cell that is either purpose-built or a worker cell extended to a size some three times that of a worker. The larva is held in a vertical cell, suspended upside down, being held by suction (molecular attraction) provided by the special food that is placed there. This could be linked to the bees' earliest connection with wasps whose combs provide vertical cells for all the brood, held in a similar way to a honeybee queen. A continuous diet of 'royal jelly' (the richest of brood foods) leads up to a seven day period of metamorphosis before they are ready to emerge as virgin queens. They can 'pipe' even before their cell is uncapped, though the sound is muffled compared with the shrill 'peep' that she can make when she gets out.

**baby workers**
Not a suitable term for the earliest stage of the honeybee's life when all eggs, unsealed and sealed brood are referred to as 'brood'. They remain as brood until they emerge as 'adults' (imagines). Newly emerged workers, 'young adults' are not difficult to spot as they are slower moving and downy-looking, lacking the shinier appearance of older bees. They spend five days longer metamorphosing in their cells than a queen and the larvae are not much different to look at than drone larvae apart from their size.
See: brood, young adult and nurse bees

**Babylonians**
They used honey as an ointment (having been mixed with butter into a paste) for treating eyes, ears and other purposes. Ritually it was concerned with magical purposes and divination for consecrating new buildings and restoring temples and for purifying the body. It was also used 'against' the eclipse of the sun. Figures of wax were also made and used for sympathetic magic, -(Frank Vernon).

**Bacillus**
A genus of bacteria which produce spores in the presence of free oxygen and are usually rod-shaped. Frequently they are pathogens and associated with a particular disease or

abnormal condition. As many appear to be endemic, it would seem that perfectly healthy bees are able to act as hosts to varieties which could, if the equilibrium of the colony is upset, lead to ill health.
See: B.larvae, bacterium and threshold

***Bacillus alvei***
A spindle-shaped pathogen associated with E.F.B. While it is in no way the prime cause, it is a common 'secondary invader'. There is also a para-alvei which behaves in a similar fashion and is said by L. Bailey to be biochemically indistinguishable.

***Bacillus apisepticus***
A pathogen thought to be responsible for 'septicaemia' a condition that can be associated with the presence of the Acarine mite.

***Bacillus brandenburgiensis***
And B.burri synonyms for 'B larvae'

**Bacillus larvae**
The causative agent of AFB White. Re-named Paenibacillus larvae. A rod-shaped bacterium the spores of which are introduced into the larvae gut by nurse bees. Here they become motile and after having been inhibited by the early bactericidal effect of the food they develop rapidly in the haemolymph and proliferate in the larvae tissue resulting in the death of the larva when it becomes quiescent prior to pupation. The spores are highly resistant to heat, desiccation and chemicals and remain viable for long periods in honey. The spores 'glow' under u/v.
See: bacteriostatic factor and gamma radiation

***Bacillus laterosphorus***
One of the most easily identifiable secondary invaders found in cases of EFB. It is a spore-forming comma-shaped bacterium.

***Bacillus thuringiensis* Berliner**
A bacterium used effectively in the control of lepidoptrous larvae. It does not appear to have any effect on honeybee brood or adults when it has been incorporated into beeswax foundation or when sprayed onto combs. Most effective against the larger wax moth – *Galleria mellonella* but *Achro grisella* is comparatively resistant to it. The warm moist conditions in the hive probably cause deterioration of the spores and crystals in the wax. Sold in USA as Thuricide. 3 tablespoons per gallon of water are used three times at seven day intervals as comb spray. Also 'Dipel', 'Brotrol Biotrol' 'Bactur' 'Sandoz' and 'Certan' Hot wax destroys it.
See: Certan

**back ache**
Discomfort if not actual spinal damage is frequently thought of as a honey-farmers disease. Lifting and loading heavy supers has caused many a slipped-disc or other consequent dorsal weakness. Lessons in basic lifting techniques and the use of suitable equipment help to prevent this trouble using the strength of the legs rather than the back and holding heavy weights as close as possible to the body are brief hints. Nevertheless many deep Langstroth users have had to give up or convert to shallower sizes.

**backcross**
A cross from a hybrid and one of its parents or between heterozygote and a homozygote.

**back end**
The end of a season especially late autumn
See: autumn

**Backyard beekeeping**
Author: W. Scott suitable for WBC enthusiasts.

**Background to beekeeping**
Author: A.C.Waine Collinwood 1955.

**back-lot beekeeping**
The back-lotters are the hobbyists or enthusiastic beekeepers, those professional,

business or other people who keep a few hives of bees for recreation, amusement or to make a little money on the side.

**bacteriophage**
An agent such as a virus, that can kill bacteria. Hence 'bactericidal'

**bacteriostatic (bacteriostasis)**
Descriptive of an agent having the property to prevent or control the deleterious growth of bacteria without actually killing it as in honey or the gut of a bee.

**Bacterium pl. bacteria**
**Bacteriophyta**
The most primitive phylum of plants. Widespread uni or multi-cellular organisms of microscopic size with a simple nucleus. Commonly known as germs or microbes; many are responsible for diseases. Types include Bacilli rod-like, motionless or motile. Spirella – curved or spiral rods and Cocci – globular (they form in masses or chains). Some are aerobic and others anaerobic.

***Bacterium eurydice* White**
Another secondary invader in E.F.B. It originates in the gut of nurse bees and is fed to the larvae. This is not always noticeably harmful but when Streptococcus pluton is present, it develops in very large numbers and the larvae often die before being capped over.

**badger *Meles meles* Brock Mustelidae** ***
A nocturnal omnivorous mammal that includes wasp and honeybee larvae in its diet. It can push hives from their stands but is more likely to demolish a wasp's nest. Some are quite large (over two feet long) with a black and white stripped face. The honey badger (ratel) *Mellivora capensis* is a honeybee pest in South Africa. A fierce animal that can cause devastating damage to hives of bees. Sadly efforts have been made in Devon to exterminate badgers by 'gassing' them, ostensibly to prevent the spread of bovine tuberculosis.
See: honey badger

**Bagster Samuel**
Author 'Management of Bees'
(Includes plate of British black bee)

**Bailey. L Sc.D Ph.D (Lond.) F. Biol.**
Retired from the Scientific Staff Bee Dept. Rothamstead Experimental Station 1980. He took a degree in Natural Sciences at Cambridge University and specialised in entomology. He worked for 18 months as a hospital biochemist and since 1951 was at Rothamstead Experimental Station doing research work on micro-organisms that infect honeybees. He has lectured all over the world and is considered one of the greatest authorities on honeybee diseases. In 1963 he published Infectious Diseases of the Honeybee' and numerous scientific works. He gave the Gooding Memorial lecture to the Central Assn. in 1984.

**bait combs**
Combs that will encourage bees to move into, occupy and remain, where the beekeeper wants them. Any disease free comb having the aroma of wax, propolis etc. will attract bees and empty hives set out with empty drawn comb (preferably some having had brood in them) are used to entice swarms to investigate and possibly move into them.. When it is required to encourage bees to move up into an empty super, combs of unsealed food or even foundation smeared with syrup may be included. Unsealed combs of brood help to hold onto an otherwise fickle cast.
See: bee-bob and baited hive

**baker's honey**
Honeys that are suitable for industrial uses or as an ingredient in other foodstuffs which are then processed. They may have a foreign taste or odour, have begun to ferment or have fermented or have been overheated. Honeys below the normal standard either as regards having a higher water content,

having been heated or possibly having an unattractive colour or odour and while edible are only considered suitable for cooking or use in industrial processes. (Defined by EEC Regulations).
See: 'industrial honey'

**balance**
A normal, healthy colony is balanced with respect to the amount and nature of its brood, the functions of its queen, number of drones and workers of various ages. The successful interplay of their various activities enables steady progress to be made in a suitable environment. Imbalance such as too many nurse bees (see 'Gerstung') too few flying bees (perhaps reduced by spray poisons), stress or instability may be associated with disease, disorder, robbing, winter depletion or other abnormality.
See: spring balance

**bald headed brood BHB**
A condition indicated by uncapped or partially uncapped brood cells showing the head of the pupae. Faecal signs may indicate that Achroia grisella (lesser wax moth) is responsible and occasionally its larger cousin *Grisella mellonella.* However bees will uncap sealed brood where it is dead, dying or parasitized and although further development to adult stage is possible, deformity of legs or wings is often noticed in the resulting adult. Dire starvation will also result in the uncapping of sealed brood.
See: A.L.561

**Baldensperger Phillipe James 1856**
Author L'Apiculture mediterranienne 1885 He tells us that Melissae were priestesses of Artemis and that the high priest was called the King Bee. He discovered yellow Saharan bees in S. Morocco.

**Balearic Island bees**
See: Spanish bees

**balling**
The smothering of a queen by worker bees. Rarely is it matricidal but in some colonies numbers of virgins, surplus to the seasonal requirement, are dispersed of in this way. Tight knots of seemingly furious workers literally hug a queen to death. The reigning queen can suffer this fate when a colony is disturbed as by manipulations too early in the season. Perhaps the bees aim at protecting their queen (or believe that she is responsible for the commotion?) A lump the size of a walnut forms around her and may even hang on for a day or two finally resulting in her demise. Blowing on them smoking them or dropping them into water may cause the breaking up of the regicidal knot and if performed soon enough it may sometimes be possible to carefully re-introduce her.

**balsam Impatiens spp Balsaminaceae *****
Half-hardy annual that spreads easily and often follows waterways. Varieties in orange, *I.capensis*, pink and yellow, *I.noli-tangere* (touch-me-not). They flower from July to the frosts. *I imphorata* (Policeman's helmet) has large flowers which encompass the bee splodging it with white pollen. *I.glandulifera* (Indian balsam) does the same in yellow. I.roylee (Himalayan) has been widespread enough to attract London beekeepers to the Welsh Harp for its late flow. An interesting feature is the ripe seed pods which, when touched, burst open and scatter their seeds. These will float along a river and sink to establish themselves on banks.

**bambara**
Sri Lankan name for *A.dorsata* or Giant Rock bee. D. Galton explains that 'bambara' was the Old Indian name for 'bee' (onomatopoeic?)
See: A.dorsata

**Banat bees**
Hungarian bees of gentle disposition. Similar to Carniolan but claimed to be a distinct race. Utilised and investigated by Bro. Adam.

**Banksia (bottle brush) *B.integrifolia* Protaceae Proteaceae** ***

A decorative Australian shrub with bright red flowers with the appearance of a cylindrical brush. They are prolific yielders of nectar which can be shaken from the active blossoms. Orange, yellow and white varieties are to be found. Golden candlesticks *B.collina*, White bottlebrush *Callisteman saliganus.*

**bar**

(a) shutting or sneck, a piece of lead (or latch) to shut up hive entrance (Butler. Cotton)
(b) Strip of wood at the top of the box, basket or skep from which bees would suspend their comb (Wildman, Isaac, Key and Cotton) First seen in Greece says Hawkins in 'Cotton'
See: anastoma

**barb** ***

A small projection which is shaped to penetrate into or catch hold of surrounding material. We are familiar with the barb of a fish-hook and know that a worker honeybee's sting has a twin series of small barbs which make it difficult for the bee to withdraw its sting once it has been sunk into rubbery material. The sting of wasps and of queen honeybees is not barbed and can be withdrawn without harm to the user.
See: lancets

**Barbeau**

Canadian who in the 19th century invented folded metal supports for queen cells and publicised his method of queen raising. In recent times the Barbeaeau cell punch is used to chop a cell containing a grub of the right age, from the brood comb for transfer to an artificial queen-cup as an alternative to grafting.

**barberry *Berberis vulgaris* and spp. Berberidacae** ***

An evergreen prickly shrub, useful for hedging that flowers Mar/May with orange-yellow clusters. These become decorative red berries so that as an ornamental plant it looks well most of the year. It likes chalk. The nectar and pollen are useful during the June gap. Pollen colours pale dull yellow, pale yellow and pale green with a 36mu gain.

**bar frame**

The first attempts to govern the exact place in which bees build their combs was the use of 'bars' or slats of wood baited with a strip of wax or comb placed across the top of the container in which they were intended to stay. When the more advanced notion of entirely enclosing the comb within a frame was thought of, the word was extended to 'bar-frame. This meant that the bees were enticed to build their comb wholly within a wooden frame.
See: bar hive and movable frame

**bar hive**

A hive in which the combs were built from or secured to wooden bars. It was noted in British literature in 1682 that the Greeks used wicker baskets with top bars.

**bark hive**

A hive made from suitable bark. Countries like Spain and Portugal produce large amounts of cork and hives made from bark of this sort of material have been used there for centuries. Thick bark has good insulative properties though it cannot be machined into the precision-made sizes that are required for movable frame hives Consequently they were fixed comb hives – alternatives to the jar, skep or log types. In Zambia and Tanzania the smooth thin bark of trees like *Cryptosepalum pseudotaxus* protected with straw is used to make hollow cylinders which are hung in trees.

**Barret G.B.**

In 1919 following Dzierzon's discovery of parthenogenesis, he confirmed the fact that by fertilising drone eggs by hand and raising normal queens from them.

**basal**

At or near the base, forming the base. E.g. 'basalar' a sclerite near the wing base.

**basalare**

A plate hinged to the pleurite associated with the fore wing root. Sclerite below insect wing base, it swings inwards on its hinge when the basalar muscle contracts. It works in harmonious opposition to the functioning of the subalar muscle allowing twisting adjustments of the wing in flight. Refer to the work of Pro. J.W.S. Pringle.

**basal joint**

The connection between the segment nearest the body and the sclerite.

**basement membrane**

A thin tough layer underlying the exoskeleton and to which muscles are attached. A membrane of specialised connective tissue beneath the epithelium as of special secreting portions of glands. Compare with basilar membrane before accepting this above definition.

**Basic Beekeeping 1978**

Author: Owen Meyer. Produce your own honey, propolis and royal jelly. Build hives and ancillary equipment. Step-by-step guide to the beekeeper's year.

**basil Ocimum spp. Labiatae *****

An aromatic culinary herb of the mint family with sweet and bush varieties. It is a tender, exotic plant and seedlings from April sowings should not go out into rich light soil until hardened off in June. Its flowers offer sweet rewards to the bees.

**Basil-thyme *Acinos arvensis***

**basilar**

Innermost layer of body wall.
See; 'basement membrane'

**basisternite**

A ventral abdominal plate or sclerite also referred to as a 'sternite.'

**basitarsus**

The largest of the tarsal segments and variously called 'palma' (mace) and 'planta' though this latter is found on the foot. The tibio tarsal joint that is the connection between the tibia and the basitarsus carries the antennae cleaner on the fore legs and the pollen press on the hind legs.

**basket hive**

See: skep

**Bath & West Show**

The Show Ground, Shepton Mallet BA4 6QN Founded 1977 for the encouragement and improvement of agricultural and commercial technology.

**Bath W.H.**

Wrote: 'Ants, bees, dragon-flies, earwigs, crickets and flies

**batik**

The use of coloured beeswax to produce 3D effect. Areas not to be dyed are waxed and coloured designs can be sequentially applied. Eva Crane tells us the technique originated in Java. Bailey described use of beeswax, paraffin and water colours in painting on Pelon (non woven fabric). Encaustic (hot wax) painting. Fine beeswax, melted and mixed with powdered pigment and applied by brush or spatula to almost any surface – Beecraft.

**Bazin G.A.**

Wrote: Natural history of bees
(1744 translation of Reaumur's work).

**B.B.J.**

The British Bee Journal Founded May 1873 by Charles Nash Abbott
Succumbed to opposition from Bee Craft etc. but re-vamped by BBKA in 2012.

**BBKA**

British Beekeepers Association

Exists to promote and further the craft of beekeeping. Founded May 16 1874
See: The many other functions:
Exam Board, Library, Sales, etc. etc.

**BBKA Beekeeping standards**
A colony: Bees including healthy brood at all stages, a fertile healthy queen and six B.S. deep or 4 other larger combs Stock: A colony and the hive it occupies.
A nucleus colony: A colony on not less than 3 B.S. deep combs
Diseases: any of the above must be free of visible pests/diseases.
The queen: The appropriate age of the queen should be stated.
Brood area: At least half the comb area to have brood at all stages, well covered with bees and total drone comb not to exceed 5%.
Frames and combs: With bee-space around frames, the combs should be fully built-out and wired. Frames securely pinned or glued and square.
Food: At least one side, adjacent to brood to have half of its area stored.
Bees: Swarms, packages by weight, queen as above and food container sufficient to ensure survival.

**BBKA Comprehensive Insurance.**
A voluntary insurance against loss or damage to hives, equipment or bees by fire, flood, theft, accidental or malicious damage and against third party claims for damage to property in connection with beekeeping activities or accidental body injury or illness caused by poisonous or other deleterious substances in food or beverages supplied by the insured member.
See: comprehensive

**BBKA News**
A bi-monthly Bulletin sent to all members of BBKA.

**BBNO**
See: Bee Books New and Old

***BDI***
See: Bee Diseases Insurance

**bean Vicia/Phaseolus spp. Leguminosae *****
A plant yielding proteinic food in a seed of characteristic 'bean' shape. Like most legumes they benefit from honeybee pollination and yield nectar and pollen and some honey-dew. Horse or Field beans 'V.faba equina' are grown 'en masse' and sensible growers pay a pollination fee to get best results. Beans will still produce seed without pollination but the pollen foragers more than make up for this. Honey is medium to dark, pleasant flavoured and the pollen pale yellow. The broad bean *V. faba* attracts aphis (as does the field bean) and insecticides may be used. *P.multiflorus*, the stick or runner bean needs cross-pollinating but yields little nectar. Beans prefer slightly acid soil pH5x5 - 6x8 and phosphorous/potash are essential.
See: field bean and runner bean

**bear *Euarctos americanus* Ursidae** ***
Black bears which may be black, cinnamon or shades of brown. They have a great fondness for the honeybee brood and will attack colonies at night, smashing a hive to get at the honey and brood. In Canada, America and Russia the annual damage done is enormous. High platforms and electric fences help and one beekeeper suggested sprinkling moth balls under the fence so that bears might sniff there and so increase the chance of getting a shock also the tying of bacon to the fences has been mentioned. A careful selection of sites helps. They are very alert to detection by man. Good climbers, fast runners and able swimmers. Their sense of smell is keen and hearing acute and they are harder to hunt than deer. Weight 90-225 kg, solitary and shy. Ursus arctos; the Eurasian brown bear. 700-800 colonies destroyed annually in Central Russia mainly by young bears – (Morse quoting 'Toumanoff')

**beard**
1.A cluster outside the entrance on hot days (Cotton)
2.barb of sting (Sydserff in Cotton) Probably

due to French word for 'beard' Bees often hang out as a cluster in front of the entrance when the stock is strong but not necessarily going to swarm. It happens especially when honey ripening calls for the entrance to be free for ventilation.

**bear fruit**

The yielding of fruit following flowering. Plants grown for their fruit will often bear masses of flowers but little fruit. The formation of well-shaped fruit results from the transfer of suitable pollen at the moment when the flower can accept it. This calls for the timely availability of suitable pollen in weather conditions that encourage its transfer, pollen tube development to the flower's ovary, fertilization and conditions conducive to subsequent development into fruit. Bees are rarely improved upon as pollinators.

**bearing**

A horizontal angle relative to a given datum line (north) giving a specific direction. The bee does not use the sun itself but the polarised light pattern which it can see, cloudy or not, as its datum. Furthermore as 'down' (the line indicated by gravity) is the datum it uses on the comb to signal the direction of food to other recruits, the direction of the 'shadows' (down-sun) is the external datum equivalent to the internal datum of gravity. However the sun moves constantly and the bee flies a zigzag course and air movements too have a considerable effect on the bearing a bee must follow. A one degree error misplaces the bee one metre for every 60m flown.

**Beaton hive**

A smaller version of the Glen hive. by J/K/ Beaton, taking 12 frames as against 15 in the Glen.
See: Glen hive

**Beck . B.F.**

Author *'Honey for Health'* 1938

**Beckley Peter**

Author: *'Keeping bees'* 1983

**Beddoes frame cleaner**

Vernon Beddoes double bladed frame scraper with smaller blade for removing frame wedge and built-in hook for removing wax from grooves in side bars. Set in polished wooden handle Reported in Bee Craft in 1983.

**bee** ***

A four-winged stinging social insect. Inc. honeybee, bumble, carpenter and others.

**bee anatomy slides**

26 b/w. Visual Publications Ltd.
The worker complete and then dissected to reveal the various organs from BBKA (still?)

**bee balm *Melissa officinalis* Labiatae** ***

Tiny white scented flowers. Lemon scented leaves – were used by skeppists to induce swarms to settle and occupy the skep. It grows up to almost a metre in height.

**bee behaviour**

A 20 min b/w film Educational Foundation for Visual Aids 33 Queen Ann St London. Doubtless superseded by now.

**bee tree *Evodia danielli, henryi, hupensis, glauca* and *velutina*** ***

Evodia means pleasant odours. Discovered by Earnest Wilson of the Arnold Arboretum in China (Shensi Hupeh) it was introduced to America in 1908. Of small to medium size though large specimens can grow to 60 feet. Its fragrant white flowers are smothered with bees in August. It is semi-ornamental (citrus family). Said to take eight years to first blooms. Also native to Korea and is semi-tropical, dioecious and also known as the Chinese bee tree.

**bee blower**

An impeller type fan usually driven by a petrol engine 1500cpm @ 3,600 rpm and weighs 48 lb. Small portable compressors (usually petrol driven are able to force a powerful jet of air along a flexible hose and out through a flat nozzle. The action of this dislodges bees from the comb face and hive walls. Used in conjunction with a light stand or chute in front of the hives, bees can be precipitated in front of the entrance and supers cleared of bees quickly and efficiently. Its use was often followed by a bout of robbing.

**bee bob (bob - a cluster or knot)**
A lump of material made to look like a swarm and suspended near the hives in an attempt to encourage any swarm that issues to settle thereon. . Its efficacy is more likely when there is a scarcity of other suitable objects for them to swarm onto. Suitable smells (extracts from the scent glands etc.) associated with the bee bob also help as bees once attracted by sight are given further encouragement when familiar smells or pheromones are present.
See: swarm lure

**bee bole (bee garth)**
An ancient hive shelter or purpose-built cavity built into the wall of a garden forming a shelf just wide enough and high enough to shelter and accommodate skeps. Sometimes an iron bar was hinged so as to swing across to be padlocked as a protection from theft. A register of known bee-boles is kept by IBRA.

**Bee Books New & Old**
Who provide and have available a very large selection of wonderful books.
John Kinross, Tapping Wall Farm,
Stathe Rd Burrowbridge, Nr. Bridgwater,
Somerset, TA7 0RY

**The Bee Boy's Song (by Rudyard Kipling)**
3rd verse:
*Marriage, birth or buryin,*
*News across the seas,*
*All you're sad or merryin,*
*You must tell the bees*
*Tell 'em coming in an' out,*
*Where the Fanners fan,*
*Cause the Bees are just about*
*As curious as a man!*
(extend?)

**bee bread**
Pollen stored in the comb. 'Farina' (Cotton) D. Galton. The use of this expression goes back to the earliest writers who doubtless saw pollen in the baskets of bees and possibly in the cells and called this food 'bee-bread'. Bee bread would of course be devoid of starch.

**bee-brush** ***
Conventionally a goose wing which presented the beekeeper with a firm grip yet whose fine feathers at the trailing edge could dislodge bees from their comb without causing them harm, was used as a bee-brush. Now-a-days modern brushes with fairly long hair laid out in a row of tufts have come into use as has soft foam-rubber attached to any suitable handle. At short notice a tuft of freshly plucked grass may serve. The aspect of hygiene should always be born in mind when more permanent brushes are used.

**bee byke**
A nest of wild bees.

**bee cap**
A cap or hood with wire gauze to protect the face.
See: cap

**beech *Fagus sylvatica* Fagaceae** ***
A tree that produces early pollen and honeydew later. Pollen a yellowish-green 30 mu.
See: Nothofagus, beech honeydew honey and bush honey

**beech honeydew honey**
Whereas the English beech does produce honeydew, in New Zealand plant sucking scale insects produce honeydew in vast quantities. The Southern beech 'Nothofagus spp. harbours over 30 species

of scale insects. The most important of these is Ultra coelostoma assimile Maskell Homoptera Margarodidae commonly known as the sooty beach scale insect or 'beech buggy'. The honey is clear and dark and does not granulate. At one time this honey fetched high prices but since then other honeys have taken pride of place, e.g. 'Manuka'.

**Bee Craft ( Ltd).**

Originally a monthly magazine by the Kent BKA it was adopted as the national Bee Journal. Obtainable by subscription or included in many cases within the subscription to the local BKA.

**The Bee Craftsman**

Author H. J. Wadey A short guide to life story and management of the honeybee 1947.

**bee culture**

This expression seems to have given way to the more widely used 'apiculture' though the excellent beekeeping journal from America is entitled 'Gleanings in Bee Culture'.

**Bee Diseases Advisory Committee**

A group of expert beekeepers set up under the auspices of the then MAFF to co-ordinate and advise Government concerning beekeeping matters and to prepare and revise such Advisory leaflets, Bulletins etc. as thought fit. Meetings occurred at frequent intervals during the seventies. See: Bee Husbandry Committee.

**Bee diseases Insurance Ltd**

Specialist member association of BBKA

**bee dress**

Apparel, attire, clothing, garb – The aim of wearing suitable protective clothing might be to permit such body movements as are necessary while handling honeybee colonies under various conditions. Some old-fashioned garb was truly quaint. Now-a-days modern fastenings and materials have made great advances. See: apparel and protective clothing.

**bee-eater Merops spp. *Coracii formes* Meropidae**

A genus of long-tailed, brightly coloured birds that feed on flying insects like the honeybee. Allied to the kingfisher. They nest in banks and seem to like venomous hymenoptera and have learned to rub the bees' abdomens against small branches to get most of the poison away. They omit this process when dealing with non-stinging insects. Varieties include : 'ornatus' and European Bee-eater *M. apiaster* also supersiliosus (rare straggler) the blue checked bee-eater and the Little Green Bee-eater *M. orientales* – 4 spp found in Iran.

**bee enemies**

See: predators

**bee-escape**

A small passage which by reason of a funnelling or conical effect, or by means of light springs. It permits the passage of bees in one direction but hinders or totally prevents their return. Bee-escapes may be built as an integral part of a clearer board or as small devices that fit into holes cut into crown or clearer boards. Care must be taken to ensure that they are not put in upside down and that they are clear of debris or gum.

**bee fair**

A fair where the major theme of beekeeping dominates. In the passing years trade stands, exhibits, displays and demonstrations have been supported by great numbers of beekeepers at the Stoneleigh Agricultural Centre. A film show, convention, meals and refreshments with ample parking facilities, offer interest, education and the chance for beekeepers from all over the country to meet and to talk. Local fairs that include a bee tent or the inclusion of a beekeeping section in the wine or flower tent are also popular summer activities.

**bee farmer**
A person who spends a great deal of his time farming honeybee colonies. The term commercial beekeeper and honey farmer are also used. It could be argued that a honey farmer's sole aim is to use bees for the production of honey and the bee farmer is more concerned with the production of mated queens, package bees, nuclei, pollination etc. but the terms are usually interchangeable.

**Bee Farmers Association The**
Open to beekeepers with more than 40 hives who are proposed and seconded by existing members. They issue a regular bulletin, circulate magazines and arrange honey jar deals, pollination contracts and other useful services.

**bee flowers**
See: *melliferous flora*

**bee fly Brachycera Bombylidae**
A two-winged fly that looks like a small bumble bee. The larvae of some species can be parasitic on solitary bees.
See: Bombylidae

**bee food**
Pollen, water and nectar that may be transformed into honey are the natural food of the bees. Nectar is processed into honey and pollen is 'prepared' and they can both be stored for future use. Honey requires the addition of water before it can be utilized. On a diet of nectar (or honey diluted with water when nectar is not available) adult bees can exist but nurse bees are only able to develop and utilise their brood food glands when pollen is available. Then they can secrete brood food for the older larvae and royal jelly for young larvae and the queen. The honey sac holds food that can act as fuel while a bee uses it to forage, to swarm, mate or just to fan.
See: artificial feeding, brood food, food sharing, nutrition, pickled pollen and royal jelly.

**bee forage**
See: bee flowers, bee plants, crops and forage

**bee garb**
See: apparel, bee-dress, dress, protective clothing, and Sherriff

**bee-garden**
A plot devoted entirely to plants that are attractive to bees and which are at the same time a source of much pleasure to the beekeeper interested in plants and gardening. F.N. Howes wrote 'Plants and Beekeeping' giving useful ideas on setting up a garden as well as thousands of useful details about bee plants.
See: bee-garden (bee bed)

**bee-garden (bee bed)**
In days gone by these were areas within a forest where 'beekeepers' were allowed to put their hives. In the new Forest, once a 'royal hunting preserve', it was only where natural enclosures existed that bee gardens could be established.

**bee-garth**
A bee-garden (apiary). In the lovely old gardens of days gone by arched recesses were built into deep walls to hold skeps that hummed with life. Amy F. E. Lioney 'The Bee Walk'
See: bee-bole

**bee gloves (gauntlets)**
Bees are likely to crawl up inside the sleeves while you are manipulating unless some kind of protective covering is used. Gloves with extra-long cuffs and elasticated ends, help to keep bees out and to prevent stinging. Smooth, kid leather gloves are still popular though cheaper rubber and plastic ones are available. Gloves also keep the hands free of propolis. Unless they are washable, gloves can spread disease. Elasticated cuffs that leave the hands free are an alternative. A good fit and comfort are important and when returning a frame

to the hive by holding the centre of the top bar one can so avoid trapping finger tips under the lugs.
See: elasticated cuff and gauntlet

**bee glue (propolis)**

This is a substance that bees collect as exudations from certain plants or failing such sources might even be collected from the tar of roads or other similar man-made materials. Some races use more than others. Bro Adam has produced a strain that uses practically none. It is a sticky substance, especially when warm but becomes brittle when cold and shares many of the properties of beeswax with which it is quite miscible. It is used to form a 'shell' around honeybee's nests and to fill small gaps which are too small for the bees but might let ants or other small predators through. It is used to reduce entrance size and received its name propolis from the Greeks who noticed pillars of the gum were constructed to prevent the entry of predators such as the Death's head hawkmoth. Colours vary through yellows and browns to red and black. It breaks down in alkalis, is insoluble in and heavier than water. It has anti-biotic properties but can harbour certain bee diseases however it can be sterilised by the use of a blow torch.
See: propolis

**Bee-go (*butyric anhydride*)**

Advertised in America as a bee repellent. When sprinkled on a cloth of absorbent material or used on a fume board after the bees have been smoked, it clears supers in a few minutes. One pint clears 300 supers. It was available from Steele & Brodie in 1983.

**bee gum A. log hive in the U.S.**

See: propolis

**bee hat**

A hat chosen for use when manipulating the hives or working in their vicinity. It would normally be light in weight and have a stiff brim to permit the fitting of a suitable veil. Straw boaters were much used at one time but strong resin-bonded fibre, moulded into suitable shapes, and light collapsible hats of closely woven material, often incorporated into a combined hat and veil, are present day alternatives.

**bee health**

The condition of balanced, normal development where a colony displays no indication of the harmful presence of pathogens or parasites. The term bee health is preferred to the word disease when the subject of bee health is discussed. So much has been said and written about honeybee diseases that it is very important to know how to maintain health in the apiary.
See: apiary hygiene, disease, hygiene and health checks.

**bee herd**

A person who watches the rising of a swarm (Keys) A person who takes care of bees, H.H. Germans use the word 'Imker' meaning 'bee-shepherd'

**beehive - beestall, skep, stock, hive.**

A domicile, chosen by man for the accommodation of a honeybee colony. Some folk use the word 'hive' when they mean a nest of honeybees, such as is frequently found in nature. Beehives may be ready-made, like a log or clay pot or purpose-built using all manner of materials to provide the bees with a weather-proof shelter in which their combs can be built. A hive has to be acceptable to the bees because to live naturally they must be free to be able to exercise their right to leave it whenever they wish.

**The bee-hive**

Author Dorothy Galton. An enquiry into its origins and history. A limited edition Northern bee books.

**bee house *****

While in the U.K. beehives are conventionally kept out in the open they are occasionally kept in specially designed out-buildings. Such buildings where hives are lined up inside with an individual entrance for each hive, are known as bee houses or house apiaries. The normal habit of storifying does tend to restrict the height of beehives in a bee house and probably for this reason together with the high cost of timber (from which bee houses are normally made) they are a comparative rarity in Britain. In places on the continent where as yet there are great forests and timber is cheaper, bee houses are much commoner. See: bee-shelter and house apiary.

**bee-hunting kit**

The age-old art of bee-hunting may be pursued in the USA using a Jurica kit which includes a beautiful redwood bee-hunting box with bee-attracting scent and full directions ABJ July 1982.

**Bee Husbandry Committee**

This was a committee of expert practical beekeepers elected by the BBKA and in 1984 under the Chairmanship of W. E. J. Hooper.

**beekeeper (apiarist)**

A person who pursues the craft of beekeeping ranging from novices or beginners to bee master or expert. People have kept bees of all sorts and for all manner of reasons but whatever they are the keeping of bees enables us to describe that person as a 'beekeeper'.
See: apiarist and bee-man.

**Beekeepers' Association (B.K.A.)**

Since 1848 when the British Beekeepers' Association was first formed, non-profit making groups of beekeepers have collectively formed working associations covering a particular area or region. They usually affiliate to the BBKA and conform to ordinary democratic practices concerning subscriptions, AGM's, minutes, meetings and the appointment of officers. There are national specialist organizations such as AEA, BIBBA, BFA, and CAB, as well as the international IBRA and Apimondia. Local groups such as DARG in Devon are also operational. See: association, beekeepers club, and beekeeping newsletter.

**beekeepers' calendar**

With space for meeting notes and beautiful pictures and monthly notes about current apiary work and plants in flower from BBKA which also produce a wealth of useful and educational material.

**beekeepers club (association)**

A club is an unspecified number of persons who co-operate in pursuing a similar interest so that any two or more beekeepers can join to form a group and entitled to call themselves a beekeeper's association. The first BKA was formed about 100 years ago. Now local groups tend to be affiliated to the central body known as the British Beekeepers' Association (BBKA) and internationally the World Organisation -'Apimondia'.
See: BBKA, and Beekeeping Assn.

**Beekeepers Guide**

Beekeepers Guide, Folly, Handbook, Manual, Encyclopaedia and other titles are too numerous to mention.

**beekeepers' journal**

The word 'journal' originally meant a newspaper or similar daily sheet.
Now it is a term used to report what might be day-to-day happenings though at greater intervals. The British Bee Journal for instance at one time came out weekly, then fort nightly and after struggling along against competition as a monthly it finally closed down. A beekeepers' journal thus implies a publication dealing with matters of interest to beekeepers and published at reasonably short and regular intervals.

**beekeepers' meeting**

The officers of a BKA will normally arrange regular meetings, say once each month. An

Agenda will be drawn up and all members sent a copy with a reminder of the time and place. A chairman will aim to cover the items on the Agenda as adequately and expeditiously as possible and under any other business (a.o.b.) scope will be given for other matters to be raised. Meetings are often arranged to coincide with talks or shows, these being the responsibility of the Secretary.

**The Beekeepers' Rule**
A circular slide rule designed by Dr. R.S.Pickard of University College Cardiff. The life-cycle of the three castes can be set against a time scale running from April to September. This enables colony management (stimulation, re-queening etc.) to be planned to achieve peak colony development or arrange for queen matings at the chosen time of the year. This may still be available from E.H. Thorne.

**beekeeping**
Apiculture or the use of honeybees by man. See: apiculture and Bulletin.

**beekeeping book**
The title of 'Beekeeping' is used so often for books that it would be unwise to attempt to list them here.

**Beekeeping at Buckfast Abbey**
Author: Bro. Adam successful beekeeper and specialist queen raiser.

**beekeeping – definition**
Managing honeybee colonies to obtain maximum population for the major nectar flow in the area and to utilise this for the storage of honey or the pollination of crops. 'Bee World' Vol..60 No.3 1979.

**Beekeeping do's and don'ts**
Author: Tickner Edwardes (famous old beekeeper). Also Beekeeping for all by the same author.

**Beekeeping examinations**
Junior, Preliminary, Intermediate, Practical, Senior, Judging, Hon. Lecturing, Fellow of BBKA. The N.D.B. (National Diploma in Beekeeping) requires a candidate to show that he is of Senior standard. America has instituted a Bee Masters exam.
See: examinations and under each heading.

**Beekeeping in antiquity**
Author: H.M. Fraser 1951

**Beekeeping in Britain**
Author: R.O.B. Manley. Well-known beekeeping specialist.

**beekeeping in schools**
Although this subject can be singularly appropriate for schools the tendency to pressurise teachers and to cut costs has resulted in far fewer schools using hives of bees as part of their educational programme (70's and 80's). When schools do use apiaries it is always wise to have at least two beekeepers on the staff as the constant movement of teachers in the scramble for promotion so often leaves a school with embarrassing, unwanted hives. Observation hives and temporary mini-nucs, have proved most educative, there was also a CSE option in beekeeping. Co-operation with CBL'S (County Beekeeping Lecturers) proved very useful when such posts were 'extant'.
See: school beekeeping

**Beekeeping in the tropics**
Author F.G. Smith 265 pp. 1960

**beekeeping in towns**
The keeping of hive in so-called built-up areas is common place and most beekeepers who are hobbyists live in or near towns. Although one's own property might not be suitable for an apiary, a diligent search often enables a nearby site to be found. The normal precautions regarding the number of other hives in the immediate vicinity (regarding disease, overcrowding etc.) should always be taken but the sheltered

micro-climate and hopefully reduced spraying risk often brings larger crops to town beekeepers than to those in the country. It is wise to hide or camouflage hives whether in town or country, because 'neighbours' might blame you for every wasp sting or mosquito bite and unfortunately pranksters and ne'er-do-wells may well be tempted to interfere with a hive made obvious by its shape and colour.

**beekeeping instructor**

A person skilled in the craft of beekeeping who has the facilities for teaching others how to successfully manage honeybee colonies. She might be one who takes beekeeping into an educational establishment, helps in disease diagnosis and who reports on and writes about apicultural matters. Various terms are used such as: 'instructor', 'lecturer' 'advisor' 'assistant' 'tutor' etc.

**beekeeping newsletter**

These mainly carry local news concerning meetings and occurrences of importance to local beekeepers. More often than not they are sent out at regular intervals (say monthly) by local BKA Secretaries to keep members enthusiastic and well-informed.

**Beekeeping Old and New**

Author: W. Herrod-Hempsall.

**beekeeping outfit**

Whereas in Britain the word outfit would be considered more appropriate as a description of the clothing a beekeeper might wear, in America the term is used to cover the property and equipment required to conduct a business of any sort. Therefore a 'beekeeping outfit' means an organisation of equipment and personnel operating in the realm of apiculture.

**Beekeeping questions and answers**

Author: C. Dadant 1978

**beekeeping shed**

An enclosed non-residential place of one storey for the purpose of storing beekeeping gear and possibly also as a place where indoor work can be carried out. The value of such a shed is enhanced by methodical lay-out and purpose-built fitments. These might include labelled hooks, shelves and other means of retaining and displaying items such as the smoker, frames, wax foundation , cupboards, drawers and a bench complete with a vice and other wood and metal working apparatus. Such a shed is a great asset at an out-apiary and particularly if it can be made bee-proof.

**beekeeping show**

When the BBJ was first published it triggered the formation of the BBKA and the first Beekeeping Show at the Crystal Palace. The BBKA has now become really famous for its enormous benefit to beekeepers. The Honey Show ranging from the National Honey Show to shows all over the country also help beekeepers to put carefully prepared honey and other bee products before the public.
See: display, exhibiting and honey show

**beekeeping (subject)**

This is the commoner expression to cover apiculture or the husbandry of colonies of honeybees. Many apicultural journals and books rejoice under the title of 'Bee-keeping' and in Great Britain anyone can keep bees without registering or having any sort of licence. While membership of a local association is highly beneficial it is estimated that only around 50% of beekeepers do belong to such groups. Most groups are affiliated to the national BBKA. Colonies that have established themselves in the wild are not thought to entitle their owners to the label 'beekeeper'. The true beekeepers have one or more hives in their care. There are hobbyists, side-liners and Commercial or Bee Farmers.
See: BBKA

**Beekeeping techniques**
A textbook for Diploma exams by A. S. C. Deans

**Beekeeping – the gentle craft**
Author: J. F. Adams 1972.

**bee-killer**
Philanthus triangulum (wasp)
See: bee wolf.

**bee lippen**
Beehive (Somerset) Anglo-Saxon leap – a basket.

**bee-louse**
See: *Braula coeca*

**bee man**
Nearly all beekeepers become known as the 'bee-man' by neighbours and those non-beekeepers who are aware that he keeps bees. Its use is in the sense – 'Ah! Here is the chap who pursues this rather unusual pastime - complimentary in the main but with a little awe, a little incredulity and perhaps a hint of an enigmatic smile.

**Bee-Mason**
A famous Sussex beekeeper (Burgess Hill). He imported large numbers of stocks in skeps from Holland in 1914. He favoured 'black and tan' an old Dutch bee. Pioneer on bee cinematography. He sent 1000 swarms to all parts of the U.K. to help re-stock following the I.O.W disease.

**bee-master**
Those beekeepers who become entirely capable and are judged by other beekeepers to be knowledgeable and experienced. Those who pass the BBKA Senior exam are referred to as bee-masters or expert beekeepers.

**The Bee-Master of Warrilow**
Author: Tickner Edwardes.

**Beemasters of the Past**
Author: Victor Dodd 1983.

**Bee matters and masters**
Author: Herbert Mace

**Bee-milk**
The food that nurse bees give to the larvae. To make this food worker bee glands are activated by the consumption of pollen and either nectar or diluted honey. In a developing colony bees eat soon after emergence and begin to produce brood food. Within a day or so and continue until their glands begin to atrophy. Over-wintering bees and bees that have been queenless can sometimes rejuvenate their glands. Each caste is fed slightly differently and workers receive a wide range of mixtures beginning with 'royal jelly' and finishing with a less rich diet.
See: bee nutrition, brood food, and pap.

**bee miller**
A colloquial expression for the 'wax moth'
See: bee moth and wax moth

**bee moth (wax)**
This refer to either the lesser or greater wax moths though other moths such as the 'Death's Head Hawk moth will invade a hive to steal honey. However the singularity of wax moths, lies in their life-cycle being dependent upon the food provided by wax and comb made by the bees.
See: greater and lesser wax moth

**bee moth *Aphomia sociella* Orchidaceae**
A common pyralid moth (Microlepidoptera) the larva of which lives in the nests of bumble bees and wasps.

**bee paralysis**
See: ABPV and CBPV

**bee parasites**
*Acarapis woodii* (dorsalis etc.), *Braula coeca*, *Melittobia acasta* and *Varroa jacobsonii*.
See: honeybee parasites

**bee pasture**

The masses of blossoms within range of the bees. The most successful foragers recruit more and more bees to specific plants so that great selectivity is shown wherever there is ample forage. At any one time a single colony will require up to an acre of flowering plants. Large trees, parks, suburban gardens, riversides, moors and agricultural crops which require pollination all contribute to the bee pasture and beekeepers generally find that it is better to position hives where good bee pasture exists rather than to try to influence the locality by growing plants for the bees (though it is always worth doing the latter).
See: *melliferous flora*

**bee pest**

A deadly or troublesome parasite, creature or pathogen, which affects bees.
See: enemies of the bees, mites, parasites and predators.

**bee plant**

Any plant that is made use of by the bee whether for the collection of propolis, pollen, nectar or honeydew.
See: melliferous flora and plant.

**bee pollinators**

Honeybees, bumble bees, *Megachile rotuda, pacifica* (alfalfa), *Nomia melanderi* (alkali bee) unidentata (two generations each year), *Osmia lignaria* (mason bee), *Fulviventris* (Spain) cotton, *Cornifrons* (to Betteville from Japan) *Exomolopsis* (not named), *Pithitis smaradila* (alfalfa) –native India.

**bee press**

Magazines, journals, bulletins, leaflets, newsletters, yearbooks etc. are printed in English and most foreign languages, by the thousand. These constitute the 'bee press' which for the most part prints responsibly and helpfully about apiculture (this cannot always be said for the general press, e.g. 'man makes 'bee-line').
See: journal, Bee Craft etc.

**bee-proof**

This is an adjectival description of something capable of excluding live bees. We would expect a veil to be bee-proof, to make our honey house bee-proof and to keep combs in a bee-proof place. This can be done by screens of the right 'mesh' when air has to be admitted, otherwise by closing all gaps so that it is impossible for a bee to gain access, as for example in a clearer board which only allows passage of bees in one direction and then remaining bee-proof against their return. So many honey crops have been lost by inserting a clearer board under supers without checking that the supers themselves were bee-proof that we have no hesitation in repeating this caution.
See: clearer board and bee-escape

**bee protein**

The aggregates of amino-acids found in the body of the bee. Nitrogenous, albuminous and often capable of producing an allergic reaction when introduced into or brought into contact with a human body. Constant doses applied indirectly to beekeepers' spouses and family can create a dangerously allergic state. Dr. Frankland states that a person may be able to withstand 100 stings but that there is enough bee protein in one sting to kill someone who has become allergic to bees. The bee protein in honey can cause acute indigestion in some people.
See: allergy, propolis dermatitis, sting and venom components

**beer**

An alcoholic beverage made by fermentation. Distinguished from ale that is flavoured with hops or other suitable flavouring. The sugar required for fermentation is derived from starch in cereal grain (malt). Honey can be substituted for all or part of the sugar. The difference between honey wine (mead) and honey beer lies in the more open fermentation (relying on a layer of carbon dioxide to exclude oxygen), the strength of the 'must' and the added flavourings, hops for beer while spices and acids might be used

with mead. The amount of honey would be in the neighbourhood of three pounds to each gallon of liquid.
See: honey beer

**bee repellent**
A substance capable of minimizing the attractiveness or even of driving a bee away from any given place. Such substances can be useful in keeping bees away from certain pesticides, to clear honey supers and in some instances to act in a subjugating fashion. Examples include: 'smoke', 'benzaldehyde', 'proprionic anhydride' the commonly used 'carbolic acid' (phenol) and 'Butyric acid'.
See: repellent

**bees (Apoidea)**
Small buzzing insects which are conspicuous during the summer and feared by some one account of their defensive weapon – the 'sting' There are some 20,000 species of bees in the world.

**Bees Act**
May prohibit or regulate the importation into or movement within Great Britain of bees and combs, bee products, hives, containers and other appliances used in connection with beekeeping or transporting bees, and any other thing which has or may have been exposed to infection with any pest or disease to which the order applies.
See: Stand-still notice

**Bees & Beekeeping**
This title has been used repeatedly: Examples – Cheshire, Deans, Morse, Pavord. Woodbury.

**Bees and honey**
Author: C.A. Carter

**Bees and Mankind**
Author John B. Free Natural history of bees of the world and the part they have played in the history of man. Entomology Dept. Rothamstead Experimental Station

**Bees and people**
Author: Ioyrish 1977

**Bees and wasps**
Author Cloudsley-Thompson – Bodley Head 1974

**bees attract bees**
See: bee bob, gregarious and swarm lure

**bees' blood**
Haemolymph, water, protein, amino acids, glucose, trehalose, fats, salts, hormones, haemocytes (corpuscles). It does not transport oxygen but it does carry carbon dioxide to a limited extent.
See: haemolymph

**bee scep (or scap)**
Hive of wicker or straw O.E.
See: skep

**bees choice of site**
In nature a swarm shows positive indications of selecting a site. This often begins by scout bees finding, beginning to clean and even mounting guard on suitable new premises. That these are not always taken up, may imply that other bees from the same colony may have been instrumental in finding a better site or that, after all, the colony decided not to swarm.
See: swarm's new home

**bees' duties**
The first three days after hatching will be spent feeding and repairing cells for the queen to lay in. From the third to thirteenth day nursing larvae. Then they follow varying house duties, cleaning, moving ripening honey, stowing nectar and pollen, sealing cells, comb building and guarding. Orientation and cleansing flights linked with removal of debris and cleansing flights. Young nurses begin by feeding brood within two days of sealing and progress to younger brood and the queen as their brood food glands develop more fully.
1-3 cleans cells,3-6 feeds older brood, 6-10

feeds younger brood, 8-12 feeds queen, 11-18 house cleaner, 12-18 makes wax and builds comb, 18-20 guard, 20 on –forage nectar-water-pollen 25 may collect propolis
See: division of labour, comb building, wax making and workers' duties

**Bees, flowers and fruit**
Author: Herbert Mace

**bee shelter**
Open shelter that allows bees to fly freely while protecting them and any manipulator from wind and rain, are not common in the U.K. though several survive from the past and are described in 'The archaeology of beekeeping' (Eva Crane). Such shelters are easier to construct, though less secure than bee-houses and offer some protection against cattle. They are rarely used for modern hives.
See: ramada

**bees killed**
Large numbers of bees may form a carpet of dead in front of the hive. This should not be confused with drone slaughter in late summer or autumn. Poison spray could be the cause and the colony may recover from such a 'knock-out' if all house-bees are intact. A dead colony will take the form of a 'silent' cluster with a heap of dead underneath if killed by starvation, most likely time, in the spring. Robbing, mice or other predators, careless use of prophylactics, suffocation (if closed or moved) and many other inadvertent or induced snags may lead to the demise of a colony. A large number of dead will inevitably be thrown out when a fine day permits a clean-up after a long spell of winter confinement.

**bee sounds**
Although bees are generally supposed to have little or no sense of hearing, they do in fact make a large variety of very distinct sounds and are clearly able to detect certain vibrations. The whirring of dance sounds, the whine of a trapped bee, moan of a queenless colony and the hiss of a queenright one, the flight notes each of the castes and the 'piping' of a queen are all examples of sounds that the bees make. Eddie Wood's Apidictor filters out the note that nurse bees overcharged with brood food make by suppressing other sounds and from its volume predicts the likelihood of swarming. Perhaps the sounds that please a beekeeper most are the contented sounds of colony ripening its stores or a swarm accepting a new hive.
See: apidictor and sounds

**bee-space**
8mm ranging 5-10mm
Since early times it has been noticed that bees do not cram every part of their nest with comb or propolis but leave clearways between their combs and the adjacent surfaces so that bees can move over the face of their combs. Langstroth was amongst the first to draw attention to the specific use of bee-space to make movable frame hives and although each race of bee may differ slightly, the clearance for bee- space is regarded as 8mm ranging from 5-10mm. Where less space is provided it may be filled in with propolis and where there is more space comb is usually built. Either way subsequent manipulation becomes more difficult.
See: bottom and top bee-space

**bee stall**
beehive O.E.

**bee stings**
A colony in normal circumstances will only sting to protect itself when this becomes necessary because it is threatened or interfered with. Although only a small percentage of the total number of bees can ever be provoked into using their stings, as the colony membership may run into as many as sixty thousand, even a small percentage can be quite enough! A human being has survived over 2000 stings though many died after receiving far less.
See: aggression, beestings, docility, swelling and temper.

**beestings**
The first milk from a cow after parturition… Not a bee word but probably derived from 'beast- ings' in the same way as 'maltings' and 'saltings'

**bee suit**
An overall or boiler suit, preferably in one piece, makes a useful 'bee-suit' as part of a beekeeper's protective clothing. Variations such as a one-piece tunic and veil, or even entire head covering, provide a light, bee-proof item of apparel which when used in conjunction with impermeable boots, gauntlets and a veil as described, offer adequate protection against be stings. Many appliance manufacturers sell suits in various chest sizes. Trouser length is not normally critical as their ends go into long boots and waist size is not particularly important provided it is adequate.

**Bees, vision, sense, language**
Author: 'von Frisch'

**Bees, wasps and allied insects (of the British Isles)**
Author: E. Step

**Bees, wasps and ants**
Author T.M. & L.T. Duncan.

**beeswax** ***
Unique to honeybees it is produced in the form of eight platelets from the wax producing glands of middle aged worker bees (mostly house bees) who masticate it to build comb. It contains 16% hydrocarbons, 31% monohydric alcohols, 3% diols, acids and esters. Melted combs are filtered to yield this relatively inert substance. Its physical properties are reasonably constant having a melting point of 142-149 F (62-65°) becoming plastic at 85°F (brood is incubated at 95°) SG ×952 - ×975. It is harder than paraffin wax (the addition of stearic acid makes it even harder and is useful when making candles. It is yielded best when groups of workers hang in festoons having loaded themselves with syrup, honey or nectar. When liquid it is very penetrative, has a similar colour to honey. It is brittle when cold has a soapy feel and characteristic aroma, its colour vary from pale yellow to dark brown. It was at one time more valuable than honey and has been used in hundreds of different ways (e.g. ear plugs, hollow tooth stopping, for polishing, water-proofing, face cream, pill coating. See: beeswax definition

**beeswax (definition)**
The melted and refined honeycomb of the bee. It has a translucency and lustre, floats in water and burns with a bright sweet-smelling flame. Soluble to a large extent in acetone, alcohol, carbon tetrachloride, chloroform or ether.

**Beeswax (excellent)**
Author: Ron Brown

**beeswax candle** ***
The temperature for 'dipping' is fairly critical 82°c – 180°F but when made the delicate texture and smooth, pale, characteristic, golden shade (akin to golden jade) has inspired men to use these for Liturgical purposes. Their clear light, sweet aroma and the virginity of the wax has been used by Christians to typify the flesh of Christ born of the Virgin Mother. The great paschal candle represents Christ, the true light. The wick is the soul and the flame of divinity absorbs and dominates the flesh and the soul. Paschal pertains to the Passover or to Easter.
See: candle, stearin, wax and wick

**beeswax – composition of**
There are more than 300 individual components of beeswax. Mainly hydrocarbons they include: methyl esters, alcohol acetates, acetoxymethyl esters and diol diacetates.
See: beeswax, beeswax properties of and waxes various.

**beeswing**

A filmy crust of tartar formed in port and similar wines after long keeping – 'beeswinged'.
See: flight wings, forewing, hind wing and hooks

**bee talks**

These may be talks of general interest about beekeeping to a non-beekeeping audience and might include the dispersal of misconceptions. Since the great wave of interest which became evident in 2009 you can talk to enthusiastic audiences who are now motivated to protect the honeybee. Talks to beekeepers will be educational and with the intention of arousing interest both in the manner of keeping bees and the improvement of apicultural activities. The BBKA are great leaders in such affairs and slowly their influence is beginning to take hold in Educational Establishments.
See: lecturer, press, speaker and speaking

**bee-tight**

Having no cracks, holes or small openings through which a bee or wasp) could gain access. This is a most essential condition when, at the end of the season, supers are put over a clearer board so that the bees will vacate them. Generally a hive should be bee-tight except for the entrance and secure ventilation. In moving hives of bees great care should be taken to avoid over-heating by using screens that are bee-tight yet permit adequate ventilation.
See: bee-proof and empty

**bee tree**

This term might be used to describe a tree that offers nectar, pollen or propolis to the bees. Also for a tree accommodating a nest of bees such as implied by the word 'gum' in the States. Space for space a tree's canopy might well offer a larger area of forage to the bees than a similar area covered with flowers. Acacia, Almond, Apricot, Lime and Chestnut are good examples.

**bee tent**

During the active season netted structures are often used to make enclosures for a temporary apiary. In Australia where a canopy of Eucalypts might yield copiously, tents are sometimes used for extraction purposes so that wet supers can go straight back on. When demonstrations are given at honey fairs, poles, ropes and canvass, with black netting on the side open to the public may form an open topped tent which provides the spectators with the facility of watching while having little fear of coming into actual contact with the bees even though the bees are free to fly over the observer's heads to their normal foraging grounds.

**bee trap**

A device, man-made or natural which detains a bee or bees, and prevents them from pursuing their normal functions. Traps are used to catch swarms, prevent drones from entering or leaving a hive, to get bees out of supers (the Porter bee-escape is an example) and to prevent a queen from passing up into the supers (queen excluder). Escape cones and other tapering slots through which bees are only likely to be able to pass in one direction, are used to get colonies out of trees and cavities (air bricks) etc.
See: bee-escape, cone escape, drone trap, and queen excluder

**beet sugar**

White crystalline sucrose derived from the starch-laden root of certain varieties of beet. It requires extensive factories and vast quantities of beet which are available under the EEC and this has cut down on the importation of cane sugar though there is still a demand for the brown (golden) cane sugars. Only white forms of cane or beet sugar are suitable for the feeding of bees. Sucrose is also obtained from the sugar cane and sugar maple.
See: sugar beet

**bee veil *****

A thin protective film of material through which the wearer can see and air can pass, yet which is quite bee-proof. Black is the only suitable colour as the fine mesh required would, in the case of any other colour, restrict the visibility by cutting down the amount of light which can pass through the veil. Gauze, nylon and cotton nettings are used and the various considerations are ease of manufacture, proofness, fit, lightness of weight, storability and ease of fitting. Many are made one-piece with a suitable hat, some even with a tunic or suit attached. It should stay-put in all conditions and should not easily be pulled off or blown against the skin. Two supporting rings (one of which may be the brim of the hat, will keep the veil clear of the face and neck while elastication at the top and bottom enable the veil to grip a hat and fit tightly around the chest.

**bee venom**

A complex mixture of toxic proteins – a vigorous allergen. Further information is given under 'bee venom activity', 'bee venom components' and 'venom'

**bee venom - activity**

According to Wm. Shipman, U.S. Navy Biology, San Diego as reported in 'Gleanings' March '79 – Phospholipase A – the allergic portion of bee venom, turns lephasine into a poison that destroys red blood cells. Mellitin (80% of venom) a strong detergent with a + electrical charge that draws itself into body cells helping the venom to permeate tissue.
See: venom

**bee venom – components**

Major ones are underlined and major allergens marked *AllergenC*, Hyaluronidase*, Phospholipase A2*, Melittin, Apimine, Cardioppep, H.C.D. peptide, Histamine, Minimine, Glucose and fructose, Water 88% Acid phosphatases

**beevert**

Drivert sugar mixed with 1% crushed pollen, used as a pollen supplement. Sheesley and Podusk 1968/9 ABC & XYZ

**beeves**

Bemyng - buzzing O.E.

**bee virus x, y**

Strains of ABPV
See: virus y found in 40% of colonies that had any trace of 'Nosema'. They are isometric particles 35mm related to virus x but commoner.

**bee walk**

Ancient rights in the forests of Europe allowed beekeepers to set-up and visit honeybee nesting sites and taxes were paid according to occupancy by the bees. Forest beekeeping was so successful that Dr. Crane opines that it probably delayed the setting up of apiary beekeeping. However there was often a conflict of interests between the forester and the forest beekeeper.

**bee-way**

The space between comb faces and between combs or frames and the end walls or surroundings. It is the maximum and minimum range of space that bees will tend to leave unencumbered, and is the space that bees require to pass back-to-back in the performance of their duties.

**The bee-way Code**

A booklet by 'DARG' 1981
*'For a safe and peaceful apiary'*
A useful booklet aimed at identifying and suggesting answers to safety problems arising from the handling of bees and establishment and management of apiaries.

**bee widow**

Used in jest to describe the wife of a beekeeper whose enthusiasm and or involvement keeps him away from his home and his wife so much that it is implied she is a widow. The condition does not exist to such a marked extent when a wife shares her husband's interest.

**bee wolf *Philanthus triangulum* Fabricus**

This is the name given to a bee hunting, solitary, digger wasp several species of which use adult honey bees to feed their larvae. Sandy soil and cracks in pavements are used for their nests and females may hunt bees at flowers and the hive entrance. They are reported as paralyzing bee's leg and wing muscles but not those of the heart and gut. Honey contents used to feed larvae while the adult finally consumes the bee.
See: *Philanthus triangulum*

**Bee World**

See: IBRA regarding quarterly issues of this journal.

**bee-yard**

A collection of hives. An apiary or place where hives are kept (U.S)

**beggar bees**

Queenless colonies will feed 'beggar bees' through gauze, especially when the beggar bees come from a queenright colony. Bees that are trapped may pass food to free bees that beg from outside. Begging can take place when there is an aperture of some sort through which alien bees can solicit food without themselves being attacked.
See: robbers

**beginner**

A person who is taking their first steps in beekeeping. A novice or one who has commenced keeping bees but has much to learn about them. Their 'status' is lost when such a person wins a prize in a honey show or gives an authoritative talk or demonstration or passes a beekeeping examination. Some prefer to retain the title 'beginner' for quite a few years. Reluctance to give up the title may stem from the fear of losing some of the thrill that is associated with first acquiring a colony of honeybees. Unfortunately anyone wearing a veil and using a smoker is deemed by members of the public to 'know all about bees'.

**Belize**

A strongly flavoured honey, one of the cheapest on the market at the time, came from this S. American country once known as British Honduras, on the eastern edge of Guatemala.

**bellbird**

The notes of this small bird resemble a bell and its melodious sounds are heard in the New Zealand and Australian bush especially where hives are kept, as for example in the southern beech honeydew area around Bell Hill N.Z.
See: honey-eater

**bell glass**

A bell-shaped glass used to cover delicate mechanisms such as clocks. Also called a bell jar. These were frequently used as supers when wooden hives came into vogue.
See: Nutts collateral hive

**bell heath Erica spp *****

It blooms from June – Sept and often overlaps and flavours the 'ling' honey. It gives rise to a dense port-wine coloured honey and this is the pink colour it often imparts to the heather honey where both plans abound. The honey can be extracted centrifugally (unlike that of ling). Its pretty pink bells have led horticulturists to develop many cultivars which range in colour and blooming time.
See: Erica spp and heath

**bellows**

An apparatus designed to produce a blast of air as a draught for an open fire (blacksmith) or to sound an organ or other musical instrument. The application of the bellows to a fire chamber enabled the first 'smokers' to be produced and this was a huge step forward in the realm of subjugating bees. In some ways a bee's 'air-sacs' act as bellows and it is possible that a queen is able to use them in conjunction with her tracheae to make the 'piping' sound. A drone needs to have his air-sacs fully expanded before coition is possible.

**belt**

Girdle round hackle of skep.
See also 'gart' (Butler)

**beneficial insects**

Honeybees are of outstanding use to man. Silkworms have over 2000 years of useful history. Cochineal comes from the Mexican 'Dactylopius coccus' and shellac from 'Lacifer lacca', locusts are considered a delicacy by some, and ladybirds demolish crop devouring aphids. Some creatures like spiders consume tons of insects – mainly harmful but they are not choosy and wasps have good and bad sides to their character. Many minute parasites have been put to use as a biological control on specific pests
See: biological control, IPM and pest

**benefits from honey**

Honey being hydroscopic, hygienic and extremely pleasant to the taste, has been used beneficially through the ages as medicine, ointment, food and sweetener. It has also found fame in the fermented form of mead and vinegar. It is very easy to digest and in fact presents sugar in a most readily assimilated form. It provides rapid warmth and energy and doubtless assists the body to 'burn-off' other foodstuffs. It contains scores of ingredients practically all of which bring benefits to the human body whether used internally or externally. It is also a great cleanser.

**benefits of nosema-free colonies**

To quote from the CEVA international pamphlet regarding the use of Fumidil
a colony is less likely to succumb to other bacterial and viral infections.
less likely to suffer heavy winter losses.
less likely to have reduced brood production.
less likely to have reduced honey yields.

**bent-nosed smoker *****

Originally the nozzles on smokers were in line with the fire chamber. This meant that to cause smoke to be blown below the horizontal the smoker had to be canted up with its base higher than the nozzle opening. Apart from the awkwardness such an angle that could lead to the blowing out of sparks together with the smoke. The bent-nosed smoker has the hinged nozzle set at an angle of about 45° or so thus pointing downwards while the fire chamber is still held comfortably upwards. Root developed the earlier smoker by adding the 'deflected nozzle.

**benty**

Ground not proper for bees (Scotland) Bonner. Chambers gives 'bent' as dry wild grass – Dorothy Galton.

**benzaldehyde $C_6H_5CHO$**

Obtained from the natural oil of bitter almonds or synthetically from toluene. It is used as a bee repellent and is more volatile (especially in a moist atmosphere) than proprionic anhydride. Exposure to air and light oxidizes it to form benzoic acid (white crystals) that should be washed off wooden parts as they are combustible. Used on a cloth or in conjunction with a fume board or chamber and bellows it has proved satisfactory for subjugating bees and especially for clearing supers at 80°F (27°C). Water and glycerine mixtures work better than pure benzaldehyde (also 90% with methyl benzoate 10% flammable) at high temperatures while neat doses are needed to do the same work at low temperatures.

**benzene ($C_6H_6$)**

Also called benzol (ole). Of considerable economic importance as so many useful compounds can be derived from it. Nitrobenzene is used for Frow mixture (treatment for acarine). Benzine (also known as benzoline) is a hydrocarbon derived from mineral oils and used for removing grease stains. It is an excellent solvent and dissolves beeswax though it is poisonous and flammable and calls for discretion where food substances are concerned. It is used as a substitute for turpentine.

**beo-ceorl**

Anglo-Saxon 'beo' – bee and 'ceorl' churl An ordinary freeman of the lowest rank. From this too we get 'beo-mothor' – queen bee.

**Berberis spp**

See: barberry

**berseem**

Pollen white barseem, bersim
See: *Trifolium alexandrinum*

**bestiary**

A treatise on beasts as written in the middle ages. Beasts of course included honeybees.

**Betts Miss A.D. B.Sc.**

born London 1884 President of Apis Club and Editor of '*Bee World*', Wrote: P*ractical Bee Anatomy 1923* and '*The Diseases of Bees*'. She favoured old English bees was a BBKA expert. Used motor cycle for going to her work as an aircraft designer (Vernon).

***Bettsia alvei***

The pollen mould. A genus in Ascosphaeracea. The other genus in this order is the pathogen that causes 'chalk brood' *A. apis*. Formerly the names *A.alvei* and *Pericystis a.* were used. *B.a* is saprophytic on pollen stored by honeybees in the cells of the comb. The mould does not attack brood. It can convert pollen into a hard lump such as one often finds thrown out of a hive when a colony expands or a swarm takes over and cleans out some old combs.
See: pollen mould

**Bevan Edward Dr.**

Wrote '*The Honeybee*' in 1827.
Its Natural history etc.
Open to members to propose and second suitable beekeepers history Physiology and management. Dedicated to Queen Victoria.

**B.F.A. Bee Farmers Association**

who maintain at least forty colonies.
See: Bee Farmers Association

**BHC hydroxychlorocyclohexan**

This is a very toxic insecticide – a chlorinated hydrocarbon. Now HCH.

**B.I.B.B.A.**

British Isles Bee Breeders Association is an educational charity formed to promote the conservation, study and restoration of our native honeybees. Founded in 1964 as Village Bee Breeders Association becoming B..I.B.B.A. in 1972 a non-profit distributing, educational body formed to promote the understanding and welfare of the native and near-native bees of these islands and to arouse interest in the science and practise of bee breeding. Apart from excellent publications another useful thing was the availability of seeds of bee plants.

**Bible**

Genesis 43.11, Proverbs 24.13, 14th Chapter of Judges, Sam.14, Matt3.3, Mark1.6, Kings 14.3, 2Sam 17. 'They compassed me about like bees Ps118:12' And they gave him a piece of broiled fish and of an honeycomb Luke 14:12 More recent translations have omitted the reference to honeycomb. And behold there was a swarm of bees and honey in the carcase of the lion. Jg14:8
The proverbial metaphor '*a land flowing with milk and honey*' is repeated twenty-one times in the bible. My son, eat thou honey, because it is good' Proverbs 24:13

**Bielby W.B**

County Beekeeping Lecturer, Yorkshire. Inventor of the 'catenary hive' Wrote: '*Home Honey Production*'

**Bienentee**

Vitaminised winter food supplement as used on the continent for the control of 'nosema'. Once available from Woodlands Apiaries.

**biennial**

A plant that lives for 2 years requiring this time to complete its cycle. It usually flowers in the second year when it fruits and dies. See: annual, ephemeral and perennial

**bier**

For carrying hives (Isaac) D. Galton.

**big fly *Mallophora rufidauda***

This predacious fly looking somewhat like a bumble bee with a reddish tip to its abdomen is a threat to bees in the Buenos Aires region of Argentina. They are swift fliers, catch honeybees using their legs as a net and then kill them and suck juices (especially nectar-laden bees) from them. Their larvae use a beetle Phyleurus vervex to shield them through the winter.

**bike (mid Eng.)**

A nest of bees, wasps or wild bees. 'Delve out the treasures frae your bike'
Robert Fergusson 1750-1774.

**bilateral symmetry**

The form of many creatures including ourselves where one side is a mirror image of the other. Compare for instance with radial symmetry. Bees too are bilaterally symmetrical.
See: radial and zygomorphic

**bilberry *Vaccinium myrtillus***
**Ericaceae** ***

A small spreading shrub that likes an acid soil. It forms ground cover in beech and pine woods and is tough enough to grow on the moors along with heath and heather. From April to June its drooping pink flowers yield pollen. The leaves are bright green and slightly toothed. Its edible black berries, called 'pick'in-hurts' or whorts, are found in quantities and on Dartmoor 'pick'n-hurts' was regarded by the natives as a holiday in July and August. The fruits are also a useful source of food for birds and wild life. Whortleberry, blueberry and American huckleberry or hurtleberrry are other names.
See: cranberry

**bindweed Convolvulus spp**

Convolvulaceae ***An herbaceous climbing weed. Greater bindweed C. sepium, can be a real nuisance in gardens. Its pure white trumpet flowers are displayed throughout the summer,. Pollen light brown. C. arvensis lesser or field bindweed, can lie prostrate, its pink blossoms being worked for their off-white pollen, though they close in the rain.

**Bingham T.F. USA 1877**

In conjunction with L.C. Root he improved Quinby's original bellows smoker by including a blast vent, a gap between the bellows and the fire box. Later the more conventional bent-nozzle was incorporated. He also invented an uncapping knife etc.

**Bio-assay**

The quantitative determination of a biologically active substance within a living organism. E.g. the procedure adopted to discover whether dead bees have been killed by a particular chemical. A common standard used is the dose required to kill 50% of the test group of insects.
(Lethal dose, or L.D. 50).
See: analysis

**biochemistry**

The chemistry of living things. The influence of biochemistry is now recognised as all pervading. Dr. A. W. Sands.

**bioflavonoid**

Found in propolis and certain plant material (esp. citrus fruits) Vitamin P. They react with animal enzymes and metabolic products to maintain normal permeability of capillaries. E.g. chrysin, techtochrysin, galangine, pinocembrine, isalpinine (mostly aglucones).
See: flavone

**biological control**
The use of living organisms to destroy or reduce pest and parasite populations. Man's influence in using one naturally harmful source to control another. This is, for example, as opposed to the use of artificially produced chemicals. The use of citrus fruit peel (due to its repellent and insecticide oil) against ants etc. is one instance and another is the use of a small parasitic wasp '*Encarsia formasa*' which has been effectively used against greenhouse whitefly (*Trialeurodes vaporarium*)'. A predatory mite has been used against spider mite. Promising candidates for the control of wax moth were '*Bacillus thuringiensis* and nuclear polyhedrosis. 'Bigger fleas have little fleas upon their backs to bite 'em. The little fleas have lesser fleas and so on ad infinitum' The ignorant premature or over frequent use of powerful chemicals can drastically upset this regime – A.L. Winfield (ADAS).
See: Certan.

**biological control of Paterson's curse (Australian honey plant)**
Four insect species were imported into Australia from France and Portugal to combat the flea beetle 'Longitarsus acnus' the larvae of which feed by tunnelling into the tap root and '*Phytoecia coerulescens*' a longhorn beetle whose larvae bore into the stem of the plant. *Dialectica scalariella* – a moth whose larvae are 'leaf miners' and Dialectica s. have been released according to Dr. Tony Wapshire and quoted in Australasian Beekeeper July 1980.

**biome**
A pocket or small part of the biosphere where the conditions have led to a relatively settled complex of fauna and flora. This may be a gradient on a mountain, a major region such as the tundra or desert. The Mediterranean biome has plants which adapt to a dry climate and includes short, stumpy trees, aromatic evergreen shrubs – 'rosemary' and 'lavender' and honeys appropriate to the region will have characteristics that do not vary much and can be identified quite easily.

**Biopoll (made in Germany)**
Trade name for a mixture of soft candy impregnated with pollen enzymes, natural vitamins and *Fumidil B*. The food is fed as normal candy in spring and it is claimed that it brings colonies forward. In 1981 this was available from Morris Beekeeping Supplies

**biosphere**
The region between levels 6000m above and 10,000m below sea-level inhabited by living things. An envelope some 20km deep enclosing the earth.

**biotic**
Life or pertaining to life.

**biotin**
Vitamin H, one of the vitamin B complex – a crystalline acid found in nectar.

**biotype**
A group of animals (or plants) with the same genetic make-up.

**birch** *Betula alba* **Betulaceae** ***
An early flowering tree Mar-Apr that can produce nectar and pale yellow pollen. The use of the word birch where 'beech' is intended often occurs in N. Z. honeydew production.

**bird**
A warm blooded vertebrate with fore limbs modified to form wings that enable most species to fly. Honey buzzards, bee eaters, and honey guides have been called 'professional' predators of bees but green woodpeckers have been known to attack hives causing damage to the structure and the demolition of the stock. Many birds are listed under their common names e.g. fly catcher, titmouse etc. Small birds like sparrows might, as a family, learn how to tap on the floorboard and entice guard bees out during cold weather only to snap them up devour the abdomen and leave other parts

behind indicative of their behaviour though rarely witnessed except with binoculars as they are shy enough to disappear when a beekeeper gets near.

**bird cherry *Prunus padus* (hayberry Scotland) Rosaceae** ***

A wild flowering cherry described by 'Colin' in BBJ) as flowering along with earliest sycamores in C. Durham.

**bird's foot trefoil *Lotus corniculatus* Leguminosae** ***

Nick-named 'Bacon and eggs' and 'Lady's slipper' and other names. It is a common native perennial and likes chalky soil and is held in high esteem as a nectar plant in Switzerland. It flowers from May to Sept. The pollen is light to medium brown, pale yellow 15mu grain.
See: Lotus

**biting**

A worker may bite through the base of a corolla to get at the otherwise unreachable nectar. This has been observed on bluebells. The drones biting seems to be confined to nibbling its way out of a cell though on one rare occasion when hives were moved too far for worker to return but dozens of drones did just that and a surprised observer was most certainly 'fussed' around head and face. Their mandibles, like those of the queen are more notched than those of workers. The queen may well use hers when fighting a rival queen. Workers use their mandibles when repelling invaders, removing obstacles (such as newspaper after 'uniting', for manipulating wax, scrabbling for pollen and feeding brood. Any tough flexible material like polythene or grape skins is too much for them though by dint of continuous effort they can wear wood away or cut expanded polystyrene to pieces. A worker too, having stung and pulled away from the victim will almost invariably attack another spot and it will claw and attempt to bite. Such biting can be felt on the lips. The mandibles of course, move sideways to meet one another.

**B.K.A.. (Beekeepers Association)**

If county orientated they may be preceded by 'K' or 'D' meaning Kent or Devon etc. because many associations though not all coincide with county boundaries. These are most useful to bring beekeepers with a special or local interest together to plan and co-ordinate their activities and to strive to improve their common lot and that of beekeeping generally.
See: Beekeepers' newsletter

**B.L.**

Bacillus larvae infection, this abbreviation is used for American Foul Brood See 'AFB', 'ABD'

**black**

Having no colour and absorbing most of the rays from light or from heat. Black is not suitable for protective clothing as bees will aggressively attack jerky movements of rough black material. (Try dangling a black ball of wool in front of the entrance!) It is a good colour for heat absorption and hive roofs in the U.K. are frequently painted black and the same 'colour' used for the outside of solar wax extractors and for veils (being more transparent). It is possible that dark bees can operate at lower temperatures than lighter bees and the apparent darkness has enabled beekeepers to work adonsoni bees at night using red light.
See: black netting

**black bear *Ursus americanus*** ***

In 1981 a survey was conducted in Florida (ABJ 1st May 1982) and black bears were considered responsible for a great deal of damage done to the apiaries of full-time beekeepers. Electric fences seemed to prove the most effective means of reducing losses.
See: bear

**black bee**

A French association of queen rearers were trying to re-select the black bee of France – *Apis mellifera mellifera*. A centre for its selection was at Rucher de Beautheil near

Coulommiers, Seine et Marne. Black (or dark) bees are found in Italy, New Zealand, N. Africa and many other places where lighter or yellow bees are also to be found. See: British black

**blackberry *Rubus fruticosus* Rosacae *****
The black fruit of the bramble now widely cultivated though in the wild an invasive, prickly shrub which covers hedgerows and can form dense clumps up to 3m high and across. Myxomatosis amongst rabbits which fed on the young shoots caused this plant to become ubiquitous. It makes a good windbreak (used with hawthorn and willow) and yields from May to the frost. Deep rooting helps even in dry spells. Many out-apiaries display a bumper blackberry crop, guarded by the bees which enable their owner to use his honey to make bramble jelly and wine. Its tough briers (stems) help to tie skeps together. Pollen grains are 20mu grey to grey-brown described also as whitish dull yellow, greenish white and grey to brown. Honey light to dark amber of good density and flavour and slow to granulate.
See: bramble

**black comb**
Comb, though beautifully white when freshly built, becomes progressively darker as it gets trampled over and especially when brood has been reared in it. Each generation of brood leaves a pupal skin behind covering its faeces, which were only evacuated after the last moult. It is then cleaned -(seemingly polished). The cells too, gradually become a little smaller. The old comb is stronger though and more heat retentive than newer comb, but it can be a source of pathogens and it is not considered an economic proposition to allow comb to be used for more than a season or two in the brood chamber. Carefully managed however, super comb might last for years. When honey is stored in the older comb it takes on a darker shade and if the comb is chewed it has an unpleasant bitter taste.
See: old comb

**blackcurrant *Ribes nigrum* Ribesiaceae *****
A shrub bearing a small, black, edible fruit. Its pollen is whitish/pale yellow. It flowers Apr-May and there are red and white varieties. Although pollination takes place when bees are absent the end flowers of trusses are much improved by honeybee visitations and heavier and better crops result from the proximity of pollinating colonies. This is due to the fact that the first flowers to open are self-fertile while the later ones are less so.

**black netting**
Black does not reflect heat or light and for this reason black netting is much easier to see through than other colours especially white. That this is not always clearly understood is probably the reason why beginners often put their veils on back-to-front. Black hive roofs absorb the winter sun's warmth and are often found on otherwise white hives.
See: netting

**black queen-cell virus**
Diagnosed in 1975 this virus disease is found in many of the bees that fall dead from winter clusters. Quite often it is associated with 'CPBV' and 'nosema' and prepupae and pupae have been found dead in combs in which queens were being reared. Such pupae were partially decomposed and had darkened their cell walls which appeared black in patches. In infected colonies the incidence in live bees decreased markedly during the summer but still 30% of colonies were found with it in the autumn. It had previously been mistakenly diagnosed as ABPV. The same anti-serum is effective for both.

**black robber disease**
Associated with 'Spring Dwindling' disease, May sickness, Bee paralysis and May Pest, although the disease is now known to be associated with a virus, The name is derived from the dark, shiny appearance that is associated with habitual robbers but

it is caused by the nibbling attacks of their nest mates and they become very dark and look as if polished. They may be hindered by guards from re-entering their colony (should they have left on a flight) which further confuses them with 'robbers'. The difference is that the robber bees will be lively and healthy whereas those with black robber disease will be unhealthy and inclined to tremble, may crawl and may have distended abdomens and associated dysentery.
See: CBPV

**Blackthorn *Prunus spinosa* Rosaceae *****

A common, thorny deciduous shrub which grows alone or in dense thickets. It does not like acid soils. Its masses of white flowers appear before the leaves in February or March. Its fruit is known as the 'sloe' otherwise it is similar to 'bullace' and plum. In suitable weather conditions it is worked by the bees for its nectar and pollen, the latter being described by Yate Allen as deep yellow with a 36mu grain. Others have called it a medium brown.

**blueberry**

Scottish name for bilberry.
See: bilberry

**blastocoel**

The cavity that appears in the mass of cells at the latter end of the period of cleavage in an egg. Cavity of a blastula (a hollow ball of cells with wall usually one layer thick).

**blastoderm**

A thickened layer formed from cleavage cells which migrate from cortical cytoplasm to surround the blastocoel.

**blastokinesis (blasto – bud Gk)**

Movement of embryo in the egg. In the final stages after some three days of incubation, blastokinesis causes the chorion (skin or shell) to rupture and enzymes released dissolve it.

**bleaching**

To whiten by chemical process or exposure to the sun. (a) This is used for making pure, white beeswax. (b) In mounting material for optical microscope work, this technique is used so that the object can be viewed by transmitted light. E.g. to render the black chitin of a bee's leg relatively transparent. This is normally done before staining or dehydrating and various chemicals such as sodium hypochlorite are used. Hydrogen peroxide, a trace of which is found in honey, is also a bleaching and sterilising agent.
See: bleaching slides. mounting and wax bleaching

**bleaching (slides)**

When material is to be bleached for microscopy this is normally preceded by the maceration and or dehydration of the material followed by the bleaching fluid. The material is then washed using several water changes. The bleaching of chitinous leg parts may take several days with careful checking to ensure the process only goes as far as required.

**blended honey**

Honey which has been mixed with another honey or sugar to achieve a more marketable product. Under present day regulations care has to be taken not to label honey in any misleading fashion. At one time sugar blends of honey were often sold as if they were composed of pure honey. Blends can be used to improve the colour, flavour and uniformity of a product. To mix the samples thorough stirring and warmth are necessary. Honey that is exported from countries with centres that take honey in from many different sources will usually be of a chosen 'blend'.

**blind gut**

The alimentary canal of the honeybee larva is not continuous until feeding has been completed and the larva is ready to make its sixth moult prior to pupating. The non-continuous gut is called a 'blind-gut'. EFB disease can cause a break through resulting in contaminated faeces being liberated into the cell.

**blind louse (*Braula coeca*)**

Also called the 'bee-louse' *L caeca* – blind'
See: Braula

**blood**

1. The fluid that circulates in animal systems and which is used to transport food and waste and sometimes gases ($O_2$ and $CO_2$ in the case of mammals) by chemical locking or diffusion.
2. It is not unknown for a bee sting to draw blood upon removal of the sting.

See: haemolymph

**blood corpuscle**

A minute protoplasmic body. Although a bee's blood is different from that of a mammal (because it does not transmit gases by locking them on to the red corpuscles) it is nevertheless a liquid which contains haemocytes (corpuscles).
See: blood, leucocyte and phagocyte.

**blood vessel**

The part of a bee that begins in the dorsal cavity of the abdomen. It takes blood into its five-chambered heart via the ostia. These valves close and the blood is propelled forwards towards the head via a convoluted tube and the aorta. The flow of blood to the brain is thus assured and back pressure then allows it to seep back through the thorax and the abdomen for recycling. The components of the blood vessel are of mesodermal origin.
See: blood and heart

**bloom**

1. Flower or blossom. Orchard tours are arranged each year so that visitors can see the masses of blossom in driving through Kentish orchards. Bees delve into the blossoms for hidden nectar and daintily arrange the pollen in their baskets.
2. The description of a fine, powdery, glaucous deposit on fruit such as plums or grapes. It can be waxy and beeswax itself develops a 'bloom' as it loses its shine. This can be seen when foundation has become old or a cake of show-wax has been left exposed to the air for some time. It can easily be refurbished by polishing with the palm of the thumb. Bloom melts at 38°C (102°).

See: blossom and wax bloom.

**blooming times**

The length of time a flower remains open and the time of the year when this occurs. Forecasts given by the East Malling Research Station each year when the various types of fruit are likely to flower. This assists beekeepers and growers in the fulfilment of pollination agreements. Early flowering plants may be quite variable as to the time of year and place of florescence. Hawthorn has begun in early April while the same plants waited until late May the following year. Generally waves of flowering begin in the Scilly Isles and travel N.E. reaching Scotland about a month to six weeks later.
See: dehiscence and florescence

**blossom honey**

Honey produced wholly or mainly from the nectar of blossoms. This distinguishes it from honeydew honey.

**blow**

To exhale with pursed lips. Should it be required to move bees on the comb, say to look for eggs, blowing is an unwise procedure as bees abhor human breath. They are better moved on their way by keeping a bare finger to two over them or if necessary by a whiff of smoke. To blow immediately alerts the bees and informs them that an animal nose or mouth is nearby and stings in that region can be quite painful. If a bee flies near the face (with or without a veil) withhold breath while it is directly in line with nose or mouth.

**blowing bees**

High velocity blowers are used to clear bees out of supers. Two cycle, knap-sack devices have proved useful for the purpose. Air velocities in the range of 160 mph have been used. This action does not in itself seem to make bees particularly bad tempered, though it can easily lead to robbing. To blow on bees by mouth is unwise. Note the old proverb 'Do not blow into a bee-hive'.

**blowlamp (blow torch)**

A small hand-held apparatus with a jet that directs a very hot flame. This useful device enables a beekeeper to apply a high temperature flame to a hive surface with the object of sterilisation. (or in some cases for drying or for the removal of paint). At one time these were primed with meths. and then fed by paraffin oil. Now-a-days butane gas or similar, is fed from a small canister. A light portable type incorporates this into the blowlamp itself and they may be fitted with 'press-button' ignition. Scorching is demanded when foul broods have been present but heat sufficient to melt wax or propolis is all that nosema spores need. Regular use of the blowlamp is a good apiary help regarding apiary hygiene.

**Blow Thomas Bates**

Born in Welwyn 1853 he founded the beekeeping appliance business that became E.H. Taylor Ltd which subsequently merged with E.H. Thornes Ltd. He invented grooved sections and was one of the first to produce the movable frame hive.

**blue**

This colour lies between indigo and green in the spectrum. Bees are sensitive to this colour and can distinguish blue-green from green or blue. It is a suitable colour for hives that are not to be camouflaged, but not suitable, especially darker shades, for protective clothing. Bees are easily trained to and attracted by blue flowers. Bee-milk surrounding young grubs appears to have a bluish tinge.

**bluebell Scilla spp *Endymion non scripta* Liliaceae ****Scilla nutans – Hyacinthus non-scripta.***

A hardy perennial bulb known as the wild hyacinth in Scotland where their bluebell is our 'harebell'. It flowers April-June and naturalises woods and shady places where it forms carpets of brilliant blue as the clusters of bell-shaped flowers disport themselves for about a week. They can grow 30cm tall and S. nutans with short white bells has been responsible for full supers when cultivated in an orchard. The honey though has the same sickly aroma that accompanies the flower. Bees will nibble through the base of the corolla when their tongues fail to reach the nectar. Pollen: white, cream, creamy white or pale yellow, with a hint of blue or even greenish with a grain of 40mu (with a single groove monocolpate). The flowers of garden varieties may have pollen of a deeper greenish blue and the flowers include pink and white as well as blue.

**blueberry *Vaccinium corymbozum* Ericaceae *****

This is a distant relative to the bilberry which is native to many parts of Britain. It is grown in the maritime provinces of N. America and benefits from honeybee pollination while providing the bees with nectar and pollen. It is advertised under several varieties in British seed catalogues. Losses of bees from CDC seriously affected the American crop in 2009 when bees known to be clear of disease were imported from Australia (according to BBC Television).

**Blue Orchard Bee *Osmia lignaria***

Widely scattered in the USA. Dr F.B. Wells discussed their breeding and use and also O.cornifrons (Japan) for apple pollination. Other non-Apis bees included Halictus, Osmia, megachile, nomia, peponapis and Xenoglossa for crops ranging from lucerne to Cucumber spp, ABJ.

**blue titmouse *Parus caeruleus* Paridae (Tom tit) *****

A colourful little bird that has even been known to nest inside the roof of a hive. It has a reputation for teaching its family to tap with its beak at the entrance of a hive and then to snap up any bee that comes out to investigate. As they are timid enough to disappear when man approaches it is worth keeping a watch on hives using binoculars when these and other creatures can often be seen doing dastardly work especially on mild winter days.
See: blue titmice and sparrow

**blue weed *Echium vulgare* *****

N.Z. name for Vipers Bugloss. A very bristly, erect biennial 30-90cm high with vivid blue flowers borne on curved spikes. A major honey plant, yielding light honey, found notably around Marlborough & Canterbury where it tends to grow along river beds and in open country. Planted at one time to give nectar when other plants dried-up it was dubbed 'Salvation Jane' but once it became rampant and troublesome the name changed to 'Paterson's curse'. However it was removed from the schedules of the Noxious Plants Act. Its honey was sent to a high class London store who re-bottled and re-named it as 'mountain blue' or some such name after which it was sold at a very high price and even turned up on Australian and N.Z. shelves in its new guise and presumably at a special price. Some bee-farmers had been quite glad to get rid of it as were certain growers who claimed it was damaging their plough shares.
See: Paterson's curse

**board**

A piece of timber sawn thin so that it has considerable length and breadth compared with its thickness. Hive walls are for example, constructed from boards. Plastics, hardboards and other materials are sometimes substituted but all must be weather-worthy. Useful for the many hive parts such as: floor boards, crown boards, clearer boards, roofs etc.
See: escape, dummy, hardboard and plywood

**BOD**

Biological oxygen demand
It is an index of the amount of oxygen required to biologically degrade a substance e.g. river effluent. It is expressed in parts per million of 02. (ppmO2).

**body**

Another word for a box, chamber or compartment that holds frames of comb and more particularly brood combs. Therefore a section of a hive might be a brood body, brood box or brood chamber. A hive box intended for honey storage would be called a super unless for 'sections' (cartons to hold about 1lb comb) when it becomes a 'crate'.
See: chamber, box, crate and super

**body of a bee**

This consists of three main parts, the head, the thorax and the abdomen. Head appendages are trophy (or mouth parts), a pair of antennae and mandibles. The thorax has two pairs of wings and three pairs of legs. The abdomen terminates with an orifice through which the shaft of the sting can be protruded. The chitinous exoskeleton encloses all the muscles, glands and viscera and is itself covered by a waxy epicuticle. The drone is by far the largest of the castes and has no sting. The queen's body equals that of the drone in length but is more tapered and the worker is relatively

**small body (liquid)**

An expression used to indicate the consistency or density in a wine such as mead. When swirled around the glass the droplets that continue to run down (after the wine has settled) gives an indication of the 'body' as does the satisfaction derived from drinking it.

**bog (quagmire)**

Ground that is usually soft, wet and spongy. It consists largely of decayed vegetation such

as moss and heather. If one wishes to keep hives there it would be best to have them anchored on pontoons. Stagnant pools frequently develop and it is difficult terrain to cross. Bees have been dropped from the air into cranberry bogs (for pollination purposes). Heather rarely yields well in bogs as they are usually low-lying and heather does best on higher ground.
See: marsh

**boiler suit**
A one-piece suit of inexpensive material worn for rough work and as protective clothing. For beekeepers it is best to use material that is light in colour and weight, which is fully zipped up the front, has pockets that do not interfere with the close-fitting of the veil and is cool and sting-proof. Example: Industrial weight, white nylon fabric. Large size thigh pockets, elasticated wrists and ankles. Chest 36-44 (larger to order).
See: bee-suit and Sheriff

**Bokhara clover (Sweet clover) Melilotus Leguminosae *****
White and yellow *.alba* and *officinalis*. A biennial that will grow over 2m in height. Honey is light, medium density with pleasant vanilla-like flavour, granulates quite quickly. Typical clover pollen – dull, greenish yellow.

**boll weevil *Anthonomus grandis***
Its invasion seriously affected the cotton industry in the USA and the use of insecticides to control it with soybean grown adjacently caused extensive bee kills.

**Bolton hive**
Described by Tarlton Rayment (Australia) as using only shallow Langstroth boxes into which frames (close-ended) are tightly wedged allowing the hive to be roughly handled – even inverted. The use of two boxes to form the brood nest earned it the cognomen of: The divisible hive. It has Lansgstroth dimensions except for its width of 13 and 7/8". One claim made for it was better control over the swarming propensity. The Bolton frame was standard shallow Langstroth with Manley type side bars.

**Bombidae**
Bumble bees, humble bees, large hairy social bees of which there are some twenty species in the U.K. Their hairs are not usually branched in the same way as honeybees whose hairs can transport large amounts of pollen. The life-cycle is akin to that of the wasp. They are good pollinators, some species having longer tongues than honeybees and therefore useful for crops such as red clover. However a hive of bees would out- number them at least 100-1 when set out in an orchard or for a crop requiring pollination. Colonies are started by over-wintering pregnant queens and increase to a few dozens or possibly hundreds and then die away to leave fecundated queens to hibernate. Usually their coloured bands are formed by their hairs rather than their chitin. They will enter hives but do little harm.
See: anodonto, bumble and odonto

**Bombus muscorum spp**
the 'carder bee'

**Bombylidae**
Flies – some parasitic on bees. They frequently resemble bees but are of course unable to sting.

**Bombylious – Buzzing like a bee O.E.**
See: bee-fly

**boost**
To help by adding strength and capability to something. When this is a honeybee colony we might 'plump', 'feed' or perhaps 're-queen'. We might also unite colonies or move strategically so that the number of their flying bees is increased, in fact there are many ways in which the strength and capability of a colony can be increased.

**borage** ***Borago officinalis*** **Boraginaceae** ***

A hardy annual used as a culinary herb and its flower for decorations in drinks. 'Borage I bring courage'. It grows 12—35cm high preferring alkaline soil. Hairy leaves and stems. Its pretty star-shaped, blue flowers appearing from April to September. It rejuvenates from self-sown seed (easy to pass on to beekeeping neighbours) but rarely in sufficient quantities to give a 'flow' and its honey is not especially pleasant (though some like it) but white with a yellowish-grey tint when crystallised. Pollen is described as almost white, white, bluish grey, fawn, light brown and greyish-yellow.
See: Boraginaceae

**Boraginaceae**

As well as borage this family of herbs includes a large number of bee-plants. The flower spikes unfurl as the flowers open and in most cases they protect their nectar from the rain and are great favourites of the bees: alkanet, blue cromwell, bugloss, comfrey forget-me-not, lungwort, Paterson's curse and viper's bugloss are some of the members.
See: under the various headings.

**borax**
**Sodium tetraborate $Na_2B_4O_710H_2O$**

Used as a flux and cleansing agent. At the rate of 1oz per 1lb of beeswax to neutralise the wax's acid content and to emulsify the beeswax when dissolved in an oil such as medicinal paraffin.
See: cold cream.

**botanical name**

The scientific name of a plant according to the system devised by Linnaeus. Such names are in two parts. The first defines the genus to which the plant belongs. The second describes the species or group of similar plants that are close enough to interbreed. So a dandelion becomes Taraxacum officinale invariably using a capital letter for the genus but lower case for the species even when it is derived from a proper name like Limnanthes douglassii. Common names tend to wax lyrical and this becomes confusing when in one region hawthorn is called 'Aglet tree' in another 'Azzy' or 'bird eagles' or 'may', so *Crataegus monogyna* or *oxyacanthoides* singles the varieties out in any country of the world.
See: classification

**botany**

The study of form and function in plants.

**botchet**

Monks of the order of Cluny gave this name to mead, the 'sweetest beverage'. It is described as a light alcoholic drink brewed with honey. A burned sack mead with a characteristic bouquet. William Lawson (1618) told how 'botchet' might be made by immersing a skep full of honeycomb in clean water. Be careful not to 'botche it! Joke

**bottle**

Although a 'bottle would normally be described as a narrow necked glass container closed by a cork in beekeeping circles a bottle takes on special guise. For instance National Honey Show bottles for 'show' must be of clear glass, punted and of round section holding approximately 26 fl.oz and without any ornamentation. Corked with plain non plastic, all-cork flanged stoppers than can be removed without the need of a mechanical aid such as a corkscrew. Honey jars were often referred to as bottles and in lands where honey never granulates it is sometimes kept in bottles.
See: bottle feeder and flanged cork.

**bottle brush genus Calistemon** ***

A tree or shrub with large, red bottle brush type flowers, found in Australia and New Zealand – a prolific yielder of nectar. Some varieties have found a home in the U.K.

**bottle feeder** ***

A jar type feeder where bees are able to take syrup from small holes bored in the lid when the jar is inverted. The bottle may sit on a plastic base that has grooves to allow bees'

access to the number of holes uncovered by rotating the bottle. They are hygienic 'contact' feeders but usually of small capacity.
See: slow and Boardman' feeders

**bottling**
The process of transferring honey, usually prepared for sale, into (glass) containers. The bottles (jars) should always be scrupulously clean and at the same temperature as the honey. A honey tank with an efficient cut-off tap is a great help. Care has to be taken to avoid the inclusion of bubbles. This can be done by minimising the free-fall of the honey as these not only spoil the appearance, cause froth to form and lead to fermentation but provide a triggering mechanism for granulation. Large scale operations use sophisticated machinery. A bottle measures volume and as specific gravities of honeys vary, it is necessary with any batch to ensure that the volume being measured accords with the weight required. With standard screw-top jars a guide is that the surface of the honey should be just hidden by the cap. When bottling mead a suitable funnel and filter are useful.
See: syphon and taps

**bottom bar**
The lowest cross-bar on a frame used in a box for holding comb as in movable frame hives. Quite often it consists of two bars through which a wired sheet of foundation can be passed. Otherwise a single wide strip of wood (or plastic) is used. Easy detachability of the bottom bar helps when a comb has to be replaced with new foundation. With this in view, nails that fix the bottom bars are hammered vertically into the side bars so that the bottom bar can be pushed out for refitting. The ¼" spare bottom bars have many uses such as supporting flimsy queen excluders when top bee-space is used and as a measure of bee-space, as a ledge under the entrance hole on a mini-nuc. etc.

**bottom bee-space**
Tiered hives have bee space above and below the combs and boxes are made so that the tops of frames are flush with the top of their box and a bee-space gap is left below. British 'National' and W.B.C. hives have this feature. It gives support to a queen excluder but causes inner covers to stick unless they are raised by a frame that is bee-space thick. An alternative is to give top bee-space
See: top bee-space

**bottom board**
Another name for the floorboard of a hive.

**bottom box (bb)**
The lowest box when more than one box is used as a brood chamber.

**bottom entrance**
Access and egress for the bees in the lowest part of the hive. As the design of the floorboard normally incorporates an entrance space so that any hive box can be set thereon and still provide an entrance' this is the most usual place for it. The debris that falls onto the floor can also be cleaned out easily by the bees. Steps should be taken to ensure that dead bees will not block a bottom entrance during the winter.
See: top-entrance

**botulism (*Clostridium botulinum*)**
A type of food poisoning caused by the ingestion of a neural toxin which can cause death through paralysis of the breathing or heart. The toxin is a by-product of the metabolism of a saprophytic bacterium '*C. botulinum*' which can only take place in a neutral environment in the absence of air of oxygen. Such an environment does sometimes exist in the human gut during the first 26 weeks of life. The bacterium can be destroyed by heat so all foods other than natural mother's milk should be sterilised by heating if to be fed to babies of that susceptible age. It is therefore recommended in some circles to keep honey,

raw vegetables and other possible sources of botulism from very young babies.

**bouquet**
The initial burst of aroma such as one gets from a newly opened bottle of mead or jar of honey, cake of wax etc. A flowering branch from a suitable 'pollinizer' transferred into a tree to supply pollen of a suitable variety to the adjacent and possibly self-sterile flowers.
See: pollination –hand

**bouton**
See: labellum

**box (container)**
1. A four sided container often of wood. Hive parts, more properly known as supers or brood chambers according to whether they are over or under a queen excluder in the hive, are often referred to as boxes – especially' brood box' as opposed to 'brood chamber'. 2. It is useful to have a portable box possibly with a carrying handle that can temporarily hold a frame of bees. Eggs, brood or a frame of bees or food can thus be moved from one hive to another.
See: body, box hive, crate, chamber and super.

**box (plant) *Buxus sempervirens* Euphorbiaceae** ***
An evergreen tree that grows in S. England and likes limey soils. Flowers Mar/May. Its honey is of indifferent flavour and pollen yellow-green. It is a slow grower, often used for miniature hedges, though it can also be found in large and interesting hedge shapes (topiary) and the close-grained wood is used for carving.

**box hives**
Hives comprising four walled boxes with internal runners to support comb lugs. The boxes may be oblong or square and usually stand on a framed floor arranged to give bees access to the box(es) via a gap on one side of the floor. The boxes fit on one another and are surmounted by an inner cover and a roof. Deep boxes are used for brood and shallow boxes for honey though this rule is not invariable. The accurate machining of wooden boxes made the advent of movable frame hives a great advance over earlier hives which were variable sized containers (skeps etc.)
See: hive

**brace comb**
Comb that is built as an addition to the ordinary store comb or brood comb, so that it forms a brace or strengthening attachment. Often found where bees have access to space to use as they like especially if the site is subject to vibration for any reason. It is less prevalent in modern hives which adhere to correct bee-space and comb spacing appropriate to the strain of bee. When two adjacent combs are joined at certain points this too would be referred to as 'brace comb'.
See: burr and wild comb.

**bracken Dennstaedtiaceae**
*Pteridium aquilinum* ***
Widespread and abundant fern which has ousted much of the heather on Dartmoor and elsewhere, changing the purple landscape to one of green. It becomes established extensively wherever there is excessive grazing or burning. It has extra-floral nectaries.

**bracket**
This was a hybrid liquor made from mead and ale. Also referred to as 'bragget' and 'bragot'.
See: braggots

**bract** ***
The small modified leaf found beneath the flowers and not infrequently the site of extra-floral nectaries. A 'bracteole' subtends an individual flower.
See: stipule

**braggot(s)**
Also known as Varii, this is an ancient drink

brewed from malt and honey. The word 'brag' meaning malt and 'gots' honeycomb. Corrupted to 'brackets'.

**bragwort**

Weak mead in Scotland (Bonner) 'bragget' (Wales) 'braggot' (Bevan) 'brawgawd or bracket.
See: John Bickerdyke 'The Curiosities of Ale and Beer' 1889). D. Galton

**brain (supra-oesophageal ganglion)**

The size of a bee's brain is relatively large compared to its body mass. The multitude of nerve endings transmitting information to the brain from the sensory organs necessitates a sizeable structure. As the bee's brain is an ideal model of our own, scientists like Dr. Pickard have been able not only to discover about the bee but to help us understand our own brains. In the bee it is bathed with blood pumped forward to it by the only blood vessel. Three component parts of the bee's brain are called 'proto' 'deuto' and 'trito' cerebrum.
See: hive mind

**brain's trust**

A meeting with selected panels of participants who compete in answering pre-arranged questions. A chairman who may be judge, questioner and adjudicator is assisted by a scorer. As a function at beekeepers' meetings this sort of event can be highly entertaining as well as educational.
See: quiz

**bramble *Rubus fruticosus* Rosaceae**

See: blackberry

**branch**

A limb, offshoot or ramification of a tree or plant. When in bloom a bee might spend its foraging life confined to one branch of an apple tree. Swarms frequently alight on a branch. County beekeeping associations are formed from a number of branches these being local groups that include mainly beekeepers within a specific area. For example the Exeter Branch together with twelve other branches formed the Devon Beekeepers Association which was affiliated to the BBKA.

**Brand's cappings melter** ***

Made in two sizes, it melts honey/wet cappings from above so that while wax melts honey sinks and remains relatively cool.

**branding**

The marking of hives with the owner's identity has become widely advocated though less widely used. A hot metal brand is used to scorch letters into the wood. Code numbers issued on request so that records can be kept and stolen hives identified though far from home. Rustling has become an all too common possibility. It has been suggested recently that postal codes might be used when branding.

**brass cone** ***

A perforated brass cone made to be screwed or nailed over the ventilation holes in a roof of the hive. The rim is about 5cm across and tappers to an opening just large enough to allow a bee to climb out but not return and it projects 3cm from the hive. If not rust-proof it can be painted and even wasps have difficulty in gaining access though the latter will do so especially if syrup is placed on top of the colony. During winter they allow moist gases that rise from the cluster to escape from the hive.
See: cone escape and cluster

**Brassica Crucifera**

A family of plants related to the cabbage. When grown for seed the yellow flowers are highly attractive to bees and yield nectar and pollen. The nectar is rich in dextrose and is inclined to granulate very rapidly indeed. Honey has a fine white granulation and those with a sensitive palate can detect a slightly mustardy flavour to it. The pollen is greenish. Members of this family are: 'broccoli', 'brussel sprout', 'cauliflower', 'charlock', 'kales', 'mustard', 'radish', 'rape'

sea-cabbage', and 'turnip'.
See: under appropriate headings.

***Brassica arvensis Sinapsis arvensis* (Charlock)**

Wild mustard producing a light yellow honey.

***Brassica napus,* Campestris**

This is oilseed rape, cole, coleseed, or colza whose widespread growth for its valuable oil has smothered fields with blazing masses of bright yellow, so conspicuous in the spring. Some varieties can be prolific nectar yielders in the right weather conditions but it granulates so quickly that many a beekeeper has had combs solidified before he could get them to the extractor. It is therefore frequently taken off before the bees actually seal it over. The white honey creams nicely and is a favourite with those who like its flavour.
See: canola and rape

***Braula coeca* (Nitsch)*** Chelorrhapha (sp. dubious) Braulidae**

A wingless fly that looks like a tiny mite. Its larvae make 0.6mm tunnels under honey cappings and the adults feet are adapted for clinging onto the hairs of bees. They stick their eggs onto honeycomb cappings and the burrowing of the larvae make thread-like irregular traces under the cappings. (This can spoil the appearance of 'sections'). Yellow-brown adults climb out to infest bees; many find their way onto the queen and they become a shiny, reddish-brown as they mature. They take food by stimulating the mouth-parts of a bee and it is likely that those on a queen benefit from her attendants. Braulae are 0.8mm and six-legged. They move rapidly out of harm's way when disturbed, though fumigants such as Folbex, tobacco smoke, camphor, naphthalene and PDB dislodge them. The disorder is named: 'Braulosis' and half-a-dozen different species have been found in various parts of the world.
See: phenothiazine and Phenovis

**Brazil**

An influx of aggressive adansonii bees detrimentally revolutionised beekeeping in Brazil in 1956 reducing their crop to only 5,000 tons, less than 1% of world production while Argentina only half her size did better – 'Mellido' (honey man). Large orchards of apples in Catalina were under pollinated and indiscriminate use of insecticides caused problems. 1973 - 4,700 tons but 18,000 in 1979. Forest tree felling, excessive grazing, fines and ground erosion has made beekeeping less; profitable in some zones.
See: Africanised bee

**Brazilian bee (stingless) *Nannotrigona* (Scaptotrigono) *postica***

There are several species of stingless bees and special hives are made for each of them. Many are far smaller than the normal honeybee and some are said to bite to make up for not being able to sting. Their honey is highly valued.

**breed**

a)To produce offspring.
b) to improve by selection, (culling, stud areas, choice of queen for producing queen cells or drones). Confusion with this term gave rise to Beo Cooper suggesting using the word 'brooding' when referring the normal production of brood. As a noun it implies a relatively homogenous group.
See: breeding, line, pure bred, race, strain, type and variety

**break crop**

A change of crop to break the otherwise annual cycle. Since the continuous sowing of the same crop year after year can lead to the increase of pests and diseases and the reduction of yields, the cycle is broken by the use of a different plant. For example the rotation of cereals may be broken by potatoes, sugar beet, oilseed rape or 'set-aside'.

**breaking dance**

The dance, performed by certain bees,

before the break up and departure of a swarm cluster. About 5% of the bees in the swarm participate in a back and forth movement and in the breaking and scenting process. The rest of the bees remain stable as if waiting for instructions. A high-pitched, buzzing sound can be heard from the cluster. The bees proceed to take-off in an orderly fashion, the cluster appearing to melt slowly away until only late returning foragers and a few stragglers search the spot for remaining traces of queen substance.

**breathing**

Insects like the honeybee have an aerating system that ramifies throughout their bodies so that oxygen makes direct contact with the cells as the blood of a bee is not constituted so as to be able to transport it. Large, paired air sacs receive air from tracheae leading from the spiracles and abdominal contractions and expansions force the air into and out of them. Leading by ever diminishing diameters the small tracheoles take oxygen to the tissues. Waste gases mingle with the air and are expelled at the spiracles. A bee is seriously inconvenienced and would eventually suffocate if its spiracles become blocked

**Breathing 'pores'**

See: spiracles.

**breeder colony**

A colony whose activities are directed to the raising of queens and or drones, as well as or instead of honey production. Such a colony should be carefully selected so that it maximises the chance of incorporating chosen characteristics into the resultant queens.

**breeder drones**

To select drones for 'air-cover' within the territory of a mating apiary a full knowledge of their grandparents must be known as it was from in those colonies that were headed by the drones grandparents, that the genetic possibilities should have been assessed. Attempts to arrange a stud area may call for isolated apiaries or a certain degree of ingenuity. Now-a-days it has been better understood that the value of cross-breeding usually outweighs the advantages of 'line breeding'

**breeding**

Whereas this term means 'rearing young' it is probably as well to use the words 'propagating' or 'rearing' when nuclei are set up with a view to making increase or producing queens in the ordinary way. This allows us to keep the term 'breeding' for use in the selective sense where by careful culling, isolation and the exercise of maximum control over genetic make-up, queens with the chosen qualities can be bred.
See: selective breeding

**breeder queen**

A queen which by virtue of its tested abilities and those of its daughters and its parent colony, is kept for the production of drones or daughter queens. Although such a queen may be well past her best from an egg-laying point of view she can be kept in a nucleus or observation hive thus acting as a gene bank and because she has outstanding characteristics she is most valuable. At U.S. auctions such queens have fetched hundreds of dollars. Bro Adam has had priceless breeder queens stolen.
See: queen classification

**brickearth**

A fertile soil found particularly on top of the gravels in the lower Thames valley. Regarded as good areas for beekeeping.

**bright**

See: light intensity

**brightness**

Honey from the hive needs careful filtering to ensure that neither wax, pollen, propolis, froth or other foreign matter is present and even then air bubbles, no matter how

minute, will detract from the honey's natural brightness. For show purposes the removal of such tiny bubbles which can cloud the honey, will need to be removed if a prize is sought. Pouring honey with a minimum of free-fall will help to prevent the introduction of bubbles in the first place then gentle warming will help to drive remaining bubbles to the surface where they may be skimmed off. Another aspect of brightness is the clear flame of a beeswax candle. Bees, who happily work in the dark, respond to bright light which draws them out to forage to defecate or even to their deaths if snow glare causes them to plunge in as if it were the sky. Shade the entrance with a slate in such circumstances and if one or two have only just buried their little bodies in shallow graves, lift them out for the warmth of your hands will help them revive and they may be returned to the colony.
See: light intensity

**brimstone**

Sulphur was used in skeppists' early days as a means of getting the honey from the bees without being stung. Fortunately 'driving' was discovered, thereafter bees could be sa ved.
See: sulphur pit

**British beekeepers' Convention**

Held annually at the National Honey Show.

**British black bee**

At the moment of penning these notes there are still a few old beekeepers whose memory goes back to the bee that was native to these islands. Although exotic yellow bees and others were imported into G.B. (even in Roman times) they were inclined to darken with successive generations until they too were akin to our old black bee. Unfortunately waves of decimation from what became known as I.O.W disease eliminated the old bee on such a large scale that it was necessary to find replacements which then swamped the country and by the First World War few if any of the genes remained in local crosses. The original British black bee was said to have made white cappings, wintered economically, had longevity and hardiness, limited swarming and the bees were dark brownish black while queens had reddish legs and they could become restless when manipulated.
See: black bee

**British standard**

Standards have been laid down by the British Standards Institute for hives, frames and associated components peculiar to the hives of this country. The BSI was founded in 1901. The standard deep frame, DN - to the trade (BSD) and the shallow SN (BSS) are found in all British catalogues. They are three inches shorter than the American counterparts.
See: 'British Standard Frame' and 'Woodbury Committee'

**British standard frame**

Designed by the BBKA's Woodbury Committee in 1882.
The deep frame is now specified as: Top bar $17^2$ long with $1½^2$ lugs at each end, width $7/8^2$ and $3/8^2$ deep. Refer to remaining standards elsewhere BS 1300 and BS1 1960
See: Hoffman and Manley

**British Wax Refinery**

St. Johns Road, Redhill, Surrey.
In 1960 this firm specialised in the utilisation of British wax and imported beeswax for many years. They also helped beekeepers by purchasing their wax and old comb. They supply wax for all manner of purposes ranging from carbon paper to cosmetics and give lectures about wax.
See: uses of wax

**broad bean *Vicia faba* Leguminosae**

This leguminous plant benefits from honeybee attention and yields nectar from which the bees process a medium to dark brown honey. Sometimes the flowers are punctured at the base when the small amount of nectar is more easily taken in this way. However many bees collect pollen from the same plants and in so doing they will effectively pollinate them.

**bronchial spasm**

A form of asthma. When provoked by, say beekeeping activity, it may be found useful to have a bronchial dilator (aerosol spray) such as Ventolin (salbutamol) or Bricanyl (terbutaline) – Dr. Riches

**brood**

Eggs, larvae and pupae of all three castes come under this heading. Eggs hatch after three days of incubation and feeding takes about 6 days according to caste. The open cells are then capped over. This gives rise to the terms 'sealed' and 'unsealed' brood. The health, age, quantity and stage of development of the brood are all aspects with which the watchful beekeeper should make himself familiar. Emerging brood is an essential asset to a developing colony and eggs or very young larvae laid by a mated queen in worker cells can be transferred to a queenless stock. See: age of brood and life-cycle

**brood - age of**

It is sometimes necessary to assess the age of brood. In the case of larvae the actual size and position in the cell (fully grown larvae stretch out lengthways) is indicative until it is sealed. To check the age visually after this necessitates uncapping and therefore killing the occupant whose eye colour changes from white to pink, to purple and finally to brown. In the case of a queen cell though, the balding at the tip and possibly any sounds coming from within, will tell us whether a cell is 'ripe'. In the case of ordinary worker brood the colour of the pupa especially the eyes, indicates the age up to the point where the capping is being nibbled open by the emerging imago.

**brood and-a-half**

The use of a system of management whereby the queen is given access to all combs in both a deep and a shallow box. When using Nationals or WBC's for instance this is a way of allowing more room for the queen to lay in and a better food supply for winter. It is over half as big again as the single deep box (1x65) and the shallow can be used below for wintering and then set on top again in the spring to allow the queen more space by moving upwards. Some beekeepers prefer to keep brood on one size of frame.

**brood body**

The chamber in which the queen is able to lay. It is usually the one placed on the floorboard surmounted by a queen excluder and with supers if any over the top. It is possible of course to place it at the top giving a top entrance but that is not usual. Other terms such as 'brood chamber' and 'brood box' are also used for this hive part.

**brood cells**

The mellifera worker cells are 5x3 - 6x3mm across, about 13mm deep and incline upwards from the septum and the horizontal between 9 and 14°. The average size of drone cells is 7x25mm across and 17mm deep. When sealed they are more highly domed than worker cells. The brood in a prosperous, healthy colony tends to form a neat concentric ovoid. Sporadic patches sometimes betray the presence of laying workers. Brood is incubated at 95 - 97°F (35°C). With the emergence of an adult bee the cell becomes imperceptibly smaller as a cocoon is left behind. Eventually it could affect the size of the bees but the comb is usually replaced before this happens. The cocoons make the cells (and therefore the comb), stronger and darker.
See: cell sizes and worker comb

**brood chamber**

The part of the hive where the brood nest is found. It may consist of one or more boxes of comb but it usually separated from the upper storage area (supers) by a queen excluder. The queen and drones in a brood chamber are given access to the entrance without having to pass through an excluder. Brood chamber size may be varied according to the prolificacy of the queen and the season though beekeepers are often quite rigid in their choice of brood chamber size using a single or double brood, brood and a half or three shallow boxes according to their fancy. The brood chamber is intended for the colony nest where the queen is free to lay and therefore not normally for surplus honey.

**brood cycle (brood rhythm)**

A worker requires 21 days from the laying of the egg to the emergence of the adult (imago), the queen 15 and drones 24 days. The queen tends to lay in concentric circles in the warmest part of the hive. This is virtually governed by the worker bees because the queen only lays in cells that have been specially prepared for her. She goes back to lay in cells as soon as they are vacated until peak laying is reached. The seasonal cycle of brood rearing follows a March to May upward curve, falls back to an August low and then tapers away to little or nothing during the non-active season from October to February. Climate and strain of bee may cause considerable variations to this pattern.

**brood disease**

A condition where a pathogen or parasite develops at the expense of honeybee brood. Despite the overwhelming presence of the threat of Varroasis, American and European Foul Brood are still very serious diseases of the brood and there are lesser brood diseases such as 'sac' and 'chalk' brood. The rarer 'stone' brood can affect adults as well as brood. Normal healthy brood has a characteristic appearance that is lost when disease is present. Symptoms include discolouration of the brood and the abnormal appearance of cells and cappings and there may be a less than pleasant odour.. There is a marked difference between the healthy, clean-looking appearance of sealed brood and greasy, slimy conditions brought about by some of the problems mentioned. See: adult bee diseases and under headings mentioned.

**brood food**

Special food produced by nurse bees whose glands are activated after they have consumed nectar and pollen (or honey, water and pollen). These glands only remain active for a limited time and their function may be governed by colony's needs, i.e. quiescence during winter but rejuvenation in early spring. Each caste is fed differently in the larval stage. Workers receive the equivalent of 125mg of honey 75-100mg of pollen during some 1000 inspections and145 feeds. The food includes a clear component (mostly protein) from brood food glands and a white substance (mostly lipid) from mandibular glands plus regurgitations from the honey sac which includes enzymes. Queens get both components whereas workers are weaned onto clear component. Drones are fed in the same way but workers get more.

See: bee-milk, brood rearing, nurse bees, pap and royal jelly

**brood food glands**

These are hypopharyngeal glands and consist of 'acini' which Dade describes as resembling miniature strings of onions and are situated over (though hypo means under) the pharynx, The glands produce a sticky, milky fluid when the bee has access to suitable food (the protein of pollen is particularly necessary) and atrophy when the feeding of brood has expended their relatively short capacity. They are rudimentary in the queen and entirely absent in the drone. Nurses allow the food to run onto their mandibles which they use to feed the larvae. The full utilisation of brood glands ages bees more than any other house activity and when the bees have completed their period as nurses the glands atrophy unless they become winter bees whose glands become dormant until spring

breeding calls for their rejuvenation.
See: hypopharyngeal gland and royal jelly

**brood frame (or comb)**
A frame containing brood. This may be of eggs, unsealed brood, sealed brood or brood of all ages. At times it may also have queen cups or even queen cells. Whereas bees are fiercely independent and colony patriotic so that they would repel any other bee (including killing any queen wantonly introduced) nevertheless they will always accept a comb of brood just as they would a comb of honey. This can be a very useful aspect for queen rearing or boosting weak colonies etc.

**brooding**
Beo Cooper's suggested term for brood rearing so that the term 'breeding' can be used for the selection of genetic characteristics via the control of male and female

**brooding temperature**
The temperature of the brood nest – about 93 - 96° (34-35°C)

**brood nest**
That part of a honeybee colony's nest which includes the brood. Its overall shape is sub-ovoid and unless there has been disease or interference it is usually quite compact. It must be kept (by the bees) at a temperature which closely approximates to 95° (35°C) -- (strangely similar to our own body temperature) and the humidity is also held at a constant figure. The queen, while actively laying will always be found within the brood nest unless she is disturbed. Most of the bees within this area will be young house-bees and their numbers will be governed by the need to control the micro-climate and the feeding requirements. Drones when present tend to be on the outer flanks. The size of the brood nest will depend on the prolificacy of the queen and tends to follow seasonal cycles unless interrupted by swarming.
See: brood rhythm

**brood pheromones**
It has been shown that the presence of unsealed brood in a colony stimulates foragers to seek pollen. The exact nature of this pheromone and its effects are still being investigated. It seems likely that as nurse bees feed the brood they, at the same time, receive material from the brood (as happens in the case of wasps).

**brood rearing**
In honeybee colonies, brood is kept in a compact mass at temperatures close to 95-96°F. Eggs, male or female; incubate for three days inside their cells. Each caste has a cell appropriate to its requirements and is fed differently. Queens are reared in scattered vertical cells, purpose-built for each individual virgin. Once an egg hatches in a queen cell it is copiously fed with royal jelly until the larva is fully developed on the 9th day when the cell is sealed. Metamorphosis continues, the virgin remaining inverted until ready to emerge on the 16th day. Workers are fed on rich bee milk in near horizontal cells and weaned onto a coarser diet, capped over on the 9th day and emerge on the 22nd. Drones are fed in a similar way to workers but food is more plentiful (this conflicts with statement given elsewhere) and they emerge on the 25th day.
See: brooding

**brood rhythm**
The ecotypes of various races of honeybees develop a seasonal cycle of brood development which is adapted to the climate and therefore the honey flows of the locality. The adaptation is not easily changed when colonies are moved to a different environment.
See: acclimatisation

**brood spreading**
The brood is normally arranged in a fairly compact mass, the queen laying in the cells as they are vacated. This makes for more efficiency and economy with regard to laying, incubation and feeding. When brood is moved away from this pattern either as

a system of management (i.e. to remove congestion) or as a means of encouraging the queen to lay over a wider area perhaps, or to extend the peak laying period, this is called 'brood spreading'. Except in the hands of an expert it is generally considered to be an unwise procedure because if cold weather supervenes the interference may lead to chilled brood

**brood trial**

The giving of a frame with some worker eggs and larvae to a colony suspected of being queenless. If queenless the bees will raise queen cells on the introduced brood, otherwise the eggs will be developed into the normal worker brood.

**broof**

A pollen substitute marketed by Thornes prepared from natural products supplemented with essential nutritional elements, vitamins, amino acids and minerals. Made into a firm dough by adding a cup of lukewarm water to a 2 lb. pack. It is claimed to stimulate brood rearing and early build-up .

**broom *Cytisus scoparius* and *sarothamnus* Leguminosae**

A hardy perennial, a bush that grows wild and blooms May – June. It does not like chalk preferring acid soils. The yellow flowers produce deep orange-yellow pollen and cultivars of various colours yield nectar as well. Pollen grains 30mu. When working for this, bees usually get the whole of their dorsal surfaces smothered with yellow. C.praecox (white) seeds were available from BIBBA. Height about 2m and scoparious. Local names include: banadle, basam, bannel, bizzom, cat's peas, genista, green broom, green wood, golden chair, lady's slipper and scobe.

**bruised comb**

Sealed honey can be 'bruised' by accident or design as for example when a beekeeper may wish to destroy the tunnels of Braula coeca or to encourage bees to use the honey in the cells for brood rearing. The flat of a hive tool accomplishes this quite readily without causing wholesale seepage. Brood can be damaged by being flattened in a similar way and this is often done deliberately to destroy unwanted drone brood. Care should be taken never to bruise or damage a queen cell in any way if it is wanted.

**brush**

Made of rosemary, hyssop, fennel and other herbs for use in hiving a swarm.
See: bee-brush

**B.S.I.**

British Standards Institution
Established by Royal Charter. BS1300 1960 Beehives, Frames and Wax Foundation. British Standards House, 2, Park Road, London W.1 First published 1946, first revision Feb. 1960...............

**bubbles in honey**

Except for heather honey which is gelatinous and retains its bubbles it is generally advisable to keep them out of honey. The process of extracting honey from the comb is certain to incorporate large quantities of air. This can be further increased if the free-fall of a running column of honey is more than an inch or so (5cm). Bubbles will include atmospheric moisture and possibly dust which will act as focal points for the formation of dextrose crystals. They may also contribute to 'frosting' (white, frost-like patches of air between the set honey and the glass). Honey which is allowed to stand after extracting will develop a thick layer of white froth and even in a warm room it takes 2 or 3 days for all the bubbles to rise and join the froth. Visual improvement of honey clouded by small bubbles can brought be brought about by heating which also delays the onset of granulation. Warmth reduces the viscosity making filtration quicker and easier.
See: heating honey

**bubble size**

In heather honey one looks to the size of

the trapped air bubbles as an indication of the true thixotropic nature of this honey, the larger the bubbles the denser the 'gel'. Also when a jar of flower honey is inverted, the speed with which the formerly upper air space forms a bubble and rises through the honey is an indication of its viscosity and therefore its water content.

**bubble test for honey viscosity**

This is explained under 'bubble size'. The speed at which a bubble will rise through honey is governed by the honey's viscosity and the temperature of the honey. When viscosities of honey are to be measured at varying temperatures the rate of a stainless steel ball passing down through the honey in a tube is a more positive means of getting accurate readings.
See: viscosity

**bucket feeder** ***

One of the most hygienic forms if contact feeder from half to two gallon in size (2-10 litre). The container has a handle that swivels out of the way, a syrup-tight lid in which a 'lapping area' of gauze or other small holes are located. Such feeders in plastic or metal are filled right up, the lids are tightly pressed home, then they are held momentarily over another container to catch the drips and a partial vacuum is formed inside, before positioning over the feed hole. If transparent the level can be seen through the sides. Remember how easy it is to start robbing if any syrup is slopped about and it is wise to have a smoker when it comes to removing the feeder.

**Buckeye hive**

An extra-large insulated hive whose outer walls accommodate Langstroth frames within.

**Buckfast Abbey**

St. Mary's Abbey, Buckfastleigh, Devon. The monks here have kept bees over many long years putting them out on the moor for additional crops of heather honey. Under Bro. Adam's enlightened guidance queen breeding apiaries have been established in the isolation of the moor and he gained a world-wide reputation for honey from the Abbey and the much sought-after strain of bee known as the Buckfast bee.
See: Buckfast-queens

**Buckfast queens**

A strain of bee bred from selected varieties at Buckfast Abbey. They were made available from Dr.J.F. Corr of Belfast, Israel and Weaver apiaries U.S.A.

**Buckthorn *Rhamnus cathartica* Rhamnaceae**

A shrub of the woods and hedgerows in S. England. It likes chalk and provides nectar and pollen in early summer. Pollen is transparent to pale yellow 16mu and the honey is dark.
See: allied shrub, dogwood

**buckwheat** *** **Fagopyrum esculentum, sagittatum Polygonaceae**

Also known as 'brank', an herbaceous plant grown for its triangular seed used for cattle food etc. Liming is possibly harmful to nectar secretion. It can be planted from late May to July 10 but is susceptible to frost and needs warmth to secrete well. It flowers August to late September but the late harvesting of seed can prove difficult. More widely grown in the USSR and USA where it produces excellent surpluses of popular, dark, mahogany-coloured, strongly flavoured honey which granulates rather rapidly and has an identifiable 'colloid' content. It is a fickle plant which is inclined to make bees cross In 'Gleanings' Dr. Richard Taylor said of it 'individual blossoms are short-lived (a day) but a field goes on looking like a vast carpet of snow as blossoms come and go for a couple of weeks. The pollen has been described as light yellow, deep yellow and yellow to brown, light yellowish green and greyish-brown, grain size 40mu.

**bud**

The characteristically shaped protuberance that be tokens the foliage or in the case of a flower bud the inflorescence, that is to follow. The size and appearance of a bud can enable the flowering date (hence potential honey-flow) to be estimated. Pink bud stage in apples comes within a few days of anthesis, lime tree buds can be deceptive often forming several weeks before the blossom and in the case of some eucalypts many months can elapse – though in each case forecasting of the likely date of flowering can be done by experts.
See: apple bud

**Buddleia (Adam Buddle)**
**Buddlejaceae Loganiacae** ***

The butterfly plant (though in recent years far fewer butterflies have besported themselves on the blossoms) An ornamented perennial that rapidly establishes itself on waste ground (London bomb-sites after WWI1) and is very attractive to insects especially butterflies though honeybees work it as well. Varieties include davidii, colvillei, variabilis and the orange-balled globosa. It can grow to a considerable size and flowers throughout the summer months sometimes helping to fill the 'June-gap'. It has a heady aroma, and benefits from heavy pruning. The pollen is light, transparent or whitish.

**Bufo**

The toad, which is known to eat bees (Esp. the 'cane toad' in Australia).
See: toad

**bugloss (ox-tongue) Boraginaceae** ***

Vipers bugloss *Echium vulgare*, (Patterson's curse/Salvation Jane Australia blue weed N.Z. Jersey bugloss *E.planagineum*, *Alkanet – Anchusa officinalis Lycopsis arvensis.*
See: borage

**buildings**

A honey house where the crop can be dealt with is probably the most important requirement in a beekeeping outfit. Hives are usually kept in the open though bee-houses or house-apiaries are common where timber is plentiful and the habit has developed. The storage of supers during the non-active season also calls for some sort of building. The position, size lay-out and cost of any adaptation or construction needs to be carefully thought out and anyone contemplating such involvement would do well to look over ones already established. Grants are more readily available to help with the purchase of buildings than beehives.

**build-up**

Each year a colony of bees tends to follow a cyclic pattern of development governed by the environmental conditions and the adaptation achieved by the particular strain of bee. After a dormant period which in Britain entails the practical cessation of brood rearing, and the formation of a winter cluster, the lengthening of the days seems to trigger off the commencement of a steadily increasing brood nest. This is the spring 'build-up'. The rate of the colony's development depends on food supply (especially pollen and water) the weather, prolificacy of the queen, and numerical strength of the colony and uninterrupted progress. The co-incidence of forager peak and honey flow is the aim of the honey farmer.

**bulb of the endophallus**
**(drone genitalia)**

An ovoid body with crescent-shaped and roughly triangular sclerotized plates in the walls and trough-like internal projections. It leads to the long ejaculatory duct and at the moment of eversion both of these pass down through the cervix which is continuous with the duct.
See: cervix and endophallus.

**bulb of the sting shaft**

An inflated part of the proximal end of the sting shaft into which venom flows when the sting is in use. It has two umbrella

valves which force venom into the shaft as the lancet is protracted. The bulb lies in a trough formed by the bi-lateral plates.
See: umbrella valves

**Bulgaria**
Folk lore and the marriage ceremony here include many references to bees and honey. The bride is given bread and honey. Honey is put onto the bridegroom's face with the words 'be as fond of each other as the bees are fond of this honey' Over 50,000 top quality queens supplied to state and private apiaries per annum by Beekeeping Experimental Station in Sofia. Queen research station at Lovech. Dadant-Blatt hives are used (big ones!)

**bulk comb honey**
Comb honey can be stored in the deep-freezer though clear honey that is not in comb will granulate in these conditions.
See: chunk honey

**bullace *Prunus domestica* Rosaceae**
The wild plum similar to the 'sloe'

**bulletin**
A short official statement covering a public event or on a person's condition or an individually chosen subject. Beekeeping bulletins are somewhat lengthier than Advisory Leaflets, the latter being free of cost while the former were sold at around 30p (1980).

**bull swarm**
The second or 'after' swarm, the first 'cast' said to be worth half as much as the original stock.

**bum (or bumble)**
To hum or make a murmuring noise like a bee. O.E.

**bumble bee *Apidae Bombida***
A large hairy, colourful bee which is social in habit. It hibernates as a fertile queen which emerges in the spring sunshine to feed and find a suitable nesting site, build a nest where it lays and incubates a small (usually eight) number of tiny workers in a manageable number of cells. Once these workers emerge they assist the mother by foraging, feeding her and encouraging her to lay in new cells which they add to the nest. The subsequent generations are normal sized workers and the nest expands possibly to as many as a hundred. Finally drones and queens are reared to mate, so that the fertile queens can repeat the cycle. Some races are plagiarised by cuckoo bees (Psithyrus) and inquilines.

**bumble bee (parasites and predators)**
A nest of bumble bees (and the bees themselves) will be more highly parasitized than the honeybee colony. This can make bumble bee queens, intended for such uses as pollination, extremely difficult to rear. Some parasites are the Nematode Sphaerularia, Bombi, the external mite: Pneumolaelaps and Kuzinia laevis whose inactive stages can be treated with kayacide. A braconid wasp Syntretus splendidus has accounted for the demise of one third of the B.pascalorum and other species. One could go on but these agencies as well a weather cause bumble bee numbers to fluctuate considerably from year to year and man using pesticides and fire, further help to decimate the bumble bee population.

**bumble bee varieties**
In Britain remaining species include Bombus hortorum, pascutorum, pratorum, ruderarius and terrestris, but up to twenty five species have been known to exist here. There are six species of Psythyrus a bumble bee queen that parasitizes other bumbles and there are inquilines (bumbles that live in company with another species) P.bohemicus and vestalis.

**bung**
Larger than the average cork or piece of wood or other material used to seal a container. In barrels (wine, beer etc.) bungs are hammered home to keep the contents

from leaking out and foreign matter from getting in. When making mead or vinegar a loose bung of cotton wool or similar porous material serves the function of keeping insects and unwanted substances out while enabling a layer of carbon dioxide to form over the fermenting liquid ('must'), until the normal fermentation valve can be inserted.

**bunt**
See Puff-ball. Butler, Thorley in Cotton. A dry puff ball can produce intoxicating smoke which used with the greatest care can cause the angriest colonies to temporarily submit.

**Bunter sandstone**
This weathers to a poor soil with no outstanding nectar source other than willowherb. However extra nutritious pollen is produced and this in turn produces bees of high longevity which itself lends to the production of good honey crops.

**burden**
The additional weight carried by a bee over and above its normal body weight. Some young house bees are remarkably strong and able to carry the bodies of drones to a considerable height and distance before dropping them. A foraging bee may have many thousands of pollen grains attached to its body as well as the pollen pellets in its corbiculae and nectar in its crop though naturally foragers for nectar would at best only carry minimum amounts of pollen. Water has to be carried, often in cold, windy conditions though it is actually lighter than nectar.

**burlap (U.S.)**
Hessian or 'gunny'. Used for sacking (especially before plastic sacks came into being). Used for wind-breaks, will serve as a foot-hold for bees collecting water. When well-weathered (almost rotting) it can be dried and burns well in a smoker, either by itself or in combination with corrugated paper, egg cartons etc.
See: hessian

**burn ointment**
Cotton: Add a lump of camphor to the sweet oil before adding wax as in his recipe for cerate. Despite doctors telling us 'dry dressings' only , many beekeepers tell of the splendid results obtained by using honey on burns and scalds saying that it takes the pain away and results in less (if any) scar.

**burr comb**
A small knob of malformed comb. Bees often build bridges of comb to fill gaps or link parts of the nest, either as a re-enforcement, ladder or such like and these irregular-shaped conglomerations of cells are variously named 'burr', 'wild' or 'brace ' comb.
See: brace and wild comb.

**burrowings**
Moles may burrow in front of hives throwing up mounds of soft earth over entrances if close to the ground, Wax moth larvae make tunnels through empty comb/ wax. Braula coeca larvae tunnel under honey cappings. Rats can burrow under hives and badgers too, the latter being able to push hives over and play havoc with the bees. The larger wax moth can even tunnel into the wood of frames.

**bursa**
A wide membranous pouch at the anterior end of the queen's sting chamber and lying aft of the vagina. At each side is a lateral pouch.

**bursal cornua**
Butt or copulatory pouch of the drone.
See: pneumophysis

**bush**
A low woody plant with many branches usually arising from near the ground. Many make excellent bee-plants. Geographically 'bush' means an area naturally covered with bushy vegetation or trees. Such areas in Australia and New Zealand include what we in Britain would call forests and there,

they contain remarkable specimens of nectar bearing plants.

**bush honey**

Honeydew honey derived from the beechwood forests in New Zealand referred to as 'birch' honey.

**Butler Rev. Charles (born 1560)**

For a few years from 1595 he was a Hampshire schoolmaster and he wrote 'Feminine Monarchy' in 1609. For the greater part of his life he was the Vicar of Wootton St. Lawrence. His daughter Elizabeth whom he named 'honey-girl' married Richard White whose grandson's grandson was Gilbert White (of Selborne fame). Further interesting details will be found in 'Hogs at the Honeypot' by Frank Vernon. A cassette 'no ordinary beekeeper' was available from BBKA.

**Butler Colin G. OBE, PHD, FRS**

Formerly Head of the Entomology Dept. at Rothamstead Experimental Station. He first discovered 'queen substance' and wrote several scientific papers and bee books. Amongst his works are: The honeybee and 'The world of the honeybee'.

**but(t)**

A straw bee-hive, skep or ruskie (Somerset, Devon and Cornwall) or a hive or swarm of bees.. 'Dhu beez uv zwaurmeen un wee aan a beet uv u bunt vur tu puut um een'. Also from Devon: Rub tha bee-butts wi' some bayne-stalks.
See: bursal cornua

**buttercup Ranunculus spp. Ranunculaceae *****

A hardy perennial (weed). Sometimes worked by bees in a time of dearth though neither its pollen or nectar are thought to be beneficial. They have been described as 'bitter' and even 'toxic'. A yellow flower of the meadow and countryside, found in many varieties flowering April-May. Thought to be one of the causes of 'May-sickness'. One variety is said to produce honey in North Island of NZ.

**button wood**

The 'Plane tree' – U.S. for sycamore.
See: sycamore

**butyric anhydride**

Unpleasantly odiferous substance used as a repellent It was marketed in N. America as 'Bee-go' for clearing supers.

**Buxus sempervirens**

See: box

**buzz**

The hum of a bee. The worker makes various kinds of buzzes as for example when flying, when entering a flower, when trapped or when about to sting. A buzz may include a rasping note or something suggestive of a summer of sunshine and flowers. A drone has a lower flight note and the queen's is different again.
See: fanning bye hive
Observatory hive O.E.

**byke**

A nest of wild bees as found in a hollow bough.
See: bike.

# C

**c./circa**
about, around or in that neighbourhood

**cabbage family Brassica**
The many useful varieties are dealt with under 'Brassica'

**cabbage tree *Cordyline Australis***
A flowering palm tree found in the antipodes and well-worked by honeybees.

**caged queens**
Queens are transported and introduced into new colonies in small cages. They are accompanied by an attendant group of nurse bees (around 8 – 10) and they should not be isolated for more than a few minutes as they cannot feed themselves effectively. The nurse bees should always be 'compatible' i.e. from the same colony as the queen. Cages used for air transport are not necessarily best for introduction though they usually are well-ventilated, have a supply of specially prepared candy preferably containing pollen, but it must not be sticky. She will live longer in the dark, is happier with her own bees but can be stored over a normal colony provided the warmth can get to her yet none of the bees are able to touch the cage for example support the cage on sticks about 1cm above the screen covering the colony). It is advocated that all the nurse bees are taken away (and subsequently checked for disease) while the queen, having been left alone for a few minutes and is then dying for company, can be introduced to the queenless bees who in turn are dying to be set right queenwise. A delay mechanism such as a piece of soft candy or newspaper delays the queen's actually mingling with the colony but allows the many tongues to explore her through the sides of the cage.

**Cajeput *Melaleuca leucadendron***
This invasive weed is credited with being a prolific nectar producer in some areas of the USA.

**cake**
A sweet baked food or shaped or moulded mass. We speak of a 'cake' of beeswax though NHS schedules now ask for a 'piece' giving a minimum thickness and range of weight, or in some classes a 'block' as prepared for sale. There are of course numerous recipes or cakes made with honey sometimes in conjunction with caster sugar. See: honey cake, ingot, patty

**calcareous**
The nature of soils, containing a substantial proportion of calcium in the form of chalk or lime. Warm, open, well-drained. Plants readily obtain water, there is abundant lime and phosphates but lack of potash, nitrates and humus. The difference between the dark soils and the whiter chalky fields together with drainage and latitude gives varying times for nectar secretion. It is always interesting to see the fields of oilseed rape coming into flower in sequence as one travels northwards.

**calcicole**
Chalk loving. Flowers which thrive in calcareous soil such as: clover, dandelion, mullein, thyme, yellow charlock etc. Opposite to 'calcifuge –member of heath family.

**calcifuge**

A plant that will only thrive in soils that are poor in chalk (calcium carbonate) – limey soils. They are called 'calcifugous' and include plants like heath, heather, rhododendron etc.
See: calcicoles

**calcium Ca**

A base metallic element forming alkalis. Calcium carbonate is chalk and in this form it is very widespread (cliffs of the eastern, southern coast). At one time clover, which grew prolifically on the Downs and similar chalky places was known as the 'honey plant par excellence'. Now-a-days we look to countries like Canada for enormous takes of clover honey.

**calcium cyanide $Ca(CN)_2$**

A deadly poison and one so dangerous that alternatives should be used whenever possible It is used as a fumigant for greenhouses. When used as an insecticide, potassium and sodium cyanides are preferred to calcium cyanamide. $CACN_2$ is another chemical of which similar use is made. The release of Prussic (or hydrocyanic) acid of which it is a salt, is lethal. Used for metal hardening and gold recovery processes.
See: cyanide

**calendar of bee-plans**

IBRA published Dorothy Hodges work and other useful lists are available and extremely useful when you get to know your district and can plan your management to achieve peak colony strength at the appropriate time of the year.
See: bee plant

**Calendula spp.**

See: marigold

**calf's dung**

When dry it was recommended for the smoker HUISH.

**Californian Bee Breeders**

Organised in 1889to serve the beekeeping industry of California. The second Annual Auction of breeder queens was held in February 1980. The price at that time for queens which were exhibited in observation hives with their brood was \$50 -\$400. Cheap at half the price – as the saying goes.

**Californian horse chestnut/Californian buckeye *Aesculus Californica***

The nectar and pollen of this plant have been said to be poisonous. This also applies to certain 'limes' in this country where bumble bees have been seen to roll about on the ground under the trees, as if 'drunk'.

**Californian poppy *Eschscholzia californica* Papaveraceae**

Various bright colours – worked by bees for bright orange pollen but not nectar.

***Calluna vulgaris* Ericaceae**

Common heather, Ling, a woody shrub that likes acid, peaty soils and is widespread on British moors. It can give heavy yields of nectar during August and September and provides a much later flow than most other plants and has become of exceptional importance to migratory beekeepers. Like so many 'fickle' honey plants it usually gives a small surplus and enough for the bees to winter on but great variations occur. Pollen has been called 'slate-grey', 'greyish brown', 'dull white','light to dark brown' and has a tetrad grain 27µ. Its honey is in a class by itself being too gelatinous to spin out in ordinary extractors and needing to be eaten in the comb or pressed out. Scottish heather honey has a world- wide reputation and it makes good mead and Athole brose.
See: Calluna cultivars, heath, ling, heather honey

***Calluna vulgaris* cultivars**

*Alba pumila*, aurea, Barnett gold, Golden feather, Robert Chapman, Silver knight, old haze, Hammondii, Peter Sparkes, Sister Anne (white, pink or purple).

**calorie**
The amount of heat required to raise 1g of water from 15° to 16°C .
See: joule

**calorific value of honey**
100g of honey has a value equivalent to 303 calories. The equivalent of 30 eggs, 6 pints of milk, 8 lbs of plums, 10 lbs of green peas, 12 lbs of apples or 20lbs of carrots.

**calyx pl. calyces**
The outer whorl of floral parts that is next to the petals. Also the cup-like head of pedunculate bodies in insects. There are four of these in the honeybee's brain and their walls encircle the Kenyon cells. Co-ordinated behaviour involving sensory feedback is associated with them and with the alpha and beta lobes of the mushroom bodies. The paired calyces form the upper part of the so-called mushroom bodies,
See: mushroom bodies, neuropile

**Campanula spp.**
Harebell *C. rotundifolia*. Perennial flowers July-Sept (Scottish bluebell) Its pollen can get into heather honey. Peach-leaved *C. persicifolia* – pollen yellow-brown, garden Campanula purple pollen in July. This is a large family seed can be found in variety in catalogues.
See: Canterbury bell

**Campden tablet Sodium metabisulphite $Na_2S_2O_5$**
It can be used to sterilise equipment or to check fermentation. As a powder or in the form of a 5ml tablet dissolved in half-a-pint of water mixed with another half-a-pint of water into which a half teaspoonful of citric acid has been dissolved. This produces a sterilising solution by virtue of the release of sulphur dioxide. Care must be taken not to spoil the flavour of mead by adding too much when using it to check fermentation.

**Canada (1980)**
Two colonies of bees went to B.C. (Victoria) in May 1858. In 1977 they imported bees and queens largely from the US 300,000 packages and queens in the one season. The number of beekeepers in 1977 was 13,400, this increased to 17,470 by 1978 and the number of colonies went up proportionately to 566,900 and presumably, judging by their exports of lovely golden clover honey to the UK this has gone on increasing.
See: Peace River and beekeeping in Nova Scotia

**Canada balsam**
A resin derived from two species of American conifer, dissolved in xylol and used for the mounting of material on permanent microscope slides. The hard resin has an R.I. of 1×56. Special clips can be obtained to hold a cover-glass over the material while the Canada balsam dries and sets.

**candidate**
A person who presents himself for a post or academic qualification (as by examination or test). A candidate may be asked, to supply credentials, to complete a form or questionnaire and possibly to pay a fee. An applicant for a job may be short-listed, that is selected for interview, from a longer list of candidates. He may be an entrant or 'sitter' for an exam; beekeeping exams include: oral questioning, the writing of answers and the practical evidence of their capabilities as a beekeeper.

**candied honey**
This means honey in solid form, granulated or crystallised – a term used in Oceania

**candles**
A candle made from pure beeswax is a beautiful thing to behold and it will burn with a clear bright flame giving off a pleasing aroma. In more recent times the addition of tallow and paraffin wax to beeswax has degraded the standard and even churches rarely use pure beeswax now. Moulds of rubber, easy release plastics of

varying shapes, using releasing agents where necessary, are one method of manufacture while dipping and rolling or pouring techniques can be used. A proper, braided wick with thickness matched to candle diameter is very important. When dipping the wick is immersed in the liquid wax lifted carefully out and then pulled taut as it dries. Subsequent dips, first in the wax and then in cold water, build up the candle. While warm and at intervals it can be rolled between two panes of glass or similar smooth material while still warm to smooth and straighten the sides. Stearic acid is used to make beeswax harder.
See: church candles, flame test, judging, showing

**candy**

Similar to 'icing' on a cake but as a winter food that can be fed by placing it directly on top of the cluster causing minimum excitement. A special texture is called for. Five cups of white granulated sugar to one cup of water, stirred carefully and boiled for a short period. then when it begins to ' cloud' pour it into suitable containers – good if one side is transparent so that progress can be followed. Ideal texture is when the candy is cold a finger nail can just scratch the surface. Trial and error will doubtless help to achieve the best result and your producing a good sample. It should not be given to a colony that has a queen excluder in place because it could tempt the cluster away from the queen. The feeding of candy is debateable because the bees should always have been given adequate food in late summer. If you have good reason to know they are short of food (maybe by 'hefting') then remember thick warm syrup might be a better alternative but this has to be carefully judged early in the year. Yellow stains on the candy makes one think in terms of disease.
See: ' Goods candy' candy recipe, queen cage candy, ragus

**candying**

An alternative way of saying that honey or sugar syrup is becoming granulated. The formation of crystals in a previously clear solution. When making candy the onset of cloudiness as it cools lets you know it is about to candy. This is the time to stir and pour into mould.

**candy recipe**

Most recipes work out at around five times as much sugar as water by volume. So use the proportions 6lb sugar to each pint of water or in metric measurements 5kg to each ltr of water. Put water into a preserving pan of enamel or aluminium and add the sugar gradually, stirring the while. Cream of tartar may be added 1/8oz with the 6lb recipe or 6×5g with the 5kg recipe. Heat slowly to avoid burning and continue to stir. Bring to the boil and continue until the temperature reaches 117°C/243°F just below the surface of the liquid. Stop and allow the syrup to cool to 40°C/105°F then stir until it appears 'milky' when it should be transferred reasonably quickly into suitable boxes (e.g. plastic margarine containers). Abbreviated: 5:1 for 3 minutes. Candy should be firm but soft enough to scratch with your finger nail.

**cannibalism**

Nurse bees will eat young larvae that hatch from fertilised eggs which are genetically male (male eggs are normally unfertilised). This is due to the secretion by the larvae which encourages the bees to eat them. Research with *A.cerana* in India has shown that the secretion is delayed and in some cases larvae are not eaten until detection at up to four days, and some apparently reach adulthood. Morse suggests that lack of pollen can lead to 'spotty-brood' caused by the eating of eggs. Recent research has shown through DNA that sibling matings by queens can lead to the laying of infertile eggs in worker cells.

**Canola**

Canadians have given this name to varieties of oil-seed rape having no Erucic acid and low in glucosinolates. By 1980 Canada had expanded her OSR production to 25% of the world total. World War II had caused a demand for the valuable steam-resistant oil from certain kinds of rape and its high protein cattle food, cooking and salad oils, margarines, cake mixes and flavourings suggest continued importance in the future.

**canopy**

An overhanging roof or cup-shaped shelter sometimes used to as a term denoting the 'dome' of stores that surmounts the bees' brood nest.
See: honey canopy

**Canterbury bell *Campanula medium***

A biennial that is popular in gardens. Many varieties from dwarf Alpines to some vigorous, tall perennials, offer colourful masses of blossom to engage the activities of bees throughout the summer months.

**cap**

1. While this word has numerous connotations it is now used by most beekeepers to refer to the cap that screws on to a glass or plastic jar for use as a honey container. Since the squat Ministry jar was standardised in World War II it has undergone several changes of shape. It is of paramount importance that the cap not only fits the jar exactly, but that it is easy to tighten and to render the hygroscopic honey completely air-tight. Caps may include a waxed wad or may have a plastic insert or be made entirely of plastic.
2. The straw cap that sat on top of skeps was the original 'super'. The saying 'If the cap fits wear it' is appropriate when we consider the fitting of helmet-type veil. It is very important that this keeps the veil away from the face and does not become detached when the beekeeper bends down or catches the brim on a blackberry bush.

See: flowed-in

**cap for skep**

A small straw extension placed above the skep COTTON. Top of a hackle, BUTLER, BAGSTER. BEVAN says 'hackle or cap'. Also 'caping' Supering with a straw cap.
See: eke

**Cape bee S.Africa**

See: *Apis mellifera capensis*

**capped brood**

After incubation of the egg and the feeding stage has passed (and this is much of the same duration for each of the three castes) the larva extends itself in its cell and begins to spin a partial cocoon. This is reinforced at the opening of the cell by the nurse bees who 'seal' or cap over the cell using wax (apparently derived from the comb as it is usually of a similar colour) and pollen together with small fibrous particles which form domed cappings that are slightly convex. In the case of queens and drones. the doming is more pronounced.
See: open brood, sealed brood, unsealed brood

**capped cells**

See: capped brood, capped honey etc.

**capped honey**

Honey comb that has been capped over or 'sealed' by the bees. If air-space is left between the cappings and the honey it will give the honeycomb a lighter appearance. The British black bee was reputed to make excellent white honeycomb. *Braula coeca*, if present, will tunnel under these cappings, spoiling their appearance with their whitish thread-like tunnels which meander along for an inch or so under the cappings. The cappings normally 'look good' unless damaged, trampled over (through being left too long in the hive), or maybe 'greasy-looking if disease is present. When honey is to be extracted the cappings need to be removed so that the honey can be centrifuged out in an extractor. The removal

is normally done with an uncapping knife or a scraping fork.
The cappings have to be drained but can be used for some types of asthma sufferers. Otherwise the wax can be washed and recovered for melting down, the sweet washings being fed back to the same colony they came from or used to make mead or vinegar. Honeycomb with a thin mid-rib makes excellent eating, though some prefer to eat it with toast or cereal to disguise the feel of wax in their mouths (indigestible but cleansing for the alimentary canal).
See: cappings

**cappings**

The separated cappings from sealed honey comb. A by-product of the honey extracting process. When the honeycomb is uncapped, prior to spinning out the honey, a layer, mainly of cappings but also part of the cell walls, is sliced off. This layer is either gravity strained or centrifuged and is broken up to make it yield most of its honey. As it is honey-wet it must be washed and dried to produce light crumbly wax ready for rendering. The washings may be fed back to the bees (from the health point of view to the colony from which they came) or used for making mead or honey vinegar. Cages have been designed to fit into supers for feeding this material back to the bees but it can cause excitement and late in the season trigger off robbing. Gravity drained cappings contain 50% honey by weight and the clean dry cappings will yield about 25% of their weight in fine quality beeswax.
See: cappings, strainer, uncapping

**cappings cleaner**

A device used to separate the honey from the wax cappings. This can be done with controlled heat or by spinning or pressing. Finally they have to be washed and dried if they are to be stored, otherwise they can be rendered in soft (or slightly acidulated) water. As honey producing outfits have become relatively numerous in some countries the successful salvaging of the tonnes of honey otherwise locked up in wax cappings has become increasingly important and many ingenious spinning devices (centrifugal) are now made.

**cappings scratcher**

A pronged device with a handle. A device for tearing the cappings off cells without totally removing the cappings is known as a cappings scratcher. When honey has to be uncapped for extraction or other purposes, it is usually sliced off but any means such as scratching the surface with a fork will have a similar effect.
See: uncapping fork

**cappings strainer/melter**

The Pratley uncapping tray is an example of an apparatus enabling honey-wet cappings to be separated into the two components. A sloping, stainless-steel tray with a deep rim and outlet for the liquid honey (and wax if heated) A hot water jacket with an electric element forms the underneath portion.
See: cappings cleaner, Pratley

**Captan**

A fungicide sometimes used on fruit even at blossom time. It appears to be somewhat repellent to bees. (Ed. unwise manoeuvre) Sometimes used with gamma BHC for seed dressings

**captivity**

Kept in confinement or restraint. Bees cannot function normally in captivity and it is a mistake to display observation hives where the bees are confined as they tend to scurry around in an excited state instead of performing their normal duties as when free to come and go as suits them. Bees may be encouraged to use a hive that you provide as their domain and we may regard them as domesticated to that extent but they are free creatures of the wild and will not willingly tolerate captivity.

**car automobile**

Ideally a four-wheel-drive motor-vehicle or

variant. A hobbyist beekeeper's car might become a portable shed if he runs a few out-apiaries. When it is intended to use a private car to move hives a hatchback or a saloon with tow-bar and trailer can be useful. Hatchbacks, small vans or shooting brakes do suffer from the necessity of smoker smell, loose bees and other embarrassments for non-beekeeping passengers. When moving hives of bees, a 'live-bee' notice on display is worthwhile and advisable and it may help to inform the police of your route and time range. Bees in an apiary 'get to know' a car and it is worth driving a short distance away before opening windows to rid the car of them.

**caramelise**

Caramel is burnt sugar. The expression to caramelise is used in conjunction with over-heated honey (see /'HMF'). Such honey can be detected by the 'caramelised' flavour which is akin to burnt toffee. Honey also becomes darker when heated. This is not likely occur under normal circumstances though granulated honey may be left too long in a warming cabinet or comb that has granulated solidly (e.g. 'oil-seed' rape) may be separated into a wax cake floating on honey which may then taste caramelised.

**carotene A pigment - found in honey**

**carbamate**

An insecticide and growth regulator certain formulations of which are sometimes used for fruit thinning.
See: carbaryl

**Carbaryl**

Carbamate insecticide, growth regulator and earthworm killer. Used as a thinner to reduce crop when over-pollination has occurred. Sevin said to be more toxic to bees at low than at high temperature (as was DDT). At least four new formulations Sevin4, Sevinol which are moderately toxic. SevinSL and XLR relatively non-toxic – all other formulations (1981) highly toxic, dangerous to bees and harmful to fish. Available forms including wettable powders. In 1982 there were 12 cases of bee deaths known to have been caused by this poison. Used on apple orchard in Kent and Essex. As of 1995 severe restrictions on use have been introduced.

**carbohydrate**

An important item of plant and animal food providing a source of energy. Chemically it is a hydrate of carbon which implies a number of carbon atoms locked into those of hydrogen and oxygen in the same proportion as in water. Starch, sugar and glucose form the supporting tissues of plants and provide an important food for animals.

**carbolic cloth $C_6H_5OH$**

Introduced by Rev.G. Raynor. Carbolic acid (phenol) was a common household antiseptic and disinfectant. Being strongly repellent to bees it was used in a diluted form to moisten cloths for subjugation and sometimes for clearing bees from supers, though due to its none-too-pleasant and penetrating smell there was a danger of contaminating the honey. It is a corrosive acid and only miscible with water when a substance such as glycerine is added. It has been superseded by numerous chemicals etc. such as Proprionic anhydride, benzaldehyde etc.
See: carbolic cloth recipe, repellents, smoke bomb

**carbolic cloth recipe**

11/2oz Calverts No. 5 Carbolic. 11/2oz glycerine and 1 Quart of water. Mix the acid and glycerine well before adding water and shake bottle well before using. R.O.B. Manley gives 2oz Calverts No.5 loz glycerine and l pint of water. Others suggest Calverts No.5 diluted to quarter strength or Jeyes Fluid @ about 10% DonWhite. Dettol might be considered a better repellent these days.

**carbon dioxide $CO_2$**

The capsules of this gas as used for

pressurising soda water siphons can be used to anaesthetise queens. However reports from Polish workers intimated that it could result in reduced longevity of newly emerged bees, reduced brood food gland development, depressed flight activity and wax gland failure. In large colonies the level kept down to 1% and fanning increases markedly if 3% exceeded. Winter clusters may reduce activity by utilising CO2 concentrations and will tolerate up to 9% but fan to ventilate at 10%. Increasing levels of world CO2 due to burning fossil fuels etc. is worrying scientist who fear an adverse effect on world climate.
See: anaesthetic, quacking

**carbon disulphide CS2**

A volatile liquid with fumes heavier than air. Toxic to humans and fumes do not disperse readily and can ignite at floor level. It is a flammable chemical used for fumigation – 1 tablespoon for each 10 frame super to kill wax moth etc.

**carbon tetrachloride $CC_{14}$, abb. CCT**

A clear liquid that is used as a degreasing substance and was once common in certain fire extinguishers. It is unlike chloroform and is toxic to humans and should not be used in a confined space. It will kill bees and wasps and it dissolves beeswax. Its advantage over petrol for the extermination of wasp nests is its non-flammability. It is harmful to fish and a well-known liver poison. It can be used to remove blemishes from cakes of wax prepared for show. A useful substance but handle with extreme care.
See: wasp

**cardboard boxes**

Boxes made from cardboard and designed to hold specific contents. Their rigidity can be strengthened with 'sticky tape' and their varied size and usefulness include swarm catching and if health is in doubt they can of course be destroyed, ('easy come – easy go'). Slight modifications such as a firm rim and entrance hole make them very appropriate together with ventilated white cloth and cord as a means of taking a swarm back to the apiary but always travel with the ventilation upwards i.e. turn the box over. If dampened the same material can be rolled into suitably sized pieces which when dry make reasonably good fuel for the smoker.

**carder bee *Bombus agrorum***

*B. agrorum* makes a nest of moss among rough grass.

**cardiac**

Pertaining to the heart or stomach. The cardiac impulse - motion caused by a rapid increase in the tension of a ventricle. Cardioblasts are cells, mesodermal in origin, from which a bee's heart is formed.

**cardo pl. cardines**

Hinge of the under lip or basal sclerite of the maxilla (mouth part).

**caress**

We may use this word somewhat anthropomorphically to describe the apparent 'feeling movements' so often made by the antennae – organs that are often referred to as 'feelers'. When a worker is seen to 'caress' a queen's body, another bee's antenna, or to examine a cell or food, it is evident that the organs of sense relating to smell and taste are being brought to bear on the subject as well as tactile sense. Such movements are therefore by way of communication or investigation.

**Carl's solution**

This solution is used for preserving material (such as dead bees) and is made up from the following: 17 parts 95% alcohol, 6 parts of formalin, 28 parts water and 2 parts glacial acetic acid which is to be added just before using.

**carnauba wax**

An important vegetable wax from the wax palm *Copernicia prunifera* grown in Northern Brazil, Chile and Peru. It is hard and used as an alternative to beeswax. Although yellowish or greenish in colour it is derived from the leaves of the plant and it contains lignoceric acid.
See: other waxes

**carnivore**

An animal or plant that eats flesh – animal tissues. Thee are plants which capture and devour various insects doubtless including bees. Hence 'carnivorous'

**carpel**

A leaf that has formed into a pistil. Several carpels may unite to form one pistil. A division of a seed capsule.

**carpenter bees Xylocopa**

A group of large, solitary bees that resemble bumble bees and burrow into rotting wood, boring tunnels in which their eggs are deposited. They swallow pollen and then regurgitate it from the honey sac when they return to the nest.

**carpet expanse**

A carpet of petals may suggest the end of a flow just as the caps of eucalypt buds may indicate that one is under way. A carpet of dead bees in front of a hive may be caused by robbing, the spraying of insecticides, build-up of disease, clear-out of dead bees after a cold spell or the casting out of drones a the end of summer.
See: strong colonies

**car polish**

May be made using 1/2oz of well-dried soap, 41/2oz beeswax, 1 pint rainwater, 1 pint turpentine. method: Finely shred soap, place in a saucepan and add rainwater then warm gently until dissolved. Put beeswax into a jar placed in a pan of water and heat gently until the wax is melted. Caution! Both wax and turpentine are flammable. Remove from heat and pour the beeswax into the turps. Then pour the warm solution of soap into the wax and turps; stirring to form a smooth emulsion – this makes about 1.8kg/4lbs of excellent polish. Apply with a soft cloth and polish with another.

**Carr Wm. Broughton**

An English clergyman and practical beekeeper whose initials W.B.C. are firmly rooted in our beekeeping as he was the inventor of a double-walled, picturesque hive by that name. He cleverly designed metal frame spacers that could be stamped out of thin sheet-metal and folded to make clips that fitted onto the frame lugs (metal ends) in a wide size for supers and narrow for brood. He also produced a section rack and uncapping knife.

**cartridge**

The fuel pack to be used in a smoker. A neat , clean, easily-ignitable roll designed to fit exactly into one's smoker has decided advantages. Furthermore it can be 'treated' at one end for easy lighting even in difficult weather conditions and the additions of certain medicants as used for *Acarine* and *Braula coeca* is also possible.

**cashiered stocks**

Stocks that were to be united or destroyed by sulphur fumes. O.E.

**Cassius Dionysius c2nd century**

In common with others said, *Leave bees one tenth of their honey, taking the rest on the day of the wild fig.* 7 July–Sacred to Vulcan.

**cast**

A group of bees that have left the hive with one or more virgin queens. 'They settled in two swarms which soon coalesced into one' GILBERT WHITE. Also called an 'after' swarm it is likely to emerge one fine day about a week after the prime swarm. As the young nubile queen requires to mate, the cast may take the form of a 'mating swarm'. Having been hived, the queen might go out to mate and the whole cast leave with her. While a cast is usually smaller than a swarm its size will vary and when several queens are present, maybe very large. They are notoriously fickle, often returning to the hive or move from one place to another. Should the cluster cover more than one 'bulge' it could be indicative of more than one queen. They may then be hived by shaking them into an empty box placed over an excluder over a box of combs containing some stores and a little unsealed brood. As the bees run down, separate virgins (some possibly being 'balled) may be sorted out. Casts are very mobile, their young queens able to fly higher and farther than laying queens. Hence they often settle in places that are difficult to get at. Food is less likely to 'anchor' a cast than some unsealed brood from a healthy colony. The drawing of comb is an indication that they have decided to stay.

***Castanea sativa***
See: Sweet/Spanish chestnut

**caste**
Social insects are differentiated into groups of structurally and functionally specialised individuals. In honeybees castes comprise the queen, the worker and the drone. Freak intermediaries are rare and such faults can only be laid at the door of queens. Such living queens are useful for research (BBBA).
See: gynandromorph

**caste differentiation**
In a normal colony a fertile female queen lays eggs of two kinds: male eggs, said to be kept free of contact with any sperm from the queen's spermatheca and female eggs which are smeared with a few male gametes (spermatozoa) from the spermatheca) by the queen and which become fertile female eggs by the time they are laid in the queen cups or worker cells. Drone eggs are laid in the larger drone cells (though they can also be laid in any cell by 'laying' workers when present). Each caste is incubated for three days and then fed appropriately according to their respective potential. Workers take 21 days, drones 24 and queens 15. Queen cells are vertical and gravity plays a part in their special development.

**castellated frame spacer**
Listed in catalogue as castellated runner or frame spacer. It is one way of securing centre to centre spacing of frames. However it prevents the sliding movement when frames are to be moved across individually or together. Also its sharp edges have to be watched out for. As there are many alternative, efficient ways of spacing the disadvantages almost certainly outway the advantages. Being totally rigid it denies the possibility of using various spacing such as when foundation needs to be drawn or wider combs produced for easier uncapping in the supers during honey flow,

**castling**
Charles Butler described after-swarms in this way in 1609 in 'The Feminine Monarchy'

**cat**
It has been jokingly intimated that cats make good 'mouse guards'. Where domestic cats are allowed to approach hives it has been known for them to try to catch bees in the air though they usually soon abandon the idea. They have been known to make themselves comfortable on a beehive roof in the sun. Bees dislike jerky, moving objects especially if dark in colour and cats are vulnerable on parts of the face and on their paws.

**catabolism kat**
The destruction or breaking down of complex organic molecules, within living creatures, usually with the liberation of energy.
See: anabolism, energy, metabolism

**catalase**
The enzyme that decomposes hydrogen

peroxide into water and oxygen. It occurs in plant and animal tissues.

**catalepsy**
The fainting or shamming of death as by a queen. This condition is thought to be due to a temporary nervous disorder. Care should be taken to close the colony and all should be well. Wax moth larvae can also do this.
See: akinesis, feigning death, thanetosis

**catalogue**
A small booklet listing items for sale with descriptions and possibly illustrations or pictures. The catalogues of most beekeeping appliance manufacturers use an alphabetical list to refer to page numbers. Sometimes prices are given alongside each item or otherwise a separate price list is provided. Annual revision and up-dating takes place and catalogues are usually available free direct or by post, many can be seen on line.

**catalyst**
An agent that remains unchanged even after it has become involved in the initiation, accelerating or retardation of a chemical or physical reaction. Enzymes are examples of catalysts and these are present in honey.

**catch**
The expression 'to catch queens' is used when nuclei or colonies are opened with a view to finding and possibly removing queens. Esp. in NZ/Australia.

**catenary hive**
The 'chain' hive. Developed by W.B.Bielby, Yorkshire. Based on the natural shape of honeycomb when built by honeybees completely free from restrictions of space. The catenary curve of the bottom board (which is continuous with the walls of the brood chamber), follows the shape assumed by a chain that is suspended at each end. Such construction gives added strength and minimises the volume of timber required. It can be built in various sizes to interchange with all manner of roofs, supers, excluders, crown boards etc. Nylon-net reinforced foundation enables top bars only to be used (frameless combs).
See: 'Home Honey Production' by Bielby

**caterpillar**
The larva of a moth or butterfly, the greater and the lesser wax moth being of great concern to beekeepers. Unlike honeybee larvae they can be very mobile and have claws and pseudopodia (sucking, clasping feet). Some complete their eating and development stages in a short time, others take years (goat moth – frequenter of rotten trees).
See: greater, lesser wax moth

**catkin amentum**
In winter this is a scaly spike that remains in its protective 'bud-stage' until environmental conditions encourage it to open out into a uni-sexual inflorescence when, as in the case of hazel and other nuts, it becomes flexible and pendulous bearing apetalous, male flowers and copious quantities of pollen which is liberated to float on the wind. Honeybees can collect and utilise such pollen even though it is not of the rounder, stickier types that they prefer. When its reproductive function has been completed it falls to the ground.
See: hazel

**catmint Nepeta spp. Labiatae**
Ideal for herbaceous border this 9" high silvery grey perennial with pretty blue flowers blooms through the summer and is beloved by bees (and cats who like to lie in it) *N.cataria* is 'catnip' and the many varieties include *grantifolia*, *mussinii* (gardens) *nuda*, *transcaucasia* and *zangesura*. The flower tubes are just right for bees but exclude flies. It will grow well on any soil.

**Cato 234-149 BC**
A very religious man, Senator and farmer. He wrote a little about bees influenced by, but taking a more business-like approach

than, Aristotle. Although he did not believe in having many slaves doubtless the 'mellarius' looked after his apiary. His writings included descriptions of the use of honey in cakes for ritualistic purposes.

**cattle**

Young calves may be over-curious and push against hives that are not surrounded by adequate fencing. Stings can penetrate their hides and cause discomfort. Milk yields have suffered when cattle were driven into a small enclosure containing a fenced apiary. Ken Stevens has had a complete apiary trampled into matchwood when a bull chased several cows (perhaps just one) through a gate that was faulty and into the apiary. In NZ where cattle range freely over large tracts hives are usually set in blocks of four so that cattle may scratch against them without knocking them over.

**caustic potash KOH**

Potassium hydroxide a strong alkali useful; for cleaning away muscles and connective tissue from chitinous parts of the honeybee during dissection. It will also dissolve wax and propolis when it is desired to make temporary or permanent mounts for viewing under the microscope.

**caustic soda NaOH**

Sodium hydroxide a white deliquescent solid which is strongly alkaline and dangerous to the skin. Not to be confused with the soda used to soften water for washing. It is used when propolis or wax has to be entirely removed from objects.
See: disinfectant, lye, soda

**CBPV Chronic bee paralysis virus**

This virus was clearly identified by Dr. Bailey and others (using the electron microscope). It has an amorphous structure and is not the same as strains of ABPV. Little is known of its natural history and the disease caused by it has no seasonal peak and may be present throughout the year. Infected bees show many varied symptoms. There may be crawlers and dysentery and bees may huddle in groups, as with acarine. Hairless, black shiny or greasy-looking bees, may tremble and have disjointed wings and bloated, distended abdomens. Healthy adults will eject such bees as they are about to die. Many will be unable to fly. It is considered that there is virtually no risk of spreading the disease by transferring bees, brood or equipment to other stocks but a colony displaying such symptoms should be requeened. The close in-breeding of queens seems to promote the trouble. It can be fatal when combined with other diseases. Susceptibility depends greatly on genetic factors. See: black robber bees, disease, viruses

**cc cubic centimetre**

This abbreviation is the equivalent of a millilitre though now 'ml' is more commonly used. A teaspoon full is about 5ml.

**C.D.B. Hive**

Introduced by the Abbott Brothers of Dublin in 1894 to the specification of the Congested Districts Board. A hive for the cottager, economical to work but able to withstand the climate. Also widely used in Scotland. A lift or 'riser' stands on the brood box to accommodate sections: when empty it can be telescoped over the lower box for extra protection in winter.

**Cecrops**

An Egyptian who wandered to Greece about 1500 BC and is thought to have introduced beekeeping there.

**cedar timber for hives**

A light, relatively weather-proof aromatic timber used for pencils and brought into use as a 'luxury' timber for hives and generally used without painting though some brands of 'Cuprinol' have been used. The English cedar *Cedrus* is not a good example because it is heavy, inclined to knots and shakes. Western Red Cedar is one of the best

and while the heartwood of *Thuja plicata* (which is not a cedar at all) provides a relatively insect and weather-proof, light, easily worked timber (expensive though). Californian redwood *Sequoia* is said to give 15 to 25 years durability, British Sequoia (red cedar) 10 to 15 years and N. American Douglas fir 10 to 15 yrs. Caribbean pitch pine and other pine woods are in common use.
See: Western Red Cedar

**cedar wood oil**
Used as a mounting medium when preparing microscope slides and as a clearer.

**celandine** *Ranunculus ficaria* **Ranunculaceae**
Lesser celandine, a yellow flower often common in woods. One of the buttercup family. The value of its pollen has been cast into doubt with some reports linking it with 'May sickness' though Whitehood describes it a good and dark orange. The greater celandine *Chelidonium majus* is not related but is a member of the poppy family.

**cell A structure containing a cavity.**

1. Biology: discrete but minute masses of protoplasm. They can differentiate to become highly variable in shape and function and are sometimes bounded by a semi-permeable membrane though the contents may disappear leaving a hard structure such as chitin in insects or wood in trees. They are the smallest units of living matter.
2. Honeycomb is largely composed of hexagonal cells though interstitial cells occupy spaces adjacent to supporting surfaces or between drone and worker cells. Queen cells are vertical, built for a once-only occupation, whereas worker cells are horizontal and built for constant use such as nectar ripening, brood rearing and storage of honey or pollen.

See: cell walls of the honeycomb

**cell base**
So that the horizontal back-to-back cells can have curved bases in favour of the occupants, the bees have contrived to build the 'septum (mid-rib) so that each cell base is composed of three 'rhombs' (diamond or lozenge shapes). Each of these form a part of the cell base on and the three adjacent cells on the other side of the comb. One can experiment with a piece of foundation by pricking a small hole in each of the rhombs which would form a single cell base and then turn it over to find a hole in each of three adjacent cells on the other side.
See: rhomb

**cell division**
See: meiosis, mitosis

**cell protector**
A simple device for the safeguarding of a queencell. When sealed queen cells are introduced into a colony which might tear them down if they were not protected in some way. Because the 'tip' of a queen cell is well reinforced with a cocoon it is relatively invulnerable. However the side walls need guarding and various methods have been employed to do this. Spiral metal coils and even sticking plaster have been used to safeguard the inmate who, upon emergence is more often than not accepted by the colony.
See: spiral, west

**cell rearing colony cell- building**
When a cell starting colony is only used as its name implies to get cells started, then the cells can be moved after 24 hours or so to a cell-finishing colony. The arrangement of such a colony will vary according to time of year, equipment available and size of the operator. A hobbyist might use one hive divided by a Snelgrove board, the bee farmer may have special prolific stocks for the purpose. Cells should only be raised in strong, well-found, healthy colonies.

**cells of the honeycomb**
Accommodation, brood, capped, deformed, drone, honey, interstitial, nurse, pollen,

queen, sealed, size of, supersedure, uncapped, unsealed walls of, worker. See under these various headings but foundation sizes can vary apart from drone and worker but also according to various races of bee.
See: cell sizes

### cell sizes

Worker base 825-820 cells per square double decimetre (2dm squared). BIBBA once distributed 700 size foundation and 923 and up to 640 are made, the latter being drone size. Worker size approximates to 2 cells per cm or 5 per inch (5.3-6.3mm) and drone 4 per inch (6.25 - 7.25mm). Sizes vary with race and bees can be accustomed to new sizes. The depth of worker cells is about 13mm the doming being considerably less than that of drone cells which approximate to 17mm deep. Queencells average 21mm in length with maximum exterior girth near the top of 10mm. (I give the following for what they are worth) Worker flats 0.195 - 0.215 and cell wall thickness: 0.22-0.25, and diagonals 1.155- 1.0

### cell starting colony

A colony that has been made ready to receive artificial grafts (or very young larvae) for the initial production of queen cells. This is done by 'plumping' or choosing a strong healthy stock. The queen and unsealed brood are removed some 4-6 hours previously and the 'prepared' frame(s) is inserted in a central gap that was purposely left empty. Feeding is carried out and free flight allowed. The cells can remain for up to ten days.
See: cell raising colony

### cellulose $C_6H_{10}O_5$

A carbohydrate and the commonest plant material being the main constituent of the cell walls that form around green plant cells. Indigestible to humans and insoluble in water. Used aa a food by ants and other creatures whose gut benefits from the inclusion of a micro fauna capable of yielding enzymes that cope with cellulose. Unlike the chitin of bees it does not contain nitrogen. In the pure form (cotton wool for instance) it will give a blue colour with iodine if first treated with strong sulphuric acid.

### Cellusolve

The trade name for a solvent used in microscopy and enabling material to be set into a permanent slide with less work than heretofor.

### cell walls

The walls of the honeycomb form hexagons so that they meet at angles closely approximating to 60 degrees. At the comb faces they are thickened so that mutual warmth is shared by brood which is, in most cases, surrounded by brood of a similar age in six adjacent cells, and joins the three rhombs forming the cell base at constant angles and through which the mutual warmth from three cells on the other side may be felt. In the brood nest a layer of larval skin sandwiches faeces against the cell base as each new bee is born. This darkens and strengthens the comb but can be the seat of infection for the nosema pathogen. New comb cell wall is <.1016mm/0.004 for drone comb it is <.1524mm/0.006.

### Celsius Swedish astronomer early 1700's

Whose name is used for the temperature scale based on the freezing and boiling points of pure water as 0° and 100° respectively. The equivalent 32° and 212° figures of the Fahrenheit scale are now on the way to becoming extinct though likely to linger on like skeps and memories of ginger beer in the minds of us older folk.
To convert from F to C multiply by 5/9 after taking away 32°.
See: Kelvin

### centimetre cm

A measure of length equal to 1/100 of a metre 39.37", 1 cm = 0.394", 1"= 2.54cm.
See: inch, fractions

**Central Association of Beekeepers**
Specialist member association of BBKA.

**central nerve trunk**
Twin cords run from the brain through the body connecting all ganglia. These are also known as twin nerve trunks or longitudinal commissures. There are two ganglia in the thorax and five in the abdomen, each innervate specific regions, the posterior ganglion being a compound that innervates all the segments posterior to it. As in other insects muscle co-ordination is not dependent on the brain and each ganglion mass appears to have independent control. over the portion that it innervates.
See: central nervous system, nerve trunk

**central nervous system CNS**
The honeybee embryo has a ganglion in each segment. These are joined together and to the brain by the long commissures (longitudinal) which form the ventral nerve trunk. Some of the ganglia fuse together forming large masses of nervous tissue in the suboesophageal and second ganglion of the adult. Each ganglion sends nerves to the adjacent parts of the body and the terminal ganglion is firmly connected to the sting and continues to function until it has been torn out. Brain and ganglia are composed of masses of nerve cells and fibres.
See: connectives, nervous system, ventral nerve cord

**centrifugal extractor**
Hrushka first invented a machine that would spin honey out of uncapped comb yet leave the comb in a re-usable condition. Modern machines may be hand-operated or power-driven and consist of a centre spindle to which cages are attached so that they may hold honeycomb and rotate within a cylindrical container. The spindle is usually set vertically though models with horizontal spindles are made. Many refinements including comb reversing, non-rusting fittings and free-running races are now used. See: tangential

**centrifugal force**
The force exerted outwards on anything that is constrained to rotate about a point or axis. The weight of a substance is increased as the speed of rotation is stepped-up and care should always be taken when spinning honey out of new combs. The wiring or reinforcement of wax foundation is done so that combs will not fly to pieces at speeds sufficient to expel normal honeys, though this does not apply to very dense or thixotrophic honeys. Centrifuges have many applications from salvaging honey from cappings, spin-drying clothes and separating pollen from honey solutions.

**cephalic Gk. the head**
In the region of the head.

**cephenes**
According to one of the pseudo-Aristotle writers sealed drone larvae, as opposed to sealed worker larvae (which became nymphs) or young drones. Butler (Feminine Monarchy) writes 'when fledged they not only serve for generation but help, by reason of their great heat, to hatch brood'. In O.E 'cephen seed' implied 'infertile egg'.
See: figulae, nymph

**ceramic**
Made from clay or similar materials: brick, pottery – widely used as a hive material before the advent of movable frame hives in the Middle East
See: hive-ceramic, hive-clay

**cerana *Apis cerana indica***
The Eastern honeybee. Generally smaller than *A.mellifera* though a group in Kashmir is reported to have worker comb as large as five and a quarter cells per linear inch.

**cerate**
Cotton says: "Melt an ounce of wax and heat an ounce of sweet oil. Do not boil but pour them together at the same temperature stirring steadily until the fluid

is in a buttery state. Stir until set to avoid separation". This is excellent for healing sores or blisters. Vernon reports that it was used in the 12th-14th centuries as an ointment base incorporating various herbs and medicaments. Lip salves still contain this base. It should have a firmer consistency than a typical ointment.

**cereal honey**

Honey-dew honey produced on winter wheat by the Bird-Cherry Aphid *Rhopalosiphum padi* also grain aphid *Sitobion avenae* on wheat and barley. It producs a pale straw-coloured honeydew. Beo Cooper, BBJ 4362 Vol Cix Feb. 1976.

**cerithus**

Alternative name for sandarche (resin) used by Pliny referring to pollen.

**ceromancy**

A word coined from the Latin word for 'wax' and 'romance'. A 'medium' tells a person's fortune by allowing them to pour liquid wax into water and then interpreting the patterns so formed.

**ceromel**

One part of wax mixed with four parts of honey, popular in the tropics for the treatment of ulcers as it does not become rancid. 'Honey & Your Health' BECK & SMEDLEY.

**ceroplastics**

The art of wax modelling. Wax soluble dyes and substances that will make beeswax easier to mould, work or release – enable a wide range of unbelievably realistic models to be made. Flowers, plants, human figures and effigies only give a brief indication of the possibilities.
See: candles, releasing agent, wax modelling, lost wax process

**Certan**

A trade name for pesticide sold for controlling wax moth larvae in stored combs or in active colonies. It is used as a water-dispensable liquid concentrate that eliminates fumigant handling and storage. 120ml plastic bottles. This bacterial insecticide is a *Bacillus thuringiensis* formation said to be safer than currently used chemical fumigants and providing a longer and more continuous protection of the stored comb against *Galleria mellonella* Greater wax moth. The bacteria in the formulation are dormant and present as spores that germinate in the wax moth's gut and destroys its digestive system. Naturally only the combs treated are protected and the moth can run riot in any adjacent combs that are untreated. Gives 12 months protection and is not toxic to bees or humans nor will it taint honey and there is no need to air combs before use.

**cervix pl. cervices**

The neck or narrow mouth of an organ. In honeybees this is the part of a drone's genitalia which follows the bulb of the endophallus in the form of a contorted tube adorned with a double fringed lobe and an arrangement of plates on both dorsal and ventral surfaces. At the moment of eversion the duct and bulb of the endophallus passes down through the cervix.
See: endophallus, bulb of

**cessation of flow**

The sudden ending of a honey flow. It may be momentary or permanent and can be caused by (a) a drop in temperature or (b) too much or too little moisture. or (c) diminished light. Bees may become 'bad tempered' particularly because more will remain in the hive causing possible congestion.

**Ceutorhynchus quadridens**

Stem weevil, oil-seed rape pest and *C.assimitis* cabbage seedpod weevil. The threat or presence of these may bring about a 'spraying episode' and it is incumbent on farmers to inform beekeepers close-by if possible.

**chains of bees**

The claws of a bee are very efficient 'hooks' and they use these to clasp one another. In this way they can hang in vertical chains, head uppermost, and so strong is their grip that a considerable weight is supported by each bee. However either a swinging of the chain or a sudden jolt can dislodge them without apparent harm.
See: festoons, comb–building, catenary hive

**chairman chair person**
The presiding officer of a meeting, committee or board. When equally divided propositions are put forward he has the casting vote and his prerogative entitles him to control the meeting by choosing speakers and making introductory and closing comments. He should be fully supported by the Secretary who prepares the 'Agenda' for meetings and deals with the minutes, correspondence etc.

*Chalicodoma muraria*
Mason bees or wall bees, as implied nest in walls. *C.pluto* is said to be the world's largest bee – 45mm/13/4" long, and rediscovered in Indonesia after having been considered extinct. Nests are built in tree trunks or termites nests and the large nests contain several queens each of which raise a few dozen bees at a time. They are not honeybees.

**chalk**
A soft, white or greyish type of limestone; the calcareous remains of minute marine organisms. 90-99% calcium carbonate CaCO3; in 'natural' waters it is kept in solution by atmospheric carbon dioxide in the form of calcium bicarbonate Ca(HCO) the loss of CO2 by heating precipitates chalk as fur in kettles and pipes where hardness is of a temporary kind. Chalk raises the pH value of soils thus causing them to be alkaline. Used for making quicklime, cement and fertilizer. Plants are sensitive to its presence some like heather hating it while clover does well on chalky soils.
See: chalk and oolite, limestone

**chalk and oolite**
Soils from cretaceous rocks laid under the influence of the sea, rich in phosphates, easily eroded, friable soil, natural drainage. Phosphate fertilisers are only available to plants when the soil is alkaline but the mineral phosphate latent is only available as plant food under acid soil conditions as when sheep have left their droppings. Nectar, attractive to honeybees, is yielded copiously in warm conditions when sheep cause a continual release of mineral phosphates to plants. A short flow calls for prolific bees.

**chalk brood *Ascosphaera apis* formerly *Pericystis apis***
A disease caused by this fungal parasite. A white mass of mycelia develops at the expense of a 3-4 day old larva drying to chalk-like remains – a sliced 'mummy' usually shows a dot where the digestive tract was (whereas similar dried pollen is layered). It is an opportunist parasite that does not spread easily as two strains of mycelium are required to produce spores though they can survive in honey for two years. 'Bailey' tells us that it rarely kills a colony and only individual ones when they are subject to stress such as 'chilling' or protein deficiency. It reaches a peak in May/June but visible signs represent only a part of the trouble as bees easily clean it out. Destruction of infected comb and 'requeening' are the usual treatments and Prof. L. Heath says that comb without bees may be cleared of spores by the use of acetic acid. The spore, germinating in the gut of a bee, cannot invade the tissues of its host unless the temperature falls slightly and briefly below 35°C. Pollen 'patties' using various fungistats have been effectively used.

**challenge**
Although bees are not normally aggressive they will attack in defence of their lives or the well-being of the colony. Examples of occasions when bees do 'challenge' are the interrogation of would-be invaders and the

queen's act of 'piping' to challenge rivals. See: interrogation, piping

**chalones**
These are the opposite to hormones and will depress or counteract the effect of a hormone. Hence 'chalonic' – depressor. See: hormone

**chamber**
Normally a room or apartment but we speak of the 'brood chamber' which means that part of the hive which the queen is allowed access to and uses for the deposition of eggs and therefore for brood rearing. Brood box is an alternative description, the word 'super' being reserved for boxes on the hive that are not intended for use in brood rearing. See: brood chamber, super

***Chamomile nobilis* Anthemis spp. Compositae**
Common chamomile not much worked by bees, possibly due to its strongly scented foliage and flowers. Has medicinal uses. Often occupies waste ground that could as easily support better honey plants.

**champagne-cork organ pit-peg**
Pit hairs and pit pegs are tactile organs buried in pits with their tips below the narrow mouth of the pit. They may also have the function of smell and taste. The champagne-cork organ is a name given to pit pegs with a similar shape to the former. See: pit-peg

**Chantry queen cage Thomas Chantry**
It uses two candy filled tunnels, one obstructed by a piece of queen excluder. Chantry devised the principle of self-release cages.

**Chapman's honey plant *Echinops sphaerocephalus***
A globe thistle growing 2-2.5m/6-8ft tall. Honeybees have been reported as slightly stupefied when gathering pollen from it in the USA. See: globe thistle

**characteristics**
Typical, distinctive qualities, marks or traits. In the realm of honeybee selection and breeding, characteristics are of the utmost importance and whether applied to the colony as a whole or to one or other of the castes, they are the means of assessing their suitability for their chosen purpose. The transmission of characteristics from parents to offspring is linked with the genetic material on the chromosomes and while there are recessive and dominant possibilities so too can there be environmental and mutagenic developments. Some examples include: disease resistance, flight/brood temperatures, longevity, overwintering economy, swarminess, temper and use of propolis.
See: excitedness, qualities of a good queen

**charlock *Sinasis arvensis***
This wild, blazing yellow flower that once, along with blood-red poppies, adorned every cornfield, has been labelled a troublesome persistent weed and suffered a corresponding fate. Yet it is still found by roadsides and sneaks its way into many an august Agricultural Establishment. It is an annual and flowers throughout the summer being known too as wild mustard, karlik, garlock and corn mustard. Its pale yellow honey like that of Brassica cousins granulates smoothly and rapidly to a mild, white set with a faint hint of mustard flavour when fresh. Pollen is described as pale greenish, light yellow and yellow, with 25-32mu grains.

**chaste honey**
Honey in by-gone days that was not quite so pure as 'virgin honey' but which was not sophisticated by any addition. From 'Hogs at the Honeypot' (Vernon) See: virgin honey

**cheating**
Bees will sometimes rob a plant of is nectar without entering the flower and pollinating

it. This is described under 'stealing nectar'. Usually insects with strong mandibles puncture the corolla near its base and from the outside. This is very likely to occur if the insect knows nectar is present but cannot reach it. However honeybee mandibles are not particularly strong for cutting plant material but they are willing to use holes already made by insects like the bumble bee. Honeybees are sometimes guilty but usually follow others.
See: stealing nectar

**checklist**

A list of actions or items required to perform a given function . These are very useful things to have, say when collecting a swarm, visiting the out-apiary etc. A hopeful alternative of trying to carry everything that could possibly be needed is not really suitable. For a swarm one might list protective clothing, smoker, fuel, means of ignition a container, ventilating cloth, string, saw, secateurs, water. The absence of keys, matches, a veil or hivetool has brought many an intended journey to an unsuccessful conclusion.

**chelate Chelicerae**

The claw-like or pincer-like parts (as of the specially adapted mouthparts of Varroa).

***Chelifer cancroides pseudoscorpions***

A small creature related to spiders and scorpions which acts as a scavenger and is said to be widespread in Britain. Their foreclaws are well developed for clinging to larger insects and a swarm of *Apis florea*, smaller honeybee, hived in Sri Lanka left quite a lot of these, where they had been shaken, after the bees had run into a box. Herod Hempsall suggested that they eat eggs and might have advantages if they we to eat Acarine mites. R.H.Brown suggests possible use to deter Varroa. See: phoresia, inquiline

**chemicals**

Substance produced by or involved in chemical processes. Drugs, sterilants, pesticides and wood preservatives are obvious examples of points at which chemistry and beekeeping overlap and the study of the nature of honey, wax and products derived therefrom also take us into the realm of chemical substances. Knowing that the student will frequently need to be able to clearly define some of these an attempt has been made to list the names, formulae and characteristics these wherever appropriate.
See: chemistry

**chemistry**

The science of elements, their laws of combination into compounds and the behaviour of both under various conditions. A knowledge of chemicals is very important to the serious minded beekeeper and Organic Chemistry covers the complex interplay of substance in living things, while Inorganic Chemistry deals with the non-living substances, many of which taint and pollute the world we live in along with organic substances.

**chemoreceptors**

Sense organs which differentiate their chemical composition. Some receptors are capable of registering a single molecule. (The drone's antennae relative to the queen's 'trail')

**chemotaxis**

is an organism's reaction to chemical stimuli.

**cherry *Prunus avium* wild cherry Rosaceae**

A fruit tree and its fleshy globular drupes. There are several wild varieties and many 'cultivars'. Most require cross-pollination to set fruit. As large numbers of hives are moved into orchards a forecast date for flowering is given and it ranges through April to early May. Large well-established trees may need up to seven stocks per hectare, the early flowering (weather wise) often restricting a colony's foraging area. The honey is dark with a strong flavour and pollen colour ranges according to type and observer from pale green to medium brown. See: gean

**chestnut Sapindaceae**

The Horse chestnut *Aesculus hippocastanum*, can yield a good supply of nectar, though cool, blustery conditions frequently limit its usefulness. Its pollen is a conspicuous brick-red. Judging by the number of 'conkers' they produce, pollination is not likely to cause problems. There is also a pink flowering water chestnut, *Eleocharis dulcis*. The sweet or Spanish chestnut, *Castanea sativa* blooms late June/July and has an unattractive smell though in France its honey seems well-sought after. It is a useful nectar plant and in Kent where it is grown in woods for use as fencing stakes it does produce good crops. See: horse, sweet chestnut

**chiasma pl. chiasmata**

1. Bundles of criss-crossing nerve fibres as those linking the honeybee's compound eyes to the brain.
2. Visible exchange of homologous segments between two out of four chromatids during meiosis resulting in the transfer of different genes to the offspring than the parent.

**chicken**

Domestic fowl. Often kept near beehives. Bees have been known to sting the birds to death (especially stinging the red fleshy comb on its head) On the other hand hives have been kept with little apparent harm to bees or fowls, inside chicken runs. Chicken wire at first presents a hindrance to the passage of bees but they become accustomed to going through it and it can be used to keep away toads, green woodpeckers and other predators. The feathers when burnt will produce a poisonous stupefying smoke, rendering the bees temporarily insensible, but are not as large as goose feathers once widely used for brushing bees from combs.

**chicory wild *Cichorium intybus***

A hardy perennial with dandelion-like leaves and whose spikes of attractive blue flowers can be seen on dry roadsides and wet, waste, gravelly places wherever the soil is limey The flowers usually grow in pairs and offer bees nectar and pollen from June onwards. The stems are rather hairy, tough and grooved. This is a useful plant because the roots when dried, roasted and ground, form the chicory that is sometimes added to coffee and the leaves when blanched make a salad. The honey is yellow with a greenish tinge and has a flavour mildly reminiscent of fresh chicory.

**chickweed *Stellaria media* Caryophyllaceae**

Bees will visit this small flower which can constitute competition in orchards in the spring and explains why good husbandry calls for a flowerless orchard sward while the fruit crops are in bloom.

**children's bee books**

These are too numerous to mention but for 8-12 year olds there is a fascinating plethora.

**chill coma**

The temperature of a honeybee must not fall below a threshold which varies slightly with different races and possibly with conditions of humidity. If below 10C/50F bees lose the power of flight whereas all power of motion is lost below 4.4C/45F. At this point they enter a 'chill coma' and recovery depends on the time spent and the depth of chill. Bees that dive into the snow mistaking the brightness for the sky can be recovered if picked up and warmed immediately.
See: comatose

**chilled brood**

This rarely occurs in nature though it is possible when a late cold snap catches a colony that has a prolific queen with too large a brood nest in the spring. It is more likely to be caused by the beekeeper himself who either reduces the number of bees relative to the brood area, or exposes the brood to temperatures below those at which it should be incubated. Beginners finding brood dead from chilling may confuse it with disease. Usually however brood at all stages will have suffered in a given area – frequently at the periphery of the brood nest. Once dead it may darken in colour but the bees should be able to remove it without harm to themselves.

**chimneying chimney-shaped broodnest**

In a natural nest of bees where ample space provides, they tend to arrange their brood nest in a vertical ovoid. Attempts to allow for this have resulted in a whole variety of frame sizes and this is not surprising because country, climate and prolificacy of the queens will all have their effect. It probably occurs quite often by chance rather than design. Consider how unsuitable the normal 'skep' or horizontal 'jar-type' hives are. Economy in the effort that the queen expends in laying, and then the workers in feeding and incubating the brood it is obvious that because warm air rises that the bees starting from the warmest part will tend to build upwards. Long- idea hives would, on the face of it, be suitable in the tropics where the need might be to disperse heat. In colder climates the vertical stacking of boxes helps bees to utilise the warmth they have. Queen excluders and gaps (bee-space) between boxes must upset the bees natural tendency to some extent. It is interesting though to find that a colony on deep Dadant frames will often build upwards with brood on just two central frames before they spread the brood outwards. Nests have been found in church steeples and, after surviving for ages, will often have enormously long vertical combs. The beekeeper today has to consider his 'back' and on many bee farms, operators have given way to using shallower frames, though with modern pallets and lifting gear much of the 'back problem' aspect has been overcome.

**China**

In ancient China, home of the sugar cane, honey was regarded as a medicine or dietary speciality. (One might say that this is how it has become in England!). Bees were kept in wooden tubs and bamboo cages. The small Chinese bee *Apis cerana* is widely kept in small hives. European bees, mainly Italian, are kept along the coast and lowland areas of N.E. and Central China. Nectar bearing plants are always blooming somewhere and migratory beekeeping is practised. The 'bare-foot doctors' make great use of hive products ('ginseng with royal jelly' is described as a double treasure). In 1977 honey exports were 15,000 tonnes. A million colonies were kept in movable frame hives and in 1980 production rose to 105,000 tonnes. Today...

**Chinese bee tree *Evodia daniellii* Rutaceae**

Referred to in 1985 Beekeepers Annual.

**Chinese slipper**

This quaint expression has been used to describe the appearance of the dried scaly remains of a larva that has died from the Sac Brood virus. The paired spiracles standing out on the small mummified remains like lace-holes. Calceolate – slipper shaped. See: mummy

**Chinese tallow tree popcorn tree *Sapium sebiferum***

It grows to 10.5m/35ft like a maple, can withstand drought and hurricanes better than oaks. Grows well in the Southern States USA; tolerant to cold down to -8C/18F. It flowers for six weeks in late spring and the bees produce a nice mild amber honey from the nectar.

***Chionodoxa luciliae***

It has blue or white flowers Mar/Apr is pretty, hardy and spreads easily providing both nectar and pollen for the bees.
See: Glory of the snow

**chiropterophilous**

Pollinated by bats. The flowers of some plants only open by night and therefore bees do not have access to them and the work of pollination is carried out by bats, moths and other nocturnal creatures.

**chitin pronounced 'Kytin'**

A colourless nitrogenous polysaccharide that forms much of the horny exo-skeleton of many arthropods including bees. It is secreted by the epidermal cells and is acetyl-glycosamine $C_{32}H_{54}O_{21}N_4$ being insoluble in water, dilute acids and alcohol. When parts of the bee are to be cleaned and clarified for mounting on microscope slides chitinous material with its associated tissue may be boiled in caustic potash and then, if necessary, bleached with sodium hypochlorite. Chitin is the only important example of the use of a carbohydrate as structural material in animals.
See: cellulose

**chitinase**

An enzyme that dissolves chitin. It is present for example in the digestive juices of a snails and therefore presumably in that of birds and other creatures able to eat and digest bees.
See: chitin

**chives *Allium schoenoprasum* onion family**

A hardy perennial yielding pollen and nectar. Height 9" and has mauve flowers in Mar-May. Where grown for seed an amber coloured honey has been harvested and the slightly onion flavour like many other 'unwanted' floral taints, disappears with keeping.

**chlorides**

Salts of hydrochloric acid. Common salt for example is Sodium chloride (NaCl). Unlike free chlorine which is a dangerous gas, chlorides themselves are not generally poisonous. They are incorporated in many bleaching compounds and form toxic insecticides such as chlorinated hydrocarbons.
See: bleaching, chlorinated hydrocarbons, salt

**chlorinated hydrocarbons**

Less accurately known as organochlorines, they are toxic, slow the rate of photosynthesis and interfere with the transmission of nerve impulses in insects, possibly upsetting Na or K ion balance. Now regarded as ecologically undesirable because of their great persistence. Biological availability and fat-solubility cause them to accumulate in the eco-system and to be concentrated in the body fat of animals and birds towards the end of the food chain. Also tend to stick to dust. *(through using them I lost my Airline Transport Pilot's Licence).* Chlordane, heptachlor, aldrin, dieldrin, toxaphene, endrin, benzene…

**chloroform $CHC_{13}$**

A toxic, organic, colourless liquid which readily vaporises. Used as an anaesthetic for humans. In excess it causes death and is used to kill insects. It can be used as a solvent for some organic substances including beeswax and propolis though less toxic substances are safer. Similar in action to carbon tetrachloride.

**chlorophyll**

Green and yellow pigments found in plants (and in honey). It occurs as discreet corpuscles within the plant cells and is vital in the process of photosynthesis as it captures light-energy. There are several closely related forms that are sensitive to different colours of light.
See: photosynthesis

**choice of**

Area, bee, colour for hive, district, hive, type of queen, race, site. These are matters over which a beekeeper has control and each will be mentioned under appropriate headings.

**cordotonal organ**

Rod or bristle-like pegs capable of detecting mechanical and sound vibration in various parts of the body of insects.

**chorion**

The soft shell covering the honeybee's egg. It is secreted by the walls of the follicles which are found in the tubules at the upper end of the queen's ovaries. The 'micropyle' is a minute opening through the chorion at the anterior end of the egg. It has an interestingly reticulated surface.

**Christmas rose *Helleborus niger* Rananculaceae**

There are wild and cultivated varieties of this perennial red, green and white Christmas flower yielding nectar and pollen when weather permits. Their vase-shaped nectaries remain attractive right on into the spring.

**Christmas tree**

Whereas in Britain this is usually a fir tree (evergreen) and hung with Christmas decorations, in Oceania where Christmas day is in mid-summer, great flowering masses adorn trees that are assiduously foraged by honeybees. In New Zealand Pohutakawa or *Metrosideros tomentosa* bears the most striking, brilliant red flowers akin to 'rata'. In coastal areas huge 'takes' of fine-grained honey are annually secured. The Christmas tree of Western Australia *Nuytsia floribunda* has cadmium orange flowers and dots the landscape of the sandy coastal plains around Christmastide.

**chromatid**

Prior to cell division, chromosomes split into longitudinally similar strands called 'chromatids'

**chromatography**

An analytical technique for separating and identifying mixtures of substances. The mixture is allowed to pass along a medium in a tube which retards each component to a different extent thus enabling it to be identified. New methods have been developed and there are many applications to beekeeping. The bio-assay of bees to determine whether they have absorbed insecticides and the identification of the constituents of beeswax, venom, pheromones, honey etc. Techniques include: gas, paper, TLC (thin layer) and high performance liquid (high pressure) chromatography.
See: analysis, assay, GLC

**chromosomal mutation**

Also 'chromosomal aberration' resulting in slightly different offspring, known as 'sports'
See: mutation, sports

**chromosome**

The microscopic threads of protein found in the nuclei of plant and animal cells, which bear the genes. They are seen more clearly at the time of cell division. The paired chromosome number is constant for cells of a given species, with the notable exception of honeybees wherein females are diploid with 32 pairs while male drones are haploid and have half that number. The genetic message consists of lengths of DNA.

**chrysalis Lepidoptera**

Gold or golden thing. The torpid, encased pupal stage of many insects. As they are born in solitary conditions, the chrysalis is a separate and complete outer case protecting the creature inside which is about to emerge as an imago. In social insects and the honeybee in particular, this pupal stage is spent in a capped cell, somewhat isolated in the case of a queen cell but in the case of drones and workers each pupa is usually surrounded by others. Although the honeybee has no chrysalis as such, it still has a cocoon and this and the last larval moult remains in the cell, accounting, together with its faeces, for the gradual darkening of brood comb.

**chrysanthemum spp. Compositae**

Single varieties yield pollen into late autumn. *C.leucanthemum* the 'oxeye' daisy is a common and conspicuous decorator of many a wilderness. Some cultivated, usually later flowering varieties, have a strong not-altogether pleasant aroma throughout the plant. These seem unattractive to bees.

**chrysin**

A plant pigment – flavonoid, found in propolis and also one of the principal colouring agents of beeswax.

**chunk honey**

The EEC Honey Regulations define it as 'honey which contains at least one piece of comb honey'. Pieces of honey comb arranged in a container. A popular way of selling in the States (USA). To obtain a good-looking product use a fully sealed comb, cut into neat shapes and surrounded by clear, light honey in a transparent container. The clear honey should be of a type or in such a condition as to retard granulation.

**church candles**

Although tallow was discovered and used for candles long, long ago, the Roman Catholic Church always used pure beeswax, its aroma when burned and its pure, steady light making it singularly appropriate. However, presumably due to expense and availability the adulteration of beeswax at first with small and then larger quantities of refined paraffin wax was permitted by successive papal encyclicals.
See: beeswax candle, candle, cerge, cierge

**chyle stomach**

The honeybee's ventriculus. (Often pulls out attached to the sting) Chyle is lymph containing globules of organised fat.

**chime**

Products of the stomach – the partly digested food. Chymification is the process of converting food into chime.

**cibarium**

The pre-oral cavity that is continuous with the bee's pharynx and is the dilated forward end of the oesophagus. Musculations enable it to be used as a pump so that liquids can be drawn up the food canal of the proboscis and pushed along the oesophagus. Its floor is the hardened hypopharyngeal plate which spreads out into a membranous sheet that connects the proboscal parts together. Dade refers to this as forming a continuous flexible, hammock-like bag.

**cierge**

A beeswax taper used in church. Cierger: to wax cloth to stiffen it (and water-proof). A tallow candle being la chandelle. Candelabrum – 'candle tree' ?

**circadian rhythm**

Biological rhythm with a period length of about one day.
See: diurnal

**circulation**

1. The movement of blood or sap. Frequently linked with recirculation when the same material might circulate many times within an organism.
2. The food or pheromones within a honeybee colony are distributed rather than circulated
3. We also refer to the 'circulation' of newsletters. journals etc. meaning the number and distribution of these.

See: circulatory system, food-sharing, queen substance

**circulatory system**

The arrangement of organs within the bee that control the movement of the blood. They include the heart and aorta, dorsal and ventral diaphragms and certain vesicles. It is an open system apart from the heart which pumps blood to the brain. Abdominal pulsations which aerate the bee also move the blood.

**circum-oesophageal connectives**
The elements connecting the brain's two ganglia – the supra and sub-oesophagal ganglia.

**citral $C_9H_{15}CHO$**
A liquid aldehyde with a lemon-like smell used in perfumery.
See: Nasonov

**citric acid $C_6H_8O_7$**
An acid found in citrus fruits e.g. lemons, useful in mead making. Also found as a trace in honey. It also forms an essential part of the cellular production of energy in the citric acid cycle.

**Citrus fruits**
Lemon, grape fruit, sweet lime, orange etc. Honey from citrus plants can be low in diastase.

**clarification**
Clove oil is used for clearing when making microscope slides. The clarification of honey is by separation and filtration. Mead or vinegar can be filtered to clarify and here are preparations to help clarification such as Isinglass.
See: clearing

**clarified honey**
Honey that has been carefully heated then filtered to remove all wax and other particles. The absolute clarification necessitates the use of the finest of filters and this normally calls for heating to lower the 'viscosity'. This, and the removal of any particles of pollen and certain natural ingredients results in that bright shining sample which so pleases honey judges, looks good on the shelf and satisfies international regulations.

**Clark Mrs. R. E. NDB**
Former President BBKA and CBL Surrey. One of the most able CBL's and most influential member of the Twickenham & Thames Valley BKA. Her inspiration and educational perspicacity inspired Major Dade to produce his most advanced 'The Anatomy and Dissection of the Honeybee'. At one time there were more holders of the NDB in that Association than any other in the country.

**Clarkia spp. e.g. *pulchela, elegans* Onagraceae**
A hardy annual growing to 60cm/2ft and flowering right through the summer. Single varieties especially interest bees yielding purple to mauve pollen.

**clarre**
A claret sweetened with mead - it was referred to by Chaucer.

**claspers**
These have been reduced to vestigial and seemingly unimportant parts of the bee though the claspers of the drone play a part in the copulation and the sensitive sting palps of workers enable them to decided from the texture of a potential victim's skin whether or not the sting is likely to be effective, ignoring tough surfaces but plunging unremittingly into softer ones.

**class**
1. A group say, of beekeeping students assembled for instruction.
2. A class in a 'honey show' would comprise exhibits that conformed to the same requirements as for example: 'light honey' two similar jars for each entry in the class.
3. One of the major divisions in the classification of living organisms including 'orders', while groups of classes form a 'phylum,'

**classification**
The hierarchical arrangement of similar groups of plants or animals. Each group shares certain characteristics with its other members. Classification starts with Kingdoms (plant or animal) and these are then split into 'phyla' then classes, orders,

families, genera and species. It should be understood that this is not a static field but that changes occur as arguments are settled and further knowledge is acquired.
See: botanical name, honeybee classification, Linnaeus nomenclature, systemics, taxonomy.

**classification of the honeybee**

Kingdom – Animalia, Phylum – Arhropoda, Class - Insecta Order – Hymenoptera (there are many other 'orders') Sub-order – Symphyta Apocrita, SuperFamily – Apoidea, Family – Apidae, Genus – Apis
See: honeybee classification, races

**claustral hive**

Invented by Abbe Gouttefangeas. It had a claustral porch (ante-chamber) and ventilating chimneys.

**clavate hairs**

Hairs thickened at one end or club-shaped as on the 'bouton'.

**claw Anguiculi or ungues**

The bee has two curved claws which are notched and pointed, on each foot. They enable it to get a grip on certain types of surface and to cling to one another's legs and bodies (as within a swarm). The size of its claws are also well adapted to the thickness of cell rims and the ladder-like structure of the honeycomb and to hanging in festoons, chains or clusters. They are essential for many kinds of activities including both fanning and stinging.

**clay**

A type of soil composed of extremely fine particles. There are many types and they are retentive of moisture expanding and plastic when wet but shrinking to become hard and cracked when dry. They lack humus. 'Boulder clay' a glacial deposit known as 'Till' is often very calcareous as is Gault a somewhat stiffer soil. There is also London clay, Oxford clay and Weald clay below the Lower Greensand. composed of extremely fine particles. There are many types and they are retentive of moisture expanding and plastic when wet but shrinking to become hard and cracked when dry. They lack humus. 'Boulder clay' a glacial deposit known as 'Till' is often very calcareous as is Gault a somewhat stiffer soil. There is also London clay, Oxford clay and Weald clay below the Lower Greensand.
See: soils

**clean floorboard**

The scraping and blow-lamping of floorboards should be done quite early in the spring (even if it means replacing the mouse guard) and whenever accumulated debris or other reason necessitates this action. As with most manoeuvres one should record this in their notes to show that this essential aspect of hygiene has been attended to. It is useful with standardised equipment to have a spare floor so that one-by-one the hives can be set onto clean boards and then each dirty one can be suitably dealt with.

**cleaning**

1. Wet supers. Ideally supers should be 'licked' clean by returning them, over an 'open' clearer board, to the colony that they came from. After a couple of days the clearer can be set to 'operate' and the dry supers can be taken and stored in a cool place. They are vulnerable to mice and wax moth. If the honey-wet combs had come from healthy colonies, they may be put back onto any strong colony (weak ones would be robbed) to be cleaned up. On weaker colonies reduce entrance size to help avoid robbing. They should be put back at dusk and precautions taken to prevent robbing such as reduced entrance size. Combs may be stored 'wet' but this method may lead to 'robbing' 'crystallisation' and 'fermentation' of the film of honey.
2. Cleaning faeces from clothing calls for the early application of a wet cloth.
3. Propolis and wax dissolves or breaks up in the presence of alkalis, these in turn however are bad for the skin. Cold propolis is brittle and can be scraped or

cracked off equipment. Both are soluble in CTC.

4. Wax – metal or plastic surfaces can be made hot, using 'almost boiling' water, and then quickly wiping over with absorptive newspaper or cloth. CTC (poisonous fumes) can be used away from the bees otherwise scraping with a suitable tool can tidy up hives and frames though beekeepers tend to overdo this as the bees idea of cleanliness is better than ours and 'exclude' can cause confusion.

See: deadman floorboard, sterilisation.

**cleansing cream**

1oz beeswax, 1oz white Vaseline, 1/2fl.oz almond oil, 1 fl.oz orange flower water, 1/4g borax, 4 1/2 fl.oz light mineral oil (or almond oil).
See: cold cream for details as to method

**cleansing dance**

Rapid stamping of the legs and a rhythmic side-to-side swinging of the body together with a raising and lowering of the body as it tries to clean around the bases of the wings using the middle pair of legs. Also called a 'shaking' dance and it can be observed at any time of the year. As soon as another bee touches its thorax it spreads its wings out on one side and curves its body to make it easier for the cleaner who works energetically using its mandibles to snip away around the other bee's body.

**cleansing flight**

Bees normally void their faeces on the wing. When a flight has been taken specifically for this purpose it is called a 'cleansing' flight. This is likely to occur after a period of bad weather when the bee has had to store its faeces in the rectum having been unable to get out of the hive. During the winter months in GB It is often mild enough for large numbers of bees to take such flights. The mood is frequently catching and sometimes when there is snow on the ground it is advisable to shield the light from the entrances (piece of slate perhaps) lest bees should be tempted to their deaths mistakenly diving into the snow which reflects to them like the sky. The front and tops of white hives frequently show evidence of cleansing flights as might nearby washing that has been hung out to dry. Offering water nearby could help but when a bee happily takes up the moisture from wet clothing the intake often prompts an outflow.

**clearer**

A device for getting bees to move out of an area inside the hive. A narrowing tunnel, cone or application of a pair or more of springs that will only allow a bee to pass in one direction As used when clearing bees from a super of honey. The words 'clearer escape' and 'exclude' can cause confusion. Perhaps the word clearer should be used for moving bees from one part of a hive to another and 'escape' for letting them get out.
See: escape, excluder, Porter, super clearer

**clearer board**

A board designed to encourage bees to pass through it from one side to the other by the inclusion of non-return valves (springs tapered to allow bee to pass in one direction only). A board may include one or more such devices and the Porter bee-escape is a well-known example. The positioning of holes and narrowing tunnels are further examples. Repellents too are made use of.
See: bee-escape, Porter, Crowther, Quebec board, tunnels

**clear honey**

Run honey or honey in the liquid form. When first processed by the bees it is like concentrated nectar – a thick, clear liquid in fact. Commercially, pure honey is either clear or set. Some nectars give rise to honey which may even granulate in the comb (brassicas) but honeys where laevulose is dominant (acacia) remain clear for a considerable period. Careful heating can restore 'set' honey to the clear state.

See: clear, creamed, crystallised, chunk' honey

**clearing**

Tissue for microscope slide presentation will often be dehydrated using alcohol, all traces of which must be removed to allow the infiltration of 'Balsam'. Cloudiness in the clearing agent indicates incomplete dehydration. Because of the mutual insolubility between alcohol and the subsequent chemicals used, xylol, clove oil, benzol, or cedar wood oil are more tolerant of water traces and also render the tissue more transparent.
See: clarification, clearing supers

**clearing supers**

Repellents, vibration (drumming), cold, one-way valves, compressed air and shaking bees from the combs are ways of getting bees off their honey comb and out of the supers. To drive back from the out-apiary with the residue of bees leaving the vehicle in ones and twos is rather anti-social, though it has been acceptable when viewed against the alternative of taking bees into the house. The ingenious, efficient and well-tried Porter bee-escape does help to avoid such embarrassments but unlike blowing, shaking or brushing, it requires two visits.
See: bee blower, empty

**cleat**

A wedge or projecting piece attached to another member to fill a gap or to prevent slipping. Not much used to describe parts of British hives. Australians use a floorboard raised on two transverse 'cleats' (wood blocks) under front and rear edges of the board. When 'cleats' – strips of wood – run front and back along the undersides of the floorboard they are useful for gripping the hive when lifters are used. An entrance block sometimes comes in for the name of 'cleat'.
See: riser

**Cleaters**

In 1630 he gave rise to the discovery that the mother of the colony was the queen.

**cleavage**

Repeated divisions of the zygote's cells, following fertilisation and forming a mass known as the 'blastula'. A series of karyokinetic divisions of blastomeres within the egg.

**cleaver**

A piece of hardwood, cut so that it will split bramble briers along their length. Each prepared brier (dried and ridded of outer layer and thorns) would be separated by splitting into three (or four) long strands. These were used as binding thread for skeps.

**Clematis spp. Ranunculaceae**

Old man's beard and Travellers joy are names for wild varieties; it is a hardy perennial that can climb to great heights (10m/30ft or more) and provides both nectar and pollen. Reports from USA tell us that the honey is extra light amber and of good flavour. Masses of small green fragrant flowers from June to August.

**Cleopatra's facial balm**

One part honey, one of milk and one egg white. Beat well and apply to clean face and neck. Leave on for half-an-hour. When it feels dry and brittle, wash it off with lukewarm water. You will feel your face tingle. Cleopatra used this formula on her entire body to keep her skin beautiful and soft. Dianne Meredith, The American Bee Journal. Feb 1979.
See: Face Lotion, Pack

**cleptolecty**

The habit displayed by some honeybees of stealing pollen from the bodies of some solitary bees.

**Clergymen beekeepers**

Many ecclesiastical gentlemen have been drawn into beekeeping. A few are listed here:

Adam Bro, Allen Yate, Brauns, Butler, Carr. Collin, Cotton, Cunning, Della Rocca, Digges, Dugat, Dzierzon, Edwardes, Filleul, Gouttefangeus, Isaac, Langstroth, Purchas, Schirach, Scholtz, Thorley, White, Wood and many others who enlightened us by passing on their knowledge of the bees.
See: bible, religion

***Clethra alnifolia*** Ericaceae
A handsome shrub native to North America, a good bee plant producing fragrant white flowers in August. Light thick honey with a fine flavour.

**clicket**
Thin boards, pierced with holes, that were placed over the entrances in winter. HH.

**climacterial number**
A critical number or period when great change is supposed to occur – CHAMBERS 1906. 9 x 7 beehives in a bee garden BUTLER Gk. klimakter – ladder.

**climate**
The generalised weather conditions over a certain area and period of time. This of course has a marked effect on the type of bee and we talk about acclimatised bees when we mean those that have been native for several generations and have become adapted to the local climate.
See: acclimatise, micro-climate

**clipped queen**
A queen whose wing or wings have been reduced by cutting to hinder her departure with a swarm. There are no hard and fast rules but virgin queens must never be so treated and young queens are rarely clipped. After one season a third of the wing may be cut away with scissors, avoiding the leading edge and if she goes into a third season possibly the other side may be cut, though short symmetrical wings can lead to limited flight.

**clipping**
A snippet or cutting as of a piece of beekeeping news extracted from the bee or national press. The clipping of a queen's wing.
See: clipped queen

**clone**
A group of individuals produced asexually from a single parent. Vegetatively, mitotic or monozygotic twins......

**cloom**
Cow dung tempered with lime, ashes, sand or fine gravel to close up skep wherever it is not tight, specially joining bottom of skep to stool. (Butler, More do you mean more? lower case in Cotton). D. Galton. Clooming – the application of a protective coating to twigs or withies. Also 'cloomed'
See: cowcloome

**close driving**
This operation of transferring bees from one skep to another was done in a similar way to 'open driving' but the upper skep (or box) was shut close down onto the full one. Any advantage gained (prevention of robbing and fewer airborne bees) may have been offset by the uncertainty as to the presence of or the assessment by visual means of, the quality of the queen.
See: driving bees

**close-ended side bars**
For super frames, side bars that are if uniform width throughout their length and which, being wider than the top or bottom bars, abut one to another to produce automatic centre-to-centre comb spacing and make them easy to uncap though they are inclined to stick together when propolised. A frame popularised by R.O.B. Manley

**closing a hive**
At the end of an operation it is important to leave the bees as undisturbed as possible. Some beekeepers mark the various hive

parts so as ensure that they are returned to exactly the same positions that they first occupied. Care should be taken to avoid crushing bees, by gently rotating bodies before lowering them into position. Roof and covers should all be left secure and as bee and weather-proof as possible. The entrance should be no larger than the colony can adequately guard yet take into account the special conditions in unusually hot weather.
See 'moving bees' for details of closing a hive altogether.
Also 'hive closure' for treatment of a hive when insecticide spraying is threatened.

**closing down for winter**
Preparing a stock to face the winter so as to minimise further attention until the late spring.
See: shutting down for winter, wintering

**cloud of bees**
Terms like a cast, swarm, mass or cluster of bees are more likely to be used by the beekeeper. However writers and ordinary folk may well consider that a large number of airborne bees moving through the air might warrant 'cloud' as a way of describing them although the interweaving pattern of such insects resembles an open network rather than a cloud.

**cloudy wing virus**
This complaint was discovered in 1977 and appears to be an infection of the trachaea. This was one of the smallest virus particles identified up to that time (17 nano-metres – 'nm'). Although the particles are physically indistinguishable from those of CBPV they are serologically unrelated i.e. different 'in vivo' from 'in vitro'. It causes inactivity and early death in adult bees BAILEY 1981.

**clove oil**
Used as an agent for 'clearing' in techniques of mounting material for optical microscope work.

**clover *Trifolium repens* – a trefoil**
White clover, was often called 'the honey plant par excellence' but in England its ability to yield prodigious amounts of nectar seem to have come to a halt. It likes alkaline, well-aerated soil. Each flower head contains 50-100 individual florets which are erect but fall when pollinated. Clovers are important for pasturage, the production of hay and for their rejuvenating effects on the soil via root nodules and their associated nitrobacter which brings about the fixation of atmospheric nitrogen. In NZ it has been suggested that for optimum nectar-secretion ground temperatures of 22-24C/70-75F are best, provided trace elements such as zinc, are present. In GB heavy flows of light honey can occur though clover is a fickle plant and conditions of soil and weather must be right. Pollen is pale yellow – Howes says it assumes dull greenish appearance when packed in the corbiculae.

**clover varieties**
White clover, wild white, Dutch and Alsike. Sainfoin has a pink flower, makes yellow wax and honey and makes good hay for horses. There are scarlet, crimson and Italian clovers, hop, zigzag and strawberry clover. Red clover (tetraploid) the second cut has shorter corollae and therefore late flows (August) sometimes occur. *Trifolium stratium* Knotted clover grows wild in the Southwest and has small patterned flowers in ovoid heads and likes dry pastures. *T.medum* zig-zag clover, is a perennial in N.England. *Melilotus alba* Bokhara clover and *M.officinalis* Yellow sweet clover are useful when established. *T. arvense* – small harefoot – is a grassy type found in the S.E. Persian, Bokhara, Honey, Yellow suckling, Calvary and Egyptian. See under various headings and 'manuring'

**cluster**
The assembly and clinging together of a number of honeybees. They have a strong tendency to do this whenever a large enough group of them are separated from their

combs and when a queen is present she has an attractive and stabilising influence. A natural swarm cluster, driven bees or a 'shook' swarm all form what is known as a 'super-organism'.
See: swarm, winter cluster

**cluster-frames**
'No-swarm' cluster frames are made of slatted plastic and provide maximum clustering space in the brood chamber with a view to eliminating swarming.

**clustering pheromone**
A constituent of the special substances released by a queen which stabilises the cluster of bees and has been called the clustering pheromone.
See: 9-O-2 pheromone, queen substance

**clypeus**
This is the chitinous frontal piece of a bee's head and derives its name from the Latin for 'shield'. It covers the area between the mandibles and where the antennae join the head.

**CNS**
Short for 'Central Nervous System' which comprises the stomogastric and peripheral nervous systems.

**coal bearing rocks**
There can be marked variations in different stratas. Soil sours quickly. Rhubarb plantations yield for a year or two and are then finished. In such places the Spring crop is of major importance.

**coast**
Coastal areas suffer from a reduction of bees' foraging area if part of this would normally extend over the sea. Land and sea breezes and exposure to strong winds can be another problem. The effect of salt also govern to some extent the type of forage to be found. Nevertheless circumstances can conspire to give good honey harvests and in particular, late crops
See: sea, seaside, shore

**cobana**
A system of comb honey production invented by Dr Wladyslaw Zbikowsky who died in 1977. Born Pennsylvania 1896, studied in Russia and Poland, began keeping bees in 1953 after retiring from medicine. He produced circular sections that fit into a frame with dimensions that fit into a super. Precision made and easy to fit foundation, time savers in fact.

**Coccoidea**
Scale insects or 'mealy bugs' small plant destroying insects whose eggs and bodies are covered with hard scale. These are spread by wood ants. Their secretions, after sucking sap from the fir trees is collected by the bees and transformed into the much sought after 'forest' honey. A symbiotic system favouring the bee and akin to honey-dew from aphids and the NZ 'beech buggy'.
See: Ultracoelostoma

**coconut *Cocos nucifera***
A tropical palm and its large, hard-shelled nut. It is a melliferous plant producing both nectar and pollen. The honeybee is a useful pollinator though wasp interference can reduce its efficiency.

**cocoon**
The lining which ensuing generations of larvae leave behind in the cell. Two factors contribute to the cocoon. First the larvae spin silken webs that interweave with the domed capping as the nurse bees, having stopped feeding, start to cap over the cells. Queens behave similarly in their purpose-built cells and make the lower tip so tough that something razor-sharp is required to cut it. After biting her way round a virgin will push the circular tip down with her head and it hangs for a while, as on a hinge, before being eaten by the bees. The other aspect of cocoons is a layer of larval faeces deposited between the 6th and 7th moults. These progressively darken the comb which

becomes tougher and cradles marginally smaller bees though long before this takes noticeable effect, the bees are likely to move off or tear down and rebuild the comb. A newly emerged bee shines up the vacated cell ready for re-use. The faeces can be a potential source of pathogens. Old combs yield nice yellow wax leaving what looks like paper-light, black comb similar to that of wasps, though tougher and two-sided. When rendering, 48 hours soaking in water is desirable as the cocoons are air-traps and prevent heat from getting at all the wax.

**co-enzyme**

A substance that accelerates or activates an enzyme. Many vitamins are co-enzymes.

**coffee *Coffee Arabica***

In coffee plantations that were buzzing with bees, 83% more seed was produced than in those where no bees were seen. Dr. Swaminathan, Food & Agricultural Secretary, India.

**cog**

A shallow box used in Scotland for a heather honey super using bars and a fairly narrow depth of about three inches or so. (*Info. uncertain*)

**cohesion**

Being united or naturally connected. The cohesion of a honeybee colony is nothing short of remarkable. The queen, the colony odour, sister-like ability to communicate, ability to identify alien creatures (including other honeybees) are all reasons why they are able to display cohesion which transforms them from a random mass of flying insects into a integrated unit – the honeybee colony.

**cohort**

Any given group such as bees of a certain age or bees performing a given tasks.

**coiled straw**

A rope of straw, formed by a chosen number of straws that could be fed through a 'gauge' and used to form a basket by making the rope into a coil. The 'rope' could be extended by feeding in more straws and the coil was bound together by thin cane (split brier). The diameter of the 'rope' or cross-section of the 'feeder' was its 'girth' and these baskets or skeps were often strong enough to bear the weight of a man
See: cleaver, feeder, skep

**coition**

Sexual copulation or the marrying of drone and nubile queen in flight.
See: copulation

**cold**

Having little or no warmth, being at a lower temperature than normal. Bees are able to withstand ambient air temperatures such as the severely cold winters of Russia and Canada provided they are well insulated within their natural canopy of honey filled wax. A strong, healthy, queenright cluster will maintain sufficient warmth to remain alive as long as it has sufficient food and is sheltered and insulated from extreme cold, wind or wet. Individual bees can become chilled and comatose before they die if there are not bees enough around them. The queen is always cherished to the last.
cold-blooded Poikilothermal Gk. poikilos – 'various', therme – 'heat'
Creatures whose temperature varies with that of their surroundings. Under some circumstances this applies to bees though under flight conditions their metabolism is speeded-up and when in a cluster with other bees they become more like a warm-blooded creature. Wasps and bumbles do not come into this category and consequently hibernate.

**cold cream**

Basically this is achieved by saponifying beeswax (with mild alkali) and mixing with a kindly oil: 1oz of pure white beeswax is shredded and put into one of two jars. Mineral oil is added and this jar will go

into a water bath ready for heating. A second similar jar will receive rainwater or distilled water to which a pinch of borax is added. Both jars are then gently heated in the water bath until the contents of both are completely dissolved. This method ensures that both parts of the mixture are at the same temperature Then taking the jar with wax which cools slowest, pour the contents of the second jar into it and stir quite vigorously to ensure a smooth creamy mix. Containerise while still warm having added perfume if you so wish. Now there are several variations possible but one of the simplest is to use the following: 28g/1oz pure white beeswax (or lightest shade that you can muster), 85ml/3 fl.oz of light medicinal paraffin oil, 55ml/2 fl.oz of distilled (or de-ionised) water and a mere 1 75g/1/16 ounce of powdered borax. Naturally the most important ingredient is some of your very best natural wax but whatever expensive oil or perfume or orange flower water you choose to substitute, you will find that you will always be asked by your friends for more.
See: plastic containers

**cold way**
This refers to the alignment of the frames in a hive so that they run parallel to the sides and all are equally exposed to any draught blowing into the entrance. Square hives allow a choice to be made in this matter while oblong hives do not.
See: warm way

**Coleoptra beetles a sub-class of Pterygota**
Forewings are hardened into 'elytra' or wing-cases, their larvae are variable some having legs and some not. There have been few reports of damage by beetles though in recent times *Aethina tumida* Small Hive Beetle has come into prominence abroad.

**collateral hive**
See: Nutts collateral hive

**Collide to bump into**
Despite the milling masses of bees that intermingle in the air around the hives and in swarms, collisions are unknown. However certain surfaces such as still water, glass and snow can cause bees to bump into them making shallow graves in the snow, making them spin round in circles in water and apart from car windscreens to harmlessly ping off glass. Should their flight be terminated by suddenly meeting your uncovered visage, then a sting is not unlikely.

**Collin Abbe**
A French priest credited with having made the first queen excluder in 1849. He wrote 'The natural history of the bees' and believed that practically no honey is lost when bees build new combs. 1879.

**colloid**
A homogeneous gelatinous substance but the term is used where particles as small as one millionth of an inch are suspended in a gas, liquid or solid. The expression 'colloidal state' is preferred. The word was derived from Greek kolla – 'glue and eidos – 'form'. In honey it tends to be a substance which does not diffuse easily through animal or vegetable membranes (difficult to filter). It may be intermediate between materials in true solution and those in suspension. Found in heather and honeydew honeys. When present it causes cloudiness and such honey may be gluey non-crystalline or may darken when heated. Smoke, mist and foam are other examples of colloids.
See: suspension

**colon**
The second or posterior part of the bee's intestine, lying between the ventriculus and the rectum and through which nutrients pass into the blood stream. See: rectum

**colony *L.colonia* – farm**
A collection of organisms living together as a separate entity and interdependent. The term used to describe one complete

unit of honeybees within their home. Foreigners often show some reluctance to use this word (this was noticed in Russia) translating beekeeping matters into English and Portuguese they have used 'bee swarms' and the Russians 'families' to describe the colony. a colony together with the hive and furniture is known as a 'stock'

**colony behaviour**

The manner in which a colony of honeybees responds to its environment. Immediate short or long-term reactions are all of interest to the beekeeper. We speak of a colony's characteristic behaviour. analysing it into parameters such as reaction to smoke, willingness to store, docility and so on and our control over these matters is as effective as our control over the queen from whom colony characteristics are derived and of course the environment in which we place them.
See: characteristics

**colony density**

According to Dr. Bailey there is an optimum density of hives for a given area. This depends on whether we are talking about 'pollination' or receiving the best honey returns relative to the surrounding forage. It is obvious that by just increasing the number of hives you will not get more honey once the optimum number of colonies is reached. Many apiaries are grossly overloaded with bees and where this is unnecessary one should consider the disease aspects and the unnecessary competition when there are far more bees than foraging conditions allow. In certain eucalyptus forests in Australia and in vast flowering masses of clover or oil-seed rape, it is hard to imagine that too many hives could be put in. At one time it was said that if every hive in the country were put into the Kentish orchards there would still be too few bees. As a rough check one hive per acre seemed reasonable but with valuable crops it is wise to overplay one's hand particularly when weather conditions only allow a small window of opportunity for successful pollination.
See: number of colonies on one site

**colony morale**

This anthropomorphic term implies that a colony may have a high or low morale, for example the zest displayed by a newly-hived swarm, a newly-mated queen or the onset of a honeyflow. Conversely the lack of vigour displayed by a queenless colony or one with disease, being robbed or in an unsuitable environment. Nevertheless some types or races of bee *A.adansonii* for instance, do show exemplary foraging capabilities and then there is colony vigour (expressed in defence) and other factors or characteristics which might well be considered as examples of 'morale'.
See: colony behaviour

**colony odour**

A colony develops a specific and individual uniqueness from its queen and from the forage brought home, the brood and food sharing. Much is put down to pheromones but however you account for it, each colony is a unit that regards all other bees or competitors, or predators as alien. Mutual identification is a necessity but smoke and other chemicals can mask this identity. When 'uniting' or 'requeening' and in the prevention of robbing, this so called 'odour' should always be taken into account.
See: hive odour

**colony rhythm**

The annual cycle of a healthy honeybee colony follows a fairly regular pattern always associated with the climate. However there is a 'balance' achieved according to the queen's ability and the beekeeper often steps in to reorganise the brood situation according to his requirements which might be to work for an early or specific kind of honeyflow. So the rhythm that develops can be learned and understood thus enabling appropriate actions to be taken at the right time. The beekeepers 'rhythm' – maybe: feed, supers on, supers off, feed? should match the environmental conditions that present themselves.

**colony strength**

Terms like 'weak' and 'strong' are applied and refer largely to the number of bees that a colony is supporting. This may vary according to the season or the actions of the beekeeper but one of the most important things is always the ability of the queen. Young? Health? Vigorous? Prolific? - colony strength will depend on these factors and of course the season's weather.
See: strength of colony, well-found

**colour**

The quality of light transmitted or reflected by a substance. Hue, purity and brightness are factors of colour which are appreciated as visual sensation. Bees can see colour in much the same way as we can except that they are blind to the red end of the spectrum while being able to see ultra violet light at the other. The colour of protective clothing, of hives etc. should not run counter to the bees' preferences. Anything resembling a bear, not merely its jerky movements but its colour and the use of white to reflect heat and black to absorb it, are points to consider. Pollen colours are not a great guide to the crops the bees are working because they are seen differently in various circumstances. They are of course the only 'colourful' things in a hive. Colour would not seem to be significant in the hive but honey, bee blood and beeswax are golden, why not paint hives in this way? A five year colour cycle was recommended for queen marking so that a queen's age would be apparent. Years ending 1-6 white or silver, 2-7 yellow or gold, 3-8 red or iridescent, 4-9 green, 5-0 blue.

**colour of hives**

White has been a popular colour but in winter a black roof is likely to absorb heat whereas a white one reflects it. In some cases it has been better not to draw attention to hives but to colour them to fit in with the background, either brown or green. From the bees' point of view they do differentiate colours and use may be made of this when raising queens or taking steps to prevent 'drifting' by using different colours at least at the entrance.

**coloured brood**

Healthy brood is always thought of as opalescent, glistening white. In the early stages of feeding a distinctly bluish tinge can be seen especially when the surplus brood food is seen against a dark comb base. Healthy brood may occasionally appear to have a pinkish or creamy tint and this can be due to pigmentation of the pollen that the nurses are eating. Diseased brood becomes discoloured so the normal white is a good guide to the health of the brood.
See: discoloured brood, pink brood

**colour of wax**

When comb or cappings are melted and filtered the clean beeswax obtained is usually of a yellow colouration. Sun bleaching or various chemical processes can render the wax almost pure white and wax that has been overheated or has too high a percentage of dross or pollen may be dark brown or even black. The most popular colours range from primrose to orange but delicate pinks and less attractive green tinges are all possible. Wax soluble dyes can also be used to impart various colours for candle making and wax painting.

**colour of the year**

The internationally agree colours for the marking of queens run as follows: Queens born year ending 1 or 6 white or silver, 2 or 7 yellow or gold, 3 or 8 red, 4 or 9 green and 5 or 0 blue. Sometimes bee magazine covers follow this colour sequence too.
See: queen marking

**colour sense**

Although colour does not seem to have any significance within the hive externally foragers are quite able to select flowers by

their colour though choice of flower would also call for their sense of smell. Because the range of their eyesight has a spectrum shift enabling them to see ultra violet but not red, it is not easy for us to realise that colours must appear different to them. For instance as brown is composed of a mixture of red and blue to our eyes, brown might well appear as blue as they are unable to see red. Also many white flowers and others have an ultra violet element which we are blind to and so would appear different to the bee. We can still use colours to help bees identify their hive or entrance and this is useful in a queen mating apiary. For protective clothing white is best and as a black ball jerked in front of the entrance will soon be filled with stings – take heed.
See: perception of colour, light, u/v light

**colt**

One of the names given to the third 'swarm' that issues from a hive. Could this indicate the naivety or 'friskyness' of a colony that swarms more than once in a season?

**coltsfoot *Tussilago farfara***

A perennial weed whose bright, star-shaped, yellow flowers appear on leafless stalks in March/April adding a welcome touch of colour to the surroundings and providing forage for the bees. It likes 'clay' and spreads easily but the small flowers belie the size of the large leaves that follow. Its pollen has been described as light to medium brown, light orange and golden and the size of the grain is 28μ. The translation of its Latin name suggests its medicinal use for driving away coughs.

**Columella, Lucius Junius Moderatus**

A Roman bee-master and agronomist (plant, soil and wind scientist) AD60. He brought a more systematic approach to husbandry and in his writing drew freely on the work of his forebears. His hints on honey extraction might guide some present-day beekeepers. He laid great store by the selection of 'race', considered that the best hives were made of cork or reeds while ceramic hives were not suited to summer warmth or winter cold. He may have been a forerunner to the builders of 'bee-boles' as he advocated the shelter of walls cut so that bees could get through them to their hives. He marked water-gathering bees so that they could be traced to their nests. He refers to the shortest day and the 'coming of the swallow' and cautions – keep a sharp look-out for hornets between 24th July and the rising of Arcturus' (bright star in the constellation of Bootes') 13th February He describes iron tools for cutting out honeycomb, mentions foul brood and possibly 'nosema' and migratory beekeeping on the Nile, drone slaughter and tearing the queen's wing.

**comb**

Honeycomb is composed of neat rows of horizontal, hexagonal, waxen cells set back-to-back in a vertical array. When melted and filtered pure beeswax is obtained but the 'dross' or remaining residue will have done much to strengthen the comb. It may, according to its age, contain cocoons, fibre, pollen husks, propolis and other materials. Several other meanings for 'comb' include a toothed wooden, metal or plastic tool for arranging the hair. On bees rows of firm hairs function as pollen combs. Many place names receive this name when it means a narrow valley or deep hollow and it is pronounced 'coom'. A fleshy, red 'comb' adorns the head of a cockerel and bees have been known to sting them.

**comb of brood**

A comb that has been used by a colony for brood rearing. Brood can imply all stages of a bee's life up to the point of emergence as an imago. They become darker and tougher than super combs only used to store honey. According to 'Cotton' a comb of brood weighs a much as a comb of honey thus giving a false impression of the amount of honey in a stock full of brood. ('Hefting might lead to a starving colony not being helped). It is essential that brood combs are not needlessly exposed to ambient air,

sunlight or wind because incubation goes on all the time.
See: hefting

**comb building**

Groups of bees ageing from 12 to 18 days hang in festoons having gorged themselves with honey (or syrup) to stimulate their wax producing glands. They will maintain a temperature just above that of the brood nest (about 35C/96F). Eight wax pockets under each bee's abdomen will produce small wax scales ('platelets'). These are lifted out by the pollen brushes of the hind leg's basitarsi and masticated until they are snow-white an pliable when, using their forelegs and mandibles, the 'mason' bees share the complex work of forming the cells and developing the comb. Gravity helps them to achieve plumb vertical combs as they extend them downwards, always keeping them parallel to the adjacent combs and leaving bee-space (ventilating space) between them. When man interferes and persuades them to build their comb out ('draw' it out) on sheets of beeswax embossed with the hexagonal cell shapes, although the bees will accept this when temperatures and food supply encourage them to do so) much foundation goes to waste because it was not inserted in the right place at the right time. The precision stamping out of cells (either drone or worker) and the alluring aroma of the newly fashioned beeswax, can tempt us to assume that it is just as alluring to the bees.

**comb collapse**

Comb becomes soft or plastic at 35C and comb containing honey in particular may collapse if the temperature goes above that figure. Despite the number of photographs taken of 'experts' holding combs out of the hive in a horizontal position it is rarely appreciated that at hive temperature combs are warm enough to sag slightly. Only in rare conditions will the bees allow temperatures to reach a dangerous level, for instance when they are interfered with, closed up during spraying episodes, or when moved without adequate 'top' ventilation or when a forest fire occurs. But, when it leads to comb collapse the bees will drown in their own honey and combs will become badly distorted. Colonies, buzzing unhappily away while being transported, will fan even more vigorously before they die if shut in a car boot and neglected for even a minute or two. Suffocation will occur before comb collapse. Always travel at all times of the year with a full ventilation screen over the combs and be sure that it is not impeded in any way.

**comb cutter**

There are variously shaped tools for cutting through comb which resists the passage of a knife and does not easily cut cleanly. The term 'comb cutter' has been applied to a shaped device that will prod out a neat piece of honeycomb (cut-comb honey). It is often made from stainless steel and is sub-rectangular and equipped with a plunger platform (like an ice-cream dispenser) which enables the precise transfer of the honeycomb chunk into the container.

**comb foundation**

A septum of 'rendered' beeswax that has been forced under pressure through 'rollers' designed to give the material the unique shape of the base of honeybee cells. The 'sheets of foundation are cut to size to fit the frames that a beekeeper has decided to use and are of varying thickness from wafer-thin for honey and slightly thicker for brood and may be reinforced with stainless wire or possibly plastic and the cell size will either be for workers or for drones. It is brittle when cold and can become 'stale' if not put into use reasonably soon after it has been manufactured.
See: foundation

**comb honey**

A market term used to describe honey when presented in the comb in its most natural, unblemished form. This can take the form of 'cut-comb' or 'sections' which were often of lime wood and made to hold a pound of

honey. It is pure honey stored by the bees when quite 'ripe' and in comb that has been newly built and not used for brood (or pollen?). It should be free from 'granulation', pollen or any foreign matter. It is sometimes put up in a jar surrounded by clear honey and may be a single piece or several pieces and is then known as: 'chunk honey'.
See: chunk honey, honeycomb, showing honey

**comb renewal**
The replacement of combs in a hive.
The reasons for removing old comb and replacing with sterilised or new comb or foundation include the following: 'hygiene, keeping wax-producing bees occupied', 'giving more space and possibly larger cells' and keeping the amount of drone comb under control. Although it is generally agreed that good bee health is easier to maintain on relatively new combs there are those who boast of having kept brood combs for tens of years. Bees in naturally chosen cavities probably tear down and replace their comb or if space allows just move on all the time to freshly built comb. Two seasons of use for brood rearing is certainly more than enough – remember the faeces left behind and the fractional reduction in cell size each time a worker is born.

**combs**
*Apis mellifera* build parallel, vertical combs shaped with rounded edges except where the comb meets its support. Consequently in nature comb shape and size will vary according to the cavity occupied. Combs two metres wide or of similar depth do occur though beekeepers usually encourage bees to fill frames of a standard size for ease of handling. The back-to-back arrangement of cells makes ingenious use of space with a minimum amount of wax yet giving maximum strength. The bulk of the comb is made of worker sized comb though 'pop' holes, interstitial cells, gaps and odd patches of drone cells are included. At certain times during the active season, entire combs of drone cells may be built outside the periphery of the brood nest. Both drone and worker cells are used for brood rearing and the ripening and storage of honey. Pollen too may be packed into either type of cell though these are nearly always within easy reach of the brood nest. Honey tends to be stored above the brood nest and cells may be extended where space allows to widths of up to 7cm to accommodate it.
See: drone comb, honeycomb, worker comb

**combs for wax recovery Melting down**
This may be done by boiling in 'soft' water (see rendering) by steam heat and pressure or more simply in a solar wax extractor. When combs are to be culled for rendering, keep them somewhere so that robbers cannot find them. Thorough soaking for a day or so helps to prevent air-traps in the cells that would reduce the effectiveness of boiling. Combs can be sliced and put into old 'tights' or stockings before going into the solar wax extractor.
See: rendering, solar wax extractor

**comb spacing**
The centre-to-centre spacing of combs is achieved by specially designed spacers, notches in the supporting ledge, 'castellated' runners, or touching shoulders on the frames. Hoffman, Manley, WBC metal (or plastic) ends are widely used and there are wide and narrow kinds. For brood a narrow spacing of 35mm/13/8" is desirable as a queen will lay in a shallow cell but not in one that is too deep (as honey cells often are). 38mm/11/2" spacing is also used. For honey combs spacing may be increased still further, though where foundation sheets are adjacent to one another these should be kept to 38mm/11/2" or less until partly started. The super frames may be spaced up to about 44mm/13/4" for cut-comb but extra wide combs of up to 72mm/3" or more have been extended and filled by the bees.

**comb sterilisation**
The sterilisation of combs increases their

effective length of life and makes for apiary hygiene. When free of bees there are two main choices: acetic acid (Glacial 80%) when stores are present and formalin when the combs are devoid of stores and bees. This is an effective way of dealing with pathogens other than AFB (which is best burnt) though in the USA high pressure sterilisation using ethyl bromide etc. is carried out. In all cases combs must be thoroughly aired before re-use. A National deep box needs 200ml of formalin or 125ml for a shallow box. 150ml formalin per 25 litres volume. Sealing in polythene aids effectiveness
See: acetic acid, formalin, fumigation, nosema, sterilisation

**comet of drones**

Visual reports of drones forming into an attenuating mass like a 'comet' are associated with the drones' natural behaviour in having located and followed a virgin queen's scent trail. This is now known to take place within a 'drone congregation area'. Having followed the trail of scent (which could be in any direction relative to the direction of the wind) the drones 'home-in' on the queen visually. The 'fleetest' reaches and attempts to copulate with the queen. Photographs of this phenomenon are available.

**comfrey *Symphytum officinale***
**Boraginaceae**

An herbaceous perennial that likes damp soil and produces nectar and pollen from May - September. Its flowers may be white, mauve or pink. Some bumbles change their buzz noticeably as they enter each flower. The faint yellow pollen has a widely oval grain of 20µ. Russian comfrey *S.resplendicum* is an 'escaped' fodder plant, now commonly established on roadsides. Apart from their use by health enthusiasts the large fleshy leaves make good compost.

**commensalism**

The balance and division of labour between plants and animals by which one or both benefit. (Fungi, algae, lichens) Such an organism is an 'inquiline' and benefits by association or sharing a home with a different species yet being neither parasitic nor symbiotic – 'eating at the same table'.
See: inquiline, symbiont

**commercial beekeeper**

In the USA someone with over 600 colonies. In recent times this expression has been superseded by 'Bee Farmer'. In Britain the BFA calls for members with over 40 colonies.
See: bee farmer, hobbyist, side-liner

**commercial hive**

The modified commercial hive is an effort to retain the outer dimensions of the smaller modified national hive while using short lugs and deeper frames so as to give a volume closer to that of the world's larger hives without a corresponding increase in the amount of timber needed. The frames are the same length as Smith frames (BS but with short lugs) and there are the customary two depths of 254 or152cm/10 or 6" respectively.

**commissure**

Transverse or longitudinal nerve fibres connecting each pair of ganglia in the ventral cord of the bee. The twin nerve trunk is composed of longitudinal commissures which link ganglia to the brain.

**committee**

A body of persons elected by a larger group to study, discover, reach conclusions and make recommendations on certain matters referred to as the committee's 'terms of reference'. The leading member is normally the 'chairman' and the size of the committee will be governed by its constitution. In BKA's members come up for re-election annually at the AGM. The officers of the association are usually ex-officio committee members, i.e. can attend and vote at meetings by right of their office.
See: committee, chairman

**common**

A tract of land owned by the local authority in trust for all member of the community. Right of access and use for various purposes is subject only to Bye-laws. At one time grazing and fishing rights were used, now-a-days especially in built-up areas, recreational facilities may be available. In country areas picnicking, rambling even temporary positioning of bee-hives, may be possible.

**common bird's foot *Ornithopus sativus***

This and *O. perpusillus* which has established itself as a weed, grows to 20cm and has yellow flowers.

**communal feeding**

In Great Britain open-air or communal feeding is discouraged. This is because it will attract bees (and for that matter other creatures) from a considerable distance and in the case of bees from other apiaries this could mean the spreading of diseases. Furthermore it usually entails feeding other people's bees. Open-air watering places are used and often baited with syrup to encourage their use. There are some aspects on the 'plus' side including economy of time and of feeders and the use of prophylactics.

**communication**

The act of or method of stimulating a wanted response in another organism. As a honeybee colony has to perform as an individual entity if it is to survive as a colony, communications between its members are of the utmost importance and are conducted by transmittable characteristics and by physio-chemical messengers to the various nerve receptors. Queen substance, the effect of brood on foragers and the awareness of intruders are examples though a short paragraph cannot do justice to so vast a subject.
See: dances

**compartment separator**

Whenever more than one queen is to be housed in the same compartment or when it is desired to create the impression of queenlessness so that nurse bees in one part of a compartment will raise and possibly allow a virgin to mate, then a division which is impermeable as regards 'touching' or 'smell' or the passing of pheromones, has to be incorporated. Vertical division boards made to fit tightly against upper cover, walls and floor, fulfil such a function. Queen excluders could be used but are permeable to 'food passing'.
See: division board, excluder

**compass**

A means of determining direction. Iron has been discovered (1983) in a granular form in rings around segments of a bee's abdomen. There are nerves leading to cells and they could be effective in sensing the earth's magnetic field like a compass.
See: light compass reaction, polarized light

**compatibility**

Ability to co-exist, to endure together. Biological commonality. When dissimilar types are crossed and form satisfactory hybrids they are said to be cross-compatible. It is much easier to unite bees and introduce new queens when both types are compatible. Imported strains may or may not be compatible with local native strains. Bro. Adam said, races or strains chosen for cross-breeding must 'notch' (prove compatible).

**competent**

To be properly qualified and able to carry out a particular function. Competency in apiculture takes some considerable time to acquire and while self-tuition may establish a degree of confidence, to be adequate in the varied circumstances confronting the apiculturist normally requires training and assessment by a qualified authority. Examinations are often looked on with unease, especially by those who do not fill the requirements but despite the inadequacy of examination methods, the setting up of appropriate bodies for this purpose has engendered a much wider degree of competency.

**competition**

A contest aimed at determining whether a person is capable of achieving a certain standard, perhaps rewarded by a prize or some advantage. Ingenious committee members have used honey tasting, cake making, and bee-bottling as well a host of other schemes to exercise talent, possibly to educate but more often to raise a laugh. Provided competitions do no harm and do not stand in the way of co-operation much good can come from them.
See: honey show

**Complete Guide to Beekeeping, The**

A book by well known American beekeeper and scientist Roger Morse. There are several books bearing similar titles from many other beekeepers.

**Compositae Dicotyledon.**

The daisy family which is the largest family of flowering plants. Their seed is often carried on a hairy parachute. Members include: 'yarrow'(milfoil), 'cat's foot', 'chamomile', 'burdock', 'Michaelmas daisy', 'thistle', 'marigold', 'sow thistle', 'chicory', 'fleabane', 'hawksbeard', 'agrimony', 'hawkweed', cat's ear', 'nipplewort', 'hawkbit', 'mayweed', 'winter heliotrope', 'ragwort', 'golden rod', 'dandelion', 'colt's foot'. See: under various headings.

**compound eye**

The bee has two strongly convex, elongated oval masses of ommatidia which form the compound eyes. Each has in the neighbourhood of 6,500 lenses and is capable of transmitting a 'mosaic' image to the brain. The drone has some 17000 lenses in all. Because of their exposed forward position on the head of the bee, with minor movements of its head the bee can see in almost every direction. The eyes are well-covered with hairs which are presumably of a protective nature. The eyes cannot be focussed but are very sensitive to movement and can differentiate shapes and colours.
See: ocellus, ommatidium, polarised light, retina, sensory transduction

**compound microscope**

Expensive, sophisticated pieces of apparatus comprising stage, mirror and optical tube with eyepiece and nosepiece for holding objective lenses. The latter is the chief lens combination which receives light from the object and transmits it along the tube in cones of a different aperture. Beware of the many cheap imitations and learn the essential basics before trying to get the best use of it. It is certainly as complicated as a good camera. Light, source and preparation of material are very important aspects of its use.
See: dissecting microscope, microscopy

**comprehensive insurance**

The BBKA comprehensive insurance policy covers loss or damage to hives, equipment and bees belonging to the insured member arising from accidental or malicious damage, fire, flood or theft. Home or out apiaries are covered and transactions can usually be made through the local BKA. There may be an 'excess' raised on some claims.

**concave cappings**

Sealed brood should be suspect if the cells are concave or flat. In the latter case they may merely have been brushed or bruised but normal cells are domed and only when the oxygen in the cell is used up by pathogens are the cappings likely to become sunken (concave), greasy-looking, cracked or broken.
See: AFB, convex

**condensation**

This occurs when a vapour is forced to deposit some of its liquid. Water droplets that form a film on a cold window pane are an example. A honeybee colony uses and produces moisture. In the active season ventilation by the bees can dispel any unwanted moisture though if it is allowed to rise and impinge on the underside of a cold metal roof, water drops may fall back onto

the bees. Ample top ventilation in winter and well-insulated roofs are proof against this.
See: frosting, humidity

**condition**

As this word is used to explain the state of anything it will be found under various headings such as 'brood condition' etc. We use the word too when disease or disorder might be more appropriate for example: 'starvation', 'dysentery', or 'excitement'. Colonies are referred to as: 'weak', 'strong', 'ready to swarm' 'well-found', 'queenright', 'queenless', 'diseased', 'healthy' etc. We also refer to weather 'conditions'.

**conditum**

An ancient Greek beverage made from wine, honey and pepper.

**condyle**

A rounded structure shaped to fit into a socket as for example the joint of a bone, or articulated joint in the chitinous leg structure.

**cone**

A circular shape that tapers perhaps to an opening or to a point. Brass cones set into a roof are used to allow bees to exit but not return. According to D. Galton, Butler used the term to describe a cluster of bees.
See: cone escape, crystalline cone of the bee's eye

**cone escape**

The metal cone escape was developed by Charles Dibborn USA in 1890. In WBC hives and others brass cone escapes are fitted to the roof to allow ventilation and bees that get over the crown board to escape. It has been used to get bees out of trees and awkward places because it provides a one way 'valve effect. Also 'conical escape'.
See: bee-escape

**congress**

International Congresses are organised every other year and held in different countries around the world. Other specialist congresses are advertised in the Bee Press from time-to-time.
See: Apimondia, tropical apiculture

**conifer *Gymnospermae Pinaceae* (pine family) Coniferae**

Mainly evergreen, cone producing trees. While they are not significant yielders of honey-dew in this country they are most useful on the Continent of Europe. As no flowers are involved (honeydew is exuded by scale insects from the sap) the flow is associated with a dearth of pollen because pollen produced by conifers is extremely copious and fine, being born by the wind and unsuitable for honeybee collection. They are useful as wind-breaks, timber for hives and for beautifying the landscape.
See: tree, forest honey

**connectives**

The ganglia of the central nervous system are all joined in a continuous line from the brain by twin longitudinal commissures which like the other paired intervening connectives e.g. the circumesophageal connectives, are composed of nerve fibres.
See: central nervous system

**conoid hairs**

These are cone-like (sub-conical) and are found on the proximal ends of the antennal segments.

**Conqueror hive**

Produced by Simmins – also double and treble 'Conquerors'. These were giant hives though their outer volume gave a false impression of the amount of space within.

**consanguinity**

Blood relationship. Said to lead to low resistance to disease, reduced viability and increased sterility, a phenomenon linked to 'inbreeding' and known as 'inbreeding recession'.

**constancy**

Bees once motivated, persist with a certain task until it has been done, or fatigue sets in, or a stimulation to pursue another course of action is received. Bees will remain constant to a single crop, where possible confining their activities to a comparatively small area. A fanning bee cannot easily be persuaded to change its occupation not even to sting. The reactions to pheromones and other stimuli therefore produce a given and usually sustained reaction, which leads to descriptions such as constancy, faithfulness and crop fidelity.
See flower constancy, crop fidelity

**contact feeder**

A feeder that is placed in contact with the bees or at least so close that they can reach the food with their tongues though prevented from getting their bodies into it (the syrup). These are far more hygienic than feeders that allow bees to jostle right up to and even into the feed. Honey tins with holes punched in the lid have sufficed on many occasions but plastic buckets incorporating a perforated screen are very satisfactory. This type of feeder is often easier to clean than the old-fashioned types with floats and awkward corners.
See: bucket feeder

**contact insecticide**

An insect poison that does not need to be eaten but kills by penetrating the body surface or blocking the spiracles.

**containers**

Honey containers should be able to withstand the weight and nature of the contents. They should be completely air-tight as honey is 'hygroscopic'. Certain metals react to form unsightly black deposits when acted upon by honey and most tins or cans are lacquered thoroughly on the inside. Only plastics that do not taint, such as food stuff types, should be used. In the 80's glass was still one of the favourite ways of displaying clear or granulated honey. Cut-comb containers, usually of light plastic with transparent lids, have become quite common. When honey is for sale strict regulations as regards labelling apply.
See: feeders, feeding, labelling

**contraction of the brood nest**

A sudden cold snap in the spring might lead to diminution in the size of the nest that the bees can properly incubate. The variations in size of the brood nest are dependent on season, weather and the condition of the colony, especially that of the queen. It is natural for the brood nest to begin to contract in the autumn though a sudden cessation of a flow, supervention of a cold spell or drought or anything that causes difficult times for the bees, might result in a diminution of the queen's egg laying rate with the inevitable consequence of a contraction of the brood nest at any time. The beekeeper can also achieve this by restricting the queen as is sometimes done before the removal of the harvest.

**contribution**

To give money, help or time to help bring something about. This is relevant in many ways in beekeeping. One might contribute help to a BKA or to a bee breeding project. Likewise one might write an article for the press. The production of nectar or pollen might be referred to as the flower's contribution.

**Convention**

A meeting or an assembly. Since 1983 a Bee Fair and Spring Convention has been held at Stoneleigh. It implies a fairly substantial number of people being attracted by suitable advertising of the place, date and purpose.

**converter clip**

A clip that can be attached to a frame with oblong section to convert the side bars to the Hoffman type of spacing. While these can be fitted to existing drawn combs it is more satisfactory to put them inside the frame with the upper edge pushed up under the top bar before it is drawn out. When the top bar is held pointing away from you the nearest bevel should be on the *right*. otherwise you will not get bevel to flat face and side bars may stick together.

**convex cell cappings**

Cells of all three castes are domed (convex), those of worker brood less so than queen or drones. When the brood is capped in normal conditions it has a regularly domed appearance. When knocked, bruised or concave it is an indication of something out of order.
See: concave

**convoluted**

Coiled or rolled up (bot.) This applies to the bee's 'aorta'.

**Cooke, Samuel**

Wrote 'Complete bee-master' in 1780.

**cooling**

A colony makes use of moisture and fanning to keep internal temperatures normal whenever outside air temperature rises above nest temperature. White or silver painted hives are often used in countries where hot sun is troublesome. Wide entrances are rarely a good plan because of robbing and predators, but in extremes bodies can be zigzagged or offset at different angles to one another so as to provide additional ventilation.
Honey: While honey should never be heated above 49C/120F when its temperature has been raised above this level it is important to cool it again as soon as possible.
House apiary: While mutual warmth is beneficial, full facilities for keeping the temperature down in hot weather should be provided. Special screens might be necessary to exclude insects and to permit ample ventilation.
Mead: the 'must' should be below blood heat before the yeast is added.
Smoke: put green grass into an over-hot smoker.
Syrup: Feed to the bees below 38°C/100F.
Wax: Cool as slowly as possible to prevent the cake, candle or model from cracking.

**Cooper B. A. B.Sc. Died 1982**

Known to thousands of beekeepers, a man of genius, dedication, conviction and leadership. One of his main aims was the restoration of our native bee. His reputation and knowledge spread over much of Europe and the world. He founded the 'Village Bee Breeders Association, now BIBBA (the British Isles Bee Breeders Association). His dynamism and fervour could not be matched. A great man who inspired others to emulate his efforts.

**Coppice**

A small wood or thicket such as one that is periodically cut to provide hurdles, stakes etc. A coppice with standard trees, leaving space between an open network of normal trees for hazels, chestnuts or other underwood, to grow and to be cut at 10-15 year intervals. Such woods are open enough to offer plenty of scope to bees having wild flowers galore including willow, bramble, rosebay, heath and other wild flowers.

**copulation coition**

When this takes place with honeybees the drone approaches the nubile queen with legs extended and turns onto its back as it grasps the queen while explosively driving his part into her. He immediately becomes inert, hanging with wings loose and supported by the queen as long as they remain locked together. Frequently the couple fall to earth but mating usually takes place in a rising air current which helps to reduce the rate of descent. Sometimes the defunct drone falls to the ground leaving the queen airborne. Usually a 'comet' of some hundred or so drones are in the vicinity of the queen.

**copulatory pouch**
**see: bursal cornua**

There is a lateral pouch on each side of the queen's bursa. They would appear to be associated with the cornua of the drone's endophallus but no positive clarification is available.

**corbicula**

pl. corbiculae pollen basket
A highly polished hollow on the outer face of the hindmost tibia of a worker's hind leg. The two form a pair which are evenly loaded with moistened (sticky) pollen grains (sometimes with propolis, wax or other materials. These are squeezed by the pollen 'press' into coagulated lumps and transferred from the auricle of the adjacent tarsal segment into the corbiculae. If they are pollen loads they are referred to as 'pollen pellets' and they are retained by fringes of curved, springy, retaining hairs that border the edges of the corbiculae and allow loads of various proportions to be safely carried. Each lump of pollen is anchored by a single spike that is positioned within the basket and also serves as a steering slide when the bee comes to push the pellet out of its basket and into a cell. This it does by pushing one leg against the other.
See: pollen basket

**cordovan**

colour of leather from Cordova in Spain
This colour in bees from Spain is a mutant that dilutes the normally black portions of the cuticle to a cordovan brown. The gene responsible is described by Morse as a single Mendelian recessive.

***coreopsis, calliopsis***

Compositae
There are many garden species with yellow, brownish or parti-coloured flowers that yield nectar and pollen in late summer. The name is derived from the bug-shaped form of the seed.

**cork *Quercus suber***

A light, porous, spongy material derived from the cork oak. Used for floats, bottle stoppers and notice boards (drawing pins will sink into it quite easily). In feeders special precautions would be needed to keep it sterile. Cork borers or cutters will make holes for tubes, fermentation valves etc and corks and cork bungs are useful in mead making.
See: cork hive, flanged cork

**cork hive**

Although the insulative properties of cork make it a suitable substance for the construction of beehives, it is only in such countries as France, Spain and Portugal that it is sufficiently cheap and plentiful to be used for hive construction.

**corma**

Barley beer strengthen with honey
A drink made by the ancient Gauls in the form of a honey-beer brewed from combs that had been extracted and therefore ostensibly less palatable than the equivalent brew made with fresh honey. The former was therefore the drink of the poor while a quality beer called 'zythus' was made for the better-off folk.

**corn cockle *Agrostemma githago***

A pink to deep purple weed once found in cornfields and worked by bees who help perhaps to prevent it from becoming extinct.

**corne honey**

Granulated but soft HH. Corned – granulated CTC.

**cornea pl. corneae**

The outer, transparent covering of each element of the compound eyes. The whole complex of such lenses on each eye. The three simple eyes have each but one corneal lens while the compound eyes have enormous numbers. They form the distal surface of each ommatidium and collectively produce an elongated, highly convex, oval surface interspersed with protective hairs. Dade says that workers have about 6,900 lenses, queens fewer and drones have about 8,600 slightly larger lenses.

**Cornelius Celsus 42 BC–37 AD**

He was called The Roman Hippocrates and Columella refers to him as an elegant writer on bees. He condemned hives made of dung as they were flammable.

**cornflower *Centaurea cyanus*, cornflower *C.moschata* Sweet Sultan**

An annual that used to be found in ripening corn. There are many garden cultivars the, pollen is pale greenish, white ,pale grey with brown tinge and pale yellow Whitehead.

**cornua**

The pneumophyses or horns that arise from the drone's genital vestibule. When a mature drone is held by the fingers a slight pressure will often cause it to evert its genitalia when the orange-tipped cornua are inflated and can be clearly seen.
See: copulatory pouch

**corolla**

The complex formed by the petals. Often colourful and of delicate texture. Bees are attracted by their shapes and colours and we see evidence of flower-insect relationships that developed over a considerable period. The petals sometimes include nectar guide lines which may reflect ultra-violet light (invisible to man) and may be shaped to form nectar tubes. Legumes have papilionaceous corollas (See APC bee). Rosaceous corollas (apple) are very 'open' and rain may wash their nectar out. The shorter corollas of second-cut red clove are more useful to honeybees the earlier flowers requiring bumble varieties with longer tongues. The corolla is often surrounded by another ring of sepals, known as the calyx.
See: cheating, nectar stealing

**corpora allata**

Small glandular globules closely associated with c.cardiaca and found above the pharynx and near the honeybee's brain. They produce neotenin the hormone that inhibits larval growth at the appropriate stages. They persist into adulthood their hormone reducing the permeability of the proctodeum while its absence increases it thus facilitating the reabsorption of water.
See: neotenin, rectum

***Corpora cardiacum***

pl. cardiaca
A neuroglandular body consisting of two small knots of tissue and situated between the cerebral ganglia and the corpora allata where the oesophagus and pharynx join. They are connected to the brain by nerves which also extend to the corpora allata.

***corpora pedunculata*** **corpus pedunculatum**

Mushroom bodies in the protocerebrum which contain small groups of nerve cells and are 'association centres' connected by nerve fibres with the optic and antennal lobes and also with the other parts of the nervous system. They perform the very important role of co-ordinating the actions of the insect according to the information received from the sense organs DADE.
See: mushroom bodies

**corrosion**

Metal deterioration due to chemical reaction. Rust is a common example and apart from nails used in hive assembly few items are made from metals that are not rust-proof. Syrup and dilute honey should always be washed from feeders, container or extractors because the corrosive activity of acids that develop when such solutions are left in contact with metal will hole aluminium feeders, leave gluey tar-like deposits in containers and extractors and both can prove distasteful and expensive. Stainless steel and Monel metal have been widely used for beekeeping apparatus since they first became available.

**corrugated packing paper card**

Two-layered , coarse, brown paper with

one ridged layer glued to a flat layer. Sold in rolls for packing up parcels. The rolls can be sawn into convenient lengths or chopped through on a chopping block so as to make cartridges that fit into a smoker. It is clean and easy to handle, can be dipped into a coloured saltpetre solution and dried for easy lighting and it is free and readily available (used to be –anyway?) Unfortunately like (egg cartons) it makes a tarry smoke that glues up the smoker which can get very hot. Precautions – don't breathe in fumes while saltpetre burns and be careful not to acquire that which has been rendered non-flammable.

**cosmetics**
Materials used for the care, beautification and adornment of the skin, hair and complexion have included products of the hive through the ages. For example, lipstick has a fair percentage of beeswax as does cold cream, and honey is used for face and hand creams, shampoos and face packs.
See: recipes, Cleopatra's face cream

**costal**
Costa is the rib of an insect wing.
Concerned with rib – see 'wing venation'

**Cotoneaster pron. Co-Tony-Aster, Rosaceae**
A hardy, evergreen perennial shrub with many cultivars in the forms of shrubs and trees. In the south it flowers about May time although it sometimes usefully fills the 'June-gap'. It yields nectar and can literally hum with bees, its pollen being of a very pale yellow with 28µ grains.
The tiny flowers become attractively bright red berries. It strikes from cuttings though its seeds need stratification. spp. include *cornuba*, *horizontalis*, *microphyllus* and *simonsii*.

**cottager hive**
A single-walled hive with many of the external characteristics of the WBC but limited to two lifts. 1884 'Modern Beekeeping'.

**cottagers**
Although any labourer, urban or rural who lived in a cottage, went by this name, a rather special beekeeping significance gravitated to it in the 1870's and 80's. Clergymen, doctors and others studied the craft so that they could encourage the poor cottager to improve his lot by keeping bees. Until that time books, and beekeeping education such as it was had been directed at better-off folk. At the founding of the Hampshire BKA in 1883 subscriptions were set at five shillings p.a. for ordinary members, half-a-crown for artisans and one shilling for cottagers.

**cotton Gossypium spp., Malvaceae**
Gives a good crop of honey in warm countries. Pollen is orange coloured; it is a quick granulating honey. The plant has both floral and extra-floral nectaries. Pink bollworm is a pest in SW USA (pheromone + insecticide) is used to control it.

**Cotton, W. C. Rev (William Charles) 1814-79**
A short and simple letter to cottagers 1837 subsequently criticised in an unnecessary way. Went to NZ and wrote a manual for the NZ beekeepers in 1848.

**cough mixture**
200 ml spirit (cooking brandy or whisky) 56g/2oz propolis. Leave for a fortnight then filter. Add 200ml glycerine and 200ml spirit. Take 2 teaspoonful's per day.
See coltsfoot

**coul-staff**
Stick through a skep held by 2 persons for carrying (Butler, More in Cotton)

**coumarin $C_9H_6O_2$**
A white crystalline substance reported as causing mild headaches to those with susceptibility. It gives the slight vanilla-like flavour to the honeys of sweet clover.

See: *Melilotus*

**County Beekeeping Lecturer CBL**

Historically the officer who was provided by the Department of Education and Science to give specialist help and advice in matters connected with apiculture. The title of CBL was based on the officer being of Lecturer grade as at an Agricultural College though Beekeepers tend to use various terms such as inspector, advisor or instructor. Inspections of bees (as for brood diseases) came under the auspices of what was then the MAFF, now Fera. Where CBL's were available beekeeping in such areas was maintained at a good level. For hobbyists, local evening day and week-end courses were arranged, often under the auspices of local BKA's but also by equipment manufacturers. Courses for teachers usually came under the umbrella of Rural Science but beekeeping was quite rare in British schools when economic circumstances caused governments to reduce expenditure on such subjects considered of lesser importance.
See: examinations and correspondence courses

**court**

A changing group of some 10 – 12 young worker bees which at any one time accompany the laying (reigning queen). Nearby bees almost invariably turn towards her on the comb and this star-shaped or daisy-petal formation (as one writer put it) is helpful, especially in observation hives in spotting where the queen is. Some of the members feed and caress her, removing any faeces or eggs that she may drop and in so doing picking up 'queen substance' which they subsequently distribute to other bees. Courtiers constantly change but some may never make direct contact with the queen.
See: attendants, food-sharing, queen courtiers/substance, retinue

**cover cloth**

A temporary working cover used while manipulating – a 'manipulating cloth'. They are usually weighted by the insertion of a stick at each end so that it resists the wind, lies taut and can be rolled and unrolled. When used in con- junction with a second one, only the frame to be taken out need be uncovered and smoked. Where a gap is left between the cloths, this can act as a barrier to the queen's passage and so help if it is required to find her. Robbing is less likely and as less heat is lost bees can be handled in slightly cooler/windier conditions. Canvas, oiled cloth or permeable material through which smoke can be puffed, may be used and cloths may be moistened or treated with a repellent. Steps should be taken not to spread pathogens and cloths should be regularly sterilised or replaced.
See: draw cloth

**coverslip**

Small squares or circles of thin glass used to cover specimens on slides for microscope viewing. Air is excluded to maximise visibility by using a film of liquid with a high refractive index, say glycerine in the case of pollen or Canada balsam for a permanent mount of a bee's wing, leg etc.

**Cowan T. W. 1874**

He was a trained engineer owning Kent Ironworks at Greenwich. Although regarded as an amateur by the commercial Manley, his British Bee-Keeper's Guide ran to thirty editions and had a great influence on beekeeping practice. He also wrote 'The Honeybee Its Natural History, Anatomy and Physiology' in 1890 and 'Wax Craft' in 1908. He was co-editor and then editor of the BBJ. He gave generously of his many talents offering a cure for Foul Brood, producing a self-reversing centrifugal extractor, introduced the radial extractor and made a rapid feeder. He held the limelight in Britain while Langstroth, Dadant and Root did the same in the USA. He left an important library in the hands of the BBKA.

**cowcloome**

A plaster made from cow dung and clay that was used for sealing hives to their stands HH. Also as a filling for wicker hive skeleton. Parget – plaster or mortar as above. Fr. 'Pourget'.

**coax**

The wide proximal segment of each leg that attaches it to the bee's thorax. It can move freely in the thoracic socket and in turn allows free articulation of the short trochanter.

**c.p.s.**

Abbreviation for cycles per second and applied to rapid movements such as an alternating electric current, wing movements or the ability of the eye to register flicker. Here are some examples: The drone's wing beat 180c.p.s. a locust 20, a worker in flight 240, noise made when stinging 300, the piping of a queen 320 – 340, while a mosquito rates 12,000 c.p.c. Hertz is now used for c.p.s. 300 Hertz being 300 c.p.s.

**Crab apple *Malus sylvestris* Rosaceae**

Hairless leaves and acid fruit, less common now than wilding orchard apples. Worked by honeybees for nectar and pollen which is pale yellow with a slight greenish tinge when packed in the corbiculae. It is useful to bees as it is found growing wild in the oak woods of southern England. Its pollen is like that of *Malus floribunda* - useful for the cross-pollination of many cropping cultivars. *(Gardeners note because hand pollination by these wild varieties can do miracles for a solitary tree in the garden).*

**craft**

Ingenuity or dexterity – a skill or a branch of skilled handwork or its professors. Whether pursued as a hobby or as a commercial enterprise or for educational purposes, beekeeping requires skill and is therefore described a craft. The official organ of the BBKA is called 'Bee Craft'.

**crafts and activities associated with beekeeping**

Batik, ceromancy, candle making, skep making, wax modelling, the making of polish, face or furniture cream, the ability to present a first-class pair of jars at Honey Show.
See: under appropriate 'headings' and uses of wax/honey

**cranberry *Vaccinium oxycoccus* Vacciniaceae**

The red acid fruit of the Ericaceae genus. Deciduous, shrubby and likes boggy ground. An Alaskan high bush cranberry *Viburnum edule* was reported as being worked by 'crawling' bees (foliage too thick to fly through) for twelve hours a day twixt May 25 and June 20 D. Tozier.
See: bilberry

**Crane Eva E, MSc PhD OBE**

As Director of IBRA 1948 – 1983 she became 'World Famous'. Under her far-seeing and brilliant motivation IBRA went from strength to strength. Some of her prestigious works are: 'A Book of Honey' 1975, and the 'Archaeology of Beekeeping'.

**cranesbill *Geranium pratense* Geraniaceae**

The common purple-flowered wild geranium found in thickets and damp places. The pollen grain is relatively large and reticulated and the nectaries are found at the base of the stamens. It is not the same as the scarlet 'bedding' geranium which is a Pelargonium and is largely ignored by bees probably on account of its odour though bees have been known to load their baskets with its bright red pollen.

**crate**

An open-work case but for bees it means a specially designed super that takes only 'sections' having 'T' piece slats, a 'follower' board and locking spring. Manufacturers list them as section 'racks'. The sections need to become fully sealed (weighing around one pound) and partly filled ones are not really

saleable. This means that unless there's a heavy flow alternative methods like 'cut-comb' are preferred.
See: cut-comb

**crawlers**

A bee that has lost its power to fly but can still use its legs for locomotion. When numbers of bees are found in this condition they are described as 'crawlers'. Some might attempt to climb grass blades or any suitable launching platform in an effort to become airborne. Diseases and conditions can reduce the ability of bees to fly. 'Crawling' has been described as a symptom of 'Acarine' and other complaints. Flight requires adequate function of all the tracheae in order that the thoracic muscles are well-oxygenated. Acarapis mite can effectively block the first pair of thoracic tracheae. It is possible of course for healthy bees to become chilled and moribund in and around the hive entrance but cold is likely to affect all muscles alike. When masses of bees are crawling – have a diagnosis made for disease as the earliest possible moment.

**crazy dance**

One of the dances performed by a returning forager intending to recruit other bees to work the same source of food. When the distance to the forage is short, as for example when a feeder is put right into the hive, bees do a very vigorous dance.

**cream of tartar**

A white powder formed of small crystals of bitartrate of potassium. A little will help to produce a fine, soft texture if added while boiling the concentrated syrup in the making of candy.

**creamed honey**

Pure granulated honey made smooth and soft by the application of warmth and whipping, stirring or vibrating which breaks crystals into a uniformly small size, As 'set' honey may be very stiff and reluctant to move, care must be taken not to overload electrical mixers by using them for this purpose. The inclusion of air bubbles will spoil the appearance of the product which should bring out all the honey's natural flavour and have a spreading texture like margarine.
See: Dyce

**Creighton Robert and Kathleen (Kent)**

Two remarkable beekeepers. A highly qualified team whose ceaseless work for the benefit of beekeepers was second to none. R. Creighton was President of BBKA and like Mrs Creighton sat on various BBKA committees. The latter was also Secretary of the Central Association of Beekeepers.

**Creosote**

In its pure form it is a colourless, oily fluid, distilled from wood-tar. Strongly antiseptic. In its crude form it is brown with a distinctive smell (although repellent to bees while fresh, that disappears when the wood has thoroughly dried out). At one time it was widely used to preserve beehives on the outside and even disinfecting them on insides. Such tarry substances are now looked upon with suspicion as they are thought to be carcinogenic. Derived from wood-tar they contain pyroligneus acid which prevents wood decay and was used to give a 'peaty' flavour to whisky.

**crescent dance**

This is one of the many dances – in most cases self-descriptive – that have been observed when returning foragers set about influencing other bees to work the food source that they had just returned from.

**cretaceous**

In geological time this was the last period of the Mesozoic era c.100 million years ago when during the period, angiosperms (seed-producing plants) became the dominant vegetation in Gondwana, in the western region of which the earliest bees may have arisen from the super-family Sphecoidea, probably from burrowing wasps.

See: geological time

**Crete Mediterranean**

Beekeeping was practised here as far back as the Minoan period (3,500 B.C.). Mythology says Zeus fed with nectar and milk by Melissa (King of Crete's daughter). Daedalus escaped with son Icarus using wings of wax. Bee of Mallia (fine golden jewel showing 2 bees holding pipe hive). Minoan coins show beekeeping. In Eastern Crete the Vraski hive was developed while in the wetter west, basket hives were set on a stand 'patichali'. Now 80,000 colonies exist, 10 per 3,000 beekeepers - 90% amateurs. Honey comes from thyme, marjoram, sage, citrus and pine and the warm climate lends itself to queen rearing and package bee production. The queens are very prolific.

**crimped wire foundation**

Wire with regular curves embedded electrically into foundation. Such wire grips the wax very well and is one of the factory-made variations in the field of foundation reinforcement which does the job efficiently, cheaply and well.
See: crimper

**crimper**

A hand-held device that will grip and ripple reinforcing wire, tensioning it and making ready for embedding the wire of the finished frame into a sheet of wax foundation.

**crimson clover *Trifolium incarnatum* Leguminosae**

An annual, flowering around June – July its honey being described by Whitehead as pale, greenish-yellow. Also called Italian clover.
See: clover

**crippled bees**

Subnormal temperatures during the pupal stage, heavy smoking (by beekeeper), overheating and parasitic interference (like Varroa) can all give rise to 'crippled' bees. Genetic defects, attributable to the queen can also give rise to unusual characteristics nearly always of a detrimental nature. 1981 Varroa mites are said to cause deformity in adult bees. An injured bee is unceremoniously cast out of the hive by its sister workers.
See: deformed bees

**crocus spp. *sativus, purpureus, nudiflorus* etc**

The spring variety blooms Feb. – Mar. and provides nectar and pollen, the latter being vital for brood rearing at that time of the year. Furthermore the bowl-shaped flower provides a shelter within which a microclimate favourable to the bee and suitable for nectar secretion is set up. Here a chilled bee might well recuperate and gather strength before returning to the hive. Pollen colours reported are: deep gold, orange/yellow, bright orange, light brown and stone colour.
See: autumn crocus

**crop**

1. The variety and mass of plants upon which a bee might successfully forage.
2. The honey take/surplus or yield.
3. The honey sac or stomach, an enlargement of the oesophagus, a transparent bag which can distend to fill a large part of the anterior end of the abdomen. Its capacity can be 100mg though 30mg are a more likely load.
See: crop fidelity

**crop fidelity**

Bees are remarkable creatures for the tenacity with which they will pursue a certain course of action and this applies to the singling out of a particular source of nectar or pollen (or perhaps honey in the case of robbing). As a rule they will have been stimulated to work a particular type and colour of blossom in a given area and will, if they are successful at foraging, try to recruit new foragers upon their return to the hive. A simple experiment can be set up with a vase of flowers of different colours (marigolds and corn flowers) and marking a bee it will be found always to go to the same colour. There are exceptions to every rule though and in rare instances, multi-coloured pollen loads have been observed.

**cross**

1. A hybrid or product of cross-fertilisation.
2. The description of the temper of bees. It is said with some truth that crossed bees can be cross.

See: cross-breed

**cross breed**

The first cross between two different strains, races or varieties. Heterozygous – a hybrid. In bee breeding this is done when combining the good characteristics of one type with the good of another, or to eliminate an unwanted characteristic. The associated heterosis or 'hybrid vigour' is a useful attribute in honey gathering colonies. It often follows when the daughter queens from a bought parent that seemed an expensive failure, seem very much better for mating with your bees. Temper may suffer though and swarming propensity too in the first cross though this may well improve when F1 and F2 are back-crossed with one of the original strains.

**cross combing**

Comb extending from one frame to another. Because of the use of wax strips or 'starters' which fail to trigger the building of combs under bars, the bees decide to build across the lines of top bars, thus rendering the set-up no more manageable than a fixed comb hive. The term 'warm-way' is best used to describe combs that are arranged across the entrance from side-to-side.

**cross-pollination Allogamy**

The fusion of male and female gametes (reproductive cells) from 'different' individuals of the species, as for example a pollen tube's nucleus reaches the nucleus of a plant's ovary or when the sperm of a drone unites with the egg nucleus. This is as distinct from self-fertilisation.

See: cross-pollination, self-pollination

**crossing-over**

During meiosis, homologous chromosomes can, due to the phenomenon of chiasmata, fail to achieve the normal linkage, with the result that genes (and therefore presumably characteristics) in the resulting progeny cause different results than in the parents. A genotype resulting from this is a 'crossover'.

**cross-pollination**

Also 'inter-varietal pollination. The transfer of pollen to the stigma of a plant of the same species though not from the same clone. Herbert Mace in "Bees Flowers and Fruit" gives compatibles for cherry, pears and plums. Lack of understanding of the significance of cross-pollination and the significance of the role of the honeybee in bringing this about, coupled with the inflow of foreign fruit where warmer temperatures gave much longer opportunities for the pollen transfer to occur have led to the gradual disappearance of large orchards in Kent and elsewhere.

See: apple types

**cross-section**

A thin slice of issue or other material, cut at right angles to the longer axis to show the characteristic detail. Cross-sections taken in three planes at right angles to one-another help to build up a 3-D image. The microtome is used for this purpose. A cross-section of a beehive could show the layered constituents; floorboard, brood chamber, excluder, super, crown board and roof.

See: microtome

**cross-sticks**

Two pieces of willow or hazel, stripped of bark and split, were pushed right through a skep at right angles to one another. Any projection was cut off. The reason for this was to give greater stability to the comb (and skep if required) at the centre. Modern skep users often forget this and wait until combs have become 'tough' before risking transportation or worse, have the combs shake loose on the journey. See: transportation of fixed comb hives

**cross-wind**

Airflow at right angles to any given direction. When a bee flies into wind its speed over the ground is reduced and vice versa its ground speed is higher when it flies down wind. A cross-wind causes the bee to head anything up to 45° away from the direction it aims to make good. When hives are arranged in a straight row there is a strong tendency for bees to drift down wind and when a prevalent wind exists hives will be found to have crops that vary along the line almost according to notes on a piano.

**cross-wire**

Frames with wax reinforced by wires running in two directions and crossing one another. Such wiring is not normally necessary or desirable as embedding is difficult so that both wires are covered with wax at the place where they cross. 'Crimping' and wave-wiring are more suitable methods where extra strength is required.

**crowding**

Techniques for producing section honey often involve deliberately 'crowding' the bees into a super (or crate) or for the purpose of queen cell raising. When there are abnormally large numbers of bees at the entrance, on the outside of the hive, or at a water or syrup supply we might use the word 'crowding'. Inside a hive it might be more appropriate to use the word 'congested' if there were too many bees. Where bees seem overcrowded it is often an indication that something unusual is imminent such as swarming. Excitement and crowding at the entrance could indicate disturbance, robbing or a returning queen. See: fanners

**crown board**

Inner cover or mat. A 'framed' and usually non-flexible cover that covers the uppermost frames allowing bee-space above them. Some are transparent, some ventilated and feed holes are often incorporated. These may be cut-out in such a way that Porter bee-escapes can be fitted (when the crown board can become a 'clearer board'). They are made from thin boarding, water-proof ply, oiled hardboard etc. and have the same shape and external dimensions as a cross-section of the hive so that the roof will pass over and so that they fit flush with the brood box or super. In NZ it was observed that the roof and crown board was sometimes a single unit. There are two good reasons for using a crown board. First the hive can be opened without the bees immediately flying forth and second the board itself may be propolised down but not the roof which would otherwise be difficult to get off. Narrow strips frame the board not only for added strength but to provide bee-space where necessary.
See: clearer board, division board, inner cover, mat, quilt

**Crowther**

A similar device to the 'Porter' bee-escape though it has more springs, providing eight ways of escape instead of two. It fits into a rimmed clearer board for use when it is required to get the bees out of a super. See: Porter

**Cruciferae**

The cabbage family with flowers on spikes, the four petals forming a cross. Honeys usually glucose dominant causing rapid granulation. It includes: alyssum, candytuft, charlock, dames's violet, rape, seakale, wall flower, wall-pepper , wall-rocket, woad and yellow rocket.

**crush**

To squeeze flat between two hard surfaces. Lots of bees suffer this fate especially at the hands of beginners who finding it impossible to hold the smoker while replacing a heavy chamber tend to lower it back into position when the bees are still swarming around the area between the two surfaces. Careful lowering with the top box at a slight angle to the lower one and gently taking the weight off and then twisting them back into line may guillotine the odd bee but minimises the risk. The use of a wedge can free the hands to use a smoker.

**crystal**

A small, transparent particle resembling glass. If of a chemical compound it will have a characteristic shape and structure. Sugar and honey crystals are typical. When crystals are found on the alighting board or in front of the entrance it can be indicative of unsuitable crystalline stores which the bees either lack the ability or the water to be able to use. Providing it is not unseasonal a feed of dilute syrup might well alleviate the trouble. In the case of crystallised honey which can be unsightly and hard to spoon out and in any case not to everyone's taste, the answer is not to clear it by warming (though a few seconds in the microwave will soon clear small samples) but to use physical energy rather than that of heat to rectify the situation. Don't risk breaking an electrical whisking device, nor breaking or bending a spoon or knife. Patience and strength is what is required. Dealing with a jar of 'tough' honey, use a strong fork. Scratch and test the surface and gradually dig and twist 'gingerly at first' until you have the surface moving. Then get bolder (if strength allows) and deliberately whip and stir with vigour, then even the hard unsightly honey will yield to produce a lovely cream with the texture of margarine. Ease of handling and improvement of flavour without any loss of important ingredients will result from this manner of dealing with the honey.
See: crystallised honey

**crystalline cone optic cone**

This is a transparent cone situated beneath the lens of the ommatidium and surrounded by pigment cells and it transmits light from the lens to the 'rhabdom'.

**Crystallised honey**

Honey that has granulated. Set honey. When this occurs in the comb the bees will often lick crystals dry and eject them from the entrance. To obtain honey from super combs that have granulated solid (as may happen with oil-seed rape) cut the comb out of the frames and melt in a honey tin or bucket kept at 60C/140F in a water bath. On cooling the layer of wax may be lifted off the now 'industrial grade (Bakers)' honey.
See: crystal

**crystallisation**

Should syrup be brought to too high a temperature, subsequent cooling by night in a feeder may cause crystallisation and this can block a feeder and create the impression that the bees do not want to take it down. In dry weather nectar and honeydew have been known to crystallise and ivy honey has solidified in the stomach of a bee.

**Crystal Palace South London 1854 – 1936**

A large edifice of glass and iron (built for the Great Exhibition 1851) )and the home of the first Honey Show until destroyed by fire. Shows were then held at Caxton Hall and various centres since. The first Great Exhibition of Bees their Produce, Hives and Bee Furniture was arranged until 1874 at the suggestion of the British Bee Journal which started in May 1873.

**cubital index**

The wing venation of hymenoptera is a useful guide to the demarcation of various species. Even within the races of honeybee there is enough difference to permit at least broad groups to be identified. The cubital index is a method of measuring a certain aspect of wing venation that enables certain conclusions to be drawn as to the possible origin of the race in question.
See: taxonomy

**Cuckoo bee Psithyrus, Aculeata**

Parasitic bees closely related to bumble bees. There are six species in GB. A female has

sharp mandibles and an aggressive array of spines and so is able to force an entry into an established nest of bumble bees when she eventually kills the existing queen and gets the workers to rear her eggs. She has no pollen baskets. Her eggs hatch into males and females and like bumbles the mated females overwinter to repeat the cycle in the spring or early summer.
See: Psithyrus

**cuckoo flower *Cardamine pratensis* Cruciferae**
Milkmaid or Lady's smock. Its soft pink or pale lilac flowers appear from Apr – June in the meadows when the cuckoo is first to be heard. Bees use it to collect both nectar and pollen.

**cucumber *Cucumis sativus* Cucurbitaceae**
Some varieties, usually ones grown in the open, need bees for pollination; others are ruined by bees (stung). Most present-day greenhouse varieties are in the latter category and painstaking steps are taken to exclude bees even one of which might spoil a great number of fruit if allowed inside. The honey is of a pale yellow to amber.

**culinary herbs**
Herbs of which use is made in the kitchen. Many of these are most attractive to the bees, when in flower, yielding pollen and nectar with characteristic flavours. Mint, sage and thyme are examples.
See: herbs

**cull**
To choose, pick or select. To separate and take away bad from good. The name given to the plant or animal so separated. Culling can be applied to the selection of colony characteristics and queens with a view to the improvement of stock. This might take the form of de-queening and uniting or the elimination, restriction or removal of unwanted drones or stock. Culling is easier than breeding and can be used as a means of stock improvement.

**cultivar**
A man-maintained plant variant (as opposed to a naturally occurring variety or sub-species) within a species which though differing slightly is less marked than a subspecies often found in the wild. A distinct group within a specie with more important distinctions than those attributable to a variety. A group made up from cultivated plants whose distinguishing features are maintained in cultivation. E.g. the apple 'Cox's Orange Pippin'.

**Cummning Rev. Dr.**
The 'Times' Bee Master. He invented hexagonal hives but promoted the Stewarton when he later came across them because they were cheaper.

**cup**
1. A queen cup or play-cell. These can be quite numerous once colonies have built up to the point where there are many drones on the wing. *(take note of this warning sign).* Their relationship to the actual possibility of swarming is somewhat obscure. Frequently they are found along the bottom edges of the combs but they are not considered to be worthy of serious attention until an egg has been placed within one – then swarming can be calculated almost to the day (egg to queen 15 days).
2. Silver cups are often displayed as a reward for wining first prize in certain classes at honey shows. Inflation caused them to become increasingly valuable and Secretaries are advised to up-date their insurance cover to compensate for possible loss.

**Cuprinol**
Six colours available all are recommended as suitable for beehives. Also Clear – a water repellent finish. Allow to dry thoroughly before use.
Not recommended: Woodworm killer and Five Star.

**currant Ribes spp. Grossulariaceae**
Black, white, red and flowering currants *R.nigrum*, *rubrum*, *albinum* are all worked by the bees and better black-currant trusses can be set when honeybee attentions add to the normal self-pollinating characteristics. Pollens reported as light to medium grey, and dull yellow, 20µ.

**custos apium**
Bee shepherd, used about the time of the Domesday Book and until the reformation. The 'custos' was a monk of some standing, e.g. the one in charge of apiary or of beekeepers.

**cut cutting**
To cut or lop off from the main body. Where colonies are kept in fixed-comb hives such as the jar hives in Malta, special knives with blades at right-angles to the handle are used to cut the honeycomb from the walls of the hive. The taking of honey is referred to as 'cutting' (we should add that it is usually accompanied by the onset of robbing).
See: uncapping

**cut-comb**
The popularity of pieces of comb drawn out by the bees themselves or from thin, unwired foundation or starters, has increased the use of small plastic containers with transparent tops to display the comb to full advantage. It is also used when honey is presented as 'chunk' honey. Honey comb from 'cogs' has been cut into pieces and sold as cut heather honey comb for many years.
See: cutting, cut-comb container

**cut-comb container**
Containers, usually of light plastic, capable of holding six to eight ounces of honeycomb chunks. A transparent, plastic lid of the 'snap-on' variety, gives a view of the contents. Name and address of the producer are to be displayed when sold on the open market. It is not wise to use comb honey that will granulate as this makes them unsightly and produces sales resistance. Heather honey in containers put straight into deep freeze were in perfect condition at least eighteen months later.

**cuticle**
The non-cellular, outer layer of the bee which is secreted by the epidermal layer and forms the hard, water-proof layer of the exoskeleton. It is largely chitinous. In the bee the waxy outer-layer is the epicuticle covering the hard exocuticle which has a soft under layer, the endocuticle.
See: chitin, metamorphosis

**cut-off**
To bring to a sudden end. This might refer to a honey flow brought to a halt by rain or the cutting of a crop of flowering plants (mustard). It also applies to the cessation of a flow as of water, honey or electricity. Special honey taps or 'gates' are made to ensure that with a quiet snap, the 'closed' position can be regained so that no post operative dripping ensues. Electric extractors usually have a speed control with a 'cut-off' or 'cut-out' position for closing down. Weather conditions such as landslides, floods or snow may lead to being cut-off from an apiary.
See: gate, clipping

**cutting**
1. When cutting pieces of comb as cut-comb, care should be taken that they just fill the container. A warm knife is not essential but ragged edges should avoided.
2. Items of news cut from newspapers and other sources. Collections of such cuttings of beekeeping interest can produce useful facts indicating the history, present state and future trends and possibilities.
3. A piece of a plant taken and planted with the intention of producing another complete and similar plant (clone). This kind of propagation is useful to spread bee forage e.g. willow.

See: cut-comb, division

**cyanides**

These give of deadly, poisonous acid fumes (HCN) when reacted with mineral acids or merely on being exposed to air. This is known as Prussic acid and is lethal to all forms of life including man. It is not only possible but very advisable to use alternative substances when wasp's nests or bees in dangerous places have to be destroyed. Such as: insecticides, petrol (or carbon tetrachloride to avoid fire hazards). If cyanides are handled at all they should be kept in well-closed containers and only used under strict supervision in well-ventilated conditions.
See: cyanogas

**cyanogas**
Used to destroy bees with apparently little harm to honey *if used as directed*! Alternatives: ETO and resmethrin

**cycle**
A round of operations or events. We have life cycle, seasonal cycle, flowering cycle and refer to regular events as 'coming round again'. While one National Honey Show is in progress the next one is already being planned. It is quite interesting and useful to keep a note of the flowering times of bee plants in your area.
The occurrence of things good and bad is described as cyclic: disease, swarming, bumper years
See: life-cycle

**cyclone**
Low atmospheric pressure (trough or low) is accompanied by anticlockwise winds in the northern hemisphere and winds and precipitation are a regular likelihood in the changing conditions associated with such a weather pattern.
See: anti-cyclone, pressure and weather

**cyclops one-eyed giants**
Occasionally bees are born with a fusion of the compound eyes at the top of the head. These are known as 'cyclopean' caused by a recessive gene which normally has a lethal effect. A drone's compound eyes join naturally at the top of its head.

***Cynoglossum officinale* spp.Boraginaceae**
Hound's tongue – it likes a sunny position in the garden. Many varieties attract bees from June to July, their tiny, deep-blue flowers can be a source of nectar.

**Cyprian bee *Apis mellifera cypria***
It resembles its Italian cousin through it is slightly smaller, brighter and has a more tapered abdomen. It has been described as somewhat transparent and yellow, bronze or carrot-coloured and the carrot colour can extend to over more than three segments. It is said to raise an excessive number of queencells, to be somewhat vicious or spiteful especially during a dearth, to be prone to developing laying workers but nevertheless to produce record honey crops being a hardy forager.

**Cyprus Eastern Mediterranean**
Beekeeping is carried out on a fairly large scale with averages of honey nearing 40kg largely from citrus, thyme and other Mediterranean shrubs. Although modern hives (Langstroths) are used cheap cement mock ups and ancient horizontal earthenwear jars can still be found. It is a fertile island and like other sub-tropical lands seems to be full of honeybee predators, hornets, swallows, death's heads and Varroa is widespread.
See: Cyprian

**cyser**
A delightful drink made from honey and apple juice. Both are capable of vigorous fermentation and they brew into a drink with a fair alcoholic content.

**cyst**
A spore-like cell with a resistant, protective wall. A pathogen in the resting stage which has secreted a tough wall around itself for protection until it finds itself in a sustainable environment for development. For example

the parasitic *Malpighamoeba mellifica* (amoeba) forms cysts which are spherical and can sometimes be seen in masses in the tubules of infected bees. Magnification x400 required.

**cytochrome**

A haemoprotein. The pink, thoracic muscles of the bee are rich in cytochrome which is their equivalent of haemoglobin. It is the substance that offers an efficient route for atmospheric oxygen to reach the indirect flight muscles and enter cell metabolism. See: flight

**Cytology**

The scientific study of plant and animal cells. Cyto – a word element referring to cells.

**Cytoplasm**

The living substance or protoplasm of a cell consists of cytoplasm. The nucleus is no part of the cytoplasm but together with it, forms the entire cell.

# D

***Daboecia cantabrica***
St. Daboecia's heath
Pink, white and silvery green varieties were given as good honey plants in Beekeeper's Annual 1984.

**Dadant & Sons Inc.**
Hamilton USA. Pioneers in the business of beekeeping supplies and presumably still going strong. Camille Pierre worked with his father Charles in authoring Langstroth's book 'The Hive and the Honey Bee' . Numerous editions followed and they took over the ABJ in 1912.

**Dadant hive**
See: modified Dadant

**Dade Major, Harry Arthur**
UK British Colonial Service, Asst. Director of Commonwealth Mycological Inst. and President Quekett Microscopical Club died July 1978. Author of the prestigious book 'The laboratory Diagnosis of Honeybee Diseases'1949.

**Dahlia spp. Compositae**
A showy, variously coloured plant which blooms from after the frosts well into the autumn. Its double flowers seem foreign to the bee but single varieties are worked for pollen which is described as golden yellow to orange.

**damage**
Damage to hives, bees and equipment can be caused in a number of ways from the natural elements of weather, fire, flood, gales, landslides and by accident, vandals and other living creatures. Ordinary wear and tear can also bring about damage which may or may not be repairable. Insurance can be arranged and is usually available through the Local Beekeepers Association.
See: repairs and insurance

**dampness**
Moisture, humidity – moist air, the effect of rain or snow. A honeybee colony is a living unit and it requires moisture and has to give off moisture in order to live. Both conditions of temperature and humidity are kept strictly under control inside their nests and it is one of the wonders of nature that despite huge variations in ambient conditions they still manage to keep their nests within the fairly narrow range of conditions providing they are not confronted with impossible tasks. Of the two, bees are more easily able to cope with temperature variations than permanent dampness. Hives should be water-proof and sites should be free of floods ad have good air drainage.
See: condensation and relative humidity

**Damson** *Prunus insititia* Rosaceae
The name of the fruit and the tree that bears it. A small, dark blue, tasty plum. It can be used with honey to make jam or a pleasant melomel. Pollen as for 'plum'.
See: melomel

**dance**
To move rhythmically with the feet and body. Since Von Frisch discovered that bees communicated the direction of sources of forage to one another within the hive by means of their antics on the comb

the matter has come in for considerable discussion. Several types of dance have been enumerated: alarm, breaking, cleaning, crazy, crescent, DVAV, figure-of-eight, joy, massage, planing, round, scrubbing, shaking, trembling, wag-tail, water-carrier, whirr, Zittertanze, etc. Each has been carefully described, the alarm for instance 'spirals or irregular zigzags with sideways abdominal wag.
See: Von Fritsch

**Dandelion** *Taraxacum officinale*, **Compositae**

Dent de lion Fr. Referring to the tooth-shaped leaves. Classed as a major honey plant despite being an unwanted, perennial weed. Its flower is worked at a lower temperature than the apple (with which it competes if the orchard sward is not kept clear) but the flower only opens fully in sunshine thus protecting itself from wet and inclement conditions from Mar-Sept. In the warmer South it may be found blooming in golden isolation all through the winter. The yellow honey has a strong flavour and is glucose dominant (granulates rapidly) with rather coarse crystals. The pollen 25-36µ, is reported variously as; deep golden yellow, orange, deep-orange, orange- yellow . It gladdens the heart of the beekeeper to watch the seemingly huge basket full carried so welcomely into their entrances during early spring
See: dandelion pollen

**dandelion pollen**

The greatest quantity is produced by the flowers between 9am and 3pm while the bees get most between 11 and 12 noon (Dr. Free). It is one of the more nutritious pollens and in delving into the flowers the foragers get smothered with it.

**danger from fungicides/herbicides**

Paraquat can harm brood. 2,4D (what does D represent here?) can often inhibit nectar secretion. Fungicides (and 'wetters) can spoil pollen and repel bees or cause them to be turned away from their own hives. Apart from the chemicals themselves the manner in which they are formulated may increase or lessen its danger to bees. Recent worries come from ultra-low-volume sprays and the micro-encapsulation when bees might pick it up as if it were pollen.
See: systemic

**dark comb**

Honey comb that has been used over and over again for brood can become quite black. It offers beekeepers a means of assessing how long a comb has been in use as brood comb. When cells are capped over or queen cells are developed on a comb , the dark colour of the comb is matched to that of the cappings. The first round of brood are much lighter than those of subsequent generations in the same comb and queen cells are darker when built on well-used comb.
See: travel stain

**dark honey**

In our 'shows' honey is divided into three classes, light, medium and dark. Official grading glasses enable judges to check that jars are in the right class. Dark honeys include: blackberry, cherry, hawthorn, heather, horse (broad) beans, plum and honeydew.
See: grading glasses, light and medium honeys

**darkness**

A new colony is said to need darkness to encourage it to start comb-building but once under way it will carry on although exposed to light. It has been queried as to whether bees can 'see' in the dark. Certainly their navigation has to be done by day but a moment's consideration tells us that no matter how dark the night a honeybee colony continues its housework as industriously as ever. An eclipse or dark cloud will send bees hurtling pell-mell back to their hives and a forthcoming shower has often been prognosticated in that way. Aggressive colonies (*A.adonsonii*) have been worked at night using a red torch light and this applies to dealing with wasp nests.

**dart**
The barbed portion of a bee's sting - the lancets - are known as the bee's dart. Also the river Dart which gave its name to Dartmoor Forest a great place for heather honey long exploited by the monks of Buckfast Abbey. You might refer to a bee's darting to the attack when enemies are near the hive.

**day**
The time between two successive nights. Unlike the weather the length of a day is entirely forecastable and it has its effect on the bee and the weather. Long days mean more daylight and possibly sunshine and thus plant growth is enhanced. In countries with a large range of latitude bees can be moved in stages northwards to benefit from the longer days during summer. Bees, though tight in the winter cluster, seem sensitive to the lengthening day and they stimulate the queen to start laying. Bees only fly by day when the sun is above the horizon.

**DCA**
Drone congregation area also called Drone assembly area'

**DD**
Disappearing disease – a good deal of work has been done on this 'condition' in the States. CDD the extra 'C' for colony is now described as a severe decline in colony population with no dead bees apparent.

**DDT Di-chloro-diphenyl trichlorethane**
A white, powdery, powerful insecticide developed in World War II. (an organochlorine). It acts as a stomach and contact poison, can cause cancer in animals and once applied is most persistent. Usually applied as a dust. Its use banned in the USA since1972 and only advised in GB when there is no effective alternative at low temperature. Dr. Roger Morse has said that contrary to public opinion it is not very toxic to bees. It has a long-term action and can enter man's food chain.
See: EPA, chlorinated hydrocarbons, phosphorous

**dead bees**
Volume about 2·75 per ml. The normal mortality of bees during the winter months often causes concern when after a spell of bad weather, when it has not been possible for the other bees to throw out the bodies, a heap of dead suddenly appears before the entrance. An average of one thousand bees may die each week.
See: expulsion, heap of

**dead brood**
Brood (eggs, sealed or unsealed) that has died through encountering a hostile environment such as chilling, over-heating, starvation or disease. When this happens the reason for it should always be acted upon to prevent recurrence and safeguard colony well-being.
See: chilled brood

**dead colony**
A colony in its hive or on its combs with all its members dead. A dead winter cluster is all too common a sight when bees have been allowed to go short of stores or harbour disease. Should such a colony be found adult bees may be used for diagnosis when nosema, acarine etc. may be found. However no chances should be taken, a match box full of the bees should be sent off for disease diagnosis, the entrance blocked after using formaldehyde to fumigate the interior or acetic acid if stores are still present.
See: nosema, taking a sample

**deadman floorboard**
A double length floorboard, a colony being placed on the front while a pile of supers to be licked clean is set on the back. A small bridge is required to make room for the edges of the two overlapping roofs.

**dead queen**
Before a queen dies in a winter cluster the bees will have made every effort to keep her alive to the very last so she is usually found

well inside the dead cluster. A dead queen continues to be attractive to workers for a considerable period after her death. When a dead queen is found outside the entrance it may be as a result of attempted requeening by the beekeeper or a failure on the part of a newly mated queen to return to her own colony. Where several dead virgins are found this usually indicates that the colony has recently swarmed or intended to do so.

**dearth**
When a dearth of nectar occurs due to the absence of forage, low temperature, strong winds, drought etc. there are many signs to enable the beekeeper to determine that this is the case. Temper of the bees can be short, robbing is constantly attempted and any sweet material soon found and water places are frequented weather permitting. Unusual sources such as buttercups and daisies may be worked. The June gap is one such annually recognizable dearth, though overlapping crops may disguise this in some localities.
See: flow, June gap

**death 'feigning'**
See: thanetosis

**deaths**
When the owner of the bees dies someone should go at midnight, tap at each hive three times and in a whisper to avoid giving them offence, tell that they must now work for their new master or mistress, otherwise they would leave their hives and never return. To quote from an old resolution 'The poor master's dead, your friend's gone, but you mun work for me… Bees, bees, bees… and the suggestion was repeated. If they weren't awakened by tapping they would pine away and perish during the ensuing year. See: funeral, superstitions

**Death's head hawkmoth**
***Acherontia atropos* Sphingidae**
Known as the 'bee-robber'; a large moth, common in warmer climates, that is capable of emitting a squeaking sound akin to the queen's piping which freezes the bees and enables it to make its escape. It has been said to station itself amongst the fanners and now bearing some of the hive bees' identity odour, ingratiate itself into the hive before using the aforementioned technique to get away. Bees in such areas build pillars of bee gum in the entrance to help prevent entry, (hence the word pro'polis) It is interesting to note that all DHH moth stages from larva to imago are able to squeak. They are found, looking almost unrecognizable, in entrance traps in warm places like Malta, Rodrigues, India. Once stuck, the bees tear them bit-by-bit removing any appendage or colourful adornment so that the transparent winged remnant is hard to identify. (Bumbles and mice can receive similar treatment. Unlike other hawk moths its proboscis is short and said to serve the insect well for piercing sealed honey caps and sucking up the honey.
See: propolis

**Debeauvoys** Fr.
In 1845 he discovered that bees would respect bee-space around a frame but it was not until 1851 that L. L. Langstroth developed the idea and put it into mass practical use.

**Deborah**
Hebrew word for honeybee (onomatopoeia?). A prophetess and judge of Israel. Bible - Judges 4,5.

**debris**
Fragments and rubbish that fall from the comb or are removed from the hive by the bees. This may accumulate considerably during winter months and calls for a sterile replacement floorboard as early as possible in the spring. The floorboard scrapings can be rolled in a sheet of newspaper for examination later. From the nature of the debris to be seen on the alighting board or ground in front of the entrance we may often deduce what is happening within the hive. Wax scales, dead workers, dried pollen, Chinese slippers, dead larvae or drones, dead virgins cappings, polished bumbles,

mummified mice and dead moths may all be seen at various times.

**decant**
To pour the clear liquid carefully off so that the 'lees'(sediment) is not disturbed or to pour from one container to another.
See: racking, syphon

**decay**
Loss of strength, wasting away, to rot or to become decomposed. This could be applied to a colony that weakens from disease or becomes a drone breeding colony, but is more appropriate for hive parts that have deteriorated due to the effects of use or weather. Western Red Cedar stands up to our climate well, most other suitable woods need a good deal of preparation and protection. Parts that might be in contact with the ground are best supported clear as even oak WBC legs will rot in time. Dry rot in the hive itself can cause long lugs to break at the frame ends.

**December**
In Britain we reach the shortest days and although severe weather usually awaits the New Year little activity is shown by colonies though when mild, some pollen may still be collected and cleansing flights will occur on mild days. As the queens will be resting there is unlikely to be much brood (if any at all) and only a few pounds of food will have been used thus far. Diseases such as nosema can cause unrest and higher cluster temperatures may cause unseasonal brooding. All apiary work for the year should be complete and an occasional check that hives are secure, especially roofs, is all that should be required. The mead that you made in October (?) should be ready for racking and last year's tasted and the previous year's enjoyed. Photographs of hives, sorting out your gear looking at how well you kept records, preparing talks, reading and writing, attending meetings. That the apiary and the hives are quiet is good. Beekeeping offers solace and this period of comatose is normal for the bees. Propolis is brittle and more easily chipped off now. Smoker, veils, overalls etc. can all be improved by repair, cleaning or servicing. There'll be snowdrops and hazel catkins soon.

**deciduous**
Tree or shrub whose leaves are retained for one growing season and then dropped for winter.

**decks**
Hives are sometimes described as having so many 'decks' e.g. 2 or 3 decks meaning that between the floor board and the inner cover there are 2/3 boxes.

**decoy hive** bait hive
A hive fitted with some comb and set out with a view to attracting a swarm. Experiments have shown that bees are more likely to occupy a hive if it is in isolation, arranged at a height of around 5m above the ground and has a capacity equal to about two Langstroth supers (one Dadant deep), National brood-and-a-half.

**de-drone**
The term de-queen is well-known and often use but it is necessary also to de-drone (that is to remove all drones) a colony when it must not be allowed to put its drones into the air as for example when nuclei are taken to stud areas for mating.
See: de-queen

**deep**
As the two main depths of frames tend to be used for every different hive type the word 'deep' refers to those items associated with the larger of the two sizes (and shallow for the other). A Langstroth deep and British shallow any beekeeper of experience would understand these terms.

**deep box**
A box that takes deep frames. Hive parts between the floor board and the crown board consist of brood chamber(s) and super(s). The brood combs are normal deep

so they would go into a deep box while the supers take shallows. The box of deep combs used for brood would be rather large and heavy to use for honey but in fact according to beekeepers' preferences either size of box can be used either as a brood chamber or for the storage of honey.
See: deep chamber

**deep chamber**
One that takes the larger comb and is most likely to be allotted to the queen and be surmounted by a queen excluder. Therefore this is where the queen lays her eggs and so it becomes the brood chamber. The depth being greater than that of the super combs implies that when full of honey it would be for most beekeepers be unnecessarily heavy. Another point to be taken into account is that between chambers there must be bee-space. To achieve this uniformly through the hive that space must be provided either above or else below the box. Care must be taken to see that each box uses the same type of spacing, e.g. either 'top' or 'bottom'. To mix these would result in either no space between boxes or otherwise double space which would lead to propolisation in one case and 'burr' comb joining boxes together when the bee-space is doubled.

**deep foundation**
Foundation is milled in long sheets that are cut to the size required and for deep frames, deep foundation is appropriate. Because of the possibility of 'sagging' wired foundation is normally used. Unless flow conditions are 'the order of the day' deep foundation is not always easy to get 'drawn out' to the bottom bar. In fact the best place to put it is in a chamber over another so that warmth rises and bees are much more likely to build the comb right down to the bottom bar.

**deep frame**
British standard deep frames are 336x220mm/14x8½" with a 432mm/17" top bar to give 38mm/1½" lugs Modified Commercial 406x254mm/16x10", Dadant 483x286mm 19x11¼", Langstroth 483x235mm/19 x 9¼".

**defecate**
To void excreta (faeces). This is normally carried out when bees are on the wing. Bees seem to have remarkably good control over this function, even for considerable periods during the winter, as they are able to retain and accumulate the faeces yet show no sign of dirtying their hives. However defecation can show up as dysenteric markings (brown or darker steaks) around the entrance if bees are under conditions of stress (such as a newly formed nucleus). Streaks and spots on washing hung out to dry are also possible when bees have settled on such clothing in order to collect water; - intake leads to output!
See: debris, dysentery, excrete, frass, faeces

**defence**
In nature bees chose sites high up in hollow trees as a defence against ants, weather and floods. The entrance, restricted with a propolis barrier in some cases, is guarded when required by potentially aggressive worker bees. Flash signals using 'alarm odour' pheromone, alert guards who may make a sweep search of the area around their nest and interrogate any moving object nearby. Buzzing, mock attacks or actually stinging, followed by attempt to bite, are part of their behaviour in endeavouring to discourage predators. *Apis florae* (the smallest of the honeybee types) are reported by Dr. Free, to form a protective curtain, shaking their bodies from side-to-side and producing a hissing sound from the shimmering mass.
See: winter defence

**defence pheromone**
The chemical – isopentyl acetate (don't use the wrong sort of nail varnish) enables bees to alert their colleagues to possible danger and there can be an immediate flash reaction which in fact only lasts for a limited period of time. (A foam blocked entrance and bash or two on the sides of a hive can cause bees to vent their anger on themselves) and a calmer period follows when they have

temporarily run out of steam. (don't try it unless you're a sadist). Noise, vibration, the opening of hives, fire, or floods may lead to the need for them to rapidly communicate with one another so that appropriate action can be put into motion.
See: alarm odour

**deformed bees**
Deformity amongst bees of the three castes can occur for a variety of reasons including genetic transfer of characteristics, disease or attack by predators (Varroa). Gynandromorphs, hermaphrodites, workers or queens with 'cloudy' wings, battered drones (especially at the time of drone evictions) are all examples.
See: crippled bees, Cyclops, gynandromorph, hermaphrodite

**dehiscence**
The bursting of fruits to liberate their seeds and also of pollen capsules in the flower's anthers in releasing their grains. When the pollen grains have matured inside the anther, the anther wall opens (usually when suitable weather conditions supervene) and the ripe pollen is discharged, this is dehiscence. Bees can physically nibble or scrabble around to cause this when the anthers are reaching maturity.

**dehumidifier**
A drying chamber in which unripe honey can be put so that it loses its moisture in a controlled fashion. It has a thermostatically controlled heat source (powered electrically) a humidity control and an adequate circulation of air may well be augmented by additional fans. Size varies, small ones being quite mobile. Settings are gradually increased starting at a mild level but progressing to a more severe level and the end of the operation. The larger the surface exposed the greater the drying rate

**dehydration**
The removal or elimination of water by heating, chemical absorption, reduction of pressure or exposure to a dry atmosphere. Silica gel, absolute alcohol, anhydrous acetone and n-butyl isopropryl alcohol are used. Some substances can be used as indicators, cobalt nitrate remaining blue when dry but becoming pink when moist. Absolute alcohol must be kept in a completely dry atmosphere.
See: dehumidifier, desiccate, honey drier

**Della Rocca L'Abbe**
Published 'Traite complet sur les abeilles' 1790. Invented hives with self-spacing top bars, side-opening doors and a super on top. He observed that bees gathered a gummy, sticky material from sunflowers. He was the Vicar-General on the island of Syros in the Aegean Sea.

**Demaree**
Named after D.W.Demaree USA 1884
A system of swarm control whereby a stock is developed until it occupies at least two brood boxes. A new brood box is placed on the floor board (after setting aside the existing boxes). It will already have been filled with disease-free empty combs except for one central space into which an active comb containing eggs, unsealed brood and the queen and adhering bees are put. An excluder is placed over this chamber and the hive reassembled having made good the empty space by moving over the frames and inserting a spare one on the flank of the box that the active comb came from. After 7 – 10 days any queen cells are destroyed and any combs without brood in the upper box are exchanged for young brood from below the excluder taking care to keep the queen below and not to separate frames of brood from one another.

**demise**
When a colony dies out it is important to try to establish the cause of its demise and not to jump to conclusions such as 'Oh, they've starved'. To leave material that is possibly contaminated for other bees to clean out can be a certain way of spreading disease pathogens.
See: carpet

**demonstration**
An explanation by way of a practical performance to show how a thing is done. A lecturer may be asked to demonstrate the Snelgrove method or the making or an artificial swarm. Demonstrations in country gardens are one of the delights of beekeeping on fine Saturday afternoons during the summer. Demonstrators should be proficient at the techniques they intend to explain and to ensure that the viewers can see and hear everything.

**De-natured sugar**
Sugar that has been rendered unfit for human consumption without (theoretically) altering its usefulness for feeding bees. The addition of fish-meal, octo acetyl sucrose, garlic powdered with vegetable charcoal, ferric oxide and sucrose octo-acetate (a bitter tasting, green compound) are some of the ingredients (chargeable to the beekeeper) added to subsidised sugar for feeding bees, Only humans could devise such systems for spoiling food stuffs. To give some idea of the efforts put in to ensure this spoilt food standard, one example is given. 0·250kg of ferric oxide (at least 50% $Fe_2O_3$) is powdered to fineness allowing 90% to pass through a 0·10mm (giving a colour dark red to brown) ... Intervention Board for Agriculture.

**Denmark**
In the 80's they imported 2,500 tonnes of honey annually. The Journal Tidsskift for Brovl was founded in 1866. An interesting beekeeping book appeared in 'Esperanto' the international language SADLY OUT OF DATE

**dense**
Thickly set or closely packed. This could apply to there being too many hives in a given region or the number of flowers on a tree. When nectar or honey are dense they have a high viscosity and density. Bees are slow and sometimes unwilling to take up very dense liquids.
See: density, number of colonies, viscosity

**densitaster**
A glass hydrometer graduated to read specific gravities between 1.220 and 1.520 at 30F. It not only indicated how thick the honey was but could be used for sampling the flavour.

**density**
The weight of a substance relative to its bulk. The mass per unit of volume. As the density of beeswax is less than that of water or honey, beeswax floats - Propolis sinks. Regarding honey, viscosity is sometimes confused with density. While the two properties are related viscosity falls with an increase of temperature whereas the density remains much the same. Plastics made to substitute for wood (in hives etc.) need to have a density roughly equivalent to that of wood. As water dissolves sugar its density increases.

**density of honey**
Density is mass per unit volume and figures are usually given as relative to that of water which has a density of 1.00 at a temperature of 20C/58F and a pressure of 1013mbs. The specific gravity of a good sample of honey is 1.4129. This will correspond to a moisture content of 18.6%. This also implies that one Imperial gallon of honey weighs 6.4kg/14.129lbs (approximately 11/2 times as heavy as water.
See: refractive index for the measurement of water content

**density of wax**
The density (or weight of a given volume) of wax varies with samples but its s.g. usually lies within the range 0.952 to 0.975. A comparison with other waxes (possibly adulterate wax) can be made by floating in water to which alcohol is gradually introduced until the wax sample sinks....
See: beeswax

**density of winter cluster**
3·3bees per cm2.
See: dead bees

**dentistry**
Blends of beeswax and microcrystalline wax with paraffin wax are used to form model gums prior to making the actual dentures.

**departure**
Setting out or leaving. Bees often fly straight into their hive entrance but nearly always hesitate momentarily on departure. When a queen goes out there is usually some evidence of excitement. The departure of a swarm from its nest is a stirring sight, with many more bees than usual swirling around and over the hive. When taking off from a temporary resting place the weighted branch slowly rises into its former position as more and more bees are uncovered by their erstwhile hive mates now eagerly scrambling skywards to reach their new home.

**deposition of nectar**
A 'house' bee depresses and raises its proboscis for 5 – 10 seconds with brief pauses, for about 20 minutes. She then enters an empty cell, ventral side uppermost and moving her head from side-to-side, paints the upper wall so that the nectar runs down. If nectar is already present she dips her mandibles in and adds her drop directly – Park

**Deposition of eggs**
See: egg-laying and oviposition

**de-queen**
While honeybee colonies usually have but one queen, to de-queen implies the destruction , elimination or removal of any queen or queens present. In other words to render the colony queenless. To do this you must either find the queen(s) or shake all bees from the hive and make them re-enter via a queen excluder.
See: re-queen

**dermis**
The innermost of the two layers of vertebrate skin. Hypodermic means to introduce beneath the skin. See: 'epidermis' in the case of the honeybee

**desensitization** immunotherapy
The protection of an individual against bee stings. It usually consists of a large number of injections starting with a minute amount and working up to point at which immunity has been reached. Freeze dried venom and venom protein, allergic extracts from honeybees and other stinging insects are now available in kit form with well described procedures from USA and elsewhere. Pure venom diluted in saline containing human serum, albumin - 15 subcutaneous injections of 1mg over 21 weeks, then at 4-6 week intervals. Rush treatment 17 injections are also possible.

**Deseret**
The former name of Utah, meaning honeybee.
See: Mormon, Utah

**desiccate**
To dry thoroughly. The removal or exhaustion of water.
See: dehydration

**desire**
In the process of 'queen raising' where starting and finishing colonies are to be maintained in a condition where they will continue to nourish queen cells, 'desire' can be created
1. by making them queenless
2 by giving bees the swarming impulse by overpopulating the colony. Strong desire enables the maximum number of cells to be accepted and to encourage optimum use of royal jelly.

**destruction**
To kill or destroy colonies found to have Foul Brood. This is normally done by Appointed Officers who burn the affected bees and comb and scorch other hive parts with a blow lamp. In the US and Oceania gaseous methods have been developed. Combs can also be destroyed by the ravages of the greater or lesser wax moth. Honey can be destroyed by fermentation.
See: ETO, fumigation, sterilisation

**detoxification**
The breaking down and neutralizing of toxic substances. In an insect's mid-gut substances which are active in detoxification are called microsomal mixed-function oxidases.
See: desensitisation

**Deutocerebrum**
That part of an insect's brain derived from the fused ganglia of the antennae. It is situated behind the protocerebrum and composed of bundles of nerve fibres. It is primarily concerned with information received by the antennae and controls their movement.
See: brain, protocerebrum, tritocerebrum

**deutonymph**
Final embryonic stage of the honeybee's parasitic mite *Varroa jacobsonii*. Deuts – those which are second in a series.

**device**
A man-made appliance, tool or implement. Beekeeping has called for ingenuity sometimes taking years to improve or develop an idea to its full and the historical build-up of items like smokers or movable frames of comb is often overlooked by those who come into the craft taking its impedimenta for granted. The word 'device' has occurred frequently in this work when the choice has been wide. For instance: apparatus, contrivance, equipment, gimmick, gear, gadget, improvisation. A device usually requires understanding and practice to make the best use of it – the smoker is a case in point. What suits one person or set of conditions may not suit another.

**devil's snuff**
Puff ball used for stupefying bees, H.H.
See: puff-ball

**dew**
It is deposited at night and is seen as a moist layer of small, clear droplets especially early in the morning. It covers vegetation and non-absorptive surfaces from which bees may collect water if conditions allow. You might see a metal roof top with a dried out circle in the middle caused by warmth from within the hive which would tell you that the roof insulation leaves something to be desired. Dew that has formed on plant surfaces that have been sprayed with pesticide could be harmful to bees.

**Dewey decimal classification**
Used in libraries to classify books in ten subject classes with further sub divisions by tens in these classes. Thus books on insects are found under 595 and bee culture 638 etc.

**Dew-point**
The temperature at which a given mass of air will begin to release its moisture. Or deposit a film of moisture (dew of condensation) on materials such as plants or glass which are at or below that temperature.

**Dextrin**
A gummy substance which is a polysaccharide carbohydrate formed as in intermediate product in the hydrolysis of starch to glucose. It has a high molecular weight, is non-sweetening, and considered to be of poor nutritional value. It is found in heather and honeydew honey possibly rendering them less suitable than other carbohydrates for winter stores.
See: starch, glucose

**dextrorotatory dextral**
Solutions of carbohydrates (like honey) possess the property of rotating the plane of polarized light passed through them. When this rotation is clockwise as viewed through the eye-piece of a polarimeter, the carbohydrate is called dextro-rotatory. Grape sugar, having this property is referred to as 'dextrose'.
See: glucose, laevulose

**dextrose** $C_6H_{12}O_6$
A simple sugar with a similar molecular structure laevulose. The two sugars form the bulk of the sugar content of honey. Also called glucose and grape sugar. The greater

its proportion in honey the more rapid the granulation It is less soluble than laevlulose or sucrose and less sweet. (when locked in glucose crystals the flavour of the honey is less sweet and only by intense stirring to 'cream' honey can this sweetness be fully released.

In honey the dextrose solution is usually super-saturated and the precipitation of its crystals increases the water content of the remaining liquid portion. This accounts for honey sometimes separating into two distinct layers. Glucose crystals are of glucose hydrate with a 9% water content while the water content of honey is 17 – 22%.

**dextrose/laevulose** comparison

Laevulose (Sometimes referred to as fruit sugar) is sweeter and less viscous than glucose which is more viscous, restores oxygen and is the first to granulate. An average of many honeys indicates a 39% laevulose to a 34% dextrose content.

**diagnosis**

The method adopted to ascertain the presence of a condition or disease from observed symptoms including the discovery of pathogens.

See: disease diagnosis

**dialects** local

Subdivision of a language. As the functioning of a colony depends on the response of one individual to another and this is termed 'communication', it has been suggested that incompatibilities can occur – as with the introduction of an alien queen – due to variations in the bee's ability to communicate, so that reading the word 'speech' for communicate, an unintelligible variation could be called a 'dialect'

See: compatibility

**diapause**

A natural rest or interruption in insect development even when they are in environmentally favourable conditions. A spontaneous state of dormancy during development or a period of quiescence during which the metabolism is greatly reduced as for example when a creature is hibernating or undergoing a dry spell etc. though it can also occur in environmentally favourable conditions.

**diaphragm**

A flexible sheet separating two compartments. A sheet of partly musculated tissue in the honeybee's abdomen. The dorsal and ventral diaphragms in the honeybee's abdomen pulsate so as to drive dorsal blood forwards and ventral blood rearwards.

See: circulatory system, dorsal and central diaphragms

**diastase** Amylase

A group of hydrolytic enzymes that split starch or glycogen into maltose. This useful substance is found in honey and as it is destroyed (or reduced) by heating, it is used to measure the quality of honey. See: enzyme

**diastole**

The phase during heartbeat when the chambers fill with blood.

See: systole

**diatomaceous earth**

A fine substance composed of the siliceous walls of diatoms and used in filtration as they have a high degree of surface area. Colloidal particles which tend to keep honeys from becoming bright are removed when passed through a cake of diatomaceous earth. There are about ten different grades available. Such filtration has been described as 'polishing' honey.

**Dichlorvos**

An organophosphorus insecticide. Dangerous to bees, and strongly absorbed by fats and waxes. A Vapona strip (containing Dichlorvos) may get into lipstick or beeswax or into foodstuffs. Bees may die when put onto foundation or into hives that have been contaminated by this chemical. In some parts of the world Vapona strips in Post Offices have 'put-paid' to queens travelling in cages. See: Vapona

**dichogamy**
The nature of plants whose male and female parts mature at different times so preventing self-pollination and encouraging cross-pollination. The value of hybrid vigour in bees has recently come to the fore as pure strains can lead to DNA problems.

**dicotyledon**
A member of the largest flowering class (Angiospermae) characterised by the presence of two seed leaves as in oak, elm, fruit trees, beans brassicas and others. See: monocotyledon

**Didymus**
Who wrote in 'The Geoponica' circa 350 A.D. He was a doctor and farmer of Alexandria. He refers to the 'Pleiades' (seven closely-grouped stars in Taurus) for the time of the year when harvests of honey might be expected.
See: Pleiades

**diffusion**
The process involved when one material moves through another from high to lower concentration. In animal or vegetable tissue the movement of substances through the semi-permeable membranes. The restoration of balance following osmotic pressure. Odours diffuse through the air. See: osmosis

**digestion**
The breaking down of food so that it may be absorbed and utilized by the organism. Food entering the bee's alimentary canal does not undergo digestive attack until it has passed the valve at the end of the honey stomach. Cells in the epithelial lining of the ventriculus then begin the process of digestion. The outer case of pollen grains may remain unchanged externally having yielded their contents via the micropylar membranes.

**digestive system** tract
A tube within a creature wherein extra-cellular digestion can take place. Food is taken in at the forward end of the alimentary canal and continue to the gut cavity into which enzymes are secreted. Here the more complicated foods (pollen in the case of the bee) are broken down. The regurgitation of nectar, with its enzymes, ensures that it will be pre-digested and the resultant honey contains simple food molecules that can be absorbed without further digestion by bee or man. From the region where digestion occurs, waste products can continue their journey along the canal to be discharged via the anus. See: alimentary canal

**Digges J.G.** The Rev
Wrote 'The Irish Bee Guide' 1904 which became 'The practical Bee Guide' Revised and reprinted at frequent intervals.

**digit**
A finger or toe or a small number.

**dilatant**
A property of some honeys which produces increased viscosity with increased shear (agitation) or stirring. This is the opposite to 'thixotropy'. The 'prickly pear' *Opuntia engelmanni* and some eucalyptus honeys have this property. It is also known as Spinnbarkeit or stringiness.
See: thixotropy

**dimensions**
Measurements in accepted units. These came in for changes in the seventies as Imperial measurements gave way to metric. The British Standards Institute had set out sizes and specifications for hives, frames and foundation. Conversion to metric was not possible by straight conversion of numbers as final figures had to be rounded off.
See: apertures, cell size, queen exc. slots, useful dimensions

**Dimethyl sulphoxide** DMSO
Used for topical application of bee venom fractions in 1979.

**Dimite**
Proprietary name for substance used for dealing with mites.

**dimorphism**
State of having two different shapes or forms as for example sexual dimorphism – a drone has about 20,000 antennal sensilla while the worker has about 6,500. Plants are descried similarly e.g. a variety having two different lengths of stamen.

**dimpled**
Having small, natural hollows. A queen cell is usually dimpled giving a much greater strength and surface area to the outside of the cell than that of a worker or drone. Smooth or stubby queen cells should not be trusted as it is differences of angle, aeration and size as well as feeding that are responsible for the difference that we see in female developed specially yet from the same egg as a worker.
See: queen cell

**Dines Arthur M.**
Awarded MBE in 1984. Author of many useful and well-thought out articles e.g. 'Honeybees from Close-up'. Assistant and then Editor of Bee Craft and President of BBKA.

**Diodorus Siculus**
Greek historian late 1st century BC who mentioned that in Roman law bees that were not enclosed in a hive were considered 'masterless' - Dr. F.M. Fraser.

**diploid males**
Normal drones are haploid but WOYKE (1967) demonstrated that the non-viable larvae of Apis were diploid males that could be reared to maturity if suitably protected but were normally eaten by workers within six hours of hatching. He went on to show that some drones were of bi-parental origin excluding parthenogenesis and androgenesis.

**Diploma in Apiculture**
Promoted in 1984 by University College, Cardiff. The one year course is primarily intended for graduate scientists and involves one year of full time study in the Bee Research Unit. 1st term: anatomy, physiology, behaviour, pathology, basic beekeeping, honey analysis, 2nd term: production of honey, wax, pollen, royal jelly, queen propagation, bee breeding, crop pollination. Visits are made to other Research centre, commercial honey farms, honey packers and fruit growers.
'Ndb'. Suggest Cut can't see any ref to apiculture on Cardiff website!

**dioecious**
A species that has male and female flowers on different plants. Hollies and willows ar examples.

**Dioscorides Pedanius**
A Greek medico AD 50 who prescribed the use of raw honey to heal cuts, two thousand years ago 'Materia Medica' Also mentioned medical uses of wax and propolis.

**Diptera** Arthropoda Insecta
Two-winged flies, some of which are occasionally confused with honeybees. Most have maggot larvae and the adults are stingless.
See: apimyiasis

**direct contact**
An agricultural chemical formulate to give maximum effect when actually brought into contact with a plant or creature, as opposed to systemics or the effects of vaporization.
See: contact insecticide, systemic, contact feeder

**direction finding**
Bees fly during daylight hours using polarised light from the sun as a compass grid. They can get back to their hive over short distances by reference to visual objects. Although they have been reported as flying by moonlight this is uncertain and by night they are attracted to any artificial light in the same way as moths, that is they fly in ever decreasing circles round any

bright source of light. A bee's compound eyes enable it to see polarised light patterns in the sky whether or not there is cloud cover. They also seem to be able to allow for the movement of the sun during the flight and while in the hive despite the fact that in the southern hemisphere it moves across the sky from right to left.

**direction of hive entrance**
The point of the compass to which the hive faces may be decided by considering a number of factors concerning bees' flight paths, effect of the early morning sun, amount of shade etc. South east is often suggested because when flight is first useful, that is an hour or so after daybreak. All directions have been compared and apart from the avoidance of direct cold winds (esp. in winter) there is no best direction. Shelter can prevent the harmful effects of winds but by facing hives in varying directions we can help to minimize drifting with its accompanying snags.

**Diphenhydramine Benadryl**
This is available as an antidote for bee-stings, in tablet form or in ointment.

**Diploblastic**
Having two distinct germ layers.

**diploid**
An organism with the chromosomes paired in the cell nucleus. Having twice the haploid number of chromosomes, from which meiosis can produce haploid gametes.
See: haploid

**direct wing muscles**
Muscles connected directly to the wing bases. In higher insects like the honeybee there are small direct muscles attached to some sclerites and to the inner surfaces of the pleurites and coxae but these only serve to alter wing pitch and to furl them. Power strokes come from the indirect flight muscles which are driven by neural rhythms (neurogenic rhythms) See: flight

**dirty**
Unclean. Any state that encourages robbers or germs. Honey, wax, honeycomb, the hive, around the entrance, a smoker – any of these things might be dirty. When honey has any foreign matter in it, be it a speck of wax or propolis or a tiny hair, it is considered as dirty. Likewise any particle of 'dross' or material other than beeswax would make that 'dirty'. 'Travel staining' of comb honey has been dealt with under that heading and dysenteric markings around an entrance suggest excitement or stress and need to be investigated. Regular cleaning of the fire chamber of smokers helps to keep them efficient. Glove and manipulation cloths should be cleaned and replaced regularly.

**disaccharide** $C_{12}H_{22}O_{11}$
A compound sugar which hydrolyses to form monosaccharides in honey thus producing a healthful, easily digested food. It is a carbohydrate with molecules too large to become digested by passing through the wall of the alimentary canal. Sucrose, one of the main sugar components of nectar is an example. Bees use the enzyme 'invertase' to transform the disaccharides of nectar into monosaccharides.
See: monosaccharide

**disadvantageous factors**
A food shortage while brooding is in progress can shorten the longevity of the bee. The use of too much smoke can cause bees' chitin to become brittle. Longevity and fecundity are usually in reverse ratio. Poor strains of bee, districts and weather can also be disadvantageous.

**disappearing disease**
A severe decline of colony population with no dead bees apparent. Carles Mraz opined it was becoming a serious problem (ABJ 1977) and due to a serious degeneration of breeding stock bees with a very short life-span were being produced. In Mexico hives containing honey and a small cluster of young bees that could be handled as if they were flies (no veil or smoker) akin to

paralysis with few dead yet scarcely any flying even during a honey flow.
See: DD

**Disc entrance**
A rotatable block or plate that can be used to give a variety of openings:
1. A variable sized opening.
2. Queen or drone excluder over entrance.
3. A ventilation screen over a closed entrance.
4. Entirely closed entrance.

These have been marketed made of plastic and special 'V' – shaped holes cut into the wall of the brood chamber to take them.

**discoloured brood**
Certain brood diseases and conditions can cause the healthy white brood to change from its normal colour and become creamy-yellow then brown. EFB should be suspected when this change of colour appears in uncapped brood, especially if the larvae are found in distorted positions in the cells. In the case of sealed brood where abnormal cappings call for an exploratory delving with a matchstick – dark brown, ropy contents should be considered as AFB (highly contagious) and immediate precautionary tactics adopted. Mandatory contact your local bee inspector.
See: coloured brood

**discoverer**
A person who finds out about something which was not previously known. The records show an impressive list of those who have contributed to our present day knowledge of bees and beekeeping. No attempt is made to list them here though many come under their respective headings elsewhere in this work.

**discovery of queen substance**
In 1961 a film was distributed by E.H.Thornes Ltd. See: Dr. Butler , 'Q – S'

**disease**
An unhealthy state or condition variously described as morbid, sickly, ill or such like terms. Bees are affected by certain pathogens many of which have the unusual ability to survive even in honey. For our convenience we divide diseases into two groups, those that affect the adult bees and those confined to the brood. Varroasis is a threat which unfortunately affects both brood and adult bees. In this, as in some other cases, the disease is named after the causative agent. The very mention of the word 'disease' is enough to put many a beginner off but rest assured, the greater your knowledge the less trouble they are likely to be.
See under the name of the disease.

**disinfectant**
A disinfectant or substance that destroys the causes of infection. A substance that cleanses by destroying pathogens and yet which can be washed away or allowed to evaporate so that no trace or smell remains. Many disinfectants such a TCP, Dettol and the once widely used carbolic acid, also act as repellents to the bees. Creosote although an excellent wood preservative (and disinfectant) is regarded as carcinogenic and although safe when absolutely dry, its strong smell is also repellent to bees. Strongly alkaline substances are used to remove wax and propolis together with the germs they may harbour. After scorching second-hand equipment wash thoroughly with 450g/1 lb. of washing soda, 225g/1/2lb bleaching powder to 4.5ltr/1 gallon of boiling water.

**dislodge**
By a sudden, unexpected movement cause either the bees on a comb or crownboard or a swarm on a branch to let go their hold and fall in the direction of the movement. The knack of not giving the bees any warning and the judgement of the correct amount of force is quite important. In the case of a swarm cluster on a branch particular care should be taken first to gently lift the branch and then to jerk it smartly downwards, preferably to strike the edge of a swarm box so that the bees as a whole flop into the box.

**dismantle**
To take apart. When a hive of bees is to be inspected certain precautions and a knowledge of how the hive was assembled, help to make the operation straight forward. According to the docility or otherwise of the bees, they should first be gently subjugated. With the amount of propolis used by most bees in the U.K. a hive tool will be essential. Gently lift off the roof, if flat, stand it upside down to receive hive chambers. Insinuate (horizontal movements – not up and down until an insertion has been made) the hive tool under the crown board so as to break the propolis seal then you can twist and apply some smoke without trapping bees. A wedge is useful to allow heavy chambers to be similarly freed, smoke applied and the gently lifted onto the upturned roof.
See: crush

**displacement crossing**
Systematic displacement crossing is used in order to remove a weakness (bad temper, susceptibility to disease and such like) from an otherwise good race 'A'. It is crossed with a race that does not display this characteristic and subsequent generations will be repeatedly back-crossed to race 'A' thereby improving the results and the inheritable traits not wanted from 'B' continually diminish. Care must be taken to ensure that the desired qualities remain preponderant.

**display**
To show ostentatiously, make visible, to give special prominence to. Insects, birds and other creatures often give a characteristic display to attract sexual partners. The honeybee queen relies on leaving a scent trail. We might display a 'honey-for-sale' notice and displays of equipment, hive products and the uses of hive products, might be put into a display at a honey show fete, fair, agricultural or beekeeping show.
See: demonstration, beekeeping/honey show

**disposition**
The mood or temper displayed over a period of time. Although some bees may remain docile at all times (when they are likely to be robbed-out) most can be aroused. Temporary arousal due to a circumstances such as interference, thundery weather or other unsettling influences, is acceptable, provided one knows when and why this is likely to occur and uses proper methods of subjugation. When a colony has a savage disposition – stings anyone who dares to approach the hive, then, in or near towns, it is socially just to move them away or requeen from a more docile strain.
See: temper

**disqualification**
The refusal to allow an article or person to qualify. Show committees usually reserve the right to refuse any exhibit or entry without having to give a reason. This power is also delegated to judge(s) though reasons are then normally stated. Disqualification might also follow if during a beekeeping examination a candidate failed to conform to the rules e.g. put his name and not his number on the paper.

**disrobe**
To undress but particularly – to take off your protective clothing. A beginner may be one of the first to take of his veil once a hive has been closed, the more experienced will only do this when at a safe distance from the hive and having checked there are no bees on his clothes. Many an over-cautious tyro has 'dressed himself up-to-the-hilt' to avoid any possibility of bees getting at his skin, only to have one crawl from under a fold and get him as he disrobes.

**dissecting microscope**
A prismatic microscope that does not reverse the image but magnifies sufficiently to enable internal and external dissections to be made (x 20). There has to be a working distance, space enough to mount and work with specimens, between the lens and the stage. Binocular patterns give useful stereoscopic vision. Dade in his 'The Anatomy and Dissection of the Honeybee'

gives useful guidance and suggestions for DIY and Owen Meyer 'On a shoestring'
See: compound, micro, microscopy

**dissecting pin**
For Acarine dissection a twin pronged pin is required. Basically two fine needles are fixed, side-by-side, so as to exactly pierce the thorax and hold it steady on a cork face for examination.
See: twin dissecting pin

**dissection**
The cutting apart of dead honeybees in order to examine the structure and relation of its parts is greatly assisted by having a suitable means of holding the body, good illumination, magnification and suitable instruments. A queen that has been replaced offers a splendid chance to see her magnificent array of ovaries. A honey jar cap is half-filled with beeswax and when firm a hole is melted in this. The body of the queen is then half-submerged and the cap filled with water. The abdomen can now be cut open like the bonnet of a car, using a fine pair of scissors. A seeker, scalpel and forceps are also useful. A wash-bottle with distilled water helps. Parts stick less under water and can be viewed with greater clarity.
See: mounting

**dissecting for acarine**
A bee is speared on its back by means of a two-pronged needle passed through its thorax at an angle that permits viewing. The head and forelegs are pushed off taking care not to mangle the adjacent breathing tubes. The bee should be canted towards the light and the tracheae can then be examined at around 15x . By lifting off the prothoracic collar a more extensive check can be made and if black or discoloured, acarine is almost certain. If needs be, the tracheae can be pulled right out and for examination at higher power put it onto a drop of water on a slide. Cover with a slip and examine at around 50x. If acarine mites are present, all stages (eggs, larvae and adults) will be seen within the tube. It is possible to look down into the tubes and at times to see live mites climb out of them.
See: acarine

**dissolve**
To uniformly disperse throughout a liquid, infinitesimally small particles of another substance. When this is possible the substances are 'miscible' otherwise they are only partially soluble or perhaps insoluble. The liquid medium is the 'solvent' and the substance that is dissolved in it, the 'solute'. Two solutions may be mixed if they are miscible sometimes an intermediary substance is used, for instance glycerine helps to dissolve carbolic acid in water. Sugar, like most substances, dissolves more rapidly in hot water. Saturation will be less at lower temperatures and sugar taken up when the solution is hot may crystallise out when the feeder gets cold at night. Wax, propolis, faeces, chitin and other bee substances will have their respective solvents.

**distal**
Describes the more distant or farther portions of an organ, being 'away from' as opposed to 'near the beginning'. A bee's claws for instance, are at the distal end of the basitarsi.

**distaste**
To dislike, to find repellent, or hate. Bees have sensitive organs of smell and taste and detect evil smells and unpleasant things yet we cannot always judge what their likes and dislikes will be. A pleasant, perfumed shampoo might put a manipulator at unnecessary risk if the perfume does not accord with their acceptance. Hive sites should be well away from stinks, incinerators and bonfires (which have the dual snag of being smelly and producing smoke). Knowledge of attractants 'swarm lure' and repellents (benzaldehyde) can be useful.

**distension**
Expansion by stretching. The structure of

the honeybee allows for certain organs to become distended. This is most noticeable in the case of a laying queen whose ovaries extend enormously and this, together with her larger thorax (allowing use of a queen excluder) and well spread legs make her relatively easy to differentiate from workers. The honey crop and rectum of a worker are also extendable though the effects of this are less obvious but can be seen when heavily laden foragers return to the hive or a bee becomes gorged with syrup.

**distinguish**
To recognise as different. Bee's ability to distinguish colours, shapes, smells and other bees are dealt with elsewhere. Some practice is required before a beekeeper can readily distinguish the queen amongst her workers. Distinguish also means sorting out the best – the best strain of bee for a district, the best apiary site, or when judging – the best exhibit. We have distinguished beekeepers, those who by service, skill and effort have achieved high reputations. Sounds and smells of the hive can be distinguished and give us useful information.

**district**
The region or locality. Knowledge of the beekeeping potential of your district can be enhanced by the use or formation of maps showing forage areas and times of blooming, unprofitable areas such as glass houses, large stretches of water, places that smell or where predators loom, the likelihood of neighbour problems or how good honey sales might be. Beekeepers in a district can work together through their local Association or Branch to benefit themselves, the craft and the public.

**diurnal rhythm**
Occurring daily. Cyclic repetitions that tend to be influenced by the day's length or whose repetition coincides with the daily rhythm might be called a diurnal rhythm. Hence the departure of foraging bees at set-times, to profit from flowers whose ability to yield nectar at certain times also following a diurnal rhythm, This is tied up with what we call the 'biological clock'. Bees develop 'habits' which can dominate their activities.
See: circadian

**diverticulum**
A tube or sac, for example the honeybees air sacs, blind at the distal end and branching off from a cavity or a canal.

**divider**
A person or thing that separates anything into parts. 1. Whenever honeycomb sections are being worked for, there is always the possibility that the comb faces will either become joined to adjacent comb faces or built irregularly rather than flat, thus detracting from their appearance. Wooden, metal or plastic sheets are used, shaped to allow bee-space between their vertical surface and the final face of the section. These give bees access but keep adjacent comb faces clear of each other. They are called 'dividers' 'separators' or 'fence separators'.
2. Division boards are used to confine bees to a particular chamber or part of a chamber.
See: division board, separator

**divides**
Or splits, an American term for making a nucleus with a view to rearing a queen or making increase by giving the divide a new queen.
See: division

**dividing colonies**
Separating a colony into two or more parts. This may be done to reduce the possibility of a colony throwing a swam or to make increase and/or to rear a new queen. Normal, healthy colonies can be separated almost randomly, provided that the brood nest is in two parts, then one will have the queen while the other part should include some eggs or very young brood. Sophisticated division calls for allowance for the loss of flying bees and the need to breed superior queens rather than have one raised poorly. Make sure that each division has

adequate food and bees.
See: splits

**division**
Used as a method of swarm control or prevention or for making increase by transferring part of a strong colony to another hive or placing it over a board with its own entrance so that two (or more) colonies are under the one roof. The part without a queen maybe left to rear one or be given another according to circumstances.
See: making increase, Snelgrove

**division board**
A vertical separator within a hive in the form of a narrow board with lugs and dimensions slightly larger than a frame so that it fits wall-to-wall and can be used as a 'follower' or means of closing a small number of frames together as when a nucleus is put into a standard sized brood chamber. Some hives are made to take such a board on the flank or flanks of the brood chamber so that its removal facilitates the lifting out of the first frame. Also used as a horizontal board to enable large colonies to be split, these are boards akin to crown boards but with entrances as for example the Snelgrove board.

**division of labour**
The integrated activities of creatures which for a time at least, are able and willing to perform a certain aspect of their corporate work. Honeybees have three castes each with a specific role to fulfil. The workers are the most versatile and generally speaking complete the work appropriate to the phase of life that they are passing through. Hence having emerged from the cells they are 'house' bees and proceed to clean cells, nurse brood, build comb, process food brought in, act as guards and fanners and then having had orientation flights they undertake foraging flights, swarming or the formation of a winter cluster. Adult worker duties may follow according to colony development and the time of year. An individual bee almost acts as if it were programmed, being very reluctant to alter its pattern of behaviour.

**DIY**
Do it yourself. Beekeeping calls for initiative, ingenuity and versatility. Although in a social community it is obviously right to utilise other people's skills to best advantage, the beekeeper is frequently forced into situations that call for the DIY approach.
See: carpentry, in-the-flat, knocked down

**D.M.P.** Di-methyl-phthalate
A strong solution of this substance was given as an alternative to Calvert's No.5 carbolic as a repellent – Deans

**DMSO** DOMOSO
A product - dimethyl sulphoxide was used to protect sperm cells so that they can be stored in liquid nitrogen. It is a base solvent for bee venom fractions and these are: PhospholipseA, Melittin and Apimine.
*Great care - it has cause irreversible ocular damage*

**DNA**
Deoxyribonucleic acid an essential compound which contains genetic codes. It is a long-chain molecule found in cell nuclei. Composed of four bases: guanine, cytosine , adenine and thymine.

**DN frames**
British Standard deep frames for National hives. DM Mod. Commercial DS – Smith

**docile**
Bees that are easily managed, hence the derivation 'docility'. Honey bees have stings and are willing to use them in defence of their colony. However many races or varieties of bees are far less inclined to sting than others. The breeding of docile bees has been very successful, though one experiment produced bees so docile that they were robbed out of existence by other bees. Young bees and colonies early in the year, tend to be docile. Of course the weather when they are manipulated and the manner of the

beekeeper also play their part.
See: temper

**dog**
Dogs often get stung, the nose being a frequent site. They get the same dose of venom that we would and having smaller bodies, should suffer more. This however rarely proves to be the case, though swollen paws and noses are not a rarity. Young dogs and puppies should be kept away from hives lest their playful nature gets them into trouble. Normally they are quick to learn and the expression 'once bitten twice shy' seems appropriate. Aerosol sprays of anti-histamine (e.g. sting relief) may be helpful.

**Dog star**
Sirius, the brightest star in the northern heavens. Significant in the ancient's celestial beekeeping calendar.

**dogwood Cornus spp. sanguinea Cornaceae**
There are many varieties of this yellow, flowered shrub which can attract bees from May – July. The pollen is yellow and caterpillars of the green hairstreak feed upon its leaves although its name comes from unfit even for dogs.

**domed cappings**
The cappings of drone brood are more domed than that of workers, which in turn have more convex coverings than cells of sealed stores. If drones are reared in worker cells the pronounced doming of the cells announces the fact that the queen is failing. Either she has become a drone breeder or a laying worker is present. The prominent cappings of these abnormal worker cells are very noticeable and may be rather scattered.

**Domesday book**
A survey of the Lands of England in record form made for William the Conqueror c.1086. ownership, extent and value of properties was given. Many entries make reference to bees and honey in those days when forest beekeeping was still commonplace.

**domesticate**
The meaning of this word lies in its origin of association with the home. As regards animals, it is used in the sense of controlling, taming or civilising so as to make them fond of the home (including the garden). To a certain extent bees can be domesticated, if only because they can be persuaded to return to their home when it is established in or by, a human habitation. However in the sense of taming or civilising the boot might be on the other foot – they might be able to teach us something!

**dominant**
A genetic characteristic that shows in the heterozygote. A character resultant from either one of two genes, say light as opposed to dark colouration. It was Mendel who first referred to dominant and recessive characteristics. When one character dominates the responsible gene has been successful but succeeding generations were outnumbered though 3 to 1 in its siblings.

**Doolittle G.M.** 1870
Father of modern Queen Rearing. In 1889 he wrote 'Scientific Queen Rearing' dedicated to Dr. S. Gallup his teacher. Founder of commercial Queen rearing and inventor of the mandril for making artificial queen cups.

**Dormant**
Temporary inactivity. A torpid state of a living organism when metabolism is minimised. This applies particularly to the honeybee colony during northern winters and is quite distinct form hibernation.
See: diapause

**dorsal diaphragm**
This is a thin membrane within the dorsal part of the abdomen of a bee. It supports the long heart and is itself supported by five pairs of fan-shaped fibres attached to the

anterior margins of the tergal plates. The diaphragm helps to form the pericardial cavity and to control the flow of blood into it. The rhythmic muscular pulsations propel blood forwards and in conjunction with the ventral diaphragm encourage a useful circulation of blood throughout the abdomen.
See: ventral diaphragm

**dorsal light reaction**
The bee's reaction to light received from above. This is 'read' by the compound eyes and in conjunction with light received by the 'ocelli' enable the bee to fly and maintain horizontal equilibrium. Dr. P.G. Mobbs tells us that a comparison by the bee of the light received allows it to make sophisticated pitch and roll movements. When the light from below is brighter (e.g. when snow has fallen) bees can get into serious trouble.

**dorsal plate**
Tergite, tergum or notum if undivided. Each segment of the abdomen is surmounted dorsally by a large back plate. These terga overlap and are connected by intersegmental membranes. They also overlap the sterna (ventral plates) at the sides and the first six of these bear lateral spiracles.

**dorsata**
Earlier name for the sub-species *Apis scutellata* – largest of the honeybees.

**dorso-ventral-abdominal-vibrations**
See: dances

**dorsal vessel**
A long slender tube that extends forwards along the centre line of the back, from abdominal segment VI. It is a blood vessel comprising pumping section or 'heart' in the abdomen and is continuous with the 'aorta' in the thorax carrying blood to just below the brain.

**double brood chamber**
When a queen is allowed access to two deep boxes this is referred to as using a 'double' brood chamber. The implications are that the queen is too prolific for her progeny to be confined either to such small frames or to a single box. Many techniques involving a queen excluder and the movement of brood such as Demaree, Snelgrove and Armitt, are possible when two equal-sized brood chambers are used. They also offer ample space for bees and for food storage with better wintering characteristics than is offered by a single small chamber.
See: brood-and-half

**double grafting**
Artificial queen cups are triggered off with an initial young larva and 24 hours later when a good bed of royal jelly surrounds it, it is removed and replaced by another carefully chosen young larva. The idea behind this extra work is to ensure that queens are raised from the very start on queen food – royal jelly.

**double hybrid**
Four dissimilar inbred lines are incorporated into one bee. The strain 'A' crossed with strain 'B' have their product mated with the product of strain 'C' crossed with strain 'D'. A resultant ABCD is thus obtained.

**double screen**
A double layered ventilation screen with a 'dead-space' in between – This should be about 18mm/3/4" and the screens separated in such a way as to prevent sagging. This enables two colonies to remain effectively separated while allowing warmth and colony odour to intermingle. Uses include helping a weak nucleus along, swarm control, requeening and queen storage.

**double-walled hives**
Serious beekeepers who want to maximize the use of timber used in hive making go in for single walled boxes. However more elaborate hives were in favour when adamant beekeepers in the 30's declared that only the WBC hive was suitable in this country. This has an outer cover composed of parts called 'lifts' while inside boxes taking normal

sized frames are kept so that there is a clear air-space (excellent insulation) between the inner boxes and the outer parts of the hive. A further architectural adornment was a picturesque gable roof and the lifts had sloping sides, said to be, though not in fact 'telescopic'. An Irish hive did go in for lifts that were actually 'telescopic' thus making a very heavy but 'warm' hive when assembled in the wintering mode. There were other advantages for this greater expenditure. They were useful for storing veil, smoker etc. when not used for bees. They insulate the inside from the sun's heat. Give good ventilation and keep the bees dry and if we go on, certainly ladies will be interested to learn, they are the lightest hives as far as weight is concerned.
WBC inners are well suited to in-house apiaries. Maybe for keeping bees up in the 'loft' too?
See: single-walled hives

**doubling**
HH said giving a 2nd brood chamber to a colony. Manley; adding a 2nd brood chamber full of good combs under the original one without an excluder. Mace: moving two colonies close together and then taking one away. Remaining hive benefits by extra foraging force and can be further reinforced with sealed brood from colony moved away.
Ratcliffe: adding a 2nd brood chamber on top the first to promote breeding (brooding) gives extra room and delays swarming. Such techniques may be used to produce a strong honey gathering colony to take advantage of a flow or if the beekeeper is very generous make nice colonies for pollination purposes. Sealed brood added to a colony being taken to the heather will increase the foraging force at a vital time.

**Downs**
Open, largely treeless, chalk hills in southern England. The nature of these when covered with clover and used for sheep grazing gave way to farming for barley and became more scrub-covered than here-to-fore especially after Myxomatosis put paid to much of the rabbit population. Where trees exist they include yew, sycamore and hawthorn and these together with clovers and other calcicoles are the native melliferous plants to which we should also add blackberry. Domesticated animals browse on them and in the exposed conditions close turf is formed and carpets of chalk-loving plans may offer forage for bees during summers that are not too dry.

**dragonfly Odonata Insecta Arthropoda**
A carnivorous winged insect whose beauty probably outweighs any mischief it causes by preying on honeybee water gatherers. They catch insects on the wing, arranging their legs in the form of a basket in order to do so. Damage to bees is local and seasonal. Larger species can be troublesome where they exist in large numbers near queen rearing apiaries especially if short of food but are always to be found near their ponds, rivers or lakes. Queens, drones and worker honeybee parts have been found scattered over the ground in apiaries. They have been reported to cause great bee losses in parts of Russia.

**drain**
To allow liquid to separate from solid material under the influence of gravity. Honeycomb is crudely reduced to honey and wax by pounding and then hanging up in a muslin bag until the dripping has ceased. More sophisticated cloths and meshes of metal or plastic are used to-day and uncapping trays are sloped to allow the honey to run away from the wax. Comb, even when cut-to-pieces, still traps a great deal of honey (about 50% by weight) and honey farmers may use heat or centrifuges to reclaim as much as possible.
See: cappings, run honey

**drainage**
The character of being able to allow either air or water to flow freely away. Sites for apiaries should ideally have good air and water drainage as neither moisture laden air nor wet ground (or floods) are good for bees.

A hollow may become boggy in wet weather or a frost bowl in cold conditions. A gentle south-facing (N. hemisphere) slope is good but steep slopes or exposed plateaux may have drainage but present other problems. Raising hives well off the ground helps drainage. Where floods are likely a firmly anchored raft might be a safeguard.

**drained honey**
The Honey Regulations of 1976 defined this as: honey obtained by draining uncapped broodless honeycomb.

**draw**
1. We refer to drawn comb, saying that the bees have 'drawn' the foundation out. Built, pulled-out, extended or developed would also be suitable alternatives.
2. The name of a means of raising money for good causes when numbered tickets are sold and at a set time a 'draw' is made to decide winners.
3. Draw in the sense of 'attract' has many beekeeping connotations. A caged queen can be used to draw loose bees together in a honey house. And a virgin in flight can draw drones towards her.

**drawback**
A disadvantage associated with what might otherwise be an advantageous situation. For example docile bees may prove relatively defenceless against their enemies, especially other bees in the vicinity. A slow start in the spring can be a drawback. Lack of the right equipment etc.

**drawn comb**
Comb built by the bees on wax sheets (foundation) that were put into frames by the beekeeper. The development of the relatively thick sheet of wax is referred to as drawing out or pulling out by the comb builders and the various stages are referred to as: just started, partly drawn, fully drawn and fully drawn and sealed. By holding a sheet to the light it can be seen that the bees do actually use the wax from the septum to help in extending the walls of the cells. A standard cell size, either worker or drone, is normal but for some races and in experimental conditions different sizes are sometimes available. Bees can be trained to adapt to slightly larger or smaller cell sizes. We should never assume that foundation is so marvellous that merely giving it to the bees will result in their drawing it out. The window of opportunity for getting foundation drawn out is not all that long in a season and much foundation is wasted by not being given to the bees at times when it is propitious for them to get to work on it. Either during a honey-flow, or by feeding syrup and by being put into a warm part of the hive (usually over the warmth of the brood nest).

**draw quilt**
Another name for a 'cover-cloth' which can be in one or two pieces. Flexible bee-compatible material weighted with stick at each side (against puffs of wind). Rolled or 'drawn' across the frames it is possible to retain colony heat and protect against attention from robbers while minimising the amount of smoke needed. They prevent light and air from disturbing the stock. They help to keep the colony under control. Honey and propolis will mark them so clean or replace regularly and remember that they can transport pathogens.

**Drenamist**
A proprietary preparation where adrenaline is supplied in an aerosol form for direct aural application. 1 pt in 1000 on doctor's prescription only.
See: isoprenaline, Medihaler –Epi

**dress**
When opening bee hives avoid wearing dark, rough, loose clothing (features common to their ancient enemy –the bear). Light weight, light coloured, smooth, bee-proof garments are called for. These can be provided by one-piece outfits though many will be content to use gum boots, a white zipped boiler suit, veil with suitable hat and if required – gloves with gauntlets.

Clothing should be free of odour. If you are a newcomer, get someone to check that your veil has black side to the front and is secure. Have spectacles, even a handkerchief inside. Pockets can contain matches (the box can come in useful) penknife, and hive tool if there's a suitable pocket.
See protective clothing

**dried honey**
Attempts to produce a dehydrate form of honey seem to have been fairly successful in the 80's. Heating honey in a vacuum for ten seconds at 115.5C/240F and then cooling has led to the formation of a brittle sheet. American firms have put this product on the market with and without other additives.

**drift**
The angle between an aircraft's heading and the track it makes good. As bees zigzag along, nevertheless holding an average heading, drift is the amount that they must allow for to get to their Von Frisch assumed target. Because of their slow flying speed relative to the wind, drift could be as much as 45° in many cases. Their return to their actual hive might be endangered in a cross wind especially if the hives are in a straight line. This is dealt with under 'drifting"

**drifting**
The entry of returning bees to the wrong hive. When hives are in a level, straight line and of similar colours, a cross wind might well be responsible for bees getting back to the wrong colony. With a load of food they are likely to be accepted even after initial molestation. This can, in some cases, lead to the hives finishing with honey crops that vary like the notes on a piano. Each one a little heavier than the next. Drones are very likely to drift and in any case are usually accepted. In such fashion diseases (especially Varroa) can be transported. Despite the pleasure of seeing a neat, orderly arrangement of hives, all at the same level, it is wiser to put them in a fairly erratic line and to vary at least the colour of their entrances to minimise 'drifting'. This is important when queens are returning from mating flights.

**drinking fountain**
A natural drinking place or the provision of a place where bees can find and use a continuous supply of water. Best in a sunny position and sheltered from winds. If this is a pond or similar water supply it is wise to set it up with 'steps' for the water collectors, in the form of moss, or Hessian – anything that gives their feet a good grip. Such 'fountains' can be baited with a little syrup to trigger them off, but open-air feeding always attracts robbers with consequent disease possibilities.
See: drinking trough, water, fountain

**drinking trough**
A gutter-shaped container in which water can be supplied. Water fountains as we usually term an artificial water supply for the bees, are commoner in the U.K. than such troughs which were widespread in the days when horses were a main means of transport. For bees, a wooden sloping side gives them a good chance of a grip without fear of a ripple sucking them in. They cannot swim and only rotate in circles until they can catch hold of something to climb out onto. Pieces of floating material may help. Where farm animals use such troughs stings could be a problem and one way to avoid this is to provide a 'baited alternative'
See drinking fountain

**drinking water**
Temperature and mineral content are important when a bee comes to collect water. Clean water is not particularly appealing to them, they have even been seen collecting from a small pool on top of a 'cow-pat' but that may have been due to the temperature! It is not easy to get them to take water in the hive so dilute syrup presents an opportunity for this and because they are 90% water just as we are, they need water every bit as much as they do nectar or pollen. Spring feeding with two of water to one of sugar stimulates brood rearing

(chilled brood will result if you do this too soon and a cold spell supervenes) and pollen collection. (Useful when bees are required for pollination though doubtful whether any supplier of hives would go to the orchards with sugar syrup!) Generally speaking unless inadequate feeding in the previous autumn has left them short, then leave well alone because toying with the fickle weather conditions is better done by the bees than by you.
See: drinking trough, water, water fountain

**drip**
You can drip feed with a slow feeder. You can drip prophylactics onto the bees. You must be careful not to drip syrup or honey around the hive as this would encourage robbing by bees and by ants. Nectar and especially honey-dew can drip from the trees at times (limes). Condensation within the hive can drip onto the bees as a result from moisture arising from the cluster and being allowed to impinge on the cold underside of a roof. Insulation and ventilation helps here.

**drip tray**
A honey-proof tray of plastic or metal upon which supers can stand either in transit or in the honey house. Metal roof coverings can also be brought to play in such circumstances as a super will fit neatly into such a tray.

**dripped honey**
Before the advent of machines for cleaning and extracting honey from the comb, the best quality liquid honey was obtained by allowing the mashed comb to drip through a filter of material. Dripped honey was then – the top quality honey.

**driven bees drummed bees, turnouts**
Bees that have been driven out of their home by the deliberate action of a beekeeper who wishes to assemble a quantity of bees. Originally, driving took place when skeppists wished to get their honey out of their skeps without sulphuring or killing the bees. It was best done in the evening to avoid robbing. Drivers would up-turn the doomed skep which was stabilised by standing it in a weighted bucket at table height. Driving irons (extra-large staples) were used on each side to fasten the lips of the two skeps together and a cloth went round the back between them Canted at an angle of 45° the hands were pummelled against the walls of the lower skep, the bees and their queen ran up into the empty one above. Standing with backs to the light, they could spot the queen more easily. When several colonies were amalgamated queens were allowed to fight it out. Strong colonies were sometimes driven with their queen to form a nucleus. Now-a-days package bees or 'shook' swarms are taken by shaking or jolting bees off their combs as driving bees is unnecessary now that fixed combs have given way to movable frames. Sent by air in gauze cages thousands of colonies are flown to the USA from places like Australia to make up for the losses due to CCD which have had serious effects in the regions where almond, cranberry and other fruits depended on them.
See: driving

**drone**
A humming sound. The sound of a drone bee is quite different from that of a worker or queen. It is one of the three castes – the 'male' bee and is normally 'haploid' takes 24 days from egg to imago and is reared in extra-large cells. After emerging (sometimes with the help of workers) it requires five days to reach maturity but its effectiveness as a male, which depends on regular flights, only lasts for a mere 40 days. It has no sting and its body functions and structure are associated with seeking nubile queens and copulation usually occurs in a 'drone congregation area' some 30m above the ground. It dies once coition has taken place, or soon afterwards having broken away from the airborne queen minus much of its genitalia. Its gene compliment is all important to the queen's subsequent progeny. Colonies produce several hundred

in the summer and few are left by autumn when a wholesale slaughter, which is often spontaneous, takes place.

**drone anatomy**
The drone is approximately twice as large as a worker. Its antennae are thicker and longer, having 13 segments compared with the female's 12. Its huge compound eyes meet in the centre and the ocelli (three simple eyes) are forced down onto the 'frons'. Its large muscular abdomen incorporates an elaborate apparatus which comes into play at maturity at the moment of copulation with a nubile queen.
See: drone assembly, genitalia

**drone assemblies**
Areas where drones assemble and cavort in readiness for mating. These tend to occur in the same places year after year influenced by a combination of atmospheric and topographical features. Studies have shown that they can be quite close to an apiary when the air happens to be relatively warm and hill or vortex thermals offer rising air which assists the copulating pair. Heights of between 30 and 50m allow the couple, whose flying capability is impaired during copulation, to lose height while still well clear of the ground. The locations where this occurs can often be 'heard' but rarely seen though drones may follow small objects like stones, thrown up towards them. The hours between 11.00 and 17.00 are favoured when winds are light.
See: drone congregation, copulation

**drone behaviour**
Drones normally appear some six to eight weeks after brood rearing begins in the early part of the year. They fly about a week after emerging. From this time on it is wise to make weekly queen cell checks. They take two weeks to mature (remember this when rearing new queens) fly in bright weather between noon and four, especially in light winds, covering two miles for exercise and to locate drone assembly areas. In the hive they are usually found on the outside of brood clusters. They generally return to their own hive but are attracted to queenless stocks and accepted when they drift to a hive that tolerates drones. They are driven out to die when there are food shortages or late in the season though some may be retained by queenless stocks. They are quite defenceless. Their late retention when other stocks have demolished their drones is a clue that such a stock might be 'queenless. Their tongues are too short to easily reach food and workers take it upon themselves to keep them nourished (or not).

**drone breeder**
A queen that has become a 'drone breeder' is one incapable of laying fertile eggs. A virgin queen may become a drone breeder if she fails to mate and a mated queen may fail due to lack of sperm (poor mating) old age or disease. A colony suffers a severe set-back if its queen becomes a drone breeder though recovery is possible during the active season, if a mated replacement queen can be successfully introduced in the failed queen's place. A colony with such a drone breeding queen at the end of winter can only be saved by the beekeeper's intervention and provided it is healthy it is best united to a queenright lot. A drone breeding queen from a good strain may be useful in a breeding apiary as her drones if properly reared will be normal.

**drone brood**
The larger drone cells containing drone brood. Bees always contrive to rear drones once the worker population has reached a certain level. However excessive quantities may occur when a queen gets into a super of drone comb or when she becomes a drone breeder or when large amounts of drone comb a present in the brood chamber. As well as being of a larger size (4 to 25mm/1" as opposed to 5 in the case of worker cells), sealed drone brood has more highly domed cappings than worker brood. When reared in worker cells due to laying workers or a failing queen the cells are highly domed, usually scattered and dwarf drones may result from them.

**drone cells**

Ranging from 6.2 - 7.2mm these are larger than worker cells four taking the space of five of the latter. They are used to store honey (easier to extract from the larger cells) and for the rearing of drones. The cells are more highly 'domed' and can easily be distinguished from worker cells. Although by using worker base foundation beekeepers try to minimise the amount of drone brood, the bees will make their own decisions tearing down and rebuilding the comb when the colony feels it is right to do so. The queen undoubtedly finds them attractive and if she escapes into a super full of drone comb she just lays and lays and lays. (Chickens happily gobble-up discarded drone combs). Possibly it is a relief for her to be able to lay in those larger cells whereas worker cells are a tight fit for her and possibly squeeze her into remembering to use her sperm pump and fertilise the egg.

**drone congregation area**

Although country folk have often heard the sound of buzzing overhead and beekeepers have looked as if expecting to see a swarm, it has only been in comparatively recent times that serious investigation has been possible. Using tethered balloons, airborne cameras and binoculars we now know that when conditions conspire to make a given spot suitable, drones will find it, and crowds from different colonies will fly up to that height (30+m) and circle around, forever watchful with their sensitive antennae and extra-large eyes, ready to plunge after any virgin queen (or anything that looks like one) and attempt to clutch hold of her for the purpose of coition. Rising air in otherwise calm conditions provide just the right conditions so that the unwieldy pair can do what nature intended and even though as the drone falls back (having virtually exploded) and the queen presumably having lost ability to fly properly, struggles loose without dropping too far and losing too much height then to the surprise of the beekeepers of old, goes on to mate with several drones.
See: mating

**drone eggs**

Whereas all the eggs that the queen lays look identical, most will have been rendered fertile by the action of the queen who acting like a hermaphrodite puts the stored drone's sperm onto them. She may be prompted to do this by the tightness of a worker cell. However a virgin queen or laying worker can also lay eggs without any possibility of being fertilized. If a fertile queen choses to do so it will be due to her inborn desire to lay in drone sized cells whenever the workers will allow this. It comes as a surprise to many people to learn that it is the queen who is 'bossed about' and becomes an egg laying slave totally dependent on her workers. The fact that drone eggs are otherwise similar to female eggs has been demonstrated by the hand fertilisation of drone eggs and so producing female progeny from them.

**drone genitalia**

Two large testes composed of sperm-forming tubules, pass their contents, on maturity, through the vasa deferentia to seminal vesicles. Here the sperm cling to the inner surface by their heads, lying in a lymph-like secretion. Mucus glands, joined by a 'U'-shaped tube where the seminal vesicles also join, become larger as the drone matures. An ejaculatory duct connects with the 'U' and permits the passage of spermatozoa at the moment of copulation.
See: drone anatomy

**drone laying queen**

A queen that only lays drone (unfertilised) eggs. This can happen at the beginning of her mature adult life through mismating or failure to mate or at the end of her days when her spermathecal contents have been used up or are no longer virile.

**drone's life cycle**

The male egg that develops into a drone is identical to any other honeybee egg except that it remains unfertilized and is haploid. It hatches after 3 days incubation, is fed

for 6 days in its drone sized cell and is then capped over. Metamorphosis takes 24 days from egg to emergence the 5th moult occurring 6 days after capping, The adult does not mature until the 38th day when it can fly and copulate with a queen. See ' drone mating' for details of that interesting affair. Success is concurrent with death. Otherwise in due course it will be killed or ejected. It can drift into other colonies as drones, despite having an alien colony odour, are usually accepted. Their life span is around eight weeks and they are dependent on the workers for food. They have short tongues and no means of defence. A food shortage in the colony might cause premature rejection otherwise there is a seasonal almost spontaneous 'turfing out' when the mating season is over and no mercy is shown.
See: drone slaughter. haploid, drone mating

**drone mating**
The drone having matured and exercised, (10 – 55 min flights at 1 – 2 weeks) flying at heights well above the foraging bees, will have located areas where other drones patrol in anticipation of finding a virgin honeybee. She will have gone through a shorter, similar process and being nubile will leave a scent trail behind her and will soon be set upon by a swarm (sometimes described as a 'comet') of drones. The drone is built for this particular purpose and uses its thicker, longer antennae to detect the whereabouts of a queen and with exceptionally large eyes, the tiny dot of a queen will soon loom up as he follows the ever increasing scent gradient (molecules of pheromone) that points the way. With special 'claspers' he will approach the queen from the rear and with his air sacs fully inflated explode essential parts of his genitalia into her. The two benefiting from thermal up-currents, otherwise flying, and perhaps falling, haphazardly as they disconnect, the queen to go on and repeat a similar match, while the drone falls uselessly to his death.
See: queen mating

**drone slaughter**
Soon after the time of year when ants swarm, heaps of dead drones are found in front of the hives. No connection exists between these to phenomena other than the time of year when the mating of flying insects is almost over. It is just a fact of bee-life that drones are only supported in colonies during the season when they are potentially useful They first appear in the spring from April on and they may be thrown out should a new queen be successfully mated or sometimes when there is a dearth of nectar following a flow.

**drone trap**
A device, usually slotted, (as for instance queen excluder slots), that will allow the passage of workers but not drones. The hive can be lifted onto a rigid excluder that sits on the floorboard and thus keeps drones and queens in the hive while workers are free to fly. Their tempers and the well-being of the colony suffers though. This method can be used to trap drones outside their hive if arranged while drones are out flying or before flying starts to confine drones to a hive if for example they are to be excluded from a 'stud' area. Purpose-built traps that may fit over the entrance are available but even when colonies seem to have a surplus of drones there is usually little point in trying to reduce their numbers by trapping. The bees themselves are able to expel drones when required and only when a drone breeding queen needs replacing should a beekeeper intervene by requeening.

**drone zone**
An area where drones may congregate in the air for the purpose of mating. Because of the height at which these occur it was quite a while before experiments were carried out to determine the exact reasons for the drones, perhaps from many colonies, to select given spots where virgin queens would also be likely to find them. It seems that when certain weather conditions prevail and at heights depending on aspects like thermal activity, the insects find that

this is conducive to the requirements of a mating couple. We now call them 'drone congregation areas - DCA's. Perhaps it is nature's way of helping to avoid sibling matings.

**droppings**

The dung of animals, frass or faecal matter dropped by bees. It is usual for the bee to void their excreta while on the wing away from their hive. In built-up areas this can lead to their soiling of articles or places where they are not always welcome. Despite the fact that it is said to make excellent manure, when a bee settles on washing hung-out-to-dry and the comfortable intake of water at a reasonable temperature presents itself, it reminds the bee that it has unwanted material aboard, then the inevitable happens. The inside of beekeepers' cars or his clothing might receive dirty looking streaks which will dry unless rubbed with a damp cloth to remove them. The excreta is composed of pollen husks and other waste. Sometimes the front of a hive is splotched in this way and this could be due to excitement or cold conditions preventing bees from flying as far as they normally would. However when 'dysentery' is the cause, then the tops of frames and even the combs can be soiled with what are called 'dysenteric markings'. A smear from the droppings under the microscope can also be used for diagnosing 'nosema'. The presence of wax moth with whitish droppings with brown specks often acts as a pointer as do the small, black, grain-like droppings of mice.

***Drosophila amelophilous***

A small black fly associated with honeybees. It lays its eggs in cappings and regurgitated food causing damp looking patches on the cappings, fermentation and dysentery.

**dross**

A term used for the scum of melted metal but used in beekeeping for the dirty residue of cocoons and other non-waxy material strained off when boiling water has been used to render comb into beeswax. or separated out in a solar wax extractor. It is called 'slum-gum' in America, is flammable and can be used as a fire lighter or dug into the ground as compost but not for indoor plants as it can become infested with flies.

**drugs organic or inorganic**

Chemical substances, often synthetic, which are used to counter an ill-effect within a living organism. Often their effects are merely inhibitory as in the case of 'fumagillin' used in the treatment of 'nosema' and Terramycin in the case of foul brood. By helping a colony to limit the production of pathogens it can give the bees a chance to restore the natural balance and so remove signs of the disease. However the use of drugs requires a thorough knowledge of possible repercussions and one should use them under instructions from well-qualified practitioners. In some instances their use as an alternative or compliment to other methods of treatment is considered a wise precaution.

**drum or barrel**

Large honey producing outfits use 25kg drums for despatching honey overseas or for long distances. They are made of metal with non-rusting linings of lacquer or other honey resistant material. Smaller quantities of honey usually go into jars or plastic tubs (buckets). For years metal 28lb tins were used but went out of fashion when the easier to clean, white, plastic buckets came into being. Wooden barrels can be used as water butts or cut down for planting bee flowers, maybe even as a temporary home for swarm.

**dry**

When a colony or box of combs is described as 'dry' it implies that they contain no stores. When in such dire straits liquid food should be sprayed, or a little carefully poured, directly onto the bees, then covered, to allow the food to take effect, and then fed in the normal way. Smoking would be quite ineffective, in fact aggravates starving bees, as smoke relies on bees gorging themselves

to bring about pacification. Dry mead is formed when less than say 1.4kg/3lb of honey per 4.5ltr/1gallon of water were used. 1.8kg/4 even 2.3kg/5 lb. produce medium to sweet mead. Weather-proof hives are absolutely necessary in our wet climate because dampness in a hive will lead to problems of ill health and rotting wood.
See: dry swarm, vapour pressure deficit

**dryer** drier
An apparatus that can be used to dehumidify, desiccate or in other words remove moisture. Unripe honey has been improved in this way. Quick drying paints contain a 'drier'. The bees themselves are busy drying the nectar so as to reduce its volume to almost one third to make their honey.
See: honey dryer

**dry swarm**
Is one that has exhausted its food and may well prove 'touchy'. Ordinarily a swarm will have been preparing to leave by 'loading-up' with food over several days before swarming. (perhaps this was one of the different sounds of a colony that E. Woods was able to pick up with his electronic swarm warning device.) If weather conditions befoul the bees intentions and hinder its finding a new home it would help them enormously to spray them carefully with dilute syrup – don't wash them off the branch but spray a little, give it time to take effect and then spray a little more. Only give what they can absorb without it dripping off them. Even an ordinary swarm not only benefits by a little clean spray of water but it can help to anchor them for long enough for the 'collector' to arrive and deal with them.

**dubbin**
A useful substance for water-proofing and softening leather. Mix 2 parts of wax with one part of Neat's-foot oil and leave to soften. Then stir until thoroughly mixed. Remember when football was football?

**Duchet, Francois, Switzerland**
In 1771 he suggested that wax was a glandular secretion.
See: Aristotle

**duct**
An open or closed tube which is intended to convey a liquid. An aqueduct is used to move water from one place to another. As parts of anatomy in the bee, ducts accept secretions from glands and control the flow of these onto delivery points such as the mandibles or the tongue. Other types of glands are ductless (endocrine).

**ductile**
Adjectival description of the plasticity of metals such as lead and copper. The drawing out of metals forming wire, nails etc. is possible because metals have this property. It means that the material remains continuous and does not break as a brittle substance would. When wiring frames the stretching of the ductile wire usually results in a characteristic musical note when the wire is plucked. Only then is it fit to embed into the wax, unless a crimper is used.

**Dugat Pere M.**
He wrote 'The Sky-scraper hive' (Gratte-ciel) describing a multi-queen system of management.

**dummy board**
A board of the same shape as a frame, i.e. with bee-space around it and with lugs so that it can take the place of a frame of comb yet not obstruct the passage of bees. Such boards can be deep or shallow. They are useful on the flank of a chamber, especially when it is desired to squash all the frames together tightly, because when opening a colony by removing the dummy the operator can more easily take out the other frames.
See: division board

**dust**
The atmosphere transports and deposits tons of material in the form of minute particles of solid matter. They can give

rise to red sun-sets, blue moons and other phenomena. Dust can block bees' spiracles and suffocate them. The use of icing sugar (powdered sugar) is used to force bees to clean themselves and thereby perhaps dislodging varroa mites.

**Dutch bees**
In 1914 when the 'Isle of Wight' disease had ravaged stocks in Great Britain, J.C. Bee-Mason introduced large numbers to help repair some of the damage. We do not hear quite so much about them these days and although their temper was uncertain and they had a marked tendency to swarm, their comb capping and hardiness was said to go well with their active nature. They were grey but had some resemblance to the Italian bee.

**Dutch clover**
Ordinary white clover, much of the seed was at one time imported from Holland. Later, varieties from New Zealand came along and the resultant hybrids with our native clover seems to have cut down on the yield of honey, clover now produces.

**Dutch elm disease**
The causative fungus *Ceratostomella ulmi* is carried by the elm bark beetle Scolytus scolytus resulting in this disease which causes extensive die-back though it rarely kills the tree. However because of the infective nature large scale destruction of infected trees has been undertaken in this country. A sweet secretion from its fungus has sometimes been collected by honeybees and it is reported that bees do not winter well on such stores.

**duties**
The co-ordinated activities of bees calls for sharing out of work in such a way that all the various tasks are adequately covered. Since some bees are more suited to certain tasks than others, it falls to the lot of those who are capable of doing a job that requires to be done, to get on with it. This does not mean that bees are not at times called upon to do work that could be performed better by others more suited but 'duties' implies the allocation of a specific job to one individual bee over any short period of time. Fanning, security, attending the queen, removing an obstacle etc.
See division of labour

**DVAV**
Dorsal-ventral-abdominal-vibrations or spirit tap dance described by Milam as performed by returning foragers and also called a 'joy' dance. In this dance one worker usually contacts another by grasping it with its front legs and vibrates its own abdomen rapidly in an up-and-down direction at a rate of about seven times a second. Prior to swarming this dance is performed on the queen herself.
See: dances

**dwarf bees**
*Apis florea* is the smallest of the honeybees but amongst *A. mellifera* dwarf drones sometimes occur when a failing queen or laying worker deposits an unfertilised egg which is subsequently nourished in a worker cell. However, poorly nourished workers or those whose food supply has been stolen or interfered with by parasites, are found from time to time. Bees intermediate between queens and workers or sometimes seen and it is unwise to rely on any queen found to be subnormal in size. Varroasis also results in deformed and undersized bees.

**dwindling**
A diminution in the number of flying bees at a time when their number should either remain steady or be increasing. Spring, summer and winter dwindling are often referred to. Causes include the sudden onset of bad weather, enticement to fly when snow is on the ground, diseases, spray poisoning, or a failing or lost queen or the prevalence of insectivorous predators.

**Dyce Dr E. J.**
In 1930 he described his process for making creamed honey involving seeding with a fine

grain and mixing at 24C/75F after killing yeasts and liquefying any coarse crystals by an earlier short period of heating. The process was patented in 1931. Honey should have a moisture content of 17·5% for winter 18%. Less than 17% makes a very hard product. He even suggested that honey should be heated and water added to bring up to the right level.

**dysentery**

The involuntary voiding of faeces or the voiding of faeces at abnormal times. The word 'dysentery' is associated with disease in humans. With bees the above conditions can be brought about by disease but the term really means that the hive is dirtied internally or externally whereas normally this does not happen because the bees void their faeces on the wing away from the hive. When 'droppings' are seen on the comb or frames in the hive or around the outside of the entrance, it is as well to have an adult bee disease check made. Brownish, black streaks may mark top bars or outer hive surfaces especially around the entrance. See: faeces

**Dzierzon Rev Dr. Johann 1845 Silesia Poland**

He proposed the Dzierzon theory including parthenogenesis e.g. in order to become qualified to lay both male and female eggs, the queen must be fecundated by a drone (male bee). At first many opposed the theory strenuously. Later it became Dzierzon's Law. All fertilised eggs are female and unfertilised eggs become males. He also invented hanging frames. C.N.Abbot in recent times covered this in his book 'Rational Beekeeping'.

# E

**earthenware**
Objects made from baked or hardened clay. Ceramics. As this technology reaches for back into man's history it is not surprising that we find hives, honey utensils etc. made from earthenware. Honey pots, crocks and jar hives are examples. In places like Malta, bees were kept in 'jar' hives (virtually like horizontal chimney-pots) until fairly recent efforts by their government persuaded beekeepers to convert to standard wooden hives and the British influence of WBC as well as other hives took precedence. As recently as the 1960's still saw bees in the horizontal ceramic hives being rowed in boats across from Malta to Gozo to collect the very special thyme honey.

**earwig Forficula auricularia Dermaptera**
Small , brown insects whose young are paler, smaller replicas. They have hidden wings and a pair of sharp pincers at the tip of the abdomen. Although they are fond of honey and will breed over the inner cover inside hives (or under frame lugs) their gut contents show that they eat outside as well as wax, honey, brood and parts of dead bees from inside. They can be quite a nuisance in the honey house, getting into any unsealed honey and possibly drowning in extracted honey if they fall into it. Upturned, hay-filled flower pots on stakes will trap them and chickens eat them.

**Eastern honeybee** *Apis cerana*
They only possess half the chromosome number of *A. mellifera* females and do not produce such strong colonies as the Western honeybee. As they are less productive too and *mellifera* can just as easily be kept in the same regions, there seems little reason for keeping them commercially but they are said to be less susceptible than ours to 'nosema'.

**eat**
To consume or devour food. When a bee appears to be feeding it is actually loading pollen into its baskets or nectar into its honey crop (sac) to carry food back to its hive mates or to pass it to house bees to put into cells for ripening and communal use, whereas when a drone or a queen take or are given food, it is for eating. When a worker bee consumes food it is passed from the crop into the ventriculus. Some predators like birds and insects include honeybees in their diet. We eat honey in a variety of ways; granulated it can be heaped thickly onto toast or if liquid it can be spread, drifted onto fruit salads, added to drinks of every sort and used to flavour butter, grape fruit in fact it has the capacity to be added to all manner of foodstuffs. In spite of the word honey going back so far in history and having been so welcome in the days before the inferior sugar arrived, it is surprising how little English people eat. Some even put it at the back of the larder or in a medicine cabinet! Australians eat ten times as much and we'd all be much healthier if we did the same.

**eating bees**
Honeybee larvae are soft and full of protein and are used as food by certain tribes (is that being 'racist'?) Such food is doubtless rich in the much valued 'royal jelly' but in fact a queen larva tastes frightful (but so do medicines that do you good). Chinese and others have ingeniously learnt how to

fry them or bake them in chocolate. All too many creatures like birds and mammals (bears – badgers…) count as 'enemies of the bees' because as the only creature that collects and stores food they are and always have been, a target for those less industrious.

**ecdysis**
Moulting, or the periodic shedding of the skin including the linings of all but the finest tracheae. It occurs during the growth and metamorphosis of the brood of honeybees and is initiated by hormones. The new soft skin that is revealed as the old skin splits, is larger and it hardens rapidly on exposure to the air.
See: moult

**ecdysone** moulting hormone
A substance concerned with the initiation of ecdysis, the hormone associated with the larval development, ecdysis meaning the moulting of the larvae. The endocrine glands are responsible for its secretion and these in turn are controlled by the brain. It governs growth and differentiation and is thought to be a steroid.
See: neotenin

***Echium plantagineum*** Purple viper's bugloss
Australian Paterson's curse or Salvation Jane or 'blueweed' in NZ, an introduction from Europe. Honey derived solely from this plant is nothing special. However, giving it fancy names and special packaging it has found its way into famous stores in London and gone back to the antipodes to be resold at prices the local beekeepers would regard mirth and amazement.
See: blueweed, viper's bugloss

**Eclipse bee-hive**
In 1879 F.A. Snell made a hive to take ten 12 x 12 American frames. It had movable sides and sections on top.

**eclosion**
Hatching from an egg or the emergence of an imago.

**ecology**
Coined by Haekel in 1869 from the Gk. kos meaning home or dwelling. Now considered as the study of the whole bio-sphere. Bees are particularly useful in ecological studies, involved as they are with so much of the environment in weather soils, plants and other creatures. The scientific study of relationships between form and function of living animal and plant communities, their energy flows and their interactions with their surroundings, animate and inanimate.

**ecomorph**
Genetically identical individuals whose different forms (morphs) are induced by environmental factors. E.g. caste differentiation in the case of queens and worker bees.

**ecosphere**
The region in which living organisms are produced and maintained and their interplay with the non-living components of their environment.

**ecosystem**
A system of living things and their environment which produce the materials and energy for continuous existence - a self-contained habitat varying from a small particle of soil to the whole world.
See: biosphere, ecosphere

**ecotone**
The border region between two adjacent ecological communities – where forest becomes grazing land.

**ecotype**
A group of plants or creatures that have adapted genetically to a particular environment while still being able to cross freely with others of their species.
Ectoderm (ecto – external or lying upon). The embryonic layer of cells that gives rise to the future external layer and the nervous system.

**ectohormone**
A hormone that acts outside the producer – pheromone.

**ectoparasite** exoparasite
A parasite, such as *Braula coeca*, that lives for at least the adult part of its existence on the outside of its host.

**edaphic**
Conditions imposed by the soil, topography or substratum rather than by the climate. Plants within a specific area will be greatly influenced by the edaphic features of that area. Edaphic factors of the soil include: texture, water content and pH.

**EDB**
See: ethyl dibromide used in USA for wax moth control.

**edible**
Suitable for eating; Most hive products are edible though in the case of wax and propolis they are not digestible. The intake of these two substances is thought to be beneficial as the wax is cleansing to the system and propolis is valued for its anti-biotic qualities. Honey is not only edible but for the vast majority who enjoy it, high beneficial. Pollen is too, though for most people special additions or preparations are necessary to make it palatable. This also applies to royal jelly while the protein value of bee grubs is not much appreciated in Europe.

**education**
The acquisition of knowledge and becoming acquainted with the worthwhile (and unfortunately, 'other') aspects of totality. Bees put humans to shame with their dedication to survival while doing much good and practically no harm. To study them is an object lesson. Research and learning about bees has taught us a great deal about them and much about ourselves. Observation hives are easier to manage than honey producing hives and every beekeepers should be in possession, or have access to one.
See: books, discoverers, inventors

**Edwards** The Rev. Tickner
Wrote "Bees as Rent Payers" 1907: 'Lore of the Honeybee' 1908: 'The Honey Star' 1913: Beekeeping for All' 1923: 'Beekeeping Do's and Don'ts' and invented the Tickner-Edwards hive

**E.E.C. Regulations**
The European Economic Community have made regulations concerning, currency, standardisation of measurements. Size of honey containers and honey standards regarding HMF content and diastase and water content.

**Eek** scottish
Eke (Keys and Bagster) Extra part added to a skep. Eke – to add to, increase, lengthen Chambers 1906. Used to describe an addition at the top of a hive usually for honey storage.

**E.F.B.** *Melissococcus pluton*
European Foul Brood a disease associated with the presence of *M. pluton* and a secondary invader such as *Bacterium Eurydice*, *Bacillus alvei* or *Enterococcus faecalis*. The bacteria are swallowed with the food and develop in the mid-gut. If the larva survives, it will be subnormal and will pass pathogens into the comb with its faeces. The disease becomes most noticeable when he amount of brood becomes large relative to the number of nurses as at the commencement of the main nectar flow. Dead larvae are usually thrown out 'piece-meal' but will dry to an infective scale if not. Death occurs about the 5$^{th}$ day just prior to capping and the larvae look dry and twist themselves into unusual positions in the cells. It is possible to have the complaint treated with antibiotics by an official of Defra. When it strikes it us during May or June. Once in the hands of Defra a 'standstill' order is placed on both bees and equipment in the apiary. *EFB is a notifiable disease, **always** contact your Regional Bee*

*Inspector.*

**EFB life cycle**

The primary causative agent is *Melissococcus pluton* a lanceolate bacterium which when swallowed in the larval food increase in number to fill the mid-gut. Some larvae die, their death being accelerated by secondary invaders such as *B. eurydice, faecalis, alvei* and others. House bees are able to cast out 'piece-meal' the dead or dying larvae thus making the disease less obvious than AFB. The remains are sometimes left to dry into a brown scale. While dead uncapped larvae may be a conspicuous aspect of the disease, pupae and adults can survive infection though bees with small abdomens may result. Treatment: to 2·5kg sugar mix Pfizer TM10 but more modern prophylactics might be available if you have the skill to tackle the disease this way. *EFB is a notifiable disease,* ***always*** *contact your Regional Bee Inspector.*

**effector organs**

Muscles and certain glands which receive instructions from sensory organs.

**effects of damp**

Bees can contend with many things but damp is not one of them and choice of site and ventilation of the hive should be arranged so that dampness does not become a problem especially during the winter months.

**effects of drugs**

Drugs are used to help combat certain conditions and diseases and are dealt with under the name of the disease or the drug.

**effects of insecticides**

Various strains and races of bees, may react differently to a respective insecticides. Temperature and formulation (manner in which the chemical is made and applied) time of day and weather conditions at the time and for a day or two following application, also have their effects. In serious cases whole apiaries are wiped out, sometimes only certain hives have a carpet of dead before the entrance. Many bees die in the field, others may cause harm to bees or brood and more rarely - honey, within the hive. It is important that details are reported whether or not any claim for possible compensation is made. In this way statistics enable the authorities to decide on what actions need to be taken to safeguard our bees. Photographs of damage (dead bees outside the hive) are very useful.

**effects of movement**

The movement of colonies from place to place is dealt with under 'moving bees', 'transportation' etc., The effect of movement on their sensory system is explained under 'sensitivity to movement'. Always remember that at any time of the year, moving colonies causes a rise in their temperature and this can lead to suffocation unless adequate 'top-ventilation' is provided.

**effects of propolis**

The wearing of gloves can help to counteract any possibility of dermatitis or the sticky, dirtying of hands. It is anti-biotic and use is made of this for gargles and other means of assimilating it into the body. Like honey it can be used for cicatrising (wound healing) as an anti-biotic, anti-fungal and anaesthetic. Mixed with spirit it can be used as a 'varnish' and it can make a good wood filler.

**effects of rain**

A fall of rain may give rise to a honey-flow or bring a honey-flow to a close depending on the state of the weather and the condition of the flowers. Although bees may work in light rain, once their bodies have become wet they cannot perform duties including flying, until they are dried off by other workers or the elements. Rain can also harm pollen and delay effective pollination.
See: rain, rainfall

**effect of repellents**

Substances such as carbolic acid which have long since been known to drive bees

away and perhaps from earliest times – smoke, have their place in controlling bees. However where the use of repellents comes under our control we must ensure that neither lives nor food sources are unwittingly made repellent by the thoughtless use of chemicals, preservatives etc.
See: repellents

**effects of smell**
The olfactory organs of bees enable them to detect and identify smells at least as effectively as we can. That does not imply that they have the same likes and dislikes. The leaves of lemon balm used to be rubbed into skeps to make them attractive to bees and some nectars appear to be more attractive to them than others even when the sugar contents are the same. Plain fresh water is ignored when a more palatable source is available and a sugar solution while practically odourless to us will strongly attract the bees.
See: attractants, olfactory sense, repellents

**effects of smoke**
Bees do not like smoke, consequently they try to move away from it. Used indiscriminately it scatters bees and causes the queen to try to get away and hide. It is useful for clearing the bees from one surface to be brought into contact with another such as replacing a super when bees might otherwise be crushed. When used to subjugate bees it masks the odour of their alarm pheromone that they may release and causes them to lower their guard. It also causes fanning and a fanning bee is a non-stinger. It makes them share and take up food, though if open food cells are not available the smoke may further aggravate them. Carelessly used it can damage chitin, brood and honey.

**effects of stings**
Human reactions to bee stings can be quite variable. There are two aspects, the pain and the after effects of being stung. The pain is akin to having a spot of very hot water flicked on your skin but the pain caused lasts longer than would the hot water. From familiarity beekeepers become accustomed to and therefore give less attention to this pain but in most cases it would be wrong to assume the discomfort is really any less even after years with the bees. A single sting (say the first one) concentrates your attention but if you are diverted by a sting or pain in another part of your body it is surprising how the intensity of the first sting seems to grow less. Though tiny, it is designed to give pain and discourage you. The effect, is another matter. Better of course if the sting is immediately 'scratched' out, but a minute hole made by the twin lancets, quickly closes and the application of 'remedies' will not have much effect. A small red dot, surrounded by a whitened circle, marks the spot. Venom is not a straight forward acid or alkali but a complicated mixture probably more akin to that of a cobra, and the changes they bring about may vary from person to person and according to reactions from previous stings. The poison activates 'mast' cells in the body and until an element of immunity is obtained, swelling and discomfort can spread from the site to far more than just the immediate area. In addition to that, once in the blood stream it can cause disturbing changes not too serious unless allergy (pain might be incidental) causes a real upheaval when anaphylactic shock can take hold and the victim, if he passes out, must then be treated in the same way as for electric shock or recovery from drowning. It behoves you to familiarise yourself with the procedure. Rest assured though, that the vast majority of people soon get used to stings and although they exercise your hidden vocabulary you can pass them off as all part of the game. We should stop there! But we must cover every single possibility so: the more severe reactions (not extreme like anaphylactic shock) can involve itching, wide-spread red blotches, malaise, wide-spread oedema, wheezing, chest constriction, abdominal cramps nausea, vertigo and vomiting or difficulty with breathing, significant

weakness, psychological disorder, decreased blood pressure, collapse, cyanosis…'He is not worthy of the honeycomb who fears the bees' stings'

**effects of vibration.**

A tap on the side of an occupied hive will receive a response in the form of a sharp hiss. A long drawn out sigh implies queenlessness. Their response to vibration tends to make them run upwards and skeppists used that feature when 'driving'. It would be true to say that continued vibration makes them unhappy but having vented 'alarm odour; its effect is relatively short-lived and when it has worn off, bees can be made to march upwards in an orderly fashion. Bees kept by the side of a railway line were reported to use far more 'brace' comb than normal. The Von Frisch dance is accompanied by body vibrations and a queen's 'piping' brings about an immediate reaction on the part of the bees. It is unlikely that they can hear the sound they make when they are flying, but their comb is a sound board for communicating by means of vibrations.

**effects of water**

Bees handle, dispose of and use water during the brooding season; yet surplus moisture can lead to troubles during winter. They can become confused when flying over water and drones have been found floating and swarms have been known to go down in the sea. Rain, real or artificial can dampen angry bees ardour and discourages robbing (try wet grass heaped over the entrances). The Sumpter method of subjugation relied on water sprays instead of smoke. If you close your bees up to avoid the action of farmers using poison sprays, be sure to provide them with water – a wick in an ordinary feeder is one idea and when moving them in hot weather give them access to water via a wetted cloth on the ventilation screen.

**effervescence**

The seething release of small bubbles. It can be caused by chemical reactions. A fermenting liquid (say mead) may exhibit this quality if pressure is released or more honey added.

**efflorescence**

The state or period of flowering or blossoming. The condition flowers are in while they are capable of being worked by the bees for nectar or pollen. Chemically it is the change of state of the surface of a crystal when a thin powdery film forms as a result of the loss of water of crystallisation. Could this latter condition account for the 'frosting' that sometimes occurs in jars of granulated honey?

**egestion**

The process of expulsion of undigested food remains. To throw out, to void or to excrete. This is the total exudation of substances including fluids from the body the opposite in fact to ingestion. Were the waste products not removed from the site of absorption they would clog the whole system. The egestion of sweet undigested carbohydrates by plant sucking insects such as aphids, lachnids etc. is called honeydew and is collected and utilised by ants, bees and other organisms.

**egg**

The ovum or initial stage in the life of an animal. Its trip from the tip of the ovariole to the oviduct takes two to three days. All eggs appear to be identical, like a tapered sausage, about 1·5mm long. If sperm is present on the egg it will become fertilised and this results in a diploid female, whereas eggs that have no contact with sperm remain haploid and become males (parthenogenesis). The egg is laid, small end down, and held momentarily against the cell base so that a glue-like substance can affix it to the wax. The eggs are then incubated at brood temperature for three days when the skin (chorion) splits as the enclosed larva frees itself. This is referred to as 'hatching', the term 'emergence' being reserved for the climbing out of the cell by the young adult (imago).

**egg** drone
The eggs of the honeybee look identical regardless of whether they are laid by a fertile queen or a virgin or a laying worker. However whereas eggs that hatch into a female larva will have been fertilised, those that hatch into a male larva will not. Eggs that become drones are haploid (one gamete) with 16 chromosomes, half the number of the diploid females. Just how the relatively few drone eggs laid by a queen become fertilized is not absolutely clear. Perhaps the queen's response to laying in the larger drone cell is not to operate her sperm pump or perhaps the workers play a part in sterilising any sperm before it can enter the micropyle of the egg after it has been laid.
See: egg, parthenogenesis

**egg** drone breeder
An egg laid by a drone breeding queen is an egg that can only hatch into a male larva because its parent was unable to fertilise it. This can be due to mis-mating or failure on the part of a queen.

**egg** embryology
1. Cleavage 14-16 hours.
2. Formation of the 'blastoderm' 30-35 hours.
3. Formation of mesoderm, mesenteron rudiments and embryonic envelope 42-48 hours.
4. Rudiments of appendages, silk glands and tracheal system appear at this time also.
5. Development of stomodial and proctodeal invaginations and neural groove 52-54 hours.
6. Formation of heart, virtual completion of development within the egg 66 hours.
7. Hatching 76-79 hours. Eggs can be stored prior to development starting and have been kept in suspension for considerable periods. Temperature variations can affect development.
See: brood temperature

**egg-laying**
Eggs are formed in the ovarioles which number around 100 and are connected to the oviduct. The ovarioles form two groups which when fully developed in the honeybee occupy much of the abdomen. From the oviduct they pass into the vagina where fertilisation can occur. The egg is then glued to the cell base having been placed in position by the queen's ovipositor. Laying workers have difficulty in placing their eggs tidily and have no means of fertilising them so only drones are derived from them. On average a new queen commences laying approximately 14 days after emergence but this depends on how soon she becomes mated.
See: laying, laying rate

**egg** laying worker
The normal worker has atrophied ovaries but in certain circumstances such as when a colony is deprived of queen substance, some workers ovaries develop and it becomes possible for them to lay male eggs. It has been reported that in the case of 'Capensis' bees male and female eggs are possible from laying workers.
See: laying worker

**eggs** other than workers
Eggs destined to become queens or drones. Workers may well have to cajole a queen into laying in queen cell cups. On the other hand she seems to find laying in drone-comb so easy that, should she get loose in a drone-comb super, there's no stopping her, and more drones will result than a colony would normally produce in a whole season.
See: egg-laying

**egg** queen
An egg destined to become a queen is identical in all respects to one destined to become a worker. It is an ordinary impregnated honeybee egg. Its incubation, like those of the other two castes requires three days. To become a queen it must be laid in a queen cup that will be developed into a vertical queen cell or perhaps in a worker cell that is modified to the shape and size of a queen cell, and be fed continuously on 'royal jelly'. Any fertile egg could

potentially become a queen and this was first realised by Dzierzon in 1845. No-one seems to have determined whether workers can carry sperm and so place it near the micropyle of an unfertilised egg.
See: parthenogenesis

**egg** sterile
A viable egg that has not been fertilised. In the case of honeybees this is an egg that is only capable of developing into a drone. In other words an infertile egg that has not been in contact with spermatozoa. Not to be confused with a 'bad' or 'addled' egg.

**Egg** unfertilised
Due to parthenogenesis it is normal for drones to develop from eggs that have not been fertilised. Similar in every other respect to a female egg it lacks the compliment of the male germ cell – spermatozoon.
See: egg drone, egg sterile

**egg worker**
An egg that develops into a worker is a fertile or 'impregnated' egg which is incubated fed and reared in a worker cell. On occasions, particularly in colonies long queenless, workers can develop their ovaries and lay eggs. Such eggs from workers are haploid and being infertile can only result in drones. A colony's viability is entirely dependent upon is ability to produce worker (fertilised) eggs.
See: laying worker, worker egg

**egress**
The means or place of going out. The entrance of a hive naturally serves as the bees' egress or exit as they go in and out by the same hole. By choice they prefer to have this at some height above the ground, possibly to make it more difficult for predators also as a protection from floods.

**Egypt** ancient
The bee was a symbol representing Kingship in Lower Egypt. Rafts were used to float hives down the Nile in early forms of migratory beekeeping. Not only was the honey valuable (doubtless pollination too) but wax had a value quite beyond anything we can appreciate today. Pictures in tombs show swarms, the cutting-out and selling of combs. In 1980 honey production reached 10,000 tonnes.

**Egyptian bee** *Apis fasciata* and *lamarckii* (of the Nile valley)
They resemble the Cyprian bee but are rather more swarmy and temper makes them difficult to handle. Their appearance is similar to that of Cyprian and Italian bees though a reddish bronze tone is detectable in their queens and the workers are of a lighter yellow with bands of white hair. They are prolific, slightly smaller than European bees and do not react as well as the latter under the influence of smoke. Laying workers are frequent too.
See: A. intermissa, lamarckii

**eidetic**
Pertaining to imagery and its persistence in the memory. Bees only fly when they can see and can find their way home again, if the hive remains in the same place that they have visualised in their memory. They are also helped by their ability to read polarized light patterns in the sky which help them with directions.
See: homing instinct, orientation flights

**ejaculatory duct**
A duct that leads from the bases of the mucus glands of the drone. It is through this that the spermatozoa pass at the moment of copulation. The openings of the seminal vesicles are brought close to the opening of the duct and as the vesicles empty semen passes into the duct followed by mucus which after coition doubles forming a 'plug'.

**ejection**
This can apply to the throwing out of crystals, dead brood, debris, dried pollen, drones, dead virgins, wax particles and workers. Much can be learnt from this kind of material regarding what has been going

on inside the hive. No unwanted material is allowed to remain in a healthy colony of bees provided they are able to throw it out. Objects like snails for instance, too large for them to handle, will be covered in propolis which becomes their sepulchres.
See: debris, drone slaughter

**ejection of workers**
Foragers may return smelling of a chemical or charged with static electricity not compatible with that of the hive bees and as a result get ejected. Bees are totally unforgiving when any ailment that is not common to them all occurs and a deformed bee, a bee with paralysis virus or any bee that fails to answer the demands the colony make on it, will be ejected. It is quite a load for one bee to carry, another yet you see them staggering away, determined to drop it as far from their nest as possible. Wholesale fighting and ejection may result from a beekeeper not having taken all the precautions necessary when uniting two dissimilar colonies.

**eke**
Originally a small straw extension added to a skep to take surplus honey, a feeder or to act as a bottom extension to the brood chamber, perhaps a ring of plaited straw but to cover eventualities such as increasing, enlarging, lengthening, augmenting or supplementing the normal skep. "We mun put an eke on that hive's full, an ther's awt' ling-bluim to come yet". Today it describes a simple, four-walled, empty box inside which a feeder can be accommodated. A shallow frame to allow candy to be applied between crown board and roof.
See: imp, nadir

**elasticated cuff**
A sleeve covering the wrist and part of the forearm. A pair of these, elasticated at each end, may be worn to keep bees from crawling up your sleeves on the inside leaving the hand quite free and therefore more dextrous for delicate bee-work like clipping or marking a queen or removing and replacing frames with maximum finesse. They are also available as a continuous extensions to the bee-glove and then referred to as bee gauntlets.
See: bee-glove, gauntlet

**elasticated veil**
A protective veil with strips of elastic inserted in a hem on the lower rim as well as the upper rim which encircles the hat. As an alternative to 'tape' elastic makes a tight, clinging fit so that the bees are effectively excluded.

**electric blower**
Devices to blow bees out of supers have been in use for a number of years, notably in Israel and the USA. When driven electrically a sensitive and variable flow of air is possible to dislodge bees from the super. Due to ease of portability small petrol driven engines are more widely used than electric. A platform or stand is sometimes put in front of the hive to take a complete super which is then blasted in such a way that the 'majority' of the bees are deposited rather firmly, in a surprised heap, in front of their entrance. It is possible that a few of the bees remain clinging to the bottom bars of the frames. Not surprisingly bee farmers do not have the interest of every single bee at heart.

**electric embedder**
Leading manufacturers of wired foundation use sophisticated devices so that one touch of the electric terminal delivers a charge which for a moment raises the temperature of the wire in the frame so that it becomes embedded neatly in the wax. DIY's may concoct a similar means of embedding using an old car battery or a mains transformer. Stainless steel wire is used now-a-days and unless completely covered by wax on both sides of the frame a tell-tale line across a sealed brood frame which the queen does not appear to have used wastes the spaces that hundreds of potential workers would have occupied. Although all too obvious, many a beekeeper overlooks this important

detail.

**electric extractor** for honey
A machine powered by electricity and designed to rotate a cage in which uncapped combs have been placed for the removal of honey. There are many examples by the equipment suppliers that are built for the hobbyist and also for large-scale enterprises. Careful earthing and normal safeguards are essential where metal machines and possibly moist conditions exist. They must be mounted on a firm base and at a height that allows the honey to be drawn off into a suitable container.
See: honey extractor

**electric fence**
Although fences that guard hives from the unwanted attention of cattle and other animals, are rarely electrified, electric fences are in widespread use as a simple means of deterring the passage of large animals. An electric pulse of high voltage is directed from a battery and through a transformer along the wire. Where bears have to be guarded against special forms of electric fencing are used with an earthing mesh set in the ground and baited to induce bears to hesitate at the wire. (Bacon and moth-balls have been used).

**electricity**
A source of energy with associated phenomena such as the production of heat, light and power – attraction and repulsion - chemical decomposition. It makes the handling of bee products speedier and simpler. Dr.R.S.Pickards investigated the honeybee brain and made "Bees, magnetism and electricity" available through the BBKA. The bee naturally picks up a certain charge of static electricity while pursuing its foraging duties and if it is interrogated by a guard bee upon its return, it can give the guard bee a shock.

**electric uncapper**
A device that uses electricity to assist in the uncapping of sealed honeycomb. The form in which current is used may vary from merely producing heat as in the case of uncapping planes, to the agitation of narrow blades which shear the cappings off.

**electro-biology**
The science of electric phenomena within various organisms.

**electronic**
Devices, systems and methods that function by some use of moving electrons. This applies to the word electric as well but whereas electricity is used to produce phenomena such as lighting, heating, chemical decomposition, attracting and repulsion, electrons by virtue of advances made with the micro-chip, are waiting to perform whatever some sufficiently understanding and imaginative person can conjure up. To name one or two techniques useful in our field: electro probe micro-analysis, electron spectroscopy, electrophoresis...
See: electron microscope

**electron microscope**
In this instrument beams of electrons are focused by electron lenses to form highly magnified images on a fluorescent screen or photographic plate. This the scanning electron microscope (SEM). A combination of static or varying electric and magnetic fields allows their capability to extend well beyond that of the optical microscope. The TEM or transmission version gives a two dimensional image.
See: microscopy

**electrophoresis**
The movement of material particles due to differences of electrical potential. The motion of colloidal particles under the influence of an electric field in a fluid.
See: paper chromatography

**elements**
Air, water, earth and fire. Atmospheric agencies or forces. Chemically: substances with only one type of atom (not composed

of molecules). The function and health of our own bodies as well as plants and bees is dependent on the fairly precise amount of certain elements that we absorb.
See: environment, honey elements, trace elements, weather

**Elm** Ulmus spp. Ulmaceae
*Ulmus minor* var. *vulgaris* the common elm tree. Blooming in February/March its flowers come before its leaves. It is rather too early for the bees though they will collect its normally airborne pollen described as yellowish-green, green, greenish, pale yellow, dull greyish, light to dark, putty-coloured with 20μ grains. Bees have been known to collect a dark sticky secretion from trees afflicted by Dutch elm disease.
See: Dutch elm disease

**Elsholtzia** spp. Labiatae
A shrub that need full sunshine and will grow to five feet but will not yield nectar and pollen until late in the year – end of Sept/Oct.

**elusive**
Not easy to find or get hold of and easily able to escape. This naturally applies to queens and sometimes to swarms. Taking the latter first, a fine spray of clean water (syrup if the bees are tetchy in a cool breeze) will help to bunch up the cluster and deter them from absconding until you are ready but hasten slowly. Queens can be clipped and marked and a box containing bees and the queen, can be taken to an adjacent site for a day or two to 'milk-off' the tiresome bees and then it should be easier to spot her. When searching through the brood chamber twice without coming across the queen, bearing in mind too that she might be on the floor then brood combs can positioned in pairs, with gaps between each pair, then sooner or later she will be found in between the combs of one of the pairs. A cast, with several queens can be hived through a queen excluder used like a colander.

**embalming**
The Assyrians and the Egyptians at times, placed their dead in honey after smearing the body with wax. H.M.Fraser.

**embedder**
A small instrument or apparatus designed to simplify the embedding of wires, or other forms of reinforcement, into sheets of wax foundation. It may take the form of a small hand-held wheel that runs along the wire but even hot screw drivers have been used. The aim though is to neatly sink the taut wire to below the surface of the wax without mutilating the embossed cell shapes.
See: electric embedder

**embedding**
Beeswax is a useful substance for embedding an insect so that part of the body protrudes while the remainder is held fast. Having done this, a layer of distilled water can then be poured on to give optimum conditions for dissecting and viewing. An old discarded queen gives this opportunity and dissection is helped rather than hindered by the water. Using small scissors to cut round and lift her abdominal, chitinous cover will display her wonderful array of ovarioles.
See: microtome, embedder

**embossed wax sheets**
Sheets of wax are passed through mangle-like rollers embossed with worker or drone-sized hexagons, the pressure and setting of the rollers, controls the thickness of the sheets, and first class foundation is made in great lengths to be cut to size as required. On a DIY basis there are various sized presses or sheets of embossed plastic which in conjunction with a mangle and sheets of plain wax, can make fairly usable foundations.

**embryo** (embryonic)
A young organism in the early stages of development. In seed plants the embryo is found in the seed. Embryo animals are found within the womb while the honeybee

embryo is within the nursery cell. Its life begins at the moment of fertilization so it is initiated after leaving the spermatheca. See parthenogenesis to consider the variation from this rule in the case of a drone.
See: embryology

**embryology**
The study of the formation and development of the early (embryonic) stages of plants and animals. The nervous system of a bee is of ectodermal origin while nerves from the sense organs originate from cells of the epidermis and grow inwards to the ganglia. Wings and legs are not apparent on embryo or larva because their rudiments are sunken into shallow pockets of the epidermis beneath the cuticle. The embryo passes through five stages of growth each finishing with a moult.
See: egg embryology, instars, respiratory system

**Emergence**
The act or fact of emerging. This could apply to the emergence of the imago from its cell (not to be confused with the word 'hatching' which is used to describe the transformation of a mature egg into a larva) It is a term that can equally well be applied to the action of a swarm in leaving the parent colony or of an individual bee leaving a hive.
See: eclosion, emerging adult

**emergency feeding**
Giving sustenance to bees either when on the point of or approaching starvation or at abnormal times of the year such as mid-winter or to stimulate the queen to lay abnormally early as for honeydew flows, pollination or queen-rearing.

**emergency queen cells**
Queen cells raised by the bees under the queenless impulse, brought about by emergency conditions that occurred naturally or were imposed by the beekeeper. The sudden loss of a queen or the giving of eggs to a colony long since queenless could cause such cells. They are not likely to produce good queens and should be replaced by more suitable cells, a virgin queen or a mated queen. Emergency cells are often built as extensions to ordinary worker cells containing larvae already too old to be developed into good queens.

**emergency swarming**
Swarming brought about by a sudden or catastrophic condition such as fire, starvation or the destruction of a colony's home. This mass exodus of the bees is also referred to as 'absconding'
See: abscond

**emerging adult**
The imago in the process of getting out of it cell. Workers nibble round their cappings protruding their antennae and possibly communicating - hive scent, queen rightness etc.- with nurse bees before they emerge. Drones perform similarly but sometimes receive assistance from workers to emerge. Queens may sometimes be held back by the workers, their isolated queen cells being surrounded by the nurse bees re-sealing as fast as the virgin tries to nibble her way out. The young virgin seems bubbling over with energy and enthusiasm to get free and as soon as the cap only requires a final push, out she scrambles and races to start demanding food from nearby nurses. At this stage the cap hangs like a flap but very soon house bees get to work, demolishing the cell and cleaning up any royal jelly left remaining inside.
See: hatching

**emerging brood**
An area of capped brood from which adult workers are emerging. As this represent the tail end of three week's nursing and incubation it is a useful bonus for plumping or strengthening a weak colony. As the emerging bees are soon able to help the recipient colony provided food is available and the minimal incubation for emergence is within the new colony's scope. Bees of this age are accepted without hesitation by any

colony and they will soon identify with it.

**Emlock**
Hive fastener used with 5/8" (1.6cm) galvanised strapping that can be adjusted quickly to fit any size hive and when pressed in locking position it makes a very secure fastening for transport. Many Australian beekeepers keep them on all the time. John Guilfoyle, Darra, Brisbane.

**empodium**
A small structure between the claws of the feet on many insects.

**empty**
Without contents. Robber bees will not hesitate to empty a super of honey when a clearer board is put on yet no precautions were taken to see that the super(s) was bee-tight. Normally a super contains occupants who would drive off any predator or interloper. As soon as their number decreases as they pass through the clearer the honey becomes more vulnerable to any unofficial visitors. Use gum, Sellotape, foam rubber, propolis or beeswax but 'seal' all gaps and do not leave in situ for more than 24 hours.
An empty feeder serves no purpose on a hive and if left it becomes propolised and mould can develop inside. Hefting can make a hive feel 'empty and action should be taken as there is no more certain killer than starvation.
See: bee-tight.

**emunctory**
A body organ that can deal with waste products as in the bee for example the malpighian tubules which act as the bee's kidneys.

**encase queen**
To cage a queen or to embed a queen in plastic or pickling solution for anatomical study. Caging within a colony could be undertaken to restrict a queen's laying or to protect her or to simply make her easier to find.

**encaustic**
The use of wax colours that are fixed with heat. Any process in which colours are burned in. The work of art resulting from this process.

**enclosure**
A tract of land surrounded by a man-made boundary. Such places often provide sites for out-apiaries. Sometimes a natural water-way may form part of the boundary when adequate consideration should be given to the possible effects of flooding.
An apiary is best enclosed for security, privacy and shelter. Dense evergreen hedges such as oleander, rhododendron or fast growing Leylandii are useful but many plants that can be easily established, maintained and yield nectar or pollen are listed in this work. Examples include: blackthorn, cotoneaster, everlasting pea, hawthorn, holly, honeysuckle (climber) ivy(climber) raspberry, snowberry, willow, yew. Sites allocated (at a price) to beekeepers in the New Forest were often situated in 'inclosures'.
See: apiary, out-apiary

**encouraging wild bees**
Bumble bee abundance depends on suitable nesting sites and the amount of wild flowers (or other suitable forage) available from April to June and of course, suitable weather. Contrary agricultural practices include the cutting down of early flowering trees and hedges, spraying dandelions and hedge bottoms. The summer burning of grassy dykesides and hedge-sides and stubbles, kill bumblebee colonies – autumn and winter burning would avoid this.
See: details on 'potash'

**endemic**
This describes conditions or diseases that occur continuously in a particular environment. Species which are native to, or as far as is known, have always existed in a particular place. Peculiar to a particular people or place.

See: enzootic

**endexine**
One of the layers of pollen coating though there is or was some disagreement concerning terminology. See: nexine

**endo**
Prefix meaning internal, opposite to ecto meaning within. Endocarp – inner layer of the pericarp (as the stone of certain fruits).

**endocrine glands** (opp. exocrine)
Glands which are ductless and therefore that produce the internal secretion of hormones. They are the hormone producing neurons (neurosecretory cells) in the CNS (Central Nervous System) whose axons terminate outside the system in neurohaemal organs, the paired thoracic glands and the corpora allata.
See: ecdysone, neotenin

**endocuticle**
The stretchable, inner layer of the insect cuticle.

**endoderm**
The future gut etc. of the embryo.

**endoparasite**
A parasite spending much of its existence living inside its host. For example *Nosema apis* whose active development occurs in the honeybee's ventriculus and which leaves the bee in the sporulated form so as to transfer itself, whenever possible to another bee.

**endophallus** penis
This evertible part of the drone's genitalia includes the ejaculatory duct and the bulb of the endophallus. Where the ejaculatory duct continues into the cervix followed by a vestibule leading to the phallotreme. See: bulb, penis, vestibule

**endoplasmic reticulum**
A double membrane to which ribosomes are often aligned. Possibly transporting away substances formed by the ribosomes.

See: mushroom bodies

**Endopterygota**
This is Division II of the Sub-class of insects Pterygota and is also called 'Holometabola' It includes bees and butterflies where the larva is markedly different from the imago.

**endoskeleton** opp, exoskeleton
The inner skeleton or inner body framework (as with human beings).

**endosternite**
An internal skeletal plate for the attachment of muscles.

**enemies**
Everything that the bees collect and produce is much desired by one predator or another. Naturally bees regard man as a predator however much he tries to befriend them. Ants, ant-eaters, armadillos, apes, badgers, bears, birds, bumble bees, earwigs, frogs, moths, mice, pollen mites, shrews, skunks, spiders, toads and wasps… Left to their own devices, honeybee colonies would spread themselves out through the forests, nesting high up in the trees. By keeping groups of hives at or near ground level we invite all manner of predators and avoidable weather, like floods, frost, bad air and minor fires.
See: bee-enemies, mites, pathogens

**energetic reward**
Dr. Sarah A.Corbet discussed the variation of sugar content in nectar at different times of the day and refers to the energetic reward as the benefit bestowed for a particular expenditure of energy. This is lowest at midday and accounts for a period of reduced foraging activity on the part of the bees around that time. An overhead sun can hinder bees navigational ability around midday though that is no problem outside the tropics.
See: energy profit, pay-load

**energy**
Work potential. It can occur in various

forms: chemical, electrical, mechanical, nuclear or thermal. Honey is an excellent source of energy in the metabolism if bees, humans and many other creatures. Energy is required to ripen honey, maintain incubation and the winter cluster, to produce wax, to fly etc. etc.
See: catabolism, trehalose

**energy profit**
This is referred to in Bumblebee Economics and is used for nurturing bumble brood. Pollinating insects forage at considerable energy expense.
See: pay-load

**English honey**
Honey that has been collected from floral sources in England. Although great variations in flavour and quality are to be found, English honey ranks with the best in the world. It is relatively expensive to produce on a marketable scale and is easily outstripped price-wise, by foreign honeys. This has led to the abuse of the 'name' i.e. cheaper foreign honeys sold under an English label. Many beekeepers here would be surprised to learn that 20% of English honey is derived from honey-dew. People generally prefer the honey of their childhood, so most like honey from English grown flowers.

**enteric canal**
Pertaining to the enteron. More often called the 'alimentary canal'.

**entomology** entomo – insect
A branch of biology and zoology that deals with insects including butterflies, moths. A subject which not infrequently leads people to beekeeping.

**entomophilous** philous – loving
A plant that requires pollinating by insects. Their structure lends itself to effective pollination by creatures such as bees as their flowers are usually of a suitable shape and size, with large perianths (flower envelope), colourful petals and large pollen grains that are often oily and sticky and occur in tetrads. Nectaries, scent and nectar guide-lines may be present and their pollen is not readily carried by air currents.
See: anemophilous

**entomophily**
The pollination of plants by insects.

**entrance**
A welcoming passage which in the case of bees is also used as their exit. Most hives will be totally enclosed apart from this and where bee-proof ventilation is provided, as in the roof. Entrance size will be governed by the strength of the colony, time of year and prevailing conditions such as heat-waves, robbing, and need for colouration if a queen needs to be specially directed back to her home. There is probably a tendency for entrances to be made too large and you only need to look at hives in warm climates to realise that protection against robbers and the ability of bees to ventilate their hives even when the opening is small, lead to reasonably small entrances being the norm.
See: bottom, top

**entrance adjustable**
Few hives are left with the same sized entrance the year round. Consequently a means of changing the size is provided, either by entrance blocks or slides, according to the hive design.
See: disc entrance

**entrance block**
An entrance reducing block. A piece of wood or suitable material that fits into the space between brood box and the floorboard offering varying sizes of entrance
1. full size with no block.
2. medium size with largest slot of the block open to bees.
3. small size when block is turned through 90° to expose only the smallest slot. A block or something similar is useful when moving bees as excitement makes them crowd to the light from an entrance and ventilation is best provided at the top by means of a travelling

screen. A weak colony might require a single bee-way entrance to avoid a pummelling from wasps or other bees.

**entrance** – bottom, American 'cleat'
Bottom entrances are conventional because it is there that bees can hand over their loads without having to carry them too far, drones and the queen if necessary have easy access and top entrances would be something of an invitation to honey robbers. However while being on constant guard against robbing, in the warmest of weather it is sometimes expedient to have openings at other levels in addition to the bottom.
See: entrance block

**entrance slide**
The famous WBC hive was designed with special entrance slides which covered every range of opening to full size down to a single bee-way.

**environment**
All the factors associated with the surroundings that collectively provide conditions for the development of growth or otherwise including the influences of position, time, weather, animals, plants, chemical, physical and electrical conditions. For example a hostile environment would make it near impossible to sustain living conditions. A suitable environment for bees would be one in which they could prosper – melliferous flowers, a pleasant climate and well-intentioned beekeepers.
See: biosphere, ecology, ecotype

**enzootic**
Afflicting living creatures of a certain species within a certain locality. Hence repeated outbreaks of chalk brood, bee-louse, acarine, CBPV, Varroa etc.
See: endemic

**enzyme**
An organic compound that is found in living cells. It is a soluble colloid and catalytic protein, the activating principal of fermentation, decay and inversion during which processes it is not used up yet speeds the action immeasurably. It is unstable and requires a certain range of temperature and pH value and its considerable influence is discontinued once the supply has stopped and it is easily destroyed or inactivated (*remember this when heating honey*). Each enzyme attacks a specific substance and its activity can be increased or reduced by appropriate activators or inhibitors. In the elaboration of honey the enzyme 'invertase' (sucrase) hydrolyses sucrose to form glucose and fructose. Enzymes are classified according to the type of reaction that they catalyse such as hydrolysing and oxidation – reduction enzymes. The noun suffix 'ase' is used for enzymes, as for example 'catalase'.
See: co-enzyme, honey enzyme', hydrogen peroxide

**ephemeral**
Lasting for a day. Short-lived animals or plants. Transitory – the life of an adult gnat.
See: circadian

***Ephestia kuehniella***
A moth which occasionally attacks comb. It is known as the Mediterranean flour moth and has pinkish white larvae.

**epicuticle** Gk. Epi – on, upon, over, above
The outer layer of a bee's skeleton.

**epidermis**
The cellular layer underlying and giving rise to, the cuticle. It is underlined by a tough, thin basement membrane.

**epilepsis**
A nervous condition in which queens will fall over on their sides and feign death in order to get out of trouble (perhaps from fright). They recover normally once the disturbance has gone. This state is also called: 'akinesis' in the case of a worker or 'thanetosis'.

**Epinephrine**
Adrenaline – useful in severe cases of stinging especially where 'anaphylactic

shock' occurs, when ***immediate*** attention is essential.
See: Drenamist, medi-haler

**epiopticon**
The middle zone of the optic lobes of insects.

**epipharynx**
The roof of a bee's mouth cavity, the upper lip, and consisting of a soft pad that lines the inner surface of the 'labrum' which is hinged to the 'clypeus' and can be brought to bear, (snuggled against), the dorsal surfaces of the upper part of the 'proboscis' thus forming an air-tight seal so that liquids can be drawn up the proboscis and into the 'cibarium'. It is raised to give access to the cibarium when food is regurgitated so that another bee may take food using its proboscis during food-sharing but is otherwise concealed by the upper lip.

**epithelium**
A layer of cellular tissue covering a free surface or lining a tube or cavity. It may form a protective or secreting layer viz. 'epidermis'.

**epochs**
Time required for making a geological series. From the Cretaceous we have Palaeocene, Eocene, Oligocene, Miocene, Pliocene, Pleistocene into modern times.

**epomphalia**
A naval ointment made from honey for the newly born – ancient Greeks.

**E.P.P.**
Abbreviation for 'Effective Pollination Period'. This term was originated during pollination research at Long Ashton Research Station and it refers to the period during which the stigmata are receptive to pollen while still allowing time for the subsequent fertilization of the flower's ovules to take place. This is flower life minus the time taken for deposited pollen grains to send tubes down to fertilise sufficient ovules to lead to the production of a perfect fruit. In cool springs the EPP of top fruit may be as short as a mere day or so.

**equalising**
Colonies are apt to come through the winter in varied conditions and the custom of using the strong ones to help the weaker ones finds favour with some beekeepers, providing the robbed stock(s) still reach their peak in time for the flow. If done with care, skill and correct timing, colonies that might otherwise have built up 'on-the-flow' instead of 'for-the-flow' can be made equal to the best honey-getters The importance of only using healthy colonies for this purpose cannot be over-emphasised.

**equilibration**
The property that solutions have of taking up or losing water so that their 'vapour pressure' equates with that of the surrounding air.
See: R.H., vapour pressure

**equilibrium table**
Honey is hygroscopic and its vapour pressure seeks to be in 'equilibrium' with any air in contact with it. That is why it is essential to seal honey in air-tight containers. If exposed to low R.H. (relative humidity) the air will draw moisture from the honey and vice-versa. *(this has been used to densify honey by drying).* The following table give RH of air against % of honey moisture content.

| | |
|---|---|
| 50 - 15.9 | 65 - 20.9 |
| 55 - 16.9 | 70 - 24.2 |
| 60 - 18.3 | 80 - 33.1 |

**equipment**
Apparatus or paraphernalia designed to fulfil specific functions. Sometimes referred to as 'gear'. Leaving gadgets and gimmicks aside, apiculture is dependent on hives, frames, and the means of handling them. There are many kinds of well-tried equipment, proved by constant use to be highly beneficial to bee or honey farming. New items constantly appear. Equipment is listed in catalogues

of bee appliance manufacturers and is on display at many honey shows.
See: equipment suppliers

**equipment supplier**
Beekeeping equipment is supplied by many firms who have been in business for many years. Some organisations come and go. Naturally prosperity depends on the state of beekeeping in that country but queens and equipment are exported by several organisations.
See: equipment manufacturers, suppliers of beekeeping equipment

*Erica carnea* **Ericaceae**
Winter or mountain heath. Although often looked upon as if it were the true heather Calluna vulgaris but the vast number of cultivars of this pretty heath which also grows amongst the wild heathers has a variety of colours and might attract bees at almost any time of the year.
See: Ericaceae, Calluna, heath

**Ericaceae**
The heath family – evergreen subshrubs. It includes: bog rosemary, bearberry, all the lings and heathers rhododendron, azalea, bilberry, cranberry, cowberry.
See: Daboecia, heath (plant), tree heath

**Eros**
The honey-thief mentioned by Theocritus but also Greek God of Love.

**Erucic acid $C_{22}H_{43}COOH$**
Found in vegetable oils, especially those derived from rape. It causes ill-health, particularly heart conditions.
See: Canola

**Erythromycin erythro – red**
An anti-biotic sold under the trade names of Erycen, Erythrocin and Erythromid.
See: Terramycin

***Escallonia langleyensis* Saxifragaceae**
This garden hybrid is a useful shrub producing often aromatic red, pink or white flowers from July onward. Bees work them well for nectar.

**escape**
An alien animal or plant which, having escaped from the place man originally intended to keep it, establishes itself and spreads in what proves to be a suitable environment. So dominating can some of these escapes become, that quite drastic measures are sometimes needed in order to attempt to rectify the situation. Examples include *Apis mellifera adonsonii* (the so-called '*killer bee*') in S. America, and Salvation Jane in Australia which was subsequently looked upon as Paterson's curse.

**escort**
A group chosen to accompany, care for and guard, someone or something of importance. A queen should not be left on her own for more than a few minutes and only then on occasions such as immediately prior to putting her into a new colony. The bees chosen to accompany her receive various epithets including 'escort'. Within the hive there are nurse bees that turn towards or attend to the queen and they would be called the attendants or courtiers. Queen cages are usually large enough to allow for up to twenty workers to accompany the queen.
See: retinue, re-worker

**Essenos**
The Greek term meaning 'King bee' the epithet of Zeus. The Essenes, a Hebrew caste, upon whom light has recently been thrown by the discovery of the Dead Sea Scrolls, were throughout to be famed for honey production.

**essentials**
For the bees this is shelter, food and forage. Man can provide these and entice colonies into semi-domestication. They are the only creatures apart from ourselves who collect, make and store food. We might take the

collection of nectar and pollen for granted and that 'stealing' some of their surplus can be recompensed if necessary by giving them sugar syrup but it is not always realised how much bees need ordinary water. Not just for cooling in hot weather but the larvae, like us, are composed of 90% water and drinking 'fountains' or places where water carriers can up-load without fear of slipping in and drowning, should always be made available.

**establish**
Bee hives cannot be placed anywhere. To install, set-up, organize, situate and find a place for hives requires some forethought and attention to what beekeepers with years of experience can tell you. So best join the Local Beekeepers Association who are ever ready to help a beginner with advice and suggestions as to how knowledge might be obtained.
See: enclosures, sites

**ester**
A fruity smelling organic compound formed by the condensation of an alcohol and an acid with the elimination of water. Di-ethyl ether is an example and waxes are largely esters of the higher fatty acids and glycerol. The saccharide esters are formed by the reaction of sugars with acids.

**ether Diethyl ether**
A volatile liquid of great mobility, heavier than air, and with a high refractive power. It is a good solvent of fats, causes cooling by its rapid evaporation and produces a vapour which acts as an anaesthetic, is highly flammable and does not mix with water. Produced by the action of sulphuric acid on alcohol. It comprises quite a large group of compounds including the elements C, H, and O e.g. $(C_2H_5)_2O$
See: ester

**ethylene dibromide EDB**
Sold as a liquid that forms a gas on exposure to air at room temperature. It kills all stages of wax moth. As it is highly toxic to humans, a fumigation chamber is necessary. 900g per 28 cubic metres or 2 lbs. per 1000 cu ft. is recommended for the sterilisation of equipment. It must not be used for combs that contain honey. It blisters skin – wash immediately with soap and water. Banned in USA in 1983.
See: PDB

**ethylene oxide $C_2H_4O$**
A colourless gas with a sweet smell. It is highly flammable. Its liquid has a low boiling point 10.5C/ 51F so it is stored and handled as a liquid at fairly low pressure. Used in US as a fumigant and insecticide for stored grain and to sterilise combs infected with diseases - nosema spores and for AFB.

**etiquette**
An evolved and accepted code of behaviour. We should be reluctant to use words designed to cover human situations when writing about bees. Nevertheless a queen establishes a certain rapport with her workers and the inter-reaction is essential if the colony is to function normally. Any break-down such as when a queen fails to 'deliver-the-goods' or when a newly introduced queen is not fully accepted, may lead to supersedure. Likewise the response of returning foragers to guard bees or one bee to another within the hive, has to be fairly precise as governed by communications of which we are only partly aware.
See: challenge, communication

**etiology aetiology**
The study of the causes of anything, as for example - diseases.

**-etum**
A suffix attached to the Latin generic name of a plant to denote an area dominated by it. Examples: Callunetum – heather moor, arboretum – tree dominated park.

**Eucalyptus spp.**
A half-hardy tree also grown in the milder

parts of the U.K. Grey Box (Gum-topped Box) E. moluccana often found in pure stands NSW.- buds Dec, flowers Feb to April, - white, poor honey producer. In Australia alone there are over 600 species and plenty grow and flower in Southern Australia especially in the southwest where forests of giant trees provide overhead canopies which hum with bees and provide nectar galore. Efflorescence can be variable and advice as the likelihood of a flow has influenced some long distant migrations. Pollen from such exotic plants has led to the conclusion that some honeys labelled, 'English' had foreign honey in them though when the spectrum of pollens is taken into account there is less room for error.

**European Foul Brood EFB**
A serious disease of the brood in which larvae normally die in the unsealed stage. This is a notifiable disease – you must contact your local bee inspector.
See: EFB

**European honey**
Up-to-date information could adjust these figures but in the 80's the highest producers were, in this order: Germany, France, Spain and Poland. Sweden has too short a season though it enjoys long summer days. In comparison Canada obtained an average of 57k per colony.

**eusocial**
Describing the complete social organization of an insect colony including:
1. Co-operative brood care by members of the same generation sharing a nest.
2. A specialised reproductive caste.
3. An overlap of generations with offspring assisting parents in the work of the colony. – The reproductive division of labour.

**eutropous opp. allotropous**
The adaptation of some insects (bees) to visit certain types of flowers.

*Euvarroa sinhai* **Sub family Varroidae**
A parasitic mite found on Apis flora and A.cerana in S.E. Asia though unlike Varroa it is not considered a danger to A. mellifera.

**evacuate**
To leave a place empty, to void faeces. In over-crowded Britain it is always possible that because of building, maybe a road, house, dam or reservoir, that an apiary has to be vacated. It is wise therefore to have a second apiary site or a potential one in mind. Bees void faeces on the wing and the term 'defecation' is used for this.

**evaporate**
To give off moisture or to turn into a vapour. Certain fumigants and subjugants used in apiculture utilise evaporation to become effective. Evaporation depends on vapour pressure and temperature. Benzaldehyde – a repellent – can be too effective if the weather is very warm. Formaldehyde may take longer to kill nosema spores in cold weather. The making of honey calls for the evaporation of water from nectar and it is interesting to note that it takes more energy to evaporate a litre of water than to raise its temperature from zero to boiling!
See: vapour pressure

**Evelyn John 1620 – 1706**
An English diarist who wrote Elysium Britanicum, a treatise on gardening with a section on bees. He also wrote in his famous diaries about his transparent observation hive.

**evening primrose** ***Oenothera biennis, stricta, missouriensis, lamarckiana*** **Onograceae**
Yellow flowers from June – Sept. A common vagrant with large flowered varieties.
Pollen light yellow with a grain over 100μ. Harwood describes pollen as bright yellow. It is a hardy perennial trailing with scented, yellow flowers opening towards evening. It will naturalise in damp places. Grows to one metre tall and it is open and visited by moths at dusk but partly open and worked by bees

for pollen by day. Oenothera means 'wine taste'.

**evergreen**

Plants which retain their green foliage throughout the winter as opposed to 'deciduous' plants which shed their leaves in the autumn and grow new ones in the spring. In some cases the new ones are not shed until the new ones are completely formed. Most of the NZ forests are composed of evergreens.
See: deciduous

**everlasting pea *Lathyrus sylvestris* Leguminosae**

The flat pea or vetchling a perennial which climbs by means of tendrils its stems becoming self-supporting. A cultivated form known as the Wagner pea has proved useful to bees and livestock while also appearing to be drought resistant.
See: vetch

**eversion**

Turning inside-out. The extension of the drone's genitalia at the moment of copulation or other stimulation, whereby the apparatus concerned with reproduction, which is normally contained within its abdomen, is explosively protruded (with an audible 'pop') so that sperm can be transferred via the penis into the corresponding female receptacles of the queen.

**evolution**

Darwin's theory of evolution suggests that all life forms are related and that changes in characteristics occurred in the course of successive generations. This can be understood more clearly by a study of taxonomy. This theory of gradual development is not shared by those who believe that the ecosphere was suddenly created.

**Examinations**

B.B.K.A. Examination Scheme inaugurated in 1882. When you commence beekeeping, it is full of interest and new knowledge is rapidly acquired. What greater pleasure than to talk about it? However unless your knowledge is 'sifted' put to the test, corrected and linked purposefully to beekeeping, over confidence and a false sense of your ability can arise. The let-downs which some beekeepers have when confronted with their inability to pass exams has caused many to try to belittle them. The quiet confidence that comes from the passing of exams and the real acquisition of knowledge helps to enhance what is after all, one of the finest pastimes. For more information contact BBKA

**examine**

To carefully scrutinise, as for example when a Regional Bee Inspector inspects colonies of bees and examines the brood for any signs of disease. Also the process of testing those who present themselves for an examination with a view for instance, of securing a beekeeping qualification.

**Examiner**

A person adjudged as suitably competent to mark an examination paper or test a candidate with practical work in an apiary or to set examination questions. According to availability and willingness an examiner will be as highly qualified in the particular field of knowledge as possible and will certainly have had to go through the rigours of the examination or test being conducted, as well as at higher levels. BBKA examiners work voluntarily and spend considerable time at meetings, forever trying to improve standards and hopefully helping students to present themselves at a standard likely to achieve success.
See: candidate

**exchange of comb**

The replacement of an old or contaminated comb by a new or sterile one. This may be drawn comb or sheets of foundation inserted when their development into full combs is likely. (The drawing out of foundation is only effectively done between the months of

May and September and when it is correctly positioned within the hive). Bees can be 'nursed' onto foundation. Say two sheets between a drawn out frame in the centre of an upper box.
See: foundation

**excipient**

Used in pharmacy to describe a more or less inert substance that can be used as a 'vehicle' for the administration of an active medicine. A spoonful of honey makes the medicine go down, in a most delightful way, Tate and Lyle probably changed the wording. But of course honey is anything but inert as it is a food and a medicine at the same time. So sugar, honey, jelly and wax are all used to help with the introduction of prophylactic treatment including bee diseases.

**excitable**

Easily aroused, soon upset. The alertness and readiness of bees to defend their domain may be excessive so that the slightest interference with them results in an immediate and unwanted response. Such bees could only be conveniently kept by very experienced beekeepers in isolated out-apiaries where their drones would be unlikely to interfere with ordinary beekeepers queen rearing programmes. Another form of excitement is when combs are lifted out of the hive for examination and the bees rush excitedly about over the comb face. The degree of excitability that a beekeeper is prepared to put up with depends on himself. Re-queening with a more placid strain should always be possible.
See: characteristics

**excluder**

A thin, slotted screen arranged horizontally right across the top bars of the brood frames, the slots being wide enough for workers to pass easily through but too narrow to permit the passage of a queen (whose thorax is slightly wider than that of worker bees). The same slots make the passage of drones impassable so be careful not to shut them over the excluder or dozen of dead bodies will result. Ordinarily queen excluders are made of slotted zinc or stainless steel sheets or from braced wires set the precise distance apart 4.14/4.24mm 0·163/0·167". The 'Waldron' is an example of a framed wire excluder. Slide your fingers across a new slotted sheet and if any sensation of a burr is found on one side, smooth it off with sandpaper. When excluders are taken off be careful not to distort them when propolis tends to hold them down and when clogged with burr comb remember they are more easily tapped clear when the wax is very cold. (i.e. during winter).

**excrement**

Excreta, faeces - if found on top bars and by hive entrances may indicate undue excitement or possibly 'dysentery'. In the latter case scrapings may be used for diagnosis under the microscope. The larva only excretes at the end of its feeding period and the faeces are trapped by the last skin as it moults. A queen's excreta are removed by workers of her 'court'. Surprising amounts of excellent manure are liberated by colonies because workers drop their faeces in flight and improve the surrounding soil.
See: debris, droppings, frass

**excretion**

The removal and elimination of unwanted products of metabolism. See function of Malpighian tubules and alimentary canal.
See: egestion, secretion

**excruciating**

Pain that causes extreme suffering. Although bee-stings are painful it is not really accurate to describe the pain they give in this way. The pain is more akin to a drop of very hot water that hits the flesh and tends to stay hot. Pummelling or massaging the spot makes it no worse and often a pain elsewhere completely negates the pain of the sting. Consequently it is not unreasonable to suppose that the pain is an induced feeling by virtue of the venom acting on the nerve endings. Subsequent discomfort of the after-effects is another matter.

**exhibiting**

1. The act of showing carefully prepared examples of honey, wax and other bee display material usually at a honey show.
2. The display of certain characteristics by a plant or creature.
See: beekeeping show, honey show

**exine (extine)**

The outer coat of a pollen grain some of which – like dandelion – are digestible. It consists of sporopollenin in an almost indestructible natural compound. The exine consists of layers of sexine and nexine or base layer, columellae or rods and tectum, a roof-like structure. The tectum may be partly or completely absent. The characteristic sculpturing of the surface may be stripy (striate), net (reticulate) etc. The ornamentation, best seen in section, may consist of borders (valla), clubs (clavae), grains (gemmae) knobs (verrucae) network (reticulum) spines (echinae) or warts (knobs) and helps to identify the grain. The inner 'nexine' has been further divided into the endexine, foot-layer and ectexine though there are disagreements as to terminology. The pollen husks from centuries past have been used to discover plants of the period and food eaten by herbivores.

**exit**

The way out. Although the opening through which bees pass to their nest is of course also their way out or exit it is customary to refer to this opening as the entrance.

**exocrine gland opp. endocrine**

Glands whose secretions are drained off by ducts. Occasionally unicellular but more often masses of tissue capable at some stage of producing a secretion that is ducted to where it can be utilised. In honeybees these glands produce both hormones and pheromones for example the mandibular glands.

**exocuticle**

The tough outer-layer of the exoskeleton. It is water-proofed by the cuticle. In composition it is mainly sclerotin but contains some chitin while the more supple endocuticle below is mainly chitinous.

**exogenous**

Originating outside the organism.

**exohormones**

Social hormones, ectohormones, pheromones. These are important in social insects and much work has been done to identify them and establish their effects.
See: queen substance

**Exopterygota**

A major division of the Pterygota. It is a sub-class of insects also known as Heterometabola and Hemimetabola in which the adults develop by successive moults from embryo forms not unlike themselves. Earwigs, dragon flies...

**exoskeleton**

A hard external structure supporting animals like crabs and tortoises. The outer layers of the honeybee's skin are hardened and water-proofed to form an exterior skeleton, similar in some respects to the human head. The layers consist of the waxy epicuticle, cuticle, epidermis, exocuticle, more flexible endocuticle and the basement membrane. The tough outer frame not only protects soft internal parts but acts as an anchor for internal muscles. Its weight limits the size of creatures that can be successful with this arrangement
See: cuticle

**expanded metal**

A sheet of metal that has been made into a network by cutting and stretching. This gives great rigidity and serves well as a mouse-guard (or excluder) provided the apertures are not too big.

**expansion**

Increasing in size and volume. Colonies can do this quite suddenly in spring and room must be given ahead of requirements or swarming is likely to result. Extensions of the hive by the addition of brood boxes or supers, if timely, will help to cope with the situation. Honey intakes are not easy to forecast but brood areas mean 3 times that area occupied by adult bees within 3 weeks! Small colonies that 'burst-their breeches' are sometimes called 'exploders'.

**experiment**

An act or operation of a tentative nature, undertaken with the purpose of testing something or discovering something unknown. This could be part of apiary management, the use of drugs, equipment, swarm control etc.

**expert**

A person who by virtue of his knowledge in a particular field is considered to be an authority on that subject. Those who pass the BBKA Senior Exam are deemed to be experts though these are not the only apicultural experts.

**expose**

To lay open to attention. Hive products are enclosed by the bees in a shell of propolis that not only keeps out the weather but makes it harder, even for small creatures like ants to gain access. It is wrong to leave wax, propolis, honey or other sweet material near the hives as this invites robbing. Losing honey to wasps is one thing, inviting a bee with Varroa or foul brood is another. When manipulating, the bees flying around will all look alike (no blazers to identify which colony they are from) so consider the use of cover-cloths and do not keep a hive open any longer than reasonably necessary.
See: dissection, opening hives

**expulsion**

Driving away, ejecting or terminating membership. Many a good and expensive mated queen has finished up dead in front of the entrance because the colony was not prepared correctly (nor the queen herself) for acceptance. A colony will throw out aliens and their own bees if they are injured or diseased. When the mating season is over even the colony's own drones are expelled.
See: expulsion of drones.

**expulsion of drones**

This occurs in normal circumstances under the following conditions:-

1. Queenright stocks normally expel their drones towards the end of summer. They are then found dead in heaps before the entrance.
2. A colony with a newly mated queen may cast out their drones.
3. During unseasonal dearths, drones may be ejected.
4. Drone larvae may also be sucked dry and thrown out as in 3.

**external cluster**

Honeybees have acquired the habit of bunching together in groups or clusters and this enables the collective identity and raising of their temperature above that of the ambient air. Anyone who braves putting a bare hand into a swarm cluster is likely to be pleasantly surprised at its warmth and softness (silky). An external cluster may form at the entrance or on the outside wall and this is often referred to as 'hanging-out' or forming a 'beard'. Also when bees leave the hive to form a swarm or cast, they are held together by the clustering instinct and sensitivity to the queen's presence.
See: winter cluster

**exteroceptor**

A sense organ or receptor that receives outside the body – stimuli from 'antennae.

**extract**

To get out by force. The removal of honey from uncapped comb is an example of extraction. Instead of allowing the honey to drip or drain from the open cells under the influence of gravity, the comb is whirled round in

such a way as to increase the effective weight of the honey and thus hasten its expulsion without damaging the comb.
See: radial, tangential and slinger

**extracted honey**

Honey that has been extracted from the combs. It may have been pressed, squeezed , scraped or expelled by centrifugal force from the cells of the comb. When filtered to remove particles of wax, propolis etc. It is referred to as 'clear' or 'run' honey. As it will be charged with air bubbles it is usual to keep it warm in a cylindrical container (with a 'honey gate' at the bottom) to allow the bubbles time to rise as froth whereupon the 'bright', clear honey can be run off into containers.

**extracting**

The act of separating honey from the comb without destroying the septum of the comb or the cell walls. It tends to be a sticky job although modern extractors and acquired 'know-how' can reduce work to a minimum. Most hobbyists probably wait until the end of the season so that only one operation is necessary. There is a general impression too that the honey will be 'riper' by the end of the season but in fact once it is sealed it is ready to extract. Heather honey is a late crop and it requires special extraction by means of a press. The work should be done in a warm, bee-proof room with care given to all aspects of hygiene, i.e. cleanable floors, working surfaces and running water. Certain elements in honey deteriorate when heat is applied, though because of its lower viscosity once it has been heated, there is always a temptation to use heat so as to aid the filtering process. Commercially, sold honey, in order to conform with regulations and put brilliant honey on shop shelves is usually heated but the hobbyist beekeeper can avoid all forms of heating and benefit from the full healthful aspects of this wonderful food and medicine. There are cold uncapping knives and honey forks and the gravitational separation of clear honey may take a little longer but eliminates the need for any heating except when preparing honey for a 'show'.
See: wax extraction.

**extractor**

An apparatus built to reduce the work-load involved in obtaining the best results from the season's harvest. For honey, a geared machine which whirls the uncapped combs around, is made to be operated manually or by electricity. There are centrifugal and radial cages in extractors that spin honey from the comb. Honey must be pressed out of combs of the thixotrophic heather honey. The cappings removed from the combs can be spun around in specially designed cages. Wax may also be extracted by steam, pressure or heat from the sun (solar wax extractor). Due to the expense of a honey extractor which may only be used once a year, Beekeeping Associations often help their members by loaning machines as do some equipment manufacturers
See: extracting & Hrushka

**extra-floral nectary**

A plant gland that secretes nectar and is positioned elsewhere other than in the flower. Though similar in structure to a floral nectary it may well produce nectar at times when the plant is not in bloom. These nectarines often yield more concentrated nectar than the floral nectary. The reason for such nectaries has included suggestions that the nectar attracts ants which then drive off other insects. Extra-floral systems are found on the 'petiole', 'calyx', corolla' and on the fruit itself. Examples include: bracken, butterwort, cotton, field bean, laurel, runner bean, sycamore and certain other trees, toothwort, vetch, yellow rattle, and nectiferous tissue is also found on folia nectaries on lesser celandine. This nectar is a different thing from honey-dew which is the secretion of insects.
See: honey-dew and leaf honey.

**exude (exudation)**

The leaking or seeping-out of a fluid. Untold numbers of plants exude sweet

secretions called honeydew, leaf honey and such like. Honeybees are usually more likely to be attracted by the bright colours and nectar guide lines of flowers, but are known to delve into gloomy undergrowth of laurel and even bracken for honeydew. It sometimes helps to almost 'train' them to some exudations such as the droplets on the whiskers that protrude from the bark of Southern beech in parts of S. Island, NZ. Moisture may trickle from a hive when nectar-ripening is in full swing.
See: extra-floral nectary, honey-dew and weep

**eye colour**

The colour of the eyes of unborn bees is a guide to their stage of metamorphosis: white, pink, purple an finally brown.
See: age of brood.

**eye compound** ***

The two compound eyes occupy most of the bee's head having large convex surfaces which enable them to see in almost all directions at once and to compare the amount of light they receive from the ocelli to stabilise their flight. Their heads can be rotated at the neck giving an even larger view of their surroundings. Each eye is composed of some 7,000 ommatidia, each having a separate lens or facet and nerve connection to the brain. The lenses are interspersed by hairs. A drone has nearly 17,000 lenses in all and the increased size of its compound eyes force the ocelli down onto the frons. Changes in pupal eye colour enable brood age to be checked.
See: compound eye, facet, ommatidium and simple eye.

**eyelet**

A metal or plastic ring for lining and reinforcing a hole. Small brass ones are made, their length equivalent to the thickness of the wood forming the side-bar of a frame. Having made a small hole in the appropriate position with a small drill or gimlet, the eyelet with its finished edge on the outside of the frame is pushed home before the reinforcing wires are fed through. They can then be pulled tight without cutting into the wood which would allow the wire to become slack.

**eye simple**

A bee has 3 ocelli or simple eyes arranged in a triangle, two over one. They are merely lenses covering elongated retinal cells that are connected to the brain by nerves. Bees are able to work in complete darkness and it is said that the simple eyes are used to register light intensity. This helps them to find the entrance, avoid bad weather (bees hurtle back home at the onset of an eclipse or the approach of a thunder cloud) and step up their activity as days lengthen and when bright weather occurs. Their ability to correlate the light from the various sets of eyes helps them to stabilise themselves in flight.

# F

### F1
This symbol is used for the first filial generation in plant and bee breeding to label the offspring as resultant from P*1* the chosen parents. When F*1* siblings are interbred their offspring become F2. The artificial insemination of honeybee queens and the breeding of pure genotypes utilises this terminology.

### Fabre 1823–1915
Popular French writer and entomologist

### face lotion
1tsp sweet almond oil and 2tsp honey. Blend and use after the skin has been thoroughly cleansed. Leave on for about 1/2 hour. Remove with soft cloth and tepid water. Apply milk astringent to close the pores and tone the skin.
See: face pack

### face pack
One third cup finely ground oatmeal (or bran, cornmeal or white flour) 3tsp. honey enough to make into a smooth paste. 1tsp rose water (or orange flower water). Blend oatmeal with honey until well mixed. If too thick and unmanageable, add a little rose water (or orange flower water) spread over clean face with the exception of the eyes and leave it on for about 1/2 hour. Relax while it is on if you can. Remove with soft cloth and warm water. A good astringent should be applied to tone the skin. Dianne Meredith
See: Cleopatra, honey mask

### facet
A smooth, flat, or round surface for articulation, an ocellus; corneal portion of insect eye. Each facet is capable of transmitting a visual stimulus to the brain.

### factors leading to successful queen introduction
1. Ensure colony is queenless
2. That it has a high proportion of young bees.
3. That there is a good flow, real or stimulated.
4. Give time for bees to become conditioned to the new queen before she is released.
5. Ensure that the queen is in the proper physiological state to match that of the receiving colony.

### facts & figures
For every pound of honey consumed by the bees – half a pound of water is release by consumption. Wedmore. It takes 10 loads of pollen to rear one worker. Volume of dead bees 2.75 per ml. To produce a pound of honey bees may have to work as many as 340 million blooms.

### faeces
Droppings, excrement – waste matter discharged from the bee's alimentary canal via the anus. Under normal conditions faeces are released in flight, well away from the hive. Undue excitement, disease or disorder may cause bees to soil the area around the entrance, the outside of the hive or even the top bars and combs. They vary in colour from yellow through brown to black and can be removed by the immediate application of a wet rag. Many practices such as supplying

water at a distance from the hive are used to encourage in-flight cleansing. The weight of faeces released by a colony is quite considerable and provides excellent manure but some pathogens survive for considerable periods in contaminated faeces.
See: soiling washing

**Fagaceae**
Monoecious trees and shrubs: Sweet chestnut, beech and holm oak.

**failing queen**
A queen whose ability to fulfil all the functions required of her, begins to decrease. This is shown by shorter laying season, more and more empty cells (cells without brood or young larvae) pollen or nectar amongst the slabs of sealed brood. Bees normally respond by attempting to replace the queen under the supersedure impulse.
See queen failure, supersedure

**faithfulness**
The fidelity of a honey bee to a particular task is one of the most fascinating aspects of beekeeping. A colony can also be faithful to a queen sometimes long after her usefulness has finished so that the bees hang on to her even when the beekeeper tries to introduce a new young fertile queen (the old one must be found and destroyed first).

**family**
A group within a taxonomical order. At one time called a natural order. In Zoology 'apidae, idae etc. In Botany rosaceae, etc. Each family contains one or more genera. Family is also an alternative name for a colony of bees.

**family tree**
A chart which records the genealogical relationship of a given queen so that its ancestry can be followed.

**famous sayings**
'The best packing for bees is bees'. On honey and wax – 'give the two noblest of things – sweetness and light'. Then: 'Upon no other thing does the honey part of the apiary depend so much as it does upon the queen. George M Doolittle 1909.
See: quotations, sayings, superstitions

**fanners**
Bees fan for various reasons. To attract other bees to an altered entrance, source of food or water, or a swarm or by way of ventilation or controlling temperature and humidity, for ripening nectar and for the dispersal of pheromones. When attraction is the object they make a backward draught, the abdomen is held high and curved to expose the Nasonov scent gland situated on the dorsal side of the penultimate segment. Once the fanning posture has been adopted the bee usually keeps it up for a considerable time, holding itself down with its claws and using its wings unhooked and vibrating at high speed. It is likely that bees draw air towards their antennae so that they can detect smells in the way that animals 'sniff'. A single bee may begin to fan assiduously, to be join by others and whole formations of fanners are seen disporting before a hive entrance, while lines of them may develop within the hive.
**See:** scenting

**farina**
The word usually refers to flour, meal, corn, nuts or starchy roots. It also describes powdery substances and is used botanically for 'pollen'.
See: bee-bread

**Fasciata Apis. mellifera fasciata (lamarki)**
Egyptian bee found in N.E. Africa which builds unusually large numbers of queen cells.

**fat**
True fat is an ester of glycerol and fatty acids. Bees are able to produce wax (a fatty substance) on a carbohydrate diet. It contains the fatty acid 'cerotic acid. Around

30% of pollen is fat. Lipide, lipid, lipin and lipoid are synonyms. Lipase is the enzyme capable of causing the break-down of fat into metabolically usable materials.
See: fat-body

**fatal**
Causing death. It is fatal (for a queen) to introduce her to a colony that already possesses one. The use of timber that has been treated with insecticide will be fatal to bees if used for hives or equipment. Slow release insecticides (such as Vapona strips hung up to kill or discourage flies) can be absorbed into fatty substances like beeswax and kill bees long afterwards. Recent work has shown that traces of unwelcome substances have crept into foundation even from the most reliable sources. Other fatal things regarding bees are starvation, damp, exposure, fire and flood and it behoves beekeepers to ensure that risks such as these are avoided as far as possible.

**fat body**
This can build up in bees especially as winter approaches and is composed of irregular masses or lobes of polyhedral cells - trophocytes, distributed throughout the body of an insect which contain various inclusions. Like the vertebrate liver they synthesise, store and mobilise lipids, proteins and glycogen. Healthy, well-balanced colonies take in seemingly endless loads of pollen in the autumn, yet there is no real evidence of this in the combs as the bulk of it is eaten and stored in the form of fat bodies in the bees themselves.

**father**
Male parent, the drone, being parthenogenetic has no father yet can beget sons who are potential fathers. Otherwise their nearest male relative is their grandfather. Sounds a bit complicated but that's parthenogenesis for you.

**Father of modern apiculture**
The title bestowed upon L.L.Langstroth due to his application of the principle of 'bee-space' which enabled movable frame hives to be brought into common use.

**fatherless drone**
As the male bee is derived from an unfertilised egg it has no male gamete fused with the female ovum and is therefore parthenogenetic or fatherless. Such eggs can be laid by a failing queen, an unmated queen, a laying worker or 'at-will' as an alternative to fertile eggs by a mated queen.
See: chromosome

**fauna**
The resident animal population of a given region, as distinguished from plants (flora).

**February**
Although winter is still with us a healthy queenright colony will sense the lengthening days and in the centre of the cluster a small brood nest will be started. From this time on the cluster lacks mobility, wears its nurse bees out more quickly, has a greater need for water and pollen and demolishes its food supply with ever growing rapidity. The return of bees with pollen loads and a gentle feel of warmth over the cluster tells us all is well. Heft hives carefully – if light a comb of sealed stores from a healthy colony may be laid flat over the feed hole. Do not be tempted to open-up unless there is absolutely no activity and you want to check for the worst. A dead colony should be sealed with acetic acid inside and a sample checked for disease. Retain mouseguards. Looking forward to next year encourage those early pollen producing plants like crocus, hazel, dandelion and colts foot.

**fecund**
Capable of producing offspring. Thus we speak of a fecund mated queen. Fecundary were hives that swarmed (Hawkins in Cotton). Fecundate – to make prolific or fruitful, to impregnate. Fecundation mating.

Fecundity – being fertile. Ample fertility, prolificacy. The condition of a queen capable of laying several times her own weight in eggs each day and of producing the largest of colonies. The physical ability of a queen to fill a large brood nest.

**feeder**

An item of equipment used for offering liquid food to bees safely and it would best conform to the following requirements:

1 having appropriate capacity
2 being easy to sterilise
3 offering access to bees in such a way that they do not drown
4 arranging suitable footholds and limited space so that bees do not fall or cannot easily be pushed into the syrup.

There are fast, slow, contact, bucket, Miller, Ashforth, metal, plastic and wooden feeders. Feeders can also be used to provide candy, pollen substitutes. Bees will drown in large numbers if they can get right into syrup and the excitement of feeding may lead to the spread of pathogens. Miller and Ashforth feeders are a complete hive chamber in themselves. Other feeders need empty chambers or deep roofs to accommodate them.
See: named 'feeders'

**feed hole**

In modern hives feed holes are normally provided in the inner cover (crown board) and flat, overlapping pieces of unchewable material such as glass or gauze are used to cover them when a feeder is not in use, so that the bees cannot get up inside the roof in which they might otherwise build wild comb or seal the roof down with propolis. For economy the hole or holes are cut so that a Porter Bee Escape will fit into them when it is required to use the inner cover as a 'clearer board'. Two such holes some inches from the centre, make it easier to position supplementary food over the cluster. A feed hole can be used not only for feeding but to give ventilation, apply a prophylactic, to allow a glimpse of development or to take a sample of bees.

**feeding**

It is necessary to distinguish between the foraging activity of a bee when it uplifts fluids such as syrup or nectar or honey via its mouth parts and feeding when nurse bees give food to larvae or drones or the queen. When a bee feeds itself food is allowed to pass the proventriculus. Up to this point in the honey stomach the contents can be regurgitated if it is necessary for the bee to pass food on beyond that point it enters the bee's digestive system.

**feeding female larvae**

From hatching and for the first day and a half, newly hatched queen and worker larvae receive similar food from the nurse bees. However it is probably best to use larvae no older than one day old when queens are to be reared from them because a larva in a queencell is given richer food from then on or conversely a worker receives a gradually changing diet until it has been weaned onto foods that have not been glandularly produced.

**feign death**

A queen, wax moth larvae and possibly unwanted visitors (hawk moth) may remain as still as death to discourage any aggression from the members of the colony. When a queen has 'cramps' or feigns death, it is likely that she will fall, apparently lifeless, to the bottom of the hive. Best to close the hive up in such circumstances, as left alone she will 'come back to life' and carry on as normal.
See: akinesis, epilepsy, stand-still, thanetosis

**fennel *Foeniculum vulgare***

A hardy perennial that yields nectar and pollen and grows to 1.5m/5ft with yellow flowers in flat umbels from July to Sept. Naturalises especially near the sea. According to Varro and the elder Pliny hives were at one time constructed from fennel stems though this was probably *Ferula communis*, giant fennel a tall ornamental apiaceous herb.

### Fellowship

The BBKA syllabus for this examination may be obtained from the Secretary. The candidate must have taken and passed the Senior and submitted a subject related to beekeeping on which he wishes to write a thesis which must be concerned with original unpublished work A successful candidate is known as Fellow of the BBKA.

### female

The sex that bears offspring. In honeybee colonies females are the result of eggs that have been fertilised and therefore diploid. Both queens and worker arise from eggs that have been fertilised and have sixteen double sets of chromosomes – thirty two in all. Their antennae each possess one less segment than their male counterparts. The queen and the worker, though two different castes, have the same basic sex.
See: queen, drone, worker

### The Feminine Monarch

Famous book by Charles Butler – a history of the bees, 1609

### femur

The third segment of a honeybee's leg. It is a relatively small segment, like the coax and it connects the trochanter to the tibia.

### fence

This may take the form of a protective screen to keep people or cattle away from the hives in an apiary. In North America single strands of electrified wire are used to keep bears from damaging hives in their 'out-yards'. Plants are used to act as a wind break or to camouflage an apiary or to cause the bees to fly at a height that keeps them well clear of footpaths etc. Where wind-proof fences are used, hives should not be set within the area of turbulence. For the fence separator in sections
See divider.

### feral

An adjective meaning wild, untamed, uncultivated, brutal or having reverted to the wild state. It can be applied to honeybees meaning that they cannot be domesticated and are quite able to survive independently of man. Feral colonies are those found in natural conditions not under the control of man. Plants or animals may become feral after having been introduced by man.
See: escape

### ferment

A chemical reaction that splits a complex organic compound into relatively simple substances. This is done by the micro-organism ('yeast' for example) and enzymes and the process is associated with the release of energy in the form of heat and gas. Fermenting stores are bad for bees. Alcoholic fermentation takes place anaerobically when sugars are converted into ethyl alcohol, and carbon dioxide (vinous) $C_6H_{12}O_6 = 2C_2H_6O + 2CO_2$. Aerobic fermentation will convert alcohol into acetic acid $C_2H_6O$ $C_2H_4O_2 + H2O$.
See: honey fermentation, fermentation valve, temperatures

### fermentation valve

A device that permits the escape of carbon dioxide during fermentation while preventing any foreign matter or oxygen from passing through to the 'must' US 'bubbler'

### fertile

Capable of reproduction. In the case of social insects fertile means capable of laying female eggs i.e. having been fecundated and having available viable male gametes.

### fertile queen

A fecundated queen or queen with viable sperm in her spermatheca. This can be put there by artificial (instrumental) insemination.

### fertile worker

An egg laying worker. This expression is used though of course workers cannot be fertilised (impregnated by a drone). Their ovaries however can develop and produce life giving eggs, e.g. drones by parthenogenesis. In *Apis mellifera capensis* bees females are also said to arise from the eggs of fertile workers.

**fertilisation**

Fusion of the reproductive germ cells (gametes) resulting in zygotes. With the plants the ovule becomes fertilised via the pollen tube. In honeybees the drone sperm nucleus fertilises the egg from the queen's ovary. Haploid drones are produced without fertilisation – see parthenogenesis. The act of being or state of being made fertile.
See: apple fertilisation, cross, self

**festoon**

A string or garland looping between two points. Chains of bees each seemingly attached by a claw to another bee's leg or elsewhere in a casual fashion yet able to support a very large number of bees. This occurs in swarm clusters and when hanging chains are formed in the process of comb building. When a swarm or cast is newly hived onto frames, such festoons can be seen when the frames are pulled apart. The curve formed by a freely hanging chain of bees is called a catenary curve.
See: catenary hive, bee chains, comb building

**Fibonacci 1170 – 1240**

An Italian mathematician, Leonardo of Pisa, formulated a series of numbers, each number in the series being the sum of the two previous numbers: 0,1,1,2,3,5,8,13,21...This has been linked with the arrangement and distribution of petals and leaves and with the drone – male parent 'nil', one male grand parent, one great, great parent, two great greats and so on.

**fibreglass**

GRP or glass resin plastic. Material woven from thousands of extremely fine filaments of glass. The massed or woven filaments are used to form insulating material and can be embedded into plastic, resin or glues to form strong, water-proof objects of various shapes e.g. feeders, bee hats, frames, hives car bodies etc.

**fibula**

Also known as 'velum' The jointed spur of the antenna cleaner, found on the inner distal edge of the tibia on the forelegs. It is used in connection with a brush lined hollow shaped to accept an antenna at the proximal end of the inner face of the basitarsus of the foreleg.
See: antenna cleaner

**fickle honey plants**

Some honey plant are much more reliable than others. Weather permitting raspberries will nearly always produce plentiful nectar. Hawthorn on the other hand can vary by up to a month in flowering and it has even been said to produce a nectar flow about once every 7 years (mighty nice honey though). Lime trees prefer warm, still, humid conditions. Wild thyme needs rain at one season while during blossom time rain would put paid to any flow. Apple and ling are another two plants that can give quite heavy flows but are sometimes poor and at other times fail altogether. Despite what is known about the conditions that produce a good nectar yields many plants are unpredictable and so the word 'fickle' is appropriate.

**finding the queen**

The use of smoke is more of a hindrance than a help when looking for the queen. Bees can be enticed to pass through a queen excluder with a view to finding her. Marking her on the thorax helps to find her but familiarity with her appearance grows with practice and experience. This can be hastened by keeping an observation hive. A dense population of badly-tempered bees makes a bad combination when a queen has to be found. A hive can be moved a few metres away and a box of beeless combs placed on the original site to give a

temporary home for the returning flying bees. After a good day of flying weather, the parent colony is easier to examine and combs may be set separately in pairs and the queen given time to get between a pair. Your eyes should then sweep the face of the comb opposite to the one you are taking out. Looked at from that angle the queen with her slightly longer legs straddling the comb can often be spotted. In sweeping your eyes around, take in the periphery first and then let your eyes move towards the centre. If she does not make herself immediately visible, go straight ahead looking at the comb in your hands, holding it vertically at a height comfortable to yourself, and inspect the comb in the manner described. She should be found between one of the pairs in this way, unless the light has driven her down onto the floorboard.
See: looking for eggs, spectacles

**finger print**
An impression of the markings on the last joint of the finger or thumb. It is a form of 'identification' and the term has been applied to the identification of the source of a sample of honey. One aspect of honey's 'finger-print' lies in the particular amino acid content and also in the types of pollen grains it contains. As Dr. Louveaux say's relative to honey, – 'Honey carries within itself its own certificate of origin'.

**finishing colony**
The colonies used to incubate queen cells that have come from a 'cell-building' colony. The greatest care must be taken to ensure that the emergence of virgins does not occur until all the other cells have either been caged or put into a mating nucleus or hive, because a free-running virgin's first task will be to immediately eliminate her rivals and what easier time than when they are encased in their cells?

**fir**
An evergreen, pyramidal, coniferous tree of the genus Abies, with flat needles and erect cones. Because their needles form a carpet that is not inductive to weeds or other growth, it does become a very suitable place for the great wood ant which builds large nests and can be very troublesome if hives are put anywhere near them. Taking bees to the heather in the South has often called for attention to care when siting hives.

**fire**
Burning or combustion. Beehives whether made of wood or straw are highly flammable. So are their contents. When we consider too, that one of the most ubiquitous and useful beekeeping tools is the 'smoker', it becomes clear that the greatest care should be exercised at all times especially in dry, windy weather, when either on lighting or extinguishing the smoker, a fire might be started. It is far better to snuffle a smoker with a twist of green grass as a plug, than to turn out the contents, even when they appear to be 'dead'. A wind can soon fan a spark into a flame and move a burning particle onto dry material some distance away. The safest plan is to plunge the remnants of the smoker contents into water. Bear in mind that while a tree can produce a million matches, one match can destroy a million trees.

**fire blight**
A bacterial disease of fruit trees, especially apple and pear, caused by *Erwinia amylovora.* The destruction of the foliage and wood is reminiscent of a tree that has been scorched by fire. The bacteria can be transported by bees but in N. America where the disease is endemic, the importance of bees for pollination is considered to far outweigh any added risk of the spread of fire blight. It also affects hawthorn, mountain ash and whitebeam. The infectious material that oozes from trees has a very low sugar content and is of more interest to flies and wasps than bees. It has been stated that fertilised blossoms become resistant to infection sooner than unfertilised ones.

**fire lighter**
When the remains of old comb (minus any

wires) are taken from the solar wax extractor while still warm, they can be compressed into suitably sized pieces to act as firelighters

**Firethorn *Pyracantha coccinea* Rosaceae**

An excellent, evergreen, nectar producing shrub, much cultivated in gardens often against walls. It flowers during the June gap and a sea of white blossom later becomes a mass of red berries (as it name implies). It will only transplant satisfactorily when young. It yields pollen too.

**fireweed *Epilobium angustifolium* UK Rosebay willow herb**

An onograceous plant with willow-like leaves and racemes of purple flowers, they are plants of the genus Epilobium. It flourishes in soils excessively rich in potash and good crops of water-white honey have been obtained where the purple stands have developed after a previous season's fire. The Peace River area in Canada (*Chamerion angustifolium)* was famous for notoriously good yields. In Britain, travelling by train or by road, the rich splashes of late summer and autumn colour, enliven the landscape and provide forage for bees, especially bumbles and other insects. After World War II it rapidly established itself on bomb sites in London and it soon takes hold on slag heaps and other newly turned over ground. A true beekeeper's heart sings with joy at such sights and the thought that she has her own bees buzzing happily away elsewhere. See: rosebay

**first flight**

'Off solo!' Circumstances can cause quite a variation in the stage at which a worker first takes to wing. Remarkably young bees are found in swarms (especially absconding swarms). The average age of the first flight during the active season is on or about the 30th day after the laying of the egg and it is likely to occur around mid-day and frequently small clouds of such bees will be seen flying, facing the hive and bobbing up and down, in what are called' play-flights'. A queen may fly on about the 21st day, and drone's flight depends much on the weather but the flights start before it is mature and on the 30th day. See: play-flight, cleansing flight, orientation

**first skeletal ring**

A chitinous surround that is open ventrally like a collar. The pro-thoracic collar. When a dead bee has been selected to be examined for signs of acarine, the head is first carefully pulled off together with first pair of legs. This exposes the prothoracic collar and it can literally be peeled off to allow the trachea to be investigated. In healthy bees they are pearly white but when the acarine mite develops in them they darken and become black.

**fixation microscopy**

When making permanent microslides, the material must be killed and 'fixed' to prevent distortion, discolouration etc. The process used is called 'fixation' and formaldehyde or certain propriety preparations are used. 10% alcohol is satisfactory for most elementary work but tends to shrink delicate protoplasmic cells. Acetic acid has a swelling action (fails to harden tissues). Alcohol/ acetic acid – 99 parts of 70% alcohol added to l part glacial acetic acid has the advantages of both but needs to be washed out with 70% alcohol. After this treatment tissues may be preserved indefinitely by storage in 70% alcohol or in formaldehyde though the latter may be used when several washings in 95% alcohol are required to remove surplus fixative

.

**fixation nitrogen**

The process whereby some plants (mainly legume) convert atmospheric nitrogen into nitrogen containing vegetable tissue (e.g. proteins). Plant proteins decay, in the presence of bacteria, to ammonium salts, or they may be converted into animal proteins first.

**fixed comb**

Comb built naturally by the bees and is

firmly attached to walls or ceiling or even developed upwards from the floor of their receptacle. Although they are always vertical and separated from one another for the passage of bees and for ventilation they are remarkably strong considering the relative flimsiness of wafer-thin beeswax. Not so long ago skeps contained 'fixed comb' and nearly all beekeeping was carried out in this fashion until the Greeks used top bars and sloping walls and later, wicker, ceramic and wooden hives. Because of the difficulty of checking that bees on fixed combs were healthy, there has long since been a ban on their importation into this country. Langstroth was the first to interpret the bee's willingness to respect bee-space provided by movable frame hives.
See: hive fixed comb

**flabellum**
The name means a fan-shaped organ. The bee uses it to wipe up minute particles of liquid – in fact it 'licks' with this part of the tongue which is also called 'bouton', 'spoon' and 'labellum'.

**flagellum antenna**
The long, flexible part of the honeybees' antennae which extends distally from the scape. Its 12 roughly cylindrical segments on the female bees (13 on the drone) are packed with sensors and can be stropped by a cleaning device formed between the tibial and tarsal joints of the forelegs. Much work is being done to identify the specific function of its various sensitive regions. It can be used to transmit or receive information and can be inserted into small spaces (e.g. to make first contact with the rest of the hive via the very first slit in the capping of emerging bees).
See: antenna, exteroceptor, sensillae

**flame test**
Characteristic colours are displayed when certan elements are put into a flame. We are discovering the make up of distant stars by such means). When a beeswax candle is given a burning test the shape and purity of the flame, absence of smoke and flicker and ease of relighting are all taken into account. See: candle purity of the flame, absence of smoke and flicker and ease of relighting are all taken into account.
See: candle

**flanged cork**
When mead is put up in bottles for display and judging it is usually specified in the schedule that all-cork, flanged stoppers should be used. These are corks have an overlapping, circular top and a tapered shank that fits tightly into the neck of the bottle. The cork can therefore be firmly gripped and twisted to assist removal.
If compression difficulties prevent the insertion of the cork, use a piece of clean string to make a path for the escaping air. The string can then be withdrawn when the cork has been driven home.

**flank combs**
The combs on the outside of the brood chamber and are likely to contain patches of drone cells unless the beekeeper has rearranged them. Beekeepers might well arrange that the worst combs (most brooded in) are put on the outside for easy replacement in the spring. Frequently you find that the cells facing the walls have less honey and rarely brood at all unless the chamber is too small. Sometimes, in the spring, a colony might start work on one side if it is warmed more than the other by means of the sun.

**flash-heater**
As the deterioration of honey due to heating is a gradual thing, dependent on the total number of calories that it has been exposed to. A short blast at high temperature having the same effect as a much longer spell at a lower temperature, advantage of this in the case of milk and other natural substances which lose some benefit after being heated. Hence the use of flash-heating of honey to so lower viscosity that it can pass through the finest of filters. Some benefits are lost of course but this is minimised by immediate

cooling, though the specific heat of honey causes it to hold on to its calories longer than many substances. Honey and wax are alike in this respect. Melting solidly granulated ivy honey in the microwave will demonstrate just how long it takes to cool with clear honey below and a solid wax slab on the surface.

**flashlight**

A lamp or a torch is an essential item when moving hives in the dark. Red light is invisible to bees and although it is not easy to see by, at least no loose bees are attracted to it as they would be to white light. If a brighter light is needed, switch it on and off so as not to attract bees to it or stand it on something away from you. Dealing with wasp nests too is best done with red light.

**Flashman G. J.**

A modest little man and lovable beekeeper who as a BBKA examiner and expert taught Ken Stevens at evening classes in Putney in 1938.

**flat**

It is not very convenient trying to keep bees in a flat. A loft might help or an observation hive perhaps. However by selling hives in-the-flat they have become easier to package and despatch. Their assembly needs reasonable care as parts can be put together upside-down. The minimal, judicial use of a smoker in conjunction with a diagonal twist of a chamber, can do much to avoid crushing bees when a hive is reassembled. Sometimes it is a toss-up as to whether a guillotined bee is much different to one that is flattened.
See: in-the-flat, long-idea hive

**flat iron**

A flat iron may be cleaned and will run more smoothly by the application of beeswax while it is hot and the iron wiped clean and then put into normal use. An American device uses a flat iron strapped to a series of cutting edges to produce cut-comb.
See: useful hints

**flavone**

The parent substance of a number of vegetable colourings also found in flowers. It is an organic compound $C_{15}H_{10}O_2$ from which various yellow dyes are prepared and it is found in propolis. It is responsible for flavour, aroma and colour of products.
See: bioflavonoids

**flavour**

A characteristic taste e.g. honey, mead, vinegar etc. Flavour is often affected by texture. For instance the same honey might appeal differently to people when presented in the comb or as clear, set or creamed honey. In honey shows in Great Britain it is usual to expect high marks to be given for flavour. No matter how well the material is presented it will not do well if the flavour is indifferent. Honey flavours vary according to their floral or other origins. Substances contributing to flavour in honeys include, the sugars, amino acids, gluconic and other acids, praline, tannins, glucosidic and alkaloidal compounds.
See: sense of taste, aldehyde, essential oils, gustatory

**flax linseed Linum usitatissimum Linaceae**

*L. perenne*, *bienne* and *cartharticum* are varieties, they are slender, erect annuals with long, narrow, lance-shaped leaves and blue flowers. Some are very good bee plants and nectar has been shaken out in spoonfuls from NZ flax in one of the driest summers on record in GB. Flax seed produces linseed oil which is used in some forms for the protection of wooden hives. Pollen is orange and dries to yellow. In NZ the plant is widespread and useful to beekeepers blooming quite early in the season.

**flexor muscle anatomical**

One of the direct flight muscles (controlled by an adjacent ganglion).

**flicker**

The bee's sensitivity to flicker is due to the

nature of its compound eyes which enable it to detect changes of motion in the visual field rather than form. In other words it does not distinguish form as clearly as the human eye but is much better able to see any form of movement (40 times faster than us).
See: compound eye, fusion flicker

**fliers**
The bees that are capable of flight. It is sometimes useful in an over-populated brood chamber when trying to find the queen, to move the hive to another nearby site so that the flying bees leave it and return to the original site. Once denuded of bees the queen is more easily found and the chamber is returned to where it came from. The same 'trick' can be used to prevent swarming or to boost a weaker colony, or to prepare a colony for working in a new environment like strawberry tunnels, or NZ beech buggy honey. To simply move a colony into a glass house would result in all the fliers heading straight for the windows. The other bees of the colony are called 'house bees' which include nurse bees and others on inside jobs because they have not yet flown.

**flight**
The act or manner of forceful passage through the air. The worker bee does not fly in 'bee-line' as common parlance would suggest. It flies in a zigzag fashion and is able to carry more than its own weight e.g. it can carry a drone. To achieve flight the thoracic indirect flight muscles need a great deal of sustenance. They cause the leading edge of the forewing to be depressed in a power stroke 250 times per second. The fore and hind wings are swung out at right angles to the thorax and locked together by a sliding fit between a ridge on the fore wings and hooks on the rear wings. thus forming a large continuous surface. The bee is dependent on flight to obtain food and water, to find a nest, to mate and to reproduce by swarming. It is capable of performing aerobatics, hovering, flying backwards, darting and zooming. The queen's ability to reach and sustain flight at levels high above the paths of the workers to copulate with several mates and return safely home is as astonishing as the miles and miles she walks during her lifetime. Several aspects of flight are dealt with under headings such as bee-line, cleansing flight, drone assembly areas, mating flight, flight muscles, navigation, orientation, play-flight, speed of flight, homing, pay load, foraging, scent trail and others.
See: rate of wing beats, flying speed, fuel consumption

**flight board alighting board**
Although many hives have no more than a bottom entrance, it has been customary (and still is with WBC hives), to have a landing or taking off board sloping up to the entrance. Famous beekeepers who have watched the return of masses of foragers during a 'flow' say that it saves the bees time as they land directly onto it instead of hovering a little uncertainly in the air in front of the hive before entering. Even fold-up versions for travelling are not uncommon. Under a number of headings flight itself has been covered but the moment when a bee ventures out of the semi-darkness and into the bright sunshine may be followed by a ritual. Its five eyes together give it an impression of its surroundings (it does not see clearly and focussed as we do). The untold number of senses on its antenna also inform it as to the weather and the sensitive units themselves (antennae) are 'stropped' in the manner of a pilot doing an outside check on his aircraft before taking to the air. Then, with that grass-hopper like agility it launches itself and is an individual, free agent. Yes, maybe it deserves to have an easily acceptable approach to its landing ground in front of the entrance after busying itself in thousands of flowers in an environment so different from its home base.

**flight path**
Upon leaving or upon returning to their hive, bees tend to fly at the best height that

will enable them to avoid making constant changes of altitude. Care should be taken when siting hives to see that their natural flight paths do not cross foot-paths or places where children, people or animals may frequently move because a collision might easily lead to a sting. Artificial barriers, like netting have been used to force bees to climb (noise abatement procedure at London Airport) before setting course, but in one case where the mesh allowed easy passage, because one bee after another hit the strands, they set to work to cut it away with their mandibles. A tall hedge, even a wall might offer both shelter and a means of getting them up and away. Clearings in woods are sometimes favoured but under trees the colonies might suffer from drops of water plopping onto their hive long after the rain stops.

**float**

To rest on the surface of a liquid or the item that does this. A small ripple might suck a bee into the water if it does not have a secure foot-hold. Bees cannot swim. If beaten down into the water by the wind there is little hope for them. Its natural instinct is to use its wings rather than legs to escape but it only succeeds in whizzing round in circles. Unless a 'float' is available for it to climb out onto it may drown, once all its spiracles are water-logged, only another bee can resuscitate it. The idea of a 'float' to allow the bees safe access to syrup in a feeder and the anchoring of bees on a raft in flooded regions are other examples of the word float entering into the realm of the bees.

**floccule**

Something that assumes the appearance of a small tuft of wool. A clump or agglomeration of pollen grains. Certain plants may yield pollen in such a way that it clings onto and even trails behind the bee.

**Flodon board**

A swarm board designed by F. Sumpter. When attached to a hive that has had a spare brood chamber set beneath it, the queen (she must be a clipped queen) will only have access to the new lower chamber but a metal trap prevents her from escaping. The swarm is therefore likely to establish itself in the lower prepared chamber beneath the parent colony.

**flood**

The submergence of land that is normally above water. Water and dampness can be fatal to honeybee colonies. A sudden, unexpected inundation of land normally free or water can catch beekeepers unawares. Careful choice of sites for apiaries should take into account any possibility of flooding. Hives, half submerged in water, have been recovered by a beekeeper with all the bees safe 'up-top' but the lower part of the combs in bad shape. For the benefit of your back and to have a marginally better micro-climate, it is best to keep hives on stands so that they are well off the ground. Tethered rafts have been mentioned and the idea of taking hives onto barges and following the oil-seed rape fields has also come to mind.

**floor**

The floor of a hive is usually a separate part (easier to clean than an attached floor) and its design allows bee-space under the brood chamber and the provision of an entrance too. In the 'honey-house' floors should be smooth, flat and hygienic. Where commercial undertakings have floors and gravity is taken into account, when honey loads arrive the unloading point is at a high level so that from there on the work is 'down hill'.

**floor board**

This has been described under 'floor' but although it can form the base of the hive, there is a growing realisation that because of Varroa and the healthy ventilation that a screen of gauze can provide floor boards have been supplemented or replaced by mesh floors. Where floor boards are used, these will collect debris that the bees do not find easy to remove. Propolised snails and slugs have been discovered when a clean-up in the spring calls for an early replacement board or thorough scraping and blow-torching the old one. With new plastic hives that have not been propolised it is wise to ensure that the hive bodies cannot slip or be pushed sideways on the floor.

**floor polish**

Beeswax is a very usable polish. In the care of bees, the substance is used to fasten everything strongly together and there is no sign of the shiny surface that presents itself when we use rendered wax to enhance the surfaces of furniture, floors etc. Rendering implies liquefying comb and passing it through a filter (piece of lint or nylon maybe) and obtaining an aromatic, translucent sweet smelling substance which for centuries made bees more valuable for this product than the honey they make. Of all the recipes that are available for making cream or polish from wax, basically softening the wax with turpentine leads to the ease of applying it to the object to be polished. Simply melt 227g/8oz of wax in .6ltr/1 pint of turpentine (both flammable) stir well and allow to cool.

**flora The Roman Goddess of Flowers**

The sum total of plant life within a given area as opposed to animals which are 'fauna'. Although beekeepers are primarily concerned with flowering plants including weeds, wild plants and those cultivated in parks and gardens and agricultural crops, the siting of an apiary should take the surrounding flora into account. Spring pollen and potential honeyflows are very important and will depend on the local flora.
Scc: forage, fauna

**floral nectary**

The part of a flower which secretes nectar. Such nectaries are to be found in various parts of different flowers. Extra-floral nectaries may also be found in the axis, carpels, stamens petals or sepals. As they entice insects into the flower to cause fertilisation they are also called nuptial nectaries.
See: extra-floral

**Florea *Apis florea***

The name given to the smallest of the honeybee species. Found in warm climates and makes but a single comb in the open, has parasites and wants to abscond if kept in containers.

**florescence efflorescence**

The period of blooming or 'anthesis'. To begin to flower or to burst into bloom. The time when bees take a particular interest in the plant.

**floret**

One of the small flowers close-packed that make up a composite flower e.g. daisy dandelion or sunflower. After being visited by a bee, each clover floret hangs down and becomes brown. The number of florets in an acre of Alsike clover may run to 400 million, red clover about 200 million.

**flow**

One of the first things the layman learns about bees in this country is that honey does not come in steadily all the time but tends to occur in flows which are often rather uncertain. A honey-flow is the actual intake of nectar into the hives and it is accompanied by great activity, an absence of robbing, neglect of the water fountain, the building of new comb, and an increase in hive weight. In its early stages the white extensions to existing cells is a good indication. At dusk the quiet buzzing of the fanners driving air into the hive on one side while bees facing the other way will be blowing aromatic sweetness from the other side into the evening air, makes for a happy feeling all round.
See: honey-flow, nectar secretion

**flowed-in**

After years of waxed cardboard liners (wads) in honey caps, they began to give way to the new method of producing a 'sealing' ring so that the wad is no longer required. This technique is described as 'flowed-in' and produces a good seal against the upper rim of the honey jar.

**flower**

The blossom of a plant and its reproductive organ. Many are attractive and useful to bees on account of their colour, odours and offerings of nectar and pollen. Parts can include receptacle, ovules, carpels, sepals, stamens, petals, styles and stigmata, but there is great variability. Considerable study is required to learn that a calyx is composed of sepals, the petals form the corolla, that together these organs are called the perianth, that pollen grains (microspores) are produced in the anthers of the stamens and at least half-a- dozen more names of parts will be found in most standard botanical text books.

**flower constancy**

Flower fidelity – a honeybee trait, the determined seeking for a particular type of flower while ignoring others that could be more rewarding. It results in the effective pollination of certain species and also the even colour of pollen loads too; this constancy enables it to transmit the knowledge more certainly to recruit bees in the hive. Multi-coloured pollen loads are rare and when bees are seen going from one flower type to another it is usually an indication of a dearth of nectar or pollen.
See: fidelity

**Flowering currant *Ribes sanguineum* Saxifragaceae**

A popular garden plant that flowers in Mar-April, its pink blossoms well-worked for pollen though the corollas of some varieties are too deep for honeybees. Bushes can grow 2m high.

**flowering plant *Spermatophyta Angiospermae***

Mostly herbaceous but also woody plants divided into the classes: Monocotyledons and Dicotyledons. Plants that bear seeds. The description of a plant during inflorescence. Bees seem to have evolved in co-operation with flowering plants and there are many examples of adaptations and specialisations where a particular plant and particular bee have become dependent on one another for survival.

**flowers bees rarely visit**

Saxifrage, Asterids (carrot, hogweed) elder, mountain ash, buttercups, dahlias, roses, chrysanthemum and double flowers. Though showy and sometimes having easily accessible nectaries. Reasons? Perhaps they have less accessible nectaries, are repellent, lack attractants or have unsuitable nectar and pollen.

**flower seed**

The significance of the seed of flowers is brought into focus when we consider the tremendous increase in seed yield brought about by an abundance of bees at the time of flowering and also the seed as a means of producing useful plants for bee forage and observation in your gardens and the environment. The huge benefit to birds and other creatures when wild flowers are pollinated leads you to realise how the whole of nature is integrated.
See: seed, seed germination

**flowers per tree**

The number of flowers and their strength is in the case of apple and other fruit trees, influenced by the weather of the previous summer and autumn. Pruning and thinning may help to rectify a trees over-production of blossom which might otherwise lead to undersized fruit and poor flower development the following year. If the canopy of huge flowering masses were spread out like a ground crop some sycamores and limes would individually cover half a hectare. When hand pollination takes place the fruit can be positioned as required but the effectiveness of bee pollination seems superior to artificial methods.

**fluid**
A substance that is able to flow or move under the influence of a force such as gravity, centripetal etc. Liquid beeswax looks remarkably similar to honey and both substances have high specific heats, that is they absorb a lot of heat as their temperature increases and give off a lot (are slow to cool) as their temperature goes down. This heat buffer is valuable in the winter when stores provide a honey/wax canopy over the winter cluster. The characteristics of honey are remarkable and put it in an entirely superior category to sugar. Sugar syrup is akin to artificial nectar and is useful in that beeswax can be built from bees fed with it. However as brood food it has to be converted by the bees into fructose and glucose.

**fluid ounce 28·413ml**
That volume of water which weighs one ounce – 20th of a pint. But looks like litres from now on.

**fluorescence fluorescent 'assay'**
The emission of light by surfaces that are bombarded by electrons. To glow strongly and characteristically. The property of producing luminosity when exposed to a bombardment by a stream of particles or certain forms of light. AFB scales fluoresce under ultra violet light. Flowers have interesting patterns including nectar guide-lines when viewed under u/v light. When a piece of filter paper is dipped vertically into diluted honey, the capillary action draws up some fluid and characteristic zones of fluorescence are exhibited – light blue topped by a white zone, under the influence of u/v light. Fluorescent assay is used with beeswax - pure white beeswax fluoresces white/blue. Ordinary British wax is variable but has a brownish-yellow and sometimes a greenish tinge while foreign honeys may run to yellow or orange. Materials are said to be phosphorescent if they continue to emit light for some time after the excitation has stopped.
See: chromatography, spectroscopy, ultra-violet light

**fly *Musca domestica* Diptera**
While commonly used for reference to the house fly it applies to all the two-winged Diptera and is also a general term for winged insects. One or two flies have larvae which are parasitic on the honeybee. As a verb it implies to move through the air on wings. A bee is able to fly instinctively once its wings have hardened and the front and rear pair are interlocked on each side.
See: apimyiasis, Diptera

**fly-catcher *Muscicapa grisola* Muscicapinae**
A member of a group of insectivorous birds which include this European species (spotted fly catcher).

**flying**
The bee's flying ability is much influenced by atmospheric conditions including light intensity. Its flying speed is not enough to enable it to battle against winds much in excess of 12mph and they are sometimes confined to their hives when 'tougher' insects like bumble bees are still able to forage. Successful sorties by queens are more likely to occur when winds are light.
See: flight, flying speed, mating, cleansing

**flying speed**
Loaded foragers return at speeds ranging from 21/26kph averaging 24kph while emptier outbound bees only achieve 20kph (11/29). They fly faster against the wind (their airspeed not groundspeed) than when with the wind. The maximum speed recorded 41kph. As wind speed approaches flying speed the large amount of drift usually causes activity from the hive to diminish. You might imagine that they could 'tack' like a ship but although they zigzag that is their security against predators rather than of tackling the wind. They can hug the ground and also even take shelter by a hedge awaiting a lull to pop over the top.
See: airspeed, speed of a bee, wind speed

**foam rubber**

Rubber that is processed so that it is one mass of bubbles. This makes it very light and spongy. Yellow or white, it is available second-hand from cushions and mattresses but if burnt produces poisonous fumes. It is useful for closing entrances (when moving hives or when spraying or other reasons call for a temporary closure of the entrance) and is best cut with a very sharp blade like that of an 'open' razor. Attached to a handle it can make a useful bee-brush and it can be used for the application of certain chemicals such as repellents, prophylactics water or syrup, or a water supply over a travelling screen.
See: unblock

**focus**

A clear and sharply defined condition of an object. The focal length of a lens is the distance at which an object is clearly defined. This can be varied by moving one lens relative to another and so providing a range through which objects can be focussed. Thousands of facts concerning the bee and its environment have become better known by virtue of using lenses to focus on things that are too small, or move too fast or too slowly to be seen with the naked eye. The bee has no means of focusing its eyes. The focussing of an eye is known as 'accommodation'.

**Folbex-strip**

From Ciba Geigy. A strip of absorbent paper containing chlorobenzilate and was used as a prophylactic in the treatment of Acarine disease. Supers should be taken off before using this chemical or honey will be contaminated. It can be used when brood is present. A strip is allowed to smoulder under a dish over the feed-hole, while the entrance is blocked with foam rubber, for one hour. This should be done at weekly intervals for at least three weeks, and one strip is used per brood chamber.
See: acarine disease

**folic acid**

One of the B complex of vitins. Certain uses have been made of this chemical – doubtless for Varroa?

**follicle cells**

These are the small cells which form a layer or continuous sheath around the egg cells in an ovariole except at the forward end where a plug of egg cell plasm makes direct contact with a nurse cell for nourishment.

**follower bee**

A bee that relentlessly stays with anyone who walks away from a hive, frequently making as if to attack but not necessarily actually doing so. They can be long or short distance types. Humorously called a 'halo'. This is a 'characteristic' than can be eliminated by breeding or requeening.

**follower board**

The name given to a certain sized board that is used in a section crate in conjunction with two spring blocks. It enables the rows of sections to be pressed firmly together and helps when they are to be removed.

**follow the sheep**

This advice was given to the beekeeper in days when wild white clover was a honey producer 'par excellence'. The sheep droppings added just the right chemicals to chalky limestone to trigger off nectar secretion from white clover. A high temperature (21C/70F) is needed, together with soil moisture and atmospheric humidity to encourage a good clover nectar flow. "Where there's sheep there's honey' - that was true 50 years ago –Frank Vernon.

**fondant**

A creamy white paste made of sugar fine enough to be eaten by bees but not as satisfactory as properly made 'candy'. If there is any chance of bees getting their wings or bodies 'sticky' there is a big risk of their spiracles becoming blocked and then only other bees can save them. This is particularly important in the case of a queen with candy as a temporary block in the end of an introducing cage.

**food**

Nectar can be used directly by the bees but for storage they convert it into honey. Only when moisture is available, either in the nectar or from elsewhere can honey be diluted sufficiently for the bees to use. This applies to crystallised stores. Pollen is eaten by nurse bees and in conjunction with diluted honey or nectar they convert it into brood food and royal jelly. Pollen is stored in cells, rammed in by the heads of workers and coated with a film of honey to protect it from the pollen mite. Bees may seem to be eating when they put their proboscises into a cell but eating has to be associated with 'food-sharing' which goes on all the time. The colonies need for water will depend on the over-all state of their stomachs. There are times however when bees need to load up. One is when a group become wax makers and they are converting what might otherwise be food, into scales of wax. Another might be in the case of 'absconding' because of a forest fire. Then a departing forager too will have taken on board sufficient 'fuel' for its anticipated 'sortie'. Bees can take liquid into their honey sac and then regurgitate it. Once beyond its ventriculus the food will be digested. From egg to pupa the food given to the brood is varied. They are weaned from rich fast developing food, in gradual stages, to the more ordinary nursing food composed from the diluted honey and pollen such as the older nurses develop. From our point of view, honey is not only a splendid, pre-digested food but a wonderful medicine too.
See: artificial food, brood food, food-sharing, royal jelly

**food attractants**

Various chemicals have been found attractive to bees and the pheromone secreted by the Nasanov gland is attractive (even to alien bees as well!) though not in the sense of a food. Sugar syrup which has no detectable smell to us, is highly attractive to bees (and wasps). Unfortunately honey can carry serious pathogens and many of the bees kept in apiaries near a firm who imported large quantities of foreign honey were found to have AFB. They can also be tempted to collect from strange sources. Ginger, milk and other substances if rendered sweet can be taken down by the bees. It offers a way of introducing prophylactics. Attempts have been made to encourage bees to pollinate crops like strawberries by feeding them syrup laced with material from the strawberry flowers. Old-fashioned skeppist used beer and lemon balm etc. to entice swarms into skeps and today, bait hives are set out with material, like 'used' combs, inside them.
See: repellents

**food chamber**

A box of stored combs given to the bees as an addition to their brood chamber. It could contain honey, or sealed sugar syrup or a mixture of both and is normally placed above the brood chamber (and without an excluder which if left between the two boxes, could by the end of the winter, leave the queen stranded below while the bees move up onto the food). In a natural nest the canopy of honey and wax in the upper part helps to contain the warmth of the cluster below. A super which contains honey that you anticipate extracting can be used as a food chamber should you be in any doubt about the amount of food you are leaving the bees. The safe amount in our climate is usually reckoned to be about 18kg/40 lb or the equivalent of 5 full standard deeps. By March the brood is taking large amounts of food and what might have seemed plenty when you 'hefted' the hives during the winter may be decreasing too rapidly and it is important not to let a strong colony go short at this time. On the other hand, if there are blocks of stores and the bees are always a little reluctant to start uncapping cells (unless they want the hygroscopic properties of honey to come into play and draw in moisture from the hive atmosphere) then the 'scratching' of sealed comb faces near the centre, will encourage them to empty the cells and if this is to be part of the brood nest – all well and good.

**food sharing**

The identification of kindred colony members is associated with the interchange of food by the passing of honey stomach contents form one bee to another. This may be 'on request' or one bee may be seen to make its stomach contents available to several other bees at once (look out for this phenomenon). The continual passing of stomach contents enables the 'hive mind' to determine its immediate needs, to play a part in establishing colony odour and literally helps to share food out to the 'communal stomach'.
See: beggar bees, epipharynx, oecotrophobiosis, food transmission

**food transmission**

Bees make antennal contact enabling them to orientate themselves towards and to communicate with one another. The receiving bee extends its tongue to take food from the mouth of the bee that is offering food. Sometimes a small group will surround the donor. This enables bees to forage as a community, recognise companions, defend the colony and organise the division of labour. Queen substance and other pheromones are circulated in this way.
See: beggar bees, food-sharing, soliciting food

**foot**

The pretarsus. The honeybee has three pairs of feet which are similar in structure and comprise double recurved hooks which are called upon to support considerable weight (as for example in a clustering swarm) a pad, which helps the bee to grip smooth surfaces (and presumably to leave a footprint pheromone) which is folded up when not in use. The pad or 'arolium' has a 'planta' is reinforced with chitinous acus, has sharp hairs and two spines. There is also an unguitractor and a manubrium. Bees can follow one another's footprints.
See: travel staining

**foot bath**

An unavoidable sterilisation treatment for the feet before a creature is able to enter its premises. Around a hive entrance the travel staining is a noticeable feature. That much of the staining comes from the antiseptic, antibiotic propolis and almost certainly serves the same function as a foot bath. This is very noticeable round the small entrance to hollow logs used by the stingless bees in Australia.
See: travel staining

**footing**

A secure basis or foundation. This is an important aspect when siting a beehive because an active colony can cause a hive to become extremely heavy and where the ground is soft or yielding, especially when a hive is resting on legs, it can lean over and eventually fall. A footing should be of adequate size, firm and level. It can help to prevent the growth of weeds in front of the entrance and help to keep the hive dry. Paving slabs can harbour ants' nests, rat runs, grass snakes etc. The best plan is to construct a sturdy, permanent stand at a suitable height (though if at the right height to avoid bending when examining the brood chamber, it may prove difficult to put the sixth or seventh super on top!).
See: site

**footprint pheromone**

Bees are able to follow quite tortuous paths previously trodden by colony members, because of a long-lasting trace of a chemical which has been called 'footprint pheromone' It is not yet certain where this pheromone is produced though it occurs all over a bee's body and is probably deposited by the feet and the abdomen. It compliments other scents produced by the bees, flowers and other sources of interest to the bees. The queen's footprint pheromone was found to be much more attractive in the hive than that of workers or drones.

See: travel stain

**forage**

The plants which supply bees with food – nectar and pollen. The crop that the bees are working on. The successful acquisition of food material as opposed to robbing or taking food artificially. Bee forage is only available during part of the year in this country. Modern farming has given little or no attention to bees in its quest for bigger and bigger marketable crops. A farmer might causally inform a beekeeper that his apiary will be surrounded by oil-seed rape and say nothing when fields of corn or rye grass reduce forage capabilities to a minimum. To-day people have suddenly awakened to the importance of this industrious little insect and now great strides are being taken to ensure that a reasonable amount of fodder is available in parks, public spaces and in fields set aside for that purpose together with corners, headlands etc. There is still a little hesitation for many who are only slightly familiar with the honeybee, to admit as others have known all along, that the honey bee is way ahead in the field of pollination. Its foraging capability and the result of its work are literally astounding. Apart from the greater interest in raising honeybee colonies than swarms of flies, bumbles or wild bees which is but one aspect, the interesting thing is the quality and quantity of fruit and seed that come from well-managed bee colonies being put in the right place at the right time. Had this been possible in Kent, the foreign competition would have had much greater opposition and flowering orchards of cherries and world famous apples would still be the order of the day instead of orchard being overwhelmed by housing estates. In its hay day, it was said that if every hive in the country were moved to Kent there would still have been fewer bees than were needed

**forager**

A field bee potentially available for, or actively engaged in, the pursuit of searching for and collecting food, water or propolis. Only the workers forage, after passing the first three weeks of their lives as 'hive bees', fulfilling all the requirements for nursing, comb building and keeping everything hygienic etc. During the really active season the hard working forager may only last three weeks. Constantly returning laden, with tattered wings and bodies almost hairless they join the many dead who surround every colony. Their bodies add to the humus, their passing akin to that of a river which remains the same though is daily different.
See: forage, field bee, hive bees

**foraging**

The activity of searching for or collecting food from natural sources. It is only possible when weather elements such as light, temperature and wind are within certain limits and is only successful when plants are actually yielding pollen, nectar or honeydew. It is fitting that only the older, more expendable, bees should perform this frequently dangerous activity that causes brood food glands to retrogress and is also incompatible with extensive wax production. In the event of a really good honey flow, bees of a much younger age become involved (brood temperature is so much easier to control in the warm weather) and the whole colony seems to be singing aloud. Queenless bees and those about to swarm forage less actively than others.

**foraging area**

The area within reach of the bees where suitable forage is available. Topography, weather conditions and other factors such as 'strain' of bee, may also have an influence. In ideal conditions bees will follow the guidance of the most successful foragers and although the immediate area around the hive will receive little attention the bees will spread out and effectively collect from sources within 500m but when large numbers of colonies compete it is found that distances exceeding 800m result in diminishing pay-loads. They can fly up to 5k/3miles if compelled to do so. However a balance has to be struck between using up

their blood fuel and the quantity that they can collect and carry home. Scout bees 'set the pace' and recruits respond according to their instructions (Von Frisch dance) and the value they put upon the returning forager's 'booty'.
See: foraging radius, fuel consumption

**foraging radius**
The distance from the hive that bees will work varies according to the quantity and attractiveness of the forage and will be governed by the wind. Water masses also govern the area covered. A radius of .8k/1/2 mile means they cover 202h/500 acres while 1.6k/1mile covers 800h/2000 acres.

**force**
Strength or power exerted on anything. It is frequently necessary to use 'force' to perform various operations in beekeeping. Whenever possible this should take the form that does no damage to equipment or to the operator. Leverage by using a suitable hive tool, well-designed carriers for lifting hives, the use of efficient and well-made equipment (honey extractors) are examples. Also a short lesson in 'lifting' as the legs must be made to do the work and not the back. Bees flight has to work with and not against the force of the wind. Perhaps they have the ability to choose to fly low into wind and higher on return. Jerks are not good where bees are concerned and patience and careful use of the hive tool, cover cloth, wedges etc. can help to eliminate the need for force. Centrifugal force has made extracting a reasonably simple job. Always advise the Police 'Force' when you are moving hives by road and have a clear label in the car **(Live Bees!).**
See: centrifugal, gravity

**forceps**
A small pair of 'tweezers' with arms that are apart when free but can be manipulated between finger and thumb so that the extremities touch. These (plural in English like trousers) help you to take hold of a very small object such as the prothoracic collar when dissecting for acarine. It has been said that a watch-maker's pair are more suitable for this than the biologist's. Don't ever be taken in by someone unfamiliar with stings who suggests removing the sting with tweezers. Get that sting out as quickly as possible by scraping it out with your finger nail.

**foregut**
The stomodeum of the larva or the anterior part of the alimentary canal in the adult including the oesophagus, honeycrop and honey stopper (proventriculus).

**foreign honey**
Many sensible countries across the world will not allow honey to come in from abroad. It is because honey can harbour bee disease pathogens from Foul Brood to Nosema and it is practically impossible to check this. The feeding of any honey not known to be free of disease spores to your bees, whether British or foreign, is to be strongly deprecated. Empty honey containers must also be sensibly handled. When it comes to quality and taste we naturally think that ours is the best. It was easy enough to scathingly say 'Oh that's just eucalyptus without realizing there are over 600 different types of eucalyptus, outnumbering all the species of trees that we have here.

**foreign matter**
Unwanted material that prevents a sample from being pure. Any organic or inorganic material or substance foreign to the composition of honey, including mould, insects, insect debris, brood, particles of wax or propolis or specks of dust. Such material may alight on the surface of exposed honey so care should be taken not to uncover clean honey in damp, dusty or otherwise unsuitable atmospheres. If tiny unwanted particles get into honey they either float on the surface or are seen through the bottom of a jar of granulated honey. Apart from colloids they are not so likely to be submerged, this makes the gravity

separation of clear honey a useful method.

**foreign proteins**

Bee venom contains many proteins and these when injected into the human body are abnormal constituents and are therefore called 'foreign proteins'. Tiny particles of matter are very likely to surround a beekeepers clothes and the air can carry particles to which some people are very allergic. These cause eyes to smart and a tendency to wheezing.

**fore-legs**

The first pair of legs which include the same segments as each of the other legs. Their speciality is the antennae cleaner on the tibio-tarsal joint. They are shorter than the other legs. The bee can take up a firm stance on its other legs while leaving the front pair free to work in conjunction with the mouthparts when building comb, scrabbling for pollen etc.

**forest**

Land covered by trees. Types include broad-leaved, coniferous and tropical and may be evergreen. Bush is distinguished from forest in countries like Australia and New Zealand where bush is natural and forest man-made. European pine forests produce honeydew while another honeydew comes from the southern beech on parts of the NZ bush. Broad-leaved sycamores and limes can be heavy nectar producers.
See: bush, Hartz

**forest beekeeping**

The early spread of bees was always enhanced by the existence of many potential nesting sites (hollow trees) in areas where weather and forage favoured their survival. Whereas honey and wax could be obtained by merely 'hunting' the establishment of 'apiary' beekeeping showed no particular advantage though it inevitably followed as the demand for wax and honey exceeded the supply

**forest honey**

Many scale insects feed on forest trees and exude sweet liquid which bees collect and transform into honey-dew honey. They feed at various times of the year and local beekeepers regularly move in to exploit the crop especially in Europe, though 'beech buggy' honey is valued in parts of the NZ 'bush' in the north of South Island. A shortage of pollen sometimes calls for the feeding of substitutes. Wasps can present a problem as they proliferate in such areas. Moving colonies which have been used to floral nectaries, can result in many sticky bees finding it hard to get used to the technique required for taking up the sweet droplets that are offered in various manners. It is probably best to send colonies deprived of their flying bees as they more readily adapt to the new environment. Lachnids (*Cinara pilicornis*) Aphidae and Coccidae and various sucking insects perform on trees such as Spruce, Larch, Scotch pine, white fir etc. *C.pectinatae* and *Lachniella costata* on Spruce depend on ants. It is not generally realised that aphids are spread by ants who farm them for their sweet exudations
The great forests of Europe have been famous for the honey derived from honeydew. Fir trees, notably spruce are full of lachnids, (sucking insects whose well-being is associated with the wood ant) and these secrete the honeydew. Beekeepers move their hives to the forest in the late summer to take advantage of the flow but there can be problems due to pollen shortage. European honeydew honey is usually dark, pleasing to the taste and it fetches a good price. Large takes of flower honey are also collected in the *Robinia pseudoacacia* forests of Hungry & Rumania.
See: bush honey, eucalyptus

**forewarning**

It is very useful to receive a warning of a happening before it occurs. This might apply to spraying or to symptoms shown by a colony itself such as starvation when grubs are being thrown out or swarming when queen-cells are being built. The presence of a mouse when wax particles litter the ground in front of the entrance and so on. Weather forecasts can be useful too.

**fore wings**

The anterior and larger of the two pairs of wings having a strong leading edge by means of which power is transmitted to the whole wing surface on the down-stroke. On the centre of the trailing edge is a rib of chitin shaped to accept hooks on the corresponding part of the hind wing. Characteristic venation enables races to be compared and identified.
See: cubital index, wing venation

***Forficula auricularia***

The earwig, a small creature often found in hives where they shelter rather than feed. However they can and do get into the honey at extracting time unless vigorously kept out. Should they drown in the honey they are found much enlarged and floating on the surface. Their 'frass' (waste akin to faeces) consists of small black particles found under the lugs of frames.

**Forget-me-not Myosotis spp. *alpestris, avensis, palustris, sylvatica* Boraginaceae**

A small blue flower that blooms May to Sept. There are biennial and perennial varieties. Its pollen grains are extremely small but are found in honey out of all proportion to the amount of nectar collected. (3-4μ). The name is similar in other languages – Fr. ne m'oublier pas and in Ger. Vergiss mein nicht.

**fork-lift truck**

A small vehicle with two parallel, horizontal arms that form a platform and can be raised or lowered for insertion under a load to be lifted or carried. They may be manually or electrically operated. As these can be transported along with beehives they have in certain circumstances proved even more useful than boom loaders, especially where only small vehicles can gain access to a site and where hives are loaded onto pallets.

**Formaldehyde H.CHO**

A colourless gas normally sold as a 40% solution in water. The full strength liquid is described under formalin and can be used for the effective sterilisation of empty combs, hives etc. The gas is released when the formalin is exposed to air in a dish, best placed on top of the frames as the fumes are heavier than air. If completely sealed all nosema, amoeba, parasites and other pathogens and all stages of wax moth will be killed in a few days, yet when completely aired the bees can make good use of the sterilised material. However such combs should be quite free of stores, which would be contaminated by formaldehyde. Do **not** breathe fumes or allow liquid to touch the skin as it is a very harmful substance. (Used in some labs. with 'gay abandon'!!!).
See: formalin

**formalin**

Used for sterilisation of non-boilable materials (antiseptic). An aqueous solution of formaldehyde. It is toxic and used in laboratories to kill and preserve plant or animal tissue. It will contaminate honey and pollen and should only be used with caution on combs that have been 'licked' clean by the bees. The combs should be aired thoroughly before re-use. Do not allow the liquid to touch the skin, nor breathe the fumes. It is however far more effective than PDB (which has now been withdrawn because of its carcinogenic properties). When stored the clear liquid may precipitate crystals of formic acid but its use is still effective.
See: formaldehyde

**formic acid HCOOH**

A colourless irritant which forms part of bee venom though it is a common substance in ants. Used in fumigants and insecticides and possibly partly effective in cases of varroa.
See: formalin

**formulation**

Many chemicals used for spraying etc. are

mixed with 'buffers' 'wetters' 'diluents' etc. so that dispersal is economic and effective. The final product will be the 'formulation' of this substance and it will usually be marketed under a trade name. There are systemic granules, wettable powders, liquids and dusts. For example the formulation of 'Metasystox' (an insecticide used amongst other things for exterminating green or black fly [aphids] on field beans) can be obtained in granular form. The manner in which a pesticide is formulated can greatly affect its toxicity to bees. Sevin can come into all 3 categories of toxicity. Furthermore despite all the care taken to make the most effective use and to minimise the danger to operatives and other forms of life (bees etc.) this can be negated by its incorrect application.

**fossa**

A pit or depression. In the bee it is a 'U'-shaped hollow lying beneath the occipital foramen in which the proboscis (tongue) is housed.

**foul brood**

Two of the brood diseases respectively known as American (AFB) and European (EFB) foul brood have wreaked havoc and seem to appear spasmodically amongst our bees. Such diseases are easily spread and can be very damaging to colonies. Consequently burning is advocated but drug treatments exist though their use and subsequent management varies from country to country. Both diseases are notifiable and once in government hands the beekeeper is slave to their orders until the trouble is completely cleared up.
See: bee inspector

**foundation**

A sheet of wax which the bees will accept as the 'septum' (mid-rib) of their comb. It can be made on home presses but commercially thin sheets of 'pure' beeswax are rolled or pressed so that, embossed with the rhombic based hexagonal pattern of natural honeycomb, they will be accepted by the bees to 'pull' or 'draw out' into comb. Mehring (Dutch) made the first foundation in 1857 and Weed and Root (USA) went on to improve it. Lamination or the insertion of stainless steel wires are used to re-enforce it (better strong when it has to go into an extractor). Although a standard cell size is made for the British bee, other sizes have been explored. Then as regards to thickness there is medium brood, thin and super-thin, the latter being good for 'cut comb' as an alternative to sections.
See: wax foundation, wiring

**foundation press**

Prior to the invention of the 'rolling mill' we used hand operated presses to make one embossed sheet of foundation at a time. They were brought in from Germany after World War II and they became more popular as the prices of foundation rose. They consist of a base tray with a hinged flap that closes down so that when filled with the right amount of beeswax at the right temperature, a sheet is produced that can be cut to size with a template. A releasing agent such as 'detergent' is needed. Instructions for making such a press were published in bee journals at that time.

**fowls**

According to 'Cotton' it is not wise to keep fowls in an apiary because once they get the taste for bees they will stand by the entrance and snap up bees as they come out. They are also inclined to jump onto hives and leave unsightly droppings behind. Other considerations are access to hives when carrying equipment and will those who attend to the fowls daily requirements take kindly to the bees?
See: chicken

**foxglove *Digitalis purpurea* Plantaginaceae**

A wild flower whose tall spikes are composed of purplish flowers, each akin to the finger of a glove. They do not like calcareous soils, flower June – Sept but are visited more by bumble bees than honeybees.

**frame**

A light, wooden or plastic structure made to hold a sheet of wax and subsequently the developed honeycomb. So that frames can be handled without damage and correct bee-space be maintained within a hive, the bees are encouraged to build their combs inside wooden (plastic) frames of a given size (see 'frame sizes'). The frame usually consists of a top bar that has projecting ends (lugs) that enable it to rest on the ledges formed by rebates within the hive wall. A complete rectangle is formed by the dovetailing of two side bars and a bottom bar (or bars). As frames are normally fitted with a sheet of wax foundation, provision for its insertion is provided in various ways such as grooved side bars and split bottom bars. The upper edge of the foundation may be gripped with either a wedge or by insertion in a longitudinal split in the top bar. The foundation has to match the type and size of the frame and grooved side bars. Five piece in-the-flat, easy to fit frames are available. BS, Manley and Hoffman frames are made in shallow and deep sizes.

**frame gripper**

A one-handed device designed to enable the user to grasp the top bar of a frame and draw it out or subsequently replace the frame into the hive. A spring loaded variety is available.

**frame holder**

Metal brackets and other contraptions have been designed so that a spare frame can be hung by the side of the hive, on a fence or other means of support. While the short term removal say of a flank comb does little harm, on the other hand exposing stores and even worse 'brood' is not to be recommended because it offers attraction to other bees and might encourage robbing. A small box which holds one frame is not a bad idea and can come in handy when it is require to transfer a frame of brood or eggs. See: frame gripper

**frame jig**

A template that will hold a frame square and steady while it is being wired or fitted with foundation. The somewhat flimsy basswood sections do call for a means of holding them square while the foundation is fitted. As dozens of similar frames are required a 'set-up' allowing for the cutting, shaping and assembly of frames, can be a useful item. Modern prejointed frames are much easier to assemble and require no jig. See: template, wiring board

**frame lifter**

A bee hive belongs to the bees and they are entitled if they wish to use propolis seals which serve them in many ways. Unfortunately although some strains use less propolis than others, the majority do 'gum' down frames so that a tool of some sort is required to release their hold and allow the frame to be taken out and examined. Initially hive tools were very robust and could lift or separate frames as well as acting as 'scrapers'. When a strong, slimmer, 'H' and 'J' type came into use, some failed to see how useful they were. Not only can this one slide into an overall pocket but the notched edge which locks underneath the top bar while a correspondingly curved portion (the tail of the 'J') neatly lifts the adjacent frame.

**frame nailer**

A rampin which has a hollow inside, magnetic at one end, so that it can accept a small nail. The nail is then positioned so that when a spring operated plunger is brought into play, the nail is forced down into the wood. Unless a glueing technique is used the nailing method does require suitable nails (gimp pins) and a minimum of four are need to affix the side bars to the top bar and then if two bottom bars are used, another four nails are required. These are best driven in so that each bottom bar can be pulled out again and future refitting with foundation will be easy. Still more nails are called for when the 'wedge' which holds the top edge of the foundation firmly in place, is hammered or pressed home with nails that also pass through the top, protruding loops of the wiring. As time goes on, more and more clever ideas come into play and reliable, electric staplers are in many cases superseding nails.

**frame sizes**

Perhaps we are a bit slow and unwilling to change from our time-honoured ways but certainly we cling and cling to the old system of measurements. This is for two reasons, it is costly to change and as they say 'You can't teach an old dog new tricks'. So here are the sizes in old-fashioned inches with conversions following. The much used British Standard deep frame has a 17"/432mm long top bar, the side bars are 81/2"/216mm for the deep and 51/2"/140mm for the shallow. The length of the bottom bars (and consequently the length of the frame under the top bar) is 14"/356mm. As we have never seen fit to go over to the almost international American sizes – our own are complicated enough, we can conveniently repeat their sizes in inches: Langstroth deep 91/4" x 19" 335 x 483mm while the shallow frame is 53/8" /137mm. There are other hive types and variations of super frames and some of the names are: Smith, Long idea, British commercial, Dadant. Manley, Hoffman which makes for confusion in ordering and stocking. Be careful to get the right size of foundation for *your* frames.

**France**

Up-to-date figures will doubtless show France as being one of the biggest honey producers in Europe but here are figures for as far back at 1970. One million colonies and 100,000 beekeepers.

**frass**

Part of the mess that might accumulate on the floorboard. The droppings of larvae especially of caterpillars including the wax moth. This will not been seen with varroa floors.
See: excrement, droppings

**freedom**

Bees become agitated if they are deprived of 'freedom'. An observation hive always looks better if its bees are able to fly and return to their entrance. When handling a queen (one that has been caged or merely during the course of manipulation) and she gets loose or takes to the air she will not fly far and you should look for her to return to the hive entrance of the spot she took off from. As much as the 'blue sky' might attract all castes of bee they are still entirely dependent on one another but to prevent them from pursuing their normal activities and therefore their 'freedom' puts them under increasing severe stress.
See: escape

**freedom from granulation**

Everyone likes to look at a beautifully clear sample of golden honey. It is attractive and appears so palatable but nature has decreed that the two simple sugars that form the bulk of the honey are locked into an agreement. They can only stay clear if the fructose dominates the glucose. What sometimes happens is that the glucose portion, ever ready to granulate, begins to do so. In crystallising it release a tiny bit of water which further liquefies the clear fructose. As a consequence we sometimes find a jar which has divided into a granulated half at the bottom while a usually darker layer of clear honey (the fructose) occupies the upper part. In the general way though, the granulation takes place with crystals of glucose forming surrounded by a film of the fructose and this can continue until the whole mass seems quite solid. The fact that the fructose is still there, and more liquid than its counterpart, can be demonstrated by vigorously stirring the honey and so smashing the crystals into tiny particles, releasing the fructose and making a highly desirable cream, the taste of which, because it has a more immediate effect on the taste buds, is delightful. So to remain free from granulation you must either have a honey like that from Acacia which is preponderantly fructose, or heat

and finely strain the honey (depriving a consumer of certain benefits) to remove any particle which could form a nucleus for starting a crystal which, once under way, is a continuous process.
See: granulation, L/D value

**free-for-all**

When carelessness has led to a wholesale bout of robbing and there is a free-for-all, not only do many bees get killed but disease may be spread and honey wasted. It is also very unpleasant for anyone to walk into the midst of the fracas. Shutting entrances down if necessary to one bee-way, throwing wet grass in front of entrances and using wet cloths or a hose may help. Oddly, once bees become satisfied by being convinced that the free-booty has been entirely cleaned up, they settle down. Leaving honey-wet extracted combs out in the open for bees to clean up (an unwise procedure) invites problems such as mentioned. But once the whole lot have been cleaned up, the last few surviving culprits, leave a warning pheromone, and other visitors soon get the message 'nothing doing'.

**freeze**

Once bees' temperature falls to the point where they are 'comatose' further chilling will bring about their deaths, but there is a short window of time during which they can be resuscitated by warmth. Bees, drawn out by the bright light shining on a fall of snow, plunge unknowingly into the reflected polarized light and melt shallow graves unless the beekeeper scoops them into his hands and warms them. Eggs too, if incubation has not commenced, can be kept in a 'no-man's land' and still, like hen's eggs, be brought into activity again. This was a method used sending very valuable bee eggs to the States. However once a bees is rendered totally immobile either by cooling or by drowning it is essential to have other bees around to help it recover. Deep freezing is a wonderful way of preventing honey from granulating because at those low temperature the action of forming crystals is brought to a stand-still. Bees 'freeze' at the sound of a queen piping and a queen will at times feign death (freeze) to get out of trouble and the larvae of wax moth 'stop dead' in their tracks when any movement would attract unwanted attention.
See: thanetosis, stand-still

**freezer**

1. Use to store honeycomb straight from the hives as it will not granulate at very low temperatures.
2. Pollen, preferably fresh, clean and dry, can be stored in plastic bags or containers.
3. Propolis and beeswax become very brittle when cold.

**French bee**

Although there are wide variations, the French bees used in England (and doubtless many came over following the Isle of Wight disease era) were reported to be hardy, prolific, and easy to manipulate. On the other hand nervousness and viciousness and a tendency to run on the combs tended to be contradicted when Wedmore observed that they were not excessive swarmers and that those from Gatinais district were among the best.

**French honeysuckle**
***Hedysarum coronarium*** **Fabaceae**

Garden varieties flower in July with red or white flowers growing to 1.2m/4' and a smaller purple-flowered *H.obscurum* is also visited for nectar.

**frequency**

The repetition of a cyclic phenomenon such as the warbling note detected by an apidictor, the micro-vibrations of thoracic muscles etc. The unit of frequency accepted in the realm of TV, radio and electrics is the 'hertz' and that is the equivalent of one complete cycle per second.
See: c.p.s., flicker

**Frisch Karl von, born Vienna 1886**

Professor at Munich University and famous for his discoveries of the dances of the bees and their means of communication. In 1944 he made the epoch-making discovery of the dance language of honeybees receiving the Nobel Prize in Physiology in 1973. Bees, their vision, chemical senses and the use of dances to indicate forage sources were all opened up by his experiments.

**frog Rana temporaria, esculena, Common and edible, Anura**

Tailless amphibian that includes bees in its diet. The giant bull frog *Rana catesbeiana* is said to occasionally eat bees US. While the cane toad is troublesome in Australia.

**frons**

The forehead or comparable structure on a bee. The queen and workers have an area that extends between the compound eyes and up to the vertex whereas in drones the extra large compound eyes force the single eyes down onto the frons thus reducing its total area as the compound eyes actually touch one another.

**front slides**

The WBC uses slides in conjunction with a porch to give a variable entrance width. They slot into guides and their lower edge is bevelled so that the lower surface adopts the same slope as the alighting board along which they can be moved. To prevent total closure the tips are cut away at the inner ends so as to form a small triangular opening when they touch. When widely open they can trip unwary passers-by.
See: sliding entrance blocks

**frost**

The white rime of minute ice crystals composed of frozen water vapour. It can be harmful to floral pollen causing digestive troubles in the bees which becomes known as 'May pest'. Frost also causes flowers to become sterile by rendering them incapable of being fertilised. In apple orchards this can reduce the size of the window of opportunity otherwise available for satisfactory bee visitation.

**frosting**

Honey on the show bench or sales shelf is severely marred if the honey is frosted, and untreated honey can become frosted quite suddenly with unsightly white flowerings. At times it even causes a gap to form between the honey and the side of the jar. Heating to expel air bubbles and bottling while warm into warm jars helps to avoid this. However glucose can crystallise either into the white anhydrous or hydrate forms. Needless to say it does not interfere with edibility.

**froth**

An aggregation of bubbles as on the surface of a liquid. The cleanest honey will produce a white creamy layer of froth after having been impregnated with air bubbles. The warmer the honey the sooner the bubbles will have risen to form froth. At ordinary room temperatures at least two days are required for honey to clear so that when bottled little or no froth reforms. Froth and scum are always viewed with disfavour.

**Frow mixture**

Frow – Richard Watson in 1918 became famous for his remedy and treatment for *Acarapis woodi* infection. The mixture was both poisonous and flammable and contained two parts petrol, two parts nitrobenzene and one part safrol oil. It was available in ampoules of 4.5ml. A modified form 5 parts ligroin, 6 nitrobenzene and 2 methyl salicylate. The treatment was best in late autumn or winter or in Feb. in the South. Contents were emptied onto a pad or thick piece of cloth about 10cm square, placed over the feed hole and then covered with a tin lid to retain the vapours in the brood chamber. The entrance should be restricted to 3cm and left for 7 days. The odours break down colony defence (identity odour) so all colonies in the apiary should be treated otherwise robbing may well ensue.

Best applied after bees have had a day of cleansing flights.

**fructose**

An intensely sweet carbohydrate - monosaccharide sugar found as D fructose in honey. Other names are laevulose and fruit sugar. A sweet sister sugar to glucose with a mirror image of its molecular structure, a hexose $C_6H_{12}O_6$ laevorotatory ketose sugar. It is a reducing sugar and more soluble in water than glucose. Fructose dominant honeys are: Robinia (acacia) sweet chestnut, Tupelo and sage (stay clear for long periods).

**fruit**

The ripened ovary of a seed plants, frequently edible. When there is a dearth of nectar bees will often take the juice from fruit that is damaged. In nearly every case the skin of the fruit is either too tough or too rubbery (as in the case of grapes) for a bee to be able to pierce the skin with its mandibles. However once bruised or damaged by a creature with sharp mandibles like the wasp, then the bee will take the juice back to its nest and recruit other foragers. The flowers of many fruit producing plants attract bees with their nectar and pollen, some more than others and with marked seasonal variations. The particularly effective pollination by honeybees warrants their widespread use in orchards.

**fruit sugar $C_6H_{12}O_6$**

A monosaccharide also known as laevulose or fructose. It is a hexose – six carbon sugar (a carbohydrate) and constitutes roughly half the sugar content of honey.
See: fructose

**Fuchsia Fuchsia magellanica Onagraceae**

A half-hardy deciduous shrub, naturalised in I.O.M. and south West. Blooms from June throughout the summer. The flower is in the form of an attractive bell and its pendant nature protects it from the rain and produces nectar and pollen, the honey is light. It likes the western seaboards and when used for hedging its vigorous growth calls for harsh cutting yet its flowers abound yielding at last a purple carpet when it is time for the bees to form their winter cluster.

**Fuchsin**

A germicidal coal-tar dye which is a greenish solid that forms deep red solutions and is used as a stain to impart a red colouration to certain preparations for the microscope It is useful for certain pollens.

**fudge**

170g/6oz sugar, .28l/1/4 pint milk,
1 tablespoon honey + knob of margarine. Bring ingredients slowly to the boil in saucepan stirring all the time. Boil for 8 minutes, allow to cool for 5 minutes. Beat until thick, pour quickly into a well-greased shallow tin and mark with a knife. When set cut into pieces. 28g/1oz of chopped nuts or fruit can be added at the beating stage if desired.

**Fuel (smoker)**

Combustible material used to form a cool smoke for the subjugation of bees. A non-toxic material that smoulders and may be used alone or in combination with various things like old, rotten hessian, corrugated paper, cardboard, rotten wood, fir cones, dried sunflower leaves, sawdust, wood shavings, and additives for easy lighting. If a cartridge is made, light it at the bottom before inserting into the fire chamber, keep your veil well clear and have spare material to hand as a refill if necessary.
See: smoker fuel

**fuel consumption**

A forager uses about 1/2mg of honey per kilometre taking off for a foraging trip with a partly filled honey crop (sac). Once this has been emptied and the bee becomes reliant on the food in its alimentary canal and blood stream then cold and wet can easily cause it to become a casualty. The food it might collect should take it home with a surplus but this balance of fuel used and forage collected will help to decide just how far away it can forage. Most effective distance 500m. Although the honey crop has a capacity of about 100mg its average load is some 30mg which would work out at about 54,000 journeys for 454g/1lb of honey. Another aspect of fuel is the amount used when moving bees to crops or visiting out apiaries and any successful honey producing organization has to take this into account. Going to the heather – this involves 4 X the single journey.

**Fuller's earth**

A fine, absorbing clay that can be used medically, as a paint extender, to remove grease from cloth or as a filter or dusting powder. As a filter it can be used to remove practically any foreign matter from honey or beeswax.

**Fumagillin**

An antibiotic derived from *Aspergillus fumigatus* and sold under the trade name of Fumidil B and used for the treatment of nosema.
See: Fumidil B

**fume board**

A panel, the same size as the horizontal cross section of the hive. Chipboard or ply covered with cloth to absorb the repellent. Used in conjunction with a cover it can form a box so that fumes are created. They can then be driven down through holes into the supers using some kind of bellows or spring mechanism.

**Fumidil B**

Abbott laboratory's name for the drug used to treat nosema – bicyclohexylammonium fumagillin. A soluble salt of an anti-biotic produced by the fermentation of *Aspergillus fumigatus*. In this form it is suitable for addition to syrup, candy or honey. Do not use at the same time as Terramycin as they inactivate one another unless a 'buffered mix is used. It inhibits the development of nosema by preventing the vegetative development of the spores whose action would otherwise cause the degeneration of epithelial cells in the bees ventriculus with the subsequent formation of enormous numbers of spores to be carried out in the bee's faeces. The recommended dose for a normal colony is 0·17g Fum B to 6.3k/14 lb. sugar dissolved in 41/7 pints of luke warm water and fed to the bees for autumn feed. Can be kept for years in deep freeze.

**fumigation**

The exposure to smoke or vapour with a view to sterilisation. Substances used include Frow treatment and Folbex treatment for acarine, acetic acid and formaldehyde for nosema, ETO for foul brood, PDB (now banned) for empty super combs against wax moth Ethyl bromide and other volatile chemicals.
See: sterilisation

**function**

Often associated with reproduction or survival in living creatures while in an organ it could be described as the special actions by which it fulfils its purpose.

**function of a honeybee colony**

To maintain itself it must be able to create and maintain warmth, secrete wax and build honeycombs to maintain and renew its labour force, to produce similar colonies in the interests of survival of the species. To provide nutrition for itself and future generations. To be able to reposition itself as necessary, to maintain health and find a suitable environment. To be able to overwinter in sufficient strength and with sufficient food reserves to enable a fresh start to be made when nectar and pollen are once more available.

**fundatrix**

Stem-mother or female of the first generation which founds a new colony –

as with Aphids. It is not always appreciated that it is ants that are responsible for the careful positioning of a fundatrix. Wood ants carry sucking insects into the spruce forests resulting in honeydew production.

**funeral**

The ceremony connected with the disposal of the dead. At the funeral of a bee-owner in olden times, the bees must have a portion of everything given to them pertaining to the funeral repast, otherwise they will die! The outside of the hives are seen hung in mourning with crepe for the deceased possessor. A swarm of bees was reported to have swirled over the cortege when the famous Hampshire beekeeper E.H.Bellairs was taken to rest at Bransgrove (1930).
See: superstitions

**fungi**

A spongy, morbid growth, mildews, moulds, smuts or rusts. A sub-division of Thallophyta. Mushrooms and moulds (mouldy pollen and chalk brood). Yeasts as used for mead or vinegar also come under this heading while certain antibiotics such as Fumagillin are derived from fungi.
See: soot fungus, fungicide

**fungistat**

Additives used in bread, dairy products etc. e.g. sodium propionate and sorbic acid used in pollen patties as a chalk brood preventative.

**funnel**

A cone-shaped utensil that tapers into a tube at the bottom which takes the form of a spout. Liquids can be poured in with little fear of spilling and directed into narrow necked containers. Lined with fine cloth or filter paper they make super filters. A large variety can be made from heavy gauge plastic bags to act as a funnel for shaking bees into a narrow observation or nucleus hive.

**furniture cream**

One of many: 8oz beeswax 1oz white wax, 2oz yellow soap, 40floz turps substitute, 30floz water and 4 tbsp vinegar. Put waxes and turps into a jar and leave in warm place (covered) for 48 hours to dissolve. Mix soap in the water and amalgamate with waxes. Add vinegar and mix. Pour cream into suitable jars or bottles. This will give a lasting polish and withstand dampness.

**furniture polish**

Beeswax is a basic polish and can be softened for application using various quantities of turpentine. A more creamy form can be made by whipping with soft water and soap. One example: .340g/12oz beeswax, 1.4l/50floz genuine turpentine, 113g/4oz soap flakes, 1l/35floz water.
See: furniture cream

**furca furcated (forked)**

Formed by internal processes of the prothoracic tergite, the 'furca' protects the first ganglion of the nervous system of the thorax.

**furcula**

The furcula is a small recurved process that arises as two branches from the base of the bulb of the sting and then unite after curving over it dorsally. It forms an attachment for muscles which play an important role in the preparation for stinging.

**furze gorse *Ulex europaeus* Fabaceae**

A low, much branched, spiny shrub with yellow flowers. Nectar is highly unlikely in the U.K. though it is a useful spring plant in NZ where it has become rampant. It is inclined to bloom throughout the year but its main efflorescence is in the spring. Pollen light brown or dirty yellow.
See: gorse

**Fuschia - Onagraceae**

A semi hardy shrub or small tree, ubiquitous in I.O.M. and S.W. Its reddish or pinkish purple flowers which are resistant to the rain can offer a feast for the bees. It can be cut back quite hard and cuttings for propagation can be taken after the first shoots appear in the spring

fusion

**fusion (flicker) frequency**

The speed at which one vision becomes continuous with the next. By comparison with humans a bee's fusion frequency is very high and indeed it is capable of noticing fast movements more perceptibly than we can.

See: flicker

# G

### gable-roofed
A roof with a triangular end wall. This type of hive roof is typical of WBC hives. Picturesque and rain-shedding with ample space for ventilators but not popular with those who work their bees commercially because they cannot be sat or stood upon, nor can other hive parts be satisfactorily set down on them or inserted into them. Also they are rather uneconomical of timber and space.
See: WBC

### gadget
An ingenious or useful implement. Beekeeping has a wealth of these devices and although it is often said 'there is nothing new under the sun' a steady stream of gadgets finds its way into the hands of apiarists always ready to make improvements or to overcome difficulties or built-in snags.
See: gimmick, implement, wrinkle

### galangine
3,5,7- tri hydroxyl Flavon an effective bacteriostatic agent found in propolis by the French scientist Dr. Villanueva.

### Galbanum
When dried used as a smoker fuel in special mouth-operated, ceramic smoker described by Columella, Palladius et al. A gum resin from Asiatic plants *Ferula galbaniflua*. It has a disagreeable odour and is a counter irritant.

### galea lacinia
The distal portion of the maxilla or outer member of the proboscis. Botanically it is the upper part of the calyx or corolla and is shaped like a helmet.
See: water balsam

### gales
Very strong winds. These are less likely to occur in the active season, in the UK when hives are extended to much higher levels than in winter. Shelter from winds without umbrageous overhanging trees, is always a good thing in an apiary but where hives are exposed, bad temper often prevails. A wire or strap can be put round the hive but it will also need guy ropes to the ground. Rocks set upon the roof are ugly and have other draw backs (trapping moisture and creepy crawlers). In winter hive configuration should be arranged so that they are no taller than is absolutely necessary.
See: prevailing wind, shelter, wind-breaks

### Gale's honey
Marketed by Joseph Farrow & Sons, Peterborough. Well-known and backed by good advertising. It is not always appreciated by those who regard foreign honey as inferior to home produce that up to ten times more foreign honey is available here than British and the promotion of honey as a product does us all good.

### *Galleria mellonella* (Fr. Cerella) Pyraloidea
The greater wax moth. A soft-looking pale, drab-coloured moth with darker brown markings over the top wings, female span 41mm, males smaller. The larvae feed on practically all hive products and complete destruction of combs can be rapid at warm

temperatures. In Tangiers where in an over-large apiary combs packed with *G.mellona* pupae had been left lying about, the moth invaded new colonies, frame by frame, presumably repelling the bees by odour, until whole colonies were ruined. It can cause absconding. The eggs are minute and hidden so that combs can be put away for winter without your realising the eggs can hatch and comb can be ruined. From egg to adult takes 49 days at 24-29C/75-85F the female moth enters hives after dark having little trouble passing guards – they can run faster than bees and can, lay about 500 eggs. New larvae run aimlessly for hours, then burrow into comb and even into the wooden frames, growing to 22mm in a silken tunnel that may extend 15cm. The grubs will run if disturbed. A pupa is formed in a tough silken cocoon that may occupy the whole cell. The moth can cause 'bald-brood' (the white bee pupa being uncovered) Spiders will attack it. Cold limit its importance in the UK.

The careful use of formaldehyde is recommended on combs licked clean by the bees. Acetic acid is an alternative.

See: *Bacillus thuringiensis*, Certan, fumigants, sterilants

**gallon**

The Imperial gallon is 4 quarts or 8 pints 1.20094 US gallon. An Imperial gallon = 4.546 litres and weighs 10lbs/4.54kg. As the s.g. of honey is 1.414 a gallon of honey would weigh just over 14lb.

See: pint

**Galton Dorothy**

Wrote 'Survey of 1000 years beekeeping in Russia' and 'The Bee Hive' and offered many extracts from old literature used in this work.

**gamete**

Mature reproductive or sexual germ cells that can fuse together to produce a 'zygote'

See: fertilisation

**gametogenesis**

The collective name for the production of male and of female gametes during the process of 'meiosis'.

See: oogenesis

**Gamma HCH formerly BHC**

A spray pesticide, a persistent organochlorine insecticide available as dusts, sprays and smokes. In 1982 there were seven known cases of colony deaths.

**Gamma radiation**

This chemical has been successfully and economically used to destroy *Paenibacillus (Bacillus) larvae* spores (causative agent of AFB) in large chambers. Commercial cobalt 60 was used in New South Wales, no damage occurred to hives which could be immediately re-stocked with bees and after seven weeks no signs of *P.larvae* infection were found. The treatment also kills nosema spores and wax moth and in 1982 work was in hand to determine the increased dosage required to control EFB. Gamma rays are the main danger caused by radioactive materials.

**ganglion pl. ganglia**

A collection of nervous tissue forming the centre from which nerve fibrils originate. In a primitive nervous system one of these will serve each segment. In the honeybee larva this tends to be true but the fusion of some ganglia during metamorphosis results in fewer ganglionic masses with some variations in size in the adult.

See: nervous system

**gap through which a worker can pass**

A bee can squeeze through a slot any wider than 2.8mm (or one that can be distorted by stretching as with some plastics. However a square hole of 3.2mm will prevent a worker passing and a round hole can be even larger - 3.6mm (mouse 9mm).

This information is necessary when forming queen excluder slots or mouse guards.

**garb**

See: apparel, protective clothing

**garden**
A plot of ground adjacent to the house where plants may be grown. As a site for hives this should not be the only place considered as there are many advantages to be had from setting up an out-apiary away from home. The washing line should not cross the bees flight path. Neighbours should not be inconvenienced. Generally it is best not to make hives too obvious and neighbour reaction can be tested by setting up an empty hive before any attempt is made to put a colony into it. A fence or screen to keep animals and children away could also be a useful construction. See: out-apiary

**Garden campanula Campanulaceae**
These include Canterbury bells and rockery plants and in July-Sept they offer nectar and pollen to the bees.

**garden escape**
A plant that began as an attractive adjunct but due to winds or other means it has escaped and grows freely elsewhere. This can apply to certain farm crops.

**garner**
To amass, acquire, gather and store.

**garth**
Archaic – an enclosed area such as a yard, garden or paddock.

**gas liquid chromatography**
One of the methods of separating mixtures into their components. The vapour of the material is carried in a steam of inert gas through a tube packed with an inert solid whose surface is covered with a layer of non-volatile liquid. A single 'shot' at the entrance comes out in a series of waves each of which is a single component of the original mixture and can be detected electrically and recorded automatically. Also gas chromatography where liquid or solid is used for differential absorption.
See: paper chromatography

**gaster**
In Hymenoptera the thorax is formed of three thoracic and one abdominal segments. The second abdominal segment becomes the petiole and is followed by the 'gaster' or hind body which is usually referred to as the abdomen.
See: alimentary canal, abdomen

**gastrulation**
The complex cell movements occurring at the end of the cleavage period. The process by which the embryonic blastula is converted into a 'gastrula' (the formation of 3 layers in characteristic order). A hollow sphere formed by a single layer of cells.

**gault**
A blue clay soil with an impregnation of phosphate nodules. Rather calcareous and somewhat stiff.

**gauntlet**
A long-sleeved, sting-proof glove. Strictly a bee glove with a gauntlet or elongated cuff to provide protection from stings and to keep bees from gaining entrance to your clothing via the sleeve. They are of soft, pliable material such as twill, a closely woven fabric and are continuous with the bee gloves and the opening through which the hand is inserted is elasticated.
See: elasticated cuff

**gauze mesh**
A thin, flexible, transparent, open material of plastic or thin wire, forming a network of threads and wires the holes or 'mesh' of which can vary in size (16 mesh would mean 16 holes to the inch). For ventilation or for the confinement or exclusion of insects, the correct mesh size is essential. It is interchangeable for many tasks with perforated zinc or plastic and gives the bees a good 'footing'. It is used for veils, in feeders and for screen boards, travelling screens, pollen traps, queen and cell cages

and many other items.

**Gayre G.R. Col**

His book 1948 gives a well-documented historical account of mead, metheglin, pyment and hippocras, as well as musum, clare and bracket. He describes some of the old utensils used for drinking them. He also produced some excellent mead commercially in the west country'.

**Gean *Prunus avium* Rosaceae**

A deciduous, rosaceous, European tree, the wild European cherry with pollen of a 'tan' hue. Many sweet cultivars have been developed from this parent. See: cherry

**Gedde John 1630 St. Andrews Scotland**

He went to London in 1675 where he put the first patent on a beehive. He wrote 'New discovery of Bee Houses' and 'English apiary' (1721) and obtained a renewal of his patent from James II for a yearly payment of 20 lb of wax to the crown. His hive may merely have been a variation of the Mew's, but Samuel Mew (presumably William Mew's son) wrote praisingly to Gedde who has had the business drive to make the hive a commercial success.

**gelatinous honey**

Honey which 'gels' or when undisturbed adopts a jelly-like composition. Such a honey is that derived from *Calluna vulgaris* better known as 'ling'. Its ability to run when agitated or stirred is called 'thixotropy' and is made use of to extract honey from the cells. NZ Manuka has similar properties.

**gene**

A unit of inheritance or character influencing part of a chromosome. When replicated, as in a cell division, it remains capable of transmitting the same characteristic. It is therefore an hereditary factor. Genes are inherited from the parents in the case of female bees but the drone can only receive those of its mother. Although a drone dies after mating its genes continue to be vitally important in subsequent colony characteristics.

**general rules**

Acid soil – dark honey, alkaline soil light honey. Fast granulation when dextrose content high giving fine crystals. Slow granulation leads to coarser crystals. When bees frequent the 'water-fountain' there is no honey flow on. Similarly robbing or attempts there at suggest no flow.

**generation**

1. Production by natural or artificial means.
2. Individuals born at about the same time.
3. Reproduction or formation.
4. We speak of the generation of eggs, brood, heat, wax and these and for other beekeeping aspects of the word -

See: eggs, heat production, 'laying' parasites, warmth, wax production

**generative organs**

These are parts of the bee concerned with reproduction and also called reproductive organs, genitalia etc. and vary with the three castes.
See: genitalia

**genetics**

The study of inherited characteristics. Male bees (drones) receive only one set of chromosomes (sixteen in number) derived purely from their mother, having no relationship to their mother's suitors. They are therefore haploid. Female bees receive paired chromosomes, thirty two in all, sixteen from each parent. They are diploid.
See: genotype, phenotype

**gen hormones**

They arise in the blood from the effect of genes. The gen hormone affecting eye colour appears identical with I-kynarenin, a substance obtained by the enzymatic oxidation of tryptophan (an amino acid).

**genital system**
See: reproductive system

**genitalia The genitals**
The organs of reproduction particularly those of the drone.
See: wedding cert.

**genotype**
The basic genetic construction of an individual organism or a group of animals with the same genetic constitution. A plant or animal whose genes are of a certain type. Genotype bees might be maintained to provide a 'gene – bank'. They will carry the desired genes without necessarily manifesting the associated characteristics though it would be hoped that these would show up in the F1 and F2 generations.
See: phenotype

**Genus pl genera**
Kind, sort or class. The generic or family name is used in conjunction with a descriptive noun or adjective, to form the binomial identification of a species. Neither the generic name e.g. Apis, nor the descriptive specific epithet 'mellifera' can by themselves identify a species, but together they form the specific name, as for example *Apis mellifera*. This identifies the species. The generic name always has a capital letter and the accompanying name is written in the lower case. When no ambiguity exists the first part of the name can be abbreviated to its capital, hence *A. mellifera*.
See: specific name

**geographical origin of honey**
Eucalyptus: Australia, Argentina and Brazil. Sunflower: Central Europe & Latin America, Lime: Central Europe & China. Pollen analysis is a guide to origin which must be established by other evidence including documented history of sample, chemical analysis and organoleptic examination. The area that honey is likely to have originated from can in many cases be detected by a careful comparison on the pollen spectrum. That is to say if one exotic type is found in company with other types found in that particular area then the chances are that no other honey samples are likely to have the same mixture. This does not take into account the effect of passing honey through containers or filters that have been 'tainted' with exotic pollens.

**geographical origin of Honey Bee**
While we are not yet certain it appears that honeybees evolved in the warmer regions of the world and then spread into cooler climes. Until man transported honeybees they were not to be found in Oceania or the New World.
See: honeybee antiquity, origin of bee

**Geological time**
An Era is a major division of time broken up into epochs. The Proterozoic or Precambrian era consisted of events prior to 570 million years ago, the Palaeozoic 570 – 225, Mesozoic 225 – 65, Cenozoic 65 to the present. Periods Cambrian through Devonian (to which first fossil insects are dated) Triassic and others to Quaternary. Epochs are main divisions of periods. Palaeocene through epochs suffixed 'ocene' to Holocene spanning millions of years and forming a geological series.
See: Cretaceous, period, epochs

**geology**
The scientific study of the composition, structure and history of the earth. As this has a direct bearing on flora its study is interwoven with any profound understanding of apiculture. The types of soil and nature of the subsoil and the forces that act upon them affect plant and animal life and so bees and bee plants.
See: geological time, soils

**The Geoponica**
Written in Constantinople circa 950 A.D. at the command of Emperor Constantine VII the twenty books inevitably include certain information on bees because the

work aimed at being very comprehensive. It was much read during the middle ages doubtless accounting for the passing on of useful and erroneous ideas, as it was largely extracted from earlier works. Individual authors include: Didymus, Paramos, and Democritus. Indications suggest that combs attached to bars were used quite early on by the Greeks.
See: Didymus

**geranic acid**
A liquid aldehyde with a strong characteristic smell. Found in scent gland secretions. Citral (geranial) $C_9H_{15}CHO$ - another source says a fragrant pale yellow liquid $C_{10}H_{18}O$ Attar of roses used in cosmetics and flavourings.

**geranium hardy Geranium nodosum Geraniaceae**
Cranesbill, a hardy perennial yielding nectar and pollen and growing to 45cm/18" tall with mauve flowers about 25mm/1" across from June to Sept. Will naturalise in open woods or grassland.

**germ**
A small cell from which a new organism can develop. A microbe, bacterium or virus.

**German bee**
Similar to the British black and heath bees but not as dark and less desirable. They tend to swarm more and have too many drones (*perhaps a desirable feature these days*), poor house cleaners and disease resisters, but gather late, winter well and make white cappings. Some varieties less desirable than others.

**germander speedwell *Veronica chamaedrys* Scrophulariaceae**
Spikes of blue flowers found in woodlands and hedge banks. Varieties of speedwell can be confusing. Pollen light grey.
See: speedwell

**German Swiss bee**
Dark, long-lived, hardy, adaptable, not inclined to swarm. Run on combs and need more subjugation that Italians and Carniolans.

**germ band**
The thickened lower portion of the blastoderm within the developing egg. It is the beginning of the embryo as it grows upwards on the sides and around the ends of the egg, finally replacing the dorsal blastoderm and forming the outer wall of the embryo.

**germ cell**
A reproductive cell or gamete.
E.g. microscpores, sperms, ovules and ova at any stage of their development.

**germinate**
The beginning of growth as from a spore, seed or pollen grain. Conditions leading to the state such temperature, moisture, chemical etc. may be very specific.
See: abrade, germination

**Gerstung F 1891-26**
A well-known student of the life-cycle of the bee. His theory of the increased number of nurse bees out-pacing the requirements of the colony and producing a surplus of brood food has had a serious place in the theories associated with the causes of swarming. The more slowly a brood area increases the sooner there is a surplus of nurse-bees and earlier start of swarm preparations.

**gestation**
The period of time between conception and delivery of the embryo. The development of the zygote into the embryo. This is a period of rapid cell division (cleavage) followed by cell re-arrangement (gastrulation) and then organogenesis (laying down the systems and organs of the adult structure).

**Ghedda wax**
Beeswax originating from southern Asia

and derived from honeybees other than *Apis mellifera*. Sometimes called 'Indian beeswax' the wax from *A.scutellata, cerana* and *florea* contains ceryl alcohol and has a different ratio of hydrocarbons to alcohols as compared with ordinary beeswax. Very suitable for making dental plate impressions. See: other waxes

**Giant honey bee *Apis dorsata*. Renamed *A. scutellata***

Also known as the 'rock-bee. It is the largest of the honeybee species and is found in the tropics and sub-tropics (middle east, India and Sri Lanka) and makes a single comb suspended high up on a cliff or rock face. Below the huge canopy of stores in extra wide comb it tapers to smaller drone cells and then to worker cells. Although, because of its size, it has a reputation for being extraordinarily fierce, it has characteristics which make it reasonably easy to handle. It rapidly withdraws from smoke and even absconds when over much is used. Natives, with primitive smokers, dangle from ropes to cut huge chunks and aim to catch the heavy lumps in their baskets. Their honey and wax is highly prized. Absconding is seasonal and even takes place to move to more shaded positions during the season. In Sri Lanka a Governor was seen being beaten by his man-servant who was only trying to drive the 'bambara' away. As a result a special cage was built inviting entry at times when the air seemed full of giant bees.

**giant puff-ball *Calvatia gigantea* (formerly *Lycoperdon giganteum*) Basidiomycota**

When ripe and full of spores it can be dried and used as toxic smoke. It was more significant in the days of the skeppists. See: puff ball

**Gift class**

Honey Show classes often include a 'gift class' where the owner pays no entry fee but donates his honey to the funds, unless being considered especially valuable, he arranges to have it returned in exchange for an agreed payment.

**gimmick**

An item such as a piece of equipment or method that is put forward as satisfactory or useful by one person yet may be described by another as a sale's gimmick'. Akin to the word gadget which suggests usefulness whereas a gimmick carries the implication of its not being widely accepted and may be clever and is probably informal.

**gimp pin named derived from upholstery**

A small nail like a panel pin, with a fairly pronounced, flat head, suitable for fastening the relatively thin wood of frames together. They may be used in conjunction with glue. Black lacquered nails of this type with suitable lengths for use in frame assembly are sold by equipment suppliers. When driven into hard material they are inclined to bend unless held rigid by pliers or a rampin.

**ginger beer**

680g/11/2 lb clear honey, 42g/11/2 oz ground ginger, 28g/1oz brewers yeast, 4.5l/1gal of water, 2 large lemons and a slice of toast. Boil water, honey and ginger for 1/2 an hour taking off scum as it forms. Cool to about 38C/100F add lemon and thinly sliced rind. Put into a vessel for fermenting and add yeast spread on toast. Cover and keep in a warm place for 3 days. Strain and bottle in strong screw-top bottles (or fitted with well-secured corks). Ready in about a week.
Gk Greek
In this work it refers to the ancient language of the Greeks.

**glabrous**

Bald, smooth, free of hair – usually applied to plants but also characteristic of an older queen which like other bees is unable to proliferate new hairs once they have been worn or pulled away. Specific plant name 'glabrous'.

**glacial acetic acid $CH_3COOH$**

A 99.5% concentration of this acid is recommended for use in destroying nosema spores. As a much lower concentration that is not so corrosive will normally be found to achieve this, ordinary commercial acetic acid is useful. It eats away or discolours metal (runners, frame wires, metal porter escapes, metal ends etc,) and attacks cement floors. It is flammable and equipment should be well-aired before going back to the bees.

**glade**

An open space in a wood or forest. It is not usual to set bees down in a forest unless it consists of flowering trees such as Robinia or Eucalyptus. A glade may offer a suitable site but access must be considered and tracks or paths which are relatively flat and free of undergrowth may seem suitable spots but there is much unseen traffic of animals too timid to show while you are around yet likely to stumble into or knock hives over when playing or being chased.

**gland**

An internal organ which acts as a factory for producing special substances of physiological significance such as enzymes, hormones etc. They may be single cells or masses of characteristically-shaped bodies. Social insects such as honeybees make especial use of their glands to function as individuals and as a colony.
See: acinus, glands and glands (plants)

**glands**

Abdominal, acid, alimentary, alkali, appendicular, endocrine, epithelial, head, hypopharyngeal, Kozchevnikov, labial, mandibular, mucus, Nasonov, postcerebral, post- genal, prothoracic, rectal, salivary, scent, silk, spermathecal, sting, sub-lingual, thoracic, thoracic salivary, venom, ventricular and wax.

**glands (plants)**

Specialised cells that can elaborate external secretions such as nectaries. Stigmata may secrete sugary substances that help to hold and stimulate initial germination of the pollen grains.

**glass**

A hard, brittle substance which can be quite transparent, gas and liquid-proof. In some cases modern plastics can be used as a substitute. In the beekeeping world many items benefit from this kind of material: honey jars, hives-fibre, crown boards, inserts, feeders, quilt, solar wax extractor and visual aids.

**glass house crops**

Cultivation of half-hardy plants in greenhouse has led to the need to pollinate flowers in an environment kept largely free of insect life. Hives of bees can be introduced to advantage provided certain precautions are taken such as not over-heating and first losing the flying bees who would hammer themselves to death flying against the glass. They must also be given ample food including pollen and suitable pollen must be at hand for the plants that are being cropped.

**glass quilt**

When the frames are to be visible via the inner cover a glass quilt in a wooden surround is often used. By using a centre strip of wood across the frame and two pieces of glass on either side a feed hole or porter escape hole can be incorporated. As glass is cold and can cause metabolic moisture to condense on the underside and drip onto the winter cluster, a layer of insulative material is best placed over the glass.
See: crown board, inner cover, quilt

**glaucous**

A waxy film that reduces water loss, greenish-blue or whitish as one finds upon a sloe. It has been used to describe the appearance of wax and possibly bees. The film that appears on wax may be polished off without detriment to the wax. As a specific name for plants 'glauca' is used.

**GLC**

Gas liquid chromatography, a technique

used mainly for analytical purposes.

**glean**
To gather slowly and laboriously. Bees are said to glean pollen when they follow the work of other bees collecting remnants without necessarily tripping the anthers or servicing the flower.

**Gleanings in Bee Culture**
An American bee magazine of world-wide interest which skilfully combines the anecdotal hobby atmosphere with education and scientific progress.

**Glen hive**
Produced at the end of the first world war by Dr. John Anderson in Scotland.
Its brood chamber holds fifteen British standard deep frames to produce larger, better colonies for the tough job on the moors. He was impressed by the prolificacy of Italian hybrid bees (after the British blacks) and designed this 15 frame hive instead of the 10 frame WBC which is similar in the double-walled aspect though the Glen hive has no plinths or projecting porches and the bees enter through a wide passage under the floor using a roller type entrance control. The legs are short.

**globe thistle Echinops spp. Compositae**
Blue globe thistle – Chapman's honey plant, a hardy perennial that grows to over a metre high, blooming from June through to August with bright blue, globular heads. It is happy on rocky uncultivated ground and is much worked by honeybees. Its pollen is light (white). As its root stock deteriorates Harwood suggests treating the plant as a biennial.

**Glory of the snow Chionodoxa luciliae Hyacinthaceae**
A spring flowering bulb with white or blue flowers Mr/Apr and yielding both nectar and pollen.

**glossa**
The bee's tongue. A hollow tube of thin, tough membrane, with a cross-section like an inverted 'C'. It has a slender, strengthening rod and is capable of being drawn backwards by adjacent muscles. It can be inflated by blood until it looks like a large bladder and in this way any particles caught in the small diameter of its canal can be ejected. The terminal bouton is a small hairy pad that enables minute droplets to be swept up.
See: glossal canal

**glossal canal**
An upper groove that runs along the worker's tongue allowing saliva (enzymes) to pass into food especially into nectar to initiate inversion of the sucrose.
It can be exposed when the bee forces blood into the ligular unfolding a plaited membrane which can then be cleaned or 'stropped' by hairs on the fore legs. Its small diameter allows it to be used as a very fine filter. (one thousandth of an inch).

**glossometer**
A device for measuring tongue length. Schools have used cards punched with holes just large enough to permit the access of a bee's tongue and placed over dishes of syrup so that the depth, when no more can be reached, is measured to give an approximation of tongue length. (e.g. compare honeybees with bumbles).

**glove**
A covering for the hand which enables one to use fingers and thumbs. Rubber, kid leather and plastic gloves with gauntlet attachments are widely used in beekeeping. They protect the hands from stings, prevent bees from getting up one's sleeves and keep the hands clean (especially from propolis).
See: bee-glove

**gluconic acid $C_6H_{12}O_7$**
Found in honey produced by gluconobacter.

**glucose $C_6H_{12}O_6$ a hexose**
A monosaccharide sugar found in honey as D-glucose and in other living cells. Other names are dextrose and grape sugar. It is dextro-rotatory and has a structure closely related to fructose. Glucose-rich honeys granulate quickly because glucose is less soluble in water than fructose. (they include rape, raspberry, dandelion and brassicas). An important energy source in metabolism and about half as sweet as ordinary sugar (sucrose). It is a reducing sugar also called aldose because one of its carbon molecules has a potentially active aldehyde group.
See: specific rotation, reducing sugar

**glucose oxidase**
In dilute solutions this enzyme reacts with glucose forming gluconic acid (the major acid in honey), and hydrogen peroxide. Exposure to light inhibits this enzymal activity.
See: honey stability

**glue**
Any adhesive substance or the act of causing two surfaces to adhere to one another. Bee glue – propolis.
See: adhesives

**glycogen $(C_6H_{10}O)n$ $(C_6H_{10}O_5)x$**
A tasteless, white polysaccharide, a carbohydrate storage product of plants and animals, animal starch. An important carbohydrate food reserve usually stored in the liver and easily hydrolysed into glucose. It is stored in the muscles and the fat body in bees and provides glucose to produce energy quickly where it is wanted. When a bee rests, glycogen is hydrolysed to glucose to provide more energy.

**gnat**
Numerous small biting two-winged insects. Not only can these penetrate most veils and give the beekeeper who manipulates in the evenings a hard time but they are said to bite the bees and make them very tetchy and bad-tempered.

**gnathosoma**
The mouthparts and antennae of Varroa mites including pedipalp and gnathosomal tube.

**goat *Capra hircus***
A horned, bearded animal reared for its milk and meat. Normally kept tethered. One attached to a water tank in a dry spell, pulled it along towards a hive where the grass was greenest but it finished up dead.

**goat's rue *Galega officinalis*, Fabaceae**
A hardy perennial, pale blue and white, grows 1.2-1,5m/4-5ft. and flowers June – Sept. Pollen light orange and grains hardly larger than the tiny forget-me-not grains. Dislikes nitrogenous soils.

**Golden Rain tree *Koelreuteria paniculata***
Pride of India. A deciduous tree US and fairly hardy in the UK flowering July – August and has attractive yellow flowers. Introduced from central China 1900.
See: laburnum

**Golden rod Solidago spp. Asteraceae**
A hardy perennial that grows 1m high. There are many cultivated varieties but at least one 'escape' grows freely on open ground, its spiked yellow flowers being well-worked by the bees as it provides pollen in the autumn. Its honey is pale yellow and of good flavour though it granulates coarsely. Pollen yellow, golden yellow. Varieties include *sempervirens* – seaside golden rod US, *chineuous* Argentina, *gigantean*, *canadensis*, *rugosa* US and *virga aurea*.
See: solidago

**Golding R.**
Hunton, Maidstone, Kent. A contemporary of Langstroth and one whose influence doubtless led to Langstroth's application of bee-space in movable frame hives. Golding had made certain improvements on the Grecian bar hive which he described in his '*The shilling Book*'. Munn, Bevan

**gonapophyses**
Chitinous outgrowths or valves subserving copulation in insects; the component parts of a sting.

**gone through**
When a colony has been thoroughly examined we say it has been 'gone through. The examination may be based on a routine inspection to check that a colony (a) is queenright
(b) has brood space
(c) has food
(d) has storage space
(e) is healthy.
On other occasions we might only check for signs of swarming such as incipient or actual queen cells or for health and when breeding and making thorough colony assessments. Notes suggest 'GT' for the above or (1/2 GT).

**Good candy Scholtz candy**
Finely ground sugar mixed into a paste with honey. Used as a candy patty or in a queen case to delay the release of a queen.
Queen cage candy was first conceived by Rev. Scholtz in 1859. Prior heating of the honey not only sterilises it but helps to produce a smoother mix which when cool remains quite firm.
See: queen cage candy

**good combs**
A comb worthy of its place in a hive. Beginners especially, are often puzzled as to what constitutes a good comb. Should I re-cycle this one? It is soon realised that new combs are virgin white and old ones tough and black with cells of reduced size. These latter are often most attractive to the bees but can easily spread disease. A good comb is a sterile one with normal sized cells, little drone comb (unless for supers) and little wasted space e.g. 'pop-holes'.

**gooseberry *Ribes uva-crispa* (syn *R. grossularia*) Saxifragaceae**
A fruit bush that flowers April/May. In France, sea-green honey gathered from the blooms of gooseberry and sycamore, is a great delicacy and said to be of unsurpassed excellence. Pollen glutinous, dull greenish yellow to greenish, brown and dark green, grains 30µ

**goose feather**
The large feathers of a goose wing were often used as a bee-brush. Now-a-days foam rubber attached to a small handle is equally effective. Feathers have been burned to asphyxiate or anaesthetize bees.
See: toxic subjugants

**gorse *Ulex europaeus* Fabaceae**
Needle furze a much-branched, low, spiny, ever-green shrub that is confined to acid soils and blooms the year-round though its fragrant yellow flowers seem at their best from March through June. Pollen is its main contribution being variously described as bright, light and deepish or dirty yellow, dull or light brown or dull or brick red orange.....It grows up to 2m high and has an invasive nature covering large areas such as roadside banks and commons and being useful for its early pollen, its spring-like stamen though, proving a little unkind to some visitors.

**G.P.O.**
In 1880 a consignment of queens arrived in Britain after 10 days in the post from Cyprus and the Holy Land. The regulations laid down by the P.O. for queen cages were 4 x 11/2 x 7/8" that is 100 x 32 x 22 mm and must have a covering of metal 24 standard wire gauge holes 1/8" dia and 1/8" apart. Compensation not payable for loss, damage or delay unless in sealed, stout manila envelopes clearly marked 'Live queen bee' with vent holes 11/2" apart at least, holes no more than 3/16".

**grading filters glass**
BBKA grading filters for honey have been made available the original ones made to BSI specifications. Two glasses held separately

and then together give the three grades of honey colour, light, medium and dark.

**grafting**
The transfer of young (12-36 hr) larvae or eggs from a natural cell to another chosen or specially prepared cell. This is to initiate the raising of a new queen and genetically suitable material is carefully (having regard to deftness, temperature and humidity) put into a colony which is strong, healthy, well-found and regards itself as queenless. The term 'grafting' was first used by Larch 1876 and the technique was successfully used by Doolittle. It is helpful to pare down the cell walls of suitably-aged brood so as to expose the young larvae and make them more easily accessible.

**grafting tool**
A tiny spatula formed by thinning and curving the end of a small piece of wood, matchstick size. Care should be taken to use material that will not readily conduct heat from the small larva, and which is easily held by the fingers and which can be conveniently slipped under the larvae. The material from which it is made should be odourless and inert. Manufacturers sell suitably curved metal tools fitted with an appropriate magnifying glass.

**grain**
There are several connotations starting with grain meaning seed. It was a measure of weight being equivalent to 50mg. The crystalline structure of granulated honey has small grains of dextrose the size of which determine whether it is fine or coarse grained. A grain of pollen has a cellulose shell and its size ranges around 25µ and its characteristic shape identifies its origin. The grain of wood is significant in that strength lies along the grain and where the grain is exposed extra attention must be given to protect it from the weather.
See: granulated honey, pollen grain, wood preservative

**gram**
The weight of one ml. of water, 1000th part of a kilogram, 0.35274oz.

**Grange observation hive**
One deep and two shallows, bottom entrance, covered by BS1300, 1960.

**Granton uncapping tray**
A long, sloping, metal tray surrounded by a 3" rim and fitted with an opening through which honey can pass. There is an underneath compartment that can hold water and an electric element is incorporated so that the honey/wax can be encouraged to separate quite quickly. Frames can be locked by their lugs into a recess so that, as they are uncapped, the cappings fall into the tray and can be dealt with.
See: uncapping knife

**granulated honey**
Honey which has become crystallized and is free of liquid. The size of the granules will vary with different kinds of honey, the smaller the grain, the smoother the texture. It is the glucose part which first begins to granulate and the greater its dominance the more readily it granulates. Long slow warming will first soften and eventually dissolve all the crystals. Honey becomes considerably lighter in shade when granulated the lightest honeys becoming snow-white and the darkest light brown. Continued stirring (more beneficial than heating) will smash crystals so that a sample may be rendered as soft as margarine but it requires determined stirring, be careful though, not to smash electric devices as the honey can be extremely resistant.
See: granulation, honey hating

**granulated sugar**
The bulk of the sugar used in ordinary households is white, coarsely ground sugar. It may be derived from beet or cane but when refined to the white condition it is almost pure sucrose and has little food value acting simply as a sweetener. Fortunately

though, because nectar is largely sucrose, ordinary white sugar makes a useful food substitute for the bees whose stomachs can handle this chemical but it is unwise to use brown sugar or golden syrup because this could cause the bees to have dysentery. Dusting with the finest ground white sugar can be part of the treatment for Varroa and for helping to unite bees. There are times in the year when dilute sugar syrup can provide essential water for brood rearing. Although we get no 'smell' from sugar syrup it is highly attractive to ants, bees and wasps.

**granulation**

The general public always seem puzzled about honey as to whether it is best clear or granulated. As explained under 'granulated honey' it is the glucose part which first precipitates crystals and its dominance will decide how quickly any given sample with transform. Temperature affects the issue, as while warm the honey tends to stay clear and so does intense cold (deep freezer) and in fact the ideal temperature for granulating is 14C/57F. The granulation process is initiated by nuclei in the form of minute bubbles or minute solid particles (say pollen) and you can help it if you wish by 'seeding'. This involves taking a small amount of a finely grained sample and mixing it with some of the honey you want to start granulating. Then stir this into the honey. That should (especially if you can keep it around 14C), allow fine (matching) crystals to form. An occasional stir for a day or so will also help.

**grape *Vitis vinifera* (In vino veritas)**

The fruit of vines. There have been no reports of bees collecting nectar or pollen from its flowers in this country. In warmer parts bees have been known to work the flowers. Bees are often blamed for damaging the ripening fruit, though a colony can starve with a ripe bunch adjacent to them because the strong flexible skins are too tough for their mandibles. Wine fortified with honey is delicious and healthful.

See: Virginia creeper a close relative to the grape

**grape fruit *Citrus paradisi***

Sub-tropical rutaceous tree with large, edible, yellow fruit and acid pulp mostly with seeds and requiring bees for pollination.

**grape sugar**

Dextrose, glucose and grape sugar a monosaccharide – hexose.

**grass**

A neatly mown grass lawn arrayed with a line of white WBC hives is a sight that might attract any non-beekeeper into making a start. The established apiarist however might look at things from another perspective. Grass and weeds should not be allowed to overgrow the entrance though bees would filter their way through even if that makes them lose temper. Wet grass can discourage a bout of robbing and a tuft might form a temporary entrance block. So too might a tuft be used to block the smoker funnel when an operation comes to an end. An over-hot smoker, belching flame can be calmed with a handful of grass tucked into the top of the fire chamber. When a nucleus is set up in the same apiary and many fliers are likely to return to the parent, then stuffing the entrance with grass – ends pointing outwards - will give them time to readjust and make the entrance normal as the grass withers.

**grassland**

Areas where grass naturally predominates due to agencies such as grazing, fire or agriculture offer poor foraging opportunities. The tendency to grow rye grass and to cut fields just before or during flowering shows a lack of interest in the needs of wild life including our bees. All too often a splendid show of dandelions or clover are victims of the chopper even when bees are busy amongst the blossoms. (gotta do our jobs ya know).

**grass verge**

When correctly sown and husbanded the grass verges that abound along our roads can be a useful source of flora bearing in mind though that even slow traffic will kill as many bees as the ones that succeed in running the gauntlet. Recent observation has shown how well 'wild' white clover predominates in council verges but to actually see bees working the flowers is another matter.

**grate**

1. The fire grate in a smoker allows air to be blown under and through the smouldering mass. It can work loose but should be taken out when cleaning is necessary and the spring legs re-aligned to help make a tight fit.
2. It is helpful to grate beeswax to make cold cream, floor polish or to soften or dissolve it in a solvent for whatever reason.

**gravid**

Pregnant - a female with eggs or pregnant uterus. The transformation of a virgin queen's abdomen into the full glory of a mated queen is a remarkable thing and this together with her sturdy, straddling legs, make her appearance fairly obvious when she stands on the comb amongst her workers.

**gravity**

The force which we and the bees feel and to some extent understand, is called the heaviness or weight and makes smaller boxes of honey less back-breaking to handle. It also gives the bees a datum from which they can signal an angle relating to the direction of the recruiter's dance. Then obeying the 'von Frisch' dances they can set off in a line, transpositioning the angle between 'down' on the combs and the angle of the scout bee with direction to take off in relative to the polarized light patterns and shadows that mark the angle of the sun. In the dark of the hive it is gravity that helped them construct vertical combs and thereby have back-to-back cells which put them into an intelligence category quite beyond other social insects and even some of ourselves! Gravity plays a part in the embryo queen's feeding and metamorphosis – the only caste requiring a vertical, purpose made cell. Excellent use can be made of gravity to render wax and clarify honey.
See: centrifugal

**grease**

A lubricant of a fatty or oily nature. It can be messy and smelly and only a type like petroleum jelly (Vaseline) can be used on metal ends, or lugs of perhaps twixt plastic chambers (when new they easily slide one against the other until the bees have had time to propolise them). It can be used as a film to exclude moisture or even to waterproof in the same way as beeswax. Its nature is too transitory for use when non-stick agents are required e.g. wax moulds.

**Great Britain**

Since 1707 this name became used for the political union of England, Scotland and Wales. England includes the Isle of Man (fine beekeeping there), Channel Islands and the Province of Northern Ireland. Britain and United Kingdom seem to be interchangeable titles so UK or GB figure under several headings in this work.

**Greater wax moth *Galleria mellonella***

This pale brown moth lives for 4 – 6 weeks. It emerges from being a pupa flies to trees and mates then almost magically, detects the presence of beeswax. Entering hives by dark it lays its eggs in crevices. It has been dealt with under its Latin name q.v. It can be quite a menace and contaminated material should never be left around but burnt or fumigated. The density of the grub's webs has to be seen to be believed and it can bind combs together especially when brood has been present into an all too solid mass. Formaldehyde and acetic acid play their part in its control.

**Great spotted woodpecker *Dendrocopos major***

Widespread in GB though the green woodpecker is the one usually associated with damage to hives.
See: green woodpecker

**Greece ancient**

In ancient Greece a girl's dowry would consist of a certain number of basket hives. Many tips regarding beekeeping come from writers of the time.
See: anastomo, kofini, vraski

**Greek bar hives**

Anastomo cofiini a basket wider at the top than the bottom in which bees build their combs from bars placed parallel across the top of the basket. The sloping sides usually discouraged any attachment to the inner walls. Inspections were therefore possible. It was one of the earliest examples of beekeepers utilizing movable combs.

**Greek bee *Apis mellifera* media var. cecropia**

*A. m. adami* was the name given to honour Bro. Adam and used for the special race of bees found still in Crete and the Aegean islands.

**Greek beekeeping**

Migratory work is an integral part of the intensive exploitation engaged in with honeybees. Colonies are taken to citrus, thyme, cotton, heather and pine. When this latter flow starts beekeepers from all directions converge onto the pine forests. *Pinus halepensis* honey comes from *Marchalina hellenica* the Lachnid sucking insects and is so abundant in Thasos island and N. Greece that it can ooze down the trunk and branches onto the ground underneath - a characteristic honey with a special flavour and reputed to be most health-giving.
See: NZ honey-dew

**Greek writers**

Ailianos, Anaxagorax 428 BC, Plato 427, Antigonos of Karystos 290, Aristomachus. Aristotle 384-322, Callimashus 230, Democritus (not the philosopher), Hecataeus of Colophon 150. Theophrastus (assistant to Aristotle) 372-287, Xenophanes 427 and Philiseus (Agrius).

**green belt**

An area of open farmland or uncultivated land deliberately left around some towns to allow for parks, recreation grounds and so on and where ordinary building – houses or factories is not permitted. These areas are often quite sheltered and can offer quite useful forage to the bees especially when certain flowering trees and shrubs are grown.

**greenfly Aphididae**

A small, hemipterous, green-coloured insect found on the underside of leaves. The insect secretes or exudes a sweet, sticky substance which falls onto the upper leaf surfaces making them shiny and attractive to bees and ants. The discarded white moults are often 'give-away' signs. Bees will collect the sweet material, especially when there has been no cleansing rain and nectar is scarce. Lime, sycamore and many other trees can become smothered by aphid secretions. Although in mating some become winged, they are usually transported and 'farmed' by ants. The surplus exudations frequently fall onto parked cars and can be very unpopular.

**green honey**

Although this expression can be used to describe unsealed, unripe honey there are honeys that are either green in colour or have a tinge of green. A green dye was inserted into cheaper sugar when the EEC instructed the 'free' issue and of course despite promises it did get into honey. Honeys with a greenish tinge include: gooseberry, kinnikinnick, sycamore, lime, phacelia, ailanthus and fields beans.
See: unripe honey, honey-dryer, dehumidifier

**greenhouse crops**

Very early fruit such a peaches and strawberries, crops that benefit from honeybee pollination, can be vastly improved if bees are use inside the greenhouse. Techniques include a good young queen, early stimulation deprivation of any fliers prior to moving hive in. Plumping of colony to maintain its balance and provision of food (pollen and nectar) and monitoring to ensure the bees are working normally.

**greensand**

A rocky soil found below the chalk, upper greensand is therefore calcareous but the lower greensand is usually acid and can produce extensive heathland. The two greensands are often separated by the Gault clay. Its glauconite content gives it a greenish-blue appearance. Clovers do well when this soil of transport originated from chalk.
See: boulder clay

**green sulphur**

This finely divided or powdered sulphur – as is used for dusting plants –has been used as a treatment for nosema. A single gram is mixed in with sterile honey but as it cause the bees to void several layers of their stomach epithelia, it is important that flying weather should precede its use or excretion within the hive could worsen the condition.

**Green woodpecker *Picus viridis*, Picidae**

The 'yaffle' can attack the thinner, weaker parts of a hive like hand-holds and knot holes and do widespread damage boring holes through which they take pollen, honey and brood. The great spotted woodpecker *Dendrocopos major* is also a minor pest that destroys a few adult bee from time to time.

**Gregarine Gregarinia**

Large protozoans with cephalonts that are oval 16-44µ. Spore producing sporonts are 35-38µ. These are parasitic inhabitants of digestive tracts and other body cavities. Fumidil B is reported a preventing infection.

**gregarious**

Living or moving in flocks or herds. When a number of bees find themselves without a queen this clustering instinct still influences them. Bees are attracted to each other by sight, vibration, heat and odour. Solitary bees frequently nest in adjacent holes making the area look as if it supports a composite nest of social insects.

**grip**

A bee can grip by means of its six pairs of claws, or pads on smooth surfaces. It needs to do this when it uncouples its wings and adopts the fanning posture. Their tenacity is remarkable as when pulling a handful out of a swarm you can feel how well they try to hang on to their comrades. The gripping also comes into force when they hug a queen to death in the unpleasant process of 'balling' a queen. A long continuous hug deprives a queen of the ability to breathe and although you might think the hugging is to protect her the action is clearly meant to be regicidal. To try interfering one can use breath, smoke or even drop the knot of bees into water. On occasions queens have been salvaged in this way, but if you try, use introduction techniques, do not hurry to re-introduce her. Never ignore a walnut-sized cluster of bees not far from the entrance.

**grooming**

Giving physical attention to prepare a creature so that it can better perform a forthcoming task. Drones and virgin queens receive appropriate attention when nubile (marriageable). A queen is constantly caressed but usually for the benefit of the colony by way of receiving and distributing queen substance. Workers attract what looks like a 'de-bugging' treatment, by adopting a straddle-like stance on the comb, their groom giving active and almost aggressive attention to parts under the wings and around the body. In the case of bees with paralysis virus, this can lead to their ejection.

**groove**
A slot or furrow channelled out often by a shaped tool. This is different from a 'rebate' as it has a 'u' shape or side walls. Where sliding panels of glass, plastic or wood need to ride along grooves this can be made smoother when beeswax is applied. Of course this does not apply to hives where the use of beeswax would gum things up. Vaseline is the better substance there and can be used to keep propolis or beeswax away. A bee also has an important groove on its tongue.
See: mouthparts

**ground**
The solid face of the earth upon which we live, grow things and rest beehives. It is always best to keep bees clear of the ground because dampness, ground frost – in fact the terrestrial microclimate is not of their choice. In nature nests are found above the ground because of the aforementioned, and of course predators. The ground under hives should always be level and firm and a covering in front of the entrance can give you information about what is being thrown out of the hive (immature brood, dead virgins, drones, crystals). Weeds should be discouraged from growing from the ground in front of the hive. Sugar can be 'ground' when needed to puff it over bees when treating for Varroa or uniting.

**groundspeed**
The actual speed of an airborne object relative to the ground. Bees adjust their speed flying harder into wind than down-wind. Wind effect means that the whole time the bee has been flying against the wind will be deducted from what would be its flight time in windless weather. The relatively stream-lined body and sensitive antennae help a bee to control itself even in windy conditions but there are limits and if a bee's average cruising speed is countered by the wind speed, then foraging is not possible.

**group**
Group one, was a classification given to highly toxic agricultural chemicals which are residually toxic for 24 hours or more and affect both crop and flowering weeds amongst the crop. A lump, cluster, swarm or mass are words used to describe various sized groups of bees. You might refer to a group of on-lookers when a hive is being demonstrated.

**growth**
Gradual increase in size. Development associated with an increase in size, as of a grub, brood nest or honeybee colony. Applicable also to the progressive impact of ideas or practices e.g. growth of a local BKA. Plant growth relative to flowering time or the uprising of weeds where they may not be wanted. The astonishingly rapid growth of honeybee larvae. The gradual development "or growth' of a colony.

**grub**
The bulky larva of certain insects including honeybees. Sometimes the term larva is preferred as grub is also the word for food. A honeybee grub requires shelter and incubation within a cell. It cannot seek food and is thus entirely dependent on the attentions of the nurse bees. It produces a pheromone that stimulates adult bees to collect pollen. A grub is white and looks like a legless caterpillar.

**Gruit**
Spice mixtures that gave mead variants their special names and flavours.

**Gruzinian bee**
A Caucasian sub race along with bees whose original homeland was in the high valleys of the Central Caucasus. Mentioned in 'Hive and the Honeybee'.

**guajillo (huajillo)** ***Acacia berlandieri*** **Mimosaceae**
An acacia shrub of the south western prairies in Texas. Its white blossoms yield

a water-white honey with a subtle milky reflection and a delicate fragrance. Some of the heaviest flows on record have come from this plant.

**guard bees**
A guard bee is one that is prepared to drive off any would-be intruder and to use its sting if necessary. This precludes young bees as older bees have less potential life than young ones and in any case the very young bee has too soft a sting and disposition to attack. Use can be made of this fact by milking off all the older bees by moving a hive well aside, when after a day's loss of flying bees, it can be used as a more docile colony for purposes of demonstration. The period of time that a bee acts as a guard bee at the entrance is limited., (as is the time 'alarm pheromone remains active). Detailed behaviour varies a lot amongst bees of different strains, the more docile simply filling the entrance when disturbed, the more aggressive flying sorties and attacking anything that moves near the hive. Guard bees at the entrance, may adopt an aggressive stance with forelegs raised and mandibles open. They are more in evidence when a nectar flow ceases and may disappear altogether during a honey flow.

**Guide bird** *Indica indica*
See: honey guide bird

**guide lines**
Avenues of direction set out by a plant or animal (including humans) to encourage a movement towards a specific idea, target or end. In so far as bees seem capable of following the same flight paths, we have still to learn about the possible significance of airborne pheromones, pollen grains or other unknown methods of laying a trail through the air, though heavy emphasis has fallen on the Von Frisch dance indications.
See: nectar guide lines

**Guinness Book of Records**
A quote in 1973 claimed that the greatest number of stings sustained by a surviving human subject was 2,443.

**gum boots Wellingtons**
A rubber boot that reaches the knee or above, impermeable to water (and to bee-stings) and helpful where snakes abound or where rough terrain or tall wet weeds have to be negotiated. Bees can become trapped twixt rubber rim and trousers, then half-crushed, they my drop down and sting your foot or ankle. Trousers and socks should be rolled in such a way as to preclude this possibility. Cool plastic insoles can be worn inside them. As with gloves, they can become contaminated with pathogens from standing on old comb or having nectar or honey dripped upon them. It is incumbent upon you, always to check that they are quite empty before putting them on and when truly 'gunged-up to have clean shoes to-hand before getting in the car.

**gums**
Early log hives in America were given this name because black gum trees had an affinity for being hollow and therefore suitable for bees. The eucalypts are colloquially referred to as 'gums' or gum trees and their wood as gumwood.

**Gusathion**
Sold under trade names of Azinphos-methyl and Aziprotryne it is an organophosphorus insecticide available in wettable powders. Very dangerous to fish, game, wild birds and animals. It had been used when plants were in full flower but in any case it is lethal to bees for up to four days after application.

**gustatory sense** *L. gustus* **– taste**
The sense of taste. Bees have been shown to be quite discriminatory and have organs of taste on their feet as well as their head. They are often quite 'fussy' as to the kind of water they will collect, seeming to prefer that with a trace of detergent (less sticky)

or mineral rather than clean water. They also have preferences for certain sugars and are frequently governed by attractants or repellents. Honey judges when giving marks for taste must find it hard to find total agreement. The taste of royal jelly is far removed from the flavour of capsules sold as containing royal jelly.

**gut**

In bees, the intestine, enteron, alimentary tract or specialised tube that extends from the ventriculus to the anus.
See: gaster, hind, large, small, mid-gut

**guttation**

The formation of drops of water on plants or the exudation of an aqueous solution by a plant. It may run down or collect on spicules at leaf bracts, or other suitable spots. This is quite common in house plants due to the plant taking up too much water (over-watering) and it is called guttation. When however the secretions are 'sweet', as is often the case when it exudes from extra-floral nectaries, it is used in the same way by bees as ordinary floral nectar to produce honey.

**gynandromorphy**

An abnormal genetic variation when an animal shows male characteristics in some parts and female in others. A sex mosaic, although rare in honeybees, various genetic irregularities can lead to the production of these freaks. Unlike hermaphrodites they do not produce both sperm and eggs.
See: hermaphrodite and deformed bees

# H

**Ha**
A hectare, a hundred 'acres', 10,000 sq mtrs, or 2·472 acres

**habit**
The characteristic behaviour shown by plants, bees or beekeepers. A disposition formed by constant repetitions of a certain action. Individual bees, being short-lived can still develop habits such as interrogating their owner's car upon each visit to an out-apiary while longer term genetic habits may be displayed by the colony as a whole.

**habitat**
The native environment in which a plant or animal is normally found. It includes the whole complex of soil, vegetation and climatic factors. The work ending 'ophilous' is used to indicate that the root describes an area in which a living thing is likely to be found, e.g. agrophilous – living in grain fields, 'limnodophilous' – dwelling in marshes, other examples will be found.

**hackle**
The whole neck plumage of the domestic cock. Pointed straw cover over a hive in winter – (More on Cotton, Butler and Keys) A rush hackle or cone-shaped weather protection of straw set over hives notably over straw skeps – D. Galton.

**haemocoele**
This originates from the first body cavity of the embryo honeybee larva. 30% of its blood is contained therein.

**haemocyte Gk. haima blood, kytos hollow**
A minute protoplasmic body which floats in the blood of a bee. Blood cell or corpulscle. See: leucocyte, phagocyte

**haemolymph Gk haima blood and *L lympha water***
A bee's blood is a fluid issue or liquid plasma containing water, protein, amino acids, glucose, trehalose, fats, salts, hormones and haemocytes which include white corpuscles and the haemolymph which is pale amber and carries food material from the digestive system to the body cells and waste to the emunctories. It has no respiratory pigment (red corpuscles) for the transportation of oxygen. The only blood vessel is a dorsal tube feeding blood to the head, otherwise it is an open system. Bees can develop blood disorders such as septicaemia, particularly from certain mites.
See: circulatory system, leucocytes, phagocytes

**hair external**
A filamentous growth from the epidermis consisting of one or more cells. Most of the honeybee's exoskeleton is covered with hairs which are hollow extensions of the cuticle formed by a single enlarged epidermal cell, the trichogen, the socket being formed by a second cell – the tormogen. They vary in length and density, many hairs having curved projections that protrude radially. They form brushes, spiny combs and act as protectors, being admirably adapted to picking up pollen and doubtless adding to the thermal efficiency of the winter cluster. Types include: clavate, bouton, conoiid,

antennal, tactile- antennal.
See: plumose, macrotrichia, setae.

**haircream**

226g/8oz liquid paraffin, ½ oz. beeswax, 114ml/4 fl.oz cold water + teaspoon borax. Warm together and drop in a little bath oil as scent.

**hair curler**

Of the various types several tubular plastic varieties with open mesh sides have proved useful for holding queens and their retinue or queen cells. The mesh should not be too large as feeding and antenna brushing is all that should be possible through the holes in the mesh. A cork, plug of candy or scrap of newspaper held over the end by an elastic band will close the ends.
See: stocking

**hairless bee**

Bumble bees look quite black and shiny should they have the misfortune to be stung and stripped after entering a colony of honeybees. A young worker is very hirsute but tends to lose hairs as it grows older and queens can become very shiny. A shiny, usually dark coloured worker, may have lost its hairs due to robbing skirmishes or it may be suffering from bee paralysis virus.

**hairless black syndrome**

Serological investigations and similar virus morphology conclude that the above is caused by CBPV (chronic bee paralysis virus) or by a serologically indistinguishable genetic variant.

**hairy willowherb *Epilobium hirsutum* Onagraceae**

Great willowherb, (codlins and cream). Large pink flowers July/August producing nectar and green pollen. A perennial that likes damp places. The elephant hawkmoth caterpillar feeds on its foliage which has a downy (hairy) covering.
See: rosebay, willowherb

**half-brother**

As the drones produced by a queen are not affected by the sperms that are essential to all the workers, the relationship between a queen's workers and her drones is half-brother and half-sister. It is important in breeding, to appreciate that drones only carry the genetic make-up of the queen (their mother) while virgins and workers will be influenced by the queen and one of the drones with which she mated.

**half-depth super**

A super that is about half the depth of the standard deep box (140mm/51/2" compared to 216mm/8 1/2"). In the case of the Langstroth with 241mm/91/2" deep boxes the 'half' size is 120mm53/8". It makes for a lighter super (deep boxes full of honey can strain your back all too easily) but there is no reason why you should not standardize on the shallow size for brood as well. Two shallows giving an even bigger area than one deep box. With Dadant, Langstroth, British commercial, Smith and WBC hives the equipment manufacturers have their work cut out keeping stock of all items.

**half-life**

The time taken for half the quantity of an unstable substance to disappear from the environment by undergoing chemical change. This is applied to radio-active materials etc. which may take a long time to become harmless. Half-lives vary from millions of years down to less than a millionth of a second. The term may be used for the time it takes for an organism to metabolically eliminate half the amount of a radio-active substance taken in or for the time required for radio-active decay to occur in half the nuclei of a specific isotope series (say cobalt 60).
See: gamma

**Halictidae**

This class of insects comprises more than 2000 described species. They are solitary bees, the female of which live long enough to see their eggs hatch and to provide pollen and nectar in the form of progressive feeding (as opposed to mass provisioning). This is said to represent one of the first evolutionary steps to the true 'social' insect *H.xanthopus* the yellow footed and *H.nitidiusculus* neat mining bee that looks like a fly.

**H & J**

The hivetool whose shape gives rise to this name and is mentioned under 'frame lifter'.

**handling queen cells**

Queen cells look robust and belie the delicacy of their contents. Great care should be taken to try to keep them in the condition that they would be in a hive. In cutting them from the comb be sure that enough of the surrounding comb is attached to avoid any possibility of causing even the slightest damage (or the bees will surely reject them) and they should be kept vertical and of course away from other queen cells. A honey jar lined with tissue paper can be made a temporary home for them and guarded against the cold. Plastic hair curlers can also be used but when the time comes for transferring them into a needful colony, then the greatest of care must be used to make them firm, vertical and to place them in a warm part of the nest. Coiled wire protectors are available which keep the sides protected while allowing just enough space at the tapered end to allow the virgin to be able to emerge and climb out. If this is done in an artificial incubator be sure that one queen cannot get out and destroy all the others. 34C/93F and 50% R.H. are the incubating conditions.

**haltere**

Flies have but one pair of functional wings. The haltere is really a modified hind-wing and possibly assists its aerodynamic qualities acting as a balancer. It is usually small and club-like.

**ham-fisted**

A person who is ham-fisted (handed) may need to wear more protective clothing than one who is adroit. Nevertheless many beekeepers who work on hundreds of hives tend to develop fists like hams and great strength is a very useful attribute when honey farming is to be the occupation. Training and the skilled use of subjugation of the bees, good equipment and understanding of their charges, helps to make a giant humble and gentle.
See: women

**hamulus pl. hamuli**

A small hook or hook-like process – in the case of honeybees wing hooks. There are some twenty of them on the leading edge of the hind wings and they engage with a stiffened spine on the corresponding trailing edge of the fore wings to form a wide enough wing to support flight while when disengaged the bee is able to fold its wings to enter cells or to fan.

**hand cream**

20g sodium alginate (or gum tragacanth) 1 fl drachm glycerine, 1fl drachm honey, 341ml of water.

1. Stir S.alginate into water and leave overnight to form a thick jelly.
2. Add the glycerine, honey and a trace of perfume.

**handhold**

A place where the hand can gain a grip to support an object or oneself. On beehives handgrips are usually 'built-in' though sometimes grooves scored out of the side walls may prove insufficient when hands are wet with honey and boxes laden. Although projections and fillets are frowned upon by many who regularly move beehives, a strip of 6mm bottom bar, strategically fixed over the groove, can improve the handgrip immensely. A loop of nylon rope knotted on the inside can also lie fairly flush against the hive and provide a satisfactory hold for all four fingers. The standard 'National' like the inner WBC chambers, would seem an ideal way of holding a box but you should consider how much extra timber this requires.

**Hanneman excluder**

An excluder that confines a queen in the bottom chamber while queens are reared in other parts of the hive. (Bulgarian).

**hanging section holder**

Virtually a substitute frame holding 3 sections in the British standard or 4 in a Langstroth. The 108mm/41⁄4" standard sections holding around 454g/1lb of honey are often supered in 'crates' with the same dimensions as the hive but in British weather many remain unfinished and therefore unsaleable. A hanging section holder, or two, is a more likely way of getting nicely filled sections and separators (plastic or metal sheets) can still be used to define the neat outline of the comb face.

**haploid**

Having a single set of unpaired chromosomes in each cell nucleus as in germ cells, (split during meiosis from the paired set of the female parent). In the case of parthenogenetic reproduction e.g. the honeybee drone, the haploid condition is practically invariable. Thus a drone's chromosome number is sixteen while that of workers or queens is thirty two - hence hapliody. Lucky, lucky me, I can live in haploidy...

**harbinger**

A precursor, something that indicates a change or situation that is to follow. Dozens of sayings regarding weather forecasts and the seasonal cycle or unfolding of events within a hive do follow a pattern which though varied is capable of some useful interpretation. For instance before a colony will swarm it proceeds through three positive stages.

1. An increase in the foraging force.
2. Drones are seen on the wing.
3. The presence of queen cells.

Many signs in the bee world enable predictions to be made when their significance is understood.

**Harbison**

The Californian who in 1857 originated the section honey-box, a 108mm/41⁄4" square of neat basswood whose four sides were dove-tailed allowing them to be pushed together. They were made so that when full and flush with sealed honey they weighed a pound. It must be more than a coincidence too that the British standard deep frame is 216mm/81/2" (2 sections) deep. Also it is convenient to fit 3 sections inside an empty shallow frame (4 in a Langstroth).

**hardboard**

Flat sheets are made from compressed wood fibres and similar materials. Unless made weather-proofed as 'oiled hardboard', it is not suitable for outdoor use but otherwise, light roofs, crown boards and other items can be made instead of the ubiquitous ply-wood.
See: ply

**hard-heads *Centaurea nigra* Compositae**

Its knob-like flower heads grow to about 1m, the pollen is grey. A kindred plant is greater knapweed *C. scabiosa.*

**hardiness**

Having the capability to survive. Being robust and vigorous. Often imported bees do not take well to our difficult climate and lacking these qualities are termed 'soft' bees. Hardy bees would be expected to have the characteristics of longevity, foraging strength and resistance to disease. It has often been said that you will find the best bees in your own back yard.

**hard water**

Water which does not lather easily with soap. The hardness is due to the presence of ions such as calcium, magnesium or iron. Soft water which does not form a scum or precipitate when used with soap is not always suitable for human consumption as the body is often dependent on the minerals present in natural water. The salts in hard water can re-act adversely when rendering beeswax and if soft (rain) water is not available the hard water should be slightly acidified. Bees are quite 'fussy' often choosing sources that seem dubious to us.

**harebell (Scottish blue bell) *Campanula rotundifolia* Campanulaceae**

A small herb with slender stems and light blue, bell-shaped flowers. A foraging bee becomes dusted on the dorsal side with pollen that is transferred to the ripe stigma of the next flower she visits. It flowers July to Sept. and is a perennial growing on hillsides, heaths etc.

**harmful substances**

Vapona (Dichlorvos, DDVP, Vaponette) is permanently toxic into beeswax. The residual effects of Dichlorvos, methyl bromide, ethylene dibromide and PDB (paradichlorobenzine) can harm bees for a very long time and can be absorbed into plastics and plastic foam. Fine dust clogs bees' spiracles.
See: toxic nectar, honey, pollen

**harvest**

The mind immediately thinks of honey but pollination services, educational benefit, research facts and pleasures are to be had as well as hive products. While many colonies are kept without taking their honey or even wax, the average hobbyist sets about the heavy, sticky and somewhat arduous task of removing and bottling honey at the end of the active season. Side-liners and bee farmers may chase crops and remove them as and when they will. Many have a strange reluctance to optimise the removal of comb for wax rendering.
See: yield, surplus

**harvesting**

Honey from quick granulating crops like oilseed rape or in the form of comb honey which must not be allowed to get 'travel-stained' should be taken off as soon as sealed. In fact for home use rape honey can be extracted as soon as a comb, held horizontally, does not spout nectar when shaken. Ordinary honeys can, for the convenience of beekeepers who only wish to extract once a year, be left until late July or August. It is even less trouble to remove in late September when the bees are not quite so alert as regards holding onto their stores but this complicates the assessment and adjustment of winter stores and the need for Varroa treatment etc. At out-apiaries blowers or hand brushing and shaking may be resorted to in order to clear bees from the supers whereas those who can avoid double journeys to apiaries can profitably use clearer boards. Don't try to clear more than two supers at a time. (who gets more honey than that anyway?) and always make absolutely sure that all nooks and crannies are sealed so that no robbing can take place because once the Porter bee escape begins to work, fewer and fewer bees remain to guard their honey. Disappointed beginners have often found that the supers which were full are almost or completely emptied the next day if other bees or wasps have been able to find their way in.
See wintering

**Harwood Alfred Francis**

A language teacher who began keeping bees in 1905 and contributed 'Bee Garden" to the BBJ and worked with acclimatised strains of American bees and experimented with Caucasians. His major work British Bee Plants in 1947 is a classic.

**Harz forest**

A central European mountain range from the Elbe to the Weser. Good honey-dew flows occur in early summer when two generations of Lachnids are found together. They feed on the sap of pines and exude sweet excretions.

**hatchability**

The likelihood of an egg's being able to change into a viable larva. This can fall to a low likelihood (50% or less) due to inbreeding and is indicated by a scattered sealed brood pattern, sometimes called a 'pepper-pot' appearance. Many beekeepers tend to ignore a preponderance of holes within an otherwise sealed brood pattern but this should only be tolerated if the queen is known to be young and prolific and the total brood count is satisfactory.
See: scattered brood

**hatching**

Movements of the larva just before hatching rupture the eggshell (chorion) about its mid point. This releases hatching fluid which disperses over the surface of the egg and dissolves the egg-shell so that no vestige of it remains once hatching is complete. Accidental wetting of the eggshell with larval food, nectar or water will automatically prevent hatching (by dilution of the membrane dissolving enzymes). The optimum RH range is 90-95%. At 80% RH, 40% of the eggs fail to hatch and less hatch at lower RH's which also damage 1-2 day old larvae and larval food dries out very quickly when the RH falls to 80%. The emergence of the imago (adult) is not referred to as hatching but as 'emergence'.

**hatching eggs**

In addition to eggs laid by the queen other creatures lay eggs that hatch within the hive. Disregarding such things as birds' nests built in the capacious interior of WBC hives, acarine mites lodge relatively huge eggs in the bees' trachea, and Braula coeca lay eggs under the honey cappings while wax moths of both varieties lay their small legs in carefully concealed crevices within the hive.

**Hawaii**

A group of islands in the N. Pacific. The first bees were taken to Honolulu in 1857. In the 1800's abortive attempts had resulted in tropical heat melting the combs. By 1939 10,000 of beeswax was exported. In winter kiawe honey is produced, macadamia nut trees bloom in summer also lehua, guava, Christmas berry and eucalypts.

**Hawkmoth Family Sphingidae**

There are several varieties of these night flying moths capable of swift flight and are mostly fond of nectar and sweet things.
See: death's head hawk moth

**hawkweed Hieracium spp. Asterales**

Widely distributed dandelion-like flowers on long thin stems. They flower May-Aug and yield nectar and pollen the latter being golden with 22μ grains. Their seeds are produced without fertilisation by a process called 'apomixis which is akin to parthenogenesis but arising from cells rather than ovaries.

**hawthorn *Crataegus monogyna* Rosaceae**

A valuable tree, useful for hedging ,when it is called quickthorn, whitethorn and other names. An effective barrier for farm animals. Its bright red 'haws' provide food for birds and small mammals though it is often cut right back which minimises flowering. There are pink and red varieties though white dominates and forms a carpet of blossom across the Downs and such places in May or June (getting earlier each year). The almond scented nectar is converted by the bees into a thick honey with an excellent, appetising nutty flavour but it is a fickle plant only yielding well in about one year out of three. It is a good show honey varying from light golden to dark amber. Its pollen is similar to apple and rose having three grooves 30μ in size and described variously as whitish, dull yellow, creamy-white and greenish.
See: May

**Haydak's formula**

A bee food in the nature of a pollen substitute which appeared in Bee World in 1977. One part of the dry mixture is mixed with two parts of disease-free honey by weight plus enough water to produce a stiff paste. Syrup can be used instead of honey. Ingredients as follows: 4 pt soya bean flour, 1 of yeast (dried Torula or other) l of skimmed milk. The addition of dried egg yolk will improve food value considerably.

**hay fever**
An allergy leading to an inflammation of the mucous membranes in the nose, around the eyes and generally in the respiratory tract. It attacks certain individuals around hay-making time and is often put down to pollen. It is possible that bee stings and or bee protein can have a similar effect or can make a person who was not susceptible, allergic to certain irritating elements. It has been suggested that 'cappings' (which contain pollen) should be eaten from about a month before the likely onset of this malady) have in some cases proved beneficial.

**hazel *Corylus avellana* Betulaceae**
A monoecious shrub, deciduous and flourishing in thickets, coppiced woods and along roadsides. Its catkins expand over an extended period through January/February and from its golden lambs' tails, prolific pollen is carried by the wind and sometimes collected by bees. It likes calcareous soils. If branches are stood in a vase over a sheet of smooth paper the pollen can be collected and made into a patty by mixing with disease-free granulated honey and fed to the bees. Pollen seems to vary from dull yellowish green, pale greenish yellow to brownish yellow. Its pliant rods can be used to make baskets and the famous 'coracles'. Small edible nuts are another outcome.

**head**
The foremost segment of the bee joined to the thorax by a short neck, normally hidden which enables the head to be rotated through a wide angle. It is largely dominated by the compound eyes which actually touch on the drone and force its 3 simple eyes (which lie between the compound eyes on the females) onto the frons. At the base of the head are the trophi with specialised mouth-parts including an extra-long tongue on workers. Two mandibles operate laterally and a pair of antennae protrude from the centre of the roughly triangular face. As this is the end of the bee that meets the world, it is equipped with most of the sense organs and the means to imbibe.
See: named parts

**head glands**
These include: hypopharyngeal (brood food), mandibula, postcerebral and postgenal.

**health**
Freedom from disease or ailments. There is a dynamic equilibrium between host and pathogen and trouble usually only occurs when something happens to shift the equilibrium point by giving an advantage to the pathogen. Avoid management techniques that require undue interference with stocks.

**health certificate**
An official certified statement that the bees are free of disease is required for the importation of bees, queen bees. Importation of package bees is not allowed. Certificate to be received by Defra before the bees arrive in this country and a copy of the certificate should accompany the consignment. Certificates are viable for 10 days only.

**heap of dead drones**
Drones may be thrown out and form a heap in front of the hive for any of the following reasons:

1. At the end of summer when after the active season they are no longer required.
2. A the cessation of a flow or during a period of dearth.
3. When a queen has mated and the colony no longer requires its drones.

**hearing sense of**
Auditory sense. Before we can be specific about a bee's ability to hear sounds it is

necessary to differentiate between what is a sound and what is a vibration Certainly bees respond to vibration and they make different sounds. The manner in which they detect vibrations is linked with overall body structure, especially the exoskeleton and sensitive hairs that transmit any movement to the brain. The comb face is also particularly suitable for the transmission of vibration as has been noticed during the Von Frisch dances.

**heath**

Barren open country that is more or less flat. A tract of uncultivated, natural land (possibly maintained as common, or beauty spot). Relatively treeless, with poor, acid soil of sands gravel or peat, supporting gorse, heath, heather, broom, bramble and other small shrubs. In the various seasons it offers forage to the bees. However it can prove to be rather dry and is usually wind-swept. Dry areas favour heath whereas the wetter heath land tends to favour ling and bracken. Ling heather is Calluna vulgaris while the bell, a similar evergreen shrub is heath.
See: heather moor

**Heath plant Erica spp. Ericaceae**

There are many cultivars in different colours blooming at various times of the year (including winter) but the wild form blooms June – September and in Sussex it gives a pink tinge to the Ling; when hives are taken to Ashdown forest. The honey is dense and of a port wine colour and its wax also has a pinkish tinge. The pollen, like that of Ling is tetrad 40µ and varies grey to pale yellow or even light brown like grey suede leather. In Dorset we find *E.ciliaris* while *E cineraria*, bell heather, is found on drier heaths while *E.tetralix* t cross-leaved heather, is found on wet heaths.
Others heaths are *E.hibernica* Irish heath, E.mackaiana Mackey's heath and E.arborea the tree heath. *E.vagans* the Cornish heath, *E.montanica* broad-leafed heath and *E.carnea* the winter heath.

**heathen**

In places where heather abounds there is a quaint superstition that it arose from the blood of a pagan. The German word 'heide' meaning heather also means pagan. Frank Vernon opines that because of their remoteness county folk in the less accessible places were the last to be converted and that this is a more likely explanation of the link between the words pagan and heather.

**heather *Calluna vulgaris* Ericaceae**

This is the true heather known as Ling. A plant of great significance to beekeepers within reach of the moors. It will grow to 2m high but the best yields are likely two to five years after the area has been burnt off (swayling). Although it is often extensive and late flowering enough to give pure samples of its unique honey which compares favourably with the Manuka (which has gained so much notoriety of late), admixtures of ragwort, blackberry, rosebay and bell heather are possible. Apart from its special nectar bees collect plenty of slate-grey pollen and propolis from the shrubs which can flower for up to six weeks though the actual flow may only last a small fraction of that time. Some reports that the bees do not winter well on this honey may well come from the fact that as they don't come back from the heather until as late as October this may well complicate the feeding (and prophylactic) issue. Each acre of heather has been said to be capable of yielding 18 lb every day for 14-20 days but as with other 'fickle' plans that are so dependent on ideal weather for heavy yields, supers are only needed once in three years and bumper crops in the south are rare indeed.

**heather beetle *Lochmaea suturalis***

Said to cause considerable damage to heather the small grubs nipping the foliage and baring the wooden stems. Rarely is permanent damage done to younger heather but older heather can be killed. In 1936 an observer reported millions of beetles as having landed on the water of Loch Awe in Argyllshire being speedily devoured by trout. Their habit is to take to the air in the spring and be carried along by the wind. Colin BBJ Nov 1981.

**heather floorboard**

Weather can be 'rugged' on the moors and a special flush floorboard which helps to keep the winds out of the hive and gives the incoming bees a porch is the speciality of this extra strong floorboard. Its entrance leads up a slope to a slot several inches back from the front of the hive.

**heather honey**

A dark, golden brown, thixotrophic, bittersweet honey with a pronounced flavour considered by many to have no equal. It is either eaten in the comb or the combs can be pressed to squeeze out their jelly-like contents. Ordinary centrifugal extractors cannot cope with this honey unless special agitators are used. Agitation of the "gel' temporarily liquefies it for pouring or extraction. Air bubbles remain trapped in the extracted honey and their size indicates the true gelatinous nature of this special product. Its water content is higher than other honeys and fermentation will almost certainly occur unless this is minimised by taking care not to include any unripe honey. Its persistent flavour makes it a favourite in cooking and heather meads are superb. Every year thousands of colonies are moved onto the moors which can be very tough on the bees and when a flow occurs the air becomes alive with bees as they dart to the blossom and thunder back into the hives increasing hive weights by as much as ten pounds a day.

**heather nectar**

Some great stretches of heather can be very disappointing from the nectar secretion aspect It yields best on a well-drained, peaty soil which has an acid subsoil over granite or iron. It does not yield well at low altitudes especially in wet bogs or light sand or gravel. Older plants are poor yielders compared with young plants 3-5 years after a heather burn. East-west valleys can offer a longer flowering period while sites over 200m above sea-level with a down-hill approach for laden bees and heather all around the hives, helps to ensure a crop.
See: other 'heather' headings

**heather press**

To squeeze out the contents of heather honey comb considerable pressure is required. Presses are made with a strong turn-screw which operates by forcing two platforms together. A nylon or scrim net is placed between these and the comb or scrapings are enveloped in this. Two-layer versions are available and the honey is led by one method or another to a gate for containerising.

**heather stance**

A place where hives can be put so that the bees can work the heather. In some parts of the country great areas of heather are available but in others only the keenster is prepared to husband his colonies in the manner best suited to the tough work the bees have to do and to drive any distance reduces any profit margin, complicates prophylactic treatment and once again is tough on the bees. Factors to be taken into account are altitude, slope relative to the sun, and the age of the heather.
See: heather nectar

**heating honey**

Raising honey temperature above hive temperatures reduces its viscosity and will facilitate filtration but care should be taken to minimise this as several ingredients suffer when honey is heated. HMF (hydroxymethylfurfuraldehyde) is as poisonous as its name is long and its levels rise with prolonged heating and conversely the valuable diastase levels diminish. Honey is a health-giving substance and certain enzymes which act as valuable catalysts in the human body may be destroyed by heating. To put the spot-light on this subject, honey deterioration takes place in exact accordance with the total amount of heat that the honey absorbs. I.e. a short hot spell is equal to a longer spell at temperatures no higher than standing barrels in the sun. So if you take honey in a hot drink, don't bring it to the boil if that is not necessary and use granulated honey which does not get full heat as immediately as clear honey. Drinks provide such a good way of getting honey into the body that this point should be carefully considered. Beeswax becomes plastic at 29C/85F so at 35C/95F (brood temperature) the comb is very soft and quite a small rise would threaten comb collapse, so take that as a yardstick for deciding how hot to let honey become.

**heat production**

Bees regulate the temperature of their hive by microvibration of their flight muscles. Increases above ambient temperature are essential for bees to be able to make wax and build comb, to ripen honey and most importantly for brood incubation and of course the survival of the winter cluster.
See: microvibration, temperature

**heat receptor**

Sensory organs that transmit the bee's sense of temperature to the brain. With slight variations between strains, colonies form a cluster to maintain warmth at a fairly definite threshold. Brood is maintained at a fairly constant level, again with racial variations. It is clear then that bees are able to register temperature levels though for details of the location of receptors and the manner in which the information is fed to the brain, the reader is referred to more scientific works.

**hectare ha**

Superficial measure of one hundred 'ares' which are the equivalent area of squares with 10 metre sides. One hectare = 2·47 acres and one acre = 0·40 hectares.

**hedge, closely planted**

A row of closely planted bushes or small trees forming a fence, barrier or boundary. Hedges can be a thousand years old. M.D.Hooper suggested 100 years for each shrub present in 30m length. Of half a million miles of hedges in 1947 a quarter of this has gone. (that was in the 80's). Unfortunately hedges need cutting to prevent them from growing tall and unkempt and encroaching on property or the clear view of drivers. Modern machines that trim hedges are pretty vandalous; leaving an eye-sore of flowerless, broken stumps. Only education can restore a compromise where all parties concerned are properly cared for.
See: grass verges hedge trimming, motorways

**hedge trimming**

As hedges usually include, or may even be made of melliferous plants they can provide useful forage and shelter affecting plants adjacent to the hedge. Bees can be seen queuing up behind a hedge in windy weather, waiting for a lull before flying over the top. Severe cutting which modern farmers and councils frequently indulge in, without relation to the flowering period can deny bees spectacular masses of welcome blossoms. Escallonia, fuschia, hawthorn, privet and willow are examples.
See hedge, pollarding, topiary

**hedgehog *Erinaceous europaeus* Erinaceidae**

An insectivorous mammal with spines, about 25cm long which rolls into a ball when startled. It will take bees from the ground in front of colonies but will also cause a disturbance at the entrance and then eat the bees that emerge. It hibernates. Getting hives well up off the ground is one way of keeping this creature at bay.

**hefting**

The hive is carefully and gently lifted at the back so that the legs or base is just clear of the ground or its support. This allows you to assess the weight of the colony by giving the half-weight. If a spring balance is used then by lifting each 'side' the two weights added together should give a fair approximation of the total weight. This can be done at regular intervals but is probably most useful for checking winter food supplies. Care should be taken to ensure that the hive is not jolted or disturbed by this action. Needless to say a little experience is required to make assessments realistic.

**Helenium spp. Compositae**

There are several colourful varieties like 'sneezewort', they make useful bee plants. In the autumn their yellow pollen is so attractive on returning bees.

**Heliotrope Heliotropium Boraginaceae**

Nectar and pollen from August on.

**Helleborus Ranunculaceae**

*H.orientalis*, *foetidus* and *lividus* are common varieties of this hardy perennial that give pollen and nectar. Growing from 30-60cm/1-2ft with pale greenish cup-shaped flowers in winter. Slow to germinate and requires good soil with plenty of compost.

**Hemiptera (Rhyncota)**
**Insecta Arthropoda Bugs.**

One of the twenty orders of the sub-class Pterygota (winged insects). This forms a large order in the exopterygote insects and includes scale insects, greenfly, creatures that are of great significance in the production of honeydew. Members characterized by the adaption of mouth-parts to sucking and they exude substances such as cuckoo spit, honeydew and waxy wool. Sub-order Homoptera: aphids, white fly and scale insects and plant and water bugs
See: forest honey

**Hemizygous**

The gene condition of haploid drones said to be hemizygous. All genes sex-linked – haplozygous.

**herb Labiatae**

1. Any flowering plant whose stem does not become woody or produce any permanent shoot systems above the ground.
2. Such a plant when valued for its scent or flavour.

Many Labiatae (dead nettle or mint family) are very useful bee plants. Many are useful for seasoning and we refer to them as culinary and medicinal herbs. Some herbs are dried and used to make aromatic smoker fuel. These include: tansy, mugwort, yarrow and milfoil. Other English herbs include: *Clinopodium vulgare* basil, catnip or catmint, Nepeta spp, lavender, lavandula spp., *Hyssopus officinalis* hyssop, *Origanum vulgare* marjoram, *Rosmarinus officinalis* rosemary, *Salvia officinalis* sage. *Satureja montana* savory, thyme, Thymus then daphne, fennel germander, lemon balm, red bee balm, sweet cicely, yellow melilot, winter savoury, bergamot.
See: US herb garden

**herbicide**

A chemical substance that is used to destroy plants e.g. as a weed-killer. There are a variety of means of application (see insecticide formulations) some posing a threat to bees, especially if they come into contact with them. There are several types and unfortunately some can render flowers that need bees, repellent to them. Another side-effect is that herbicides can impart a smell to the bee or perhaps a static charge, that may cause it to be rejected by its own colony when it returns. There is also a possibility of harm to the brood. A further possibility is when bees collect surface water.

**heredity**

The passing on of genetic characteristics by parents to their offspring. Heirship, inheritance.

**hermaphrodite**

Bisexual, having both stamens and carpels on the same flower, or in the case of an animal producing both male and female gametes. In this respect it differs from a gynandromorph or freak bee. Once a queen

has successfully mated she might be termed a hermaphrodite as she carries gametes of both sexes.
See gynandromorph

**Herodotus c.5 BC**
A Greek historian, commonly known as the 'Father of history' wrote that Egyptian beekeepers put hives on boats sailing up the River Nile in winter and then in the warmer spring allowed the boats to drift slowly down river again.

**Herring H. T.**
An expert in making wooden former trays for producing thin sheets of foundation and using plastic moulding sheets so that when passed through the rubber roller of a mangle produced excellent sheets of wax foundation.

**Herrod-Hempsall, William**
Appointed as Instructor in Beekeeping at Swanley College 1898 became Advisor to the Min. of Ag. During World War II. (His brother Joe edited the BBJ)
Hons. Lecturing 1914 and wrote several books including 'The Anatomy, Physiology and Natural History of the Honey Bee' and was described in the US as 'The Doyen of British Beekeeping'. He was very attached to the WBC hive.

**Herzog excluder**
A German-made, framed, wire excluder. Unlike the Waldron it provides correct bee-space while offering maximum freedom for movement and ventilation.

***Hesperis matronalis***
Dame's violet, sweet rocket – a hardy perennial that grows to 60cm/2ft with sweet-scented mauve or white flowers. May – July. Excellent for naturalizing in shady places.

**hessian burlap (US)**
A strong, coarse cloth of jute or hemp much used for sacking agricultural produce before the advent of the plastic sack. Also used as windbreaks while it was relatively cheap material. In a weathered, half-rotted form it could be used when dry to fuel the smoker a it smouldered steadily and made an acceptably cool and non-irritant smoke.
See burlap

**Heteroptera Insecta Hemiptera**
A sub-order of bugs which are characterised by their wings being closed flat over their backs. Bark bugs, stink and squash bugs etc.

**heterosaccharide**
These are not strictly carbohydrates though on hydrolysis they yield glucuronic or galacturonic acids that are derived respectively from glucose and galactose whence: pectins, mucilages, gums and chitin.

**heterosis**
Hybrid vigour – the improved characteristics of progeny over parents. Where two genetically different lines are crossed, the offspring frequently display greater vitality than either of the parents. This is course very important where line breeding has selected certain characteristics but the resultant queens produce less vigorous bees than their hybrid daughter's progeny. In fruit and seed production too, there are advantages where cross-pollination occurs even when crops are self-fertile. It begins to fade out in the F2 generation.

**hexagonal**
A six-sided polygon (in the beginning six bees must have got together to arrange this, eh?) with 120° angles between them. Six equilateral triangles can be assembled to form this shape but it can be displayed in two fashions

a) with point uppermost – the vertical mode (as is usual in honeycomb) or
b) flat side uppermost.

Bees utilise this shape because of its economy and suitability the side walls being shared and a roundness being effected by a thickening of the entrance to the cell, the top rim.

**hexapoda**
Classification of the six-footed Insecta class.
hexose
A class of sugars containing six atoms of carbon – a simple sugar or monosaccharide, glucose and fructose are examples and form the bulk of the sugars in honey. It can be condensed to form starch.
See: laevulose, dextrose

**hierarchial system classification**
The arrangement of taxonomic groups in an ascending series of inclusiveness.

**Himalayan balsam *Impatiens glandulifera* Balsaminacea**
Water balsam, Indian balsam or Policeman's helmet are referred to as *I.glandulifera.* It has established itself so well around certain areas of water that some beekeepers move their hives to exploit the late August honey flow. The pollen is white and as seeds ripen their pods they have the fascinating property of exploding audibly when touched. dusting the back of bees with off-white flour. Balsamina is a cultivated garden variety which has double flowers and is of little use to bees.

**hindgut**
The posterior portion of the alimentary tract. In the larva the proctodeum is in contact but has no connection with the mid-gut. That part of the gut which is posterior to the mid-gut or ventriculus.

**hind legs**
These are the largest and strongest of the three pairs. They are highly specialised in the case of workers having identical joints to the other castes but forming the pollen press at the tibio-tarsal joint and the pollen basket on the outer face of the tibia. The claws are correspondingly strong and the legs when spread permit the bee to lift and use the fore legs for various purposes. The queen, having to spend much of her life walking, has exceptionally strong legs which help to make her stand out on the comb face and straddle adjacent cells when inserting her abdomen to lay.

**hind-wing**
This is a similar but smaller veined structure, to the forewing. It lies under the forewing when the bee folds its wings over its back. This arrangement allows workers to go head-first into cells. When extended for flight, the leading edge has a row of some twenty hooks which engage in a sliding fit onto the chitinous portion of the forewing's trailing edge adjacent to the hooked area. Sometimes called the secondary wing, it trails after the power stroke and disconnects when the bee engages in fanning.

**hints useful**
Accident, ankles, clean floorboard, elusive, finding queen, firelighter, flat-iron, flashlight, foam rubber, free-for-all, funnel, hair curler, handgrip, lighten beeswax, liner, looking for eggs, ply, pond, pour, release, security, sneeze, speaking, speck, spin, stocking, straighten, strong colonies, tape, tie, unblock and Vaseline.

**hippocras**
A sweet mead made by adding spices to pyment (which is made from pure grape juice and honey).

**Hippocrates**
The famous Greek physician who was reputed to have lived to the great age of 111 having eaten honey all his life. Known as the 'Father of medicine', his system gave rise to the Hippocratic Oath used by doctors throughout the world, too few of whom advocate the health-giving properties of honey to-day.

**hiss**
A short sharp sound like that of a goose or a snake. When queenright and a colony has its hive rapped, it emits a sudden burst of sound described as a 'hiss' which dies away as rapidly as it started. Queenless bees tend

to continue the sound in what is referred to as a queenless 'moan'. So when a hive of bees is given a tap, a brief non-attenuated 'hiss' is a normal healthy response.

**histamine**

An amine $C_5H_9N_3$ derived from the amino acid histidine. It is an organic base that can be released from the tissues as a result of an allergic reaction to the penetration of bee venom. When freed, it causes dilation of the capillaries, small arteries and venules, which lead to a general fall in blood pressure, while the stimulatory effect on the nervous system, gastric secretion and bronchioles can promote constriction leading to bronchospasm in allergic individuals. For many who are not particularly allergic, stings will still cause turgidity in the immediate neighbour-hood. Anti-histamine taken orally or in the form of sprays and ointment can often give effective relief and it has been suggested that a couple of aspirin type tablets taken an hour before stings can be useful in many cases.

**histology**

The scientific study of organic tissues especially their microscopic structure.

**hitch-hiking**

The use of one organism by another for transport – not parasitism.

**hive**

We hive bees in beehives. They are artificial or man-chosen containers specially designed to accept them. In NZ bees in the feral state are still referred to as a hive rather than a nest. A colony of bees when in a hive is called a 'stock' and for the management the hive needs to be made in sections for easy manipulation. A floor, box of combs, a queen excluder perhaps then a box called a super then a top cover – the crown board, surmounted by the roof. It has to be a weather-proof shelter and as the bees have to be free to come and go, the contents must be acceptable to them. See: hive furniture etc.

**The Hive & the Honeybee**

This book was first published by Langstroth in 1853. It ran to over 50 editions was subsequently revised by C.P.Dadant in 1890 and it has continued to be extensively revised and reprinted ever since.

**hive assembly**

Hives are not standardized as much as they should be in UK. in fact you would think that each new beekeeper has decided to bring out a new type. World-wide there is no doubt that the American Langstroth hive is the most prevalent but they are not widely used here. As their measurements need to be fairly precise because bees will only allow free movement of their combs if they are correctly spaced, it is best to buy from accredited manufacturers. For ease of transport and economy packs of material are sold, in-the-flat, to be assembled by the purchaser but those prepared to pay can always buy them made up and ready to use. DIY calls for close attention to the accompanying instructions because it is easy to build them up inside out or upside down and if nailed-up or glued, it is not easy to correct such faults. US called this type of package 'knocked down' instead of in-the-flat.

**hive baited**

A hive set out in a likely position to attract a swarm and treated inside so that the interior is attractive enough to encourage the swarm to seek out, occupy and develop its nest there. A decoy hive like this used to be treated with beer or lemon balm in the days of skeppists, but now a healthy used comb or two is a better bet. By choice swarms locate their new home well above the ground and in a space with a volume equivalent to a brood and a half is best.

**hive – bar**

Before foundation was invented the hive furniture consisted of nothing more than bars across the top for the bees to attach their combs to. Perhaps the underside of the bars would be waxed to encourage bees to follow the line of the bars but nothing could guarantee that they would not build across the bars rendering inspection impossible. Sloping side walls following the natural curvature of the combs helped to prevent the combs drawn down from the top bar being fixed firmly against the side walls. In this year 2010 there has been a sudden desire to let the bees run riot with their comb building but for sensible economy there is everything to be said for going along with the established trend and using movable frames and bee-space to full advantage and regarding hive bars (top-bars) as things of the past.

**hive bee**

Any bee that can be persuaded to stay-put in a container might well be described as a hive bee. Honeybees exist in four types and these are the ones best known as 'hive bees' though solitary bees, bumbles and stingless bees if semi-domesticated could rightfully come under this heading. Apis mellifera bees remain hive-bound (and so hive bees) until they reach the time when they start foraging and performing duties outside their hives.

**hive box**

As hives became sectional, comprised of various chambers, bodies or boxes, they took on names to differentiate the particular purpose that each section was intended for. So brood chamber, super, section box, could each be identified. You might have two brood boxes (chambers) or several supers and the term box would not identify what part of the hive it belonged to but once the parts for brood are named and those for honey become supers, then the words hive box are not particularly appropriate.

**hive carrier**

A device that makes the lifting and carrying of a hive relatively easy. Ingenuity has led to a host of different appliances for this job. Strong, light and collapsible are terms that describe some of the better models. Some grip the hive firmly, others depend on the hive having runners underneath for the attachment of handgrips. The use of pallets and fork-lift trucks has made large scale commercial beekeeping extremely efficient.

**hive ceramic**

Clay hives were widely used in warm countries where this material was common-place but in Europe the tendency was to use straw until the movable frame came along. A ceramic hive would ordinarily be set down on the ground more like a chimney pot so that combs would be of a manageable size. The ends were easily protected and could allow an entrance at one end. Often the sides were decorated with engravings and large cactus leaves often used to shield them from the heat of the sun. They were not easy to transport and in Malta where Pio Sant having the right to exploit the thyme honey on Comino, the smallest of the islands had to fix them firmly in his rowing boat to keep them upright because if rolled the bees would get quite furious. At harvest time, long rods with a blade at right angles would be used to secure the combs of honey.

**hive-clay**

Before the advent of the movable framed hive which made wooden hives more convenient straw 'skeps' were widely used and could be make quite easily by farm worker and cottagers. Outside Europe, in warmer climes straw gave way to baked clay and ceramic hives of various shapes held naturally built comb often protected by cactus leaves from the heat of the sun. Sometimes the hives were blocked together in a self-sheltering structure while at other times long, cylindrical tubes, akin to our chimney pots, were laid horizontally upon the ground.

**hive clips**

Modern hives can function without any specific need for clips. Yet the all-important spacing of frames, though automatic using Hoffman frames or Manley, is done by the use of attachments, like the 'metal end'

(plastic often used now) which clip onto the lugs and set the space between frames. Another form of clip is a device like a staple which can be driven in to fix hive bodies so that they remain solid and don't separate when migratory work or just moving hives is contemplated. The 'follower' in a section rack will also have a spring, clip-like device for squeezing the rows of section boxes together.
See: metal ends

**hive closing**

Subsequent upon taking the hive apart to carry out an examination, it is necessary to re-assemble it so that everything is left in its identical order or re-arranged to plan. Although the term 'closing' might be used for this shutting down operation it does not imply closing the entrance. On the other hand there are occasions when total closure such as when guarding against spraying operations or moving hives to another site call for complete and secure closure. Much ventilation is done via the entrance so as soon as this is closed extra ventilation must be given. Without it, vibration will cause the colony to become excited, temperature will rise and great harm will befall. A full sized gauze screen is the answer and this would ordinarily take the place of the crown board, allowing a moist cloth to be laid on the gauze, if bees are moved in really hot weather. Mesh varroa floors give bottom ventilation.
See: hive closure

**hive closure**

When a warning that spraying with dangerous chemicals is received, you can either move hives away altogether, or if necessary close them while all the bees are at home, i.e. after dusk and before flying starts the next day. Success depends of the weather experienced but hives should be given a complete ventilation screen instead of the inner cover (crown board) and a constant supply of water by means of seepage from wet foam rubber or some such means. The whole entrance should be blocked completely so that it is dark and does not cause bees to crowd around it trying desperately to get out.
See: Wardecker waterer, overheating

**hive cloth**

Or cover cloth – This is an important piece of equipment for a number of reasons. Bees don't like suddenly being exposed to the light and brood especially should only be exposed to the open-air for as short a time as possible. Also the smell of an open hive attracts unwanted attention from other bees, wasps, wax moth and other predators seen or unseen. If two hygienically clean cloths are used in conjunction with one another, it should only be necessary for a short burst of light descending on the bees while a single comb is lifted out for examination. Keeping them covered means that the vital heat that they have produced by way of incubation, wax making, ripening nectar and so forth is not lost. Cloths should be designed to suit their purpose having a piece of dowel stapled to one end to counter the effect of any wind that might blow them loose and permitting the whole thing to be rolled in stages as the examination proceeds. The material used will inevitably get sticky from honey and propolis and you will have to devise a means of regularly cleaning them. Canvas (deck chair canvas?) or hessian but whatever is used should not let light through and be a suitable material to be compatible with wax, honey and propolis. Obviously nothing hairy. It is a fact that few beekeeper discover (and that is just as well) that when a hive is opened right up and just left exposed to the light and air, they become docile!! Needless to say this would not be wise in fact downright stupid, to do in cold, windy weather conditions. The sort of situation where this can be used is when a colony is to be opened up time and time again for demonstration purposes to a crowd of onlookers, then leaving it open is an option but here too, cover cloths might form a compromise.

**hive collateral**
One of the earliest ingenious efforts to give a hive architectural qualities.
See: Nutts collateral hive

**hive colour**
White is a popular colour for hives yet it renders them conspicuous and at times this can be a disadvantage. It reflects the heat of the sun which might be useful in summer but not during winter. Some folk have gone in for black roofs to allow some winter benefit from the sun. The bees seem indifferent and bright colours are often used and certainly help to minimise 'drifting' also helping newly mated queens return safely. You might suppose that as trees are their natural home that this should be the colour used. It has been said that bees can clearly distinguish blue-green from either blue or green and that it is a colour they favour.
See: colour of hives

**Hive Dadant**
In some respects this is like a overlarge Langstroth but in use they do not seem at all compatible. They take 12 combs which are almost 50mm deeper. The roof is just over 51x48cm/20x19" with a crown board 51x47cm/20 x 181/2".

**hive decoy**
A hive without bees set-up with the object of attracting a swarm. Old, but healthy, comb in a two box hive which is raised to a level well above the ground (3-4m possibly) and about half-a-mile from the apiary optimises the chances of success.
See: bait combs/hive

**Hive density**
For fruit pollination 5 colonies per hectare. As regards apiary siting in areas with a reasonable amount of forage you might go up to twenty hives. For migratory beekeeping where there are acres of moor for heather or when field beans, oilseed rape etc. are visited, then the number can be increased. Countries with the largest number of colonies (density) in Europe in the 80's were Czechoslovakia, Greece and Switzerland. Russia, China and the U.S.A. run into millions. Despite large scale bee farmers (whose bees tend to follow the crops) hobbyist beekeepers outnumber those of the commercial set-ups but with the greatest density around towns. It has to be realized though that once a certain limit has been reached, say twenty, putting in another five will increase competition but will not increase the honey take by 25% as might be expected, so it is wise to run several out-apiaries rather than put too many colonies in one location.

**hive double**
A twin-hive, one colony set above another will benefit the upper one by mutual warmth. This can sometimes be used to help a weaker colony or when a nucleus or queen mating extension is made above a separating board.
A book called the 'Sky scraper hive, outlined a method of running several colonies one above the either, each of course with their own queen. Attempts to have two queens in one hive have included using two queen excluders with a shared super area between them but this did not lead to the expected increase in the honey crop.
See: skyscraper, multi-queen

**hive entrance**
The opening through which the bees pass into and out of the hive. Whereas special holes are sometimes made for this purpose, it is customary to design the floorboard so that when the brood chamber sits upon it a bottom entrance is provided. This is convenient because it allows the drones to come and go. Top entrances have been tried by some and of course when a second colony or swarm control method calls for it, an upper entrance can be arranged though it is wise to have that second entrance facing in the opposite direction to the lower one. The size of an entrance can be varied according to the weather and colony strength and of course in winter we usually put a mouse

guard over it.
See: direction, bottom entrance, mouse guard

**hive escape**

A brass cone or tube may be set up for ventilation and while allowing bees to pass out, prevents re-entry or the entry of the uninvited (robber bees or wasps). With such an upper escape it can happen that drones get trapped up top and proceed to block the excluder with their dead bodies.
See: Porter

**hive fastening**

When cattle or large animals might brush against or knock over hives it might be best to arrange them in touching blocks of four. When it becomes necessary to fasten them, as when transporting, to prevent one body sliding over another you are faced with two alternatives. Either to lock them together (as with angle brackets or staples) or to strap them together so firmly that movement is almost impossible. Vibration on a journey can easily work a hive loose and although the bees usually cling to the outside of their hive it can be a bit off-putting if it happens in the family car. Remember to inform police and label the car LIVE BEES when taking bees on the highway.

**hive feet**

Hive legs might be more appropriate and these are found on the W.B.C. hive and they are angled for greater stability. They do not last for many years if they are set directly onto the ground. Dampness sooner or later makes them deteriorate. For this reason it is not unusual to place them on a concrete slab or firm dry base. Most other hives are made so that they are best put onto stands,
a) to give a better working height
b) for clearance from wet growth, the ingress of ants and a much better micro-climate than exists at ground level.
Much of the 'robbing' done by ants is unseen and often very serious, so steps should always be taken to stop them from developing their nests 'right on top', so to speak, of the hives.

**hive fillet**

Flanges or plinths around the base of chambers to overlap slightly for weather protection. They are not as widely used now-a-days because two flat surfaces are soon partly propolised by the bees and therefore weather proof. However the design of such plinths incorporated a bevelled surface to shed the rain and a longitudinal groove on the underside to ensure that water cannot seep backwards towards the hive.

**hive fixed comb**

Prior to Langstroth the bees in skeps or boxes were free to build their comb 'willy nilly', that is to say join it onto roof and sides and to build brace combs together here and there. A cane was often pushed horizontally right through a skep to keep the combs steady. At brood temperatures wax is unbelievably soft and while it is not interfered with, can support a hundred times its own weight with honey. The examination of such colonies was difficult if not impossible, queens were rarely seen and the health of brood could not easily be ascertained. Despite their accuracy of building vertical combs and ensuring bee and ventilation space between them, somehow the parallel nature of the combs was randomly twisted and no two skeps would have had quite the same pattern. To cut out fixed combs could be a messy process and special knives were made and used. The long-idea type of hive, like the chimney pot type laid horizontally on the ground in Malta, required especially long handled knives with the blade set at right angles to the handle. Although strong enough when naturally fixed in their nest, once pieces of honey comb were removed it was not easy to handle them because to apply enough finger strength to hold the weight caused the fingers to dig into the comb or allow it to break away altogether. Only by placing smallish pieces on the flat of the hand could the honey filled comb be supported without damage.

See: movable comb

**hive floor**

Following the same periphery as the hive a flat board rimmed on three sides so that it leaves bee-space plus under the lowest combs and also determines the size of the entrance. It has become current practice by 2010 to allow for Varroa treatment and so gauze screens are often substituted for the solid floors. However when solid floors are used they become receptacles for any unwanted material that is too heavy for the bees to cart out of the hive. Odd things that might become smelly and are too heavy for the bees to eject, are covered with little sepulchres of propolis. They can get to look very untidy when exposed to view. This does not imply that the bees cannot cope with their own sanitary arrangements but it makes the average beekeeper decide to scrape and blow-torch the floor, especially after the winter months, and give them a nice 'clean' one to set about propolising and carrying on again as usual.

**hive furniture**

A empty hive is just as acceptable to bees as one that is furnished, but for the beekeeper's benefit the contents are arranged to suit his or her beekeeping methods of management. Consequently as time rolled by and ingenious and helpful inventions were made, a whole string of inclusions have become possible. Of course accurately made frames are the first essential, then methods of spacing them and keeping the queen in what is termed the brood chamber, call for articles made of metal or plastic such as queen excluders, metal or plastic spacers, crown boards or inner cover, dummy boards, clearer boards and feeders, section crates, and whatever else the equipment manufacturers can come up with. Honey can become a very expensive item to produce!

**hive insulation**

The retention of warmth by a beehive is normally due to the thickness of the walls. A single wall of say seven eighths inch thickness will protect the bees against our normal winter cold while allowing some of the sun's warmth to penetrate. A crown board can be further insulated and when the early brood nest's warmth can be felt by your hands you might be tempted to cover it with a nice thick chunk of polystyrene. Double-walled hives like the WBC have an airspace allowing the inner walls to be thinner and lighter than those of a single-walled hive. The air space between outside and inside walls persuading many beekeepers to agree with the inventor W.Broughton Carr that this was undoubtedly the best hive for the British climate. It worked! They are good hives, but expense and manoeuvrability have to be taken on board so such hives are becoming rarer except in many cases by beginners who are won over by the elegance of the sloping walls and gabled roof, entrance porch and variable slide entrance. Gosh, I'll have to get one.

**hive large**

The Conqueror and Glen hives and even the WBC look larger than their contents would imply. Of course, any hive, made for vertical stacking, can be extended to make it look enormous. In melliferous countries where they might look askance at anything smaller than a deep Langstroth box, when six or seven are piled onto the one hive (and full of honey) it is necessary to climb onto transport to form a platform so that another 'super' can go on. When nectar does come flooding in (let's hope it will this year of course) instead of piling more and more supers on, it is best to take one off when fully sealed, extract the honey and immediately return it. Instead of being angry with you the bees will relish a new box of wet combs and it spurs them on to greater efforts. It becomes harder to examine the brood once a hive is loaded with heavy supers but 'Bigger da man,bigger da chest' but watch your back.

**hive leaf**

A hive arranged with frames that can be

opened out like the leaves of a book. This lay-out was one that the famous Huber used to make his invaluable observations.

**hive legs**

Projections from the base of a hive either to form a stand or from the floor itself. They can set the brood chamber at a more convenient height for observations though you should bear in mind that with supers to go on, lifting boxes on high can call for a tall beekeeper or the use of a step. Because of the importance of these when it comes to a hive that can double or treble its weight once the bees have got cracking, they must be strong and stable. Being the part most likely to suffer from damp and so begin to rot, they should be adequately treated with wood preservative, perhaps greased, or stood in containers of oil which also helps to defy the ants.

**hive lift**

Perhaps this word came into play because in the case of the famous WBC the outer walls comprise telescopic (they aren't really) walls that are 'lifted' off before an inspection of the real hive parts can be made. They are wider at the bottom than the top and fillets along the inside lower edge prevent them from telescoping when in use. It may be sacrilegious to suggest it but anyone converting from WBC to National will find the lifts, cut down and fastened to the inner walls, make boxes practically compatible with the national. When stacking WBC lifts (which can come in two depths) place them diagonally to one another, and by the way, they can all too easily be knocked corner-wise out of square. Internal brackets can strengthen in this respect.

**hive log**

The first move from 'hunting' for hives and using man-made ones doubtless came from finding bees in hollow trees and going from there to encourage their using hollow logs. They could be transported, hauled up into trees and emptied and re-used, perhaps to attract swarms as well.

**hive long-idea**

A hive taking one size of frame in horizontal array and if required, using vertical queen excluders. The roof is often one-piece but several crown boards enable section by section to be dealt with. Lifting is minimized and natives have made similar hives carving out logs, which can be strung by wires between trees to out-do the proliferous ants.

**hive maintenance**

Because of the weight of hives and their relatively flimsy structure, they can easily be damaged, the careless use of a hive-tool can add to this. Where damage occurs its early rectification is advisable because weather can step in to further worsen the trouble, bees will glue up gaps with propolis, and unwanted cracks or crevices are beloved by the wax moth and other small creatures. In weather-proofing a hive, care must be taken not to use substances repellent to bees or even poisonous. The advent of plastic hives was at first frowned upon but after many years of excellent service they have become accepted. However they can be damaged by heat, even ultra-violet light and slugs and snails love to cling to the inner walls. Whenever hives are free of bees the opportunity should be taken to clean and sterilize them but air them well before re-use. In NZ it was observed that the dry wood of their Langstroths after months in the sun could benefit by a dip in hot, liquid paraffin wax which sank into the grain and after draining and setting gave a long lasting protection which bees readily accepted.

**hive manufacture**

The mass-production of well-made hives has made home production less attractive. It is important that hives and frames are made to very close limits. The ease of assembly from in-the-flat has reduced cost and become common place. New things like glued staples and plastics of all sorts have come into their own and made life a lot simpler for the modern beekeeper. Standardisation has rather gone by the board as so many choices have been made available. In many countries they just refer to a hive and it automatically means the same type. Here though, its what type do you want and the beginner rarely has a clue. Smith, Dadant, Langstroth, National, Modified National, Commercial, Glen. But what about this one and the resplendent WBC is pointed out, standing white and proud on a neat patch of green grass; 'Yes that's it says the beginner (unless he or she has been to classes).

**hive materials**

Beekeeping was an occupation which even the poor could aspire to and all sorts of material were used. To quote Eva Crane: mud, straw, cowdung, fired clay, woven cane, woven wicker, coiled straw, hewn wood, bark, cork, shells, coco nut, plant material, gourds etc. and to these we might add: cement, GRP, hardboard, plywood, glass, Perspex and plastics. Wood is still the predominant material with plastics perhaps fast catching up. Do the bees care?

**hive mates (nest mates)**

The honeybees that are accepted as members of a single colony. In some cases (perhaps more often than is generally supposed) both mother and daughter might be functioning in the same colony. Or a number of virgins (kept apart by the bees) may be ready to leave for mating. Drones of course can come and go if they want, because colonies will only reject them at times of dearth or at the end of season. Colony identity, built up from their queen's pheromonic distribution and the actual smell of the working colony enables them to reject any outsider (other than drones). This means that the members of a colony share a positive bond and by virtue of their changing physical capabilities they work 'happily' together sharing the responsibilities which go along with being a honeybee.
See: identity odour

**hive mind**

The corporate actions and behaviour of a colony of honeybees has often been anthropomorphically described as due to the 'hive-mind'. Although bees' brains are individually quite small They are able to communicate and share experiences and apparently make collective decisions. We now know that chemical and glandular influences may well govern many decisions.

**hive model**

Model hives, made as exact replicas but to half or quarter scale, make attractive displays and are useful for educational purposes. Perspex, wood and ply make suitable materials – a quarter scale WBC is large enough to accommodate a bumble bee colony. Small hives for purely ornamental purposes can add charm to the small garden or for interior decoration.

**hive - movable frame**

A hive which by virtue of having bee-space all round the frames of comb, allows them to be removed or repositioned. This was either difficult or impossible with fixed comb hives. The Greeks were thought to have been the first to use this principal in their bar hives and Langstroth the first to fully appreciate the implications of bee-space and to apply it to methods of mass producing movable frame hives, gaining in the process the title of 'Father of Modern Beekeeping'. See: movable frame

**hive multi-storey**

The basic winter configuration of a hive is often just a single storey affair with its one chamber, roof and floorboard. With the coming of spring and development of the brood nest it behoves you to consider giving them more space and this can be done by adding another chamber, either deep or shallow. To prevent the queen from moving upwards as she is bound to want to as warmth rises and she always lays in concentric ovals steadily upwards if allowed to, a slotted layer (through which

workers but not the queen can pass) is inserted. This is sometimes omitted but because beekeepers want to keep honey stores separate from brood and pollen an excluder is used to confine her to the lower part of the hive. As the season progresses and more space is needed both for brood and for storing and ripening nectar, extra 'bodies' (chambers) are added and this can then be described as a multi-storey hive, but as this situation is normal the term multi-storey hive should perhaps be kept for hives that are extended upwards either to accommodate an extra queen rearing portion or even, like the 'Skyscraper hive, be divided into separated parts, each with their own entrance and their own queen. Very tall hives can be unstable and precautions should be taken to ensure that they are safe from toppling.
See: skyscraper hive

**hive – national**
The British national hive follows the revised specification of the British Standards Institution. It is 460mm/181/8" square, takes BS frames and has bottom spacing. This is quite significant because where top spacing is used as in Smith, Langstroth and other hives, the two cannot be mixed or it results in double bee space or no bee space at all, between chambers. No space means propolis and double space, burr comb. Each box takes eleven frames, spaced 38mm/11/2" centre to centre. The side walls are of single thickness while the end walls are double. There is a modified version with single walls giving better hand holds and saving timber. Canadian red cedar is widely used for its construction.
See: national hive and modified national

**hive number**
Hive identity is useful for note keeping and diagnostic tests but you can give your hives names if you prefer but it does help, once more than two hives are run, to clearly know the history and actions required for each one. Another aspect is the number of hives in a given apiary, area, or country.
See: hives per hectare

**hives observatory**
The value of an observation hive is completely overlooked by the majority of beekeepers. Once you have made one it will dawn on you how much you can learn from being able to actually see the bees at work. Their tidiness, the queen and her daisy-petal retinue, and perhaps most important, the condition of the honey gathering stakes, i.e. onset of a honey flow. They need only contain two or three combs and can be used to store or rear a queen. Both sides of the combs should be visible and it is best to keep them covered to conserve heat and allow them to work in the dark. A feeder is very necessary and a clear tunnel which must be prevented from becoming blocked, should lead out of the shed, house or wherever the hive is positioned, to the outside world. Keep it in a position where the sun cannot touch it.

**hive odour**
Well the hive itself should have no odour (red cedar may smell attractive but that does not mean bees share the pleasure). But colony odour is another thing, bees have an extraordinarily acute sense of smell. Like dogs they can single out any one from a dozen different odours and that is why colony odour becomes a life-saving aspect for bees, who are thus able to distinguish friend from foe. The specific odour is acquired because the bees share a queen, share food substances, share the smell of their brood and share the smell of incoming stores. Smoke upsets their whole regime and way of life and their 'guard is down' when they have to struggle to keep out of smoke or blow smoke away. Care should always be taken when odiferous substances are brought near the bees and whenever a queen is marked make absolutely certain there is no trace of smell before she is returned to her bees. The airing of combs treated with acetic acid or formaldehyde before use is important. The air within the brood nest is 'special'. So special that it is a surprise that

no-one seems to have drawn out a sample for sale(!) by means of a filter at the end of a tube incorporating a pump.
See: odour hive, identity odour

**hive paint**
Although red cedar is claimed to weather into a respectable colour and to remain weather-proof, it often gets painted all the same. You must of course be very careful what wood preservatives or paint you use because so many include insecticides these days and bees are, after all, insects! Even when non-poisonous they might well be repellent. Varnishes are particularly suspect. Modern acrylic paints can be useful. Bearing in mind that bees are not colour blind (except to red) any colour chosen should take into account the reflection of heat and light, the durability and ease of application. White is likely to remain the copied colour but there is a school of thought that camouflage is better. Why not jazz it up though and have all the colours of the rainbow? On WBC flight boards a sprinkling of sand on a coat of wet paint, gives the bees a nice foothold when running back after foraging.
See: hot dip

**hive plans**
When using plans bear in mind that you must either work in old British measurements or in the metric system. Plans help you to get a better understanding of how the whole thing works. Useful too for scaling down and making smaller models. During the course of time the world of bee literature, magazines etc. have given superior plans not only of hives but honey extractors, solar wax extractors, smokers, observation hives and all else. BBKA have always offered a wealth of useful ideas and material including plans galore.

**hive plastic**
When plastic hives were first introduced, the conservative British beekeeper did not readily take to the idea, and many excellent hives seem to have disappeared. In the 70s it was possible to buy snap-together (like Lego) supers, in white plastic from America, they were only thirty bob (£1.50), but importing agents and the customs trebled the price. They were exceeding long lasting. So easy to assemble. Another type was of brown plastic with the density and properties of wood as regards cutting and sawing. They too lasted the test of time and were quite inter-changeable with their wooden national counterparts. Now frames and all manner of beekeeping items are made in plastic which is light, clean and accurate and bees propolise them never noticing tother from which.

**hive preservative**
Plastics can be washed, sterilized and scraped and offer a sensible alternative to wood which needs preservation. New liquids constantly appear on the market and the ever-relevant warning 'watch out for harmful substances' that can be incorporated without notices making the fact abundantly clear. Creosote, once used with gay abandon, is now frowned upon and Cuprinol, Solignum et al have come along in its stead. The hot-dip for tired timbers is not likely to catch on unless you have lots of space in a big garden and can bring a bulk of second grade paraffin wax to 100C; then you might want to engage in this useful procedure. Tongs are used to submerge hive bodies for one minute, then a few seconds draining followed by a second or two's 2nd dipping. They remain sound, neither warping, cracking or there we go again – its flammable and you may well think a wood fire is a cheap way of heating but not if the lot goes up!

**hive products**
Although honey is perhaps the most obvious commodity produced by bees and made use of by man, wax comes a close second and almost every substance and item that comes from the hive seems to have been put to some good use. Hence bee venom is marketed as is propolis, pollen, royal jelly brood and bees.
See: harvest

**hive protection**

Routine maintenance to ensure that roofs remain sound, that splits or loose knots have not opened gaps and that boring insects, woodpeckers and other creatures have not caused damage – these things help to check hive conditions. Shelter from severe winds and weather, siting so as to give good drainage and rendering the hive inconspicuous to avoid rustling or vandalism also help. Hives may be branded with the owner's identification number and comprehensive insurance is offered by BBKA.

Wire netting and flapping plastic help to ward off woodpeckers and mouse guards over the entrance from Nov to March is also a wise precaution.

**hive records**

A beginner might be slow to start serious note taking. Perhaps because the significant points only become clear as time progresses. But even with just one hive it is a very good habit when gloves are off and things are fresh in mind to note the time, date, weather and actions taken or observations made. Also looking ahead noting items that will be required when next the hive(s) is inspected. To leave cards inside the roof as so many do, is not as sensible as having a book, set aside for the purpose and making columns so that facts are listed accurately but as briefly as necessary. Recommended shorthand like fb for floorboard X for queen excluder and Q for queen can be vastly increased as words become more familiar. Eggs, Queen, Sealed brood, when present with number of combs covered all make good reading afterwards and help the beekeeper to remain on top of the job. A kind of log book.

**hive roof**

The removable topmost cover of the hive. Unless hinged and this is rare, it should not be referred to as a 'lid'. Flat roofs can still be slanting to allow rain to drip off at the back while still being suitable for sitting, even standing on and upturned are useful to temporarily contain things. The gable roof, though architecturally pretty is not as handy. Apart from being water-proof a roof should be provided with ventilation and give space enough to accommodate a feeder and in some cases even a veil, hive tool and such like though an empty WBC might serve this purpose !!?? If the roof is lightweight it must be designed so that the wind cannot lift it off. Heavy roofs have their place and can be more robust when used to stand on when you want to pick the adjacent blackberries and all that. Shallow roofs are rather pointless and in NZ the idea of combining roof with crown board was catching on. Propolis and a slight overlap on two sides doing the job of holding it in position. An overlap on four sides would preclude the easy use of a hive tool for lifting it clear. If bees can get up into the roof they may well glue it up or build comb and make it tricky to get off.

**hive runner**

A strip of metal or plastic along the edge of the rebate on which the lugs of the frames rest, is so helpful. How and why some people have taken to those sharp, castellated runners (in fact they're not runners at all) must be because they have not heard of Hoffman frames which lock frames very securely together yet can still be slid along the runners when manipulations take place. Runners allow only a thin strip to make contact with the lug so that propolisation is minimal. If used in conjunction with metal ends these should be aligned to touch the upper edge of the runner. The other hive runners are two strips of wood along the bottom edge of the sides of the floorboard.

**hive scale**

It is not a common practice to weigh hives. Nevertheless if the weight is known it can be very helpful and its diurnal variation gives a good indication of the colony's progress and well-being. Some people have specialised in doing this even mounting the hive on a scale and many useful facts come to light. It has been suggested that knowing the weight

variation can help to indicate the likelihood of a swarm. And the swarm's departure should make a sudden drop in hive weight. With a water pendulum and microswitch a warning device could be arranged eh? If a spring balance and hook latched under one of the runners on one side gives a measure which is added to a reading taken on the other side, that will give the hive's weight. The accuracy of 'hefting' can also be checked in this way.
See weighing hives, scale

**hive scraper**

Hive tools are made so that as well as breaking propolis seals they can lift frames and act as small scrapers. Larger paint scrapers are suitable for wide areas like floor boards. A curved end help to scrape by pulling towards you but you should be reminded that propolis and burr comb should be minimal if proper bee-space is adhered to and it really isn't necessary to keep hives open while every top bar is spotlessly scraped. Of course it is necessary to avoid a build up under excluders especially the sheet type. Scraping almost invariably upsets normal bees and if after a few days they will have built it all up again would you be winning or losing?

**hive site**

A site that is suitable for a hive will be one in a healthy area where good forage is available and the bees are useful to the community at large without being any sort of embarrassment. Shelter, security, good air drainage, access and overseeing by someone responsible - if you find all this then you're lucky. More and more hives are being kept in towns and out-apiaries are often associated with having to climb over fences, walk across tricky areas, climb hills and to have the key to the padlock and face the wrath of anyone whose nosey dog was stung through being allowed to run where it pleased. But despite everything the mood of the public has really had a rightful boost and pleasant associations can be built up (honey helps) and long lasting friendships made.
See: drainage, enclosure, out-apiary

**hive skyscraper**

A book with this title was top of the pops at one time and involved using queen excluders to separate many colonies all heaped one above the other in a towering array. All with their own entrances rather like a long idea hive with adjacent colonies separated horizontally. Ventilation is highly important because waste gases tend to move downwards and the secret of bee economy is to have the right temperature and humidity in their nest. It is not really a sensible proposition to have so many spare supers that you can afford to go on and adding them until you need to stand on a ladder or platform to reach the top. Uneconomical because it is only in glut years which like the roulette wheel can come in groups but more often go ages and ages before the extra supers are required. So a better plan is to take the view that a bird in the hand is worth two in the bush and get that super off when the honey is sealed and extract and return it within 24 hours. You'll find the bees are lining up, waiting for it and it gives a marvellous kick to their morale as you will see from their increased activity.
See: skyscraper

**hive – Smith**

A hive which Langstroth might have been proud of because if follows his line of thought in wasting no timber. Four walls, eleven frames with half inch lugs and top spacing. The same British standard frame minus 50mm/2" of lug with corresponding shorter sides to the hive yet holding just the same amount of comb. They are commoner up north and have to be asked for specially here. Nice to use and though not as light as the WBC they are lighter than other hives. Probably the best of the British standard hives.
See: Smith hive

**hive spacing**

The spacing of hives within an apiary should never be too close even though touching

hives might be considered a way of sharing warmth. But colonies are all enemies of one another and fiercely protect what is theirs. It is true that foragers can 'drift' into other hives and this is all too likely if they are closely spaced, but from the health point of view and ease of handling it is best to not only keep them well apart but forget all about regimented rows and try to make them look different by irregular spacing and for your convenience being able to manipulate one without standing in the flight path of another. There are times when it is helpful to move a hive to one side swapping positions (to prevent swarming) or to lose flying bees for some reason, so give yourself space to do this kind of thing.
See: hive density

**hives per hectare**

When growers come to the conclusion that honeybee pollination is a very economical and useful thing they look to beekeepers to have the knowledge as to how many hives per ha are recommended. Few if any know. However the agricultural authorities in New Zealand brought out a list which is surprising seeing how it varies with different crops:
Almonds 5, apples 2, blackberry 2, buckwheat 4, cherries 2, kiwifruit 8, cranberries 20, gherkins 6, blackcurrants 5, lavender 2, macadamia 4, onion seed 30, peach 2, prunes 5+, pecans 2, raspberry 4, sunflowers 2.

**hive stand**

In the case of the WBC and some other hives this is incorporated into the floor board. Twin stands may be economical but can cause 'drifting' and the easier spread of disease. Stands may be portable or remain fixed but the materials used for legs or posts would need to be proof against rotting or frequent replacement will be necessary. A stand should offer a firm base for the hive, level, though a slight incline towards the entrance is beneficial.
See: stand, trestle

**hive standard**

In Britain we probably have more different hives than seems to be the case in other countries, whether by choice or the power of the first mass producers, is another matter. Although here, the most widely used hive is probably the British standard 'national', there are a plethora of hives. To take one, the WBC; it is listed as the 'standard telescopic hive' that is not accurate as the sloping lifts of the outer part are definitely not telescopic. Variants of the WBC in the form of the Burgess and CDB hives are truly telescopic and by inverting the lifts they go over one another forming a double or treble outer wall. Conversion to metric sizes might bring even more complications but, spare a thought for the manufacturers who are called upon to stock such a variation. Efforts to get beekeepers to agree on one specific type have fallen on stony ground.
See: CDB, telescopic hive

**hive Stewarton**

The name was derived from the Ayrshire town in which the originator lived. It was one of the first attempts to make a bar-hive so that combs could be removed and replaced. However it was octagonal so the frames nearest the centre were larger than those towards the flanks. It was a beautifully made hive with a pinnacled roof and the carpenters were kept busy making them for the many orders that rolled in.

**hive strain**

It is sometimes necessary to impose enormous leverage to loosen a well-propolised frame that has remained in-situ for a long time, or to separate boxes. The snapping off of the unnecessarily long lugs is not uncommon and hive corners can deteriorate when hive tools are misused in the effort to separate hive bodies. Leverage may be applied to the top bar at points other than the lugs and once here is room other frames may be levered sideways before lifting. Diagonal strains should always be avoided and when transported methods of fastening should not impose undue strains

upon the timber.
See: strain, race, variety

**hive strapping**

When hives are to be moved the various parts need to be firmly clamped together because although under normal operating conditions parts are firmly held together by propolis, The vibration and jogging of moving can easily break this seal and if this happens as you are pushing the hive into the car, that can be a trifle embarrassing! Rayon, steel, nylon and other tough tapes are used, with special tensioning devices like buckles for pulling the strapping taut. The point at which the strap tends to dig into the wood should first be protected.
See: moving, staples

**hive straw**

Straw skeps were at one time made in all sorts of shapes and sizes and often made with relatively little skill. Now if skeps are made the whole process has become an art where the right kind of straw (wheat straw) which has been cut by hand is passed through a ring after being rolled up in damp cloth for 24 hours before use. A really well made skep is firm enough to sit on but having written all this, bear in mind that although the antiquity of the idea is appealing, the use of straw is not good when diseases are taken into account and a cardboard box serves the purpose better because it can be destroyed as opposed to using a skep.
See: skep, straw hive, straw

**hive substances**

Here are substances used or made by the bees: beeswax, bee-milk, honey, nectar, pollen, propolis, royal jelly, venom and water.

**hives under trees**

On the face of it keeping a hive under a tree seems a reasonable idea. But, there are snags. Weather-wise, long after rain has fallen, dripping goes on plopping onto the hive in irregular spells according to the wind, while the same wind can move the roots and disturb the hive in that way. The falling of the tree and lightning strikes are rather less likely to occur but twigs and branches fall and over all it is an umbrageous mass which may cut off daylight such as entices the foragers out early in the morning. A nest in the hollow of a tree does take advantage of the tree's shelter.

**hive to**

To hive a swarm you have to decide whether they are to be shaken down in front of the hive (preferably at dusk) or into the hive (as in inclement weather). In the first case a sloping board leading up to the entrance and covered with a clean, white cloth allows you to see the queen perhaps and any debris left on the cloth can be taken for examination. If you want to further encourage the bees to enter – and some may well be facing the wrong way, be brave, take a handful an flick them remorselessly onto the entrance. Definitely no smoke. In the second case, an empty box is placed over the one containing the combs. You might see fit to put an excluder over the frames if you want to find the queen or queens. Go ahead and heave the bees into the empty box. Heave, slap and knock for all you're worth, as many bees will be clinging to the inside of the box you were carrying them in, and why not, they thought it was home and may even have started trying to build comb. Cover, then on the following day, replace the empty upper box (throwing any bees clustering in it in front of the hive), with a crown board and roof unless another box of combs is obviously necessary as in the case of a big swarm. Q.E.D.

**hive tool**

A specially designed metal tool of considerable tensile strength, hand-held and capable of being clipped on to the back of the smoker, slipped into a purpose-made, reinforced, pocket of your overall or hung from a nail in the shed. Quite a number of these are lost. In the long grass, left where they shouldn't be – anyway a bright colour helps but it pays to develop the habit no matter how busy or occupied you are to always note where you have put it. If you arrive at an out-apiary without a veil, or without gloves, or without a smoker, make sure it's never without hive tool. This indispensable object can scrape, lever cut, tap and most importantly, act almost like a key when opening a hive is necessary.

**hive top**

Hackles were used in the days of our great grandfathers to shield the skeps from severe weather. A hive roof is a more substantial part these days and is strong, preferably flat, weather proof and provides ventilation over the crown board. It must be proof against wasps or robber bees and should be wind-resistant and reasonably easy to take off and to replace. In spring a frosted roof might develop a clear patch in in the middle indicating warmth rising up from the cluster below. If you stand your smoker there, remember don't knock the hive and alert the bees before you have even started.

**hive twin**

Side-by-side brood chambers each with their own queen. There are one or two advantages but several snags. On the credit side there is less space occupied and mutual warmth can be beneficial. But drifting, robbing and the need for a special roof and difficulties manipulating, weigh rather heavily against the advantages.

**hive ventilation**

E. B. Wedmore wrote book on this subject in 1947. Modern investigations have shown that the bees are very capable of arranging their own ventilation by fanning in the summer. In winter the cluster assumes characteristics more like a mammal and the rising waste gases are best disposed of by top ventilation that also prevents condensation. Top packing which hinders ventilation should not be used until bees are sufficiently developed to be able to indulge in fanning.
See: ventilation

**hive WBC**

A hive comprising a floor with integral stand, porch and alighting board. On this platform stand thin wooden boxes that hold ten normal-spaced frames or eight wide-spaced. It uses bottom spacing and entrance slides fit into the porch. Over the thin inner walls are outer lifts. They are tapered, narrow on top and wider at the bottom, thus producing a dead-air, insulated space around the inner boxes. They also have an inner plinth at the bottom so that the lifts slightly overlap each other thus shedding the rain. The roof is gabled. The outer walls can be painted any colour but white is most popular whilst a black roof aids the colony in winter.
See: WBC hive

**hive wide idea**

This is another way of saying 'long-idea' hive. An elongated, horizontal box containing a large number of frames and a vertical queen excluder. The size is limited though supering can be arranged. The roof too, is a long one, hinged at the back. It does away with heavy lifting and may have sloping walls so that, using bars instead of frames, natural shaped unwired combs can develop. Not a hive that you would want to transport.

**hive wooden**

The majority of hives today are still made from wood. Softwoods need painting but unpainted red cedar hives are popular (if expensive). Heavy hardwoods, although durable, make for expense too and heavy lifting. As the bees natural habitat is frequently a hollow tree, which they clean and line with propolis, wood is a natural material that has good heat insulating properties, being workable and allowing moisture to find a natural balance.
See: wooden hive, plastic hive

**hiving**

The act of getting bees into a hive. The domestication of bees has necessitated transferring them from their wild or 'feral' state and succeeding in getting them to occupy a domain arranged by the beekeeper. Scents, superstitions and useful tricks have been used to bring about the successful hiving of bees. However bees will not stay in hive they consider unsatisfactory (e.g. contaminated with insecticide or smelly), casts and swarms can be fickle and leave after emptying the feeder!

**hiving package bees**

Package bees will have suffered the stress of traveling and on arrival should be sheltered in a cool, dark place having brushed the gauze with a pint of tepid syrup. Hiving should be carried out the next day if weather conditions permit. No replacement workers will be born for three weeks, by which time the unit will be at its lowest ebb if no 'plumping' is undertaken. The hive should be set up with foundation and such frames of drawn comb and disease-free stores as can be made available. The queen is checked in her cage and set for the candy-timed release. Shake bees from the package onto the frames, put the queen cage between entire top bars and close the hive. When the queen is in full lay, about a week later, add healthy emerging brood if available. Feeding should be continuous until all foundation has been drawn out.

See: package bees

**hiving a swarm**

The establishment of a swarm in a bee hive. The hive should be made ready, clean and fitted with suitable frames either of foundation or disease-free comb. To hive in broad daylight could be accompanied by the swarm becoming airborne and perhaps absconding or moving to a difficult location. At dusk therefore, the bees can be thrown down onto a white sheet placed on a sloping board leading up to the entrance. Usually they will run in by themselves but keep an eye open for the queen. Once she is in the bees are likely to follow but should they appear to be reluctant and some are running or pointing in the wrong direction, physically take a handful and throw them onto the entrance. If smoke is used it will block the useful effect of the scent that many fanners near the entrance will be emitting so be vary sparing and keep it well away from the entrance. Sometimes the swarm which seemed extra large, may well be a collection of casts. In which case several queens will be amongst the throng and it is handy to have an empty matchbox handy so that they can be captured as you only want one inside. See: taking a swarm.

**hiving sough (pronounced Sow)**

The sound that the bees make before they swarm. There is a reduced activity and masses of bees will be loaded with stores (which might be responsible for the slight change of their buzzing note) and giving out this rather melancholy, continuous hum – it is often heard the evening before the swarm issues.

See: queenless moan

**HMF**

Hydroxymethylfurfuraldehyde – the longest word in this beekeeping dictionary. Poison too, yet it is a substance found in honey which increases measurably as honey is heated. (used along with diastase content when passing EEC Standards). Its level should be below 40mg/kg. It is formed by the decomposition of fructose in the presence of an acid when heated. The average content in honey is 1·24mg per 100g. Tests are able to differentiate between honey mixed with invert sugar, long stored honey and heated honey.

**hobbyist beekeeper**

90% of British beekeepers are hobbyists. At one time their main object was to produce that marvellous compound – honey – for themselves and as a well-received gift to friends and associates. By 2010 a sudden awareness on the part of the public that bees were getting scarce and that many of our luxury goods depended on them, e.g. almonds, apples, avocadoes, blackberries, cherries, kiwifruit, gherkins, blackcurrants, lavender, onions, peaches, raspberries and sunflowers. A great number of people decided either to take up beekeeping or at least to be totally sympathetic to the need for bees. Hence instead of regarding beekeepers as 'odd bods' they decided to join their ranks and discover the educational, economic and social benefits from indulging in this totally useful activity. Long may the bees go round. See: bee farmer

**Hodges Dorothy**

1974 saw a reprint of 'Pollen Loads of the Honeybee' a beautifully illustrated book that has become a classic on pollen grains. Perhaps, as you will see by the faithful reporting of pollen colours and sizes through this work, that there is a great variation in opinions as to pollen colours. Take a pellet of pollen, use it on white paper to make a mark as if a line of chalk, label it, then check and see how different it becomes in a day or so when the colours will have changed. Interesting too if you have a u/v source to see what colours the bee sees.

**Hoffman frame**

This type of frame in all the various sizes, has a shoulder which not only gives precise centre-to-centre spacing but locks the frames together so that supers can be carried long-ways up (held closer to the body) or safely placed facing any way in a transport. The wide shoulder is achieved by widening the top of the side bars. Not only just a fatter bar but a bevelled edge on one side which engages with a 'flat' on the other side, minimising propolisation yet firmly distancing one frame from the next. It is most important when assembling these frames that you get the bevel on the correct side. Looking down onto a top bar held sideways, the pointed (bevelled) side of the Hoffman points forwards on the right side, while on the left it points rearwards. In other words the direction of the two pointed sides would give an anti-clockwise rotation. If this isn't clear, don't guess because if the side bars are put in the wrong way round, then you get flat on flat, point on point and this nullifies the reason for having Hoffman (grooves for holding foundation should be on the inner face). The sharpness of this bevel may not always be as acute on cheaper frames, get them from a good firm. See: Manley Diagram?

**Hogs at the Honeypot**

Author Frank G. Vernon the story of Hampshire beekeepers by the author of '*Teach yourself Beekeeping*'.

**hogweed Heracleum spp. Apiaceae**

They include 'giant hogweed' *H.mantegazzianum* more spectacular than useful. A hardy perennial with white flowers in large umbels up to one foot diameter from June to Sept. Some folk have skin that is sensitive to this plant, children use the hollow stems as pea-shooters and the umbels provide a feeding and mating ground for insects.

**holding down cage**

A cage that can amply accommodate a queen, designed to gently hold her against the comb while she is marked. An open mesh of light netting covers the cage which, when she is 'trapped, bears against her sufficiently to restrict movement while not harming her, yet leaving the operator free to apply marking paint. (The cage has a sharp side enabling it to be pressed into the comb). Pressure is then released so that when the paint has dried thoroughly, she can be allowed to go back with the bees.

**Holland**

Beekeeping is taken seriously here and much help was given to the U.K. when skeps of Dutch bees were brought in to replace our devastating losses during the period when I.O.W. disease was rampant. They have two hive designs: the double walled Simplex and single-walled Spaakast, both using a frame akin to the British Smith frame. The annual Bee Market at Veenendaal has been visited frequently by groups of British beekeepers. See: Dutch bees

**holly *Ilex aquifolium*, *opaca* (American) Aquifoliaceae**

Britain's widespread, evergreen tree with glossy spiny-edged leaves (less spiny higher up the tree), fragrant white flowers and conspicuous red berries that have made it popular in Christmas songs and for decorations. It grows slowly and is a good hedge plant flowering during May and into the June gap between earlier fruit and the clover. It does well in a wet climate and certain cultivars are amongst the best American honey plants. Honeys are described as light amber with a pleasant flavour, golden with attractive bouquet, and the pollen as light green or greenish yellow with a 25µ grain. The honey has also been described as oily, dark and smooth.

**hollyhock *Althaea rosea* Malvaceae**

Native to China but now a tall garden plant that will grow to more than 2m with flowers opening higher and higher up the stem. Single varieties mostly in pastel shades, look most attractive with bees scrabbling for pollen in their wide open flowers that continue until they are cut down by the frosts. The pollen has been described as light to orange; Yate Allan says pale yellow with a 90µ grain (one of the largest). Howes refers to bees looking as though white-washed and a bumble fell several inches before recovering itself. The seeds are easily collected and should be sown outdoors in early summer but treat as a biennial because the root stock deteriorates.

**holm oak *Quercus ilex* Fagaceae**

An evergreen tree whose young leaves are spiky like holly. It flourishes in exposed positions, even salt-laden winds fail to harm this rugged plant, and it can be used for hedging. At times they hum with honeybees, their long dangling catkins offering more for certain than just pollen.

**Holometabola (endopterygota)**

Insects that have a complete metamorphosis. The class Insecta is divided into three groups: Ametabola, Hemimetabola and Coco cola sorry Holometabola. There are nine orders within Holometabola and these include Hymenoptera (bees, wasps and ants) Lepidoptera – butterflies and moths, and Diptera (two wing-d flies).

**Holotype**

The specimen selected to designate a taxonomic new species.
See: paratype

**homeless bees Nomada spp.**

These are similar to cuckoo bees but they parasitize species of solitary bees. One example is the wasp-like Nomada which parasitizes the mining bee Adrena. There are more than twenty species.
See: Nomada

**homeostasis**

A state of equilibrium between organisms and their environment usually referred to as the 'balance of nature' or the balance of functions and chemical composition within an organism. It does of course steadily change due to climatic and other influences and in any case it is always prone to cyclic variations.

**home yard**

US. What we would call a base or home-apiary as opposed to apiaries away from home (out-apiaries) established either on a temporary or semi-permanent arrangement.
See: out-apiary

**Homer c.850 BC**

Famous Greek poet who wrote Iliad and was the reputed author of Odyssey. In those early days writing was more likely to mix fact and fancy but dealt with mythological, symbolic or mystical uses of bees and honey (as in festivals) rather than scientific matters. Nevertheless, Aristotle who wrote extensively on bees some 500 years later would have been familiar with his descriptions.

**homing**

The return flight that brings a bee back home. The sense of being aware as to which way it must return after flying away from its base and negotiating an area of flowers, not only implies having a sense of direction but being able to use it even after being trapped for some time at a distance. Doubtless its special eyes, which can read direction from the lines of polarized sunlight, help but if a bee flies out of a car window, somehow or other it must make it to a hive or die. Drones and queens too have this remarkable gift of knowing their way back. Compared with life in the hive, the outside environment can only be interpreted using their senses of smell, vibrations, light-orientation and familiarity. Do the myriads of radio type waves really upset them in any way? Can they register shapes, colours, shade and shadows and pick up miniscule traces incumbent on the air. Their survival depends entirely on this particular ability.
See: direction finding, homing instinct

**homing instinct**

The inborn and acquired characteristics that enable a bee to leave its nest, forage and return directly to its entrance (or immediate hive surroundings). To all intents and purposes it must take on board an entire map of what it sees (what its brain registers). Put a colony onto an entirely new site and bees unhesitatingly set off in the new environment and as assuredly find their way back. They are only confused if the new site is within flying range of their previous abode. In the case of a swarm the old memory sheet must be wiped blank. How, Oh how, do they do it? Colonies should be moved no more than a few feet during the active season. Over three miles for safety otherwise. If spiders can spin enormously long webs can bees make enormously long, invisible trails in the air?

**homoiothermic**

Warm blooded. Usually applied to living things that maintain a constant temperature above that of their normal surroundings. This does apply to a colony's brood nest but individual bees can only do this collectively or they become comatose at below 10C and moribund at lower temperatures.
See: poikilothermic

**homologous**

Having a corresponding form, position or function or having the same structure or origin but being used for a different function. Different creatures might have organs with a similar position or function in the embryo though not necessarily in the adult. The similarity being attributed to a common ancestry in evolution.

**Homoptera Insecta Hemiptera**

A sub-order of wingless and four-winged insects characterised by the wings sloping at an angle over the body, They include: frog hoppers, leaf hoppers, aphids, plant lice, white flies and scale insects.

**homozygous**

A plant or animal having genes in the corresponding loci of a pair of chromosomes, or in other words bearing similar alleles in relation to a particular character. Hence homozygote. A high level of homozygosity leads to uniformity but prolonged inbreeding leads to loss of vigour and a scattered brood pattern with bees. Sometimes though this is due to diploid male eggs which are eaten.

**honesty *Lunaria annua* Brassicaceae**

A hardy biennial that grows to 1m/3 ft with pretty, sweet-scented purple or white flowers from April through May. A plant for the bee garden, self-generating and useful. It forms flat, transparent, papery, silver-white satin seed pods that can be dried and used for decorations. They yield both nectar and pollen, the latter being almost colourless, The grain having three grooves and sized 25µ.

**honey**

The fluid viscous or crystallised food

produced by honeybees from the nectar of blossoms or from secretions of living parts of plants, or other secretions found on them, which honeybees collect, transform, combine with substances of their own and store and leave to mature in honeycomb. It has a characteristic flavour, colour (usually golden) appearance and odour according to its source and the crystallised form can be whipped into a creamy texture.
See: honey definition, misdescription, showing honey

**honey acids**
Acetic, benzoic, butyric, citric, formic, gluconic (see gluconobacter), glucose-6-phosphate, glycolic, isovaleric, lactic, malic, maleic, oxalic, pyroglutamic, pyruvic, succinic, tartaric, d-ketoglutaric, b-glycerophosphate and 2- or 3-phosphoglyceric, phenylacetic, proprionic, valeric and amino acids. The most prevalent acid is gluconic. Whew!
See: finger print

**honey adulteration**
Adulterants or materials that can be used for fraudulent addition to honey include: syrup made from cane or beet sugar, invert sugar and corn (maize) syrup, 'isomerised' high-fructose corn syrup variously called Isoglucose, Isomerose, Flo-sweet, Cornsweet and HFCS.
See: adulteration of honey

**honey analysis**
Many conventional tests can be applied to honey to ascertain its constituents both as regards quality and quantity. The advance of science has made a detailed study of specific ingredients possible. Rex Sawyer BSc. has contributed in no small manner to one aspect now known as melissopalynology or the study of pollens in honey. Dr. Eva Crane's book 'Honey' goes into enormous detail on the subject.

**honey (ancient)**
Corn, crystallised in comb (Butler). Live –run from 1st swarm or ordinary from old stock (Butler). Maiden – virgin (Sydserff in Cotton). Stone – crystallised in comb (Cotton). Virgin – run or 1st swarm (More in Cotton, Keys: Bonner: Huish: Duncan) Today: clear, granulated, crystallised, soft set, comb, cut-comb, chunk, blended, pure, light medium, dark and named according to source (plant or place).

**honey anti-biotic component**
Its antibiotic qualities come from is high sugar content, low pH value and hydrogen peroxide which is formed thus:
Glucose, water, oxygen, gluconic acid, hydrogen peroxide
$C_6H_{12}O_6+H_2O+O_2=C_6H_{12}O_7+H_2O_2$
See: honey stability, inhibine

**honey arch**
See: honey bar, canopy

**honey aroma**
The aroma of honey is naturally associated with that of the flowers from which its nectar was collected. Sometimes this is more dominant than others but due to the presence of volatile oils which can easily pass away in time or be driven off by heating, it is only when a well-sealed jar of honey is first opened that an initial burst of fragrance is fully appreciated. Heather honey has a persistent aroma (reminiscent of the moors) that will even withstand cooking as in a baked cake. Other honeys can still be identified from mead or marmalade made with their use. There are plants whose unwanted contribution to the aroma is not so welcome, sweet chestnut can offend though in France it is held as a delicacy. Privet, ragwort and one or two others, but an unwanted aroma in an otherwise nice honey, will gradually disappear after a spell in store.

**honey badger (ratel)** ***Mellivora capensis***
A powerfully built animal with short legs

and a thick coat it can break hives in much the same way as bears. In topical Africa hives are not set on the ground but hung from the branch of a tree by a rope or wire and a forked stick is wedged between the angles of the branches. Badgers are said to be able to empty a hive at night by repeatedly holding their tail in front of the entrance, this enables the badger to carry large numbers of bees away and it finally takes the almost unguarded combs. Kigatura also told us in 'Bee World' that they form a protective coat by allowing sun to bake wet mud that they have rolled in.
See: Kenya top-bar hive

**honey baked apples**

Wash and core the apples leaving part of the core in the bottom of the apples as a plug. Fill the cavity with honey, using as much as the tartness of the apples requires. For variety add a bit of lemon juice, or a few cinnamon candies. You may stuff the cavity with raisins and dates or other fruit combinations. Bake as normal; delicious.

**honey bar**

When a canopy of sealed honey has been built up over the brood this can act as a barrier that discourages the queen from moving up – always providing that space still remains below for her to lay in. This barrier of honey is sometimes referred to as a 'honey bar' and is used by some beekeepers as an alternative to a queen excluder.
This can be misleading though because bees can readily turn a comb of food into a comb of brood.

**honey bear**

White, plastic, squeezable, hand-sized bears which become brown when filled with honey. They have an easily removable nozzle on top which is snipped to give the desired flow of honey and can be cut off by releasing pressure on the sides. The honey used should be clear and not too thick. It can be warmed or if needs be the honey diluted.

**honeybee *Apis aculeata, dorsata, mellifera, cerana* and *florea Apoidea***

In that order of size. There are many sub-species and in the case of the commercial honeybee: races, strains and lines. It is a highly social insect with many unique features including specialisation and its spectacular honeycomb. Its main food is honey and pollen though it collects all manner of sweet secretions (aphid and plant), water and propolis and other similar materials. They include three castes: one fertile queen (though exceptionally by man's contrivance more) a few hundred drones during the active season and 50 to 80 thousand workers in the case of *A.mellifera*. The microclimate of the nest is controlled and invaders are unwelcome. A highly specialised flying insect.
See: stingless bees

**honeybee classification**

Kingdom – Animalia, Phylum- Arthropoda, Class – Insecta, Sub-class – Pterygota, Division – Endopterygota, Order – Hymenoptera, Sub-order – Apocrita, Super-family – Aculeata, Family – Apoidea, Sub-family – Apidae (four tribes of which one is Apini), Genus – Apis, Species - mellifera (formerly mellifica), Variety – Caucasia.

**honeybee parasites and predators**

Since Varroa has raised its ugly head it tends to have swamped out many of the diseases that went before but of course they are still with us. The small hive beetle is another that has recently come into view. Each of these will be found under their respective headings.

***'Honey Bee Pests, Predators and Diseases'***

By Professor of Apiculture at Cornell University this authentic work and 'The Complete Guide to Beekeeping, Comb honey production' and 'Bees & Beekeeping' have been co-authored by Ted Hooper.

**honeybee pollinator**

The reason why they are so significant is that honeybees have bodies and habits which fit them admirably to most pollination tasks. Their hairs, although not at first as conspicuous as the bumble bees' are individually branched and ideal for picking up pollen grains. A bee might be foraging for nectar but its body becomes covered with pollen and on returning to its nest-mates, this gets spread around. The pollen intended for their own use is sterilised and packed into pollen baskets and plays no part in pollination being stored safely in cells for the use of nurse bees. Though not quite as ubiquitous as they once were, they still account for the seeding of millions of wild plants which in turn benefit other insects, birds, small mammals and help to beautify the countryside. They submit to management and can be positioned so that vast areas of important crops can receive the benefit of their efforts. Compared with bumbles and wild bees, the honeybee can be vastly increased in numbers if needs be.

**honey beer**

Requirements: 680g/11/2lb honey, Juice of 3 lemons, 28g/1oz ginger, 4.5l/1gal. water.

1. Take 2.25l of the water, add ginger and boil for 1/2 hr.
2. Stir honey and lemon juice in to the remaining cold water.
3. Add the boiled water and ginger to the above mix.
4. Check temperature (not above blood heat) add yeast spread on toast 24 hrs.
5. Strain through muslin (nylon) and allow to stand until sediment settled.
6. Bottle and it will be ready to drink in 2 or 3 days.

**honey bottle**

A pot, a bottle, or a jar. Honey containers. Honey's texture and nature necessitates it being kept sealed and in a wide necked container. During World War II a jar was designed to conform with these requirements and to hold one pound of honey (454g). It has stood the test of time though new types of caps, or lids have come into use. There is a smaller 1/2lb size but glass has in many cases given way to plastic. However glass is washable and re-usable. The threaded neck, standard for so long, is now shared by a partner of a non-interchangeable design. Regulations increase as the years go by and all manner of additional facts now have to appear on the labels. Beebase has a leaflet available.

**honey brassica**

The yellow flowers of the brassica family can yield nectar profusely but being glucose rich it granulates rapidly. It does not do to leave it too long in the hives. The honey is of a light colour with a pleasing flavour and slight mustardy after taste which often calls for blending with another type of honey.

**honey bread**

When making bread use honey at the rate of 6% of the weight of flour. The hygroscopic nature of honey keeps the bread fresh for a longer period than would plain sugar.

**honey butter**

Blend a cup of clover honey with three quarters of a cup of butter or margarine. Store in the fridge when not in use.

**Honey buzzard *Pernis apivorus***

A European bird with brown plumage and white streaks underneath and sting-proof feathers that protect it from the stings of wasps (or bees) which it digs out of their nests in the same way as it does ant pupae. It will also dispose of lizards and frogs - a few pairs have been known to breed in the south.

**honey cake**

The use of honey can improve the colour, texture, flavour and keeping qualities of a cake. Heating tends to drive away some of the natural aroma of the honey though less so when heather honey is used. 454g/1lb of honey is roughly equivalent 340g/3/4lb of sugar and 1.125l/1/4 pint of water in recipes.

**honey canopy**

Under natural conditions bees arrange their brood in compact, concentric ovals with pollen adjacent and a dome of sealed honey overhead and perhaps on the flanks. This dome or canopy is a good insulator having a low thermal conductivity and high specific heat. The cells of honey are often elongated before sealing minimising air-space between the combs. The canopy also tends to act as a barrier to the queen keeping her in the warm 'pocket' below. A reserve of 5-10 kg should always be present with another 20kg for overwintering in the UK. 'The singing masons building roofs of gold"; Shakespeare.

**honey cappings**

Honey-wet wax cappings if allowed to gravity drain for 24 hours or so after thoroughly mashing them so that no pockets of wax harbour quantities of honey, still retain 50% of honey by weight. 8 lb of honey-wet cappings will produce a 'must' when added to one gallon of water for mead making; i.e. 1.82kg/4 lb honey to 4.5l/1gal of water. The clean wax can be rendered into first class wax

**honey carton**

Waxed hardboard cartons printed with decorations are sold holding weights of honey in the region of a pound to a kilo. Screwed caps or press-in lids are alternative ways for sealing but the contents cannot be seen and so rely on the labelling. They are light and strong but cannot easily be re-used.

**honey cells**

Each hexagonal cell is separately sealed and so accessible one at a time. Both drone and worker cells are used for the storage of honey and they normally form a neat pattern alongside and over the brood nest. Once sealed, bees show a reluctance to open cells, though damage a few and they will soon be emptied. The hygroscopic nature of the honey is turned to advantage in the spring when by exposing it to the hive atmosphere it absorbs water which they are likely to be short of.

See: ripening

**honey characteristics**

These are looked for when you are putting honey up for show. Clear honey should have flavour, aroma, density, colour brightness and normal L/D value i.e. glucose/fructose ratio. Granulated honey follows suit with attention to appearance and texture as well. Comb honey must be well-sealed with light cappings and no trace of wax moth or Braula tunnelling or travel staining.

**honey chemistry**

Honey is a supersaturated solution of simple sugars with enhanced flavour due to the extra ingredients that come from the plants and the bees. The chemistry has been closely studied and is a very complicated subject and lists of ingredients and average quantities have been published. The 24 acids have been given under 'honey acids' but for the other 140 different ingredients you could not do better than refer to Doctor Eva Crane's works.

See: honey sugar alcohols, honey anti-biotic

**honey chunk**

Honey containing one or more carefully cut pieces of honeycomb. The honey used should not only be clear but preferably from a high fructose variety resistant to granulation as the beautiful appearance of the honeycomb should not be obscured by any cloudiness and naturally the lighter the shade of honey the better. The clear glass container should be elegant and display the contents to full advantage.

**honey class**

The colours of different honeys range from water-white to practically black. Three divisions of colour; light, medium and dark are used to separate honeys for show purposes and special grading glasses are used by the judges to fail (as they love to) any entry that has been put into the wrong

class. Apart from clear honey other classes include creamed, granulated, cut-comb, sections, a comb suitable for extraction. With a precise 'Pfund grader' we get: extra light, light, pale, medium and dark.
See: grading glasses

**honey clear**

Honey is a clear liquid but it can and will granulate. Furthermore once the beekeeper gets hold of it, all sorts of bits and pieces can get into it to obscure its clarity in one way or another. These include, particles of propolis, wax, air bubbles, even hair or wood and it usually takes some time after the extracting process for a bulk of honey to settle down and release the majority of these other substances by gravity allowing them to rise in the form a froth (or scum) at the top. This method of separation goes a long way towards making it a table food and delicacy but the public always searching for better than best, prefer to see a gleaming, crystal clear, translucent specimen and the way to go about that is take advantage of the characteristic of viscosity. Honey is slow to absorb heat but as it does so the viscosity goes down. Long before extremely high temperatures are reached it becomes so runny that it will pass through the finest of filters. Diatomaceous earth provides a wonderful filter but cheaper methods prevail. Once all air bubbles are cleared and only pure 'slightly changed' honey remains, then it does look beautiful. Clear, golden, bottled sunshine...bye, bye enzymes, diastase, valuable trace elements.

**honey colour**

When in a country like Australia you find a large beekeeping outfit displaying a vast range of honeys from water-white through all shades of amber to tinges of green and pink, mahogany and so on to black, you realise that judging honey by colour is not in itself an identifying feature though when other features such as aroma, flavour, density and taste are considered then most experts can go a long way towards suggesting the origin of a single-source honey. As well as different plants, soils too affect the colour; alkaline soils tend to produce light honey, while the same plant on acid soils gives a darker honey due to the uptake of oxides of iron and manganese.
See: grading, honey light/medium/dark, spectrum

**honeycomb**

Comb is the unique material upon which the lives of bees totally depend. Bees use it in many ways from ripening nectar into honey, rearing brood in three types of cell, worker, drone and queen, storing pollen, storing honey and clustering in during winter and as a signalling board – basically as a home. This is honeycomb. When we talk about comb honey, cut comb, supers of comb, sections, we are then referring to our food, the way we look at the bees' products. How they ever got together to produce this amazingly intricate structure has puzzled scientists and is an object worthy of study. It commences by younger bees (say 5-15 days), huddling together after gorging themselves with carbohydrate and developing their wax producing glands. The wax is formed in platelets which they hook out from the four pairs of wax pockets and proceed to masticate into a workable white substance which then becomes communal property as the so-called masons among them set to work constructing the rhomb-based, back-to-back, hexagonal cells of the vertical, parallel combs. When melted and filtered they yield what we call beeswax.
See: comb foundation and septum

**honeycomb moth**

Honeycomb is soon destroyed by wax moth larvae, if it is left unprotected by bees, at temperatures where the wax moth life-cycle can occur. The moth is naturally born to be able to evade bees and often gets into hives at night. Her eggs are very small and barely noticeable and combs put away (even sections once out of the hives) can become alive with the grubs which savage the comb, build thick masses of web, drop their black faeces and unless dealt with by the bees can even drive the bees away and take over. Gaseous treatment becomes essential. Strong colonies help as does making sure no unprotected comb (or pieces) are left around. See: Galleria

**honeycomb section**

Basswood strips that could be folded and dovetailed into four and a quarter inch square boxes, became known as 'sections'. Using extra-light foundation or strips of wax the bees can be persuaded to put one pound of honey into them. There is no pretending that the bees like these tiny boxes, preferring long, deep combs themselves. However by crowding bees and putting the sections on when a flow is imminent, it is possible to get them drawn, filled and sealed. A box with same peripheral dimensions as the hive but filled with rows of sections is referred to as section crate.
See: Cobana, section holder

**honey composition**

As honey is often brushed aside as just being sugar it is worth looking at the difference. Ordinary white sugar is pure sucrose, sucrose and nothing else. Dissolve it in water and it reaches a saturation point at 66%. Now to start off with, honey is 80% sugar so the bees have developed a useful trick for getting their food supply preserved and more packed into the same space as would be possible with sucrose. The uniqueness of honey has foiled all attempts to copy it and although several poor substitutes are available, they are no rival for pure honey. The multitude of ingredients belie the fact that it is mainly water, fructose and glucose but apart from a brief reference here look elsewhere for honey ingredients. It has minerals almost equivalent to those in our own blood, amino acids, essential oils, enzymes; like the structure of honeycomb, it took scientists ages and ages to work out even part of it.
See: honey composition, ingredients etc. etc.

**honey condition (can be impaired by)**

Excessive moisture, air-bubbles, froth or scum, specks and wax particles (foreign matter), damage by heating, chemical taints (smoker) and fermentation.

**honey constituents**

'A Comprehensive Survey of Honey' by Eva Crane, MSc PhD, lists 181 substances known to be present in honey. By percentage, fructose 22-54%, water 13-26, glucose 20-44, sucrose 0-7. This leaves about 7% for pollen, aroma constituents, enzymes, vitamins and minerals.
See: honey ingredients

**honey contamination**

The spoiling of good honey. Substances used for subjugating the bees can easily contaminate honey. Admixtures of cheap sugars or the addition of other ingredients is also possible. Heat, moisture and fermentation are other aspects of items that will spoil honey if allowed to do so. The greatest care should be exercised to keep honey in the condition that the bees achieved, avoiding the addition of any foreign matter by any means whatever.
See: honey condition and definition

**honey cream**

Heat clover or similar light honey to 54-60C/130-140F and mix thick cream at a rate of two parts honey to three of cream. Blend carefully and store in wide mouthed container in the fridge. From Honey & Your Health – Cecil Tonsley.

**honey creamed**

Granulated honey that has been warmed then whipped or stirred vigorously until the crystals are uniformly small and the texture is akin to butter. It should have an attractive appearance and be free of air bubbles while having a well-pronounced flavour. It is a popular form as it can be handled more easily than liquid honey which may run or drip.
See: honey cream

**honey crop harvest**

That part of the honeybee's stores that are removed by the beekeeper. In addition to its own day-to-day requirements a colony might develop 100kg of surplus in a complete year. A beekeeper in the U.K. is usually quite happy if he can take a reasonable amount of this he can leave sufficient food for the bees to over-winter. Migratory work or other reasons might make it wise to remove honey before the end of the season, but regardless of all else, always make sure that the bees have the recommended 20kg, even if this has to be arrived at by feeding them up to 50% of sugar (in the form of thick syrup fed in good time). Sugar is a pretty useless food for us, but for the bees it is the only substitute that their specialised gut can take.

**honey crop sac**

The part of a honeybees' alimentary canal which acts as a storage 'chamber' from which the contents can be regurgitated if required. It has gone by various names – crop, sac or bag and it size and likely contents are described under honey sac.

**honey crystallisation**

The ratio of the two simple sugars in honey vary according to source. It is the glucose part which is less sweet and is inclined to precipitate crystals. In clear honey some form of 'nucleus' or starting point, is required to set a crystal forming. Once this begins it is a continuous process and will finally result in a mass of glucose crystals surrounded by a thin film of fructose. Temperature plays a part. For instance one of the latest flows comes from ivy which is rich in glucose and temperatures are low in late September when it goes on yielding, The result is that bees have been found with their honey sac clogged with crystals. The only recourse is to feed dilute syrup. If the first crystals are small, similar ones may follow and 'seeding' is a manner or triggering off a smooth granulation and 12C/54F is the ideal temperature for this. Once granulated the honey becomes noticeably lighter in colour. For 'showing' be sure to examine the jar from underneath because the tiniest speck of material will be visible there. A short spell in the micro-wave will change inedible comb honey from its gritty, crystalline nature into a cake of wax floating on a sea of clear honey. Such heated honey is known as Baker's honey.
See: creaming

**honey crystals**

When large crystals form this is due to the glucose portion of the honey and they can be unsightly and reminiscent of sucrose when eaten in this condition. Some of the best honeys granulate like this. There are two ways of improving the situation, one is to warm the honey (38C/100F should be sufficient – more heat causes valuable aspects of the honey to diminish or disappear) or otherwise very determined stirring will whip it into a creamy texture akin to butter. This can be done with a fork in a pound-sized jar but do a layer at a time unless you are very strong. In a 30lb (approx. 13.5kg) bucket however gentle heating over-night say with a 25Watt bulb (this was with old type incandescent bulb-other heat sources may have to be found) in a disused frig or other sealed unit, will make it soft enough to stir with a strong kitchen utensil. 'Seeding' is another approach. It implies using a sample of smoothly grained honey. First, using as little heat as possible, clear the honey to be seeded. Then when cool, some can be put into a convenient container, bowl or wide-necked jar so that stirring is easy. Add some of the smooth sample and stir until it is well mixed and can be poured into the cool, clear sample. Half-a pound mixed into a 30lb bucket and then kept at as near 18C/65F as possible, giving an occasional stir as it gets under way, then both taste and texture will be found most acceptable.
See: glucose

**honey dark**

Colour is in no way indicative of the flavour. Honey-dew honeys can be dark because where sweet secretions are gathered from leaves. This is where the 'soot' fungus is found and its black colour gets into the honey. Plants grown on acid soils also tend to give darker honey. Plants known to produce dark honey are buckwheat, blackberry, cherry, hawthorn, heather, broad bean and plum. If customers are put off by the colour the best thing is to do a swap with a friend who has light honey and do a bit of blending.

**honey definition**

Poets, scientists, encyclopaedists and government departments have all had a go at defining this wonderful, natural substance. You will find our definition under 'honey' qv.

**honeydew**

Sucking insects of many varieties such as aphides bite into plants and in order to gain sufficient nourishment themselves, produce a surfeit of carbohydrate which they forcibly eject as a spray which falls onto the upper surfaces of leaves, twigs etc. (e.g. lime trees). Rain can bring such a 'flow' to a halt. In countries like NZ where a rather special insect nick-named the 'beech buggy' copiously ejects droplets of clear honey-dew which make the bark of the Southern Beech look as if covered in fur. The droplets cover everything below, including street signs which all become black from the activities of the soot fungus. Although clear when first collected, that honey-dew gradually darkens when exposed to air and light. Forests in Europe have annual migrations of beekeepers all anxious to get the very special honey-dew from the pine trees. Where honey-dew is the main crop bees might require help with pollen. Wasps, that can overwinter without hibernating, become an absolute pest in their millions when so much honey-dew is available. The honey made from such sources is rich in colloids and these improve the health-giving qualities. Bumbles have been found rolling on the ground inebriated, under Tilia tomentosa and other late flowering limes due to the honeydew containing melezitose. There are very rare instances where this symbiotic link with unusual plants and their sucking insects can lead to poisonous honey. Much more honey-dew honey is produced than we are generally aware of.

**honeydew insects**

These are very numerous and are mentioned under several headings. In Britain aphides and Lachnids operate on lime and pine trees, in Europe Psyllids, scale insects (Coccoidea), aphids (66 species), 20 Lachnids, 9 Chaitophoridae, 15 Callaphidadae, 10 Aphididae, 4 Thelaxidae and 3 Pemphigidae all sucking away at the phloem sap – mainly on fir trees. See: scale insects, sucking insects

**honeydew symbiosis**

Hemiptera (Rhyncota), *Buchnaria pectinatae* yields on silver fir, *Cinara pinciola* on Sitka spruce, (*Picea sitchensis*) in Scotland –honey with good flavour which crystallises to a pale colour (similar comments made about oak) *Cinara acutirostris* on *Pinus nigra*, Lime trees *T. petiolaris* due to melezitose and mannose, Marhalina (Greece), Ultracoelostoma NZ.
See: baum, cereal, wald and tree – honey.

**honey dilatant**

Honey which has the property that enables it to be drawn out into long 'threads'.

**honey dispenser**

Plastic squeeze bears and other containers that can be filled with liquid honey and used to release a small, controlled stream of honey as required. The honey runs more easily when warm or when dilute and steps should be taken to prevent thickening or granulation of the honey inside the dispenser. Jugs with cut-off flaps and sealable tubes are other alternatives.

**honey drinks**

As honey blends so well, flavour-wise, and physically, with so many drinks both hot and cold, the list of these would have to be endless. Lemon and milk can be served hot or cold sweetened with honey. Honey dissolves more easily in hot liquids. It can be used with spirits. Drambuie, a famous liqueur tastes like whisky and honey, dry wine responds fruitfully to the addition of honey, tea, coffee fruit juices and milk drinks can all be flavoured to advantage. Warming or prolonged stirring may be necessary to ensure that a layer of undissolved honey is not left at the bottom of the drinking vessel. Drinks offer a very satisfactory way of getting this valuable food substance into the body.

**honey dryer**

Honey containing more than the regulation allowance for moisture can be dehumidified by blowing warm dry air through the supers. Special rooms (Arizona or 'hot' rooms) with dehumidifiers and special ventilation to carry off moisture laden air.
Special electrically heated straps have been successfully used around containers and used in conjunction with fan heaters. Care must be taken not to over-dry the honey.
See: ripeners

**honey eater oscine Family Melphagidae**

Birds of some seventy species found in Australia. The tongue is like a small paint brush and mops up thin fluid with ease. Pollination is helped by brushing pollen from flower to flower with the wings while they search for food. Example the Eastern Shine-bill. The bill and tongue is adapted for getting nectar out of flowers.
These birds are strictly speaking nectar eaters (not honey), should we point this out?

**honey electrical conductivity**

The resistance of honey to the passage of electricity and vice versa the ability to conduct electricity varies over a wide range. Heather honey is on the high conductivity side of flower honeys but honeydew honeys generally have a high rating which tends to be linked with their unsuitability as food for overwintering colonies in cold climates.

**honey elements**

Magnesium - Mg, Sulphur - S, Phosphorous - P, Iron - Fe, Calcium - Ca, Chlorine - Cl, Potassium - K, Iodine - I. and Sodium - Na.
See: elements, trace elements

**honey enzymes**

Complex, organic substances originating from living cells and capable of producing, by catalytic action, certain chemical changes such as digestion in organic substances. They are present in various nectars and the bees add enzymes, particularly invertase to turn the sugar content of the nectar into the fructose and glucose of honey. Diastase is used to measure the degree of heat to which natural honey has been submitted and its level must not fall below a certain threshold if it is to be allowed entry into EEC countries. Glucose oxidase has been shown to play an important part in honey's inhibitory role. Peroxidase, lipase are all involved.
See: enzyme

**honey Erica**

Erica is a large family of plants but the 'heath' is the best known yielder of a dense honey with the dark reddish colour of port wine and with a pronounced and pleasant flavour. It can be extracted centrifugally (whereas the ling honey is thixotropic) but it is thick and dense. The nectar flow comes in late summer and can extend into the heather flow often imparting a decidedly pink colour to the resultant heather honey.
See: bell, Erica, thixotropy

**honey Eucalyptus**

There are so many trees of this species (more than all the European trees put together) that to assume set characteristics for their honey could not be accurate. As Australian honey is blended for export much of which arrives in the UK is of medium colour and with a characteristic flavour. The word eucalyptus has become associated with the oil derived from the leaves of some species but this of course, has no connection with the flavour of the honey. Quite a number of these trees are now grown in the UK but only yield, if at all, in warm conditions.

**honey exhibiting**

Ever since beekeepers formed groups or associations there have been those who wished to compete to produce the best tasting hone, the nicest looking comb and so on. When you take the trouble to exhibit honey so that it is carefully examined by discriminating judges and put on full view to the public as well as the critical eye of beekeepers, then experience is gained which is for the benefit of the craft and helps to maintain high standards and to sell honey to the public.
See: showing

**honey extracting house**

A place arranged so that supers of honey can be moved in with a minimum of lifting, the honey extracted and filtered into containers and the wet supers returned to the bees or put into storage. Temperature control is needed and the whole place should be well-lit with bee-proof screens on windows and doors to the open. Much thought has been put into the lay-out of such a room or 'house' and a work plan must be facilitated by an arrangement of the various pieces of apparatus so that all movements can be carried out in a continuous sequence. All surfaces should be smooth, hygienic and easily cleanable. The heavy supers will be at the starting point, their uncapping must allow for the collection and processing of the cappings, the extractor should be adjacent and everything arrange so that the movement of honey is 'downhill'. The extractor needs a firm base and must be at a height that allows for the filter and containers to be put directly beneath the honey tap. Replacement filters should be to-hand and all containers should be scrupulously dry and clean. The type of extractor, and other pieces of equipment will be dependent on the size of the whole operation. Before constructing a shed or house for this purpose the viewing of one or two existing establishments is advised.
See: honey house

**honey extractor**

Since Hrushka first designed a honey 'slinger' great strides have been made both with the extracting apparatus and the items to be extracted. The majority of machines today hold the uncapped combs vertically in cages that can be rotated by hand or electricity so that the honey is expelled within a cylinder, and runs down the inside walls to a honey gate or special snap cut-off tap. As the speed of rotation increases so too does the force of expulsion (centrifugal) and combs reinforced by wires can withstand considerable force, provided they are held against the mesh of the cage. High revolutions or a gap between comb and cage can cause damage to the combs. Some extractors will have a reversing device but most will necessitate the turning round of the comb so that both sides are emptied. The operator should wear suitable garb (to catch your tie in the works might prove fatal!). The extractor size has to conform to the size of the frames, British Standard, Langstroth etc. Some extractor types include: centrifugal, hand or power operated, radial, reversible, tangential and press.

**honey fermentation**

Honey is a supersaturated solution of sugars and on average contains around 18% water. When this moisture content exceeds 19% and some honeys may do so, then fermentation is possible. Between the temperatures of 12-38C/55-100F live yeast is always present, in a highly sugar-tolerant form, unless it has been killed by heating. Honey also has the characteristic of being 'hydroscopic' meaning that when exposed to air with a vapour pressure greater than that which exists at the surface of the honey, it will absorb moisture. Because of this in the case of honeys liable to fermentation, a layer of fermenting honey may be noticed, with a 'winey' smell. Unless the honey has been

disturbed this top layer may be quite shallow and the honey below still normal. Heather is a honey liable to fermentation on account of its higher water content and sometimes a glucose rich sample of normal honey may begin to crystallise and in so doing yield some of its water to the remaining fructose portion which will remain clear and might ferment. When this condition occurs it is not uncommon to find a jar half-and-half with crystallisation at the bottom and clear honey on the top. When it is desired to allow honey to be fermented into mead the highly sugar-tolerant yeast (which does no favour to the subsequent mead's flavour) is destroyed by raising the temperature of the 'must' (1.3-2.2kg honey to 4.5l water) to 38C and then yeast of a known variety added.
See: fermentation, mead, vinegar

**honey filter**

Although honey is pure and clean, once it has been handled by a beekeeper it is likely to contain particles of wax, propolis and other material with which it has come in contact. Such solid particles can be filtered out first using a coarse filter to remove larger bits of wax etc. which might block a finer filter and can take place fairly rapidly. Then the honey can be passed through a straining cloth or finer filters according to the beekeepers needs. The finest filters slow up the work unless the honey has undergone some degree of heating. Most commercial bee farmers will be more interested in the importance of appearance so most shop honey will have lost some of its valuable ingredients though must conform to the standards regarding HMF and diastase. Once extracted a honey tank or 'ripener' so-called will after a day or two allow fairly clean honey to be run off from below while the scum (froth) will stay on top and line the sides of the tank as the clean honey is run off.
See: showing

**honey flavour**

Your very first taste of honey may bias you in favour of that particular taste. Otherwise peoples' tastes vary considerably, some like a mild or bland flavour and others something exciting. Mishandling, such as by heating, can drive away the honey's natural, distinctive, pleasant flavour but categories are used to describe flavours thus:- extra delicate, delicate, mild, medium, strong and extra strong while specific terms such as minty, mustardy, spicy etc. will also be found.
See: aldehyde, ester, flavour, gustatory, taste, organoleptic

**honey flow**

An observable increase in the amount of honey being stored. This initially shows as white extensions to the cells in the upper part of the comb (known as drawing), then by much uncapped honey which gradually becomes sealed stores. According to their timing, flows may be described as early, main or late. A specific flow such as a clover flow would be related to a crop that yielded to the exclusion of other sources. The pre-requisite of a good flow is good weather, nectar bearing flowers and strong colonies of bees. 'Migratory' is the term given to beekeepers who move their colonies to areas that are expected to give a good flow – such as heather. It is not easy to quote figures for the speed at which supers can be filled but from foundation to a completely sealed super (BS shallows) has occurred in a day or so while world records would be up as high as two deep Langstroths in little more than a day or so from Eucalypts or Willowherb.

**honey, tree or forest**

In Britain we get tree honey from Sycamores, Sweet Chestnuts, Acacia, Limes but as to forests, only where a reasonable amount of heather exists could we speak of forest honey. Yet in Europe an annual migratory move to the forests is normal and it comes (apart from acacia), more from sucking insects than from the flowers of trees. The Australian forests are an example of enormous amounts of honey from the great, overhead canopies of the Eucalyptus. Doubtless in Africa and South America the

forests may produce flows but here only certain arable crops, orchards, the moors, parks and gardens are our honey producers.
See: migratory beekeeping

**honey fudge**
A sweet candy made from ingredients including honey. Here's one recipe:
454g/1 lb. of granulated sugar, 57g/2oz unsweetened chocolate, 1/4 teaspoon salt, 250ml/1 cup of evaporated milk, 2 tbsp. butter, 1 cup of nuts.
Method: Boil the first four ingredients together for 5 minutes then add 62ml/1/4 cup of honey and cook until soft ball stage, that is 115C/240F. Add 2 tbsp. butter and allow to stand until lukewarm. Beat until creamy, add the cup of nuts and pour into a buttered tin, cutting into pieces when firm.

**honey fungus *Armillaria mellea* Basidiomycetes**
Bootlace fungus. Its only link with honey lies in the colour of its spore-forming toadstools frequently seen growing on old tree stumps. It is an edible fungus but difficult to eradicate as it spreads underground with black 'bootlace' threads causing great damage to woody tissue killing trees and shrubs. The mild South West seems to favour it.

**honey gate**
This is an alternative name for the tap used to control the flow of honey from a tank, extractor or other large honey container. Set in the bottom of the tank they are designed to allow positive, easy opening and closing with full and intermediate positions so that high or low rates of flow can be established, yet readily cut-off without subsequent dripping. One such is of plastic with a honey-seal insert and all that is required is to cut a 1 and l eighth" hole in the side of any plastic container suitable for honey and screw the gate into position. Leave room at the bottom for tightening the large, plastic nut.

**honey grading**
On the international market honey has to conform to certain standards of purity, density and quality and is then graded by colour according to the pfund scale or more recently by the honey colour comparator ranging from 1 – 140mm. - into water-white, extra white, white, extra light amber, light amber and amber to dark. To those of us who know and value good honey it comes as a slight surprise to learn that the highest prices are paid for the light honeys.

**honey grading glasses**
These can be used by judges and others to classify which class a honey should be placed in. They consist of two squares of tinted glass. One glass has an amber shade which indicates the darkest end of the light class. The second glass indicates the darkest shade in the medium class. They need to be used against a good light source with a white background. A small wallet is provided to safeguard the two grading glasses.

**honey granulated**
Most clear honeys will granulate, some quicker than others. Glucose-rich honeys go first. Granulated honey usually has a lighter shade than the clear honey from which it was formed. Classes for granulated (or set) honey favour an even, fine grain with light colouration. No liquid honey must be visible. Foreign matter disqualifies and judges usually look through the base of the jar where specks of wax are often overlooked by the exhibitor. Froth or uneven granulation would lose points.
See: honey crystallised, creamed

**honey guide bird Indicator spp. Indicatoridae**
Found in southern and tropical Africa, Asia and East Indies. About 200mm/8" long it is a parasitic bird like the cuckoo laying in the nests of other birds that use holes – woodpeckers, bee-eaters, kingfishers. It makes a harsh rattling call and flies from tree to tree ahead of its collaborator (large

mammal/man), becomes silent and when the mammal has moved closer, continues and so leads it to a bees' nest and when the nest has been exposed will eat pollen, brood and combs, having the ability to digest wax and preferring these to honey. Varieties include *Indicator indicator, I.predotiscus* and *I.melichneutes robustus.*

**honey hand cream**

Mix 115g/1/4lb softened lard with two egg yolks; 1tbsp honey; 1tbsp ground almonds and 1tbspn rose water. Add a few drops of almond essence and work into a stiff paste. This is an early Victorian recipe and was said to work wonders for chapped hands.

**honey harmful**

If a bee should collect some harmful honey it is unlikely to make it back to the hive and if it does the chances are it will be rejected. However there have been reports as far back as Xenophon and recent proof that 'tutu' (Coriaria spp) could yield a toxic honey-dew. Thus beekeeping is forbidden around the area of the Bay of Plenty in NZ where the plant is found. Headaches and nausea have been attributed to the eating of the poisonous honey.
See: honey-dew, nectar, toxic honey

**honey harvest**

Honey successfully taken from the bees. In exploiting bees man also husbands them, helps them to avoid problems and to find suitable forage. Efforts to increase national output receive many set-backs due mainly to variations in the weather. World honey production however, steadily increases. Bees are always ready to take back honey so never risk leaving the slightest crack or hole through which bees (or wasps) could get to the honey you are trying to take off.
See: harvest, honey crop

**honey heather**

Also described under 'heather honey' this famous product is a world-beater when it comes to quality and flavour. In its pure form it remains a 'gel' with imprisoned air-bubbles throughout though some modern methods of spinning the honey out with a high speed centrifuge, can result in an amber honey that is bubble free. It has a higher water content than other honeys which needs to be taken into account when storing as it can lead to fermentation by the highly sugar tolerant yeasts that are present in all honeys. In the form of comb honey it looks and tastes delicious. It out rivals Manuka, a similar honey, and though selling at a far lower price still earns the right to be sold at higher prices than ordinary flower honey. Its retentive aroma of the moors is resistant to heat in so far as cakes, mead or other cooking processes do not take away the obvious presence of the honey.

**honey-home**

A hive moulded with integral skin from structural polyurethane. It was claimed to be lighter, better insulated and to have a harder surface –giving immunity to woodpeckers and to last longer than the more conventional Western Red Cedar.

**honey house**

A building used for the extraction and handling of honey. As honey is heavy to handle it helps when loaded transport can have level and easy access to the uncapping machine. Sizes may vary according to the scale of the operation -government help is given in some countries where even a 500 colony farmer may be handling over 100 tons. Lay-out is important so that the honey is kept moving: uncapping, extracting, filtering and containerising. Warmth and humidity control and he exclusion of flying insects are essential. Hygienic cleanliness, good water supply and conformity to local regulations are further considerations.
See: extracting house

**honey hydrometer**

A hydrometer is an instrument for measuring the water content of a substance and a honey hydrometer is graded to cover the range of specific gravities (densities)

of various honeys. Made of glass or plastic it is inserted into the surface of the honey where it remains upright and finds its own level against which the specific gravity and thus density of the sample can be read - S.g. honey 1.414. So 4.5l (1gal) of honey weighs just over 6.3kg (14 lb).
See: density, s.g., viscosity

**honey hygroscopic**
Sealed in the comb and kept in the hive with the bees, honey has a remarkably long life. Some races leave an air-space under the cappings (which then look beautifully white) and this will have a set vapour pressure and help insulation. Once extracted and sealed it has an even longer life provided the water content is reasonably low. Once uncovered the surface layer will absorb moisture from the air whenever (as is usual) the vapour pressure (relative humidity) of the air is high. Conversely honey can be 'dried'. Not only does this property have a detrimental effect if it leads to fermentation but it is extremely useful when applied to the skin or wounds because it can draw inter-cellular fluid into itself and being antiseptic is safe for this purpose. Manuka honey has gained fame in this respect, partly because it comes from the Tea Tree and especially because, like heather, of its gelatinous texture which helps when applying it to the body.
See: honey moisture equilibrium

**honey ice-cream**
The flavour of ice cream can be much enhanced by the addition of honey while it is being mixed. Ordinary ice cream can, in the opinion of many, be given an even better taste when a little honey is drizzled onto it. A combination of ice cream and honey with ordinary cream, banana, nuts and such like, can lead to the production of exciting, pretty, healthful sundaes.

**honey judge**
A person chosen by virtue of qualifications, to act as the arbiter at a honey show. The BBKA have an examination for judges and initially anyone intending to take it should enter in the various classes of shows and act as a judge's steward whenever this can be arranged. The National Honey Show is used for this purpose but local shows too can offer valuable experience. Judges and stewards wear a white knee-length coat and have a kit including a torch, tissues, magnifying glass etc.

**honey judge's exam**
The BBKA exam qualifies one to call himself a BBKA Honey Judge. A precursor is the holding of a certificate higher than the Preliminary, evidence of being a successful exhibitor and having served as a steward in at least two large shows. The exam is a combined practical and oral test where the candidate will be expected to have a wide knowledge of bee products and appliances and to be aware of common faults and mistakes, to be conversant with show rules and to understand methods of allocating awards.
See: grading glasses, honey hydrometer

**honey judging**
At British honey shows, apart from comb honey, classes call for two standard squat 454g (1lb) jars. They must be identical, clean and filled correctly and correctly graded in light, medium or dark classes. Granulated honey must have no liquid, nor liquid honey any granules. Appearance, aroma, density, flavour and conformity to the requirements of the schedule, enable the judge to arrange the winning entries, his own personal opinion finalising. Honey is handled with scrupulous care with the aid of a steward whose opinion may be sought. Judges are often asked to range far beyond honey and wax to cakes, sweets, wines, - in fact ambitious schedules have classes that give almost unlimited scope for material connected in any way with beekeeping.

**honey keeping qualities**
Ripe honey can be kept in air-tight containers for a long, long time. Some darkening and either partial or complete crystallisation is likely to occur with age

depending on the temperature at which it is kept and the relative amounts of sister sugars that are in it (L/D ratio). Honey that has crystallised (granulated) can be cleared again by warming as that reaction is reversible but an hour at 48C (120F) is better than trying to hurry it. The specific heat of honey is very high meaning that it takes a long time to heat up and to cool down again. The laevulose: dextrose ratio is the decisive factor as regards granulation and there are instances where the dextrose (glucose) portion begins to crystallise and the crystals in forming release a tiny amount of water. This goes into the laevulose (fructose) portion and can turn it from the ripe condition to one where fermentation might occur. In any case when this separation does happen the dextrose crystals lie at the bottom of the jar while a clear layer of the sweeter fructose is found at the top, half-and-half.
See: honey fermentation

**honey label**
EEC regulations stipulate that honey labels must bear the name and address of the person responsible for it, i.e. producer, packer or seller. The description of the honey can be simple or specific e.g. Pure English Honey or Exmoor Heather honey but the contents must be mainly from that source and not contain honeys gathered elsewhere. In other words the label must truly state what is in the container. The size of the lettering also follows a set scale depending on the size of the container and the weight must be clearly stated in grams as well as other avoirdupois.
Consult Beebase for the latest regulations.

**honey leaf**
Honey derived from material collected by the bees from the leaves of plants. This expression has been occluded by the widespread use of honeydew to cover sweet secretions other than from floral nectaries.

**honey light**
Honey within the range from water while to medium amber. The actual line of demarcation being determined by the use of an official grading glass. Light honey is popular on the show bench and the grading for export is based on lightness. While light honeys a can have a pronounced, delightful flavour, darker ones often taste better. A light honey can display chunk honey better and granulates with a nice looking white grain. Very light honey should be kept separate from dark honeys which would spoil its appearance.
See: light honey

**honey liquid**
Runny honey or clear honey as expelled from the honeycomb. Honey can of course be eaten in the comb or in the creamed or granulated form. Liquid honey is considered to be brighter and clearer when extracted from combs that have never contained brood. Liquid honey has variable traces of pollen, wax or other substances in it. Sources may be indicted by the pollen. As it is a supersaturated solution of sugars granulation is always possible. Its density, viscosity and L/D ratio will vary within reasonably narrow limits.

**honey loosener**
When thick honey in worker comb is to be extracted it is best if this can be done while the honey is relatively warm. In the case of heather honey which comes in a jelly-like state agitation by pronged gadgets or microwave might help with the extraction though it is more normal to use sections or cut-comb or else to use a heather press to extract it from the combs.

**honey marketing**
Whereas overseas honey which outnumbers British honey perhaps tenfold or more, is imported, it is handled by large organisations such as Kimpton Brothers Ltd. The marketing of our own honey is usually done by the producer. Few are able to indulge in large-scale advertising but as only four thousand tons of superior British honey is likely to find its way onto the

market no great difficulty exists in selling it.

**honey mask**

Whip together egg white, honey and barley flour. From Australasian Beekeeper.
See: face pack

**honey mineral content**

The marvel of honey from the point of view of health is the value of its trace elements which in conjunction with enzymes enable all parts of the body to benefit especially because of it generally easy digestibility. The number of parts per million give no idea of the catalytic effects that such small quantities of matter can have. Don't knock any of these out by heating.
See: honey elements

**honey mint sauce**

Half fill a jar with clear honey and warm it slightly – too much heat destroys the flavour of the honey. Stir in as much chopped mint as it will take. This will keep very well and it is delicious with roast lamb.
See: honey mineral content

**honeymoon honey-sweet, moon-month**

Mead was drunk at weddings for the state of the moon in the days of King Alfred the Great (8th century). That period of time when newly-weds spend their moments in happy and harmonious relationship. That sweet, transient period when, as at the full moon, the tide of love is full. The moon was thought to be a cup of honey and the term honeymoon has become associated with fullness and sweetness.

**honey moth *Achroia grisella* Pyraloidea**

Lesser wax moth. Their larvae shun daylight, feign death. The larger wax moth *Galleria mellonella* is sometimes seen hovering around hive entrances at dusk.
See: wax moth

**honey mouth**

The bee's stomach mouth or honey stopper.
See: proventriculus

**honey's optical properties**

Density, fluorescence, opacity, optical density, optical rotation, and refractive index.
See: polarimeter

**honey organic**

To claim that honey is organic implies that the beekeeper has much greater control of the bees' movements than is reasonably possible. Nevertheless efforts to exclude artificial inorganic substances from honey are to be complimented.

**honey pasteurizer**

Most germs are killed by honey or cannot live for long if surrounded by it. However botulism has been shown to survive and though much is lost by flash-heating honey the pasteurising process is used, especially when super-bright honey with a long shelf-life is required. Pasteurizing is achieved by passing honey through a narrow gauge tube where it is flash-heated and rapidly cooled so that it does not darken noticeably.
See: heating honey

**honey physical properties**

When ripe honey is freshly stored in the comb it is a viscous, sticky liquid – a super-saturated solution of monosaccharides. The main monosaccharides are the two simple sugars (hexose sugars with a molecule half the size of sucrose) and they have names according to their ability to rotate a plane of polarized light: laevulose to the left and dextrose to the right. This enables a distinction to be made between honey and sugar. Dextrose is also called 'glucose' and is responsible for the granulation aspects and is not as sweet as its sister sugar laevulose also named fructose. The average water content is 17% and the specific gravity (density) is 1·414. Water weighs 1kg per litre, honey c. 1.4kg. When honey is heated its 'specific heat' makes it slow to absorb heat and then to release it, making honey a useful baffle against temperature changes. Heating however has a marked effect on its viscosity which decreases until the liquid is so 'runny' that it will go through extremely fine filters. It has characteristic aromas which vanish when heated while its very special flavour continues to survive. Other interesting features are its electrical and thermal conductivity and refractive index.

**honey pigments**

Pigments that give various honeys distinctive colorations are very often derived from the plant's pollen. Saffron from Crocus sativus.

**honey plant, equipment**

The various kinds of apparatus (devices) on the market which uncap, extract, filter and containerise honey become more and more sophisticated. All levels of beekeepers are catered for from hobbyist to honey farmer and there is keen competition at International Beekeeping Congresses for manufacturers to have their particular make of equipment adjudged as best. Gold or silver awards are proudly won and Britain has its fair share of credits but there are some good firms abroad.
See: honey processing apparatus

**Honey possum *Tarsipes spenserae***

A marsupial that seems to feed on nectar.

**honey potential**

When a given crop is examined to determine the maximum potential honey return per hectare, useful information emerges, though variations will occur according to place and season. For instance in Scotland, oil-seed rape has been compared with ling heather as yielding up to 50-250kg /110-550 lb per hectare.
See: pollen potential, yield of honey per acre

**honey press**

The squeezing of honeycomb to force out the honey is rather wasteful of both honey and comb. Honey presses are not generally used for flower honey but for the thick (gelatinous) thixotropic heather honey, which will not budge from the comb when submitted to normal extraction processes, so a honey press is necessary. One type has a strong turn-screw which enables enormous pressure to force two teak draining plates together. The comb, or waxen mush, is enclosed in a filter cloth of strong material such as nylon. The honey is extruded, complete with air bubbles, for containerization.

**honey processing**

The bees process nectar into honey which is straight away edible but in order to get it into containers, contamination from air bubbles and processing apparatus calls for special treatment to increase shelf-life and make it visually attractive to the customer. While a 'rough and ready' treatment by means of gravity separation in a large, narrow receptacle will separate the froth from the clear honey which can be tapped off after a day or so of settling, further steps are necessary to improve its appearance. Of course the beekeeper has the benefit of being able to eat the honey in the relatively unspoilt condition where the removal of the froth means that most of the bits and pieces (comb, propolis, bee's wing or leg) will have been cleared away. However to conform to regulations and expectations more elaborate steps are taken. Now the object is to make the honey 'look' nice. Advantage is taken of the fact that the honey's viscosity diminishes according to how much heat is applied. It is not necessary to raise the temperature much before it becomes liquid enough to pass through ordinary straining cloths of muslin, nylon etc. but to drive off all air bubbles (which incidentally precipitate granulation if left in the honey) further heating may be required to pass it through the finest of filters and produce that brilliant, translucent, golden quality which makes it a winner on both the show bench and the super market shelves. The colour and smoothness of granulation will depend on source and even dark honey's benefit by becoming lighter when granulated. A further step in the process can be to cream

the honey. Vigorous stirring is one way, but gentle heating and stirring is easier and makes a lovely, smooth-as-butter texture which has an enhanced flavour by virtue of the tiny crystals going straight to the taste buds in your mouth.
See: creaming, extractor, filter, processing apparatus, seeding, uncapping knife

**honey production**
There are of course many excellent reasons for keeping bees as well as for honey production. Anyone who succeeds in making a living out of bees in the UK will not do it on the sale of his honey alone but will have to engage in many associated pursuits associated with beekeeping. He is likely to buy and sell other people's honey and perhaps to sell bees, queens and equipment, to market all hive products and their derivatives and maybe to engage in pollination work. To get honey from bees year in and year out requires much skill and hard work. Competition with say, the Chinese, Russians or Americans doesn't come into the picture. At home you would need to run several apiaries, to engage in pollination contracts and migratory beekeeping where you chase the crops as and when they become available. Why not combine it with holidays for instance having them aboard a barge on the canals?

**honey pump**
A heavy duty pump capable of driving honey through pipes. Where honey farmers handle large volumes of honey such pumps are essential to get the honey from the spot where it has been separated from the comb and cappings. Such pumps require warm honey (due to the high viscosity of cold honey) and thermostatic arrangements are associated to ensure constant delivery from the pumps.

**honey quality**
Density, appearance and flavour can be readily examined but when it comes to nutritive properties that is not so easy to identify. For instance merely being light in colour puts it up to the top for market prices, yet all the while a new gimmick name or title sends the popularity and price of honey soaring. At one time Salvation Jane (blue weed) an Australian honey which beekeepers had been feeding back to the bees, suddenly took on a new significance and was being sold at Fortnum and Masons as a delicacy. Then along came the Southern beech honey-dew honey and this went to the top, more recently honey from the Tea tree (Manuka) went soaring because of its medicinal use in hospitals. Honey has been a natural food substance since the very first man tasted it. Its very name is welded into literature and history and makes any product more attractive if it is entitled to say 'with honey'. In truth honey is a valuable food and medicine and the reasons for this have been clearly given in this work. Beekeepers and those who have eaten honey all their lives will tell you that one honey is probably just as good as another, always assuming that it has not been tampered with.

**honey recipes**
As honey blends so well with all manner of foodstuffs from flavouring ice-cream to basting ham, it is obvious that the endless lists of recipes, several of which are included under their various headings, cannot be given under one heading but one simple one is: 'eat it'.

**honey regulations**
The EEC Council Directive has spewed out specifications galore on things like labelling and importation requirements, perhaps one whose significance might be mentioned here is the level of HMF in imported honey must be less than 40mg/kg and 'diastase must have an activity of not less than 8.
See: sale of honey

**honey ripe**
The ripeness of honey applies more than anything else to its water content. We tend to assume that honey which has been capped over by the bees is certain to be ripe and this is usually so. However honeys are sometimes

found to be watery and to have higher than normal moisture content even though extracted from sealed comb. The opposite can also be the case where honey that has not been capped over is ripe. A reasonable 'rule of thumb' is to hold the comb horizontally (don't do that in the ordinary way especially with brood) and shake it. If the honey stays in the comb and nothing drips out, then it should be ripe. (This is important with oil-seed rape honey which needs to be extracted as soon as possible to avoid granulation in the comb). Artificial methods of extracting moisture from honey below the required density are making considerable headway.
See: dehumidifier, green honey, ripeners

**honey ripener**
At first, large containers holding extracted honey were termed 'ripeners' but they were really just honey settling tanks. Now that devices are available on a commercial scale for removing moisture, these might more correctly called 'ripeners'.

**honey ripening**
It has been found that in order to produce a kilo of honey three kilos of nectar must be collected. This necessitates the evaporation of huge quantities of moisture by fanning. Enzymatic action begins to invert sucrose into simple sugars the moment it is passed down into a bee's honey stomach. Manipulation with the proboscis, moving it from cell to cell and constant aeration by fanning concentrates the simple sugars so that 80% of sugars go into solution. (a sucrose solution becomes saturated at 66%). Once the density of the honey is right, bees cap the cell over. For the feeding of adult bees and brood, nectar is more suitable than honey. Honey is more stable and economical of storage space. Methods have been devised for blowing warm air through supers to encourage ripening.

**honey run**
This is just another way of referring to liquid honey that has been extracted from the combs and is free of granulation.

**honey sac, crop**
The enlarged, musculated, distal end of the oesophagus. The word 'crop' doubtless compares it with a bird's crop – part of the alimentary canal where food can remain for a while before being digested or in the case of bees before being regurgitated. Many terms such as honey stomach, pot and bag are also used to describe this part of the bee's anatomy. It connects to the ventriculus where digestion takes place via a valve which is capable of sifting pollen and is strongly musculated.
See: honey stomach

**honey seeding**
The triggering off of the granulation process by stirring some fine grained honey into the clear honey and subsequently keeping it at the temperature best suited go granulation – around 15C. One way of introducing the 'seed' is to mix some of the smooth sample with a reasonable quantity of the honey to be seeded. Stir the two together until they are well mixed and then stir this into the waiting honey. Although only the top layer will benefit from it initially, it will work its way down but you can hasten and assist this by giving the mass a stir or two every day or so. See: seeding honey

**honey settling tank**
Often called just 'honey tank' and it is a means of quickly getting the crudely extracted honey into a suitable, temporary container. A straining cloth or other filtering device can be put over the tank but this would retard the speed of entry. A cut-off tap at the bottom allows honey to be drawn off provided you have the tank on a level which gives space for 30 lb. tubs or whatever container is next in the process, to go under the tap. It is possible to put other filters between tank and containers.
See: honey ripener, honey tank

**honey show**

An Exhibition of Apiculture. A room or hall is laid out with tables for the pre-arranged show which may cover a local area, county or a larger region. A schedule covering each of the classes, will have been drawn up and competition entries and fees paid by the appointed date. A judge or judges will have been booked and prize cards set out by them as appropriate. Where trophies are awarded these will be handed out, usually by a notable person, at the end of the show.
See: disqualification, honey show schedule, points, Crystal Palace

**honey show schedule**
Rules governing a honey show will ordinarily be set-out in a small pamphlet entitled 'Honey Show Schedule'. This guides – the judges, show secretary and exhibitors by giving details of classes and their requirements, fees and last date of entry. Also the time, place and date of the show and the names and qualifications of the judge or judges.
See: honey show

**honey sign**
This can refer to a sign that is probably to be displayed out-of-doors with suitable wording to inform the public at large that the displayer has honey for sale. Unless it is protected by glass it needs to be weather-proof (paper dipped into liquid wax becomes weather-proof). Manufacturers list a variety of these in their catalogues.

**honey skin treatment**
250g pharmaceutical lanolin, 125g raw honey and 125ml sweet almond oil. Melt honey if granulated, heat lanolin in a water bath (bain-marie), add almond oil and stir until well blended.
See: Cleopatra's facial balm, face pack, face lotion

**honey slinger**
Centrifugal force has been used in a variety of sophisticated ways to expel, throw, spin or sling honey from honeycomb. The earliest attempts were little more than a can attached by cord to a pole in such a manner that the tin with honeycomb inside could be swung around through 360°. The first commercial equivalent has been credited to Major Hrushka. British bee books contain illustrations of simple slingers that were used in earlier times.

**honey spectrum**
Starting with water-white rosebay willowherb, raspberry, brassicas, charlock, acacia, clover, lime, holly by now a golden amber, chestnut, dandelion, sainfoin clove, sycamore, willow, field bean, heather and blackberry, then port wine heath, cherry, hawthorn, horsebeans, plum, buckwheat and the dark honeydews and honey that looks quite black yet is natural and tasty.

**honey spinbarkeit**
German word describes the characteristic of some honeys which have a stringiness or spinnability.

**honey spoon**
Non-slip spoons with a small projection that prevents them from sliding down into the honey. Stainless steel is used in preference to other metals. These are listed in the catalogues of leading appliance manufacturers.

**honey stability**
Bees have rendered nectar more stable by converting it into honey. Honey derives its stability by its high sugar concentration, low pH value, its osmotic pressure and the presence of hydrogen peroxide. When water is allowed to dilute honey (as when bees use it for feeding larvae), the glucose oxidase present attacks the honey's glucose forming, gluconic acid and hydrogen peroxide.

**honey stickiness**
The viscosity of honey at normal temperatures causes it to become quite 'sticky'. This property is made use of by the bee to deter small creatures like the pollen

mite, by 'pickling' pollen under a layer of honey. Some beekeepers prefer to store their supers honey-wet as a precaution against wax moth.
See: viscosity

**honey stomach**
Also called sac or crop. Explained under 'honey crop' it is a transparent bag used for the temporary storage of fluids or for their transportation. Food on route for the stomach proper (ventriculus) must pass through the honey stomach. Although its size would enable it to hold 100mg of fluid, normal loads of nectar, water or honey rarely exceed 40mg. Bees about to swarm begin to load their crops several days before departure and weigh 4000 to 454g as opposed to emptier best that weigh about 5000 to 454g.

**honey stopper**
The stomach mouth, proventriculus or proventricular valve It is at the end of the foregut though it looks like a projection of the ventriculus into the crop It is a valve consisting of four musculated, triangular lips which are lined with fine, closely-set hairs which enable the bee to filter pollen and pass it on to the stomach while withstanding the pressure imposed when the muscles that surround the crop are used to drive food back up the oesophagus.

**honey storage**
The ability of the bee to store honey makes it unique and has greatly assisted its survival. Recognisable honey has been taken from the pyramids. Even in the comb it has a long life and hives that have remained undisturbed for years have had combs of honey that were quite black. Although honey slowly darkens and tends eventually to deteriorate with age, properly ripened honey sealed in dry air-tight containers and kept at reasonable temperature, can remain good in storage for many years. Self-stacking drums and tins of metal or plastic are used though retail sales usually call for glass jars or other small containers.

**honey strainer**
Manufacturers list many of these ranging from conical funnels of perforated metal to honey tank inclusions, grids, platforms and other surfaces through which honey can pass leaving behind any unwanted particles of wax or other material. The finest particles are however most often filtered out using fine meshed cloth by amateurs. The professionals have large scale apparatus which can remove particles as small as pollen grains.
See: filter

**Honey-sucker**
A type of bird. It is very likely that this name is commonly used for the various birds e.g. humming bird, which have learned to use their bills to sup nectar from the corollas of flowers.
See: honey eater

**honey suckle Lonicera spp Caprifoliaceae**
The deep corollas of wild varieties suit moths with their exceptionally long tongues rather than bees There are numbers of cultivated varieties known to be visited by bees, some exotic ones flowering quite early in the year.

**honey sugar alcohols**
These are numerous and inter-related to ketones, aldehydes, acids and their esters. They are: benzyl alcohol, dulcitol, ethanol, furfuryl alcohol, inositol, methanol, ribitol and sorbitol A high sugar alcohol content in nectar tends to discourage bees from collecting it. Fermenting honey is injurious to bees.

**honey sugar content**
Different honeys will have variable sugar contents. Generally fructose is dominant with glucose a close second. Other sugars will often be present but rarely exceed 4% and this will mainly consist of sucrose. As

bees are capable of cracking sucrose into the simple sugars, one might assume that the slight trace of sucrose is deliberate on their part.

**honey, sugary**

This lay-man's description of partial or complete granulation stems from the notion that honey should always be clear. Any crystals seen in it are assumed by those who are unaware of the properties of honey to be sugar of the sucrose variety. This faulty reasoning, which has to be taken into account at sales outlets, has caused producers to 'pasteurise' honey so that it has a long shelf-life, remaining clear indefinitely.

**honey, surface layer**

Despite moisture being taken up by the surface layer of honey on certain occasions, it is slow to diffuse into the rest of the honey. Naturally you might assume when you lift the cap off such honey, that the winey smell means that the whole of the contents are ruined. Not so only the first centimetre or so need be removed to allow the normal honey to be safely devoured.

**honey take**

A beekeeper might refer to his year's harvest as his honey 'take' How much have you taken off someone might be asked? The honey take is at the choice of the beekeeper and account must be taken of the store of honey left for the bees to winter on and that should amount to 20kg even if this has to be made up from sugar syrup.

**honey tank**

A large, temporary storage for honey. At one time these were made from metal that stood up to honey but now-a-days the ubiquitous plastic has come into use for most storage items. The deeper the tank the better chance of the froth rising to the top, leaving relatively clear honey below and with a honey tap at the bottom, this can be drawn off. As the level goes down, more and more of the froth lines the walls, so that most of the clear honey can be taken out. The froth, a creamy mass (mess) is mainly air-bubbles though any bits of wax, propolis or such like will be there too, but taste it, is very nice, and probably a healthy thing to eat. Honey becomes positively corrosive if left in a metal container that is not constructed out of suitable metal or lined with such. It becomes black and will have absorbed substances that are undesirable. Gentle warmth, say 27C/80F will facilitate the rising of air-bubbles.

**honey taster**

A small object shaped to pick up a controlled amount of honey and with a suitable handle It gives the user the opportunity to taste the sample.

**honey thermal activity**

A block or canopy of sealed honey acts as a blanket over the winter cluster.
On opening a hive in the spring, if all the bees are up top, this could imply not only have they lost this blanket but they might well be short of food. The thermal conductivity of honey is low because viscous honey behaves like a solid. Things like this have all added to the bees miraculous ability to survive for millions of years.

**honey toffee**

2 cups honey, 2 of sugar and 2/3 of water and 1/4tsp of salt. Bring to the boil and continue until a droplet falling into cold water becomes a hard glassy bead. Pour onto generously buttered platter, then 'pull' it until it assumes the colour of flax. Cut into pieces and wrap each in waxed paper.

**honey toxins**

Although extremely rare the analysis of suspected samples has shown the following: acetylandromedol, andromedotoxins, desacetylpieristoxinB, gelsemine, hyenanchin and tutin.

**honey tree**

Just as a honey plant would be looked upon

as one visited by bees so we might refer to a honey tree in this respect. Our useful trees in Britain in order of flowering sequence are: Sycamore, the stone fruit trees, Horse Chestnut, Holly, Robinia, Sweet Chestnut, pine and Lime. Exotic trees are becoming more and more prevalent, such as varieties of Eucalyptus but none of these have the same significance as do foreign examples which yield annually and copiously and are regularly migrated to by the European, Oceanic and American beekeepers
See: forest and bush honey

**honey turbidity**
Turbid means opaque, muddy looking or containing particles of extraneous matter. In honey this might mean a cloudiness that cannot be dispersed by ordinary means of filtration. Air-bubbles are often the culprits but these can be removed by applying warmth to the sample. Colloids (common in honey-dew honey) remain to lessen the attractiveness but to improve the health giving qualities of some honeys.
See: colloid

**honey twin-spin**
The electrically operated table-top honey extractor with controllable speed that will extract two B.S deep; or two shallow combs without any need for change over half-way as both sides are extracted at same time and it is said to be easy to clean.

**honey types**
Blended, clear, chunk, comb, crystallised, dark, granulated, light medium, soft set or named according to plant source or place, e.g. Scottish Heather Honey, Maltese Thyme, Australian blue flower, Chinese Mountain Honey or Canadian Clover - but the statement on the label must be true.

**honey uses**
As honey is not only a foodstuff but a stable, easily applied liquid or soft-set solid, its uses are innumerable especially as man has been familiar with this substance since the very earliest times. It is benign on the external or internal skin. It changes the character and flavour or all many of food items. It can be used (unusually) as anti-freeze, shampoo, temporary glue, for making mead or vinegar, as an ointment or skin cream...

**honey vinegar**
This is linked with mead making because before the acetic acid of vinegar can be produced it has to go through the fermentation process of forming alcohol (vinous stage) which in turn becomes acid on exposure to air, little more than a pound of honey to the gallon (against 2-5 when making mead).
Here's one of the many recipes:-
Boil 1lb honey in a gallon of water to kill harmful micro-organisms and to improve the colour add 1/2 pint apple cider or fruit juice. Minerals may be deficient so add a very small amount of potassium bicarbonate (or tartrate) and ammonium phosphate. When the temperature is low enough add fresh yeast (blood temperature or below) as a cake or on fresh toast Now expose to air (best fermentation takes place between 27-38C/80-100F) put into a shallow container. To assist maximum exposure to air shavings from oak, ash or hazel may be added. By way of yeast 'mother of vinegar' or sediment from a previous brew can be used for this second process. Cover with muslin, cheese cloth or material which keeps insects out but allows air to pass through Keep temperature constant by leaving in a warm place, skim and when liquid becomes clear, siphon off leaving the dregs, add one third volume of strong honey vinegar and leave for 3 months before bottling.

**honey viscosity**
This property combines density and stickiness and varies according to temperature When cold honey becomes thick and even a low density honey becomes particularly sluggish after half an hour in the frig it will respond to rising temperatures by becoming less and less viscous A curve drawn showing reduction of

viscosity against rise of temperature shows us that at 28-45C /80-90F an optimum point is reached where further increases of temperature make less difference For settling, and for containerising then, this is the best temperature to work in.

**honey vitamins**

Although these are only present in small amounts in honey, owing to the high digestibility factor they could be more significant than their small quantities suggest: B (thiamine), vitamin B2 complex, Riboflavin, nicotinic acid (niacin), B6 (pyridoxine), pantothenic acid, and vitamin C. Occasionally H and K.

**honey-wet**

To describe a comb as being honey-wet means that it has had as much honey extracted as possible by ordinary means but still has a film of honey adhering to the faces of cells in the comb The bees are only too happy to 'lick' this off but it can cause robbing if done carelessly One beekeeper who made a point of stacking all the wet supers in the open air and allowing a free-for-all finished up with Foul Brood. Not only combs but equipment too can be honey-wet and because of its hygroscopic nature the wet honey attracts moisture from the atmosphere and this makes a corrosive mixture that is harmful. The complete washing of extractors, trays, containers and anything metal especially, should always follow their having been used for honey. On wood it can stain and attracts dirt. Although honey calls for careful cleaning up it is in itself a remarkable cleansing agent.

**honey whipped**

Complete granulation can result in very hard honey The closely knit crystals can be separated by heating or by friction The normal process is to warm the sample and this needs to be done slowly as honey takes a long time to absorb heat and to apply too much in order to hurry things will spoil the honey Once soft enough various types of spinning, whirring or mixing tools can be applied but be warned it is easy to damage tools unless the tenacious nature of honey is allowed for. For those who dislike applying any heat, vigorous stirring will achieve the same creamy condition which comes from the smashing of crystals into their tiniest size The final product of the treated honey results in a texture akin to butter and the flavour is enhanced too. Honey can of course be whipped with cream, ice-cream, butter etc.

**honey yield**

The amount of honey that it has been possible to take from the bees and utilize or the amount of nectar that various plants can yield. Those capable of producing up to 500kg per ha (500 lb per acre) are: white dead nettle, sage, sweet clover and black locust. The years that can be called good (or even glut) years were 1798, 1818, 1858, 1868, 1908 which indicate the ups and downs of English honey takes...

See: crop, harvest, honey potential, surplus

**hooks hamuli**

The row of hooks on the leading edge of the hind wings of a bee. They are chitinous and intricately curved to make a sliding lock on the corresponding chitinous ridge along the trailing edge of the forewing. From being folded on their backs the forewings are drawn forwards, engaging the hind wings on the way, and allow the powered down stroke to be transmitted to the whole surface of the two wings on each side of the bee. This locking feature allowed by the hamuli, not only gives the bee a large area of wing, but can be disconnected as easily as they were linked up and this makes fanning possible (with the wings disjoined) and entry into cells with the wings neatly folded on their backs. Hooks are used for many purposes, to suspend a smoker, in the bee shed, and elsewhere but on beekeepers protective clothing Velcro has stepped in to fill the bill.

**hoop**

A circular band or ring of wood, plastic or metal. Skeppists made good use of wooden hoops to strengthen the top edges (and sometimes the bottoms) of skeps enabling them to be attached to one another or to give added protection at the bottom and some used them to attach 'bars' to.

**Hooper W. E. J., N.D.B.**
Known as Ted. Retired beekeeping lecturer at Writtle. Chairman A.E.A Exam Board, NDB Board, and served on BBKA Board and Bee Husbandry Committee Author of one of the best bee books or modern times '*Guide to Bees and Honey*' and with co- author Roger Morse, '*The Illustrated Encyclopaedia of Beekeeping*'.

**hop *Humulus lupulus* Urticaceae**
Although bees do sometimes visit the flowers for pale yellow pollen and for honey-dew, the cultivated plant can be a menace to bees as it is sprayed regularly with insecticides and no other flowers allowed to compete on its territory.

**hop clover *Trifolium procumbens* Leguminosae**
Hop trefoil – a little yellow flower that resembles hops, flowers May-Aug.

**horehound *Marrubium vulgare* Lamiaceae**
White horehound is a useful perennial bee plant and the dried leaves were once use to make infusions for the relief of coughs. Its small white flowers can be seen from July-Sept and its nectar makes an amber to dark honey Also black horehound *Ballota nigra.*

**horizontal**
That which lies in the plane of the horizon. Feeders may waste syrup or capacity if not set horizontally on the hives. Where hives are inclined slightly forwards a slip of material could beneficially be placed under the front side of a feeder. A solar wax extractor should be set so that the glass is at right angles to the sun's rays at mid-day which, according to the time of year is at about 30 degrees to the horizontal. Brood combs (when warm and the wax is no longer firm), should not be held horizontally as they may sag and the slightest movement of the brood's surroundings is to be avoided, apart from pollen or nectar falling out. Another aspect of the horizontal and similarly vertical arrangements is the shape of the hexagonal cell which can have two vertical sides and points uppermost (the correct way) or two horizontal flats and no vertical sides. Make a drawing and see for yourself.

**hormone**
An organic substance of glandular origin which activates or controls specially receptive organs when transported to them within the body. Minute quantities when moved within a creature can cause a noticeable effect.
See: juvenile, pheromone

**hormone inhibitor**
Plant growth depends on hormones and conversely on hormone inhibitors. Two of the latter are abscisic acid and xanthoxin. Animals too are influenced similarly by hormones and their inhibitors.
See: growth hormone

**hornbeam *Carpinus betulus* Betulaceae**
In April the yellowish-green catkins are occasionally visited for pollen. It has been reported as producing honey-dew collected and worked by bees in July.

**hornet (European) *Vespa crabro* Aculeata**
Our largest wasp, tawny yellow with pale-brown velvety antennae, head and face yellow with a butterfly-shaped patch between the antennae. It is a social insect and as large as Apis sculata the giant honeybee of the tropics. It nests in hollow (often rotten) trees and contrary to popular notions it is not aggressive unless interfered with. It will eat other wasps but has become rare though some nests are still found in the Southwest. Should they attack colonies of bees, a few dozen can demolish a small colony within an hour. Varieties include:

orientalis, mandarinia and mongolica. They have trouble with them in Japan.

**horse**

Although horses have quite thick skin they do suffer if attacked by bees and have in fact been stung to death. Their coats are less protective than those of bovines. As with humans they are most likely to be attacked if, while hot and sweating, they move rapidly by or knock into hives. Where possible hives should be positioned so that the bees' line-of-flight' takes them nowhere near the horses, furthermore apiaries in fields used by these animals should be well fenced.

**horse chestnut *Aesculus hippocastanum* Sapindaceae**

A tall deciduous tree with handsome erect racemes of white and pink flowers with coloured centres. They bloom may to June, have inedible nuts (conkers) which were at one time ground as a treatment for horses (hence name) and originated from the Balkans. The glossy fruits earn it the name of 'buckeye' in the States It grows rapidly on moist soils but needs plenty of space. The pollen has been called deep orange brown, bright and brick red, pink and deep pink with an 18µ grain. A red variety A.carnea blooms a fortnight later. The Indian Horse Chestnut A. indica blooming a fortnight after that giving a deep red pollen The Californian buckeye is suspected of stupefying foragers. They often give splendid floral displays which are associated with little or no honey. The nectar's sugar content is high 70% and the honey is described by Burtt as very dark .
horsemint Monarda punctata Lamiaceae American horse-mint a perennial which is the principal source of Thymol which has been advocated as a useful addition to syrup for winter feeding. Its honey is said to have a minty flavour.

**horticulture**

The science and craft of cultivating orchards and gardens, greenhouse plants etc.

**host**

An animal or plant that provides nutrition for certain parasites A living organism that harbours parasites For instance the honeybee with regard to Acarine, Braula and Varroa. Also more pleasantly, a beekeeper who makes his apiary available for an demonstration or meeting.

**hostathion Triazophos**

A Trade name for the organophosphorus insecticide used aerially on rape causing heavy losses to honeybee colonies in 1978. It is listed as being highly dangerous to bees and is used against bladder pod midge after petal fall, a point of time not easily defined. No access for livestock to treated area for one week. No harvesting of peas for 3 weeks and other edibles 4 weeks.

**hot dip (NZ)**

For hive preservation a suitable tank is required which will take a wax mixture heated to 135C and hive parts immersed therein for 10 minutes. Not to exceed 150C though a second tank at that temperature is useful. The flash point is 200C. It sterilises, expels moisture and the timber absorbs material which extends its life and durability. It can be painted immediately. The composition of the wax is 70% paraffin wax as a base, 10% microcrystalline wax for tensile strength, 20% polyethylene for hardness etc. Do not use wood resin, linseed oil or petroleum oils. The polyethylene improves viscosity, flexibility, moisture repellency and paint adhesion and it can be painted directly; Use (V-A) vinyl acrylic or acrylic gloss paint. Other recipes – 5 min dip at 160C followed immediately by acrylic paint or a 10 min dip a 150C when hives have to be sterilized after AFB.

**hound's tongue *Cynoglossum officinale* Boraginaceae**

One dictionary describes this plant as a troublesome weed. It is a perennial and its flowers are maroon or claret-coloured, they appear in early summer. It likes the sea and is to be found around sand-dunes, waste places and along roadsides whenever it escapes cleaning up operations. Bees gather some nectar and pollen from the plant whose greyish, hairy leaves are shaped like a dog's tongue

**house apiary**
A structure built or used for the accommodation of hives. In a lean-to a row of hives along one wall or two rows in a gable-roofed shed with entrances to hives on the outside with a coloured shape to identify each one. Windows should be shuttered and have a bee-escape slot over the tops Cheap, light hive parts such as WBC inner bodies can be used and only light top coverings are required. The beekeeper can manipulate during any weather in the active season and everything can be locked-up securely. Two tiers are possible for over wintering. Beehouses are common near forested areas in Germany, Switzerland and Austria. There is a good pamphlet on them by J Spiller

**house bee**
An adult bee that has not yet taken up duties outside the hive. During the active season the average time spent as a house bee is generally considered to be about three weeks. A bee's life is extended considerably if it remains longer than the average time inside the hive. This applies to overwintering bees who having formed part of the winter cluster, go on to rear brood in the spring and of course to the queen who having mated, continues her duties for several years.

**house plant e.g. Solanum capsicastrum**
A plant that is kept indoors, usually as a decoration or to shelter it from unsuitable weather. If for instance a plant like the winter cherry needs to be pollinated, this can be done by hand using a small brush to transfer pollen, otherwise a brief spell or two out in the sunshine might encourage a bee to find it.

**Howes F. N., D.Sc (G.B.)**
Author of 'Plants & Bee-keeping'. A dictionary of useful and everyday plants and their common names It has been of invaluable use for improving details about plants described in this work.

**Hoy's Octagonal Box Beehive**
Richard Hoy wrote knowledgably about bees in1788 and this hive was produced while he was running a warehouse for honey in Piccadilly. Octagonal beehives had been used, discussed and patented from the middle of the 17th century. Samuel Hartlib, William Mew, Dr Wilkins, John Gedde, John Evelyn, Christopher Wren, Moses Rusden, Thorley and others used them but by 1756 Stephen White and then Thomas Wildman moved on to square boxes.

**H.P.A.**
The Honey Producers Association changed its name to B.F.A., Bee Farmers Association in the 60's. Members were expected to run over 40 stocks and to be recommended by an existing member.

**HPLC**
High performance liquid chromatography, one of a group of analytical methods used in laboratories.

**Hrushka Major von de Francesca**
Austria 1865, who had noticed honey ejected as his son carelessly swung some honeycomb he was carrying and went on to produce the first centrifugal honey extractor – a hand-spun, single frame apparatus.

**huckleberry Gaylussacia Ericaceae**
Dark blue or black berries and one of a number of similar plants (vaccinium) that produce luxury items of nutritious food and whose crops are greatly increased when honeybees abound.
See: bilberry

**hum**
A low continuous, pleasant droning sound

such as emitted by a flying insect such as a bee. The sound that comes from a contented colony or when a swarm is on the march into a new hive.
See: hiss, moan, warble, humble bee

**humble bee**
More often called a bumble bee. An O. E word with onomatopoeic connections. To a child it might be assumed to come from the insect's mouth but as we know it is the sound of the rapidly moving wings that produce the hum and it varies as a bee changes pitch on entering a flower.

**humectant**
A substance that retains or preserves moisture as opposed to hygroscopic which means absorbing or attracting moisture – a valuable quality that honey has and makes it useful as a moisturiser.
See: demoisturiser

**humeli**
A concentrate of honey boiled in water that was used by the ancient beekeeper of Crete and other parts to feed their colonies.

**humidity**
The condition of the atmosphere with respect to its water content. The actual amount of water-vapour held in a given volume of air is described as its absolute humidity or vapour concentration. When warm air is cooled its relative humidity increases and this accounts for condensation in the hives when spent air from the cluster is able to impinge on a cold non-porous surface. During winter well-ventilated roofs are a help in the connection.
See: dampness, condensation, vapour pressure

**humus**
The dark, fully decomposed material found in soils of both animal and vegetable origin. It is an important ingredient in soil fertility, supplying moisture and valuable trace elements both of which can be highly significant in plant growth and nectar secretion. Humus is in the colloidal state like clay and forms a clay-humus complex.

**Hungary**
Their famous scientific adviser Orosi Pal made great contributions to beekeeping knowledge. Here we find 250,000 hectares of acacia (Robinia pseudoacacia – not native) forests, the largest area of this tree in Europe There is an 8 – 12 day difference in the blooming times from south to north and this enables beekeepers to migrate with their bees to sites further and further north (just as Norwegian beekeepers follow wild raspberry along the coast) There is also some chestnut, lime and raspberry, while cultivated crops include rape, sunflower, mustard, crimson clover, alfalfa and sainfoin (Burgundy hay). The government gives considerable help to beekeepers A special race of bees, the Carnica variety is widely used and a popular hive is the great Boczonadi a German Lagerbeute or long-idea hive with 24 frames (variations run from 15 – 36). Colonies with Varroasis and Foul Brood have to be destroyed though compensation is paid.

**hunger swarm**
A swarm which occurs when every bee in the colony leaves (absconds) its former home on account of food shortage or famine. The hive is left quite clean-looking. The trait is common with mellifera's cousins in warmer lands where monsoons or other climatic changes make a change of area essential.
See: abscond

**Hunter Dr John, FRS 1792**
A UK surgeon, born in East Kilbride, who invented an observation hive and the division board and explained the secretion of wax and function of the spermatheca. His 'Observations on Hive Bees' was posthumously released in 1861.

**husbandry**
The careful and thrifty management of

a farming enterprise. Prudent economy. A group under the auspices of the then Ministry of Agriculture were responsible for the publication of Advisory Leaflets and to give advice on apicultural matters as required.

**hut**

A small, humble dwelling quite apart from the house. Such a place is useful for temporary accommodation at an out-apiary and when light, water and services are available it can, if suitably laid out, be used as a honey extracting house.
See: house apiary

**HWP**

Honey working party of COPA (Council of the Agricultural Organisation of the EEC.
See: CONBA

**hyaline**

From the Greek, meaning glassy or transparent – bee's wings.

**hyaluronidase**

Along with another enzyme – phospholipase this combines with a protein, apitoxin, to form the venom mixture.
See: venom components

**Hybla honey**

Hybla was a town in Sicily and honey from its wild thyme was said by the ancients to have a soft and delicate fragrance second only to Hymettan honey.

**hybrid**

A plant or animal arising from the union of gametes from different genera, species, sub-species - or even clearly marked varieties within a species. Dadant gave – a scientific cross between previously tested and selected inbred lines. In beekeeping the term hybrid is usually applied to the offspring from parents of known lineage; while 'mongrel' is kept for the offspring of parents who themselves have come from mixed sources. Except where it is possible to maintain a pure genotype (e.g. by isolation) bees are likely to become hybrids though the usage of the term mongrel would be applied.

**hybrid vigour heterosis**

In 1887 Charles Darwin showed that under natural conditions, cross-pollinated plants will crowd out and overwhelm weak and sickly plants that are produced by self-pollination. This discovery led to the widely accepted theory of hybrid vigour. It can also be shown to exist in the animal kingdom and is of course applicable to honeybees.
See: crossbreed, heterosis

**hybridize**

To cross varieties (see hybrid) or to breed hybrids. By usage this term has come to be used in more varied ways than was originally intended.
See: hybrid

**hydathode**

An epidermal structure that is specialized for the secretion or exudation of water.

**hydrogen peroxide $H_2O_2$**

A colourless unstable liquid that is soluble in water. An oxidising agent found in nectar and honey. The liberation and accumulation of this substance is by way of the honey glucose oxidase enzyme system. It is a bleaching agent and anti-biotic. It is heat and light sensitive. Fully ripened honey contains less than dilute honey.

**hydrolyse**

To subject material to hydrolysis. A number of enzymes hydrolyse various disaccharides to monosaccharides thus: $C_{12}H_{22}O_{11}$=2($C_6H_{12}O_6$) by the addition of a molecule of water.
See: hydrolysis

**hydrolysis**

The chemical decomposition that takes place when a compound is turned into other compounds by taking up the elements of water. The interaction of substances with water, e.g. the addition of a sugar molecule to a water molecule in the inversion process forms two molecules of the simple sugars. Starches and sugars are broken down into simpler more digestible sugars, such as glucose, in the process of digestion. Hydrolysis is the reverse of condensation.

**hydromel**

The French and the Roman name for mead Pliny states: 'Add honey to rainwater and keep in the sun for 40 days from the rising of the Dog Star (Sirius) – this would be more likely to produce vinegar!
See: Pliny, mulsum

**hydrometer**

A small apparatus or piece of equipment used to determine the specific gravity of liquids. A graduated stem is plunged into the liquid and it finds its own level against which the s.g can be read off. Because of the wide range of s.g.'s special hydrometers are calibrated and used for certain groups of liquid, one for wine being unsuitable for honey.
See: s.g.

**hydrophilic**

Being water soluble.
See: lipophilic

**hydroxymethylfurfuraldehyde**

HMF a toxic substance found in honey and which increases in amount when honey is heated. Its level should be below 40mg/kg and this one of the yardsticks chosen for the identification of acceptable honeys under EEC Regulations. Adulteration of honeys with invert syrup will show up as excess HMF in the Fiches test, as it is a breakdown product of glucose and found in chemically inverted syrup.
See: HMF

**Hygnius Julius 63 BC – 11 AD**

He was referred to by Columella regarding the tradition of the fabulous origin or bees.

**hygrometer**

An instrument for measuring atmospheric humidity There are numerous kinds, some based on the cooling effect of the evaporation of water (the wet and dry bulb thermometer) others on the turgidity of certain materials (human hair, Seaweed) in various conditions of humidity.

**hygroreceptor**

A sensillum that detects the presence of moisture. These are said to be found on the eight distal segments of the antennae and are capable of perceiving an RH difference of 5%.

**hygroscopic**

Capable of absorbing or yielding water The sensitivity or affinity of a substance for water. Honey is hygroscopic and must be harvested when ripe and stored in air-tight containers. The drying out of honey into a toffee-like substance or its dilution to the point of fermentation depends on the relative vapour pressure of the honey and the surrounding air. The hygroscopic nature of honey enables it to remove moisture from a burn or open wound thus keeping it clean as it heals and preventing a dressing from adhering to the wound.
See: honey moisture

**hygroscopicity of honey**

The high viscosity of honey prevents rapid diffusion of any moisture absorbed at the surface When exposed to air with a relative humidity lower than the honey, drying out will cause a thin film to form on the surface. A thin film of honey exposed to dry air becomes very 'tacky'. Honey with a 17·4% moisture content is in equilibrium with air at 58% RH

**Hymenoptera Insecta Arthropoda**

Small to medium insects, many of them social – over 6000 British species. The order of insects that includes the honey-bees, also taking in wasps and ants and creatures having two pairs of membranous wings and grub-like larvae. The order is divided into sub-orders, Symphyta (sawflies) and Aprocrita (the remainder).

**Hymettus**
Hymettan honey famous since the days of ancient Greece, collected from Mount Hymettus in S.E. Greece near Athens 3370 ft

**hyper**
Used as a prefix to mean above or in excess (exaggeration). A hyper parasite is an organism that lives parasitically on another parasite. Hypersensitive to bee stings means showing a greater reaction than most people; also Latin super and supra opp. to 'hypo' with which it is easily confused.

**Hypericum spp Clusiaceae**
Commonly St John's Wort but many wild and cultivated varieties often known by this name. It keeps bees busy in many a garden, collecting orange pollen.

**hyperparasite**
An organism which parasitizes another parasite Some viruses do this. Where the virulence of a harmful parasite can be minimised by the introduction of a hyperparasite, this can be a useful alternative to crude chemical or other methods of control and is termed 'biological control'.

**hypo Gk**
Under or below. Latin infra, infer and sub; opposite to hyper – over.

**hypoallergic**
Does not cause any allergic reactions in human beings. This has been used officially to describe beeswax which is widely used for medicinal purposes and in cosmetics.

**hypodermis**
The cellular layer lying beneath the cuticle or the surface epithelium. Beneath the epidermis.

**hypopharyngeal gland**
Brood food gland. Paired ducts 11/2 times the length of the bee's body to which 1000 or more berry-shaped bodies are attached. The pharynx is able to suck-out the secretions They are absent in drones and vestigial in queens. They produce royal jelly.

**hypopharyngeal plate**
A hardened plate lying on the floor of the cibarium with its forward lobe inclined downwards.

**hypostoma**
The basis of the mouthparts of Varroa mite which comes into contact with the tegument of the bee.

**hyssop *Hyssopus officinalis* Lamiaceae**
A fragrant perennial herb with white or blue flowers that grows to 45cm tall from June- Oct. F. N. Howes suggests that it makes a nice quartet for bees with Catmint, Lavender and rosemary and tells us that bees revel in the blossoms, helping themselves to nectar and pollen. A useful herb yielding fragrant leaves for inclusion in salads etc.

**hysteresis**
The after effect of pollination by bees In physics hysteresis implies a change that has taken place physically and which is later reflected in the condition that results from such energy transference.
See: abrade

# I

**I.A.A. heteroauxin**
Indole acetic acid – significance thought to be relevant in the realm of pollen and nectar secretion. It is a growth promoting hormone $C_{10}H_9O_2N$. Substances that will promote cell enlargement are termed 'auxins'.

**IBAP**
The Intervention Board for Agricultural Produce. Set up under the EEC and responsible amongst other things, for the statistics relating to beekeeping and the arrangements for distributing EEC subsidies to beekeepers as in 1980 and 1981.

**IBRA**
International Bee Research Association set up in 1949 to improve beekeeping practice leading to higher colony yields and better pollination services. Under the leadership of its Director – Dr. Eva Crane – vast improvements in information storage and retrieval took place and regular publications of 'Bee World', 'Apicultural Abstracts' and 'Journal of Apicultural Research.
Use the link www.ibra.org.uk/ to access the latest information and download articles.
International Bee Research Association,
16 North Road, Cardiff, CF10 3DY, U.K.

**Icarus mythology**
Son of Daedalus (Athenian) who fled to Crete having made wings of wax and feathers. Icarus flew to near the sun, the wax melted and he fell into the Aegean. His father, at a lower altitude reached Cumas (Italy). Icarus - name acquired by the International Pilot's Association. Shouldn't have used bees wax!

**ICBB**
International Commission for Bee Botany set up by the International Union of Biological Sciences in 1950. Working groups included Honey research, Nectar secretion, Honey-dew, Bee protection, Mediterranean bee flora, and pollination.

**Iceland Poppy *Papaver nudicaule* Papaveraceae**
A poppy of the Arctic regions having pretty flowers in shades of yellow, orange, salmon, rose, pink, cream and white as well as bi-coloured varieties. Bees collect pollen ranging from yellow to quite dark.

**Ice plant *Sedum spectabile* Crassulaceae**
It draws butterflies like a magnet flowering in late summer to a height of 30cm. It is a robust plant which puts up with all types of soil, heat and cold, even shade. Other plants such a Mesembryanthemum also share the name of ice-plant.

**icing sugar**
Confectioner's sugar, a finely ground white sugar which sets into fondant, frosting or icing when mixed with a small amount of water. Though dearer than white granulated sugar it offers a quick way of making up a suitable plug for a queen cage though it holds little nutrition for a queen unless heated honey and or pollen are also mixed in. More recently it has been used as a treatment for Varroa. When dusted with the powder the bees immediately set about cleaning off their bodies and this is supposed to knock Varroa mites off as well. An even finer powder is produced for wine

makers.
See 'Goods candy'

**ideal weather conditions**
Conditions affecting nectar secretion are often dependent on previous weather that has influenced the growth and development of a plant. No matter how good your system of management or ability of the bees, all will depend on the occurrence of a nectar flow in suitable flying conditions. Bright, still weather with temperatures in the 20C's coupled with sufficient moisture in the soil, can produce a honeyflow when bees abound.
See: nectar secretion

**ideal wood preservative**
It should protect wood against fungal decay and woodworm, minimise swelling and shrinking, allow wood to breathe thereby helping the bees to keep healthy and free of disease and allow the bees to be introduced a few days after the treatment. It should not taint the honey, seal the pores of the wood and prevent it breathing, drive the bees away, attract robbers, require a number of weeks drying before a treated hive can be used, nor contain harmful insecticide which although not intended to kill the bee has the power to do so.
See: hive preservative

**identifying**
Hive, queen or swarm. Records are an essential part of beekeeping progress. With them goes the need to be able to identify a colony and its queen. Hive numbers may include apiary and colony identity, 4G meaning colony No. 4 at Grange apiary. Queens are more likely to be given a colour to identify their year of birth or this may be done by a system of wing clipping. Swarms are not easy to identify unless the queen is specially marked or they are kept in sight from the moment they issue.

**identity odour**
The honeybees of a colony intercommunicate to warn of danger, absence of a queen, or need to emphasize the direction of their efforts. The information is passed between colony members who in turn are able to recognize one another. Strangers are immediately recognized as such and dealt with accordingly. A beekeeper can artificially overcome this lack of willingness to accept anything which does not bear 'colony odour', either by caging until the new odour has been acquired (as in the case of a different queen) or in the form of brood which is never rejected.
The manner in which a colony develops this identifying feature is partly understood. For instance queen substance is past around the whole colony and food sharing, grooming and other social mannerisms like the balance in the activities between the various glandular developments and appropriate physical ability permitting them to work in unison and to mesh into a composite whole. Exceptions are the drones and while a colony is supporting drones, those from other colonies can drift and their sense often guides them to where their potential is most wanted. Robbers, like any skilful predator, can overcome this important defence which otherwise specifically identifies a colony as a single unit.

**IgG**
Immunoglobulin G also M, A, D and E. The five major structural classes of gamma globulins found in the gamma fraction of the protein element of human blood. Allergy or immunity developed as a result of the introduction of a foreign antigen to the body, is dependent on the body's production of gamma globulins.

**ileum small intestine**
The anterior part of a bee's hind-gut which adjoins the ventriculus.

*Ilex aquifolium* **Holly Aquifoliaceae**
A common evergreen shrub with many cultivars and male, female and hermaphrodite varieties exist, it blooms in May/June, its small fragrant white flowers are quite inconspicuous but secrete nectar

freely and are much appreciate by the bees during the 'June gap'. The male plants bear none of the bright red berries, the source of song and story around Christmas time. The blue butterfly larvae feed on it leaves and its wood is used for carving and marquetry.
See: holly, holm oak

**Iliad Ancient Greece**

A Greek epic poem ascribed to Homer telling of the siege of Troy and he refers to honey as being the food of Kings.

**Illingworth, Leonard, Foxton, Cambridgeshire**

A well-known bee-farmer and linguist – a pre-war notable who wrote reviews of French and German books for BBJ and Bee World and was Secretary of the Apis Club.

**image**

A mental picture or representation. The overall impression that the public has of any group or activity that they might pursue and its likely effect on society as a whole. Fortunately after years of being kicked into the grass it has dawned on those who have a great effect on public opinion that bees are really important. So much so that the beekeeping associations who have plugged and plugged in vain in the past are now being listened to and 'out-of-the-blue' there is almost an excess of keensters and not enough bees to go round. We might say that the 'image' of beekeeping has been well a truly polished up 2010. Perhaps honey will follow suit.

**imago pl. imagines**

The final and sometimes the shortest phase of an insect's life. The perfect or adult insect. In beekeeping this means the worker, queen or drone as it emerges from its cell. The word 'adult' or princess in the case of a queen are more usual terms.
See: sexual maturity

**immunise**

To render a person, plant or animal resistant to a certain disease or disorder.
See: bee venom, atopy and immunotherapy

**immunity**

The ability of a creature to resist the effects of pathogens. To have no susceptibility to a disease or disorder. It can be induced, acquired or inherited.
See: allergy and immunoglobulin

**immunoglobulin**

A globulin protein that is found circulating in the human blood and tissue fluids. There are five of these anti-bodies and some are capable of combining with and thereby preventing further penetration of certain toxins such as bee venom but there are others that may trigger the release of histamine and prostaglandins. Prof. Lessof, Dr R.C.Riches and others have done valuable work in helping those beekeepers who have proved allergic to stings, especially those who are only stung occasionally (spouses).
See: allergy, IgG, immunotherapy, immunity

**immunotherapy**

Desensitisation – The curative treatment required to bestow upon a living thing the ability to resist disease or disorder. For those who suffer general symptoms or extensive local swelling from bee stings, minute doses of venom can be subcutaneously injected at regular intervals, the dose being gradually increased until the recipient can tolerate 100mg - the equivalent of two stings. See: bee venom, sting treatment

*Impatiens glandulifera roylei*
**Balsaminaceae**

Water or Himalayan balsam. *Impatiens* implying 'impatient' because the seed pods when ripe cannot wait to burst open. Bees disappear into the flower and come out with their backs dusted with the white pollen from overhead stamens. They flower late in the year August/September providing a good crop for those within reach of the

water-side areas favouring this plant.
See: Himalayan balsam

**imperfect queen**

A queen that is less than perfect and has some recognizable deficiency. While a queen's appearance as regards her size and completeness is very important, queens should be judged by the size and compactness of their brood pattern and by the characteristics of the colony they produce.

**implement**

1. An instrument, tool or utensil. This covers a multitude of items in the field of gadgetry though perhaps none more important than the hive-tool and smoker.
2. To implement an order or act as of Parliament. The enforcement of the law.

**Importation of Bees, The (Prohibition) Order 1979**

This came into force May 25th 1979 banning imports of bees from countries known to be infected with Varroa and restricted imports from other countries to queen bees with a small number of attendant workers.

**Improved National**

The original National hive modified to provide more efficient hand-holds but otherwise square, bottom spaced and taking eleven standard frames like the National.
See: modified national

**impulse swarming**

Starvation, over-crowding or unsuitable home. These could be reasons for a sudden departure of a swarm. Normally there is much prior preparation within a colony before they swarm. Bees load up with food and queen cells are started.

**inanition**

Emptiness from lack of nourishment, starvation. This probably kills more bees than disease though fortunately it is not catching. It is a certain killer however and it beholds beekeepers to make sure that at all times there is a reasonable food reserve within the hive. Lovely strong stocks of bees have been found dead in the orchards, having been taken there with the brood chamber full of brood but when weather intervened, despite being surrounded by blossoms, the bees simply died.

**inbreeding**

Inbreeding fixes the genes making them homozygous and where pure 'line' strains have been produced in isolated mountain valleys, it has produced bees with 'fixed' characteristics but as recessive characteristics can lead to unwanted or harmful development 'out breeding' is used to restore what is lost or they deteriorate as compared with the vigour of hybrids.

**inch fractions**

Fractions of inches run into many places of decimals when converted into millimetres. The following are round off to the second decimal place but beware multiplying up without reverting to 1 metre equalling 39.75 inches. Despite metrification, many older books whilst still valid as information give imperial dimensions. Due care should be taken when converting!

**incline**

To tilt, lean or slope. Where hives are placed on a slope it is as well to cut level platforms sufficiently large to enable a manipulator to have ample room to handle them. A solar wax extractor needs to face south and be inclined so that the double glass panel is at around 30 degrees to the ground so that the sun's rays strike it at right angles. Some bees are more inclined to swarm than others.

**incompatibility**

The inability to exist in harmony. This could be pollen which is incapable of causing fertilization or a queen that is not compatible with the colony.

**increase**

To make greater in any respect. Apiculturists speak of 'increase' when they desire to increase the number of colonies. This can occur through swarming but whether planned or not it comes from the utilization of additional queens.

**increasing forage**

The distribution of seeds, cuttings and bee plants themselves. The utilization of tips, dumps roadside verges and other non farmable areas. The recovery of land from desert or sea. The use of containers – decorative pots and stands, hanging baskets and window boxes. Re-seeding after destruction by fire, flood or other catastrophe.
See: out-apiary

**incubation**

The maintenance of temperature and humidity in order to allow a newly born creature to survive and develop. In the case of honeybees the optimum temperature of 36C is required (survival minimum 32C) This has to be maintained for three weeks so that an egg can be converted into an imago (or adult insect). Because queens are born in individual cells it is possible to take these out, once sealed, and incubate them artificially by maintaining the same vertical posture, temperature required and humidity. The incubation of some predators, namely Varroa, wax moth etc. does offer a tiny window of opportunity in the war against them if their weakness against temperature variation could be exploited.

**index**

The alphabetical indices at the ends of books are exceedingly useful for reference purposes. An index might also be a sign guiding you to whatever you might choose to investigate as for example the cubital index in the venation of a bee's wing that can help to separate one race from another or the determine the amount of say, moisture in the atmosphere.
See cubital, RH

**India**

A land where honey-hunting dates back through millennia and where the four varieties of honeybees are found, the giant honeybee, eastern and western and the small florea. The Indian Bee Journal appears in English. Honey and wax are much prized commodities.

**Indian BeanTree *Catalpa bignonioides* Bignoniaceae**

A N. American (but widely naturalised) tree of elegant shape with attractive bell-shaped blossoms, flowering July and August with clusters of whitish flowers amongst large leaves. As the flowers are 4cm across bees get right inside them and they have also been observed to work extra-floral nectaries under the leaves.

**Indian bee *Apis cerana* (formerly *indica*) Apidae**

Also known as the Eastern honeybee. It is not as easily domesticated as *A mellifera* because it tends to abscond rather easily. Slightly smaller than the Western bee and still quite a good honey producer though far less than *A.m.* It also evolved to the extent that it is more resistant to certain parasites such as Varroa.

**Indian summer**

A sort period of warm, dry, sunny weather often settles in despite summer having passed. It can occur as late as November and may well be accompanied, especially in the South, by a mini honey-flow from the remaining flora notably ivy and it helps to give the bees a final chance to seal all stores and have cleansing flights. Ivy honey granulates so fast that dilute syrup helps to give bees water with which to re-liquefy it.

**indigenous**

Native as opposed to having been introduced or exotic. Originating in and characterizing a given place. It is applied to both fauna and flora and very often indicates a robustness or survival capability for the conditions found in that area.

**indirect flight muscles**

Longitudinal and vertical indirect flight muscles. The alternate flexing of these two bands of muscles produces a thoracic vibration which powers the wings and in winter produces body heat. The frequency is determined by the resonating thorax and not the nervous system which only regulates and maintains muscle activity.
See: flight , rate of wing beats, direct wing muscles

**individual**

A single, separate, indivisible entity.
Each bee is an entire individual and can support life for a limited period on its own. Honeybees however are soon 'unhinged' if kept alone and can only continue to fulfil their normal span of life in company with at least a reasonable number of compatible bees. It has been said that drones that are isolated, live longer with their heads off. Beekeepers tend to be individualists and many prefer to work very large outfits rather than enter into partnership.

**indoors**

In a house or building. Bees are rarely kept indoors except for observation hives and building such as house apiaries or bee houses. Provided their nests are adequately ventilated bees can be trained to use remarkably long tunnel approaches to their home. Absconding swarms are eventually likely if they are perpetually forced to walk a long way before getting into their nest or out into the open.

**Industrial honey**

When honey is fit for human consumption but of a quality below that accepted for the normal market, it can be designated 'baker's honey' or industrial honey; for use in cooking or other processes where flavour or colour are not of particular importance. The actual threshold standards are laid down by the EEC Honey Regulations. Examples are honey with a moisture content greater than 23%, honey with any foreign taste or flavour, honey which has begun to ferment or effervesce, honey that has been heated to the extent that destroys or inactivates natural enzymes, excess HMF etc.

**industry**

Involvement in a task or systematic work or labour. Bees having embraced this way of life millions of years ago are constantly held up as an example to man especially as their industrious efforts are unselfish and improve rather than pollute the ecosphere.

**infection**

The impregnation or invasion of a living thing with disease producing germs.
As honeybees suffer from several bee-transmitted ailments, apiary hygiene should always be practiced in such a way as to avoid any possibility of careless infections. Pointers include re-queening, replacing time-expired comb with foundation, sterilizing used equipment before giving to the bees. Diagnosis to ensure any disease is checked forthwith. Gloves, hives, honey impedimenta can all carry spores of disease and therefore be infectious.

**infestation**

The act or state of introducing destructive pathogens or parasites which are capable of surviving and increasing or overrunning colonies. Whereas acarine mites die very quickly when separated from their hosts, Varroa is said to survive for days. Robbing, drifting, accepting stray swarms, can all bring about infestations.

**inflorescence**

The actual flowering, blossoming or coming into bloom. The whole flowering part or the characteristic arrangement of flowers on the axis (stem).

**infuriate**
Cause to display what appears to be extreme anger. With the vast majority of honeybee colonies it is quite easy to do this. Once they have been triggered off to release 'alarm odour' a colony can seemingly 'go berserk'. Harsh knocks on their home, certain chemicals or smells, rapid jerky movements by large hairy objects – all these things can be guaranteed in most cases to infuriate the bees and excite them to attack.

**ingestion**
The act of taking food into the body for digestion. An adult bee does this via its mouth but in the larval stage the whole body lies upon a bed of liquid food absorbing it from the underside while breathing from the upper. Care must be taken when moving small larvae when 'grafting' to put the larva down in the same position, food-side down.
See: ingurgitate

**ingot**
The resultant piece of metal secured by pouring it liquid into a mould. Beeswax responds too, to being set in a shaped mould and candles have been made this way and 1oz blocks (ingots) marked with word 'beeswax 1oz have doubtless been changed now for the nearest equivalent metric size 3g.

**ingurgitate**
To swallow in excessive amounts.
We frequently use the word regurgitate when talking about worker bees but they have a 2-speed intake ingestion process.

1. Where small quantities of liquid are lapped-up using the labellum and the glossal groove to take the substance in.
2. Where the proboscis is formed into a tube in which the tongue operates as a pump in the process of ingurgitating.

**inheritance**
The transmission of biological characteristics from parents to offspring.

**inhibine**
It has been shown that by enzymatic reaction glucose oxidase leads to the formation of H2O2 in honey. It is a biologically active material (bactericide) and it is found in both nectar and honey. It inhibits the growth of micro-organisms. Although it is heat and light sensitive some factors are more stable than others but it is rendered ineffective when honey is heated. The anti-biotic activity is further enhanced by the acidity and osmotic pressure of honey. This is significant when considering the use of honey on wounds and 'Manuka' is not alone in achieving these results.

**inhibit**
To restrain, hinder, arrest or check an action or process. Fumidil B is an anti-biotic which, when fed as recommended to honeybee colonies, inhibits the vegetative development of Nosema spores. This leads to a drastic reduction of spores though it does not kill existing ones. With infection minimized the colony can out-grow the disease. Other diseases can be inhibited by the careful use of drugs.

**inhibitor**
A substance that can be used to inhibit the development of a disease, disorder or condition. Inhibitory substances usually take the form of drugs but an anti-enzymic substance may be found to inhibit the effect of specific enzymes. For example the human alimentary epithelium is protected against digestive gut secretions, (lack of this presumably leads to ulcerations?).
See: inhibit, drugs

**inner bodies**
Where a hive has an outer protective wall the chambers within are known as inner bodies. These are kept out of the weather and shielded from cold winds and are made of light timber. The WBC's with such light inner bodies are often favourites with lady beekeepers and make a light component hives in a bee house.

**inner cover**

Sometimes called a 'crown board' though many substitute materials are used to act as an inner cover of the tops of the frames. Glass 'quilts', canvass sheets and other means are used, their main object being to let you take the roof off without immediately exposing the bees; it is also useful to have a feed hole incorporated. In the spring beekeepers often put added insulation on top when the early brood nest is under way and the weather can be cool.
See: mat

**innervate**

To stimulate via nerve impulses. To convey nervous energy or the development of nerves from peripheral areas into the central nervous system.

**inoculum**

Source of contamination e.g. pathogen; some other material – your hands after handling foul brood combs.

**inorganic**

A substance that has not required any living process for its production, Minerals are inorganic while carbon is organic. Inorganic chemistry is that branch dealing with the formation, structure and properties of compound of elements other than carbon (apart from CO2 and carbonate salts).

**inquiline Latin – tenant**

Cohabitation with another species, not parasitic though possibly getting a share of the food. *Braula coeca* is sometimes described in this way. Many other creatures have been attracted to the warmth and security of a beehive: birds have nested in the top of WBC hives. Earwigs, spiders and wood eaters and even slugs have been numerous in some hives, less so when the colony is strong and dry. Cuckoo bumble bees are another example. See: commensal, symbiont

**inquisitive bee**

A bee that seems to be curious or trying to gain knowledge as of the whereabouts of a possible attacker, forage or food supply or a suitable site for a swarm.

**insect hexapod**

The largest class in the order Arthropoda: small invertebrate creatures having a pair of antennae, three pairs of legs, head, thorax and abdomen and in many cases two pairs of membranous wings. The class 'Insecta' includes amongst other genera, bugs, earwigs, flies, greenflies, butterflies, ants, wasps, and bees. There are about 22,000 species in the UK.
See: bees

**insecticidal fumigants**

Be extremely careful when handling any of these: calcium cyanide, chlordane vapours, chlorinated hydrocarbons, ethylene bromide, methyl bromide, naphthalene, ethylene oxide, Phostoxin, formaldehyde and Vapona. Many have come and gone because of their virulence. Until fairly recently PDB – paradichlorobenzine, was happily passed around until it was found to be carcinogenic. Beeswax (comb) is extremely likely to pick up any of these contaminants and most are 'residual'.

**insecticide**

A substance that kills insects. Unfortunately these can easily get into the hands of irresponsible people. All over the world chemicals of all kinds come in for periods of testing and there have been many occasions when bees suffer. Toxicity and residual action are used to establish groupings to indicate their poisonous effects. The matter is in government hands but crops have to take priority over insects and it is not all that simple to arrive at a satisfactory balance.
See: spray mortality

**Insecticide formulations**

The precise chemical mixture and examples of the form in which it is applied. There are stickers, spreaders (usually oils) and wetters, while micro-encapsulation can cause similarity to pollen syndromes as in methyl parathion: D – dust - controlled droplet application. EC – emulsifiable concentrate. F – flowable. G – granular, MA – concentrate application at mosquito abatement rates. S – solution. SP – soluble powder. ULV – ultra low volume 70-120µ. WP – wettable powder. See: formulations, spray

**insecticide toxicity**
For all information on this subject visit or contact the HSE (Health and Safety Executive) website www.pesticides.gov.uk/farmers_growers_home.asp.

**insectivore**
A creature that devours insects. These include many birds, mammals and reptiles. Examples include: bats, green woodpeckers, shrikes, swifts, honey birds, shrews, toads, moles, hedgehogs and other insects.

**inseminate**
To sow or to inject seed into something. To make a plant or animal fertile. With queens artificial or instrumental as opposed to natural insemination.

**insemination**
The sperm masses from several drones are discharged into the vaginal pouch. Each new drone is said to pull away the remains of any traces of genitalia from earlier suitors. The spermatozoa is at first held in the distended lateral oviducts but muscular contraction by the queen forces them into the vagina where they are stopped by the valve-like fold and directed into the spermatheca. See: sperm

**insoluble**
Knowing what can and cannot be dissolved is helpful as regards getting substances into solution (catalysts) and separating substances like a mixture of honey and beeswax. Cappings, wet with honey, can be washed clean with water. The sweet water can go back to the bees, be made into a fizzy drink, or mead or even into honey vinegar, but the wax needs to be melted before it can be passed through a suitable cloth filter, to extricate the dross and leave you with a fine golden liquid (for all the world like hot honey) that can go into moulds and then fulfil one or more of the hundred and one jobs it can do.
See: solvent

**inspection**
Looking into or examining. An inspection is an implication that something is being critically examined for readiness or correction. It might be a list of items required to perform a certain operation, or the visit of an inspector to ascertain that certain standards are maintained. As regards a colony it could be a routine examination or cursory check via a glass quilt or hole in the crown board. Weekly checks (inspections) during the season for queen cells, or to ensure sufficient space is there for laying or for storing. In fact active beekeeping is a series of inspections to control and check progress.

**instar stadium**
A period of post-embryonic growth (eating and development) between moults.

**instinct**
Innate behaviour or an inborn pattern of behaviour as opposed to learned behaviour. An automatic co-ordinated response to environmental or physical condition at a particular time. The instances where bees are totally influenced by their nervous systems include vibration of wings, response to alarm odour, dislodging pollen from corbiculae, nurse bees attraction to open brood etc.
See: intelligence

**instrumental insemination**
Of a queen requires delicate instruments, a high degree of skill and appropriate anaesthesia ($CO_2$). One advantage is the

complete control and choice of sperm but as opposed to this although selective breeding is possible, the resulting queens fail to be as robust and egg-worthy as those naturally mated. A microscope, small syringe and means of holding open the queen's vagina (by-passing the valve fold) are required.

**insulation**

A non-conductor to prevent or reduce the passage of electricity, heat or sound. Honey-filled honeycomb is a wonderful heat insulator and a surplus of such food over their winter requirements is another encouragement to their survival. As honey warming cabinet the insulation provided by an unwanted refrigerator, with a 25 watt bulb inside works a treat. A solar wax extractor which utilizes the sun's heat via a double-glazed panel, needs to be painted black and well insulated. Insulated side plates (PVC) for an observation hive will assist the small colony to continue to rear brood. As bees keep their brood at our blood temperature, although they supply their own insulation, some help from us at times is valuable.

See: solar, winter packing

**insurance**

Indemnification against actual or possible loss by payment of an annual premium. Several schemes are of particular concern to the beekeeper. BBKA and local branches offer cover against injury, damage to property or harm caused by deleterious matter in food , stings or disease. Details will vary in different localities and types of beekeeping operation.

**integument shell**

The word is derived from the Latin 'integumentum' meaning covering. It is used to describe the seed coat and also the external protective layer of Arthropods'. A bee's integument is its exo-skeleton, it is multi-layered and waxy, keeping body fluids in and exterior liquids out, though it is through the integument that contact poisons are able to pass. It is also vulnerable at certain points as for example the spiracles (10 pairs), the fumes of petrol for instance almost make a bee (or a wasp) fall apart.

**interchanging**

This leads us to queens, colonies, floorboards or anything else that can be transposed. First – floorboards. These have largely been replaced by screens since the advent of Varroa but as they collect whatever rubbish the bees are not able to clear out for themselves, it is good practice to change them and this can be done by having a spare so that it can replace one, which is then thoroughly cleaned and sterilized and then used for the next hive and so on. This can be done in the spring though the words 'spring cleaning' are not really appropriate for beekeeping. Hygiene is a permanent necessity when dealing with bees. Next colonies, there are occasions when one colony can be swopped for another, one instance is in good flying weather when a weak colony is transposed with a strong one. This could hold the strong one from swarming while helping to strengthen the weak one. Such manoeuvres should only be performed when there is a flow on and bees are more readily accepted as they return to strange hives. Queens are rarely swopped and both would need to be carefully initiated into their new colony. In any such exchanges bee health (absence of disease) is an absolute priority.

**intercommunication**

The means whereby two or more creatures in a group are able to have their behaviour influenced by a stimulus transmitted by one and received by the another. Within the honeybee colony there is much intercommunication, in fact their well-being entirely depends on it. It can be chemical, physical, visual or take forms which we as yet dimly comprehend.

See: alarm odour, queenlessness, swarming impulse

**interesting facts**

Queen larvae are provisioned ten times

oftener than worker larvae. It takes 600 bees and 30,000 miles of flight to produce a pound of honey. A bee with an empty crop flies with its rear legs extended backwards. A colony needs 4 million pollen loads per year. There are 450,000 scales to one pound of beeswax. The top cells can support x 1320 of their own weight. A larva receives 10,000 visits. There are 10,000 journeys to a pound of honey. In the hive no bee ever starts and goes on to finish any particular job. A queen walks 275m in laying 2000 eggs. Well-fed bees are 4000 to the pound while hungry ones come out at 5000 to the pound. A flying insect is up 10 degrees warmer than the surrounding air. A termite queen can lay 40,000 eggs a day. A pound of wax needs half a million wax scales. The groove in a bee's tongue is only one thousandth of an inch diameter. The thickness of a cell wall is about four thousandths of an inch. A forager uses about 1/2mg honey per kilometre. Incubation requires 34C/93F and 50% relative humidity.
See: dimensions, temperatures, useful dimensions, facts and figures

**interference**
Action taken which hampers the state which would otherwise prevail. In a beekeeper's early days he/she might well be inclined to open up without any good plan or reason and leave the bees worse than if they had been left alone. Ask yourself is my manipulation going to hinder or help?

**interior**
The inside. Bees work perfectly well in the dark of the hive and there is no question that they go on working right through the night. It is of course necessary that some light penetrates to encourage foragers to set off on their journeys. In the hollow of a tree, bees will propolise the interior to make it water-proof, for reasons of hygiene (it is anti-biotic) to prevent the smell of their attractive stores from wafting around and to help exclude tiny predators. Their cleverness compared with wasps of making vertical combs and thus being able to use both sides, is just one of the things that makes honeybees unique. The environment inside their nest virtually puts them into a world of their own and it goes back millions of times longer than man's incursions.

**intersegmental membrane**
This is composed of thin flexible layers of cuticle enabling movements of the harder chitinous segments to which it connects. It can fold and is waterproof. The distension of the intersegmental membrane allows the queen's abdomen to expand and elongate as it fills with eggs.

**interstitial cells**
Cells of the honeycomb built in the interstices or spaces such as between the larger drone cells and the worker cells or where the comb is attached to a surface such as a top bar, roof, floor or wall. Such cells may have three, four or five sides as distinct from the normal hexagonal (six-sided) cells.

**intestine**
The digestive portion of the alimentary (food) canal from stomach proper – the ventriculus – to the anus.
See: small and large intestine

**in-the-flat**
English version of the American 'knocked down'. Frames are usually sold in this fashion - a top bar, two side bars and two bottom strips which together form the bottom bar. These can pressed home to form a complete rectangular frame held fast by nailing or glueing. Hive parts pack and travel more easily in-the-flat and can be assembled in the same way as frames.

**intine**
The endosporium or inner membrane of a pollen grain or of a spore. A thin, clear semi-permeable layer that is composed of two layers – nexine and sexine.
See: angiosperm, pollen grain

**introductory cage, queen**
A small, ventilated cage with a small compartment for candy, that can be used to introduce a new queen to a colony prepared for her reception. Many such cages are used to send queens, together with around a dozen attendant workers, through the post. It is however advisable to transfer the queen to a new, clean cage without the workers for the final introduction. Such a cage should be slim enough to fit between the top bars, to have a mesh that allows the queen to be fed and touched without being harmed and to have a small tunnel through which she can pass into the colony once the candy block has been eaten sufficiently.

**intruder**
A creature that makes an unwelcome invasion of a honeybee colony. Intruders usually seek food and some are able to do this almost unmolested while others may be ejected with a struggle, often causing serious loss of life to colony guards and others who become involved. Bumble bees are rather big for the workers to handle but once stung they are stripped of hairs and other appendages and thrown out, to lie like black, shiny beetles, onto the forecourt. Wasps may be killed, though it takes more than one bee to do this, and they are tough, furtive and can become a real menace if they signal their success to their sisters. Others include: ants, earwigs, mice, slugs, wood lice and moths (wax).
See: enemies, hive mates, identity odour, predator

**invade**
To overrun or infest, to enter and spread harm by invading. Apis adonsonii – the African bee
– escaped from what was intended to be 'safe keeping' and invaded South America and then went on to threaten wider and wider circles where the climate suited it. Weeds and wasps have travelled from England and become widely established in places where they were not wanted. Bees, being beneficial insects, are not considered as invaders unless they occupy buildings, electrical sub-stations and other places where they are not only unwelcome but may put safety at risk.

**invagination**
The introverted, drawn in or folded back on itself – the formation of the gastrula in the embryo.

**invention**
The origination of a device or the conception of an idea and its implementation. Beekeeping has advanced into modern apiculture on the strength of inventions. While bees and honeycomb still closely resemble what earliest man discovered, our means of utilizing them to our advantage has advanced enormously and happily will continue to do so. Here are some of the advances made:

| | |
|---|---|
| 1675 | Greeks used open baskets with top bars. |
| 1768 | T. Wildman made box hive with wooden bars, |
| 1792 | J. Hunter - division board. |
| 1796 | Keys slides between bars. |
| 1843 | Kretchmer lst foundation, |
| 1851 | Langstroth bee-space and movable frame hive, |
| 1857 | Mehring improved foundation, |
| 1857 | Harbison 4 piece section, |
| 1862 | California – solar wax extractor, |
| 1865 | Hruschka lst honey extractor, |
| 1870 | Wheeler – metal end, |
| 1870 | Quinby – bellows smoker, |
| 1874 | Queen excluder, |
| 1876 | Cook –basswood |
| 1879 | Grooved one piece-section, |
| 1890 | Cowan – radial extractor, |
| 1891 | Porter bee-escape. |
| 1961 | Dr. Butler – queen substance. |

**inventory**
A written record of goods or possessions – a detailed list of items in stock. Only by keeping some sort of inventory can one order each season's requirements and proceed with their beekeeping on an orderly fashion. Living from hand-to-mouth causes problems for beekeepers and suppliers alike.

**inversion**
The act of inverting something or turning it upside-down. The hydrolysis of carbohydrates such as sucrose, results in 'inversion' as the rotation of a plane of polarized light is reversed in the inverted fructose/glucose solution so formed. This is brought about by the bee's enzyme invertase. In meteorology a reversal of the normal lapse rate gives rise to stable conditions often marked by a tell-tale haze layer. In summer this can and often does, lead to honey-flow conditions.

**inversion of queen cells**
A queen cell left in an inverted position will be torn down by the bees. The effects of gravity with the cell in its normal vertical position, are important to a queen's proper development. Skeppists were aware of this and turned skeps upside down as a means of checking swarming when this was required.

**invertase**
A plant or honeybee enzyme that splits sucrose into the sister sugars with a mirror-image chemical structure – dextrose and laevulose. This process is known as inversion and the resultant sugar mixture 'invert sugar'. It is produced in the hypopharyngeal (salivary) glands of workers.

**invertebrate**
A spineless animal, an animal whose body structure does not include a backbone. For example a bee.

**invert sugar**
A hygroscopic mixture of fructose and glucose. Sugar that has been inverted by honeybee enzymes or by the action of man, involving the use of heat and acid. Over-feeding in the autumn results in much effort by the bees to invert it and can have a detrimental effect on the bees - a physiological drawback. See: inversion.

**in vitro**
Experimental biology in isolation from the whole organism – (in-glass). It was useful in the diagnosis of Bee Paralysis Virus before electron microscopy enabled more detailed studies to be made.
See: in vivo (below)

**in vivo**
experiments observing behaviour within living organisms.

**ion**
An electrically charged atom, radical or molecule formed by the gain or loss of one or more electrons. A charge of electricity can exhibit an imbalance of positive and negatively charged particles. Thunderstorms can do this and bees become tetchy in such conditions, one returning with a reversed charge can cause bad temper at the hive and even rejection on account of it.
See: electrical charges

**I.O.W. disease**
A bee disease that reached epidemic proportions several times between 1905 and 1919 was first reported on the Isle of Wight and spread to the mainland literally obliterating the extant native bee, nostalgically referred to as the 'British black'. Many foul brood ridden apiaries were wiped out by this 'new' disease which undoubtedly included what are now known as 'acarine', 'nosema' and other diseases (especially viruses). The native bee was gradually replaced by more resistant exotic strains but numbers never regained their former high levels. Many of the suggested remedies and cures would have made the bees ill anyway and it is significant that it happened at a time when skeppists were changing their habits by going over to movable frame hives.

**iron Fe**
Metal that is magnetisable, ductile and malleable. Acted upon by honey it becomes black ferric tannate but alloyed with carbon it gives us varying qualities of steel from which many items concerned with apiculture

are made. Alloyed with both carbon and chromium, it gives us stainless steel which is resistant to honey water corrosion and is used for honey tanks, extractors, hive tools etc.
See: compass, Monel metal

**iron pan**
This occurs on soils like Millstone grit and Keuper Marl and gives rise to plants like ling and willowherb. It is a thin red layer some inches below the surface which is impervious to the drainage of excess water. The soil becomes sour and is ideal for ling. It can be disturbed by deep rooting plants such as trees, bracken and bramble. These conditions favour less prolific bees.
See: Keuper marl

**irradiate**
To expose to radiation. Some experiments have been made to deal with the pathogens of EFB and AFB by means of radiation.
See: gamma

**irritability**
The state of being easily irritated. Bees and wasps are sometimes if not often, of this nature. However much has been done to develop more docile strains of bee and to teach beekeepers that whacking smoke all over them is not the best manner of subduing them. Careful handling and giving a colony 'time' to accept your presence – perhaps a preliminary puff at the entrance then a minute for them to get the message, help, as do cover cloths and general, confident, reasonably slow actions. Suit your reactions to theirs and when wise to do so, close them before you become too involved and try again when it has become more propitious
See: irritation, temper

**irritation**
Excitement to anger, to inflame or to cause an uneasy sensation in a living organism, Both bees and beekeepers can become irritated and of course either might be the original cause. Difficulties arise when the whole colony has become irritated as when they have been alerted by alarm odour. Avoid knocks and jerks and only expose some of the combs to broad daylight, cool air, (and other bees). Always have a smoker going properly before you start and a means of setting it down so that you are not kippered. Judicial use of cool smoke will involve gently puffing across the top of the frames (never down into) and working slowly and calmly using your hive tool to best advantage. An experienced beekeeper seems to have a magic charm but really it is because he or she knows just what they are about to do.

**iso Gk. equal**
Hence: isohyet. isobar, isotherm , isosceles etc.

**iso-amyl acetate pheromone**
Alarm odour discovered in 1962. When released it has a far reaching effect on surrounding bees. However although the commotion caused is usually enough to drive off a predator, especially when a few stings have been delivered, it does not have a long duration. Strangely, though few put it to the test, bees that are shut up and hammered, vent their spleen on themselves and after a reasonable pause can be opened up to find them busy fanning and perhaps loading their honey stomachs. The inrush of bees waiting outside seems to offset the mad rush of those near the entrance to get out. The principal can also be observed when bees are 'driven' or when they are transported.

**isohyet**
Lines joining places of equal rainfall. These usually mark bands of flora or distinctive countryside as for instance in the SW of Australia.

**isomer**
Having same formula but different molecular structure like: fructose/glucose.

**isomerose syrup**
Although it is predominantly fructose and glucose it is difficult to keep in solution and its crystals are hard and difficult to liquefy. It can be distinguished from honey-dew material by mass spectrography which can differentiate the carbon elements in honey.

**isopentyl acetate**
It has a very long formula and is produced by the bee's sting apparatus. It is highly volatile and only lasts for a short time but when released it alerts the bees to attack and reduces the number of departing foragers.

**isophene**
A phenocontour or line through places showing the limits of the frequency of a particular variant form. Lines for example drawn to show the spread of the so-called 'killer' bees in S. America.

**isopleth**
Lines joining places of equal sunshine, Useful to attract holiday folk.

**isoprenaline**
A drug that helps counteract the release of histamine especially when stings are likely to affect a person's breathing.

**Israel**
40 hives per 1000 inhabitants, a nine month long honey season with melons, orange, grapefruit and mustard with re-afforestation including eucalypts and hives rented out for pollination.

**Italian bee**
The varieties that have been widely propagated and distributed world-wide tend to be a light yellow bee, very prolific but inclined to be swarmy.

**Italy**
A yellow, docile strain of bee has been widely used in warm countries such as America, New Zealand and Australia. In Britain they have been found to 'eat themselves out of house and home' but do well in warm seasons and it has been said that subsequent generations crossed with local bees are more stable and economic in our climate. Their queens are lighter and easier to spot.

**ivy *Hedera helix* Araliacea**
A ubiquitous late-flowering, climbing, woody plant with ever-green, shiny leaves. Its flowering defies the first of the frosts and an Indian summer can mean a small surplus, the bees being willing to go on even making wax in the South. Pollen is dull yellow, greenish yellow to brown with a 25μ grain and the honey rapidly granulates into a rock-hard, but smooth crystallised lump with a sweet, some say slightly bitter, taste. The concentrated nectar can sometimes be seen on the flower head and it tastes quite sweet. It does seem to smother trees and ubiquitous is an appropriate word. Bees with solidified nectar in their crops have been found dead. A feed of dilute syrup is handy at such times and a need to watch out in the spring when brooding calls for increasing amounts of water while bees wrestle with ivy crystals.

# J

**Janscha Anton 1734 – 1773 Slovene (Yugoslavia)**

He observed and recorded the mating flight thus ascertaining that honeybees mate outside their hives and with several drones and display the mating sign when they return to their colony. It is suggested that he was familiar with parthenogenesis much earlier than Dierzon. On morale he said 'everything becomes cheerful and brisk in the presence of the queen' and he was also responsible for queen introducing cages.

**January**

Although the queen's egg-laying will have become minimal, with the slow but the obvious increasing amount of day-light, the bees respond to it and the consumption of stores increases. As the days lengthen, the cold strengthens. This is when good husbandry in the autumn begins to pay off. The mouse guards keep that little rodent from sneaking in while the cluster is tight and stealing valuable stores and leaving objectionable faeces. If freakish weather permits foragers to seek out those flowers like mahonia, hellebore etc. they may cleverly dodge the mouse guard which could trip their pollen loads off. You cannot be active but you can be observant. No particles of wax (sign of a mouse) no dead grubs (sign of premature brood rearing) but cleansing flights in those short moments when the duty pilot gives 'clearance' and even a glimpse of pollen can be backed by a feeling of warmth when you put your hand above the cluster, if the bees are right up top that could indicate that they are using the last of their stores. Did you put by a spare comb of sealed stores? These things should tell you all is well and help you to be patient, after all Feb is a short month.

**Japan**

A great importer of honey. More valuable though is the honey from Apis cerana japonica which is kept in hollow logs or boxes. The vast stretch of latitude from south to north gives migratory beekeepers a season of changing opportunities. 60% of the beekeepers were professionals. In tropical parts and elsewhere hornets can be troublesome. Even strong colonies can be in trouble trying to defend themselves.

**Japanese/Chinese candle tree *Triadica sebifera* Syn *Sapium sebiferum* Chinese tallow tree**

These are to be found in Britain and reputed to be good nectar plants.
No ref for this under Japanese only Chinese!

**Japanese hornet *Vespa mandarinia***

These have figured in several good films and are colossal giants as compared with ordinary honeybees. Even the masses in strong colonies have to work hard to withstand an attack from these hornets. But if they can repel the first one or two they might cope, if several get away with carting bees off to their nest though, then sisters follow sisters. Mitsubachi Kagaka 'B' jrnl.

**Japanese knotweed *Polygonum cuspidatum* Polygonaceae**

A deep rooted weed brought in from Japan which has rather overrun its welcome here. It is absolutely rampant and is found all over the place. It has been said to have kept bees busy but anyone attempting to grow it would be very unpopular as great efforts are being made to control or exterminate it including getting special predators from Japan. There are other varieties flowering for six weeks from mid June with rosy-red flowers on terminal spikes but they don't necessarily confine themselves to wet ditches and ponds as was once the case.

**Japan wax**

Of Japan tallow, a pale yellow somewhat hard lustrous material harvested in China and Japan from the fruit berries of Rhus succedanea.
See: Madam Tussauds

**Japonica *Chaenomeles laganaria* Rosaceae**

Japanese quince, a shrub with crimson flowers often as early as Christmas but it will flower on until June. Bees mainly collect its pollen'

**Jar**

A glass, earthenware or plastic container with a wide mouth. Honey jars are often called pots or bottles. In Britain we have the standard British squat jar in one pound and half pound sizes. Good quality, clear, flint glass is commonly used without any tinting which would give false colouring to the honey. See: honey jars

**J.A.R. The Journal of Apicultural Research (IBRA).**

**jargon**

Speech or writing peculiar to a profession of group such as beekeepers. The stock had over-wintered and was showing three seams. Play-flights increased not long after re-queening and although they had chimneyed, supering, followed by a flow, soon brought them to peak strength. No need for plumping though a cast had been newspapered onto the brood earlier on and this permitted tilting to obviate any piping. Supersedure occurred, a nearby congregation area giving the virgin full scope. A lingy stretch finally ensured a better-than-average take.

**Jar hives**

In Malta and other Mediterranean countries hives are extended horizontally in the form of a wide tube (akin to our chimney pots) instead of vertically as is the custom in cooler climes. They are made of baked clay and the initial part is ventilated at one end the other receiving an extension or being covered by a vertical board allowing a small entrance. The initial section is like a jar and this type of hive is known as a jar hive.

**jasmine *Jasminum nudiflorum* Oleacea**

Winter jasmine (native of China) in the same family as lilac and privet. Bees will work it, though its early flowers, which appear before the leaves, come in December and January and rarely coincide with suitable weather. Not to be confused with American yellow Jessamine (*Gelsemium sempervirens)* a climbing vine native to coastal plains that blooms in early April but whose fragrant blooms spell death to the honeybee.

**jaw**

In the honeybee itself the mouth has no jaws but mounted on the lower side of the head above the tongue are mandibles. These are hard, moving parts that work sideways in opposition to one another. They let the bee chew wax and certain fibres and can even nip mildly so that a sensitive area like the human lip, can just feel the pinch. Guard bees hold the mandibles open in an aggressive stance when confronted by intruders at the entrance.
See: biting, mandibles

**Jerusalem pine (or Turkish) *Pinus brutia***

One of the honeydew yielding pines of Greece to which beekeepers annually migrate.

**jig**
A pre-formed device that offers a shaping service for the specific purposes such as the assembly of frames or section boxes or for drilling holes or marking material in a standardised fashion. Although a jig may take skill and care in making it can save considerable amounts of time and help to improve accuracy, possibly use materials more economically.
See template, wiring

**jittery flight**
Ribbands description of the fitful flight displayed by visiting robber bees which he opined was due to their different scent and consequent attacks on them that followed recognition. In addition their legs stick out rearwards to adjust for their empty honey sac.

**job**
An individual piece of work. In the hive, bees tackle tasks appropriate to their development and the needs of the colony. Despite the neat, tidy array of cells and their contents it is surprising to discover by observation that an individual bee rarely starts and finishes a job but that they are so integrated that one inevitably carries on where another leaves off. So jobs get done even though the workers will have changed or added to one another's efforts to get any piece of work finished.

**Johnson's organ**
The organ of Johnson, present in the pedicel (basal joint of the flagellum of each antenna). It consists of a radial arrangement of chordotonal sensillae (sensory cells) that surround the nerve trunk running through the antenna. The tips of the sensory cells reach the inter-segmental membrane round the joint above the pedicel and are said to be sensitive to antennal vibrations and to give the bee a sense of its speed of passage through the air (airspeed).

**joint**
Segment of an articulated body such as the joint of a bee's leg, femur, tibia etc. The corners of wooden hives are dove-tailed, lock jointed or butt jointed, a joint in an article made of wood is where two different pieces are held together.

**joule**
A unit of work or energy. It is defined as the work done when the point of application of a force of one Newton is displaced through a distance of one metre. Equal to 10 ergs or 0·74 foot-pounds. The MJ (mega-joule) is one million joules.

**journals**
Some journals in English: American Bee Journal, Apiacta, Apicultural Abstracts, Australian Bee Journal, Australian Beekeeper, Bee Craft, Bee World. Bee-keeping, Canadian Bee Journal, Canadian Bee-keeping, Gleanings in Bee Culture, Indian Bee Journal, Irish Bee-keeper, Journal of apicultural Research, New Zealand Beekeeper, The Scottish Bee Journal, The Scottish Beekeeper, The Speedy Bee, South African Bee Journal and the Indian Bee Journal.

**joy dance**
A bee places its front legs on the body of another bee and makes five or six shaking movements. Yipee!

**judge**
A person appointed to decide on the winner and order of merit of competitors in a contest such as a honey show. As such a show may be wide ranging, including educational displays, wax, wine and honey, judges are frequently specialised in one or other of these lines.
See: honey judge

**judge's steward**
A person appointed to assist the judge at a honey show (competition). He or she will be expected to know just what a judge is

looking for in the allocation of awards and of anticipating or carrying out requirements from the supply of clean tasting spatulas to the giving of opinions when asked. Experience in working as a judge's steward is essential for anyone aspiring to take the BBKA examination for judging for which the Senior examination is a preliminary.

**judging candles**
Each candle should be of equal size and length. Tips well-formed, wax free from extraneous mater, the sides parallel and symmetrical and smooth or if patterned – sharp. Some tackiness is expected and tall fat candles are preferred to small thin ones. They should be bleached rather than dyed. The wick should be plaited and positioned quite centrally. In burning the flame should match the size of the candle and be bright. A little guttering is acceptable. The liquid wax in the well of the candle should not be excessive. On being blown out smouldering should last ten to fifteen seconds. Relighting should be easy.
See: candles

**judging honey**
Initially, especially in large classes, elimination points are looked for, these include honey type correct for colour grading or texture, then both jars are checked as identical and both correctly filled. Surviving jars will have the cap of one of the pair loosened slightly. Then the judge savouring its aroma will remove the cap for as short a time a necessary to test density and flavour. The sample will then be carefully examined for any foreign particles on the surface, suspended or on the bottom. Cap and wad will be seen to be clean. Final sorting of the award winners will be carried out by flavour, appearance and using the steward's confirmatory senses if required.
See: honey judge, judging

**judging wax**
Beeswax should have a smooth texture, pleasing characteristic aroma, be absolutely clean, free of blemishes whether cracks or marks, conform to the requirements of the class as stated in the schedule. Thickness and total weight are within limits and the appearance of the piece whether polished or not will then be assessed by the judge so that awards in order of merit can be made. Colour, cleanliness, uniformity of appearance, freedom from cracking or shrinkage, aroma and freedom from adulteration or chemical treatment will be looked for.

**July**
In many areas the season for honey collection and colony development peters out this month. Not in the south west nor for heather goers. Sealed honey can come off at any time, though usually one extraction is enough for most beekeepers. Be on guard to stem the bee's desire to rob – prevent it; cures can be troublesome and may spread disease and cause many bees to waste their energies if not their lives. Heather colonies will need: plumping, young queens and early removal of surplus with plans for their special supers. Other colonies need not be rushed into honey extraction. If feeding is required consider whether beekeeping is not costing you too much. Carry out any prophylactic treatment. Start making winter plans. Keep supers bee-tight. Manipulations by evening or in the rain may give bees an extra day or so foraging. The number of first class foraging days can, in some summers be counted on the fingers of one hand. If you open them up on any of these days it would cut their potential by a very sizeable amount.

**jumbo hive**
A hive with Langstroth periphery but Dadant depth brood frames – 10 with 35mm spacing. It uses either Dadant or Langstroth depth supers. Originally it was called the Quinby hive when it took eight 11¼" frames that were 18 ½" long.

**jumpiness**
Bees that make jerky, fitful movements and are likely to leave the comb being held. Beowulf Cooper described bees in this way

when there were midges about and these were 'biting' the bees.

**June**

Young queens and regular queen cell checks should have prevented or minimized swarming but don't let-up. Freakish weather calls for vigilance. Always be ready to feed or unite. Nuclei, split-board techniques and other methods of making increase and rearing queens can be pursued but remember 'making increase' may run foul of making honey. Queens mated this month are the most useful for heading colonies destined to go to the heather. Keep ahead on space and get foundation drawn, next month might be too late. Look out for other sites or good forage spots for putting hives temporarily or permanently.

**June gap**

Honey flows have habits of developing whenever local flora is available and the weather is right. As a general rule though, a lull occurs when the spring flow diminishes and the main flow has not yet set in. There are plants that fill in this lull, which has acquired the name of the 'June gap'. Mustard, field beans, holly, raspberry and other plants will yield nectar in this period but it is not uncommon to find colonies working hard in excellent weather but losing weight.

**Junior exam**

BBKA exam syllabuses can be obtained from the Examination Secretary. Schools, Scout groups and similar young persons are catered for. A paper is set with twenty questions covering elementary aspects of beekeeping and requiring only short answers and in addition the candidate is examined orally for up to ten minutes.

**Juvenile hormone (neotenin)**

A modifying agent that favours the development of larval structures and opposes adult differentiation. High and low levels of this determine queen/worker differentiation during the third day of larval development. It has also been found to affect the reaction of the larva to gravity, queens remaining upside-down during metamorphosis.

# K

**Kashmir bee virus**
Found in Apis cerana. Dr. Bailey identified a similar strain of virus in Australia affecting *A. mellifera*. Kashmir 'A' the Indian strain and Kashmir 'B' the Australian. It kills adult bees in six days and it can kill larvae. Confirmed in Queensland , N.S.W. and S. Australia. Some bees are virus resistant while other strains are very susceptible.
See: over-crowding

**kataphase**
This is the final stage in cell division from the formation of chromosomes to actual cell division.

**keel**
Lateral parts of a papilionoid flower (Pea-Leguminosae) where united petals form a boat-shaped structure (carina) below the wings and enclose the sexual parts. The wings cover and conceal the keel but these are forced apart and the bee may receive a blow from the staminal column which this action releases. Lucerne, a crop with this type of flower gets various responses by the bees and bad temper at the hives may be one of these. Dr. Adey tells us that foraging honeybees have been shown to deposit a substance which sometimes deters other foragers.

**Kelvin**
The basic SI unit of temperature equal to one degree centigrade, the unit of the Kelvin scale, Symbol K.

**Kemlea Bee Supplies**
Starcroft Apiaries, Catsfield, Battle, Sussex. Suppliers of bees, foundation and all manner of products.

**Kenya**
Several groups of honeybees characterised by the specific geographical distribution controlled by natural barriers include: Apis melifera litorea, monticola and rubica. A. Litorea a small yellow honeybee is found along the hot, tropical, coastal strip up to 500m. A. Monticolae a large black bee occupies the tropical, cool, forested mountains between 2400 and 3100m. The Kenya bee that occupies the Savannah plains is similar to A.m.scutellata (giant bee). A wasp Philanthas diadema captures bees on flowers while Paloras latifrons takes them at the hive entrance. Safari ants (Iridomyrmex spp) are also predators. Traditional methods using hollowed out logs are giving way to the KTBH hive as this is less wasteful of timber. From Bee World No.2 1984 by K.I.Kagatura.
See: KTBH

**Kenya top-bar hive**
It has sides that slope outwards 115° from the base (bees do not attach comb to the sloping wall). It is hung between posts on wires greased at various points to deter ants. A Langstroth frame with median cross bar affords support to combs in the absence of foundation. The ingenious honey badger is said to learn to tip a suspended hive so that combs spill out.
See: honey badger

**Kerr Robert**
He made a hive in 1819 that was superior

to anything that had gone before. Kerr, known locally as 'Bee Robin' had taken the octagonal shape favoured by William Mew (1652) and using wood and cabinet maker's skills produced a bee-space hive which enabled manipulations to be made – a worthy fore runner to the Langstroth.

**ketones**

Organic compounds containing the carbonyl group for example acetone $CH3COCH3$ where the first group is repeated after the CO. Camphor used in making plastics. Ketone bodies, compounds related to acetone, are found in blood, urine etc.
See: aliphatic

**Keuper marl**

This forms a light, loamy soil free from pebbles (sometimes an iron pan develops). Honey is usually darker when derived from plants growing over an iron pan. Good crops of light honey can be obtained under almost any system of management but considerable modification of practice is required to fully exploit iron pan areas. In the Midlands it is often overlaid with soil of transport (mudstone) when iron pan easily develops and only spring honey or blackberry become useful or else by a quartz pebble layer which provides some of the best nectar areas in the country.

**Keuper sandstone**

Found around Birmingham and Liverpool. On this type of soil limes produce a heavy crop every year. A large brood chamber and prolific bees are best here but as Bunter sandstone is found adjacent to Keuper sandstone one often finds adjacent districts that demand entirely different bees.

**kex**

Hollow stems from Heracleum sphondylium hog weed or elder, used as feeders in skeps or for watering (Butler, Keys). Stems also used as candle moulds. The dry stalks of hemlock or other umbelliferous plants also 'kecks'.

**Keys John**

Wrote 'Practical Bee – Master 1780 and 'The Ancient bee-master's farewell' 1796. He used slides between bars – contemporary of Anton Janscha.

**Khalifman I.A.**

Author of '*Bees*' which gained the Stalin prize in 1951.

**kidney**

A bean-shaped, glandular organ which is composed of tubes which condition the blood that passes through extracting waste and balancing levels of substances in the blood. A bee's kidneys take the form of loose Malpighian tubules which ride in the blood.
See: Malpighian tubules

**killer bees**

A scaremongering title glorified by the 'Press' yet originating from reasonably ordinary honeybees *Apis mellifera scutellata* which required careful handling but had escaped from a consignment in the U.S. (from S. Africa) and upset many of the bee-farmers who needed to work with more manageable bees.
See: Africanised bees

**kilogram, Kg**

A unit of mass and weight equal to 1000g and the weight of a litre of water, (2·20462lbs).

**kilometre, Km 0·62137 statute miles (roughly 5/8ml)**

A unit of length and common measure of distances 1000 metres or 3200ft

**kinesis Gk. movement**

Random movement depending on the quality of the stimulus.

**king bee**

Charles Butler. The largest and most important bee just had to be 'King'. The discovery that it laid eggs caused some re-thinking. Still to-day, in some Arab countries, it is still usual to refer to the queen as the Sheik. Many a potential king tumbles from the sky in a DCA.

**king bloom**

The first flower to open at the end of a fruit truss. When this is well pollinated it often produces the largest fruit on the truss (blackberry). In the case of apples Gloria Hoffman (Michigan State University), Red delicious she said could get by with fewer bees if the king bloom opened when conditions were ideal for pollination.

**Kingdom**

The highest category (most inclusive) of the taxonomic hierarchy. One of the great divisions of natural living objects. In taxonomy all living things are divided into two Kingdoms: Animal and Plant. It is possible that a third Kingdom may be erected when more is known about such substances as viruses etc. See: systemics, taxonomy

**knapsack haversack, rucksack**

A leather, plastic or canvas bag or case carried on the back. Such an item can be extremely useful for carrying beekeeping gear as there are usually several pockets and this facilitates finding things that are wanted. Where gates have to be negotiated it is good to have both hands (and arms) free while carrying one's gear on the back. Cover cloths, gloves, hammer and nails, hive tool, matches, metal ends, porter escapes, queen cages, queen marking kit, scissors, secateurs, smoker fuel and veil are all good contenders for space

**knapweed (hardheads) Centaurea spp. Asteraceae**

A perennial that likes sea-cliffs and was once common in meadows and pastures. It flowers from June to August. Resulting honey is said to be thin, golden and to have a sharp flavour. Its pollen is reported as greenish yellow to light brown and pale yellow with a 28μ grain.

**knock-out**

To render senseless or to destroy. Unfortunately there are times when in order to save one form of life another has to be destroyed. a convenient way of dealing with live bees in a matchbox when sent for disease diagnosis or other purposes that call for fresh, dead bees, is to stuff a little Kleenex or other absorbent tissue into the slightly opened end of the box. Then using a small piece of similar means of delivery, wet the absorbent tissue with a suitable killing agent such a amyl acetate and close the box for a minute or two. Alternately place in a frig until dead.
See: ammonium nitrate, amyl acetate, anaesthetic, taking a sample

**knotgrass**

A low growing weed with very small greenish flowers.
See: Japanese knotweed

**knowledge**

Acquaintance with facts and truths gained by sight, experience and experimentation, especially learning from other's mistakes. Beekeeping Associations have been set up to disseminate information and suggest some of the best sources of apicultural knowledge. See: stimulate interest

**Koran**

The sacred scripture of Islam. Honey is a medicine for the body and the Koran is medicine for the soul. The Koran includes one book entitled the '*Bee*' Sura 16. The bee is said to produce medicine for men. '*Honey wherein is healing for mankind*'. The Mohammedan concept of paradise is a place with '*rivers flowing with honey*' A man came to Mohammed saying 'my brother lies ill'. He was told to give him honey and water. The man returned to say his brother was no better. The same advice was given a 2nd and 3rd time whereupon Mohammed said '*Tell your brother his stomach is lying*'. This done and more honey taken the brother recovered.

**Kozchevnikov gland**

A cluster of cells in the sting chamber which probably produce pheromones associated with queen substance.
See: isopentyl

**Kratchmer Gottlieb Germany**

He was credited with having made the first 'foundation'.
See: Mehring

**krupnik**

A Polish drink made from whisky boiled with honey which had to be drunk hot in the winter.

**KTBH**

The Kenyan Top Bar Hive. A movable comb hive based on the same principle as the ancient Greek basket hive. Coffin-shaped and angled to a narrower base to discourage comb attachment to the side walls. Special queen excluders in a trapezoidal frame can be used and the hive is usually suspended by two wires at each end to keep ants at bay.
See: Kenya top-bar hive

**'k' winged**

Workers are sometimes found with hind wings that protrude rigidly outwards at right-angles to the body. The wings separate and jut from the thorax forming the shape of a 'k'. This has been attributed to the presence of the Acarine mite but is also found with healthy bees and can be associated with grooming though a complete explanation has yet to be given.

**kairomones**

Odours, pheromones produced by one species but for the benefit mainly of other species that perceive them. Doubtless this aids commensals, inquilines and other symbionts.

# L

**l**
Litre from which 'ml' milliliter etc. US liter. No full stop should be used.

**label**
A relatively small attachment, often of paper, which bears a description of the contents and details such as producer, packer or distributor. Under EC and recent regulations for consumer safeguard positive requirements are now in force covering accuracy of description, quantity, ingredients, size of lettering and so on.

**labelling**
The affixation of labels to items such as jars of honey for sale should always be done thoughtfully as the label may encourage or discourage sales. Legal requirements demand: name of product, weight of contents, name and address of supplier and country of origin if imported. Size of letters, 1 oz. not less than 2mm, 2 or 4 oz. 3mm, for 8,12, 1, 11/2, or 2 lb. - 4mm while exceeding 2 lb. no less than 6mm. Permitted abbreviations: pound lb., ounce oz., kilogram kg,
gram g and there should be one type space between the figure and the unit used – 1 lb. Special rules usually apply to labelling in Honey Shows. A leaflet is available from 'fera' The Food and Environment Research Agency.
See: marking on containers, lettering, SI

labellum Latin – diminutive meaning a little lip. A small hairy disc at the distal end of the glossa (tongue). It is also called the flabellum which means a fan-shaped organ. Bouton or spoon are other names used. The bee is able to 'lick' with this part of the tongue either to wet a surface or to wipe up small particles of liquid using the curved hairs which assist this function.

**labial glands**
Snodgrass entitles the glands which are ingrowths of the labial segment 'labial glands'. They include post cerebral, thoracic, salivary and mandibular glands. Used in chewing, alkaline up to pH 8·5 but produce no sugary or fatty material.

**labial palp**
One of a pair of long, slender, articulated segments extending distally from the prementum and closing over the glossa to form the tube of the proboscis by aligning with the corresponding maxillae which swing under the tongue.
See: palp – labial

**Labiatae labiate: having lip-like parts**
Dead nettle family, often aromatic herbs. They include: thyme, horehound, mint, basil, yellow archangel, deadnettle, catmint, balm, selfheal, clary, savory, woundwort, betony, vervain, germander, and sage.

**labium**
This means lower lip and it includes the inner members of the proboscis. They are the postmentum, prementum, labial palps, two paraglossae and the complete tongue. It is articulated to the middle of the lorum.

***Laboratory diagnosis of honey bee diseases* - A classic**
Book by H. A. Dade Queckett 1949.

**labour, division of**
Being social insects the honeybee colony has developed into a society where the work is shared out according to ability and availability. In a general way the division of labour in a normal colony during the active season follows the age pattern of the bees. They hatch, clean cells, eat then feed older brood, then younger brood and the queen, older brood again and then make wax, act as house bees, guard bees, foragers for food and water and then collectors of propolis. The chain of divisions may be broken of necessity – when a queen dies or when they swarm or winter ensues.
See: polytheism

**labrum**
This means upper lip and it is the sclerotized flap that is hinged to the lower part of the clypeus. On the inner side is the epipharynx.
See: labrum, epipharynx, sclerotin

**laburnum *Laburnum anagyroides* Genisteae**
A tree that lights up the suburbs with its golden chains hanging in clusters until they carpet the pavements with their spent petals. No reports of bees on it and it is said to be poisonous – the black seeds certainly are.

**Lachnids**
E.g. *Lachniella costata* which is found on spruce, *Cinara pectinate* on firs, *C. acutirostris* on *Pinus nigra* in southern Europe. also *C. brauni, shimitshecki* etc. Plant sucking insects that produce honey-dew. It is becoming evident that much more honey-dew honey is produced than was generally supposed and many sources such as laurel, holly and various fir trees go unnoticed because when bees disappear into the dark green foliage they are lost to sight.
See: honey-dew

**lacinia pl. iae**
Mouthpart, flap or hinge. These are found on the inner surfaces of the maxillae between the large stipites and galeae. Although the laciniae are quite small lobes on the honeybee they are larger in less specialised insects.

**lacquer**
Resin or cellulose ester dissolved in a volatile solvent. The resultant varnish used to be baked onto the lining of honey tins as it produces a protective semi-durable coat on certain metals thus preserving the metal from the corrosive effects of honey and water. Its place has been taken over for honey containers which now use hard, clear polyurethane varnishes or just plastic buckets. Propolis can be dissolved in suitable solvents (alcohol) and varnishes made in this way were said to be used on violins made by such famous people as Stradivarius.

**lactic acid**
A hygroscopic, syrupy liquid found in sour milk and other substances and small quantities are found in honey. Its chemical formula: $CH_3CH(OH)COOH$.

**ladder**
Bees hanging onto one another, making wax (Sydserff in Cotton). *A festoon*. The open cells' edges make a vertical ladder for the bees whose hooked feet can easily pass over them.

**laevorotatory**
Anti-clockwise rotation. The physical properties of substances capable of rotating the plane of polarized light to the left or anti-clockwise. The honey sugar fructose is called levulose on this account.
See: dextro – right hand rotation

**Laidlaw, Dr. Harry H. Houston, Texas 1907**
His dedication and enthusiasm for beekeeping have won him a great reputation in the American beekeeping industry and

throughout the world. He made important contributions to bee research, particularly in bee genetics, queen breeding and artificial insemination and in the means of halting the northern advance of the Africanised bee from S. America.

**lake**

A large inland area or large pool of usually fresh water. Islands in such lakes sometimes offer opportunities for isolation breeding though where water surrounds and covers what would be the foraging area, food can be a problem. Still water can be dangerous to bees as reflected light can cause them to regard it as 'sky' and swarms have been known to settle around a queen and drown. In a dry area a lake-side site can be very beneficial though access must be considered. As always where relatively calm water exists there comes the chance to keep hives on rafts and thus move them to follow the forage.
See: waterways

**lamina externa**

The galea of the trophi - shield of the mouthparts? The lamina interna forms the lacinia of the trophi.

**laminated**

Composed of layers beaten or bonded into a plate or sheet. Foundation is usually made out of one-piece beeswax but laminated forms using a basal, plastic layer of other materials are made.

**lancets**

Two of these spears work in co-ordination with the shaft of the sting. The inner surfaces slide against one another and ride in grooves in the shaft (stylet) while the outer surfaces each have upwards of half-a-dozen barbs which open out into the wound when the bee pulls on the sting but close and allow the lancet to enter more deeply as the sting is forced further in. They are made of tough 'chitin' and are of a reddish hue.
See: barb

**Langstroth frame**

Like most frames it has a deep size and a shallow size though in heavy honey producing countries the deep size is often used to the exclusion of the shallow. The top-bar is 19" long and the side bars 9" deep LD, and shallow side bars 5" LS, bottom bars 17" (483, 229, 127, 432mm) thus giving ample overlapping lugs at either end of the top bar. It is usual to have Hoffman type side bars.
See: Hoffman

**Langstroth hive**

Although one of the first movable frame hives it is a remarkably efficient. Economical hive using a minimum of timber to give maximum volume. The shorter lugs save two inches of wood and yet fulfil their purpose adequately. Being oblong has hidden advantages, when empty one chamber can be put inside another, and when carried, using Hoffman frames, a box can be held in the more comfortable position against the chest because even when combs are horizontal instead of vertical they are firmly held in position. Some beekeepers are starting to realise that by working with just one size of frame manipulations like swarm control and queen rearing become easier. Nevertheless, probably because deeper frames were required in the brood chamber while honey supers were lighter with shallow frames – the Langstroth follows suit with two frame sizes. Boxes have a periphery 19"x 16 1/4", deep frames 9", shallows 5" and top bars are 19" long. 10 frames to a box and another advantage is top spacing (metric dimensions given in entry above).
It is the most widely used hive in the English speaking world and is totally adequate in all climates.

**Langstroth, Rev Lorenzo Lorraine USA 1810 – 1895**

He patented the first commercially produced hive using bee-space and it has perhaps become the most widely used hive in the world. In 1853 he published 'The Hive and the Honeybee' and annual revised editions followed. In 1861 the American Bee Journal carried an eye-witness account of the queen's mating.

**language**
Beekeeping terms in several European languages were edited by Dr. Eva Crane and published by IBRA. Like music and mathematics, languages seem to put us apart from other creatures. Nevertheless communication uses other forms than speech and the social life of the honeybees gives ample demonstrations of the manner in which they share and act upon information. Pheromones and in-built characteristics arising from 'genes', allow the bees to use their sensitive organs in such a way that an intelligence comes forth, the use of which must put us to shame because they are highly beneficial to the eco system and have survived through ages which we are unlikely to be able to emulate. Their dances, comb building, complete submission to duty and skilful use of the earth's resources returning benefits reciprocally, leaves us linguistically short of words.

**Lanius spp.**
A predacious bird such as the shrike which feeds on insects amongst other things.
See: shrike

**large intestine**
The colon or rectum of the bee which follows the small intestine and terminates at the anus.

**large-scale**
This indicates carrying out a particular activity in a big way. Bee-farming on a large scale would imply keeping hives in thousands with many off-shoots such as hive manufacture, fulfilling pollination contracts, queen breeding, package bees, honey production and making the best use of all hive products.

**larva pl. larvae**
The active insect stage between ovum and pupa. The grub, embryo or baby bee. It is white, helpless and confined to its cell, must be incubated within a narrow range of temperature and humidity. It is that stage of a honeybee's life from hatching out of the egg to the pre-pupal stage when sealed in its cell. a newly emerged larva is 1·6mm long 0·1mg in weight and legless.
See: grub

**larval colour**
In the case of a normal, healthy larvae they are described as glistening, pearly white. The significance of this fact lies in the recognition of healthy brood. Pollen and other causes can occasionally lead to a creamy or even pinky tinge but anything other than the normal healthy colour should be investigated by an expert as soon as possible. Chilled brood can be grey or black and glutinous brown indicates the serious AFB condition where a match stick might pull out a gluey brown 'rope' – very infectious.

**larval faeces**
Each larva deposits two layers of light-brown cocoon material in its cell and its faeces are sandwiched between them particularly around the cell base. Should these be impregnated with pathogens such as nosema spores or amoeba cysts, the spread of such diseases is a distinct possibility. This is an important reason for replacing brood comb after it has been used for several generations of brood. During the winter faeces are withheld until a short flight is possible because bees normally defecate on the wing away from the hive. Should they soil clothing, marks should be wiped off as quickly as possible with a damp cloth. Dysentery is a condition where streaks are left on the outside of the hive and on top bars. Unless such material is shown to contain pathogens it is usually a passing condition, brought on for instance by feeding unsuitable material.

**larval food**
This is called brood food and is produced

by the brood food glands of nurse bees who feed on a diet of nectar or diluted honey and pollen. It is a creamy secretion which at its richest is known as royal jelly. Queens and drones are well supplied with the richest food while worker brood receives a slowly changing diet with a gradual weaning onto nectar and pollen. The feeding stage of all castes last about six days.

**larval growth**

This is remarkably rapid and the weight and size of the larva increases more and more quickly until full size is reached after almost six days of feeding. It is governed by the hormones AH and JH and the controlled and careful nursing that include 1300 visits and a steadily changing diet.
See: brood food

**larval skin**

A larva sheds its skin six times during the feeding process. As the old skin splits and its place is taken by a larger skin that has developed underneath, it is dissolved by enzymes and re-absorbed. Pearly white and slightly translucent, its appearance is a good guide to its health.

**late**

Month-by-month details of work to be carried out and the four seasons covered under their various headings. Time and tide wait for no man and you should always try to keep ahead of the bees requirements. Feeding should be completed in good time to allow for the proper 'ripening' before sealing but this takes much energy out of the bees whose strength needs to be conserved for over-wintering. Mouse guards might trap a mouse inside if put on too late.

**lateral**

From, at or pertaining to the side. Lateral movement would mean sideways movement. A lateral extension – one arising from the side. A side elevation as in a plan, would be the lateral view. Some hives like Nutts and Long idea, make use of lateral as opposed to vertical development. Frames are then arranged in juxtaposition. Lateral thinking - solving problems by associated ideas, imagination as opposed to step-by-step reasoning – useful in certain situations.

**laurel *Prunus laurocerasus* Rosaceae**

A small ever-green tree. Its flowers are visited in April by bees for nectar and pollen but its main attraction is its extra-floral nectaries under the leaf surfaces. Laurus nobilis Bay, has small yellow flowers in May/June which are well-worked by the bees and its leaves are used as a flavouring.

**lavender Lavandula spp. Lamiaceae**

A small ornamental old-world shrub that prefers light, chalky, well-drained soils. Its pale-purple, fragrant flowers yield nectar and pollen from June/July and on to the frost. It can be dried to perfume drawers of linen and in Victorian times was burned to perfume rooms. It yields an essential oil used in porcelain painting and veterinary medicine. Its honey is famous, being described as golden or dark, granulating smoothly and having a most pleasant taste and fragrance. The glorious blue in cultivated rows has been used to grace many a French and Austrian travel brochure. See: Narbonne and sea lavender.

**lawn**

An area of grass that is kept mown to encourage short, dense grass for recreation or a framework for flower borders. While little by way of forage is found on bowling greens, it does not take long for the ubiquitous clover, dandelion and other flowering plants to reach dehiscence on some lawns which often benefit from watering when all else is dry.

**lay flat**

To rest in the horizontal position. When a comb of brood (not queen cells) is laid flat over the top bars, the bees will continue to rear the brood if they are strong enough to sustain incubation and will use the stores as opportunity permits. If left the bees will eventually rebuild the comb in the vertical plane.

**laying**

A queen should begin to lay about twelve days after emergence if weather conditions are suitable for mating. She confines her activities to areas where the cells are kept at appropriate brood rearing temperatures (about 35C/95F) and in cells that have been cleaned and polished and marked with the pheromone that stimulates her to lay. Occasionally a queen will begin laying more than one egg in a cell but this usually rectifies itself once she has settled down. A large number of eggs in one cell are usually indicative of laying workers. A queen always seems very keen to lay in drone cells (and should never be given access to drone super combs) but not so keen, unless constantly chivvied, to lay in queen cups.

**laying queen**

A mated, laying queen may well be described as the heart and soul of a colony. While she can lay an adequate number of worker eggs and yield pheromones that inhibit queen rearing, the colony behaves normally. A queen may be present without laying as in period of dearth and some winter months. Or she may not have mated or become a drone layer, or she may be laying as well as a daughter queen in the same colony.
See: drone layer, mated queen

**laying rate**

A queen's egg laying rate is not only governed by the nature of her food and the amount that she is given but also by the number of cells that have been prepared and are at the right temperature. The rate will vary according to the prevailing conditions such as the ambient air temperatures and the inflow of nectar and pollen. Furthermore younger queens will lay faster and more continuously. A rate of one every seventeen seconds having been observed though this would mean 5000 eggs a day whereas 1000 a day is a more likely average over the peak laying period in the UK; some races of course, being much more prolific than others.

**laying worker**

Although workers are imperfect females and cannot mate, their vestigial ovaries tend to develop at the same time as their brood food glands, especially when they are unable to dispose of their brood food but are inhibited from doing so when abundant 'queen substance' is available. Races vary in their willingness to permit laying workers and small patches of drone brood above the excluder can be tell-tale signs. Only Capensis varieties have been known to lay diploid eggs, laying workers in Europe laying only the haploid eggs that become drones. A colony that becomes, or is made queenless, soon has laying workers. A cell may receive batches of scattered eggs though laid by different workers. Such colonies are not easy to requeen except by uniting. A Taranov board can be used to get rid of them as they will not fly. Give such a colony a frame of emerging brood a day or two before attempting to re-queen.
See: re-queen

**LD value**

Lethal dose. Measured in micrograms, a chemical expressed as a percentage of the number of contacted bees that die. e.g. LD25 means it will kill 25%.

**L/D value**

Used to indicate the laevulose/dextrose ratio of these sugars in honey. Not to be confused with LD – lethal dose.

**leaching**

The effects cause by the constant percolation of water through anything. The gradual impoverishment of surface soil due to rain dissolving the soluble materials and allowing these to percolate to lower levels. The 'leachate' is the drained water with its higher concentration of salts. Acid rain can leach calcium ions from the soil and leaves of plants upsetting the normal production of enzymes.

**leaf-cutting bee Megachile spp. Aculeata**

A genus of carpenter bees that builds the walls of its cells from pieces cut out of leaves of plants especially roses. The long row of cells resembles the appearance of a green cigar. Such bees are loathe to sting. Megachile – 'big lips'. Most are parasitized. Bred in Canada and the States for the pollination of alfalfa.
See: *M. rotunda*

**leaf hive (Huber)**

A specially designed hive where supers could be set upon a brood chamber which had frames of comb that were hinged and could be opened out like the leaves of a book for examination or observation. It was also referred to as the 'book' hive.
See: Huber

**leaf honey**

The term honey-dew seems to have supplanted the expression 'leaf-honey, yet especially in dry summers, all around are leaves glazed with a sweet, sticky covering that is often worked by the bees. Leaf honey is therefore honey derived from the leaves though in fact this is almost certainly the exudations of underleaf sucking insects and is therefore more properly known as honey-dew honey.

**leaflets**

Most beekeeping organizations produce leaflets giving condensed information on a variety of aspects. BBKA, BIBBA, IBRA and a whole number of other sources help to keep us informed and up-to-date. Examples include: Show rules, Trees for bees, Queen introduction, Taking a sample, Swarm control, Honey and so forth.

**leaking feeders**

Feeders that rely on air-pressure to hold their contents can suffer when increased temperatures by day cause the air-space over the syrup to expand and this forces syrup onto the bees. Lever-lid and plastic honey buckets have this fault though well-insulated roofs minimize the problem as does putting warn syrup onto the hives. Other causes of leaking are ill-fitting lids or minute holes as are inclined for form in aluminium feeders, especially if they are not regularly washed out.

**learner**

Many people learn about bees, even having lessons in handling, yet do not take up beekeeping. As confidence and knowledge is only acquired by fairly frequent, albeit intermittent practice, it behoves those whose wish to acquire these attributes, to join a local association and arrange with beekeepers to have practice at manipulating colonies from time to time.

**learning by bees**

Adaptive changes of behaviour that occur as a result of experience. There are five kinds: Habituation (weaning), Conditioning (trial and error) Latent and Insight (adaptive re-organisation of experience). To what extent, if any these processes can be transmitted from one generation to another e.g. comb building, is of great scientific interest.

**leather**

The prepared skin of animals can incite bees to waste their stings on it, especially the non-shiny, non-polished kinds. Be careful not to wear anything of the sort when working with bees. Normally shiny belts are all right but a loose tag might invite attention and cause bees to throw away their lives.

**leather polish**

For brown shoes or upholstery: 3oz/84g beeswax; 1oz/28g white wax; 1oz/28g Castile soap; 1 pint/.571 turpentine; 1 pint/.571 rainwater.
Method
Shred wax into a jar and add the turpentine. Place the jar in a water bath and heat gently

until dissolved. Care as its flammable. Boil the water and stir in the shredded soap. When solutions are quite liquid and still 'warm' pour the soapy one into the wax solution and stir well. Pour into jars, shake before use.

**lecture**

A lecture as opposed to a talk, usually aims to instruct and educate and implies that the lecturer is qualified and able to impart some of his/her knowledge to some of the audience. People are sometimes too modest or lacking in confidence to give a lecture even when qualified to do so. Confidence comes from careful preparation, a sensible assessment that the material offered is sound and likely to be of use to the many in the audience, and the pleasing assurance that when it is a beekeeping audience they love to hear about their favourite subject and are usually quite tolerant.
See: lecturer

**lecturer**

A regular member of the teaching staff at a College or university. A BBKA exam is open to those who have passed the senior examination. It takes place before a public audience, lasts 30 – 40 minutes and two examiners are present. At the end of the lecture ten minutes is devoted to questions from the audience or examiners. This followed by an impromptu talk lasting between ten and twenty minute on a subject relevant to beekeeping, chosen by the examiners and on which question arising have to be answered. A successful candidate is known as a BBKA Lecturer. Qualified beekeepers at Agricultural Colleges were once employed as lecturers.
See: bee talks, Hons lecturing, speaker

**lee**

1. The sheltered part or the side turned away from the wind. An apiary site that is sheltered by a rise in the ground or by a wall, fence or thick hedge is in a favoured position. Hence leeward and windward sides.
2. When making mead or vinegar the material deposited at the bottom is given the name of 'lees'.

See: dross

**Lee James Uxbridge**

His father Robert started a well-known firm in 1862 producing beekeeping appliances. James introduced the plinthless hive in 1898, a time when most hives used external over-lapping plinths which tended to make them difficult to separate (no point of entry for the hive tool).

**leg**

A body appendage used to support and perambulate it. The bee has three pairs of these segmented appendages and the front pair can be used in the fashion of arms when scrabbling for pollen, or building honeycomb or tussling with invaders. They are shorter than the middle pair and possess antennae cleaners while the hind legs are the longest. Each is segmented - starting with the coxa, then the trochanter, femur, tibia and tarsus (the latter having five divisions). All six feet are similar and have strong pairs of claws and pads for walking on smooth surfaces. The legs of the worker are highly specialised for food collection, use in conjunction with mouth parts, for cleaning the antenna and for general work within the nest. They allow the bees to cling together as when balling or hanging in a swarm, making comb or rarely 'to ball the queen'. The drone's legs are designed – like his whole body – for mating (lacking hive tools), while the queen's legs are highly important as they help her to deposit eggs accurately, to cover miles and miles on the combs and are not only strong but make her stand out on the comb and when necessary, to do battle with her sisters.
See: fore, middle, hind legs

**legality**

There is no regulation forbidding the practice of apiculture on private property in the UK. This is not so in some countries. It is required that no nuisance is caused

to other members of the public. You are expected to take all sensible precautions when transporting colonies of bees (mark your car 'Live Bees') and inform the police. Permission must be sought and often fees paid to position hives on National Trust, Forestry Commission or Water authority property. There are many regulations governing the preparation and sale of honey. Indemnity against harm from stings or cleanliness of honey is often included in an insurance policy taken out.
See EEC Regs, labeling

**leg honey**
See sandarac (Butler). Possibly too the meaning of pollen carried on the legs having the appearance of sandarac.

**legume (pod) Leguminosae, Dicotyledons, Angiospermae.**
Plants of the taxonomic family Leguminosae, the pea family including: bird's foot trefoil, broom, clovers, gorse, laburnum (poisonous), lucerne, medick, melilot, peas, restharrow, rue, sainfoin, vetches etc. The word also means pod, seed vessel or vegetable. Their root modules harbour the nitrogen fixing Azotobacter. The plants need bees for pollination and many give good honey crops. As soil improvers they succeed without need of artificial fertilizers.
See: Azotobacter, pea-flower family

**Leguminosae flower**
Five petalled flowers, the upper is a broad petal called the 'standard', side petals called 'wings' and the two bottom petals (often united) are known as the 'keel'.
See: pea flower family

**lemon *Citrus medica* var. *limonum* Rutaceae**
A yellow acid fruit of a sub-tropical tree. Its flowers are visited by bees which improve crop yields while they benefit by early nectar and pollen. Grown along with oranges and grapefruit in NZ gardens. Lemon juice is used in a variety of drinks which are much enhanced by the addition of honey. Honey, lemon and glycerine can make a sore throat more comfortable.

**lemon balm *Melissa officinalis* Lamiaceae**
Bee balm – a native to Mediterranean area of Europe but spread as a cultivated herb. Its essential oils contain elements similar to some that are found in the Nasanov gland secretions and before aspirin an infusion was used for the relief of headaches. Reference to its attractive power to bees were made in Virgil's 4th Georgic and frequently since. It was crushed and used to entice swarms into new homes.
See: bee balm, Melissa, mint

**length**
The linear dimension, size or time of anything measured from end to end. This is obvious enough when the ends are clearly defined but can lead to trouble with frame sizes according to whether the lugs are included. Depth, width and thickness together with length enable the full dimensions of a hive section to be given but beware of ambiguity and confirm with a sketch if necessary. Day, time, tongue, festive seasons, and lengths of winter are given under these headings.

**length of day**
As this increases with higher latitudes the shorter active season is sometimes compensated by more flying hours in the day. As long days aid plant growth and nectar secretion, bees do well when moved north, though wintering difficulties usually force beekeepers to winter their bees further south. This is not noticeable in these small islands but in America and Japan advantage is taken of northern honey flows and southern breeding programmes.

**length of frame**
The standard British (BS) frame is 432mm/17" long and can be used with long or short lugs to give 366mm/14" or

406mm/16" comb lengths. When uncapping the longer shallow is as easy to deal with as its shorter companion. American frames only use short lugs giving more comb space and they are 482mm/19" long and comb length 423mm/17".

**length of life**

Worker including life as brood – 6 weeks (summer) up to 6 months (winter). Drone including life as brood 3 – 4 months (they sometimes overwinter). Queen including life as brood, 3 – 5 years (exceptionally up to 7 years) they can be superseded or replaced by the beekeepers. A colony which might supersede instead of swarming and go on indefinitely.
See: age of queen, worker brood

**length of the active season**

This is mainly dependent on the hours of daylight and temperature and varies by as much as a month between extreme areas in G.B.

**length of time to mate and lay**

A queen usually makes her first cleansing and orientation flights 3 – 5 days after emerging. She makes a serious effort to secure mating when she is 5 – 14 days old. At 30 days she is likely to fail altogether and if kept, is likely to become a drone breeder. Upon return from her matings, her ovaries develop and she lays within 2 – 3 days.
See: mating

**length of tongue**

Extending from the labial prementum down to the labellum or tip of the tongue it varies with different races of honeybee from 5-7mm. This is longer than many bumble bees though shorter than some. The tongue length of honeybees can govern the plants that they can effectively work for nectar. Plants like honeysuckle with extra-long corollae are out of reach though bees may nibble through the base on some flowers. Red clover is borderline, the second cut being within reach of most honeybees.
See: tongue length

**length of winter**

Although this is entirely forecastable in terms of daylight, typical winter weather is hard to identify for Great Britain and as yet even harder to forecast. However the honeybee season closes down in October and no nectar of any significance is likely to be available again, until at the earliest, the following April. If we label this the bees' winter it lasts for seven months and all colonies require a food reserve of 20kg to take them through the inactive period.

**lens pl. lenses**

A disc-shaped transparent object with curved opposite surfaces that cause light rays to converge and allow focusing. It (they) may merely collect light and transmit the fact by nerve impulses to the brain as with a bee's simple eyes. They may form an image as on the human retinue, camera film or screen, or they may be used to magnify or reduce an image. When two or more lenses form a unit this is known as a compound lens.
See: compound eye, microscope

**Lepidoptera Insects, Arthropoda**

An order of endopterygote insects, butterflies and moths. to a small extent honeybees may compete with these creatures for food. The death's head hawk moth does 'steal' honey but may lose its lives in the process (it is a migrant moth to G.B.). The two wax moths 'greater and lesser, are pests to be reckoned with especially when you are storing empty honey comb. They have egg, caterpillar and chrysalis stages before becoming an imago.
See: moth

**lesser celandine *Ranunculus ficaria* Ranunculaceae**

A native perennial with heart-shaped leaves and yellow flowers that blooms from March to May and is worked for yellow pollen which microscopically appears rough with 3 grooves.

**lesser wax moth *Achroia grisella***
Like the greater kind it is drawn to wax comb as if by a magnet. A strong colony might keep it well in bay but any comb not protected can be riddled with webbed tunnels all impregnated with droppings.

**let-alone beekeeping**
Since the advent of movable frame hives it has become possible to manipulate colonies at will. Many come in for regular inspections often in the form of week-end beekeeping. On the other hand let-alone beekeeping either implies leaving the bees to their own devices or reducing the number of examinations to an absolute minimum. For example two annual manipulations involving 'supers-on' and 'supers-off'. This makes checks for disease, comb replacement and requeening difficult if not impossible. It is unlikely that let-alone methods of beekeeping can be reasonably adopted if one is surrounded by other beekeepers.

**lethargic**
Apathy or the drowsy suspension of activities. This term might be applied to bees or people especially when they are not pursuing the course of action which you feel they should. A colony might appear lethargic if it were in the throes of some disorder – parasitic or starvation or possibly prior to swarming.

**lettering**
Size on honey jars less than 50g 2mm letters, 500g 3mm. Full details of labelling can be obtained from Fera.

**leucocytes**
Colourless, white blood corpuscles, concerned with the destruction of foreign matter including disease pathogens (micro-organisms) They can become phagocytes.
See: phagocyte

**level**
Although apiary sites may benefit by not being on the level, hives themselves should be level in all respects other than a slight forward incline to assist the removal of debris and discourage the entry of water. A saucer of water can act as a spirit level. An area of level ground large enough to enable the operator to stand and set out equipment while manipulating is also a good thing close to a hive.
See: horizontal, spirit level, vertical

**levorotatory laevorotatory**
Left-hand or anti-clockwise rotation. Applied to translucent substances that can rotate a plane of polarized light so that it enters at one angle and comes out in another plane which is inclined anti-clockwise to that of the entry plane.
Honey which fructose dominant rotates slightly anti-clockwise.
See: dextrose

**levulose laevulose**
Fructose or fruit sugar, one of the chief components of honey. A sugar solution that rotates the plane of polarize light anti-clockwise – to the left.

**liaison bees**
Bees acting as go-betweens such as scouts endeavouring to lead a swarm to a new home, or house bees that meet returning forages and help them by off-loading and distributing their booty. Bees whose loyalty is beginning to shift into a new direction as when a new queen is introduced or uniting takes place.
See: beggar bees

**lias**
Marine sediments as exemplified in the Lower Jurassic rocks of N.W. Europe, a kind of limestone laid under the influence of the sea and similar to chalk but weathers to a stiff, pale grey, retentive clay. In heavy rain it is too wet and cold for clover to secrete, while dry weather causes the clay to crack

and dry quickly. Only in occasional warm, humid summer can clover yield well. A wider variety of flowers do well on this soil and trees (sycamore etc.) call for a non-prolific bee able to gather an early harvest without turning the crop into brood. Ling in Westmoreland yields on this limestone due to an overlying acid layer.

**library**
A place where reading material is available and can be read. While excellent libraries of bee books have been set-up, many local associations run small lending libraries. These are worthy causes as there is often as much pleasure to be derived from reading about bees as actually keeping them and much knowledge can be acquired in this way.

**Libya**
N. Africa where a long strip of land borders the Libyan desert. Their bee is Apis intermissa. Wax moth can be a serious problem and where re-afforestation takes place it is not uncommon to find too many hives on top of one another on a single site.

**lid**
Roof or cap, the outer cover or closure for a box or container. It may be hinged or removable. Honey jars use lids that can be screwed on tightly so as to seal them reasonably well from the effects of atmospheric moisture. Past bee-masters have intimated their displeasure when beekeepers have called a hive roof a lid.
See: cap

**life cycle**
The development from egg fertilization to the mature adult stage or the continuous cycle as of a honeybee colony. An organism's pattern of life from birth to death, including the production of gametes. Each of the three castes of the honeybee has a very different life cycle, while the colony has but a seasonal cycle. The speed of completing a cycle is significant in the case of pathogens. Where bees can outbreed parasites for example the infection may diminish or disappear.
See: length of life, drone, queen worker, worker brood

***The life of the Bee***
A classic by M. Maeterlinck – an enjoyable read. Several books same title.

**lift**
An outer hive section onto which the roof fits. It has to be 'lifted' off to uncover the inner hive. The lowest lift of a WBC (sometimes called a 'riser') incorporates a porch with entrance slides.
See: hive lift, riser

**light**
Visible energy received from the sun or that which enables objects to be seen. Bees are governed to a large extent by light intensity and only fly away from the hive when the sun can be used as a navigational aid. The bee's inability to see red light and the significance of ultra-violet and polarized light are explained under the appropriate headings. The increasing length of daylight and light intensity in the spring, causes nature to stir. The queen is stimulated to start a new brood nest. Entrance flight activity is often directly proportional to the brightness of the light (even luring them out when snow brightness2 occurs). When trapped they always try to escape via the light.
See: photo positive, polarized, shield entrance, u/v

**light compass reaction**
Human eyes cannot see the patterns of polarized light emanating from the sun without the aid of special glasses. Bees can however, and they use this information to give them a datum line from which directions can be angled. 'Down – sun' (in the direction of the shadows to us) links with 'down' in the hive and angular bearings to left or right of down are interpreted into similar flight angles from the down – sun,

polarized light indications. Thus the bee uses sun-light as a compass pointer.
See: homing, compass, polarized light

**lighten beeswax**

Beeswax can be bleached almost white by the use of hydrogen peroxide (H2O2), using 10ml per pint of water and bringing to the boil.

**light honey**

Honey whose colour falls within the light class: acacia, apple, charlock, clover, holly, brassicas, raspberry, rosebay. Further descriptions may be used as light honey covers one third of the range.
See: colour comparator, honey spectrum, pfund

**lighting a smoker**

Practice makes perfect but it is possible to burn your fingers (or gloves), burn a hole in the veil or even set fire to the forest. It helps if this can be done out-of-the-wind. Use perfectly dry, even 'primed' fuel and use the fire chamber to shield the match or lighter flame. Light the bottom end of the smoker cartridge or place 'easy-to-burn' (crumpled newspaper, wood shavings) to get ignition and initial heat then add one of the many things, like sawdust, dry leaves, dry rotten wood, old sacking, dry dung – there are hosts of smoulderable substances, but if the smouldering material can break loose, plug the smoker with green grass. It is best to get it well-alight by using blasts from the bellows to help, then when a puff or two tells you that it is in good working order, it can be rested in a vertical position. Keep it down-wind, avoid breathing the smoke it is extremely bad for you, and avoid getting the smoker (and therefore the smoke) too hot. Use as sparingly as possible, the bees don't like it, and it is harmful to their chitin and especially to the brood. Keep some spare fuel handy and never take chances with the remnants when you have finished with the smoker. PUT IT OUT! Stamping is risky when there's a wind, better to keep some water available to ensure this is done really effectively. Alternately plug the smoke aperture with a twist of grass, a cork or similar plug.
See: smoker

**light intensity**

Work inside a beehive is continuous during the active season but by day, light intensity has a marked effect on foraging. In conditions that are otherwise suitable, dull or gloomy skies reduce the number of foragers while bright skies encourage what is often called 'furious activity' at the hive entrance. An exposure metre will give a reading, though bees are sensitive to aspects of light such as the colour temperature and ultra violet light.
See: light

**ligroin**

A petroleum spirit – mixture of paraffin hydrocarbons – with a boiling range of 70-120C. Highly volatile and sometimes included in treatments against certain parasites.
See: benzene

**Ligurian bee (Italian)*Apis ligustica*, a variety of Italian bee from the region of Liguria, NW Italy.**

**Ligustrum vulgare privet**

This plant can grow into a small tree and produces masses of flower that give off an over-powering (non-too pleasant) odour and which are avidly worked by the bees, some beekeepers take off supers to avoid its inclusion, but honey tainted with the smell does improve if kept for a while. The plants large black berries are inedible to man There is a Japanese variety L.ovalifolium an evergreen with less conspicuous flowers that are 'alive' with bees in July.

**lime *Citrus aurantifolia***

The sweet lime with small greenish-yellow, acid fruit of a small tropical tree, and as so often with fruit names, the name of the tree itself. Famous for its thirst quenching lime juice and makes a nice beverage when sweetened with honey.

**limestone**

Rock composed of calcite carbonate CaCO3, the calcareous remains from a past geological age when the chalky skeletons of marine creatures were laid down. Although hard, it is partly water-soluble and has a pronounced influence in maintaining alkalinity in over-lying soils while permitting excellent drainage.

**lime tree Malva spp. Malvaceae**

The linden tree, a major honey producer in parks, gardens and tree-lined avenues and streets of European towns. here are many varieties including Korean, Chinese, Japanese, American, Russian. A large tree will offer a canopy of blossom equivalent to a whole garden. In London blooming begins around the 3 week of June but varieties go well on through the summer. They are sources of honey-dew too which aphis produce and which becomes covered by the soot fungus that develops on the glistening upper surface of the heart-shaped leaves. It is collected along with the honey-dew making a very dark, almost black honey. Some have even jibed – it's like axle grease. In America the 'basswood' is used for making the 41/4" square honeycomb 'sections'. When pure the honey is not very dense, has a greenish tinge and a slight peppermint taste. The pollen' colour has been called yellow, pale-yellow, greenish yellow and orange and the grain is 25µ. It is said to yield well on Keuper-Sandstone but its flowering often coincides with blustery weather, Plants may take fifteen to twenty years to bloom. Some of the late varieties have a bad name on account of carpets of so-called 'drunk' honeybees and especially bumbles that form a carpet beneath the trees causing better grass to grow there. However either the nectar or the pollen seems to cause a paralysis of the bees flight muscles particularly in dry years on dry soils but this might be due to the honey-dew. It is a tree associated with happy lovers; 'The night air is full of heady perfume, The nectar literally fell in showers and lovers walked beneath your linden trees'. Seeds need to be scarified and softened with $H_2SO_4$ then scattered thickly 1cm deep following a period of severe drought.
See: linden

***Limnanthes douglasii* Limnanthaceae**

A hardy annual variously called 'cheese and eggs/ poached egg plant/ butter and eggs and meadow foam'. It self-sows and makes a nice rockery plant 15cm tall and forms a mass of yellow/white flowers which are worked well by the bees for both nectar and pollen. Autumn sowings give plants that follow crocuses and the flowers are sweetly scented. Pollen is colourless or very pale-grey with a green tinge, grain 22µ. They do best in damp humus in full sunshine. Sow Sept or Mar.

***Limonium vulgare***

A herb with bluish-purple flowers which grown on coastal, salt marshes and cliffs, further details given under sea lavender.

**Lindauer, Prof. M Munich**

An avid researcher whose work included activities of bees, comb building, communication, dances, nurse bees, organs of gravity, orientation, swarming behaviour and water storage amongst other things. He wrote 'Communication among social bees' 1961.

**linden**

The German name for lime tree or basswood. Late frosts can damage the blossom. Leaves are heart-shaped and the pendulous, yellowish cream-coloured flowers often fill the air with perfume. Some varieties are inclined to drip with honeydew and this can be unpopular in towns where cars and pavements are affected. Tilia americana is used for basswood sections and the silky finish makes the timber useful for 'turning' while the shavings make excellent smoker fuel. *T.periolaris* and *tormentosa* on

dry soils and especially after dry periods produce nectar and pollen that is toxic to bees as the 'mannose' in the nectar causes paralysis of the flight muscles. *T. oliveri* and other large beautiful Oriental trees are harmless in this respect and free of honeydew.
See: lime tree

**line**
A pedigree resulting from known lineal descent, lineage, genetic continuity or unbroken links in a family tree traceable through several generations.
See: breed, phenotype, race, strain, sub-species, variety

**line breeding**
A strain, race or breed produced when parentage of queens is controlled so that the offspring display characteristics that are continuous with earlier family traits. Artificial insemination has given us better control, but queens so treated are more suitable for the 'banking' of genetic material than the production of honey-producing colonies, though advances are still being made.
See: in-breeding

**liner**
Bees line their nest cavity with a thin layer of propolis re-enforced where necessary to fill cavities and keep out small creatures and the weather. Honey and syrup containers can be re-enforced with plastic bags. The ubiquitous food tin (can) can also be selected for size and inserted with both ends removed, into the fire chamber of the smoker and, being replaceable, extends the life of the smoker and aids cleaning.

**ling *Calluna vulgaris* Ericaceae**
Anglo-Saxon 'lig' means fire and the moors in full bloom have an autumnal beauty enhanced by its characteristic colour and aroma. The common heather is a perennial shrub extending over quite wide areas (especially moors) in Great Britain. Due to its late flowering (Aug-Sept) and tendency to suppress other flowering plants, bees are rarely kept permanently on the heather but are moved to it during August where it acts as an additional crop for most beekeepers. They are then brought home again for wintering in less rugged quarters. In choosing sites you should be cognisant of access, shelter and large areas of *C.vulgaris* and the fact that it yields better on higher ground. It results in a gelatinous honey which refuses to come out of the comb unless pressed. To get the best out of this crop you need to start preparing quite early in the season so that a young (preferably June mated) queen and adequate numbers of flying bees will hatch out during the six odd weeks at the heather (plumping helps) and really strong stocks are best suited for the job.
See: heather/honey/nectar, *C. vulgaris*

**lingel**
Shoemakers thread, rubbed with beeswax.

**linkage**
The association of two or more genes so that 'linked' characteristics are transmitted to the offspring. If for example colour and temper were linked, it would then be possible to cull, i.e. to select temper by choosing those queens and drones of the appropriate colour.

**Linnaeus 1707 – 78**
The Latinized name of Carl von Linne, the Swedish naturalist whose 'Systems Naturae' (1735) established the basis of binomial classification of the plant and animal Kingdoms. The tenth edition in 1758 led to its acceptance by the whole world.
See: classification, nomenclature, systemics, taxonomy

**linolenic acid**
The main polyunsaturated (present in some varieties of oil-seed rape). Unsaturated fatty acid $C_{17}H_{29}COOH$ occurring as a glyceride. Also linoleic acid $C_{17}H_{31}COOH$ found in all

the drying oils and especially in linseed oil. See: erucic acid

**linseed oil**

Derived from the seed of flax. A drying oil used in paints (though water-bound paints seem to be getting priority these days) for wood protection, linoleum etc. It is sold as raw or as boiled linseed oil. It can be used directly onto timber and when dry polished with beeswax.

**lint**

A soft linen material, one side like gauze but the other scraped into a soft woolly down for dressing wounds. This material has proved useful as a filter for beeswax. Liquid wax is poured through from the fluffy side of the lint, the mass of small fibres trapping any solid particles and allowing the clean wax to pass through.

**lipase**

An enzyme produced by certain organs of the digestive system which converts oils or fats in fatty acids and glycerol.

**lipophilic**

Wax soluble. Some Acaricides and Varroa treatments get into honey and especially (due to repeated applications) into wax. Hydrophilic – water soluble.

**lipsalve**

2 tablespoons beeswax, 85 g/3 oz. almond/olive or essential oil. Warm until dissolved and then stir until cool.

**liquid**

A state in which a substance's molecules are free to move amongst themselves but do not tend to separate as do gases. Stable substances will adopt solid, liquid or gaseous states according to temperature and pressure. Liquid honey loses its free flowing nature when its molecules are locked into crystals. Beeswax which becomes liquid at 62 – 65C looks remarkably like honey.

**liquid honey**

Is a viscous, translucent solution of sugars. When clear and golden it has been euphemistically called 'liquid sunshine'. Some honeys like 'acacia' remain liquid and clear for long periods but when the glucose portion begins to granulate in a honey, it first becomes cloudy and then solidifies. Only by removing all bubbles and solid particles however small, can a sample be kept clear. This is done by heating and passing through fine filters, a process that causes honey to lose some of its vital ingredients.

**liquid paraffin**

A clear, comparatively odourless, oily liquid distilled from petroleum (used as a laxative). Pharmacies stock heavy and light qualities. Light oil is able to dissolve beeswax and is used in certain recipes.
See: cold cream

**liquefaction**

The process of liquefying a substance. A more appropriate term than melting for bringing honey from the granulated to the liquid state. Basically heating 60-70C (140-158F) and stirring is required. A well-insulated cabinet (defunct deep-freezer) fitted with a low-powered lamp, can be used to liquefy a 30 lb. bucket of granulated honey. Be careful not to heat for longer than necessary because certain valuable ingredients go to waste when heat is applied.
See: liquefying granulated honey

**liquefying granulated honey**

As a temperature well below boiling water suffices to bring granulated honey into the liquid state, a water bath is often used if a large enough container can be found. The temperature of a greenhouse or airing cupboard may well go much of the way towards softening and clearing granulated honey. Commercially and for the handyman a thermostatically controlled box like a disused 'frig' is a great help 60 – 70C is enough.
See: liquefaction

**listen**
To allow the mind to appreciate sensations received via the ears. Rarely as effective a way of learning as seeing or performing, but one that can fill the mind with relaxation and wonder. The humming of bees as on a fine summer's day when you stand beneath a flowering lime tree. The sound of a swarm, bees working in a super or the piping of a newly hatched virgin, the subdued roar of a colony humiliated by having been inspected, the queenright hiss and queenless moan. The disappointment of absolute silence when a hive is struck but winter has claimed its inmates. We can listen and observe and in doing so increase our knowledge.
See: sounds

**listless**
Languid, indifferent and seeming to lack either energy or enthusiasm. When a colony appears to be listless, acarine or some other disease or disorder should be suspected. Starvation, damage from spraying, whatever the cause, when it is noticed by comparison with other stocks, the trouble should be looked into right away and steps taken to rectify.

**literature**
Printed matter of all kinds, books, pamphlet, journals, leaflets in fact a whole mass of material. Although books have been mentioned especially old writings, there has been little attempt to keep up-to-date though several headings cover some of the ground
See: books, poems, rhymes, sayings

**lithosphere**
The layers of soil and rock which comprise the earth's crust. Sometimes honey shows have put up a map covering the locality and samples of honey from various places have been found to have interesting variations. Where information has been available various soils and their influence on beekeeping have been given but they are far from comprehensive.
See: soils

**litmus**
A blue powder obtained from lichens. It is impregnated into paper and used to determine whether a liquid is alkaline or acidic. Blue – alkali, red – acid. See: pH

**litre**
A unit of capacity that is the equivalent of 1kg at its maximum density. In certain fields it is fast replacing the gallon which is the equivalent to 10 lb. water.
A useful approximation is 13⁄4 pints to the litre. A small 'l' signifies a litre which works out at 1·76077 pints or 0·21998 gallons. As petrol has gone up in price to at least five times its earlier value a litre now costs what a gallon used to (2010).

**little bee, dwarf bee *Apis florea***
This is one of the four honeybees that live in the tropics and sub-tropics. Its honey is much prized but it is not easy to expand them commercially because they abscond and build a single comb in the open-air.
Its cell is similar to but a good deal smaller (10 to the inch) than *A. mellifer's* honeycomb and it suffers parasites some almost a tenth the size of itself. They are reasonably docile to handle.

**little blacks**
The name given to bees suffering from paralysis virus and they are also called 'black robbers' However bees that have given over to robbing become hairless, due to many a tussle, and look a greasy, shiny black.

**livestock**
The larger animals usually kept on a farm and some of which can become quite skittish when kept near hives. Apiaries should be located so that the bees' flight paths are clear of such animals. A horse is more badly affected by stings than a cow though cows can go off milking when stung. It is unwise to tether a goat too near a beehive. You could refer to a stock of bees as being

livestock.

**living**

To make one's living from beekeeping is fraught with the possibilities of failure especially in the UK where weather is so uncertain and costs are high. Often families who are all keen to work together and love bees, go on to expand and by covering all the many aspects of beekeeping not merely producing and selling honey but all hive products, fees for pollination, and the encouragement of hobbyists by supplying equipment and selling honey for them. Quite often too, apiculture becomes a means of research, education and subject for writing and when someone has a good foundation so that making a profit from the bees is not essential, then it may supplement their living costs without their feeling distraught if their bees fail them.
See: bee-farmer, hobbyist and side-liner

**lizard Lacertidae**

We have to consider these pleasant little creatures as enemies of the bees. World-wide there are many species and the loss of bees due to their depredations must be relatively small here. Their elongated bodies, scales and tapering tales are well-camouflaged, but their stationary posture and sudden turn of speed make them invulnerable as far as bees are concerned.
See: enemies of the bees

**load**

That which the bee carries back to the hive. A loaded bee flies with hind legs well forward under its body while an empty one stretches them well out behind. Liquid can only be carried in its honey crop (sac), pollen can be carried on body hairs but the pellets go into corbiculae (baskets) on their rear legs as does propolis but only specialized bees go in for that. A bee can just about carry a dead drone over a short distance. When transporting hives certain precautions are necessary to avoid losing bees en route and to fasten them securely and by co-operating with the authorities as regards route and facilities.
See: loading

**loading**

1. hives for transport. Hives are best fitted with overall ventilation screens and stacked so that they fit tightly against one another yet with their screens exposed to the airflow. The platform of the transport should not be uncomfortably high and two men co-operating can do the work better and in less than half the time that one would take. Mechanical and electric devices using pallets are the order of the day when it comes to bee farming.
2. When loading uncapped combs into an extractor their weight should be symmetrically balanced so that undue vibrations and stresses that might lead to damage do not occur. Spinning should be done slowly, the combs reversed and faster spinning to finally empty first one side and then the other.

**loam**

Loose soil composed of sand, clay and organic matter and having fertility.

**lobe**

A roundish extension or division of an organ or segment. A lobe on the posterior edge of the bee's protergum, covers the first spiracle situated in the membranous wall of a cavity. It is fringed with a thick brush of branched hairs. Until the bee is about 5 days old and these hairs have hardened, the acarine mite is able to gain its entry to the first pair of spiracles.

**local locality**

Applying to the area surrounding where you live. Beekeeping Associations are organized locally, regionally and nationally. Bees forage over the local area and the potential will depend on the available flora. If there is insufficient forage due to the cement jungle, large expanses of water or other reasons it is worth considering setting up an out-apiary or engaging in migratory beekeeping.

**location**
A place. Bees' lives depend entirely upon their ability to be able to fly from and then to relocate their nest. Beekeeping success if largely tied up with having your apiary in a good location from the bees point of view. To change their location calls for an understanding of how far bees can be taken and still 'find their way home'. A general rule is not to move them more than a metre or so, otherwise to move them more than three miles. Even from that distance drones will find their way back in good weather conditions. The memory of a colony in this respect lasts as long as the bees flying life. On return from six weeks on the heather it is reasonably safe to rearrange the positions of the hives when they are returned. During the winter it is not a good idea to disturb bees by moving them although a careful re-organisation within an apiary is possible. This rule might be usefully broken when, knowing that the flying bees will return to the spot they have got used to, a swap of hives can help to weaken one and strengthen another if done in good flying weather. Fighting rarely ensues if the bees are coming back with nectar on board.

**lock slide**
A locking device that enables two chambers to be fastened together. A pair of metal brackets are attached to the upper and lower edges of adjacent boxes so that a 'v' shaped locking slide can be tapped in between them. Once these are in place the two chambers are held firmly together. The v-shaped piece is as easily removed again by tapping it out.

**lock spring**
This is a wooden block into which a spring has been fitted. It can be pressed down between the 'follower' and the end wall of a section crate to force all the sections up tightly together and make the whole box rigid and correctly bee-spaced. Their release facilitates the separation of the section boxes.

**locus**
The position in a chromosome relative to other genes in which there is always one, and only one, gene of the same kind, or one set of genes allomorphic to one another.

**locust false acacia**
A thorny, branched, white flowered tree. Also the 'carob' and honey locust. In Malta an early flow from the flowering carob is later followed after the mid-summer dearth, by a honeydew flow from the pods.

**loess**
Soils formed from the fine silt or dust dropped or carried by wind and rain. Usually yellowish and calcareous in Europe and Asia and the Mississippi valley of U.S., North China and Russia. A loamy deposit that weathers to a brown or even reddish hue. Its fine texture encourages excellent root growth.

**loganberry *Rubus loganobaccus* Rosaceae**
First grown by J. H. Logan in 1880 California. A fruiting cane with the habit of producing long shoots like the blackberry but prostrate. It likes rich, deep loams and follows the early flowering pattern and good nectar yielding characters of the raspberry from which it was derived. Its pollen is greenish white and is available over a long flowering period. Frequent visits by bees result in a good crop of well-shaped fruit.

**log hive**
Man probably learned to domesticate honeybees by moving logs in which they had nested. African natives still set-out empty logs that have been 'baited' hauling them up into the trees for protection. The use of gum trees resulted in early hives being called 'gums'. We read in 'Gleanings' that Grandpa's bee gums stood on a makeshift, slap-dash bench in a grove of long-leafed pines in Eastern Carolina. Man's skill as a carpenter has led us to use lighter structures

than the logs and to the use of movable combs.
See: hive log

**Long Ashton**
A research station where R.R. Williamson and others instigated much work on pollination.

**longevity pron. lonjevity**
1. The actual length of a creature's life.
2. A great length of life. It is usually assumed that a queen capable of a long effective life, produces long-lived workers. Are longevity and prolificacy inversely proportional? The British black was said to live eight weeks in summer while Italians only lived five. Longevity of workers links better with a variable climate like ours with its shorter and less certain flows. When considering longevity full account must be taken of acclimatization. In this respect local bees are best. The length of foraging life must have a positive effect on a colony's honey-gathering potential.

**long-idea hive**
One that has a level and horizontal arrangement of combs to minimize lifting and to keep the whole hive weather-proof and compact under one roof.
See: wide idea

**longitudinal flight muscle**
When these masses of thoracic muscle contract the roof of the thorax is raised and thus the wings go through one power stroke. In order to keep these muscles fed with the extra oxygen they require, the largest spiracles and trachea are adjacent (the spiracles face forwards slightly) and cytochrome which gives them a pink coloration also assists in the rapid transfer of oxygen.
See; flight muscles, vertical

**long-tongued bumble bees**
*Bombus hortorum* and *B. agrorum*. Those knowing little about bumble bees might tend to assume that they all have longer tongues, work harder and are possibly more plentiful than honeybees. It would require 400 bumbles per acre to pollinate Kentish orchards in the spring if they could be persuaded to stick to the job of working fruit blossom.

**Lonicera spp Caprifoliaceae**
Honeysuckle useful as a bee-shrub and requiring little attention once established in a bee-garden.

**looking for eggs**
This is one of those 'skills' that takes a little while to acquire. First it is necessary to know exactly what an egg looks like and which part of the hive they are likely to be found in. It will be in the area where the seams of bees are thickest. Then the brood pattern must be interpreted, for instance a gap in the middle of sealed brood may well have cells with eggs in them. The oblate patches of brood will be graded so that as you get to younger and younger larvae, eggs are then likely to be seen alongside. You should hold the frame vertically at 'eye-level' with the light back over your shoulder. Shiny cells awaiting eggs take on a slightly different appearance once the queen has backed into them – get to know this.
See: finding the queen, spectacles

**loosestrife, purple *Lythrum salicaria* Lythraceae**
A marsh plant with spikes of purple flower frequently found along river banks.

**lore**
Traditional knowledge on a particular subject, sometimes anecdotal, usually of a popular nature. Tickner Edward's '*Lore of the Honey Bee*' is a book that has given pleasure to thousands.

**lorum pl. lora**

Thong or leash. This is the part inside the head to which the postmentum of the proboscis is articulated. It is a 'v'-shaped sclerite that yokes the distal ends of the cardines to the postmentum.

**loss of**

The failure to keep, have or get. To suggest a few of the many ways a loss can occur in beekeeping the following are suggested. Loss of brood, cast, crop, dignity, drones, forage, foragers, honey, life, queen, site, sting, stores, swarm, temper, water gatherers, wax, weight.
See: rustling, vandalism

**lost wax process**

*Cera perdita, Cire perdue.* Most ancient Greek, Etruscan and Roman bronze figures were made by this process using beeswax. A casting process whereby a wax model was encased in plaster, then the wax was melted out of the mould and replaced by liquid metal. When ordering statues etc. the stipulation was ' it must be the genuine article and not with any fault patched up with wax'. Without wax. Sen cire, that is where the English expression 'sincere' came from.

**lot**

A collective term for the bees of a colony. Americans use the term 'back-lotter. We might refer to a bad-tempered colony as a touchy 'lot', bees without a queen are a queenless lot. Perhaps the word is more often used in a derogatory sense to describe a colony, one lot might rob another lot.

***Lotus corniculatus* bird's foot trefoil**

A clover-like plant common in dry pastures. Heads of yellow flowers attract bees from June onwards and it has a deep tap-root rendering it less susceptible to drought conditions.
See: bird's foot

**Loudon, J. C. Born 1783, Cambuslong, Lanarkshire**

In 1840 he wrote *'The cottagers' manual'* and recommended that a 'niche in wall' (bee-bole), should be fitted with an iron bar against theft.

**louse pl. lice**

A small, wingless parasitic insect highly adapted for clinging to the host, often flattened and difficult to dislodge. Feet and mouth parts specially adapted to suit the conditions presented by the host. Braula coeca is called the bee louse but has paled into insignificance in the face of the more predacious Varroa.

**lower lias**

An alkaline soil that is suited to certain of the clover crops and with the right conditions of moisture and temperature can produce an excellent nectar flow.

**lubrication**

Beeswax is an excellent lubricant and may be used for a whole variety of items to make them slide more easily when one surface has to pass by or through another. Cotton, a flat iron, thread, wood augers, zip fasteners, belts, drawers, saws, curtains… It remains water-proof and strangely - non-sticky.

**Lucerne (alfalfa) *Medicago sativa* Fabaceae**

Medick. A European forage plant, a hardy perennial with large prostrate, purple flowers from June through to September. It is often cut for fodder before it blooms. It will naturalise in grass, especially in chalky soils and is widespread on waste ground. For best pollination results fresh colonies should be exchanged every few days during the flowering period (presumably because bees tire of being bashed every time they visit a flower). The early season honey is light but becomes amber by autumn and is mild flavoured granulating into a fine, hard, white grain. The pollen is greyish-brown, light cream with a 30μ grain.
See: alfalfa

**lug**

A projecting piece by which anything is held or supported – (not the ear). The extended top bars of frames form lugs which support the frame and offer a ledge for the fingers to fit under when the frame is held (that does not imply that the frame cannot be held single-handedly by the centre of the top bar). The extra length of the lugs on British Standard frames, which almost act as handles, require four extra inches of wide timber on every box just to provide this 'handle'. One use for this extra-long lug is to hold one end firmly with a punch like grip and then clout that hand with the other fist (clenched) when a sudden jolt is required to dislodge the bees from a frame.

**lumber**

Timber, wood made into boards or planks as prepared for market. In the UK the word timber is commonly used (lumber less often) as it also means junk and we do say 'to get lumbered' (meaning to have to put up with something or somebody).

**lumen**

1. The derived SI unit of luminous flux, a measure of light intensity. One lumen per square metre is known as a lux.
2. In biology it is the cavity within a tubular organ or the central cavity of a plant cell.

**luminescence**

Light that we 'can' see emitted from substances bathed in ultra-violet light, the form of light humans 'cannot' see. Light akin to incandescent but occurring at room temperatures. Widely used now-a-days on uniforms worn at night. Light from a non-thermal source.
See: fluorescence, microscopy

**Lunaria spp. honesty**

A cruciferous herb with purple flowers and semi-transparent, satiny pods.

**Luneburg Heide**

More recently famed because it was here that the German capitulation occurred in May 1945 at the end of World War II. A great heath in Lower Saxony where Calluna and juniper abound and has been the source of much honey and always receives regular visitations by beekeepers for its late summer honey.

**lupin *Lupinus polyphyllus* Fabaceae**

Not considered particularly useful to honeybees as they are scarcely heavy enough to trip the flower. Well-worked by some bumblebees though, yielding pollen from bright yellow to maroon and reddish brown. In the 80's *L. augustifolius* and albus were strains genetically engineered as protein producers but although useful for pollen are in no way comparable with oil-seed rape though many hectares have been grown.
See: tree lupin

**lye sodium hydroxide NaOH**

A solution resulting from 'leaching'. Used boiling to sterilize hive parts when it cleans off wax and propolis and washes germs away. However it is extremely caustic and its misuse can lead to hours of suffering as it is very harmful to the skin.

**lymphocytes**

Small colourless corpuscles found in the blood and lymph of humans and which play a significant part in dealing with immunity or allergy to sting venom, bee protein generally and other allergens that you are likely to encounter when beekeeping. Along with plasma cells they are present in the reticulo-endothelial system and are capable of synthesizing immunoglobulin antibodies. You shouldn't have asked.

# M

**m**
Metre not to be confused with mile 'mi'.
15m = 15 metres.

**Mace, Herbert George, F.E.S. Essex, B 1882**
He took up beekeeping in 1908 and became a famous author and lecturer. Some of his works include; *Bees Flowers and Fruit*, *Bee Matters and Masters*, *A Book About a Bee*, *Adventures among Bees*, *Some other bees*, *Modern Beekeeping*, *Beekeeping Associations*, *Bee Farming in Britain*. He also designed a scale, bee box and stand. He favoured 'blacks' and kept 6 – 8 hives for honey and experiment but was expelled from BBKA by council in 1929.

**Macedonian bee A.m.cecropia Kiesw**
From Southern Yugoslavia & N. Greece – belongs to Carnica race.

**maceration**
To soften or separate parts of anything by the use of a liquid using heat if necessary. Its application to beekeeping could be many fold. Two examples include the washing of honey from wax cappings prior to making mead, vinegar or feeding the sweetened water back to the bees. Its use in the preparation of material for microscope slides when an improvement in the clarity of parts can be achieved by soaking in KOH for several days followed by washing with several changes of water.

**machine**
An assembly of man-made parts or system intended to perform some kind of work, for example a honey extractor, foundation mill, special machines for producing beehives or other beekeeping equipment. Machines are intended to give greater accuracy, more economical use of labour (automation) and in some cases 'mass-production' than ordinary hand labour and individual tools would permit.
See: apparatus

**McIndoo N.E.**
In 1914 he did considerable work on honeybee antennae, sense of smell etc. and on olfactory sense, scent producing organ and sense organs of mouth parts 1916.

**macro**
Long, large, great excessive in duration or size, also 'mega' as opposed to micro. Macrocosm – the Great World or Universe.

**macrotrichia hairs**
Bristle-like structures (setae) or projections found on the body and wings of insects. Although frequently referred to as 'hairs' they are extremely varied and numerous and range from sharp spines to brushes, combs and other highly specialised arrangements. it is noteworthy that the hairs on a honeybee are much more branched and capable of pollen retention than those of bumble bees.

**Madame Tussauds**
The famous London museum where models are three parts beeswax and one part Japan wax.
See also: lost wax process

**madwort**

1. ***Asperugo procumbens*** Boraginaceae
2. Sweet alyssum *Lobularia maritima*. Honey scented spring and summer 10cm tall masses.
See: *Alyssum maritimum*

**Maeterlinck, Count Maurice**

Belgian scientist and poet 1862 – 1949 who received the Nobel prize for literature 1911, La Vie des Abeilles (Life of the Bee). The English translation by Alfred Sutro sold over 100,000 copies. He also wrote 'Life of the ant'.

**magazine**

A periodical publication, usually with a titled paper cover, containing articles which may be illustrated. Bee Journals for various countries make interesting reading.
See: journal, publication

**maggots**

Larvae (Sydserff in Cotton, Keys, Hiush and Bagster). Legless larvae as of bees and flies.

**magnesium Mg**

A light, ductile, silver-white metal that burns with blinding light. Base for magnesia. Important element in soil connected with nectar secretion.

**magnetite $FeOFe_2O_3$**

A very common black iron oxide which produces a magnetic field (such as the earth's) which has been shown to affect the bees' ability to align combs correctly and also to affect their navigational ability. Magnetic material is thought to accumulate in certain parts of a honeybee's body and magnetic crystals have been found in the proximal end of a bee's abdomen. It is also found in some humans.
See: Honeybee brain, magnetic field

**Magnification**

The use of an instrument such as a magnifying glass or more complicated array of lenses designed to increase the apparent size of an object thus making smaller details of its structure clear. A ratio of image to object size 20x means an apparent increase of size to twenty times as for acarine dissection.
See: optical, electron microscope

**maiden**

Pertaining to or befitting a girl or unmarried woman. It could be used for a virgin queen but seldom is.
See: maiden honey/swarm

**maiden honey**

Honey derived from a swarm that has come from a swarm of the same summer. Much prized in olden times.

**maiden swarm**

A swarm that issues from a colony that began as a swarm of that same season.

**mailing cage**

A small, ventilated, bee-proof cage for the transportation of bees by public carriers and which is usually designed so that it can be used for queen introduction.
See: queen introduction, shipping cage

**maintenance**

To keep repaired and in good condition and effectively to keep in existence.
Honeybee colonies are self-replacing and under the watchful eye of a knowledgeable beekeeper go on indefinitely, though their hives do not. As hives are normally kept in the open and exposed to the elements, they need to be made of durable material such as plastic or wood which if necessary can be painted and repaired. Western Red Cedar has been a popular timber in this country because it is light, fairly weather-proof and insect repellent. Regular oiling, cleaning, repainting, blowlamping, strengthening and checking will be necessary to keep all your equipment in good order. The winter provides an opportunity to give a thorough going over things like apparel, smoker, stored comb and of course, to keep your

knowledge up-to-date.
See: hot dip, paint, repairs

### making comb

Bees construct a unique form of comb. The European mellifera requires darkness to start off but will continue once combs are started. Scutellata and florea the 'giant' and the 'little' bee make their comb in the open air. If you are disposed to insinuate your fingers into a swarm you will find that it feels 'silky' and 'warm'. In a cavity, bees of the wax producing age will huddle together and raising the temperature a little, begin to hook out the wax plates that develop in the (eight) wax pockets on the underside of their abdomens. This they masticate it into a workable, white material and using other necessary reinforcements such as propolis, pollen husks, fibres (even the purest newly made comb will give us some of this material when melted and filtered). They do, what has amazed even the cleverest humans since they first investigated a bee's nest, develop arrays of hexagonal cells, arranged back-to back in vertical, parallel rows. Typical of our honeybee, they work together, never competing, so that each cell might be the result of many different bees, each playing a small part in its creation. To align virtually cylindrical cells back-to-back would puzzle even a clever school girl (and her teacher) if it entailed rounding off each cell at the bottom. By pushing a needle through the three rhomlike triangles that form the base of a cell (using a sheet of foundation to do this) it will be found that the holes made appear in three different cells on the opposite side. So all those millions of years ago these almost magical insects discovered this trick which is beyond the wasps, ants and other social insects. Making comb then is not only unique but truly remarkable.

### making honey

Commencing with nectar, having a varying sugar content of around 33%, which they suck up from floral and other sources, they say to themselves. 'Ah this is nice and runny and easy to lick up and push down into our special stomachs' but it won't keep if we try to store it away like this'. So as with pollen, they begin to 'treat' it and once it has been unloaded (sometimes via house bees) into a cell, the 'process' of ripening it begins. Enzymes which they added to it on the way home, together with a great effort of fanning on the part of the house bees, the liquid is concentrated until the sugar content is pushed up to around 80% and in the course of this, the large sugar molecule of the sucrose, (which neither they nor we can immediately digest) is cracked from being sucrose to form two simple, easily digestible sugars, fructose and glucose. Once again they did not have the benefit of science to help them develop this life-saving means of storing food away and staying usable the whole winter through, but now we know. Yes, sucrose is $C_{12}H_{22}O_{11}$ a big molecule, but using an enzyme they not only remove some water ($H_2O$) but split it into the two mirror image, simple sugars of fructose and glucose. (also called laevulose and dextrose) which are hexozes each with a formula of $C_6H_{12}O_6$ - smaller molecules that can pass directly into the blood. Being more concentrated yet still containing all the beneficial ingredients that came from the nectar it occupies less cell space, but has to be diluted before the bees can use it. So as is the case with honeycomb, honey itself is a wonderful unique thing.

### making increase

Producing additional colonies. This can be done during the active season when new queens can be reared. Quite often because ordinary colonies can produce queen cells advantage can be taken of these to make increase while at the same time forming a method of swarm control. The situation can always be precipitated by taking the existing queen away (putting her temporarily in a small nuc. perhaps) when the bees finding themselves queenless, go ahead and make queen cells. Where queens are marked so that they can be found easily, the subsequent steps can follow reasonably easily. To search for the queen in a strong stock and

when heavy supers have to be lifted off can encourage an arbitrary splitting of a colony so that there is no need to actually find the queen. Once the colony is divided into two parts, the one without the queen will build queen cells while the other will have eggs and young brood. Many systems of swarm control work in this way either by using a nucleus or a Snelgrove board but the size of the original colony, time of year and honey flow conditions will require the use of knowledge and experience to decide how many new queens will be given the chance to emerge and mate, always bearing in mind that queen cells must be separated or the first virgin to emerge will destroy the others. The collection of a swarm offers the chance of gaining a colony but care must be taken to safeguard the health of your bees by putting the swarm somewhere other than next to your existing bees until a health check can be carried out on them. Although a queenless colony can be persuaded to take eggs, a queen cell, a virgin, or a mated queen care must be taken (i.e. to ensure no existing scrub queen, or queen cells are present) to introduce the new potential material so that it is willingly accepted. Special queen cages are available for this purpose.
See: divides, nucleus, increase, Snelgrove

**making candy**
Pure, white sugar and water can be candied to make a smooth, easily scratchable block of candy. Using one cup of water to six cups of sugar, the mixture must be stirred and boiled for several minutes. As it cools, the clear liquid will become cloudy, then you must be ready with containers so that it can be given a final stir and then be poured to cool in the chosen receptacles.
See: candy

**making wax foundation**
It should be said that because the techniques and long-time skill of equipment manufacturers who have strong rollers that act like a mangle and press thicker sheets of wax down to wafer thin sheets of foundation (they wire them efficiently if required too); that you should always compare the cost of purchase with the time and trouble that making your own foundation would incur. Hand presses can be bought, and plastic sheets embossed with honeycomb cell shape too for use with your own mangle, but it takes a lot of skill to turn out usable sheets that are not too thick and then templates are required to cut them to your chosen various sizes.

**male**
Belonging to the sex that begets young and that produces the means of fertilization. The male honeybee is called the 'drone' a caste that is only tolerated for that part of the season when mating is possible.

**male gamete**
The electron microscope studies of the honeybee spermatozoon show that the acrosomal complex, or head with its nucleus, is quite an intricate structure having perforatorium, axial filament (or rod) and galea. The tail includes the axoneme (the axial thread of the flagellum) and two asymmetric mitochondrial derivatives.

**malformation**
Faulty or anomalous formation especially of living things.
1. Of honeybees the deformity caused by Varroa or a harmful genetic fault.
2. Of fruit the impoverished and distorted shape of many fruits such as apples and strawberries, can be laid at the door of inadequate pollination or insufficient honeybee visits.

**malic acid**
An acid found to a greater extent in apples, especially the unripe, than other fruits. It is a colourless crystalline, dibasic hydroxyl acid C2H3OH(COOH)2 - a component of the citric acid cycle. Used in flavouring and for ageing wine.

**Mallophora big fly**
It catches bees while flying, bites them in

the neck and sucks their lymph. There are *M. ruficauda*, *robusta* and *bigotii*. Bees have become too scared to forage.
See: big-fly

**mallow *Malva sylvestris* Malvaceae**

The common mallow is a mauve, perennial weed that flowers from June to August. It produces nectar and pollen the latter being a pale mauve, pale yellow to white and has a large grain, 72μ, like its relative the hollyhock.

**malodorous**

Sometimes bees make honey that seems to humans to have a rather off-putting smell. In many cases such honey, left to mature, even in enclosed containers, gradually loses the unwanted aroma and consequently becomes quite palatable. Examples include honey derived from privet, ragwort and sweet chestnut.
See: toxic honeys

**Malpighamoeba formerly *Vahlkampia mellifera***

The pathogen that gives rise to the disease in adult honeybees known as 'amoeba'. It is most easily diagnosed when macerated malpighian tubules are examined at high power 400-500x, when the spherical cysts can be seen.

**Malpighi, Marcello 1628 – 94 Italy**

Whose name became associated with many anatomical parts, including the kidney tubes of the bee. He was the first to use a microscope to study anatomy.

**malpighian tubules**

Long narrow tubes, some one hundred in number, which lie in the abdominal blood stream from which they extract nitrogenous waste. This is passed into the ventriculus at the position where they lead into the alimentary canal just after the small intestine When amoeba disease is present they are host to a parasite Malpighamoeba mellificae.

**Malta**

Although these Mediterranean islands are small 95mi2, beekeeping has gone on for centuries using horizontal, ceramic, tubular hives of 'jars'. The Government encouraged the conversion to wooden hives and 1000 or more are kept. They have to be kept out-of-sight and out of the sun. Education of children did not discourage them from being bee-haters and spiteful vandalism was not uncommon. The name Malta comes from 'melita' meaning honey. There is an early and late season flow with a dearth during the hottest part of the summer which is usually very dry. The bee is dark and slightly smaller than A. mellifera and imports from Italy seem to steadily revert to the native bee. When driving around it is almost impossible to see any beekeeping, the hives are usually hidden and sometimes locked away in caves.

**maltose**

A reducing disaccharide. White crystalline malt sugar $C_{12}H_{22}O_{11}.H_2O$ is formed by the action of diastase on most starches. It is found in quite significant quantities in honey. It is less sweet even than glucose.
See: sweetness factor

***Malus floribunda***

A decorative apple whose pollen is both prolific and useful for cross-pollinating many of the cultivars in current use for cropping. M. floribunda can be grown in containers, have its flowering time influenced by temperature control and then placed temporarily and directly into orchards or be used as bouquets (small bunches of pollen active flowers set out amongst the blossoms awaiting pollination).

***Malus pumila* formerly *Pyrus malus***

The apple tree. Many of these are self-sterile (notably Cox which requires Early Victoria, Worcester Pearmain or other suitable pollinizer) Most benefit by cross-pollination when a good set of fruit is required.
See: apple, malic acid

**mammal**

Warm blooded (self-regulating body temperature), back boned, and females suckle their young. Viviparous – brings forth young. badger, bear, whale, man. The winter configuration of the honeybee colony in the form of a composite cluster approximates to a mammal in so far as its temperature control is concerned.

**man**

Man, a far more recent arrival on this planet than the honeybee, has seen fit to exploit its characteristics in various ways. While enabling many more colonies to exist in chosen areas than would be possible under natural conditions, he also seeks to improve the bee to suit his purposes.
See: apiarist, beekeeper, beginner, expert, mistress, mellarius, mellitarii, rustling, vandalism, women

**management**

The control and direction of anything. The management of honeybee colonies would dictate the type of hives used, strain of bee, location and manner in which stocks were manipulated. The management of a beekeeping enterprise would govern production of the chosen end-products, marketing them and general business and financial arrangements associated therewith. There are a plethora of books on the subject.
See: husbandry, manipulation, system, tape recordings

**mandible**

A jaw-like, biting organ. A pair of these are hinged at the side of the mouth and they open and close sideways. The musculation and two-point articulation on each mandible cause them to turn inwards and backwards as they are closed. They are fed with a secretion from the mandibular glands and are grooved so that a spoon-like delivery of food can be made. They are used to create an aggressive stance when acting as guard bees but they are too smooth to be damaging being more suited to manipulating wax than cutting anything tough. Mandibulate – having mandibles.

**mandibular gland**

These are single, lobate sacs lying under the genae just above the mandibles. Rudimentary in the drones, they are large in workers and very large in the queen. A short duct opens in the membrane at the root of the mandibles and secretions run down a groove which leads to the mandibular spoon. 10-hydroxydeconoic acid is found in brood food. In workers 2 heptanone, an alarm pheromone, is also produced and when stinging the mandibles also nip the victim so that it is 'marked' as well as stung. The secretions excite other workers. This gland is also associated with colony identity which prevents workers but allows drones to be accepted by other colonies. Queens or workers must have their colony identity masked before they are likely to be accepted by another colony.
See: Q's mandib. glands, colony odour, pheromones

**Mandibular pheromone**

10-hydroxydeconoic acid and Heptanone-2-one are pheromonic secretions from this gland. The latter is used when combs are built and it excites other workers whenever deposited. It is a volatile, strongly-smelling constituent possibly used as a mild deterrent for instance to tell nurses that a larva has just been fed or that a worthwhile food source has been cleaned up. See: 2-heptnone (under 'two'), aliphatic ketone

**mandril**

A tool used for moulding wax into the shape of a queen cell cup when producing queen cells for queen rearing. A small piece of dowelling, rounded at each end, about 9·5mm dia. Less used now that there are satisfactory plastic queen cell cups.

**manipulating cloth**

Now that we have become 'extra' disease conscious a 'cover cloth' that goes from

hive to hive is sometimes frowned upon. However within a small apiary where there is interchange of material between hives, keeping the aspect of hygiene in mind, a suitable, flexible cloth that can be weighted with a 10mm piece of dowelling, is a most useful commodity, allowing less smoke to be used and keeping the bees under good control, while excluding robbers and keeping unwanted light out of the chamber. When two are used in conjunction with one another it is possible to move across the frames only exposing one at a time. Bearing in mind the disturbance caused by an inspection of the combs, which should be minimized and not maximized as so often happens when the bees are left uncovered and combs are taken out and stood beside the hive, the use of a cover cloth is highly beneficial.
See: cover cloth

**manipulation**

The control and management of stocks including the handling of bees and their combs and the arrangement or rearrangement of the hive and its contents. Manipulation is usually done for a specific purpose such as checking progress, their health and need for more space and that stores are sufficient. Also perhaps to re-queen a colony, carry out swarm control and to exchange old combs for foundation and for migrators to take stocks away to more distant crops or pollination.
See: management, techniques, systems

**Manley, R.O.B.**

Who died in 1977 was a well-known and very successful bee farmer. In addition to all the helpful advice and encouragement that he gave during lecturing, he wrote 'Honey Farming' in 1946 and 'Beekeeping in Britain' in 1948. He also produced the very sound, close-ended super frame which makes uncapping easier, provides an inner wall within the hive and holds frames firmly together for traveling. Its dimensions are: top-bar 27mm x 69mm deep with side bars 41.5mm wide and with two bottom bars.

**manna lerps – dried honeydew**

The anal secretion of the aphis (*Cocci maniparas*) which lived on the Tamarisk trees – Albert Hind. Crystallized honeydew such as the manna of the Hebrews which is produced by a coccus (sucking insect) on the Tamarisk bush in the Sinai desert. The exudates of the flowering ash Fraxinus ornus in S. Europe.

**mannitol**

Found in manna, a white sweetish crystalline carbohydrate alcohol with optically different forms $C_6H_8(OH)_6$. A polyhydric alcohol related to sugars – Vines & Rees. Used as a dietary supplement and nutrient.

**mannose**

A hexose sugar found in the nectar of certain plants, notably some forms of lime (*Tilia tormentosa*). Bees cannot properly metabolize this sugar and in the partially metabolized state it inhibits the vital metabolism of the normal laevulose and dextrose sugars, thus causing a fall in the bee's blood sugar level. Many are found dead or dying (bumble bees seem even more prone) under the trees, sometimes they are referred to as 'drunk'.
See: lime trees

**manubrium**

Meaning 'handle' this word is used for several anatomical parts on other creatures as well as bees. A plate on the dorsal side of a bee's foot which bears five or six bristles and is attached to the arcus. As the manubrium is pulled down (as when the claws fail to find a grip on a smooth surface) the arcus spreads and the arolium is spread.

**Manuka honey *Leptospermum scoparium***

The Red Tea Tree from the leaves of which Capt. Cook on his arrival in NZ made 'tea'. A small, erect, evergreen shrub native to NZ with a small white flower which grows in masses usually with an alternative source or two which prevent the otherwise thixotrophic nature (like heather) from causing beekeepers to have to tear the comb down to the mid-rib to get the thick honey. It is distinctly flavoured and has recently come to the fore on account of its use in hospitals to treat wounds (though other less expensive honeys are just as effective). Its pollen is a 'muddy' white.

**manuring**

The application of chemical or organic fertilizers. Although manuring can have a marked effect on the copiousness of a nectar flow (hence the attractiveness of a plant to honeybee pollinators) or conversely it can retard nectar secretion, comparatively little attention seems to have been paid to this. Clovers need soils whose lime, phosphate and potash levels are adequate if flowers are to be worked.
See: stimulation of nectar

**maple Acer spp. Sapindales**

A tree valued for its wood and its sap. There are many different species in N. temperate zones. *A. platanoides* – the Norway maple, *A. saccharum* – the sugar maple. The sugar content of the sap is too low to interest honeybees though floral nectar is useful in early spring when they swarm with bees if the weather co-operates. They flower Mar/April and the pollen is reported as dark brown, pale yellow and light green-ish yellow with a grain of 25µ.

**Maraldi, Jacques Philippe 1665 – 1729**

A third generation Frenchman out of Italy. Astronomer and mathematician. He used a one-frame observation hive in the garden of the French Royal Observatory in Paris in 1687 and wrote 'Observation sur les abeilles' in 1712. It contained the first accounts of many features of bee-life which are now taken for granted and he demonstrated the geometry of comb building.

**March**

This can be a 'touch and go' month if colonies are short of food and brooding has increased using up all the food available. Provided they are well supplied, there is warmth above the nest and pollen is being taken in, things are best left alone. But nothing kills as surely as starvation so where serious doubt exists (check by hefting) then it is not too early to give them some tepid, dilute syrup (pound to a pint) but don't give them more than they can take down overnight. Candy is an alternative but it gives the bees much work to do to use it and water is often their main requirement. Beware gales, floods and late snowfalls. Clean floorboards can be given with a minimum of disturbance and the entrance can be down to three or four inches. If you have not already taken the mouse guards off, they can certainly come off now.

***Marchalina hellenica* (Genadius)**

A sucking insect which secretes honeydew and feeds on Pinus brudia in East Mediterranean countries. In Thassos beekeepers move their bees in on the flow during September and October.

**marigold *Calendula officinalis* Calenduleae**

Many of the single varieties make good bee plants but this one is a golden flowered plant with strongly scented foliage. There are corn and marsh marigolds.

**maritime**

On or near the sea. Maritimum is attached to plant names to indicate that they are found at the coast. E.g. *Eryngium maritimum*. Temperatures near the coast tend to be influenced more by sea temperatures which only change slowly as the year goes by, sea winds tending to be cold in the spring but milder in the autumn.

**marjoram the mint family**

*Origanum vulgare*, heracledicum, and marjorams are hardy perennials yielding nectar and pollen and growing up to 60cm tall. Wild marjoram with its masses of purple flowers and long stems, favours calcareous soils and blooms from July. The general flavour of the honey is improved by its inclusion though it is rarely found as a mono-crop. Pollen is medium grey to medium brown. The nectar's sugar content can be as high as 70%. It is suitable for the herb garden.

**marked queen**

Easier to find and colour coding can give her age. A dot on the dorsal surface of her fused segmented thorax is reasonably impermeable but use quick drying paint and make absolutely sure she is free of smell before being returned to her bees.
See: colour of the year

**market garden**

An area given over to the cultivation of fruit and vegetables. In size it lies between a farm and the small holding. Time can rarely be spared for the running of honeybee colonies and also as pollination can fall foul of greenhouse requirements especially when the growing of cucumbers calls for the exclusion of all insects.

**marketing**

The act of buying or selling in the market. There are local, national and world markets for honey, bees, wax and other hive products. Several countries have Honey Marketing Boards which aim to help the honey farmer. Although honey is a well-known product its sales are directly proportional to the scale and effectiveness of advertising and marketing techniques. Beekeepers are not always good promoters of their own produce.
See: EEC Regulations, labelling, size of letters

**marking**

Equipment, queens, exam papers, hive entrances...Bumble bees mark their territories. Honeybees mark their hives and soil around the entrance. Excited young nuclei will do this and dysentery may cause it. Washing, hanging out to dry, may also become marked as bees collecting water frequently decide to 'off-load' at the same time. Rustling of stocks has prompted the 'branding' of hives.
See: defecate, branding

**marking on containers**

1. Honey for sale should show quantity by weight in Imperial and metric units and 1, 2, 4, 8, 12 oz., l lb., 11/2 and multiples of l lb. are permissible.
2. The name, trade name and address of producer, packer or seller.
3. A description should be in one of the acceptable terms, i.e. honey, comb honey, heather honey etc. The minimum height of figures used on 1 lb. jars is 4mm. Units of weight lb. oz., kg or g.

There is a Fera information sheet available.
See: labelling, lettering

**marl**

1. An earthy deposit composed of clay and calcium carbonate. Clay soils which often have a high lime content, some having originated from fresh and others from sea-waters. Sandy marls will improve clayey soils and clay marls will improve sandy soil.
2. The covering of wicker with 'fat earth' or clay was called 'marling' and such techniques were employed for the production of fixed-comb hives in early times.

**marmalade**

When honey is used instead of sugar, each kilo of honey should be taken as the equivalent of three quarters of a kilo of sugar and the amount of water reduced by 11 fl oz for every kilo of honey used. The resultant marmalade will be dark and, for those who like honey, have a very appealing flavour. A tin of 'Marmade' reads: to make 6 lb empty contents into a saucepan together with 3/4 pint of water and 4 lb sugar. But if 5 lb of honey are used to replace both water and sugar, a very nice product is obtained.

**marrow** *Curcurbita peop* **Cucurbitaceae**

This large fruit which is eaten as a vegetable, bears distinct male and female flowers which need pollinating. Various insects will do this but in one's own garden it is simpler to remove a suitable male pollen donor (pollinizer) keeping it in water for a while ,and then use it directly like a brush onto the female flower. The family include gourds, pumpkins, cucumbers etc.

**marsh**

An area, usually low-lying, where due to poor drainage or the impermeable nature of the sub-soil, it remains water-logged. At various seasons it may be merely wet-land while at other times it could be completely under water. Such places may well support bees from the point of view of foraging opportunities, though transportation and siting would have to take certain problems such as flooding, into account. The specific name of 'palustris' is used to indicate that a plant or creature is to be found in the marshes. Near coasts salt marshes with their own particular flora like sea-lavender, will sometimes be found.
See: bog

**marsh marigold** ***Caltha palustris***
**see marigold Ranunculaceae**

Looking like a large buttercup, this flowers in damp locations offering nectar and pollen to the bee. It is also called 'king cup'.

**Martin, John**

In 1684 in Bienenbuchel published in Germany, he correctly stated that beeswax was not gathered from flowers but produced by worker bees. It had formerly been asserted by Aristotle, Charles Butler and others that it was a floral exudation.

**mason bee** ***Chalicodoma muraria***
**Megachilidae**

These small solitary bees construct their nests of clay in cavities where mortar has become loose or similar places. Journalists frequently take up the theme in their newspapers saying that bees are destroying buildings.
See: solitary bees

**masque of Reaumur**

This is to be found at the distal end of the drone's genitalia and looks like a mask where the anal opening, genital vent and associated markings give it this characteristic appearance.

**mass**

1. A body's resistance to acceleration (proportionate to weight).
2. An aggregation of articles or creatures forming an indefinite shape. Bees tend to be attracted to bees and form clumps, but when spread out, as on a flat surface, they would be described as a mass. Sometimes when a colony has been disturbed the whole face of a hive will be;'black with bees'. When a swarm is hived they spread out in a mass before converging on the entrance. They can be encouraged to form chosen shapes by writing letters or making images with honey and letting them lick it up.

**massacre of the drones**

This is covered under drone slaughter and slaughter of the drones. A dearth of nectar at a time when colonies are short of stores may also lead to the ejection of drones, in fact white pupae sucked dry are often a sign of near starvation in the spring. Drones may also be dispensed with when a queen has been mated. Wasps will often take drones or ailing bees from outside a hive without actually making an entry.
See: drone slaughter, slaughter of the drones

**massage dance**

A bee, on the comb, is said to begin by bending its head in a peculiar way with its mandibles open wide and its tongue partly protruded. Neighbouring bees who notice this become excited and turn towards her using their antennae and forelegs to touch her sides from below, to climb over her and

use their mandibles too, in what appears to be a cleaning operation. The attending bees clean their antennae from time-to-time and the bee performing the so-called dance opens her mandibles and extends a dry tongue which she strops while splaying the proboscal parts for attention by the workers also helping her. A bee treated like this eventually quietens down and moves away.
See: dances

**mass crawling**
A situation where hundreds (maybe thousands) of bees are seen crawling on the ground and on the front and around the hive. They may be unable to fly. This was a common condition at the time when the Isle of Wight disease was rampant. It may well be due to a parasitic invasion of their main thoracic trachea by the acarine mite which can get into the spiracles of young bees. The first pair of spiracles face forwards slightly and are essential for flight.
See: acarine, Isle-of-Wight

**mass feeding or provisioning**
The provisioning of a cell with sufficient food to see the larva through its complete development from its breaking out of the egg until it becomes adult. This is not the way of the honeybee where nurses make progressive feeds, weaning the larvae in the process. On the other hand solitary bees are left with a supply of food which they eat before pupating.
See: progressive feeding, nursing

**mast cell**
The presence of these cells in the connective tissue leads to turgidity and inflammation when they are stimulated to release histamine (and other chemicals) by an allergen such as bee venom. When stung it is the over-release of histamine which causes swelling, irritation and discomfort.

**master beekeeper**
Someone eminently skilled in the science (art or craft) of beekeeping. A person who has mastered the art – mastery.
See: bee master

**masticate**
To chew. The jaws or mandibles of workers are suitable for breaking pollen from the anthers and for shaping and fashioning wax. Paper and wood can be torn or chopped like other hive debris, but plastic is flexible material and is not easily dealt with. The stronger mandibles of the wasp enable them to chew wood into papery substance for making their nest. When bees remove the wax scales from their wax pockets, labial secretions are added to the wax as it is masticated and formed into pure white comb.
See: eclectic melting point

**mat**
This word is used in some countries (NZ) to imply the inner cover but we usually refer to it as crown board. It is an inner cover over the frames, like a second line of defence under the roof. Where a flexible cover is used it is likely to become glued down with propolis and in tearing it off the bees can become aggravated.
See: crown board, inner cover

**match**
The safety match that can be struck to provide a small source of light and flame is frequently used to ignite a smoker though now that cheap lighters are plentiful many use these. They have another use. The thin, square cross-section strip of wood is just right for putting under each corner of a crown board to give additional ventilation. It can also be used as a probe when delving into a diseased looking larva which if dark in colour and suffering from Foul Brood can be twisted and withdrawn stretching a ropey string which is an unfortunate sign of AFB and should be reported immediately. When discarding matches remember the saying: 'one tree can make a million matches, but one match can burn down a million trees'.

**mated queen**
There may be a short period of several days between a queen's return from successful mating flights and her commencing to lay. True success is gauged not merely be the appearance of eggs in worker cells but the uniformly flat cappings of normal sealed worker brood. A mated queen co-ordinates colony activities boosting morale (by welding their behaviour patterns into a purposeful drive) suppressing laying workers, and inhibiting queencell raising until her ability to produce queen substance in sufficient quantity leads to swarming or supersedure.
See: fecund, laying queen

**material**
The substance from which an object is made. Old fashioned materials have stood the test of time but more exacting requirements such as rapid flight, travel in outer space and other recent pursuits have led to the development of plastics, alloys and other substances many of which have been incorporated into hive making and the production of beekeeping equipment. See: hive materials, timber

**mathematics**
The science devoted to the measuring of relationships and properties of amounts whether expressed numerically or in other forms. Maths has been linked with Physics in the subject index. Pocket calculators make otherwise difficult problems easily solvable provided the elusive 'common sense' is used as well. Conversions from our time-honoured tables, into metric are quite simple using such aids to mathematics.

**mating**
Copulation of queen and drone. A drone cannot evert its genitalia unless he has become mature and is airborne with air sacs fully inflated. When ready for mating they tend to patrol where rising air (thermals) help to slow the fall of the linked couple whose flying ability is temporarily restricted (!). Queens, some 5 to 15 days after emergence from their cells, release a pheromonic scent trail about 12m/40ft above the ground and the drone's extra sensitive antennae can pick up molecules from a considerable distance away (from where a queen can be little more than a dot in the sky). It almost goes without saying that this is likely to take place when the wind is light and during the warmest part of the day. Using his claspers the drone will mount on top of the queen's abdomen and an astonishingly violent (they say 'accompanied by an audible 'pop') delivery of his genitalia into the queen's open sting chamber, fulfils the purpose for which he was born. He falls back leaving the queen to carry on, the mucus plug he left to retain the semen, apparently proving no obstacle to the accompanying drones awaiting their turn. Tumbling earthwards he will at least have left his DNA to live on after him. Just how many drones follow suit and how many additional flights she might take, are things that vary with circumstances, but the contents of a queen's spermatheca, in which she can nurture sperms for several years, show that anything between 5 and a dozen or more fathers are possible. The relatively small Drone Congregation Area probably helps the good lady to find her way back to her eagerly awaiting sisters and she may well be wearing her 'wedding certificate' in the form of genitalian parts still visibly attached to her.
See: drone assemblies

**mating hive**
A hive made for the purpose of allowing a queen be prepared for her wedding flights and to allow her successful return. It can be quite small and provided it can comfortably house the small group of bees required, together with sufficient food then the nature of the container is not particularly important. Comb is essential and ample bees to maintain warmth and feeding may be necessary if the queen is to be allowed to lay for a while. You have to bear in mind of course that a small lot of bees like that are

vulnerable to robbing so keep the entrance small. As she is not accompanied you will be aware of the danger a queen is exposed to once she sets off for a drone congregation area. A colourful entrance will help her to re-locate the hive.
See: mating, nucleus

**mating sign**
A returning, newly-mated queen may well have the tell-tale signs of the drone's genitalia still partly protruding from the tip of her abdomen until the workers within the colony have removed them. The sign that she has been successfully mated follows when eggs in worker cells confirm this.

**mating swarm**
A swarm which usually absconds and includes a nubile queen or queens and a considerable number of workers whose fidelity has been transferred from the parent colony to a new queen. They often emerge from over-large nuclei or colonies that were caused to raise emergency queens. Robbing or disturbance can also cause this phenomenon.
See: absconding, cast

**maturation**

1. Having become mature. Dade says queens mature 21 days after the laying of the egg, workers 36, and drones 38.
2. The production of eggs or sperms from oogonia and spermatogonia.
3. Mead should be kept for a year before drinking and maturation can continue for another year or two, in fact a poor tasting mead often improves with keeping.

See: mature drones

**mature**
The achievement of complete natural growth or development. To ripen or to become ready for mating. This applies to: drone, honey, mead, queen, pollen, seed.
See: mature drone.

**Maurizio, Dr. Anna P.**
A Swiss scientist who was Vice President of IBRA and made enormous contributions to the sum total of world beekeeping knowledge with very special discoveries in the realm of pollen, bee plants etc. Her books include '*Das Tract pflanzenbuch Nektar und Pollen*'.

**Maury yeast**
One of the strains of yeast considered especially useful for the fermentation of mead. The natural yeast that is found in honey is not suitable for producing a fine tasting mead but there are strains of wine yeast that do very well.

**maxilla pl. maxillae**
One of the mouth parts; They are appendages behind the mandibles. They are composed of stipites, galeae and laciniae and the vestigial maxillary palps. Each maxilla is articulated to the cardo and lorum, the cardines articulating in turn with the postmentum so that the maxillae lie on either side of the central components of the proboscis. The articulations serve to close the maxillae over the glossa and together with the labial palps form a sucking tube.

**May, Crataegus, Hawthorn**
Local names include: aglet tree, azzytree, bird eagles, bread and cheese, cuckoo's bread, cheese tree, hag, hagbush, hagthorn, heg-peg bush, hipperty haw, holy innocents, ladies' meat, may blossom, may tree, moon flower, pegall bush, quick, quickthorn, scrog bush, and whitethorn.The old saying, 'Cast not a clout 'till May is out' may refer to the end of the month or the dehiscence of may. The end of the month would be more positive because it is a fickle plant whose flowering times varies by up to a month and its nectar only becomes useful when a temperature of around 21C/70F occurs and this has meant a good yield only once in 5 – 7 years.

**May month**

Now the season is under way and sycamore and fruit, oilseed rape and other plants can yield a surplus when the weather permits and colonies are sufficiently advanced to benefit from it. It is often said that the colonies build up on the strength of the flow (in other words no surplus because all is eaten). Space must be given ahead of the bees requirements both for brood and for honey. This is a good time to get foundation drawn out and the months when this can be done are all too few. Colony checks should be made regularly now but keep them brief, efficient and purposeful. Carry out swarm prevention and control in accordance with plans you have made and keep records. Strong well-found colonies pay dividends so be forever watchful as regards hygiene, health and condition of the queen.

**may pest**

Or May sickness, a condition of adult bees which has sometimes been described as paralysis ? and at others, pollen blockage of the rectum where young bees are found running on the alighting board or around the entrance with their colons blocked with pollen husks. Pollen from buttercups, ivy and frost damaged sources has been held responsible. Other names have been used such as hairless bees, spring dwindling, little blacks, bee paralysis and black robber disease. It is really a condition rather than a disease.

**mb millibar**

A widely used unit of pressure. 0·001 bar. We are familiar with isobars the lines drawn linking places with the same atmospheric pressure on a map and indicating highs and lows. With present day accuracy it behoves a beekeeper, knowing the state of her local flora, to take weather conditions into account with regard to various aspects of beekeeping management.

**mead**

A drink made by fermenting honey in water. It was at one time made with barley and was more like beer than wine. The name is now reserved for honey wine but all sorts of variations such as the adding of spices (melomels) has given rise to a host of names that will be found under 'drinks'. Meads can be dry, medium, sweet, still or sparkling and need about a year to complete their working and then continue to improve over the next few years. It is an ancient drink and local dialects have their own words: maethe, meath, meathe, meeth, meth. Germans use Met and Honigmet, Russians – med, Czechs medovina, French hydromel and middle English mede and medu while Welsh medd. See: mead making

**mead making**

Use sterile equipment and warm the 'must' to a temperature that kills any yeasts present (no need to boil as really hot water does the trick). The amount of honey varies according to taste: dry still .9-1.5kg/2 – 31/2 lb, medium sweet to rich desert 1.8-2.7kg/4 – 6 lb to each 4.5l/gallon of water. This is the 'must' and it is vital that it is completely sterile – add 70g/21/2 oz cream of tartar and one heaped tsp of yeast some of which can be started in a small amount of the 'must' 24 hrs ahead. Other additives can include l tsp citric acid, 1/4 tsp grape tannin, or 1/4 cup of strong tea, 2 tsp wine nutrient, One 3mg vitamin B tablet. Bung the container with a porous plug (cotton wool?) during the initial rapid fermentation period and then use a fermentation valve or lock. Rack whenever 'lees" (deposit) accumulates. When decanting, the short 'breathing space' is actually beneficial though air must otherwise be excluded. The layer over the wine would normally be carbon dioxide. See: cyser, hippocras, hydromel, melomel, metheglin pyment

**meadow**

An area of grassland bounded so that it may be used for the grazing of cattle or for the production of hay. At one time a rich mixture of grasses and flowering plants offered bees useful forage. Man's control of the land has gradually changed the quality and quantity of meadows in the UK. Crops are often grown and either cut or sprayed before the bees have had time to accustom themselves to the source. The specific name 'pratensis' implies that a plant or animal is to be found in the meadow. There is a move afoot in 2010 to regain some of this 'wild' type native flora for the benefit of various insects and other creatures.

**meadow sage *Salvia pratensis* Lamiaceae**
A wild sage and pleasant culinary herb that likes dry, limey soils though it is not likely to be found growing widely enough to give a nectar flow. It grows to two feet and produces both nectar and pollen.

**meadow sweet *Spirea ulmaria* Rosaceae**
Meadow wort or Queen of the meadow a perennial which flowers in dense masses from July to September preferring locations that are damp. Its yellow flowers yield bright yellow pollen. The plant has aspirin-like qualities and in Anglo-Saxon times was used to flavour mead.

**measurements**
And measuring – the extent, dimension or quantity of anything e.g. of length, height, width, depth, volume, temperature, weight, relative humidity, direction of movement, speed, power, cycles per second, colour, light intensity, weather characteristics, cost etc. When making bee equipment remember 'measure twice and cut once.
See: dimensions, figures, temperatures, sizes, useful dimensions

**mechanical smoker**
A smoker operated by a fan that can be switched on or off. Normally a hand-operated bellows is fitted but clockwork and electrically operated fan-type smokers have been made for years especially in Germany. The Konigs Vulvan is an example with a lever-operated fan.

**median segment**
The intermediate or middle segment as of the abdomen, thorax, leg etc.

**Medic *Medicago lupulina* Fabaceae**
Yellow trefoil – it likes an alkaline soil and blooms from May to August offering nectar and pollen. Also known as 'black medic' the flowers being followed by distinctive black seed pods.

***Medicago sativa* Alfalfa, Lucerne**
Good crops of honey are taken when bees work this plant for seed production in Canada and the US. It is grown in Britain mainly for cattle forage and hay but good honey crops are seldom to be had. The lighter its honey the greater its density.

**Mediterranean flour moth *Anagasta kuehniella***
It feeds on pollen in the hive and burrows through combs causing damage in the US. It is controlled like wax moth but poor ventilation and warmth encourages the moths.
See: grain itch mite, Indian meal moth, predators

**medium brood**
One of the thicknesses of foundation. Heavier than fine or extra fine and thick enough to allow reinforcement and for its use to make plumb vertical combs in the brood chamber. One sheet of medium foundation is approximately the equivalent to the wax that an ordinary beekeeper can salvage from one comb.

**medium honeys**
Honeys whose colours range between the light and dark classes or the blend of light and dark honeys. Produced from horse chestnut, dandelion, lime, sainfoin, sycamore and willow.
See: light and dark honeys

**medlar *Mespilus germanica* Rosaceae**
This small fruit tree is not very popular

these days though both wild and cultivated forms are good bee plants. It flowers in May and yields both nectar and pollen. The fruit resembles an open-topped crab apple and is picked when tobacco brown. Harwood writes '*When in blossom the medlar is a charmingly pretty sight, the wide-open flowers have the effect of a cloud of butterflies just alighted*'. It originates from the Caucasus and some are now found wild here in the S.E.

**meer**

Mead made with buckwheat honey

**meeting place**

A drone congregation area is a place where queens find suitors and this is up in the air at a height of around 20m. When beekeepers meet a suitable room may be a hall or other building that is not too far from the majority of the members, which can be temperature controlled and which enables cars to be parked nearby. Other refinements include the availability of audio-visual aids a telephone and of course the hire cost is another relevant aspect.

**mega**

Used as a prefix to mean great in number or size. One million times a given unit, or ten to the power of 6. See: macro

***Megachile rotundata* Leaf-cutter bee**

A solitary European bee bred in Canada for the pollination of alfalfa where it is said to rival honeybees in ease of manipulation and propagation. 4·8mm straws are inserted into drilled blocks, Soda straws 5·5mm inside dia and 107mm long are also used. They are more sensitive to variations in light levels than honeybees but are good alfalfa pollinators and are now used in a number of countries.
See: leaf-cutter bees

**Mehring, Johannes 1857 Ger.**

He made foundation but without walls, not as strong as the subsequent Weed foundation but he used carved wooden plates to form sheets of wax embossed with hexagons upon which the bees would build comb.

**meiosis**

This kind of cell division leads to the production of gametes, i.e. haploid germ cells from the diploid material (fertile components). The phases begin as for 'mitosis' but at prophase there is a significant difference in the behaviour of the chromosomes. A text book on biology should be consulted for details of the various phases and possibilities regarding the distribution of genes.
See: synapsis

**mel, meli L. honey**

Hence: melezitose, meliphagous, melipona, Meliponinae, meliponins, melilotus, melissopalynology, melissomelus, melittobia, mellarius, mellifera, melliferous, mellifica, mellifluous, melligo, mellisugent and mellivorous.

***Melaloncha ronnai* Phoridae**

Said to paralyse host in which it pupates. A member of Phoridae which normally scavenge dead bees.

***Melanosella mors apis***

A fungus that causes melanosis, the blackening and malfunctioning of the queen's ovaries. It also attacks oviducts and spermathecae and is sometimes found in the intestines. There is no known treatment.
See: melanosis

**melanosis**

Caused by a primitive organism, a fungal disease that blackens and destroys the organs it attacks. It is believed to be transported by the blood and produces a morbid development of dark pigment in the queen's egg cells (ovaries) or in the poison sac and rectum of queens. When infected, queens soon become non-layers and are superseded. Melano means black.

**melezitose**

A trisaccharide sugar. When winter stores contain this sugar (possibly from honeydew) the high dextrin content renders them unsuitable for safe wintering in the British climate. For example the overloaded and irritated rectums might lead to dysentery or the escalation of Nosema. It has been found in honeydew produced by aphis on lime trees.

***Meligethes aeneus***

Pollen beetle pest on oils-seed rape.
See: pollen beetle

**melilot Sweet clover Fabaceae**

A clover-like fabaceous herb with compound leaves and clusters of fragrant flowers.

**Melilotus spp. Fabaceae**

Melilot, sweet clover, Bokhara, white and yellow melilot. A tall, hardy annual or biennial that does not like acid soil and produces long slender racemes from June to September. It can grow as large as small bush, likes limestone soil, hot days and cool nights. Yellow sweet clover is overlapped by the white *M. alba*, which flowers a little later. The honey is light greenish, of good quality and has a vanilla-like flavour and granulates quickly. The pollen colours reported are light brown, dull ochre, dull greenish-yellow and its grain is 18µ. Seeds germinate best if frosted.
See: clover

**Melliphagous**

Gk. phagein – to eat. Feeding upon honey (humans?)
See: mellivorous

**meliponiculture**

The husbandry of stingless bees.

**melipona Apidae**

Stingless bees found in the tropics and warmer sub-tropics. The Yucatan Maya have exploited the stingless bee Melipona beechi since prehistoric times. They make blisters of wax the size of walnuts; these are all joined one to another and are full of honey. The bee is about half the size of a honeybee and its honey is preferred. Hollow logs are used with a plug at each end and an entrance hole arranged in the middle. Pollen pots are also taken out and eaten. Another stingless bee *Lestris melitta liao* robs *M. beecheii* whose enemies also include army ants and flies. The honey is watery with a delicate flavour and a pH3·8. If a bee is accidentally killed it is folded in a bit of leaf and buried. In Brazil various stingless bees are domesticated and specially designed boxes are sold to house them.
See: meliponines

**meliponines Apidae**

There are several hundred species of these tropical social bees. The stingless bees Trigona and Melipona are native to Central and Southern America. They resemble solitary bees and wasps in that they mass provision their young. Cells are then sealed. They can bite and exude a sticky irritating substance. Gourds, clay pots and hollow trunks, sometimes purpose-built wooden boxes are used for them. Yields are small and honey water content may be high. *M. quadrifasciata*, *rufiventris*, *marginata*, *scutellaris*, *beecheii* and *fulvipes* are some of the varieties found.

***Melissa officinalis* Lamiaceae**

Bee balm, lemon balm, a hardy perennial that grows to 75cm, has small white flowers, likes a light soil and sunny position and blooms from July to September. The leaves have a lemon scent and were used by skeppists to settle swarms and can be used for flavouring. Back in the time of the Greeks it was written about as having the power to attract swarms. Recent scientific work has shown that there are several fractions in the Nasanov scent secretions that are similar to those in lemon balm.

**Melissococcus pluton**

Revised name for the causative agent of EFB, as Streptococcus are not true anaerobes and will grow in the presents of oxygen. Furthermore Melissococcus has a DNA composition outside the Streptococcus range.

***Melissodes agilis***

A useful pollinator of sunflowers used in Utah.

**melissopalynology**

The science of pollen analysis in honey. One use is to identify and classify pollens and other small particles (spores, yeasts, soot and plant particles) so that the honey's plant source or sources can be determined. Microscopic analysis of honey has become more efficient in recent times. In Britain the work of Rex Sawyer has set new standards of education and accuracy.
See: pollen identification for beekeepers

**melittin**

A protein which can damage blood cells and release histamine. It is a main constituent of bee venom and it causes local pain, inflammation and largely accounts for the local and general inflammation.

**mellarius mellitarii**

A Roman slave whose task it was to look after the hives of bees.
See: apiarius

**melliferous flora L. mel – honey and ferre –to carry**

Honey plants, those with flowers that produce nectar. Although it would not be possible to incorporate even a sizeable fraction of the plants that are useful to bees, every effort has been made by reference to well-known books and lists to include most melliferous plants common in Great Britain and many of significance in other countries. By giving every reported colour of pollen it is hoped to show that a spectrum within a certain range exists for many plants. Colour is also influenced by the bee's 'wetting' of the pollen and the fact that the colours are not always fast and often change under the influence of light.
See: bee plants

**mellifluous mellifluent**

Flowing with honey, sweetly flowing.

**melittoplis**

A commensal found in hives, not thought to be a pest, probably oophilous and smaller than Varroa jacobsoni and Tropilaelaps, brown, ovate and flattened dorso-ventrally.

**mellivorous L. mel – honey,**

vorare – to devour meliphagous Gk. phagein – to eat.

**meloja**

Spanish name for the syrup made by concentrating the liquid obtained from the washing of honeycombs.

**melomel**

Honey wine incorporating fruit juices other than grape, apple or mulberry.
See: cyser, morat, pyment

**melon**

Experiments showed that the proximity of hives not only increased the yield but that the prolificacy was increased by 23%.

**melting point**

The temperature at which a substance changes from the solid to the liquid state. It should be born in mind when handling combs that as wax becomes plastic, then soft at a considerably lower temperature than its melting point of 63C/145F. there is not much gap between that and 34C/94F at which point combs begin to collapse as might occur when bees are hot and excited (say in traveling). A strange phenomenon is called the 'eutectic melting point' when two substances with a higher melting point blend and become liquid at a lower temperature. This applies to beeswax and the use of additives can make it easily 'workable' at low temperature (ear-plugs?).

**member**

A person who belongs to a designated group such as a beekeepers' association. As beekeepers form groups to improve their relationship to society and to further their own individual and collective aims it is wise to take advantage of these benefits. For example bee equipment can be made more readily available and maybe at a discount. Also education and hobby involvement are provided via Field Days Congresses and so forth. Member is also used for parts of the body, in the case of bees we normally say 'appendage' See: organizations

**membrane**

A thin sheet of pliable or elastic plant or animal tissue, connecting or separating regions or covering surfaces. The intersegmental membrane of bees allows for the distensions or accommodation. The wings are described as membranous.
See: osmosis, intersegmental membrane

**memory**

The ability to retain or recall previous experiences. Bees are creatures of habit and once trained or accustomed to a certain pattern of behaviour, tend to repeat it while their body processes allow and the need for action exists. Bees have a time and spatial understanding which is used for foraging purposes. Much of the navigation ascribed to angles of the sun becomes a matter of memory and bees can be shown to make a visual rather than dead reckoning return to their entrance. Swarming seems to obliterate memories of their former home though the memory is retained long enough for them to return if their queen does not make it. Robbing can become an ingrained characteristic as can bad temper when some colonies are constantly interfered with. Bees in an out-apiary often show every indication of remembering the apiarist's car.

**Mendelism Mendel 1822 - 1884 Austrian biologist**

At an Augustinian monastery at Brunn in Moravia he carried out plant breeding experiments over 25 years. The theories of heredity put forward by Abbe Gregor G. Mendel and the law he discovered relating to the inheritance of characteristics, is one of the hall-marks of man's progress in the realm of genetics. It did not surface in the scientific world until 1900. 'The offspring of parents with different characteristics will exhibit these in a definite ratio'.

**mentum**

Chin – part of the labium bearing movable parts.
See: pre, post mentum

**Mephitis spp. skunk**

A small, striped, fur-bearing animal which seems fairly immune to stings though a strong colony will usually drive them to another one. It appears unlikely that they can feel any pain. They will pull mouseguards off and are most active at daybreak and just before dusk. Fortunately not a British predator.

**meront**

The stage in the development of nosema as the planont having fed on the contents of the ventricular cell divides into two and two again forming four daughters or meronts.
See: planonts

**Merops spp. Bee-eater**

An insectivorous bird which has a long slender bill, some with brilliant plumage. In the Sudan, where it is called the carmine bee eater, it is reported as having killed many bees. *M. apiaster*, the European bee-eater not only eats bees but insects that eat bees. *M. superciliousus* the blue tailed and *M. orientalis* the little green bee-eater. Bumble bees too feature prominently in their diet and stinging insects may be 'bee-rubbed; against a perch.

**mesenteron**
The embryonic stomach. It forms around the yolk of the egg and becomes the forerunner of the ventriculus.

**mesh**
One of the small gaps provided by the criss-crossing strands of a net. Fabric, gauze, nylon, wire and other materials may form a mesh which permits air or liquids to pass through yet restrains particles or objects larger than the gaps (holes) formed by the network of strands. Widely used in filters, strainers and screens to make bee veils or insect barriers. Queen cage mesh is usually 3-4mm.

**mesoblast**
The mesoderm or middle layer of an embryo. Composed of cells that have moved from the surface of the embryo into the interior. It is the germ-layer demarcated within the embryo after gastrulation but before derivative tissues have differentiated.

**mesoderm**
In the embryo this is the name given to the third original cell layer which forms between the entoderm (endo) and the ectoderm. Meso – middle and derma – skin, it gives rise to the muscle, blood etc.

**mesodermal originations**
Ovary, testes, vasa deferentia and accessory glands are thought to be of mesodermal origin while the ejaculatory duct and vagina are of ectodermal origin.

**mesothorax**
This follows the prothorax. In this region the plates are all firmly fused together but there is a scutal fissure between it and the scutellum. The forewings are accommodated in a gap between the tergite and pleurite. This gap is covered by a tough but flexible membrane which is continuous with the forewing. The tegula overlaps the wing root and there are provisions in the edges of the pleurites for the articulation of the wing sclerites, and the middle legs are articulated from the pleurites.
See: scutum

***Mespilus germanica* Medlar (formerly *Pirus g.*)**
This picturesque tree with crooked branches grows wild but is rare now. It is still of some consequence to beekeepers in Portugal. Like the medlar its fruit is not edible until on the point of decay.
See: medlar

**mesquite *Prosopis glandulosa, juliflora* Fabaceae**
A shrub or small tree regarded as an invasive weed (though it is a prolific source of honey in some south western parts of the US, Mexico etc). Super rich bean-like pods make good fodder. Also listed as honey locust a thorny tree, Mentioned under Maltese honey.

**metabolism**
The total effects of a body's use of food or the processes involved in the maintenance of life. The human body has been described as a metabolic whirlpool. This implies that a living creature is constantly re-organising compounds either by catabolism to break-down or simplify material, or by anabolism which concerns itself with the synthesis or building process. For example the break-down of food is accompanied by the release of warmth and energy, whereas the synthesising of amino acids and protein is essential to body building.
See: blood sugar, mannose

**metal**
An electropositive element and good conductor of heat or electricity, which will actively combine to form salts. Bees prefer wood to metal, largely because of heat transfer but doubtless due to chemical and electrical reactions. Metal has the advantage of being capable of forming sheets, being moulded or stretched into wires.
We also speak of broken stones forming a

macadamized or 'metal' road. The absence of these in some counties leaves huge areas of bee-forage 'untouched'.
See: metals

**See: deadman floorboard, sterilisation.**

Metal divider

Used in 'section' crates. A thin sheet of metal shaped to make a clean separation surface between the faces of adjacent sections, yet cut-away to allow the easy passage of workers between one row and another. They can be obtained three or four sections in length according to whether British or American hives are used. They are for 41/4" square sections. Wood or plastic can be used but thin metal is light and reasonably durable. Thorne's catalogue describes them as tinplate spacers, intended to prevent brace comb and to ensure attractive level cappings.

**metal ends**

Invented by W. Broughton Carr, small clips stamped from thin sheet metal, were folded to slide onto the frame lug and provide centre-to-centre spacing of near to 11/2"/38mm. A wider one, nearer 2"/51mm was also made for wider super combs. Although these can be staggered or alternated with clipless lugs to give narrow spacing, beginners especially are advised to keep them out of the brood chamber. Rusting, sharp edges and lack of rigidity have been overcome by replacing them with plastic ends of a similar pattern although these are not quite as versatile. Hoffman side bars offer a very reasonable alternative as the spacing is automatically built-in.

**metal feeder**

These are rapidly being replaced by plastic counterparts, which are easy to clean and quite acceptable to the bees. The over-all, large wooden feeders, Ashforth and the like, have advantages for heavy feeding and also fit in with the hive design.

**metal queen excluder**

Slotted, zinc sheets could be stamped out and were quite popular but they tend to become propolised in national hives and have burr comb built under them in hives with 'top-spacing'. Here again, plastic has again begun to take over but the sturdy Waldron with rigid, parallel wires fixed in a wooden frame, although more expensive, is very durable and long-lasting.
See: excluder, queen excluder

**metal roof**

The roof of a hive must be fairly robust as well as leak-proof and for this purpose metal does seem an obvious choice. The weight is also important though when correctly designed a wooden roof surmounted by a sheet of metal can be made so that it is wind resistant only being 'liftable' by a direct upward movement. Insulation under the metal is important to minimise temperature changes and the possibility of condensation, though roof ventilation should take care of that.
See: roof

**metals**

Metallic elements are crystalline when solid and display lustre when freshly cut. Soil minerals affect nectar secretion. Extractors and honey tanks must be made of stainless steel or similar honey resistant metal because a mixture of water and honey will attack lacquer and tin linings. Bees invariably try to cover any metal within the hive with their 'gum'. Smokers – an important part of most beekeeper's equipment -must be copper or a metal which can stand up to regular scorching from the burning smoker fuel and hive tools must be as light as possible compatible with great tensile strength and ability to take a fairly sharp edge for scraping purposes. Monel metal was used to reinforce beeswax foundation but even the best foundation often fails to completely embed the wire and the tell-tale rows of empty cells make an uncomfortable looking pattern sometimes,

on otherwise beautiful slabs of brood.
See: metal roof, minerals

**metal work**

The craft of using metal to make or maintain objects and this is widely used for the benefit of beekeepers: honey extractors (enclosing metal or plastic cages) queen excluders, metal spacers, hive tools, roof covers, bee escapes, ventilators and smokers are just a few examples.
See: wood work

**metamorphosis**

The changes in form and structure that occur as the larval stages of an insect progress toward the imago. Some insects such as earwigs are born looking rather like pale, miniature adults. Honeybees change from egg to ancestral looking larvae, these undergo further gradual changes during pupation and eventually mature as an imago or perfect insect. The complete change of form is the metamorphosis.
See: age of brood

**metaphase**

The stage of meiosis and mitosis when the chromosomes are arranged on the equator of the spindle of the cell nucleus.

**metascutellum**

The scutellum of the insect's metathorax. It forms a bump across the back of the thorax. The part of the queen generally used for 'marking'.
See: scutellum

**metathorax**

The third of the four segments that compose the thorax, situated between the mesothorax and the propodeum. Its plates are relatively small and almost hide its two spiracles though these can be seen more easily on newly emerged bees.

**meteorology**

The scientific study of the atmosphere with related phenomena, particularly the weather and climate. The Met Office (former Meteorological) is a government department which issues weather forecasts, warnings etc. Gales and floods are sometimes usefully prognosticated but long term UK weather seems to be anybody's guess and even freak dry periods like the two months normal rain in 15 months of 1983/4 came without warning. Satellite 'actuals' are now making forecasts considerably more reliable.

**methelglin**

A dry mead to which spices have been added. A sweet version with spices is known as Sack metheglin.

**method**

A system, or form of management or orderly manner of going about things. For instance analysis, filtering honey, mead or wax, swarm control or prevention, queen introduction, uniting or filling feeders or bottles.

**method of destruction**

When a colony has had the misfortune to get Foulbrood, destruction by fire is the normal means of ridding the material of infection. The hive entrance is sealed when all bees are inside and 140ml/1/4pint of petrol is poured through the feed-hole and immediately sealed. A hole, deep enough to take all the frames, combs and bees is dug. A small fire started in the bottom and the contents of the hive, ensuring that all bees, comb, frames and loose material are put onto the fire in such a way as to obtain complete combustion. The hole is then filled, the area sprayed with Jeyes or similar disinfectant and the hive and parts completely scorched inside. Finally scrubbing of all parts before re-use, with hot lye, using rubber gloves to protect the hands.

**method of filling bottles**

Once clean mead (siphoned, filtered etc.) has been poured into a clean bottle using a suitable funnel until it is within 2cm of the top, some difficulty may be experienced inserting the cork due to compression of air above the surface of the liquid. This can be overcome by using a short piece of clean string placed so that one end enters the bottle top and an airway is provided alongside the cork as it is forced home. The string can then be pulled clear as hand pressure is kept on the cork.

**methylated spirits meths**

Ethyl alcohol denatured to prevent its use as a beverage. It contains methyl alcohol (wood alcohol) pyridine and methyl violet dye which gives it its colour. It will dissolve wax and propolis and is used for cleaning purposes. It tends to harden the skin if used to clean the hands.

**methyl bromide $CH_3Br$**

Kills all stages of wax moth but it is a very volatile substance with toxic fumes and the best results are in a gas-tight fumigation chamber.

**methyl salicylate**

A synthetic chemical – methyl-ortho-hydroxy-benzoate – Oil of Wintergreen. It was used by allowing evaporation from a small bottle with a wick protruding, inside the hive, as a precautionary measure against the migration of acarine mites.

**metre**

3·28084 ft or 39·37 inches. The symbol 'm' a basic S.I. unit of length. Approximately one ten millionth of the distance from the pole of the earth to the equator (like to check it?).

**metric abbreviations**

They are:-
atm atmosphere, cl centilitre, ca centare, cg centigram, cl centilitre, cm centimetre, dag decagram, dal decalitre, dam decametre, dg decagram, dl decilitre, ha hectare, hg hectogram, hl hectolitre. hm hectometre, Hz hertz, kg kilogram, kl kilolitre, km kilometre, KW kilowatt, l litre, m metre, t metric tonne, mi mile,
mg milligram, ml millilitre, mm millimeter
t tonne,
a year or atto, d da. Plurals the same.

**metric prefixes**

T tera - one million million, G giga - one thousand million, M mega - one million, k kilo - one thousand, h hecto - one hundred, da deka - ten, d deci - one tenth, c centi - one hundred, m milli - one thousandth, mu (μ) micro - one millionth, n nano - one thousand millionth, p pico - one million millionth, f femto - one thousand million millionth and a atto - one million million millionth.
See: systéme

**metric system**

Systéme International d'Unités.
A decimal system of weights and measures. The gradual introduction of the metric system which began with coinage, has led to widespread changes (aided inflation too as prices rounded upwards) and repercussions not the least of which have been repercussions from the insular British. Such were the errors in translating 2 lb to 1 pint for winter feeding that the government sent out a corrective circular 7 kg for 4 litre is not far out though in any case one kilo of sugar will go into a pint though the resultant syrup is very thick.

**Mew, William 1655**

Rector of Eastington in Gloucester left the design for an octagonal wooden box with windows that could be storified or 'nadired' and henceforth supers instead of bottom extensions (ekes) could be used. Although not a publicity seeker many became interested in the hive and much has been written about it. Samual Hartlib, William's son Samuel Mew, Dr.Wilkins, John Gedde, John Evelyn, Christopher Wren, Moses Rusden and Thorley. It gave way 100 years later to a square box designed by Stephen White then Thomas Wildman. See: Mews and Hoy.

**Mexico**

The number of beekeepers and yield of honey in 1980 was similar to ours. The honey from the Yucatan peninsula is amber to extra light with 19% moisture content. In 1981 the World Beekeeping Congress took place in Acapulco.

**M.G. wax extractor**

A cylindrical container capable of being heated underneath. A spout with a funnel leads water down into the bottom. Waxy material (combs etc.) to be rendered are put into the container with enough water to go over the inlet. The waxy material to be rendered (having been previously soaked for 24 hours) is put into the container and a circle of Hessian clamped over the top. It is allowed to simmer until the wax has melted, then boiling water is slowly poured in via the funnel until liquid wax is forced up through the Hessian. This spills into a circular trough and is ducted via an outlet spout into a suitable mould. Boiling water is added until no more wax come up through the hessian.

**mi**

Abbreviation for statute mile (1760 yards) Not to be confused with 'm' metre, or 'ml' millilitre.

**Michaelmas daisy Aster spp. Asteraceae**

Starwort. A perennial Aster beloved of bees in the late autumn sunshine. More likely to provide pollen than nectar. Honey is amber with a pronounced flavour that improves with age. There are over 400 varieties. Pollen described as yellow, transparent and almost colourless, the grain 25µ.

**micro Gk. micros**

Very small. The word is used to form the diminutive as in microbe but also to mean enlargement as in the cases of microphone and microscope. Also to prefix units one millionth of the stem such as microgram.

**microbe**

A microscopic organism or germ such as a bacterium, frequently pathogenic but also responsible for natural decay, soil improvement, fermentation etc.

**microclimate**

The special climate that exist in a particular habitat which makes it different from that of the surrounding environmental conditions. Small-scale climate such as within a flower, hive, garden or greenhouse etc.

**microfiche**

Miniature information storage on a sheet of microfilm usually 10 x 15cm (4" x6") capable of accommodating and preserving a considerable number of book pages in reduced form. Also called 'fiche' A micro reader x48 is required and most public libraries have these.

**micromillimetre 0·000039 inch (µ)**

One millionth part of a millimetre – a millimicron. A micrometre is a micron.

**micron pl. micra**

Unit of length – a microscopical measurement 0·001 millimetre, as used for the measurement of pollen grains in this work. It is symbolised by the Greek letter µ . Also millimicron which equals 10 Angstroms.

**micropyle**

1. A small pore, covered only by a very thin membrane in the chorion of the egg of the honeybee. It is positioned at the apex of the egg i.e. the end that is not attached to the cell base and it is through this aperture that the male germ cell is able to penetrate via the spermatozoon. (Do nurse bees play any part in controlling this?)
2. An entry point at the apex of an ovule for the admission of the pollen tube in a flower.

**microscope**

A lens assembly positioned to obtain the best resolution while increasing the apparent size of the object viewed. Thus rendering visible objects too small to be seen with the naked eye. These vary from hand lenses to dissecting microscopes that rectify the object and 'compounds with an eye piece sometimes binocular. These devices have now advanced to electron and scanning electron microscopes that can photograph at powers vastly in excess of the optical microscope.
See: microscopy

**microscopical analysis of honey**

Gives geographical covering and botanical origin of honey. About contamination by brood, dust, soot etc. its yeast content and other micro particulars. Pollens that are over-presented include Myosotis Forget-me-not, Castanea sativa Sweet Chestnut. Under presented: citrus, Lavendula, Rosmarinus, Salvia, Robinia, Tilia, Medicago, Epilobium and Curcurbiteae.
See: melissopalynology

**microsome**

Cell particles of minute size comprising granules of protoplasm found in living cells. They contain a number of enzymes (ribonucleic acid) and are functional in the realm of protein synthesis.
See: mitochondria, ribosomes

**microspore**

Bot. The cell from which a pollen grain develops in flowering plants.
Zoo. Spore of A.F.B., nosema etc.

**microsporidia**

Very small spore-like elements produced asexually by monads. Developed by a spore mother-cell owing to meiotic anomaly.

**microtome**

An instrument designed for cutting extremely thin slices of material for microscopic examination. Material is normally either frozen of embedded in paraffin wax, the wax is then removed by a solvent. Microtomy – the cutting of very thin slices.
See: cross-section

**microvibration**

Small rapid spasmodic movements of the flight muscles are an example of microvibration which is essential to honeybee survival.

**microwave oven**

An oven that heats by means of microwaves. Provided that no metal (such as honey jar caps) are included such ovens can be very useful for softening or liquefying solid (granulated) honey. To soften - one minute at full power (600 watts) and a little longer to clear.

**middle 'C'**

A musical note near the middle of a piano keyboard. The sound a swarm is said to produce as it starts moving to leave the hive. The note can be replicated by pressing a moist cloth against the glass of an observation hive and moving it firmly downwards when the 'squeak', as long as it lasts, will cause all the bees to stand quite still. This is the sound a virgin queen makes (called 'piping') when she is about to leave the hive and if it causes nearby bees to 'freeze' then her way of departure is cleared.

**middle legs**

On the honeybee this pair of legs is shorter than the hind legs but longer than the forelegs. They bear no special tools other than a single spine on the distal end of the tibia. Other insects (wasps and bumbles) are richly endowed with such spines while this remaining pair on the honeybee have not been observed to serve any particular function. Possibilities might include extracting wax platelets or dealing with pollen (?).They do help to stabilize a honeybee when it uses its antennae cleaners of the forelegs or the pollen sorting apparatus of the rear legs. The claws are similar to those on the other legs having ample strength for hanging in festoons or clusters.

**mid-gut**

The larvae develop three unconnected parts as their alimentary canal; the anterior stomodeum, the mesenteron (mid-gut) and the proctodeum. Whereas the stomodeum grows as an epithelial tube to join the mid-gut, the hind-gut (proctodeum) does not connect with it until the larva has finished feeding and is about to pupate.

**Midnite**

A commercial hybrid bee derived from a scientific four-line cross using Caucasian, and Carniolan strains. First released in 1967. Said to be gentle and willing to work at low temperatures.

**mid-rib**

The septum or centre line of the comb. Although it is built in the vertical plane overall, it is composed of rhombic cell bases. Three rhombs, each inclined at a slight angle to the vertical, form the concave cell base on one side while serving one third of a cell on the other side. When foundation is used it will be noticed that the septum is thinned down considerably by the bees. However successive broods of pupating larvae thicken the septum by adding skin moults with layers of faeces sandwiched between, so darkening and strengthening the septum with what are called 'cocoons'.
See: cocoon

**mid-summer**

The moment when the sun is on the tropic of Capricorn. This is when we have our longest day but mid-summer's day is June 24th and the solstice is about 21/22nd when the sun is at its highest. The solar wax extractor will be at its best and maximum foraging could occur if the weather is right. Sea temperatures lag behind the sun so still continue to rise leading to warm on-shore breezes in the autumn.

**mignonette *Reseda odorata*, lutea Reseda-ceae**

French pronunciation. A hardy annual that has racemes (spikes) of small whitish, fragrant flowers with prominent reddish-yellow or brownish anthers and blooms from June to August. Found wild on the moist, chalk soils and grows wild on chalk hills in the sun. Bees can be seen going backwards and forwards anxiously collecting nectar and pollen and re-visiting again and again. Its height 15 -30 cm and pollen grain 20μ, described as transparent yellow, fawn, brown, orange dark yellow and dull red brown.

**migratory beekeeping**

Columella describes how the inhabitants of Achaia imitated the Egyptian custom and took hives overseas to benefit from the wonderful forage on the Attic peninsular. Solon writes about caravans of bees and bees on rafts in 600 B.C. and the floating of hives down the Nile was practised in very early times. The Romans moved hives down on the Po and the Rhine. Migratory beekeeping tends to be practised more in countries where a wide variation of latitude enables successive honey flows to be secured; e.g. Japan and Norway. Large scale movements to orchard areas for fruit tree pollination to acacia (Robbinia pseudoacacia) forests in Europe and to the great tracts of heather, are regularly practised. A 1400 mile trip in 26 hours was said to have had no adverse effect upon the bees!
See: moving hives

**mildew**

A whitish coating which may be caused by microscopic parasitic fungi. Also describes mould or even 'bloom' which is not pathogenic. It can be found on damp walls and over-wintered combs in a hive, also old pollen and dead brood,. It is surface

deep and disappears when dry conditions supervene.
See: chalk brood, pollen mould

**milk and honey**
Honey blends with all manner of foods and drinks but with hot milk it is specially favoured as it makes a tasty nourishing beverage suitable for young and old alike. The Bible reference to a 'Land flowing with milk and honey' tells us how long, long ago the value of honey was appreciated.

**milk test**
For Foulbrood. A suspected scale of AFB is ground up and intermixed with two drops of warm, fresh, whole milk. Transfer to a microscope slide and rub thoroughly for up to ten seconds. It will, if genuine, coagulate the milk. EFB remains take 80-120 seconds to coagulate with milk. Precautions must be taken to sterilize everything used, keeping contaminated material away from the bees.

**milk vetch *Astragalus glycyphyllos* Fabaceae**
A perennial climbing plant with green zigzag stems that sprawl along the ground for approx. 60 - 75cm. Sea vetch flowers a dingy yellow. Its name comes from its alleged ability to increase milk yield of goats.

**Milkweed *Asclepius syriaca*, Pleurisy root Apocynaceae**
Butterfly weed *Asclepius tuberosa* in US. A superior nectar yielding plant, yellow to orange with deep roots. May be planted in pastures as cows will not eat it; highly attractive to bees though not first bloom (blooms twice). Also A. asperula and speciosa.

**Miller, C.C. Dr. Marengo, Ill US Died 1920**
Wrote '*Years among the Bees*' Ohio, Medina. Introduced 'T' supers. a feeder, queen introduction cage, a queen rearing method and declared 'Breed from the Best.

**Miller feeder**
A box-type of feeder that covers the whole hive and gives access to the bees via a central slot running right across the feeder. Usually made of wood and lined with water-proof paint. The central slots virtually divide the feeder into two reservoirs, each capable of holding up to a gallon of syrup. The painted walls of the slots are covered with gauze or similar material to give bees a foothold and a panel of glass, ply or gauze covers the slots to confine bees to that region. A cover such as a crownboard is put on to prevent bees getting into and drowning in the syrup. Some beekeepers leave them on to provide top-ventilation in the winter.
See: overall feeder.

**milli**
Prefix meaning division of the following word by an element of one thousand.
E.g. millibar, unit of atmospheric pressure widely used in meteorology.

**millilitre ml**
A metric unit of capacity 0·001 litre, previously called a cubic centimetre (c.c.).

**millimetre mm**
A metric unit of length – one thousandth of a metre.

**millimicron**
One millionth of a millimetre.
A micrometer.

**millstone grit**
This type of soil is found on the Pennine range and hills of north Devon. In contrast to chalk and oolite it is derived from sandstone laid under desert conditions with little plant food in it composition. Deficient in magnesium which deters clover secretion. It makes a light, well-drained soil of a hungry nature and readily forms iron pan.
See: iron pan

***Mimosa wattle* Acacia spp Fabaceae**
A fragrant evergreen tree that belongs to

tropical or warm regions but manages to survive in the Channel Islands and the southwest. There are about 550 species in the sub-tropics. *A. dealbata* is the silver wattle of Australia. Many are valuable for their astringent and gum-yielding properties (e.g. gum Arabic) while some yield tannin. Much of the scrub is wattle. Although the fragrant flower with its fluffy yellow balls has been sold for years on the London streets in January, (hot-house grown), its pollen is windborne and the beads of honeydew that form on indoor plants do not seem to materialize on those plant that survive outdoors.
See: Robinia

**mineral**

An inorganic compound that occurs naturally and has characteristics and composition that is fixed. Sulphur and gold are examples of minerals made of only one element but most are found in combinations of two or more such as salt NaCl, calcium carbonate $Na_2CO_3$. The minerals found in bees, honey or collected pollen, can be a useful indicator of the presence of particular minerals within their foraging area.
Dr. E. Crane calls them 'prospectors'.
See: mineral content of honey

**mineral content of honey**

The minerals found in the bee are remarkably similar to those found in human blood. They affect honey colour as do other factors. Dark honeys are richer in minerals than lighter ones. Some chemical elements found in all honeys include:
I (iodine),Si, Al, Fe, Cu, Na, K, Mg, Zn, Ca, Cl, P, S, and Mn.
See: honey minerals

**minerals in pollen**

Calcium, iron, copper, magnesium, phosphorous, sodium, potassium, aluminium. manganese, and sulphur.
See: minerals of honey

**minicosy**

Insulated cover for small queen raising nucleus colony as advocated by BIBBA.

**minim M**

The smallest liquid measure – a drop.
A mediaeval unit of volume ·00061 cu in
60 minims equal 1/8 fl.oz, 3·5ml (drachm), or one small teaspoonful.

**mining bee Andrena Aculeata Apidae**

A solitary, hairy bee that burrows into dry earth to make its nest. Restricted to the Columbia basin. Often gregarious. A few Halictus have developed a worker caste and become sub-social. The food mass on which the female lays her egg is a solid, dough-like, dry ball of nectar and pollen. Andrena can be parasitized by Nomada.

**mini-nucleus**

A small nucleus such as one might use in a 'stud' area, comprising the smallest viable unit with, at the outset, a ripe queen cell, virgin or nubile queen. Usually very well insulated against heat loss, having a small but adequate food supply (often in the form of candy) and having an easily identifiable but small entrance.
See: baby nucleus, nucleus

**mint Mentha spp. Lamiaceae**

The aromatic, perennial herb that blooms from July through September. Its amber honey has a fleeting minty or pepperminty flavour. There are untold varieties and most of them yield nectar freely. *M. acquatica* yields nectar with a small vitamin C content, *M. piperita* is used as peppermint in medicine and confectionary. *M. sachalinensis* is the common or garden mint (peas mint or spearmint). The pollen is pale yellow almost transparent.
See: horsemint, lemon balm, pennyroyal

**miodomel**

The Polish monks of St. Basil made a superior mead flavoured with hops, excellent for digestion and a remedy for gout and rheumatism.

**mirror**

Water, glass and other flat surfaces that reflect light can confuse bees and they have been known to fly into window panes though their flying speed and tough, chitinous covering allows them to escape serious harm. Drones do not go out to collect water, yet having found one or two floating in a pool it was assumed that in nil-wind conditions the surface had given the same polarized light pattern as the sky and they must have plunged to their deaths instead of entering a drone congregation area above. It has been said that flashing the reflected light from the sun onto a swarm has caused it to leave its clustering position.

**miscible**

Capable of being mixed in all proportions, e.g. water and alcohol. When two liquids are not miscible an intermediary substance into which either will dissolve, may be used to assist mixing. (water, glycerine and carbolic acid).

**misdescription**

Misrepresentation. Fraudulent, misleading labelling including the wording or illustrations used. A term like English honey may be used only for honey made by bees in England. One can be more specific giving county or region and even floral source if the honey was derived mainly from that flower, e.g. Devon Heather Honey though this could still be labelled 'English honey'.

**mistletoe *Viscum album* Santalaceae**

A parasitic perennial plant that grows on a variety of trees and has yellowish flowers on leathery stalks and white berries. Whitehead classes it as useful for both nectar and pollen and states flowering time as March/April. A relative of this parasitic plant is Nuytsia floribunda (Western Australia/NZ) which begins as a creeping vine but eventually kills its host tree and takes over to become a forest giant. The red flowers of the canopy can be seen dotted about the NZ landscape and it is bountiful in producing enormous quantities of nectar. White 'rata' honey.

**mite Arachnida Acari**

A small arachnid with a sac-like body many being tiny parasites. Bumbles tend to have a more varied fauna than honeybees, the latter sometimes playing host to the Acarine mites and the larger Braula coeca or bee-louse. The pollen mite is another highly destructive of stored pollen. There are enormous numbers of similar parasites.

**mitochondrion pl.dria**

Minute, self-replicating, semi-solid bodies or organelles found in the cytoplasm of eukaryotic cells containing protein, fat and many enzymes. A mitochondrial sheath surrounds the spiral thread of the spermatozoan body. A mitochondrion is larger than a microsome or lysosome and is associated with respiration and energy production.
See: microsome

**mitosis**

The indirect method or process that takes place in normal growth of cell division when reproduction is not to follow. It is explained in four phases: prophase, metaphase, anaphase and telophase. For details consult a standard text book.

**mixed equipment**

Although the need for standardization is constantly emphasized and moves towards having one recognized standard are always being mooted, there are really quite a host of different hive types. Bearing in mind the need to interchange or to change over, here are one or two points where a fortuitous similarity of size exists between two different types: Dadant and Langstroth have the same 'length' of frames, The Dadant width is close to the National, Inner WBC width is the same as Langstroth so WBC inners will sit on Langstroth floors. Smith shallow frames fit crosswise in Langstroth shallow boxes. Two Langstroth shallow boxes will take Dadant deep frames. WBC 'metal' ends may have outsides bent forward through 180 degrees and then they will touch adjacent Hoffman shoulders. WBC bodies reinforced at the sides by wood cut from the lifts, practically fit onto nationals.

**mixed pollen loads**

Honeybees are usually faithful to one kind and colour of flower on any particular trip (so both baskets match), In periods of dearth bees can be seen searching from one kind of blossom to another but to see more than one colour of pollen in a pollen pellet is rare though it is sometimes seen on bumble bees' legs. See: oligotrophic

**mixture**

A blend or combination of substances. This may apply to a solid, liquid, gas of any sort of material. We mix our syrup by stirring sugar into hot (bro. Adam always had his monks use 'cold') water. A mixture may need filtering to remove one or more of the ingredients. Mixed pollen loads are unusual on honeybees yet within the hive pollens of various colours are found cheek-by-jowl. A remedy may take the form of a mixture, e.g. 'Frow'.

**mnemonic**

A memory aid. In practical beekeeping simple rules may well be transformed into mnemonics, e.g. always put the brood nest foundation between the food and the brood. The international queen marking colours might lend themselves to a rhyme, white, yellow, red, green, blue – you work one out.

**mobile site**

Caravans or portable bee-houses have been used as apiaries in some countries. A barge might well provide a chance to move hives along water-ways. Rafts were used by ancient Egyptians to take advantage of the gradually changing foraging conditions along the Nile. It would be interesting to learn whether bees first taken to America or Oceania were given any chances for flight while on route.

**model hive**

According to publicity and technical ability to churn out hives by the thousand, several hive types have come to the fore. Starting with the early straw skeps, strange contraptions followed and then more serious constructions went to Langstroth in America and the white, so-called 'telescopic' WBC here. As the emblem of the 'bee' has been used so much, advertising also makes use of whatever hive strikes the artist's opinion as being 'typical' or 'model'. In another sense, a model hive would be one especially made for education or artistic purpose, usually on a smaller scale. Such hives might go along with the garden 'gnomes'.
See: hive model

**moderator**

Arbitrator or mediator. An officer who presides over the conduct of say a public forum, legislative body and so forth. The BBKA appoint a 'moderator' who oversees the examination arrangements including the setting and marking of examination papers, conduct of examiners etc.

**modified commercial hive**

Or National Major hive. 181⁄4" (464mm) square with bottom spacing. The deep frame is 16 x 10" (406 x 254mm) with a 171⁄4" (438mm) top bar, the shallow frame being 6" (152mm) deep. See: hive commercial

**modified Dadant**

Although Americans have for the most part gone over to Langstroth hives, some people prefer the even larger frame and hive – the modified Dadant, which has a periphery of 20 x 181⁄2" (508 x 470mm) – roughly Langstroth length and National width; which holds eleven Quinby depth frames 111⁄4", shallows 65⁄8". Quinby liked 11⁄2" centre to centre frame spacing and the external size will take 11 of these or 12 of the narrow (13/8"). The brood area is 3740 sq. inches 2.41 sq. metres, top spacing is used.

**modified national hive**

181/8" (461mm) square with 11 frames or 12 at 1³/₈" spacing. As with the national each box comprises eight pieces of wood which assemble to offer hand-holds like the WBC and a weather strip along back and front. Frames are flush with the top of each box so that relatively flimsy queen excluders can be used. Bee-space is provided underneath the frames. Like other square hives combs can be aligned either 'warm' or 'cold' way.

**modified Snelgrove board**

The Snelgrove board is normally fitted with a centre opening and three pairs of entrances, one above the other on three sides of the board. Modification reduces the number of entrances, but all such boards have at least one pair of entrances so that the colony, separated into two parts, can be manipulated to increase the number of foragers in the lower part while permitting a new queen to be raised in the upper part.

**modify**

To alter with a view to improving something. Even in countries where the Langstroth hive is the standard, ubiquitous hive, there is always someone wanting to make some variation in the shape or size (especially depth of frames). Britain has never achieved the same widespread acceptance of any one hive pattern, though one of the most widely used is the modified national.
See: metal work

**Moir Library c/o Music room**

Scottish BKA a Central Library that contains a good selection of bee books. George IV Bridge, Edinburgh EH1 1EG.

**moisture**

A liquid having the ability to render things moderately or slightly damp. The wetting element is most commonly water but any other liquid may have the same property. The degree of moisture is referred to as the 'moisture content' and the amount present in the air leads to conditions that cause mist or fog and that are called 'oppressive' or 'bracing' etc. We refer to 'wet' supers meaning that the empty cells have a coating of honey which is no more than a thin film left behind after extraction.
See: relative humidity, vapour pressure

**moisture content of honey**

Under Schedule II of the Honey Regulations heather and clover honey should not have a moisture content of more than 23% and other honeys 21%. The moisture content varies the viscosity and specific gravity. Honey with a 14% moisture content has a s.g. of 1·4453 whereas at 21% it falls to 1·3966. At 21% moisture content there are 14 lb honey to a gallon, (a gallon of water weighs 10 lb).
See: equilibrium, specific gravity

**moisture equilibrium (honey)**

Honey with a 17·4% moisture content is in equilibrium with air at 58% relative humidity so it would lose moisture to air at 50% RH but absorb moisture from air of 60% RH. Equilibrium tables show: at RH50 at equilibrium with honey at 15·9% water content, 60 -18·3, 70 - 24·2, 80 - 33·1.
See: hygroscopic, moisture content

**molecular attraction**

This phenomenon can be seen when a liquid is caused to 'ride-up' the surface of a container, or when two objects floating in water are pulled towards one another by the force. It is a force of attraction acting on atoms and molecules of a substance other than gravity.

**molest**

To interfere with something with a view to

stealing, annoying or injuring. The bees' own defence stands them in good stead as regards discouraging most would-be molesters, In late summer the only UK molester (there are bears in some countries) is the wasp which can be a nuisance. Other predators can be put into this category but are dealt with under 'Predators'.
See: predator, vandalism, wasp

**Monel metal**
An alloy of nickel, copper 67/28 with 5% other metals. Produced in Canada. It is a silvery white, rust-free and corrosive resistant alloy used widely for honey storage tanks and extractors. Monel wire is recommended for wax reinforcement. 67% nickel, 39% copper, 1·4 iron and 1% manganese, by Monel Nickel Co.

**mongrel**
Hybrid offspring of repeated crossings, often used in a depreciatory fashion. An animal or plant whose parents were of dissimilar origins. Applies to the vast majority of UK honeybees to-day.
See: crossbreed, degenerate

**monocolporate**
Used to describe a pollen grain having one furrow and one pore.

**monocotyledon**
A plant which arises from the seed with only one primary shoot. One of the two sub-classes in Angiosperms. Grass is a typical monocot.
See: dicotyledon

**monocropping monoculture**
An agricultural term used when large areas are cultivated with the same crop. Unless special precautions (including the use of sprays, artificial fertilizers etc,) are taken, damaging pests are encouraged by monocropping and yield may seriously decline. From the beekeeping point of view crops like clover, alfalfa, and oil-seed rape have given bees a bonanza when grown in vast quantities. Alternatively barley and rye grass have made areas a 'no-go' place for the honey farmer. Monoculture (2010) has suspect side effects injurious to bees caused by lack of variety in forage.

**monocular Visual-aids**
This applies to microscopes with one eyepiece. While binocular instruments allow 3-D effect and are better for dissecting, a monocular has the advantage when one eye can be opened to see down the tube and the other for making a drawing of what can be seen.

**monoecious**
The plant equivalent of animal hermaphrodite. Having unisexual flowers, yet bearing the flowers of both sexes on the same plant, i.e. stamens and pistils appear on separate flowers on the plant. Hazel is an example where the male catkins are very different in position and appearance from the small red female flowers. This characteristic encourages cross-pollination when the two sexes mature at different times. Compare with dioecious.

**monosaccharide $C_6H_{12}O_6$**
The simplified form of carbohydrate e.g. glucose and fructose. Two monosaccharides can join to form a disaccharide in fact are building blocks for the formation of more complex sugars. Monosaccharides cannot be split into simpler sugars. Glucose is a hexose (6 carbon atoms) with an aldehyde group (CHO) while fructose has the same formula but contains a ketone group (CO). They are reducing sugars.
See: disaccharide

**monostrain**
Colonies with similar genes and therefore characteristics, as for example those raised from sibling drone mothers. The setting up of a monostrain area, where all queens are sisters (and therefore all their drones are siblings too) is a pre-requisite to forming

a 'stud' area. Provided such an area is reasonably isolated from other drones, a high percentage of true matings can be expected.

**monsoon**

An atmospheric pressure system that has a cyclic repetition resulting in a complete reversal of prevailing wind from one season to the other. It occurs in tropical areas and its onset can bring several months of heavy rain, thunderstorms and hurricane-force winds. Bees acclimatized to such regions e.g. Apis scutella, cerana and florea, abscond to avoid the almost certain destruction that would ensue if they remained domiciled in one place.

***Monstera tuberculata***

A tropical plant producing fruit. An Australian variety produces edible fruit but you need to know just when to eat it. *M.deliciosa* (Swiss cheese plant).

**monstrosity**

A living organism that exhibits an abnormal form or structure, deviating greatly from the natural or normal form. A morphotype. In flowers the conversion of anthers into petals, thus forming what are known as 'double' flowers, although monstrous can be pretty, yet usually of little use to bees compared with single flowers.

**montan wax**

A hard, white bituminous wax obtained from lignite and peat and used for polishes, candles and in insulation.
See: waxes

**mood**

The behaviour and reaction of a colony at any given time. Most colonies do show variable responses to manipulation and governing factors apart from the basic temperament of the strain are: queenrightness, whether there is a flow on, how skilful the operator is and the weather conditions at the time.

**moon**

The earth's satellite, responsible for tides and our month. An ancient Germanic belief supposed that the moon was a huge cup filled with honey and that honey fell to the earth upon the oak and the ash. Artemis, the moon-goddess was often symbolized as a bee. Honey coloured moon – honey nice moon month. Madhukara (Indian mythology) meant honey giver.
See: honeymoon

**moor**

A tract of open ground with an acid, peaty soil, often well above sea-level and usually dominated by heath and heather with allied under-shrubs and coarse grasses. It is usually rough terrain unsuitable for most types of farming but the home of many game birds and deer. Beekeepers seeking an additional crop of honey from the heather in August and September migrate to the moors but heather is fickle and a boost to winter stores hardly offsets the cost of petrol for the double journey. Despite that those near enough tell of fantastic 'takes' when the condition of the heather and weather conditions allowed bees to revel in glorious, sweet smelling honey flows.
See: heath, heather nectar

**morale**

Used anthropomorphically in the sense that a colony may work with zest or seem to 'loaf about'. Zest is very noticeable when a swarm first occupies a new home or after a virgin has mated and started laying. We judge their morale by flying activity, increase of hive weight and general conformation to the pattern of behaviour expected of a healthy, queenright colony according to the season.

**Morat**

Mulberry melomel.
See: mulberry

**Morator aetatulae**

Pathogen causing sac brood.

**More, D**

Wrote '*The bee book history and natural history of the honeybee*'. Ideas for interesting gardens, including a chapter on bee-gardens.

**moribund**

In a state of destitution. Unfortunately colonies are sometimes found in this state, even on the verge of extinction. It can be due to unchecked disease or heavy robbing by human or other predators. The aim of a good beekeeper is to ensure that colonies are kept healthy and never have a level of stores below a sensible threshold (10kg in summer and 40kg in autumn) so that a moribund condition never occurs.

**Mormons**

Followers of a religion founded by Joseph Smith USA. The honeybee symbolizes the two tenets of the Mormon Church 'Industry and co-operation'.

**morphology**

The study of form and structure (as opposed to function) of animals or plants, language, or features of the earth.

**mosaic vision**

The visual image transmitted to the bee's brain by the compound eyes. Numerous images, each separately pictured in the lenses of the ommatidium, are transmitted via the optic nerve to the brain. The acceptance of this vast spread of information helps the bee to detect movement quickly but prevents the bee from concentrating or 'focussing' into one three-dimensional image as does the facility of 'accommodation' in our own eyes.

**moth**

An insect that flies outside daylight hours and so offers little competition to bees though it is unusually fond of honey. The 'Death's head' in particular is mentioned along with the larger and lesser wax moths who feed on wax and damage honey combs.

**moth eaten**

Having been devoured by the larvae of moths. In most cases strong colonies can control wax moth pests. However when combs are removed from the hives, except in very cold conditions they will be prey to the ravages of wax worms. The webs will interconnect adjacent frames and render the whole mass useless and fit only for burning. The use of fumigants (acetic acid or formaldehyde) and careful storage of comb when it is away from bees pays dividends.

**mother**

The queen honeybee. Frequently in fact usually, all the bees in a hive are the progeny of the one queen. Capable of laying a million eggs and living for several years. Mother is a term favoured by Russian writers in preference to queen.

**mother of vinegar**

An agglomeration of stringy, gelatinous or mucilaginous material that forms in vinegar and is largely a concentration of the Acetobacter aceti which can be used to start a new batch of vinegar.

**motherwort *Leonurus cardiaca* Lamiaceae**

A common weed with clusters of pink or purple flowers. It blooms from July on and can grow to 1m, yields nectar and white pollen.

**motile**

Capable of spontaneous motion ~ used for microorganisms etc.

**motor nerve**

Peripheral nerves consisting of nerve fibres of motor neurone which connect with the effector organ and conduct impulses from the central nervous system to the body muscles and glands. Vibration of the thoracic wing muscles occurs as a result of such motor nerves.
See: association fibres, central nervous system, sensory nerves, nerve trunk etc.

**motor neurons**

Nerve cells concerned with the regulation of movement. The extremely rapid vibrations of the thorax to produce 'piping' notes, fanning or flight, are a result of the resonance of the drum-like thorax (fused segments) powered by the thoracic muscles.

**motor ways**

As these have formed part of the 'concrete jungle' (over three million acres of farmland are paved over annually in America - 1·4 ha [1985]) much valuable bee-forage has been lost. Planting on the roadside verges can be helpful though bees that cross motorways are very vulnerable and many finish up in radiator matrices or flattened on windscreens or imploded by the sudden pressure drop as they are sucked over streamlined cars. When bees are to be transported on motorways the police should be informed and one's route carefully planned and chosen.

**mould**

1. A downy, furry covering associated with decay as well as the name given to saprophytic fungi. Any superficial growth of fungal mycelium.
2. A hollow shape or matrix used to give shape in the making of candles and other wax products, the making of wax foundation and in the moulding of plastic beehives etc. Releasing agents such as silicone, glycerine, grease or soap are used so that the moulded wax does not refuse to leave its mould.

See: candle, chalk brood, mildew, pollen mould

**mouldy comb**

Pathogenic mould can exist in the form of chalk brood and the very rare 'stone brood'. Other moulds, usually associated with neglected brood or pollen and dampness, can in the ordinary course of events be cleaned up by a strong colony without harm to themselves. Mould is less likely to be found if only strong colonies are kept in sound, dry hives and on sheltered sites.

**moult**

The shedding of an outer layer of skin that has outlived its usefulness. Most larvae develop in stages throwing aside (or moulting) old skins at intervals, six moults in all. Chitin is unstretchable so when growth expands to the point of splitting the existing coat a new larger one is ready underneath. Honeybees not only produce a loose, larger skin under their existing one but by enzymatic action are able to absorb the previous skin. This also applies to the chorion of the egg. Greenfly are often first detected by the appearance of white moults.

**moulting hormone**

Ecdysone secreted by the endocrine glands. JHI (juvenile hormone I) JHII and JHIII have been isolated and are known to be terpenoids (as are certain scent gland secretions). Moulting hormone has been isolated and identified as crystalline, silky, fibrous needles that melt 235 – 237C absorb u/v and infra-red light and are optically active – molecular weight 300.

See: prothoracic glands

**mountain**

Elevated ground larger than a hill and usually rising to a peak. The mountain foothills and slopes can offer both shelter and a water supply for bee plants. Such are the areas best suited for 'bush' honey on the north facing NZ slopes. However the lower temperatures render the higher regions poor to useless for bees and also act as a barrier. Use has been made of this for isolated queen rearing, as for example in some Austrian valleys. Heather can be useful on lower mountain slopes and moors. Montana is a specific name. Monticola is used to describe plants or creatures of the mountains. Mountain slopes that face the sun often give rise to flora that is highly beneficial to the bees (see NZ beech honeydew).

**mountain ash *Sorbus aucuparia***

A small tree having pinnate leaves, small white flowers and bright red or orange berries. Several Australian eucalypts use this name.
See: rowan

**mounting**

Drone copulation. Height of hives – storifying and supering, Temporary and permanent slides.
See: mounting slides

**mounting slides**

There are many techniques for the preparation of material for viewing by increased illumination and magnification. When the material can be made translucent, a sub-stage mirror will reflect light up through it. Stages include: fixing, dissecting, macerating, bleaching, staining, dehydrating, clearing and mounting according to requirements. Short cuts can be taken but the best results come from the careful pursuance of each of the relevant stages. Temporary mounts can often be made using a 30% aqueous solution of glycerine. Permanent mounts in a medium such as Canada balsam take longer as successively stronger strengths of alcohol, will be followed by absolute alcohol, so that there is no water present to cloud the balsam. Modern proprietary substances can make things easier. Acetone is good but anhydrous types for final stages are expensive. Cellusolve is not suitable for animal tissue. Make friends with a laboratory assistant.

**mouse Mus spp. pl. mice**

A field mouse *M. sylvaticus* is a small rodent that will feed on honey, pollen and bees. While the bees are active they can easily scare a mouse away or sting it to death. However after they have formed a winter cluster, a mouse can move in and establish a nest. Surrounding itself with dry grass, torn newspaper and other rubbish the mouse and its family are almost invulnerable and can survive the winter without difficulty. Depredations include the offensive smell created by urine and droppings and bees never like re-using combs damaged by mice. The use of full entrance 'mouse guards' is strongly recommended from November to March inclusive. If a mouse is stung to death within the hive, the bees being unable to move it, proceed to strip it of hairs and flesh and may get the skeleton out, if not then they embalm it with propolis. Skeletons have been found, brown with propolis and looking like a piece of cord, on the ground in front of hive. Ashes and pine needles scattered around the hive helps discourage them. Another interesting animal known as the 'honey mouse' *Tarsipes spenserae*, is well endowed for climbing shrubs to feed on nectar. It has a prehensile tail, long snout and tongue and is a marsupial.
See: mouse guard, shrew

**moused**

The description of a colony that has been invaded by a mouse (or mice) during the winter. Unfortunately this often means the colony's demise because the mouse has eaten the food that would have enabled the bees to survive. The smell and comb damage may well dissuade new bees from accepting such a hive and at best such a colony is likely to be poorly – don't let it happen.
See: mouse guard

**mouse guard**

Once nights become frosty it is wise to fit a device across the entrance which while allowing free passage to the bees makes entry by a mouse impossible. A mesh of ungnawable material, with slots no wider than 9mm. Slightly larger holes are still effective because the mouse's skull is wider than its height. Confronted with such an obstacle a mouse will not go in to feed, sensing that it would encounter difficulties on the way out.
See: queen excluder

**mouth buccal cavity**

In the bee it is at the distal end of the

cibarium through which food passes to reach the alimentary canal.
See: mouthparts

**mouthparts**

The bee's mouthparts are highly specialised and include the mouth and surrounding organs. These are the proboscis which lies below and the labrum and mandibles above. A ventral plate at the mouth opening is called the hypopharynx and leads to the cibarium or food chamber which is musculated and acts like a pump passing fluids back into the pharynx which tapers into the oesophagus. Beneath the ventral plate lies the salivarium at the proximal end of the tongue. Under the labrum, which is hinged to the clypeus, is a soft pad – the epipharynx – which closes over the proboscis to make an air-tight joint.

**movable comb**

Natural comb or comb built on foundation that can be moved without having to cut it free. As bees always attached their combs firmly to the roof and adjacent walls of the hive, it was not until Langstroth invented movable frames that they, and the combs built inside them, could be moved about at will. Earlier Greek hives used tapered baskets with top bars from which combs were suspended. The fragile nature of the comb to top bar attachment called for great care in handling.
See: anastoma, Langstroth, movable frame

**movable frame**

Modern hives are designed to take advantage of the fact that where frames are surrounded by exact 'bee-space' they can be lifted out or removed as required. Only when this distance is varied will the bees either build brace comb if it is too large or propolise the gap if it is too small. Long ago Greek hives were using this principal but it was not until Langstroth in America brought modern beekeeping into line by mass producing hives with correct bee-space that its commercial significance was appreciated.

**movement**

All movements near the hive should be smooth, steady and on the slow side. Rapid, jerky movements invite attention and coupled with the exhalation of human breath give the game away – you're a predator.
See: swipe, transporting bees

**movement of entrance**

Even a small alteration of the position of the entrance will set returning bees fanning to warn each other that something is amiss. Should a hive be rotated through ninety degrees the bees will get used to the new entrance but will continue to land in the old place and then walk round to the new position. Moves during the active season should not exceed a few feet unless the hive is taken right away to a new site at least three miles distant. Shorter distances could mean that bees who have learnt their way around the old area will go back to the old location. If a re-arrangement within an apiary is vital, then move the hive in short day-to-day stages. In winter when flying has virtually ceased a hive can be moved if extreme care is taken not to cause any disturbance to the cluster. Best during a really cold snap.
See: moving bees

**moving bees**

Disease can easily be spread in this way so consider both the health of the bees being moved and that of the area they are going to. Prior inspection of the new site is advisable and it should be at least three miles away. Transportation involves vibration which in itself keeps the bees under control (journeys with open entrances are not unusual in Australia) but the excitement raises their temperature dangerously so steps must be taken to give them more than adequate ventilation en route. This calls for a travelling screen instead of the crown board and the facility to spray them with clean water if necessary. If entrances are to be made bee-proof, use foam rubber because gauze or perforated zinc will cause then to crowd around the entrance trying to escape. It is always helpful to advise the police of

your intended route and to label the vehicle 'LIVE BEES'. Secure strapping should take account of the stresses likely to be applied (what would happen if one rolled over?). On arrival don't put the roofs back on until the last moment and count the number of entrance blocks removed to ensure no hive is left blocked up.
See: hive fastening, over-heating, tie

**mow**
To use a device to cut the grass short. When this has to be done near the hives it can be very irritating to the bees. Not only does the noise upset them, and some machines have an aggravating effect but the smell of cut grass has an adverse effect on their temper. The fact that the operator may be perspiring and moving jerkily close to the entrances is fraught with the possibility of stinging. Solutions include:

1. Not to cut the grass in front of the hives in flying weather.
2. Move the hives back a metre or two until the job is done.
3. Keep the area in front of the hives free of grass by use of gravel, bark, tarmac or such like.
4. Use weed-killer instead of grass cutting.

**mucilage**
A sticky secretion produced by certain plants. It sometimes has the property of absorbing water and swelling. Some forms may be collected and used as propolis by the bees or it may cause pollen grains to adhere to the hairs of a bee. Sometimes the pollen forms a conglomeration that sticks to and trails behind the bees.

**mucus glands (accessory glands)**
A gland that secretes mucus. Part of the drone's genitalia. Dade describes them as large, club-shaped sacs filled with mucus and joined at their bases to form a large -U- where the seminal vesicles are connected by short narrow tubes. The mucus glands increase in size as the drone matures. Mucus is a viscous suspension in a fluid acting as a lubricant. Irritation by pollen grains can cause a morbid secretion in some humans.

**mud**
Earth that has been made so wet that it becomes soft, slippery and somewhat sticky. When parking cars at Agricultural Shows or Field Days or when pulling trailers of hives into orchards or other apiaries or when engaged in migratory beekeeping, attention should always be given to the state of the ground. Conditions can worsen and suitable precautions should be taken as hives of bees are no ordinary load! Sacks can be laid under wheels, gravel, ashes or logs may be used to help one out of slippery situations. Where possible one should anticipate trouble early in the proceedings because constant revving and skidding may embed the wheels deep into the mud and render extrication more difficult. Special tyres, four-wheel-drive or a tow from a tractor (be sure you have proper attachments) are other means of overcoming problems of this sort.

**Mulberry *Morus alba, nigra, rubra***
**Moraceae**
A sturdy tree that bears greenish white flowers from May – July. Its fruits, much desired in days past, are red, white or black according to the species. Useful for both nectar and pollen. Wine from its fruit when made with honey has the distinct name 'morat' – mulberry melomel. The leaves of the white mulberry which does not grow well in the UK, are used to feed silk worms.

**Mullein *Verbascum thapsus***
**Scrophularaceae**
Aaron's rod, Hag's taper – Great mullein, a late flowering biennial, 2m tall, a chalk lover found in dry sandy soil, pinewoods and banks, that secretes nectar sparingly and provides brick red to orange pollen. The fluffy, white coating on the leaves used to be scraped off to make candle wicks. Hundreds of species can be found world-wide.
See: Aaron's rod

**mulsum**

Mull or mules, a drink the Romans made by blending mead and wine. It was mentioned by Varro and Pliny XI referred to mulsum as 'honeyed wine'.
See: Oenomel

**multi**

Word prefix meaning many, e.g. multicellular – having many cells'

**multiple matings**

The Volcano experiments by the Rutner Bros. in Italy showed for the first time that a nubile queen may mate with from 3 – 11 drones. This had been suspected for some time as queens had been seen to leave the hive repeatedly for mating and the contents of a newly mated queen's spermatheca indicated that several drones must have been involved. Our present knowledge of drone assembly areas has quite demolished earlier notions and the number of drones varies but on occasions it can reach quite unexpectedly high numbers
See: drone assembly, congregation areas

**multiple race crosses**

Races A and B are crossed and F1 crosses bred from them in order to reduce the decrease in productivity and vitality. F3 and F4 generations may be crossed with race C and at another point with race D and so on. Thus inbreeding is prevented from causing degeneration.

**multi queen casts**

While most casts have but a single, virgin queen, it is quite common for casts to emerge within minutes of one another and for two or more to settle with mutual contact. Casts with more than one queen are usually larger than ordinary casts (thus appearing similar to a large swarm) and often have bulges, each with a queen. It is not uncommon to find queens that are being 'balled' in such composite casts. If hived into an empty box, over an excluder over a box that contains a frame of unsealed brood and honey, it should be possible to select one good queen to run in below and for any others to be disposed of as thought fit.

**multi queen colonies**

Apis mellifera colonies are expected to have only one queen, however there are supersedure types where two queens – mother and daughter live in harmony for a considerable period of time, each with their own brood nest (albeit a smaller one for the old queen). By intervention a beekeeper can arrange to have more than one queen by the imposition of queen excluders or screens but such larger colonies do not do justice by getting more than double the ordinary crop and it is not easy to winter colonies in this formation.
See: multi queen hive, Snelgrove

**multi-queen colonies**

It is possible during the summer to have two or more queens, separated by excluders, yet sharing the same colony odour, the workers having access to all parts of the hive including the supers. Where enormous colonies are required, all under one roof, this system makes it possible though there are manipulative problems. Once accustomed to doing so, the bees work in harmony sharing supers and in the right flow conditions can amass a large surplus. However in many cases large numbers of workers remain faithful to their own queen and precautions should always be taken when the queen lay-out is changed. This might entail caging a queen and the use of newspaper as in uniting. To turn a hive into a multi-queen hive, an extra queen excluder is needed and provision for bees to have their own direct entrance to the part if the hive where their queen is.

**mummy**

Mummies, the remains of dead larvae, are cleaned out by the bees but may be found in the 'dross' when floor boards are given a clean, especially after the winter. The difference between dry mouldy pollen and the dead remains of a chalk brood victim

lies in the structure of the pellet, the ones that were once larvae show signs of segmentation.
See: Chinese slipper

**muralis**
A specific name added to a genus to indicate its existence in walls,

**murrain**
Plague or pestilence, usually highly infectious but more likely to be used to describe diseases of cattle but used by some to cover bee-diseases such as foul brood.

**muscle**
A discreet bundle of contractile fibres in an animal's body. Tissue composed of highly contractile cells, sometimes called muscle fibres. Cardiac muscle being different in characteristics to the thoracic indirect flight muscles. The function of muscle is to produce pressure or movement. Honeybee muscles are contained within the exoskeleton anchored by apodema and phragma. Only the thoracic muscles show a pink colouration due to the presence of 'cytochrome'. The bees' muscles include: compressor, dilator, extensor, flexor, indirect flight, of the leg and thorax, retractor, heart, diaphragmic and sphincter
.

**museum**
An educational establishment designed to store, care for and exhibit items of artistic, historical and scientific interest. Displays of bee and allied insect material are often augmented by the installation of an observation hive . Care should be taken when this is done to ensure that it is in a place where the direct rays of the sun do not fall on the glass, yet where it is well-lit, at a suitable height and regularly attended to by a knowledgeable beekeeper.

**mushroom bodies**
The paired Corpora pedunculata named Pilzhutformiger Korper by Dietl in 1876. Stalked bodies, largest in the worker bee and smallest in the drone, found in the bee's brain. They are neurosecretory cells and have alpha and beta lobes that have discriminatory capability with regard to sight and smell and probably temperature, humidity and touch. The alpha lobes being implicated in the learning process.
See: calyx, Kenyon cell, neuropile, neurosecretory cells

**music inspired by bees**
Flight of the Bumble bee, Rock song 'Birds and the bees' 1965, blues 'King bee' 1964, Apple trees and honeybees and snow-white turtle doves. Bee madrigal. Although bees seem to have no objection to most forms of music they do not seem to pay any attention to it unless it causes their nest to vibrate.

**must**
The prepared but unfermented liquid to be used for the making of new wine. The sweet liquid obtained by washing the cappings after all the honey that can be drained of by gravity straining has gone. Its sugar content can be tested for gravity with a hydrometer (or an unbroken fresh hen's egg which should float and show above the surface an area around the size of a 50p coin. If less add a little more honey.

**mustard Brassica spp.**
An annual that flowers within ten weeks of sowing and is a summer plant. There is wild mustard and charlock, white mustard Sinapis alba and black mustard S. nigra. The honey has the faint hotness of mustard when fresh and a strong aroma, it granulates rapidly but it blends well with other honeys. Supers can fill fast when it is grown for seed. It has a coarse granulation and pollen is light yellow. The plant likes a deep, moist, fertile soil.
*Like oilseed rape – '*
*Its blazing yellow calls*
*The bees from nearby fields,*
*E'en apple trees*
*They plunder its never ending sweets*
*For lightness the honey clover beats*

**mutagen**

Any reagent or force (radiation) that can change the genetic material of a germ cell or somatic cell. This may result in daughter cells with changed characteristics or it may cause carcinogenic or teratogenetic (birth) defects. Russians under G.A. Avetisyan have used chemical agents and gamma rays to improve their selection methodology with Carpathian bees.

**mutant**

An organism that displays a transmittable characteristic which has not previously occurred amongst its ancestors. Some apparent mutants are probably due to recombinations of genes (chiasmic) rather than an actual mutation or change in the genes ability to govern cell identities.

**mutation**

The random alteration of genes in the reproductive cells which may occur suddenly or gradually. Genes are normally 'copied' but when modified by a mutation the changed structure will be repeated. If a particular mutation protects the organism in a hostile environment then the mutant has a good chance if survival. In breeding mutations can be propagated selectively resulting in 'new' types.
See: chromosomal mutation

**mutilation**

To make a thing imperfect by damage or the removal of a part. Bees have no means of proliferating tissue other than in the normal cycles of glandular activity and digestion. Consequently when any damage occurs to a bee such as a split or tattered wing, a missing part (appendage such as segment of a leg or antenna) or fracture of its chitinous exoskeleton, it has no way of regaining its original condition. If this is sensed by its nest mates, such a bee is likely to be ejected.

**Muttoo Rajendranath**

He started publishing the Indian Bee Journal in 1939 as the official organ of the All India Beekeeping Association and he became known as the 'Father of Indian Beekeeping'.

**mutual recognition**

Pheromones account for many behaviourisms within the honeybee colony and mutual recognition pheromone is the name given to one such. Queen substance as well as food sharing contribute to the colony 'odour' thus ensuring that members of the colony are recognized and not treated as intruders when they return to the nest. The smell of otherwise innocuous substances (perhaps a change in their static electricity charge?) could lead to the destruction of this identifying smell with the result that their nest mates might not accept them.

**mycelium**

The vegetative (thallus) part of a fungi or the mass of 'hyphae' as seen when a fungus is vegetative as in the white, almost fluffy appearance in the early stages of chalk brood.

**Mycocidin**

A mycostatic substance developed in the US for the control of *Ascosphaera apis* (chalk brood).

**mycology**

The fungi found in any particular area and the name of the branch of Botany that deals with fungi. H. A. Dade was a mycologist at Kew Gardens, London.

**mycosis**

Parasitic animal diseases caused by fungus. Stonc brood Aspergillus flavus, a very rare disease, would come under this heading as would Chalk brood which is also caused by the fungal parasite *Ascosphaera apis.*
See: chalk and stone brood

**myogenic rhythm**

The muscular auto-contraction mechanism that vibrates the thorax and allows the neurogenic rhythm of the flight muscles to be translated into regulated wing movements.

**mythology**

The science of myths (traditional stories from preliterate societies). Beekeeping is so ancient that it is interwoven with early mythology, as for example the wax wings of Daedalus and his son Icarus or the burying of an ox to produce a swarm of bees. It involves many mistaken beliefs but indicates early man's interpretation of his environment.See: moon and stars

# N

**Nadir (opp. zenith)**
Lying beneath. Used to describe an extension given to a beehive underneath the existing parts. The word 'eke' is used for the same purpose, though it is more often used to mean on top of the hive.
See: eke

**nadiring**
Adding another box under brood combs –More, Cotton, Keys, Bevan, Bagster. Nadir – the point in the heavens that is diametrically opposite the zenith, the lowest point of anything. Bevan suggested that old stocks should be supered (storified) and that swarms should be nadired. It is likely that bees use the nadir as opposed to the zenith when performing the Von Frisch dance.

**nails**
A relatively slender piece of metal with one end pointed and the other widened to form a head which can be driven (by hammer, rampin etc.) firmly into wood thus making a permanent fix by joining two pieces together or allowing glue to set while the two pieces are held firmly together. Great specialization resulting in steel, galvanized, brass and other types of nail, fits nails for specific purposes as does their shape and size. Equipment manufacturers sell special black, lacquered frame nails and packs of roughened galvanized nails, for hive assembly.
See: rampin, gimp pin, screw etc.

**nailing**
The fastening of two or more pieces of material, often wood, together. As opposed to screwing, the pieces can be prised apart again. The surface on which the nailing takes place should be firm and the material to be nailed set firmly upon it. In the case of frame assembly noise can be reduced by the use if a rampin. Nails appropriate to the task should be used and every nail carefully and strategically placed. Blunt nails are less inclined to split wood and compressing a small circle by hammering the head onto the wood facilitates the subsequent entry of the point without splitting.

**nailing tool**
A hammer of appropriate size is normally used to drive a nail home. A blunt nail is less likely to split the wood than a sharp one. Furthermore the surface into which the nail is to be driven should either be compressed slightly by striking the nail upside down onto the wood so that the head makes a depression in the right place or a small receiving hole may be drilled. A nail can be steadied by gripping the shank with long-nosed pliers. The rampin is a useful device for inserting small nails. With the advent of special glues, and stapling machines it is possible to avoid nailing in certain cases.
See: nail, rampin, screw, staple and glue.

**nano**
A prefix for one thousand millionth. Micro is one thousand times larger. A nanometer is a convenient size for measuring virus particles. See: iridescent virus.

***Nannotrigona (Scaptotrigona) postica***
A Brazilian stingless bee.

**nanometer**

One thousandth of a micron (0·001 mu). A prefix implying 10 to the minus 9 or decimal eight noughts then one. One thousand millionths of a metre. A very small measure of length which should not be confused with a nautical mile.

**Narbonne honey**

Famous honey sold in big London stores. Narbonne was a port in the south of France well-known in Roman times as doubtless was the rosemary flavoured honey gathered there.
See: rosemary.

**narrow spacing**

The centre to centre spacing of brood combs is 1" and this is known as narrow spacing and metal or plastic clips or Hoffman side bars, are used to give this spacing, sometimes 1½" is used. In the supers however it is not unusual to find wider sizes according to whether Manley, Hoffman or wide British are used. Honey combs can reach a considerable width, even in extreme cases almost 3" but the main object there is to achieve flat sealed comb faces that can be uncapped easily. The large Rock honeybee makes its single comb extra wide at the top for the storage of honey then tapers to lower brood cells below.

**Nasonov gland** ***

The transliteration of the Russian name has resulted in other spellings. It is a crescent-shaped gland that lies under the anterior part of tergite seven and passes its secretions (pheromones) into the air via the transverse scent canal and the application of fanning. It comprises seven terpenoids: (z)-citral, (E)-citral, geraniol, geranic acid, nerol, nerolic acid and (E,E)-farnesol. Workers may release this secretion to 'mark' and render sources that are useful but have little smell of their own.
See: essential oil, scent gland, scent gland secretions and terpenoid.

**National Beekeeping Associations**

B.B.K.A., Scottish B.K.A., Welsh B.K.A., Irish B.KA., IBRA, CBA, BIBBA, AEA, BFA, CONBA et al.

**The National Bee Unit**

The Food Environment and Research Agency, Sand Hutton, York YO41 1LZ
Tel. 01904 462510

**National Beekeeping Centre, Stoneleigh, Kenilworth, Warwickshire, CV8 2LG**

**National Diploma in Beekeeping**

Candidates will be required to show that they have passed the BBKA Senior exam or its equivalent. It is a two day examination with a viva voce part complimentary to two, three hour written papers.
All scientific aspects of beekeeping and allied subjects are covered including work in the laboratory and the use of microscopes. Lists of recommended books are available. A successful candidate may use NDB after their name.
See: NDB Hons.

**National Honey Show & BBKA Convention**

It originated in 1921 when Kent and Surrey BKA's had the first show at Crystal Palace in 1923. The National Honey Show was constituted and continued until fire destroyed Crystal Palace in 1936. On The third week of October each year until 1984 it was held at Caxton Hall, London and became a Mecca for beekeepers from all over the country. It was the centre for the BBKA's Judge's examination and included a honey show, trade centre, lectures, meetings (such as BFA and AEA) and there is a rest room and refreshments. It created a truly national occasion for the enjoyment of beekeepers from all parts of the country and other parts of the world.
The National Centre has now been well established at Stoneleigh.
See: NHS publications and points

**National major hive**
This hive uses the same sized floorboard as the national hive but the area of comb has been increased to approximately Langstroth size by the use of the shorter Langstroth length lugs and deeper frames 16 x 10 and 16 x 6".
See: modified commercial.

**National minor hive (Smith hive)**
This hive has the advantages of the Langstroth in that it only uses four walls per chamber, top spacing and short lugs, yet it takes British Standard frames with shortened lugs.
See: Smith hive.

**native**
Belonging to by birth and by birth of parents, in the same place. Indigenous origin – not foreign. An animal or plant indigenous to a particular region. Native bees are those which have lived and thrived in one area for many generations.

**natural selection**
The survival of such animals and plants that are best able to adapt themselves to the environment and the elimination of those which are not. Charles Darwin 1809-82 established the principal that the normal processes of evolutionary change eliminated species that could not satisfactorily adapt to their changing environment while favouring the survival if those fit enough to do so.

**natural swarm queen cell**
A queen cell that has been produced by the bees under normal swarming conditions, as opposed to being raised under the supersedure impulse or due to the interference by the beekeeper or by the colony's suddenly trying to rectify a queenless situation.
naturally crystallised
Granulated honey entered in honey shows is specified in this way with a rider 'or soft set' sometimes added. This differentiates it from creamed or clear honey.

**navigation**
The faculty of keeping one's bearings so that a journey may be undertaken and result in successful homing. Von Frisch's investigations and discoveries awakened many readers who at first expressed disbelief that bees were able not only to navigate but to communicate the ability to recruits in the hive. Their ability to see the sun whether hidden by cloud or not, as a compass, restricts them to day-time flight. Dr.Melvin Kreithen tells us that bees are sensitive to small changes in the earth's magnetic field as for example as a result of sun spot activity. Allowance for the sun's movement (though it moves right to left in the southern hemisphere) and the deflection by the wind, are seemingly compensated for, though bees do drift and presumably some get lost. Distance may well be measured in terms of energy expended and scent trails may also help foragers. The map-like appearance of colours and shapes probably plays a part too.
See: Von Frisch.

**Northern Bee Books**
Mytholmroyd, Hebden Bridge, HX75JS
Have a wonderful variety of bee books new and old.

**NDB Hons.**
The honours for the National Diploma in Beekeeping is akin to Fellowship of the BBKA in that subjects have to be approved prior to the compilation and acceptance of a thesis which must be original unpublished work. The successful candidate will be entitled to use NDB Hons after his name.
Details available on website:
www.national-diploma-bees.org.uk

**neat's foot oil**
A pale yellow oil derived from feet, skin and bones of cattle. It blends well with beeswax to make leather soft and water-proof.
See: dubbin.

**neck**
The head of a bee is rounded sufficiently

to overlap the short neck and the thorax to which it is joined tends to almost conceal it. The neck permits full rotation of the head giving the bee an extremely wide range of vision with only a relatively blind area to the rear. Although thin it conducts the blood vessel leading to the brain as well as the nerve cord, alimentary canal and oesophagus etc.

**nectar**

An aqueous secretion from nectaries containing sugars, small amounts of salts, proteins, acids, enzymes and aromatic substances originated from the phloem sap and collected by foragers as food. The sugars are mainly sucrose though fructose, glucose, oligosaccharides and sugar alcohols are found in some groups as the nectars of various plants can differ considerably. The sugar and water contents can be very variable depending on the plant, temperature, soil type and condition.

**nectar concentration**

The percentage of sugar in nectar tends to vary diurnally and from plant to plant and is a balance between secretion and evaporation. Longer corollas offer more protection than open flowers. Here are some concentrations: apple 40%, plum 15-20%, bean 22%, cherry 20-25%, lime 32-35%, sunflower 37%, Lucerne/clover 40%, raspberry 46%, dandelion 50%, borage 53%, Robinia 55%, willow 60% marjoram and horse chestnut 76%. At high concentrations >50% honeybees sucking speed declines and they will gather more from a nectar with 45% than one with 55%. This also applies to the rate at which a syrup feeder can be emptied.
See: sugar content and energetic reward.

**nectar flow**

Flowers arrive in their due season yet inclement or unsuitable weather at the time of blooming can stultify nectar secretion. Plants such as buttercups, clover, hawthorn and rosebay may gladden the eye yet offer nothing to the bee if the wind is cold. Supers have been filled by blue bell honey in the spring and by ivy in the autumn but low temperatures minimise such possibilities. When nectar is abundant this is a 'nectar flow'; when bees work it then we get a 'honey flow'.

**nectar glands**

Plant organs that produce nectar, more commonly referred to as nectaries. They may be floral or extra-floral. Quite often they are to be found at the base of the flower and positioned so that the pollinators have to brush past pollen covered stamens to get to them. Some plants may have nectar secreting glands that produce nectar on other parts than the flowers.
See: nectary.

**nectar guide**

Lines, spots and other devices that contrast with petal colour and sometimes reflect ultra-violet light. They guide insects instinctively to the nectary. They are also known as 'sap-sots'. The flower shape itself is often a nectar guide, shape, colour and geometrical arrangement helping a trained insect to locate the nectary. Bees are particularly sensitive to these patterns while beetles for example may stumble somewhat haphazardly around a flower centre.
See: soya bean.

**nectariferous plants**

Plants capable of yielding nectar are often marked 'N' in bee-plant lists. This may come from floral or extra-floral nectaries or both.

**nectarivorous –**

nectar sipping as applied to certain insects, birds and bats.
See: bee plant, melliferous and nectarivorous

**nectar load**

The amount of nectar carried back to the hive in the bee's crop. While the sugar content may vary from 20-60% the maximum load in unlikely to exceed 70mg

(that is 85% of the bee's weight) and 20-40mg is the average (Dade). Bees make 7-13 trips per day each lasting around 30 mins with 4 min intervals. Up to 24 trips a day have been recorded. Foraging shortens a bee's life considerably though not as much as brood rearing. About 10mg is retained for fuel carrying the normal payload.
See: deposition of nectar, honey stomach and payload.

**nectar quality**
It has not been shown that any particular combinations of sugar are better for the bee, though nectars vary in this respect. Nectars with more sucrose are usually more attractive and attractants and repellents can be present though not in themselves important foodwise. The nutritional value of pollens has been studied and comparative lists made. Honeydews have been generally classed as bad for wintering and heather has occasionally been viewed in the same light.
See: nectar concentration and toxic nectar.

**nectar secretion**
Factors which affect the quantity and quality of nectar include: soil and atmospheric humidity, the type of soil, use made of fertilizers, vigour of the plant, temperature, vapour pressure, wind, time of day and year, length of day, sunshine, climate and microclimate. On light soils best results may come after a cool night and rainy spell with a day temperature of 80F while on a clay, high night temperatures are helpful. Sainfoin yields better on boulder clay than chalk, limes on Keuper sandstone. Many plants have threshold temperatures for secretion: clover 73 sycamore 85 hawthorn 70 while micro elements Mg Fe Mn Cu Zn etc. in base solutions increase nectar secretion.
See: yield of honey per acre.

**nectary (sugar valve)**
A secretary gland composed of a group of modified plant cells responsible for the production and display of nectar, frequently within the flower though the number and position vary.
See: nectar guide/secretion and hydathode

**nectary deep**
Many flowers have nectaries that are too deep for the honeybee or are only accessible when they are well-filled with nectar. Sometimes a bee will nibble through the base of the corolla although this in no way precludes visits which effectively pollinate by pollen gatherers. Blue bell and field beans are examples. In others, once the nectar reaches a level the bee can touch it is able to draw out most of the nectar (due to molecular attraction). In the case of red clover the flowers after the first cut have corollae of a more suitable length for the honeybee.

**nectary extra-floral**
Because flowers are scented and conspicuous we tend to think that it is only here that plants offer nectar to the bees. In fact much nectar is made available by plants via nectaries outside the flowers and these may give a flow long after the flowers have finished.
See: extra-floral nectary.

**nectar yield**
The habits of different flowers are variable with regard to the time of day, quantity and quality of nectar that they secrete. Dr Ingrid Williams reported that with Swede rape flowers yield more nectar if it is collected more often (this also applies to honeydew yield in NZ beech). Relative humidity can affect concentration of the nectar but for any given plant it stays within a certain limit tending to become less concentrated as the day wears on.
See: honey potential.

**needle - (skep)**
A large needle about the size of a 4" nail, made of metal, bone or plastic is used for feeding the binding cane through the skep walls as the coil of straw rope

progresses, thus tying the consecutive rows of rope together. The binding cane was traditionally split bramble cane and other suitable alternatives are now available.

**neglected brood**
Brood that has died through receiving insufficient attention from the nurse bees. There are many things that could cause this condition which rarely if ever exists in strong colonies. Onset of cold, lack of intake, CDD.
See: addled brood and drone neglected brood.

**Neighbour George & Sons London 1862**
Father and son Alfred were equipment manufacturers at the time when box hives were beginning to take over from straw skeps, having an office in London and an apiary in Hyde Park. In 1844 George sent a Nutts hive with bees to NZ. They made Woodbury bar and frame hives with short lugged 13 x 7¼" frames. The improved Cottage and Single box hive. Links continued to sales at Gamages's London store and to the formation of Robert Lee & Sons. Alfred often accompanied Thomas Nutt on visits to patrons in the vicinity of London.

**Nematoda**
Unsegmented thread-like, thin, whitish round worms that usually dwell and mate in the soil. Some forms are parasitic and while rare in the honeybee, wasps, bumble and solitary bees are often their natural hosts. For example Neoplectana carpocapsae are known to infect wasps but it has been shown that it can kill honeybees.
See: parasites, pests and Sphaerularia.

***Neocypholaelaps indica***
Found in Sri Lanka, India & Nepal. Not thought to cause economic damage to honeybees though visibly dependent on adult honeybees which carry it to pollen stores on which it feeds.

**Neo-Epinine (Isoprenaline)**
Used in inhalers for hay-fever sufferers. An anti-histamine which in some cases acts as a treatment for bee stings.

**Nepeta spp. *** *Catmint, catnip.* *Nepeta mussinii gdns. Glechoma hederacea* Europe.**
See: catmint

**nerolic acid**
Terpenic acid cisform – one of the terpenoids which together with a number of other substances go to form the secretion of honeybee's scent glands.
See: Nasonov.

**nerve cells**
Bundle of fibres capable of transmitting sensory stimuli and motor impulses. See under these headings: CNS central nerve trunk or commissure, fibre, ganglion, motor, optic, peripheral, stomogastric, sympathetic, nervous.

**nerve cells**
These are bundles of fibres that are capable of transmitting sensory stimuli and motor impulses.They have a nucleus, axon, dendrite and neurofibrils. Motor nerves may terminate in synaptic knobs. Masses of these and their associated nerve fibres go to form the bee's brain, ganglia and connecting commissures.
See: CNS, central trunk, commissure, fibre, ganglion, motor, nervous system, optic, peripheral, stomogastric, sympathetic etc.

**nerve fibres**
Nerves are composed of thread-like fibres which together with nerve cells form the complete nervous system. A nerve fibre has an axis cylinder surrounded by a sheath and bundles of these fibres form nerves which may be bound up in a connective tissue sheath.
See: association fibres and nerve cells.

**nerve trunk**

The central nerve trunk or twin, longitudinal commissures of nerve fibres extending from the brain and via the eight ganglia to the sting See: central nerve trunk.

**nervure**

A rib-like structure supporting the membranous wings. Although circulation of blood through the wings is vestigial, the nervures of the wing contain valves and pulsating vessels so that blood follows a definite course through the hollow collateral channels, the hind wings proving more simplified than the forewings. Cutting of the wings does not lead to bleeding because the gap is immediately sealed by a cap of leucocytes.
See: venation.

**nest**

A place taken and furnished by an animal or colony for the initiation, care and safety of its young. In the case of bees – the brood nest and home for the entire insect colony.

**nest bees**

House bees. Those with special duties to perform before they advance to foraging . Nurses and other bees whose work tends to keep them in the brood nest area.
See: nest mates.

**nest-brood**

A honeybee colony tends to keep all its brood in a fairly compact ovoid which is flanked by pollen and surmounted by honey, nursing including feeding and incubation is thereby made most efficient.
See: brood nest.

**nest cavity**

A swarm prefers a cavity with a volume of about 40 litres. Bees usually choose an airy but draught-free cavity in a tree or similar hollow but will also use caves. They tend to look for nest sites well above ground level where they are safer from certain predators and floods.

**nest-mates**

All the bees of a colony that are accepted as members. sharing as they do colony 'odour'. Exceptions occur when a queen has been replaced or when 'drifting' occurs or of course when a beekeeper has intervened as for example by 'uniting'.
See: hive mates.

**nest (mouse)**

A mouse may attempt to make its nest inside a hive when, after a cold spell of weather. the bees have been forced into a tight cluster and it gets an opportunity to gain access.
See: mouse, moused and mouse damage.

**netting**

A uniform mesh of cotton, string, nylon, metal or plastic. Used in many ways for the benefit of apiarists. The veil utilizes black mesh, fine enough to exclude bees but not air (or gnats!). Ventilation and traveling screens are made with a suitable mesh but if plastic is used an odd bee might force its way through and what is more, be forever careful not to get the smoker or anything hot near it. A veiled bee-suit can be put on a cool wash inside a special bag inside the washing machine.
See: gauze and perforated zinc.

**neurohormones**

Neuro – pertaining to nerves. Certain neurosecretory cells in the region of a honeybee's brain are functional as endocrine glands in the production of hormones and neurohormones and can cause changes in the electric potential, affecting nerve activity and communication.
See: neurosecretory and endoplasm.

**neuromere**

A segment that corresponds in length to the extent influenced by a pair of spinal nerves. In the larvae there are twenty neuromeres grouped 3 - 3 - 14. (head, thorax abdomen) and the nerves originate from the cells of the epidermis and grow inwards to the ganglia.

**neuropile**

Those areas of the brain where neuronal filaments and dendritic fields are densely interlaced but neuronal cells are few or absent. Two cup-shaped regions are called the medial and lateral calyces.

**neurosecretory cells**

A gland-like nerve cell such as the endocrine cells located in the pars intercerebralis of the brain which secrete a hormone therefore neurohormone.

**new comb *****

Comb, freshly built by the honeybees, is light (almost pure white in colour) and as suitable for the processing of nectar, the storage of honey or the rearing of brood. When melted and filtered very fine samples of wax can be obtained. In certain circumstances such as when bees have built comb while on a diet of sainfoin honey, the comb will be saffron yellow and from heather a slight pink. New comb is usually free from pathogens.

**New Forest**

A famous wooded area interspersed with heathland (S. England). Bees have been taken there for the heather for centuries but in later times the Forestry commission make the same charge for a hive as for a caravan, driving away many who would like to get a late honey crop.

**newsletter**

A duplicated sheet or sheets sent to members of any group or organization such as a local or countrywide B.K.A. Its contents are arranged to keep members informed about the activities of their group and happenings of interest locally, at home and abroad. Monthly letters are most popular and are sometimes included with a nationwide magazine such as Bee Craft. Increasing postal costs were becoming a problem but these are being alleviated by the use of e-mail.

**newspaper**

1. In 'uniting' it is safer to dequeen the lower stock and set the queenright bees over them with a sheet of pricked newspaper in between. 2. A queen may be introduced in a hair-curler (perforated, plastic cylinder) with a pricked single layer of newspaper fastened over the 'open' end.
3. When more than one super has to go on each hive at an out-apiary an appropriate delay can be built in by leaving a sheet of newspaper between them.
4. Wrap sterile newspaper around supers to be stored for winter, include a small wad of cotton wool soaked with acetic acid and seal with sticky tape. Further, insert in black plastic refuse sack if desired.
5. Pin newspaper over the entrance for a temporary closure when a prophylactic vapour is to be used. The mandibles of bees enable them to chew newspaper into a state akin to that required for 'papier mache'.

See: article, papering, comb storage, cutting and uniting.

**newspaper 'uniting'**

One of the more reliable ways of causing two colonies to join forces without fighting is to dequeen the lower one and place the queenright lot over sheet of newspaper (with small slit cut in its middle) on top. The slit encourages a bee to cut it with its mandibles and mutual contact begins to occur slowly and by the following day a chewed away circle in the centre would be found while the returning bees do not immediately contact the queen who is surrounded by her original workers. Papier mache-like snippets of paper sometimes marked yellow on the edges, are ejected from the entrance. No need to disturb them for several days.

**Newton**

An SI unit of force required to give an acceleration of one metre per second to a mass of one kilogram. Equal to 100,000 dynes See: specific heat and joule.

**New Zealand N.Z.**
Lying on almost the other side of the world to Spain these islands cover a north-south stretch of over 500 miles. Population 4 million They suffer the effect of being on a 'fault line'. – See mention under several headings. A great country for honey and beekeepers.

**nexine**
The unstructured inner layer of exine (the outer coat of the pollen) it has an inner layer called endexine and a foot layer on which columellae (pillar-like structures) stand. See: exine, intine, and pollen grain.

**niacin**
See: nicotinic acid.

**Nicander**
Of Colophon about 150 B.C. wrote books which dealt with bees and his authority almost certainly influenced AElian, Columella, Pliny and possibly Virgil. His books were lost in the great fire at Alexandria though quotations via other authors exist.

**Nicotiana spp. *****
The tobacco family Solanaceae
Tender perennials requiring rich soil and full sun (which could be put to much better use!). Best treated as annuals. The long corollas of most are of little use to bees though there are many different varieties. Fragrant and in many bright colours. Howes says honey when obtained is strong flavoured and dark – akin to that of buckwheat.

**nicotinic acid**
Niacin. Derived from the oxidization of nicotine $(C_5H_4N)COOH$
A vitamin found in honey and the first vitamin to be isolated in the pure state. The component of vitamin B (anti-pellagra factor) essential for growth.

**nidificate**
To build a nest.

**night**
During the winter little variation takes place in a honeybee colony between night and day. In summer brood rearing, wax building and honey ripening gives the bees plenty of work to do by night. Activity within the nest seems to be unaffected by darkness. The shorter the night the longer time the potential foraging during the day. Bees can be manipulated at night using a red light. See: moving bees.

**night classes**
Evening classes – more popular in winter than in summer. Useful for beekeeping instruction though it is necessary to organize a minimum number of students, a suitable place and a qualified teacher. County Councils will do this if numbers warrant, though fees go up and up. BKA's can of course organize their own.

**nigrosin**
A chemical dye ranging from blue to black, that is made by heating nitrosophenol, aniline and aniline chloride. It is used for pigments and shoe polishes. Sometimes used to form a dark background against which Nosema spores can be seen and photographed.

**Nile (Ancient Egypt)**
See migratory beekeeping and the early use of the Nile for exploiting the seasonal flooding and movement of the river.

**nine**
Hormonic chemicals frequently begin with 9. For example a report indicated that 5000mu.g per day of 9 oxotrans
2 – decenoic acid were produced by a queen (from the mandibular glands.

**nip**
To pinch or give a small bite or attempt to do so. When a bee commits itself to an all-out attack, one needs protective clothing. Should it be able to plant its sting it will try to disengage by tearing out its sting and immediately re-attacking clutching the victim's flesh with its claws, pressing the tip of its now useless abdomen against the one attacked and using its mandibles to 'nip'. This can be felt if such a bee is held against the lips. See: biting.

**nitrate**
The radical $NO_3$, a salt used in the form of potassium and sodium nitrate for fertilizing the soil. Its application encourages leaf growth and is counter-productive when used on clover (from the beekeeper's point of view that is) as it reduces nectar secretion. Leaching into rivers has caused pollution problems and an excess of nitrates in food is dangerous. Whack it on if you want to win prizes in shows!

**nitric oxide (NO)**
A colourless, dangerous gas – under suspicion as affecting the earth's atmospheric radiation shield. One of the oxides of nitrogen that are produced as a gas when saltpetre is burned, though it rapidly oxidizes to $NO_2$ (the dioxide) in the air.

**nitrobacteria**
These are found in the soil and are associated with leguminous root nodules and are concerned with the fixation of nitrogen and therefore soil improvement. See: Azotobacter and nitrate.

**nitrobenzene (oil of Mirbane) $C_6H_5NO_2$**
Is a pale yellow, oily liquid that is strongly refracting. Melting point 6C boiling point 211C. It has an intense odour of bitter almonds and is derived from benzene. It is poisonous by inhalation and by skin absorption producing chronic intoxication; it also attacks haemoglobin and harms the lungs at 1 ppm and vaporizes readily. Used as one of the constituents of Frow mixture, an acaricide and well-tried remedy for Acarine disease. Also referred to as a bee repellent.

**nitrogen (N)**
A colourless, odourless gas that comprises four fifths of the volume of the earth's atmosphere. It is of great importance in the form of nitrates as a soil fertilizer. Nitrification, or the fixation of atmospheric nitrogen, is effected by certain bacteria 'nitrobacter' which are found in the nodules of many leguminous roots. See. 'N' fixing.

**nitrogen fertilizer**
Nitrogen is taken out of the atmosphere by plants which die and subsequent degeneration of the tissue puts nitrogen into the soil where it can be re-absorbed by living plants. Where there is a deficiency of nitrogen due to leaching, over-cropping or other reasons, artificial nitrogen fertilizers may be used. The application of these must be timely, or leaf growth may minimise flowers and so, nectar production. See: leguminosae and potash.

**nitrogen fixing bacteria**
Symbiotic bacteria of the nitrogen fixing kind inhabit nodules on the roots of some plants notably Leguminosae (Papilionaceae). They belong to the genus Azotobacter and Clostridium. Organic nitrogen becomes ammonia, then nitrite, then nitrobacter further oxidizes the nitrite to nitrate. These are examples of chemosynthetic organisms. See: botulism, clover and leguminosae.

**nitrous oxide $N_2O$**
Laughing gas. It is sometimes produced by burning 'saltpetre' in smoker fuel to produce anaesthesia in the bees. Other oxides can also be formed and one of these is rather harmful. Nitrous oxide itself is invisible and sweet tasting.

In so far as you can taste 'smell'.
See: potassium nitrate.

**Nitzsch**

The scientist whose name is attached to the mite Braula coeca Nitzsch which he discovered in 1818. It is more conspicuous than the Varroa mite and tends to gravitate to queen. It tickles bees causing them to regurgitate food but apparently does not harm to brood.
See: Braula coeca.

**noise**

Having kept bees on the threshold of Heathrow airport, exhaust deposits from aircraft were certainly more significant than the effect of noise. As their sense of hearing is more in the realm of appreciating vibrations, a colony near a railway track might build more brace comb. The vibration of a queen's 'piping' is transmitted via the comb and Von Frisch dances also take this kind of 'noise' as a means of communication.
See: cps, sounds, and vibration.

**nomenclature**

This includes giving names to taxa and the system of names produced. The scheme which is universally adopted is the 'binomial system' devised by Linnaeus. Each species has two names: the generic and the specific. The generic name which always uses a capital letter, tells us which genus it belongs to and hybrids will have 'x' inserted between this and the second name which is in lower case and tells us the species of the organism. This is followed by 'L' to indicate Linnaeus and the original author's name may follow. E.g. *Cucurbita pepo ovifera* the vegetable pumpkin. See: Linnaeus, systemics and taxonomy.

***Nomia melanderi***

A soil-nesting, alkali bee found In N.W. USA, reared for alfalfa pollination. It is a non-social, gregarious, ground nesting species which prefers soil that is sub-irrigated. Also called the APC bee.
See: alkali bee.

**Nomocharis spp. Liliaceae**

It is a Himalayan bulb which likes peaty soil in semi-shade, its nodding pink flowers have deep purple eyes. There are many varieties in beautiful colours.

**non-floral nectar**

See flower honey, extra floral nectaries, leaf honey, honey-dew and glands (plants).

**non-reducing sugars**

Sucrose and trehalose. All the monosaccharides are reducing sugars.

**North America**

European honeybees were shipped to North America from 1621 onwards – E. Crane. Weather, nesting sites and forage encouraged their spread. Later in the 19th century direct shipments were made to the west coast and in such regions as British Columbia the bees spread rapidly – from 'Archaeology of Beekeeping'.

**Northern Bee Books**

Ruth & Jeremy Burbidge

**Norway**

West side of the Scandinavian peninsular. In 1978 there were 4,500 beekeepers with about 60,000 colonies. A good deal of honey comes from wild raspberry and the extra-long days of higher latitudes help to produce an annual average of some 1,200 tons, 90% of this is handled by a Honey Co-operative. Carniolan bees are reared near Asker and other good plants include rape, clover and blackberry. Its western seaboard is over 1000 miles long and coastal indentations probably triple this number.

**nosema**

A protozoan infection by a spore forming micro-organism Nosema apis Zander. Enormous numbers of spores multiply in the bee's ventriculus and are voided with the faeces. The disease adversely affects performance, inhibiting the production of brood food and shortening the life of queens and workers. It causes supersedure and decreases honey production. There can be as many as 30 to 50 million spores per bee, and old comb may well have live spores sandwiched in the cocoons. The disease appears to be closely associated with the presence of certain viruses including black queen cell or bee virus 'y'. Treatment calls for replacing brood comb, requeening and Fumidil B can be fed as an inhibitory agent. See: ETO, Fumidil B and sterilization.

**notes**

The keeping of notes helps with forward planning and lists of items required or jobs to be done and can be an important aspect of beekeeping. Two things render note taking at the hive a little tricky. First, every pen, microphone or booklet runs the risk of sticking to propolised gloves or fingers. Second, even stout cards and certainly thinner pieces of paper become illegible when the bees have nibbled holes in them. Various systems of simple signs can be used: a block, having six faces could show – food, foundation, space, new queen, super clearer, prophylactic. Shorthand: F foundation, Q queen, X excluder, Fo food, S super, BC brood chamber, Fe feeder, saves space. With clean hands and a moment to contemplate, the computer is at your service – start a Beehive file. Dates, in particular, can become vitally important. How long for drones to mature or queen cell to be ripe?

**notice**

An announcement published or displayed for all to see. Meetings, directions to an apiary. These need to be clear, neat and apposite. An apiary and hives whenever moved to places where ownership might be usefully made known should bear details of name, address, phone number, E-mail address….. but please always make it quite clear which person has been allocated the responsibility for removing notices when an 'occasion' has been finished and the notice needs to be taken down. Notice also implies 'observation'. The enjoyment and education to be derived from being observant is endless.

**notum**

Tergite or tergum. Not Tweedle dum and Tweedle dee but as Dade says these are tergites and are segmental sclerotic plates on the dorsal surface of the abdomen and together with the sternites (the abdomen having no pleurites) form the segments of the abdomen.
See: tergite nourishing cream
Melt ½ oz. beeswax in a basin by warming in a waterbath. Stir in 4oz honey and ½oz sweet almond oil. Stir until ingredients are thoroughly blended. Remove from heat but keep stirring. When almost cool, add a little perfume (if desired) and pour into a wide-necked, screw-topped cosmetic jar.

**November**

All hives should now be set up so that the bees can remain quiet and unmolested until spring. Floods, gales and freak weather must always be followed up by checks that all is well. This is a good time of year for looking over records to improve next year's beekeeping, for repairing and refurbishing equipment, making an inventory, getting your order in before prices rise and planning your winter's reading. Are your hives 'branded'? Have they been rendered as inconspicuous as possible? Are they labelled with your name, address and telephone number or e-mail address?

**novice**

One who is as yet unskilled – a newcomer to the craft. The novice's prize at the honey show is won by someone who has not been awarded a first prize at a honey show before. See: beginner.

**N & P**

Abbreviation for nectar and pollen.

**nubile queen**

The perfect female honeybee, sexually

mature and at the stage when mating is likely. Unlike workers and drones a virgin seems lively from the moment she scrambles out of her cell.

**nucleic acid D.N.A.**
Deoxyribonucleic acid has a wide range of structures (long chain compounds) and plays a vital role in the formation of genes, being self-replicating. Composed of purines, pyrimidines, sugars and phosphoric acid.

**nucleus (beekeeping)**
Plural nuclei, commonly nuc or nukes.
A small unit of honeybees set up by the beekeeper for the purpose of: 1. storing a queen, 2. nursing a queen` 3. building up into a colony, or 4. use as a mating nucleus (box). The size can vary, the smallest being a mini nuc and the largest a full-sized colony. Such a nucleus would not be relied upon to gather its own food but would be fed, especially before during and after a new queen begins to lay. When setting one up in your own apiary, allow for the fact that most flying bees will return to their former home. A nucleus must contain either a mated queen or the wherewithal to produce one, i.e. female eggs or day-old brood, a virgin queen or a queen cell at some stage of development. Remember they are vulnerable to robbing especially when wasps are around.
See: queen raising nuc. and nucleus (Biol).

**nucleus (Biol & General)**
The central part or beginning around which additions accumulate. The granulation of honey can be speeded up when numerous nuclei are present – these are often in the form of minute air bubbles, particles of pollen etc.
See: nucleus (beekeeping).

**nucleus hive (nuc)**
A hive purpose-built or adapted to accommodate a small colony of bees, usually for the purpose of getting a queen mated or for the temporary storage of a queen. Use can be made of a standard hive with a 'dummy' board confining the bees to a small number of standard frames (often half-length so that when set end-to-end they can be placed on top of an ordinary hive. Provision for feeding and a small entrance are important attributes.
See: mini-nucleus and nucleus.

**nucleus mating**
A small colony of bees used for the purpose of getting a queen or queens mated.
Such a colony would have but one brood box and the means of feeding it.
Queen breeders sometimes set up rows of these on posts in stud apiaries perhaps in isolated locations. Because of their small size they are vulnerable to robbing, but on the other hand they are relatively easy to make up and to move to areas where robbing is unlikely.
See: nucleus and mating.

**nucleus mini**
A small non-standard hive possibly using half-length frames so that they can be aligned end-to-end and be put into an ordinary hive for getting them drawn out and filled with food perhaps. When propagating large numbers of queens, small nucs comprising in some cases little more than a polystyrene flower pot, are given a queen cell, food supply and just sufficient bees with a view to getting the queen mated. They are easy to transport to a stud area (or place where non-sibling drones will be flying).

**Nuka Hiva**
An island near Nuka Oa in S. Pacific Fr. in the Marquesas twixt NZ & US.

**number**
See Dimensions, facts, number of cells per frame, statistics, colonies on one site, temperatures, Fibonacci, the 3 – 6 – 12 worker stages 21 days egg to imago.

**number of cells**
Allowing for the wooden surround and assuming worker base at 840 cells per

decimetre squared (counting both sides of the comb) Deeps work out:
Dadant 9200, Langstroth 7400, British commercial 7600, Standard British 3300. These figures are rounded to the nearest 50. It will be seen that a British commercial is roughly equivalent to a Langstroth and a 14 x 12 to a Dadant. A British brood and a half lies between Langstroth and Dadant.

**number of colonies**

Although these are figures for 1977 they may still be of interest: Argentina 1000 colonies with an average yield of 22kg, Australia 374 (?) 40, Chile 525/15, France 1030/9, Ger. Fed Reb 1100/18, Italy 750/9, Japan 326/19, Mexico 2000/30, Switzerland 295/7, U.K. 243/15, USA 4315/19. Russia 1000/19. Hive numbers are in 1000's ?? but yields are in kilos. (more accurate up-to-date info would be useful).

**number on one site**

The number of hives that can reasonably exploit an area to full advantage cannot be laid down without regard to the foraging available, however at the high end where crops like oil-seed rape or heather abound, a twenty metre spacing of hives over the whole area would theoretically give the best returns. In the U.K. apiaries that exceed 25 hives may find that yields do not increase as numbers of colonies are increased. Very large apiaries will also have additional robbing and drifting problems. See: colony density, optimum and over crowding.

**nuptial flight**

Wedding flight. The flight of virgin queens (and ants) that leads to fertilization. Little more than three weeks pass between the laying of the egg and a virgin's departure to seek a drone congregation area. It has become clear that in order to fill her spermatheca with sperms enough to last her life-time, a queen may require not only to copulate with quite a few drones but also may well go off on several mating flights. Once she has commenced laying, a condition that many nurse bees must have been anxiously awaiting, she settles down to the one and only duty. Her unbarbed sting will have been sheathed and is now significant as an ovipositor.
See: fecundate and mating flight.

**nurse bees**

A young bee of from 3 to 15 days of age which by virtue of its developed brood food glands is able to attend to the feeding requirements of the queen, brood and drones. In a straight forward colony development, bees of this age will perform their duties until their brood food glands begin to deteriorate. They then proceed to other household duties, wax making, comb building grooming, taking nectar from returning workers and duties labelling them as house bees will lead to play flights and then to foraging. In the autumn the development of the 'fat-body' by the consumption of pollen, will set then up as winter bees which will enabling them to produce brood food and become nurse bees in the early spring.

**nurse colony**

A colony set aside for queen raising. Whether starters, finishers or temporary places for queen storage they would be called nurse colonies as opposed to the honey producing colonies. Some methods such as the Snelgrove system, enable colonies to be used for honey production and queen raising at the same time.
See: nurse bee, trophocyte.

**nursery cage**

A cage that confines a queen cell or queen so that it can be kept at hive temperature and receive food without any fear of the destruction of its contents. Queens are kept in batteries in nursery cages and ripe cells can be allowed to yield their virgins in such cages. Being bee-proof, with ventilation and correct space for the queen's emergence, these facilitate the production of virgins ready for introducing into queenless colonies. As they will have had no contact with bees their 'aroma' is neutral and they are relatively easy to insinuate into a suitable stock.

**nursing duties**

Although the work done by the colony as a whole in maintaining the warmth and relative humidity of the brood nest is part of the nursing process, the actual feeding where continuous attention including the weaning of workers is carried out by the nurse bees. The adult worker emerges and immediately begins to nourish and develop its hypopharyngeal (brood food) glands by eating honey, water and pollen and the rich brood food it produces is then fed to the newly hatched larvae (or perhaps the queen). The gradual change in the condition of these glands leads to their ability to wean larvae by feeding older and older larvae as the glands deteriorate. The possible ability to revitalize these glands is shown when older bees are compelled to feed young larvae and the ability of bees that have over-wintered to start a new brood nest indicates that this must be normal though one doubts whether bees reared in that way are 'up-to-scratch' See: nurse bees.

**nutrition**

The act or process of putting food into a living organism. The actual food or nutriment. The process of converting food into living tissue. The storage of food within the body, the study of food value etc. The honeybees are able to derive their entire nutritive requirements from pollen and nectar though at times when insufficient nectar is forthcoming, water is collected and during periods of dearth their stores of honey are called upon. The queen's requirements are quite special because she needs to be constantly delivering eggs and so 'royal jelly' is her diet. The workers cannot build new replacement tissue as regards wear and tear though throughout their lives glandular changes and developments take place so that their requirements are carbohydrate to allow then to produce energy and warmth. Their food requirement in repose is given as 0.7mg per hour and 11.5 in flight. See: alimentary canal, artificial food, digestive tract, brood food, enzymes, fat body and royal jelly.

**nutritional value of pollen**

Ranking highly are: crocus, chestnut, clovers, poppy, top fruit and willow. Then come: dandelion. cotton-wood: elm and maple, followed by; alder and hazel. Finally pines, evergreens and wind-borne pollens. Mixtures of pollens are more beneficial than that of a single species and pollen from the comb will have undergone a fermentation process which increases the digestibility of pollen proteins. Bees longevity is based on pollen. Dry, old pollen is almost worthless as its digestible albumen will have disappeared. Much has been learnt about the amino acids, enzymes and other pollen ingredients.
See: alfalfa and P production.

**Nutt's collateral hive**

Described in Thomas Nutt's book 'Humanity to honeybees' in 1832. It consisted of three boxes set side-by-side. These could be swivelled apart and had feeding drawers beneath each one. The central portion was the stock box and was surmounted by an octagonal cover, including as did each of the three boxes, an inspection window. It could hold a bell glass and the boxes at each side could also be extended in that fashion. They were 'cabinet made' and the handsome roof of the octagonal central portion rose to a neatly domed knob for handling. Nutt also designed an 'Inverted hive' and an Observation hive.

**Nuytsia floribunda**

A relative of the parasitic mistletoe which begins as a vine but grows to kill its host and becoming a giant free-standing tree. It is found on the coastal plains of W. Australia and its glorious cadmium yellow flowers attract hosts of bees around December/ January and it has become known as the

Christmas tree. Its honey is green but of an indifferent flavour. In NZ a variety with similar parasitic habit is known as 'rata' and its brilliant red canopies stand out as they dot the bush around Christmas time – a veritable feast for bees who produce a white, smooth-grained honey with a delightful taste. See: Manuka and Pohutakawa.

**nylon**

A strong, resilient, thermoplastic, synthetic polyamide that can be extruded into fibres, made into rope, material sheets and castings. It has become an enormously useful substance in many aspects of beekeeping. Nylon rope handles stay flush against hive sides yet last for ages.. For extractor bearings, filters and strainers and hive strapping – it has stepped in and replaced both wood a metal devices with a clean, silky-smooth, durable alternative.

**nymph**

Pupa or chrysalis. The young stage of an exopterygote insect. It resembles the adult having the same kind of mouthparts and compound eyes, is wingless or has wings incompletely developed. It is sexually immature. Although there is no nymph stage in endopterygota (which includes honeybees) this term is sometimes used by beekeepers to describe the pupal stage of the brood after the first moult and was used by Butler, Huish and Bevan to describe pupae. Pliny the Elder wrote 'The common bees are called 'nymphs' when they begin to assume their perfect form.

**NZ Abbreviation for New Zealand**

Mentioned repeatedly for its many useful additions to beekeeping know-how and ability to produce one special honey after another.

**N.Z. beech honeydew**

A special substance produced from the southern beech tree in the northern half of South Island. The forest (bush) literally drips honeydew where only wasps or bees can take advantage of it as roads have not yet arrived. Towards the end of the season migrations to suitable sites are undertaken by some honey farmers mainly on the eastern side of the main divide, north of Mt. Somers and on the west coast north of a line drawn between Bell Hill and Atarua in the Grey River valley. There are varieties like mountain black, red and hard beech trees, standing 20m high and clothed with the fur-like silvery tubules from which the glistening drops of honeydew swing. It covers sign posts and the sticky sweetness is immediately invaded by the soot fungus which often obliterates the letters of nearby road signs. As with lime tree honeydew here, some of this fungus finds its way into honey which darkens on that account though light itself can darken honey.
See: honeydew.

**NZ beekeeping**

Lady Hobson and Rev. William Charles Cotton took bees from NSW (Australia) and later from England in 1842. Earlier the ship 'James' embarked at Mangunga, Hokianga, March 13th 1839 when the Rev. Bumby (missionary) and Mary Anna Bumby (according to Isaac Hopkins) took two straw hives. Later two hives of Italian bees were reported in NZ Herald Aug 13, 1877. By May 31 – 1980 there were 5,217 beekeepers with 19,450 apiaries and 233,810 hives (not so different from UK though honey crops were vastly different) 9000 queens were exported. Unfortunately the advent of Varroa has interfered with what was such a timely source of queens – available in March!

**NZ flax *Phormium tenax***

Perennials with clumps of sword-like leaves. Copious nectar can be had in July the bright orange pollen becoming a duller yellow once stale. In suitable conditions spoonful's of nectar may be shaken from the flowers. It grows wild in the damper regions. It grows and sometimes produces long flowering racemes in the south west of the UK.
It needs mulching and sheltering from

cold winter.

**NZ forage**

Rewarewa, (*Knightia excelsa*) also called honeysuckle has bright red flowers and pleasant honey in early summer. Blackberry, thistles, gorse (*Ulex europaeus*. Butter cup honey has been reported (probably king cups). The wettest areas on the west of the divide still give excellent crops particularly in years when the rata vine flowers widely. Some of the largest operators are in marginally warmer but drier west of S. Island while conditions can border on the subtropical in the north of N.Island. The area around Nelson is famous for its orchards.
See: NZ honey.

**NZ honey**

Although the bulk of New Zealand honey is collected on the Canterbury Plains with a large proportion of clover, honeydew from the north of S. Island's foothills could increase. In some areas the rata vine yields very heavily while blueweed, flax, kamahi, Manuka, blackberry, rewarewa, thistle, pohutakawa and other plants yield well in moist warm conditions that so often prevail. Blueweed grows along river beds and open country around Marlborough and Canterbury. Landslides, forest fires and flash floods do cause problems (quakes?).
See: beech.

# O

**O.A.C. Strainer**
This is special type of honey strainer used and designed by the Ontario College of Agriculture.

**oak Quercus spp. *** Fagaceae**
A tree whose leaves give rise to honeydew and its flowers pollen in April/May. The timber is strong and useful for hive stands. Rather too heavy for hives, though doubtless many a hollow oak has been populated by a feral colony and it has also found its place in history for barrels, casks and helped to win many a battle in the days of sailing ships.
See: cork oak.

**objective (lens)**
The lens or lenses that first receive the rays of light from an object and forms the image that is to be viewed or photographed in a telescope or microscope.

**oblique bands**
These are found on the drone's genitalia, on the walls of the bulb of the endophallus.

**oblong hives**
Whereas square hives have the advantage of being turned so that combs face either 'cold' or 'warm' way, oblong hives have the advantage of storage when empty because turned sideways one body will fit inside another. Some older versions of the WBC were oblong but to-day it is the Smith (British size) or the Dadant and Langstroth (US size) that are oblong.
See: under those names.

**oblong plate (sting)**
This is one of three pairs of plates associated with the functioning of the sting mechanism. They are immovably fixed on the side of the bulb and stylet of the sting and the quadrate and triangular plates, which are articulated, are also connected to them. A ramus extends from the oblong plate and the lancets can slide along it forming a tube through which the venom is pumped.
See: quadrate and triangular plates.

**observation hive *****
A hive that takes standard-sized frames, but ideally displays both sides of them via transparent walls. Very early hives were made of mica or talc and other transparent materials. Sizes may vary from a cross-section of a hive with deep, shallow and section layers, to hives that cover a whole wall, exposing eight or more full depth frames to view. It is rarely worth over-wintering bees in such hives which are best made up in May and if fed continuously to restrict laying or if brood can be added or taken away as required, can be kept going until they are united to a normal stock by the end of September. Any wooden frame that will hold a frame and has glass or Perspex sides will suffice. A tunnel or tube can lead from top or bottom to give bees access. Swivelling, double-glazing and the incorporation of feeding devices are also useful. The queen can be marked as can bees of certain ages. A magnifying glass, torch and note book should be to hand. Where possible keep on the north side of a building.

**obstruct**

To make entry and exit more difficult. Closure of hive entrances is dealt with elsewhere but it is sometimes necessary to partially block or divert bees. Examples are the leaning of a slate or pane of glass over the entrance to confuse robbers or to shield the entrance from snow-glare. When weeds and unwanted herbage is allowed to grow up and smother an entrance the bees are usually remarkably capable of cutting or finding their way through. The dampness and harm to the hive base is enough to suggest that this fault should not be encouraged. When wire netting is placed over the entrance to keep off birds, cane toads or other predators this too is a form of obstruction.

**occipital foramen**

The opening in the occiput through which the organs inside the head are connected to the thorax.
See: occiput.

**occiput**

The plate of the exoskeleton forming the back of the head on insects.

**occupy**

To invade or take possession of something. Bees will find a new home when they swarm. At times a vacant hive is occupied. Bait hives are often set-out for this purpose. Certain predators, commensals, inquilines and parasites may also occupy a hive or the bees that are in it. A colony is more likely to be robbed out than it is to leave its hive to try to join another queenright colony, though this has been known to happen.
A beekeeper will set up his or her hives and occupy an apiary.
See: bait hive and swarm lure.

**ocellus pl. ocelli**

The simple eye, of which the bee has three, situated two over one on the frons. They are simple lenses and are thought to be used in the dark and poor light conditions of the hive and to measure light intensity. They bring about dorsal light reaction during flight so that there is a stabilization of movements in the 'pitch and 'rolling' planes. The comparison of the amount of light entering the simple eyes as compared with the two compound eyes leads to a sophisticated control of pitch and roll. Ocelli is also the name given to the eye-like spots on some butterfly wings.

**octagon**

A figure having eight sides. The regular octagon (sides and angles equal) is the cross section of an ommatidium. It was considered by many, like Evelyn, to have a special significance to the bee and there were octagonal bee houses and hives.
See: octagonal hive.

**octagonal hive**

Mew designed an octagonal hive and Gedde, White, Kerr and Cumming all went along the same lines.
See: Mew's octagonal hive.

**October**

At this time it is neither wise to move colonies nor to feed them though in the south feeding can go on as long as weather conditions are conducive to the syrup being taken down and sealed. Hives are best sheltered from the elements and reduced in height while precautions need to be taken to keep roofs in position. Strapping or encircling with wire is so much tidier than using huge rocks on the roofs. Inner covers give a second line of defence. Mouse guards should go on once the pollen intake is minimal. A pregnant mouse may attempt to establish itself whenever the bees are forced to cluster continuously for more than 24 hours so don't get caught out. Apiaries should be arranged so that little more than cursory checks are needed until spring. Store empty supers by stacking and bee-proofing. They can be covered with sealed newspaper (wax moths don't like it) and prophylactics such as acetic acid also help to deter moths or the hatching of their multitudinous eggs that are there, but hard to see. This is a month when the clocks go back but regardless of the time, actual conditions will often vary from a standard assumed for any one part of the country.

**odd**

Strange or abnormal behaviour by a colony of bees should always be investigated. For instance a great activity could be due to robbing, while inactivity could mean that some harm has befallen a stock. A knot of bees in front of the hive on the ground may have a queen in it. A burst of flying while snow is on the ground may lead to many dead (use tile to shade entrance). An untouched feeder when they are short of food may mean that they cannot get at the syrup for some reason. You might almost say, oddities are not uncommon in beekeeping.

**Odonata** ***

Dragonflies – they come under the sub-class of Apterygota. Carnivorous as nymphs and adults. Fierce enemies of other insects. They can fly along with their legs hanging down beneath them like a net, ready to swoop up their prey. Watering places are frequented by these and several other predators.
See: dragonfly.

**Odontobombus (odonto – tooth or teeth)**

Pocket making bumble bees (group of at least nine species)
See: Anodontobombus.

**odour**

The property affecting the sense of smell. A scent or aroma which usually has a most pleasant nature wafts from the entrance when honey is being ripened. A honeybee is well endowed with olfactory sensilla and they play an important part in identifying their own colony members, determining that the queen is present and in foraging. Pheromones may be thought of as odours but they have additional chemo-physical effects. It should be remembered that the bee's sense of smell does not always coincide with ours. They differentiate between water and syrup and rarely like our skin creams and shampoos.

**odour – alarm**

A pungent smell as of 'pear drops' when released by the bees. It is pheromone that gets an immediate response from other bees but interestingly although immediate in effect it is of relatively short duration. Chemical analysis show a whole spectrum of parameters that come into play and with incredible speed as a whole colony is put on the alert. As they tumble out you can almost read their saying '*where is it, let me get at it!*'.
See: alarm odour and iso-pentyl.

**odour – colony**

Although each colony working on similar food sources will appear to have the same sort of odour, one has only to put a few bees from one hive into another to see that their alien odour is instantly recognized The day-to-day smell of the bees will be much affected by the crops they work whereas the colony itself with its queen and all its stores and comb, will have a highly individual smell from the bees point of view.
See: colony odour and hive odour.

**odour of nectar**

As each flower and nectary will have its own characteristic smell, nectar can often be distinguished by this property. When a colony has been working on a flow all day, the moisture laden air being driven out of the hive as they go about ripening it, can be a pleasant thing to experience and sometimes identify the source of the forage. It has been described as a 'heady perfume' and fields of clover, avenues of limes and of course 'heather' will each provide you with an identifying odour.

**odour hive**

The composite smell of the colony itself is important for their own identification. To the human nose, waxen comb, honey, propolis and bees, do give a total smell that is pleasant and quite distinctive. Superimposed upon this will be the effect of any current flow, especially when nectar is being ripened. Certain conditions that can occur due to mice, dampness, disease or other unwanted phenomena may spoil the normal smell and give the beekeeper a warning indication that something is amiss. See: colony/hive odour.

**odour of honey**

At the moment of unscrewing the cap, a jar of honey should yield a pleasant, characteristic odour which will give a good idea as to the delicacy of its contents. Honeys with strong flavours will have more dominant odours as a rule than those whose flavour is more subtle. Where unwanted aromas as from privet, sweet chestnut, ragwort etc. are present, such honey will nearly always improve with keeping. The 'honey is just sugar brigade'(No names no pack drill) have obviously no sense of smell (or taste one presumes).

**odour of mead**

The subtle but significant bouquet produced by a good mead is best savoured at the moment when it has been freshly poured. Should any trace of unwanted aroma exist it is unlikely that the flavour will be of the best though as with honeys some unwanted aromas gradually disappear.

**odour of wax**

Wax has a sweet, characteristic and distinctive odour. This comes out when beeswax candles are burnt as opposed to those made from paraffin wax. The sweet smell of new foundation should not beguile you into assuming that it will attract bees in the same way that honey does. Swarms put in boxes, only partly filled with frames of foundation, have often filled the empty space with their combs ignoring the foundation. Showmen have been known to keep a prized cake of wax in honey to retain its odour. It is possible to get special perfumes suitable for impregnating wax so that in burning the aroma is liberated and wax will 'store' subtle aromas. When ordinary beeswax is heated to melting point the aroma given off is highly attractive to bees and wasps.

**oecotrophobiosis**

Social cohesion in the insect colony as exhibited by one food sharing household.

**oenocyte**

One of the two elements of the fat body. Special cells found in and around the fat body (which is highly significant when it comes to overwintering bees.)

**Oenomel**

A drink from ancient Greece used by the Romans, wine mixed with honey, something combining both strength and sweetness.

**Oenothera spp.**

It prefers poor, well-drained soil and has a long flowering season
See: evening primrose.

**oesophagus**

A fine, thin-walled tube leading from the pharynx to the honey stomach. It passes from the head and through the thorax, widening to form the honey sac in the anterior part of the abdomen.

**off-course**

That bees get lost and 'drift' into other colonies has been well established. In fact bees are very good at 'searching' and only need to pick up an olfactory clue which the radiating lines of foragers probably leave in all directions, to find their way home again. The numbers of bees that return to a former site when only moved a mile or so, as in the fruit orchards, (road distance between sites may have been over 3 miles) is remarkable. Some bees can fly in windier conditions than others and there is no doubt that they become skilful at using the shelter of hedges etc. to avoid the effect of heavy gusts.

**official**

1. Authorised or authoritative as for example official information.
2. A holder of post or office. We are familiar with Secretary, General Secretary, Chairman and President and so on but new posts often need to be created to meet new times. Publicity, audio/visual aids, apiary manager and so on, each branch of activity having a person (official or officer) in charge.

**offset**

Or 'split'. A nucleus of bees taken from the parent colony with a view to rearing a new queen or making increase or perhaps as a means of swarm prevention.

**offspring**

The generation of bees produced by one parent. We might expect a queen's offspring to be of uniform colour but in fact there often seems to be a mixture. Some are say, black and tan and others dark. One reason for this is that the layers of sperm within the queen's spermatheca are getting used one at a time (see patrilineage). Older bees tend to become darker and shinier, drones do not always share their sister's colour and the temper of a colony can change. The significance of the drones not necessarily being related to their sister's fathers should always be taken into account when breeding projects are planned.

**oil**

Typically unctuous, viscous, combustible substance. Old engine oil in tin trays can be used as a deterrent or trap to prevent ants from getting into hives when placed under the legs of hives though bees can get into oil and it is fatal to them. Paraffin is slightly repellent to ants. Some oils are under the headings: oil of almonds, linseed, paraffin, safrole, turpentine, wintergreen and benzaldehyde. For specific purposes grease such as Vaseline jelly are used for lubrication and the anti-propolis treatment of hive furniture.

**oil of almonds**

An artificial, synthetic oil known as benzaldehyde which can be used as a repellent and has been used with various success according to ambient air temperature in the clearing of supers. Almond oil which is extracted from bitter almonds is a relatively colourless balm that can be used with beeswax in the preparation of skin creams.
See: nourishing cream and benzaldehyde.

**oil-immersion**

An oil-immersion lens is essential when the highest possible resolution and magnification are required with a light microscope. A special objective lens is used in conjunction with an oil such as cedar wood oil, having the same refractive index as glass filling the space between the objective lens and the cover slip. It is better to use a water soluble substance – 50/50 honey and distilled water – because there is less chance of dissolving lens cement or material on the slide.

**oil of safrole**

An oil with a highly invasive and unpleasant smell used in the Frow treatment for acarine.

**oil of wintergreen methyl salicylate**

This has been used as a preventative for Acarine. The idea is that as the blind female mite that causes the disease finds its way into either of the first pair of thoracic tracheae of a young bee, then a confusing scent such as that of this. Oil may well cause the death of the mite due to its inability to find a satisfactory home. Best used in warm weather.

**oilseed comparisons**

Three useful bee plants: Soya bean 37·9% protein 18% fat, Rapeseed 20·4% protein, '43·6%' fat', Sunflower 16·8% p. 25·9%. Rape is easier to grow in temperate regions – there are well over fourteen varieties best

sown before mid-Sept and eleven spring varieties.
See: OSR, soya bean and sunflower.

**oilseed rape pests**

Pollen beetles eat pollen but do little damage, seed weevil lay in maturing pods and eat a few seeds, pod midge eat pod wall and usually all seeds from the pod are lost.

**ointments**

Honey is sterile and hygroscopic and is useful as an ointment though sticky and therefore needs to be covered. It can be useful in dilute forms. For bee stings: Benadryl, Anthisan and Thephorin etc.

**old bees**

There are many things that can shorten a worker bee's life but apart from mishaps, summer bees working tirelessly are lucky to last any more than 6 weeks from emergence to death. Old bees are the first to sting (wise -because why waste a younger life?) will have become darker and relatively hairless, their wings show signs of wear being somewhat tattered and ragged at the edges. We call them 'industrious' but that hardly does credit to the enormous amount of effort they put into each of the phases that follow the maturing of their different glands while 'house bees' and the unbelievable effort they put into foraging visiting hundreds and hundreds of flowers to fill their corbiculae with just two pellets. Nectar, water and propolis are collected, often in dangerous conditions and when a tired old bee comes in with its last load we would be amazed if we only knew what a selfless effort it had made for the colony. Workers that over-winter, ever ready to perform their duties, are forced to shelter patiently while short days and lower temperatures keep them hive bound. Turning food into warmth energy, their lives may well extend to six months, a month for every week of their summer sisters. Drones have but one duty (if called upon) and their lives may be terminated by mating, or an change of heart on the part workers when a queen has been mated or a dearth occurs, or by the natural clear out at the end of summer. So we might give those 2 or 3 months at the most, unless queenlessness allows the colony to tolerate some of them through the winter. Queens, the hardest workers of them all, are still robust enough to outlive many generations of their progeny and although likely to start failing in their third year can go on for one or two more. Their early majesty slowly gives way after laying tens of thousands of eggs, their gait and fading beauty make them easier to pick out and to pity.
See: old queens.

**old comb** ***

Comb darkens as it is trampled on and as generations of brood leave their faeces and cocoons behind. One might say that it is beloved by the bees, the queen seeming to prefer laying in it but hard, dark old comb should have no place when mankind wishes to exploit the bees to their advantage. One melted down comb produces enough wax to make a sheet of new foundation. Pathogens from wax moth grubs to nosema spores revel in old comb. Equipment manufacturers give a fair price for it and it is very satisfying to see the sun's rays liberating the pure wax from the old comb and dross as it passes through those tights you become tired of, in the solar wax extractor your husband/wife so skilfully made.
See: black/dark comb and comb renewal.

**old man's beard** *** ***Clematis vitalba***
**Ranunculaceae**

A tough, rampant climber that spreads over hedges and does its best to fell very tall trees if left to it own devices. It likes alkaline soil and grows from cuttings (it's a devil to get rid of) but its white or greenish flowers smell sweetly of vanilla and Howes tells us that it can buzz with bees as it yields nectar as well as pollen, the nectar appearing as droplets on the filaments of the flowers. There are many cultivated varieties - at their best in mid-summer.

**old queens**
Despite living so much longer than any of her progeny it is not, in modern beekeeping practice, a good idea to give space to old queens. Note Americans are known to have paid hundreds if not thousands of dollars for a breeding queen with much sought after characteristics. An old queen lays later in the spring and fewer eggs than a young queen, she is more likely to swarm and will, through contact with untold numbers of progeny (and visitors) be potentially less healthy. They are not as reliable as younger queens to winter well, in fact, all-in-all, it is wise to manage colonies so that queens do not have to enter a third year. In appearance (she will have suffered from various mishandlings by errant workers) wings a bit shabby, gait slow, yet having grown fond of her, some beekeepers find it hard to think of replacing her. Fiendish though it may seem her frozen body is useful for purposes of dissection.

**Oldfield Dr. Josiah**
'Honey for Health' An American best seller. He was one of the many totally convinced on the usefulness of bee products for human health.

**old red sandstone**
There are three types, lower, middle and upper. Reports on the middle old red sandstone state that it can produce an early crop which might restrict the queen's laying unless large brood chambers are used enabling the colony to build up naturally and harvest a later crop in the supers.

**old virgin**
Once a new queen has become several weeks old without mating, in other words becomes an old virgin, she is unlikely to become a good queen. The spermatheca contains a fluid which is displaced as the spermatozoa enter but in the old virgin this fluid begins to coagulate. Eggs laid by older virgins will only produce drones (dwarf drones if laid in worker cells) as they will not be fertilized.

**olfactory**
Concerned with the sense of smell. The bee has a highly specialised ability to find and act upon those parts of its environment that are appropriate to its success. Recognition of its hive mates, discovery of food sources or a new home, awareness of danger, queenlessness or the queen's presence are a few of the aspects of colony life that require a special olfactory sense. In the case of the drone, despite having extra-large compound eyes, it is heavily dependent on its olfactory capability (enhanced by an extra-long antennae) to engage with a queen when patrolling up aloft. The touching of antennae seems to be more meaningful than merely recognizing a nest mate. It sounds strange but bees must 'sniff' in this way.
See: odour.

**olfactory receptors**
The antennae bear large numbers of sensillae many of which are concerned with the sense of smell and taste. They seem to take the form of specialized hairs, pegs or plates. However the bee is certainly able to recognize odours using other parts of its body. Considering the fact that life goes on in the honeybee's nest in total darkness, the olfactory capability must play a great part.

**oligosaccharides oligo –**
few or little. Non-reducing sugars.

**oligotrophic**
Providing insufficient nourishment.
The name for insects that only visit a few species of plants.

**ommatidium pl. ia**
One of the thousands of similar units that form the bee's compound eyes. Externally the tiny lenses can be seen and numerous hairs arise amongst these, presumably by way of protection. Each ommatidium has a crystalline cone under the normal lens and the cone is seated amongst pigment cells which prevent light rays crossing from one ommatidium to another. A rhabdom,

or continuation of the cone, passes down into the basement membrane where it connects to nerve fibres. Each rhabdom is surrounded by retinula cells which in turn are surrounded by other pigment cells. Each rhabdom has eight retinulae around it. The opposite twist on two of the long visual cells allows the bee to determine the sky's direction of polarization by comparison.

**omphacomel**

Pyment or wine made from grape juice and honey – authority Palladius (Dr. H. M. Fraser).

**one-hour (method of queen introduction)**

A colony is deprived of its queen. The new queen is caged with her own workers until ten minutes before introduction. Then the attendants are removed and the queen spends ten minutes alone before being put between centre frames in her cage that is sealed with a candy plug through which a matchstick hollow has been pushed.

**onion *** Allium/cepa Liliaceae**

Honey produced from this biennial does not necessarily carry the distinctive (and perhaps unwanted) flavour of the onion itself. The genus includes: chives, garlic, leeks, onions and shallots whose flowers are all useful to the bees and offer nectar and pollen over quite a long period. The juice from the leaves was thought by some to provide a useful antidote to bee stings when rubbed on the affected part.

**oocyte**

An egg before the formation of the first polar body. A cell that undergoes meiosis and thereby forms an ovum.
See: oogonia

**oogenesis**

The formation, development and maturation of the ovum, or the production of female gametes at meiosis.

**oogonia**

The primary female germ cells which form in the narrow end of the queen's ovaric tubules and enlarge first into oocytes and then eggs or nurse cells.

**Oolitic limestone**

On this type of soil clover crops well and yields nectar freely when temperature and plant varieties are right. Lucerne for example, when left to flower and not prematurely cut for hay, needs temperatures in the seventies F.

**Oolite – limestone with a texture as of fish-roe.**

**oophilous**

A creature that eats the eggs of arthropods.

**opalescent honey**

Honey with a milky iridescence or cloudy honey. Some honeys even after filtering, may still restrict the passage of light and do not present a bright, clear appearance. This can be due to colloids and often applies to honeydew honeys.

**opaque honey**

Honey that is impenetrable to light. Honey is generally clear when first sealed over by the bees. It is however natural for it to precipitate crystals when stored, unless it has been carefully filtered and heated. Crystals and/or air bubbles render the honey opaque. The presence of colloids (as with some honeydews) will also tend to make it opaque. Honey that is clear and bright may win prizes at shows and sells well but from the healthy food point of view opaque honey is every bit as good if not better.

**open**

1. To open up a colony means to remove the roof and inner cover, or to separate chambers, so that manipulation of the frames and an examination of the hive contents can take place.
2. Flowers can normally admit a bee as soon

as they are open, and in the case of apple and other fruit the sooner pollen transfer is effected the better.
3. An 'open' class at a show means open to all-comers.

**open brood**
Uncapped or unsealed brood. A Queencell would be unsealed until capped over and 'open' once the inmate has emerged. Brood in fact, from the egg to the moment of sealing.

**opening**
A space through which something can pass. Splits, knot holes, holes due to poor maintenance and woodpecker damage – any of these may provide openings which some of the bees will not hesitate to use as entrances, nor wasps or robber bees for that matter. Openings through which a predator might pass should be kept to a minimum so that the bees can adequately guard their colony. An opening in a wood could provide a site for an apiary but it is best not to keep hives under trees.
See: entrance, gaps and slots.

**opening hives**
Have all appropriate equipment to hand. Know what you are about to do. Subdue (wait a full minute after smoking entrance) before opening. Make use of a hygienic cover-cloth, as this excludes other bees, keeps valuable warmth in and shields them from the glare of unaccustomed daylight while minimising smoke required and keeping many of the bees from taking to the air.

**open-topped hive**
Whereas this expression probably followed Langstroths development of movable frames as opposed to the former skeps that had to be up-turned for even a cursory examination, it also suggests 'top spacing'. Reasons for the use of 'top' as opposed to 'bottom' spacing would seem relatively unimportant but whichever you choose do not mix the two in one hive.

**operculum**
An organ serving as a lid or cover such as the lobes covering the first thoracic spiracles.

**opercles**
The French word for 'cappings'.

**optic**
Pertaining to the eye as an organ of sight or the visual function of the brain or to instruments which function by means of sight and light.

**optical density**
The degree to which honey will allow light to pass through it, is opacity or how opaque it is. This would not necessarily be associated with honey's density (mass per unit volume) or its viscosity but would depend on colour, colloidal suspensions, amount of pollen, air bubbles or crystals.
See: density and s.g.

**optical properties of honey**
See density, fluorescence, opacity, optical density/rotation, polarimeter, polarized light and refractive index.

**optical rotation**
The angular rotation of the plane of polarized light when it has passed through an optically active substance such as honey.
See: polarized light.

**optic cone**
The crystalline cone of the ommatidium which is one of several thousand similar component parts which go to form the bee's compound eyes.
See: ommatidium.

**optic lobes**
Great bundles of crossing and re-crossing nerve fibres (chiasmata) connecting eyes to the post cerebrum. Collectively the opticon forming the greater part of the

protocerebrum
See: optic nerve and opticon.

**optic nerves**

The large mass of nerves leading from the bee's compound eyes to the brain and collectively termed the optic lobe.
See: optic lobe.

**opticon**

The inner zone of the optic lobes.

**optic rod**

The rhabdomeric extension which leads the light passing through the crystalline cone of the ommatidium into the nerve fibres.

**optimum**

The most favourable amount, suitable size, quantity or temperature. We might refer to the optimum size of the colon for example when it comes to wintering. Although very large colonies, (obtained perhaps by uniting two very strong colonies) will winter successfully, they may consume more stores pro rata than a four frame nucleus which can also be wintered in mild conditions. Lying between these sizes is the normal forty to fifty thousand bee colony which with a good and suitable environment will winter healthily and economically.
See: No. of colonies and right size of hive.

**optimum temperature**

The most suitable temperature when handling bees is above 60F (16C) and for extracting honey 90F (35C), for clover secretion 70F (25C). These are optimum temperatures. See: temperature.

**oral**

Uttered by word or mouth, using speech, pertaining to the mouth. We speak of an 'oral' to describe an examination of a candidate by word of mouth as opposed to the writing of answers to questions.

**oral plate**

The oral plate of the honeybee is also known as the cibarium, or pharyngeal plate. It is a hardened plate on the floor of the cibarium, the anterior lobe of which bends downwards.

**Orange *** *Citrus aurantium* (and vars) Rutaceae**

Seville (bitter) *C.amara*, Grape fruit *C. decumana*, Tangerine *C. nobilis*. Globe-shaped, reddish-yellow fruit. New flowers appear while the last year's fruit is still ripening in December and it yields nectar and whitish-yellow or deep-yellow pollen, 25mu grain. The flavour of the honey is influenced by, amongst other things, methyl anthranilate.

**orchard *****

2-7 million flowers per acre (c5 -20 per ha) of enclosed land devoted to cultivation of fruit trees. Often the practice is to grow only one type of fruit as the period of flowering is relatively short. During this time bees will be useful for the purposes of pollination and may gather some surplus honey. However to keep bees on the site is not normally desirable due to the absence of flowers after petal fall and the use of sprays. Bees pollinate best when carefully prepared colonies are brought into the orchards just as the first blossoms are bursting. Snags from spraying can occur when adjacent orchards carry different fruit. Hand-pollination was used for early flowering nut trees and where non-fruiting male trees are required, plants kept in containers can be moved in at blossom time so that the whole area is full of productive trees (kiwi fruit). Cross-pollination gives better crops even when trees are self-fertile and tables of matching pollens are available, humble plants like crab apples and *M. floribunda* can help and in private gardens bouquets from adjacent gardens might help if branches, tied in vases, are inserted amongst the needful blossoms. The importance of the right sort of pollen at the right time cannot be over-emphasized.

**order**

Despite the apparent shambles in a hive, the result of the workers labours always appears

to be orderly. In fact where any disorder appears it points to abnormality such as a failing queen, the effects of robbers, predators, disease or the weather. Orders for bees or beekeeping equipment should always be given in good time. An order can be of a body or society of persons or a command of a court or a judge. It can also be a parliamentary instruction governing certain aspects of the laws of the land. In taxonomy too, it is a group consisting of families, forming a part of a class.
See: party parliamentary etc.

**organ**
A complete part or member within an animal or plant performing its own particular function and having its own structure, e.g. antenna, heart, sting.
See: organism.

**organic**
Derived from living things. In chemistry – appertaining to carbon compounds e.g. hydrocarbons. Pathology: affecting the structure of an organ. Pathology: of the body organs.
See: organic honey and inorganic.

**organizer**
One who arranges harmonious and united action, bringing about interdependence an co-ordination. After many years amongst beekeepers it has become very clear that quite a lot of them have no wish to be organized. However committees, education authorities and those with the brief of having to shepherd other's activities, have to appoint organizers who fortunately are always to be found.
See: adviser and instructor.

**organism**
A single entity made up of tissues, organs and organ systems. These vary from the very simple to the very complex. Any form of animal or plant life, a living thing composed of mutually dependent parts. An array of protoplasm capable of carrying on life processes. This includes unicellular bacteria. Virus particles, which are infinitely smaller, may well be at the limit between that which is a living organism and that which is not. Ability to feed and reproduce are characteristics of most organisms. We have these characteristics as do bees.

**organization**
A structure composed of interdependent parts. This applies to a honeybee colony, a beekeepers' association or to an equipment manufacturer. To systematize or give character or to arrange things to achieve a certain end. Hence 'organizer'
See: organizer.

**organochlorine**
Insecticides etc. Organochlorides can be highly injurious to living systems and some of them have a very long life. One might say that at times they were used with 'gay abandon'.

**organoleptic**
This word is used to describe a human analysis or measurement of a certain property e.g. the taste of honey or wine.
See: - gustatory and taste.

**organophosphorus (esters)**
They block the enzyme acetyl cholinesterase which maintains the organization and transmission of nerve impulses. Generally bio-degradable, but insects may develop resistance. It is more toxic to insects than mammals and it can be carcinogenic though it is said to be short-lived, degrading into harmless end-products.

**organza**
A fabric mesh made from a mixture of silk or nylon with cotton. It has been used as a coarse filter for honey, its tough quality and consistency of mesh making it long-lasting and suitable.

**Oriental poppy *** *Papaver oriental* Papaveraceae**

They make large, single flowers in May and June which are excellent for both nectar and pollen which is a deep, royal blue (black Harwood) and they like hot sun and poor dry soil. A popular border plant but it tends to sprawl.

**orientation**

To turn or move so as to achieve the desired relationship with surroundings. Ability to travel so as to reach a target e.g. forage or nest. The bees' directional flight seems to be measured in terms of angles relative to the sun's azimuth (direction), at a particular time. In fact the direction of the sun whether visible or not governs the polarized light patterns which cover the sky and the bees are capable of sensing this. Flying becomes erratic if attempted by artificial light or if confused by reflections from glass, snow or water. A dark cloud or an eclipse sends them scuttling back home.
See: homing instinct, navigation, nadir and von Frisch

**orientation flights**

Play flights – they take place around mid-day when hives suddenly seem to be surrounded by activity as young bees in various stages of initiation, zigzag and hover facing the hive while drone activity is also at its height. It does not seem to interfere with the return or departure of the foragers who impatiently plough their way through the milling mass. The first flights serve to accustom the bees to the outside appearance of their home and this eidetic image doubtless remains locked in their memory banks and is ready to serve them if it is necessary for them to correct any homeward drift by visual reference to the appearance of the area in the immediate vicinity of their home.
See: homing instinct

***Origanum vulgare* - marjoram**

A useful herb beloved of the bees.

**origin of the honeybee**

The honeybee is thought to have originated in South Asia and to have moved thence to Africa and Europe. This was some eighty million years ago, about the time when the upper chalk was deposited. Bees that became fossilized in amber and comb in marble, help us to draw certain conclusions. It is likely that their ancestors came from wasp-like creatures whose tongues became specialised for delving deep into floral nectaries and whose oesophagi enlarged at the stomach end to form a storage and carrying crop. They will have developed, along with honeydew producing and flowering plants through 10 – 20 million years. True honeybees probably arose in the Old World in the Oligocene. A mellifera is believed to have originated in the African tropics or sub-tropics around the end of Tertiary but its fossil records are too incomplete for any definite verification.

**-ose**

A name ending for chemicals, especially carbohydrates like sugars: sucrose, glucose, fructose, laevulose, dextrose and mannose.

**osier *** *Salix viminalis* Salicaceae**

Willows, specially 'coppiced' to get long green stalks, or withies a willow with long narrow leaves the flexible twigs of which were much used for basket making. It was once used for the wicker framework used to support clay or mud covering early types of beehives.
See: cloom and harl.

***Osmia lignaria* spp.**

The blue orchard mason bee increased the yields of apple and almond orchards quite significantly when used by the Logan Laboratories (which house U.S. National Pollinating Insects Collection). *O. cornifrons* (Japan) is about two thirds the size of a honeybee and it is useful at low temperatures for apple tree pollination though pollen tube growth is slow at temperatures too low for the honeybee. *O. coerulescens* is especially useful on red clover. *O. pilicornis* has foxy-red fur on its thorax and first two abdominal segments and was re-discovered in S. England after being considered extinct.

**osmosis**

The tendency for conditions of 'ion' density in two miscible fluids to equalize on either side of a semi-permeable membrane. These are widespread in nature and liquid is caused to pass through by osmotic pressure. This pressure causes water, or the stronger solution of dissolved particles, to pass through a membrane into the weaker solution. The distribution of water in living things is greatly influenced by osmosis – communication between cells - the secretion of nectar etc. It allows plants to absorb water against the force of gravity.

**OSR Oil-seed rape**

There was a boom in the growth of oil-seed rape reported in Ontario when new varieties became available that were low in unwanted substances such as: erucic acid and glucosates which contained undesirable, fatty acid which has unwanted pungent flavours or aromas. Since Britain joined the E.E.C. hectarage shot up. However blazing yellow fields offer nothing until an equable temperature of 60 – 70F (15 – 25C) maintains. Turnip rape is usually spring sown in Canada and Sweden but here it is mostly swede rape sown in August.
See: break crop.

**ostium pl. ostia**

A mouth-like opening or valve. These are for example five pairs of ostia in the muscular walls of the bee's heart. These are one-way valves allowing blood to enter the heart chambers as they are dilated at diastole but which close so that in contracting at systole the blood is propelled anteriorly.

**ounce**

A unit of weight standard as one sixteenth of a pound. Its metric equivalent is 28·349g or 1g equals ·03527oz. A fluid ounce is the volume of water that weighs one ounce.
See: fluid ounce.

**out-apiary**

A place away from home or base where some beehives are kept. The reasons for an out-apiary might include: securing matings with desired drones but excluding others, overcrowding at the base apiary (more than 25 colonies rarely yield well on ordinary sites), use of special facilities or forage, spreading a large operation over a good range of foraging areas. Associations and honey farmers should always be on the lookout for suitable sites where out-apiaries can be established.
See: apiary, hive site and site.

**out-doors**

Out of doors, in the open. This is where bee hives are normally kept so that the bees can range freely over the surrounding countryside (or foraging area). Some cover may be given to hives in the form of an open shelter or house apiary, but in the ordinary way hives remain permanently in the open exposed to all weathers.
See: house apiary.

**outfit**

This is a word used by the Americans where we might say 'firm' or 'undertaking'. It can refer to apparel but is more suggestive of 'equipping' a person to do something.
See: bee gear.

**outstanding**

Superior. This might well refer to a race or type of bee. When selections are made for breeding bear in mind that it is the genes of the drones and queens and not just the best honey yielder, because robbing, drifting, hybrid vigour and other factors can bias choice whereas longevity, disease resistance, docility and other transferable factors need to be taken into account.

**outyard**

While Americans use this term for an

out-apiary the English word 'yard' means something much smaller and usually surrounded by buildings close by.

**ovarioles**

Ovarian tubules that collectively form the ovaries of the queen (or laying worker). In a normal laying queen the paired ovaries each comprise one hundred and fifty or more such tubules or ovarioles. They begin anteriorly as a thin thread which increases in diameter as cells are budded off. These differentiate as they pass along the tubule becoming true egg cells, nurse or follicle cells. The eggs increase in size at the expense of the other cells to reach maturity at the position where the now wider tubules open into the lateral ducts of the oviduct. See: oviduct.

**ovary** ***

1. The female gonad or gland that is responsible for the production of (a) eggs, and in the case of honeybee queens (b) hormones which effect sex characteristics. The queen when in full lay has two huge ovaries which extend her abdomen and comprise some one hundred and eighty ovarioles (tubules) each.
2. It also means the future seed box or fruit of a flowering plant

**overall**

A loose protective garment comprising stout trousers with an extension covering the chest and having shoulder straps. Useful when doing work in the honey house or bee shed but not when working with the bees which requires the cover that a boiler suit or bee suit provides. We refer to the 'overall' size of a thing e.g. National hive 18" overall perimeter.
See: boiler suit.

**overall feeder**

A feeder built to go onto the hive as a complete box with the same periphery as the hive itself. Usually made from wood (though plastic versions are available and painted to make them waterproof. A slot running right across the feeder gives a wide access to the bees. It is a rapid feeder. Ashforth type has slot at one end so slight tilt on hive helps bees to completely empty it while the Miller has the slot across the centre allowing the slot to be positioned over the cluster.

**over crowded**

A colony is overcrowded when any part (particularly the brood nest) has an excess of colony members. Such overcrowding may induce swarming especially early in the season. Overcrowded supers are a pre-requisite for the production of honeycomb sections. Bees will often collect on the exterior walls of a hive or hang like a 'beard' before the entrance when overcrowded in warm weather. Viruses spread more easily when bees are overcrowded.

**over heating**

When brood nest temperatures rise above 35C brood may be damaged or die and this can become likely when bees become excited due to confinement as when they are being transported. Bees die at 50C but even in the hottest parts of the world bees rarely die in natural conditions provided they have water and are adequately ventilated. Comb collapse is another matter because it is already soft and fairly plastic at brood temperature. Honey has a high specific heat which helps bees avoid sudden fluctuations. However sustained over heating can cause a colony to drown in its own honey when combs give way. Honey begins to lose some of its natural characteristics the moment its temperature is raised. Even slight heating will cause it to lose some of its diastase and aromatic qualities. Once 120C is exceeded continued heating will lead to darkening, a rise in HMF content and caramelisation.

**overkill**

Killing or destroying a larger slice of the environment than is good or intended. No matter how careful man is in aiming death-dealing actions at identifiable and often minute targets he is apt to apply far stronger lethal and damaging doses than are strictly necessary in order to achieve his purpose. Accident, ignorance, carelessness and thoughtlessness can lead to overkill.

**over pollination**

Although under pollination is perhaps the more likely and serious fault, in an effort to secure a good top fruit set during dubious weather conditions, orchards were sometimes over-saturated with honeybee colonies. Should unusually good weather supervene, then an overset will result unless the colonies are removed when the work they are intended to do has been done. Over pollination where too many flowers are set, calls for 'thinning' (a costly operation) or fruit may develop that is unmarketably small.

**overstocking**

Having too many colonies in too small or unsuitable an area. It is generally considered in the UK that no more than 25 hives can be kept on one site if you have honey production in view. Obviously this is not true when productive sources such as heather or oilseed rape are considered. The number of hives should be limited according to the nectar potential unless deliberate overpopulation is provided for certain pollination requirements.
See: hives per acre and density of hives.

**overturn**

To upset or turn something over onto its side, roof or to cause a hive to fall over. This may be caused by large animals, gales or a falling tree or land slide. Hives standing on soft ground may overturn when the weight of supers increases. Careful selection of the apiary site and where necessary the locking together of all hive parts, or setting them side-by-side in blocks (usually with the entrances facing in different directions) as well as providing firm bases, will help to avoid this kind of catastrophe. Heavy rocks or stones on top of a roof may help but look ugly and can damage a hive.

**overwintering**

The requisites for safe wintering include a strong healthy stock of bees with a good queen, a sound sheltered hive with adequate, suitable stores and the entrance should be protected against mice. In exceptionally cold climates (Canada/Russia) an external covering may be put right over a hive though it is often easier to destroy the bees and start afresh with package bees each spring. A colony properly set up in September should survive with only occasional, external checks by the beekeeper until more space has to be given in the spring.
See: honey thermal conductivity and wintering.

**oviduct**

The posterior ends of each mass of matured ovarioles open into two lateral oviducts. These are broad and short and meet to form the median oviduct which opens into the vagina. They serve to transport and to nourish the eggs. As the name implies eggs are ducted through these from ovarioles where they originate, to the vagina where they may be fertilized by the queen (unless intended to be drones) before being laid.

**oviparous**

Creatures that lay eggs which are matured and hatched externally from their bodies. This of course applies to the honeybee. Their eggs require three day's incubation, yet, surprisingly, they can lie around for a considerable time like chicken's eggs before the hen begins to incubate them. It has been reported that eggs out-of-the-hive for forty days, proved viable when put back in the brood nest. Buckfast strain eggs were sent to America and successfully reared into queen bees there.
See: viviparous.

**oviposit**

The queen inspects a cell and if satisfied that it has been prepared for her, moves a few cells forward to straddle a position that enables her to insert her abdomen and

glue a tiny, sausage-shaped egg standing perpendicular to the base of the cell. Workers enter cells head-first so the queens sliding backwards gives the rim a different appearance which enables an observant beekeeper to find the area which she was last using. During the few moments that she takes to lay, she is surrounded by half-a-dozen or so attentive bees. Constant egg-laying requires constant feeding, so there are frequent pauses when the attendant nurse bees lavish their royal jelly upon the queen.
See: laying queen, drone eggs and oviposition.

**oviposition**
The laying and positioning of eggs. The final, neat, concentric ovals of brood arise from the queen's activities. She is attracted to that part of the nest where 'house bees' have prepared the cells. Not only are they clean and polished (it is quite easy for a beekeeper to see where it is next intended she should lay) but they are kept warm and will have been labelled in the bee's pheromonic language 'lay-here'. Because of her shape and size it is not possible for her to lay eggs one-by-one along a row but she must criss-cross the comb, a routine that requires strong legs and causes her to cover greater distances on foot than a workers flies in a lifetime!
See: oviposit.

**ovipositor**
A special organ at the end of the queen's abdomen which enables her to deposit eggs. In the case of the honeybee queen Dade tells us that clipping off the shaft of her sting does not interfere with the operation of egg laying.

**ovoid**
Egg-shaped – a solid oval. The brood nest is often described as having this shape.

**ovoid appendage**
This is adjacent to the mucous gland on the drone's genitalia.

**ovule**
A small egg or rudimentary seed. This structure is found in seed plants at the base of the style and is the part that develops into a seed after fertilization.
See: ovum.

**ovulate**
To lay an egg. Workers and unmated queens can produce eggs that may be developed into drones. The queen's ovulation is controlled by the worker's feeding her and the accessibility of suitable cells.
See: oviposition and reticulation.

**ovum**
The unfertilized cell, egg or seed containing a haploid nucleus. The female germ cell. The honeybee's egg is laid perpendicular to the cell base so that while embryonic development occurs within, air is free to circulate all around it. To become fertile an ovum must be fertilized
See: egg.

**ownership**
The criminal 'rustling' of stocks has led to the use of branding techniques. Ownership implies having the right of possession. This can be difficult sometimes if a swarm flies off. Ancient law insisted that the swarm remained the property of the original owner provided he kept it in sight. The changing of ownership in the past was usually accompanied by ritual-like behaviour. Crosses on trees where bees' nests were discovered are still to be found and respected. Bees were never paid for in anything as mundane as cash. There is still a peculiar, mistaken impression by some that the bees and so the beekeeper, are getting something for nothing. Who owns a hive that has been left in an old orchard? It could be a liability!

**oxalic acid**

A white crystalline dibasic acid $(COOH)_2 \cdot {}_2H_2O$. This substance was once put forward as a means of improving the colour of beeswax but to-day it has come into its own again in the realm of Varroa treatment.

**oxidization**

To combine with oxygen (to get rusty). An oxidizing agent is one that brings this process about. opp. 'reduction'

**oxygen $O_2$**

Found as a free gas constituting one fifth of the atmosphere. Responsible for life and combustion. To absorb oxygen as a chemical process is oxidation.

**oxymel**

The eyes of bees were burnt and compounded with honey to form oxymel a famous preparation useful for all sorts of eye disorders in mediaeval times.
Also the name of a beverage using vinegar, water and honey.

**oz.**

Abbreviation for an ounce.

# P

**P1**
The symbol used to denominate the first parental generation from which breeding experiments start. P2 would be the grand parents and so on. Offspring become F1, 2 and so on.
See: parents.

**pabulum**
That which nourishes plant or animal. Nourishment of the mind. Fit for food.
See: pap.

**pachystegia** ***_Olearia insignis_**
**Compositae**
A vigorous shrub with strong leathery leaves and downy shoots. It grows to 1m producing masses of large white flowers with yellow centres (2½ - 6cm) in late summer.

**package bees**
A combless package of bees sold by weight, about 4000 to the pound or 9000 to the kilo. Enclosed in a well-ventilated, gauze cage with a small feeder and the queen secure in a smaller cage through which the bees can feed her.. 18C is optimal for such bees. Thousands of such colonies are sent by road and air and Canadian beekeeping has been largely dependent on supplies of these from the southern USA. The advent of Colony Disappearing Disease caused Americans to buy Australian bees to help make up the deficit. Because Varroa, Foul brood, Nosema and other diseases are widespread in some foreign countries, Canadians buy from healthy sources and use the precautionary Terramycin and Fumidil B. In 1981/2 UK imports were only permitted from a few sources but moves took place to staunch this.
See: hiving package bees, shipping cage and shook swarm.

**packing bees**
Concerning wintering it has been said that the best packing for bees is bees but packing may take the form of insulation over the combs or around the hive. As packing can become wet and certain kinds are teased to pieces by the bees, it behoves you to consider what kinds of packing , if any, are suitable according to the circumstances.
See: package bees & pkg for winter

**packing hives for winter**
For external packing black is favoured as it absorb heat from the sun. Excellent non-heat conductive materials have been produced though consideration should always be given to toxicity, resistance to moisture, toughness to resist bees' attempts to remove it, cheapness and ease of handling. Since the advent of Varroa many beekeepers winter the bees over screens. Ideas on insulation over the crown board vary but top ventilation is important.

**packing sections**
These can be individually packed into cardboard boxes but when packed en masse no movement within the package is permissible as damage to the comb surface ruins the contents. Biscuit tins that hold layers of eight are useful. The above refers to the 4¼" square sections holding approximately one pound weight of honeycomb. Other sizes and shapes of sections call for varied packing

arrangements. They can be deep frozen and if they go in with clear honey they will come out that way even after a long period. While on the hives it is usual to ensure that they are packed warmly and 'packed' with bees

**packing tins**

The inexorable change in packing arrangements with ever more use of plastics makes many of the older famous 28 lb tins and smaller sizes out-of –fashion. Honey drums with special linings are probably still used for exports but as countries like China and Russia come into the picture it is likely that as with plastic honey buckets they too might convert to plastics.

**pad**

A cushion-like wad of some soft material used for the protection, comfort or the application of a volatile substance. Small rectangles of absorbent material are used for the application of Frow (acarine remedy). Where fumes are required to permeate a chamber from the top, the feed hole of a crownboard will take a small pad and once the fluid has been poured onto it, an air-tight container (bowl) can be set over it so that the fumes go below. See: wad

**Pagden system**

Put a swarm on the site it originated from after moving the parent colony aside. Originally contrived for fixed frame hives but quite suitable for hives to-day. See: Pagden.

**Pagoda tree *** *Sophora japonica* Leguminoseae**

Native to China. It is a deciduous legume liking alkaline, gravel, mineral soil having pinnate leaves with clusters of greenish-white flowers which are worked for nectar and pollen for several weeks in the summer.

**pain**

A distressful sensation of being hurt – an excessive stimulation of nerve endings or even trunks. The pain of a bee sting has been described as 'excruciating agony' or "*I hardly felt i*t" so pain varies with bee, season and individual and is rarely considered in isolation but in association with the buzzing noise and vibration, apprehension of what might happen and fear of receiving more stings. Even a really painful sting scarcely exceeds the pain of a severe pinch, yet when stung, the site of the sting can be pummelled without further pain and in healthy tissue there is no subsequent bruising.

**paint for beehives**

External quality paint or wood preservative (taking care to ensure no insecticide has been incorporated) can be used but the wood needs to continue to 'breathe'. White or red cedar are colours commonly used and from the appearance point of view, a repaint every year or so helps. For water-proofing feeders a dip in molten paraffin wax or the use of black, glossy Bitumos or the like. Always ensure that the smell of paint has completely disappeared before offering the hive to bees

**palaeontology**

The study of early life-forms as they existed in geological periods. The early stone-age fossils laid down 570-225 million years ago.

**Palestinian bees**

Bees of the Holy Land having many of the characteristics of Syrians though being rather more nervous and not very good tempered. The queens are prolific layers being long and slender and producing slightly smaller workers than European bees. As a race they are inclined to produce more laying workers especially after becoming queenless

**Palladius Roman writer AD 350**

Diverging very little from the Aristotle – Pliny – Columella tradition, he wrote in a condensed and practical fashion. There should be two or three entrances in a hive, no larger than the size of a bee; for narrow entry will thus obstruct harmful creatures. On the other hand they give bees alternative ways out if an entrance is besieged. He mentions the preserving of fruit in honey, the positioning of an apiary, bee plants, types of hive, in fact quite comprehensively.

**pallbearer bee**

A worker be that carries a dead bee from the hive Quite often such a bee will carry the corpse, even one as large as a drone, to some height and distance before dropping it. It has been observed that the same bee was seen to remove a dead corpse one week after first having been observed performing the same duty.
See: age-mates

**pallet**

A portable platform for storing or moving uniform objects – hives or boxes of honey jars. Used in conjunction with a fork-lift truck these are labour saving items that make the handling of various items quick and efficient and within the scope of one man.

**palm** *** *Salix caprea* **Salicaceae**

Goat willow a dioecious shrub (small tree) whose flowers arrive before the characteristic, tapered leaves. Reproduces easily from seed and colonises waste ground (likes damp woodland, scrubs and hedges). Also a tropical or sub-tropical variety, evergreen, colourful and picturesque, with branched trunks and a crown of palmate leaves. They make lovely flower arrangements blooming along with daffodils.
See: sallow and willow

**palma**

Mace refers to the forelegs basitarsus as the palma – the word is associated with palm as of the hand.

**palp pl. palpi**

An organ used to touch, guide or feel gently and with sensitivity, usually distal. Insects use them for feeding or for tactile or olfactory purposes. They are highly specialised and in the honeybee examples are the labial palps and the sting palps.

**palp–labial**

The pair of labial attachments articulated from the prementum. They are hairy, segmented and help to form the proboscal tube by combining with the maxillae. By comparison with the maxillae the labial palpi are quite slender.

**palp maxillary**

While the appendages of the mouth parts are well-developed on the wasp and the cockroach they are vestigial on the honeybee.

**palp sting**

The sting sheath is comprised of two soft extensions of the oblong plates about the same length as the shaft. The bee can use the sensitive palps to determine whether or not the material is suitable or worthy of an attack.

**palynology**

The scientific study of living and fossil pollen and spores. The analysis of substances to discover the identity of the pollen grains they contain and to learn more about their source of origin. Melissopalynology is the application of the above to the study of honey.

**pamphlet**

A small, unbound, printed publication somewhat larger than a leaflet. They sometimes run to a number of pages. Larger versions of advices in special covers are called bulletins. Special lectures are often reported in pamphlet form – BBKA, BIBBA etc.

**Pan** ***

An Arcadian God of mythology, half-man, half-goat son of Amalthaea, a goat nymph. Reared as a half-brother sharing milk and honey diet in Pan-guarded flocks, herds and beehives. Alder bark pipes cemented together with beeswax provided his musical instrument, the pan-pipes. A divinity whose protection Theocritus thought should be sort for beehives. Fraser.

**panel pin**
A small nail with a narrow head, used to attach ply or hardboard panels to a frame as when making crown, clear and 'split' boards, the assembly of frames etc.
See: gimp pin

**panicle**
A cluster of loose groups of flowers forming a composite raceme in which each branch has another raceme.

**pantoporate**
A word used to describe pollen grains having many pores. The 'beet' family Chenopodiaceae has pollen that is pantoporate.

**pantothenic acid $C_9H_{17}NO_5$ short version**
An oily acid (hydroxyl) found in plant and animal tissues. It is essential for cell growth – a component of the vitamin B complex and royal jelly is rich in this substance.

**pap**
Brood food – a mixture of honey, pollen and water fed to the older grubs by the youngest nurse bees – (Schofield).

**Papaver spp. *** Papaveraceae**
Showy sun-lovers, mostly short lived flowers with large seed capsules with intricate compartments that assist seed dispersal.
See: Californian, field and Iceland – under 'poppy'

**paper**
Mankind probably learned how to make paper by studying the behaviour of wasps when nest building. The vast qualities of paper that we have today and the fact that it can be stiffened or glazed by beeswax give ample room for thought and experimentation, However newspaper can fill a significant role in beekeeping. If used to start a smoker, crumble a small piece because a flat, half-burnt piece over the grid of the fire chamber hinders the passage of smoke – best lit before donning gloves and veil as these are flammable. Broadsheets provide amply large pieces for 'uniting' bees when two layers, pricked in a few places, separate two alien lots. It can deter wax moths when sealed around dry super combs. Corrugated sheets are clean to handle and will cut to shape (a roll can be sawn through) for use as a smoker cartridge though beware of unwanted ingredients ranging from anti-flame material to toxic chemicals.

**paper chromatography**
Mixtures are analysed by using special porous paper marked with indicators. The identity of the substance can be checked by the rate at which the compound in solution moves across the paper. A small spot of the solution of the mixture is placed near one end of a strip of absorbent paper. This end is placed in solvent mixture (often water and a mineral acid and butanol). This solvent soaks its way along the paper strip, causing the mixture's components to move with it. Because the component substances are different, they move at different rates and this enables them to be separately identified. This technique is used to isolate and identify the various substances, especially the sugars and allied carbohydrates in honey.
See: analysis, assay, chromatography and electrophoresis.

**papering**
To paper one colony onto another, in other words to unite two colonies together using the newspaper method. This is a common expression in NZ. At a distant out-apiary it might be more profitable to 'paper' two colonies together in the case of queen failure or other non-disease fault than to have to requeen.

Colonies united thus are –'newspapered'.
See: newspaper method.

**Pappus**
A mathematician who studied the shapes of honeycomb when he lived in Alexandria around AD 379 Dr F.M. Fraser.

**para**
A prefix indicating: aside, beside, beyond, amiss or near or something that protects or stops (guards against). Chemically a compound with benzene incorporated in its formula. Para dichlorobenzene PDB was much used against wax moth until found to be carcinogenic.

**paraffin oil**
A mixture of liquid hydrocarbons said to discourage earwigs but very liable to taint honey. Burns with a smoky flame though far less volatile than petrol. Boiling point in the range of 150 – 300C. Light, medicinal, liquid paraffin, clear oily liquid used as a laxative, dissolves beeswax and can be used for making skin creams.
See: cold cream and paraffin wax

**paraffin wax**
White wax or candle grease. $CnH_2n+_2$ – a white, translucent solid which melts in the range of 50 – 67C. The 58 – 60 is quite suitable though more expensive grades up to 67C can be obtained. (Beeswax melts between 62 – 65C). White wax is available in granular form for easy handling and weighing and it mixes in any proportions with beeswax in the making of candles, polishes etc. It is unsatisfactory in beeswax foundation and when burnt it has a smokier unpleasant aroma whereas beeswax candles smell sweet and look good. It is almost as inert as beeswax resisting most chemicals. Can be used as a lubricant (zips, drawers, curtain rails) and has long been used in liquid form to brush onto sections and frame lugs to discourage the application of propolis. In NZ it is used for a hot wax dip but the flash point of 400C has to be taken into account.
See: candles, hot dip and white wax.

**paraglossa pl. ae**
A pair of short lobes on the inner side of the labial palps. They form an important part in the function of the proboscis as they partially surround the base of the tongue allowing saliva to pass onto the tongue and effecting a tube-like seal when the proboscis is used for rapidly imbibing liquids. A pretty sight which a bee forthrightly displays.
See: salivary syringe.

**parallel radial extractor**
This consists of 3 banks of parallel combs arranged to rotate in a vertical plane around a horizontal spindle. It is like a normal radial extractor except that the rotating spindle is at ninety degrees to the vertical.

**paralysis**
A disease characterised by the crippling of power or activity. Honeybees have been observed to suffer from conditions that for want of deeper knowledge (which we are now obtaining through the use of the electron microscope) have been dubbed paralysis. Many of the poisons given to the bees in an effort to effect cures have led to apimyiasis, acute and chronic viruses which are being shown to link with other better known diseases, nosema for example.
See: black robber disease, ABPV, CBPV, hairless and viruses.

**parameral plates**
Two mirror-image plates such as the twin plates that form part of the sting structure and more especially those which flank the valve-like plates on the drone's copulatory organs.
See: quadrate plates.

**parameter**
A variable which is considered as a total spectrum so that the effects of other variations may be looked into. This mathematical term is used when discussing

say, nectar secretion, when one of the parameters would be the weather. That in itself is a variable but it is a parameter that can be set into perspective by other variables such as the soil conditions or type of plant.

**parasite**
An organism that lives in or with another organism, benefiting by way of food, shelter or often both, but not always to the detriment of the host, which is rarely killed. Types: 'obligate' – can only live on its host, 'facultative' – can earn its living in other ways, 'klepto' – one that robs another of its food.
See: honeybee parasites, commensal, inquiline, symbiont and parasitoid.

**parasitoid**
A parasite which eventually kills its prey (ichneumon).

**parent**
The father or mother. In breeding those denoted P1 and their first generation of progeny F1. The raising of queen cells for careful selection of the P1 generation and in the breeding of queens, the P1 has to be the result of painstaking selection of their forebears.
See: parent colony.

**parent colony**
The original colony. Splits, divides, nuclei, swarms, casts and other offsets are related to the colony from which they originated and this is referred to as the 'parent colony'. When a large swarm emerges from the parent colony it may become smaller than its swarm. The parent colony of necessity gets a new queen, has all the brood, the established comb and the safe nesting place.

**park** ***
An expansive area, usually landscaped and set aside for public use or benefit. Large National parks in some countries cover areas the size of Wales. Flowering plants, herbaceous borders, trees and shrubs make useful forage for bees in some instances. Sometimes referred to as 'gardens' e.g. Kew Gardens, or a park might include gardens. The public usually have free access though parts may be kept private as for example the setting up of a small apiary.

**Parliament**
The legislature of Great Britain, responsible for statutory instruments, orders, edicts and bills. For example: the disease of Bees Order 1962, Bees Bill 1979 and so on. Ministries become Departments, Coalition predominates now (2010) what next?

**parsley** *** *Petroselinum crispum*
A garden herb used to garnish or season food. Cross-pollination was found to double seed output in USA.

**parsnip ***Pastinaca Umbelliferae**
A good bee plant when allowed to flower from July to ;September. Its honey is mediocre and of light amber shade while the pollen is a greenish-yellow.

**partheno**
Words prefixed by this element have the meaning of 'virgin' quality or without fertilization.

**parthenocarpy**
The formation of seedless fruit without fertilization (from an unpollinated flower) as in the case of banana and pineapple but also to a lesser extent with some types of fruit grown in the UK. Fertilisation provides a stimulus for the production of growth hormones which result in fruit formation. There are possibilities that hormones can be formed as a result of other stimuli.

**parthenogenesis**
Reproduction without fertilization. The development of a new individual from an ovum that has not been fertilized. This happens in asexual generations of aphis and

is the normal way in which honeybee drones are produced. Such individuals are usually 'haploid'.

**parthenogenetic females**

This is an unusual characteristic relative to 'ploidy', whilst males are usually produced parthenogenetically, on rare occasions in South Africa for instance it is well known that queens can be raised there from laying worker eggs. It is due to the variation in the 'ploidy' of this variety of *A. mellifera*.

**partial acceptance**

The smooth running of a colony depends very much on a harmonious relationship between the queen and all the other bees. When a new queen is imposed on the bees by the beekeeper the initial antipathy must be fully overcome or a situation can arise wherein those bees that have accepted her frequently have to drive off other bees that make as if to harm her. To what extent her work and safety is impaired in such circumstances as this, will depend on the queen's compatibility and ability to conform to the colony requirements. Supersedure is very common in such circumstances.
See: acceptance

**Passion flower *** *Passiflora caerulea***
**Passifloraceae**

Tendrilled, woody climbers with short-lived summer flowers whose unusual structure has been linked with the Passion of Christ. The spiny centre – the crown of thorns, and the petals are said to represent the nails and the cross and 12 twelve apostles though two are missing (Judas and Peter). The edible fruit produced requires a good summer to ripen it.

**pasteurize**

To expose to a high temperature to destroy certain micro-organisms and to arrest fermentation. The temperature of 140F (60C) is often used with honey, not to pasteurize it but to clear granulated honey or to liquefy it sufficiently to pass more easily through filters.
See: heating honey.

**pastime**

An activity undertaken to make time pass in an acceptable fashion. If this becomes a regular activity it is presumably quite pleasurable though beekeeping not only involves pleasure but a great deal of work and so becomes a hobby or 'flank occupation'.
See: hobby.

**pasture**

The grass eaten by grazing animals and the ground suitable for this by virtue of a cover of herbage. Meadows and standing leys contributed enormously to the forage in days of yore. Sadly, modern methods of agriculture in GB seem to have given scant attention to the bee. Rye grass, spraying and cutting before full flower, are aspects of loss to the honeybee. In 2010 some attempts are being made to rectify this against a World food shortage.
See: meadow.

**Paterson's curse *** *Echium plantagineum***
**Boraginaceae**

Blue weed. It varies from viper's bugloss in that it has four stamens that project beyond adjacent parts while Paterson's curse has only two and it's corolla is red but changes quickly to purple blue. It naturalises well along with *Trifolium subterraneum* both providing valuable cattle forage. Its dark blue pollen has a high protein content and egg-shaped grains, while its honey is light and of good flavour. In 1979 attempts were made to eradicate it but were restrained in 1980. The leaf-miner *Dialectica scalariella* and flea beetle *Longitarsus echni* from the Iberian peninsula were to be biological controls. (they also attack borage, comfrey and heliotrope).Honey derived from this plant can contain Pyrrolizidine alkaloids.
See: biological control and Salvation Jane.

**pathogen**
A foreign inclusion (parasite) within an organism that can cause disease. Hence pathogenic. Within its limits a pathogen will evolve in the way that best secures its chances of survival, harmfulness towards its host is immaterial to it unless its chances of survival are changed thereby – Infectious diseases of the Honeybee Dr. L. Bailey

**pathology**
The science of the nature, causes and remedies of diseases together with the entire study of diseases.

**patchali**
This was a stand used by early beekeepers in western Crete to support hives clear of the ground which was often wet.
See: Crete.

**patrilineage**
Bees with the same father (from the same batch of sperms) constitute a 'paternity group' or patrilineage of full sisters while bees in different paternity groups of the colony are related as half-sisters.

**pattern of behaviour**
Individual bees are slow to alter their pattern of behaviour acting almost as if they had been programmed to perform a certain series of actions. Examples of this include:
(a) their persistence in returning to the same departure point for days after the entrance has been moved,
(b) a fanner will 'budge-over' if pushed with a finger to continue fanning ' never to attack.
(c) A forager who has lost pollen pellets at the entrance, still goes through the process of trying to slide both pellets into a cell.

**payload**
The weight of a returning forager minus the weight when it left the hive. The profit making part of the cargo transported. Like a small aeroplane a bee burns off fuel as it flies, but in the case of the bee, the fuel comes out of what would otherwise be the payload. It is interesting to consider the specials needs of a water carrier.
See: energetic reward.

**Pchelovodstvo (Apiculture)**
The official Russian journal on beekeeping. Russia has over ten million colonies of bees. The Russian word for 'bear' means honey lover.

**Pea *** *Pisum sativum* Leguminosae**
Of the many wild and cultivated varieties annual garden peas offer little to honeybees. However the everlasting or flat pea is quoted by Howes as being worked industriously for nectar. This is slow to establish itself but could form a useful part of any barrier at an out-apiary.

**Peace River**
Flows from the Rocky Mountains in Eastern British Columbia, north east through Alberta to the Slave River. It goes through some of the heaviest honey yielding bee pasturages in the world.

**Peach *** *Prunus persica* Rosaceae**
Native to China it is an early flowering tree requiring a warm habitat. Hand-pollination is often carried out as the blossom in March is too early for honeybees to be reliable agents for pollen transfer. The nectar too is not very attractive and bees are easily lured to contemporary sources such as dandelion. The pollen is yellow and honey delicately almond-flavoured while the fruit is sweet and juicy
with a textured pinkish-yellow skin, too tough for bees to puncture though they will take juices should a peach fall to the ground or have its skin pierced by a bird or a wasp.

**pea flower family *** Leguminosae**
Having papilionaceous flowers (with irregular corollas). Many of these are excellent bee-plants. Their seed pods split into two halves while the characteristic

flowers lack radial symmetry having a central 'standard', two 'wings' upon which the bee usually alights and two bottom petals, often joined, known as the 'keel pressure on which often triggers off a slap-on-the-back for the bee as the stigmas are tripped. Members include: beans, baptisia, bird's foot trefoil, broom, clover, goat's hue, gorse, greenweed, French honeysuckle, Judas tree, laburnum, lupin, pagoda tree, pea, rest harrow, Robinia and vetches.
See: Legume and leguminoseae.

**peak brood cycle**

Peak refers to the top of a curve, drawn to show the amount of brood in a colony on a time base. Under natural conditions (with 'near' native bees) a colony will adjust its development according to the weather and length of day but whether brought on with 'fits-and- starts' or by steady week by week improvement it will reach a peak and at that moment swarming becomes a certain possibility. If it –'platforms' out and replacement bees take over as others die, this can lead to a bumper crop but is more likely to occur when queens are young.
See: peak colony.

**peak colony**

Just as an individual bee's life will tend to follow the development of its various glands, so the colony as a whole is balanced by having the right proportions of bees of various ages. It was natural at first to think that as there was only one, long-lived, superior bee - the queen- that she must be in-charge. We now know that an interweaving of the various functions and the use of pheromones and food sharing all go to making up the colony's progress, always governed to a great extent by the weather. So although we can go 'part-way' towards governing colony development we cannot say this about the weather.

**peak colony strength**

This occurs when the maximum number of foraging bees are available. According to the type of bee and the district, this condition may continue for some time but once the effect of reduced egg-laying begins to result in a falling off in the number of foragers the colony will have passed its peak.
See: honey/nectar flow.

**peak egg-laying**

The moment when a queen reaches the point of maximum egg deposition. This will be governed by the strength of the colony, time of year and condition of the queen. Young queens are potentially heavier egg layers than older ones.
The curve of egg-laying usually has a flattened peak so that a good queen will lay at a high rate for days or even weeks.
See: colony peak.

**peak foraging**

This occurs when weather and flow conditions are right provided the colony concerned has reached the optimum number of foragers i.e. having a good queen, good health and no carnage through spraying. Our sometimes fickle weather can result in this happening in bursts.

**Pear** *** ***Pyrus communis*** **Rosaceae**

A tree that continues to go on yielding fruit into old age. Like the peach it is an early flowerer – March/April and the content of its nectar's sugar is low and variable and therefore unattractive. There are literally dozens of different varieties. The sweet, edible fruit has a characteristic shape, being rounded but tapered towards the stem. Its pollen has been described as: whitish, pale green and greenish-yellow. It is often grown against walls helping to form a local micro-climate. When obtained the honey is said to be dark with a strong nutty flavour.

**pearly-white honey/wax**

Dr. Eva Crane mentions in 'A book of honey' that the honey derived from 'morning glory' has this quality.
F. N. Howes says the Campanilla of Cuba produces honey equal to that of Lucerne or sage and the comb built from it is pearly

white yielding a wax as white as tallow.

**peck**
British Imperial dry measure of two gallons or a quarter of a bushel. Equal to 9·10 litres (7·57 US).

**pecten (rastellum)** ***
A comb-like assemblage of stiff hairs or spines. It assumes great significance as the distal edge of the hind legs tibiae. The left leg's pecten is combed through the pollen brushes on the basitarsus of the opposite hind leg. The pollen thus combed out falls onto the auricle (part of the 'pollen press') from whence it is squeezed upwards into the pollen basket.

**pectolase pectozyme**
An enzyme preparation that is sold for helping to break down the fruit in the preparation of a wine 'must'. Useful with pyment, cyser and other melomels.

**pedicel** ***
1. The stalk of an individual flower, fruit or inflorescence.
2. The proximal segment of the antenna which includes the organ of Johnston.
3. The slender link between thorax and abdomen though this is normally called the 'petiole'.

See: neck and organ of Johnston

**pedigree**
Ancestral line or line of decent. A valuable breeder queen will not only have a pedigree (past parentage) but will have evidence of having produced superior daughter queens (progeny).

***Pediculoides ventricosus***
The grain mite which is found in Australia. Its larvae leave tunnels under honey cappings which look remarkably like those of the bee louse – *Braula coeca*. The mite sucks juices from immature bees in capped-over cells.
See: Mediterranean flour moth and predators.

**pedology**
The science of the genesis and classification of soils – origin, characteristics and uses.
peduncle
The main stalk of a cluster of flowers, flower head or inflorescence. Also the small footstalk joining leaf to branch. In the case of insects like the bee it is an alternative name to 'petiole' for the waist (stalk) between the thorax and the abdomen. It is also used for a stalk-like bundle of nerve fibres.
See: petiole and pedicel and waist.

**peeping (or piping)**
Cotton referred to this long ago. It is a sound that only a queen can make. Normally she behaves more like a 'slave' of the workers but when a virgin emerges from its cell it can make this unusual squeaking sound which causes every bee to pay homage and stand still. She can thus move at will and if allowed to do so, will proceed to sting virgins that are still enclosed in their cells, their location being given away by their responsive 'squawks'. The sound is a clear note approximate to middle 'C', which can be heard well away from the hive. The Deaths Head can make this sound too.
See: piping.

**pelargonium** ***
Plants of the geranium family. Tender evergreen sub-shrubs which like sun or part shade and well-drained soil.
See: geranium.

**pellicle**
A silk-like or filmy protective covering. Sometimes remnants of the pupal skin adhere to young bees long after they have left the cell.

**penicillin**
An anti-biotic produced by moulds of the genus Penicillium, a green fungus. Also the

name for products made synthetically from Penicillin.
See: Penicillin waksmanii.

**Penicillin waksmanii**
Can be pumped by high pressure spray into combs – in situ. Used by Americans as a comb 'cleaner'. Claimed to break up AFB scale residues thus simplifying the work of the bees in removing scales. The hard black scales are full of AFB spores and the above procedure is not recommended.

**penis**
The male organ which plays the major role in the transfer of sperm from the male drone to the female queen. In the honeybee this organ is known as the endophallus as at rest it lies entirely within the abdomen and plays no part in the excretion of faeces. It comprises the long ejaculatory duct which leads into the bulb, the cervix follows and then the vestibule from which two cornua (horns) arise and it ends at the phallotreme. When a mature drone copulates a violent contraction causes the penis to evert and plunge into the female orifice.
See: bulb, cornua, endophallus, phallotreme and vestibule.

**pennyroyal 8***** ***Mentha pulegium***
**Labiatae**
A perennial herb with hairy leaves the source of a valuable oil, its small blue flowers are useful to bees throughout the summer months as are the other mints
See: mint.

**Penrose uncapper NZ**
A fully automatic uncapping machine which takes the frame by the lugs, uncaps both sides and has it ready for extracting. Gold medal at 25th Congress, Grenoble 1975.

**Peppermint** *** ***Mentha piperita***
**Labiatae**
Like pennyroyal it has downy leaves, yields an important oil. Purple or white flowers and the amber honey which improves with keeping as initially it has a noticeably minty flavour.

**pepper-pot**
Describing the appearance of brood where too many empty cells or cells of pollen or nectar or young brood are scattered amongst sealed cells. Queen replacement is suggested by this as the gaps can indicate a failure of the original egg to mature.
See: scattered brood.

**Peptide 101**
The pharmacological part of the venom reported to be ten times more effective than cortisone in the treatment of some kinds of arthritis and patented under this name.

**perambulate**
To walk about and to inspect. The bees most economical method of transporting itself and any load is to fly. Within the hive this is hardly possible so perambulation is the method by which movement occurs. The queen covers astonishing distances, - up to 100m per day when most active, but food sharing and the early off-loading of incoming material helps to minimise worker perambulation. A queen does not put her eggs into cells one-by-one consecutively in a straight line but zigzags around looking for cells she has not laid in. Six strong legs are an essential requisite.
See: walk.

**perception**
The awareness of the external world via the senses. Bees are able to use much the same senses as man though their need to specialize has made them far more sensitive than we are to certain aspects while less so in others.
See: appreciation of colour, light, olfactory, tactile.

**perennial**
Enduring, continuing indefinitely.
This can apply to a colony of honeybees.
A plant that grows for more than two

seasons. A herbaceous perennial is a plant whose leaves die down in the autumn but whose roots send up new shoots in the spring. Woody perennials continue to make growth thus becoming larger every year. Length of life will vary according to species, soil condition and climate.
See: annual, biennial and ephemeral.

**perforated cappings** ***
This refers to brood that has been sealed and subsequently torn open by the bees. It rarely occurs but when it does it is usually symptomatic of the serious disease known as Foul Brood (AFB). It can take the form of 'bald' brood but here the cappings are not so much perforated as eaten away by wax moth larvae. Robbing can show up when honey cappings are torn raggedly open. Normal sealing and unsealing is done neatly with a circular hole.

**perforated partition**
A division or separating screen through which air can pass or bees can pass food and queen substance. Usually vertical and enabling the queen to be confined to a particular comb or combs. Sometimes used to encourage the building of queen cells in the queenless part.

**perforated screen**
Gauze, plastic, perforated zinc and other materials are used in division boards or as completely framed screens for confining bees to the hive (give water supply if done to hold bees back during the active season) for travelling or for separating a colony into two or more parts. Bees are inclined to fill holes in the mesh with propolis and when not required for ventilation they are best covered with polythene which can be pulled off leaving the screen free of propolis

**perforextractor**
This device available from Thornes, enables heather honey to be extracted without destroying the combs. Firstly the comb is uncapped in the conventional manner and laid flat on a timber base board and held by metal clamps. The perforextractor, a block with evenly-spaced steel needles, is then pressed firmly all over the comb, the needles penetrating through to the board, continue over the whole face of the comb. The agitation renders the honey temporarily fluid so that it may be extracted immediately in a tangential extractor. Avoid using any pollen-clogged combs as they tend to break up.

**perfume**
It is unwise to assume that bees have the same reaction to perfumes as we do. In fact many embrocations, shampoos and skin dressings actually aggravate bees. Lemon balm and 'swarm lure' have been suggested as attractive to bees but when in doubt just aim for cleanliness. The smell of cut grass, perspiration and other items mentioned can cause the response from the bees to be less than pleasant. Peppermint which is not objectionable to the bees, can be used to disguise other smells as for example colony odour when uniting. Bees will scent-mark food sources that are lacking in aroma to help nest-mates find them
See: aroma and smell.

**peri**
Meaning about, around or beyond – periphery and perimeter.

**perianth** ***
The outer envelope of the flower enclosing both stamens and carpels. The external floral whorls, the calyx and the corolla.
See: botanical dictionary for variations mono/dicotyledons etc.

**pericarp**
The cell wall of an ovary after it has matured into a fruit such as nuts, pomes, drupes etc. and sometimes consists of the layers epi, meso and endo-carp. Its development is initiated and stimulated by the original action of pollination.
See: 'hysterisis'.

***Pericystis apis, alvei*** **-**
now referred to as *Bettsia alvei* genus Ascosphaeraceae – the pollen mould.

**periodicals**
Magazines, journals, newsletters, bulletins and other publications that are issued at intervals. Some might start as a weekly, become a monthly and finally a quarterly

**periopticon**
A ganglionic mass associated with the nerves of the eye and lying beneath the basilar membrane – the zone of the optic lobes.

**peripheral nervous system**
This is concerned primarily with the innervation of the sensory cells in the integument. The sensitive nerves serve the outer regions and extend inwards to the central nervous system. The network of nerve endings where the sense organs around the exo-skeleton collect information to pass to the brain for processing and co-ordination.
See: C.N.S. motor nerve and stomogastric.

**peristalsis**
Alternate waves of contraction induced in tubular organs such as the alimentary canal which can move the contents along and expose them to various processes. Peristaltic pumps have been developed to drive honey through tubing without incorporating air bubbles.

**peritrophic membrane**
peri – round, trophe – food. The thin lining to the ventriculus through which digestive juices and enzymes (secreted by the cellular layer of the ventricular wall) pass into the food and through which the products of digestion pass into the blood via the stomach wall. Thin membranes forming a delicate, cylindrical covering around the food mass within the ventriculus.

**permeability**
The degree to which a membrane will allow a molecule of a given kind to diffuse or pass through it. The composition of an organism and its secretions are very dependent on the variability of its biological membranes.
See: osmosis

**Permian rock**
Sedimentary deposits of the sixth and last period of the Palaeozoic era. Soils that were formed vary but where an excess of potash occurs and magnesium is deficient willow herb flourishes and provides a useful late crop for migratory beekeepers.

**perpetual**
Lasting for ever. Plants that go on flowering right through the growing season. These are rarely the heavy nectar yielding plants that produce positive 'flows' but are useful as steady suppliers of nectar and pollen. Trees and hedge plants come under this heading. In the S.W. hedges of Escallonia and fuschia are commonplace and their vigorous growth requires hard cut-back. Bees work them from early summer to the frosts.

**Persimmon *** *Diospyros kaki* (Japan) Ebenaceae**
A tree with plum-like fruits that are sweet and edible when thoroughly ripe. (Rich red and orange). Worked by bees on Japanese hillsides, grown in Australia and certain varieties are grown here in the UK but mainly for the attractive foliage. They like hot summers and the fruit (about the size of an orange) stays on the trees in the autumn, long after the leaves have fallen and it is extremely bitter until ripe.

**Perspex**
A clear acrylic plastic used in sheet form instead of glass. Its heat conductivity is less than glass and this makes it useful when used as a material in contact with the bees. However it scratches rather easily. For transparent crown boards or flat pieces to go over feed holes and possibly for observation hives, it can be useful. When cutting use a sharp, fine-toothed saw and avoid splitting.

Rough edges can be sand-papered. Try making an observation hive.

**perspiration**

A product of man's sweat glands which does not befriend man to bee (neither does breath). Remember this when mowing, concreting or working near the hives. Those who are inclined to perspire when working at the hives might see fit to use an elasticated band of towelling (such as tennis players wear) inside their veils.

**pests**

Any life-form that competes with man for shelter or food that is destructive, noxious or troublesome, a nuisance such as being a hazard to health. Many creatures are favoured in one respect and disliked in another. Wasps kill millions of caterpillars and houseflies but plague beach users and house wives. Honeybees depend on the education of the public to prevent their coming into the 'pest' category. Fortunately the attitude to bees and honey has undergone a remarkable change for the better of late (2011)

See: predators, weeds, pesticide and enemies.

**Pfund grader**

This is a device for measuring the colour of honey so that it can be allocated a class or grade. Whereas dark honey may sell as readily as light in the UK, in America and on world markets, top grade honey is the lightest and dark honey may well get fed back to the bees.

***Phacelia campanularia*** ***
**Hydrophyllaceae**

Other speciesinclude: tenacetifolia, viscida, whitlavia etc. A hardy annual with bright blue flowers that grows to 9" and has a long flowering period. It can be planted in orchards to improve local fauna or between rows of potatoes. It can be ploughed in as green manure and also used to entice bees to red clover and Lucerne. Best sown in late September for spring or in June for flowers eight weeks later Aug-Sept. Does well as bedding like catmint or for edging. The honey is light green with a fine flavour. Pollen from P. tanacetifolia is navy blue, drying to a lighter shade. P. campanularia is cream, light brown or greyish yellow. It is worked all day by the bees.

**pH value**

The degree of acidity or alkalinity of a substance is expressed as its pH value. This is a useful means of assessing soil types and many plants are sensitive to this characteristic. pH7 is neutral and the scale goes down to '0' for acidity and up to 14 for alkalinity. The concentration of hydrogen 'ions' increases or decreases ten times for each unit of pH change. The pH values of honey range between 3·2 and 4·5 showing its positive acidity.

**phage Gk. phagein – to eat (thus destroy)**

A word ending –suffix- meaning eating or devouring. An agent (bacteriophage) that causes the destruction of a micro-organism.

**phagocyte**

A blood cell capable of killing and digesting bacteria. They discharge anti-toxins and anti-bodies (agglutins) making bacteria more readily ingestible by other phagocytes. Not all leucocytes are capable of becoming phagocytes and lymphocytes cannot act in this way.

**phallotreme**

The drone's genital opening at the end of the vestibule. The penis valve.

**phallus**

The undifferentiated embryonic organ out of which the essential sex organ develops, the clitoris or penis or symbol of the male reproductive organ.

See: phallotreme.

**phanerogam**

A plant that has conspicuous flowers and true seeds – the opposite to cryptogam which has no apparent reproductive organs. No longer in technical usage.

**pharate**
Instar within previous cuticle prior to ecdysis (during morphosis).

**Pharmacia (GB) Ltd**
In 1980 pure venom preparations were available from this firm, as used for immunotherapy against bee stings. The treatment is for minute doses of venom to be injected subcutaneously at regular intervals, the dose being gradually increased until the recipient can tolerate 100mg of venom – the equivalent of two full stings.

**Pharmalgen**
Vaccine available 1980 and containing only pure wasp or bee venom for treating patients who are hyposensitive to stings, It could be had on NHS prescription.

**pharyngeal glands**
The hypopharyngeal or brood-food glands.

**pharynx**
This is the bee's gullet. It is continuous with the cibarium and thence the oesophagus.

**phenol (carbolic acid) $C_6H_5OH$**
Caustic and poisonous, used in resins, plastics and disinfectants. Highly repellent to bees even when very diluted.
See: carbolic cloth

**phenology**
The study and recording of cyclic happenings in the bio-sphere, such as flowering times, bird migration or plant or animal changes that are annually repeated at varying times due to seasonal or climatic influences. The study of isophenes.
Short for phenomenology.

**phenomenon**
As this word is used to cover the unusual as well as the unaccountable for many people a swarm of bees might well be described as such. There is something humbling and educational about a swarm of bees, though often when its capture requires expeditious and ingenious activity there is insufficient time to go into details with onlookers.

**phenotype**
A group of individuals whose hereditary characteristics (genotypes) have interacted with the environment to produce members of similar habit or appearance but not necessarily of the same genotype. A genotype remains stable whereas a phenotype changes continuously, thus a genotype will act jointly with its environment to develop into a phenotype.
The result of biological phenomena having altered (particularly appearance) a genotype.
The sum total of all characteristics.
See: genotype.

**pheromone (defence)**
One of the most significant aspects of a honeybee's defence mechanism is the establishment and recognition of colony odour. When the guard bees are alerted to attack a potential or actual aggressor, 'alarm-odour' also comes into play. When released it has a marked immediate effect, but does not persist.
See: alarm/colony odour.

**pheromone (alarm)**
Isopentyl acetate – released by scent gland and 2 heptanone from the mandibular gland. Alarm odour act as a repellent to foragers.

**pheromone (aggregation)**
Much work has been done in the 70's on honeybee pheromones and the trace chemicals responsible for this aspect of behaviour have been reported on by various scientific workers. In honeybees 9H2 is one responsible element.
See: pheromone (clustering).

**pheromone (clustering)**
Bees from a colony will tend to form into groups when they have no combs to cluster

on. The attraction is partly visual as can be demonstrated by their collecting on anything that looks like a mass of bees. It has been discovered however, that a pheromone has been exuded by the bees and that this acts as a powerful urge for them to assemble in one cluster. The dominance of the queen in this respect is notable. See: aggregation.

**pheromone (foot-print)**

As its name implies this trace chemical left behind by a bee when its feet touch certain surfaces, can help other bees to follow the track taken. This is significant relative to hive and nest entrances and the constant tramping of the bees towards their flight hole often develops into a visible mark. See: footprint pheromone.

**pheromone (forage marking)**

A volatile pheromone left at a food source by foragers. It induces others to alight. That from the dorsal abdominal surface is more effective than that from the thorax.

**pheromone (mutual recognition)**

The ability to identify an invader is a corollary of the bees of a colony being able to recognize one another. Since senses other than visual come into their own in the darkness of the hive, pheromones play a large part in the ability of the bees to mutually recognize one another. Hence the name 'mutual recognition pheromone' is used to label this aspect of their pheromonal behaviour.

**pheromone (sex)**

Released by an airborne nubile queen.

**pheromone (trail marking)**

Odour marks or trails used by termites, ants and bumble bees.

***Philanthus triangulum* Hymenoptera**

A wasp, little more than ½" long and slender. The bee killer, bee wolf, digger wasp. Found in Europe and N. America. Honeybee pest in Japan. It likes sandy soil and cracks in pavement. Parasitic wasps may provide biological control. Recent reports of the bee-wolf say it feeds almost exclusively on honeybees, taking them at the flowers but they become relatively inactive at ambient air temperature below 18C.

**Philippines (7000 islands)**

Heavily forested mountains. A. mellifera brought by colonists in 1639. Today they usually fizzle out due to Varroa while the other 3 races of honeybee survive.

**Philiscus of Thasos**

Called Agrius – kept bees in uninhabited country and wrote book (unfortunately lost) – said Pliny.

**phloem sap**

Nectar is largely derived from the phloem sap which is the watery fluid circulating through a plant and carrying food etc. to the tissues.
See: photosynthesis.

**phoresy (phoresia)**

Phoretic hitch-hiking, the manner in which mites like Varroa are able to get onto bees. The use of one organism by another for transport without parasitism. Apis florea carry a scavenging creature (large for its size) which symbiotically does useful work of clearing unwanted material.

***Phormium tenax***

Widespread in areas of heavy rainfall in NZ.
See: NZ flax.

**phosphate**

Usually sodium phosphate but any salt or ester of phosphorous – the 'P' element in NKP fertilizer. An important buffer in cell sap absorbed rapidly by actively growing plants.

**phospholipase**

The main allergen in bee venom. It causes failure of cellular functions and releases

histamine from cells. A haemo-toxin that destroys blood cells. It combines with lipids in the cells to form a potent toxin. Found in bee and snake venoms.
See: hyaluronidase and venom components.

**phosphorous (P)**
Solid, non-metallic element. Yellow phosphorous is poisonous, flammable and luminous in the dark. Red phosphorous less flammable and poisonous. Organophosphorus esters (OP) used in many insecticides.

**Phostoxin Aluminium phosphide**
Sold in tablets in a re-sealable canister. then released they give off poisonous phosphine gas that has a strong carbide smell. One tablet placed over a dozen supers kills all stages of wax moth. It is widely used in grain mills, warehouses, etc.

**photography**
The production of images by the chemical action of light and other radiant energies on sensitive surfaces. Photographs, transparencies were most useful to illustrate talks or articles. The advances that have taken place since this paragraph was written have now opened up a complete new world of facilities combining digital cameras with computers and means of display.

**photomicrograph**
A photograph taken through a microscope. As the bee is so small, yet highly intricate and specialized, the study of details of its structure is greatly enhanced by this use of Bees sometimes fail to use plastic photomicrography. This has now been taken to greater depths by the use of the electron microscope.
See: SEM and TEM

**photopositive**
Showing a reaction to light. Bees become photopositive when they begin to forage, previous to which they are accustomed to the dark of the hive – photo-response (Berthold, Benton 1970)

**photosynthesis**
The synthesis of complex organic materials by plants using carbon dioxide, water and inorganic salts. Sunlight is the source of energy and a common catalyst is green pigmented chlorophyll. It is the plant's method of storing energy from the sun. Sunlight, water and carbon dioxide create sugars in the chloroplasts.

**phragma pl phragmae**
Ingrowths of the exoskeleton which take the form of strengthened plate edges. These stiffen the shell and provide anchorage for the muscles. The longitudinal wing muscles for example are attached to the first and second phragmae of the thorax.
See: apodema, scutellum and tentorium

**phylogeny**
The development of a race or of a type of plant or animal

**phylum pl phyla**
A race of organisms descended from common ancestors. This is a major group in the classification of animals and the equivalent of a division in plant taxonomy.

**physical properties of honey, nectar, sugar syrup, wax etc.**
See: headings by these names

**physics**
The science dealing with matter and energy in its various states and processes. Closely linked with mathematics and listed with that science under SI – subject index. A physicist is a scientist who specialises in physics.

**phytoinhibitory Phyto – plant.**
Inhibition of plant growth. The phytotoxic activity found in propolis extracts were not due to galangine but an aqueous extract of

propolis inhibited the sprouting of potato tubers and the growth of lettuce and rice grains. Russians say, also cannabis sativa.

**Pi pl pis.**
The Gk for the ratio of the circumference of a circle to its diameter. It will not work out to an exact number though it has an exact value that approximates to 3.141592 To calculate the speed at which a comb is rotating around the extractor; 60 x 2 R x rpm = distance per hour.

**Piana Gaetana Italy.**
He established a large queen-rearing enterprise at a time when Italian queens were being demanded over much of the bee keeping world. After his death in 1937 his sons, Guilio and Gian continued to expand it. For many years they also published: *L'apicoltore d'Italia.*

**Piast**
Beekeeper who became king of Poland in AD 824.

**(PIB) BIBBA code for 'pollen in amongst brood'**
This is an indication of a break in the laying cycle or bees that take advantage of the queen's brood being scattered. It may in some cases be a genetic trait.

**Picea sp Pinaceae**
Hardy evergreen trees that dislike air pollution. The many varieties include some beautifully graceful and handsome specimens and although their pollen is powdery and wafted by the wind (of little use to bees), they do in many cases, give rise to the hosting of sucking insects that yield honeydew.
See: spruce

**piece-meal, piece by piece**
When anything can be broken up into small pieces and carried by individual bees out of the hive, this is the way that they go about keeping their hive clean. A small piece of paper dropped into an observation hive, provides an interesting lesson. Larvae, dead from EFB, creatures like mice and bumbles, too large for a single bee to handle, newspaper and expanded polystyrene – these things come out of the hive 'piece meal'.

**Pickard R. S. Phd BSc.**
One time head of the Bee Research Unit, Dept of Zoology, University of Cardiff.
He wrote 'Honey bee biology with man in mind'. He has lectured to the CAB, on 'The honeybee brain and bees, magnetism and electricity'. June '76 and Feb '77. At Cleppa Park he ran a field station where queens were specially bred.
Senior lecturer in Neurophysiology.
See: Diploma in Apiculture

**pickled brood**
See: sac brood

**pickled pollen**
Pollen stored for future use by the bees. Fresh pollen is far more nutritious than pollen that has been stored for any length of time or than any substitute. Nevertheless they have to go for long periods when brood has to be fed, yet there is either no pollen available or the weather is quite unsuitable. Just as nectar is being processed from the moment it is collected, so too is the pollen. The 'treated pollen' is rammed into cells and first covered by a film of honey. If it is to be kept for a longer period, say at the onset of winter, more honey will go on top and it will be finally sealed over, looking for all the world like sealed honey. Either in the temporary 'sticky' condition or covered right over, this pickled pollen is less vulnerable to the pollen mite and does not ferment in that condition.

**Pierco foundation**
Plastic foundation (sheets embossed with suitable hexagons), is made by Pierco. Inc. Upland, California, USA. It is covered with a thin layer of beeswax. In 1982 successful results were achieved.

See: Arnaba and Stapla

**piginent**
A spiced wine with honey

**pigment**
Natural colouring matter in plants and animals. A dry substance usually pulverised, which when added to a liquid medium that holds it in suspension becomes paint, ink or dye. Some pollens are so rich in pigment that the comb, honey and adjacent wood in the hive becomes stained. In the case of sainfoin this is saffron yellow. Pigments and their significance to bees and their honey follow

**pigmentation (chitin)**
Much of the chitin on a honeybee is black. This is a good colour for heat absorption. Many bees from warmer climates have yellow bands; in fact their abdominal segments can appear almost transparent when they are viewed against the light. Legs too, where heat absorption is not a good thing, are often in lighter shades on some bees. The colouration of bumble bees is largely from the pigmentation of the body hairs rather than their chitin. Queens have been described as having 'red legs'
See: next entry

**pigmentation (eyes)**
The compound eyes are only able to transmit the thousands of individual images to the brain because of the special pigment cells that surround each crystalline cone and its rhabdomic extension.

**pigments**
See: honey pigments

**pilosity**
Hairyness. Pilose means covered with hair, soft hair, furry or downy.

**pinching**
The destruction of any developing queen cells, by squeezing the cell between finger and thumb. Queen cell pinching is a time consuming process and is not likely to be a deterrent to swarming, though if thoroughly carried out (a process that causes a major disturbance within a colony), it can delay the departure of a swarm.

**Pine Pinus sp. Pinaceae**
They have needle shaped leaves in clusters and produce no nectar and the pollen is not suitable for bees. However crops of high quality honeydew honey are obtained when wood ants are able to nurture certain 'lachnids' at the trees expense. The timber is used for hives (though not so much in the UK where red cedar is the vogue).
See: Scot's Pine, Pine honeydew, tree honey

**pine honeydew**
Baumhonig, Waldhonig, Tree Honey. The European pine forests yield great quantities of honeydew. This is brought about by various sucking insects which are themselves distributed and protected by wood ants. The phloem sap that they take up passes through special filter chambers in the insects and they then exude the sweet excess which is collected and transformed into honey by the bees. In many countries where beekeepers move their hives to the forest they reap the late but welcome harvest; Austria, France, Germany, Greece, Spain and Switzerland and others.

**pingers.**
'They (the bees) shoot up as soon as the crown board is removed 'ping, ping, ping' on the wire mesh of my veil' – Rex Boys BBJ

**pink brood**
Pigmentation from certain nectars and pollens can bring about a pinkish tinge to otherwise normal, healthy larvae.
See: coloured brood

**pink honey**
Heather honey often receives an inclusion of the mahogany coloured Erica honey which gives it a slightly pinkish hue.

**pink wax**
Possibly due to the same pigment as above, the wax rendered from the comb worked by the bees at the heather often has a delicate pink colouration.

**pint**
One eighth of a gallon, 20 fl oz. or 0.56825 litres. A widespread Imperial measure until the advent of S.I. in the UK (Only 16 fl oz. in the USA) A litre equals roughly 1¾ pints.

**pipe**
A tubular conveyance for fluid or gas. When used as an entrance to the observation hive clear plastic tubing can be useful, though it should have at least ½" internal diameter (preferably 1" or 2.5 cm) and it needs to be roughened on the inside or to have a cord inserted if the bees have to walk 'uphill' Condensation within the tube may cause problems and a small hole or two drilled near the lowest point may help surplus moisture to escape.
See: plastic piping

**pipe**
Queen cell (by common people) (Butler) DG A queen cell might well have been likened to the inverted bowl of a smoker's clay pipe

**pipe cover queen cage**
A domed, circular protective screen strengthened at the rim about 5cm across. It can be pressed into the comb face and so enclose the queen over some food with or without attendants. Similar in appearance to an old fashioned smoker's pipe covers. It is a suitable thing to use when uniting bees and a valuable queen has to be safeguarded. The attitude of the bees to the queen can be judges by their behaviour on the surface of the screen. Should they show signs of aggressively trying to get at her, delay setting her free. After a day or two the top of the cage may well have been joined by the bees to the adjacent comb face so take especial care when prising the two combs apart that you do not pull the cage open inadvertently!

**piping**
A sound peculiar to the queen when there are rivals in the offing. It has a strikingly pure tone of 320 – 340 cps (Woods1950) and it is produced in short, intermittent bleeps. Cotton called it peeping and seeping was another term used. Virgins sometimes reply with a dull squawk or quack from inside their cells (323 cps) both sounds are thought to cause bees to delay the release of more virgins. It is a substrate-borne sound; Prof Free says 'queen produces the piping sound by contracting its flight muscles with its wings folded. This gives a frequency vibration of 435 – 493 cps (double that produced when the wings are spread), at the same time she presses her thorax against the comb surface and communicates the vibration to bees in contact with the comb (and of course queen cells)
The comb is thus used in the manner of a tuning fork.
See: substrate

**Pirimicarb**
A carbamate insecticide which although harmful to livestock, has a short persistence and may not harm bees as much as some similar formulations. Available as dispersible grains, wettable powders or smokes.
See: Aphox

**Piriton (chlorpheniramine)**
A bee sting remedy to be taken in tablet form 4 times per day. If a beekeeper is sensitive to 'beedust' one of these antihistamine tablets should be taken one hour before opening a hive, but for some the substance may be excessively sedative.
See: Triludan

**pistil**
The female part of a flower consisting when complete of an ovary, style and stigma. Flowers vary a lot with regard to

the positioning of these organs. Once virile pollen has fastened on to the stigma at the distal end of the pistil pollen is then able to send its tube down through the style and the male nucleus of pollen fuses with the nucleus of the ovule to initiate seed production. That is fertilisation.

**pistillate**
A flower, frequently of a dioecious plant, such as kiwi fruit, that bears only female reproductive organs – does not produce viable pollen.
See: staminate

**pit**
A fairly deep hole in the ground made for some purpose such as the disposal of contaminated bee material. Any smell of wax, propolis, honey or traces of bee pheromones, will go on attracting bees and if the material is diseased, this can be dangerous. Consequently any pit that is dug to get rid of such items should allow plenty of space so that a deep layer of earth can completely cover everything afterwards. Needless to say when burning AFB, combs etc. the pit should be quite dry.
See: pit hairs and pegs and method of destruction.

**pitch and roll**
To pitch means to alter a thing's attitude in the fore and aft line (to climb or dive) To roll implies a change in the lateral plane i.e. left or right wing down
See: dorsal light reaction

**pitching**
A swarm settling –term used by Sydserff in Cotton DG. A swarm would have been described as having pitched upon a branch or a post for instance.

**pit hairs**
Sense hairs of the antennae which are buried in pits with their tips just below the narrow opening to the pit. They are fine bristles arising from the cuticle.

**pit peg**
Also referred to as a champagne –cork organ (due to their shape)These too are found on the antennae, particularly towards the tips and are, like the pit hairs described above thought to be associated with the sense of smell and possibly taste.
See champagne pit hair

**placebo**
A neutral substance or method that is either used as a control or for its psychological effects in the treatment of a state, disease, or condition. A medicine that has a psychological rather than a physiological function
When a small child is stung adult sympathy and attention may be far more important than any actual substance applied.

**placoid sensillae**
Plate-like organs found on the sides of the antennae which face outwards, especially on the eight terminal segments. In many insects these detect pressure changes and could be associated with flying speed or the checking of wind speed before attempting to leave the hive.
See: sense plates

**plan**
A drawing made to scale so that a design can be converted in to a hive, feeder, special floor or other piece of equipment. Also a scheme of anticipated work aimed at completing a certain activity successfully. BBKA BIBBA and others have made useful plans available for mini-nucs, solar wax extractors and so on.
See: planning and preparation

**plane**
A flat or level surface that contains all straight lines connecting any two points on it. We speak of an optically plane surface or a plane of light like the following;

**plane of polarised light**
The state in which the rays of light are

transmitted through a substance and confined to a single plane (see above) The measurement of the angle of this plane as the light enters, and again when it leaves any translucent substance tells us something about the nature of the substance. For example the difference between sugars such as fructose, glucose or sucrose.
See: polarised light, Polaroid.

**Plane tree platanus**
There are many varieties and hybrids. The London plane trees coped in the days of smog and atmospheric pollution by the fashion in which they shed their bark. They have ball shaped fruit clusters and large leaves with pointed lobes.
See: sycamore

**planing**
Single bees or rows of them perform rhythmic dances with heads directed downwards and tongues half unfolded, looking as if they were sweeping or polishing. Observed by Mehring in 1866 and cited by Morgenthaler in 1931, mentioned by Ribbands.
See: dances

**planning**
Operations with bees require fore thought and preparation. Having decided what condition a colony is likely to be in and what items of equipment might be required so that it will need no more attention until the next planned visit we need a plan. We can draw up a list of items that will be wanted and set out the various things to be done in a sensible and logical order. By such means many a snag can be avoided and the best possible use made of the time spent. Planning one's work and working one's plan is a great aid to successful beekeeping.
See: plan and preparation

**planont**
A wandering stage during which a threadlike polar filament leaves a Nosema spore to penetrate between the cells of the wall of the ventriculus where they usually enter epithelial cells.
See: meront

**plant**
Any living organism that is a member of the plant kingdom. Its cell walls are of cellulose, it will lack power of locomotion and have no nervous tissue but be able to synthesise inorganic substances. Herbs are the non-woody plants that die back to the ground each year. Shrubs merge into the smaller trees and these are woody, forming more permanent structures above the ground, the biggest trees being the largest and oldest living things on the planet. Trees may be evergreen or deciduous. Herbs may be annuals, biennials or perennials.
See: plant kingdom.

**Planta Dr A. von Swiss 1884.**
He showed that bees use pollen with the wax when capping over the brood cells.

**plantar**
L= sole of foot, the first tarsal joint of insects

**plant- honey extracting**
Extractors that take large numbers of frames(21 or more) are usually of the radial type, using a vertical spindle. No matter how large the primary extractor is, it is often found useful to have a secondary unit, albeit smaller as this enables continuous extraction without any pauses while one is being loaded or unloaded. Special apparatus for dealing with large quantities of honey-loaded 'cappings' are also important, as are filters and the means of keeping honey 'on the move'. Cleanliness and ease of operation are important.
See: honey house

**plant kingdom**
One of the main divisions of the living world including : algae, bryophytes, pteridophytes, seed plants and fungi.

**plants (slides)**

Important to honeybees. 80.Ohio with cassette or lecture notes N. American flowers though many grow in UK L and botanical names. From BBKA

**plants (slides)**

Pictures for tracing. 44 Encyclopaedia Britannica with captions. Plants and trees in colour with drawings alongside to assist in making visual aids. From BBKA

***Plants and beekeeping* (book)**

Dr F N Howes. Publisher: Faber and Faber, 3 Queen Square, London WC1N 3AU. Reprint 1979. Available from NBB.

***Plants for bees***

A calendar of bee plants by Dorothy Hodges, author: *Pollen loads of the Honeybee.* From IBRA

**plant – wax rendering.**

The apparatus required for getting clean beeswax from old comb, wax scraps or cappings is varied. All depend on bringing the wax into a liquid state so that it can be separated from the slum gum (dross) and filtered before casting into sheets or ingots. Hot water, steam, solar heat together with centrifuges, filters and presses are combined or used individually in various ways.
See: solar wax extractor, wax extractor and rendering

**plasmolysis**

Contraction of the protoplasm in a living cell when water is removed from it by exosmosis

**plastic**

Concerned with or pertaining to moulding or modelling. It refers to the physical property of permitting flow or deformation rather than any chemical property. Wax becomes plastic (soft) at 85 F (30C). The introduction of synthetic plastics is slowly revolutionising beekeeping equipment. Lead by Thornes, British manufacturers now make so many useful items that a list would not be possible here. Several items follow though. Some of the thermoplastics: PVC and poly-(ethylene, propylene, and styrene)

**plastic bag**

Flimsy polythene bags of various sizes and thicknesses have many uses in beekeeping. Waterproof and airtight, they can make clean linings or hold scrap comb, and a roll of them (Snappies) can be slipped into the apiary box. When opening a queen cage to release workers this can be done with the hands inside a plastic bag so that no one can escape. Notices or notes can go in so as to remain bee and weather proof. The list is endless.

**plastic container**

For honey, cold cream, beeswax, in fact it's a 'you name it' situation for a multitude of plastic containers that come in all descriptions. 50 ml white Min. of Health plastic cold cream containers can be obtained from A. W. Gregory Ltd. Glynde house, Glynde Street, London. SE41RY and other suppliers. Plastic honey bottles, squeeze tubes, honey tanks and the rest are made by others.

**plastic disc entrance.**

This has been dealt with under disc entrance but is another example of a very versatile object that has become possible due to the weatherproof, mouldable properties of plastics.

**plastic equipment**

Many new items have come on to the market, all the better for being made out of plastic. A short list is given and one well designed article Steele and Brodie – a capping strainer in heavy duty polythene. The holes in the cross bar accept frame lugs, while the deep, easy to clean container will hold a considerable quantity of cappings and is easy to handle. Cages, escapes, excluders, feeders, frames, gloves, helmets, porters, screens, sections, spacers, taps and veils all now come in the plastic form. Naturally there are occasional snags. Black stains can form in feeders and some types of plastic seem repellent to bees and often they prefer wood.

**plastic excluder**

Like other metal parts used in the hive the slotted zinc queen excluder now has its counterpart in a less flexible plastic form. Bees do not seem entirely happy with them (any more than they are with metal excluders) and it is advisable to ensure that they are quite smooth and to get the bees used to them (i.e. put them on top of the frames for the bees to lick and make acceptable before forcing the bees to pass' through the slots' ). Framing is still a good idea as it ensures correct bee space and makes the excluders easier to handle including putting them on or taking them off. Especially using hives with top spacing. See: queen excluder.

**plastic feeder.**

The use of hygienic, easy to clean material is fine for feeders. The bucket type is another obtainable from suppliers such as Steele and Brodie and others. It is of large capacity but cannot be refilled on the hive. When placed over a feed hole it provides its own bee space so that the bees can get at the ring of small holes.
See: bucket feeders.

**plastic foundation**

Attempts to make entire combs from plastic have met with some success but the bees still differentiate between the two substances and seem to prefer their own. With strong colonies in' flow conditions' some very satisfactory results have been obtained. The plastic sheets and laminates covered with a thin wax film are improving all the time. Thornes and other suppliers offer several kinds.
See: Arnaba, Pierco, Ready comb, and Stapla.

**plastic frame**

Frames and sections made from plastic instead of wood have come on to the market since the 70's. One firm who advertise them as requiring no nails is Foster Magneto Co.

**plastic hat**

Strong, lightweight helmets are available with adjustable internal fittings and a firm brim for supporting a veil. Tough, durable plastic will stand up to hard wear and tear and they are cool enough if there are ventilation holes and a good airspace between the head and hat. They do however take up more space than the collapsible hat.

**plastic hives.**

Tough closed-cell foam (isocyanuric) GRP (glass reinforced polyester) are claimed to have 30 years durability and to require no painting yet can be scraped and washed clean. In America Kelleys have produced quite a good one in Langstroth dimensions while in Britain, Honey Home have made a good substitute for the national. The high cost of an initial female mould makes it highly necessary to standardise as the cost tumbles when very large numbers of the same type can be sold. Heat, ultraviolet light and other environmental conditions can cause them 'pit', distort and deteriorate but many are still very useable after 10 years and fulfil all the demands made of them. The smooth non-absorptive walls can harbour moisture and encourage slugs and creatures that are rarely found in wooden hives. The likelihood of timber scarcity and price variations probably suggest that plastic hives have come to stay.

**plastic honey tank**

Generally speaking these are lighter and easier to wash and to handle than metal tanks. It is easier too, to give them a tapering shape for easy cleaning. Whereas a honey/water mix can be most damaging to metal containers, plastic seems to be a very suitable substance to keep in contact with honey and to clean afterwards. They can take a special filter or a cloth can be tied over the top. Various shapes and sizes are made.

**Plasticore**
A sheet of strong film punched with tiny holes and coated on both sides with pure beeswax. The holes have the effect of riveting the beeswax through the film. Two large holes are left in the bottom corners for the bees to use as pop holes as there is no way that the plastic can be chewed by their mandibles. The finished sheets are bound at each end by metal strips for firm and easy insertion into frames. Fresh wax and a good flow enable the bees to draw Plasticore into very superior combs.

**plastic piping**
Flexible transparent piping (tubing) is very useful to provide entrance tunnels for observation hives and can be obtained from various sources.

**plastic spacer clip**
The cost of a wooden frame plus a clip to convert it to Hoffman sidebars is probably more than a Hoffman frame. However as a lot of people have existing frames and wish to convert them to Hoffman spacing these clips will do this. They are however best inserted inside the frame under the top bar before the foundation is put in. The old fashioned 'metal end' has been replicated in plastic form. It is not easy to enthuse over them.

**plastic veil**
Black plastic screens bound in tough canvas make good veils provided the hot smoker is kept well away from them. The bees take more kindly to the smooth plastic used than to many forms of black netting material.
See: Sherriff

**plate**
A thin flat layer or scale - a plate-like organ. E.g. parameral plates, pleurites etc.

**Plato**
Greek philosopher 427 – 347 BC
Known as the 'Athenian Bee' A great admirer of Socrates. His eloquence to have been assured when, as an infant, a swarm of bees settle on his mouth. This seal of honeyed eloquence was repeated on several notable members of subsequent generations. 'Clever little bees'

**play flights**
Young bees make short flights keeping close to or within sight of the hive entrance. They come out in the middle of the day especially in warm conditions and can be so numerous as to create the impression that the colony is swarming. The activity does not as a rule last for long and bees hover facing the entrance and orientating themselves to their surroundings and possibly cleansing themselves. The visual images recorded in their brain programmes them for an accurate return to their home when they subsequently forage.
See: first flight, orientation.

**pleach**
To plait or interlace branches or twigs as in the structure of ancient hives. Which were then cloomed, or in the making of a hedge or windbreak.
See: wattle

**Pleiades (Seven sisters)** ***
A conspicuous group of stars in the shoulder of the constellation of Taurus, the Bull. A cluster of over 2000 stars of which six or seven are easily visible to the naked eye. Seasons and time for various actions in the beekeeping world were always related to the natural calendar of the skies.

**pleiotropy**
The various effects from a single genetic factor. Interaction between a species and its living conditions, e.g. acclimatisation, 'polymorphism'.

**pleurite pleuro (the side)**
The two lateral plates which together with the dorsal tergite and ventral tergite form the outer casing of the thoracic segments

**pleuron pl pleura Gk pleura 'rib'**
An external lateral piece of a thoracic segment referred to more precisely by Dade as pleurites. (seat of the forelegs).

**Plexiglas**
US trademark for a transparent substitute for glass – polymethyl methacrylate.
See: Perspex.

**plinth.**
Normally used for a thing made of stone but also used for the peripheral projection that allows for the rain to run off at the base of a beehive lift. Many of the early wooden hives were made with a plinth or a 'fillet' of wood in the form of a bevelled strip, cut to allow it to just overlap the box or floorboard on which it was set. The underside would be channelled so that the rain would drip off before touching the hive walls. A major snag was the difficulty that ensued when a beekeeper tried to insert a hive tool, to separate propolised boxes as the slit between them would be overlapped by the plinth.

**Pliny the Elder A.D.23-79**
A Roman naturalist and prolific writer who believed that honey should be included in one's daily diet to promote health and longevity. Like his contemporaries he had rather fanciful notions about the economy of the hive. 'Honey was engendered from the air especially when Sirius is shining – heavenly origin (long fall?). All insects retire for winter from the 'setting to the rising of the Pleiades' 11 Nov – 23 Mar. He was an exhaustive extractor of information from earlier works and doubtless used a little imagination too. Red and black pollen depended on the direction of the wind. He mentioned the location of forage areas by scout bees. Likened bees to dolphins as feral creatures. Pliny the younger explained his uncle's indefatigable methods. Wrote of the 'honey industry' in Britain during the Roman invasion.
See: mulsum and stars

**-ploid**
A word element in cytology and genetics indicating the number of sets of chromosomes, thus we get haploid and diploid (2n), tetraploid (4n) etc. Female honeybees are diploid. Gametes have haploid arrangements producing diploidy by fusion. Drones remain haploid freaks.

**plum *** *Prunus domestica* and spp. Rosaceae**
A small, hardy deciduous tree bearing oblong, stoned fruit and flowering early in spring . Often grown around orchards to form a wind-break and help the establishment of wild pollinating insects Pollen is reported as light, deep yellow , pale green, yellowish green, and deep yellow, the grain is larger than that of cherry. Many varieties are self-sterile but in any case cross-pollination is beneficial. When the flower opens its stigma is immediately receptive and continues so, thus eventual self-pollination is possible even if insect visits fail. The nectar is weak and bees are easily attracted to other sources. The E.P.P. can be very short in a bad spring.

**plumose (plume)** ***
Feathered or feathery. It has been used to describe the honeybees' pollen collecting hairs. However 'feather' tends to imply small processes arising from the stem of a bee's hair in opposite directions but in the same plane (like a blade). This is not so in the honeybee as the processes protrude in all directions and form ideal pollen traps, much more so than in their cousins the bumble bee.

**plumping**
Increasing the strength of a colony by giving it combs of eggs, sealed or unsealed brood or better still emerging brood according to the

colony's need and capability of nourishing it. This is a useful technique for producing early strong colonies for queen raising, pollination or early flows. Plumped (Pd).

**plum swarm**
The first and therefore the most valuable swarm. The first migration of bees from a hive.
See: prime swarm.

**ply**
To repeatedly apply oneself to a given task. Bees will ply to and fro, visiting a source of food carrying their loads back to the hive.

**plywood**
Laminated sheets of wood glued together with alternating thicknesses are available, as are various qualities and marine or external quality ply have been used for hive roofs. These are lighter and warmer than metalled ones. Care must be taken to ensure that ordinary ply is not used where weather or moisture from the bees may cause the glue to become ineffective and also to check that the glue does not contain any insecticide.
See: hardboard.

**pneumophysis pl. ses.**
Bursal cornua or horns of the drone's genitalia. These cream coloured, orange tipped expansions are made interestingly and clearly available when a mature drone is held by its thorax (rather firmly) when it will evert its genital apparatus.

**poach**
To encroach on another person's property. Although few beekeepers would ever consider putting their hives on anyone else's property without first seeking permission, it has been done both by accident and design aimed at bringing some benefit to the 'dumper' it may be classed as poaching. Although it is not legally necessary it is very wise to label hives that are moved onto someone's land, with the owner's name and address.
See: branding and rustling.

**Pochote *Bombacopsis quinata***
A wood that is used for beehives in Costa Rica and which withstands 40 years or more of use but has now become rather expensive.
See: red cedar.

**pocket**
A pouch-like receptacle, hollow or cavity. Special pockets in bee-suits are useful for the immediate availability of a hive tool etc. Breast pockets can allow a channel between them which can allow bees to get through where the elasticated hem of the veil might lift and let them through. The corbiculae – pollen baskets - have been referred to as pockets.

**pod midge *Pasyneura brassica***
A pest that is found on oil-seed rape. Efforts to control it can necessitate the use of chemicals that are harmful to honeybees. Law suits have resulted from this.

**Poems**
*A beehive's hum shall sooth my ear.*
*A Wish 1763-1855*

*A play o' bees in May's*
*worth a 'noble' the same day.*
*A play of bees in June's purty soon.*
*A play of bees in July's nod worth a butterfly.*

*A noble was an obsolete gold coin while 'play o' bees was a swarm.*

*As bees bizz out wi' angry fyke*
*When plundering herds assail their byke.*

*At a month's end ....*
*But I who leave my queen of panthers*
*As a tired honey-heavy bee*
*Gilt with sweet dust from gold-grained anthers*
*Leaves the rose chalice, what for me?*
*From odours of the chaliced centre*
*From the amorous anthers' golden grime*
*That scorch and smutch all wings that enter*
*I fly forth hot from honey time*
Swinburne

*A wonder here, for in this tiny jar*
*The essence of a thousand acres rest*
*Left but the cap of all this summer bliss*
*The peerless captive greets you with a kiss'*
G.G.D. Desmond.

From the *Bee Boy's song* by Rudyard Kipling
*Marriage, birth or buryin'*
*News across the seas,*
*All you're sad and merrying'*
*You must tell the bees*
*Tell 'em coming in an' out*
*Where the Fanners fan*
*'Cause the Bees are just about*
*As curious as man!*

*The bees, rejoicing o'er their summer toils*
*Unnumbered buds, an' flow'rs delicious spoils*
*Sealed with frugal care in massive waxen piles,*
*Are doom'd by man, that tyrant o'er the weak*
*The death o' devil's smoor'd wi' brimstone reek*
Burns

*Car, la Bretonne*
*Elle est mignonne,*
*Elle est gentille,*
*Comme la grenadille*
*Ou le ble noir* (buckwheat)

*The careful insect 'midst his work I view*
*Now from the flowers exhaust the fragrant dew,*
*With golden treasures load his little thighs*
*And steer his distant journey through the skies.*
John Gay. Rural Sports.

The collected poems:
*'Bees are Black with Gilt Surcingles*
*Buccaneers of Buzz'*
Emile Dickenson 1830 – 1886

*Excellent herbs had our fathers of old*
*Excellent herbs to ease their pain*
*Alexanders and Marigold, Eyebright, Orris and elecampane,*
*Basil, rocket, Valerian, Rue (Almost singing themselves they run)*
*Vervain, Dittany, Call-me-to-you, Cowslip, Melilot, Rose of the sun,*
*Anything that grew out of the mould*
*Was an excellent herb to our fathers of old*
Rudyard Kipling

*Fish the beck for miller's thumb*
*Near the honeybee's homeward hum*
*See the dragonfly's startled flight*
*Hear the wise old owl at night*
A.L. Hind

*Have your skep ready, drowse them with your smoke*
*Whether they cluster on the handy bough,*
*Or in the difficult hedge be nimble now*
*for bees are captious folk*
*And quick to turn against the lubber's touch,*
*But if you shake them into their wicker hutch,*
*Firmly, and turn towards the hive your skep,*
*Into the hive the clustered thousands stream,*
*Mounting the little slatted sloping step*
*A ready colony, queen, workers and drones,*
*Patient to build again the waxen thrones.*
Sackville-West 1892 – 1962

Henry V Act 1 Scene 2
*Obedience for so work the honeybees'*
*Creatures that by rule in nature teach*
*The act of order to a peopled kingdom*
Wm. Shakespeare 1564 -1616

*His helmet now shall make a hive for bees.*
*Polyhymnia 'A farewell to Arms'.*

*How doth the little busy bee*
*improve each shining hour*
*And gather honey all the day*
*From every opening flower!*
*How skilfully she builds her cell.*
*How neat she spreads the wax.*
*And labours hard to store it well*
*With the sweet food she makes.*
Isaac Watts 1674 – 1747

*I saw the gorse on linnet heath,*
*The foamy blossom of the hawthorn trees*
*The pearly trace of spider's webs ,*
*I heard a labour song of the honeybees*
ALH.

*The Lake Isle of Innesfree.*
*Nine bean rows will I have, a hive for the honeybee.*

*And I live alone in a bee-loud glade.*
Wm.B Keats 1865 – 1939

Meditations.
*What is not good for the swarm is not good for the bee.*
Marcus Aurelius 121 – 180 A.D.

*The pedigree of honeybees*
*does not concern the bee*
*A clover any time to him is aristocracy*
Emily Dickinson

II Penserosa
*Hide me from day's garish eye while the bee with honied thigh*
*That at her flowery work doth sing,*
John Milton 1608 – 1674

Pretty Words
*'Honeyed words like bees,*
*Gilded and sticky, with a little sting'*
Elinor H. Wylie 1885 - 1928

*Sing a song of honey*
*Honey from the white rose, honey from the red*
*Is it not a pretty thing to spread it on your bread?*
(It continues nicely)
Barbara Euphan Todd

*Six shining panels gird each polished round. The door's fine rim, with waxen fillet bound. White walls so thin with sister walls combined.*
*Weak in themselves a sure dependence find.*
Evans

*'So roses grow on thorns,*
*and honey wears a sting'*
Dr. Isaac Watts.

*Sweeter than honey and the honeycomb.*
Ps 19, 10Ps

The Tempest Act V Sc 1
*Where the bee sucks there suck I,*
*In a cowslips bell I lie.*
*There I couch when owls do cry.*
*On the bat's back I do fly*
*After summer merrily, merrily, merrily shall I live now.*
*Under the blossom that hangs on the bough.*
Shakespeare

Tennyson *'Eleanore'*
*Or the yellow banded bees ,*
*Through half-open lattices,*
*Coming in the scented breeze.*
*Fed thee a child, lying alone*
*With whitest honey in fairy gardens culled;*
*A glorious child dreaming alone*
*in silk soft folds upon yielding down,*
*With the hum of swarming bees*
*Into dreamful slumbers lulled.*

*There's a whisper down the field,*
*And the ricks stand grey to the sun'*
*Singing Over then, come over,*
*for the bee has quit the clover*
*and your English summer's done*
Rudyard Kipling 1865 – 1936

Virgil 4th Bk of Georgics 50 BC.
*The work goes on like wild fire, the honey smells of thyme.*

*Hundreds of quotations like this come from this well- known writer who must have known and studied bees as much as any man.*

*Here's when the swarms are eager of their play and loathe their empty hive and idly stray.*
*Restrain the wanton fugitives and take timely care to bring the tyrants back. The task is easy but to clip the wings of their high-flying kings.*
*At their command the people swarm away.*
*Confine the tyrant and the slaves will stay.*
*Whether thou build the place for the bees with twisted osiers or with the barks of trees make but a narrow mouth*
Virgil.

Then from Kubla Khan – Coleridge:
*Weave a circle round him thrice and close your eyes with holy dread ,*
*For he on honey-dew hath fed*
*And drank the milk of Paradise.*

**Pohutakawa** ***Metrosideros excelsa***

A New Zealand coastal tree with brilliant red flowers similar to the rata. Capable of yielding a very fine honey around Christmas time.

**poikilothermal**

Having a body temperature that fluctuates with that of the surrounding environment. It applies to reptiles and fish. While to some extent it also applies to honeybees because they have the ability to increase their temperature by the microvibration of their thoracic muscles when they act jointly within a cluster. Yet in the cold and alone their tenure of life is short .
See: cold-bloodied comatose and homoiothermal.

**pointsettia** *** ***Euphorbia pulcherrima***

A native shrub in Mexico and Central America. Popular as a house plant in the UK especially at Christmas time when its brilliant scarlet bracts contrast with it green or variegated leaves. It exudes fair-sized globules of nectar (as dense as honey) though there are rarely if ever opportunities for the bees to work it.

**points**

N.H.S. (Honey show) points are awarded according to the scale of lst 6, 2nd 5, 3rd 4, 4th and VHC 3, HC 2 and C 1. VHC = Very highly recommended.

**poison**

A substance capable of impairing health or causing death. An undesirable substance that causes physical or chemical body processes to behave abnormally
See: harmful substance, spray poisoning, toxin, venom and virus.

**poison contact**

Relating to bees a contact poison is one that becomes effective when a bee actually touches it or vice versa. Such a poison would not give off poisonous fumes but would be most effective as a dust.

**poison gland** ***

Also known as the poison or venom gland. It is a long, forked tubule with tips that are both slightly expanded; the single proximal tube widens to form a large club-shaped sac in which the secretion of the gland is stored, Dade.
See: alkaline gland, venom gland and venom components.

**poisonous honeys**

Honeys from the yellow jasmine (U.S.) and mountain laurel, have been reported as poisonous when eaten from uncapped cells but wholesome when capped. Poisonous honeydew can occasionally arise under freak conditions with the 'tutu' in New Zealand. Although the authorities are aware of the exact cause and take every precaution to ensure that no bees can get anywhere near foraging distance, it is still of interest to note that seriously poisonous sources do exist. As poisons that are likely to affect humans are also likely to harm the bees, they do in fact act as a 'buffer' between the plants and ourselves. Pyrrolizidine alkaloids are found in certain honeys but do not assume serious significance. See: toxic honey and toxic nectar.

**poisonous plants**

A plant which when even a small part is eaten (including roots, seeds etc). could cause the consumer to become ill. Chemically alkaloids or glycosides are responsible for most of the potent poisons. In belladonna atropine (alkaloid) and in foxglove, digitoxin (glycoside) are present but do not seem to get into the honey when nectar is collected from their flowers. Several poisonous plants like privet, ivy, ragwort, rhododendron and yew are not usually troublesome and have less than pleasant taste and aroma. These plants are dealt with under their respective headings. Some pollens , namely buttercup and scabious have been reported as harmful in certain circumstances.
See: andromedotoxin mannose, toxic nectar, toxic honey and 'tutu'.

**poisonous sugars**
Galactose, arabinose, xylose, melibiose, mannose, raffinose, sacchyose and lactose. Pectin, agar and many gums are toxic or can hydrolyze to toxic sugars.

**poison sac** ***
The female bee's venom storage chamber. Poison from the venom gland passes into the widened portion of the gland which takes the form of a club-shaped sac which tapers to the narrow duct opening into the bulb of the sting.
See: venom sac.

**poison spray**
These include compounds with varying toxicity to bees: Organophosphorous, chlorinated hydrocarbons (organochlorines) dinitro and inorganic compounds. Methods of application include: wetters, powders, systemic, granular and UHV See: pesticide, insecticide and spray.

**poke**
A sac or bag of woven material such as muslin or linen. Quite large ones were used for straining honey. FV suggested the size of a pillow-case, conically shaped, with a bar that enables it to be supported from the mantelpiece so that honey can drip into a bowl in front of the fire. Forefunners to the more modern conical metal or plastic sieves. See: 'honey poke'.

**Poland**
2 million colonies average holding 10 - Nation of beekeepers who regard bees along with sheep, pigs as significantly important creatures.

**polar filament**
The slender apical end of egg-tube of insect ovary.

**polarimeter**
Polariscope an optical instrument for measuring the rotation of the plane of polarized light as for example when passed through an optical compound such as a sugar solution. It enables levulo and dextrorotatory sugars to be identified.
See: dextrorotatory, Polaroid and polarized light.

**polarized light**
Ordinary light consists of electromagnetic vibrations travelling in every possible plane at right angles to the light path. In polarized light the vibrations are non-random and confined to one plane. It is produced by passing ordinary light through a transparent, plastic film 'Polaroid' or through a specially made calcite prism (Nicol prism) Many substances like sugar solutions have the property of rotating the plane of polarized light by a specific and characteristic amount. This affords a means of identification as for example in honey analysis. The axis of rotation of the light from any point in the sky is always perpendicular to the plane of a triangle connecting the observer, the sun and the point in the sky at which he is looking.
See: light-compass and Polaroid

**Polaroid**
Trademark of a material formed by a sandwich of orientated iodo-quinine crystals between two protective sheets of glass.
It has the property of transmitting light rays, the vibrations of which are confined to a single plane. Honey and nectar have positive properties in this respect Polarising crystals include calcite or Iceland Spar.
See: dextro and laevulo rotation.

**pole star Polaris**
Gives us true north and though faint it has no bright stars adjacent. Hive entrances are often related to direction, the sun, shade or shelter from prevailing winds. South is sun's mid-day direction or point the hand hour of your watch towards the sun and south is half way between that and 12. Every 400 years or so the two poles interchange. That will affect us more than it will the bees.
There is no Southern equivalent in the southern skies.

**pollarding**
The treatment used to encourage a dense growth of new shoots to arise from trees at the same level i.e. the lopping off of the crown or topmost branches. Sometimes trees such as limes are cut back to the trunk presumably to prevent offending branches from obscuring vision, dropping honeydew or needing more frequent attention.

**pollen** ***
The fertilising element of flowering plants. The microspores produced and discharged by the anthers. Looking like fine powder to the naked eye it is composed of minute grains each with a size, shape, colour and composition distinctive to the particular plant. Once locked onto an appropriate stigma, in the right environmental conditions, it germinates to produce the male reproductive cells. Gathered by bees it is sterilized in the course of storing to prevent pollen tube growth in the comb and used as food providing them with their all-important protein supply. Try a few grains on a glass slide in weak sugar solution, magnification will give a good idea of what would happen to unpickled grains.
See: pollen grain, exine, intine and tape recording.

**pollen analysis (palynology)**
The identification of pollen grains in honey and the discovery of the plant that produced them. See: melissopalynology

**pollen barrier**
The first brood and suitable weather conditions give a hearty stimulation to young bees to collect pollen and this tends to be packed solid in frames adjacent to the brood. Although it is used at a fabulous rate when conditions are good it can act as a barrier as the queen can only lay in warm, empty, prepared cells. So early in the year it is probably unwise to interfere though you might hear much said about 'pollen clogged combs'.

**pollen basket**
The corbicula – absent in drones and queens and cuckoo bumbles. The long, curved, retaining hairs have been observed to enclose pollen masses the size of a worker's head though such loads are only encountered with strong young bees and when pollen is plentiful and much needed.
See: corbicula.

**pollen beetle *Meligethes aeneus***
Pest on oil-seed rape – buds and flowers fail to set and fall from the plant leaving podless stalks. The beetle is more abundant on spring rather than winter rape and more so at the edge of a crop than the centre. Farmers are frequently forced into using pesticides perhaps only for the headlands.

**pollen brush** ***
Scopa or sarothrum. This is the inner surface of the hind-leg's basitarsus. Use a hand lens to look at this on the worker. The transverse rows of hairs, each a very effective 'rake' become loaded with pollen which the 'rastellum' combs from the opposite leg onto the shelf of the 'press'. The middle legs are also endowed with hairs and like the hairs all over the bee's body they are especially suitable for retaining pollen grains' (far in excess of the bumble's capability)
See: pollen grains and packing.

**pollen canopy**
This can be noticeable during the brood development period when because pollen is always stored adjacent to the brood nest (or between the brood nest and the honey) it tends to form a canopy over the nest. It is a layer of variable thickness according to the size of the brood nest available to them under the honey canopy. This double canopy of pollen and honey has a considerable influence in helping the bees to control the microclimate of the broodnest.
See: pollen cells.

**pollen cells *****

Cells in which pollen has been placed for temporary or long-time storage. The variegated colours can be seen forming a canopy over and filling the flank combs alongside the brood nest. They are not usually seen within the brood nest unless the queen is failing or the bees have swarmed. Worker cells are used in the main, though drone cells are sometimes used. Pollen may be hidden under a normally-sealed layer of honey. Blocks of pollen like this are usually within the area of the brood nest. Honeycomb may be spoilt by an occasional cell of pollen but having adequate brood nest size helps to obviate this. Combs for extraction are harder to uncap when areas of pollen filled cells are encountered and isolated cells of pollen quite ruin the value of comb honey (despite its nutritional value to humans). It is often possible to detect pollen hidden beneath honey by holding the comb up to the light. Brood chambers of adequate size help to prevent pollen from getting into combs intended for honey surplus or through the excluder.

**pollen cellulose coat**

Each grain of pollen is enclosed in a small cellulose capsule which is relatively indestructible and of characteristic size and shape. Pollen from plants that lived in ages past has helped us to identify the flora of those times. The husks are usually plentiful in honeybee faeces. The cellulose coat has membranous germ spores through which osmosis allows the contents to be digested. See: exine and pollen digestion.

**pollen – classification of**

The arrangement of pollens into associated groups. Size and shape of the grains together with the number of spores or grooves has enabled positive classification techniques to be established. See: melissopalynology, pollen grain and pollen terminology.

**pollen clogged**

When filtering honey cloths can become pollen clogged and filters need to be changed when necessary. Bees have been found abnormally (excitedly) on the alighting board or in front of the hive and this has been ascribed to May pest or pollen clogged 'colons'.

**pollen collection**

Foragers may collect small loads of pollen incidentally while working for nectar but those after pollen alone start with an adequate food supply to fuel them for the return journey. In flowers the anthers may be brushed against or actively nibbled. The hairs of the first tarsomere clean the head while the middle leg's basitarsus clears pollen from the thorax and passes it to the hind legs. The hind legs broad basitarsi brush pollen from the abdomen. Honey regurgitated from the crop is used to moisten the pollen and the combined harvest collects on the basitarsus from which the 'rastellum' combs it onto the 'auricle' of the 'pollen press' from whence it is squeezed into the corbicula where it forms a compact mass (pellet)

**pollen collector**

Younger foragers make the best pollen collectors and in marginal weather conditions this can be a testing experience as the loads are about one fifth of their body weight. The loads require an enormous number of flower visitations (50 to 1000). The size of the pollen loads can be used to ascertain the development of new foragers in the spring. As brood pheromones stimulate foragers to become pollen gatherers (the feeding of sugar syrup seems to do likewise) these, like the amount of brood, govern the number of bees recruited to pollen gathering. Not every bee necessarily becomes a pollen gatherer while some may work on this task throughout their foraging lives.

**pollen colour**
From the beekeeper's viewpoint pollen is beautifully displayed both on the bee's legs and in the honeycomb. The colours vary according to the plants visited but so can the colour from any particular source. This has been shown throughout this work where each observer's readings have been faithfully reported. In many case a spectrum for each plant could be drawn up. The variations can be caused by the condition of the pollen, the bee's wetting of the grains to secure fixing and sterilizing the grains, and just as petal colours can vary so too can pollen grains. As pellets frequently fall off as the bee scrambles into the entrance, it is not hard to make a collection of grains that would otherwise go to waste. Using them as you might use a small piece of chalk, make lines on a white sheet, link them up if you can with plants you see the bees are working. The next day it will become obvious that some colours are not 'fast'. The bright orange of flax became a dull yellow in less than a day. It seems that just as honeys can go from extra white to black, through tinges of green to mahogany etc. so too can pollen. See Dorothy Hodges pollen colour chart. Compare: grey – apple and raspberry, black – anemone and poppy, yellow –dandelion and sunflower, blue – viper's bugloss, crimson – horse chestnut, navy – phacelia, indigo – mallow, turquoise – rosebay, brown – clover and heather, purple – campanula.

**pollen comb**
The hairs associated with pollen collection could be ranged from soft retaining hairs, like those surrounding the corbiculae, to the branched hairs that sweep up pollen and are found all over the body and middle legs to the stiffer combs of the hind leg's inner basitarsi and then the very pronounced 'rake' of the distal tibial edge – the pecten. See: pollen brush.

**pollen compatibility**
The suitability of a particular type of pollen for its duty in bringing about fertilization of the recipient flower. If the 'pollinator' that 'pollinizes' a flower does so in the right conditions and with compatible pollen it should lead to fertilization. To be viable the pollen must not only be in the right condition (this can be influenced by humans) but it must be compatible. Its shape, size and physical condition enables it to chemically respond to the condition of the stigma and, wasting no time, send forth its pollen tube to grow right down into the ovum. This you can see, if fresh (not sterilized) grains are put onto a microscope slide with a 30% sugar solution how these tubes can grow! The appropriate chemical 'match' is essential. Crab apple or Malus floribunda pollen may successfully fertilize a cox but the fruit will carry the genes of both.

**pollen composition of**
Protein, fat and carbohydrate can be found in fairly even quantities in some pollens but variations are enormous and 20% protein 8%fat, 30% sugar and 12% water with ash and undetermined residue is an average sort of breakdown. Elements like nitrogen, sulphur and phosphorous, calcium, potassium, vitamin A,C,D,E,M,B1, B2, B3, B6, B12 amino acids, reducing sugars, vegetable oils, biotin and rutin together with digestive enzymes go towards completing the picture.

**pollen consumption**
Some bees only 2 hours after emergence begin to eat pollen. By 4 days most feed heavily on pollen reaching a maximum consumption at 5 days by which time they are actively engaged in brood rearing. Pollen consumption then diminishes between the age of 8 to 13 days as brood glands retrogress and protein is transferred to its flight muscles and stored in the case of winter bees in their 'fat body'. A colony uses upwards of 75 lb (35 kg)of pollen each season though this figure has to be related to the prolificacy of the queen.

**pollen digestion**
Active proteolytic enzymes with the capability of pollen cleavage are found in the

gut of the worker. Pollen protein content varies from 10 – 40% and 3 – 6 day old bees are likely to break up and digest the most grains and will feed a greater or lesser number of larvae proportionate to the pollen intake.
Extracted from ABC & XYZ.

**pollen dispenser**
Hive inserts designed to encourage departing foragers to pass through selected pollen, have been used in an attempt to improve pollination of top fruit. A trough is provided near the entrance, replenished from a 'hopper', containing 'ripe', compatible, viable pollen, mixed with diluent such as lycopodium powder, to overcome the pollen's natural stickiness.

**pollen gatherer**
A bee whose primary concern is the successful collection of pollen, though some nectar gatherers may also return with relatively small pollen loads. Pollen gatherers do not fly as far from the hive as some of those collecting nectar. Bees are stimulated to gather more pollen by brood produced pheromones and by the feeding of sugar syrup.
See: collector, pellet and pollen load.

**pollen (going in)**
When bees are seen to be taking pollen into the hive, large loads in spring are a good indication that young bees are on-the-wing and that brood rearing is progressing. This and warmth over the brood nest can be linked to foraging activity to assess the queenrightness and colony strength so that if 'hefting' also confirms adequate stores are present, the bees need not be opened up unnecessarily early to confirm these things.

**pollen grain *****
A microspore of a seed plant, dust-like particles known collectively as pollen. They usually contain two nuclei, the generative nucleus and the pollen tube nucleus. The individually, cellulose packaged, male gamete as developed to ripeness in the capsule of a flower's anther. The pollen mother cells develop in the sporogenous tissue and when ripe that disintegrates and the pollen is released. The pollen grain's structure; that is its shape, size and surface characteristics (important when latching onto the stigma), are usually positive enough to permit identification by microscopic techniques. From tiny forget-me-not 3-4mu to Rosemary and willow herb 100mu.
See: Angiosperm, exine, intine, stamen, sporopollenin, pollen load and pellet.

**Pollen identification of**
For beekeepers by Rex Sawyer College Cardiff Press.

**pollen - identification of**
The identification of pollen grains gives us information as to what plant and possibly when and where it grew. The presence of any particular pollen in honey does not emphatically tell us the source as some pollens are ubiquitous and some honey sources have little or no pollen. Geographical origins of honey are linked however not to one pollen type but to the spectrum of pollen so that manuka, cabbage tree, kamahi and cat's ear, tend to put the source in Oceania with a strong possibility of NZ.
See: pollen signature, palynology and melissopalynology.

**pollen intake**
Factors affecting this include race of bee, time of year, weather conditions and pollen availability. There are two occasions when more than the usual amount of pollen may be seen to go into the hives. The first is whenever there is an expansion of the brood nest (a) in the spring (b) when a new queen has started laying (c) when stimulative syrup feed has been given. The second is in the autumn when bees are busy getting into a physical state for the winter which will include having bees able to feed early spring brood. This is done by enlarging the 'fat body' by their eating large amounts of

pollen. This sight often surprises beekeepers who wonder where all the pollen is going to.

**pollenkitt**
German name for the coating of sticky material that cover entomophilous pollen grains. It means pollen glue and covers the exine of most angiosperm pollen.
See: anemophilous.

**pollen load**
The total removable load of pollen packed into a returning bee's corbiculae. Two pellets formed from masses of pollen grains (almost invariably of the same type) that have been moistened to render them sticky by the forager. Various figures have been quoted. Dr. Butler stated about 248 flowers visited, Ribbands 32 dandelions, Vansell 100 pears 84. Average weight 11mg but hard maple as much as 29mg. Trips usually take 6 – 10 mins but up to three hours has been reported! Another figure 24 mins. average. It takes about 10 loads to rear a worker larva; bees average 10 trips a day. A colony will visit as many as twenty million flowers for a day's pollen and use about 100 lb. in a season.
See: grain, pellet, pollen basket etc.

**The Pollen Loads of the Honeybee**
This book is a classic in the realm of pollen identification by colour and was reprinted in1978 London IBRA.

**pollen mite *Mellitifis alvearius***
These small mites seem to be ubiquitous and any pollen combs left unguarded (as in a shed) are soon occupied by the mites and the pollen is turned into dry dust which can fall out of the cells like dust. The mites do little or no harm to an active colony which covers its stored pollen with a film of honey to render it immune to attacks from the mites. Pollen in combs away from the bees is in fact cleaned out by the mites whereas it would otherwise, in damp conditions, become a mass of mould which to beginners looks like 'chalk brood'.;

**pollen mould *Bettsia alvei* or *Pericystis alvei***
Pollen is found in honey itself and this effectively preserves it. When stored for the use of the bees it will have been protected from mites by a sticky layer, ready for use by the nurse bees. Mould is always likely to form when the bees have not covered the pollen with a film of honey and damp conditions aggravate it. When dry, bees have little difficulty in cleaning either the mould or dry lumps of pollen and the pellets are found thrown out after a spring clean-up.
See: pickled pollen.

**pollen nutrition**
Late scientific investigations have shown how pollens can vary but an old record – Mauricio (1960) gave (a) highly nutritive: fruit, willow, corn, white clover then (b) less nutritive: elm, dandelion and (c) fair: alder, hazel and (d) poor: pine. More recently high value given to crocus, chestnut, clover, poppy, top fruit and willow. A mixture is better than a single source but old dry pollen is worthless.
See: nutritive value of pollen.

**pollen pack**
The lump of pollen or pollen grains, bunched together by the bee and variously called, load, pellet granules etc. Mixed pollens are rarely found in honeybee loads as they are faithful to one kind of plant when collecting, bumbles though do sometimes have mixed colours in their corbiculae. If not already sticky the grains are moistened by regurgitations from the honey sac so that the whole mass of grains adhere together in a single lump. A characteristic routine is followed, sometimes even when a bee has perchance lost part of its load; one leg forces the pellet out of the retaining hairs on the other side and off the spine, right into a cell. A house bee will then ram the pellets flat with her head. A bee finds it difficult to fly if one pellet is dislodged from one of its baskets.
See: pollen.

**pollen packing**
Many flower structures force the bee to brush past pollen-covered anthers giving it little room to go through the movements that are necessary to transfer the grains into its basket. Much of this therefore goes on whilst the bee is actually in the air. Poppies are good plants to observe this while they hover over the flowers. Doubtless large numbers of grains are scattered around and there is probably a useful 'trail' of pollen let loose in the orchard as the bees wing their way back. The arrangement of hairs on the pollen basket facilitates a rearward movement of each pollen pellet and once they have been slid off into a cell house bees will use their heads as battering rams to pack it economically into the cell. At this stage the individual grains will have been given a 'treatment' by the forager so that there is no tendency for the pollen tubes to grow out of the grains as this would make the pollen useless for consumption and also take up a lot more space. When packed a shiny film covers them.

**pollen pellet** ***
One of the two symmetrical (balanced) lumps of pollen that the bee packs into its pollen baskets. They can become dislodged (by using a pollen trap) or accidentally and once knocked off the bee, do not seem to be recovered. Two pellets would average about 11mg when fresh. When pollen is sold it is usually in the form of pellets.
See: pollen load pack etc.

**pollen pickled**
Pollen stored by the bees and not intended for immediate use. It differs from pollen on the flowers in that the bee has added enzymes to prevent pollen tube growth in the comb and a sticky layer to give it some protection from the ubiquitous pollen mite.

**pollen ploidy**
The number of sets of chromosomes is fixed for any particular pollen and the fertilisation of a flower is dependent upon its receiving compatible pollen of the right ploidy.

**pollen potential**
A plant's pollen potential will depend on the timing relative to weather conditions and pollen can be spoilt by the rain or frost. Some plants are noted for their pollen and although clouds can be discharged (and are collected from the ground and eaten in some lands) airborne pollen is not an easy pick-up for the bees that prefer the rounder stickier types. The bees themselves are not necessarily experts at choosing the most nutritious pollens and at times suffer from poor pollens, for example when surrounded by areas not yielding good pollen e.g. honeydews and mono-crops of the wrong variety and the bees strange choices include collecting tar instead of propolis in Malta, needing help to offload propolis and waiting for the sun to melt wax on a hive roof so that they can collect it in the same manner as propolis. The fact that pollens can pass viruses on to the bee and act as vectors in allowing poisoned pollen to go back to the hives are reminders that we must pay heed to the bees vulnerability which comes to light when the unknown reasons for CDD are explored.

**pollen press (anatomy)**
The tibio-tarsal joint of the worker bee's hind leg is an adaptation that greatly assists pollen collection in their work. The distal edge of the tibia bears a ledge which by the flexing of the joint bears on a corresponding toothed shelf on the proximal end of the basitarsus. Along the inner edge of the tibial ledge is a row of stiff spines 'the rastellum'. This combs pollen grains from the pollen brush of the opposite leg and they are held by the retaining hairs of the auricle and squeezed between the tibio-tarsal surfaces to form a compact mass that is pushed into the corbicula on the outer surface of the tibia. The angle of the teeth on the auricle together with the retaining fringe of hairs,

ensure that the flexing of the joint directs the pollen mass into the pollen basket.

**pollen production efficiency**
Malus (apples) Golden Hornet 100, Hillieri 46, Winter Gold 45, Alden Harmensis 28, John Downie 8, Profusion 6. Other species Egremont Russet 35, James Grieve 29, Golden Delicious 27, Millers Seedling 21, Lord Lambourne 14, Worcester Pearmain 12.

**pollen rake** ***
The pecten or rastellum found on the distal edge of the hindmost tibia, forming part of the pollen press. This rakes through the 'scopa' or pollen comb of the opposite leg. See: rastellum.

**pollen release**
This occurs when the dehiscence of the stamens takes place. Flower development and favourable weather conditions will cause the capsules of the anthers to split and pollen is shed. Wet or frost may delay dehiscence. Plants are quite varied in their floral mechanisms so that pollen is more plentiful and easier to collect from some flowers than others. Broom for example flicks open like an uncoiled spring daubing the honeybee with yellow pollen over most of its dorsal surface. Other flowers will only yield pollen at certain times of the day. In one's own garden a supply of useful pollen can be greatly encouraged by taking branches of buds that are bursting into flower and putting them into water 'indoors'. The amount of pollen that is then released far exceed anything that can happen in the open. The flower can then be used like little brushes to apply pollen to the stamens that actually require it. Frost and applications of spray can be highly detrimental to pollen.

**pollen reserve**
It has been shown that brood rearing is often related to pollen income or to their pollen reserve when new pollen is not available. Not all races of bee are acclimatized sufficiently to have set up reserves of pollen to cater for periods when unsuitable weather conditions prevail. Different races vary enormously in their enthusiasm for pollen collection and storage. It has been postulated that a reserve of 500 square inches of pollen filled cells is desirable, preferably within the cluster. See: pickled pollen.

**pollen signature**
The pollen spectrum obtained by an examination of the bee's stores at any one time, allows us to identify the sources of forage worked by the bees. Whereas a single pollen's identification will not necessarily fix a honey's origin, the identification of several pollens (collected in one area at one time) in a honey will give a positive enough identity to allow the use of a 'pollen signature' (or fingerprint). In this way plant sources and the geographical point of origin can often be determined.
See: pollen identification and melissopalynology.

**pollen slide**
Pollen may be dropped onto a slide with a smear of glycerine and viewed directly through a cover slip. For a more permanent slide proceed as follows: avoid contamination by allowing a flower to open indoors. Wash out pollen into a watch glass filled with ether. Decant and wash with fresh ether. Dry, then transfer to a slide and spread out. When drying keep the temperature below 40C. Keeping them in this temperature range allow grains to swell. To stain add alcoholic fuchsin solution to melted glycerine gelatine and use this with degreased preparations of pollen as above. Use 0·2 - 1·5ml of alcoholic solution per 10ml liquefied glycerine gelatine.

**pollen sources**
Bees usually pick up some pollen while they are foraging for nectar. They will carry some on their body hairs and possibly some will be put into their baskets or rejected.

However the pollen gatherer is always to be found when the brood nest is developing and plants used include: blackberry, broom, coltsfoot, clover, crocus, daisy, dandelion, fruit trees, gorse, hawthorn, hazel, heath, heather, horse chestnut, ivy, lime, Michaelmas daisy, poppy, thistle, willow, yew etc.
See: under various 'pollen' headings.

**pollen storage**
Although freshly gathered pollen is infinitely better for bees than that which has been stored, the bees 'treat' pollen as they collect it then it is rammed into prepared cells. Then it is coated with a thin film of honey or if it is to be capped over with wax for later use a layer of honey, after which it is sealed with the same appearance as a comb of sealed honey, though always in the vicinity of the brood nest..
See: cells, canopy, freezer, pollen barrier, reserve, substitute and supplement.

**pollen substitute (patty)**
Dissolve 3½ lb sugar in 16 floz boiling water. Remove from heat and slowly stir in 1 lb soya flour seeing that the ingredients are thoroughly mixed. While still fairly hot spread in ¼" layers on five 10" square pieces of waxed paper. Lay over the brood frames. Americans say F-200 followed by Nutrisoy were the best soya flours to use. Another recipe: 3 parts expeller soya flour to 1 part skimmed milk (not thought desirable by some) ½part fish meal and natural pollen. Add a little sugar syrup and mix into a dough-like material. Place patty on wax paper and lay over the top bars of the frames directly above the cluster.

**pollen supplement**
Natural pollen or a mixture of this with substitute materials intended as a stimulant to brood rearing or to make up for a deficiency of natural pollen as for example when honeydew flows are worked. Fresh pollen mixed with soya-bean flower proved useful. All emphasize the benefits of adding fresh pollen if possible; this is perhaps when one colony can be used to help another. Never feed 'dry'- stand in a small quantity of water overnight. Warning pollen can carry disease pathogens.

**pollen terminology**
The exine (external pollen coat) is described according to its architecture. The intine or inner layers are sometimes divided into ectexine and sexine. The classification of grains has led to the use of terms that translate the number of pores or furrows into Latin.
See: exine, intine and pollen grain.

**pollen transfer**
The action that results in pollen being moved from the part of the flower where it is produced – the male element – to the part of a flower where it is needed for pollen tube growth and subsequent fertilization of the ovary. When the transfer is effected by bees they are referred to as pollinators, the plants that yield the pollen are the pollinizers. Many agents such as wind, water, gravity and animals of different kinds move pollen in this act of transfer.

**pollen trap**
Traps can be used over the entrance to dislodge the nutritious pellets from the foragers as they enter the hive but the shortfall causes more bees to divert their attention to pollen collection and often temper suffers. The trap might consist of a mesh designed to obstruct the passage of a bee so that if it is carrying a fair-sized pollen load, one or both of the pellets will be dislodged and fall into a tray covered against the weather and the attention of bees trying to recover the pollen. Plastic meshes are kinder to the bees than most metal ones and devices are used that incorporate a sliding panel so that a harvest can be taken in such a way that any one colony does not suffer unduly.

**pollen tube *****
A tubular process that grows out of a pollen

grain when it finds itself in compatible circumstances e.g. once it has become attached to the appropriate stigma. It is the expansion of the intine which emerges through one of the germ-pores (pits) in the exine. The pollen tube's growth down the style to the ovary depends on the new plants stimulatory effect governed amongst other things by temperature and humidity. There are two nuclei, the tube nucleus and a generative nucleus which moves down the tube and fuses with the ovule. Grains can be made to sprout in vitro but when packed in honeybee cells, pollen tube growth is inhibited by enzymes applied by the bees. Such pollen has therefore been rendered sterile. The numerous branched hairs of the honeybee carry fertile pollen which makes them such valuable pollinators.

**pollen viability**

The capability of a pollen grain to fulfil its pollination role. To do this the donor plant must be compatible with the recipient. Nature helps by marking flowers with colour, odour and taste, taking advantage of the fact that honeybees can detect these differences and are usually faithful to a certain type of flower on any one trip. Both the individual character of the pollen itself and the female structure of the flower only become functional within a relatively narrow range of time, temperature and humidity. Poor weather, especially frosts, but also rain, the use of pesticides and damage from pathogens, can hinder or prevent effective pollination.

**pollinate**

To transport and effectively position suitable pollen onto the stigma. Hence pollination and the agent of transfer is called the 'pollinator' while 'pollinizer' refers to the plant bearing the donor pollen.
See: pollen transfer, hand and cross pollination.

**pollinating agents**

Factors that take part in the transfer of pollen to its rightful target: Insects including honeybees, moths, flies and other bees, bats and beetles, birds and mammals, water, wind and gravity. Pollinating insects and flowering plants developed in the Cretaceous period 120 – 130 million years ago. None are more numerous and controllable than the honeybee. Bats, beetles, flies moths, birds and mammals, and then gravity, water and the wind are all able in their different ways to help plants with this dependence on external factors. Pollinating insects and flowering plants developed in the Cretaceous period 120 – 130 million years ago.

**pollination**

A useful service carried out by a number of agents especially the honeybee. It entails the transferring of viable pollen from the anther to the stigma. As the timing and density of the honeybee population can be regulated, widespread use of honeybee colonies is carried out all over the world. Much of this has been taken for granted and lack of information about compatible pollens has led to the collapse of very useful food sources 'to wit Kentish cherries. Most plants for instance benefit from cross-pollination even when they are declared to be self-fertile. Intervention such as hand-pollinating peaches rather than miraculously expecting bees to forage actively so early in the year are not generally appreciated. To hand-pollinate garden fruit trees where the fruit is developed on the best side for ripening and picking, comes as a surprise to many.

**pollination agreements (contract)**

Identification of participants, rental price, time of delivery of colonies, number of colonies, positioning, operation and maintenance, protection against insecticides, removal of colonies, care to avoid stinging, payment etc.
See: pollination contract

**pollination by bumble bees**

Unlike honeybees there are dozens of different species of bumble bees some

of which with their longer tongues are extremely useful pollinators, on red clover for instance. The snag is that their colonies have to be re-started each year and are relatively small. The size of the queens makes them conspicuous but their hairs do not hold pollen grains like the branched hairs of a honeybee. They can work in cooler temperatures than honeybees but at lower temperatures the pollen rarely works efficiently.
See: pollen viability.

**pollination – close**
This is when the anthers of a flower actually come into contact with its stigmas and this leads to successful fertilization.

**pollination of crops**
Bees will do more effective work if they can be moved in and grouped in the centre of a crop at the moment when it first comes into flower. The yield of American cranberries shot up as the number of colonies increased. Czechoslovakian figures gave 3 per ha gooseberries, 8 red currants,
6 blackcurrants, 8 strawberries and
2 raspberries. The size and quality of fruit also improved. A strawberry requires many bee visits to form a perfect fruit and this applies to kiwi fruit and others. Weather at the time, size of plants and the bees' requirement for pollen – all affect the issue.

**pollination – cross**
Cross pollination is the transfer of pollen from the flowers of one plant to another of the same species but derived from different seed. Generally speaking plants seem to benefit by cross-pollination even when self-fertile because there is a certain vigour about the pollen tube growth and subsequent fertilization. More and better seed is produced and with man's intervention earlier crops can be obtained. (strawberries and runner beans).

**pollination fee**
The price paid by the grower for each full colony delivered to a chosen site, usually for one month. The Bee Farmers Association act as agents to arrange contracts for their members covering colony numbers, dates and positioning.

**pollination – hand**
Brushing one flower against another may suffice but the use of portable bouquets. small brushes, application by aerosol and other methods of spreading the pollen are numerous. The donor plant may only make its pollen available for a short time though this can be enhanced by taking budding plants into a greenhouse or other warmish location. Tales of whole Chinese families setting to work to make up for the dearth of insects caused by the careless use of insecticides are rivalled by stories of old Kentish fellows who went into the cob nut orchards, hammering their backs against the trees and coming out looking like Chinamen.

**pollination – self**
A flower which uses its own pollen for fertilisation. This could be a hermaphrodite flower allowing pollen to fall onto its own stigmas or the pollen being moved within the flower by some agency.
See: pollination – close etc.

**pollination value**
Honeybees have established a regime in which they are rewarded for the service that they provide to all manner of plants. Man, going first of all for the honey, was not slow to realize the value of wax and went on to discover the enormous difference they made to food crops and then to all the wild flowers that make life possible for hosts of creatures. Only recently in this country, have we come to realize what beekeepers have known all along, that honeybees are a valuable and essential part of the environment. To set values on their worth is second to ensuring their survival.

**pollinator**

A living creature which actively transfers pollen to a needful and recipient stigma. Not unnaturally, folk have an aversion to being stung. This immediately reduces the number of people with time for bees. There was always a positive tendency not to understand and admire what the beekeeper was doing but to poke fun whenever possible. No journalist can resist what he or she thinks are witty comments when just plain information would be more valuable. But the bee is not just a buzzy thing that can sting;! It has survived far longer than mankind is likely to. It is remarkably well equipped (its body hairs can hold 60,000 grains) it can look after itself. It stores one of the finest foods. Asks nothing for the service it gives, helps to make us healthy is good for the environment. Yes, there are other pollinators but there simply isn't anything that compares with *Apis mellifera*.

**pollinium**

An agglutinated mass of adhering pollen grains which together with caudicle and adhesive disc form the pollinium. Bees can occasionally be seen returning with a trailing mass of pollen attached to them. Compare with mucilaginous substance on mistletoe and threads binding fireweed and other pollens.

**pollinizer**

A donor plant that provides the pollen to be utilized by another plant of the same species. Plants like Cox's Orange Pippin apples produce very marketable fruit when pollinated by other less marketable varieties (even Malus floribunda – an apple grown for its decorative blossom). The second variety is the pollinizer. There is of course no need to occupy valuable food producing ground when pollinizers can be grown in containers and moved in alongside the needful crops when required. The pollen from a pollinizer can be used for hand pollination – or of course added to pollen supplements for the bees.

**pollution**

Making anything unclean or dirty. The polluter should pay. With ever increasing costs facing the public spirited beekeeper, it would seem only fair that those who cause damage to his charges should provide adequate recompense. Such a statement arises as a result of the use of certain agricultural and other chemicals.

**police**

When moving bees it is wise to inform the police of the nature of the load, its route and the range of time. Reciprocally they will advise you of any difficulties you may encounter. Label your transport clearly 'Live Bees' so as to be readable from 20m.

**polish**

Beeswax and propolis can take on a beautiful shine, Once a film of wax adheres to wood the more it is rubbed the better the shine. It can improve leather making it supple and water-proof and plastics too are easier to dust once treated.

**polyandry**

Having more than one husband at a time. This applies to the honeybee queen and refers to the female's consorting with several drones.

**poly – more than one.**

**polyethism**

Behavioural division of labour within a colony of insects.

**polyflora**

Honey from sources other than mono crops, e.g. oil-seed rape, or honeydew.

**Polygonum *Polygonum maculata* (persicaria) Polygonaceae**

Smartweed – a good bee plant that spreads easily in some regions. It can impart a not altogether pleasant flavour to the honey though this very dark honey has been described as spicy and it is quick to granulate.

**polymorphism**

The occurrence of different forms of organs in the same individual at different periods of life, e.g. the silk glands of the larvae becoming the salivary glands of the adult. Also the occurrence of different forms of individuals within the same species – soldier ants as opposed to workers. This can be due to environmental factors (feeding) as opposed to genetic origin. Passing through many different forms or stages.

**polyploidy**

When the cell nucleus of an organism has three or more times the basic haploid number of chromosomes. It is rare in animals but significant in *Apis mellifera capensis* where laying workers are said to be able to produce females by parthenogenesis and it is found in some pollen types.

**polypropylene**

Used for hive strapping and honey containers a synthetic plastic polymer of propylene. Similar to polythene but of greater strength.
See: polythene.

**polysaccharide**

A carbohydrate group of nine or more bonded compounds synthesized from monosaccharide molecules such as starch, cellulose etc. Important structural and energy-rich material found in organisms. Either insoluble in water or forms colloidal solutions, non-reducing and do not have a sweet taste.
See: chitin.

**polystyrene**

A clear plastic polymer of styrene. In its white expanded form it makes excellent insulative layers and packing material but the bees can and will nibble it to pieces. It can be used as a hard-rigid plastic for moulded articles.
See: polythene.

**polythene**

A plastic polymer of ethylene, transparent sheets of which are used for piping, containers and packaging. Polythene bags have become so useful (and troublesome) and many people have their own 'sealers'. Pollen can be deep-frozen for future use in such bags. They can be used in-situ as feeders. They can line leaky containers or act as liners to honey tins. Bees can get trapped in them, conversely as bees slip on them they can be made into a funnel for getting bees into awkward slots like observation hives. Ingenuity leads to a multitude of uses for this remarkable material.
See: plastic

**polyurethane**

A polymer of urethane used for making rigid foam and products including varnishes and paints, also used for insulation, decoration etc. The tough, clear varnish is useful for lining feeders. When using for beekeeping articles, dry thoroughly, wash before use and avoid breathing fumes while working with it.

**polyvinyl chloride**

PVC a colourless thermoplastic resin produced by the polymerization of vinyl chloride. It has resistance to water, acids and alkalies. In the form of sheeting it has replaced many materials such as Hessian (burlap) for sacking, packaging etc.

**pome**

False, fleshy fruit developed from the receptacle of the flower and not the ovary – the part that we eat in fact.
See: drupe.

**pond** ***

A hollow in the ground, man-made or 'natural' which contains water. Garden

ponds are places where water collecting bees may come in large numbers. Drowning is a distinct possibility and can be avoided if mossy, sloping platforms of material on which the bees can get a grip are provided and where the sun can lift the chill off a little. To the bee, water is a glutinous fluid and a ripple can pull a bee into the water where cold if not suffocation causes their demise. Lifted out, a bee that has not been in too long can be resuscitated when held in a warm palm. A hive, placed on a raft, is safe from ants but don't fall in when removing supers!
See: water fountain.

**pop-hole** ***
Combs which are always built parallel and vertical often develop 'pop-holes' through which bees can pass from one comb to another (good places where a queen can hide from you). With strict bee-space, to avoid propolis, bees may leave gaps at the bottom of frames,' – don't expect perfect uniformity to your requirements.

**poplar** *** **Populus spp. Salicaceae**
These tall trees are often used as windbreaks. They produce grey-green, light green, or dull yellow pollen with a 16mu grain in March or early April. There are balsamic secretions collected as propolis rich in abietic acid. They grow rapidly producing light, soft wood and food for Lepidoptera including the poplar hawk

**poppy Papaver spp. Papaveraceae**
There are so many varieties. Gaucium flavum the yellow, horned poppy is a hardy annual with showy flowers, pink, bright yellow, red and variegated. They produce little or no nectar and their pollen is dark, bottle-green to almost black. Grain size 20mu, average height 55cm. The field poppy (Flanders – 'till the poppies bloom again'..) The longest seed pods of any British plant are found on the yellow, horned poppy found on coastal shingle. Californian, Iceland, Oriental, Corn and the opium poppy (somniferens) Afghanistan ? The promising buds can be carefully unwrapped to produce a wrinkly but almost normal sized bloom – interest for the kids.
See: note on pollen packing.

**popular smoker.** ***
The name given to the smallest (unpopular) smoker, It stands upright on its base with the nozzle in line with the fire chamber. Its contents do not last long and pulling off the nozzle can burn your hands –gloves? The hinged nozzles are more satisfactory and deliver smoke at the right angle.

**porch** ***
Pretty but unnecessary. The elaborate WBC incorporates a porch and useful entrance slides. Coupled with a sloping alighting board which truly the bees seem to like, you could almost fall for this oldy-worldy kind of beekeeping.

**pore**
A minute hole or potential passageway. Such as the micropores of a pollen grain or a bee's egg (enabling fertilization to take place). They serve as a means of absorption or transpiration or for the extrusion of the contents as when a pollen tube develops.

**porous**
Having many pores or interstices such as foam rubber or lint. Permeable by air, water or the like. The ability of such substances or material to 'hold' liquids makes them useful for the application of water or prophylactics. Brood cappings and the cappings of sealed stores are slightly porous and comb can 'weep' and brood cells are not entirely airtight.

**portable bee house**
Caravans arranged as bee houses or house apiaries set on wheels. Bee houses are in themselves uncommon in the UK largely due to the high cost of timber. So few people have taken the trouble to set up a bee house that can be moved on water or by road, though there is much to be said for the idea

as the movement of colonies is essential for pollination and optimising the honeycrop. It is more widely used abroad.

**portal**
A word used more in olden days to describe an opening or entrance. In so far as it permits ingress and egress it is perhaps more appropriate than entrance.

**Porter bee escape** ***
An ingenious device designed by E. C. Porter in 1891. It consists of sheet metal folded to form a flat tube (now widely replaced by plastic) through which the bees gain access from the upper side. Their egress is possible via one or other of two pairs of fine springs that open when pushed by the bees but only in one direction. The lower part of the flat tube is detachable by sliding and this allows for adjustment and cleaning and forms ends that are open to let bees that pass through the springs go to the underside of the clearer board.
See: clearer board.

**Portugal**
The abundance of cork has led to its use for hives, though for modern honey farming the more durable wooden hives are used. A round of tree trunk, looking for all the world like a piece of tree trunk, except for a cap to protect it from rain and a white band around its middle to distinguish it from an old log, could be seen in many places. A 300 mile north-south stretch of coast bordering the Atlantic must have a marked influence on their climate and the opportunities for migratory work. Honey production is some three times that of the UK and they only import a fifth as much.

**pose** ***
To adopt a characteristic attitude as when a bee challenges another, fans, invites grooming or indicates humble submission to an aggressive fellow bee. When a colony is uncovered in winter, bees may initially be reluctant to leave the cluster and attack but large numbers will immediately adopt a tail-up gait even protruding their stings which may even be adorned with a droplet of venom. Any such exposure by the beekeeper, should be of very short duration and for a specific purpose such as checking the seams of bees for subsequent pollination strength.
See: seam of bees.

**post**
A strong piece of timber or metal used as a vertical support. Nucleus hives can be perched on top for ease of operation and for the flight and return of a queen while protecting them from ground borne predators. Chicken or barbed wire, strung between posts can make a fence to keep cattle or sheep away when a migratory site or out-apiary is set up or netting perhaps on an allotment. Laid horizontally they can raise hives off the ground and act as runners.
See: bee bob.

**postage stamps** ***
Many stamps have illustrated bees and hives and though rare in the UK (despite the importance of beekeeping) many countries, especially while hosting Bee Congresses, have made special issues portraying beekeeping.

**postal cage** ***
Specially designed cages, often interlocking, have been in use including for air travel. They are always ventilated but without fear of stings for anyone handling them, have a supply of 'candy' and often a means of using them as an introduction cage. The contents would of course include a queen together with around ten young or suitable workers (one's that are able to feed and attend to the queen). E-mail or telephone calls have brought queens from far and wide 'post haste' but the ugly problem of spreading disease has brought restrictions from some countries and a consequent reduction in the trade. Recently 'Hawaii' due to the small hive beetle.

**post-cerebral gland**

A much branched pair at the back of the head also called the occipital salivary gland. It lies behind the brain of the bee, its acini having a rather characteristic shape. Their duct joins the median duct that carries thoracic salivary secretion and then passes under the pharynx. Well-developed in the queen and workers but atrophied in drones. See: acinus.

**post code**

On its arrival few people realized the terrific (useful) affects it would have. Previously, one might ask, Could you tell me..? however did we find our way through the tortuous countryside. Now as a navigational aid it is quite superb. It has been suggested we use it for 'branding' hives.

**posterior**

Coming after or situated at or relatively near the hind end. Caudal. A bee carries its sting posteriorly.

**posterior intestine – 'rectum'**
**opp. anterior.**

See: dorsal

**postgenal glands**

Small sac-like glands that are situated behind the inner walls of the fossa. Little is known about them but they comprise a flat mass of delicate cells which open through a minute canal in the inner face of a supporting plate near the mandibular glands. Almost certainly associated with pheromones. Homologous in some wasps, bumbles and other hymenoptera.
Perhaps more up-to-date info is available.

**postmentum (submentum)**

A small triangular segment which is articulated between the lora at the proximal end of the inner members of the proboscis. It is attached distally to the prementum.
See: prementum and proboscis.

**post nuptial**

After mating the drone descends somewhat clumsily to the ground only living for about as long as a worker that has lost its sting. A queen is often seen to trail part of the drone's genitalia as she returns to her colony. Once freed from the drone's encumbrance her interest in his fate is seemingly non-existent and she probably intends to mate again with other suitors.
See: wedding certificate

**posture**

The position or condition of anything at a particular moment. The stature of a guard bee with head and forelegs raised up from the other legs and mandibles menacing. Or the stance of a beekeeper examining a frame of bees. One's attitude. Or the unusual tilting of a heavy hive that threatens to fall over.
See: back ache and pose.

**pot**

This word has been tied up with beekeeping through the years and was used for beehive and of course honey pots were famous before glass 'jars' came into use.

**potash**

Potassium carbonate as derived from wood ashes is one of the ingredients that form fertilizer applications. Potash boosts the sugar content of clover nectars (and other legumes) and also materially assists in the ripening of seed for harvesting. N.B. Potash should not be given prior to spring grazing on soils where hypomagnesaemia is a problem.

**potassium (K)**

A silvery white metal which oxidizes rapidly in air and reacts violently in water to form alkaline potassium hydroxide. Similar in many ways to calcium. Various salts of potassium are useful in beekeeping and honey contains more potassium than any other mineral.
See: cyanide, potash and potassium nitrate.

**potassium nitrate $KNO_3$**
A white crystalline compound known as 'saltpetre'. Used in explosives, matches etc. Use one pound to a gallon (2 oz. to a pint) of water to produce a dip to wet the ends of smoker cartridges so that they light more easily. Stain with dye but get rid of fumes before using the smoker on the bees.

**potherb**
Herbs used in cooking either as foods in themselves or for seasoning. Many of them flower attractively and provide provender for both man and bee.
See: herb.

**pound lb.**
A unit of weight or mass equivalent to 16oz or 0·45359 kg. Syrup feeds are suggested as sugar 1:1 for stimulative feeding and 2 to the pint (1·6kg per litre) for thicker winter feed. A note at the time of switching to decimals was 'and it took a little while longer before inflation pushed honey through the one pound a pound barrier. The pound has taken a good pounding!

**pour**
To cause a fluid to pass out of a vessel and usually to aim it at a target area.
A beekeeper might fill a bucket with syrup for transportation to an out-apiary. If the bucket is full and a strong wind is blowing spillage will lead to waste and worse still robbing. A vessel with a spout like a watering can could help. When pouring honey into containers avoid any unnecessary 'free-fall' as this will incorporate air bubbles. Where possible allow honey to run down a slope rather than fall in a stream through the air. To handle bees when it pours is possible with an umbrella but not pleasant.

**powder**
Fine loose particles like dust. Pollen is often referred to in this way but bees collect, -'handle' it and pack it as pellets in their cells. The pollen mite, when given the opportunity, will convert stored pollen into powder that will fall like a shower out of old comb if held flat. Some prophylactics can be applied in powder form and flour for uniting or icing sugar for Varroa are two examples. Once a bee's spiracles get blocked with powder it is in some difficulty unless helped by other bees.
See: debris, frass and pollen mite.

**powerful**
The drone is a big powerful bee yet his entire power is devoted to flying and mating. The increased noise and activity of their flights around the hive in the middle of the day can give the impression of anger and in fact once when a colony had been moved just too far for workers to return to the old site, a mass of drones did fly at the face and actually pounded the beekeeper, making themselves a nuisance. Being non-aggress due to their lack of armament they are prey to wasps, spiders, birds and other predators. Their autumn turn out is performed with only the merest recalcitrance being shown. The application of power for extracting, carpentry etc. is an important aspect of beekeeping.

**pp.**
Abbreviation for pages, postage and packing - See spp.

**ppm**
Parts per million – this degree of measurement is used to set levels on the permissible amounts of toxic materials permitted in water, foods, honey etc. also of material in the atmosphere. The Icelandic volcano recently brought jet flying to a standstill the danger level being ppm of airborne dust. A drone is sensitive to incredibly small amounts of queen substance when flying at congregation levels.

**PQN peak queen cell number**
This is the maximum queen cell possibility at any given time, calculated by adding together the number of queen cells with brood in them, plus the number of virgins

and mature queen(s). From this figure an assessment of a colony's swarminess can be made. From lecture by Beo. Cooper, while Director BIBBA.

**Practical examination (BBKA)**
Candidates must have passed the preliminary examination and are required to have kept bees successfully and have a thoroughly sound practical knowledge of the craft and the skill to apply it. A successful candidate is known as a Practical Beekeeper and is qualified to enter for the examination in Judging. It is a practical examination covering all aspects of beekeeping and is aimed at those with wide skills but who are disinclined to take a written examination. Still valid?

**practices**
Dr. L. Bailey suggested that beekeeping practices that lead to the aggravation and spread of diseases include: Disturbing colonies unduly, hindering their normal development, keeping colonies queenless, moving brood or using contaminated combs and giving them unsatisfactory food.

**pratensis**
Specific name used to imply that a plant or creature is usually found in the meadow, e.g. *Trifolium pratensis* – red clover.

**Pratley uncapping tray** ***
Made in tin-plate or stainless steel, this is a water-packed jacketed tank (deep tray) which can be connected to the mains electricity supply and used hot. It has a trough for heating knives and an uncapping rest ( including hole for the frame lug) and the honey and cappings sliced from the comb fall onto a (heated if desired) sloping platform which quickly melts and separates honey from wax and allows the mixture to form a cake on top of the honey.

**Praying mantis** ***Mantis religiosa***
A carnivorous insect. There are several mantids that have been reported as eating honeybees though they seem just as partial if not more so, to flies. While at rest they fold their front legs as if in prayer. They are quite rare in the UK.

**precipitation**
The deposit that settles out of a liquid. This can be due to foreign matter or a lowering of temperature if the liquid was saturated. Syrup that is made up too thick at high temperatures may deposit crystals that prevent the bees from gaining full access to the syrup. When wine or mead is made the 'lees' precipitates out and clearer wine can then be racked or siphoned off. Proprietary substances may be obtained from the pharmacists to encourage the clearance of wine by precipitating the elements that cause it to cloud. A judge always inspects the underside of a jar of granulated show honey, as so often small specks can be seen there. From the meteorological viewpoint anything that descends to the earth from the skies is regarded as precipitation whether hail, rain, sleet, snow, volcanic dust or strange creatures.

**precursor**
Something that indicates a future happening. Drones are a precursor of the swarming season (swarms or queen rearing). The bursting of flower bud - a possible flow, the hasty sudden return of foragers – a shower in the offing.

**predators** ***
Creatures that pillage, plunder, rob or prey on other animals or exert a wasteful influence on them. Small predators of the honeybee include: ants. flies, bumble bees, chelifers, dragon flies, earwigs, frogs, hornets, the wax moths, lizards, moths, mites (specialised) spiders, toads, and wasps. Amongst birds and mammals are badgers, green wood-peckers`, kingfishers, man, mice, rats, shrews, shrikes, squirrels, sparrows, swallows, tits and sea gulls. There are many other exotic predators: apes, bears, honey guide birds, sphinx moths, opossums, praying mantis, raccoons, ratels, skunks, and

yellow bee pirates. How do bees survive? You've guessed it- their stings.

**preliminary exam**

(up-date)
The syllabus for this BBKA exam was available and although it comprehensively covers a wide range of beekeeping it is first principals that are looked for rather than a detailed knowledge. The candidate will be asked to manipulate a colony and to have a reasonable knowledge of its contents. Also subjects dealt with orally will cover bees natural history, disease, poisoning, harvesting and to know the names of hive parts and equipment. The passing of the preliminary is a necessary first step to all subsequent BBKA exams.
See: intermediate.

**premature**

Arriving before it is due. The emergence of brood is remarkably punctual though delays can occur due to temperatures, prematurity is less likely. A 'pulled' virgin is the name given to a virgin that is persuaded to climb out into the beekeeper's hands.
See: pulled virgin.

**prementum**

The base of the median labium of the proboscis. It is a long cylindrical plate joined to a smaller, triangular postmentum. It bears the distal appendages: paraglossae. glossa and labial palps.

**preparation**

Actions taken in accordance with a plan to ensure that you have the knowledge, equipment and other necessities to perform a given task. In few fields is the need for careful preparation more necessary than in beekeeping. Moving hives for example requires careful, thoughtful preparation.
See: plan and preparation for exams.

**preparation for exams**

Nothing substitutes for experience with the bees. However as all the varying conditions take endless time, much benefit and pleasure can be derived from the widest possible reading. It is usually helpful to work within an group or, if needs be with just one other, so that discussion, question and answer can take place. Writing answers to questions should always be part of your study methods. A happy, relaxed approach is most helpful.
See: taking exams.

**preparation of microscope slides**

Sometimes material, like a wing, can be put straight onto a spot of water on a slide and held flat by a cover slip and viewed until the water evaporates. Otherwise material may have to be dissected, sliced, preserved, fixed, stained, bleached and hydrated before setting and mounting in a medium. Fixing can be done with 70% alcohol 99parts with 1 part glacial acetic acid. Stains include: nigrosin, basic fuchsin, methyl green or blue etc. Dehydration requires increasing strengths of alcohol up to absolute or anhydrous acetone. Clearing makes use of xylol, clove oil, benzol, cedar wood oil (clove-oil is more tolerant of water traces) Mounting in xylol-balsam mixture permits evaporation of xylol. Wax can be used to seal the edges of the mount when quite dry.
See: sectioning, preserving, maceration and bleaching.

**preparations for swarming**

These are of course made by the bees rather than the beekeeper.
See: swarming preparations.

**prepare**

To make ready. Beekeeping is rather dependent on having made things ready. To be caught out 'living hand-to-mouth' can lead to disaster. Under the various headings preparatory steps will be found dealing with candy for bees, comb for honey extraction or rendering, for exams, for a flow, handling bees, the heather, pollination, queen introduction, requeening, syrup feeding, transporting, uniting and wintering.

**preparing show wax**

Use the whitest cappings (from a solar wax extractor if possible) and filter through a fine white fabric of flannel (use lint fluffy side up) using a minimum of heat. Use a porcelain or glass mould that has been carefully polished with yellow soap. Support the mould in a large vessel of soft (or slightly acidulated) water. When the wax has developed a crust, scoop some water gently onto the wax, cover and cool slowly. The wax will float up when cool. Polish cake with the ball of your thumb. Keep wax in light honey to retain its aroma.
See: beeswax and render.

**preservative**

A substance that can be used to prevent decay, decomposition or fermentation. Certain preservatives are more suitable than others for preserving timber. Formaldehyde is used to preserve plant or animal tissue so that it can be examined at a later date.
See: hive/wood preservative and pentachlorophenol.

**preserve**

To keep in a good healthy or sound condition. Honey does not need preserving merely closing in an air-tight, sealed container. Honey itself can be used for preserving fruit etc. Pollen can be preserved in combs in plastic bags placed in deep freeze. Honeycomb sections likewise.
See: wood preservatives.

**preserving fluid**

For the safekeeping of material to be used for subsequent microscopy, a solution known as Carl's solution can be made as follows: 17parts 95% alcohol, 6 parts formalin, 28 parts water and 2 parts glacial acetic acid added just before using.

**President**

A person appointed to preside over an organized body of people. As the status of such a group is enhanced by such leadership it is not uncommon to find those with great names invited to the post. Generally in beekeeping circles officers by dint of generous and efficient service, rise through being Secretary or Chairman to Vice-president before serving a term as President.

**press**

The collective of printed publications and that section of the media that deals with them. Unfortunately the press often misinforms the public on matters related to beekeeping. As languages were from the earliest times formed around activities such as ours, expressions like 'stung into action', make a 'bee-line' or honey, honey(to sweeten anything) gives reporters access to an easy line of jokes which seem to please them more than the facts. There is of course the 'bee-press' which tends to be read by those keen enough to realize its value. Associations would do well to appoint a press or public relations officer.
See: bee-press, heather press, journals and magazines.

**pressed honey**

Honeys other than the thixotrophic heather honey are rarely pressed out of the combs but spun out by centrifugal force. When honey is expelled from the combs by pressure, a great deal of force is required because although the comb is really quite flimsy it still has surprising strength when efforts are made to squash it. Once passed through straining cloths and air bubbles allowed to rise, honey is much the same whether pressed, squashed or expelled centrifugally.

**pressure -atmospheric**

Atmospheric pressure forms patterns whereby air tends to move from high to low pressure areas. However the rotation of the earth causes anti-clockwise winds around 'lows' in the northern hemisphere and clockwise around 'highs'. The rotations are reversed in the southern hemisphere. Generalisations are not possible but lows are associated with unsettled, changeable and often windy weather, while highs may bring thunderstorms, fog or winter overcast but usually light winds and pleasant conditions that can set in for considerable periods. A link between weather, flora and time of year can be useful for judging likely bee-activity. The wonderful advances made in the display of meteorological conditions on the TV give us all a much better idea of what is going on.

pretarsus The bee's foot or 5th tarsomere. See: foot.

**prevailing wind**

In Britain we get plenty of winds from the south-west but many aspects of wind shelter would include the topography, strength and temperature of the wind and position of natural wind breaks. Wind can play havoc with queen mating and the bees' maximum, working flying speed is of nothing, once winds or gusts reach speeds of over twenty five miles an hour. The build-up of snow will also be relevant whenever a wintering site is chosen.

See: cross-wind, wind and wind breaks.

**prevention of disease**

To maintain healthy stocks a good knowledge of diseases is essential Carelessness and ignorance are probably the main causes of the spread of diseases. Apiary hygiene should always be foremost in your mind when systems of management are thought out and applied. Replace queens and brood combs before they reach the third year and be cautious and thorough when dealing with your own or second-hand equipment, swarms of uncertain origin etc. Always sterilize combs and hive when they have been emptied of bees.

See: fumigation and sterilization.

**prevention of drifting**

Despite all the accounts of bee's navigational ability to forage and return to their own hive, it is quite evident that the last few yards are almost certainly dependent on visual references. Distinctive colours and shapes help returning bees and the heights and direction entrances face also influence them. To place hives in a neat row may look nice and suit you but it is abnormal from the bees' point of view and drifting may well occur, so much so that a steady cross-wind during a honey flow can result in a crop spectrum ranging regularly along the row according to the wind. The importance of individual hive entrances comes to the fore when mated queens return from their wedding flights and the spread or avoidance of disease by making it as easy as possible for colonies to locate their own entrance, are also points worth bearing in mind.

**prevention of fighting**

Many beekeepers fail to appreciate that robbing is going on until fighting occurs Safeguard against both by keeping colonies strong, with entrance size matched to colony strength and the likelihood of robbing. Always 'play safe' when uniting colonies by taking all sensible precautions – see 'uniting'. Imitation rain or wet grass heaped over the entrances may help to quell a struggle. If one particular colony is taking the brunt of attacks, either move it three miles or more away or interchange its position with one of the attackers.

**prevention of robbing**

Bees are inveterate robbers and there is no faithfulness between colonies, only by keeping up colony strength coupled with control of entrance size can robbing be avoided. It should be realized that much robbing goes on unnoticed. Furthermore once a bee has taken to robbing it will continue to do so and even recruit others. Avoid unnecessary exposure of colonies when manipulating (use cover cloths). When feeding, do so at dusk, only giving what the colony can put away in a reasonably short time. Avoid spilling syrup, do not leave comb lying about. Never expose honey-wet combs to let the bees 'clean up' as this will attract possible disease from afar. A sheet of glass in front of the entrance will deter robbers and a robbing session can be quelled by hanging a wet sack over the colony or wet grass over the entrance. Try not to get into a position where such actions are needed.

Robbing by ants and wasps is another aspect, oil, grease and borax can be used against ants but wasps, if they are encouraged by your having fed sugar syrup or a strong nest is near can destroy a colony or weaken it and even harm the queen if they get away with a prolonged attack. Close entrances to a one or two bee-way and find and destroy nearby wasp nests. A funnel, petrol and dusk can reach an underground nest. The fumes and liquid are quite fatal to them, no need to burn your eyebrows by dropping a match in for good measure
See: prevention of fighting.

**prevention of swarming**
Despite the fact that swarming is a perfectly natural phenomenon and that bees seem to prosper when allowed to swarm (new queens are reared under ideal conditions), methods to control or prevent swarms are numerous. The use of young queens (never taking them into a third year) and using a strain not given to excessive swarming, keeping ahead of them with brood and storage space, all help. Clipping a queen's wing may hinder a swarm's departure. The annual need to produce a first class queen can be linked with swarm prevention. Intricate, involved methods with specialized equipment are not to everyone's taste, but transferring written instructions into purposeful activity affords an interesting experience.

**preventative**
An agent or measure that helps to ward off disease. Although there are many prophylactics that are used to reduce the virulence of disease, to control or eliminate disease preventative measures do not involve drugs but management aimed at keeping colonies healthy. Replacement of comb with sheets of foundation, replacement of queens and the cleaning and sterilization of equipment are examples of preventative measures.
See: prevention of.

**Priapus son of Dionyous**
Like Pan he was also a guardian of beehives. Deity of gardens and vineyards the personification of male procreative power as a God. Half-goat, half-man –replete with macrophallus.

**prick**
A stab by a tiny point such as a pin or a thorn. When such a point is as small as a bee's sting it would scarcely be felt were it not for the inclusion of pain producing venom. A prick or two is all that is needed when newspaper covers the end of a queen cage (say hair curler) and in the layer of newspaper when two alien colonies are being united. Heather honey can be released by pricking the cells.

**prime**
To prepare or make ready for a particular purpose. A colony may be primed by 'plumping' with another's brood to increase its foraging strength or getting sections worked. When filtering, a dry cloth can impede the passage of honey until the under surface has been 'primed' by touching it with a saucer or such like. The first swarm is called the 'prime' swarm if smaller ones follow they are 'casts'.

**primer**
When a fine grained sample of honey is stirred into a liquid sample to encourage it to precipitate granulation we would call that a 'primer' and the process 'seeding'.
See: seeding and starter.

**prime swarm**
The first swarm that a colony throws in a particular season. It is usually a large one and includes the colony's laying queen and a mixture of bees of various ages. In text book weather conditions, prime swarms issue in May when the first of the replacement queen cells have been sealed and the mature drones are on the wing. Top swarm and plum swarm are other names.

**princess**
The non-reigning female member of a royal family. The fanciful name given to a virgin queen (it was 'prince' in earlier times).

**princess tree *** *Paulownia tomentosa***
Foxglove tree with large erect inflorescences of pale violet, bell-shaped flowers that the bees crawl right into. Hardy when mature though herbaceous when young. It likes the sun and a good, well-drained soil. Similar to the Catalpas.

**privet *** *Ligustrum vulgare* Oleaceae**
Garden privet is an evergreen shrub with small, white, tubular flowers. It imparts a slightly undesirable odour to honey which when pure is thick and dark. The plant is very attractive to bees in June and July and beekeepers often take supers off rather than get their good honey contaminated. However the honey will improve with keeping. Hedges are usually clipped short but large sources are available whenever privet is allowed to develop to it full stature. Its pollen is pale green, 24mu and it likes calcareous soils.

**prize**
A complimentary token of one's skill, luck or enterprise in satisfying the conditions laid down for its award. Honey Shows allot prizes for winning entries and in so doing encourage the presentation of honey to reach the highest standard.

**probability**
The likelihood of an occurrence. Weather is forecast on the basis of probability, in some circumstances the probability of honey flows can be foretold. Disease and swarming might also be assessed from the point of view of probability. In statistics '0' means it will not happen while '1' means it is certain to happen. Therefore the closer a figure of probability is to '1' the more likely it is. See: statistics.

**problems**
*Is the colony queenright?* see Q.rt test.
*How to cope with a granulated super*
– see 'rapid granulation.
*How to get disease diagnosed*
– see 'sending a sample'.
*How do I deal with bad tempered lot*
– see 'requeening'.
*How do I find the queen? see 'queen finding'.*
*How can I transport my hive?*
– see 'moving bees.
*When to use thick or thin syrup?*
– see 'syrup strength.
*What food reserves should my bees have?*
– see food reserves.
*Dead bees in front of hive –See 'dead bees'.*
*can I use creosote,* – see creosote.

**proboscis *** pl. proboscises**
Not permanently functional as it is in most insects but temporarily improvised by bringing together free parts of the maxillae (galeae) and labium (labial palps) to form a non-rigid tube inside which the glossa can be moved back and forth. When not in use it is drawn up behind the head by swinging on the suspending cardines. When protracted the broad galeae (front) and labial palpi (rear) form a tube. When the muscles are relaxed it is retracted, but two strong muscles can erect it into the working position. The tongue moves rapidly back and forth while its flexible tip is swung around with an agile 'lapping' motion. The long, hairy tongue has a cross-lined appearance due to the presence in its wall of hard rings bearing hairs separated by smooth membranous intervals which enable the tongue's length to vary. Relaxation of two strong muscles enables the elasticity of a rod in the tongue to straighten it. The food canal connects with the mouth proper while the sucking apparatus is contained within the head,. Blood pressure allows the tongue to be inflated for stropping and clearing. Tongue length varies 5·9 - 7·1mm.

**process (anat) A natural outgrowth,**
projection or appendage. As a means of improving we would say a systematic

series of actions performed with a view to achieving certain changes. When dealing with hive products, as honey is a unique food substance and wax is a unique material capable of a thousand uses, processes associated with these open up a complete new realm of activity with as much fascination and usefulness as beekeeping. See: extracting, rendering and recipes.

**processing micro slides**
Bleaching, clearing, dehydration, fixation, maceration, mounting, preservation, sectioning and staining.
See: these titles and preparation micro/ pollen slides.

**processing honey**
Bees process nectar into honey. Honey removal, uncapping, extracting, settling and containerizing are all part of the beekeeper's processing of honey.
See: honey extraction.

**processing wax**
Old combs, scrapings or cappings can be rendered into clean wax. The best quality is likely to come from washed honeycomb cappings. It is important that only soft or acidulated water is allowed to touch the comb and old comb should be sliced and allowed to soak for several days before being boiled. The hot liquid wax is extremely fluid and can pass through a variety of filtering agents. Even burying in sand and boiling in barrels or tied in weighted 'old' tights or melted in the solar wax extractor. Temperature controlled ovens, micro-wave ovens, steam heat and hot water methods call for many ingenious types of apparatus. The stable nature of wax (apart from ravages by wax moth) allow wax to be kept for unlimited periods.
See: beeswax, rendering, wax extracting plant and solar wax extractor.

**proctiger**
The last segment situated within the sting chamber on females and bearing the anus. It is connected to and comes away with the sting. Drones have two 'claspers' on either side of the proctiger which is on the final and partly visible abdominal segment.
See: anus, scent and wax glands.

**proctodeum**
The embryonic hind gut.

**produce**
Anything that is brought into being. The most fundamental language tells us that bees make honey. When exploited by man he is regarded as the 'producer'. In the past wax was more durable and useful than honey. Due to the utter cleanliness of the bees every scrap of material from the hive has been proved useful in one way or another. Not only does the bee farmer produce services like pollination but honey, wax, propolis, pollen and royal jelly all of which have countless uses and applications, leading to products such as: mead, vinegar, candles, cold cream, dubbin, polishes and lubricants
See: hive products.

**production of beekeepers**
Agricultural Colleges would seem reasonable locations for the training of beekeepers yet as they close down at the end of June the most useful beekeeping part of the year eludes them. The production of beekeepers in the UK has been somewhat haphazard. In schools it was sometimes squeezed into Rural Science but tended to get lost as was the case with horticulture where beekeeping was looked at as quite an insignificant aspect of plant welfare. Beekeeping Societies began in earnest with the BBKA but everything was done on a voluntary basis with little or no help or encouragement from the Government. Not until 2009 did somebody wake up! Better late than never…

**production of bees**
Queen bees by air mail, package bees – in parts of the world where weather conditions favour and economic situations

allow, enormous strides have been made in producing the fundamental requirements of the beekeeper, namely bees! Partly due to our weather and it might be said inertia on the part of the government the UK has one of the lowest percentages of beekeepers. Consequently we import ten times more honey than we produce and fruit and nut production have leached away so that most of these things come from abroad too. Steps taken to avoid the importation of disease have caused a further decline in our ability to have more bees and more beekeepers. The influence of weather can be appreciated by the southern states of the US having bee farmers galore while in Canada there are honey farmers galore.

**production of honey**

Large scale honey production calls for reasonably cheap hives and good honey flows. These conditions are found in Canada, USA, Australia, NZ and S. Africa (where mainly Langstroth hives are used). China, Russia and India each have their own stories to tell. The UK probably has the dearest wax, wood and fuel which have been at the crest of inflation and are all vital in the realm of modern beekeeping. Many a thriving enterprise has fizzled out because of the vicissitudes of British conditions.

**production of judges**

Honey Show judges came into being as a result of the shows themselves. Exhibitors in their efforts to become more and more successful sooner or later ask for or are given the opportunity to act as a judge's steward. From such activities they can learn the appropriate skills required and then with the accepted qualifications enter for the BBKA Judge's Exam whereupon they are officially listed, if successful, and are then in great demand up and down the country.

**production of lecturers**

As beekeeping is such a specialised subject there is no course of training strictly appropriate to the production of qualified beekeeping lecturers in this country. The BBKA Hons. Lecturing and National Diploma in Beekeeping examinations ensure a fairly high grade of lecturer even though they do not carry the weight of University qualifications.

**production of pollen**

Plants produce pollen in the form of male germ cells leading to fertilization. That it provides the honeybees with a source of protein is fortunate. The bees' storage of pollen is proportionate to their needs unlike nectar which they are willing to store until further orders. If the beekeeper wishes to produce pollen as a crop it is not usual to raid the combs but to take steps to dislodge the pollen pellets from the legs of returning foragers. Traps can do this but are an interference with the bees' economy and are best used for short spells. Trays containing the freshly collected pellets are checked and the pollen can be kept for long periods in deep freeze, otherwise, once stale, much of its food benefits disappear. There are other ways of getting pollen but the wonderful efforts of the bees have led us to stealing it from them.

**production of propolis**

Colonies vary in their propensity to collect and use propolis. It usually takes the form of a gum that is not easy for them to collect and beekeepers usually scorn their efforts, scrape it off and throw it away. It comes in various qualities and colours but the bees give it a lasting quality adding their antibiotic in the process. Used to water-proof and sterilize their nest it is found in parts of the hive to suit their convenience and not easy as a rule to harvest. Their tendency to clog mesh has been used as a method of allowing a supply to be taken from them. It has a pleasant, enduring, balsamic smell and where sizeable amounts of clean propolis can be slipped into a jar, it is worth doing so because it offers non-beekeepers an opportunity to appreciate a characteristic, hive-like smell which along with beeswax gives the recipient a sensation of that redolent aroma of bees and their nest.

**production of queens**

More queens are produced naturally every year than we can ever use. Like sycamore seedlings that sprout in the gutters – individually useful but the majority go to waste. Although every beekeeper, knowing queen cell raising is a perfectly normal feature, could incorporate queen rearing with normal hive management, many, failing to realize what is needed to turn the contents of a queen cell into a laying queen, let the opportunities go by. Small queen mating boxes only need a cupful or so of bees, a supply of food, some comb (say the size of a section) and a queen cell from a healthy colony which bears the characteristics looked for e.g. prolificacy, health and energy –for docility we look to the drones. Artificial grafting or temporary dequeening to stimulate queen cell production, are alternatives to waiting for swarm cells to be built. Whatever method is undertaken every new colony needs a new queen and spare mated queens are an asset. When produced commercially they are rightfully expensive and will lack some of the virtues that bees in your own locality, in your own apiary, possess
See: queen propagation.

**production of royal jelly**

Although bees naturally produce royal jelly, to encourage them to produce more, calls for the use of a queenless colony. The finest quality of royal jelly is produced by strong healthy colonies. It has to be removed from the queen cell, freed of its embryo and put into a preserving medium immediately. Alcohol and freeze drying are used and commercially there will be secrets untold. Its value to humans is doubtless exaggerated but as a money spinner it has found an accepted market. Advertising helps. White and creamy, with a sharp taste of abietic acid which seems to disappear from the capsules available from various pharmacies.

**progeny**

Offspring, issue, descendants. In an ordinary colony of honeybees all members are the progeny of the one queen mother. A queen has to be judged by the effectiveness of her progeny.

**programming**

1. To arrange the order and nature of events at a beekeeper's meeting, field day, week-end course etc.
2. To cause a machine to follow a particular performance.
3. To feed a set of instructions written in suitable language into a computer to enable it to perform logical operations on data.
4. To arrange learning processes –'programmer'.

**progressive feeding**

The manner in which nurse bees feed the young. When food is taken to the young larvae, stage by stage, as opposed to the mass provisioning that occurs when an egg is placed on a bed of food and the larva left to fend for itself. Regarded as an advance from the solitary insects to the social insects method of feeding their young. Honeybee nurses keep what amounts to a continuous watch, feeding the larvae at frequent intervals governed by pheromonic control. Once the nurse bees begin to put food into a cell the young larva is virtually floated on its food. On the third day less food is given and when this reserve is eaten nurses switch over to progressive feeding or feeding at intervals. In all there are about 10,000 visits, something like 1300 a day and as many as 3000 bees may be concerned with the nursing of one larva, taking some ten hours of attention. Queens receive ten times more visits than workers.
See: mass provisioning.

**projecting microscope**

In the days when transparent microscope slides were projected onto a screen, often in a completely blacked-out room, things could be quite cumbersome. Now-a-days much of this type of display has been replaced by 'power point' use of lap-top computers, static television etc.

**projector**

An apparatus that uses light and lenses to project and enlarge an image. Such advances have now been made in the realm of visual and audio aids that fantastic electron microscope pictures are commonplace on telly and in the classroom that a complete new world has been opened up, showing things never seen before and hastening and improving the learning process. What unbelievable marvels these would have seemed in the past.

**Prokopovich Peter J,**

The great Russian beekeeper said to have used movable frames as early as1841 and to have described self-spacing sections with 4 bee-ways. His very useful hives opened at the side not at the top. One of the earliest inventors (c.1806) to use frames for supers, though there was no clearance between these and the hive walls and bees were free to build comb as they wished in the lower portion of the hive. He lived in a forest area and doubtless some ideas followed the traditional forest beekeeping.

**proline**

Common amino acid found in protein and in honey.

**prolific**

Fruitful, capable of producing abundant offspring. A prolific queen needs a larger brood chamber than one who is not prolific. One noticeable difference between beekeeping in Britain, Canada and NZ for example is the small size of British colonies. Large colonies require a prolific queen and the right environmental conditions. Where massive takes can be counted upon this type of queen comes into her own but this rarely applies to the UK.
See: fecundity.

**pronotum**

The tergite of the prothorax or prothoracic collar. It bears two hair-fringed lobes that cover the first pair of spiracles. In dissection for acarine the collar is peeled off with a circular motion to expose the full extent of the first pair of thoracic spiracles.

**pronucleus**

The haploid nucleus of a sperm or egg prior to fusion of the nucleus in fertilization.

**propagate (make increase)**

To cause plants or animals to multiply by any natural process from the parents. Making nuclei for increase, splitting. dividing and hiving swarms, would all come under this heading.

**properties of wax**

Amorphous solid material, insoluble in water, melting point 62-65C having become friable at 49C (120F). It is resistant to the passage of heat. In its liquid form it has great penetrative qualities being extremely fluid and it possesses a golden colour, looking much like honey. It has a high specific heat, remaining hot for a considerable period of time and contracts upon cooling leading to 'cracking' unless the process is carried out very slowly. The solid feels smooth (silky almost soapy) and has a pleasant, characteristic aroma. Its colour is almost a translucent yellow but it can be bleached white. It floats on water having an s.g. of 0·952 - 0·975. It is soluble in acetone, hot amyl alcohol, benzole, benzene, carbon tetrachloride, chloroform, dioxin, ether, isoproptyl ether, paraffin, turpentine etc. and has an acid number 16·6 - 20·7, RF 1·4398 - 1·4451 and is hypoallergenic. It saponifies with alkalis and emulsifies well on boiling with a little water and potassium carbonate.
See: adulteration, beeswax, colour of wax,

other waxes, strength of, tests for, propolis and temperatures.

**properties of honey**

Honey is a supersaturated solution of simple sugars, its average 's.g'. is 1·415 @ 60F (15C) and its water content 17 -21%. 100g produce 303 calories. It is hygroscopic and sticky,: heather honey and Manuka are thixotrophic, Colour varies from extra white through amber, gold, green, pink and mahogany to black. Where dextrose dominates granulation occurs rapidly but the solid form can be reverted to liquid by heating. It is a stable substance with long-lasting qualities if kept sealed in air-tight containers. Its refractive index (RF) is 1·49. A piece of filter paper saturated with honey produces a characteristic fluorescence under U/V light. It is a unique substance produced only by honeybees. After 78 years using honey instead of sugar the author claims that it is not only a very pleasant and nutritious food but a medicine and ointment too. The honeybee is grossly undervalued unless its many 'unique' and special qualities are appreciated: honey, comb structure, wax, branched hairs, controllable life cycle and contribution to the benefit of its environment.

**properties of propolis**

An aromatic substance collected from trees etc. such as poplar and horse chestnut. It is brittle when cold, heavier than water and melts at about 150F (65c). It dissolves in molten beeswax, ether, chloroform, carbon tetrachloride and partly in alcohol, turpentine, petrol and ammonia. It is sticky, water-proof and tenacious at hive temperatures. It is anti-biotic and used as a gargle etc. It has a sweet, balsamic aroma which clings to it (useful in a potpourri) Colours vary according to source but yellowish-red or brown to black can be found. Composition: 55% resins and balsam, 30% beeswax, 10% aromatic oils, 4% pollen, and bioflavanoids, phenolic acid and minerals.

**prophase**

The initial stage in meiosis or mitosis during which chromosomes appear within the nucleus of a cell and undergo pairing as the nuclear membrane begins to disappear and meiotic division proceeds. In meiotic division it occurs in 5 stages as a thread thickens in phases: lepto, zygo, pachy, diplo-tene and diakinesis.
See: meiosis and mitosis but use official text book for verification.

**prophylactic**

A substance which defends or protects from disease, a chemical or drug used for this purpose. As our knowledge of diseases grows so does the list of suggested prophylactics – sulfathiozole, FumidilB, terramycine (TAFSP), Frow, Folbex. Phenothiazine, Penicillium waksmannii, have probably all become old-fashioned.
See: varroa.

**proprionic anhydride**

A colourless, volatile chemical used as a repellent to remove bees from supers etc. It is used in a 50/50 mixture in water.
See: bee-go, benzaldehyde and carbolic.

**propodeal spiracles**

The two symmetrical, musculated spiracles of the propodeum are large and effective as they play great part in the 'breathing' or the refreshment of the bees' muscles with oxygen so it is not surprising that they exist close to the great indirect flight muscles of the thorax. They lead into the large thoracic air sacs. These are of course very important in the drone who can only copulate when its air-sacs are fully inflated

**propodeum**

The fourth or distal segment of the thorax. It has the same structural characteristics as the abdominal segments and is therefore thought to have been a part of the abdomen on an earlier ancestor of the honeybee. It has no pleurite and bears spiracles that are musculated and larger than those of the

abdomen. Its large convex tergite completes the shape of the thorax and tapers to a narrow fit around the petiole.
See: alitrunk.

**propolis Gk 'Before the city'**
A gummy, resinous, aromatic hive product collected by the more mature foragers from such sources as poplar and horse chestnut trees etc. It is brittle when cold but 'tacky' and water-proof at hive temperatures. Colonies collect variable amounts – around 150 – 200g p.a. It is soluble in wax, ether and chloroform and is saponified by alkalis. Used by the bees to line their nests, fill gaps and strengthen combs. It keeps in odours that might attract predators. It can be yellow, red, brown or black. Maltese bees collect tar from the hot road surfaces in the absence of trees. Caucasian bees use more than Russian dark forest bees. Far East bees use little and cerana, florea and dorsata none. It is not stored neatly but in streaks and lumps. It can harbour disease spores despite having pharmacological properties. 1 in 2000 beekeepers are allergic to it. Once used to varnish violins. Some bees build pillars of it to protect the entrance. Balm, bee gum and bee glue are other names for it. Used with a hot putty knife it makes a good wood-filler. The Greeks gave it the name meaning 'before the city'
See: properties, uses of propolis and propolis collection.

**propolis collection**
It is collected during the warm hours of the day by the more mature foragers. The workers may return to the hive for sustenance without unloading. It is collected from leaves, buds and the bark of conifers. Colonies may collect up to 200g per year though some races are more avid collectors than others. In suitable conditions the gum is loosened by the mandibles and carried in shiny (pollen is never shiny) lumps in their corbiculae. Once in the hive, the bee need help to unload it and failure to offload sometimes results in the bee having to go out the next day to get it warm enough to transfer. A piece left in the hot sun on a roof, will enable an observer to see how a visiting bee deals with it.

**propolis constituents**
Over 40 different substances have been identified in propolis and these include 30% waxes, 55% resins and balsams, 10% ethereal oils and 5% pollen. Some excellent reports have been published in IBRA's Research News.
See: properties of propolis.

**propolis dermatitis**
Some beekeepers have suffered from dermatitis of the hands as a result of contact with propolis. It is thought to be more severe when associated with resins from poplar trees. The use of a barrier cream such as 80 Betnovate /Locoid, rubber or plastic gloves and washing the propolis off quickly with methylated spirits after handling are ways of minimizing the risks. Tridan tablets taken orally and applications of steroid cream will control skin rash caused by propolis but prevention (wearing gloves) is better than cure.
See: allergy.

**propolis gargle**
Put 10g of finely cut propolis into a bottle. Top with 30/40g of 96% alcohol. Place in a dark place for 2 days shaking periodically. On the 3rd day add water to increase the total to 100g and filter. To gargle, add 5 – 10g of the propolis solution to a ½ glass of warm water. Do not use stronger solutions.

**propolised**
To have become covered with or stuck down by propolis. Gloves and other items of equipment become sources of pathogens once propolised. A propolis seal may be helpful to repel or keep out invaders but can actually impede ventilation. Bees have been known to suffocate when their ventilation had been propolised in an observation hive. Care must be taken to dislodge a crown board (insert flat of hive tool and twist)

in such a way as to preclude any jolting or jarring also of a frame that has been 'glued down' by the bees.

**proposition (remit)**

A clearly worded suggestion that is put forward to be voted on at a meeting. It has to be proposed by one member and seconded by another. Other members may suggest amendments which if carried by a majority vote, obviate the necessity for voting on the original proposition. Changes of rules or actions to be taken are dealt with in this fashion. The Chairperson may have a casting vote.

**propupa (pre-pupa)**

The stage after the sealing or 'capping over' of a cell, prior to the pupal stage.

**prosopis spp.** *** **Mimosacaea**

Agerba bean or Mesquite a drought tolerant plant which can reach 18m down to reach water. It provides nectar and pollen in semi-arid regions. Pods, rich in sugar, can be ground and used as flour. Useful for shade in desert areas.

**prostaglandins**

Substances that should act as mediators at the site of injury usually by such action as the release of histamine. When an atopic person is susceptible and over reacts, aspirin is a potent antagonist to the formation of prostaglandins and Dr. Richards has recommended that two aspirin tablets taken one hour before exposure to the bees will give a suitable form of protection.

**protandrous**

Flowers such as the dandelion in which the anthers mature before the carpels thus discouraging self-fertilisation.

**protection against**

Covered under individual titles such as; animals, flood, predators, starvation, snow and wind.

**protective clothing**

This is a general term covering the kind of attire worn to prevent bees from using their stings on the wearer's flesh. It would probably include a zipped boiler suit in smooth, white, bee-proof material, gum boots, gauntlets and a veil with a stiff brimmed hat, all these items should be robust enough to be re-usable, yet entirely bee-proof and reasonably light in weight. They should be held clear of the face and tied, elasticated or zipped to keep the bees out, regardless of the attitudes and contortions that the operator might adopt. The veil should be black for good transparent vision.
See: under individual items and Velcro, zip and Sheriff.

**protein**

A complex organic compound with high molecular weight, synthesized from amino acids by all living things. In foods this is the tissue-builder and for bees pollen is their only source of protein.

**prothoracic collar** ***

This is the first skeletal ring and it just fails to meet on the ventral side of the thorax and so has been dubbed 'collar'. It is usually necessary to remove this if a thorough inspection of the first pair of thoracic trachaea is to be examined for Acarapis infection as the spiracle and largest diameter of the trachea is covered by the lateral flanges of the ring.

**prothoracic glands**

Two strips of tissue on the ventral side of the thorax join in the median plane but degenerate ten days after the larva's final moult. They are responsible for the production of moulting hormone (ecdysone).
See: moulting hormone.

**prothorax**

The first of the three segments that together form the thorax of the bee. It bears the fore-legs and it is these and the prothoracic collar which are removed to expose the tracheae which Acarapis mites may inhabit.

**protocerebrum**
The largest of the three component parts that make up the brain. It consists mostly of the great optic lobes of the chiasmata of the compound eyes and processes visual and sensory information and controls complex behavioural sequences.

**protogynous (see: protandrous)**
Flowers whose carpels mature before anthers making self-pollination unlikely.

**protonymph**
The penultimate embryonic stage in the life cycle of the honeybee's parasitic mite *Varroa jacobsonii / destructor*

**protoplasm**
A complex semi-fluid substance capable of spontaneous motion and regarded as the physical basis of life. The substance within and including the plasma-membrane of a cell but not secretions or ingested material. See: cytoplasm.

**protozoan pl. protozoans**
A phylum composed of unicellular animals. These include several that are pathogenic to bees, their well-defined nuclei distinguishing them from bacteria. e.g. *Nosema apis* and *Malpighamoeba mellifica*. Many are useful in consuming dead organic material.

**proventricular valve**
Proventriculus or 'honey stopper'. The musculated valve which terminates the fore-gut and controls the movement of food into the ventriculus

**provisioning (mass)**
A female mother bee (solitary) will provide food en masse for her young. For example a ball of pollen glued together with nectar, will have an egg laid upon it in a safe cavity furnished by the mother who will then have no further interest in the nest.
See: mass feeding.

**provisioning (progressive)**
The frequent application of brood food, which in the case of worker larvae, varies steadily including weaning and a final cut-off that triggers the sealing of the cell, is a complex process governed by signals released in the brood cells by pheromones.
See: progressive feeding.

**provoke**
To arouse to sudden anger. Stormy weather, especially when the negative and positive ions are frequently changing can along with a sudden knock, human breath and other factors, lead to the arousing of an immediate, unkindly response from the bees. The art of good beekeeping is to avoid provocation of this nature.
proximal
At or near the point of origin or attachment as opposed to distal. The proximal ends of the antennae are attached to the bee's head.

**Prunus spp. *** Rosaceae**
Almond, apricot, blackthorn, cherry, damson, peach, plum and many other good bee plants.
See: individual headings.

**pseudoscorpion**
*Chelifer cancroides* found on *Apis florea* in Sri Lanka – probably useful as they eat wax moth larvae and many varieties of mites (Varroa?)

**Psithyrus spp. Cuckoo -type**
There are some species of this cuckoo-type bumble bee in the UK. The females are not capable of making their own nests and are built with aggressive mandibles, spines and the means of forcing an entry into a legitimate bumble bee's nest. Here they kill and usurp the queen whose workers then begin to tend and nurture the cuckoo's eggs. Each species tends to have a specific host and is of similar appearance.

See: cuckoo bee.

**psychrometer**

An instrument that enables the temperature and relative humidity to be measured in a few seconds.
See: wet and dry bulb thermometer.

**Pterocephalus Dipsacaceae**

A tufted, cushion-forming rock plant which likes calcareous soil and sunny position. Ideal for a dry wall. Very hardy, propagated by division, a pretty plant that gets attention from the bees.

**Pterygota**

This sub-class of insects is composed of 'winged' insects and includes the honeybee. It is further divided into two major divisions. Div 1 Exopterygota (alt. Hemimetabola) and Div II Endopterygota (alt. Holometabola) there are twenty orders in this sub class.

**pubescence**

Arrival at puberty. Covered with soft down or hairs as found in newly emerged bees. Each of the three castes look decidedly pubescent when they first emerge from their cells. Hardening of wings, legs, sting and the constant contact with other bees, gradually removes this young-looking appearance and a harder, shinier look results.
See: pellicle.

**publications**

Copies of books, periodicals, recipes and the like, reproduced by typing, printing, photocopying and other methods (CD's) for the distribution to sections of the public.
See: advisory leaflet, bulletin, journal, magazine, periodicals, NHS publications, reprints, quarterlies, yearbooks etc.
Beekeeping organizations include: ADAS, BBKA, IBRA, BIBBA, Bee Craft, NHS, DARG, CAB and many counties and almost every country have their respective Bee publications.

**public relations**

The promotion of 'good-will' among the public. In 2010 a sudden surge of interest almost rocked the existing organizations, beekeeping manufacturers and the like after years when beekeeping education in schools was lacking to the extent that many grew up afraid of bees and quite unaware of their usefulness to mankind. Toppling the balance has had repercussions, mainly good, but landed us rather unprepared for such a burgeoning of interest. At one time nearly every county had a County Beekeeping Lecturer but these melted away presumably because of a government policy to save money and because of the scant appreciation put upon this 'little side-issue' of beekeeping. Ken Stevens left his post of CBL at Bicton College in 1980 under the restricting influence of Maggie Thatcher

**puff-ball *Calvatia gigantea, pulverentules* Lycoperdon**

There are many varieties of this fungus which is found in meadows and woods and which when ripe and dry acts as a stupefying fuel if burned in a beekeeper's smoker.
It can grow to five feet (1.5m) across and is called by names such as: the larger mushroom, bunt, burt, devil's snuff, maximus. frog's cheese, fuzz-ball, puckfist, punk, mully puff, tuewood etc. The ball-like fruit body yields clouds of spores when the mature puff-ball is broken but slices of the white ball can be cut dried in the sun and used in the smoker.

**puffer**

Alternative name for a smoker. A hand-held apparatus that can be squeezed thus closing bellows (which are sprung so as to open again) that drive air through a chamber in which smouldering material has been placed. The resultant smoker under the force of draught from the bellows is then directed by a nozzle joined onto the fire chamber. The nozzle may be hinged or merely a slide fit.
See smoker.

**'pulled' virgin**

Immediately prior to the queen's emergence from her cell there is a period during which she is both adult and mobile within the sealed cell. When such a queen is released by the beekeeper she is known as a 'pulled' virgin. Virgins still lacking the full intensity of their final colouration have been released and still matured apparently satisfactorily. Normally there is no point in doing this however, in certain circumstances, where to give a faulty queen cell would lead to certain ruin, there is an advantage in seeing the fully developed queen's emergence amongst the bees.
See: queen – pulled.

**pulvillus pl. pulvilli**

A cushion-like pad or process on an insect's foot. Dade tells us that the arolium is the name of the pad on the dorsal surface of a bee's foot. A pulvillus is an adhesive pad between the claws of some insects. When the claws cannot take hold the pad comes into play, enabling the bee to settle onto, and if necessary to climb, on smooth surfaces like glass.
See: aroleum and foot.

**pummel**

To repeatedly beat with the hands. Drumming with the hands can act as a driving or subjugating force on the bees. Frequently used in the days of 'skeps'. When it is desired to wash the honey-wet cappings there is no method any more effective than working with the hands. This helps to get most of the wax broken up and the action is very kind to the hands. Pummel only comes close to the word required as stirring and squeezing with the fingers are part of the process described.
See: driving.

**pump**

An apparatus designed to propel a fluid (liquid/air) A necessity where there is a a large turn-over of honey as on a honey farm. Honey is approximately one and a half times as heavy as water and much more viscous. Pumps that are to be repeatedly used for long periods must be able to stand up to the job. One such, the Jabsco is a one inch, high quality pump with a neoprene impeller. It is self-priming and pumps honey without aeration. All honey pumps do better with warm honey; at 80F (28C) the viscosity is very much reduced.

**punch**

A tool for piercing, perforating or stamping material. A specially designed punch, capable of being used to cut a cell through the comb in such a way as to leave cell and egg or very young larva intact this was thought to be an improvement on methods consisting of lifting young larvae into artificial queen cells in the process known as 'grafting'. A punch for driving in gimp pins when assembling frames has a magnetic holding device to steady the gimp pin.

**punk**

A preparation that will smoulder. Mainly US – decayed wood, when dry can be used. Some trees yield more suitable punk for use in the smoker than others. Some particularly fine material came from monkey-puzzle Araucaria araucana. It should be clean, easy to light, cut and handle.

**pupa**

After the fifth moult a creature very much like the imago is formed. It is however soft and white and has only small pouches that will become wings. This is an 'instar' which lasts eight days in the case of a worker during which time colour develops the eyes going through white, pink, purple to brown. By uncapping a cell of worker brood, its age can be assessed from its eye colour. Queens pupate for five days while drones require eight.
See: pupal changes.

**pupal changes**

After spinning its cocoon the larva becomes quiescent, contracting its head within the larval head and antennae begin their development. Thoracic and abdominal appendages become evident.

1. The head and the appendages are well developed.
2. The head is now directed ventrally.
3. The mesothorax has greatly increased in size.
4. The first abdominal segment is now fused with the thorax.
5. The remainder of the abdomen is separated from the rest of the body by a deep constriction. This takes 8 – 9 days for a worker or drone but only 5 days for the queen.

**pupal skin**

The final moult allows the young adult to leave its pupal skin as a cell lining. It actually sandwiches its faeces between the previous cell lining and the pupal skin. As a result of these layers, brood comb becomes progressively darker, tougher and more likely to harbour pathogens. Nurse bees polish and flatten the cell once its occupant has left so that it can be re-used for honey, pollen or brood or as a cavity for an overwintering bee.

**pupate**

To become a pupa. Honeybees become pupae after the 5th moult. Apart from undeveloped wings they display all the adult's external features.
See: pupa and pupation

**pupation**

That stage of metamorphosis when the larval prepupa changes into a pupa. Many comparatively rapid alterations in body form and development occur during this stage. These include remodelling of the head and mouthparts, telescoping of the abdominal segments as the sting begins to shape.
See: pupa.

**pure**

Quite clean and free from extraneous matter. Several examples follow. When we say pure water we may well mean 'quite fit to drink' from our point of view bees often choose 'impure' water as being more satisfactory to their requirements while our ideals on purity may go on to include chlorine, fluorides and other dubious compounds. Pure honey or pure wax will imply no adulteration.

**pure line**

Obtained by selective inbreeding resulting in bees that are homozygous for all genes. In other words the genetic characteristics have become stereotyped.

**purple loosestrife *** *Lythrum salicaria* Lythracaea**

An herbaceous perennial native to Euroasia. It likes moist conditions and is adapted to boggy or moist land by banks of streams, ponds and marshes, though it will grow in flower beds. The seed requires very moist conditions for germination but the plant spreads easily and can choke out native vegetation. The seeds are said to float for several days and then sink. Plants grow up to 2m high and secrete best under humid conditions from June to September. The honey, scathingly referred to as like amber motor oil US, has a greenish tinge, is thick and heavy-bodied with a distinctive but not too strong a flavour. Pollen is dull yellow, pale green, yellowish green, dark green with two sizes of grain 25 and 30mu. See yellow loosestrife and loosestrife.

**purslane *** *Portulaca oleraceae* Portulacaceae**

Half-hardy annuals with red or yellow flowers open from spring to the frosts. Honey slightly amber, good body and with distinctive, delightful flavour. Used as a potherb and salad plant.

**pursuer**

A bee that goes after you and shows every sign of being determined not to give up the effort of trying to sting you, Also known as 'followers'. BIBBA decided it was a characteristic that could be 'bred' out of bees.

**pvc**

This synthetic material can be made into sheets of varying thicknesses from flimsy to tough resilient material much used for sacks, greenhouses etc. It is airtight and waterproof but its effective life is governed by exposure to ultra violet light and it is destroyed by heat while combustion yields unpleasant (poisonous) gases.
See: polyvinyl chloride

**pyemotes herfsi Cudemans**

The adult is a shining, pearl-like ectoparasite of the honeybee. It is round bulb-like 0·3 - 0·4mm dia.

**pycnometer**

Measures the moisture content of honey more accurately than the refractometer.

**pylorus**

Pyloric sphincter, a thickening of the walls of the intestine in the region where the posterior portion of the ventriculus is joined by the malpighian tubules and enters the small intestine It regulates the passage of digesting food into the intestine. Its inner surface is bristled encouraging rearward movement.
See: sphincter and ventricular valve.

**pyment**

Omphacomel – a sweet wine that is made using grape juice and honey. With added herbs it becomes hippocras. Fruit juices with honey form melomels but special names are given when the fruit juice is apple (cyser) or grape (pyment) or mulberry ( morat).
See: omphacomel

**pyracantha** *** ***P. coccinea*** **Rosaceae**

An evergreen shrub with clusters of bright red fruits (there are yellow varieties). Useful to bees as it flowers during the June gap. A common garden plant, although tough when established it can only be transplanted when young.
See: firethorn.

**pyrethrin**

Two compounds by this name are obtained from pyrethrum flowers and are used as contact insecticides. pyrethrin – 2 $C_{22}H_{28}O_5$ and pyrethrin – 1 $C_{22}H_{28}O_3$.
See: pyrethroids.

pyrethroids

These are synthetic compounds related to the pyrethrins. They are devastatingly or highly toxic to bees though unstable (deteriorate rapidly) to the action of air, moisture and alkalis. Their toxicity to mammals is low because of rapid detoxification by body enzymes. Insects have shown an ability to become more immune to their toxicity but it appears that they can have a repellent action on bees.

**pyridoxine $C_8H_{11}O_3N$**

It forms together with closely related pyridoxal and pyridoxamine Vitamin B6 and its deficiency causes dermatitis or anaemia

**pyrrolizidine alkaloid**

Can cause acute and chronic poisoning in animals affecting the liver and it has been found in trace quantities in some honeys.

***Pyrus communis*** ***

The common pear cultivated by the Ancient Greeks it originally came from W. Asia. It grows wild and is a thorny tree with bitter-gritty fruit but is used as root stock for cultivated and garden varieties. Flowering in April its pollen is useful but the nectar is low in sugar. It is susceptible to fire blight and to mistletoe.

# Q

## QC

Abbreviation for queen cell under BBIBA's code words. QCC capped QC, QCE - QC with egg QCL – QC with larva and QCM – QC emerged, QCM – queen clipped and marked, QNS – queen not seen, QPBX – queen put below excluder, QQ – more than one queen present. For note keeping their code system has been carefully thought out.

## quadrate plate

Part of the sting mechanism. It is pulled forward by the contraction of the protractor muscle attached to its posterior margin. The movement is imparted to the adjacent triangular plate. The bilateral symmetry of the sting is fulfilled by two of these plates, one the mirror image of the other. Three plates together with dorsal anterior region of the sting, give rise to alarm odour.

## qualification

A state of preparedness which befits one to fulfil some function or office normally prescribed by fairly rigid conditions and the satisfaction of which is set out on a certificate. – Certificated. A qualified beekeeper would be one who has passed one or more of the various examinations in apiculture.

## qualities of a good queen

A queen's qualities must be judged by the characteristics she imparts to her progeny. Bro Adam separated these into groups

(a) those of fundamental importance
(b) secondary
(c) those of technical/aesthetical value.

(a) Fecundity, industry, resistance to disease, disinclination to swarm
(b) Longevity, wing power, keen sense of smell and self-defence, hardiness, wintering ability, spring build up, thrift, self-provisioning, comb building and the disposition to enter supers readily
(c) Good temper and calm behaviour, freedom from brace comb and propolis, attractive cappings and a keen sense of orientation.
See: characteristics, longevity, wing power etc. and profitable bees.

## quantities of nectar

238 lb per acre per day from alfalfa. 100 or so kg per hectare is a good average for most plants.
See: honey potential and nectar yield.

## quarry

An open excavation from which stone, chalk etc. is obtained by various methods. Disused quarries or disused parts of active ones have vehicular access and shelter and can, if in a good foraging district, offer a useful site for an apiary. Vandalism and the proximity (as a rule) of plenty of large stones, has to be taken into account. Chalk quarries develop a micro-climate which can lead to a DCA (drone congregation area) forming above them.

## Quebec board

A bee-escape board utilizing a framed board with a central 2¼" hole. On one side of the board two triangles of ¼" square section bottom bar are formed ¾" apart, with the outer lengths 12¼" long. Ends of bars are cut at 30 degree angles and set to form ¼" gaps at corners. The triangles are covered with ¼" mesh or gauze forming narrowing tunnels beneath.

**queen**

The honeybee queen mother. The fertile female of ants, bees, wasps and termites. The reproductive female in colonies of social Hymenoptera – Henderson. The one bee upon which the well-being and capability of the whole colony depends. Good queens are never too highly priced. You pay the highest price by not having a good queen. Success in most cases demands success in your own queen raising. See: queen bee.

**queen accepted**

A new queen that has been introduced successfully to a colony. Acceptance implies that she is able to live in continuous harmony with the rest of the colony and this includes the laying of fertile eggs. Partial acceptance where the queen's performance is much below her potential due to the non-cooperation by some members of the colony and downright rejection, are possible alternatives to the above when a colony is given a new queen. Subsequent supersedure can indicate a delayed rejection.

**queen activity**

Activity when a queen is present in a colony suggests that she is fulfilling her role as the egg layer and this results in warmth being created as brood is incubated and pollen loads being brought in under the stimulation of brood pheromones and the need for protein to produce brood food.

**queen age of**

Queens rarely (accompany a) swarm in their first year unless congestion or other causes combine to bring this about. A queen that is given full scope for egg laying is best replaced before two seasons have past, lest she should fail during the winter months resulting in a queenless colony in the spring. Three year old queens will be swarmier than younger queens but if it is thought fit to hang on to a queen because of her excellent record, then it is probably best to scale down her egg laying by putting her into a smaller nucleus or observation hive. Naturally, the bees themselves react to a queen that is failing (dependent as they are on receiving enough 'queen-substance') by superseding her with a new queen, but if honey crops are the main target of your beekeeping it is wise to incorporate queen raising into your normal management, i.e. start the new season with the intention of producing new queens for at least half of your colonies. See: queen introduction.

**queen anatomy**

The main visible difference between a queen and a worker is the length of the queen's body which extends well beyond her wing tips. When in full lay her abdomen is characteristically large. Her thorax too, is just that much larger so that sheets or screens of 'queen excluder' can prevent her passing through slots which allow workers through but hold back both queen (and drones). Her tongue is shorter but her substantial legs are longer, making her stand out a little on the comb. It is useful to concentrate on the queen's appearance from your earliest start in beekeeping. Handling too is easier if practice is first obtained by handling the harmless drones. Her unbarbed sting is very, very rarely used on humans and she can be comfortably held between your lips, keeping her sting for possible rivals (queens of the same age will fight and their anatomy is such that only one can plunge the final sting which destroys her opponent but leaves her unscathed) such as the unborn while in their sealed cells. She can 'pipe', that is make a sharp 'beep beeping', noise which causes surrounding bees to 'freeze' on the spot. When she does this, rival virgins give their presence away by emitting a 'muffled' pipe from within their cells.

**queen attendants**

A constantly changing retinue of nurse bees endowed with the richest brood food – 'royal jelly' move towards the queen, and as she makes the necessary pause to deposit an egg, it is quite usual to see a ring of bees facing towards her which leads to the expression 'daisy petal' for the shape that they adopt on the comb. Many anthropomorphic expressions, like retinue, attendants and court are used but the important thing is that the reciprocal receiving of 'queen substance' by those 'favoured' bees is then eagerly sought by other colony members. It is this, in fact, what leads to

(a) the knowledge that they are queenright and
(b) it is a significant part of the colony's identity 'smell'.

Unlike many creatures a queen bee needs at least a thousand bees if they are to form a successful colony though for short periods in a queen cage a small retinue, (preferably of her own young bees) can sustain her. When a beekeeper removes a queen, in a very short time the colony begins to show signs of distress with bees running hither and thither. On releasing her she almost bull-dozes her way back in while the mood of the workers is displayed by a positive hum (almost akin to a cat purring) with many adopting a fanning posture. When requeening by use of a delayed entry cage, it is important to put the queen into a clean cage alone, as her attendants would emphasize the fact that she is 'alien'. See: queen introduction and re-workered.

**queen 'balling of'**

A queen, dead or alive, will always attract worker bees. A queen cage left on a hive roof will soon be surrounded by bees. Let loose amongst alien bees the queen would not be stung but smothered. This can sometimes occur when due to unwanted disturbance (say very early in the year) the bees act as if to protect her but go on to form a tighter and tighter knot (the size of a wall nut) which they will maintain, sometimes for days, until the poor queen is killed. This so-called 'balling' is often missed, especially if it occurs in the grass in front of a hive. Taken in hand the bees in such a 'lump' will scatter and the queen can be rescued. To avoid this with a valuable new queen you will do well to carefully follow the instructions for 'queen introduction'.
See: balling and queen introduction.

**queen bank**

An artificial arrangement for holding a number of queens in close proximity yet each with its own attendants confined to their own small compartment with 'Good' candy or other suitable food. No contact 'bee-to-bee' should be possible between the compartments. They are best kept over a strong, queen-right colony benefiting from such environmental factors as arise from that arrangement. It has been found possible to nurture a score or so of queens for from four to six weeks provided the food supply is maintained.

**queen bee**

Parent of the honey gathering colony. No colony can be better than its queen, that is to say all genetic characteristics are passed on from her to the colony. Her egg laying capability is far in excess of that of queen wasps or bumbles. She can suffer diseases like other bees. Special food – 'royal jelly', is fed to her by the bees with highly developed brood food glands. Normally only one queen is found in a colony though temporarily daughter/mother combinations (which continue laying, though a comb or two away from each other) are possible. She is larger, her abdomen extending well beyond her wing-tips. She may live for several years. She holds her colony together and excites its activity. Queens can be marked on the thorax and have forewing(s) clipped, without unduly impairing their performance in the nest.

**queen bumble bee**

She is considerably larger than the other castes and mates without ascending to the heights honeybees fly to when copulating and then goes into hibernation for the winter. Awakening in the spring, she must then find a nesting spot, make a nest and commence to lay and incubate a small amount of brood, foraging the meantime until the first of her young can take over colony work. Cuckoo bumble queens can sometimes usurp certain varieties. See: bumbles and Psithyrus.

**queen cage**

At times it becomes necessary to cage a queen. When this is done a small, well ventilated cage is provided which can also accommodate a half-dozen or so suitable workers (her own progeny or young workers who are able to feed her). Many designs according to whether she is to be trapped within a colony yet continue laying or for transportation when several cages might be locked together. When used as introduction cages, as well as a small food compartment, they are made for a slow release and usually slide open and shut. Slim enough to go between two combs, their mesh is such that bees can reach through to her with their tongues and antennae but cannot harm her.

**queen cage candy**

One quick way of making this is to use very fine icing sugar mixed with enough honey to make a firm, handleable mix. This serves two purposes to sustain her (the addition of a little pollen may help but her real food is 'royal jelly') but also to allow the colony's bees to eat their way slowly through to her. Care must be taken not to make it too sticky on the one hand nor too hard for her to 'lick' on the other. Be sure to use honey which you are certain contains no pathogens. A guide to making this is to warm honey and then stir in about 2½ times as much icing sugar to make a firm paste.

**queen cage posting**

Cages should be 4 x 1¼ x " with a secure mesh (3 – 4 mm) and labelled 'Live Queen Bee' Postal regulations in 1980 required that the cage be enclosed in a stout manilla envelope labelled 'Live Queen Bee' with vent holes no greater than 3/16ths" and at least 1½" apart. These were the names then extant: Adam, Benton, Brice, Butler, Jay Smith, Ideal, Latham, Miller, Meadows, Pieyre and Worth. See: holding-down, queen cage and hair-curler.

**queen catcher**

Manufacturers have made glass types that allow an undistorted examination of the queen to take place. Others akin to a match box (boxes smelling of matches are not recommended) which partly open can go over a queen and then be slid shut. There is a bull-dog clip type made of queen excluder and devices allowing the queen to be sucked up onto a screen. By hand, a queen should be approached from the rear, with moistened fingers and taken hold of by the wings, then offered up to the middle finger of the other hand (which she will grasp with her legs) the thumb of that hand can bear down on the legs holding her firmly while leaving your other hand free to manipulate marker or scissors. Always allow ample time for marker paint to lose its smell and dry before releasing a queen amongst the bees.

**queen cell**

These are purpose built and not re-used. Almost reverting to the primitive style of the wasps who use vertical cells, an acorn-sized vertical cell, isolated from its adjoining cells, usually stands well out from the comb. It begins as a cup (before active use), once a female egg (fertilized) has been laid in it and incubated for three days, the larva is continuously fed on the richest brood food and the cell walls extended downwards as the larva grows over the next five days. Its walls are dimpled, the cell stands out from the comb face and can be easily seen once the surrounding bees have been moved although at times irregular comb may hide them. Although the porous walls are vulnerable to attack from rival virgins or workers who have changed their minds, the tip when it is finally sealed is strongly re-enforced by the cocoon that the inmate spins. At this stage the queen is in a similar condition to having become a chrysalis. When ripe the tip is nibbled bare by the workers and when ready to emerge the virgin gets to work with her mandibles, possibly the only time she uses them, and clips her way round until the end of the cell can be pushed open in the nature of a trap door. When valuable cells are handled this should always be done with the greatest care never jolting, crushing or inverting them. These cells can be triggered off by rendering a colony queenless. It is also possible to make artificial wax cups in the process of 'grafting' (inserting a day-old larva into one and putting it into a queenless colony)

**queen cell cage**

Queen cells with living occupants need incubation just like any other brood. Cages permit a safe, rigid attachment of a cell so that it remains vertical and is clear at the tip so that the virgin can emerge when ready. Gauze keeps bees away so that several cages can be set up in a colony acting as a nursery and at the same time the environmental conditions are the same inside and outside the cage. Conical, open, coiled wire cages with an extended arm for insertion into the comb are available. When cutting out a cell embedded in the comb, take every precaution to avoid damaging it in any way as the bees will certainly refuse it if you do. See: pulled virgin, spiral cell protector and West cell protector.

**queen cell – emergency**

The word 'emergency' implies that the bees have had to hurry the production of a queen as might be the case when their own queen suddenly and unexpectedly disappears. In these circumstances instead of a carefully unfolding development taking place resulting in a queen cell being furnished with a fertile egg and going through the normal three-day incubation period before feeding with royal jelly is started, the cell may be developed straight out of a cell in the worker comb. It will look short and stubby and rarely produces a good quality queen. See: queen cell scrub.

**queen cell –extra long**

Unusually long queen cells are not uncommon. As with any queen cell deformity no faith should be put in such a cell. They have been known to contain a drone, or perhaps dead workers and to result in poor queens. It is only wise to use queen cells that are right in every way.

**queen cell –natural destruction**

If a queen mutilates a queen cell or puts her head into an occupied cell it soon becomes empty. Queen cell protectors are used when queen cells from one colony are put into another to avoid their being torn down. When a change in conditions occurs that does not seem favourable, bees can 'change their minds' and tear down queen cells. Once a colony has successfully achieved the production of a new queen she or the workers will make short work of any remaining queen cells.
See: spiral cell protector.

**queen cell –new**

A new queen cell is usually of the same colour as the surrounding comb. However the position and shape are useful evidence that a cell is a new one. Quite often a cell is built in isolation – even fixed to the frame or a wire! Always close to the brood nest however. Cells buried in the comb or built amongst drone brood are not to be trusted
See: queen cell old.

**queen cell - old**

An old queen cell is one that no longer contains a viable inmate and is usually smoother (lacking the well-defined dimples of a healthy new cell) and somewhat darker than a new or active queen cell. The darker shade comes from 'travel stain' as it will have been much climbed over by bees. A cell may appear new when the cap has been sealed back again after a virgin's emergence but generally speaking old cells are partly demolished though not necessarily moved away altogether after use. A queen cell is never re-used.

**queen cell – opened**
An empty queen cell, intact in every way except for an open tip with or without a hanging flap, is indicative of a recently emerged virgin which may or may not be in the hive. The possibilities are that she has left with a cast, that she is out on a possible mating flight, that she has been 'balled' by the bees - but if she is around in the hive 'piping' is likely. This is a shrill, intermittent bleep which can be heard from well outside the hive. It is not long before such a cell is reduced in size or torn down.

**queen cell protector**
Where the beekeeper wants to go against the bees natural inclinations and to introduce an unborn virgin into the hive a sealed queen cell may be introduced provided steps are taken to prevent the bees (or possibly the queen) from damaging it via the vulnerable side walls. The cocooned tip is relatively invulnerable but the sides (which can be covered with micro sticking plaster) must be protected.

**queen cell – scrub**
This term is used by beekeepers to describe either a queen cell that is likely to produce a scrub queen or an undersized or distorted queen cell. Circumstances that give rise to this phenomenon include lack of nutrition, nursing by over-aged bees, undue haste in the hurry to become queenright after sudden queenlessness, low temperature and other abnormal conditions. See: queen cell emergency and scrub queen

**queen cell – supersedure**
A queen cell that has been deliberately and carefully built by a queenright colony that does not intend to swarm but to rear a replacement queen. Should a queen begin to fail as regards the colony's need for queen substance or egg laying capacity, the bees may during the active season take steps to replace her by the construction of perhaps one or a small group of queen cells near the centre of the nest. Swarm cells are often scattered around the periphery of the nest and are much more numerous. The renewal of a queen in this way often goes unnoticed by the beekeeper. Only when queens are marked and careful notes are kept is an unplanned replacement likely to be seen. To what extent bees try to get the best of both worlds by rearing a few queens as if to swarm and then changing over to supersedure is not entirely clear. What is certain is that some strains are more inclined to supersedure and go through several seasons replacing their queens without swarming. It is not easy to breed these supersedure type of bees as unlike swarmy strains they are not willing to build a lot of cells and if their eggs are given to other colonies to develop into queens, crossing these back to the drones that carry the supersedure trait presents several problems.

**queen classification**
Untested: sold soon after they begin to lay in their mating nuclei. 99% of all commercially raised queens are sold 'untested'. Tested: after purity of breeding is determined by seeing what the queen's progeny are like. Select tested: queens are placed in producing hives and their progeny tested for genetic traits. If these are found to be satisfactory they can then be sold as selected tested queens. Usually 1st year as queen mothers, select tested queen mothers are used by queen rearers to produce 24 hr. old larvae for the production of queens for sale. Breeder: The most desirable and expensive. Kept until daughter queens tested to determine that genetic traits are as required. Usually two or more years old.

**queen clipping**
The cutting off of part of the queen's wings. Great care must be taken not to harm the queen but part of the trailing edge of the fore wing can be done by an expert without

serious detriment to the queen. A queen is taken by the wings and held so that she grips a finger of the opposite hand with her feet. Then the thumb is brought down firmly onto her legs leaving the first hand free to use a fine pair of sharp nail scissors. Virgins need their wings to mate, first year queens rarely swarm but second year queens may well be given this treatment. See: clipped queen.

**queen – colour of the year**

The internationally agreed colour to denote the year of birth of a queen is as follows:
year ending 1 or 6 white or silver,
2 or 7 yellow or gold,
3 or 8 red/iridescent,
4 or 9 green,
5 or 0 blue.
Colours are only really important when queens are sent or received from abroad.

**queen courtiers**

As the queen walks over the surface of the comb nearby bees will turn towards her forming. This can be seen in an observation hive as out in the daylight she is more likely to be trying to hide herself. The pattern formed is often referred to as a daisy pattern and such anthropomorphic terms as: attendants, court, retinues are used.

**queen cramps**

Under such circumstances such as handling or interference by the beekeeper, a queen will sometimes 'feign death' by curling up, falling or keeping quite still. She might even produce a kink in her body and the only evidence of life is a slight abdominal breathing. Later when encouraged to do so by the attention of her bees she will return to her duties as if nothing had happened. See: akinesis, action inhibition, balling, catalepsy, catatonia, epilepsis and thanetosis.

**queen cup**

A waxen, inverted cup (much the same shape as that which holds an acorn) is fixed so that it can be developed downwards. They are often quite numerous as swarming time approaches and sometimes called 'play cells'. They are of course an indication that the colony might have swarming in mind but are not to be taken too seriously until an egg appears in them. Artificial ones can be made with a suitable piece of dowelling (mandril) coated with wax and plastic ones are available It is best that bees should have a chance to lick them over before being put into use.

**queen drone breeder**

A queen whose eggs can only become drones as she cannot fertilize them. If reared in worker cells dwarf drones will result. This condition can arise in two ways. The first is when a virgin, the colony's only hope, fails to mate and she becomes a drone breeder and second when a good laying queen begins to fail and worker eggs are inexorably replaced by drones. In the latter case the colony is doomed unless it is given eggs or very young brood. It is a situation to be guard against by always having young vigorous queens Highly domed and scattered cells of brood in worker comb are a 'give-away' signal that this has happened. Even if this is discovered while the colony is still reasonably strong, the nurse bees will be few if any and the best plan rather than giving them the task of trying to rear a new queen, is to unite it to a queenright nucleus or stock. They may remain strangely loyal to an old useless queen so be careful when trying to introduce a valuable new queen to them. This means finding and destroying the old queen first and giving them a frame of emerging brood so that here will be some young bees to take an interest in the new queen.

**queen erratic**

A good, healthy, laying queen usually walks in a stately fashion about the brood nest. Jerky or furtive movements are more likely to indicate that the queen is unmated or unwanted. Excitable queens may give rise to excitable bees. A queen may also be described as erratic if her egg laying pattern does not conform to concentric circles but is patchy. Sometimes lines of broodless cells mark reinforcement wires so don't blame the queen but those who wired the frame.

**queen excluder**

A device invented by Abbe Collin in 1874. The principal is that a sheet of wood, wire, metal or plastic, through which workers may freely pass, is used to confine the queen (whose thorax is larger) to a certain part of the hive. This area becomes known as the brood chamber. The lower part of the hive is most commonly assigned for this, so that the queen (and drones) are free to use the exit if required. The parts of the hive from which she is excluded are normally reserved for honey storage and known as the supers, though with some methods of swarm control brood is put over the excluder. When drones are trapped over the excluder they become agitated and will die blocking the slots in their effort to escape and fulfil their function. Some beekeepers prefer not to use excluders but then brood is found in combs otherwise intended for honey storage. Slots or holes should be 0·165" (4·2mm). See: QE framed.

**queen excluder (framed)**

While flexible zinc or plastic sheets of excluder are widely used on hives with bottom spacing i.e. top bars are flush with outside walls – they become firmly glued to the top bars and can suffer from distortion and aggravate the bees when removed. Framed excluders can be twisted off (like crown boards) and remain rigid and do not sag when used over top-spaced frames. They are more durable, efficient and cleanable. The slots through which workers can freely pass but which impede the queen need to be fairly accurate at 0165" or (4·2mm). See: queen excluder slots.

**queen excluder slots**

Manley considered a slightly smaller slot more suitable (perhaps he had the odd queen slipping through!) 0·16" . With new sheet excluders check whether the slots have a light burr on one side as the bees can be put off if these are not smoothed off or at best put on the upper side. Where possible slots are best arranged to run across the top bars.

**queen drone breeder**

A queen whose eggs can only become drones as she cannot fertilize them. If reared in. worker cells .dwarf drones will result. This condition can arise in two ways. The first is when a virgin, the colony's only hope, fails to mate and becomes a drone breeder and second when a queen that has functioned well eventually begins to fail and worker egg are inexorably outnumbered .by drones In the latter case the colony is doomed unless given eggs or very young brood. It is a situation to be guarded against by always having young vigorous queens. Highly domed and scattered cells of sealed brood in worker comb are the 'give-away' signal that this has happened. Even if this is discovered while the colony is still fairly strong there is likely to be a shortage of bees still in a state to be able to rear a queen. and the best plan would be to unite them to a queenright nucleus or stock. They may remain strangely 'loyal' to an old useless queen so be very careful when trying to introduce a valuable new queen to them. This means finding and destroying the old queen first and giving them a frame of emerging brood so that there will be some young bees to act as nurses to the new queen.

**queen failing**

When a queen fails during the active season it is quite usual for the bees themselves to supersede her – that is to use one or more of her eggs to raise a replacement queen. Should failure occur during winter when no drones are available and the weather unsuitable for mating, then the colony's survival is in peril, You should guard against queen failure by using young, well-mated queens. Do not be misled into thinking the bees know best. They are strangely faithful to a queen who is past her best and the honey gathering potential is often lost

because a poor queen is tolerated.
See: drone breeder, queen - drone breeding, queen failure and supersedure.

**queen failure**

The reduced efficiency of a queen to lay fertile eggs. This may be total, as in the case of a drone breeder or partial as when scattered holes appear in the slabs of sealed brood and yet the bees are slow to start superseding. Total failure is usually disastrous in the winter but often rectified by the bees in the active season when drones are available.

**queen food (larvae)**

The royal jelly fed to a of a queen 3 days old contains as much as four times the quantity of pantothenic acid as that of a worker larva. The mandibular secretions of nurse bees in a queenless colony have on average eight times more pantothenic and 115 times more biopterin as the secretions of nurse bees from queenright colonies. As there are no apparent changes in hypopharyngeal gland secretion, it has been thought that in queenless colonies nurse bees specialize in feeding either queen or worker larvae, hence queens get mandibular and hypopharyngeal secretions whilst workers only receive the latter. Prospective queens finally attain twice the pupal weight of a worker.
See: brood food, royal jelly and worker jelly.

**queen glabrous**

A queen having a surface devoid of hair or pubescence, in fact smooth, shiny and hairless. Such a queen is likely to be old and bears the signs of constant attention from nurse bees; her wings might be slightly tattered too. Experienced beekeepers can usually estimate her age as youth and age become widely separated.

**queen imported**

Since the advent of Varroa the importation of queens which had become quite common, came to a close for a while until we were found to be riddled with Varroa ourselves. Our one-time native 'black' bee has long since disappeared and queens, nuclei and package bees have been imported 'ad-lib'. Depending on source they may take time to accommodate themselves to our erratic climate. It is not so much the queen herself who might adapt but the generations of her progeny. There is much to be said for 'hybrid vigour' and as drones rarely come in from abroad an imported queen's daughter virgins will mate with local drones giving us hybrids. The first cross might be an improvement on the parents but ill-temper is not uncommon.

**queen injured**

It is commonly supposed that an injured queen (like an injured queen cell) will immediately be replaced by the actions of the bees during the active season. Yet queens have been known to go on laying, apparently normally, for years even with such faults as a 'kink' in the abdomen, part or all of a leg missing and of course they survive careful clipping and marking. However it is not wise to take chances with a queen. The whole well-being of the colony depends on her – you need the best.
See: queen cramps and supersedure.

**queen introduction**

Despite the bees apparent adoration of their queen they will instantly treat a stranger as 'alien' and stranger means anything but one of their own colony. Consequently to introduce a new queen to a colony requires
(a) that the colony is queenless and has been stimulated by a feed and
(b) that the new queen is in an acceptable condition to be attended to by strange bees. Ensuring that the colony concerned is queenless is the first requirement but the timing of this is important. To attempt to just swop one laying queen for another can work in some circumstances but it is better to de-queen and give the bees six hours to fully appreciate their loss. After twelve hours they may well have begun to build queen cells from the former queen's brood and this could lead to rejection of the new

one. So between and say, six to ten hours of queenlessness, is to be aimed at. The new queen may be in a special cage with a small group of young bees attending to her. So she will not be yearning for food. Remove her attendants or put her on her own in a new cage and give her ten minutes which will be enough to have her keen to receive workers attention again. Then, with as little disturbance to the colony as possible, insert the caged queen between the top, centre bars of the brood chamber. The cage will have holes through which antennae and tongues can reach her and in accepting the strange bees' attentions, both will begin to acquire the same odour. A delay, before they are allowed to make close contact with her, will have been arranged by having a 'candy' block or perhaps strategically placed newspaper strip (maybe with a pin hole in it) barring the entry of bees. Provided she will not suffer by becoming cold or not getting any attention, the delay period can profitably be extended to beyond the half-hour or so. Apart from a very careful check on the cage having been vacated, leave the colony well alone for five days or so. Her acceptance will be indicated by a brood nest having been started. This gradual introduction when, at first just one bee gets into her cage and then another, leads to mutual acceptance which is so important. Let loose too soon, she will be smothered by the bees in a strange phenomenon which is called 'balling.
See: acceptance.

**queen introduction (safe)**
C. P. Dadant suggested that the most certain method of queen introduction was to cage the old queen between two combs near the centre of the brood for two to four hours. She is then removed and the new queen is put into the same cage in the same position after which the hive is not opened for at least two days.

**queen larva**
The cell of a queen larva is larger in diameter than a worker cell and is open end downwards. The larva has a greater respiration rate than a worker larva and is fed continuously on 'royal jelly'. They hatch from the female egg after three days of incubation and are sealed with a surplus of royal jelly on the 9th day from laying.

**queen laying**
The egg-laying rate of queens is all too often copied from one book to another and astonishingly high figures of 2000 eggs a day are quoted. As the number of adult bees resulting from such an egg laying rate would approximate to over 80,000 bees it will be appreciated that queens rarely reach such high figures .The laying rate is dependent on the colony well-being as much as the queen and follows a curve from January to peak around May and diminish to zero in the autumn.

**queenless**
Literally a colony that has no queen. A colony without a queen or hope of raising one themselves we would describe as queenless. It is necessary to differentiate between a stock that is potentially queenright, i.e has either eggs, very young brood, a queencell or a virgin and a queenright colony which is one with a mated queen. A colony may have a failed queen. In this case she must be removed so that the colony is queenless before requeening is attempted. Colonies long queenless are apt to give rise to laying workers.
See: orphan. queenright and test comb.

**queen life cycle**
The egg is diploid, fertilized and incubated for 3 days. The grub is fed on royal jelly as the cell is extended vertically downwards. Capped on the 9th day, the 5th moult on the 12th while metamorphosis completes by the 16th. The adult virgin emerges, matures by 21st and makes exploratory flights. Mating may occur within a few days but can take a week or two (though failure to mate after 3 weeks will almost certainly result in a queen capable of laying only drone eggs) She may mate with several drones returning to the hive after each She

lays in a day or so and does not leave the hive again (generally) except with a swarm. This rarely happens in her first year and in nature queens may live for four years or more. Supersedure is an alternative to swarming and in some colonies co-existence of mother and daughter can occur before the older queen is eventually phased out. See: drone breeder, mating, queencell and royal jelly.

**queen – long lived**
Long lived queens are said to give rise to long-lived workers. Yet to keep queens into a third year implies that they have not been worked hard enough and that swarming is no problem. To check on longevity bees should be marked and records kept rather than trying to sustain a queen for seven or eight years just to see how long she might live. The length of life of a queen, as of a worker, depends on the amount of energy she has been called upon to expend. Breeder queens should be kept in small well-found colonies to lengthen their useful lives. See: queen age of.

**queen's mandibular glands**
These are responsible for the production of the special pheromone 'queen substance' and she can produce around 10,000 mg per day. Its constituents include an unsaturated fatty acid 9-oxo-E-deconoic acid and 9-hydroxyl,des-trans-2-enoic acid. The interplay of these two chemicals in the numerous roles performed are still a subject for clarification. They include drone aphrodisiac, cluster stabilizer, queen cell and worker ovary inhibitor and colony cohesive factor. See: queen substance, 9-0-2 etc.

**queen marking**
The difficulty so many beginners have in finding a queen has led to splashes of colour being daubed over the area approximating to the curved dorsal surface of her thorax (scutum). Marking is also useful to help identify the origin of a swarm, to label an individual queen and to mark her with the international colour of the year. Where paint is used it should be bright and quick drying (take care no smell accompanies her back to the nest) Glue and discs are also used and marking outfits and pens are available. A queen can be caged for marking. See: holding down cage, isotope tags, manganese dowex beads, queen marking outfit.

**queen marking outfit**
There are special queen catchers, marking cages and other aids. Thornes for instance say their cage provides the gentle way of holding the queen for marking and examination, while providing an applicator, glue and coloured discs or five colour of safe quick-drying paint.
See: queen marking and colour of year.

**queen marking paint**
Nail varnish usually upsets bees due to the irritating solvent. Equipment manufacturers sell suitable quick drying paints. Due to the rapidity of drying it is usually better to use a rod rather than a brush. Easy applicators are available.

**queen – mated**
A fully mated queen is one who within the first few weeks of her adult life has been effectively mated. Mating may necessitate several flights and to be carried out successfully reasonable weather and a fair number of mature drones on the wing are essential. Once the mating period is over she remains in the hive and under the stimulus of being fed by the nurse bees she begins to lay fertile eggs. If she fails to do this she may become a drone breeding queen. See: fertile, laying and queen drone breeding.

**queen mating**
A nubile virgin may make exploratory flights when 3 to 5 days old. She is dependent on light winds and reasonably good weather as she usually mates high in the air where mature drones are flying in what is known as

a drone congregation area. She flies during late morning or early afternoon and weather permitting usually succeeds by the 8th or 9th day. Up to ten or more drones will willingly sacrifice their lives in this process.

**queen mating station**

This is an apiary set aside for raising queens When isolation is possible and a drone stud is set up, selective breeding is possible.

**queen mediocre**

As success in beekeeping is largely dependent on the quality of the queen a less than satisfactory or mediocre queen should be culled and replaced by one more likely to have the desired characteristics. Mediocre queens not only hold back the colony they lead but are responsible for the production of drones who may well continue to distribute the unwanted characteristics through nubile queen for several generations. Bro. Adam said that 'a mediocre queen is one of the costliest things in beekeeping'

**queen mother.**

A colony having but one queen is composed of offspring from the one parent. She is thus not only queen but queen mother. For a time when first mated a new queen may operate with her brothers and sisters. When intervention by the beekeeper causes a colony to be requeened, she may lead alien bees until her progeny gradually replaces them. Under supersedure there may be a queen mother and a queen grandmother both laying for a while. Multiple queen hives are artificial arrangements to suit certain forms of management.
See: multiple queen hives, re-queening and supersedure.

**queen-new**

A colony that has just superseded will have a new queen. The virgin that mates from a cast will then act as the new queen to that colony. A beekeeper may see fit to change the queen of a colony. The new one may not necessarily be a young one though it is rarely wise to re-queen with anything but a young one. A new queen can come about through the beekeeper removing the existing one during the mating season and allowing the bees to requeen themselves or by giving the de-queened hive a few eggs, a queen cell, or a virgin or a queen from another colony, one newly produced or a 'bought' queen.

**queen – non prolific**

The potential egg laying rate is a variable factor based on a queen's age, race and the condition of her colony. A healthy young queen may well be less prolific than the average and such a queen would be more suited to the beekeeper who is content with a single brood chamber of BS frames. There is the school of thought that links the non-prolific strains with longevity though this is best judged from the progeny rather than the queen mother. Most queens respond to ample space and good conditions and bees that only use two Langstroth shallows for brood in England may need two deeps in Canada. NZ or Australia. Nevertheless where winter economy and the ability to succeed in our difficult climate calls for a relatively non-prolific strain, careful breeding from such strains would be useful, though one imported queen might frustrate many years of effort by allowing drones from the prolific strains to intermingle with those more carefully selected in the stud area surrounding a mating apiary.

**queen of the meadow *** Filipendula spp. Rosaceae**

In Anglo Saxon times the plant was used to flavour mead. Its sap contains the same chemical group as aspirin and was used to calm fevers or induce sweating. It likes marshes and fens and partial shade. Prone to mildew. The flowers are creamy white or yellow in the wild but various shades of pink are possible in cultivars.
See: meadowsweet.

**queen off colour**

Queens do not suffer temporary periods

of ill health. They are either good queens or not, though their capability is usually a reflection of the colony's well-being. There is never really any question of treating a queen to rid her of disease. Whenever a colony has disease, in addition to such treatment as is appropriate the colony should also be re-queened.

**queen old**

After two seasons of laying a queen enters the realm of being an old queen. By this time she will be very slow on the comb, her wings are likely to be 'tatty' at the edges and she will have been worn very smooth and shiny by the bee's constant attentions. The older a queen the more likely she is to swarm, to be superseded and to carry mites and pathogens. Apiary hygiene calls for replacement of queens before they become old. The oldest queens will be those put into observation hives or held in nuclei for selected egg production.

**queen – ovaries**

A queen honeybee has an astonishing ovarial array. Comprising two collections of tubules (ovarioles) which become so massive when fully charged with mature and developing eggs, that they cause the abdomen to expand and elongate. Each collection includes in the region of 300 egg tubes functioning at the same time. Starting at their tips which are joined together anteriorly, egg cells and nurse cells form and move along the tubes. The eggs steadily increase in size nourished by the nurse cells which continue to absorb food through the walls of the tubule. As the egg approaches full development the nurse cells (and follicle cells which surround them) disappear and the haploid eggs are passed one-by-one into the oviduct.

**queen piping**

Normally when a colony has 'thrown' its prime swarm there will be queen cells, the number varying according to race of bee, climate, weather and so forth. Although sisters in these cells, will each regard the others as rivals. So much so that the first one to emerge might well dispatch the others in their cells – if the workers allow. At this stage the virgin is capable of 'piping'. She makes a penetrating 'beep' 'beep' which can be heard several feet away from the colony. The sound is peculiar to queens and has the effect of causing surrounding bees to 'freeze' on the spot (this can be demonstrated in an observation hive by stroking the glass in such a way as to cause a squeak). Causing the other bees to freeze gives her easy passage either to depart on a mating flight or to attack her potential rivals. This is made easier because they are still trapped in their cells and (foolishly) give their position away by giving a responsive squawk – the clear piping sound being muffled. Occurring in high summer when 'after swarms' or 'casts' are not uncommon, it is sometimes possible to find several dead virgins on the ground outside the entrance.
See: piping.

**queen – prolific**

A queen laying eggs in excess of 1000 a day. As explained under 'non-prolific' the laying rate of any queen is governed by the colony itself and the conditions at the time but some are able to build up into far larger colonies than others. The factor of longevity also comes into the calculation and prolificacy is thought to be associated with shorter active lives and a quicker 'turn-over' of bees. Prolific bees tend to convert most of their stores into bees unless the peak fortuitously coincides with a really good flow.
See: queen non-prolific.

**queen propagation**

Normally a colony that is either queenless or has been made to re-act as if it were queenless, will raise queens from worker eggs. As genetic variations call for the same care in breeding honeybees as in any other form of stock, the queen over whom we have a great deal of control until she mates, should always come from carefully selected parentage. As good queens are the most significant contribution to successful beekeeping, they should be raised under the best possible conditions – time of year, strength and health of colonies with mating apiaries flooded with suitable drones. Even then we must be disciplined enough to reject all but the queens whose progeny confirm that the genetic make-up is what we wanted. See: production of queens.

**queen – pulled**

Or pulled virgin. A ripe queen cell will have its tip nibbled down to the cocoon by bees of the colony. It then appears balder and of a lighter shade than the rest of the cell. The queen herself may also be heard moving about within and signs of her mandibles nibbling on a circular path around the cell tip may also be noticed; her antennae may even be protruded (just checking you might think). At this stage the cell cap may be sawn carefully open with a razor-sharp blade. The queen will need to be patiently coaxed out and is then known as a pulled queen. Not having made contact with the other bees, she is at this stage fairly easy to introduce into an alien but queenless colony.

**queen raising**

See under grafting, cell-building, starter and finishing colony.

**queen raising nuclei**

The production of really first class queens is obviously best done in strong well-found colonies. This coincides with swarming preparations when good queen cells are literally two-a-penny as late spring glides into summer. As a swarm prevention technique for example though, a queen raising nucleus may be set up. Ensure that drones have been raised three weeks before the queens. Not less than three frames from a strong healthy colony should have an ample supply of bees shaken in. By taking it directly over three miles away so as to retain all flyers then fewer bees are needed but the loss of flying bees back to parent if kept in the same apiary can be reduced by stuffing the entrance with fresh grass, ends pointing outwards. Do not position in full sunlight. Leave for two days with adequate ventilation and food, then ensuring first that the nuc is queenless, and removing any queen cells they may have started, introduce a carefully raised queen cell. Syrup at 1:1 strength should be fed before during and after the mating of the queen. Do not place in the heat of the sun. and keep the entrance small, coloured and marked. Virgins are not easily found and you could be searching while they are aloft. If bees start taking pollen in that is often a good sign that she is mated and laying. A test can be performed by inserting a small triangle of eggs into a centre frame and if queen cells are built on it as opposed to normal workers being developed then your virgin may have been hi-jacked by a bird (seagulls fly at DCA heights in some coastal areas). See: nucleus.

**queen – rat-tailed**

This seemingly uncomplimentary description is that of a 'scrub' queen. One that has an abnormally thin abdomen even when she is in full lay. It is not wise to allow such a queen's genes to be broadcast so replace her with a good one, as to cull her would encourage her bees to raise another queen with the same built-in deformity and her drones to carry the fault on high.

**queen rearing**

The arrangement of honeybee colonies so as to encourage the production of selected queens as and when required. Colonies belonging to beginners often get out-of-hand, swarming indiscriminately. Then, untold new queens are reared naturally and tend to pollute the DCA's which more careful beekeepers will have established. Good queens are so important for good health and productivity that regular replacement with first class local stock should form an integral part of good apiary management. Time. effort and money spent on stock improvement is rewardingly spent. See: queen propagation.

**queen rejection**
Left to their own devices bees eventually 'ball' or supersede the queen. Rejection, lying in front of the entrance, a dead queen (regardless of cost) is likely to have been caused by the unwitting beekeeper. To make a colony accept a new queen can vary from the unbelievably simple to the unbelievably difficult. Precautions involve preparing both the queen and the recipient colony for the event. Feeding the expectant colony can help to put them in the right mood. Rejection can come from the bees simply not having been conditioned to identifying themselves with the new (alien) queen and it must be remembered that if the new queen is a laying queen she needs to be nursed over this 'short' period of change-over. She must be 'crying-out' for attention and the colony just ripe for her reception. 6 to 10 hours of queenlessness will achieve that but 10 minutes alone for the queen. A cage protecting the queen but allowing tongue and antenna probing can usefully delay full scale contact. Short cuts using smoke, water or icing sugar dust should only be risked if you have several back-up queens available. It is her queen substance the bees must want and she must dearly want their welcome attentions. Rejection thwarts what should be a smooth running and important part of apiary management. Apart from the intake of pollen don't rush checking her acceptance. Before looking for eggs a seven day pause is advisable.
See: queen replacement.

**queen replacement**
Establishing a new queen as the 'head' (hub) of the colony. A new queen is primarily required to safeguard the considerable investment that hive, back-up equipment and activities present. Sometimes this is done to achieve docility. Whenever disease rears its ugly head it is wise to re-queen. She, the longest lived and most exposed bee, is as likely as any to carry pathogens. If the superiority of a queen persuades you to keep her longer than usual, remember she can have her egg laying scaled down by being put into an observation hive or nucleus.
See: supersedure.

**queen retinue**
The dictionary lists retinue as 'a body of retainers in attendance upon an important personage. A colony's being and survival depends absolutely and entirely upon their queen. Like so many amazing things in the world of honeybees, her influence is little short of magical. Only a small group of workers can reach and touch the queen at any one time yet her influence spreads rapidly throughout the whole colony. That group is forever changing, made up from bees having something to offer the queen and whose lives are only one twenty sixth of an average queen's life.
See: attendants, court, daisy petal pattern etc.

**queenright**
This term is applied to a colony that has a mated queen. A colony in a normal state. A colony with a queen capable of laying fertile eggs. Tapped, the hive will utter a short overall burst of buzz and when queenless the sound attenuates into what is called a 'moan'. Remove a queen and the bees soon begin to notice and search for her, release her after a while and they positively purr.

**queen rival**
A rival queen is one who could take over another's duties. Princesses are born rivals but bees usually take care to keep them apart. That a queen is armed for that one particular fight if necessary is demonstrated by her barbless sting and body that can only be pierced by a rivals sting when one has secured a coup d'état position. Laying queens rarely come face to face but if they do a fight is bound to ensue.
See: balling, supersedure, queen replacement.

**queen's egg laying capacity**
Bearing in mind that this can vary according to the queen's age, the race of bee, the

weather conditions and the state of the colony, there is still an optimum egg laying rate for a queen. If she is fully acclimatized she will have a seasonal egg laying curve. Most queens are capable of laying somewhere around one million eggs in a life time, but probably lay fewer than 250,000 even when pushed to the full. A colony of 50,000 bees requires a laying rate of an average of 1000 eggs a day. Each worker egg needs the queen to operate her sperm pump and an unfertilized egg in a worker cell is soon eaten by the bees; new queens might, here and there, lay more than one egg in a cell (perhaps initial enthusiasm or learning how to use her ovipositing apparatus).

**queen selection**

Where queens are propagated from carefully chosen stocks, selection of those that appear to have the right characteristics and have mated with the right drones can lead to stock improvement. The bees themselves are always willing to 'select' when confronted with more than one queen as in the case of swarms that combine or of multi-queen casts. It is not certain that they always know best. Man's interference with bees is often aimed at exploiting them to bring out the best in them and for him or herself the most profit.

**queen – shape of**

Whereas a queen should be judged by her progeny and the size and shape of her brood nest rather than from her own actual appearance, there are many characteristics such as an unduly pointed abdomen, atrophied appendages and other obvious deformities that would in most beekeeper's eyes, render queens unacceptable.
Conversely, many a beautiful and apparently well-formed queen might prove inefficient or even worthless.
See also weight and size.

**queen short**

A short queen would not only be more difficult to find but is far less likely to make a good queen than a normal sized one. Virgin queens get smaller after they have been out of the cell for a few days (trim and ready for action) but should always enlarge to normal size once they get into full lay. It is the breadth of a queen's thorax which is significant relative to her inability to pass through an excluder and not her length.
See: queen cell scrub.

**queen sterile**

A sterile queen is incapable of laying fertilized eggs – a drone breeder in fact. This occurs when a virgin has been unable to fly, find drone partners and mate or sometimes when after a period of successful egg laying a queen becomes incapable of laying other than drone eggs.
See: drone breeder.

**queen –sting of**

The lancets of a queen's sting have only undeveloped barbs (barbette) She is therefore able to penetrate the walls of a sister cell and sting the unborn inmate. On rare occasions a queen will be confronted by one of a similar age and a fight will ensue. Nature has arranged that the victor remains unscathed by only one being able to secure the right position that allows her to pierce the other with her sting. Her unbarbed sting is easily withdrawn. The venom in her poison sac remains effective for a fairly limited time and the sting of an older queen has little or no effect. Her sting is of course more strongly attached to her body than is the case of workers.

**queen storage**

An individual queen can be kept alive for several weeks in a small cage provided there are around a dozen attendant workers and a supply of food in the form of soft candy. Batches of such cages have been kept for a month or more over the feed hole (for warmth) of a strong colony in early spring though there must be no possibility of contact between the attendants and the bees of the colony nor must there be between the bees in adjacent cages.
It is very difficult to overwinter with more

than one queen in a colony yet queen storage offers a possible way of keeping spare queens.

**queen stubby**

A queen with a shorter abdomen than normal. As with any other visible malformation of a queen it is most unwise to tolerate such queens unless as an absolute emergency and then over the shortest possible length of time. Remember that a laying queen is passing her faults on through her progeny.
See: queen scrub.

**queen substance**

The discovery by Dr. C.G. Butler that many aspects of colony behaviour were controlled by the release of a pheromone manufactured in the queen's mandibular glands, triggered off widespread research. Modern methods of quantitative and qualitative analysis cause a constant up-dating of the chemicals involved. 9O2 and 9H2 are certainly main components. The following are examples of effects of the queen's pheromone 'queen substance'. The nubile queen's attraction to drones via the scent trail, the swarm cluster stabilizer, the inhibitory effect against queen cell building and laying worker ovarial development and the stabilization of colony morale.

**queen tested**

Queens that have been checked as laying eggs that produce suitable adult bees. In the ordinary way a queen that has mated and begun to lay is ready for sale. However such a queen is untested. A queen that has been checked as being responsible for a good brood pattern and satisfactory progeny is naturally more valuable (costly).

**queen -undersized**

A self-explanatory term that has been described under: scrub, stubby, rat-tailed etc. No sub normal queen should be allowed to head a colony because her progeny will be capable of passing on the same defects which could go on to affect the well-being of other colonies in the area'. An undersized queen may well slip through a queen excluder.

**queen unsatisfactory**

As soon as a queen shows any sign of being unsatisfactory steps should be taken to replace her and that, to put it bluntly, entails destroying her. So many of us love bees and to bring this about gives one a sad, guilty feeling. A suggestion is to put her in a small cage in the freezer!

**queen vigorous**

The vigour of a queen is displayed by her ability to head a colony that overwinters well, is disease resistant, builds up rapidly in the spring and which produces a good harvest. She may appear very slow and sedate on the comb, her vigour is to be judged though, by her achievements as outlined.

**queen virgin**

A virgin queen can emerge from her cell on the fifteenth day from the laying of the egg. However queencells are usually built vertically and well separated so that workers can encircle them and where necessary confine and protect the unborn virgin until such time as the colony is ready to release her. Unlike workers, virgins are very active from the moment they climb out of their cells. They are mature on the 21st day and become stale if not mated within three weeks.
See: virgin queen and mating.

**queen wasp *****

She is considerably larger and slower flying and less inclined to sting than her workers. They usually overwinter in this country in hibernation and emerge sluggish in the spring to search for a site to build and lay in a small inverted, papery cup of a nest. Due to competition for sites, predators and indifferent weather, only one out of every 100 succeed in producing a functional colony. It is not true therefore that every one killed either in autumn or spring will be one less nest.

**queen – weight of**

The weight prospective queens attain is twice that of a worker larva. It has been found that queens weighing 180 – 200mg received 96% more acceptance compared to far fewer over and under this weight. Different races vary though in weight and final ovarial development. It does not follow that the largest, heaviest queens are necessarily the best but the use of undersized queens should be avoided. See: weights and queen accepted.

**queen young**

A young queen is a queen that is in its first year of lay. Even if she has mated late in the year and therefore starts season number two as a young queen she will not be a young queen by the end of that season. Furthermore from the colony's point of view, replacement of such a young queen is more safely undertaken at the end of her second season rather than at the beginning of her third.

**quell**

To suppress or put a stop to some unwanted activity such as robbing. To say 'don't let it start' does not help once it occurs. A really serious bout of robbing calls for drastic action. Water is a good subjugant if a hose is to hand or a bucket and wide-necked plastic jug can help to simulate rainfall. Wet sacks, reduction of entrances to one bee-way, grass piled over entrances help and removal of the culprits if this can be discovered are a few suggestions. Needless to say smoking is absolutely useless.

**questionnaire**

A list of carefully thought out questions posed to obtain opinions and thus gather information which may be used for statistical or other purposes. Much useful research can be done in this way though one is always dependent on the help and consideration given by those to whom the questionnaire is sent.

**quickthorn** ***

Hawthorn, when used as a hedge it is usually neatly trimmed and bears minimum flowers as a result. It is a fickle plant though only rarely yielding its delicious honey. See: hawthorn.

**quiescent**

At rest, inactive, almost motionless. This applies to a honeybee colony in the cold of a British winter when bees form a tight and virtually motionless cluster unless disturbed.

**quilt**

A piece of material used in place of a crown board or perhaps in spring, in addition thereto. Flexible and bee-proof it may include a flap to allow feeding but should be stout enough to withstand nibbling. A closely woven material which will certainly become propolised on the underside and upon the removal of this aggravates bees so crown boards are used as a rule by preference.

**Quinby Moses (U.S.A) 1810 – 1875**

I 1866 he introduced the larger size of frame which was later modified and used in the Dadant hive. His hive held nine or ten 8 ½" x 11¼"frames (spaced 1½") In 1883 he wrote Mysteries of Beekeeping giving details of queen rearing and in 1870 he improved the bellows smoker with an unventilated fuel chamber that had bellows at the side. See: Jumbo

**quince** *** ***Cydonia oblonga*** **Rosaceae**

A hard, yellowish, acid fruit on a stout, branched shrub. A source of nectar and pollen from Jan – May. The fruit is bright yellow when ripe and is used to flavour apple tarts and for making into jelly. Varieties range from white to pink and red and are spiny. Very hardy but requires a very warm summer to ripen fruit.

**quiz**

To examine or test. This often takes the form of groups of beekeepers from one branch competing with those of another to achieve the best results answering questions posed by the quizmaster. A good inter-branch activity that can promote study, competition and amusement.
See: brains trust.

**quotations**

Upon no other thing does the honey part of the apiary depend so much as it does upon the queen. George Doolittle 1909.
"Taking them vore" – a Devon expression for sulphuring bees in the olden days when skeps were used.
See: but(t), poems, rest harrow, sayings, snippets and superstitions.

# R

**rabbet (rebate)**

A cut, groove or recess in the edge – a step shaped cut in a piece of wood, as is used in brood chambers and supers to provide a recess, normally reinforced by a 'runner' to support the lugs of the frames.

**raccoon Procyon spp.**

An arboreal, meat-eating mammal; common in North America and it includes bees in its diet.

**race**

A geographical sub-species. A variety with relatively fixed characteristics composed of phenotypes and probably genotypes. The many races of our honeybees – Apis mellifera are listed under their Latin names. See: line, races strain, type and variety.

**raceme (Bot.)**

**racemose – having racemes**

A simple , elongate inflorescence having a common axis bearing stalked flowers that open in order from the proximal to the distal end, i.e. upwards on rosebay and downwards in sycamore.

**races**

A geographical sub-species, an identifiable group, class or kind with fixed characteristics that are used to distinguish races of honeybee: chitinous plates on male sex glands (penis valves), cubital index, number of hamuli, width of metatarsus, shape and size of wax glands. Mellifera can be divided into national or geographical names – European, Oriental and African. See: line, race, strain, type and variety.

**racking**

The process of drawing off wine that has been working by siphoning or decanting from the 'lees' (dregs) into a clean vessel. This should be done whenever there is a sediment as it is here that unwanted bacteria may develop and spoil the flavour. See: decant and siphon.

**radial honey extractor *****

In the barrel of this extractor, (as opposed to the tangential type), the frames are disposed radially about the central spindle so that many frames can fit into the same space that fewer would occupy in a tangential extractor. Each frame is spaced widely between top bars while the gap narrows until quite a small clearance exists between the bottom bars which are nearest to the central spindle. There is no need to reverse the frames as both sides of the combs are emptied at the same time. Whether shallow or deep, the same number of frames can be extracted (unlike tangentials which only take half the number of deeps as do shallows. Hybrid forms where frames are between tangential and radial distribution are also made and screens that can convert radials to tangentials are advertised by Thornes. See: reversible and tangential.

**radial symmetry**

Repetition of shapes in a plane around a central point or axis. This applies to many flowers though not to Leguminosae for example. Although bees can work the whole area surrounding their hives for miles, many factors such as sources of forage and wind, combine to remove any radial symmetry from the pattern of foraging. See: bilateral symmetry and zygomorphic.

**radius**

(a) The distance from the centre of a circle to the circumference.
(b) a wing vein.
(c) The foraging area used by a particular colony is expressed as a circle of a given radius with the hive at the centre.
(d) From the radius of a rotating cage or mechanism (honey extractor) we can calculate the speed of movement when we also know the number of resolutions per minute and therefore the increased force of 'g' imposed on the comb and the honey.

**raffinose $C_{18}H_{22}O_{16}$**

Trisaccharide sugar formed in small quantities when sucrose is broken down into simpler sugars and new sugars of high molecular weight are synthesized by group transference. See: melezitose.

**ragged cappings**

Once honey has been capped over, the bees do not readily uncap cells unless there is no other food available. Should predators, (including honeybees from other colonies) begin to steal honey, then cells are torn open in an untidy fashion, giving the appearance of 'ragged cappings' and allowing bits of the cappings to fall as debris onto the floor and towards the entrance. Signs of mice, together with their droppings, and other robbers can be spotted in this way as some of the debris will be seen at the entrance, Empty honeycomb then, with some of the cappings still in place but with ragged holes torn, tells us that the stores have been robbed.

**Ragus candy**

This is a proprietary brand of candy made from pure, white, refined, cane, invert sugar and water and sold as being eminently suitable for wintering. It contains no mineral or organic matter, no acid inverted sugar, no commercial glucose and it is claimed that it remains firm in the hive and will not under any circumstances liquefy.

**Ragwort *** *Senecio jacobaea* Compositae**

An old world, perennial herb with irregularly lobed leaves, that has become a troublesome weed. The yellow flowers are daisy-like and it is a relative of groundsel, producing a rank, bitter, yellow honey with undesirable odour. Pollen is deep or dull yellow. It withstands drought and is frequent on neglected pasture. Its foliage is the common food of the cinnabar moth and where it grows in sufficient quantity, it can get into and spoil heather honey as it flowers from June to October. The honey does the bees no harm and is quite satisfactory for wintering. Honey contaminated in this way can improve with keeping. In Scotland it is known as 'stinking Willie'.

**rails**

Horizontal bars that may rest on the ground or be supported by posts can be used to set hives at a good working height and be clear of the poor micro-climate at soil level, weeds and predators. Hives can be slid along them but should be locked on in some way to prevent them from being pushed off the rails. Manipulating equipment spares, boxes of combs etc. can go under the rails while working at the hives.

**rain**

Water drops that fall from cloud. It usually confines bees to their hives, washes nectar from flowers, affects soil temperature and causes leaching. It can wash honey-dew away. The timing of rain is important as it may make or mar a crop. *A Maltese* beekeeper said that rain is essential to produce nectar bearing thyme flowers but ruins secretion if it occurs while they are in bloom. Referring to *Apis florea*,
Dr. Free states 'When it is raining the bees of the 'curtain' face upward with wings partially spread nd raise their bodies at an angle of 15 degrees to the comb face and

thus form overlapping layers, rather like tiles on a roof. Providing it is not cold little harm comes from manipulating bees in the rain'. (some folk use umbrellas).
See: rainfall.

**rainfall**
The amount of rainfall is a significant aspect of climate, weather and consequently flora. Whenever it is uncertain so is the success of beekeeping. In Britain we have dry summers, wet summers and mixed-up summers. The onset of a honey flow not only requires warm temperatures but a cessation of rain (unless it falls at night), with sufficient soil moisture and plants that have had enough to achieve full development. Rainfall tends to increase across Britain as one goes west.
See: cessation of flow, isohyet, moisture deficit and rain.

**rain water**
Although this can be contaminated it is 'soft' and after simple filtering can be used to render wax which does not respond well to hard water. Arranging for water to collect for the bees use, should allow for heavy falls and for 'floats' to allow bees to get a 'footing' without fear of being sucked in and drowned.

**ramada**
A long, covered arbour under which rows of hives are placed to protect them from the sun in warmer climates and in arid conditions. Similar straw-covered shelters are often pictured over rows of skeps set-up on a bench in olden times.
See: bee shelter.

**Ramadan**
Muslim fast rigidly carried out from dawn to dusk on the ninth month of their year. Subsequent feasts make great use of honey.

**ramify - ramification**
To divide, spread out or extend into smaller and smaller branches thus forming a network or reticulation, e.g. a bee's air sacs ramify into tracheoles. The activity of beekeeping eventually causes a person to branch out into many aspects of life that at first sight may appear to have little or no connection with the bees.
See: reticulate.

**rampin** ***
A hand-tool for holding a small nail by magnetism and which by means of a spring-loaded mechanism, drive a nail squarely home when compressed. A useful device for inserting nails (gimp pins) when assembling frames without the noise or need for hammering and striking one's fingers.
See: frame nailer, staples and glue.

**ramus**
A branch as of a plant vein or bone. That part of the sting known as the ramus has its proximal end fixed to the distal end of the triangular plate. It is flexible and runs along a semi-circular track finally becoming the lancet which projects rearwards. The two rami slide along but are locked onto the shaft and the space between the three parts is filled by the bulb of the sting.

**range**
We speak of the bee's range of flight, range of pollen or honey colours, the extent of a colony's foraging area, the range of colours seen or sounds made or registered by the bee and so on. The limits between which a certain action or variation is possible. Also mountain ranges which have as marked effect on the environment.

**Ranunculus** *** **Ranunculaceae**
The plant family which includes buttercups and generally has five yellow petals and sepals, a large number of stamens and the pistil composed of many segments known as carpels.
See: butter cup.

**Rape (oilseed) *** *Brassica napus* (Swede rape) Oleifera**

*B.campestris* (turnip rape), cole, coleseed, colza. Our entry into the EEC soon brought about a huge increase in the hectarage of this plant. It not only produces protein-rich cattle food, but margarine and has many other uses including motor fuel. Its honey is light and so rich in glucose that it granulates rapidly. Beekeepers often extract just before it is sealed in order to avoid having combs clogged with set honey. It likes deep, moist, fertile soil and there are winter and spring sowings. Its seed yields 20% protein, 43% fat and cattle fodder. The pollen is light yellow and its proximity to orchards and other crops can result in the under-pollination of these, as the bees seem very attracted to the rape.
See: OSR and oilseed comparisons and rapid granulation.

**rapid granulation**

Certain honeys, especially those gathered from the brassicas- rape, charlock, mustard, cabbage, dandelion, ivy etc. granulate so quickly that they can solidify in the comb (or even in the bee!). This can occur during extraction and while filtering and once an extractor is lined with a granular film it will trigger other honeys to follow suit. Once a super has granulated solid it is probably best to cut out the comb and micro-wave it so that a layer of wax can be separated from the spoilt but still usable honey.

**Raspberry *** *Rubus idaeus* Rosaceae**

A plant growing in the form of canes (wild kinds sometimes prickly) having small juicy drupelets. In Scotland where it is cultivated in large areas producing water-white honey with a delicate flavour. The nectar is 21 – 31% sugar and the honey granulates rapidly. In Norway the very long west coast provides migrating beekeepers with a long season of forage. As it flowers during June it is useful to fill what is often a 'gap' in the mid-season flow. Pollen reports come as light, mid-grey, white, greenish white, very pale yellow, dull yellow and its grain is 23mu.

**rastellum**

Pecten or pollen rake. Its position is on the inner distal edge of the tibia of the hind leg, It takes the form of a stout comb being composed of a row of wide, pointed spines which clean the accumulated pollen from the opposite basitarsus causing it to fall onto the auricle.
See: pecten and pollen rake.

**rat *** Rattus spp.**

A long-tailed rodent that can be destructive to stored equipment, make burrows under hives and will pull out loose entrance blocks. They are too large to get into a hive via the entrance but could be troublesome if hives and equipment are kept in places frequented by them.

**Rata *** *Metrosideros robusta, salucida***

A NZ forest giant which yields copious amounts of nectar, one tree being able to supply several colonies of bees. It begins life as a parasitic vine but eventually kills its host and becomes a free-standing tree. Dotted around the landscape its blazing red makes it especially noticeable around Christmas time. The honey has an attractive flavour and granulates quickly.
See: pohutakawa.

**rate**

The speed at which anything moves or progresses. Wing beats are governed not by the nervous system but by the resonance of the thoracic box which pulls on the muscles, causing their contraction to produce myogenic rhythm. The rate of colony build-up should be related to an anticipated flow (early OSR and sycamore). The rate of spinning an extractor should begin slowly and increase as the combs empty. Rates can be slow, normal, fast etc.
See: direct and indirect flight muscles.

**Ratel** *** *Mellivora capensis*
**(honey badger)**
A badger-like carnivore found in Africa and India which is inordinately fond of honey. It leaves tell-tale scratch marks around the inaccessible Trigona nests.

**raw honey**
Unripe honey i.e. honey with a high water content that has not been properly matured. Combs that are mis-handled by holding them horizontally while manipulating during a honey flow may drip raw nectar/honey. Similarly when shaking bees from combs with cells of unripe honey, much food is thrown out with the bees. See: dehumidifier.

**Raynor Rev G.**
He introduced the use of carbolic acid as a subjugant, read paper on the Ligurian bee to BBKA in 1880 and was the first inventor of a very popular reversing tangential extractor and a Raynor queen introduction cage. First importer of American machine for making first class foundation.

**rear**
To bring up, raise or cultivate. This term is used when bees 'rear' brood or when the beekeeper rears queens, drones etc. Care should be taken to differentiate between the terms 'rear' and breed or propagate. To breed queens would imply having some control over the genetic structure if only by selection. Propagate would suggest the production of daughter queens, while rearing would apply to any method producing new queens. See: breed, hindmost and posterior.

**Reaumur Rene A. F. de**
A French physicist 1683-1757 who discovered the function of the spermatheca. 1734 Memoirs pour servir a l'histoire de insects. He made one of the earliest observation hives gave his name to the 'masque de Reaumur' on the drone. Thought that mating occurred within the hive.

**rebate**
Rabbet, cutaway – a groove or recess on the edge or surface of a board or such like. This is a carpentry technique used for joints and for the ledges on which the frame lugs rest. Some beekeepers who make their own hives set up a machine so that the rebated corners and the recess for the ledges (runners) that support the frames, can be carried out sequentially at the same time.

**receptacle (bot)**
Apex of the flower-stalk bearing flower parts. The cup-shaped surface from which the flower parts arise. Also the word for a place or container that can hold something. The cells of the honey comb are receptacles.

**receptacle seminis (spermatheca)**
The receptacle for the live spermatozoa in the highly specialized queen honeybee. See: spermatheca.

**receptor**
A sensitive part of a creature that detects various stimuli and by virtue of its effect on the nervous system influences the behaviour of the creature.

**recessive**
A recessive gene has no effect on the genotype unless homozygous. These always give way to the dominant equivalent characteristic but when there is no dominant gene to supersede it then the recessive characteristic is displayed. In-breeding can bring about this detrimental change. opp. dominant.

**recipes**
Several are given: beer, candy, car polish, cough mixture, cold cream, face cream/lotion, floor polish, fudge, furniture cream, ginger beer, hand lotion/cream, hair cream, honey skin treatment, leather polish, linoleum polish, lip salve, mead, propolis gargle, strawberry jam, vinegar, water-proofer for boots.

**reciprocal hybrids**
When crossing a black drone with a yellow queen the result may be different from the reciprocal hybrid obtained by crossing a black queen with a yellow drone. Artificial insemination makes such comparisons possible whereas isolation apiaries are much more of a problem to find.
See: genetics.

**recognition odour**
The exact nature of this quality that enables a bee to recognize its fellow colony members and to be so recognized, has been made a little clearer now that we know more about pheromones. When uniting, over-riding odours can be used but ordinarily the sum total of queen substance, common food supply and the smell of the combs, propolis etc. all combine to provide this odour.

**recognition of**
disease, foes, own hive, a nubile queen, queenlessness, shapes, smell, sounds, forage, source of food, (honeydew), queen, beekeeper's car – see under respective word.

**record**
(a) A written expression to preserve facts,
(b) Highest or most extreme degree reached. E.g. hive or weather records, record for the highest crop of honey.
See: hive records and notes.

**recruits**
A newly acquired individual who follows a certain pattern of behaviour. A newly recruited forager that works a particular source of food having been excited by the dance of a successful scout bee. The act of recruiting.
See: scout.

**re-crystalize**
Honey that has been cleared of granulation (usually by warming) will be slow to re-crystalize but will eventually do so.

**rectal ampulla**
Responsible for the absorption of protein.

**rectal glands**
Also called –rectal pads – partly chitinized and arranged around walls of the expandable rectum. They are able to absorb and store salt and play a part in the recycling of water (important when overwintering).

**rectum**
The colon or large intestine, the part of the alimentary canal that lies between the small intestine and the anus. The rectum is capable of distension (in the same way as the honey crop) and it is not possible for both of these to be fully distended at the same time (hence marks on neighbours washing when bees collect water).When confined to the hive for long periods in the winter faeces accumulate in the rectum until a cleansing flight becomes possible. Work has shown that the rectum acts as a water reservoir the permeability of the proctodeum varying under the control of hormone from the corpora allata.
See: corpora allata.

**recycle**
To put through the cycle again. Bees produce wax and honey and both have to be recycled. The danger of disease pathogens being transferred in this way must be taken into account. Nosemic honey can be boiled and re-fed but sugar is hygienic and cheaper. Wax can also be produced on a sugar diet and recycling suggests bees could be given the opportunity to produce more than they would naturally.

**red**
A colour that is not detectable by the eye of a bee. Red light may therefore be used to manipulate a hive at night (white light would draw bees from the hive. Flowers that appear red to us might well contain other colours visible to a bee. Presumably orange would appear yellow to a bee but as their visual perception takes in the ultra violet

(which is invisible to us) they get an entirely different impression of colours which they can identify and distinguish.

**Red cedar** *** ***Juniper virginiana Thuya plicata***

North American trees whose timber is reddish, fragrant, soft and light in weight. It is easy to work while being fairly weather resistant. Though one of the more expensive timbers used for hives, it is generally the one specified in the U.K. It is not very suitable for frames and its aromatic nature does not endear it to the bees. If untreated it is inclined to crack or split when it is old. There is an Australian variety *Adrela toona* Meliaceae. See: pochote.

**Red chestnut** *** ***Aesculus carnea***

A common decorative hybrid between horse chestnut (which flowers two weeks earlier) and the red 'buck-eye' from N. America – its candles of flowers which are well-worked by the bees are pink or dull red and it has scaly bark. The timber is easily shaped but is very light and weak.

**Red clover** *** ***Trifolium pretense***
**Leguminoseae**

A perennial with pink flower heads which can bloom right into August and can fill supers with fine, water-white honey. The tetraploid variety gives more nectar but has a longer corolla. Though bumbles are its natural pollinators because nectaries are 10-20mm long and honeybee tongues rarely exceed 7mm but they can if the nectar depth reaches 3mm, suck the nectary dry due to capillarity. A second flowering after an initial cut reduces the corollae to 8-12mm. As recommendations are to take hay from red clover in 2 or 3 cuts bees can then work the strip most effectively for say 10 days giving a higher seed yield and uniformly timed set. The pollen is variously reported as dark/pale yellow, and yellow with a grain of 32/40mu. In dry weather bees may desert the crop or steal nectar from the base of flowers. The honey is similar to that of white clover though it granulates faster.

**Red currant** *** ***Ribes rubrum***
**Saxifragaceae**

An April flowering cane shrub bearing small red edible fruits. It is related to the white and black currants. Its pollen has been described as whitish yellow and dull or pale yellow with a grain of 20mu See: currant.

**Red deadnettle** *** ***Lamium purpureum***
**Labiatae**

A square-stemmed annual found on wasteland, roadsides etc. flowering from February onwards. F. N. Howes describes its pollen as a beautiful dark orange colour while R. O. B. Manley says pink and other sources - Dines, say red but the variations may be due to different habitats. See: white deadnettle.

**Red-hotpoker** *** ***Kniphofia aloides***
**Liliaceae**

Erect spikes of glowingly flame coloured trumpets – the torch lily. It grows to 3-8 ft. and flowers June-Oct. Not all varieties have corollas with sufficient rotundity to admit the honeybee. The colourful, pendulant, tubular flowers should be dead-headed and the plant crowns protected in winter.

**red pollen**

Pink, reddish brown, brick and crimson pollens come from Mignonette, Horse chestnut, Sainfoin, London pride, Red deadnettle, mullein etc.

**Red squirrel** *** ***Sciurus vulgaris***
**Sciuridea**

It is an arboreal rodent once common in the U.K. but ousted by the grey squirrel. It has been reported as including bees in its diet.

**reducing sugar**

A sugar with potentially active aldehyde or ketone groups. The main honey sugars – laevulose and dextrose – are both reducing sugars, whereas sucrose which under the proposed honey standards should not exceed 5% in honey, is a non-reducing sugar. A reducing agent causes another substance to undergo reduction and is oxidized in the process.

**reduction**
Chemical reactions may add or take away oxygen in the process. Reduction implies loss of oxygen while the reverse is oxidation.

**reduction division**
Meiosis. This refers to cell-division in the course of reproduction. The stages that a cell passes through to produce four haploid gametes. It is the halving process that takes place prior to fertilization. Honeybees are a special case in that the drones are haploid before cell division takes place.

**reed**
A tall aquatic grass with hollow stem. Phragmites communis for example. Used for thatching and early beehives.

**referee (umpire)**
An arbitrator to whom a dispute is referred or person elected to give an appropriate reference for a candidate presenting herself for examination, employment etc. One prepared to give a reference.

**reference**
A reference book is a dictionary or encyclopaedia to be consulted rather than read continuously. An indication of the part of a book where information, clarification or proof may be found. Drawing the attention of an authority to a matter requiring decision, settlement or consideration.

**refining**
To reclaim wax from old comb or to improve the quality of the reclaimed wax
See: wax rendering.

**reflex**
An involuntary response. An automatic and practically immediate response evoked by a particular stimulus. Pressure on the dorsal side of the bee's thorax results in the raising of the wings. This is a mechanical reaction though, and reflexes normally referred to are of a nervous reaction.

**refractive index**
The ratio of the perpendicular distances of a ray before and after refraction or the ratio of the velocity of light on the substance to that in a vacuum. A simple refractometer enables this to be measured using quite small amounts of honey . This enables us to determine the water content and specific gravity of the honey. The s.g., moisture content and therefore refractive index are variable c. 1·48 - 1·50. (Honey is one and a half times as heavy as water).
A refractometer is an optical device allowing the user to determine the moisture content of a solution like honey. It determines the refractive index thus: specific gravity, density, viscosity and water content. These are related and at a given temperature one can be used to discover the other.
See: s.g. and R.I.

**refugium**
Any area which by virtue of certain fortunes has enabled it to become a refuge for plants or animals which have subsequently been unable to spread out of it (presumably because the immediate environment – sea or suchlike, is unsuitable for them).
It was thought for instance, that the British Black bee which was wiped out in Great Britain, might have been found in a niche in Tasmania

**regicidal knot**
During the 'balling' of the queen a clump of tightly packed bees surround her. It has been referred to in this way (reginicidal?)
See: balling.

**registration of beekeepers**
The administration and cost of any scheme to register all those who keep bees would

be neither easy nor cheap. Arguments have waged for a considerable time as to whether beekeepers should have to say officially where their bees are and how many colonies there are in each apiary. In some parts of the world registration is compulsory.

**regulation**

A Government order having the force of law. The act of regulating and controlling. The microclimate of the brood nest, control of the queen's egg-laying, the number of queen cells built, these and other things are regulated in a honeybee colony and have been described as the 'hive-mind'. Investigations however show that automatic response to stimuli, the effect of pheromones and the normal cyclic development of individuals and the colony are the real regulators.

**regurgitation**

The use of a worker's muscles to deliver the contents of its honey crop (sac) into its labium or tongue and thence to other bees or into a cell is regurgitation. Food sharing where two or more bees may share the regurgitation of one that has brought in fresh food or has crop contents that will benefit them, goes on continuously when a colony is active. The act of regurgitation is part of the honey ripening process.

**re-infection**

A second infection that follows recovery from the first one. Old combs are able to carry pathogens especially of nosema and foul brood.

**re-enforce**

This general term meaning to brace or strengthen, is used in beekeeping with certain particular references such as the re-enforcement of foundation to prevent sagging or collapse in the extractor. Also the re-enforcement of mead, a wine which under natural conditions does not exceed 14% alcoholic content. The number of young bees or foraging bees might be re-enforced by 'doubling', 'plumping' or otherwise increasing the normal strength of a colony.

**rejection**

The discarding or throwing away of an item. Bees will eject any creature considered to be alien – even a splendid laying queen although they are queenless, unless both queen and colony have been put into a mutually acceptable state. Drones are rejected when the 'hive-mind' considers fit. A good beekeeper rejects poor queens or combs that have become unfit due to age or performance. Predators and foreign material also suffer this fate if the bees are capable of dealing with them. Certain bees that are injured or diseased may also be rejected by a colony. See: cull and throwing out.

**relative humidity (r.h.)**

The ratio expressed as a percentage of the water present in the atmosphere to the amount required to saturate it at the same temperature. A given parcel of air becomes wetter as it cooled and drier as its temperature rises. As brood nest temperature is fairly constant around 34C the r.h. is around 40% though some scientists have quoted higher figures. When 'grafting' or keeping brood out of a hive the r.h. should be well above this figure as desiccation is most harmful. See: equilibrium and humidity.

**release**

To let go or free from confinement. A delay in the release of a queen from an introduction cage is dealt with under 'queen introduction'. A queen excluder will not only confine a queen but drones too and unless drones are released the excluder becomes clogged by drones that have died in their effort to struggle through. Once transported and bees have been set down on their new site, no attempt to block upper ventilation (putting a crown board or roof on) should be made until entrances are opened and this is best done after they have been allowed to settle after the buffeting of transport. Their release should take place only on the actual site they are to occupy.

**releaser**

A stimulus that sets off an instinctive reaction, e.g. the release of 'alarm odour' by a single bee (releaser pheromone). When an ingot of beeswax is to be released from its mould this will have been made easier when a releasing agent has been used.
See: releasing agent and G.R.P.

**releasing agent**

Name given to the material used when a moulded substance such as plastic (g.r.p.) or beeswax has to be released after setting or curing from its mould. In the case of beeswax when moulded into foundation or candles a solution made from water can prove quite satisfactory. Aerosols of silicone are available and can be used to spray a mould before filling with beeswax and there are special linings that are now widely used.
See: wax moulding.

**religion**

In the quest for establishing the values of an ideal life much has been written. It is reasonably certain that all religious writings include references to the honeybee in some way or another. We find bees, honey and wax play important parts in mythology and in ceremonial acts. Islam's heaven has rivers of honey – the food of the Gods. The early Christians used only pure beeswax candles, the bees were emblems of purity, chastity and hard work. See: Babylonians, Bible, Koran, Talmud, Rig-Veda, Mormon, Rash Hashanah, Ramadan and clergymen beekeepers.

**reluctance**

Having no desire or being unwilling to do something. Bees may be reluctant to use a feeder, to accept a new queen, to unite with another colony, to occupy a super or to enter a new home. The coaxing or governing required to overcome reluctance that the bees show is part of the skill of good beekeeping.
See: partial acceptance.

**remedial effects**

Nothing is as likely to do more for a healthy colony than a good, new queen. Extra space can remedy overcrowding. Foundation can remedy loss of activity due to idle bees. A good honey flow can set a lot of things right.
See: drugs, new comb, prophylactics and sterilization.

**remedy – antidote**

Something intended as a cure for disease or disorder. To make good, heal or rectify. Old-fashioned remedies for bee stings ran into dozens. For instance contact with wet earth, cold metal, something hot, onion juice, blue bag, soda, vinegar, ammonia.....Modern proprietary preparations include useful aerosols containing anti-histamine, pain-killers etc.
See: diseases, recipes and sting treatment, removal of propolis
Or spilt beeswax from clothes etc. – use folded newspaper, filter paper or tissue and a hot smoothing iron. Caked utensils can be warmed with hot water and then wiped out with absorbent newspaper before finally cleaning. Freezing to make propolis brittle when it may be hammered out is another method.

**remunerator**

It was the name for part of the famous 'Remunerator and Preserver' hive promoted by the Western Apiarian Society (Isaac). The lower part was the Remunerator and the upper the Preserver with the bees etc. The Remunerator or 'nadir' was taken for honey.

**rendering wax**

Causing comb to separate into liquid beeswax and dross. The process of transforming natural comb cappings and scrapings into clean ingots of wax. Pressure and heat are required and highly sophisticated apparatuses and techniques

were used to produce the super clean pharmaceutical wax. Simple apparatus can however extract a fair proportion of good quality wax from old comb. Water should be distilled, de-ionized or acidulated. Hard water requires a dessertspoonful of white vinegar to each gallon of water.
See: steam and solar wax extractors and wax rendering.

**rent**
Colonies that are put out for pollination purposes should be suitable for the service and a rent is usually agreed upon in the form of a pollination fee for a given period of time. Commercially this is arranged by the Bee Farmers Association. When a site is made available for the accommodation of hives, private arrangements can be made and a donation of a couple of pounds of honey per annum is considered reasonable.

**repairs**
Renovation and the temporary or permanent repairs to hives and equipment will be an on-going process. At out-apiaries many beekeepers carry a hammer, pliers and nails to carry out on-the-spot repairs. Regular overhaul of equipment during the winter months pays dividends in economy and efficiency. We are fortunate in having so many useful products such as glues, fillers, paints, plastic brackets, buckets and all sorts of easy-to-use items. It goes without saying that repairs are so much easier to do when free of bees.
See: damage.

**repellent**
Something that repels or drives back, which is repulsive or causes distaste or aversion. Smoke, disinfectant, the smell of hands and several chemicals formulated to assist with the clearing of supers – benzaldehyde, bee-go and others, come under this heading. Protective clothing is better than trying to apply repellent to the body.
See: bee repellent and smoke bomb.

**reprints**
Leaflets, pamphlets, bulletins and other means of giving abstracts or condensed information are listed under respective headings. Many Societies are willing to disseminate information and guidance by means of reprints (photo-copies). Notably BRA, Bee Craft, BIBBA,
Central Assn., BBKA, NHS, DEFRA etc.
See: publications.

**reproductive isolation**
Normal reproduction of honeybees is initiated by matings in the free air when any drone may mate with a queen. When conditions imposed by the beekeeper lead to the isolation of queens from other drone sources this is referred to as reproductive isolation.

**reproductive organs (drone)**
Testes, vasa deferentia, seminal vesicle, mucus glands, the ejaculatory duct, bulb of the endophallus, cervix and the horns which arise from each side of the vestibule – finally the phallotreme (genital opening).
See: drone.

**reproductive organs (queen)**
The two ovaries comprise some 300 tubules or ovarioles. Each leads into its oviduct and these join to form the median oviduct and just below the point where this goes into the vagina, a duct from the spermatheca enters. The vaginal orifice has two lateral pouches alongside and opens into the bursa.

**reproductive system**
The organs collectively associated with the procreation of the species in both males and females. Also referred to as: genital system, genitalia and sexual organs.

**Reptilia Vertebrata**
Reptiles form a class of cold blooded vertebrates which includes orders such as Squamata: snakes and lizards and other bee-eating creatures. They are not usually serious enemies of the bees in

Britain so it is rarely necessary to take any steps to protect colonies from them. See: lizard

**requeen**

To establish a new queen in a colony. This usually involves removing the existing queen and getting the colony into a mood of acceptance before inserting a new queen initially protected by a specially constructed cage. Regular annual re-queening is carried out in America when colonies are worked hard on pollination circuits. For best honey production it is best to have a queen that is less than three years of age. When a colony has been queenless for some time (and possibly developed – laying workers) it is best to give them a comb of emerging brood two days before requeening them. Uniting also offers a method of re-establishing queenrightness. See: introductory queen cage, laying worker, uniting.

**requeened (rqd)**

When a new queen has been satisfactorily introduced in a colony either by cage, direct introduction or uniting to a queenright nucleus or other method. It is important to check that she has been fully accepted and is laying normally.

**re-queening (by queencell)**

This can work if done during a honey flow; a ripe queen cell is inserted between the top bars (under a queen excluder – if present). Provided the bees do not destroy it, then shortly after emergence the virgin usually kills the old queen though sometimes the two queens will exist together for a while. This does not seem to provoke swarming. The sides of such a cell can be protected by 'micropore' – it must of course be kept vertical.

**research**

Diligent, systematic inquiry in order to discover facts by the scientific study of a subject involving critical investigation. In recent years research has been sparsely financed but suddenly there has been a 'wake-up' call (2010) and various organizations have been generously allotting money to research project associated with bees in general. There is still a tendency to under-play the role of the honeybee perhaps indicating a residual reluctance to accept the fact that the honeybee is quite unique and exceptionally well equipped to cope with all that is asked of it.

***Reseda odorata*** ***** Resedaceae**

Having erect racemes of sweet smelling flowers it can be grown in succession to flower all summer. It likes rich, alkaline soil and will grow well in pots. Vars. crimson fragrance, machet, rubin and red monarch. See: mignonette.

**reserve**

That which is kept over or held for future use. While this term might be used for the amount of honey a beekeeper holds in stock (or sugar or jars) one significant use of the term is in relation to the honeybee's food 'reserve'. For winter the food reserve might vary from twenty kilos in Europe to forty kilos in Canada, but because long-lasting damage can be done by causing a colony to live hand-to-mouth, a permanent reserve of at least five kilos is a good beekeeping rule. Reserve might also mean the price that a seller is willing to accept at an auction. Areas reserved for certain aspects of ecology include nature reserves

**reservoir**

A body of water collected in an artificial lake. Apart from offering nothing by way of forage over a large area, to the bees they are not at all safe, for bees can easily be 'drawn' in and drowned should they attempt to take water from it.

**reservoir bees**

Circumstances can exist when a certain number of bees in a colony, have honey sacs that contain either water or very dilute honey. Also known as 'tanker bees'. Scientists have established that such bees exist because water is not stored as such. However it is possible that bees might be landed with unwanted loads of water yet be unwilling to discharge it at the end of the day.

**residual effect**

The effect of using a chemical may be momentary or it may continue, possibly with decreasing virulence, for some time. Insecticides can be 'volatile' when fumes or released vapours kill, 'systemic' when the poison is absorbed by the plant, or 'contact' when insects that touch the active poison suffer its consequences. The duration of the toxicity after the insecticide has been applied is its residual effect. Naturally those with residual effects are more dangerous and more likely to take their toll of bees. Certain chemicals like DDT, because of their indestructibility and involvement in ecological cycles, have been banned in responsible countries. Certain materials (plastics, foams and wax) can absorb the toxic qualities of substances and care should be taken when Dichlorvos, methyl bromide, PDB (now banned) ethyl dibromide and such like are used.

**resin**

A non-volatile solid or semi-solid, organic substance. An amorphous substance that exudes from plants. Some forms may well be collected and used as propolis by the bees. Cowan mentioned it in 'Wax Craft' for incorporation into beeswax when large moulds call for greater strength. When added to wax polishes used on floors, resin helps to reduce the danger of slipping.
See: stearine and Canada balsam.

**resistance**

Immunity. Having the ability to withstand infection. This may be a natural trait, some bees being more resistant to certain diseases and conditions than others, or it may be imparted to them by the use of drugs or prophylactics (inhibitor substances). Fumidil B, Sulphathiazole and Erythromycin.
See: drugs.

**Resmethrin**

This has been described as a relatively safe chemical for killing bees where it is uneconomical to overwinter them or presumably when disease treatment is not considered advisable. Available in aerosol form giving a spray with 2% Resmethrin. A ten second burst from a new aerosol was sufficient when applied under the inner cover onto clustering bees.

**resolution**

The act or process of making the correct number of close lines visible or of distinguishing between the individual parts of an image. In microscopy the actual magnification and the amount of light are usually of secondary importance to the ability to see minute details accurately.
See: microscopy.

**resonance**

Having a prolonged, subtle, stimulating effect beyond the initial impact. The comb of the bees is hardly looked upon as a resonant substance, yet it is capable of transmitting vibrations – 'substrate sound' and the queen hugs the comb when she vibrates her body in the act of 'piping' and this has an immediate and 'freezing' effect on the bees. Workers can be seen hugging the comb as they do 'waggle' dances. Drumming (pounding the sides with your fists) is also used as a means of driving bees upwards.
See: piping.

**respiration**

Inhalation and exhalation, the process by which energy is liberated within the body of an organism and so becomes available for use in vital activities. It is nearly always an oxidizing action using oxygen from the atmosphere and producing waste products such as carbon dioxide and water. In plants

energy is released from stored foods – the reverse of photosynthesis but in bees the expansion and contraction of the air sacs enables air to be drawn in and led via the tracheoles to the body cells direct. In insects blood has a relatively small expiratory role and is less concerned with the carriage of oxygen than that of carbon dioxide. Internal tissue or cell respiration can be aerobic or anaerobic.
See: respiratory system and R.Q.

**respiratory movement**
Heaving of the chest and diaphragmic movement in man. Slight telescoping of the abdominal segments over the inter-segmental membrane in the bee. Bees at rest and while in the hive, show little sign of respiratory movement though breathing always continues even in the cluster in the dead of winter. However a bee that rests momentarily while foraging or swarming will be seen to rapidly pulsate its abdomen.

**respiratory system**
The tracheal, respiration system originates from tubular ingrowths along the sides of the body, the external openings of which are the 'spiracles'. In the adult ten pairs of spiracles lead, via tracheae, into tracheal sacs (large air bags which are bi-laterally symmetrical and joined by ventral, transverse commissures. The sacs perform as bellows and the rhythmic abdominal pulsations serve to compress and dilate them using the blood as a hydraulic medium. Oxygen is taken to the tissues via tracheoles which are ramifications from the larger trunks and the air sacs.

**response**
The reaction of an organism or part of it, to stimuli. This could be the bee's sting, its own behaviour or that of the whole colony or a part of it. Anticipation of a bee or a colony's response to conditions that have just prevailed or are prevailing is of great importance in apiculture. Response to smoke, to a new queen, to a honey flow – these are examples.

**rest**
The work of a colony seems unending, yet individual bees appear to remain stationary and do nothing. You might say that a colony of bees 'rests' during the winter. In fact much of what goes on is hidden. The various glands are brought into play successively and such changes – such as the production of wax, or brood food may occur better when a bee appears to be resting than when it is active.

**rest harrow *** Ononis repens Leguminoseae**
A wild plant with pink pea flowers, Its name comes from its tough, woody roots and stems which used to hinder the plough. From June to August its delicate, rose-pink flowers, shaped like butterflies, are there for the bees. 'It maketh oxen whilst they be ploughing to rest or stand still'.

**resting cell (cytology)**
A resting nucleus implying that whatever other activities it may have, the cell is not in the process of dividing. When metabolism has been reduced to a minimum, many cells that would otherwise be active will be resting.

**restive**
Restless, uneasy or impatient of control. A colony that is difficult to control by ordinary methods of subjugation might be described in this way. It could be due to their basic temper, atmospheric conditions or other disturbances. Birds and animals that disappear when the beekeeper arrives can give rise to a colony's uneasy state.
See: restless.

**restless**
A colony that is restless will indicate signs of uneasiness as for example when made queenless or when being attacked, disturbed or robbed. The restlessness may continue for days or until the colony has been restored to a normal condition. Perhaps the site is too windy or unwelcome smells from a nearby

factory or incinerator could be responsible. During winter dysenteric droppings may occur.
See: queenless moan.

**restriction of the brood**

Reduction in effective egg-laying by the queen to a point of cessation. The normal expansion of the brood nest takes place progressively under ideal conditions with healthy bees and a vigorous queen, suitable weather and an adequate supply of food. Sudden adverse changes in any of these or interference by the beekeeper, or lack of space, may lead to a restriction of the brood.

**reticulation**

Latticed or covered by a network such as the nervation of a leaf or the wing of an insect. A close-up of the chorion of an egg, the exine of a pollen grain or the surface of a stigma may also be marked by distinct patterns which may be granulated or reticulated – e.g. ivy pollen. See: ramify.

**retina**

The inner, nervous, light-sensitive membrane of the eye which receives images that can be transmitted to the brain. The bee's retina is composed of several thousand structured units – the ommatidia - and while the same light receiving and nerve transmitting properties are there, the sight of the bee is interestingly different from our own. Bear in mind that in addition to the compound eyes there are three simple eyes and all of these integrate to enable bees to perform their duties.
See: sight.

**retinue**

A body of retainers in attendance upon royalty. The queen's retinue changes in number and composition. The queen herself can often be brought to notice by the bees nearest her who form a 'court' facing towards her. Anything up to a dozen young bees, some will feed her, lick or seemingly caress her. Her bouts of searching for suitable cells (cells clearly prepared by the house bees) and laying, are interspersed with ten minute breaks during which time bees of her retinue will feed her. To lay over a thousand eggs per day means that they probably exceed her body weight.
See: attendants, court, escort and queen substance.

**retinula**

The innermost element of the ommatidium, composed of elongated, pigmented cells.

**retrogressive**

Backward moving, a term used in genetics to indicate degeneration or the assumption of characteristics as possessed by earlier and usually lower type of development. It normally indicates going back to a previous or less satisfactory stage – the opposite to progressive.

**return**

Although it is obvious to a beekeeper that bees must return to their nest or they are likely to perish, many of the general public are surprised to be told this is so. Growers have sometimes asked that hives should be repositioned a hundred yards or so to suit their convenience and have to be informed that once accustomed to a site the bees must be allowed to stay or else be moved right away. The number of foragers that do return to a vacated site will depend on (a) the foraging force and (b) the distance they are moved. Bees deprived of their home in this way fly searchingly around and while they are not aggressive, anyone who has asked for the hives to be moved may take some convincing. Even a month later a colony returned to the vicinity will still include a few bees that go back to their old site. This home-finding ability is vital, the whole existence of the colony depends on it

**reverse side**

When holding a comb up at eye level and then requiring to reverse it so that the other side can to lay the comb flat – even

momentarily. (hold the top bar vertical and swivel comb through 180 degrees). It is not the end of the world to lay a comb flat but bees build vertical combs in the dark. To expose them to daylight and worse still to turn cells intended to be near horizontal, into a position where nectar or pollen can fall out, or even the queen, is best avoided. A warm comb, though held firm by its embedded wires, can sag very slightly destroying the accurate cell depth by causing cells on one side to lengthen and on the other to shorten. Certain liberties can be taken but when handling queen cells, even though no apparent harm comes from exposure or altering the inmates normal position – you want the most important bee to be as perfect as possible, take no chances.
See: horizontal and vertical.

**reversible extractor**

A tangential extractor with frame-holding cages that can be rotated through 180 degrees so that both sides of the combs can be extracted without the necessity of having to lift out and physically reverse the combs as would otherwise be done.
See: radial and tangential.

**re-worker**

When a queen has been transported in a small cage and her escort is changed this is known as re-workering the cage.
The importance of having young, well-fed workers to form the retinue comes into play when after travelling it is desired to store the queen for a while before introduction.
The terms re-queening and re-droning are also used.
See: attendants, cage and queen attendants

**R.H. -relative humidity**

Saturated air has an R.H. of 100%. Any smaller percentage tells how far away from saturation a given sample is. The air over a sealed sample of honey will be around 60%. Brood nest R.H. is around 40%. The lower the R.H. the faster evaporation takes place.
See: relative humidity.

**rhabdom (rhabdos – rod)**

A transparent rod surrounded by retinal cells with outer pigmentation that runs from the base of the crystalline cone beneath the lens of the compound eye facet, down to the nerve fibre at the proximal end.

**rhabdomeric microvilli**

These important parts of the bee's compound eyes enable it to distinguish the down-sun direction, even beneath overcast and this they relate to the gravitational datum – 'down' – on the comb and they measure azimuth bearings to and from the hive accordingly so that the angle subtended between the down-sun direction (or the down-comb direction) and the intended flight path (or direction of dance on the comb) is the same.
See: von Frisch.

***Rhamnus cathartica***

A small, thorny shrub which bears clusters of black berries and whose foliage feeds the brimstone caterpillar.
See: buckthorn.

**R. H. B. Ron Brown**

R. H. Brown, OBE, BSc Distinguished beekeeper and author of the award winning '*Beeswax*' and also 1000 yrs. Bkpg in Devon.

**rheumatism**

A complaint commonly affecting human joints. There is a widespread belief amongst older folk that stinging by bees either prevents or ameliorates rheumatic conditions. As rheumatism seems to take many forms, some of these may well be less virulent where a regular intake of bee-venom is customary. There are examples of life-long beekeepers who have had little trouble with this complaint.

**Rhizobium**

A rod-like bacterium found in the root nodules of leguminous plants and responsible for the fixation of atmospheric nitrogen. When clover was included in the meadows nitrogen was continuously put back into the soil by their nodules and the bees played their part in pollinating the florets. Modern agriculture seems to have chosen barley, rye and other crops where nitrogen is applied artificially, therefore the sustenance for bees has become minimal or even totally absent. Apart from chemicals and radio waves which also have their effect on bees, this change in agriculture must contribute to what is happening to our bees today.

**Rhododendron spp. *** Ericaceae**

There are over one thousand species from giants to dwarfs. This ornamental shrub likes acid, lime-free soil, moderate temperatures and high humidity. It blooms from January to September. Some forms have spicy scents and several with the finest fragrances must be cross-pollinated to produce seed. They can be unbelievably beautiful with their contrasting colours, intensely coloured nectar pouches deep in the flowers with honey guide lines very evident. Reports from Scotland have suggested that toxic nectar and harm from sticky white pollen has been a problem. It is an evergreen and its huge woody stems in the case of very old shrubs, have supported a canopy under which hives could be kept. Where it establishes itself, as for example in Snowdonia, it can become a dominating plant that eradicates all else with its dense foliage. Mosses, toadstools, bluebells, hazel and wood sorrel, normally found under trees, have all been obliterated.
See: rhododendron honey and Xenophon.

**rhododendron honey**

The works of earliest writers were favoured by being read over and over again. Xenophon's story of the soldiers who ate honey said to have been derived from the rhododendron, has therefore become extremely well-known. Many of the soldiers suffered toxic symptoms and some species of the plant do contain 'andromedotoxin' one of the world's deadliest poisons. Bumble bees work these long trumpet-shaped flowers and suffer no worse than when they work some of the Tilia. Perhaps Xenophon's soldiers ate from bumble nests? In any case beekeepers in this country have little to fear if rhododendrons grow nearby and counties with many wild varieties have not reported any sickness or fatality as a result of eating the honey.

**rhodomel**

Palladius tells us that this drink was manufactured from roses and honey as its name in Greek implies.

**rhomb - rhombus**

An oblique, angled, equilateral parallelogram. The bases of honeycomb cells become almost rounded due to their being formed from three diamond-shaped (rhombic) plates. Take a needle and pierce each of the three plates on a sheet of foundation. then turn it over and you will find that the holes have been formed in the base of three cells on this side. Hence a cell on one side is centred between three on the other. This unique and mathematically ingenious arrangement puts the honeybee in a class by itself! It enables back-to-back horizontal cells (compare with the wasp) to be built, using their unique beeswax, with maximum economy of material and mutual warmth and convenience for nursing, storage of pollen and ripening of honey.

**rhymes**

See poems, poetry, sayings, and superstitions.

**Rhynchota**

rhynchos – a beak or snout Hemiptera *R.p.luxurians*, *decaisneana*, *ambigua*, *hartwigi*, *viscosa* and simper florens. Honey dew is produced by these plant insects (frequently positioned and farmed by ants) whose mouthparts comprise four pricking bristles that move against each other and puncture plant tissues. The liquid intake of young larvae may greatly exceed their body weight per hour and they secrete 'honeydew'. The

newly secreted liquid is almost colourless but becomes brown and tends to solidify on being exposed to air. It is also rapidly invaded by the 'soot fungus' and this can impart a blackness to the honey
See: NZ honeydew, Hemiptera, Coccoidea and leaf honey.

**R.I.**
See refractive index.

**Ribbands C.R. M.A. Sc.D**
Principal Scientific Officer at Rothamstead Experimental Station who wrote, amongst several other enlightening works - 'The behavioural and social life of honeybees 1953.

***Ribes grossularia*** *******
**gooseberry Saxifragaceae**
There are red, black and white currant varieties of this flowering bush and when they flower in April the fruiting varieties do better when bees are around. Though honey is rarely obtained pure it is said to be of medium colour and of excellent flavour. The pollen is greenish-yellow.

***Ribes nigrum*** **- blackcurrant**
Grown for its small, black, edible fruit. It can be propagated easily from cuttings. The long flowering racemes become more needful of pollination as the distal flowers open. Where pollination is inadequate it is quite usual to get small clusters of currants in the earliest flowers which are partly self-pollinating. Where grown commercially spraying is likely. Blooming occurs during the dubious weather in April.

***Ribes sanguineum*** ***** flowering currant**
Howes tells us they are popular in gardens, blossom early in the season (April) and are good bee plants. Some have corollas too long for the honeybee to reach the nectar but plants are well-worked for pollen.

**riboflavin $C_{17}H_{20}O_6N_4$**
A crystalline, orange-yellow pigment – a factor in the vitamin B complex (B2 and G). It is concerned with cellular oxidation (essential in the diet) and required by man and bee. With H4O6 ending along with it, it provides the co-enzyme necessary for tissue metabolism. It is present in honey, milk, fresh meat, eggs, fresh vegetables and is necessary for growth.

**ribonucleic acid RNA**
A chemical messenger and intermediary between the structure of D.N.A. and protein. It is found in all living cells, mainly in the cytoplasm and plays an essential role in protein synthesis.
See: ribosome

**ribosome**
Spherical granules or microsomal particles containing up to two or more cells of RNA. Found on nuclear membrane and membranes of the endoplasmic reticulum. They take part in the synthesis of proteins. Dr. R.S. Pickard describes them as the molecular work benches upon which enzyme proteins are constructed by linking together sequences of amino acids.

**Rickettsia spp. Chlamydiales**
Small, gram-negative, bacteria-like organisms with bacterial type cell wall. Intracellular development requiring living tissue for their development and reproduction. Have been found in bees causing problems with the haemolymph.
See: spiroplasmosis.

**riddle - riddling**
To use as a sieve. A. M. Sturges uses the word to describe the shaking of bees onto an excluder placed over a box of combs with a view to finding the queen when most of the bees have run down below.

**RIE**
Rocket immunoelectrophoresis a technique for the detection of pathogens in infected bees or from their food source.

**Riem Johann 1770/82**
He was the first to report and explain the behaviour of laying workers.

**right size of hive**
Langstroth reckoned that 2000 cu ins should be the interior size. The bees and beekeepers ever since have been trying to prove him wrong. With the advent of wooden, movable-frame hives it became possible to vary the size occupied by the colony from its minimal winter dimensions up to the size required for record honey takes. As regards frame size and number of frames per box the only right size will be the most commonly used size.

**Rig -Veda 2000 – 300 B.C.**
The Vedas comprising four books of wisdom regarding Hinduism. They contain much advice as to how honey and bees should be used. Honey is often looked upon more as a medicine than a food by the people of India. In Hindu mythology the moon was referred to as Madhukara – honey giver. Vishnu, India and Krishna were called Madhava or honey-born. Oldest document among the sacred scriptures of the world's living religions.

**riparian**
Referring to things that dwell or grow on the banks of rivers or other areas of water. The right to fish.

**ripe honey**
This is normal honey which has been completely transformed from the original nectars having a minimal water and sucrose content. It is usually derived from sealed honeycomb though even this, at times, may contain unripe honey. Honey can be ripened artificially. The fascinating thing about bee-ripened honey is its storability.
No other creature, apart from man, has rivalled this method of safeguarding their summer's work.
See: dehumidifier and unripe honey.

**ripener *****
At one time this word was widely used to describe a honey settling tank. In fact to ripen would at least imply the removal of the unripe honey's water content. Unless therefore some additional means of extracting water vapour is provided, honey tank is the best word to use for this piece of apparatus. Once honey has been extracted from the comb and filtered, many air bubbles will have been trapped and only slowly rise to the surface. The settling process should therefore be allowed to continue in a warm atmosphere for a day or so before the clear honey is run off from below the froth. Small bubbles in honey cause cloudiness, can cause 'frosting' and may trigger off granulation.

**ripe queen cell**
A queen cell in which a virgin is ready to bite her way out. The tip of such a cell will have had the wax removed and the cocoon will be visible giving it a somewhat bald appearance. Sounds may emanate from it and possibly through any small slit, cut by the virgin may give a view of her near the tip, an antenna or tongue might appear. Such queen cells can be safely transported provided they are not squashed, chilled or put anywhere so that the queen might emerge and get away.
See: pulled queen.

**riser**
Alternative name for a lift used with the C.D.B. hive. An outer wall often filleted in such a way that it can be used either way up. One way up it increases the size and capacity of the hive, the other way up it overlaps the lower body thus giving an extra wall to retain heat and act as a protection against the elements. Countries where Langstroths are widely used, call the three strips fixed around the edge of the floorboard on which the brood box stands, 'risers'.
See: cleat and runner.

**rival queen**

A queen in the same hive as another queen of similar age or where a younger queen is not living in harmony with her parent and is likely to be superseded by her. On occasions when supersedure takes place the daughter will receive more attention than the mother who then slowly ebbs away.
See: queen rival.

**river**

A natural, flowing stream of water that cuts its own course across the land to reach the sea. River basins will have their own flora. Rivers often flow through valleys which remain well-irrigated and sheltered, though winds can be funnelled along them.
See: migratory beekeeping and waterways.

**RNA ribonucleic acid**

It occurs in several forms, for example: messenger, ribosomal and transfer DNA.

**road sides**

The waysides or borders of the road. Similar areas are the central reservations or the strips that separate dual carriage ways and the like. These can offer useful forage for bees though traffic must always receive priority with the result that spraying, untimely cutting and collisions between bees and vehicles are hazards that must be considered before describing them as benefits or areas to be especially planted with nectar and pollen producing plants.

**roar**

A deep, prolonged sound of distress, rage or excitement. When a colony of bees is caused to 'buzz' by getting them all to fan, this is called a roar. It occurs when a colony is transferred from a tree to a hive or when a stock has been over-manipulated.
See: queenless moan.

**robbed**

Unless special precautions are taken it is all too common to find a newly made up nucleus empty in a day or so. Honeybees will work hard and deliberately to render a colony in its vicinity defunct by stealing all its food. This may be nature's way of spacing colonies out so that they have free range over their surrounding territory. At the present time when much attention is given to the need for docile bees, it should be understood that a truly docile colony puts up little resistance against robber bees. Strong, queenright colonies, are usually able to hold their own against robbers but lack of morale due to queenlessness or weakness can encourage determined bees to rob them to extinction.
See: robber and robbing.

**robber**

Bees are inveterate robbers and once accustomed to this nefarious pastime are slow to change back to normal foraging activity unless a heavy 'honey flow' supervenes. Robber bees can become hairless and shiny and quite often black. (A colourful bumble bee will be rendered black and shiny if it is overcome by bees when in the act of robbing). 'Empty" robber bees approach the entrance furtively with hind legs outstretched and their flight includes rapid darts from side to side unlike the wearier and more direct flight of returning foragers. Bees suffering from paralysis virus may be pushed about by healthy bees and begin to assume the hairless state or robbers.

**robbing**

This is often encouraged by the beekeeper especially when too many colonies are kept on the one site. Although wasps, ants and many other insects will rob honeybee colonies, beekeepers tend to reserve the expression 'robbing' for the stealing of honey by one honeybee colony from another. There is little sign of robbing during a 'honey flow' but whenever weather conditions permit flight during a period of dearth, there is a tendency for robbers to investigate other weaker hives. If they succeed further attempts are sure to continue and this can lead to a spread of

disease (robbers may be from other apiaries) and the decimation of weak or small lots. Keeping strong colonies with entrance size matched to their strength is a safeguard.
See: ragged cappings.

***Robinia pseudoacacia* *** Leguminosae**

Native to N. America, it was introduced to Europe in 1640 and has spread to become the main honey producer in Hungary and Romania. It is a medium sized, deciduous tree whose wood is hard and durable hence the name 'post locust' though it has a variety of names: common locust, acacia, black, white and yellow locust, false acacia, pea flower, locust tree etc. It likes high temperature when it secretes heavily with a sugar content as high as 60%. The light honey is pale yellow with a mild taste (popular on world markets) and its rich laevulose content makes it slow to granulate. Its long, hanging, white chains of flowers only last for 10 – 12 days in May to June. Pollen is grey to pale yellow and has three grooves 25mu. It does not seem to be a heavy yielder in the UK. True acacia is a mimosa though acacia honey is invariably that of pseudoacacia.

**rock bee Apis scutellata**

Also known as the Giant bee. It is the largest of the four honeybees and is found only in the tropics. It builds a huge single comb high on rocky cliff faces.
See: Apis scutellata.

**rock cress *** Arabis spp. Cruciferae**

This is an early flowerer – March, and Howes says it gives an abundance of pollen and nectar. The hairy rock cress *A.hirsuta* is sometimes common in dry rocky places.

**rocking**

The manner in which a bee 'hears' is to interpret vibrations rather than listen to airborne sound. Consequently dances which convey information frequently involve body movements which translate vibrations into visual or substrate sound. Rocking describes the vibratory body movements that have been observed in some of the 'dances'.
See: dances.

**rodent Mammalia Rodentia**

Mammals with teeth adapted to gnawing. Field mice can be especially troublesome to bees while colonies are in winter cluster unless the wise beekeeper has taken preventative measures in the form of mouse guards at every entrance. Rats have been known to burrow under and nest under hives. Both the above would be prepared to eat honeycomb and mice readily nest inside combs whether or not they are occupied by bees. They stand no chance against bees during the active season.
See: mouse, shrew, rat and red squirrel.

**rogue bees**

Whereas a whole colony may have a disposition to rob where weaker colonies give them the opportunity, rogue bees are those which may even cause damage within their own hive, possibly eating eggs or in other ways. Bees accustomed to forays with a view to robbing usually become black and shiny though this must not be confused with similar symptoms as for 'black robber disease' likely to be caused by a virus.
See: beggar bees.

**Roman writers**

Aelian Ad200, Athenauueus 2nd century Ad, Casius Dionysis lst century, Cato 234-149 BC, Columella AD60, Cornelius Celsus 42 BC – 37 AD, Didymus Ad 350, Diodorus Siculus late lst century, Julius Hyginus 63BC ' 11 AD, Pappus AD 379, Palladius AD 350, Paramos4-500Ad, Pliny AD23-79, Senaca AD 65, Varro 116-27BC, Virgil 70-21BC.

**Romania**

Bordered by 4 countries and the Black Sea. One of the biggest honey producers in Europe. False acacia is one of the main sources as it is in Hungary. A large quantity of pollen, royal jelly and propolis is also produced.

**roof garden *****

The use of flat roofs for this purpose can provide bees with what is often sheltered forage in towns. They are usually out-of-sight from below except when the aspect along the front is exploited as when decoratively adorning some form of advertising. It has been known for hives to be successfully managed in roof gardens but like the top of garages can be windy and may have access problems.

**roof rack**

A metal rack that fits onto the top of a car's roof for carrying luggage and other articles. Beehives have been transported in this fashion though any height would make them unwieldy and dangerous especially in windy conditions. Better, whenever possible, to have hives as low and near to the road as can be arranged.

**Root A. I. (USA)**

He blazed the way for practical apiculture and in 1883 he wrote The '*ABC & XYZ of Bee Culture*'.

**rope hive lift**

Two complete loops of rope which when pulled tight are twice the width of the hive (length of each piece four times the width of the hive) can be set on the ground and once a hive has been lifted onto them, two stout poles can be slipped through the loops, parallel to the hive sides. Two people, one in front and one behind, may then lift the poles. Then as the poles are raised, the looped rope bites into the sides of the hive, holding it steady and at a reasonable height so that the lifters can carry it, though care must be taken to keep the poles level and for the one in front to warn when any step or obstruction is encountered.

**ropiness *****

Used to describe the condition of the putrid brood that is symptomatic of the disease known as American foul brood, When a match stick or similar object is twisted in the remains of a grub that has died from A.F.B. then a brown, gluey mess will adhere to it as it is withdrawn, forming a short 'rope'. These remains dry into black scales which fluoresce under U.V.

**rose spp. *****

Single varieties yield golden pollen in abundance but though the flowers have such a delicate aroma, it is usually a surprise for those new to beekeeping to learn that little or nothing is ever heard of rose honey. The pollen can be pale greenish-yellow with a grain size of 25mu.

**Rosaceae Dicotyledones Angiosperrmae**

Trees, shrubs and herbs having five petals and sepals arranged alternatively in an open circle resembling the flower of a rose. Almond, apple, bramble, cherry, cotoneaster, damson, dropwort, geum, hawthorn, laurel, rowan, strawberry, and whitebeam are examples. The flowers tend to be very exposed to rain which often washes the nectar away and soaks the pollen.

**Rosebay willowherb *****
***Chamerion angustifolium* Onagraceae**

It flourishes in soils excessively rich in potash and has aggressive creeping roots. A tall, showy, gregarious plant with bright, pinkish-purple flowering spikes, also known as 'fireweed' and stands of it are seen from July onwards. Attempts have been made by some to have it listed as a noxious weed. Reports of huge takes have come in from areas in Canada where it rapidly establishes itself after forest fires or where trees have been felled. It forms masses of fluffy seeds which are easily dispersed by the wind. It is best propagated from the roots. Its honey is water white and of high grade. Pollen saxe-blue to greenish blue and the large grains with three conical projections are 72mu and their viscous threads often adhere to the bee due to the sticky viscin of these threads. Heavy yields are also reported from Siberia. Species include: angustifolium, hirsutum, montanum, obscurum, palustre, parviflorum and tetragonum. Other members of the

Onagraceae family include Fuchsia, Clarkia (Godetia and evening primrose (Oenethera)

**Rosemary *** Rosmarinus (dew of the sea) spp. Labiatae**

An evergreen menthaceous shrub with pale mauve flowers. Native to the Mediterranean region but it grows well in this country. Pure rosemary honey is obtained in some countries e.g. Spain and France. In the UK it blooms April – June and likes warm situations and light soils. It yields a fragrant essential oil. The honey is light amber and bears the subtle characteristic aroma and flavour. The pollen grains are amongst the largest.
See: Narbonne.

**rose-water**

Made by steeping – distilling rose petals in water. Referred to in *'beeswax skin cream'*.
See: Recipe.

**Rosh Hashanah**

The Jewish New Year of which honey is one of the symbols. A two day Jewish holiday. Family members dip bread or apple into honey – a symbol of hope for the coming year. Eaten after a Jewish prayer calling on God to cause a good and sweet year to descend upon them.

**ros melleus**

Ancient words for honeydew. Literally it means honeyed dew. However it equally well could be translated *'nectar'* since the ancients believed honey to be showered onto flowers and leaves from the heavens. Incoming nectar was called ros melleus regardless of its origin.

**rotation**

1. A honey extractor should not be rotated too fast as the increased gravitational force will break combs. A retaining screen should be in contact with the uncapped comb face and faster spinning is possible once half the honey has been expelled from the inner face (not applicable to radial).
2. The rotation of crops used to give bees a fair crack-of-the-whip. Now the use of artificial fertilizers has led to constant replanting of nectarless crops. Root crops used to be followed by cereal then legume and wheat or potatoes, beet, oats then clover

**Rothamstead**

Bee research has been carried out here since 1934 and the MAFF Bee Dept. was once housed there. In 1985 beekeeping research was discontinued. Dr, Juliet Osborne's letter in BBKA News No. 183 June 2010 indicates that some re-establishment has taken place.

**rotten**

Bees abhor anything rotten and it is a sign of serious disorder when rotten brood is discovered. Foul brood has been known since earliest times and the type called American often has a putrid, gluey smell associated with it. Rotten sacking (largely replaced by plastics these days) when dry, is clean smelling and still makes good smoker fuel. The splayed legs of WBC hives are given to rotting after long use.

**round dance**

While humans form a revolving movement the bees' round dance is formed by a single bee and is closely followed by nearby, would-bee recruits. It is a circular dance performed when the source of food is close to the hive. Syrup feeders over the bees can cause this phenomenon. An octagonal dance has been reported during which the bee makes at least three types of movement.
See: dance.

**Rowan *** *Sorbus aucuparia* Rosaceae**

Mountain ash. A small European tree which is deciduous and likes acid soils. Its greenish-white flower with warm, creamy anthers form masses in May/June have a powerful musky odour. Bees do not work them much but the honey is said to be of medium colour and the pollen grain a deep yellow (golden) 25mu. It develops pretty, red berries and together with its feathery foliage it is much used for ornamental planting. Seeds lie dormant for 18 months before sprouting and they are widely spread by birds but most seedlings are soon destroyed by sheep or deer and only a few survive in the scree of crags or waterfalls. The autumn tinting of the leaves is spectacular and the berries can be made into jelly.

**rowl**
The circular 'withe' on which skeps were woven

**royal jelly**
A white cream which soon darkens and changes character on exposure to air. This specialized brood food contains large amounts of hydroxylated, fatty acid (E)-10-hydroxy-trans-2-decenoic acid and nucleic acids RNA and DNA together with sugars, proteins, vitamins and amino acids, cholesterol and water. Chromatograms show numerous peaks. It is found in fresh royal jelly and in the worker's mandibular glands. It has pH 4·8. and it is estimated that a queen, who is fed continuously on r.j., will consume 300mg of royal jelly to achieve a birth weight of 200mg. The x-lipolic acid is said to accelerate the metabolism in the human liver. While it is evidently the ideal food to enable complete development of a perfect queen its significance as a human food is open to question. To wit, it has a very distinct and non too pleasant a taste yet none of the preparations I have received from Pharmacists have this flavour

**R.Q. respiratory quotient**
The ratio between the volume of carbon dioxide produced and the volume of oxygen used. During flight when the bee's most active metabolism occurs, the RQ is considerably helped by the cytochrome of the indirect flight muscles.

**rubber *Hevea brasiliensis***
A tree native to S. America but widely grown for its milky juice or 'latex' that is 'tapped' from the trees and is a major source of commercial rubber. In India the bees are able to collect a great deal of nectar from its flowers but no pollen. There are however, extra-floral nectaries which are well-worked especially by *Apis cerana*.

***Rubus fruticosus*** *** **Rosaceae**
Blackberry or bramble. There are many wild and cultivated varieties. The wild kinds grow in profusion on almost any kind of soil. It is often abundant on hedges and prefers places that are not too shady. The tough, wiry stems were once stripped of thorns and split as bindings for straw skeps and thatch. The honey has a coarse flavour, rather low sugar content and granulates slowly into a dense mass.
See: blackberry and bramble.

***Rubus idaeus*** *** **Rosaceae**
The raspberry which is very attractive to all manner of insects yet even more and better fruit can be obtained by moving hives close and saturating the area with honeybees. Light honey. Beware of insecticide sprays.
See: blackberry and loganberry.

**ruby honey**
One source of rich, ruby or port-wine coloured honey is the heath – bell heath and in the south, heather honey and wax derived therefrom have a tinge of this colour.

**ruche**
Old French bark of a tree, beehive made of barks. Medieval Latin rusca and French for hive 'la ruche.'

**ruderal**
Growing in waste ground usually brought into being by man's mining, quarrying, tipping etc. From the beekeeping point of view where sites for hives are not easily found, not only do such areas as this offer more space but after a year or two provide a useful biome. Debris and rubbish may

hinder access and security from vandals must be taken into account.

**rudimentary**
Biologically, vestigial, emaciated, in an imperfectly developed condition, at an early stage of development, checked in its early growth. A worker's ovaries or the drones mandibular glands would be referred to as rudimentary.

**rules**
Principles of conduct laid down for the running of associations, honey shows etc. Alterations are normally made by majority decisions at AGM's.

**run**
To move rapidly or at a fast pace on foot or in the case of a fluid to flow. Bees are not very fast on their feet and are outstripped within the hive by the speedier predator – the wax moth. Braula make very rapid movements over bee's bodies and are quite difficult to catch. Heather and Manuka honey which set into a 'gel' but will run after being agitated.

**run honey**
Clear honey – doubtless this name goes back to the time before honey was spun from the combs but was run out after the comb that had been crushed. It is to distinguish honey freshly separated from the combs from honey in the form of honeycomb.

**runner**
Strips of angled metal or plastic that are used to reinforce the wooden rebate in the hive wall that support the lugs of frames. They also permit easy movement of the frames when it is required to slide one or more frames across the hive body. Such movement is not possible when the runner is notched or castellated which is sometimes done to provide frames with correct spacing from centre to centre. They are made to fasten lugs for carrying purposes.
See: crawlers

**Runner bean *** *Phaseolus multiflorus***
**Leguminosae**
Stick bean or scarlet runner. Honeybees can be encouraged to work these by 'scenting' but bumbles nibble holes through the base of the corolla and hive bees follow. It has been said that when hive bees take to the runner bean flowers it may imply that all else must be finished. The honey and pollen are similar to that from field beans. Pollen is collected between 8 and 10 in the morning but nibbling of the corollae can occur in late afternoon.

**running honey**
The speed of movement of liquid honey is linked with its temperature and viscosity. A long free-fall is inadvisable as this will certainly incorporate unwanted amounts of air bubbles. Always arrange for it to run along a slope so that the free travel through the air-space is minimized. It runs silently – many devices are made to ensure that a flow is cut-off smartly when a certain level is reached. Once on the floor the work of thousands of our little friends is wasted as is the spouse's temper.

**rural**
Characteristic of the countryside. In the days of skep beekeeping the craft was mainly a rural pursuit. Gradually, doubtless due to the higher cost of wooden hives, and the artisan type of user – apiculture has become more widespread in and around towns both in the UK, Europe and America.

**rusca**
Latin for 'rush' see present day ruskie. Also ruchae – baskets of honeycomb.
See: Columella

**Rusden Moses**
An apothecary appointed Bee Master to King Charles II at the instigation of John Evelyn. In 1679 he published his book *'A further discovery of bees'* treating of the nature, government, generation and preservation of the bee, with the

experiments and improvements. He denied that the queen was feminine and proved it by dissecting a queen to show the 'testes'.

**Rush Juncus Juncaceae**

The stems of this plant were dried and used as an alternative material to wattle, wicker, rue or wheat straw for making hives (skeps) in earlier times when movable, wooden frame hives had not come into general use. The soft, common rush was preferred as being less brittle than sedge rush stalks. They have the advantage over straw of having no 'ear'. Cut late in July or August, in long lengths, tied into sheaves and stored in a dry place until needed.

**Russia**

A huge country reaching from Europe in the west to China in the east. Bees are utilized in every possible way for pollination under glass as well as in the open for the production of venom, pollen, propolis, and royal jelly as well as honey and wax. 10 million colonies (doubtless many more now) special breeding farms with enormous numbers of queens and package bees. Forest covers 40% of the country, lime, rosebay, raspberry, wild angelica, etc. Daily weight increases of 30kg not unusual. In desert oases fruit trees, cotton, hemp, pumpkin and legumes make beekeeping profitable. Certain races are world famous and some are of course 'winter-hardy'.

**rustling**

It is an unfortunate fact that bee 'rustling' occurs. Hives should be identified by branding (post codes suggested) and kept as inconspicuous as possible. Complete hives with full supers and even nuclei have been taken. Convictions are relatively few and the trouble is found in many countries all over the world. An easy 'prize' if precautions are not taken.

**Ruttner Prof. Frierich**

Lunz, Austria an internationally well-known beekeeping scientist and authority on races of bee and be breeding.

**Rye-grass *Lolium perenne* Graminaceae**

Perennial fodder grass. Unfortunately for beekeeping this nutritious grass which is of no use to the bees, has become widely sown in lays for the development of stock. It provides an impoverished habitat for wild life. The old meadows with their mixture that always included clovers, seem to be things of the past, though in recent times common sense is beginning to reign.

**rye straw**

This material was preferentially used in Scotland, mid and northern England for the making of skeps. It was considered to be finer and softer as well as longer and tougher than the common rush. Wheat straw was more prevalent in Irish skep making.

# S

**sac**

A bag-like structure in animal or plant as for example one containing fluid.
Also Sack, bag or pouch.
See: sac brood, air, honey, poison and tracheal sacs.

**sac brood**

A brood disease caused by a virus *Morator aetatulae.* The larvae die after spinning their cocoon and pupating. It is also called 'pickled brood'. The bees uncap the cells and these can be seen after the normal brood has hatched. The prepupae becomes a watery sac which is flabby, a light yellow at first and then darkens to dark brown. The sac-like corpse is easily removable though it may dry into a silvery scale known as a 'Chinese slipper'. The dead larval sack lies lengthways in its cell. Most infection occurs during summer and nurse bees become infected themselves. Although this cause no sickness it stops them from further nursing which of course is all to the good. The virus cannot survive dry nor in honey for more than a month and not beyond 10 minutes at 58C or momentarily at 80C. Requeening is recommended.

**saccharin $C_6H_4SO_2CONH$**

Derived from toluene and is four times as sweet as sugar. It is used as a substitute for sugar. Hamlyn gives it as a white crystalline powder 500 times sweeter than cane sugar-used as a calorific-free sweetener.

**Saccharomyces spp. (yeasts)**

*S.epilloidous* – a selected yeast that produces very little acetic acid. One that is used for 'musts' with a low sugar content - (below 15%).

**saccharose**

See cane and beet sugar and sucrose.

**Sucrose**

a disaccharide that is found in nectar, and seems to act as an attractant compared with other sugars. White sugar used by beekeepers for feeding bees. Also called ordinary sugar, cane and beet sugar.

**sacking**

A stout, coarse, woven material made from hemp, jute or the like and used widely in former days before the advent of plastic sacks in the eighties. Useful for wind-breaks and for smoker fuel if slightly rotten and containing no chemicals. Such material is best soaked and then left in the open for some time before drying, cutting to size and using in the smoker.
See: burlap, Hessian and honey poke.

**sack mead**

A sweet mead which can be drunk with a desert. 4 to 4½ lb. of honey per gallon of water.

**Sack**

strong light-coloured wine.

**sack**

metheglin A more highly spiced sack-mead.

**sacrificial combs**

This expression has been used in conjunction with the treatment of Varroa infection. Brood rearing is limited to only a few combs so that Varroa can be concentrated there. These combs are then removed and destroyed. Drone comb is very suitable for this. The reduction of infestation in this fashion is an important support measure to improve the chances of success when using chemotherapy treatment. Varroa appears to prefer drone to worker larvae.

**safety**
Although the vast majority of people suffer little more than temporary discomfort from bee stings and most beekeepers by dint of perseverance and familiarity come to be quite nonchalant where bee stings are concerned, it behoves us to remember that the uninitiated can be very scared of the insects and that for a small proportion of the public a sting can be downright dangerous. Be circumspect and insist on the use of bee-proof clothing being worn and checked by any newcomer to the hives.

**safflower *** *Carthamus tinctorius* Compositae**
Native of the Old World bearing large orange-red flower heads producing oilseed. Sometimes intersown in alfalfa areas to provide protein-rich pollen (alfalfa pollen lacks sufficient protein) Bees collect its pollen in the mornings returning to alfalfa in the afternoons. It also supplies nectar which blends well with the alfalfa. It assists soil reclamation by lowering the water table and increasing salts in the soil. It does well in India though it will grow in S. England. See: bee drives and alfalfa.

**safrol oil (safrole) $C_{10}H_{10}0_2$**
Obtained from the oil of sassafras which is extracted from the root of the North American tree *Sassafras albidum* (the Ague tree). A constituent of the original Frow mixture used as a volatile fumigant against certain bee parasites. Melting point 11C, boiling point 233C – it is also found in camphorated oil and is used in perfumery, soap and flavouring.

**sag**
To bulge away from the vertical. The prudent beekeeper takes all steps possible to retain the upright, plumb vertical line of the honeycomb, by keeping foundation taught and hives quite level. A diagram would show how a sagging mid-rib leads to overlength cells on one side and correspondingly short ones on the opposite side of the comb.

**sage *Salvia officinalis* spp. Labiatae**
A perennial shrub. The blue flowers of this culinary herb provide excellent bee forage. Britain's climate is usually too cool and moist for the plant to do of its best. The greyish-green leaves are harvested and dried for seasoning. The light, mild honey remains clear as it has a high laevulose content. It is a great favourite in the States where black, purple and white varieties are common. A brownish-yellow pollen has a grain size of 35mu.

**Saharan bee *A.m.major***
See : N. Africa

**sainfoin ** *Onobrychis sativa, vicifolia* Leguminosae**
Saint foin (wholesome hay) a European fabaceous herb 'esparsette'. A native perennial that does well on boulder clay and chalk, withstands drought having a penetrating tap root. Common and giant forms with rose-pink flowers that bloom from May to August. It likes a limestone subsoil and yields at a lower temperature than other clovers. Once grown as a fodder especially for horses. The honey is a lemon yellow and a bright yellow pigment colours the wax and even the frames; the honey is not quite as dense as that of other clovers. The pollen has been described as: reddish, brown, yellow and slate grey and the grains are 30mu. It is still to be found on the limestone grassland of southern England and Wales. With the ever increasing interest in equestrian pursuits it would be fine if plots were once again grown for the hay which horses like so much.
Oh, for yesterday again!

**St. Ambrose St. Ambrosius**
***Patronius Apicultorum,***
340 – 397 AD. He was renowned for his sweetness of speech, arising it is said, from a swarm of bees touching his mouth when he was in the cradle (as was Plato). He is the Patron Saint of the old and venerable craft of beekeeping and December 7th is the Saint's holiday. He was Governor of Emiglia and Liguria (N.Italy) and then Archbishop of Milan. He wrote 'Hexameron' and many hymns and devised the Ambrosian Chant, an arrangement of church music. His skeleton lies in his old church in Milan. Saint Ambrogio. One quote: The fruit of the bees is desired of all, and is equally sweet to kings and beggars, and it is not only pleasing but profitable and healthful, it sweetens their mouths, cures their wounds and conveys remedies to inward ulcers.
See: ambrosia.

**St Daboec's heath *****
***Daboecia cantabrica* Ericaceae**
A short evergreen shrub of W. Europe available from nurserymen.

**St. John's Wort *** *Hypericum perforatum***
***Hypericacea androsaemum* (Tutsan)**
The golden flowers (like stars) are worked for their orange pollen from July to September. Tutsan is a hardy perennial with pale yellow flowers from July to September and the flowers are followed by conical fruits changing from green to red to purple black. Suitable as a shrub or hedge. Evergreen – it should be pruned hard in spring. Aaron's beard (H.calycinum) were obtainable from BIBBA.

**Sale of :**
bees, bee produce, equipment etc.
That beekeeping is mainly a hobby in G.B. is obvious from the fact that a hive costs the equivalent of 100 lb. of honey, the frames, wax and other essentials a further 100 lb. and then only by skilled beekeeping is the outlay going to be recompensed over the next six years. Although English honey can be sold for twice the price of imported honey, wax, petrol and timber (essential for honey production) cost twice as much here as in countries where the cheaper honey comes from. There are opportunities for the sale of healthy bees and well-made equipment but it is akin to angling or golf in that the purchasers can only expect pleasure rather than profit from the investment. The sale of bees and equipment is also restricted not merely by the high cost but the lack of opportunity and shortage of educational facilities, though recent changes are taking place in this regard sale of honey produced in the U.K. You are encouraged to 'register' if you wish to sell honey to the general public and then your facilities will come under inspection. The numerous Food and Drugs acts covering the sale of honey have increased still further since our entry into the EEC. Equipment manufacturers can help you with correct labelling and honey shows help you to achieve high standards for presenting your product.
See: honey regulations.

**Salicaceae *** Angiospermae Dicotyledons**
The willow family. Willows , sallows and poplars. Deciduous trees or shrubs, dioecious, the female plants bearing catkins and both producing nectar. Early flowering and the balsam poplars are well-worked for propolis.
See: salix, sallows and willows.

**saliva**
Fluid produced by glands and discharged into the mouth. The mixed secretions of the postcerebral and thoracic salivary glands of the bee. The thoracic secretion is stored in two small sacs from which ducts run to the median duct in the head and thence to the salivarium.
See: salivarium and salivary syringe.

**salivarium**
A pouch under the hypopharynx which is largely overlapped by the paraglossa and at the bottom of which is an opening leading

to the salivary syringe. In the larva this is the spinneret of the silk glands.
See: saliva, salivary canal, gland and syringe.

**salivary canal**
The common duct of the salivary glands which leads to the salivarium.
See: saliva, salivary gland and syringe.

**salivary gland**
There is a salivary gland in the larva which has no counterpart in the adult. Adults have a head salivary or postcerebral gland which shares a common duct with the thoracic salivary whose secretion is stored in two small sacs which are ducted to join with the postcerebral secretions in the head.

**salivary syringe**
A small pocket in the prementum into which the common duct of the salivary glands open. It is musculated and can eject saliva into the cavity on the labium at the root of the tongue where, due to the overlapping paraglossae, it is conveyed around the base of the tongue into the glossal canal underneath. Here it can be mixed with incoming food or used to moisten solid food.

**Salix spp.**
A very large genus of shrubs and trees with rough, pliable branches used for wicker work. They thrive in damp places and have narrow, lanceolate leaves on willows and broader more elliptical leaves on sallows. They are favourite alternative foods for many lepidopterous larvae. Rosebay willowherb derives its name from its willow-shaped leaves.
See: sallow, willow and windbreak.

**sallow *** Salicaceae**
The broad leaved willows including the Pussy willow famed for its golden, male catkins which hum with bees scrabbling for pollen in the early spring (from March to May). Both male and female flowers secrete nectar under favourable conditions.
They are easily propagated from cutting.
The pollen is yellow.
See: willow.

**salt NaCl**
Common salt is sodium chloride a crystalline compound used for seasoning and preserving. Bees have been found to prefer slightly salted (-1%) water than fresh, yet quantities exceeding this amount become increasingly toxic. Rectal glands absorb and store salt and this could have an effect on the recycling of water. When 'baiting' a source of water as at a bees' water fountain, sucrose is the most suitable additive.

**saltpetre (nitre) $KNO_3$**
Nitrate of potash. When used as a 'severe' subjugant one recipe calls for 3oz per ½pint (75g to 250ml). Add to paperhangers size (glue) at 2oz to the pint and mix into a thick smooth paste. Apply to corrugated paper which should be lit from the top. (Put smoker nozzle against the entrance hole, puff and leave for one hour. (?!!) The Editor cannot go along with this.
See: potassium nitrate.

**salts in honey**
Phosphates and compounds of calcium, iodine, iron and magnesium are present in very small amounts as are other trace elements.
See: salt.

**Salvation Jane *** *Echium lycopsis* Boraginaceae**
Australian bee plant closely related to viper's bugloss, introduced to S. Australia by Surveyor General Paterson in honour of his wife Jane. It was favoured as Salvation Jane saving great herds of cattle in times of drought. However it grew too vigorously in NSW and cattle suffered from alkaloids hence its variation to Paterson's curse. It has a better reputation in NZ where it is known as blueweed. In S. Australia much honey comes from it as it is widespread, persistent and highly productive – often the dominant species.

See: blueweed, biological control and Paterson.

**salvemet**
Ancient German remedy made from honey and crushed bees, used by women on St. Catherine's day and having a beneficial effect.

**Salvia spp.*** *S.pratensis* Labiatae**
Meadow sage – there are broad and narrow leaved, red and purple sages. The scarlet garden varieties are rarely productive as bee plants though *S. superba* is reported as useful. Sage – *Salvia officinalis* is a hardy perennial. Varieties yield nectar and pollen, grow to two feet high and produce whorls of purple flowers from early spring until late summer. It naturalizes on dry chalky soils. Howes tells us it can form a four foot high bush of blue flowers amongst purple bracts besieged by bees and butterflies from June to October.

**sample**
A small part of anything intended to show the average quality, style or condition of the general amount. Could apply to many things in beekeeping: bees, queens, honey, wax or by-products such as these. See: pamphlet 'taking a sample' and spray poison.

**sandarac**
This is a dry, friable, almost transparent, tasteless, yellowish resin imported at one time from Mogador, Morocco CTC. Derived from a large NW African conifer Callitris quadrivalvis. Used in incense and for making varnish. It can be used as an additive to beeswax for certain purposes.

**sandstone**
Bunter, Keuper and old red. A rock formed by the consolidation of sand. The grains are held together by a cement of silica, lime etc. Some with an embodiment of quartz, can be extremely durable, other readily disintegrate as they weather.

**sap**
The watery fluid transported upwards by the xylem tissues of vascular plants from the root system to the leaves. It carries soil nutrients (e.g. dissolved minerals) to plant cells for food production. The water is then lost by transpiration. Cell sap is a non-living fluid in the vacuoles of the living cell: it contains variable amounts of food, waste materials, inorganic salts and nitrogenous compounds. It keeps the stems and leaves of herbaceous plants turgid.

**saponin *L. saponis* - soap**
A plant glucoside, the aqueous solution of which froths when shaken forming colloidal foam. It is used in detergents, emulsifiers and foaming agents.
See: cold cream and saponification.

**saprophyte (saprozoic)**
An organism that obtains organic matter in solution from dead or decaying tissues of plants and animals. It is often a secondary invader giving special symptoms such as smell or colour with a disease like EFB.

**sarothrum (scopa)**
Pollen brush or comb.
See: pollen comb and scopa.

**saturated**
For winter feed thick syrup is fed. This gives a close to saturation point where no more sugar can be dissolved in the water. In the case of sucrose this is 66% (14% less sugar than honey contains). Both the air and the ground can become saturated with water. Sites should have good air and soil drainage.
See: r.h.

**savoury *** *Satureja montana, hortensis* (annual) Labiatae**
A culinary herb of the mint family well worked by the bees but mainly for the yellow pollen. It is a strongly flavoured herb and the flowers in June/July are pale and rather insignificant.
See: field scabious and herbs.

**saw-cut top bars**

Used on frames that are designed to take sheets of foundation where the weight of the subsequent comb has to be supported from the top bar. Two methods are in general use to hold the foundation in place and the long saw cut from lug to lug is one. The other is to use a wedge when a continuous strip is tightly wedged and nailed under the top bar usually in a manner that traps the loops of the embedded wire. Both methods can harbour wax moth. The split top bar gives slightly more comb area but the wedge is easier to use.

**sayings**

Bees do nothing invariably. The best packing for bees is bees. Follow the sheep. It takes bees to make honey and it takes honey to make bees. A cat in gloves catches no mice. Neighbour. If a swarm settles on dead wood a death will take place in the family within one year. Let the bees tell you. The native basswood trees were loaded with blossoms the colour of butter. Out of the strong came forth sweetness – biblical reference used on Tate & Lyles golden syrup – motif since 1885. Suddenly all were queenright. Telling the bees. Taking them vore (Devon expression for sulphuring the bees in olden times). Upon the queen depends everything in beekeeping – Elija Gallup. On no other one thing does the honey part of the apiary depend so much See: poems and superstitions.

**scabious *** Scabiosa spp. Dipsacaceae**

Wild and cultivated forms are worked by bees mainly for the pale yellow pollen 50mu. Some varieties are more useful for nectar than others. A hairy willowherb which bears dense compound heads of blue flowers in July and August has been described as harmful.

**scale**

A small, flat, plate-like structure.

1. We describe the sub-ovals of wax produced by workers as 'scales' of wax.
2. The dried remains of larvae which have died from AFB. Such scales are black and found at the base of the cells and they contain millions of spores of the disease and they are tough and long-lived and they fluoresce under u/v light.. Blow-lamping or burning are the only methods of safely destroying them.
3. An apparatus used to weigh a package of bees, hive, honey or wax etc.

See: hive scale, spring balance, wax scale and weighing.

**scale insects Hemiptera, Homoptera Rhyncota**

Small plant pests which suck phloem sap via modified mouthparts many yielding honeydew. Most females have a waxy or resinous scale or shield covering their bodies. Many are attended to and even transported and farmed by ants as their sweet secretion 'honeydew' is a useful food. They include: the plant lice: green, white and black fly, Lachnids, Coccids, Scolypopa, Cinara, Ultracoelostoma and other sucking insects like leaf and tree hoppers and gall insects.

See: honeydew honey, NZ honeydew and sucking insects.

**scalpel**

A small, straight blade attached to a convenient handle so that tiny, accurate cuts can be made. It should form part of a kit of dissecting instruments. Handles with changeable blades are available. Stainless steel is best where corrosive liquids may be encountered.

See: microtome.

**scanning electron microscope**

A beam of electrons are focused onto an object and by scanning it and bouncing the results onto a cathode-ray screen, a three dimensional image can be formed.

See: SEM.

**scap (scep)**

As skeps pre-date the era of consistent

spelling it is not surprising to find several variations of the word 'skep'.

**scape *****

The basal joint of the antenna. It is inflexible though it can be partially rotated at the socket where it enters the head just above the clypeus. It occupies about a third of the antenna's total length the remainder being the flexible flagellum. The workers are longer than that of the drone though this is reversed in the case of the flagellum. See: antenna and flagellum.

**scarlet clover *** *Trifolium incarnatum* Leguminosae**

There are early, medium and late varieties but they usually bloom before other clovers. The flower heads become somewhat prickly with age and can give cattle digestive troubles.
See: clover.

**scattered brood**

An abnormal brood pattern where sealed cells are interspersed with gaps. A healthy, normal queen lays eggs in concentric ovals and an even patch of sealed brood results. The scattering of holes or younger brood is sometimes referred to as 'pepper pot' or shot brood. It may be due to disease, a failing queen or a queen whose eggs lack 'hatchability'.

**scent**

A distinctive kind of smell. As a noun it describes minute particles of matter, often volatile oils, which are carried better in moist than in dry air, and register via the olfactory nerves as smell. It is also used as a verb. The bees' scent glands enable them to 'scent' sources of forage. Important pheromones of course, do far more than affect the sense of smell; they evoke a particular type of response, as for example in the case of alarm odour.
See: aroma, odour, smell, scenting and sniffing.

**scent gland (Nasonov)**

The white or cream-coloured crescent under the tergite of the penultimate segment of the worker. It is on the dorsal side and displays the gland as the distal tip of the abdomen is bent downwards, though the abdomen itself is raised. This arched posture is invariably accompanied by active fanning so that the effects of the pheromone pass into the air and can attract other bees towards them, and it frequently triggers off a similar reaction on their part. The gland cells discharge their product via minute, individual ducts opening into a pocket at the base of the tergal plate. Jacobs reported that single gland cells are scattered all over the bee but concentration of 5/600 cells in one site enable the giving out of odour to be regulated.
See: Nasanov, scent gland secretions, scenting and sniffing.

**scent gland secretions**

Geranic acid is transformed into terpenic acid, geraniol is transformed into primary terpenic alcohol, (z)-citral, nerolic acid, cisform (terpenic acid) nerol and farnesol. These are terpenoids.

**scenting**

Worker bees are able to do this by flexing the tip of the abdomen and secreting Nasanov pheromone which is then propelled rearwards by the motion of their wings. Incidentally just as they can initiate a scent, so they can 'sniff' by using their wings to blow air backwards thus drawing air from in front of them over their antennae.
See: scent gland.

**scent marked**

The queen and workers are able to 'scent-mark' that is to leave a lingering trace of chemical, so that bees are attracted to it, repelled by it or caused to perform in a particular way. When a sting is used, the spot is 'scent marked' as is the cell of a larva just fed. The cell of a larva just fed is identified so that other nurses go to another cell, a swarm will find its queen and continue to display much interest in the spot where she alighted or walked on. Flowers and entrances can be marked. Items of equipment that seem perfectly wholesome to us become more acceptable to the bees when they have had the opportunity to 'deal with them'. This often calls for the covering up of unwanted odours and then giving the material a 'smother' of bee scent.

**scent pheromone**

This is released by the Nasanov gland and contains geraniol, citral, nerolic and geranic acids. Also the mating pheromone of a nubile queen on a mating flight though of course this is a mandibular gland secretion. See: Nasanov and mating.

**scent trail**

As a nubile queen flies out for potential mating she releases a pheromone once she is above 12m. The molecules she leaves through the air are known as the scent trail and are most likely to trigger a response from drones when between 30 and 60m above the ground in a drone assembly area. Some confusion has been caused by land lubbers who failed to appreciate that an airborne queen may fly in any direction and that the drones would therefore approach her 'up-scent' (towards her) along her line of flight which may be in any direction relative to the wind. It is possible too that workers can leave some kind of scent trail.
See: scent pheromone.

**Schirach Adam G. Rev (Germany)**

Between 1760 and 1770 he showed that queens could be reared from worker larvae.

**Scholtz**

Quote in 'The Hive and the Honeybee' as having been reported by Langstroth as the first to devise the use of 'candy' in queen cages. Honey is heated and mixed with about four times its volume of sugar 150F with continuous stirring.

**School beekeeping**

Bees kept for educational purposes. This might well include the encouragement of solitary and bumble bee nests or a small colony, observation hive or even an apiary of several honey gathering colonies. It would usually be taken on as an extracurricular activity and require having a teacher who is sufficiently qualified together with at least one other member of staff with interest and confidence. When County Beekeeping Lecturers were employed by the Ministry of Education many schools received encouragement and help but when a certain prime minister came along - one by one they disappeared. '*Apiculture for Schools*' was a booklet made available to Devon schools in the 70's.

**Schwirrlauf**

Landauer gave this name to the whirring run made by the bees about to swarm. They run across the combs with partially open wings making a continuous sound as of a piping queen (about 250 cps) They run in a straight line, pausing every inch or so to make a high pitched buzz. It is a behaviour pattern that is passed on rapidly from bee to bee.

**science**

The systematic application of methods based on organized understanding; some of the sciences having a direct bearing on apiculture are listed: aetiology, agriculture, apiology, astrology, astronomy, biochemistry, biology, botany, chemistry, cytology, embryology, entomology, geology, histology, mathematics, melissopalynology, meteorology, microscopy, morphology, mycology, mythology, palaeontology, palynology, pathology, phenology, palynology , pathology, phenology, physics and pedology

**Scilla spp. *** Liliaceae**

Bell-shaped flower that blooms in the spring having slightly smaller flowers than the wild bluebell and so, on occasions, being able to

offer nectar more easily to the bee.
See: bluebell.

**sclerite**
A hard chitinous plate or spicule forming part of the exoskeleton. In the bee, articular and axillary sclerites are associated with the wing. See: axillary sclerite, spicule and wing.

**sclerotin**
A tanned protein which together with chitin, goes to make up the exocuticle or outer surface of the bee's exoskeleton.

**scolophore**
The chordotonal sensilla of the auditory apparatus of an insect. Such organs are found in the legs and head of the honeybee. Each scolophorous organ is composed of a chain of three cells with the sense cell below. A fine process leads from the nerve cells. through the two cells above, as far as the cuticle. They may register the effects of muscle-tension on the body wall

**scopa pl scopae**
The bee's pollen brush. The collective rows of closely-set hairs that cover the broad, flat, inner surfaces of the basitarsi of the hind legs of a worker honeybee are used in the process of pollen collection. In various solitary bees such as Halictus and Andrena they have scopae, large pollen brushes, on the proximal part of the femur and pollen is deposited there and on the tibia. Female Megachile and Osmia have ventral abdominal bristles forming scopae which serve for pollen collection and transport. See: pollen brush.

**scorch**
To apply sufficient heat to dry and sterilize enough to shrivel paint and darken wood. This method is widely used on hives, especially their interior, to avoid the spread of predators, to kill disease spores and to render equipment suitable for honeybee occupation.
See: blow lamp.

**Scotland**
Conditions in the west are milder and wetter, even resulting in the existence of sub-tropical species of plants in a few especially sheltered places where the weather associated with the Gulf Stream is felt. Heather suits the later development of colonies and some crops such as cultivated raspberries are quite extensive. The mountainous terrain makes the environment quite varied and as a general rule flowering times tend to lag up to a month behind SW England especially at the start of the season. See: Scottish BKA and Beekeeper.

**Scots pine (fir) *** *Pinus sylvestris***
A tough native tree with prickly cones and needle-like leaves. It is sometimes said to be a source of honeydew and is grown for its useful timber.

**Scottish BKA,**
Bee Journal, Scottish beekeeper etc.
Details of addresses to follow.

**scout bees**
As in-nest workers change over to foraging Lindauer discovered that about one in seventeen went out and located their first food source without having followed any bee dance by experienced foragers. These are 'scout' bees, though any successful returning bee may dance and recruit others to the same food source and in between collecting trips, groups working the same plant species are said to congregate preferentially together. Bees are always scouring and searching every corner of their territory, especially when nectar is not plentiful. At swarming times scout bees seek-out and even arrange guards, on empty cavities. See: dances, recruit and wag-tail.

**scramble**
To move hurriedly and urgently in a disorderly fashion. The bees order and

uniformity often develops from an initial chaos or apparent shambles, as when a swarm is in the air or after being thrown in front of a hive. At first they move in any direction, then slowly as one or two lead the way and as fanning begins, more and more follow suit, pointing their bodies towards the entrance, begin to fan and move steadily (indeed some run) towards it. The fanning releases 'here's home' pheromones which incidentally, can also pass to bees which do not belong to the colony.

**scraper *****

A person who scrapes or a tool for so doing. Hive tools as well as being designed to help lift propolised frames out, have edges bevelled and relatively sharp, to act as scrapers. However a wide paint-scraper can be used when large surfaces like floor boards, crownboards etc. need to be scraped before blow-lamping. When wax and propolis are cold (as in winter) they are more brittle and less inclined to stick. The hive tool can be given a resting place on a smoker, hung on a nail or lost in the grass.

**screen**

A screen may be a surface used to reflect light, as when visual aids are utilized, or as a surface which controls and limits the passage of particles, insects etc. A queen excluder is a type of screen and wind-breaks may be used to screen or shelter hives. A colander is a type of screen and queen excluders can be used in that fashion. Wire or plastic netting may be used to screen (protect) hives from certain predators and netting to keep bees in yet allow air circulation or to divert their flight paths over the heads of the unwary.
See: projector, riddle, sieve, travelling and ventilation screen.

**screening**

Bees may be compelled to pass through a 'queen excluder' in order to find a queen or to separate out the drones.
See: screen.

**screen travelling or ventilation**

A framed sheet of material, which while bee-proof, permits the free passage of air. Because of the heat produced by a colony excited by transportation or confined during flying conditions (as when shut in while the area is sprayed) a screen, fitting right over to the top bars, with the same periphery as the hive, is really essential to prevent the over-heating of brood or even the collapse of combs which become 'plastic' at temperatures not far above brood levels. Similar screens, placed under the brood box have recently come into play to allow varroa mites to fall through yet be unable to climb back and these are placed under the brood chamber.

**screw-eye spacers**

A screw-eye can be screwed into the side bar so as to space the frame from the wall and to give correct centre-to-centre frame spacing. It is probably best to use two on each frame. One say, top right and the other bottom left but they must be screwed in accurately to achieve the right spacing. It is a laborious and not inexpensive method compared with such methods as using Hoffman shoulders on the side bars.

**Scrophulariaceae *****

The figwort family, herbaceous plants including: figwort, toadflax, cow-wheat, bartsia, Aaron's rod, speedwells. Flowers July/August and happy in any soil and it spreads luxuriously near water.

**scrubbing dance**

This is seen when bees are cleaning the hive or using propolis. They make rocking or wash-board movements as antennae, tarsi and mandibles are used in rapid, sweeping movements by young bees in what are believed to be cleaning, scraping and polishing activities on hive surfaces.

**scrubbing solution**

Dissolve 1 lb washing soda, 1 gallon of hot water, add ½ lb bleaching powder, stir and

allow to settle. Pour off clear liquid into another vessel. Apply freshly made with a brush. Wear rubber gloves and wash items with clean water afterwards.
See: sterilization.

**scrub queen**
Or queen cell. The word 'scrub' is used in the sense of something undersized or inferior. Whereas a good queen cannot be judged by her excellent appearance, on the other hand one with a poor appearance should not be risked. Therefore scrub queencells or scrub queens should be replaced by superior ones. See: queen-scrub.

**scum**
A film of extraneous matter on a liquid. The froth, bubbles and impurities that rise to the surface of a liquid such as honey or beeswax when liquefied. The removal of any such blemish is essential if honey is to gain a prize when presented on the show bench. In the bottling of honey it is wise to minimize the formation of scum by carefully filtering and then allowing it to stand in a warm (90F) place for a day or two before running off. Take all steps to prevent the inclusion of air bubbles. See: skim.

**scutellum**
A shield-like part, a small plate. The transverse, prominent roll of chitin situated across the posterior edge of the scutum of a honeybee. The scutal fissure lies between these two transverse, dorsal plates. An interior extension of the scutellum is a U- shaped band to which the longitudinal, indirect flight muscles are attached, their contraction causing an arching of the thoracic roof.

**scutellata**
A local race of honeybees in North Africa and now the name given to the former: *Apis adonsonii.*

**scutum**
A strong, chitinous plate covering most of the thorax. This is the spot that is normally chosen for affixing any mark required on a queen. It is strongly domed and adjacent to a prominent roll which goes across the thorax behind it and called the scutellum. Between these two plates lies the 'scutal fissure' which plays a vital part in allowing the flexing of the thorax under the influence of the indirect flight muscles. It is the size of the thorax which prevents the queen from getting through a queen excluder.
See: flight.

**sea**
The mass of salt water covering much of the globe. Plants growing on coasts are often given the specific name 'maritimum' e.g. *Eryngium maritimum* – sea holly. Colonies kept near the coast will have part of their foraging area reduced and as the sea temperatures and on-shore breezes have a marked influence on the local climate, care should be taken in siting apiaries as for example by providing shelter from winds. There is also a distinct possibility that Drone Congregation Areas might coincide with areas where gulls might interrupt the flight of queens. World temperatureas are largely governed by sea temperatures.

**sea bindweed *** *Convolvulus soldanella***
It likes cliff tops and has pale to deep purple flowers. Each flower lasts but a day but the plant continues to bloom all through the summer. It is not a great nectar yielder but is likely to be visited when there is a lack of more suitable sources in the area.

**sea cabbage *Brassica oleracea* Cruciferae**
A seaside plant mentioned by 'Colin" in the BBJ. It has typical yellow flowers like the cultivated cabbage.
See: brassica

**sea-green honey**
Gooseberry, sycamore and lime are amongst the plants that produce green or green-tinged honey.
See: gooseberry etc.

**sea holly *** *Eryngium maritimum* Umbelliferae**

It likes shingle. Its roots can be eaten, candied with sugar and orange water. Mainly a southern sea-side plant with powder-like blue flowers and prickly leaves, growing up to two feet tall. Worked by honeybees and can boost nectar intake when sufficiently widespread.

**sea lavender *** Limonium spp. Plumbaginaceae**

*L. vulgare* grows on muddy shores and salt marshes, sometimes covering large areas with lavender or pinkish flowers, blooming from July and right through to November. It is like statice and has deep roots enabling it to cling to windy places. In Essex beekeepers have moved colonies to benefit from a late crop. Its pollen is golden yellow 38mu. Jeff Rounce says no pollen from it. Its dried flowers are used as indoor decorations. At Saltdean (Brighton) a large area disappeared (probably deliberately removed).

**sealed honey**

Honey that has been ripened and covered over with a thin, slightly porous capping of wax, pollen and fibres. Bees often show a reluctance to seal honeycombs and similarly to uncap them once sealed. The work of sealing and opening up is always done neatly. An exception is when robber bees take honey when ragged, torn cappings and debris that falls willy-nilly are tell-tale signs of what has been happening. When partially capped supers are removed it is sometimes wise to spin out the unsealed honey (which may ferment if kept) for immediate use, or to use it to make mead or vinegar.
See: travel stain, weeping and white cappings.

**sea level**

A.M.S.L. Average mean sea-level. Calculated as the average tide level Newlyn, Penzance, Cornwall. It is the level that the sea would assume if there were no waves, tides or swells. Heather is not thought to yield well at sea-level while other crops say honeydew in NZ only occur between sea-level and 850m. Apiaries at sea-level (or below) may be more prone to flooding than those on higher sloping ground. Heights are often expressed as above ground level as for example DCA'S.

**seam of bees**

A relatively dense packing of bees between the top bars of the frames. Each packed row of bees is known as a 'seam' and the position, compactness and number of seams tells us a great deal about the state of the colony, likelihood of the queen's position, pollinating strength etc. It is an indication of the possible size of the brood nest and the clean healthy appearance of the bees, warmth arising and the signs of pollen going in give us a good indication of the colony's strength and potential. For early pollination work the number of seams is paramount. Eight in America but in our spring four or five seams in a national's brood chamber in March is about right.

**sea pink *** *Armeria maritima***

Narrow leaves that grow in tufts and dense heads of pink or white flowers. It can tolerate infertile terrain and a high proportion of salt in the soil. Like sea-lavender the dried flowers can be used for flower arrangements.
See: thrift.

**searcher bees**

Bees are frequently seen to be searching as around empty hives, hollow trees or other possible sites, or where syrup or honey has previously been exposed. There is little doubt that they are able to convey the notion of what they have discovered to other bees who might well be persuaded to accompany them to any site where a reward is likely. This characteristic is also noticed amongst wasps and the understanding of the word social, as applied to insects, is apparent - their ability to act as a society bestows upon them the sense of a community, working together, as opposed to their having individual lives.

**sea rocket** *** ***Cakile maritima*** **Cruciferae**

A succulent herb whose bright white, pink or pale lilac flowers certainly supply pollen. Sometimes it forms a distinct line of vegetation along a beach about the 'drift' line (brightening up the otherwise often untidy area of flotsam and jetsam). It copes with being buried under sand and its seeds are spread by the tide. Because the plants conserve every drop of water it is unlikely that nectar would be yielded.

**seaside**

Synonymous with coast. As Britain has such a long coast line and towns by the seaside must include many beekeepers, it is safe to assume that many hives are kept within flying distance of the sea. In Malta bees forage from one island to another even though they are about two miles apart. However conditions at seaside places are fickle, once in a while giving bumper crops, usually mediocre and all too often poor. Where possible an apiary should be established as far as conveniently inland and sheltered from wind.
See: coast and shore.

**season (active – dormant)**

Although the year is naturally divided into four periods, spring, summer, autumn and winter, as far as bees are concerned they have two seasons; March to October when they are active and November to February when they are relatively dormant and existing on stores collected during the active season. This applies to Britain anyway but as a general rule the life-cycle of a honeybee colony fits into this framework.
See: months by name and seasonal hints.

**seasonal hints**

Unlike the farmer who may sow in spring and harvest in autumn beekeeping depends on the quality of its queens. Wintering with good ones is the basis of success the following season. After a successful winter then comes the spring build-up, supering, swarm control and harvesting. The season is dealt with by entries under the name of every month.
See: months, harvesting, spring management, supering, swarm control and wintering.

**seat**

An object for sitting on. When manipulating the combs of a brood chamber, normally the lower part of a hive, a considerable time may be spent with a bent back and that is not good or comfortable for the spine. A seat such as a deep hive roof can come to the rescue and it is often a good idea to use roofs that are suitable for sitting on at the side of the hive where combs can be held at eye-level yet the back held straight. A strong, deep roof has many advantages, it can hold items, offer ample feeding space, give extra winter insulation and even maybe stood upon.

**secateurs**

One-handed pruning shears with comfortable handles so that more strength can be applied to sever stout, woody material such as twigs and small branches. When it comes to taking a swarm, secateurs have the advantage over a saw in that an entire branch, swarm and all, can be parted from a bush or tree without jerking and dislodging the swarm as sawing would almost certainly do. Also twigs and leaves can be cleared from under a swarm allowing it to be jerked off a heftier branch and into a suitable receptacle.

**secluded**

Screened from view or hidden. A suitable place for an apiary as from the point of view of privacy and shelter it is better for the bees and the operator. Although people are coming round to the view that bees are useful creatures, it still makes some folk a little 'edgy' to see an active beehive anywhere near small children or animals.

**secondary invaders**

This term is used particularly when the establishment of a primary pathogen which causes disease or disorder is then accompanied by other bacteria that do not in themselves cause disease but are merely associated with it. EFB is a case in point though links between the success of certain pathogens with the presence of certain viruses are being discovered.
See: saprophyte.

**secondary wing – hind**

The smaller wing of the pair on a honeybee has hooks on its leading edge which lock into a shaped-rib in the rear centre of the forewing. It is not driven directly by thoracic movements but follows the forewing which takes the power stroke and then trails to reposition for the next. When not in use it is conveniently folded under the forewing upon the honeybee's dorsal surface (thorax and abdomen)
See: hindwing.

**second-hand equipment**

Equipment that has been owned previously or used by someone else. Beekeepers often give-up in a hurry and some extraordinarily good bargains can be obtained.
However certain risks are involved. The manufacturers will normally make-good any faulty item or deficiency but this is rarely possible in the case of used equipment which might be home-made, non-standard or even riddled with the spores of disease. It is wise to assume that any equipment bought needs sterilizing before bees are put near it and that hives of bees should be carefully checked for good health.

**secretary**

An officer responsible for recording the minutes of meetings and dealing with the general correspondence of an association. Together with the Chairman and Treasurer they form the nucleus of the Committee or officers of any formally united group of beekeepers or others.

**secretion**

The release of substances from a plant or body and the material, other than waste, yielded. A substance elaborated from the cells, especially from glandular cells. It implies work on the part of cells in passing the substance, also called a 'secretion' through the plasma-membrane. We talk of the function of wax-secretion by bees of certain ages. Compare with diffusion.
See: excretion, osmosis.

**secretion of nectar**

Plants vary enormously in their ability to secrete nectar. Some may yield so small a amount that bees may have to lap it up with their labella or cut through the corolla to reach it. Others may produce nectar that drips to the ground. The humidity, light intensity and all these things, can affect nectar secretion.
See: extra-floral, nectar secretion, yield, nectary, quantities.

**section** ***

1. A small frame fitted with a starter or a sheet of wax foundation and intended to be filled with honeycomb. One of the most widely used was the 4¼" square, basswood section that holds around one pound of honey. Designs and materials have come into play of various shapes and containing various amounts of honeycomb. J.S. Harbison originated a 'section honey-box' in California in 1857 and this had large 5 x 4 sections nailed at the corners. He was called the Father of Californian bee-culture. In 1876 A.J. Cook (USA) came up with the basswood section and in 1879 James Forncrock made the grooved, one-piece section.
2. A thin slice, as cut on a microtome, for detailed study under the microscope.

See: Cobana, packing sections and cross-section.

**section crate**

A wooden box with slats arranged so as to support rows of three BS or four US 4¼" square sections, giving bees access to the bee-ways. With a similar periphery to the hive, it takes the place of a shallow-frame super. The sections are held closely together within the crate together with a 'follower' board and spring blocks.

**sectioning (microscopy)**

In order to examine the cellular structure of sections of organic tissue, very fine slices (cut directly or by use of a microtome) are selected and prepared. To support tissue while being cut it can be embedded in paraffin wax or frozen. Wax is not miscible with alcohol so place the material in 50/50 cedar wood oil/alcohol and when it has sunk to the bottom transfer to pure cedar wood oil. 75, 50, 25% then pure oil. Wax can be removed using xylol. Follow by staining if required. Modern scanning techniques can do this now.
See: microtome.

**security**

Safety from interference or danger. The security of hives, because they are kept out in the open in all weathers, is important. There are so many considerations such as weather, access, animal interference, vandalism and so on that you must always be circumspect in setting up an apiary. It is possible to insure bees and hives and as values increase this becomes a growing need. Emlocks, house apiaries and other defences are well worth looking into.
See: comprehensive insurance, out-apiary and rustling

**sedge Carex spp. Cyperaceae**

Some varieties of this rush-like reed, which grows in wet places, have been used for making earlier types of beehives. E.g. *Cynosurus cristatus* and *Carex mariscus.* Unlike rush it has a solid, often three-sided stem.
See: coiled straw, clooming, skep and straw.

**sediment**

Material that settles naturally to the bottom of a liquid, i.e. matter that has not been taken into solution and is too heavy to be supported by the liquid unless agitated. It may be separated from the liquid by siphoning or filtration. When, in order to clarify a liquid or to extract more of the solid or semi-solid material that would normally become sediment, precipitation is encouraged by the use of further substances or by centrifuging.
See: clarification and finings.

**seed**

A structure relatively resistant to hostile conditions, usually small and rounded with a protective coat food reserve and the germ centre from which a new plant develops. The ovule after it has been fertilized and matured. A mature fruit which contains an embryo which potentially can germinate under the appropriate conditions. It results from the propagation of the ovule. The relatively small propagative part of a plant from which a new crop can be grown.
See: seed germination.

**seed germination**

Some seeds, especially those of many wild flowers go into a period of dormancy which has to be broken before they can germinate. Methods to cause this include: chilling – stratification at temperatures below freezing.

**Scarifying**

scratching or pricking of seeds. Exposure to light or air . Submerging in water (boiling water with some Australian flowers) or burning of material just over the surface. Pollen grains may also require a particular set of circumstances to facilitate pollen tube growth.
See: abrade.

**seeding honey**

The mixing into a cool, clear honey of a fine grained sample (ivy, OSR etc.) with a view to encouraging a fine grain throughout when the honey finally sets. The element is called the primer or starter. Such a method can be used when the honey otherwise forms a coarse or unsightly grain. Gently heating will clear an unsightly granulation and when cool about one twentieth part of smooth grained honey is mixed in. A temperature of around 54F should then be maintained until the honey sets.

**seed weevil Lariidae or Curealionidae**

Winter rape crops have increased weevil infestations according to the number of years rape has been grown on the farm (not so for spring rape) . One of the numerous beetles that harm crops and stored food.

**seeker (equipment)**

This is a needle or bodkin-like probe that is useful when dissecting and it becomes necessary to delve into tissue or objects or to hold with one hand while the forceps are brought up with the other hand.

**seep**

The slow passage of a liquid through anything porous, cracked or only partially retentive of liquid. Some liquids have great penetrative power and leaking feeders, honey tanks and containers can all present problems. Some wooden feeders show no signs of leaking until carefully checked. Then it may be discovered that a colony thought to be taking down several litres of syrup, is sharing it with the whole neighbourhood. Liquid beeswax will find its way out of any faulty container. Beeswax can be used to seal containers in certain conditions though it cannot withstand heat or vibration when cold. Water is sometimes allowed to seep out onto a surface from which water gatherers can collect it.

**segment**

One of the parts into which a thing naturally divides. Each of the sections of an arthropod's body. Hence segmented. We refer to the segments of a honeybee's leg coxa, trochanter down to the pretarsus. The scape is a segment of the antennae. See: segmentation.

**segmentation**

Primitive forms of both Annelida and Arthropoda display segmentation wherein a repetition of pattern produces a number of joined units. When these are of a similar nature this is metameric augmentation. The honeybee larva displays this feature with only the end, head and posterior segments -showing any marked differentiation. In the adult bee the segments of the head are fused together, those of the thorax joined together, while each abdominal segment is separated by softer inter-segmental membrane. The appendages of the adult consist of articulated segments.

**selected, tested queens**

Queens that have been used in producing colonies and their progeny tested for genetic traits. If these traits are found to be satisfactory they can then be sold as selected tested queens. Usually first year as queen mothers select tested. Queen mothers are used by queen rearers to produce 24 hr. old larvae for the production of queens for sale. See: queen classification.

**selection**

Exercising of choice. While natural selection is demonstrated by the ability to survive, the choice of the best characteristics to suit man's particular requirements calls for the careful choice of what he considers best and often the ruthless rejection of genetic sources considered less desirable. The importance of selection in the realm of breeding stock cannot be over-emphasized. See: selective breeding.

**selective breeding**

The breeding of plants or animals choosing desired characteristics. Because the natural mating of honeybees takes place somewhat randomly as regards the actual drones that copulate with a queen, selective breeding can only take place when some control can be exercised over the male bees. Isolating mating apiaries or flooding the appropriate areas with chosen drones are means utilized.

Finally the rejection or culling of variants enables selective breeding to result in the steady improvement (?)of stock.
See: cull, genetics and reproductive isolation.

**selfing**
Self fertilization – autogamy. The utilization, by a plant, of its own pollen for the purpose of fertilization in its own flowers.
See: self-pollination.

**self-heal *** *Prunella vulgaris* Labiateae**
A hardy perennial that will grow up to 30cm. Although its purple flowers yield nectar and pollen the long corolla often prevents the honeybee from reaching the goodies. Perhaps bees could be taught to deposit a little first and then, due to capillary action, remove the lot? The plant was at one time accredited with great properties for healing.

**selfless**
Without concern for oneself. Individual bees are quite selfless and give their all to their colony. A colony on the other hand is very possessive, may be ready to steal and will not readily part with anything except pollinating services which it gives efficiently if unwittingly.

**self-pollinating**
The successful transfer of pollen from an anther to the stigma of the same plant. It supposes that the plant in question is self-fertile. This does not mean that cross-pollination is impossible, in fact in the vast majority of cases better results are obtained when cross-pollination also occurs.

**self-spacing**
When applied to frames this means that they are designed so as to give the correct spacing from centre-line to centre-line when they are made to touch one another. In Britain we seem to have a 'hang-up' on this subject of frame spacing which does not seem to exist in the large honey producing outfits of the world. Different races of bees will, under natural conditions, space their combs slightly differently. With *A. mellifera* in the UK 1 " seems to be the accepted spacing though wider spacing may be used in the supers. The brood chamber spacing is claimed to reduce swarming and the number of drones produced.
See: frame spacers

**self-sterile**
A plant that is incapable of producing viable seed unless it is cross-pollinated. A flower whose own pollen is incapable of performing the function of fertilization even when the pollen reaches its otherwise receptive stigma. For example in apples - -- the Cox's Orange Pippin.

**selling**
To pass on to a purchaser for some consideration. It is not surprising that in most countries individual hobbyists can undercut market prices. This is one good reason why those who become serious beekeepers often buy honey to resell. Undercutting is of course harmful to the interests of those whose living may be largely dependent on producing and selling honey. Any adverse action where bee farmers make beekeeping more costly or precarious for hobbyists is a source of immediate concern.
Factors to be considered are:

1. producing an item for which there is a market (expensive advertising is for the big boys).
2. Keeping a high and reliable standard.
3. Setting a sensible price so that a continuous supply can be maintained.

See: labelling and insurance.

**SEM**
The scanning, electron microscope – capable of a resolution down to 7nm. Used for the study of pollen grains and various structures of the honeybee etc.
See: electron microscope.

**semen (sperm or seed)**

The sexually, reproductive, impregnating fluid as found in the testes for example. Products of the male reproductive organs or seminal fluid. In the case of the drone it includes enormous numbers of sperms, and a fluid that enables them to be motile. It has been proved possible to 'freeze' (and then store or transport) drone sperm.
See: sperm.

**seminal duct vas deferens anat.**

A coiled tube or duct leading from the testes to the seminal vesicle in the drone and whose function is to transport semen.

**seminal pump anat.**

A muscular pump adjacent to the spermatheca in the queen honeybee, who by mating, has incorporated in her body a supply of fertilizing fluid and can now control its release so that she can lay fertile eggs at will for several years.

**seminal vesicle anat.**

The vesicular seminalis an organ that stores the sperm in the mature male. The sperm vessel. In the queen the spermatheca also does this keeping it viable by glandular secretions.

**Seneca Marcus Annaeus 55BC – AD39**

He suggested that bees might browse on flowers and change the sweet nectar by some intermingling and property of their breath, into honey to be preserved and stored.

**sending a sample of adult bees**

The regular examination of samples, typical of the bees in a colony, should be carried out annually or whenever adult bee health is in any doubt. It is considered that a matchbox full (about 30+ workers) is sufficient. They are more easily 'taken' in a larger glass or plastic container and can be rendered immobile in a refrigerator and then transferred without food into a matchbox, labelled with colony number and the name and address of sender. Sometimes these can be examined locally otherwise send to Government Research. Checking for virus particles requires electron microscopy.
See: DEFRA.

**sending a sample of comb**

Where brood disease is suspected Fera should be approached but where it is necessary to send a piece of comb, neither honey nor dead bees are particularly useful. A comb preferably containing brood only, should be marked by embedding a spent matchstick in the comb pointing to a suspicious cell. This should be wrapped in grease-proof paper (not in an unventilated plastic bag) put into a cardboard box or other suitable postal container. Then with colony number and owner identification it should go to the appropriate authority. (Fera?) Take precautions in the apiary until 'all-clear' is given.

**sending a sample of suspected spray damage**

A photograph should always be taken at the moment damage is found so that labelled with time and place the sample would be corroborating evidence. A free service should be available but it is always important to supply as many facts as possible. How and what was sprayed. Your co-operation is nationally helpful.
See: spray poisoning.

**senility**

Weakness, wearing away or inability to function adequately on account of old age. The life led by many beekeepers helps to retard senility but senility of queen bees must be guarded against by regular, careful re-queening, though some queens might be allowed a prolonged spell of life in a smaller unit so that their genetic material may be used or investigated.

**senior course**

See: Senior Exam and BBKA courses

**senior exam**

Syllabuses for Senior exam can be obtained from the BBKA Exam Secretary. A candidate must have passed both parts of the Intermediate exam. (now modules) The ground is covered to a deeper extent and includes wider aspects such as historical knowledge. Part I of the exam is taken in the candidate's apiary where any aspect of bee work from clipping a queen to taking a sample for disease diagnosis may be called for. Also a thorough inspection of the apiary, honey extracting equipment and microscope work will be performed. Part II is a more advanced version of the Intermediate. Successful candidates may be referred to as Master Beekeepers.
See: lecturer, judge and NDB.

***Senotainia tricuspis* Meigen**

An endoparasite of honeybees. The female attacks bees as they leave the hive and deposits one or two tiny larvae on the intersegmental membranes between the head and the thorax. The larvae penetrate to the thoracic muscles where they develop into the second instar feeding on haemolymph. The bee dies and the larvae then feed on solid tissues before moulting to their third instar when it moves into the abdomen and devours its contents. It may enter another bee and grows to 8 or 9mm and goes into the ground to pupate, hatching after about 14 days or overwintering. No alarming depopulation of colonies has been recorded.
See: apimyiasis.

**sense of availability of food**

The bees' sense organs are all designed for the ready recognition of food. When one colony in an apiary is being fed during the active part of the day, when one colony from the heather is put amongst the others or when a colony has been thrown off guard by the use of some volatile material, the other colonies are quick to sense the possibility of easily gained food. Bees will always exploit the sources that are easiest and profitable.

**sense of colour**

The bees are responsive to various colours in a similar way to ourselves except that they have a shift on the spectrum which enables them to appreciate ultra violet but are blind to red. Black and white are clear to them helping them to spot a returning beekeeper and showing a marked dislike of anything jerky and black. See appreciation of colour, colour, light and u/v.

**sense of danger**

Bees, having so many enemies that would steal their stores or damage their colony if they were able, are forever on the alert guarding their homes against intruders. Sudden noises, vibrations or bursts of daylight within their homes will cause an outburst of 'buzzing'. If any of them exude alarm odour (a pheromone) the message is rapidly passed around the colony. They can become incensed and violently attack any moving creature which has been marked as a target by being stung. In attacking they are selfless, unhesitatingly giving up their lives in an attempt to save their colony, its food and their queen.

**sense of form**

Shape and movement are aspects of vision which in some ways they share with us. Though they are far more sensitive to movement and get an all-round view of their environment albeit they cannot focus in the way that we can. Knowledge of this enables us to render ourselves less conspicuous to them by moving slowly or keeping almost still, when a human might look like a tree or a post. Distinctive shapes at or near their nest can help bees to identify their hive thus reducing drifting. The manner of returning bees will alter when they return to find you standing by their hive – instead of diving straight in to the entrance they will sometimes hover around 'checking up' and may 'buzz' you.

**sense hair (anat.);**
A sensory hair such as those found on the antennae. As a hair stands proud of the surfaced from which it arises and its surface area is considerable, sense hairs are well-suited to the work they perform such as the gathering of information and its transmission to the brain. They are however somewhat vulnerable to damage.
See: sensory hair and sensilla.

**sense of hearing**
It has long been argued as to whether or not bees have any sense of hearing. Certainly by our standards the number of sounds they make are somewhat limited. Nevertheless a queen's piping has an instant and remarkable effect on a colony though this may well be due to vibration, as much as or more than. to the sound which a human ear detects. They appear to have no equivalent to our ears though many parts or the whole of their skeleton might act like a tympanic membrane. It could be said that bees react readily to vibration but do not show a high level of response to airborne sounds. Your breath will have far more effect on them than your voice. Their buzz may be merely a by-product of their thoracic wing vibration.
See: c.p.s. sounds and vibration.

**sense of location**
A bee, leaving the hive for the first time, turns round, hovers and faces it recording details in its memory bank which appear to be obliterated if at some time later it leaves with a swarm. Its entry to its colony remains visual even though the von Frisch navigation sense enables it to return from a foraging expedition. Thus once a bee has left a hive by any aperture – knot hole or gap between hive bodies, it will always return to that spot. Should the opening be no longer available it has to learn how to walk from there to any other opening into the hive that it is able to find. Unlike homing pigeons bees have to completely re-orientate when moved to a new location. This leads to drifting, when several hives are positioned on a new site. Should a flight from the new site bring a bee into an old area that it recognizes, the chances are that its memory pattern will attract it unerringly back to the spot where its original entrance was. Bees taken away from their base in a matchbox can only find their way back if they have a visual knowledge of the spot where they are released. They appear however to be able to identify and follow any other honeybees recent track through the air and can make sweeping searches.
See: eidetic.

**sense organ**
The larger sense organs such as the antennae, compound and simple eyes are found on the head and convey information to the bee's brain. Upon these organs and on other parts of the exoskeleton we find minute sense organs known as sensilla all of which reinforce the information sent to the brain. Advice concerning the environment, especially as regards food, other individuals and so on, are transmitted from the 'receptors' via the nerves. Upon reaching the brain the information is processed and acted upon.
See: receptor and sensillum.

**sense peg**
Similar to sense hair but having a short stout peg instead of a bristle. These are sensilla and many thousands are found on the bees' antennae

**sense plate**
Sense plates consist of hollows in the cuticle on the outer surface of the final eight segments of the antennae, appearing densely on the terminal segments. Each is capped by a thin sensitive plate level with the surrounding surface and connecting with the sensory cells. We know more about their form than their function though clearly they are able to read information which the forward position and flexibility of the antennae give them. They have been shown to respond to odour. Sense of loss. A honeybee colony that is being robbed is more concerned about the interference than

the loss. If the robbing is done stealthily enough the aggressor may get away without being attacked. The beekeeper in fact is just as likely to be stung when giving the bees food as when taking their honey. However for a day or two after any large amount of honey has been taken the bees can be very tetchy. They are far more aware of the loss of forage or a queen.

**sense of queenlessness**

The constant movement of bees within the hive, their food sharing and grooming, all help in the distribution of pheromones. Queen substance is one of these and while it is present the bees function as a normal working unit. Should a bee question the presence of a queen (only a fraction of their number can see or touch her in a day) it will 'sniff' by drawing air towards its antennae while making a small buzzing movement with its wings. Failure to achieve a satisfactory response will lead to many bees behaving thus and setting up what is called the queenless roar. Should a queen be removed temporarily, the bees race around looking for her even hurrying over the outside walls of the hive. Gently released at the entrance after ten minutes, a seemingly 'happy' purr develops as more and more bees fan the message to the whole colony. Should an accident occur that deprives them of their queen, they remain unsettled and inclined to sting, queen cells are started but conditions are not normal again until a new queen begins to lay when a great surge of effort to re-establish normality occurs.

**sense of the queen's presence**

Having observed the uproar that occurs when a colony is first made queenless, it is not surprising how readily bees perform their various functions when all is well. Yet there is a constant recognition of the single queen's presence even when the colony is enormous, and some bees go through their lives without meeting the queen,. This is brought about by the transfer of queen substance from bee to bee. It was found when in hiving a swarm their queen was temporarily caged, that on release she first ran away from it, then stopped, angrily it seemed, to demand food which was readily given her. The bee that fed her soon had a ring of bees begging for a 'taste' and then each of these bees developed circles of others happy to be given that all pervading pheromone - 'queen substance'.

**sense of sight**

Bees can determine the position of the sun by the polarized light pattern in the sky even when the sun is obscured by cloud. They can only navigate when the sun is above the horizon. Their eyes see ultra-violet light but are blind to red (a red torch can be used to handle them – or wasps- by night). Shapes are meaningful to bees though they lack our visual acuity being unable to 'focus' their eyes. Their 'flicker frequency' is much higher than ours and even the slightest movement is immediately detected by them. Their two compound eyes can, with the aid of head movement, cover most of the sky, with just a small 'blind' area rearwards. The three simple eyes give them sensitivity to light intensity which influences foraging. (the passage of a dark cloud over the sun can send them all scuttling back home). See: retina.

**sense of smell**

A bee's sense of smell is thought to equal or probably to exceed our own. As the senses of smell and taste are close or sometimes interwoven, it may not be possible to easily determine whether a bee is 'sniffing' or tasting. Little attention has been paid to the fanning posture of bees which enables them to draw aromas towards their antennae just as humans use their noses for sniffing. The act of scenting on a food source could easily be confused with sniffing if such behaviour is ever experienced. As bees are quick to differentiate syrup from water and to know when honey wet combs have been put on a hive for 'licking-up' and to spot hives that have just returned from the heather, there is every reason to suppose that 'smelling' is a normal activity that has a marked effect

on honeybee behaviour. The mouthparts and the antennae certainly play a major role with regard to their sense of smell.
See: olfactory.

**sense plate**
Sense plates consist of hollows in the cuticle on the outer surface of the final eight segments of the antennae, appearing densely on the terminal segments. Each is capped by a thin sensitive plate level with the surrounding surface and connecting with the sensory cells. We know more about their form than their function though clearly they are able to read information which the forward position and flexibility of the antennae give them. They have been shown to respond to odour

**sense pore**
These are amongst the smallest of the sensillae and are found in great numbers over the surface of the antenna. They expose the tips of sensory cells.

**senses of bees**
Alarm, association with other colony members, availability of food, colour, direction, form, hearing (auditory), location, sight, smell (olfactory), touch (tactile), and vibration. These are some of the bees' senses and further details are given under appropriate headings.
See: alarm odour and sense organs (sensilla).

**sense of loss**
A honeybee colony that is being robbed seems to be more concerned about the interference than the loss. If the robbing is done quietly enough the aggressor may get away without a single sting, (whether a wasp or a beekeeper!). One is in fact more likely to be stung when feeding the bees than when taking the honey. However for a day or two after any large amount of honey has been taken the bees can be tetchy. They are far more readily aware of the sudden loss of a queen or of forage.

**sense (taste)**
Bees are able to differentiate between different strengths and types of sugars and mixtures thereof. Density of solution also plays a part and their sense of smell would also doubtless come into play. Von Frisch tells us that bees can distinguish between: sweet, sour, salty and bitter.
See: sense of smell.

**sense (touch)**
There are many ways in which bees are able to demonstrate their sense of touch. The sensitive antenna are often called 'feelers' though the range of sensations that they can determine go beyond the sense of touch. We speak if their caressing a queen. When a bee stings it also 'feels' the surface (checks over) before the lancets are forced into the material to be stung. Bees often climb over one another when entering the hive and clearly recognize their relatives in that way.
See: tactile sense.

**sensillum pl. sensilla**
Small organs of sense. Scientists continue to discover more and more about the structure and function of the bees sensilla. All the various interpretations of a bee's surroundings are transmitted to its brain via these minute, interesting organs. A bee is endowed with senses of smell, taste, touch and sight, and via vibrations at least, hearing. The various groups of organs are labelled according to their position, shape or sense that they appreciate, e.g. tactile sensilla.
See: pegs, pit hairs, plates, placoidal, scolophore, sense hair and sense of loss.

**sensilum campaniform**
Bell-shaped sensilla.

**sensillum chordotal**
Rod or bristle-like receptors associated with sound or vibration.

**sensitivity to movement**
Quick or jerky movements are unwise when we are near the hives because bees

are much more sensitive to movement than we are. Continuous movements to us may be broken into components by the bee which is able to differentiate between plant and animal movements. They also allow quite accurately for the movement of the sun when navigating and for any tendency to drift from their chosen track. Animals that are habitual predators (e.g. wax moth) have learned the trick of defence involving remaining stationary when interrogated. See: flicker and fusion frequency.

**sensory hair**
Hair sensilla on the organ of Johnson have been reported as capable of perceiving vibration such as the body waggles communicating distance of a source of forage. They also take the form of innervated elastic bands stretched between the walls of the tibia and said to be sensitive to vibrations akin to the piping of a queen. See: sense hair.

**sensory receptors**
'Proprioceptors' register relations between one part of an insect's body and another. Exoreceptors may register changes outside the animal.

**sensory transduction**
The change of one form of energy, say light rays, into another. In the case of the bee this occurs in the retinula cells which are nerve cells that transmit information received as light into nervous energy that is interpreted by the optic lobe of the brain.

**sepal *****
A leaflike subdivision of the calyx, lying outside the petals of a flower. They form a protective casing which surrounds and shelters the flower bud.

**separator**
A thin panel or plate of plastic, wood or metal which separates rows of sections in a section crate. It allows passage-way for bees but prevents them from joining one comb face to another and helps to produce a clean, flat surface to the honeycomb. Where these are not used, comb faces can be spoiled by brace comb or by comb faces bulging to up to three inches thick while the adjacent comb is diminished correspondingly. It is also called a 'fence' and is sometimes confused with a honey 'extractor'. These are listed in manufacturers' catalogues as 'dividers' when associated with sections. See: divider/extractor.

**September**
Whereas honey removal may be delayed until this month, winter feeding can no longer delayed and thick warm syrup should be applied so that all colonies are up to 20kg minimum well before the end of the month (and before the –ivy' flow).When bringing colonies back from the heather, their honey will be all too attractive to the other colonies (so restrict entrances). Uniting, final hive configuration and siting ready for winter should only leave mouse guards and final checks remaining to be done. Do not give wet supers to weak colonies and arrange feeding or clearing of wet supers until nightfall.

**septicaemia *Bacillus apisepticus***
A disease of the bee's blood that can cause death. It has been associated with 'acarine' but there is no certainty yet as to the cause or the cure of this minor complaint. Pathogens include *B. apisepticus* and *Pseudomonas apisepticus*. Later work suggested that there are several different causes of the complaint in which the blood changes to a chalky white. It causes muscle degeneration and can kill a bee in 20 to 36 hours after inoculation. Bees that die often have a putrid odour, muscles decay and body segments fall apart.
See: spiroplasmosis.

**septum**
A partition or wall separating two masses. A dividing wall or membrane. The mid-rib or foundation or the continuous plane of the cell-bases.

**sepulchre**

Tomb or grave. Used by some writers to describe the enclosure of propolis sometimes found surrounding a once living creature that has found its way into a beehive. Having been stung or suffocated to death by the bees it then, being too large to pull out of the hive, is covered over with hygienic propolis to prevent germs or unpleasant smells from polluting the hive atmosphere.

**serosa**

A serous (watery) membrane such as that which surrounds the young embryo within the egg. The outer larval membrane of insects.

**service tree** ***
***Sorbus torminalis/domestica*** **Rosaceae**

A perennial - producing clusters of white flowers in May/June and attracting bees to both nectar and pollen.

**servicing**

The action of a bee in transferring the appropriate ripe pollen grains to the receptive stigma in such a way as to lead to the subsequent fertilization of the ovule. The term is more commonly used when a male animal mates with a female.
See: pollinate.

**seta pl. setae L. a bristle**

A stiff hair or bristle. There are many of these on the honeybee. The anchoring hair in the bowl of the corbicula for example.; this is curved and points forwards thus helping to direct the pollen pellet as it is dislodged from the basket into the cell. The handling of pollen makes great use of the various setae.

**setbacks**

A check to progress. Bees are able to catch up again and offset most setbacks though the total reward gained by the beekeeper may well be diminished. When most of the flying bees have been 'knocked-out' by spraying, several days later the colony may appear quite normal again; naturally honey will have been lost though. Chilled brood may also be offset by heavier subsequent laying and a failing queen gives way to a supersedure queen whose laying soon brings the colony to peak strength again.

**set honey**

Naturally granulated honey as opposed to clear or run honey. A term used in the trade rather than by the beekeeper.

**settling**

1. Bees are spoken of as settling on a flower or hive; perhaps alighting would be a more appropriate word. To settle implies resting. Bees do this of course but on the whole have a remarkable capacity for work.
2. We leave murky liquids to settle, i.e. to clarify, leaving the sediment at the bottom.
3. A hive may lean over as it becomes heavier due to its settling into soft ground unless it has been firmly supported. 4. We speak of settling with regard to paying our beekeeping subscriptions and also of arguments and disagreements as being settled.
See: alight.

**settling tank**

It is a convenience to have a fairly large tank into which filtered honey may be run in order to settle. Not that any foreign body should be there to descend but so that air bubbles may rise to form froth and thus leave clear honey that can be run off via a suitable honey valve at the bottom of the tank. The one hundred weight size was popular and it incorporated a filter at the top and a honey gate (valve) at the bottom. As well as considerations of durability, cheapness and capacity, we have to consider how stable they are and how easy to clean thoroughly.

**sex**

The determination of sex in honeybees is

unusual in that parthenogenesis accounts for the production of the males (drones) while normally fertilization is responsible for females whether workers or queens. In most creatures there is a sex chromosome but with honeybees the sex is determined by 'ploidy' a single (haploid) set of chromosomes in the males and a double set (diploid) in females giving them 32 chromosomes.

**sex characteristics**

Special characteristics that are displayed by one sex but not the other in the same species, or characteristics which collectively determine whether a creature is classed as male or female.
See: sex differences/linkage

**sex differences**

Some of these include the following: male honeybees are haploid, have an additional segment on their larger antennae and specialized genitalia. They lack long tongues and their legs are not adapted for pollen collection nor do they have stings. Females form two castes, the queens and the workers are both diploid and have stings. Queens have fully developed genitalia while worker ovaries are vestigial. Worker legs and mouthparts are highly specialized for social life with their colony.

**sexine**

Part of the pollen coating.
See: exine.

**sex linkage**

A link between a given characteristic on one or other sex. Sex determination in honeybees is by chromosome number the males haploid and females diploid.

**sex of eggs**

A honeybee's sex is determined at the mature egg stage. The sex of the egg depends on whether or not it is fertilized. If a drone's sperm has entered the micropyle and fertilized the egg, it becomes female and is capable of being reared into a worker or queen according to the size of cell, type of food etc. If the egg is not fertilized it develops parthenogenetically into a normal drone if fed and reared in a drone cell. Such eggs can become dwarf drones if reared in worker cells.
See: drone egg, laying worker and female egg.

**sex organs**

The genitalia or organs of reproduction. These are only fully developed in the queen and drone and although some workers may lay eggs they are not capable of mating.
See: reproductive organs, drone and queen.

**sex pheromones**

The sex attractant that becomes effective when the nubile queen reaches a height of at least 12m, though mating situations would be more likely between 30 and 60m. The 9-0-2 component of the queen's mandibular gland pheromone is the major sex of A.mellifera queens which usually mate in drone assembly areas. It also appears to be the attractant in the case of A. florea and scutellata. Dr. Free suggested that confusion between the three tropical species does not occur because their drones tend to fly at different times of the day.
*A.Florea* 12 – 1430, *cerana* and *scutellata* at 1615 – 1715 and 1800 -1845 respectively.
See: mating and scent trail

**sexual attraction**

Queens may well be attracted to areas in the sky where drones congregate and conversely drones can detect, are attracted by and will follow a nubile queen's scent trail.
See: mating.

**sexual cell**

Cells do not in themselves determine the sex of the inmate though each caste is reared in a cell appropriate to its particular needs and in a normal well-regulated colony the appropriate cells are used for male or female eggs according to their size and in accordance with the colony's requirements.

**sexual maturity**

Drones reach maturity thirteen days after emergence (38th day from egg) while the queen matures on the twenty first day from the egg. When raising queens one should always be well ahead with the drones first. See: genitalia and reproductive system.
sexual reproduction
Reproduction involving the fusion of haploid gametes into the diploid zygote. This occurs in the reproduction of honeybee females (workers and queens) but the drones are produced from infertile eggs.

**s.g. Specific gravity.**

Density or weight per unit volume.
See: specific gravity.

**shade**

Generally speaking it is rarely necessary to provide shade for hives in the UK as the sun never gets high enough in the heavens to produce the heat of sub-tropical or tropical countries. In fact it is often preferable to keep hives in the sun,'. However swarming with some strains has been tentatively linked to lack of shade so some beekeepers arrange to ward off the sun between 1 and 3 p.m. by strategic positioning under shrubs or such cover.
See: ramada and snow.

**shake**

When it is required to get bees off a comb to put them into another hive or to see better whether there are eggs or for any other reason, they must be 'jerked' off. This assumes that they have had no prior warning – a half-hearted shake will leave plenty clinging on and require a much more positive jerk than would have been necessary if the first move had been very decided and without warning. Needless to say a firm grip has to be taken on the frame. When removing a swarm from a branch the importance of giving them a decided 'jerk' without any warning is paramount and when finally hiving them, because they will have clustered together with claws locked into partners, it will be necessary to treat the mass more like glue and give them an especially determined shake, perhaps followed by another or a beating of the container. In all such procedures the presence of a queen or especially of queen cells, calls for a more circumspect approach.
See: shaking bees.

**shaking bees**

When bees are shaken from the combs or hive it is highly likely that some will take to the air, they are no more likely to sting on this account and most will fly back to their entrance and begin 'fanning'. A 'shook' swarm implies separating a large number of bees by shaking. Removing honey by shaking bees from the combs is usually helped if a soft brush is at hand to clear those that are left (especially along the bottom bars of the frame. When transporting bees, the shaking and vibration causes their temperature to rise and combs can even collapse if steps have not been taken to provide ample ventilation via a top screen. Brood, too, can be damaged by over-heating and moving bees very early in the day and perhaps spraying the top screen with clean water, are means of reducing the effect caused by the excitement of being moved).

**shallow**

Since the advent of box hives with movable frames it has been customary to have two sizes - deep and shallow. Ostensibly so that the queen has use of the deep brood combs while the shallow combs are used for the heavy honey. In fact either can be used for both, the queen, (bees themselves) might seem to prefer very deep combs (and no queen excluder) but in the interest of avoiding back-breaking work, bees can be persuaded to do a very good job on shallow combs. A brood chamber consisting of two or three shallow boxes offers versatility and interchangeable, one-sized frames.
See: sizes.

**shampoo**

Using honey (well diluted) and a free-lathering soap is a good substitute for shaving cream or shampoo. Work into a lather and rinse well and it will leave the hair soft and glossy. Or use expensive shampoo first and then the inexpensive spoonful of diluted honey, well rubbed into the scalp and left for a minute before using the shower and putting your tongue out to tell you when the honey has completely done its job. Lovely hair after such treatment.
See: shaving cream.

**shaving cream**

Select a free-lathering soap and cut into shavings. Dry thoroughly in an open oven. Run through a fine sieve or grind in a hand-flour mill. Mix 2 parts of powdered soap and 4 parts of honey with one part of soft water. Mix thoroughly and allow to thicken to the right consistency. Work up to a lather on the face (or head if used as shampoo) and rinse adequately. Helps to give the hair 'body'.

**shed**

A small outhouse for shelter or storage. To make the best use of a shed is to have an orderly arrangement that allows for the best utilization of space and the rapid retrieval of items needed. A shed used in connection with beekeeping allows for boxes of combs to be stacked with ventilation screens at the top and bottom, raised sufficiently to keep the stack aired. A window and work bench offers opportunity for winter jobs, cleaning etc. and for the making up of frames in the early spring. Boxes sealed with a pad moistened with acetic acid help eliminate wax moth. Remember that even when no traces are visible the dozens of minute eggs will hatch out if conditions allow. Tools, bee-gear, smoker, fuel etc. can be usefully housed. Fire is an ever present danger – take no chances in this respect.
See: bee house, hut and house-apiary

**sheep**

These animals get along with bees except that they are inclined to displace hives by rubbing themselves against such objects. Although their wool protects them they can be more vulnerable after shearing and in any case it is wise to put adequate fencing around the hives to that they cannot come into contact. Old time beekeepers used to say 'follow the sheep' this was associated with the benefits that accrue from their droppings on calcareous soils where clover's nectar secretion was augmented.

**shelter**

Exposed sites, say a high plateau or hillside where a valley funnels the prevailing wind, are to be avoided and sites with natural shelter should be chosen whenever possible. The micro-climate in a sheltered, south-facing valley may be several degrees warmer and allow more flying hours and produce better crops. Shelter in the form of trees or hedges, can break the wind force and artificial barriers too can be helpful.
See: hut, shed and wind-breaks.

**Shepherd tube**

A device used to 'pipe' bees from an upper brood chamber that has been raised over the lower one as in some systems of swarm control. The queen and brood are put below but all brood with queencells are put on top. A ½" tube connects the upper box to the lower entrance. A new queen raised upstairs, mates and moves in below. Induced supersedure one might say.
See: Flodon board.

**shield entrance from**

(a) bright light when it is snowy.
(b) robber bees when circumstances call for this.
(c) From cold or fierce winds.

Snow glare can attract bees out and away from their winter cluster to almost certain doom. A slate or board place so as the shield the entrance will cut down the reflection from snow.
The strength of a colony and the time of year should dictate when a smaller entrance should be used. In a determined wasp attack

it may be necessary to close down to a single bee-way but when bees are robbing within an apiary a pane of glass can help detract them or if necessary a wet sack or wet grass will make access difficult for robbers. A rare but wholesale bout of robbing may call for all-round reduction of entrance sizes and where necessary the use of a hose to have the effect of rain. Shelter from winds calls for wise siting and the use of natural barriers or hedges, shrubs etc. An open type of fence allowing limited penetration is preferable to one that can cause turbulence even leading to snow drifts.
See: snow glare.

**shim**

A thin slip or wedge such as used in machinery to make parts fit or the name given to slates, tiles or other pieces of material to level hives. Well-worn extractors can possibly benefit from the insertion of a shim of suitable thickness though in this time of durable hard plastics this is less likely but gears sometimes need re-aligning.

**shipping cage** ***

Package – a framed box of bee-proof mesh to carry a given quantity of shaken bees. The queen is included in a small separate cage and a temporary provision of food is made for her and the bees. The design of such cages gives them strength enough to be stacked. They are vulnerable to temperature extremes and insecticides and of course they need prompt and careful treatment by the recipient.
See: package bees.

**shirty**

A colloquial term used to describe the attitude of a colony which is not very tractable.

**shock**

A sudden disturbance or commotion. Perhaps we are too concerned with the outward demonstration of resistance by the bees to our interference rather than the shock that is given to the colony and of its repercussions for them. Stress has been referred to as a cause of disease because it lowers the bees' resistance and healthy resilience. A beekeeper might walk away from a re-assembled hive feeling that he has done a good job. If smoke alone sets thousands fanning and has its harmful effects on their stores, brood and comb (even their chitin deteriorates when heavily smoked) he may be in error in assuming that all will return to normal in a short time.
See: setbacks.

**shoe polish**

Beeswax is a unique and unbelievably useful substance which these wonderful little pollinators produce. Amongst so many other uses shoe polish is one of its many applications:
Take 1oz shredded wax, ½oz shredded soap, 4 tablespoons of 'real' turps. Pour the turps over soap/wax mixture and leave for 24 hours. Then add a little boiling water to make creamy and lamp 'black' if required. Floor polish, car polish and other water-proofing and cleaning agents can also be made.

**shook swarm**

A mass of bees with a queen, produced by shaking them from their combs. As the age range of a mixture of bees is rather random it does not have quite the same characteristics as a natural swarm. Bees sold by the kilo (about 11000 bees) are now called package bees. Consider this weight factor when you collect a swarm in a cardboard box.
See: package bees.

**shore**

The flattish strip of land adjacent to the beach which is washed by tides. The fauna and flora will have become established to the rather special conditions. In the main bumble bees probably do better there than honeybees but there are many seaside beekeepers and their bees may well work the flowers in season or even collect water and minerals that present themselves. Crops of honey vary with the season – amount of rainfall, absence of over-supply of rain. Wind and the co-incidence of gull's flight with DCA's can suggest better sites even if only slightly inland. See: biome, coast, river bank and seaside.

**shortage**

A deficiency – a well-found colony will respond to natural conditions whereas a colony with a shortage of food, bees of the right age, a good queen or some other important necessity, will be unable to take advantage of its environment and may even succumb to it. A shortage of food is sometimes indicated by white grubs, that have been sucked dry, being thrown out of a hive. A shortage of nectar will, in fair weather, result in large numbers of water gatherers at ponds, water feeders and other sources as during the breeding season, much water is required. A shortage of nectar bearing forage may give rise to the bees working unprofitable flowers and often flying from one variety to another. See: fidelity and reserve.

**shot brood**

Sealed brood with over-many empty cells and thus conspicuous 'holes' in the brood pattern, indicate the failure of the egg or its subsequent larva, as it may fairly be assumed that the queen originally laid in every cell. A holey brood pattern may be called 'pepper-pot' or patchy and suggests the need for a queen replacement.
See: drone layer and spotty brood

**show**

A display, exhibition or competition which calls for an organized, educational array – so useful in the realm of beekeeping where common ideas, advancement and improvement are so beneficial. These occur in beekeeping circles in many parts of the world.

**showcase**

A transparent case used to display its contents – possibly to show off a carefully prepared piece of wax to advantage. In the realm of shows the showcase is more likely to be used to hold a comb of sealed honey considered suitable for extraction. It should be amply wide to take the frame while allowing some space between the comb face and the glass.

**showing**

The following items have been included in this work: candles, comb honey, honey, mead, vinegar and wax.

**showing candles**

Three similar candles are to be out up for display and the judge will light one, extinguish it, and then relight to check ease of relighting and allow for two hours burning. The proper wick should be used, tips well-formed, wax clean, and sides smooth, parallel and symmetrical. Tall, thick bleached candles are preferred. The flame should be clear and bright and of a size that matches the candle and it should not gutter or smoke. The melted wax in the well of the candle should not be excessive or allowed to run down and the smouldering time when puffed out should be 10 – 15 seconds.
See: judging candles.

**showing comb honey**

(1) A section should be evenly capped over on both sides up to the clean wooden (or plastic) surround and displayed in a suitable box or case that show the comb.
(2) Cut-comb must show no sign of granulation or loose honey, the edges should be neatly cut to suit the container which in itself should be appropriate to the size specified.
(3) A comb for extraction should be well filled and sealed and easy to uncap without touching the frame. In all but the latter case, traces of Braula or travel staining would default.

**showing mead**

When mead is entered for a show it is normally required to be displayed in 20oz, punted. clear-white, glass bottles with a flanged cork and the exhibitor's number 2cm from the bottom. It must show no signs of working and be free of sediment. There will be classes for dry, medium and sweet meads though sparkling meads and melomels may be permitted in some classes. Appearance, bouquet, flavour, body, alcoholic content, brightness, clarity and cleanliness all contribute to enhancing the exhibitor's chances of winning. A good mead maybe half empty when the judge has finally decided.

**showing vinegar**

Honey vinegar is not often given a place in shows but when there is a class for this wholesome liquid it is displayed in a similar way to mead – in a clear glass 20oz bottle. It should be bright, clean, free of sediment and have the typical strong taste of honey vinegar. Working should be quite complete and no hint of wineyness should exist.
See: honey vinegar

**showing wax**

Properties looked for include: colour, appearance, aroma, texture, transparency and tenacity. Schedules normally specify weight and thickness of sample. Commercial wax may be broken to display smoothness of texture and candles may be lit.

**shrew *** Sorex Insectivora Soricidae**

This ferocious little glutton is mouse-like and short-lived. The pigmy shrew *S. minutus*, seems able to get through some types of mouse guard. In addition to being able to kill and eat torpid bees, it has 'scent' glands and can produce an odour which is so repellent that it can drive a weak lot of bees out to their deaths. It is nocturnal and has a sharp snout. The disturbance they cause can lead to dysentery and nosema.

**shrike *** Lanius spp. (Oscine)** ***Laniidae collurio, elegans, excubitor, minor***

This bird is a great insect-eater and records of its predaceous nature date back to Virgil. It has a strong hooked and toothed bill and fixes its victims on prickles around the nest in the form of a 'larder'.

**shrub**

A perennial plant that has no trunk but branches freely from ground level. There are however some shrubs that will in certain conditions produce a trunk and grow into a small tree. A branched, woody plant that does not develop to the size of a tree – below 5m.:
Willows, hazels, cotoneasters, pyracantha, buddleias, snowberry, blackthorn, laurestinus, berberis, tamarisk, sumacs.

**shrubby honeysuckle Lonicea spp.** ***standishii*** **Caprifoliaceae**

Wild, climbing honeysuckles need long-tongued moths to exploit their corolla tubes but this garden variety suits honeybees which sup nectar and collect pollen that is said to vary from greenish-yellow to light brown. *L. fragrantissimo* blooms January to February with creamy, white flowers before its leaves. A bushy relative to the well-known climbers.

**shut**

Closed or obstructed. The opening and closing of gates leading to apiaries may add to security but can hinder a vehicle's approach to the hives. The 'country code' must always be followed in such matters. Hive closure includes slides that allow bees into extracted supers to clean up and may then be shut so that the bees can only leave via the clearer(s) are useful when at the end of summer, robbing is to be vigilantly guarded against. A door or window that is not shut might encourage bees or wasps to enter where honey-wet are exposed. The shutting in of bees to avoid spray damage needs skilful judgement and handling and is not as satisfactory as moving them to as new site.

**shutting bees down for winter**

Successful wintering depends on having a strong healthy colony with a good queen not past her third season. Also adequate stores (20kg) up to half of which may be fed as sugar syrup ignoring weight of water), and the feeding completed in good time by the end of September depending on season and the region. A weather-proof hive on a site sheltered from winds, flood, cold air pools and interference, and a carefully fitted mouse guard over the open entrance will enable the beekeeper to say that her bees are shut down for winter. Needless to say 'shut' does not refer to the entrance which may be reduced and should be secured by a mouse guard.

**S.I.**

Systeme International d'Unites. Of the wide range of measurements to the power of ten one is picked out as an example – k – kilo 103. Gram, litre, metre etc.
See: Systeme.

**siblings**

Sibs. Offspring of the same male and female parent. A sibling species or pair will have similar or even identical appearances yet can have distinct genetic make-ups and may never interbreed, This can apply to drones or virgins but it should be remembered that because of parthenogenesis a drone is only a half-brother to virgins reared in the same colony.

**side bar**

The end bar of a frame. They are usually dove-tailed (mortised) into the top bar so that the ends of this form the lugs. The side bars govern the depth of the frame and they can control the distance from comb centre to centre if the width of the side bar is such that it touches either the wall or the adjacent side bar, as is the case with Hoffman type frames. The lower end of the side bar is cut away so as to take the bottom bar which may be a single strip or two separate pieces so that the foundation can be passed between them.
See: frame.

**side effect**

1. Any effect other than the one desired, that is caused by the use of a drug. When a drug performs precisely according to our wishes we say it is 'specific'. Unfortunately bees suffer from drugs and chemicals which though applied with the best of intentions often do harm.
2. Wind can have a side effect on the flight of bees. Although they fly 'zigzag' they are still purposeful in their direction and of course this must allow for wind effect.

In a Drone Congregation Area all bees are carried along in the same current of air. However the queen's ability to return to her colony has to take account of the displacement caused by her mating activities.

**sideliner**

Sideline is an additional line of business or goods. A sideline beekeeper is in a class between the hobbyist and the full-time bee farmer. The Bee Farmers Association's rules of membership require a certain number of hives to be managed. Many beekeepers work along commercial lines, i.e. aiming to make a profit, although they cannot afford the time or perhaps the investment to operate on a full-time commercial basis. These sideliners would usually have other full-time occupations.

**side rail**

An American term for the floor-board 'runner'

**sieve**

A utensil with a mesh or perforated screen which enables certain objects, material or liquids to pass through while restraining others. Funnels for sliding bees into packages or mini-nucs may be fitted with a drone excluding screen (sieve). A pollen trap may be fitted with a tray to receive pollen pellets via a mesh that keeps bees out but allows pellets through. Wax cappings can be separated from their washings by means of a sieve which can take the form if a nylon cloth, stocking or something similar. Stockings filled with cappings or comb and weighted with stones in a boiler can act as a sieve or filter to allow the clean beeswax to rise to the surface (use acidulated water). The stocking idea is also useful in solar wax extractors.

**sight**

The faculty of vision. Bees are clearly able to see and to transmit information about what they see to the brain. However there are some very distinct differences compared with ourselves. Instead of having a retina their nearest equivalent is the retinula of the compound eye. Their eyes detect colour, shape, form, movement and light polarization but they cannot form sharp images by accommodation as in our eyes. They also have three simple eyes which are light sensitive.

**sign**

A hint or indication, a suggestion as to an existing situation or one that is to follow. The state of the flora, likely weather, beginning of a nectar flow or dearth, the depredations of enemies, these and hosts of other things are usually evident to those with experience and understanding of enough to ''read the signs'.

**signalling behaviour**

Bees have numerous ways of signalling to other bees, primarily of course to members of their own colony. Mainly the use of pheromones as in scenting, fanning and marking. A queen may 'pipe' and bees may solicit food from one another or encourage another bee to groom them. The queen lays a scent trail while foragers seem able to drive away competitors.

**signs inside and outside the hive**

The whole gamut of behavioural patterns and the various conditions of a colony are displayed to the diligent observer. Interpretation of sounds, smells, weight of hive, activity in front of the entrance, debris, floor board scrapings or anything passing through a bottom screen, the state of the weather and flora, the number of seams of bees, the appearance of the winter cluster (through a transparent crown board), the feeling of warmth over the brood nest, the reaction to exposed sweet material or attention to a water fountain, these are signs that can be read from outside the hive. Healthy brood patterns, queen cups, queen cells, ragged cappings, Braula, wax moth, need for more space etc. are looked for inside the hive.

**silent robbing**

Robbing that carries on assiduously without being obvious. That is to say the more noticeable aspects such as fighting, ragged combs, debris at the entrance are lacking. It is robbing that takes place with few visual signs rather than with less noise as the word 'silent' might imply. See: intruder, robbing, beggar bees.

**slicone**

Compounds made by substituting silicone for carbon in an organic substance. Aerosols of silicone are useful as for releasing beeswax from moulds. Silicones can also be added to wax polishes. Oils, greases, resins and synthetic rubbers come from silicones. The compounds are characterized by thermal stability, water repellence and physiochemical inertness.

**silk glands**

Present in the larvae and later becoming salivary glands. They are in the form of a pair of long, kinked glands extending through most of the larva's body and uniting to form a common duct discharging into the spinneret of the labium.

**Silurian rocks**

Trees grow at a high altitude on these soils but flower later so most beekeepers work for spring honey. The honey produced wins many first prizes at shows apart from heather honey as this becomes mixed with blackberry and other late nectar producing plants that grow at these higher levels.

**Silver lime *** *Tilia petiolaris***
Also called the pendant, white lime, it flowers later than vulgaris (europea), mongolica and some others and is renowned for its toxic nectar that can kill wild bees and result in large numbers of drunk-looking bumbles crawling in the grass beneath such trees. Honeybees can be affected enough for scouts to 'recruit' foragers to such a source.

**silver swarm**
An old name for a swarm that emerges after August. Squib was also used.

***Silybum marianum* *****
A hardy annual giving pollen and nectar and growing from 2 to 5 ft. Large, rose-red flower heads in July – Sept, growing from large rosettes of spiky mottled or white, veined leaves. Well worked by bees.

**Simbles**
Hesiod, an early Greek writer referred to drones as living in 'simbles' – these would have been wicker hives, later to become the clay simbles as used in Attica.

**Simmins hive**
The fore runner of the Modified Commercial hive. Simmins Samuel (GB) wrote 'Modern bee farm in 1887 and produced the Conqueror hive.

**simple eye *****
Ocellus – these are three simple eyes (ocelli) situated on the vertex and lower on the drone as his large compound eyes meet at the top of the head. They are used to compare the light received by the compound eyes and this helps to stabilize the bee in flight. Each ocellus consists of a lens over a layer of simple, elongated retinal cells which are connected to nerve fibres. They are thought to be of use to measure light intensity to warn them of thundery rain in flight and enable them to decide whether to set-off from the entrance and help them to find their way out of the hive. Recent researches show that they play a part in navigation.

**simple microscope**
A magnifying glass or hand lens as compared with a compound microscope which uses a system of lenses.

**Simplicity hive**
The original name for the national hive.

**Simpson J. M. A. (Cantab) Ph, D (Lond.)**
Retired from Scientific Staff at Rothamstead Experimental Station in 1978 after 30 years of loyal and valuable service.

**simulate**
To create the impression that a certain state of affairs exists despite the fact that in reality it may not be so. Bees under the influence of a honey flow will rear brood, draw foundation into comb, accept a new queen or unite more easily to another colony. It is helpful therefore if a flow can be simulated by the feeding of sugar syrup so that these activities are not at the whim of the weather. We may simulate queenlessness by caging or temporary removal of a queen to get queen cells started.

**Sinapis arvensis (*Brassica arvensis*) *** Cruciferae**
Charlock or wild mustard once described as a troublesome weed in cornfields. Now sprayed into insignificance though seeds under the surface remain viable for years. A useful bee plant, though like other members of the Brassica family, its honey granulates very rapidly.

**sincerely**
This word is believed to have originated from 'genuine'. Roman marble statues were genuine if they had not been patched with

wax. They were therefore 'without wax' that is 'sine/cere'

**single species honey**
This term is reserved by Louveaux and others to mean: honey produced by bees allowed to forage on one flowering plant only, under experimental conditions that prevent access to any other source of food.

**single-walled**
As opposed the double-walled hives like the WBC, single-walled hives are composed of boxes, open at top and bottom, standing on a floor board and surmounted by a crown board and roof. The four walls of the national hive for example are of " western red cedar protecting the bees while the outer surfaces are exposed to the weather. See: double-wall and walls.

**sinus**
A space, cavity or dilation within an organism. The dorsal and ventral sinuses of a bee are those spaces formed between the respective diaphragms and the outer skeleton of the abdomen. These sinuses play their part in the open-flow movement of the blood. The dorsal sinus houses the heart while twin inner cords run through the ventral sinus. See: dorsal/diaphragms.

**siphon** ***
A tube taking the form of an inverted U which allows liquid to be drawn upwards out of a container and over into a receptacle at a lower level. It stabilizes when the levels are equal. As mead can be drawn off without disturbing the sediment (lees) this is a useful way of separating the sediment from the clear liquid when the fermentation is complete. Decanting, as an alternative, may well lead to a disturbance of any sediment.

**siphuncle**
A tube-like structure in aphids – a dorsal tube formerly called a nectary. Aphids secrete honey-dew from these – they are a pair of tubes situated near the tip of the abdomen.

**Sirius** *
The dog star in Canis Major and the brightest star in the heavens. In Britain it is visible well above the horizon during summer and especially around the Mediterranean. Orion's belt (three evenly spaced stars) inclining from top right to bottom left, points downwards towards Sirius. In the antipodes the same three stars point upwards to Sirius. As a heavenly body, fixed in relation to the other stars, it has figured as a significant item in man's natural calendar. Used for a calendar indicating to the ancients when supers should go on. See: stars.

**sisters**
Female offspring with the same parents. The queen may share a colony with her worker sisters until she outlives them and is then surrounded by her own progeny. It is assumed that all the workers of a colony are identical sisters though 'drifting' or interference by the beekeeper (plumping, diverting flying bees or requeening) might vary this rule. As a queen mates with more than one drone, workers in a colony may be half-sisters and of course drones, having no father, are only half-brothers to workers of a similar age. See: patrilineage and sibling.

**sites**
The location or place where anything is, For example apiary sites. To site is to place a hive or apiary in a certain location. Sites should be carefully chosen; they may be temporary or permanent. Movement from one site to another is migratory beekeeping. It is wise to have reserve sites for the isolation of swarms, for the movement of nuclei or to move if it becomes necessary to leave an existing site. The location of a site will have a strong influence on the potential honey crop. In countries like Australia good siting and migratory beekeeping is of the utmost importance.

See: drainage, enclosure, footing,- and hives under trees.

**sitology sito – food**

The science of food, diet and nutrition. The importance of the effects of groups of food substances on the human, comes into consideration when we realize that honey contains over 180 more ingredients than sugar. Furthermore large honey producers who tend to play down the value of fresh honey compared with honey that has been processed for selling, should compare a fresh apple with a baked apple.

**situation**

A place in relation to location or conditions in relation to circumstances. Success in beekeeping depends heavily on having hives in the right place at the right time.

**size of brood chamber**

The majority of beekeepers seem to want to work with a certain size of brood chamber. Only a comparatively few accept the principal, so easily applied with modern hives, of allowing the brood chamber to respond in size to the type of bee and the stage of development of the colony. Overcrowding of the brood nest while the colony is expanding in the spring often leads to swarming. At other times restriction of the brood is best for the bees. Brood chambers vary according to hive type but may also comprise one or more boxes which may not be of the same depth (making comb interchange difficult).
See: brood-and-a-half, double brood and dimensions.

**size of entrance**

In nature the opening through which the bees gain access to their nest is likely to be of a size dictated more by chance than judgement. Survival however is more likely when their cavity is situated in relation to the entrance so as to give protection from the weather. Bees can and do, restrict the opening by using propolis. British hives tend to use the full 16 x 1" summer entrance with the use of entrance blocks (cleats) or slides to give smaller sizes as required to retain warmth, keep out wind and discourage robbing.

**sizes**

Worker 13 – 17mm head to tail
(compare cell size)
Wing span 18mm
Each wing 8mm
Thorax 4mm
Tongue 6 – 7mm
Honey sac 20mm3.
Egg mm
See: frame/cell sizes.

**skeletal system**

The hard skeletal framework internal or external which protects and supports a creature's softer and more vulnerable organs. At first thought it may seem a little odd that a bee wears its skeleton on the outside. However when we compare it to the structure of our own heads, in this respect at least it is similar.
See: exoskeleton.

**skep *****

A basket-shaped hive made from grass or straw rope or plastered wicker-work.. Having originated in Germany they became widespread in Europe up to the turn of the century when the gradual change over to wooden, movable frame hives was accompanied by the Isle of Wight disease. A skep sits on a suitable board which should provide the entrance as cutting a hole in the rim of the skep would weaken it. Old Dutch skeps had their entrance hole about half-way up. The basket is of course inverted so that the dome keeps out the weather while the rim acts as a firm base. A well-made 16" skep weighs about 5 lb. Sometimes Gipsy pot-type caps of sedge were added and sometimes ekes. The materials were cheap and plentiful and as the bees suited themselves as to how they arranged their combs inspections were tedious and combs (and usually bees) were often destroyed.

Attempts to return to this primitive style are pursued by some non-progressive beekeepers today but with a view to 'helping' the bees. The word 'Skeppa' means half-bushel in Scandinavia.

**skeppist**

A beekeeper who kept his colonies in skeps. A hundred years ago most beekeepers were loathe to change to 'wooden' hives. One famous beekeeper of that time was Pettigrew. When British bees were suffering from the ravages of a spectrum of diseases and conditions known as I.O.W. disease, Dutch skeps with resistant bees, helped to save the day though it was thought to have spelt the end to the old British black bee. See: skeppist's terminology

**skeppist's terminology**

Bee bole, bramble, butt, cane, cross-sticks, driving irons, eke, feeder (horn), hackle, needle (skep) ruskie, rush (common). Rye straw, skep, skep stand, skewer, skip, stall, sulphur pit, wheat (straw). Other names for skeps include: scep or scap, butt and ruskie (Scotland). For a stand, skeps were often stood on a round slice of tree trunk or three-legged stool, the latter giving better protection from mice.

**skewer**

A long pin of wood or metal used as a skewer. A securing needle which was forced down through the walls of two skeps when 'driving' was to be performed. The upper skep was positioned on the lower upside down skep, as if it were hinged and the skewer driven through so as to hold the skeps rigidly at this point without impeding the upward progress of the bees. See: driving irons/bees.

**skid board (dolly)**

A low truck or platform with small wheels used for moving heavy supers about in the honey house with a minimum of lifting.

**skills**

Abilities developed through practice, knowledge and aptitude. Many different skills are called for in the craft of Apiculture. It takes two years to make a 'smoker boy'. Like controlling any animal – dog, horse or creature whose ways and reactions must be understood to get the best results from them, bees require the understanding that only comes from years of intelligent practice. To 'read' their intentions and anticipate their behaviour, linked with the use of sensible, efficient equipment, marks the good beeman or woman. Taking a swarm, honey extraction, honey judging, judging when to do what, these are skills that take time to acquire.

**skim**

To remove any floating material from a liquid surface. When viewed against the light so that a liquid honey surface appears shiny – even a speck or hair can be seen if present. Provided any such blemish is minimal, a piece of waxed, water-proof paper (as used on the top of jam) can be set down onto the honey surface. Once it has made contact and stuck to- the surface it can be lifted off, using a spoon and saucer, so that the blemish comes away on the thin film of honey adhering to the paper, but don't let drops fall back into the honey.

**Skimmia *** Rosaceae**

A compact, evergreen. Low-growing shrub with dense panicles of fragrant, creamy-white flowers that burst from pink buds in March and April. They tolerate atmospheric pollution and carry their brilliant red berries through to the winter but their young leaves are susceptible to frost.

**skin lotion**

1 dessert spoonful of honey, 3oz glycerine, 1½oz lemon juice, ½oz red lotion, 1 oz. alcohol and 2 oz. rose water.

**skip**

This pronunciation persists when older beekeepers speak of a skep. Keys spelt it that way. Dictionaries tend to use 'skep' for iron boxes such as were used to raise coal or minerals and as attached to a crane.

**Skunk *** Two genera Mephitis & Spilogale Mustelidae**

Small, striped, fury, bush-tailed N. American mammals. A skunk can consume ¼ to 1 lb. of bees per night and keeps at it until it has cleaned a colony out, then it moves on to the next and repeats the act. (Not CDD - ?!). It ejects faecal fluid when attacked or worried. They are carriers of rabies and have a diet similar to a house cat. Poisoned eggs have been used to kill them where troublesome in Canada. They go for colonies at ground level but can be trapped. They leave tell-tale scratches on the earth in front of a hive and regurgitate marble-sized clumps of bee parts. Keep strong colonies raised above the ground and sink boards with nails hammered through just below the ground in front of the hives. They can pull mouse guards off and act at day-break or just before dark.

**skyscraper hive**

An exceptionally tall hive, often containing two or more queens each separated vertically and having entrances at various levels. Pere Dugat wrote a book with this title and gave full details of his method of getting several queens with their own brood chambers all working to the mutual benefit. See: hive skyscraper.

**skywards**

Up into the sky. To fly to higher levels. This is the route taken by queens and drones when they leave a hive - the exception is when a queen is surrounded by bees when she leaves with a swarm. Contrary to a layman's belief drones are not to be found lazing around in the sun or resting on or visiting flowers and queens are even jostled by non-swarming bees if they do not climb to mating altitudes. On rare occasions when a queen is seen on an object outside the hive it is either because of intervention by the beekeeper or she is resting after attempting to move away with a swarm.

**Sladen F. W. L. (GB)**

In 1902 he wrote an account of bee's scent organs, bred golden bees near Dover, wrote '*Queen rearing in England*' 1905 and in 1911 described how pollen is collected and the part played in the process by the auricle. He was also an expert on bumble bees.

**slaughter of the drones *****

Towards the end of summer when swarming and mating are done with, the bees will have garnered their stores and be preparing for the winter months when drones would require feeding yet offer little in return. The drone's useful life would not extend through the winter in any case. Earlier they were fed and groomed but now they are neglected and harried. They become weaker and are driven out, prey for ants or birds and the workers do nothing to protect them. Gradually, in queenright colonies, hundreds of drones may be found in a dead heap in front of the entrance. A queenless colony and occasionally others will retain some drones.
See: drone slaughter, massacre and dead bees.

**sleeve**

1. That part of a garment that covers the forearm. When manipulating bees it is probably better to have sleeves rolled up if they cannot be elasticated or covered by gauntlets because bees that get inside will inevitably crawl upwards (the same applies to trouser legs.
2. Slides may be kept in plastic sleeves.
3. Sleeves or plastic covers as used for bank and identity cards and so on can be used for bee-notes in the hives so that bees cannot nibble them away.

**slide**

1 A slip of good plate glass with ground edges and corners, of uniform thickness (about 1 mm) for displaying material for viewing under a microscope.
2. A black and white or colour transparency (photograph) mounted in glass for the projection of an image onto a screen. Two by two inch frames holding images on 35mm film were useful when visual aids were in their infancy. Microscope slides are usually 3 x 1" and special projectors were available or there are attachments for use with 2 x 2 projectors or permanent mounts for use with the projecting microscope. Now by use of the computer screen much of the above is out-dated.

See: entrance slide.

**slide rule**

A device where one graduated scale can be moved relative to another. A beekeeping version set positive factors like times of incubation against dates so that forward planning for queen raising, getting colonies to peak strength for pollination and other forms of management can be arranged. To a large extent calculators (being so small and convenient) have taken their place and mini computers are getting smaller and smaller.

**sliding entrance blocks**

A pair of these, bevelled to match the slope of the alighting board are used on WBC hives. They fit nicely under a ridge that forms part of the porch. They can be opened out or slid inwards across the entrance, and a small niche is cut in the end preventing them from completely closing the entrance. They can be removed altogether thus offering entrance sizes from one bee way up to full width. They are also called front slides. They can be painted any colour. Avoid tripping over them though, when extended.

**slinger**

A simple, original form of honey extractor. A single honeycomb or piece of one, was uncapped and lain flat at the bottom of a container. A cord was then tied to the container (or its handle if suitable) and then swung around, possibly from a pole so that the centrifugal forces imparted caused the honey to leave the underside of the comb and collect in the bottom of the container. See: Hrushka.

**Sloe *** *Prunus spinosa* Rosaceae**

Blackthorn. A small, sour, black, (though often covered with 'bloom') plum-like fruit and the shrub that bears it. It can be used to make a pleasant melomel.
See: blackthorn.

**Sloth bear *Melursus ursinus***

It lives in India and Sri Lanka and passes its day in caves or concealed in bushes. The tongue is exceptionally long which enables it to scoop up termites and ants and its claws are prodigiously strong, efficient tools for digging out the nests of bees for this bear loves honey.

**slotted floor board**

Slots in a floor board allow clustering space and the easy removal (by bees) of debris. The Killion board is an example. Such slots are usually over a floor so that no timber is saved. Security and protection from draught precludes the use of a floorboard that is truly slatted. Such were the thoughts until Varroa came along and now, it seems, open mesh floor boards are the order of the day.

**slotted separator**

The section divider separates adjacent rows of sections but is cut away in vertical slots so that bees may pass from one row to another at the edges of the sections. Other separators fitted with slots are used either to give clustering space and help to discourage swarming, or for ventilation purposes or to confine the queen or separate her from another part of the hive.

**slotted queen excluder**

Queens are kept out of the supers (due to their slightly wider thoraces) or confined to boxes intended for brood chambers by sheets of worker permeable material. The openings in such sheets are intended to prevent queens from passing through (and incidentally drones too). Slots are used, there being long and short varieties with carefully machined slots less than 4·2 mm wide. Holes are rarely used but parallel wires or plastic strips are used, the Hertzog and Waldron excluders being examples. The spacing then has to be 4·11mm.
See: gap, useful dimensions, framed and zinc excluders.

**slow feeders**

A feeder designed to give only a few bees at a time access to the syrup. Older jar-type feeders even gave the option of using one or more small holes (up to ten) by rotating the jar on its base. Slow feeders are meant to stimulate egg laying and to improve colony morale, whereas rapid feeders, while useful for winter feeding, might cause blocking of potential brood cells.

**Slow paralysis virus**

This was first discovered in GB when infected with preparations of this virus adult bees died after about 12 days, typically suffering paralysis of the anterior 2 pairs of legs for a day or two before death.

**Slow release units**

Insecticides that release their vapours over a long period of time. The Environmental Protection Agency has cautioned against their use in rooms where food is prepared or where infants or aged persons are confined. Some of the names given include: DDVP, pest strips, Vapona, Sectovap, Mafu, Flitox and Cooper flystrip. Chemicals include: umagillin – 2, and 2 – dichlorovinyl dimethyl phosphate. Queen bees in cages have succumbed due to the proximity of a 'fly-killer' and foundationis also easily contaminated.

**Slug Gastropoda Pulmonata**

A gastropod, elongated, no shell, slimy, and slow moving. This destructive garden pest curls up when disturbed. It has a slimy body akin to those of snails. When they are found in hives it is usually on walls where dampness has been encouraged through lack of ventilation. Some smooth plastic walls seem more vulnerable than the more customary wood, and many large slugs seem to be able to enjoy the comfort of such shelter without the bees being able to eject them. Ashes round a hive helps and they hate salt.

**Slum gum U.S. – wax residue**

The residue left behind when the clean wax has been separated from the comb during the process of 'rendering'. It is generally referred to as 'dross' in the U.K.

**small claims**

A booklet published by H.M.S.O. entitled '*Small Claims in the County Court*' covers small claims (up to £500.00) with little cost to the beekeeper – as for example for Spray Damage, so wrote G. L. Mills in Bee Craft, May 1983.
See: compensation.

**Smaller wax moth *Achroia grisella* (Fabricus)**

This little creature can destroy comb unless precautions are taken. In the hives strong colonies can keep it in check, though sections and comb honey may be spoilt. Once combs are bee-less, nosema treatment (formaldehyde for empty comb or acetic acid for combs with stores) is wise as the former kills all stages while the latter helps to control it. Supers should be covered so as to exclude wax moth until they can be put on the hive again. The moths lay a lot of eggs which are not easy to see. Acetic acid can be used to protect overwintered empty combs.
See: Achroia grisella and lesser wax moth.

**Small hive beetle *** *Aethina tumida***

Although known about for a long time it

is only recently that this creature and the damage it can do to a colony, has become of general concern. In 2010 there have been no reports of its presence in the U.K. but we have been warned to look out for it and to report it to Fera immediately.

**small intestine**

That part of the honeybee's alimentary canal that comes after the ventriculus (stomach) and along to the rectum. Differentiated from the larval hind-gut. In most creatures the small intestine is concerned with digestion and the large intestine with the preparation of faeces by the removal of water. This latter function is performed by the rectum in Apis.

**Smear**

A thin film of liquid such as that deposited on a microscope slide for the purpose of examining a ground-up sample of bees for the diagnosis of diseases such as nosema and amoeba.

**smell**

Noun: A good sense of smell is a characteristic of the honeybee and a useful attribute for the beekeeper. It can be used by both to warn of danger and in humans it can be a source of comfort and pleasure and of course displeasure.
Verb: To test by the sense of smell or to give out an odour. Characteristic smells are a feature of beekeeping: honey, beeswax, mead, a hive interior or the ripening of nectar along with many others.
See: olfactory.

**Smith hive**

The National Minor designed by that great Scottish beekeeper Willie Smith. Its great advantages lie in its simplicity, lightness and ease of construction. It utilizes British frames with an inch cut from their unnecessarily long lugs. The hive is similar to a Langstroth being single-walled and having top bee-space and the shallow frames will fit into a Langstroth super that has been rebated along its sides.
See: hive Smith.

**smoke**

Visible vaporous products of combustion. Water vapour with carbon in suspension. A dense, cool, non-toxic smoke is what beekeepers try to produce when subjugating and controlling large numbers of bees. Smouldering material which produces smoke without sparks when puffed is satisfactory provided the resultant smoke is as clean and as sweet smelling as possible. Bonfires are best kept well away from hives though well-diluted smoke is not unduly harmful. Types of fuel and the lighting of smokers are dealt with under other headings.
See: effects of smoke, smoker fuels and grass.

**smoke bomb**

An aerosol, small enough to go into an overall pocket, containing hardwood smoke concentrate. When the operating button is depressed, a cloud of 'artificial' smoke is delivered. This is useful in so far as it is immediately available, requires no lighting and is cool. However surfaces touched remain repellent for a while and it is not wise to use it on supers. Although expensive every split second of use is controlled and there is no unwanted wafting of smoke straight onto the beekeeper.
See: smoker cartridge

**smoker cartridge**

A roughly cylindrical roll of material specially cut and kept dry for ignition and putting into the smoker. The material used should be clean and easy to handle: corrugated wrapping paper, rotten sacking (Hessian/burlap). It can be treated with a dyed substance for easier lighting provided this is not likely to harm the bees. Saltpetre with a coloured dye has been used as a liquid drip which when dry enables the cartridge to be set-alight more easily. Material should fill the fire chamber allowing only minimum space for the passage of air If too loosely packed it can cause hotter smoke and burn through rapidly. See: cartridge and potassium nitrate.

**smoker fuels**
Materials that can be ignited and will then continue to smoulder giving dense, cool smoke when required. These include: rotten sacking (Hessian), rotten wood (punk) an some kinds of sawdust provided they are held back by grass or a less penetrable block for fear of puffing sparks into the hive, corrugated cardboard and other packing materials such as pithy egg cartons(first trampled flat), fir cones, dried sunflower leaves and stalks, cotton waste, but all materials should be dry and free of chemical contamination. A positive flame is required to start things off and as the length of time it can smoulder is limited it is wise to have spare material to hand so that the smoker can be quickly replenished. See: smoker cartridge

**smoker (function)**
Impelled by hand pressure on the bellows and directed by a nozzle, clouds of cool smoke should be delivered by a smoker as required. Once well alight it should be possible to temporarily lay it on its side without its going out. All too often the smoker is left to stand upright, belching wasteful, harmful smoke in a choking column polluting the atmosphere. The cartridge or mass of material to be smouldered should packed fairly tightly but arranged so as to permit the air that is pumped in to pass through and emerge from the nozzle as a cloud of smoke. For safety (avoid burning a hole in the veil) it is wise to get the smoker going first and this is best done out-of-the-wind , light the 'bottom' of the cartridge before inserting into the fire chamber. Give an exploratory puff or two before carefully applying it. It is harmful, especially to brood and does the chitin of the bees no good. A waft or two to clear bees when replacing boxes – try to avoid belching it down between combs or puffing smoke at bees that do not require it. To wit use it as little as possible. Nevertheless don't be caught out by finding the bees difficult to handle and having no smoker going.

**smoker (history)**
In 1865 Moses Quinby made a simple smoker through which tobacco smoke was blown. In 1870 he made one that incorporated a fire chamber and bellows, but if fell to T. F. Bingham in 1877 to further improve Quinby's smoker by arranging the bellows alongside the fire chamber and leaving a gap between them so that the fuel was free to smoulder. By 1879 A. I. Root introduced the more modern smoker with a more sophisticated nozzle. See: smoker cartridge, fuel, function, structure etc.

**smoker (structure)**
This has become one of the best-known tools of the beekeeper. While aerosol and automatic types are available the conventional hand-held and operated device is still the most widely used. Basically it consists of a fire chamber with a hinged nozzle cap. At the side, near the base of the fire chamber, is an opening into which a small jet of free air can be pumped from an adjacent bellows box and a tube. The air passes under a perforated grid above which there is material which when ignited will continue to smoulder. Once ignited and put 'burning end down' into the fire chamber from the top the nozzle is locked on the smoker is functional

**smoker types** ***
give ancient Egyptian and early Italian illustrations...

**smoking bees**
Smoke is usually thought of as a subjugator but its most important use is probably to get bees out of the way so that they are not crushed when boxes are re-assembled and sometimes to move them away from a certain place. Otherwise the action of smoke is two-fold. It masks their identity odour

and makes them fan and also causes many to put their heads into cells to take up food lest the smoke suggests a forest fire or merely to get away from the smoke.). Once fanning, a bee is completely occupied and shows no immediate desire to change its posture and least of all to sting. If your object is to subjugate (make them less inclined to resist your interference by buzzing you or stinging) then use the smallest amount necessary to engage the colony's attention, then wait a minute or two for it to take effect before opening up. To minimize the need for smoke, move carefully and slowly, use a cover cloth to keep out the light and keep in the warmth and don't blow at the bees but at the air around them. Smokers may be a source of conflagration!

**smothering**

To deprive of the air that is necessary for life. Bees can be inadvertently smothered by shutting them in during hot weather e.g. by enclosing a swarm in a container without adequate ventilation. An observation hive can easily have its entrance blocked and a colony being transported become covered, whereas jogging and enclosure can cause excitement and a rise in their temperature and great need for fresh air. In an unfortunate condition such as fatal disease, wait until all are home, pour some petrol through the feed hole and seal up. If, because a queen is alien or some great disturbance causes the bees to cluster tightly around her, she must be released from their hugging (even dropping the walnut-sized lump into water) or she will be suffocated and this can take days. We call it 'balling' the queen. When drastic circumstances call for suffocation, such as fatal disease, wait until all the bees are home, then[pour some patrol through the feed hole and seal them up

See: balling

**snail Mollusca Gastropoda**

A small, slow-moving creature with a shell. It likes the damp. Snails are rarely any trouble to bees, though on occasions, especially when hives are on damp sites and are not well ventilated, they gain access, probably for protection from the weather. Should their 'armoured tanks' foil the bees attempts to send them on their way, then it is quite usual to find them enclosed in a propolis sepulchre. Where wet slabs give them the chance to lay their dozens of eggs, they can suddenly become numerous. Merely carrying them off to a new site won't do, Mark them and see. They can return from considerable distances!

See: slug and sepulchre.

**snake *** Squamata Serpentes**

Snakes cause little trouble in Britain though there is always a possibility that a beekeeper might inadvertently get a bite from an adder, perhaps when taking or bringing hives from the moors. Gum boots help. They do not exist in NZ.

See: lizard and cane toad.

**sneeze**

A spasmodic emission of breath via the nose. A sting near the nose can bring on a fit of sneezing. A form of allergy occurs when a person tends to have a 'runny' nose while operating with bees .As a veil hinders the application of a handkerchief to that part of the anatomy it is probably a wise precaution for vulnerable people to keep a handkerchief available inside their veil.

See: wheeze.

**Snelgrove board**

A rimmed, horizontal, division board with a central hole that can be quickly and easily covered with gauze or perforated material which prevents the passage of bees though enables them to communicate as though of one colony. One side of the rim, which takes the form of a batten around both upper and lower sides of the board, is untouched, while the other three sides each have an adjacent upper and lower entrances (3 pairs of entrances) cut so that they can be left open or shut.

**Snelgrove L. E., MA, MSc**
H.M. Inspector of Schools. President of Somerset BKA, Lecturer and Judge. Produced the Snelgrove board in 1931. Wrote '*Swarm control and prevention*', '*Queen rearing*' and '*A method of re-queening*'. (water method).

**Snelgrove method of swarm control**
Build a colony to double brood and super by the end of May. Re-arrange so that the lower chamber has no brood except a little unsealed on central comb. Put queen into this box and cover with queen excluder, place the super over the excluder and the box of brood goes on top. 3 days later the Snelgrove board is inserted under the top box. It has a gauze screen over the central hole and a top side entrance on the board is opened. 7 days later this entrance is closed; its partner below is opened as is the upper entrance on the other side of the board. In a further 7 days' time the upper entrance is closed and its partner below opened and the remaining upper entrance at the rear of the board is opened. The new queen is thus able to fly and mate from the upper entrance at the rear of the hive. Feed upstairs lot.

**SN frames**
BSS for shallow frame for national or WBC, SM for modified commercial, SS for Smith. See: DN.

**sniffing**
A bee can draw air towards its antennae and so sample something that is ahead of it. It is as reasonable to say that a bee can 'sniff' as to say it can hear. It has neither ears nor nose yet it registers vibrations (sounds) and odours. See: scenting, fanner and sneeze.

**snippets**
1 A buzzing with bees like a bean field in bloom.
2. An the wild honey's still in the gum tree bough.
3. The native basswood trees were loaded with large healthy blossoms the colour of butter.
4. Bees gain experience with a particular type of plant improving skill in entering and manipulating its peculiar maze of floral structures and thereby increasing their foraging efficiency.
5. Eight little scales of wax looking much like half-posted letters – reference to workers with plates of wax protruding from their pockets. Tickner Edwards.
6. He also tells us that his allowance to each half-dozen monks at dinner was a sextarium of mead – several gallons.
7. From the point of view of wild-life, rye grass provides a very impoverished habitat.; Prof.N.W.Moore.
8. The journey to a flower uses up fuel and a visit is only worthwhile if the flower provides more fuel than the visit costs. Dr.Sarah Corbet.
9. The relationship between man and bees – and their honey and wax – has been a long one. Their mysterious and apparently disciplined community within the hive fascinates us – from '*The Archaeology of Beekeeping*' Eva Crane.

See: Mythology and Sayings.

**snow *****
Originating as white flakes it can fall to produce a carpet of various thicknesses and deep drifts. It is frozen water vapour. It can attract bees out of their hives due to the brilliant reflection of sunshine from its surface. Bees are confused by it because they fly with most light on the upper part of their eyes and the similarity to light from the sky causes them to dive into it, melting small graves. Hive entrances should be shielded from the bright light reflections that a carpet of snow throws up. Snow heaped on or over hives acts as a good wind shield and insulator but a prolonged smothering of the entrance could lead to suffocation. Picked out from the snow quickly, they respond to the warmth of one's hands and can be put back. See: snowglare.

**Snowberry *** *Symphoricarpus albus, racemosus* Caprifoliaceae**

A hardy shrub that flowers from June to August and is constantly worked by honeybees. It spreads easily via suckers, grows from 1 – 3 ft high and can form a thicket. Its small, bell-shaped, pinkish-white flowers develop into large white ornamental berries. (Hence its name). As well as nectar it yields whitish-yellow pollen with a 35mu grain.

**Snowdrop *** *Calanthus nivalis* Amaryllidaceae**

A small drooping, white, bell-shaped flower arising from a bulb which blooms for several weeks from January to March. It is widespread throughout the U.K. and great, white masses dangle in accompaniment to the breeze in the woods of western Scotland, one of whose beekeepers described its pollen as orange though bright yellow, greenish yellow and even brownish red have been reported. Temperature usually discourages nectar secretion but the charming plant obliges sometimes making nectar and pollen available to the bees.

**snowglare**

Reflection from the bright snow can utterly confuse the bee. The hobbyist beekeeper with a hive or two not far from the back door, may well shield the hive entrances from the light reflected by the snow. Heavy slates are suitable for this purpose. The bee farmer will have to rely on the careful siting of his apiaries and the arrangement of the hives within them.
See: snow

**social insects**

A family of insects living together and exhibiting some degree of mutual co-operation with a parent. They form a highly organized, perennial community in which there is co-operation between successive generations living together in a common abode or shelter. The more advanced forms of social insects are found in two Orders: Isoptera and Hymenoptera. Both seem to have developed independently. The termites being only distantly related to the Hymenopterous bees, wasps and ants. Large colonies are formed within which there is division of labour that is based on a caste system. Egg laying is normally confined to a single perfect female known as the 'queen'. The two orders differ as regards males. In Hymenoptera these only exist to mate and either die or are killed off before the onset of winter.
See: eusocial and slaughter of the drones.

**Socrates hive (Greece)**

A two-queen Langstroth hive with two entrances aimed at producing a huge foraging force.

**soda $Na_2CO_3 + H_2O$**

Washing soda – carbonate of sodium.
A white, crystalline solid which dissolves in water softening it and making it alkaline. The use of a tablespoonful per gallon is recommended for washing propolised metal or plastic parts, though the use of spirit is kinder to metal surfaces. Do not use with aluminium feeders, nor with water to be used for wax rendering as it would spoil the texture of the wax. Acidulated water should be used for that purpose. The bicarbonate $NaHCO_3$ is used in cooking and medicine. Soda water is merely water charged with carbon dioxide.
See caustic soda.

**Sodium Na**

One of the elements – a soft metal that oxidizes rapidly on exposure to air.
Commonly present in salt (NaCl).
A common base for many useful compounds e.g. Chile umagilli $NaNO_3$.

**sodium metabisulphite**

10g to a litre for sterilizing also used to arrest fermentation

**soft bees**

Bees whose qualities lack the necessary

robustness for survival and to do well in our climate. These occur when queens that have been mass-produced in warm, overseas climates are established in colonies here.

**soft fruit**
Soft, stoneless fruit from the canes and bushes that include: blackberry, currant, gooseberry, loganberry, raspberry, and strawberry. All benefit from the services that honeybees render as they scrabble for pollen or nectar.
See: under respective headings.

**soft soap**
A soft liquid soap made from vegetable oils, glycerine, oleic acid and KOH with distilled water. Used in certain recipes with beeswax and helpful for skin disorders.

**soft set**
A term used for crystallized honey which has been softened, stirred and re-set – or totally liquefied, seeded, and re-set. It should be completely crystallized, with a smooth, spreadable texture and show no liquid or air-scum.

**soft water**
Water with little or no dissolved salts of calcium or magnesium. This is the best kind for use in rendering wax, obtaining a good lather but not necessarily the healthiest to drink.
See: de-ionized water.

**software**
This includes all the manuals, flow charts, information stores, programmes, routines and explanations of how to use the hardware which it makes functional.

**softwood**
Wood of the coniferous tree, such as pine. Light in texture and easily worked it is usually a fairly quick-growing type of timber, radiata pine being widely grown and used for beehives in NZ. As opposed to the hardwoods: oak lime etc. which are from the broad-leafed flowering trees and are mostly heavy woods – harder to work and the grain is uncertain – they are also more expensive.

**soil**
Earth which is created by erosion or disintegration of rocks which becomes impregnated with organic material and supports vegetation. The type of soil, its moisture and temperature have effects on nectar secretion. Chalk and limestone favour wild flowers in mid-season, while acid soils, especially at higher altitudes, favour the heathers or late summer. Light soils drain better and are usually warmer earlier in the year, though dry conditions may mean a cessation of flow. Heavy soils that retain moisture can yield well, late in the season as, for example, with late second crop red clover and mustard.
See: influence of soil and soil types/fertility

**soil fertility**
The ability of 'earth' to support plant growth. It also affects nectar secretion. Certain plants have fairly specific soil requirements. An example is ling heather which secretes best between 600 and 1000 feet ASL, though lower on the west coast of Scotland. It likes shallow, acid soil especially on igneous and metamorphic rocks – whinstone and greenstone, however there is little nectar if roots are wet hence the preference for higher ground.
See: Leguminosae, nitrogen fixing and bean.

**soil moisture deficit**
This is the difference between moisture quantity in the soil and its capacity. In agriculture its measurement determines whether or not irrigation is required, or how much rain would be required to return conditions to normal.

**soil properties**
Texture, structure, porosity, water-holding capacity, temperature, nutrients, pH and ease of tillage. Elsewhere reference has been made to the soils of transport. These depend

on the nature of the transported material as opposed to 'sedentary' soils that are formed in situ and are therefore dependent on the nature of the parent rock. Sandy soils are coarse with large pore spaces and are easily aerated since the drainage is rapid. They have low water holding capacity, have low retention of ions and leach rapidly. They are seldom acid, will till easily. Clayey soils are fine, clod easily and have small pore spaces. They have a high water holding capacity are cold when wet but heat more rapidly when well drained. Can be very acid and tillage difficult.

**soil types**

Brick earth, bunter sandstone, calcareous, clay loams, boulder clay and clays (fine textured) gault, greensand , loams and silt (medium textured) loes, lower lias, marls, old red limestone, oolitic limestone, peaty soils, Permian rocks, sands and sandy loams (coarse textured) Silurian rocks.
Wild soil groups: tundra, desert, podsoils (heathland, coniferous) brown earths, laterities, blackearths, prairie, steppe, savannah, mountain and high plateau, oases, tropical and 'mangrove' swamps, rendzinas, loes (wind borne) alluvial(water borne) gley and bog soils.
See: downs, moorland, heath and soil.

**Solanum *** *S. jasminoides* Solanaceae**

A showy, climbing shrub with creamy, white flowers. It is tender and needs warm wall. Family includes nightshades, tobacco and include the house plant 'winter cherry'. This latter can be put outside for pollination or possibly hand-pollinated to produce the pretty cherries.

**solar energy**

The energy from the sun. It is used to separate wax from old black comb in solar wax extractors but a host of other beekeeping applications could follow the new practical methods of converting the sun's rays into directly usable energy.
See: solar wax smelter.

**solar wax smelter (extractor) *****

Mentioned in California in 1862. An inclined tray in an insulated box with a removable (or hinged) double-glazed cover. It is aimed at the mid-day sun and kept free of shadows. Waxy material including equipment can benefit and be made sterile by the heat produced. Burr comb, cappings, rejected comb, scrapings – any such material – can be put into old tights or stockings or the liquid wax allowed to run through a perforated barrier. The insulating air layer between the double sheet of glass enables it to retain its heat even when the sun goes in for short periods. The wax runs into a suitable mould No smell – no waste of energy. More beekeepers should use these. They can be made large enough to take queen excluders and are best painted black. In G.B. Glass should be angled at around 44 degrees to the horizontal.

**soliciting food (begging)**

The behaviour which involves the angle of approach, characteristic extension of the tongue and the use of antennae to bring the desired reaction from another bee which will open its mandibles and make contents from its honey sac available to the soliciting bee (sometimes one or two).
See: beggar bees.

**Solidago spp. *** Compositae**

A tall autumn plant bearing masses of small, yellow flowers. In the U.K. we are familiar with *S. virgurea* which becomes smothered by bees and other insects in late summer. There are however very many varieties especially in N. America where they are valuable honeybee plants. Where surplus honey is taken it is described as yellow, thick and of fine flavour. It crystallizes with a coarse grain.
See: golden rod

**solid food**

Candy, fondant or an occasional pollen supplement or substitute, or pollen itself. Solid sugar feeding is problematical as fine dusts can enter the bee's spiracles and larger crystals tend to be thrown out of the hive unless moistened. Syrup is the normal way of feeding though patties are used. In British winters liquid feeding is generally considered inadvisable and in emergency candy is given. Candy causes less excitement and does not ferment but can spread diseases like nosema.

**solidify**

To become solid compact or hard. 1, Honey will granulate into solid form, the ones from brassicas (OSR) being rich in glucose do this all too rapidly. 2. Beeswax when liquefied contracts as it cools and should be free to break away inside the mould (use releasing agent) and cooled slowly to avoid cracking.

**solitary bees**

There are at least 227 varieties in the U.K. and they are sometimes referred to as 'wild' bees. Examples include: Megachile the leaf cutter, Halictus, Osmia the mason or miner, Andrena the carpenter, Anthidium manicatum or wool carder. These are the non-social bees that live on nectar and pollen yet are of a solitary habit as regards nest building and the provision of food for their young. Only parents of a few varieties live to see their offspring.

**Solomon's seal** ***
***Polygonatum multiflorum*** **Liliaceae**

A hardy perennial growing one or two feet (30 – 60cm). Useful as an early summer plant yielding both nectar and pollen – May to July. Its berries are poisonous.

**solstice**

Twice a year the sun reaches its highest declination when it is over the tropics. The dates are 21 June and 22 December, when respectively we experience the longest and shortest days. Temperatures lag behind these dates so that hopefully it is still getting warmer at the summer solstice and conversely cooler from 22 Dec in the northern hemisphere. The longer foraging day in the northern latitudes gives unexpectedly large honey crops in areas that are very cold during the winters. Solar wax extractors (unless swivelable and rotatable), should fact south and be inclined at about 30 degrees to the horizontal to catch maximum sunlight.

**solubility**

Ability to dissolve in a liquid. Honey and sugar are easily soluble in water. This helps as regards feeding colonies and washing honey extracting equipment. Propolis is slightly soluble in alcohol and can be removed with methylated spirit. From equipment it can be dissolved by the use of an alkali. Wax is soluble in ether, alcohol, turpentine and other oils like paraffin and carbon tetrachloride.

**solute**

A substance that is dissolved in another substance – propolis will dissolve in beeswax.

**solution**

A mixture of two substances where the molecules are evenly spaced but no chemical reaction occurs. When sugar syrup is made the sugar is the solute, the water is the solvent and the syrup is the resulting solution. Honey is a supersaturated 'solution' of sugars in water. Simple sugars can reach a far higher saturation point than compound sugars, hence the cleverness of bees to store their food in monosaccharide form. The higher the temperature the greater the concentration possible and conversely sugar may crystallize out in a cold feeder at night.

**solvent**

A liquid that is capable of dissolving another substance. Water is an extremely useful solvent and should be used cold to wash honey wax accumulation away. Hot water would melt the wax and spread it whereas honey is quite soluble in cold water. There are a number of liquids that can be used for dissolving wax and these include acetone and CTC while others are listed under 'properties of beeswax.

**somatic cells**

Cells which take part in the formation of the body part of a plant or animal (as opposed to germinal cells) becoming differentiated into various tissues, organs etc.

**somite (melamere)**

A body segment of an articulate animal.

**Sooty fungus**

Leaves covered with the sweet secretions of creatures like aphids, become are invaded by a black mould. A similar black mould can form on the syrup-wet walls of plastic feeders. The mould can cause a black discolouration in honey derived from such surfaces and the exceptionally black 'cart grease' honey collected from bees near London lime trees is an example of this. See: sooty mould – link with honey dew.

**sooty mould fungus**
***Aerogennothereca elegans*** **– Hughes 1967**

It smothers the trunks of trees infested with honeydew producing insects. The mould is entirely nourished by the honeydew. Particularly widespread on the northern half of S, Island NZ
Also 'fumago' of the fungi 'imperfecti group. See: sooty fungus

**sorbitol $C_6H_8(OH)_6$**

A white, crystalline powder and isomer of mannitol. A sugar alcohol that is not sweet yet is nutritionally useful to the bee in that it is slowly metabolized to give energy (unlike the sugars used to power the flight muscles). Isomeric with mannitol. Used in resins and varnishes and as a substitute for sugar and for making ascorbic acid.

**sore throat medicine**

Mix 1oz marshmallow root with 1oz honey in 4/5ths pint water. Gargle well with the liquid several times a day.

**sororicide**

The dispatch of a successful virgin queen's sisters as in making rare use of her sting she goes from queencell to queencell killing the captive occupants. It could be applied to workers balling virgin queens developed from the same queen's eggs.

**Sorrel *Rumex acetosa* *** Polygonaceae**

A perennial found in meadows and moist pastures. Its pollen has been described as very pale yellow and as greenish Height 25 – 50cm.
See: sorrel tree and sourwood.

**sounds**

A swarm issuing or entering a hive. Bees (workers, drones) on the wing. A trapped bee. A stinging bee (the yell of a beekeeper stung). A piping queen. A queenless (moan) or queenright (hiss). Colony. Onomatopoeia. See: substrate borne sound, bee sounds, hiving sough, listen.

**Sources**

of disease, infection, knowledge, nectar, pollen, propolis, honey, forage, water, water sources. Covered under these headings.

**Sourwood tree *****
***Oxydendron arboretum*** **Ericaceae**

Early honey plant producing water- white honey (western N. Carolina). Surplus usually confined to the foothills and mountains of the State. Piedmont country from western Virginia to western Florida. Honey is a very pale, straw-colour and has a mild but pleasing flavour – another report: a light honey with a strong minty odour and taste.. White bell-shaped blossoms, purplish-red leaves in the fall. Other names: sorrel, lily-of-the-valley and elk tree. Name comes from the acid leaves and it favours acid soil.

**South Africa**

Honey production in the 80's was 5,000 tons. Big crops from the Eucalyptus forests (Natal, Zululand and the Transvaal).
Cape fruit utilizes pollination services

and there were 1500 hobbyists. There are two indigenous honeybee sub-species A.m.capensis and scutellata. The latter produces aggressive, hard working bees suited to a tropical habitat. Capensis is native to the south western Cape. It is a black bee, less productive but easier to handle than scutellata. Able to produce females from its laying workers. Journal South African Bee Journal.
See: southern bee-eater.

**Soya bean** *** *Glycine max (soja)*
**Leguminosae**
A valuable oil-producing plant. The flower is pentamerous, asymmetrical, purple or white, it has u/v nectar guide lines and there are some two million flowers per hectare. The nectar has a sugar content of 37 – 43% and is transformed into a light lemon honey with flavour described as indifferent to excellent. The oil is used as food and in the manufacture of soap and candles.
See: soya bean flour.

**soya bean flour**
Out of several varieties the type with less than 1% oil and with Urease enzyme and anti-trypsin element removed, proved most acceptable to bees for use as a pollen substitute.

**Sp plural spp.**
Species following generic name e.g. Apis mellifera (mellifera being the species).

**spaced dummy.**
An ordinary dummy board with staple driven in about 2" from each corner of one side of the board. They should protrude ¼"so that they automatically space the board this distance from the side wall. Such an arrangement is recommended in a Smith hive when eleven 1 " frames are used.

**spacers**
The most commonly used frame spacer is the Hoffman shoulder as fitted on the side bar of frames. Metal (plastic) ends designed by W.B. Carr are widely used in the U.K. Brackets, clips, dowels, screw-eyes, nails and cut away lug supports are all used in various shapes and forms to secure the required centre-to-centre spacing of the combs in a hive.
See: frame spacers, Hoffman, metal end, and Yorkshire.

**Spain**
Famous for cork, orange blossom and rosemary honey. Abbe Della Rocca, ancient rock paintings, Migratory beekeeping takes place to collect the late honeydew.
See: Spanish

**Spanish bees**
Medium brown with many grey hairs. Large, prolific, slight swarmers but rather aggressive. The Balearic Isles bees are very different – small grizzly and they build twisted and irregular combs and have no commercial value.

**Spanish chestnut** *** *Castanea sativa*
**Fagaceae**
See: sweet chestnut.

**spark**
A small burning or smouldering particle such as fly from a match that is being struck or possibly from a smoker that is being puffed. All too many fires have been started by sparks falling or being blown into dry flammable material, somewhere hives were burnt and also heath and forest fires. Although smelly to cork up a smoker nozzle with a plug of grass, this method means you take all your sparks with you and the partially burned smoker fuel is often quite easy to re-ignite. Grass can also be used over the hot fuel to prevent hot smoke and or sparks from being ejected into a hive.

**Sparrows** *** ***Prunella modularis***
**Passeriformes**
The hedge sparrow is a thin-billed, insect eating song bird, not related to house or tree sparrows It has been seen to pester

bees by tapping at the entrance and then whisking off any bee that investigates the disturbance. Beekeepers rarely see this though they might find mutilated bees around the entrance, because the bird is so timid that it can only be spied at the hive by inconspicuous bird watches. Passer domesticus the house sparrow is also a destroyer of insects.

**spatula**

A small implement with a thin handle but spoon-like blade. For honey tasting a glass rod with one end flattened after heating is quite suitable as honey can be cleaned easily. A spatula may well be used when taking small quantities of powdery substance from narrow topped bottles – yeast, sodium metabisulphite, Fumidil B, borax etc.….

**speaker**

A person who speaks formally before audiences. Generally speaking, the public are quite interested in bees as a subject though they may not like to get near them. This offers golden opportunities for enthusiasts to give talks and help to enlighten those who might otherwise remain ignorant of the very special place and extreme usefulness that should be attributed to beekeeping. Many social groups: W.I., YFC, Lions, Rotary, Toc H and lots of others, regularly need speakers. Use of visual aspects via computer projection etc. and samples of honey, wax and model hives and carefully prepared talks help to supplement a new speaker's confidence.

**speaking**

The act of expressing oneself by the aural use of language. When speaking to an audience however small, it is necessary to speak more slowly and clearly than when carrying on an ordinary conversation. Use of dialect, the vernacular or tendency to hurry, should be avoided and one's voice should be directed to the back row. While notes are useful they should be in the form of headings, diagrams or reminders so that one does not appear to be reading though it is always possible to ask the audience's permission to read a given passage – 'Listen to this….' Be humble too, the audience usually knows best.

**specialization**

The pursuit of a special line of work, study or activity. To become adapted to a particular environment. To modify or utilize an organ for a special function. To adopt a structure that only permits certain limited function or functions.
See: specialize.

**specialize**

To pursue a relatively narrow aspect of a subject. Many people are beekeepers or have a knowledge of beekeeping but some 'specialize' in mead making, migratory beekeeping, queen raising, or some other specialist pursuit. Normally people only become specialists in a few branches of knowledge and are quick to identify anyone who professes to be familiar with it but is not. Few of the general public appreciate the specialization that beekeeping demands and regard anyone with a veil and a smoker as an expert.

**species abbreviated sp. Pl. species – spp.**

The smallest and only natural group within a genus. An assemblage of plants or animals with a very large number of characters in common, maintained and distinguished from other such groups by constant features. Natural populations that are potentially interbreedable. Frequently they are productively isolated from other such groups. Grouped within a Genera our honeybees belong to the group within Apis being called mellifera.

**specification**

A statement of exactly what is required. A list of materials needed to complete an object such as a floorboard, honey warming cabinet or any of the items that appear in journals for the DIY beekeeper.

**specific gravity s.g.**
The relative density of a substance compared with water. The ratio of density at a given temperature to the density of water at 4C. With s.g. of water as 1.00, beeswax averages at 0·963 and honey 1·414. S.g. at a given temperature and pressure enables us to determine other physical properties of honey as both moisture content and refractive index are linked with the s.g. One scale of measurement is the Baume scale.
See: honey density, moisture content and refractive index.

**specific heat (or heat capacity)**
The heat required to raise a given mass of a substance through one degree centigrade. Now expressed in 'joules' per kilogram per 'kelvin'. For liquid honey this is about 0·6 while granulated honey is higher. Because of its high specific heat beeswax when liquefied remains hot long after the source of heat has been removed.
See: joule, Kelvin, Newton and melting point.

**specific name**
Under the binomial system of classification each species has two names – generic and specific. The generic name identifies its family group and the specific name the species. Specie names are derived either from adjectives or nouns and always use the lower case even if derived from a proper name e.g. Portugal becomes lusitanica and should agree with the gender of the generic name Rosa alba for instance and.... The foreign aspect purpurea for purple and coccinea for scarlet can be off-putting but more acceptable than English so we may not know that sativa means cultivated and that pratense of the meadows. Several other common ones are explained under their respective headings.
See: genus.

**specific rotation**
Optically active substances like honey sugars have this property which depends on the concentration of the substance in solution in grams per cm cubed, the length of the path travelled by the light used and the temperature. Dextro-rotation is indicated with 'a+' and laevulo-rotation with 'a-'.

**speck**
A small particle differing in colour or texture from its surroundings. A piece of honeycomb may appear blemished by a speck of darker wax or tiny dot of propolis. A speck seen through the glass at the bottom of a jar of honey will cause it to be rejected as a non-prize winner. Specks can also be caused by frass or tiny droppings of creatures like ear-wigs and wood-eaters. When the debris left behind by wax moth, Braula, pollen mites or mice can be identified, this is a step towards being able to do something about them.
See: frass, skim and dysentery.

**specimen (uma.)**
A representative part (or example) of a substance, plant or animal. Not used in relation to honeybee colonies though an individual predator might be referred to as a fine, large, or nasty little specimen.

**spectacles**
It is not easy to don spectacles once a veil is in position so if you wear them make sure you have the right pair on before getting into your veil. Looking for a queen or evidence of her presence (eggs), calls for good or suitably corrected eyesight. Beware of conditions such as a poorly ventilated veil that could cause lenses to steam up and also of perspiration either wetting the glasses or making them slip off. It has been known for a poor-sighted person to be 'buzzed' near a hive and in swiping, to knock his specs. Into the grass. The ensuing embarrassment of searching , on hands and knees, while continuing to be stung, does not bear thinking about.

**spectrum of nectar sugars**
In rare cases sucrose can be the only sugar

present and in most cases it predominates. In Cruciferae sucrose is almost absent. In Castanea sativa, Robinia pseudoacacia and Trifolium pretense there is more fructose than glucose and they are therefore slow to granulate. With Taraxum officinale and Brassica napus glucose predominates and granulation is rapid.
See: nectar.

**spectrophotometric method**
This is a technique where solutions and reagents ($H_2SO_4$ and phenol) are mixed and the absorption of sugar can be calculated from the resultant colour using a standard curve. The development of such methods of assay, using new techniques and apparatus is forever on the move. The Journal of Apicultural Research has included many original papers covering such methods for making comparisons between parts of spectra.

**spectroscopy**
The study of a substances internal structure by the interpretation of a spectrum of light passed through it. Prism instruments and interferometers make this possible. The science of the use of the spectroscope and spectrum analysis. Mass spectrography has been used to differentiate between isomerose syrup and honeydew honey and X-ray spectroscopy for tracing the sources of trace elements in plants.

**spectrum pl. tra**
A broad sequence or range of related qualities e.g. complete range of colours as dispersed from light.
See: parameter and spectra.

**speed of the bee**
A loaded bee surprisingly flies faster than an empty one. Probably it gets more air miles per micro-litre this way. Out-going it could vary 11 – 29 kph averaging 20kph (just over 12 mph) while a loaded home comer would fly between 21 -25 kph averaging 24kph (16 mph). Their airspeed is faster when flying against the wind than with it. Work discontinues when winds exceed 24kph though they can, as many people have discovered, put on a burst of 40kph. Cross-winds not only reduce their ground speed but give large angles of drift.
See: airspeed and flying speed.

**Speedwell *** *Veronica spicata* Scrophulariaceae**
A perennial that grows to 35cm and offers a bounty of nectar and pollen in its clusters of ultramarine blue flowers on show from July to September. There are over 20 species, dry and water varieties. Speed you well, I bring victory – picking the flower will cause a storm A hairy creeping plant with upright flower stems.
See: germander.

**spelling**
Words that can easily be misspelt include: advisor, defecate, supersede, nuc (nucleus), repellent, Nasonov, movable, storey, honeys, propagate, pollination, thixotrophy, pollinizer with apologies for inconsistency of spelling in this work especially as American 'ize' endings come in so often.

**sperm (spermatic fluid)**
The male fertilizing element, reproductive cell or gamete. It has a minute body with a long vibratile tail. In the drone they pass down from the testes along the vasa deferentia into the sperm vesicles where they are temporarily stored with heads buried in the soft cellular walls of the vesicles. In the mating season spermatozoa are sent down through the ejaculatory duct in a secretion from the mucous glands and the mass fills the bulb of the penis. On transfer to the queen they are nourished, and maintained until required, in the spermatheca. A queen is able to keep them viable for several years. It is fairly certain that not one but several are released onto each egg that is to be fertilized.
See: male gamete, semen, spermatozoan and insemination.

**spermatheca (receptaculum seminis)**
Surrounded by a temperature resistant tracheal envelope it initially holds 5 – 7 million spermatozoa and sustained 36C will destroy fertilizing ability though not motility. Brood nest area is 36C but can drop to 32C. It has two glands and a pump, a spherical sac which, during the early part of a queen's life, can absorb hundreds of thousands of sperms (capacity said by Dade to be 7 million) and to be able to keep sperms viable for several years.
See: spermathecal contents, duct, gland, pump and valve.

**spermathecal area**
This includes: bursa, oviducts, sperm glands, sperm pump and valve, spermatheca, vagina and valve fold.

**spermathecal duct**
This duct leads from the twin glands to a median duct incorporating a pump and valve and thence into the vagina. It is instrumental in allowing the spermatheca to become charged with sperm during the queen's matings and then comes into play allowing the queen to fertilize eggs at will, a valve fold in the vagina bringing eggs into contact with the duct where it enters the vagina.

**spermathecal gland**
Two branches of this gland are looped over the dorsal surface of the spermatheca and their common duct joins the spermathecal duct above its opening into the spermatheca. Its function is initially to attract the migration of sperm by chemotaxis so that the spermatheca can be charged and in the laying queen the gland produces nutrients that keep the stored spermatozoa viable.

**spermathecal pump**
This lies external to the spermatheca as part of the spermathecal duct just below the spermatheca and incorporating a valve so that sperm can be drawn from the spermatheca and delivered under control as required into the vagina where the valve fold can offer eggs up to the opening of the duct.

**spermathecal valve**
The valve is closely associated with the pump at the 'S' bend in the spermathecal duct (below its connection with the spermatheca) The pump draws a small quantity of sperm into the 'S' bend and the valve enables the pump to force sperm down the duct towards the vagina.

**spermatogenesis**
The name given to the process of sperm formation. The origin or development of spermatozoa or the production of male gametes in the testes
See: oogenesis

**spermatogonium**
One of the primitive cells that give rise to spermatocytes (male cell at maturity)

**spermatophore**
A capsule of albuminous matter containing sperm.

**spermophyta**
A primary plant division of seed bearing plants. For non-honeydew nectar these include most useful bee plants.

**spermatozoon pl. spermatozoa**
An active gamete found in the semen. Those of the honeybee are found in the testes of the drone but move into the seminal vesicles as the drone becomes mature. During mating they are transferred to the queen where they move into and are stored in her spermatheca for availability when she lays female eggs. It is said that she places two or three sperms on each egg. They have the form of slender threads, about 0·25mm long and propel themselves by flagellating their bodies in a film of moisture. When in natural storage in the seminal vesicles of the male or spermatheca of the queen, they lie in swathes, side-by-side. They are fairly robust organisms and have been keep frozen

and transported for A.I. purposes by plane across the world.
See: male gamete.

**spermiogenesis**
Development of the spermatozoon from the spermatid.

**sperm storage**
Long range bee breeding programmes either call for the expensive maintenance of large numbers of queens or the long-term storage of semen from different lines. Unfortunately although the storage of sperms by the freezing method (-196C) in liquid nitrogen, results in the survival of a good proportion was not successful as some were damaged or killed and genetic damage attributable to this has shown up in some granddaughters of the queens inseminated with sperm stored in this fashion (1981). Doubtless needs up-dating

***Sphaerularia bombi***
Nematode parasite of bumble bees that develops inside a queen (up to 100 nematodes) which usually returns to their hibernaculum to die. The dissemination of the juvenile nematodes into the soil makes ready to infect new queens when they subsequently enter hibernation.
See: Nematoda.

**sphincter**
An annular band of constrictive muscles that can control the size of an orifice on a hollow organ like the oesophageal/ventricular connection or body orifice such as a spiracle.

**spice**
A pungent aromatic vegetable substance. Spiced meads, such as metheglin used: rosemary, ginger, coriander, aniseed, lemon or orange peel, mace, nutmeg, cinnamon, juniper, elder flower and cloves.
See: herbs.

**spicule**
A minute pointed or needle-like body such as the central hair in the bowl of the pollen basket.
See: corbicula.

**Spider Arachnida *** Araneae**
Friend or foe? Eight legged, wingless, predatory arachnids with two-part bodies (cephalothorax and abdomen). They will catch insects in their webs and wax moth are as vulnerable as bees but their webs can cause a great sense of annoyance even when bees are able to break away from them. The webs are very tenacious on bees bodies. There are so many varieties of spiders and doubtless they account for a large number of bee deaths. In double-walled hives there is usually one well-fed, obvious looking spider, sometimes referred to as 'the duty spider'. When killed it is immediately replaced by another. They do of course eat wax moths.
See: Arachnida.

**Spiller John C.**
A Somerset beekeeper who began about 1880. He was a builder's merchant, Lecturer and Judge and a great prize winner for honey. In 1952 he wrote '*House Apiary*' having used bee houses since 1913. He invented thick aluminium dividers to improve the capping of sections.

**spin**
1 To rotate rapidly. The forces produced by spinning were first used to extract honey by Hrushka. When an extractor is not evenly loaded or before there is sufficient weight of honey in the bottom to help hold it down, spinning may well cause the machine to wobble violently. As the honey gate is at the bottom it is not always convenient to bolt it to the floor. Cradles are available and 'T' shaped support platforms . Start by spinning slowly and increase speed after much of the honey has been expelled.
2. Bees tend to spin round when they get into water.
3. A larva spins its cocoon using the mandibula salivarium as the spinneret, which later become thoracic salivary

glands for making silk. 4. Bees will clasp one another in combat and frequently begin to spin around when they are locked in their struggles.

**Spindle tree** *** *Euonymus europaeus* **Celestraceae**

An attractive tree that grows to 20ft (6m). It likes chalk and can tolerate shade. It is most beautiful in autumn when laden with bright pink 4-lobed capsules which are ½" long and hang in clusters on short leafy shoots. The flowers are small, greenish-white, inconspicuous with 4 petals. The leaves taper to a point and are toothed round the margin. It is a host to scale insects and caterpillars. The wood is often used to make 'spindles' and high quality charcoal for artists.

**spine**

A stiff pointed process or projection. Many parts of the bee are arrayed with rows of stiff hairs answering the description and serving as brushes and combs for the cleaning of the body or movement or transfer of pollen or wax. Wasps being more aggressive have more.
See: spur.

**Spinnbarkeit**

Literally this means spinable but has been used to mean stringiness or fibre-like quality. The formation of long strings when honey is forced from the cells or a rod is dipped into the honey and moved quickly away. The property is also referred to as dilatancy which means that viscosity increases with stirring.
(opp. Thixotrophic)
See: dilatant.

**spinneret**

The larval silk glands unite in a common duct which opens into the spinneret as the larva's labium and is for spinning a silky thread to form the cocoon. Honeybee larva merely help to line their cell capping in this fashion, rendering it tough enough for bees to walk over and in the case of queen cells, to resist stings. The spinneret is found on the labium of the larva and becomes the salivarium in the adult.
See: salivarium, silk glands and spin.

**spinning**

On completion of the eating stage, the larvae turn lengthwise in their cells, heads outermost and set to work spinning a partial cocoon as the nurse bees help to develop a strong, porous capping over the opening. Whereas the larvae of many insects spin cocoons that completely surround their metamorphosing bodies, a honeybee merely helps to strengthen the inside of the cell covering (capping) by spinning fine silky threads of considerable strength. It is reasonably difficult (requiring a razor-sharp blade) to cut through the tip of a queen cell.

**spiracle**

A small opening (breathing hole) in the chitinous exoskeleton at the proximal end of the tracheae through which respiratory gases pass. In all there are ten pairs. Each of the three pairs on the thorax are different, the first pair facing slightly forwards being well placed to receive a 'ram-jet' effect in flight and having hair-fringed flaps. The second pair are small and inconspicuous being partly hidden while the third pair are almost identical with the seven pairs of the abdomen. Each of these is musculated though the last pair are on the internal sting apparatus. The first thoracic pair are vulnerable to invasion by Acarapis woodi (predatory mite) 'during an adult's first five days' of life after which the fringes of hairs stiffen. Larvae lie in their food only breathing from the spiracles on the upper surface of their bodies and care must be taken when 'grafting' to ensure that the small larvae do not drown by being placed, wrong-side-down onto a new bed of food.

**spiracle – abdominal**

The seven abdominal pairs of spiracles are similar to those on the propodeum. They are musculated (thus preventing access by the acarine mite) and can be opened and shut at will by the bee.

**spiracular plate**

This is an atrophied plate of the 8th segment and is weakly attached to the sting.

Spiracle – propodeal

The largest pair of spiracles in the bee are those of the propodeum (by origin an abdominal segment) and they play an important part in respiration.

**spiracles – thoracic**

There are three pairs of thoracic spiracles. The first pair are the places where acarine mites can gain access to young bee's trachea. The second pair belong to the metathorax and are small and hidden by the membrane in a notch between two pleurites. The third pair are the largest on the honeybee and are found on the sides of the propodeum. These are musculated and similar to abdominal spiracles.

See: breathing pores.

**Spiraea spp.**

A perennial herb with dense heads of small, cream coloured flowers. It likes wet ground, is sweet-smelling and regenerates easily.

See: meadow sweet.

**spiral cell-protector'**

A device shaped to accommodate a queen cell and made from coiled wire. The narrow bottom opening is just wide enough to expose the tough tip of the cell which is opened like a flap when the virgin emerges and yet which resists any attempt by the bees to pierce it. The cell's side walls are protected by the wire, yet open to allow incubation. A metal slide covers the top and a projection at the top end of the wire permits holding or fixing to the comb. As an alternative, 'micropore' tape has been wrapped around queen cells leaving the tip free as mentioned.'

See: West cell protector.

**spiral thickening**

The strengthening of breathing tubes, where two way passage of gases is required without use of muscles, is effected by spiral thickening. In the bee's trachea, spiral strengthening holds the tube open all the time and the actual thickening takes the form of a continuous, coiled, spiral cage of cuticle upon the chitinous wall, When dissecting the first pair of thoracic spiracles, it is seen that this permanent distension of the tubes makes a perfect home for Acarapis families and the taenidia can be pulled out like a spring. Human wind pipes are reinforced similarly.

See: taenidium.

**spirit level**

An instrument with a flat base and which incorporates a glass tube in a window showing a bubble in the fluid medium. It is graduated so that when the line cuts through the centre of the bubble, the instrument is level or if fitted with a tube at right angles to that one, 'vertical'. Judging hive levels by eye can lead to unnecessary error and once bees have accepted a hive and built combs, it is quite upsetting to then alter the level of the hive, In the absence of a spirit level, any container in which water level can be seen may be utilized.

See: level and vertical.

**Spiroplasmosis**

Caused by Spiroplasma an organism about which little is known (1978). Bees contaminated by the pathogen develop the spiroplasma in their haemolymph and usually die within 3 – 7 days. Tulip poplar and Southern magnolia were two of the plant sources of this trouble ABJ 1978. Phylogenetically it is classified as intermediate between bacteria and viruses.

**spiteful**

As this usually means malicious and associated with the desire to annoy or hurt another, it should not be applied to bees whose object in threatening us or using their stings on us, is to make us go away and thus protect their property.

**split**
A man-made division of a honeybee colony – hence 'split' board. Snelgrove designed an elaborate board with six individual entrances so that a colony could be split, allowed to rear a new queen and yet return foragers to the parent colony at intervals.
See: division board, Snelgrove, splits, split board and tilt.

**split board**
A board used horizontally to make a split or division of a colony without the need for any additional equipment. It is a flat sheet of suitable material such as ply, hardboard etc. framed all round except where a gap on one side forms an entrance for the part of the colony set upon it.
See: division board and Snelgrove.

**splits**
Colonies that are strong in bees may be 'split' into two or more parts. While there will ordinarily only be one queen in one of the parts, provided the other part(s) has sufficient food, bees and the wherewithal to raise a new queen (virgin, queen cell or eggs) new colonies can be formed. Warmth and economy of equipment can be effected by putting splits over division boards e.g. Snelgrove, though the making up of nuclei enables easy transportation to drone stud areas for selective matings.
See: division board, dividing colonies, making increase/ nuclei.

**splitting**
The separation of one hive box from another calls for the 'skilful' use of a hive tool because in the majority of cases the bees will have glued the two parts so firmly together with propolis that getting them part could seriously 'jar' the bees. As 'splitting' implies rending or cleaving, it will be seen as a suitable term bearing in mind that the propolis seal has to be broken. A quick , two chamber separation for queen cell checks is known as splitting and this calls for a lifting and levering of one box on the other so that queen cells can be seen if present in the lower part of the upper box. It is a quick but not infallible method of checking.
See: division, splits and 'tilt' method

**sponge rubber**
Also called 'foam' rubber. This was available in clean, convenient sheets which could be neatly cut into strips using a very sharp knife. As upholstery containing this material was found to be a deadly fire hazard (toxic fumes arise when it burns) it is no longer ubiquitous. However the multitude of uses it can be put to are worth mentioning. The spongy material can hold liquid and it has uses for applying certain cures. Its flexibility and ease of being torn into appropriately sized pieces mean that it can block a hair curler used as a queen cage, stop up a hole, crack or crevice through which you have no wish to let bees pass, pack round inside a honey jar to contain something delicate (queen cell, piece of comb, bee specimen etc.) and block entrances for travel or to totally enclose a hive for some reason . Use pieces of reasonable size so that they do not become loose and always remember the need for sterility.
See: foam rubber.

**spore**
A walled body containing the wherewithal to produce an adult individual. A minute cell that becomes free. Sometimes unicellular. A reproductive body as of fungi, bacteria and Protozoa. For example the spores of wild yeasts which if not sterilized can spoil mead and spores that can cause American foul brood sporulations that can have a serious effect on beekeeping. A sporicide is an agent that kills spores.

**sporopollenin**
The substance comprising the outer layer or coat of a pollen grain – exine. It is one of the most stable (resistant to the elements) materials known in the organic world. It is

formed by the oxidative polymerization of carotenes and carotene esters.

**sport**
A somatic mutation. In biology this means a genetic change or mutation, or an offspring with a characteristic not found in its parents. It is used to describe a new generation that does not run true to type. As the male parentage is so difficult to control in honeybee matings it is not uncommon to find offspring that do not run true to type though this does not imply that the new queen is a sport.

**sporulate**
To produce or release spores

**spotting**
In the sense of searching for and finding we use the word spotting a queen or her eggs or spotting a get-away swarm.
Washing, hung out to dry, may become marked when bees taking up moisture decide they need to defecate and this leads to spotting.
See: find the queen and worker egg.

**spotty brood**
Sealed brood that is spoilt by occasional holes which may be empty, contain nectar or have young larvae in them. These virtual gaps in a smooth area of sealed brood indicate that for one reason or another the original egg in that cell has failed. The finger of suspicion points to the queen who may have laid an egg with the wrong ploidy, she may be failing or might be diseased. It is always wise to cull such queens and replace with vigorous, healthy stock though if disease is the cause then appropriate action should be taken.
See: pepper-pot/shot brood.

**Spp**
Abbreviation for species.

**spray**
Finely divided liquid particles that can be blown through the air. Many chemical poisons developed for military purposes, included chlorinated hydrocarbons, phosphates and carbamates etc. some of which have been adapted and developed for agricultural use. Farmers are knowledgeable and careful. Unfortunately deadly sprays get into the hands who are either unknowing or irresponsible with seriously harmful effects on the most useful of insects – the honeybee.

**sprayguard entrance**
A device that allows returning bees to get back into the hive but prevents others from leaving. It permits limited ventilation and has a series of one-way (inward) bee traps incorporated and the whole thing is fitted in place of the normal entrance block. It is intended for temporary use while local foraging has been made unsafe by the use of insecticide. When bees are thus confined they must be provided with water inside the hive.

**Spray mortality**
The likelihood of mortality is ranked high or low according to various parameters. First as regards timing: High if done at mid-day, when warm, sunny and windless. Low if in late evening when cold windy or cloudy. Then as regards flowering; High at full flower and when attractive weeds exist. Low when there is no flower and crop is clean, Bee's flight; High when flight path goes over crop but low when foraging is away from crop. Pesticide; High when liquid, high bee toxicity, when there is fumigant effect or when persistent, but low when granules are used, liquid toxicity low or a harmless formulation is used.

**spray poisoning**
When a colony suffers from a pesticide application, send fresh sample of 200 dead bees in a ventilated, stiff cardboard box to Fera. It is most helpful if as many of the following details as possible can be given: Nature or trade name of spray, crop treated, hectarage, material responsible, method of

application, colonies affected, other colonies known to be affected. Label samples clearly with date of collection and sender's name and address. Photographic evidence is especially helpful.
See: pesticide poisoning.

**spread brood**
To increase the natural volume of brood by physically interfering and moving brood, usually with a view to encouraging the queen to lay in the area opened up. As such action might easily cause chilling of the brood it can go seriously wrong should unsuitable weather ensue. Judgement calls for experience and skill. Reversing two brood chambers can also lay itself open to the chilling of brood.

**spread diseases**
The nature of disease is such that it exists because it is able to spread from one living organism to another. The manner in which this occurs depends on the structure and function of the pathogen. Knowledge of these factors enables us to limit disease spread. Viruses, spores, bacteria, germs, mites and other parasites can usually be destroyed by heat, fumigation or the use of prophylactics. Used equipment, hive products, queens, bees, swarms and the beekeeper himself, can all act as vectors in the spread of disease. Be exacting rather than trusting.
See: apiary hygiene.

**spread honey**
A little honey, being sticky, can go a long way. Wiping with a damp cloth can leave stretches stickier than ever. Fortunately it is quite soluble in cold water and with a little circumspection a garment or utensil can be washed thoroughly clean. Cleaner often than it was before. To produce a 'spread' on toast or bread it is easier when the honey is not too thin. The viscosity of runny honey can be markedly increased (temporarily) by a short spell in the refrigerator.

**spring**
The season that follows winter and leads to summer. It is more clear-cut in temperate latitudes but varies in relation to the mass of land and the surrounding sea. The southern hemisphere of course has seasons reciprocal to ours. Bees do not wait until the spring to stir themselves from their winter cluster but, according to the increasing length of daylight, will start developing a brood nest. This requires bees capable of bringing their brood food glands into play and the demand for water to dilute their honeys stores becomes paramount. Some can be obtained by their opening cells and allowing the honey to absorb hive atmospheric moisture. This can have the appearance of 'incoming' nectar. The overwintered bee's fat body comes into action but fresh pollen at this time is worth its weight in gold.
This is where the behaviour of both bees and beekeeper in the previous autumn count for so much. An apparently strong lot can go down almost overnight if they run out of food, so don't let this happen. 2 parts sugar to: 1 of water, feed tepid, directly into the brood nest at dusk, can save the day. Watch their requirements for space too.
See: Mar, Apr, May.

**spring balance**
A weighing scale. Although lever types are also common, the spring balance is lighter and in many cases more easily adapted to the weighing of beehives. The passage of years has brought in so many more accurate measuring devices. However, a record of hive weights can prove a useful guide to the choice of best sites and a year by year comparison of weights – most interesting. To those unable to assess by 'hefting' a hook and spring balance used by tipping up one side and adding it to the reciprocal reading from the other side can be a good guide. Accurate readings can show the departure of a swarm, say when a super is required or feeding necessary but most beekeepers will never have weighed a hive…

**spring block**
A pair, clamped behind a 'follower' board in a section crate will squeeze the sections up tight, The small wooden block has a projecting spring which can be forced twixt follower and wall.

**spring development factors**
Warmth, food supply, pollen, flight, light, sufficient ventilation and humidity and most of all a 'good queen'.

**spring dwindling**
Old, over-wintered bees can die of rapidly when early spring duties involve external work. Those with experience enough to go out for pollen or water can be easily overwhelmed by the elements. Following aspects like CDC, a setback such as a big loss of bees, will hold any colony back if not actually lead to their destruction.

**spring feeding**
Where spring feeding is resorted to and it is usually much wiser to have over-wintered with a good surplus of usable stores, but 'needs must'. A colony will be as happy to get the water as the sugar so use dilute syrup (pound to pint). Super soft candies are a temptation but tepid syrup put over brood nest at dusk and well wrapped to retain colony warmth can avoid a tragedy but care is needed as feeding encourages more brood and unless the cold relents more feeding may be called for.

**spring management**
A clean floorboard (scraped and blow lamped) should be given during a dry spell in February or March. The mouseguard can come off once pollen is going in (though this might be delayed where night frosts are still continuous). Keep entrance size reduced until colony advanced. Learn all you can from external observations like, warmth over brood nest, pollen going in in good quantities, no suspicious debris being thrown from entrance, a tap on side responded by a hiss and not a moan and take both weather and flora into account. The queen must not be restricted for space, nor bees hurried or delayed by lack of or over-anxious setting up of super space. The flight of first drones warns that swarming is now possible. A young queen and space just ahead of the bees should minimize swarming risks. Once you have drones think about queen rearing forthwith. Excellent queens can be reared with little or no interference with your potential honey crop. See: supering.

**Spruce *** *Picea alba* Pinaceae**
The Christmas tree or Norway spruce. These conifers produce vast amounts of quantities of highly prized honeydew in European forests. This is part of a symbiotic process in which insects (such as lachnids) pierce the trees and produce sweet secretions that are collected by bees. 'Migration to these areas may lead to hives deficient in pollen which can be rectified by giving them combs of pollen from normal colonies as the flow goes on until well into autumn.

**spun honey**
Honey that has been thrown out of the comb or from cappings by the process of spinning. Early extractors were called slingers or spinners.

**spur**
A sharp, stiff, chitinous outgrowth on the legs of certain insects. Only one conspicuous and possibly vestigial pair are found on the tibia of the middle leg of a worker whereas wasps and cuckoo bumble bees seem to have them sticking out aggressively from their legs. Botanically the name spur is also given to a process of a petal or sepal which functions as a nectar receptacle. See: extra-floral nectary.

**spur embedder *****
A device used to embed the wire into wax foundation. One model has a brass wheel with spurs, the tips of which are grooved to guide the tool along the wire.

A convenient wooden handle supports the wheel at the end of the arm. It is sometimes found useful to keep the metal warm by dipping it into hot water before use. See: Woiblet

**Spurge *** Euphorbia spp. Euphorbiasceae**

A plant whose greenish-yellow flowers are not particularly conspicuous. There are many species e.g. Caper spurge E. lathyris which flower in May/June and provide a dark brown pollen.

**square**

Having right angled corners. When this feature of beehives is marred by lack of care in handling boxes, accidents etc. it should be rectified. Steel or plastic corner brackets will strengthen and bring sides back into line so that frames can be aligned truly parallel to the side walls and bee space adhered to.

**squat**

To sit in a crouching position. To adopt this position while examining the brood combs can prove uncomfortable for most people. A strong , flat-topped roof may provide a temporary seat or the chamber being looked into raised by standing on an upturned roof. Such methods help to avoid leaning forward with back bent for any length of time. When examining combs the common fault of holding them flat, should be avoided. Combs are built vertical, should be kept vertical and held at eye-level with the light shining over one's shoulder so that eggs and all else can be clearly seen. Minimize twisting when holding something heavy.

**squatters**

Those who settle on land or building without obtaining the owner's permission. One could look upon a swarm as behaving in this way when it goes off and finds a home on someone else's property or in a spare hive left for the purpose. Hives left empty should either be sealed during the swarming season or left with healthy combs properly spaced and aligned. Many a good swarm has moved into a hive and built its comb around unsuitable frames, on-the 'skew' and then been quite difficult to re-domesticate. Beekeepers should accept responsibility for collecting swarms and not allow them to move into properties and embarrass the residents. See: stray swarm.

**squawk**

Perhaps this word describes a sound somewhere between a squeak and a croak. At all events it is sometimes used when referring to the responsive sound made by virgins still in their queen cells when a piping queen on the comb challenges her rivals. If not interfered with by the bees or beekeeper, the squawk may well lead to the captive virgin's demise as the challenger can kill with impunity, stinging through the vulnerable sides of their cells.

**squib**

Or 'silver' were names given to a swarm that issues after August.

**Squirrel *** *Sciurus carolinensis* (grey), *vulgaria* (red)**

Although squirrels do not normally interfere with bees, where hives are kept in or near trees, the creature (which does not hibernate) can cause a disturbance to the quiescent stock while it is clustering by dealing with nuts and fir cones while sitting on the roof or alighting board of a hive. See: red squirrel.

**Sri Lanka**

Formerly Ceylon here all four varieties of the honeybee are found. In one spot, on a location known as Lion rock, a wire cage had been provided for protection against possible forays by A.scutellata, after a notable person had been 'attacked' upon the rock. Predators abound. The highlands are surrounded by coastal lowlands and swamps and there is much jungle. Honey is regarded as a special, almost 'mystic, food.

**stadium pl. stadia**
The stage of growing that a larva passes through between moults (ecdyses).
See: instar.

**staggered spacing**
When precise, close spacing of frames is required, for instance when new foundation is to be drawn into worker comb, we may set spacers such as the ubiquitous metal (or plastic) end, so that each alternate pair overlap the ones adjacent. Once comb has been started one can revert to normal spacing. Close ended frames like Hoffman cannot be used in this way.
(diagram?)

**syrup**
When sugar is dissolved in water with a view to feeding bees it is called 'syrup'. Dilute (pound to a pint) is used for spring feeding or when bees can make good use of the extra water. When fed to bolster their stores as for winter feeding the denser solution is appropriate. At 2 ½ lb. to one pint saturation point is reached (c.66% sucrose). 2 : 1 is normally recommended and this has a s.g. 1·32. Honey is 1·414. The quickest method is to bring the required amount of water to the boil and then stir the sugar in until completely dissolved.

**stagnant**
Water or air that is not free to flow and collects in the open. Those who know the bee is one of nature's cleanest creatures and whose food products are eaten without hesitation, are sometimes astonished to see the sources from which bees will take water. In truth bees do not seem overly fond of absolutely pure water and are probably sensitive to ingredients that could make such water less acceptable to us. Their sense of taste and smell and their ability to filter and sterilize anything they imbibe means that the question of their using anything stagnant does not arise.

**staining of washing**
Bees can be attracted to washing hung out to dry, particularly in spring when safe sources of water are so valuable to them. Warm from the sun, with lower tension due to minute traces of detergent and offering an excellent foothold - no wonder if they receive insect visitations. Bees take the water into their honey stomach and it is not surprising that this might trigger off the desire to void their faeces. Soiled material should be washed with clean water as soon as possible. Those who keep bees in built-up areas may consider making special water fountains for their bees.
See: water fountain.

**staining combs and frames**
Unless bees are confined to the hive for six weeks or more the marking of combs and hive parts should not occur in healthy colonies. Dysenteric signs are possible when diseases such as acarine, nosema and amoeba are present as irritation coupled with the need to void faeces can lead to the release of faecal matter within the hive.
See: cleansing flight and defecation.

**staining hive exterior**
When a hive becomes marked with dysenteric droppings a sample of bees (and possibly a scraping of the faecal material) should be sent off for diagnosis. Undue excitement as for example when a small nucleus is made up, can lead to the area around the front of the hive being marked. Obviously this is more apparent on light hives than darker ones.

**staining – microscopy**
Staining techniques used to colour certain types of tissue so that they can be examined with greater clarity depend on whether water or alcohol stains are used and conditioning is needed.
Background staining to give contrast is used, such as the illumination of white nosema spores against a black background (nigrosin). As mentioned

stains must be soluble in the medium so as to preserve or fix the slide material. Basic fuchsin, methyl blue or green etc. See catalogues of staining schedules.

**stainless steel**

A hard steel alloyed with chromium and proof against rust and the attacks from honey which when dissolved in water brings about the discoloration and degeneration of most other metals. Therefore, although expensive, most long-lasting forms of equipment for dealing with honey or wax (and where plastic materials are not suitable) are made of stainless steel. For example, honey extractors and tanks, uncapping trays and wax extractors.

**stalk**

The main stem of an herbaceous plant. Dried stalks of cereals and rushes were much used in the past for skeps, caps and even hollow kex for feeding bees (as we now use drinking straws). Some solitary bees make their nests in the hollow stems of plants and artificial blocks cut with hollows of the same dimensions have been successfully used to move solitary bees into areas for pollination. See: APC bee.

**stamen pl.stamena**

The pollen bearing part or male reproductive organ of a flower comprising a stalk or 'filament' that bears at its tip the pollen producing anther. In the developing stamen the pollen mother cells divide twice making four cells. At one stage there is reduction division (meiosis). The grains are then in the form of a tetrad at which time the architecture of the external pollen coat is formed.

**stamina**

Ability to withstand or endure disease, long periods of effort, abnormal conditions etc. Strength of physical constitution. To quote Bro, Adam 'All-out fecundity usually, though not invariably, means offspring are short-lived and lacking in stamina. See: qualities of a good queen.

**staminate**

It describes a flower, frequently of a dioecious plant such as holly, which bears male stamens and pollen but does not produce seed or fruit.

**stamp**

To bring the foot down sharply onto the ground. This might be done to exterminate a predator or to break a piece of material for use in or on a hive. To stamp near hives is likely to cause the bees to investigate. Foundation is put through rollers to 'stamp' or emboss the shape of honeycomb cell bases. Postage stamps bearing bee pictures rarely appear on our stamps but do quite frequently on stamps of other nations.

**stand**

1. A platform devised to keep a hive clear of the ground. This has several advantages (a) it improves the microclimate of the entrance.
2. Keeps entrance clear of weed growth.
3. Gives the beekeeper a better working height.
4. Protects the floorboard from damp. Also used to describe a mass of flowering plants for instance a 'stand of rosebay willowherb'.

**standard**

The uniformity of dimensions that is called for by the figures set out by the British Standards Institute. That includes: hives, frames, honey quality etc. Botanically the central petal of a pea-flower. A tree or shrub not supported by a wall.

**Standard abbreviations**

In 1978 BIBBA issued cards setting out encoding and decoding letters to cover a variety of beekeeping aspects so that accuracy and brevity could be obtained. Labelled RC7 it was intended for use with bee breeding record cards.

**stand-still**

(Freeze) – The constant movement of bees inside their hive can be brought to a temporary stand-still when a vibration set up by a queen's 'piping' occurs. This can be simulated by causing a squeak on the glass of an observation hive when bees will dutifully stop moving. It is linked with the possibility of a queen being able to move through an otherwise jostling mass. She also has the ability to feign 'death' by adopting an attitude resembling that of a dead queen only to recover again when conditions return to normal. Another stand-still situation is when Fera imposes a an order on an apiary where foul brood has been found, which forbids the movement of bees or beekeeping material into or out of the apiary during the period that treatment and sterilization is carried out. See: akinesis and freeze.

**staple**

Large, rustproof, coppered staples that do little damage to a hive are made to be driven in at angles to one another to secure one chamber to another for transport. The use of staples in general, has advanced to the use of stapling guns of various sizes to make driving them home accurate and simple. See: stapling gun

**stapling gun**

In the 70's devices came onto the market enabling staples of various sizes to be fixed rapidly and accurately, This made them available for all manner of beekeeping uses. Glued staples can help to make strong, weather-proof joints.

**stars**

Distant, celestial bodies that were used to check seasonal happenings before calendars and clocks became available, Columella makes many references to stars. Bees were sometimes thought of as stars around the honey pot (moon). Pliny an ancient writer remarked that when Sirius was shining brightly it betokened a honey-flow. See: aqua mulsa, Columella, hydromel, Pleiades, Pliny the elder and Sirius.

**starch $C_6H_{10}O_5$**

A white, tasteless solid – linked units of D-glucose. The common carbohydrate produced in plants and stored in seeds. A translucent jelly once used for stiffening clothes. It is convertible into sugar and enters into various forms of food. Starch-gum is another name for dextrin. A starch granule consists of a core of amylase enclosed in a rind of amylopectin. The former can be hydrolysed completely into glucose while the latter has first to be phosphatized by phosphatase before it can be hydrolysed.

**starter or primer**

1. A thin strip of comb or wax fastened into a frame or section to encourage the bees to build their comb where the beekeeper wants it, yet without the need for a full sheet of foundation. The subsequent 'natural' comb has a softer septum than that which is built on foundation.
2. Finely granulated honey used to 'seed' a sample so that it will crystallize well.
3. A small quantity of the sweet liquid with yeast enough to trigger off fermentation with a view to adding to the 'must' before fermentation has become well-established.

**starter colony**

A queenless, vigorous colony in which grafted larvae are put so that queen cells are initiated. After some hours (usually 24) the grafts are transferred to a cell-building colony. Such a colony can be made by rendering an existing colony queenless or by removing bees and brood from several colonies and placing them in one hive. A frame of newly grafted cells is given every three days and finished cells removed on their 9th day.

**starting beekeeping**

The conscious acceptance of responsibility for a colony of bees. Fera has pamphlets

'Advice to intending beekeepers'. In 2010 the sudden rush by so many people to take up beekeeping has stretched educational facilities to the limit. It is hoped that this 'dilution' can be absorbed and adequately coped with. Any kind of 'impatience' instead of pondering may lead one into trouble. Siting, type of hive, access to proficient help, patience and perseverance - it is not easy to put that in a nut-shell.

**starvation**

Death caused by the deprivation of food. Colonies are most vulnerable when the brood nest has been developed and inclement weather occurs in early spring. March is notorious for colony demise unless a reasonable food reserve has been assured. Food reserves should never go below 15 pounds (the equivalent of 3 full, standard, deep, British, national frames), and wintering in Britain calls for 40 lb. A colony would 'feel' light long before starvation occurs. 'Hefting' should help to avoid such a crisis. See: massacre and starved bees.

**starvation swarm**

Rarely, in the UK at any rate, do bees that are short of food leave their hive 'en masse' on that account though when this occurs it is called a starvation swarm. Honeybees in some countries are known to swarm seasonally when a period of dearth is about to overtake them. When death comes to a colony through starvation most of the bees will be found with their heads tucked into the cells, there is no energy left for them to swarm. See: starved bees.

**starved bees**

Bees dead from starvation have, on examination have less glucose and fructose in their thoraces thus enabling them to be distinguished from those killed by insecticide or cold. Many die with their heads in the cells and the queen will be found amidst a small group of those who tried to the last to save her. Be careful though, because even the slightest movement remaining should encourage you to sprinkle them with tepid syrup and bees have been brought back from the brink of death this way.

**static (electricity)**

An electrical charge can develop, usually through friction of any substance that is not earthed. A free flying bee may return to the hive with a different charge from that of its nest mates. It is often noticed that colonies are extra 'touchy' on days when there is a lot of static about. Returning bees may even be regarded as alien on this account. See: electrical charges.

**statistics**

The interpretation of numerical data and facts derived from these. Like other sciences it has its own technical terms, for example $P \geq 0{\cdot}001$ means a probability of less than one. It is useful in research and in discovering the most likely way to achieve chosen goals. Here are one or two examples: With a crop capacity of 100mg a honeybee only carries a nectar load of 20 – 40mg and so 12 to 24,000 journeys are required to a pound of nectar and 10 – 50,000 for a pound of honey. From this we could say, statistics suggest that honeybees must make thirty thousand journeys to produce a pound of honey. Then again, with a water load of 20 – 60mg a bee would require 8 – 24,000 journeys for 1 lb of water, that is 10 – 3,000 for pint and 14 – 42,000 for a litre.

**statutory instrument**

A statutory rule or order such as laid before Parliament, an instrument being an agency by which something is effected – a formal legal contract or document setting out rules or conditions. A Bill becomes an Act when it has passed through its first and second reading in the House of Commons, then Committee and Report stage before a third reading all repeated in the House of Lords. Finally it becomes

an Act when Royal Assent is given.
See: Parliament.

**steal**
To take without right, or to insinuate something unobtrusively into a new place. The latter applying to re-queening and the former to the removal of honey from bees especially when there has been no compensatory care and attention on the part of the beekeeper.
See: robbing, rustling and stealing nectar.

**stealing nectar**
If we consider nectar as an insect's reward for servicing a flower then, when a 'back avenue' approach is made and the corolla's tube is nibbled through from the outside without any pollen transfer, then this is referred to as 'stealing'. Where the corolla is deep or difficult to enter, bumbles and other insects may cut a hole to obtain the nectar. The mandibles of the honeybee have great difficulty with flexible plant material but will use a hole already made by a bumble bee. Honeybees have been seen to act in this fashion on field beans but careful observation showed them to be outnumbered by those that entered the flowers legitimately for pollen.
See: begging, cheating, cleptolecty and robbing.

**stearin $C_3H_5(C_{18}H_{35}O_2)_3$**
A glycerine ester of stearic acid. A soft, white, colourless, semi-solid which is the main constituent of the more solid fats such as mutton suet. When making objects from beeswax, candles for example, it is useful to add a small quantity of stearin (about 10%) This is frequently done when beeswax and paraffin wax are used together. It gives the candle an opaque appearance, helps to harden the beeswax and causes a slight shrinkage which helps when releasing from the mould.
See: beeswax candle and resin.

**Stem weevil *Ceuthorhynchus quadrilens***
OSR pest

**step comb**
Connects comb or frame to the one above, or the floorboard to a bottom bar. Correct beespace should minikise this sort of wild comb also known as 'burr' and 'brace' comb.

**sterilant**
An agent used to render anything free of living germs or pathogens. Oxyfume 12 utilizes ethylene oxide as a gaseous sterilant. The fumes of formalin can be used to treat supers when free of stores. Any of the strong disinfectants or temperatures of 0 to minus 5F kills all stages of wax moth in 2 hours or 115 – 120F in 1½ hrs. Gamma radiation, acetic acid, scrubbing solution, hibitane, umagil tablets, ETO, formalin, umagillin are further examples of substances used.

**sterile**
1 Not fertile – unable to reproduce sexually. Barren.
2. Aseptic – free of living organisms. Hence sterilize and sterilization.
   See: sterile egg.

**sterile egg**
An addled egg. One that fails to hatch. Normal eggs hatch after three days of incubation within their cells. Unfertilized eggs produce drone larvae and fertilized eggs become female larvae. When the development of an embryo begins within the egg but a larva fails to hatch, this is a sterile egg and in most cases it is due to a genetic fault on the part of the queen though desiccation or chilling may also be a cause.

**sterilization**
Any form of sterilization while bees are present on combs would be of a prophylactic nature such as the use of miticides.
Once equipment is free of live bees, the opportunity should be taken to lightly scorch woodwork – there is no need to blacken or damage the wood –but metal

parts can be boiled and plastic washed or fumigated. Fumigants will depend on facilities, acetic acid may be used against nosema or formalin when there are stores present. As burning is still considered appropriate for American foul brood in this country heavy atmospheres of ETO are less likely to be used.
See: method of destruction, nosema, comb sterilization.

**sternite**

A sternal sclerite or ventral plate of an arthropod segment. The ventral plates of a bee's abdomen are the sternites which overlap one another in streamline fashion and are in turn overlapped by the tergites.

**Stewarton hivern**

Ayrshire hive. Designed, used and sold by Robert Kerr a Scottish cabinet maker 1755 -1840 consisting of tiered, octagonal boxes with fixed top bars having sliding strips between them, the centre comb being longer than those on the flanks. This became the first reference to body boxes, a clear forerunner of to-day's movable frame hives.
See: Kerr.

**stick (verb)**

To fasten two surfaces together so that they adhere to one another. To a small creature like a bee water proves much stickier than we can imagine consequently good footholds should be present when water is provided in the open air so that the water cannot over power and pull the bee in. Honey's sticky nature (and propolis too) call for immediate remedies and a wet cloth and certainly a supply of water are more than useful when dealing with hives. Handles, even the steering wheel of the car,
---- it is unbelievable how easily sticky honey is passed on from one to the other. The gum-like tenacity of propolis, not noticed until new equipment has received a coat, demands that a sturdy hive tool is to hand, otherwise separating hive bodies can be near to impossible.

**stigma**

1. That part of a flower's carpel that is receptive to pollen at the apex of the pistil, often sticky. The female receptacle for pollen.
2. Also imputation attaching to one's reputation e.g. no stigma having had foul brood in one's apiary.
3. A thickening on the costal border of the wings near the apex.

**stimulate secretion**

To arouse interest or provide effort or action. (a) interest in beekeeping. (b) bees to forage. (c) queen to lay. (d) colony to develop. (e) pollen collection.
As follows:

(a) Education via the media, Honey Shows, County Agricultural Shows, and personal help and co-operation.
(b) Bright light, good weather, strong colonies, using a strain with a strong inclination for storing,
(c) Length of days, dilute syrup feeding, a healthy young queen, space for her to lay in,
(d) A vigorous healthy queen, adequate number of bees, good comb, a suitable environment, stimulative feeding or a nectar flow.
(e) brood pheromones, the feeding of dilute syrup, a suitable environment.

**stimulation of nectar secretion**

Both raspberry and alsike improved when aqueous borax solution was used. Red clover benefited when 2kg boron with 30kg Phosphate and 40kg fertilizer were used per hectare. Also spraying with succinic, ascorbic and citric acid and ammonium molybdenate, While potash manuring decreased the nectar secretion of red clover, cotton benefited from superphosphate, and Alfalfa gave more nectar when treated with herbicide.

**stimulative feeding**

Syrup feeding in conditions that approximate to a nectar flow i.e. reasonably

weak (½ : 1 or 1 : 1) sugar syrup fed directly to the bees slowly but continuously while the stimulatory effect is required. Usually done to cause colonies to make an early and or rapid start in the spring and to encourage mating and subsequent laying. It has been noticed that such feeding seems to encourage foragers to bring in more pollen than they otherwise would.
See: spring feeding.

**sting *** (terebra) U.S. 'stinger'**
A sharp pointed, venom bearing organ consisting of lancets, barbs and venom canal. The action of causing a sharp, distinctly noticeable pain. The worker's sting lies in the sting chamber and consists of three pairs of plates, articulated so that they lever the lancets and accessory parts causing a puncture in the victim's skin through which venom is pumped. A venom gland ducts poison into the bulb of the sting and it passes through the shaft formed by the two barbed lancets which ride on the sharply pointed stylet.
See: remedies, sting components and function.

**sting (wound)**
The area affected by the sting of a bee. Despite the minuteness of the actual lancets and quantity of venom, humans generally pay great heed to the warning that bees give us by stinging. Most people who do not keep bees and many who do, remember all the details of their first sting. Once the sting has been scraped out, a small red dot becomes visible at the point where it penetrated and a pale weal forms as a circle around it. A small dot of blood may be drawn and a small whitehead may subsequently form. Depending on the individual degree of immunity, the area may swell and the affects may reach other parts of the body. Irritation may occur but the release of histamine which causes the swelling can worsen to the point of disfiguring the face or making a limb feel extremely uncomfortable. This may take hours to occur and can be quite worrying. It does not help at all to be told that in the very, very rare cases of death following a sting, that always takes place within a half hour because the sufferer might feel so bad that they want to see a doctor. Anti-histamine helps but that would have better been taken before the sting. After 24 – 48 hours the effects usually rapidly diminish regardless of the treatment received.
See: sting antidote.

**sting antidote**
Proprietary medicines are usually sold to cover honeybee, bumble bee, wasp stings and even mosquito bites, despite the fact that the venom from each is quite different. Avil, Benadryl, Piriton, Drenamist, medi-haler, Epi isoprenal and Becotide !!! Beekeepers who are stung regularly over years and years benefit from better health than most. (The writer is over 94 and testifies to that). Unfortunately there are those whose reaction is bad enough for them to declare themselves 'allergic'. Naturally they try to keep away from bees or wasps but this is not always easy. Anyone who fears getting anaphylactic shock should obtain an adrenaline injector from the Doctor. The number of 'old wives' remedies are so many and varied, often proving contradictory that to list them would be quite pointless. 'He is not worthy of the honeycomb who fears the bee's sting'.
See: sting 1st aid.

**sting at rest**
When not in action the worker's sting is entirely retracted within the sting chamber with the shaft turned up so that its base is concealed between the oblong plates and its distal part ensheathed between the two projecting lobes (sting palps). The unbarbed sting of the queen functions as an ovipositor but can be used when she confronts a rival though her venom deteriorates as she ages.
See: sting function.

**sting cavity**
Chamber or pouch – the aperture at the distal end of the abdomen in which the sting

apparatus is held. It is suspended in the membranous wall of the chamber – ready for use!

**sting components**

Glandular system, venom and reservoir, the sting shaft and associated tissue. The shaft consists of three components which taper together to form a sharp, hollow point. The stylet with track-and-groove connections, along which the lancets can move back and forth. See: sting at rest and venom components.

**sting (first aid)**

The immediate 'scratching out' of the small venom apparatus (which appears as a white, pulsating sac). A blunt finger nail, drawn firmly and quickly over the skin, should slide the sting out without squeezing in any more poison. To pick or pull out the 'thorn-like' sting would cause more poison to enter the wound. The tiny puncture is filled with antiseptic venom designed to give you a very distinct message and the hole closes and is not receptive to other material so the application of skin treatments is really a waste of time while oral applications such as aspirin or anti-histamine aerosol or tablets are the best approach. See sting treatment. A person may be disturbed to find that the swelling and throbbing continues into the second day but it will generally subside after that whether or not further remedial action is applied.
See: allergic, hypersensitive and treatment.

**sting function**

The sting of a worker bee is protruded by a complex plate and muscular action. Venom is pumped through a central poison canal, two lancets slide back-and-forth along a track-and-groove connection to the stylet, barbs prevent the lancets extraction from the wound and the sting shaft penetrates until the venom sac touches the victim's skin. The bee, in trying to make a second attack, tears itself away from the entire venom apparatus and returns to bite, claw and worry. Its life is now curtailed and it is unlikely to return to or be of further use to the colony. For this reason younger bees are disinclined to sting is because they have their whole life ahead of them.. The apparatus, even from a recently killed bee, still functions. Should a bee fail to pull away it rotates on the skin and twists out the barbs if allowed to do so.

**sting glands**

Two glands, the venom gland and the alkaline gland, are associated with the sting. The alkaline gland is a whitish, strap-shaped organ whose secretion is poured into the sting chamber. Dade refers to another conspicuous gland of unknown function, attached to the ventral gland of each quadrate plate. Lubrication of the sting is a possibility.
See: alkaline and venom gland.

**stingless bees**

Numerous varieties exist and many are worked for their honey and wax. Most are found in warmer countries – especially Brazil. There are: Meliponini ' Melliponida. Trigona iridipennis and hockingsi, Nannotrigona and in Africa Apotrigona. Some can bite you others are reasonably passive. Special hives are sold according to type and their honey and wax is often highly valued. One is remarkably similar to but much smaller than A. mellifera and its minute pollen baskets fill up as they work for all the world like their larger cousins.
See Melipona/Brazilian.

**sting palps**

Two long, soft, finger-like lobes are attached posteriorly to the oblong plates of the sting. The bee determines by 'feel' whether or not the flesh of a potential victim can be penetrated. Glass, tough rubber, and plastic are not attacked whereas anything leathery or soft, especially hairy, is liable to initiate the stinging reflex.

**stings (effects of)**

A sting usually elevates the blood plasma cortisol level two or three times above normal for about ten days. Then the level returns to normal. If stings are received oftener than at ten day intervals the cortisol level stays high giving protection against allergic reactions. More than one sting a week usually enables the receiver to develop immunity. Lack of certain vitamins, especially B5, is said to increase the possibility of sting reaction. Royal jelly is said to be rich in B5 – pantothenic acid.

**sting reaction**

The normal response to a bee sting usually depends on whether the body has encountered that particular kind of protein before. There may be local redness, pain sensation and swelling with the possibility of a more general reaction in the form of skin rash or asthma. An allergic reaction might follow this pattern: 1st sting ‘ swelling, 2nd whole limb swollen, 3rd Generalized swelling, nettle rash and in some cases asthma, 4th collapse and unconsciousness, 5th serious collapse with incontinence and unconsciousness while a further sting might bring immediate collapse and possibly death. Anaphylactic shock arrests the heart and immediate help is essential by chest compression on a flat surface.
See: allergy, antidote, stings and venom.

**sting treatment**

The immediate scratching out of the small venom apparatus. A blunt finger nail drawn firmly and quickly over the skin suffices to scratch out the sting without squeezing in any more poison as would occur if tweezers, biting, picking or sucking out were employed. The tiny puncture is filled with antiseptic venom remember, so go ahead at once and ‘scratch’ it out. Subsequent application of a proprietary anti-histamine by aerosol onto the spot and the swallowing of a recognized anti-histamine tablet will also help. Where none of these are available no lasting harm will occur unless the victim is ‘allergic’. The normal person may be disturbed that the swelling and throbbing should go on into a second day but it will subside after that, whether or not further remedial action taken. Where only swelling occurs a beekeeper may adopt a stoic attitude and eventually develop immunity. Help can be had either by taking two aspirin tablets or an anti-histamine tablet before going to the apiary.
See: anti-histamine, immunotherapy and sting antidote.

**stipe pl. stipites**

The proximal segments of the maxillae. They are hinged to the cardines and swivel the maxillae over the glossal complex so that in conjunction with the labial palps the tube that surrounds the tongue and completes the proboscis, is formed.
Stipule One of a pair of small leaf-like appendages at the base of a leaf-stalk.
See: bract .

**stir**

1 A colony of bees is often referred to as having been ‘stirred-up’ after quite normal manipulations have been carried out. It is a fact that bees will go quietly about their business if they are not disturbed, opening up and taking frames out is an entirely different matter and a colony will then pay much more attention to passers-by or others in the vicinity.
2. Sugar syrup should be prepared by bringing water to the boil and then stirring in the sugar until dissolved.
3. When ‘seeding’ honey a fine grained sample should be stirred into a similar amount of the honey to be seeded and stirred to equal consistency before being stirred into the top of the bulk to be seeded.

**stock**

1 A loose term to describe a colony complete with its hive. A colony kept by a beekeeper for a particular purpose as opposed to a colony in the wild.
2 The entire collection of goods or equipment such as the reserve supply held by manufacturers of beekeeping equipment.

**stocking**
A foot and leg covering. The strong flexible ones that present-day ladies wear are very useful as filters either for honey or beeswax. For example in a solar wax extractor the old comb cut into pieces and placed in a stocking not only enables the clean liquid wax to run through but offers the means of removing the spent comb or dross. The stocking can be held open by a non-flexible sleeve (piece of piping?) so that the chopped up comb can be easily passed into it. Boiling water extraction is also possible by inserting stones (or similar non-soluble, weighty objects) in with the comb so that it stays at the bottom of the container and allows the clean wax to rise to the top.

**stomach**
The ventriculus is the bee's stomach but there is also an enlarged posterior end of the oesophagus called the honey stomach, crop or honey sac. Digestion proper takes place once food has passed through a valve (proventriculus) into the ventriculus. See: pylorus, proventriculus and malpighian.

**stomach mouth**
The proventriculus or 'honey stopper'

**stomodeum**
The larval or embryonic fore-gut which begins as an invagination of the ectoderm. A small pit appears at the front of the mesenteron, deepens to become a tubular ingrowth which eventually becomes the fore-gut and is continuous with the rest of the alimentary canal.

**stomogastric nerve system**
The sympathetic nervous system. It is connected to the brain by the frontal commissures and the stomogastric nerve extends posteriorly along the dorsal wall of the pharynx. It is concerned with the innervation of the digestive circulatory and reproductive systems.

**stone**
A piece of hard material (rock) ubiquitous and useful as it is heavy, durable and to some extent portable. Its various beekeeping uses would overfill this page but here are some: To add weight to prevent a roof from being blown off in a gale. As pebbles to give bees a foot-hold when collecting water. To weigh down a porous container when rendering wax (see stocking). To raise the level of wine (sterilize first) when the gap between it and the fermentation valve is too big. To be used for signs – 'this hive needs re-queening', 'feeding' etc. For making a base for a hive to stand on. To enable tyres to gain a purchase in muddy ground.
See: soil.

**stone brood**
A mycotic disease and standard mycological techniques enable pathogens Aspergillus flavus, Fumigatus nigra or mycosis to be identified. It has also been called 'new bee disease', black brood and bee pest. Sealed and unsealed larvae turn white and fluffy, then pale greenish as spores develop and then become as hard as stone. Little is known about it in the U.K. but the spores are said to be capable of causing injury to the eyes and lungs of humans. All stages of brood and adults can be affected. Aspergillus flavus is yellow-green and causes the disease when a highly virulent form attacks a colony with low resistance. A.fumigatus is grey-green. A. niger presumably black. Low temperature and undernourishment make bees susceptible. Common in Europe and N. America but rare in the U.K.

**stonecrop *** *Sedum acre* Crassulaceae**
A perennial with a low habit, fleshy leaves and surviving on stonewalls or rocks in exposed conditions with little soil. Harwood refers to a 'cloth of gold'. Whitehead to a bluish pollen and Hayes gives interesting details of cultivated varieties as well.

**Stoneleigh**
See National Beekeeping Headquarters.

**stoneware honey crocks**
These and earthenware (ceramic) crocks were popular before costs and the disappearance of craftsmen made them impracticable. However there are still avid collectors of such items – see cover to Eva Crane's 'A Book of Honey' and many and various are the forms of pottery used to enhance honeyed contents as the old crafts become popular again.

**stopper**
A plug, cork, bung or anything that enables a container to be opened or resealed. Honey is usually put into screw-top jars and fitted with a cap that has a waxed wad or other method of making an air-tight seal when screwed down tightly. Substances used as prophylactics, for sterilizing and other apicultural affairs each need containers with appropriate stoppers. Ground glass stoppers are used for highly volatile substances such as acids and strong alkalis. Plastic is used sometimes, even for acetic acid, when, once again, a wad is incorporated in a screw cap.
See: fermentation valve, honey stopper, cap and wad, and flanged cork.

**stopping (filling)**
Beeswax and propolis can both be used as a wood filler (stopping) and to seal, temporarily, leaks in hive chambers, clearer boards or even feeders. At times, liquid beeswax can be used to make an air-tight sea.
See: sandarac.

**storage**
To put things away neatly and securely for future use. Of all living creatures honeybees exemplify this capability. When we extract and store honey, containers should be absolutely air-tight. Combs without bees become fragile in the cold and prone to attack from wax moth. Supers can be sterilised (see fumigants) and stacked in the open where they remain aired or indoors if space allows but mice and wax moth (all stages) must be kept out. Queen excluders, travelling screens, plastic sheets and newspapers will help. Queens can be stored for up to a month in summer, but must have water, food and attendants and be kept at hive temperature and be separated from alien bees.
See: queen bank and queen storage

**storify**
To super or to add additional layers to the hive as the colony develops and stores are accumulated.

**stores**
Food, preserved and sealed economically in honeycomb cells for use in the immediate or more distant future. It is a characteristic, unique to honeybees, to be able to subsist entirely on food gathered by their now deceased sisters. Such stores not only ensure their survival by supplying a source of energy and warmth but provide a heat-retaining canopy.
A host of descriptions are associated with their stores: ample, inadequate, minimal, poor quality, ripe, sealed, unsealed, unsuitable, surplus, winter etc.
See: honey, nectar, pollen, pickled pollen, winter and reserve stores.

**storing**
Of honey, pollen, royal jelly and equipment. The putting into reserve of food material. When bees are storing food there is an intake of food (nectar, honey and pollen) from outside the hive and the colony's weight is increasing. A characteristic displayed more strongly by some colonies of bees than others.

**stormy fermentation**
When a 'must' is first prepared and primed with a working yeast at a suitable temperature after an initial lull, the rate of fermentation can become quite vigorous. This has been called 'stormy fermentation'

and until it has settled down so that an ordinary fermentation lock can be used, it is best to close the orifice with a sterile, lightly packed ball of cotton wool until the fermentation can be safely controlled with a normal air-lock.

### STP

Standard temperature and pressure. As many attributes vary in accordance with the changes of temperature and pressure (humidity for one), a standard datum is necessary. In chemistry 0 degree centigrade and 760mm of mercury are used. When tables quote, say the specific gravity of honey, they give 1·4129 at a temperature of 20C and 760mm pressure. Where water is concerned zero centigrade is its freezing point but it is heaviest at +4°C.
See: moisture content and specific gravity.

**straighten**

A buckled, zinc queen excluder should be cleaned and completely flattened by passing it through a mangle because to work properly it should be quite flat. It is best to frame them with ¼" lathes so that this gives bee-space and helps to prevent buckling. When making beeswax candles a wick can be stretched straight while still warm after dipping in the hot wax and the subsequent candle can be rolled between flat sheets (such as glass) as dipping continues. Metal ends can be cleaned, opened and flattened in a vice (not so the modern plastic ones).

**straight line**

Although neatness and tidiness would suggest aligning hives precisely in a straight line this is likely to lead to 'drifting' where bees enter an adjacent hive instead of their own. The expression 'to make a bee-line' implies that bees fly in a straight line. If they were to do so they would be more vulnerable to their many predators. It is not difficult to observe that they actually fly in a zigzag fashion which gives them a constantly changing view of their surroundings. A bee can dart from side to side too– its dexterous flight capability is truly remarkable.

**straight nosed smoker**

Although such smokers are still available the invention of the bent-nosed variety made the control of smoke direction much easier.

**strain**

1 To filter and so cleanse a liquid by passing it through a mesh in the form of a sieve, filter cloth or other method. This applies to substances like honey, wax and wine.
2. In genetics it can mean a type within a species, having definite features distinguishing it from others of the species. This can be a descendant from a common ancestor showing similar characteristics. Breed, race, line, ecotype etc.
3. To apply undue force. To withstand the strains imposed when transporting hives they should always be securely strapped.
See: line, race, type, variety, eco/geno/phono type.

**strain conformity**

With decisive characteristics of honeybee strains, such aspects as the cubital index, hair length, colouration and width of tomentosae all go to help the identification of a particular type. One looks for things like gentleness, spring development, swarming propensity (or lack of the latter).
See: cubital index and tomentum.

**strainer**

An implement that processes liquids by passing them through a mesh fine enough to remove impurities. Plastics and perforated material including sheets or fine particles are used when mead, vinegar, honey or beeswax are purified. A dense liquid like honey calls for an initial coarse filter followed by subsequent finer ones. As impurities run from particles of wax or propolis, pollen, and tiny air bubbles, it behoves one to use such methods as are appropriate to the arrival at a stage acceptable. For example for the shop shelf or show bench the aim is to produce a clear, bright, 'nice looking' sample. This invariably means taking advantage of honey's viscosity by heating which so reduces its viscosity that it will pass through the finest of filters and be clear of air bubbles.. Good bye though to some of those trace elements pollen and possibly valuable ingredients which abound in fresh honey. See: filter and stocking.

**straining cloth**

A finely meshed piece of nylon, cotton or suitable material. There is a standard type of straining cloth. Speed of passage through the cloth depends on its mesh, the density and the temperature of the liquid and the standard or purification already achieved and finally desired. A cloth may need to be 'primed' by first pouring a little warm honey through or 'touching the underside' to get the flow started.
See: filter and stocking.

**straw**

The dry, jointed, hollow stem of grasses. Certain types of cereal grain, after threshing, are particularly suitable for skep making where a certain flexibility helps, as opposed to that used for thatching – (long stemmed wheat or rye grass). Merrist widgeon is one such and its long, unbroken straw, though brittle when dry can be kept wrapped in a moist cloth overnight. Cutting from the ground is best done manually because modern harvesters tend to buckle the straw. The stems 'en masse' are also called straw, it is used as fodder, stored in bales and is flammable

**strawberry ***Fragaria vesca Rosaceae**

A reddish, fleshy fruit and the small, stemless perennial plant that bears it. The fruit is cleaner when grown on a bed of straw. Each flower is a composite requiring several insect visits if a fully developed strawberry is to result. Even in low polythene tunnels honeybees have been used to improve the crop in regard to quantity and quality (many varieties are given to malformation) and bees are especially valuable for early varieties. To avoid bees simply bashing away at the polythene to get out, it is best to lose the flying bees first before setting the hive down in a prepared hollow. Although most varieties give little nectar the pollen is described as pale-yellow to dull brownish-yellow, the grain size 20mu.

**strawberry jam**

4½ cups of prepared fruit (about two quarts of ripe strawberries). 1 box (1¾ oz.) powdered fruit pectin and 7 cups mild flavoured honey. Crush ripened berries – one layer at a time. Put fruit into a 6 – 8 quart pan. Mix pectin well in. Stir to boil on high heat, stir in honey and bring to the boil for two minutes. Stir constantly. Remove from heat and skim with a metal spoon.

**Strawberry tree *** *Arbutus unedo* Ericaceae**

A small, evergreen tree bearing strawberry-like fruits. Howes tells us that it only grows wild in southern Ireland and has pinkish-white, pitcher-shaped flowers and blooms from October to December. It does better in Mediterranean regions though bees have been seen working it when autumn weather conditions were abnormally fine. Its honey is lemon coloured but the taste is somewhat bitter and has a characteristic odour. Common in NZ and cultivated in some gardens in S.W. England.

**straw skep *****

A basket woven from straw and intended for catching swarms or hiving bees. As straw was a common enough material before the advent of movable frame hives, cottagers who kept bees would constantly make new skeps for the new season's swarms. They were stood on a base board covered by a straw stook or stood inside a 'bee-bole' for winter. Size and shape varied though they tended to be domed or conical, had entrance holes and sometimes interwoven handles or feed holes. Honey harvests implied killing the bees until someone developed the technique of 'driving' bees from a full skep into an empty one by pummelling on the sides of the lower skep.

**stream (creek)**
A small, natural watercourse. This might provide the bees with water but could flood. Creeks in Australia and New Zealand are given to occasional 'flash' floods. It would be never be wise to set hives on the bed of a dried out stream. Another consideration could be the difficulty of crossing the stream. Sometimes they provide boundaries, protection for cattle and useful irrigation. Useful for washing honey coated items.

**streamlined**
Shaped to offer the least resistance to the flow of a fluid. A bee's body not only allows it to get into cells of a minimal size, to develop power of flight so that it can support weights greater than its own body weight but the overall outline of the bee is remarkably streamlined. The heavy rear legs act as a counter balance to the shifting centre of gravity of a bee as it becomes loaded. Empty robber bees fly with their legs outstretched.

**strength of beeswax**
Every ounce weight of comb is capable of holding 40oz of honey even when suspended from the ceiling or any flat surface to which the bees have attached it. Furthermore bees may cluster on such comb and in all, at hive temperatures it is estimated that beeswax can support up to one hundred times its own weight. The wafer-thin cell walls (1/1000th"), the tenacity of wax, and its water-proof nature, make this pollinating insect one of the most remarkable. Materials paper, cloth etc. dipped into liquid wax become strengthened and water-proof. Heat changes its character and it is usefully flammable.

**strength of a colony**
The smallest colony likely to have any chance of survival would have to include several hundreds if not thousands of healthy workers as well as a queen. Colonies at peak development will include 50 – 80 thousand workers according to the race of bee and the climate. The size, numerically then, is what is referred to as the strength of a colony. Hence adjectives include: weak, backward, healthy and strong when colony 'strength' is to be quantified. Counting the number of 'seams' of bees is one way of checking.
See: strong colony.

***Streptococcus faecalis***
Found as a secondary invader in larvae infected by EFB – also referred to as S. apis. The causative agent of EFB is *S.pluton* which has lanceolate cocci.
See: S.pluton.

***Streptococcus pluton***
This was the previous name for what is now called Melissococcus pluton which in company with *B. pluton* and other bacteria are the pathogens associated with EFB. It has been treated with Terramycin and is killed after 10 mins. At 79°C.
See: *Melissococcus pluton.*

**Streptomycin**
An antibiotic that has proved effective against certain bacterial diseases. Streptomycin sulphate and dihydro streptomycin sulphate are water soluble prophylactics useful in the treatment of EFB. Fera are responsible for dealing with this disease in this country.
See: Terramycin and Melissococcus.

**stress**
The reaction of some animal, plant or thing that has abnormal conditions or pressures brought to bear upon it. Physical or mental deformation caused by strain. In the case of honeybee colonies environmental conditions vie with the actions of beekeepers in imposing stress. Here are a few stressful conditions, the significance of which is that they may lead to an unbalanced pathological state, hence disease or disorder: prolonged confinement, frequent unseasonal inspection, transportation, robbing, the attentions of predators, exposure to unfavourable climatic influences etc.

**strigilis *** L. curry-comb**
The Latin also meant a curved blade used for scraping the skin.
See: antenna cleaner, strop.

**strong colonies**
Powerful- capable of fulfilling the tasks expected of them. A strong colony is one that has reached maximum or near maximum size for the time of year. A simple calculation of colony size (in terms of bees) in November and again in March will tell us that some 20,000 bees will die in the five months, that is almost 100 each day, so strong colonies may well have clear-outs of dead bees that seem quite horrifying after periods of winter confinement. Be sure that entrances do not become blocked – have a sample examined.
See: carpet and strength of colony.

**strop**
1 A cleaning or refurbishing motion that involves sliding, wiping or sweeping, as for example when a bee uses its antenna cleaner or forelegs to clean antenna or tongue.
2. The sharpening of a razor on a leather strap (strop) – use beeswax on the leather!

**stud**
This word adopted from horse breeding, means the maintenance of selected males (drones). There can be only a limited control of drones (who are free to fly to drone congregation areas) and for pure matings it is essential to take all possible steps to exclude stray or unwanted drones from an area in which queens are to be mated. Such areas may be achieved by isolation or by re-queening all hives in a given area with siblings so as to produce a mono-strain area. In Austria those who pay to move nucs to such areas are fined for every drone found in those nucs. These areas may be called 'stud' areas.

**student**
A person who studies a subject as for example in preparation for sitting an examination. A great deal of preparation for bee exams has to be done on one's own as courses are relatively few but books are many. The ordinary College year does not harmoniously blend with the study of beekeeping as even if done full time, holidays break in just when the best opportunities occur for work in the apiary. A majority of students look forward to advancing from their studies to a full-time job in the same sphere. Britain is very limited in beekeeping opportunity while courses and jobs are much better linked where Bee Farmers are always looking for labour e.g. Australia, Canada New Zealand and USA. A course followed by splendid opportunities is run for example in Queensland, Australia.

**style *****
That part of the pistil of a flower that supports the stigma. Although slender it is through this that the pollen tube must grow to reach the ovum for fertilization.

**stylet**
That part of the sting shaft along which the lancets reciprocate and protrude. It is rigid, sharp and slender. The lancets make a sliding fit onto intermittent 'rails' on the stylet, combining with it to form the venom canal and business end of the sting apparatus.

**Stylops Stylopsiptera**
A minute parasitic insect which rarely parasitizes honeybee larvae but the females remain in the host, which may, as a consequence, be deformed. The total damage to colonies is said to be 'slight'.

**sub**
A prefix meaning under (like hypo) e.g. subclass, subdue, subcommittee, subsist etc.

**subalare**
A small plate hinged to a pleurite and connected by a tendon to the subalare muscle. The 2nd axillary sclerite associated with the wing root mechanism is attached to the subalare.

**sub committee**
A secondary committee appointed out of the main committee. Whereas the work of a full committee is of more general nature special sub-committees are appointed to investigate matters of a more specific nature with a view to reporting back their conclusions to the general committee.

**subdue**
To control by superiority, to bring under subjection. When mankind first discovered that he could handle fire and smoke while other creatures could not this gave him certain advantages and they have been exploited in connection with the bees. Few creatures can withstand the full fury of an attack from a strong colony of bees, yet when subdued bees become a different proposition. To subdue bees means to temporarily put them into a state where man's manipulation of the colony is tolerated and then for the bees to return to normal. See: subjugation.

**sub-family**
A category in classification between a Family and a Genus, with names ending – inae.

**sub-genus**
A sub-division within a Genus or below a Genus but above a species, (rare).

**Subject index**
As this work is intended to take the form of a dictionary, yet to be of maximum use to beekeepers and all kinds of beekeeping students in particular, as well as cross-referencing to assist with 'associated' ideas when a word has been chosen, all the words to do with that subject are alphabetically listed under that subject with page numbers alongside(?). This should help with the discovery of words that were previously unknown or did not come to mind. A guide at the front is also given which lists the titles chosen for subject headings

**subjugation**
When applied to bees this means the use of chemicals (including smoke), vibration etc. to keep the bees reasonably quiet during manipulation. Bees of different races and strains may vary in their response to subjugation. The weather, a dearth or flow – each will have a marked effect on a colony's behaviour when steps are taken to subjugate it. Smoke, water, vibration, repellents, exposure to light and air, continued interference, cold and anaesthetics have all been used to subdue bees. However it is always important when using any method to assess the effect it is having before assuming that it is reasonable to continue with that particular method on any particular occasion. Brood, chitin, and colony morale may all be damaged by inappropriate actions.
See: anaesthetize, effect of smoke, potassium nitrate, subjugants, smoking bees and toxic subjugants.

**subjugation by smoke**
Too much smoke is very harmful while too little may be useless. Smoke has more effect when bees have access to open, liquid food and can be useless if they have not. An orthodox approach is to don protective garb and ensure that the smoker is well-fuelled and responding to the stimulus of pumping the bellows. A 'drift' of smoke across the entrance clears enough bees out of the way to permit touching the nozzle against the open entrance. Two full puffs diagonally across the floor board goes up into the combs, but wait at least two minutes for the full effect. Insert your hive tool between the hive chambers smoking through narrow gaps on either side at each corner. Once the upper part begins to lift don't let it down as bees would crush. Ensure lower frames stay put, blow smoke over the top bars

exposed (allow wind to assist). Put cover cloth on and only smoke seams as uncovered and the lugs of the frames when returned. Docile colonies need little or no smoke See: smoking bees and potassium nitrate.

**sublimation**
To change directly from solid to gaseous form. This applies to certain crystals such as PDB (now thankfully banned) camphor and other anti-parasitic devices (treatment of Varroa). They are often very dependent on humidity and temperature.

**submentum**
Part of the trophy, better known as the postmentum.

**submerge**
To immerse or cover with water. As a means of subduing a queen so that her reactions are quite submissive (and in this state a queenless colony is more likely to accept her). One way of doing this is to put her into a small, clean (fumeless) box and with a small gap to allow water to enter, submerge the box by hand, in tepid water for a second. The box is then withdrawn, shaken very gently and the queen is then ready to be introduced into the queenless colony where the youngest bees are likely to be.

**submerged antipathy**
When alien bees are to be encouraged to join one another or to accept a new queen, if adequate precautions are not taken it is sometimes possible to have every sign that all is well only to find, subsequently, that there is turmoil or a queen (even after some initial laying) is superseded. To avoid such occurrences always make absolutely sure that both parties are in a condition where they badly need and want or will accept the other. Don't rush such matters and refrain from being impatient. See: acceptance.

**submissive**
To become humbly obedient.

Communication between nest mates is essential and an incorrect response may have dire consequences, To overcome such situations bees will sometimes become entirely submissive rather than invite aggression. Tales are told of whole colonies joining forces with and helping raiders from another colony to clean out their own stores.` A bee that has 'drifted' into a colony by mistake will 'get its tail down' and allow itself to be 'frisked' by the alien guards. Sometimes they are accepted. Queens, when introduced to new colonies, must mesh in with their requirements and this is best achieved by rendering them submissive either by keeping them on their own for 15 minutes or wetting them with dilute syrup or other means. When a colony that has been thoroughly aroused and alarm odour has affected every one of them, they 'run-out-of-steam' after a reasonably short time because that pheromone is transient and all adopt a buzzing roar so that neither kicking the hive nor other interference alters their behaviour. See: beggar bees, submerge and thanatosis.

**suboesophageal ganglion**
The ganglion which is close to the brain but below the oesophagus. It sends nerves to the mandibles and the proboscis and is linked by paired connectives to the thoracic nerve cord.

**suborder**
Taxonomically this is a group of related families, lower than an order and higher than a family.

**subscription**
A reasonably small payment such as that intended to cover cost and postage of a magazine or periodical or cost of membership to an association (also referred to as 'dues'). It may be covered by a banker's order or be covenanted so that the recipients may claim a proportional rebate from the Inland Revenue. One may subscribe to a beekeeping journal.

**subspecies Ssp.**

Taxonomy tries to divide groups into species and within these can be found varieties. A sub-species lies in between the two and forms a somewhat ambiguous rank. Races or types that are morphologically distinct but interbreedable. The sub division may be geographical or ecological. For example similar insects that have evolved differently as a result of living in different areas or habitats and which can interbreed. A different group within a specie with more important distinctions than those attributable to a variety. They tend to disappear through interbreeding.
See: line, race, specie, strain, type and variety.

**substances bees collect**

Normally nectar, pollen, propolis and water are the substances collected by honeybees. However if honey or wax are exposed they can and do collect (or rob) both of these. Other strange materials have been carried back to the hives by foragers including: coal dust, paint, tar and sawdust and they sometimes display an unaccountable interest in petrol pumps the fumes of which are totally fatal to them.

**substances harmful to bees**

Poisons such as arsenic and poisonous gases, wax soluble volatile substances like petroleum, fine powders especially insecticides, too much smoke, evil smelling substances –creosote and nail varnish. Vaseline (also known as petroleum jelly) appears to be harmless to bees.
See: harmful substances.

**substandard**

Less than satisfactory. When a colony's performance has to be described this way – 're-queen'. Equipment from manufacturers with a reputation will usually be excellent and replacement is likely should an item be substandard. Beware of cheap alternatives as beekeeping demands the best.

**substrate**

The substance upon which an enzyme or ferment works. The ground or solid object upon which animals walk, e.g. honeycomb. That which something is incumbent upon. The substratum or subsoil.

**substrate borne sound**

Vibrations detected by the bee's feet, may not be audible to man but are significant to the bees. The queen's piping is a combined use of a substrate 'platform' that helps to transmit or resonate the sound (vibrations) that she creates. Other communications between bees we have yet to discover.

**subtend**

A chord or angle formed between two lines. At the hive the direction of the sun subtends the same angle relative to the direction of food as the vertical subtends to the direction of the food collector's dance. This is usually taken to mean from the upright on the comb but is more likely to be angled relative to downwards which the direction bees are sensitive to in the darkness of the hive. (The datum being the line of gravity whereas we have been assuming they 'read' upwards as the direction of the sun). Polarized light patterns act as their guide 'not the actual sun'.

**successful mating**

To achieve this not one but many virile drones are required. In a drone congregation area a queen not only finds drones (they find her) but gets some protection from their numbers (she can for example be vulnerable to sea gulls) Observation of exactly what goes on up there is not easy but what is certain is the number of sperms that finally reside in the queen's spermatheca. From its capacity it is clear that enormous numbers of drone's sperm could be accommodated and these she nourishes and continues to use for years to come. Actual numbers of matings vary from one drone (unsatisfactory) to a dozen or more, though much higher figures have been suggested and reported.

**Succinic acid $(CH_2)_2COOH_2$**
A white, crystalline, soluble acid which can be made synthetically though it occurs naturally in amber. It is used in lacquers, dyes, perfumes etc. Significant in the Kreb's cycle (citric acid cycle) and is of metabolic importance. It stimulates nectar secretion

**sucker *****
A shoot that arises from the base of the stem or directly from the root of an established plant. These can give rise to new plants and form a useful means of spreading bee-plants around the district.

**sucrase**
An enzyme that hydrolyses sucrose into fructose and glucose as in the transformation of nectar (sucrose) into honey (fructose and glucose).
See: enzyme and inversion.

**sucrose $C_{12}H_{22}O_{11}$**
Although sugars cover a wide range of organic substances, white, crystalline sucrose is what people generally regard as sugar. To say honey is just sugar (as some misleading scientists have done) belies the truth that honey is very different from sucrose). However most nectars are solutions of sucrose but they are inverted by the bees into two simple sugars: fructose and glucose. It still persists in honey around 2 – 5% (virtually a 'trace). It is a sweet-tasting carbohydrate food with a compound molecule that has to be 'inverted' (split) before digestion can take place. Commercially it is cane or beet sugar and chemically it is known as saccharose, a disaccharide sugar which can be synthesized under respective names. Sucrose along with trehalose is a non-reducing sugar.

**sucrose octo-acetate**
A bitter tasting chemical that was once used to de-nature sugar at the beekeepers' expense when EEC regulations granted us an issue of sugar to help when in Europe a poor season had left so many colonies in a parlous condition.

**suffocation – smothering**
Killing by depriving of air. Colonies can be suffocated should their entrance become blocked (especially observation hives) or nuclei with small entrances. Travelling with bees in warm weather makes them particularly prone. Also the careless handling of swarms, if cut off from air during transport, can suffocate in a very short time. Balling is the manner in which bees might suffocate a queen.
See: balling, smothering and ventilation screen.

**sugar**
Sanskrit carkara (originally applied to grains of sand). Sugars are soluble in water and have a sweet taste. They range from monose to decose. A sweet, crystalline substance of plant origin. A fermentable carbohydrate. It can be synthetically produced in many forms and there are simple and compound sugars. Sucrose is the only form suitable for feeding to bees and this should always be dissolved as white crystals in the appropriate quantity of clean water. The enzymes that bees carry in their saliva can invert the sucrose (as is found in nectar) into the simple sugars: fructose and glucose. Brown sugar, molasses etc.: overload the bee's gut and would almost certainly cause dysentery.
See: carbohydrate, monosaccharide, poisonous sugars, sandarac and various other headings.

**sugar beet *Beta vulgaris***
A variety of beet cultivated for its white roots from which household sugar is extracted. Quite large factories are required to get the finished product and quotas are allocated under the EEC regulations. The price is artificially fixed and British beekeepers have never been given the benefit of low prices though the survival of colonies frequently depends on fairly heavy sugar feeding. The 1980 position

was that the authorities appear willing to subsidise beekeepers if their own respective governments are willing to agree to it.

**sugar candy**
Candy for bees needs to be soft and devoid of anything harmful. It is not difficult to make if exact measurements of water and sugar and correct timing is used to consolidate.
See: candy.

**Sugar cane *Saccharum officinarum***
The chief source of sugar, this tall grass is grown in certain warm countries (e.g. Cuba). Sugar from the beet is not chemically distinguishable but whichever is used for the bees it should be refined and white as anything additional to sucrose and water can imperil the bees' safety during the winter when gut contents should be benign and in no way irritant. Bees can produce wax on a diet of sucrose.

**sugar chemistry**
Ptyalin, a salivary enzyme, breaks starch into maltose (disaccharide formed from 2 molecules of glucose).

**sugar content of nectar**
This varies from plant to plant and while sucrose is the dominant form of sugar in nectar favoured by honeybees, a large variety of other sugars, principally fructose and glucose are present but the strength of solution also differs tending to vary constantly according to conditions in any given plant but as variable as the sugar spectrum itself throughout the range of bee plants worked. Examples of sugar contents are: plum 15%, bean 22, lime 32, sunflower 37, white clover 40, raspberry 46, borage 53, black locust 55, and marjoram 76.
See: nectar concentration.

**sugar – de-natured**
Sugar deliberately spoiled at the purchaser's expense so that it is unsuitable for human consumption (and can ostensibly spoil honey should it perchance be robbed by bees and put in the wrong place) but can be fed to the bees. Fish meal, bitter chemicals and persistent dyes have all been conjured up as ways of making beekeepers lives more difficult. Bees never appeared to be Government's favourites. Hopefully some are beginning to see the light
See: de-natured.

**sugar for bees**
Although many beekeepers try to refrain from feeding substitute food such as cane or beet sugar, the survival and well-being of queens and colonies in the number required for agriculture make such feeding essential. It is however always reckoned that at least half the bees' provender should be natural food. Thick 2 : 1 sugar syrup is used for winter storage and fed before October while 1 : 1 or weaker syrup may be used during the active season to stimulate egg-laying. It is undesirable to feed liquid food from November to March, even if the bees can be persuaded to take it, so candy is used. Variations occur but feeding other solid sugar is not recommended as a method to be used by beginners.

**sugar spectrum**
The range of different sugars in nectar varies even more than those in honey where the fructose glucose ratio is given as a measure of the honey's likelihood of granulating quickly. In nectars the respective relative quantities of sucrose, fructose, glucose and other sugars, may well account for the attractiveness of a nectar even when the overall sugar content is low.

**sugar syrup**
Sucrose solution. Density can be varied from 5 sugar to one of water by weight or 1 : 1, or in the case of thick syrup 2 : 1 or saturation level. All manner of methods to liquefy the sugar so that the sugar is free of crystals at all temperatures it is likely to encounter are used. Bro. Adam was said to insist that all sugar used at Buckfast Abbey should

be stirred with paddles in cold water until dissolved. Some half fill a bucket feeder and top-up with water on the site. Water first brought to the boil and then the sugar stirred in, is probably as good a way as any, though a rough 'rule-of-thumb" winter feed, is level sugar in a heating pan, fill to the same level with boiling water and stir until dissolved.
See: artificial food and feeding syrup.

**sugar tolerant yeasts**
Several species of these have been found in honey including Zygosaccharomyces, Saccharomyces, Nematospora, Schizosaccharomyces, Schwanniomyces and Torula. They will work at very high concentrations of sugar when the temperature and water content of honey allow. They are unsuitable yeasts for mead making.
See: fermentation

**sugary honey**
The glucose fraction of honey is easily triggered into crystals. This not only makes clear honey unsightly (and causes the uninitiated to assume that it has had sugar added) but leads to further granulation. Gently warming for as long as is necessary to return the crystals to clear liquid is the best answer to the problem. Coarse granulation is best countered by re-heating and 'seeding' with a fine grained honey.
See: seeding.

**sulcus Sulcate**
Having a groove or furrow For example a pollen grain might have long, narrow grooves and be termed 'sulcate'.

**sulphates**
The salts of sulphuric acid, the battery acid. When neat alcohol is used in the final stage of dehydrating material to be used on a microscope slide, copper sulphate (in the anhydrous form) will turn blue if any water is present.

**Sulphathiazole $C_9H_9N_3O_2S_2$**
A sulphanilamide derivative – a drug used in the treatment of pneumonia and used at one time to destroy bacteria responsible for AFB which it can only do when the bacteria is in the vegetative stage.

**sulphur (brimstone) (S)**
A pale yellow, non-metallic element found widely in nature and used for the preparation of sulphuric acid (one of the strongest acids) from which many useful salts including sulphates are derived. It melts at a fairly low temperature, burns to form choking sulphur dioxide and occurs in several allotropic forms, flowers of sulphur, green sulphur which has been used in a honey mix for bees to cure ailments like nosema. Useful in agriculture as disinfectant and in insecticides. It converts softwood into a hard durable product and is trioxide will form acid rain.

**sulphuring**
'taking them vore' Old Devon. A barbaric custom existed which entailed holding the lightest and heaviest skeps (that were to yield the year's harvest of honey) over a sulphur (brimstone) sprinkled fire. The poisonous fumes (sulphur dioxide) killed the bees and doubtless flavoured the honey.
See: driving/sulphur pit.

**sulphur pit**
Before the art of driving bees came into play it was customary to kill the bees with fumes of sulphur dioxide in order to take their honey. Medium weight skeps were usually kept for overwintering and to provide the next year's swarms. A red fire was prepared in a small pit, then by sprinkling a minimum of sulphur on the embers and then holding the skep over the pit, the bees could be asphyxiated without undue harm to the combs and their honey. Earth could be used to prevent fumes or bees escaping and some skill was required not to overheat combs nor taint the honey.
See: sulphuring, driving and sulphur.

**Sumac *** Rhus spp. Anacardiaceae**
There are many varieties of this shrub and they bloom from May to August. The nectar is exposed and available to many insects. It likes hot clear, days and can produce a flow when sufficiently abundant during a heat wave. The honey has a transient bitterness and poison ivy is one of its species. Howes makes a special reference to Chinese sumac *R. potaninii* which swarms with bees in May or June.

**summary**
This often follows a well-written article and an abbreviated extract or short description (especially scientific articles). – A short comprehensive digest of facts or statements.

**summer**
Normally the warmest season of the year when the sun reaches its highest altitude. Following spring when colonies will have burgeoned forth after their winter quiescence swarming will have to be curbed or controlled and regular queen cell checks are usually considered essential though supersedure strains are said to exist. Supers should always be given before they are needed but it is a small fault to be early though a more serious mistake to be late. Once sealed they can be replaced with empties but never leave a colony with less than 10kg of food. Queen rearing should be a part of apiary management unless one has a very good source from which to get new queens.

**summer dwindling**
An unexpected diminution in the number of bees during the summer. If known to have been caused by swarming, it is understandable, If there are many dead in front of the entrance, spray poisoning should be looked into. Queen failure and disease are other possibilities. During a normal summer a colony that does not swarm should increase to peak size and maintain that condition until the queen's laying is reduced towards the end of summer.
See: spring dwindling.

**sun solar orb**
Our nearest parent star. All forms of energy – heat, light, radiation etc. come from the sun and activate plants and other living things in the biosphere. In Britain colonies do best when they receive winter sun and are often faced towards the morning sun to arouse foragers early. The total amount of sunshine is also significant as regards the honey crop. The bees can always 'read' the direction of the sun from the appearance of the sky whether or not there is overcast due to their ability to 'see' polarized light. A solar wax extractor should face true south and be inclined so that the mid-day sun's rays strike the glass at right angles. See: bleaching, solar energy, light, ramada, shade, solar extractor and sunlight.

**Sunflower *** Helianthus spp. Compositae**
A hardy annual that grows seven to eight feet high. Single varieties flower from July to the frosts, the large, yellow-rayed heads seed freely benefiting from honeybee visitations (Argentina reported 66% increase in seed yield, 5.8% oil content and 12·5% energy/germination). The honey is golden-amber, medium coloured and has a characteristic flavour and is very smooth on the palate while granulation is soft though slightly coarse. Its flowers last for 20 days. Recommended to use 1 or 2 hives per acre (2 per hectare). Its yellow pollen and gum is collected and the dried leaves and stems make good smoker fuel. The oilseed is used for cooking, margarine and confectionary. Shelled seeds are nice to eat and most useful as bird food. The cost return is lower that OSR and insecticides are used against sunflower moth and seed weevil.

**sunken cappings**
Under ordinary circumstances the bees dome cells at least slightly whether they are used for food or brood. However cells can be flattened or depressed by knocking though

the bees can rectify this in the case of honey. When sealed brood is invaded by bacteria as in the case of disease, the capping may then be drawn in, producing a sunken, concave or cracked appearance.

**sunlight**

The light of the sun when reflected from snow can entice bees out to their deaths. Its direction helps recruits to find food and bees are flying creatures of the day because of their reliance on the polarized light pattern which the sun creates. (when the sun is overhead there is usually a lull in proceedings). After a dull period the sun intensifies flight activity and it influences nectar secretion. It is lethal to brood though little seems to be done by most beekeepers to shield unsealed brood when they are manipulating. It is especially damaging if allowed to fall on the glass of an observation hive. Package bees or confined bees such as a swarm must be protected from the sun. It can raise the temperature of dark surfaces rapidly and black is used for solar extractors which trap heat by the use of double-glazing. It can be used to bleach beeswax. Bees do not seem to react in any way to the moon.

**super**

A chamber usually holding shallow frames intended for storing honey though usable for brood as well. Holding shallow combs is lighter of course than a brood box. It is customary to put these above the deep frames of the brood chamber, but being lighter there is some sense in using only supers and thus standardizing on one size of frame. Bees clearly prefer very deep combs but in nature they often build shallow ones to occupy limited space and it is much lighter work for beekeepers to use shallow frames. Perhaps the term came into use thinking of superstructure (the upper part of a ship). In the days of skeps or top bar hives, section crates, ekes and other upper extensions were used. The super is versatile in so far as it can go above or below a deep box and can be fitted with a variety of frames – Hoffman, Manley etc. See supered and supering.

**super clearer**

A close fitting board, framed to arrange bee-space below and ample clearance for any extended combs on supers placed above. One or more holes are fitted with devices – like the once popular Porter bee-escape or cut so as to encourage a one-way movement of bees from the super(s) being cleared to the chamber below. When used it leaves the honey combs unprotected so it is essential to ensure that there is no way in for other bees or wasps. Many a crop has been lost through over-looking this point.

**super family**

A number of closely related families within an Order and suffixed oidea - A good oidea! Ranking below an Order and its sub-divisions but above a family.

**supered**

See: supering
A colony that has been given additional storage space for honey.

**super glue**

A very strong type of glue that forms a tight bond within a few seconds. It can be used with beeswax.

**supering**

This term was long ago used for storifying, tiering-up or the adding of supers. The first super is not just to accommodate possible honey but to allow for the dramatic expansion in the number of bees that occurs because the cluster spreads out and hundreds of new bees are emerging every day. Therefore unless a colony is weak and being fed (when a healthy uniting should be considered) get the first super on sooner rather than later. Further supers call for a knowledge of local flora, understanding of the weather and peeps under (or through) the crown board. An observation hive can be useful in this and other connections. For

simplicity a new super can just go 'on-top' but sometimes bees need a little persuasion. Foundation should be put in the warmest spot available and where frame size allows, a small swap such as removing two frames from below, replaced by 2 of foundation with a drawn comb between them and the two drawn ones taken out put together, centrally, in the one above. This makes a bridge to 'draw' bees up. For cut-comb production use 'starters' alternated with drawn comb.
See: sections and starters.

**super-organism**

A colony of social insects, a term first coined for an ant colony by Wheeler, the various functions of the colony making it akin to the organism of an animal.

**supersaturated**

A solution that has been taken past the point of saturation. At high temperatures a fluid can hold more particles than when cooler. Air can be described in this way as regards its ability to hold water. Such conditions are unstable in so far as the surplus may well be liberated in the form of fog or precipitation Honey is a supersaturated sugar solution and will therefore tend to granulate being 80% sugar against a maximum concentration of sucrose in water of 66%.
See: r.h.

**supersedure cell**

A cell that is to produce a supersedure queen is not noticeably different from any other kind of queen cell except that a colony only builds a few cells when supersedure is to follow and those cells are built in close proximity, near the centre of the nest, and not spread-out but a knowledge of the colony's characteristics in this respect is important.

**Supersedure characteristics**

A. C. Waring of BIBBA listed the following:

1. Few cells are raised.
2. The old queen continues to lay.
3. Queen cells are well-built.
4. No antagonism between queens
5. Both queens continue to live in the same colony.
6. There is no 'piping' as when virgins appear in swarming colonies.

Other secondary characteristics include:

1. Slow build-up.
2. Minimal mating excitement.
3. Mating in presence of brood.
4. Great pollen collectors.
5. Longevity of both queen and workers.
6. Low peak queen cell number.
7. Thrift.
8. Local mating habit.
9. Toleration of drones,
10. Non-migratory drones.

**supersedure swarm**

The dividing line between swarming and supersedure is never very definite though a few queen cells near the centre of the brood nest suggest the latter and a large number of cells around the periphery and elsewhere the former. Weather, the food situation and the race of bee can cause the 'cookie to crumble' this way or that. In other words, while swarms are not normally associated with straight forward supersedure they can occur associated with this impulse, as the virgin leaves with a cloud of bees, ostensibly to mate.

**superstitions**

Fear of the unknown; It is very unlucky if at swarming time bees alight on a dead tree as it betokens a death in the family. Candle colours white or red – love, gold-power, blue contentment, green – growth or wealth. In Ancient Crete the head of a horse or jackal or a broken clay pot or egg shells were suspended from a tree or pole to protect the apiary from ' the evil eye. In Dalmatia the bridegroom's mother presented the bride with a spoonful of honey but as soon as the bride opened her lips to receive it, the spoon would be withdrawn – Cecil Tonsley. In Brittany, Westphalia and Lincolnshire betrothals are announced to the bees and hives are decorated with red and white ribbons. The new couple must introduce

themselves to the bees. In Germany and Austria it was customary at weddings to dress the hives with red cloth. Drones were said to be honeybees that had lost their stings and grown fat. To Ancients the moon was a honey-pot and the stars, bees buzzing round it.
See: black hatred, deaths, funeral and heathen.

**supraoesophageal ganglion**
**supra – above**

The bee's largest centre of nerve tissue – the brain, situated beneath and between the antennae and compound eyes. It comprises two protocerebral lobes, two deutocerebral lobes and the small tritocerebrum.
See: protocerebrum.

**Surculose Bot. *** mimosa**

A plant that produces suckers. Mimosa (or more correctly Acacia) and Tree of heaven are examples. When new plants appear to spring up at some distance from the stem it is not wise to try to take them out until their connection to the parent has been severed and they have established themselves in their own right. Mimosa also known as 'wattle' is a common Australian plant but needs the mild SW (Channel Isles) to 'hang on' in the U.K.

**surface layer of honey**

Although honey is hygroscopic and will attract water from the surrounding atmosphere, when the moisture equilibrium allows, normally only the surface is exposed and the moisture gained is only slowly diffused throughout the honey, in fact only the first centimetre or so is affected in the early stages and, quite often, honey with a faulty container only begins to ferment as a result of airborne moisture in the top layers.

**surface tension**

Although no film exists the surface layer of a liquid has unbalanced molecular forces and these can cause properties akin to a stretched elastic membrane. In a tube we find that mercury bulges to form a convex meniscus while water creeps up at the edges so that its meniscus is concave, the pulling effect that the surface of water may have on a small creature like a bee is enough to engulf it. Clothes and wicks, filter paper and other materials whose structure forms the equivalent of tubes are able to draw liquids up (molecular attraction) and feeders sometimes make use of this fact.
See: stick.

**surge**

A sudden or steady forward movement. When handling swarms bees will sometimes begin to surge in a certain direction. Like sheep or birds it is often a case of follow my leader but pheromones and fanning play their part. Get enough bees fanning and success is yours.

**surplus**

The amount in excess of what is used or needed. Surplus is a word used in beekeeping to cover the amount of honey that may legitimately be removed without causing a colony to suffer in any way. The true average surplus from an apiary is the total amount of harvest (having given true regard to the above definition) divided by the number of colonies overwintered. So if ten colonies overwinter, become fifteen and produce 500 lb of honey, the average surplus for that year would be 50 lb.

**surroundings**

Those conditions and objects that effect any particular thing or undertaking. The immediate environment. This is important when considering a honeybee colony not merely as a super organism but one with 'arms' that reach out into the surrounding territory. Apiary sites need to be chosen with circumspection.
Survey of 1000 years beekeeping in Russia
A book by Dorothy Galton 1971 obtainable from NBB.

**survival**
The continuance of life. In the history of the world some species have become extinct others have survived. Survival requires the ability to remain alive and successfully beget offspring. In nature we see the survival of the fittest. Man's interference often leads to survival of such plants and creatures that he chooses to be most useful.

**susceptible**
Impressionable, readily subject to an influence. Some strains of bee are more susceptible to certain diseases than others. Conversely we find that some bees are less susceptible and able to withstand diseases. Over many years bees tend to stabilize in their own environment. The introduction of alien parasites or transfer of bees to new environments can show up susceptibilities. Queens imported into Britain from areas where acarine disease is absent may well prove highly susceptible to the disease here.

**suspend**
To hang by attachment to an upper surface. Bees are quite capable of suspending their combs which when full of honey weigh thirty times the weight of the wax (brood is quite heavy too). They do interbrace them and use side-walls as support. These are not necessary when combs are built upwards from a base but bees are probably less inclined to build upwards as predators can gain easier access and there is nowhere for debris to fall nor hanging bees to cluster. Queen cells are sometimes hung away from the main comb. In Africa hives are suspended from greased wires to ward of ants. Flying activities are suspended at nightfall.
See: suspension.

**suspension**
Particles that pervade a fluid medium resisting the tendency to rise (as does scum or froth in a liquid) or to precipitate. Colloids in honey for example, will resist filtration yet cause cloudiness that may worsen with heating.
See: colloid.

**suture**
A junction between two hardened plates of the exoskeleton. The hinged line of closure on a bivalve shell. A seam where to adjacent parts join – these can be seen between some of the fused segments of the bee's thorax. Compare with sulcus.

**swaling**
An Exmoor term for muirburn. The burning of the furze and heather is also done on Dartmoor in order that the grass may spring up and afford pasturage for the cattle. Only those in charge of beasts have the right to swale and its practice is confined to the months of March and April, In England and Wales burning is prohibited between April lst and Oct 31st except in special circumstances as in Northumberland where the dates are 16th April to end of October. Burning may not be started between sunset and sunrise and 48 hours' notice must be given to those adjacent or with an interest in the land.

**swallow *** *Hirundo rustica Passeri formes* Hirundinidae**
Long-winged birds with swift, graceful flight. This summer visiting bird spends a great deal of time flying low and catching such insects as it might encounter. A solitary, flying queen would be vulnerable though she may well have some cover from workers until she approaches the height at which drones begin to congregate thus giving her safety in numbers. Sea gulls also gain this privilege but their webbed feet preclude landing on trees so wooded surrounds to an apiary help. Back to swallows; which may well swoop down and take 'returning' bees by the edge of level crops like rape, field beans etc.

**Swammerdam Jan Dutch 1637 – 1600**
He was first to show by dissection, soon

after the first microscope was invented, the sexes of the castes and that a queen could be raised from a worker egg. His discovery of the queen's ovaries proved once and for all that she was not the 'King" bee as many had continued to think up to that time. He also wrote: 'Biblia Naturae" of 1669 and Miraculum Naturae 1672.

**swamp**

A lowland region permanently saturated with water . It has its own particular flora and there are occasions when bees (possibly using hives on rafts) can be useful there. Cranberry bogs in U.S.

**Swarfega**

The proprietary name for a hand cleanser. It takes the form of a soft, green jelly that is applied to the dirty (propolised) hands and rubbed in before water is used. It will help to rid the hands not only of propolis but oil, grease, paint, tar, creosote, resins, waxes and rubber compounds.

**swarm** ***

Prime or top swarm. It is used to mean a great number of things in motion in the air but it is important to understand that in beekeeping a swarm is composed of the colony's queen and a large number (usually about half) of its workers and as likely as not some drones. They will have voluntarily left their home which becomes the parent colony, to set up a new home and to virtually form a new colony. Their issue is accompanied by excitement (sometimes on the part of the beekeeper too) and they seem to almost tumble out in a crazy sort of rush – sometimes having to urge the queen to join them in a swirling aerial mass. The old mother queen tires and settles, the bees surround her, perhaps dangling from a branch in a sizeable lump but scouts find a new home and they depart thence.
See: absconding, after, artificial, bees choice of site, casts, hunger, lost, mating, migratory, natural, ownership of, starvation and swarm's new home.

**swarm box**

A box or container of cardboard, wood or plastic in which a swarm can be taken and transported ready for hiving, It is a convenience if it can be held in one hand so as to leave the other free as when climbing a ladder. It should be light yet providing ventilation. In the interests of hygiene a disposable cardboard box is better than a skep which could be contaminated. Swarms can vary in size and it is advisable when a report is received that the size of the swarm can be estimated: e.g. melon, football.

**swarm catcher**

A person who attempts to or succeeds in securing (catching) a swarm. A device such as a closable net at the end of a long pole or a box set up as a bait hive so as to entice a stray swarm into it.

**swarm cluster** ***

This may be of many shapes and sizes according to the original strength of the parent colony from which it issued and the disposition of the object upon which they settle. Swarms probably average 2 – 5 lb. – up to 20,000 bees and they can combine to form enormous masses. An uneven lumpiness often suggests the presence of more than one queen. A queen's stabilizing influence will hold the cluster together but it is due to her pheromones and does not readily pass at once to every bee consequently you have to allow time for this to happen as one by one they start fanning and passing the message "the queen is here'. A spray of clean water or rain will cause them to tighten in a collectable cluster. They will chose a place out of the sun and although surprisingly subdued once they have collected together there will be scout and foraging bees coming and going but they will usually have 'a place to go' in mind and before long off they go.

**swarm control**

Unlike methods of swarm prevention, swarm control assumes that a colony has

initiated the first phase of swarming ( to wit, started building queen cells).
It outlines steps that must be taken to successfully use the swarm and deal with the parent colony including actions to attempt to prevent the issue of casts. Methods such as: Hebden, Pagden and Heddon based on making artificial swarms are just a few of hundreds of systems that have been explained, practiced, cursed, praised and used to advantage. The characteristic of swarming is always there and it behoves the understanding beekeeper to make good use of this by way of rearing young queens and taking into his or her own hands how long a queen should be allowed to reign.

**swarmed**
A colony that has recently thrown a swarm will be temporarily queenless and may have sealed brood and tell-tale queen cells. These latter may range from unsealed cells with queen larvae in them to sealed or recently emerged cells and in the later stages cells that have been torn down. One or more virgins may be free in the hive after swarming. The sound of 'piping' suggests this. Dead virgins may be found outside the entrance. When doubt exists as to whether the colony has a virgin or means of producing one, remember she may be out on a mating flight and if eggs fail to be seen then give them some eggs from another colony. This will either trigger a queen to lay or will be developed into queen cells or worker brood according to whether or not they have a queen.

**swarm names**
The first natural swarm to issue with the original mother queen is referred to as a 'prime' swarm; other names include: maiden, plum or top swarm and 'flight' and a play of bees. Swarms caused by abnormal conditions include: absconding, artificial, desperation, hunger, lost, migratory, starvation and driven or shook swarms. Casts or after swarms which usually issue about a week after the prime swarm, sometimes have more than one queen. Their significance to skeppists led to the use of special names the meaning of which have been lost. All were colts but third issues were a smart, tail, foolish or spindle; fourth issues squib or silver and a swarm after August was a wing, cuts, castlings, bulls, hubs and wheels.

**swarm of Bees in May**
This well-known rhyme handed down from generation to generation implies that an early swarm could be put to good use and would have the best chance of producing a sizeable crop. The silver spoon linked with June suggests a reduced but nevertheless useful swarm. July swarms were not looked upon with favour and are likened to a pig's eye or a fly. August with disgust and you'll always remember the one in September.
See: swarm names.

**swarming fever**
Ants and termites display what appears to be a mood of intense excitement as they spew forth mating swarms. So do honeybees. Honeybees will swarm for a variety of reasons including mating but generally as a part of their cyclic development. When the time comes it is associated with a number of physiological changes. Bees load up with food a day or two before-hand once queen cells have been started, the queen is fed less to lighten her weight, and although foraging continues and ordinary colony life carries on, there appears to be much idleness and bees loaded with food and often with wax pockets full, remain at home even spread out on the outside of the hive – basking in the sun one might say. All this in ideal weather but the latter can make their plans go awry.
See: swarming preparations.

**swarming habit**
The seasonal manner in which a large proportion of the colony leave their existing home with their queen and seem to ignore their former possession and set up anew. The strength of this trait, always characteristic to some extent in honeybees, varies with race and strain.

Great efforts have been made to reduce this tendency to a minimum in bees bred for honey production. However other characteristics are also sought thus the swarming habit persists while other useful traits that go with it (vigour and health) are bred into the bees. .
See: supersedure.

**swarming impulse**

This is the urge to swarm as opposed to the inborn trait of swarming. In fact the translation of the inborn trait into positive action. The swarming impulse may be sharpened by seasonal factors, the age of the queen, the suitability of the environment, including actions taken by the beekeeper. When the swarming impulse is strong we approach the state described as swarming fever and a single colony may throw a swarm followed by many after swarms (casts). If it is believed to be genetic (on the part of the queen) as opposed to environmental it would be wise to cull such a queen to stop this characteristic from continuing.

**swarming preparations**

Although the production of mature drones does not imply swarming, it is an essential precursor. Similarly the building of queen cups – often called 'play cells'. Once eggs appear in these (about seven days before the issue of a prime swarm when weather allows) swarming becomes likely. At this stage the queen is given less food, lays less and becomes lighter. About half the adult bees (depending on race) begin to give up foraging and increase body-weight and hive temperature putting extra food into their honey crops and in many cases developing wax in their wax pockets and a mood, sometimes called 'swarming fever' becomes prevalent. Circumstances even at this stage can develop to cause them to discontinue preparations such as cessation of a flow, deterioration of the weather or action by the beekeeper but otherwise swarming becomes highly probable.
See: swarming fever/habit.

**swarming season**

The season of the year during which bees are inclined to swarm in Britain is mainly May and June though often extends into July. Weather conditions can sometimes lead to April or August swarms but the bulk of the swarming occurs in May/June.

**swarming strain**

The variability of the display of the swarming habit has led us to classify certain races or sub-species as inveterate swarmers. A strain of bees, therefore given to unusually frequent swarming (when colonies swarm every year for instance) would be called a swarming strain. They may be very prolific and useful to bee farmers. For honey production though less swarmy bees are more desirable.

**Swarm lure**

A pheromone produced synthetically and marketed by Steele & Brodie. Created to make bait hives more effective by increasing their attractiveness to honeybee swarms.

**swarm management**

The utilization of swarm(s) that have issued to make new colonies for honey production etc. A swarm might be de-queened and united to another stock requiring a boost. It is probably best to first hive the swarm and de-queen it when it has been checked as all right for uniting to another colony. Two swarms can be put together but once again it is probably best to hive them separately first and then eliminate the least desirable queen before uniting. Another of the many possibilities is to move the parent colony aside and hive its swarm on the original site and then use the parent for further queen raising.
See: Demaree, prevention of swarming, swarm control, Snelgrove.

**swarm prevention**

Methods of management aimed at preventing the initiation of swarms, i.e. the construction of queen cells for swarming.

Young queens, the use of adequate space for the queen, her brood and for storing are aspects of swarm prevention. Special hives and systems, devices, tricks and carefully thought out operations have all been written up and written-off. The bees will still swarm though.
See: Demaree, prevention of swarming, Snelgrove and swarm control

**swarm's new home**

It has been noted throughout the ages that once a swarm has established itself in a new home it does not return to the parent even if it runs short of food while the parent still has plenty. One reason for this is that the new colony will soon develop its own identification mark in the form of 'hive odour' and another lies in the fixation of their new starting point relative to bearings into and from the former one. The almost strange blotting out of former memories can be appreciated when we see how quickly bees accustom themselves to a new setting when moved right away from their original site
See: bees choice of site.

**Sweden**

Population over 8 million with 20,000 beekeepers who get excellent co-operation from the government and have a national honey packing unit. Good crops of heather honey are secured but the heather is mostly scattered in glades amongst the pine forests.

sweet

Having a pleasing taste – applicable to sugar, honey etc. Whereas sugar is uniformly sweet there can be a considerable variation in honey flavours, those rich in glucose being less sweet than those where fructose is dominant. Heather honey is less sweet than most honeys. English vocabulary incorporated both the words sweet and honey to refer to something especially 'nice'. In most cases (apart from sarcasm) it tends to be complimentary.

**sweet alyssum** ***

Lobularia is its correct title but see *Alyssum maritimum*.

**sweet chestnut** *** ***Castanea sativa***
**Fagacea**

A tree that bears edible nuts enclosed in a prickly bur. It is also known as Spanish chestnut and it is valued for timber, likes light sandy soil and grows to 500 ft. and lives up to 500 years. It is coppiced and grown in large numbers to make fencing stakes. Supers can fill with light, yellow honey in July and although it has an unwanted, heady aroma this disappears with keeping. As regards pollen colours reports range from greenish-brown to very pale yellow and the grain is 13mu. C. pubinervis is the Chinese sweet chestnut and the Japanese get good crops of honey from this tree.

**sweet cicely** *** ***Myrrhis odorata***

A hardy perennial that grows up to a metre tall. It has a useful place in the herb garden and yields both nectar and pollen.

**sweet clover** *** ***Melilotus alba***

Bokhara, honey lotus, it has slender racemes from June to September and pale greenish honey.
See: melilot.

**Sweet lime *Citrus medica* var.** *limetta*
**Rutaceae**

A greenish-yellow, acid fruit and its tree. Like the other fruit bearing citrus trees it flowers early when insects are few in number. The introduction of hives vastly increases the likelihood of flowers being pollinated during the effective pollination period. When honey is said to come from the limes one refers to a green-tinged honey coming from the lime tree (linden) Tilia spp.
See: lime.

**sweetness factor**

Bees' preference in nectar is reported as first sucrose, then glucose followed by maltose and finally fructose (the sweetest of honey sugars). Scientific ratings for various sugars are: sucrose 100, fructose 175, glucose 66 and maltose 30.

**sweet rocket**
An ultra-hardy , evergreen, branching plant.

**swelling**
An area enlarged by becoming turgid or puffed-up. In the early stages of beekeeping this is an all too common result of bee stings. It is brought about because one fraction of the venom triggers off the release of histamine unless one has immunity, natural or acquired. The grossness of such swellings can cause limbs to ache and throb and what can be more upsetting disfigure the face by puffing out lips and closing eyes. When early man was first stung and was asked 'what hit you?' it must have caused considerable apprehension to be told – 'a tiny flying insect'. Such swellings usually take thirty six hours getting worse before slowly returning to normal.

**swift *** *Apus apus* Apodidae**
Long, narrow wings and short tail, an insect-eating bird which works in flocks and is a regular summer visitor in this country. Doubtless many bees are included in their diet when they fly near apiaries. They rarely alight on the ground and have difficulty in getting airborne again if they do. A low-flying queen is likely to be most vulnerable when not surrounded by workers or drones. See: swallow.

**swimming**
Bees never take voluntarily to water though tired foragers and queens have been reported to alight on water. Once the surface tension is broken a bee vibrates its wings until it becomes chilled, its spiracles blocked or it reaches a surface that enables it to climb out. Unfortunately they seem to be attracted to swim-pools and are often found dead or drowning. Water gatherers are at risk especially when air temperatures are low as a ripple will draw them in and their subsequent movements are inclined to be haphazard in a medium which to them would be as sticky as treacle to ourselves. Bees found apparently bloated and dead may recover if lifted onto a warm, dry surface.

**swipe**
To make a rapid sweep through the air with one's hands. This is the all too obvious reaction when the uninitiated is 'buzzed' by an insect. That swiping actually attracts bees can easily be demonstrated by standing near an open hive and simply swiping. Workers bees make any such error all too obvious. A queen flies slowly and a deft person might swipe her into his hand but practice on drones first. We must look like trees to other creatures ' move like one when bees are around.

**switched**
When two colonies of unequal strengths are interchanged the weaker one gains bees while the stronger one loses some. Why should one do this? Well it is a technique that is useful and time is short but a colony is on the point of swarming. By swapping it with a weaker one the trouble can be temporarily cured.

**Switzerland**
A landlocked, European country with 6 million inhabitants and some 25,000 beekeepers. Pine honey dew is produced when 'buchneria' invade the trees and honey comes from Alpine rose (rhododendron), bilberry and raspberry.

**swivel**
To rotate or turn or twist in either direction. An observation hive displays its contents more effectively when able to swivel. Solar wax extractors are too heavy to face any direction than south though daily swivelling could be advantageous. It is not good to swivel from the waist when lifting heavy weights like supers – move your feet, shuffle round. Frequently when one object is

difficult to insert into another a slight swivel helps.

**sycamore** *** ***Acer pseudoplatanus***
**Aceraceae**
A tall, shady and ornamental, deciduous tree which is grown for its wood and readily spreads by saplings that spring up in the vicinity. Its long flowering period helps to defy the often boisterous windy conditions but the long, yellow-green tassels are easily torn off by strong winds. It not only provides copious nectar which comes from the flowers and the extra floral nectaries while further supplies of honey dew also come from the associated aphid secretion. The resultant honey is amber with a greenish-tinge – the pollen greenish-yellow, grey or bright yellow and the grain 35mu.

**Sydserff Robert**
Wrote '*Treatise on bees*' 1792 ' much referred to by Cotton. Used the word 'barb' for the sting.

**syllabus**
The subjects to be studied for a particular examination or course.
See BBKA exams.

**sylph**
A fairy or mythical being though the word is also used in place of nymph referring to a pupa. Not an appropriate word for describing brood at any stage. Esperanto word for brood disease 'nimfopesto'.

**sylvatica**
Or sylvestris, a specific name implying that the plant or creature is to be found in the woods. (Pennsylvania).

**symbiont**
A partner in symbiosis, a condition in which living organisms live in mutually beneficial partnership e.g. flowers and bees, which are actually dependent on one another for existence.
See: commensal and inquiline.

**symbiosis**
A compatible association or relationship between two dissimilar organisms to their mutual advantage. A classic case is the association of wood ants and lachnids to provide honeydew from the spruce, collected and transformed into honey dew honey by the bees and removed and eaten by the beekeeper.
See: clover and nitrobacter.

**symbolic use of bees and hives**
Beehives and bees were often chosen to signify industriousness as the foundation for prosperity by English business houses. Other symbolic meanings were 'perseverance, loyalty and thrift'. Over 66 heraldic coats containing honeybees have been issued by the College of Heralds in London (1968).

**symmetry**
When form, size and arrangement corresponds. A flower may have bi-lateral or radial symmetry. There is an appeal about the orderliness and tidiness of symmetry. Uniformity attracts many and repels some. We should guard against the symmetrical lay-out. Of hives. Honeybees usually space their nests well apart. When set close together there is over-population of the foraging area and bees get confused and may enter the wrong hive taking diseases with them, causing possible fighting and upsetting hive records.

**sympathetic nerve system**
That part of the nervous system known as the autonomic as opposed to the parasympathetic system. It applies to the segmental nerves supplying the spiracles.

**Symphyta**
This sub-order together with Apocrita forms the order Hymenoptera of the class Insecta.

**symposium pl. symposiums, sia**
A meeting or conference called to discuss

some topic. A collection of opinions put forward or account of such a meeting.

**symptom**
A sign or indication that gives evidence of a condition or the onset or effect of a disease or condition. Learning the symptoms of bee diseases helps to prevent, control or identify them and is an essential to keeping healthy colonies.
See: syndrome.

**synapse**
The terminal branching of dendrons and axons which serve to give contiguity between two nerve cells. The region of contact between two or more nerve cells where transmission of impulses occurs from the axon of one neurone to the dendrite of another.
See: CNS.

**synapsis**
The joining together of homologous chromosomes from each parent during meiosis.

**Syndrome**
A number of symptoms which taken together are a useful guide to, if not a positive identification, of a disease or condition.

**synergist**
A substance which is most effective when in company with another specific substance. For example in alarm odour isopentyl-acetate and Z-11-eicosen-1-ol. The word synergistic means co-operative while synergic indicates operating together as in the case of muscles, hormones etc.

**syngamy**
Sexual reproduction – the fusing of gametes (fertilization).

**synonymous**
Having the same or equivalent meaning, or expressing the same idea e.g. apiculture and beekeeping.

**synopsis**
An abbreviated outline of a subject. E.g. give a synopsis of the methods of swarm control.

**synthesis**
The construction of a complex substance from basic substances. We frequently use this word when a substance like a pheromone has been artificially created. Often the result leaves a lot to be desired but occasionally it is better as it may obviate certain side-effects that the real thing has.
See: swarm lure.

**Syriac Book of Medicines**
It includes 300 recipes in which honey is used – the Liturgical language of eastern Christian churches.

**Syrian bees**
Syria at the eastern end of the Mediterranean has two varieties with similar appearance but one being as vicious as the other is docile. Resembling the Cyprian and Italian bees they have a good reputation as workers. They swarm a lot, casts have many queens but they do not winter well.
See: Apis mellifera syriaca.

**syringe** ***
A device with a thin nozzle and a piston or pump which can take up a liquid and then eject it with some accuracy in any chosen place or direction. Useful for cleaning, washing or applying measured (especially small) quantities of drugs or other liquids.

**syrup**
Although syrup is any thick, sweet liquid used in cooking, syrup to beekeepers is a solution of pure, white sugar in water. Other sugars and additives are used at one's own risk. Time has shown that without harming their excretory system bees can utilize sugar syrup to produce energy and wax. 8 lb. to 4

pints gives one gallon of syrup which is the equivalent of 10 lb. honey according to Ted Hooper. For winter feed: 2 lb. to 1 pint, 8kg to 9 pints (5 litres), 1 cwt or 50kg to 7 gal. See: Fumidil B and dissolve.

**system**

1. Organs or parts of organs concerned with the same function. The collective parts forming a complex or aspect of a functioning unit or assemblage.
2. A method or arrangement for management, e.g. the Demaree system.

**Systeme International d'Unites**

(tera, giga etc but cannot show10 to power of 6 etc.

**systemic**

Pertaining to or affecting an entire plant or body. Generally distributed throughout an organism. Systemic compounds (including insecticides) are often far safer where bees are concerned because the poison is applied to the foliage, seed or soil and normally only kills creatures that bite or suck from the plant itself. Nectar is not affected, though there was some doubt in this respect about the compound Schradan.

**systemic compound**

A compound which when applied to foliage or sometimes to seed or soil, is moved to other parts of the plant.
See: systemic and translocated herbicides.

**systemics**

A branch of biology concerned with the recognition and grouping of living organisms. It comprises classification and nomenclature. A study of systems.
See: classification, Linnaeus , nomenclature, taxonomy.

**systole**

The rhythmical phase of the human heart-beat that propels the blood from the heart in response to the contraction of the heart muscle after each dilation or 'diastole'.

# T

**table spoon**
One level tablespoon = 15ml or ½ fluid oz. or 4 teaspoons.
See: teaspoon.

**tablet**
A small, flattish piece of compressed medicinal powder with precise dimensions and weight. Examples: Benadryl (diphenhydramine – 50mg). Piriton (chlorpheniramine maleate – 4mg. Useful for 'oral' admittance whereas powders are less easy to transport, measure and apply.

**Tachinidae** ***Parasitic diptera***
Rondaniooestrus apivorus is one of the enormous family of flies associated with honeybees. Reported as causing apimyiasis.

**tactile**
Concerned with the sense of touch. For example 'tactile hairs' on the antennae.

**taenidium pl. taenidia**
The spiral thread of cuticle in the bee's breathing tubes (tracheae). It holds the tube open whichever way gases pass.

**TAFSP**
Terramycin Animal Formula Soluble Powder. Extender patties used in the treatment of EFB were made using 4g of the above with 133g of granulated sugar, adding 67g of vegetable shortening such as margarine.

**tail-board**
A hinged board at the back of a truck or trailer, that can be removed or lowered to make loading easier or to act as a ramp.

**tailgate loader**
Hydraulic, battery-operated platform that raises or lowers hives from ground level to the deck of a truck yet folds away under the deck when not in use. NZ Beekeeper Dec. 77.

**tajonal** *** ***Viguiera heliantoides***
In Yucatan thick beds of yellow flowers blanket roadsides and clearings deep into the jungle providing nectar and pollen for the bees.

**take**
One refers to a honey 'take' meaning the surplus honey or the crop that the beekeeper has decided to take from her bees. The gathering up or collection of a swarm is also referred to as to 'take' a swarm. 'Take' is also used to describe the total amount of money obtained for honey sold at a meeting etc.

**take off**
To leave the ground and become airborne. Bees can do this in-a-flash. Initially when leaving, after contact with other bees inside the hive, they brush their eyes and antennae, to free them of any particles in a sort of 'pre take-off" drill during which they probably make an assessment of the environmental conditions. The small 'gust' made when a bee departs a flower is probably significant as regards spreading the floral pollen around. We take mouse guards off when pollen begins to come in, in the spring. When a swarm takes off there is a great commotion at the hive and again when it

takes off from a branch which may lift when relieved of the weight.

**taking a sample**

The removal of live or dead bees that belonged to a colony so that their examination can lead to an assessment of that colony's health by diagnosing any disease present. Consideration must be given to the size of the small parcel. Some 40 workers in the case of disease, 200 for suspected viral infections or poisoning by spray. To avoid disturbance one can set a jar over the feed hole and having trapped enough bees, put them in the refrigerator and thence to a matchbox. See: spray poisoning.

**taking a swarm**

Once a swarm has settled it must be coaxed or forced into a suitable receptacle so that the bees can be safely hived. Hiving is best done at dusk to avoid the chance of their becoming airborne. Some ingenuity may be called for but there are useful tips. A spray with clean water helps to settle them or hold them until the beekeeper arrives. Avoid blotting out their signals which the use of smoke can do. Only use smoke to mask the scent left behind on the branch once the bulk of the bees have been secured. Bees can be shaken down (give as little warning as possible) brushed down or driven up by vibration, or if absolutely necessary scattered by the use of smoke. A white sheet, placed on the ground as nearly as possible underneath the spot they'd settled on can become the base for the upturned box that you shook the swarm into. A stick or stone can be put under the rim to make an entrance. Normally they're docile and a handful can be thrown in front of the stone to get 'fanning' started and so attract airborne bees to their temporary home. Be sure to give them adequate ventilation once in your car. Put the open side of the box upwards with the cloth (netting helps) over it to keep the bees in. Their myriads of claws enable them to cling together so that the mass assumes the quality of treacle. So be prepared to shake 'hard' when tipping them onto a board to run up into the hive. See: hiving a swarm and dislodge.

**taking exams**

There is a lot that can be learnt about the technique of taking exams. Naturally one should, by sheer application to the task, be able to keep one's cool. Time must be spent carefully studying the question so that all parts of the answer are strictly relevant. As there are a set number of questions to be attempted and a set time (every minute of which should be used even if it means reading and re-reading all that has been written) a simple division tells the candidate how much time should be spent on each question. This can usually be worked out for previous papers before going in for the exam.

Try not to get carried away with enthusiasm in answering a favoured question only to find that you have less time than you need on the others. Deduct reading time. If four questions are to be dealt with in two and a half hours settle for a strict no-more-than-half-an-hour, on each question leaving space to elaborate and time to read back over everything. Unless something is actually written down or drawn on the answer paper no marks can be given, so if your best has been done but there is still one question you find baffling – have a go – even expanding on the question might trigger off something helpful. Get comfortable, have suitable equipment and space. Be careful not to waste any time looking around at others and wondering how they are getting on. See: courses.

**talk**

To express oneself by articulation. A talk on the subject of beekeeping might be 'off-the-cuff' (impromptu) if the need suddenly arises and the speaker is fluent and well acquainted with the subject. When asked to give a talk, find out how many listeners and what kind of audience you might expect. Careful planning and preparation helps as does the wise and sensible use of other people's work in the form of audio-visual aids etc. See: lecture.

**tall**

Hives are kept as low as possible so as to offer little wind resistance during the winter. They grow taller as supers are added and in a good year it might even be necessary to stand on a hive roof or truck to reach the uppermost boxes. When a check of the brood chamber is required but several heavy supers are sitting on it, then have sufficient cover cloths so that the supers can be removed and covered against robbers. A tall beekeeper should beware of knocking his veil off as he passes under branches or obstructions. It is wise for all beekeepers, especially tall ones, to have some basic training in 'weight-lifting' i.e. use leg strength to reduce strain on the back. When you hear about very large takes be prepared to accept them 'with a pinch of salt'. ('tall' stories).

**tallow**

A hard or solid fat, off-white and tasteless, obtained from parts of animal's bodies and used in foodstuffs for dressing leather, making soap, candles and lubricants. It is sometimes used in conjunction with beeswax.

**Tallow tree (Chinese) *****
***Sapium sebiferum***

Can grow to 50 ft. in a suitable sub-tropical environment. It has little yellow, catkin-like flowers in spring and poplar-like leaves which become red/yellow in the autumn. Although it is a fast grower the wood is hard. Unfortunately the tree is semi-hardy in cool climates. The seed coating yields candle wax.

**Talmud**

A Rabbinical thesaurus. Commentaries on the oral law made by Judah. The fundamental code of Jewish law. B.Batra 18A there is a warning about the unsuitability of brassica honey for bees. Bees could not be given water on Saturdays or holidays though hives could be protected from the sun or rain. Honey is recommended for ulcerated wounds.

**tamarisk *** *Tamarix gallica, anglica***

An ornamental shrub or small tree. An invasive plant reputed to give good honey flows in some regions (described as a dark brown honey). It thrives on the coastal areas where its feathery foliage and yielding branches are often used as a wind break. Hebrew Tamarisk – broom). In areas where the weather is mild it can bloom until Christmas and a honey-like substance is exuded from the cut stem of *T. gallica*.

**tamping**

To squash down with a series of light strokes. Bees use their heads to ram or 'tamp' pollen so that it is packed tidily and solidly within the cells. Bees returning to the hive with pollen usually press the pellets off into a cell and leave to get more. The tamping is done by house bees with an average age range of 11 – 20 days.

**tangential honey extractor *****

A machine with cages arranged tangentially to the direction of rotation enabling the contents of a comb on the outside to be expelled with extra force as the speed of rotation is increased. With this type of machine it is best to spin slowly at first as the inner faces of the combs retain their honey until the combs are turned round and their other sides face outwards. If all the weight on the inside and is pushed too firmly outwards combs can be fractured at speed. See: radial and reversible.

**tanging**

Ringing or tinkling. From Roman times the technique of making a noise either to draw attention to a swarm or to get it to settle, has led to such practices as beating a metal pan with a key and even firing shot or ringing bells. Beating upon metal to make a din Herrod Hepsal. Tanging with a pan, basin, candlestick or such-like instrument of brass – Butler.

**tannin (tannic acid)**

A inhibine and. astringent vegetable compound found in oak bark, nut galls etc. Tanning is the art of converting skin into leather. It was also used as a yeast nutrient and to clear wines.
See: honey yeast nutrient.

**Tanzania East Africa formerly Tanganyika and Zanzibar.**

They have a beekeeping training Institute and in 1980 the honey production was 9500 tonnes.

**tap (USA –faucet)**

A device for controlling the flow of a liquid. Because of the nature of honey to drip, honey taps are called 'gates' and are designed to cut off the flow cleanly and at once. The word is also widely used in other ways e.g. to tap a female thread, or to tap a beekeeper's knowledge or electric current. You may tap on the side wall of a hive to learn from the responsive sound whether a hive is queenright or queenless. (hiss or moan).
See: queenless moan.

**tape recordings**

Sound recordings were made on magnetic tape but developments in the electrical world give as many ways of recording either sights or sounds. Even the wearing of a veil does not hinder the tiny devices that are available for doing the job today though watch out for propolis on the microphone). Making and keeping records are valuable aids which allow us to benefit by forward planning and from further study to act as reminders or as a means of education.

**tapes**

Thin strips of material useful for fastening. All manner of tapes are available to help with beekeeping practices. Examples are adhesive, double-sided, reinforced, rayon – some are exceedingly strong and there are electrical warming tapes. For sealing in an air-tight fashion, closing gaps, holding something fast for a moment or more permanently. Don't leave sticky side exposed where bees can get stuck on it. – what did we do before all these ingenious ideas came into force?

**tap strainer** ***

The filtration of honey would be slow if there were any kind of filter or restriction in the honey tap (gate) itself. Consequently metal and plastic filters which can be used with meshes of increasing fineness, to keep the honey flowing and yet remove those particles of propolis, wax and such-like that accompany the honey coming from the extractor. Of course the properties of honey will allow an automatic gravity clearance if suitable containers allow it to stand in a warm environment for a day or so. Fine strainers will clog unless a coarser one has first taken out most of the larger particles. The introduction of air bubbles can be reduced by minimizing the honey's 'free fall'.

**Taranov board**

Devised by the famous Russian beekeeper of this name, this is a sloping board that leads up to the hive entrance. It has a clustering space underneath and it is positioned so that only bees that can fly are able to bridge the 10 cm gap left between it and the hive. It has various uses but in principal the bees are shaken onto it and only bees that can fly get back into the hive. A shook swarm can be made or all the bees shaken from the hive and the queen and young bees established elsewhere as a method of swarm control. It has been suggested as a means of separating workers from other bees.

**tarsomere**

One of the five divisions of each tarsus. The first and largest is the basitarsus and on the hind leg this bears the pollen brushes on the inner surfaces, then three small tarsomeres join to the distal tarsomere which is the foot or pretarsus.

**tarsus**
The distal group of five tarsomeres on each leg of the bee.

**tartaric acid $C_4H_6O_6$**
Cream of tartar. An organic acid found in four isomeric forms, one is dextrorotatory and is a colourless, crystalline substance obtained from grapes. Used to make cream of tartar with fine grain and soft texture. When a little is boiled with sugar it produces a fine candy for feeding bees.

**Tasmania**
Southernmost Island of Australia 26,215 sq. mi. Mountainous and much forested. Bees were first successfully imported in 1831 and spread during the next 50 years. Then Italian bees and movable frame hives came into use. Now there are over 1000 beekeepers and 800 tonnes of superior honey is exported every year. Blackberry honey which is mild and light and has a fine grain vies with that of clover and the unique leatherwood. There are two species of Eucryphia only found in the wet western half and its honey is extra light and aromatic. The 3 – 4 month season is short by Australian standards. Some descendants of the old British Black bee are still to be found there but they are said to be less productive and swarmier than the Italian and Caucasians though they winter more economically.

**tassel (bot)**
A pollen bearing inflorescence as of sycamore.

**taste**
The sense invoked by savouring the flavour of a substance in the mouth. The sensation perceived when something comes into contact with the tongue. Sweet, sour, salty, bitter...
See: flavour, honey flavour and gustatory sense.

**taxis pl. taxes**
A suffix used to indicate an organism's movement with or without growth following the stem which indicates the cause. e.g phototaxis (light) geotaxis (gravity) but also heat, sound, touch, water etc.

**taxon pl. taxa**
Hence taxonomy. The formal unit used in classification e.g. Order, Family, Species.
Gk. tax: order, arrangement
See: taxonomic endings.

**taxonomic category**
Groups of the same taxonomic level

**taxonomic endings**
Botanical and bacterial: order – ales, suborder – ineae, family – aceae, tribe –eae and subtribe - inae. Zoological: superfamily – oidea, family – idea, subfamily – inae, tribe – ini.

**taxonomy**
A branch of biology concerned with the practice of classification. The taxonomy of the honeybee uses characteristics such as the cubital index (a measurement of wing blood vessels), the number of wing hooks, width of the metatarsus, shape and size of wax glands and the shape of the chitinous plates of the penis valves.
See: classification, Linnaeus, nomenclature and systemics.

***Taxus baccata* – yew** ***
Evergreen conifer with thick, dark green foliage and fine grained elastic wood. Beautiful fleshy red fruit with poisonous stones. It produces clouds of pollen but it does not seem useful to bees. The oldest yews are amongst the most ancient of trees in Britain.

**Taylors**
Founded by Thomas Bates Blow in 1880. By

its 100th year it had become E. H. Taylor Ltd famous for first class equipment and then was incorporated into E. H. Thorne (Beehives)Ltd in 1984.

**teaching aid**

Something that helps a student to learn. This may be used by a teacher or be self-operated as in the case of programmed learning, videos etc.

**teasel *** *Dipsacus fullonum***

A hardy biennial that grows to 5 f. A prickly leafed herb with thistle-like ovoid, purple flower heads which produce masses of fluffy, white seeds. It has structures by the leaf bases which are cup-like and collect rain water around the stem. It yields both nectar and pollen on many ovoid, bristly heads from June until August and the honey is described as light and pleasantly flavoured though somewhat thin. It naturalizes in damp places. The dry, stiff, brush-like tassels are used for the teaseling of cloth and its brown characteristic shape is often used to enhance flower decoration.

**teaspoon**

One level teaspoon = 5 ml or one sixth fl.oz Abbreviation tsp. See: tablespoon.

**technique**

A way of accomplishing something or the method of performances. See: manipulation, management, system.

**tectum**

Roof of a pollen grain which often has architechtural ornamentation. See: exine.

**Teeswain**

A notched runner (support for the lugs of frames) used to provide precise centre-to-centre spacing of the frames. Sometimes referred to as 'castellated spacing'. It cannot be varied and prevents the easy sliding of frames across a chamber. Has very 'sharp' edges. Normally for brood chamber use where the dead square sharpness locks frames securely for travelling.

**tegmentum pl. tegmenta**

Roof. The 9th abdominal tergite of male insects forming a pair of leathery fore wings serving as protective covering for the hind wings of certain insects.

**tegula**

To be arranged like over-lapping tiles. The large scale that overlaps and protects the root of the forewing. The abdominal segments tend to overlap in this fashion.

**telescopic hive**

A hive built with lifts that can fit over one another (as when the inner boxes are reduced for winter) and conversely can be expanded by up-turning the lifts so that they fit snugly on top of one another. The CDB hive is one such and though WBC's built by Burgess also had this feature (later models did not have this telescopic capability).. Each of three lifts would in turn fit over the one beneath so that when it was telescoped for wintering, or in summer the upper super could be given a two-walled lift to keep a box of sections warm.

**television TV**

A powerful medium for communication on a very wide basis. Bee matters including films are covered from time to time and can be purchased for association or private use. Computers have now brought in unlimited numbers of aids.

**Tellian bee *Apis mellifera intermissa***

Found in Maghreb from Morocco to Libya. A bad tempered bee, small and dark. A yellow bee of the Sahara is said to be a sub-race found in oases.

**telling the bees**

This was an ancient ritual whereby the bees were told when their master was dead. Failure to do this was believed to result in

the death or departure of the bees. It would seem likely that the bees' survival had often been dependent on their master. Once he went it was not long before they followed if given no attention. A variation of this ritual was to tell the bees of any important family event and this notion is carried forward in '*The Bee Boy's Song*' by Rudyard Kipling.
See: superstitions.

**telophase**
When the chromosomes having reached their appropriate poles, most species enter the final stage of meiotic and mitotic cell division during which the nucleolus re-appears and the nuclei revert to the resting stage.

**TEM**
Transmission electron microscope which gives only a two dimensional view of the object but permits staining techniques to be used. The elemental analysis of honey (including heavy pollutants like lead) is carried out by x-ray transmission electron microscopy.
See: electron microscope, and SEM.

**temper**
The showing of an outburst of anger. The characteristic of displaying their reaction to humans. Brazilian Africanized honeybees were found to be more aggressive when the temperature and humidity were high. They will then sting more and pursue any attacker over a greater distance. Descriptions include: fierce, unmanageable, tiger-like, vicious and aggressive.
See: anger and disposition.

**temperatures**
Brood nest 32 – 36C (90 – 97), Sustained 36C fatal to brood, optimum laying 30 – 35C (86 – 95) Bees die at 49 – 50C and life is endangered below 45F (5C). They become comatose and lose ability to fly at minus 4C (25F) Outside of cluster shell 7C while at the centre 24·5C. Hive working temperatures 15 – 26C ( 60 – 80). Swarm 95F (34C). Winter thoracic 69 – 80F (15 -26C). Bees are immobilized at about 10C and die after 2 or 3 days. but after 3 hours at minus 3C and one hour or less at minus 4C. Flights for bowel clearance and water collection 8 – 10C. Foraging depends on race and line but 13 – 43C. Normal hive duties 20C. Winter cluster forms at 14C. Wax secretion 35C (95F) it becomes plastic at 30 (86F), soft 35C (95F), melts 61C (142F) best candle dip 82F ( 18C) comb collapse 63C (145), while propolis is brittle when cold, plastic 30C (86F), soft 35C (95) and melts at a higher temp. than wax – 65C (150). Hot dip paraffin wax (150C) Fermentation possible 52C (70F), yeast functions 10 -25C. Granulation of honey favourable at 50 – 64F (10 – 18). Nectar is secreted by sycamore 12C, apple and hawthorn 20C while clover needs 22C.

**temperature control In the hive**
Fanning (including the use of water) or the adjustment of metabolic processes including micro vibrations of the thoracic muscles within a chosen and prepared environment to adjust the nest temperature to the normal level despite variations in the ambient air temperature. The packing tightly together or opening out of the cluster and the internal use of food help a colony to subsist during the non active season.

**tension**
The stretch or strain that pulls a thing tight. Wire used to strengthen wax foundation so that it does not subsequently sag and so that it gives support to honeycombs that will withstand rapid spinning in the honey extractor, is usually pulled very tight so that when plucked a high note is emitted. Various techniques ranging from bending the side bars inwards and releasing once the wire is fixed or anchoring and then pulling the wire to the correct tension are used. Crimping does away with these techniques and easy-to-use devices are available to do that. The bees can detect when a manipulator is 'tense'.
See: wiring frames.

**tenth segment**
On both sexes the tenth segment is reduced to a small conical lobe bearing the anus. In the female it is entirely concealed in a chamber at the end of the abdomen that contains the sting.

**tentorium**
The chitinous bars and bridge which form a framework (tentorial bridge) within the head of an insect. The framework associated with the cavity of the brain.
See: phragma.

**terebra – actual sting**
An ovipositor modified for stinging.

**tergite**
The dorsal sclerite of an insect's abdominal segment. The backplate. The tergite of the prothorax encircles the neck like a collar and is swivelled off when one views the first thoracic trachea for mites (acarine). The largest thoracic tergite forming the strongly domed scutum is useful for 'marking' with paint (or attaching a small sticker) without injury to the insect. It is never wise to mark a queen until she has completed her mating flights. Abdominal tergites overlap one another rearwards and are joined by intersegmental membrane, They also overlap the ventral sternites.

**tergum**
A back plate, the cuticular plate on the dorsal side of a segment on an arthropod (bee).
See: tergite.

**terminal filament (of ovariole)**
Each of the queen's 150 or so ovarioles begins at its anterior end as a thin thread or terminal filament. These adhere together and are attached to the ventral side of the heart at the anterior end of the abdomen.

**terminology**
Words or terms used in any branch of science, art or craft. Whether American, old, new, quaint or technical – every attempt has been made in this compilation to include and explain words used in any form of beekeeping literature in the English language.

**termites *** Isoptera**
Pale-coloured, soft, mainly tropical, social insects. Also called white ants. An order of exopterygote insects resembling true ants though independently evolved. Reported as a problem where wooden beehives are kept in tropical areas. They require to stay out of the daylight and to travel in tunnels. These can be very extensive. They are able to use the sun's polarized light and the moon for navigation.

**terpenes $C_{10}H_{16}$ and $C_5H_8$**
Derived from cyclohexamine and the camphors. Monocyclic hydrocarbons which occur in essential or volatile oils. They include: turpentine, eucalyptus, aniseed and cinnamon oils also attar of roses and oil of wintergreen.
See: essential oils, flavour and Nasonov.

**terrain**
The surface of the earth particularly the natural features of the land. As the type of terrain is of fundamental importance to beekeeping, wherever possible the potentials of each specific type have been covered under the respective heading.
Hence: bog, bracken, bush, chalk, coast, Downs, forest, grassland, hedge, lake, marsh, meadow, moor, mountain, orchard, pasture, refugium, roadsides, site, wasteland, waterways and wilderness.
See: foraging area.

**Terramycin**
An antibiotic oxytetracycline hydrochloride (OTC). TM – 50D is twice as strong as TM – 25 and five times as strong as TM10. It has been recommended that this drug, which is used for the treatment of EFB should not be fed at the same time as Fumidil B as they tend to inactivate one another. Some

formulations have corrected this by altering the pH factor. However as Terramycin is unstable in honey and syrup (degrades within 6 weeks or so), it is best fed as a mixture with icing sugar as dust – 1 : 5 sugar. It is stable in the dry form and can be mixed with sulphathiazole. This information is from American sources as a figure of 3·25g in syrup at 3 – 4 day intervals has been quoted but its action is inhibitory. See: Canadian experiment under Oxytetra

**territory**

An area over which an animal, bird or insect operates. Some like the robin jealously defend their territory mainly against their own species. Honeybees seem to be able to largely dominate visits to melliferous sources (except against their own species) within about three miles of their hive or nest. Mating can bring in drones from considerably further. They can operate over a much wider area than workers and queens only slightly less so. Bumble bees seem less versatile with regard to foraging and tend to work the territory in waves.

**terror**

Intense, overpowering fear. It has to be admitted that some people are genuinely terrified of bees and it is essential to take this into account when dealing with bees that could be encountered by other members of the public. A bee has so scared a car driver that the centre reservation of the motor-way was crossed. The mind boggles at the unpleasantness of certain possibilities. Public education has been almost entirely in the province of the beekeeper.

**test**

The procedure adopted to ascertain the properties, qualities or condition of something. A complete analysis of honey and beeswax has been offered by various sources. A trial or exams are alternative meanings of 'test'. To put someone to the test means to check their ability to do something. Because few consider all the ramifications of beekeeping until they become involved it is necessary to grade tests so that skills can be continually developed. No-one knows it all. See: exam/trial.

**test comb**

This is a comb containing normal healthy brood, especially eggs, that is put into a colony suspected of being without a queen. A queenless stock will almost certainly set about building queen cells over some of the brood and a queenright stock will just develop them into normal brood. Prior to this a feed might trigger off some egg laying. As combs are not always interchangeable a 3 cm triangle containing eggs can be cut out and inserted into a comb in the queenless lot.

**test for beeswax**

To check the authenticity of a sample of beeswax we require to know its specific gravity, melting point, resistance to acids and its solubility in certain liquids and its flammability. When an alkali is added no glycerine is liberated as would be the case with other fatty acids. It does not react with chlorine or bromine. Comparison with a similar piece of known beeswax helps diagnosis: check floatability smell when burnt, texture, colour, aroma, acidity. Float in water and add alcohol until it sinks. Its acidity can be discovered by dissolving the wax in warm alcohol when blue litmus should take about 15 mins to turn red.

**test for honey**

Specific tests for diastase and HMF content form part of the EEC import regulations. So much is now known about honey that numerous tests have been devised to measure its density, optical rotation, colour etc. See: pfund, polarimeter, honey judging and properties of honey.

**test for insecticide contamination**

Bees and contaminated material put into a sealed polythene bag with air enough to keep bees alive for some time. If material is not contaminated the prolonged life of the bees gives a check on the material being tested. A control bag - i.e. the same amount of air in bag with bees but without suspect material.

**test gauge**

A 'feeler' that enables you to check the width of a gap such as queen excluder slots, bee-space etc. Bottom bars can be useful though they are usually ¼" whereas five sixteenths approximates more closely to bee-space.

**tested queen**

A queen that was mated and checked as successfully laying worker brood. The test can be extended to verifying that her offspring have the desired characteristics.

**testicular tubule (testiole)**

The testes of the drone are composed of bundles of tubules in which spermatozoa a produced and matured.

**testis (testicle) pl. testes**

The male sex gland. In drones just emerged from their cells the testes are relatively large, white, bean-shaped bodies that occupy a large amount of abdominal space. By the time they are mature (about 13 days after emergence) they have become small, greenish-yellow scraps of tissue their contents having passed into the seminal vesicles via the vas deferentia.

**tests**

May be carried out for: cane sugar, colloids, foul brood (milk test), honey flow, h.viscosity, honeydew, need for water, purity of wax, queenrightness, setting of honey sweets, ultraviolet light, Varroa mites etc.

**tethered queens**

In the interest of observing the mating behaviour of drones, nubile queens have been tethered with fine nylon thread to hydrogen balloons so that they fly at a height where copulation might occur. As the queen's scent trail would not be free to move normally through the air, as its point of origin was so much more restricted than if she were free. It was originally supposed that drones always flew up-wind to find nubile queens but in a moving air masse the queen might leave a scent trail in any direction relative to the actual wind,

**tetraploid**

Having four times the haploid number of chromosomes in the nucleus – polyploidy.

**texture of wax**

The smooth, silky feel of beeswax goes with its aroma and translucent nature to render it rather special (unique) and candles, models and other beeswax artefacts have been described as having a jade-like quality. Provided alkalis, including hard water, are kept from the wax its texture is non-granular and non-greasy. When we used the word texture in connection with beeswax we are referring to that special and essential quality that is conveyed by touch and its pleasing non-granular appearance to the eye.

**thanetosis**

Adopting the pose of 'rigor mortis'. The habit of feigning death (e.g. on the part of the queen) presumably with a view to causing possible predators to lose interest. Wax moth larvae and certain other creatures display this phenomenon. Thanatos ' Greek mythological personification of death.
See akinesis, epilepsis, catalepsy and action inhibition.

**Theophrastus**

Encyclopaedist – Greek philosopher. Referred to several times for his writing about beekeeping matters in Dr. F. Malcom Fraser's "Beekeeping in Antiquity.

**theory**

A coherent group of propositions that account for known facts as opposed to a hypothesis which may be no more than conjecture. Plants are observed to give rise to honeys described as dark, medium or light. When reports are received that blackberry honey from one quarter is medium and that of another light we may theorize as to whether this is due to the type of soil (acid or alkaline) or the rate of secretion. Certain honeys, notably from honeydew, darken on exposure to air. It is my theory therefore, though it would have to be proved by further investigation, that light honeys are obtained where 'flows' are rapid and the honey is processed and stored quickly.

**thermal conductivity**

This is very low when honey is cold but increases with temperature rises and decreases as the water content goes up. Quoted as 123 at 2C, 143 at 71C when water content is 21%.
Thermal conductivity is the number of calories flowing across a 1cm cube per second when opposite sides of the cube have a 1°C temperature difference. Wax is also a slow conductor of heat and combined with honey their properties assist the winter cluster if a good canopy of sealed stores is present.

**thermal vaporizer**

Certain formulations of insecticides are available for insecticidal slow release units or thermal vaporizers. In 1977 they included: Vapona, Sectovap, Mafu, Flitox, Cooper flystrip etc.

**thermometer**

An instrument that measures temperature. There are now available sensitive 'probe' types that enable readings of cluster temperatures or other interesting facets of beekeeping thus helping us to further understand the ways of the bee.

**thermostat**

An automatic device that controls temperature by means of a relay activated by thermal conduction or convection. It has many uses including warm tapes for liquefying honey that has granulated in large containers, for keeping fermenting mead at the best temperature and for controlling a cabinet where wax is filtered or solid honey cleared or Queen cells incubated.

**thesis**

A proposition that advances an original point of view. There are accepted ways of setting out a thesis as required for certain aspects of research or the upgrading of an educational 'Degree"

**thiamine**

Vitamin B or aneurin. A hydrochloride of an organic base. A white, crystalline solid of the vitamin B complex essential to all forms of life – required by the nervous system. It is found in honey.

**thin layer chromatography**

This is identical in principal with paper chromatography except that the medium instead of being paper is a thin layer of absorbent, inert substance (such as kieselguhr) coated onto a glass or plastic base.

**third party**

All paid up members of BBKA (through affiliation from local BKA's) are automatically covered for claims against them for injury, damage to property or damage due to deleterious matter in food, including claims from stings (but not nuisance caused by bees). Maximum cover is 500,000 though beekeepers must bear the first £100.

**thistle *** *Carduus & Cirsium* spp. Compositae**

A prickly-leafed weed with characteristic flower shape. Fluffy, white seeds spread by the wind. Flowers mostly purple. Bees work many varieties of wild and cultivated thistles and gain quantities of nectar and pollen. The honey is graded as light and of excellent flavour while pollens are whitish and sometimes described as transparent. Milk or holly thistle – *Silybum marianum*.

See: globe, field and Arabian thistle and teasel

**Thixotropy**

This is a physical property of a substance in which it becomes gelatinous (jelly-like) when left undisturbed but returns to the fluid state when shipped or agitated. An isothermal reversible gel-sol-gel-sol transformation induced by shearing and subsequent rest. Heather honey (Calluna vulgaris) and Manuka have this property and Karvi honey (Carvia callosa) from India. See: dilatant.

**thoracic salivary gland**

Large, paired salivary glands each with their own reservoir are found in the thorax, their ducts leading forward to a common median duct shared by the salivary glands of the head (post cerebral). The thoracic glands are developed from what in the embryo were two long, kinked silk glands.

**thorax**

The portion of a honeybee's body that lies between the head and the abdomen. It is largely filled with the indirect flight muscles though the aorta, oesophagus and nerve chord pass through and there are thoracic salivary glands. It also includes three pairs of spiracles and air sacs. The appendages include three pairs of legs, one pair to each segment and after the prothorax and two pairs of wings. The scutellum of the queen is reasonably bald and impermeable and offers an ideal place for placing identification marks.

**Thorley Rev. John**

In 1744 he produced 'Melissa Logia' and observed that the wax scales were carried in the wax pockets and also made several anatomical discoveries, e.g. the honey stomach. He was opposed to sulphur killings and in 1774 he wrote 'An enquiry into the nature, order and environment of bees.

**Thornes**

One of Britain's and Europe's leading suppliers of all things appertaining to beekeeping. E.H. Thorne (Beehives) Ltd, Beehive Works, Wragby, Lincoln, LN3 5 LA. Edgar Henry Thorne started producing hives and equipment in 1912. His son Leslie Thorne established one of the most go-ahead firms for the supply of beekeeping equipment and in 1984 incorporated the whole of the well-known firm E. H. Taylor Ltd.

**threshold**

A place or point of entering or beginning. A level of quantity, temperature or concentration at which a particular occurrence begins. For example there is a threshold limiting value (TLV) for maximum concentration of pollutants, say carbon monoxide (CO) of thirty parts per million. Bees become aware of pheromones when their concentration reaches a certain threshold level. The 'alighting board' could also be termed the threshold. Likewise the area next to the entrance, it is here that tell-tale signs are often seen, the casting out of brood when near starving or of drones, dead virgins or intruders that have been stung to death.

**thrift *** *Armeria maritima***
**Plumbaginaceae**

1. Sea pink is common on cliffs and salt marshes. Called 'plant of sympathy' it likes the sea and forms grassy mats and their carpets of rose-pink, honey-scented flowers arrive from April to August providing nectar and pollen for the bees. Pollen is dark yellow with a 40mu grain.
2. Thrift when it applies to economical management, frugality to permit successful wintering, yet vigorous growth when appropriate, is an important honeybee characteristic which along with stamina, wing power and self-defence go to make the most successful colonies.

See: sea pink.

**throw**
This word has been used to describe the behaviour of a colony in producing a swarm or cast. Quote ' a hive may throw several casts'. The word hive being less suitable than 'colony' in such circumstances. It is true that the queen is often seen to leave with some reluctance after many bees have taken to the air, in other words perhaps she is 'thrown' out. A beekeeper might throw a swarm in front or into a hive (shake would perhaps be a better word). Honey is thrown (spun) from a comb by centrifugal force. A beekeeper could be thrown by problems a colony might present him.

**throwing out**
Much information can be gathered from the throwing out of materials which bees discard. Particles of comb might suggest robbing or the presence of a mouse. Crystals might mean granulated stores that the bees are unable or unwilling to utilise. General debris may accumulate on the floor board after the depredations of the pollen mite. Dry, white larvae may indicate the colony is near starvation if they are worker larvae or merely a change of need (status) if drone larvae. Surplus virgins may be cast out after swarming fever has subsided. The autumnal massacre of the drones is a well-known phenomenon. The ejection of particles of newspaper almost chewed into powder will follow the newspaper method of uniting. Predators and articles offensive to the bees but too large for them to remove, will be rendered hygienic by coating or sealing under propolis.

**Thuja plicata**
The tree that produces Western red cedar and native to N. America. It has deep-green foliage, grows to almost 200ft (60m) and is used for ladders, poles, boats and weather boarding as it is said to last for a life-time out-of-doors without chemical treatment. The wood is scented (not necessarily the bees' favourite smell) and the name cedar was given because of its fragrant foliage. It is widely used as a hedge shrub as it makes a dense evergreen barrier and stands clipping well.
See: Junipers and western red cedar.

**thunder**
A rumbling, reverberating noise that can often be heard from afar. It emanates from the vacuum caused when a charge of lightning passes through the air. When overhead it can cause an ear-splitting 'crack' which can be upsetting to humans and presumably bees. At any rate in the dry weather bees can show signs of being unusually tetchy though this may well be associated with an abnormally high static charge in the air and to which they are particularly sensitive. Heavy showers are often associated with thunder and this fact together with darkening skies would also lead to surliness on the part of the bees. It is often supposed that bees can anticipate a storm and an apiary may suddenly become alive with hosts of returning forgers that are regarded as harbingers of nearby bad weather.

**thyme *** Thymus spp. Labiatae**
A hardy perennial that grows to 15cm. Nectarwise this is a rain-sensitive plant producing a mass of bluish-purple flowers in June and July. The plant is dried and used for seasoning. Mediterranean countries can produce almost pure samples of this fragrant light honey (including the famous Mt, Hymettus honey) with its mild, spicy flavour. It likes well-drained soil and a sunny position. Varieties include: T. pulegioides – larger wild thyme, chamdedrys, serpyllum, vulgaris (6" small pink-whitish flowers July/ September and capitatus. Pollen is very pale-green to brown or dull yellow ' grain 25mu. Thymol is also produced from this plant.

**Thymol $C_{10}H_{14}O$**
A crystalline phenol. If enough is added to prevent the syrup fermenting then the quantity is enough to damage the bee's gut. An antiseptic that will preserve a sugar solution against mould or fermentation. A. C. Waine suggests buying 2% solution in

surgical spirit and adding a spoonful (5ml) to a pint. 9 ml to a litre if bees need feeding. Ron Brown recommends 4g crystals in 12oz medical bottle surgical spirit. Then level teaspoonful of this solution to every quart of thick sugar syrup fed after end of Sept. Generally additions to sugar syrup are not encouraged. The best advice seems to be get good 2:1 syrup into the colonies in ample time for them to store and seal it for winter.

**tibia pl. tibiae**

The third segment of the leg between the femur and the tarsus. On the worker's hind legs the outer face constitutes the corbicula or pollen basket. The tibio-tarsal joint on the worker forms the antenna cleaner on the first leg. The pollen press on the hind leg. The tibiae on the middle legs each carry a spine. As several pieces of significant apparatus are to be found between this and the basitarsus on worker's legs the expression tibio-tarsal joint is used frequently to describe this articulation of the legs. See: tibio-tarsal joint.

**tibio-tarsal joint**

The front and hind legs of workers are equipped with special apparatus in this area. The antenna cleaner on front legs (in common with both other castes) and the pollen press of the hind legs. See: basitarsus.

**tie**

To fasten with rope, cord or string. It is often necessary to tie things together, to keep bees in for transport, or to secure a hive against wind or predators. It should be remembered that ordinary rope stretches when dry but becomes taut when wet. Tying a bee hive for transportation can be done with rope which will give an excellent hand-hold but every strand must be drawn as tight as possible and only counted on for a few hours as the ropes soon work loose. Staples, cleats, tapes are more commonly used. See: Emlock.

**tiering**

In nature combs are more often than not built downwards from the point of attachment and honey is stored only at the top of the

central combs and a canopy is formed by increasing the stores until the flank combs may be full. The beekeeper however, wishing to increase the capacity of the hives for stores without disturbing the brood nest is inclined to pile extra boxes on top rather than below or horizontally. Most modern hives are tiered or 'supered', the extra boxes for honey going on top. See: long-idea hives and supering.

**Tilia spp.** ***

The Basswood, linden or Bee tree which has yellowish or cream-coloured flowers and more or less heart-shaped leaves. There are many plants of this genus. It is often planted in avenues and has fine grained wood T. American a shade tree used too for making honeycomb sections. It flowers June - Aug yielding best at temperatures 66 – 70F. The honey is light green and 'minty'. *T. vulgaris* is widely planted in parks and 'gardens

and is vulnerable to aphid secretion which in turn attracts the soot fungus and in the London area has made honey very dark. Over 20 species, some late flowering and one or two producing toxic nectar (bumbles often seen crawling as if drunk on the grass under the trees). *T.oliveri*, *petiolaris* and *orbicularis* the weeping, silver lime. See: lime.

**till**

A soil formed by glacial transport, a mixture of clay, sand, gravel and boulders.

**'tilt' method**

This is a way of checking for queen cells by lifting one side of the upper brood box as if the boxes were hinged and looking for queen cells by puffing a little smoke in the split so formed, Care must be taken to prevent supers when present from sliding off. It is necessary to have two or more parts for the brood and the method relies on the queen cells having been built or started where they will be visible when the upper part of the hive is tilted away from the lower part. A useful idea as it saves time and disturbs only slightly though it is not always reliable. though.

**timber**
Wood in the form of trees intended to be milled for joinery. The resultant planks that emerge at the saw-mill. Although we often find that the word wood is used when timber would be more meaningful. Places where timber is grown often provide good sites for permanent or temporary apiaries. Timber is now sold in metric sizes in the form of planed or sawn.
See: lumber, pochote, softwood, timber for hives, wood and western red cedar.

**timber for hives**
Prime, clear Western Red Cedar is the soft, light, easily worked, weather-proof timber specified for most hives in G.B. It is however expensive. It has to be imported, mainly from Canada. NZ and Australia widely use radiate pine; Ponderosa pine and deal are also used.
See: hive materials. The harder wood is used for frames.

**time**
A non-spacial continuum only but dimly understood by humans who tend to assume that events occur in irreversible succession.
See: day, geological time, length of and time lag.

**time lag**
The time needed from the moment when an event is initiated to the actual occurrence. Six weeks from egg laying peak to foraging peak. Sixteen days from the sudden demise of a queen in the active season to the arrival of virgins. You must allow for time lags by planning ahead. For example a swarm is initiated many days before it emerges. There is a time lag between the initiation of queen cells and the laying by the resultant queen and then the hatching of her progeny. Adult drones take at least a week longer to mature than queens.

**time table**
A schedule for a planned sequence of events. This becomes significant when Field Days, Week-end Courses, queen rearing and many other aspects of beekeeping have to be committed to paper or computer.

**tin**
A can or air-tight container of metal (tin once widely used) or plastic, ubiquitously cylindrical but formed in other shapes to suit particular circumstances. The coming of plastics rapidly replaced honey tins providing the same stacking possibilities, ease of sealing and cleaning, Honey Farmers use 300kg steel drums with lacquer linings and special trolleys, dollies, hoists etc. are made to handle them.
See: drum, honey tin and packing tins.

**tin smoker barrel**
The fire chamber or smoker barrel of the smoker, like the exhaust of a car, has to undergo oxidisation on the outside and coking-up with carbon on the inside. Under the onset of these conditions, accelerated by heat, a resistant type of metal is advisable – tin and copper are easily workable into the shapes required and answer quite well to the requirements mentioned. A smoker with its fire chamber and also presumably its nozzle will be made out of suitable long-lasting material.

**Tinsley Joseph**
Lecturer in beekeeping at West Scotland Agricultural College. Wrote: Preparation of honey and wax for the Show bench 1908. Wintering of bees 1928, Beekeeping up-to-date and the Rearing of Queen bees 1944. He made useful researches into queen mating and wintering. Expert BBKA and SBA and 'Judge'.

**tin snips**

Small hand-operated shears for cutting sheet metal. Roof coverings, mouse guards, metal ends (spacers) and hosts of other apicultural items that are developed from sheet metal, perforated, slotted or plain, call for the use of snips.

**tips**

Hints and useful pointers abound throughout this work but are too numerous to be listed separately.
See: cautions and hints.

**tissue**

The substance or cellular material of organic matter. An interwoven mass. Usually fairly soft. An aggregate of cells and cell products which form a definite kind of structural material in an animal or a plant. Paper tissues are useful (as is foam rubber) for the temporary blocking of holes or cracks to prevent bees from getting where they are not wanted. An empty 1 lb. honey jar lined with tissue can be used to transport a delicate item like a queen cell.

**tit (titmouse) *** Parus spp. Paridae**

This small bird causes the greatest damage to bees during the winter months when it lures bees from the hive by tapping with its beak near the entrance. A bee that goes to investigate is promptly devoured. They will collect dead bees from in front of the entrance to feed their young in summer.

**toad *** Bufo spp. Bufonidae**

The common toad is a frog-like, tailless amphibian, a reptile that can take a bee in a flash, especially at the edges of ponds where bees go for water. B.marinus is troublesome in the USA and in Australia where this toad was introduced to keep down sugar cane pests it has developed a great liking for bees so hives must be kept well off the ground or behind chicken wire.

**toad-flax Linaria spp. ***
Scrophulariaceae**

A showy, two-lipped flower, yellow or orange. There are wild and cultivated species providing the honeybee with useful minor sources of nectar and pollen in June. Varieties can grow to lm and the pollen reported as medium yellow to pale yellowish green.

**tobacco Nicotiana spp.
Solanaceae**

Flowering tobacco. Many species have long-tubed, starry flowers that open in the evening. *N. tabacum* is cultivated for its leaves (using best quality soil for harmful purposes). Nicotine, a poisonous alkaloid derived from the tobacco plant is used as an insecticide and in some forms of control where Varroa is present.

**tomato pl. oes *** Lycopersicum esculentum Solanaceae**

A tender herb grown for its fleshy fruit (yellow or red). Bees have to be persuaded to work its flowers presumably because they have no attractant (or have a repellent) but better crops can be obtained with bees.

**tomentum pl. tomenta**

A band of fine hairs on the middle three abdominal tergites of workers. Closely matted hairs, pubescence – with hairs pressed close to the surface.

**ton**

A unit of weight: Metric tonne = 1000kg or 2205 lb. A British ton = 1016kg or 2240 lb while a US ton = 907kg or 2000 lb.

**tongue**

The ligula or glossa A freely movable oral organ, A long, inflatable part of the bee's proboscis. Drone and queen have relatively short tongues. The worker's has a length of some 5mm and a canal about one third of the thickness of a human hair with a slender rod running through its length. It is thickly covered with hairs and the distal end bears a

spoon or bouton.
See: glossal canal and groove, and short tongued bumbles.

**Tonsley Cecil C FRES, BEM**

One of the best-known and respected recent British beekeepers.
He was Editor of the British Bee Journal, wrote Honey for Health and many leaflets on honey. In turn he became Chairman then President of BBKA a Lecturer, Judge and Bee Farmer.

**tool**

An instrument designed to assist a person to carry out certain kinds of work. One very useful and necessary example is the 'hive-tool'. The construction of hives, frames and equipment of various sorts, together with maintenance, calls for the use of various tools according to the material (wood, plastic, metal) to be worked upon. Brief details of some that are required for dissecting: manipulating and handling bee materials are given under respective headings.
See: appliance, device and gimmick.

**top bar**

The strongest (supporting) bar on the frame which carries the whole weight when the comb is filled with honey. The two ends become lugs that rest on the runners and can take WBC or various types of spacers. They can be either saw-cut or have a wedge running along the underside to hold foundation in place. They are supplied in two widths for economy or exact bee-space. A recent interest has been aroused in using only the top bar so that the comb can assume its natural 'catenary' shape and do without side or bottom bars but such combs are vulnerable when handled and are wasteful when crushed in the course of extraction of little use in extractors!

**top bee-space**

The height of runners is arranged to allow the bee-space to be formed at the top of the chamber (as opposed to the bottom).
It has many advantages but must never be interchanged with bottom spacing because either 'no' bee space or else 'double' bee-space would ensue. It is standard in the Smith and in American hives allowing freedom for the passage of bees in winter, better ventilation and frames stay put when the upper cover (crown board) is raised.
This can be merely a flat board and gives space for bees whereas in the case of bottom spacing the tops of the frames and hive walls are 'flush' allowing no space for bees which can immediately spew over the edges when the crown board is taken off for inspection.
Useful too in observation hives or nuclei which can utilise both top and bottom bee-space. It also gives space for prophylactics such as Apiguard and makes access to feeders easier.

**top crossing (genetics)**

A superior male (pollen, plant or sire) is identified and crossed with a large number of individuals in the breeding population. A valuable method for rapidly increasing the frequency of desirable characteristics within breeding populations. Bee breeding has problems regarding the application of such techniques but the experts are gradually overcoming this.

**top entrance**

A means of access and egress for the bees at or near the top of the hive. Bees will accept an entrance in almost any position close to their nest but for the convenience of beekeepers it is more usually placed at the front of the floorboard. However as a flight hole for drones, protection against falls of snow, in multi-queen hives and over split boards and to provide additional ventilation or to integrate with queen excluders, top entrances are used.
See: bottom entrance and disc entrance.

**top feeder**

Feeding bees with syrup is usually done from the top using various types of feeders but when an over all feeder takes the place of a super it allows for a considerable amount of food and there are two well-known types: Miller and Ashforth the former having a feeding portion at one end and the Ashforth having a feeding strip running across the centre.

**top fruit**

Tree fruit as opposed to a bush or canes. Pomes. Their honey is often dark and honeybees are nearly always the most useful and effective pollinators. Almond, apple, cherry, pear, plum, quince, medlar, peach and nectarine are examples.

**top packing**

As heat rises and can escape from the top of the frames they are normally kept covered by a 'crown board'. Bees packed between the combs are referred to as 'seams' and these together with the canopy of food form an insulating cover. However extra packing can be put over the crown board and this would be called – 'top packing' and of course it is covered by the roof. It is rarely necessary in this country but beekeepers are often tempted to put extra insulation there in the spring when the brood nest begins to expand. If such packing is used it is essential that it remains dry and does not consist of material which the bees could get at and chew nor should it impede ventilation.

**top heavy**

Likely to overbalance because the greater weight is on top.
Hives stacked vertically on a firm base are unlikely to suffer through being in this condition but on spongy ground a slight tilt may develop until the hive falls over. Also when travelling with hives they should be securely fastened so that they can withstand severe jolts and application of brakes.

**topple over**

To tumble over and fall. Should a hive containing bees do this, trouble can be reduced if the hive is quickly re-assembled. Frames tend to fall out once a box is upside down but a stick or a board promptly placed under the top bars may help to lift it upright frames and all. When carrying bees a smoker should always be kept ready in case of such an eventuality.
See: top heavy.

**top spacing**

This refers to bee-space and is covered under 'top bee-space'. Combs must be surrounded by bee-space in all directions both vertically and horizontally. Langstroths and Smiths use top spacing other British types usually use bottom spacing.

**top ventilation**

It is important that the products of metabolism – waste gases, moisture etc. are free to be carried away and neither accumulate nor lead to condensation and mould. Roofs should have ventilation holes to allow for this. It has been suggested that crown boards should be raise by a minute amount (as on match sticks) but leaving a feed hole 'open' may well lead to the propolisation of the roof or become sealed with propolis if covered with gauze. Where bottom spacing is used as on National hives, lack of space at the top (as top bars are flush with the side walls of the hive) calls for rimmed crown boards or in winter the laying of bottom bars (¼" square section) across the frames which can improve ventilation and give bees a passage across the frames. An eke is also required when certain prophylactic treatments are necessary.
See: ventilation.

**torch**

A portable light produced by the flame of flammable material (including beeswax?) wound about the end of a stick and ignited. A red torch gives some light for us to see in the dark but is not visible to the bees. As moving bees often calls for waiting until dusk to shut them up, bear in mind that a torch is often a necessary requirement.
See: blow lamp and flashlight.

**torment**
To agitate or upset a great deal. Unfortunately a little knowledge is a dangerous thing and some youngsters have been known to throw stones at hives or shoot airgun pellets at the entrance when their owner is not in the vicinity. Vandalism can be covered by insurance but wherever possible hives should be kept in an apiary where they are reasonably secure and if possible inconspicuous. Education of the young (and those older) also helps

**torn**
Pulled apart by force. Bees will tear and remove offending material – newspaper polystyrene etc. but when honeycomb is damaged you might well suspect robbing. A tear in a bee suit could lead to trouble and a damaged queen excluder can render it ineffective. If AFB strikes then cappings over affected brood may be sunken or torn.

Torres Luis Mendez de Spain 1586
He was the first to give a description of the queen bee as a female which laid eggs. The queen was the mother. – 'the leader bee in the hive lays eggs' He wrote 'Tractado breve de la cultivacion y cura de las colmenas.

**Torulopsis - a yeast plant.**
Torula and zygosaccharomyces are yeasts found in honey and will bring about fermentation at quite high sugar levels within a 50 – 80F temperature range. Bear in mind that when fermentation occurs in the top layer it is very slow to diffuse into the honey below so check whether only a small layer at the top is affected.

**touch - sense of**
The ability to ascertain the nature of something by feeling it. Bees have highly motile and sensitive antennae and mouthparts. Also the sting palps are able to determine whether or not a surface warrants stinging. Frequently, as with our own senses, the sense of touch is reinforced by, or related to, the responses from the other senses. Hence when a worker touches another's antennae with its own, the sense of feel and smell together with other behavioural responses go to make the action akin to our conversation.

**touchwood**
Tinder – wood that has died and lost its moisture, usually by the action of fungi. Such dry wood will as likely as not smoulder when ignited and dry rotten wood when available in clean, handy pieces makes useful smoker fuel.

**Townsendia spp.**
Aster tribe, a N. American hardy alpine with large, daisy-type flowers in summer against tufts of narrow, green leaves

**toxic**
Harmful, poisonous – for toxic substances in honey see 'HMF' and other toxic materials with applications to beekeeping are listed under their headings – creosote, carbolic, formalin etc.

toxic honey
Quite often in dry years, when honeydew is plentiful there are reports of bees suffering from toxic nectar. Mannose (reported by von Frisch in 1930 as accumulating in the bee gut as it cannot be broken down. Plants implicated include: Azalea pontica, R.aboreum, buttercups, laurels, some limes.(*T.petiolaris, tomentosa, orbicularis, argentea*). Substances confirmed as andromedotoxin-acetylandromedol etc. but higher molecular sugars also either prove repellent or poisonous (bumbles are more prone to suffer) yellow jasmine Gelsemium sempervirens and mountain laurel, poisonous ivy, kalmia, nightshade, belladonna alkaloids and NZ honeydew from *Coriaria arborea* where aphid *Scolypopa australis* is found many reports from instances abroad. Melezitose found in honeydew from limes caused high mortality as from *T.platyphyllos.*

**toxic pollen**
Plants whose nectar is suspect usually have pollen that comes under suspicion too.

Those frequently mentioned as being harmful include: buttercup, Ranunculus, rhododendron, foxglove (Digitalis) and scabious. There is unease too about the effect of certain insecticides and chemicals on the pollen.

**toxic subjugants**
Substances that can be burnt in the smoker and which render bees temporarily immobile include: dry, ripe puff ball, ammonium nitrate, chicken feathers and saltpetre. Benzaldehyde has been used as a repellent (for super clearing) and carbon dioxide when A.I. is carried out on queens. Anything that has the effect of immobilizing bees almost invariably causes them to be ill-disposed upon arousal.
See: smoker fuels.

**toxic sugars**
These are mainly polysaccharides (compounds of high molecular weight) formed by the condensation of mono-saccharide units. They are either soluble in water or form colloidal solutions. The breakdown and synthesis of sugars such as occurs in essential stages in the respiratory processes of bees, requires the presence of an appropriate enzyme. Some such are: galactose, lactose, raffinose, stachyose, rhamnose, mannose and melezitose.
See: toxic honey.

**trace elements**
An element found in minute amounts in an organism enabling it to maintain certain essential processes. In plants the content of trace elements is roughly equal to that of the human body - 0·02% of the dry matter. Honey and other hive products contain many of these trace elements (1) essential: Fe, Zn, Cu, Ma, Mn, I. (2) minor: N, V, F, Br, Cr, Al, Sr. (3) unessential: Si, Li, Ru, Ag, Cs, B, Mg.
See: elements and honey elements.

**trachea *** pl. tracheae**
A spirally reinforced, breathing tube which conducts air from the body opening (spiracle) to the air sac and the transverse commissures. The bee's breathing tubes resemble the human windpipe. They are epithelial tubes held permanently open by a cuticular lining of taenidium which forms continuous spirals that prevent the walls from collapsing and allow the efficient intake of oxygen and discharge of waste gases. They lead from the spiracles to airsacs which are connected by similarly constructed transverse commissures. The tracheae lead from the air sacs attenuating into tracheoles. The anterior spiracles are associated with the queen's piping and are the seat of the acarine mites.
See: spiracles, trachea, taenidium and tracheal sac.

**tracheal sac**
Longitudinal expansions of the trachea. A tracheal sac acts like a bellows and is connected to the spiracula trachea and also connected by transverse commissures with the bilaterally symmetrical opposite member. Sacs surround the brain, are large in the rear part of the thorax and largest in the abdomen. From the sacs smaller branches ramify to all parts of the body. The inflation of the drone's air sacs is essential at the moment of coition with the queen.
See: air sac

**tracheoles**
The ramified breathing tubes which conduct air from the air sacs to the tissues. They have no spiral thickening as do the tracheae. They form a network and are open at the extremities where they contain a little of the surrounding blood plasma through which gaseous exchange takes place.

**tracheoles**
The ramified breathing tubes which conduct air from the air sacs to the tissues. They have no spiral thickenings as do the trachea. They form a network and are open at their extremities where they contain a little of the surrounding blood plasma through which numerous gaseous exchanges take place.

**trailer**
A small vehicle towed behind a large one. These can be a useful asset when hives are to be moved as for migratory beekeeping. They keep bees from the car, provide adequate ventilation and can make for easy loading and a tail board can be an advantage. Tyres should be correctly inflated and in good condition and a 'Live Bees" notice should be clearly displayed.

**trail odour**
The queen especially but also workers and perhaps drones, leave a 'follow me' trail through the air. This is thought to be detectable by other bees of the same species, identifiable and possibly even colony specific. It could account for what bees do when inadvertently let loose by a travelling beekeeper or why they so easily locate other hives. But in the case of the queen much work has been done and the pheromones involved identified, synthesised and made use of. Swarm lure is another instance of airborne signalling.
See: pheromone and scent

**trait**
A distinguishing feature, quality or characteristic. With bees we refer to traits such as - disease resistance, swarming propensity, urge to store, tendency to rob and so on

**transfer**
To convey something from one place to another. Bernard Shaw was quoted as saying, 'All problems could be solved if we could always have the right thing in the right place at the right time'. Ownership of hives or products can be transferred from one beekeeper to another. When bees were transferred from one skep to another this was called 'driving'.
Characteristics can be transferred from parent to offspring. Feral bees might be transferred into domestication. A colony can be moved a short distance without bees becoming lost but they can be transferred to a site more than three miles away and become accustomed to their new site.
See: moving bees and transportation.

**transferring tool**
When queen raising with artificial queen cells a small instrument called a grafting (or transferring) tool, shaped to help transfer a young larva from its bed of royal jelly and into the new cell can be bought or made or a small brush might be used for the purpose. Conditions should take into account the temperature and humidity and the heat conductivity of the surfaces involved. Also you have to be aware that a larva has a right and wrong way up. Its under surface will be clogged with food and the upper surface must be kept upwards so that it can continue breathing.
See: grafting and grafting tool.

**transition cells**
A cell in the honeycomb which is neither a queencell nor hexagonal. They are found between worker and drone cells, where cells touch the frames or sides of the hive and where 'pop-holes' occur. Such cells are irregular in size and depth and may have three, four or five sides. Their necessity arises from the need for rigidity while switching from one size of cell to another.
See: interstitial.

**translocated herbicide**
A herbicide which when absorbed into the plant via the leaves or roots, moves within it and finally kills it.

**translucent**
Allowing some light to sine through. Beeswax in thin sheets and honeycomb share this property giving beeswax a jade-like quality and honeycomb to be examined for pollen content or type of honey contained.

**transpiration**
The diffusion of water vapour into the air by evaporation from living plant tissue. The exudation of vapour through pores or stomata. Leaves wilt to prevent this occur-

ring to excess. Large trees give up hundreds of gallons of water to the air. It is not a physical function of the cells but purely the effect of a wet surface losing water vapour to an unsaturated atmosphere. Plants vary in their structure, many having a water-proof cutinised or suberised cell layer to reduce what would otherwise be a widespread loss. Successful land plants compromise by having perforations in the protective layer. See: respiration.

**transportation (of fixed comb hives)**
Cylindrical hives, 'jar' type, can only lie horizontal in their original position as any rotation would cause combs to slant with a severe risk of comb-collapse. Apart from marking the uppermost side, wedges should be fixed so that in transport the combs remain vertical. Log hives are similar. Skeps, suitably reinforced with cross-sticks can be carefully inverted and transported upside down the domed end being held in a WBC lift or something similar. Mouths of such hives can be securely covered by nylon mesh. See: moving hives and transporting bees.

**transporting bees**
Prepare hives a day or so before hand ensuring that they have adequate food, that parts are effectively fastened together with strapping, staples or other locking mechanisms and use full top ventilation screens over the combs. Use adequate size square section foam rubber strips with short cord tied to one end (for easy removal after they have been finally located on their new site). Firmly block the entrance with the foam once all bees are back after dusk. Ideally move in the early morning having previously advised the Police and marked the vehicle 'LIVE BEES'. Have veils and smokers with you, align frames in the direction of travel. Tie or wedge hives firmly and expose travelling screens to the moving airstream. In hot weather screens should be sprayed with clean water now and again. On arrival, set hives down to allow them time to recover from the jolting, ensuring spacing and levelling is as required and if entrances are to be replaced by crown boards – do them one by one and returning roofs but leave the removal of entrance blocks until last.

**travellers' joy *** Clematis spp.**
Ranunculaceae Old man's beard C.vitalba. Wild clematis and cultivated varieties produce a lot of greyish-yellow pollen which is sometimes collected by bees. White flower heads and feathery fruits – they like calcareous soils. and chins. (smile),

**travelling box**
A special, bee-proof, ventilated box which is strong, light and easy to secure, open and close, and carry. Designed to accommodate four or five frames (such as a nucleus) and fit to be despatched by rail or road transport. They should take a set number of frames so that they are held together firmly and the entrance must be sealable. Sometimes the queen is enclosed in a small cage for subsequent release.
See: package bees and queen cage.

**travelling cage *****
This can be for package bees, a queen and attendants, or perhaps for a swarm. Ventilation and bee-proofness are important attributes as are the need for food and moisture,. See: package bees and queen cage.

**travelling screen**
A framed screen with the same dimensions and shape as the hive periphery. It should be strong enough to fit, bee-tight, on top of the frames instead of the inner cover or crown board and to allow another hive to be put upon it while still preserving ventilation and allowing strapping or other means of fixing. While perforated zinc, plastic or similar materials are usable they must be able to resist heat, distortion and tearing. A strengthening cross-bar is sometimes advisable. Ability to spray in hot weather conditions is helpful.

**travel stain**
The surface of honeycomb though originally white becomes progressively dark-

ened due to the constant passing of bees. The marks left by the conglomeration of footprints made by bees. When virgin white honeycomb is sought it should not remain long in the hive once sealed because of this travel staining. Bees can mark glass and the entrance to their hive. See: dark comb.

**treasurer**

One entrusted with the funds of an association, to whom dues are paid and who meets agreed expenses. The treasurer is also required to keep account books and to give oral and written explanations of the accounts at the AGM. they are inevitably a member of the committee and executive council together with the Secretary and Chairman.

**Trebizond**

Medieval empire of Asiatic Turkey. The inhabitants of Trebizond have been recorded as paying taxes to the Romans in equivalents of wax. It was close to this place that in 400 BC Xenophon's soldiers, who ate local honey, suffered according to quantity, behaving as if drunk or even becoming mad. However they all seemed to recover in a day or so ' The Persian Expedition.

See: Xenophon.

**tree** ***

A tall woody perennial with a stout self-supporting main trunk which carries branches above clear of the ground. The large flowering canopies of many trees produce more flowers per acre than anything herbaceous. Top fruit, hawthorn, sycamore, horse and sweet chestnut and robinia are all useful bee plants. The timber from basswood is used for sections while pine and cedar are used for hives. Large quantities of honeydew can also be exuded from trees. One that can normally grow more than 5m high. A flowering tree can produce an acre of forage on far less ground and offers a refuge for birds, a colony or a swarm.

**tree heath *Erica arborea* Ericaceae**

A shrubby heath, worked well by bees and hardy in the mild climate of S.W. It has white flowers from Mar-Apr. These and other varieties, some will tolerate lime, a few grow bigger than shrubs and they cannot withstand harsh winters.

**tree honey**

This is derived from early sources such as: pear, plum, cherry, apple, hawthorn, and sycamore. Rather dark, slow coarse granulating honey with an excellent flavour – it has been called 'wood-honey'. It may also be used to describe the secretions of insects that feed on the phloem sap of trees: lachnids, Lachnella, costate, Cinara brauni, acutinostris, pectinate, shimitshacki, that frequent firs in S. Europe. Such insects are distributed by wood ants and the honey is highly prized and large numbers of beekeepers migrate annually to the forests. Pollen is usually lacking in some environments. See: honeydew, forest honey, bush honey, wild honey.

**tree lupin** *** ***Lupinus arboreus* Leguminosae**

A perennial border plant with white, yellow and purple flowers in June – August. It yields nectar is drought resistant is fast growing and is a luxuriant hardy, semi-wood shrub that likes good soil.

**tree mallow *Lavatera arborea***

Malvaceae A hardy biennial that grows to 5 ft. yielding both nectar and pollen. It is a robust plant bearing large, rose-pink flowers from June – August and it thrives in warm spots near the sea.

**tree medick *Medicago arborea***

Yellow flowers in spring and useful to bees.

**tree of heaven** *** ***Ailanthus altissima*,**

glandulosa Simarubaceae A handsome Chinese tree with vigorous growth to 20m (65ft) with pinnate leaves and greenish flowers in July or August. Its honey has been described as having an unpleasant after-taste but improving with keeping. A pale, greenish-yellow or greenish-brown honey. Its wonderful red colouration in the autumn is reason enough for including it in the garden where it is grown for shade and suckers freely.

**Trees and shrubs valuable to bees**
A book By M.F. Mountain London IBRA 1965.

**trees significance**
200 species of young trees were available from Rockhampton but the number of indigenous species is quite low in GB and honey from tree sources accounts for less than that from other sources while in Australia for instance some 80% of their honey comes from hundreds and hundreds of different species.

**tree wasp *Vespula sylvestris norwegica* Aculeata**
They hang their papery nests in trees or bushes instead of holes or cavities like other wasps. Possibly their greater vulnerability makes these wasps more aggressive than other types including the hornet.

**trefoil – bird's foot *Lotus corniculatus***
The trefoils are plants having compound leaves each with three leaflets – the clovers. Pods spread like crow's foot. Its widespread nature is clear from the seventy names given to it.

**trehalose $C_{22}O_{12}$**
A white, crystalline disaccharide, non-reducing sugar. Present in high concentrations in bees' blood and consumed when flight muscles are used to raise temperature to keep above the threshold of immobility. Also found in yeast and fungi.

**trestle**
A supporting frame. Australians use a four-piece transportable trestle to support hives. They are simply constructed with two horizontal beams that fit into two end pieces that provide locking cross-beams to hold the horizontals when splayed, thus providing a stable trestle capable of taking considerable weight.
See: hive stand and stand.

**trial**
A series of tests or experiments. The observation of the behaviour of bees or colonies in a particular environment. The use of bees for experimental purposes calls for time and space at places with research facilities – sadly lacking in 1985.
See: test.

**triangular**
Six triangles will fit into a hexagon (strength). A 'T' - shaped base for an extractor prevents wobble. Triangular parts of a bee include the plate of the sting and the scale/hairy patch –penis.

**tribe**
A group of animals which are classified as ranking between a family and a genus. Any aggregate of people united by ties of descent from a common ancestor and sharing the same leader.
See: classification, family and genus.

**trichogen (tricho – hair)**
A cell that produces hairs or bristles in insects.

**trickle**
A slow, irregular leak which if left unchecked can cause a lot of trouble. Honey moves silently and care needs to be taken to ensure that an unobserved trickle (or tap left on) does not occur. To see the bee's efforts go to waste is unforgiveable and a sticky mess can be troublesome to clear up. Water 'fountains' to provide open-air places for bees can use the trickle method of allowing a continuous trickle of water to leak onto a sloping board.

**tricolporate**

Pollen having three furrows and pores (Dicotyledons).
See: pollen terminology.

**Trifolium - the clovers**

Good bee plants and soil enrichers. Examples T. alexandrium, hybridium, incarnatum, pratense (red clover), repens etc.....

**trigger – off**

It is frequently found that when a comb of eggs is given to a colony known to have a virgin who appears to be slow to start laying, that this triggers her off and she begins to lay on adjacent comb. Probably nurse bees and house bees responding to the presence of the new brood put in by the beekeeper, prepare adjacent cells (feed the queen) and induce an otherwise tardy queen to begin laying. When bees get excited and their temperature rises, panic sets in and things can go from bad to worse – another example of one situation triggering another off.

**Trigona stingless bees (Moka)**

Found in S. Africa. They nest in rock-like soil, can be detected by the small whirling cloud like a mini-tornado over their tiny chimney. They yield up to 2 or3 bottlefuls of delicious non-cloying honey and they can be domesticated.
See: Brazilian.

**Triludan**

An alternative antihistamine to Piriton which for some people can be excessively sedative
See: Piriton

**Trimoline**

A commercially produced invert sugar

**triploid**

Having three sets of chromosomes. When this unusual structure occurs in honeybees it is possible for meiosis to produce haploid material. Thus even by parthenogenesis workers can be derived from eggs that have not been fertilized.

**triploid pollen**

The ploidy of pollen can have a great influence on the compatibility of pollen in pollen transfer and such plants are usually sterile or more difficult to pollinate.

**tritocerebrum**

The smaller part of the bee's brain. It sends nerves to the labrum, the frons and the fore-gut but is not easily distinguished from the other components as it integrates with the circum-oesophageal connectives that run into the sub-oesophageal ganglion.
See: brain, deutocerebrum and protocerebrum.

**trochanter**

The second proximal segment of each leg preceded by the coxa and followed by the femur. Like the coxa it is a relatively stubby and unspecialized segment similar in shape and structure to those on the legs of a crab.

**trophallaxis**

In return for feeding, the larvae furnish secretions for the nurse (or possibly other bees). 7% glucide and 1·4% amino acid. (Israel). Reciprocal food sharing, a form of communication which keeps colony members aware of a colony's requirements – for instance water requirement etc. The interplay of pheromones between larvae and adults is still being investigated. Wasps may nip a larva and cause it to secrete a watery-looking liquid, said to have the same proportions of sugar and protein as those found in human milk.

**trophi**

Mouth parts including mandibles and maxillae. The proboscis is normally retracted but the associated muscles can extend it and the tongue can be stropped and inflated to clear unwanted material from the filter-sized canal.
See: mouthparts, proboscis.

**_Trophocyte tropho_ – nourishment**
Nurse cell that nourishes the adjacent egg within the ovules of the queen.

**trophy**
A cup or plaque such as those awarded at Honey Shows for the greatest number of points (most prizes) in certain classes. It is customary to hold such a trophy for one year; sometimes a miniature is awarded as a 'keepsake'.

**tropical apiculture**
IBRA sponsor conferences on Apiculture in Tropical Climates which are held in various countries at the invitation of their governments, e.g. India 1980 and Kenya 1984. The tropical crops as reported by Dr. Free include: avocado, cashew, citrus, coconut, coffee, cucumber, mango, okra, pumpkin and piments.

***Tropilaelaps clareae* spp.**
A small ectoparasitic mite discovered in 1961 whose importance was overshadowed by Varroa but which has infested bees in the Philippines, Hong Kong, Vietnam, India, Malaya and Java. It acts in a manner similar to Varroa causing disfigured or even killing bees and has made the introduction of honeybees more difficult in tropical Asia.

**tropism**
The movement or response to the influences of external stimuli.
See: thixotropy.

**truant swarm**
**(absent without permission)**
A swarm that gets clean away so that its former owner does not claim it and therefore its origin may well be obscure. It will have the distinct possibility of being able to spread disease. In built-up areas such swarms have caused trouble and expense by setting up colonies in unwanted places. Also referred to as an errant swarm.

**truck**
A sturdy, motor vehicle with the rear end designed for easy loading. Shooting brakes, utilities (utes), hatchbacks etc. are all put to use in the transport of beehives which are best moved with an airflow passing over the ventilation screens.

**'T'-shaped**
Material in the shape of a letter 'T' or having a 'T' shaped cross-section. For example pieces with an inverted 'T' section are used to support sections in crates and also for the sturdy construction of a base for a honey extractor.

**tube**
A hollow cylindrical length of material able to convey fluids or gases and function as a passage. Closed ends as for example test tubes in a laboratory can be used for displaying or for storage. Strength and flexibility and the multitude of various substances used for their construction make them ubiquitous.
Funnels taper into tubes and in the form of siphons the control of fluid substances of all sorts become easily possible. For observation hive entrances, the prevention of robbing and the filling of feeders without splashing, candle making and so on ad infinitum.
See: tunnel.

**tui *Prosthemadera novae-zealandiae***
In NZ it is sometimes known as the 'honey-eater' and is known to frequent places that attract bees such as the honeydew of the bush country. It has white tufts under the throat and has been nick-named the parson-bird (Maori). Its melodious chirping often indicates to 'honey prospectors' where there are good spots in the Bush.

**tulip tree *Liriodendron tulipifera***
**Magnoliaceae**
Lilly tree with tulip-like flowers with greenish-yellow petals with orange internal colouring. They have a grey bark which becomes orange with age. They come into

flower when about 15 years old and in the Eastern United States they are variously called yellow poplar, white, blue or tulip poplar, whitewood cucumber tree, saddle leaf, fiddle tree, hickory poplar..... They produce aphid honeydew too. The honey is dark and initially reddish though it matures to a darker hue. It blooms for about 3 weeks following the last frosts – useful for build-up. Its wood which is soft and easily worked is used for cabinet making.

**Tunisia**
A North African country pop. 7 million with 40% agricultural workers, sandwiched between Algeria and Libya and having a long Mediterranean coastline. The Tunisian bee A. m. intermissa is smallish, dark and hasty-tempered.

**tunnel hives**
These were mentioned by Pliny the Elder as earthenware and wicker pipe hives but he may well have been referring to the 'jar' hives such as were in recent use in Malta. The references were to hives in Southern Italy though their shape has been described as rectangular, presumably meaning their cross-section and likely therefore to have been made of wood. Long knives with projecting edges at right angles made the cutting-out of comb a bit easier – were described by Columella. Forerunners indeed of the modern hive tool.

**tunnel**
A usually horizontal passage that allows movement underneath or through something. Narrowing tunnels which allow a bee-way at the end but confront any bee that tries to enter from the other direction not only a tiny aperture but sharp projecting corners. These are used for clearer boards and obviate the necessity for springs, they are usually positioned at the corners. Bees have been found useful in ground-level plastic, strawberry tunnels. The tunnels made by wax moth larvae and Braula mites can spoil combs and make their presence known.

See: Braula, Quebec board and wax moth.

**tupelo *** Sour gum USA Cornaceae**
This tree is valued in the States as a prolific yielder of mild, amber honey with a greenish tinge, flavour 'exquisite', rich in laevulose-d its high fructose content inhibits granulation a It likes hot summers and does not do well in the UK. Varieties bloom from March through June. It likes deep swamps (swamp tree) and river bottoms and some varieties make strong, tough timber while others give soft, light wood. Tall tupelos rise from the still, brown waters of the Mississippi's deep water swamps.

**turbid (containing sediment)**
Opaque or cloudy due to particles held in suspension. Mead, especially when disturbed while lees are present may have this appearance. When allowed to settle the mead can be siphoned or decanted off but may require further filtering.

**turgid**
Fully expanded due to water intake – swollen or distended but not with air. When in use the honeybee's tongue might be described as turgid but it can further inflate the organ with air in the act of cleansing. Plants lose turgidity when they 'wilt'. When some folk are stung the release of histamine causes turgidity making the region swollen and painful.

**Turkey**
Mostly covering the Anatolian plateau (Asia minor) with two million colonies of bees with a tenth of these in the Izmir region. The Langstroth hive is favoured and honey comes from clover, sunflower, vetch, cotton, sainfoin, pine honeydew (from SW coast) and heather. Huge wax processing plant and enormous number of queens produced.
See: Anatolian.

**turn**
To make anything move around a central point, pole or spindle – to rotate, twist or

wind. Many flowers turn to face the sun. Fr. turnesol 'sunflower'. A solar wax extractor should point towards the mid-day sun. Honey is expelled by rotating combs in an extractor. Bees help flowers to 'turn' into seed or fruit. You can examine a comb held vertically if you turn so that the sun shines over your shoulder.

**turpenoid**

A main group of essential oils (oils which volatilize completely when heated) found in plants and while they may merely be end products of metabolism, many perform special functions such as attracting insects or resisting fungal attack. The Nasonov gland secretion contains seven of these essential oils.
See: Nasonov and terpenes.

**turpentine**

Oil of turpentine is derived from the terebinth, a tree that grows in Mediterranean regions and the thin volatile oil of Pinus palustris. It is a common solvent in slow-drying paints (water based paints are tending to take over today) and will dissolve beeswax and propolis and is used for making beeswax polishes. Also called 'turps'. There is also white spirit or turpentine substitute.
See: white spirit.

**Tussilago *** (Coltsfoot) *Tussilago farfara***

A widespread weed that blooms early in the year and can be useful for pollen, It has medicinal properties.

**tutu *Coriaria arborea***

A tree found in NZ and said to be responsible for the production of poisonous honey. Although widespread it is only in a very local region that a certain leaf hopper Scolypopa australis, feeding on its berries has produced honeydew with a toxic content (tutin). The NZ authorities ban beekeeping whenever the honeydew is likely in that area and every precaution is taken to ensure that all NZ honey is of the highest standard.

**tweezers (forceps)**

A small metal (plastic) pincer-like implement used for plucking or for handling small objects. Quite useful when dissecting bees. A sting should always be 'scraped' out at once and not removed by tweezers which would force more poison into the wound.
See: forceps.

**twelve apostles**

Strangely enough you can often count twelve workers as the number that comfortably surround a queen on the comb.

**twin dissecting pin**

A small two-pronged 'fork' which can be made by setting two stainless-steel needles on a small handle. This allows an object, such as a bee's thorax, to be pinioned so that it will not swivel when worked upon. No.8 sewing needles set one sixteenth of an inch apart in a wooden handle and bound on either side of a 1·6mm wire with tape or frame reinforcing wire covered over with valve rubber will suffice.

**twin hive**

Side-by-side hives have been possible since vertical walls as in Langstroth hives (and others) were adopted, set alongside each other with their own queens but given common supers over the excluders. Special roofs are needed but manipulation is tricky.

**two-deck**

An American expression to indicate the number of boxes forming a hive's brood chamber – 3 deck etc.

**two queen colonies**

When by accident or design a single colony contains two laying queens. This is not the same as having two queens separated by a board and having their own entrances but under the same roof as in the Snelgrove system. Nor is it brought about by placing a weak colony over a ventilated screen on top of a stronger colony. A genuine two-queen colony may well use an excluder and super to keep the queens apart but the progeny of each will be free to intermingle and all contribute to the common store.

**type (Genetic)**

This is a group that embodies characteristic qualities – it may be a genus, specie or subspecie. It could be a strain, breed or a representative specimen used to symbolize the group. Hence: typify, typical etc.

**tyre**

Heavy rubber outer casings of air tubes on car wheels. When spent heaps if these appear as 'weights' to hold plastic sheets over farm materials etc. They can be taken away from tyre specialists for the asking. They can make hive stands or cut in two water troughs. Too ugly for roof weighting though.

**tythe bees**

10th part given to parson (end 16th cent,) Southern in Cotton according to Dorothy Galton.

# U

**ubiquitous**
Occurring or assuming to be, everywhere – omni present, e.g. yeast spores, wax moth, pollen mite. Although it is possible to ensure that any given spot is rendered free of such organisms, relaxation of hygiene leads to almost certain re-invasion. When in certain weather conditions the air is full of pollen grains they can be described as ubiquitous.

**U.K. beekeeping**
The number of beekeepers and the interest in beekeeping reached a peak during World War II. Post war governments only regarded beekeeping as a very minor adjunct to farming. and the number of colonies steadily declined. By 1977 we had only a quarter of the number of stocks that existed in equivalent European countries. At that time 32,000 beekeepers were in the 1-10 group, 3140 had 11-39 colonies, 340 had 40-100 and 135 had over 100 ' about 36,000 beekeepers in all with less than 200,000 stocks. At that time honey production was around 1,500 tonnes while 14,000 were imported.

***Ulex europaeus*** ***** gorse Leguminoseae**
A low. spiny. much branched shrub with bright yellow flowers which seem to flower the year round. In warm, dry weather it can burn fiercely and gorse fires have destroyed much heather. A popular bee plant in NZ where it provides an early boost to colonies.
See: gorse.

**Ulmus sp.**
The elm a tree planted for shade and ornament but unfortunately prone to Dutch elm disease. It has coarsely toothed leaves on one side longer than the other.
See: elm.

**Ulster Beekeepers Association**
up-date

***Ultra coelostoma assimile*** **Homoptera Margarodidae**
The second instar female nymph is responsible for producing honeydew on NZ beech trees. First instar nymphs are crawlers, adult females do not feed and males are not thought to be present. After crawler stage the insect occupies a crevice in the bark and secretes a waxy 'test' – its long proboscis penetrates to the phloem sap and excess nutrients are exuded in the form of honeydew droplets at the end of the hair-like waxy tests. This gives the trees the appearance of a 'fur-like' covering and droplets fall, if not collected and duly the 'soot fungus' blacken surroundings.

**ultraviolet light – UV**
A part of the spectrum that we cannot see yet which is visible to the bee. Nectar guide lines in flowers are often displayed in this colour being obvious to the bees but invisible to us. It is used in fluorescent assay, the long wave 3650A being most reliable for beeswax though 2537A wavelength is useful too. Honey too gives a characteristic fluorescence. It will register on suitable film but care should be taken as it is harmful to human eyes. American foul brood scales iridesce when activated by UV.

**ultraviolet honey purity check**
When a piece of filter paper is dipped

vertically into pure honey, capillary action will draw up some of the honey into the filter paper. In ultraviolet light this will show a characteristic blue fluorescence tipped by a white zone. Its intensity is greatest when the water content is low. Artificial honey does not have this property so this provides a honey purity check.

**umbel (parasol)**

A flat or rounded inflorescence having a single point of attachment. Umbelliferae (Dicotyledons), a large family including giant hogweed. Plants belonging to the Umbelliferae display this type of flower. Parsley and carrot flowers also form this 'sunshade' umbel. Other family members are: Fool's parsley, Angelica, garlic, hemlock (poisonous), pignut, sea holly, fennel, and parsnip.

**umbrella valves**

These are within the bulb of the sting and their function is to force venom from the bulb into the shaft as the lancets are protracted and then collapsing as they return. The name is derived from the shape which resembles the moving parts of an umbrella.

**unassembled**

This refers to hive parts sold in separate pieces for assembly i.e. 'in-the-flat' or knocked down.

**unblock**

To remove a block. Entrance blocks should only be taken right out when stocks are strong enough to defend themselves against other bees and predators, and when the active season is in full swing. When moving bees and foam rubber blocks have been used to render entrances bee-tight, a piece of string tied to the end of the foam and Sellotaped to the front of the hive helps when it comes to unblocking and reduces the likelihood of someone's boot treading on the string and opening up the hive as it is being moved.

A curved piece of wire can haul dead bees out of a blocked entrance should winter dead have blocked the entrance.

See: entrance block and foam rubber.

**uncap**

To remove the cap. Its significance in beekeeping lies in the slicing or carving off of the sealed waxen covering that bees put over their cells of honey. The individual cappings form a sheet (unfortunately not always quite flat) the removal of which enables the operator to spin the honey out in the extractor. This process is known as 'uncapping' and may just take a thin slice, wet with honey on the underside though in some circumstances a deeper cut is necessary when more honey and some of the cell wall is also removed.

**uncapped**

Uncapped honey cells and brood cells are undergoing attention from the bee. The uncapping of honey can take various forms. You uncap when you slice off a layer containing the cappings so that the open cells allow honey to be spun out in a whirling extractor. Robber bees can uncap honeycomb in an untidy way leaving untidy debris whereas normal uncapping is done carefully and progressively. Brood is uncapped by the nibbling of the inmate and nurse bees so that the material is eaten. A queen cell is uncapped by a virgin queen as she nibbles round and once she has removed the greater part she pushes it open in the form of a flap, as if on a hinge, and promptly climbs out.

A worker may climb in and eat up the remaining royal jelly and for a short while the cell will otherwise remain intact but then they will be eaten away leaving little remaining as each cell is built for and only used by a single queen.

**uncapped brood**

A cell containing an egg or grub is referred to as 'unsealed brood', the egg being incubated for three days and the larva being fed for another six before capping over takes place. there follows a period when metamorphosis takes place over a period of twelve days by which time the adult bee is ready to emerge. When cells are found with a pupa exposed this is likely to have been caused by wax moth larvae and is called 'bald brood'. Once a sealed brood cell is damaged or prematurely uncapped the larva inside dies.

**uncapped honey**

Once honey has been capped over bees tend to be reluctant to uncap it though if damaged or uncapped by the beekeeper bees lose no time in removing the honey. A second inflow of nectar often calls for the adjacent cells to be extended giving an uneven surface to the comb. It is quite ingenious the way bees treat each cell as an individual honey pot and come spring, when water is hard to get, they will uncap a few cells and the hygroscopic nature of the honey allows it to absorb moisture from the hive atmosphere. This can give the beekeeper the appearance of there having been an inflow of nectar. If robbers have been at work the uncapping of cells is always 'ragged' and bits of capping fall as debris.

**uncapped queen cell**

As a queen cell is always vertical and opens at the base we can tell when emergence is likely because workers who monitor the cell will have begun to remove some of the wax giving the tip of the cell a bald appearance as the cocoon becomes exposed. Queens are always controlled by the workers despite the obvious layman's idea that have given her the name of 'queen' as opposed to mother. The cell is usually 'proud' of the comb so that nurse bees can surround it and at times they confine an unborn virgin, even re-sealing the cap as she tries to cut her way out. Normally though her eager nibbling away at the tip of the cocoon results in her being able to push the cell cap down where it may hang there like a flap unless it falls away, while she departs with alacrity, keen to begin her life which should exceed that of all the other bees in that it may even last for several years.

**uncapping**

The process of uncapping, performed when run honey is to be extracted, is carried out in a variety of ways. Honey farmers would of course use special machines with vibrating blades for speed and ease of operation. The majority of beekeepers use knives or planes or even steel uncapping forks to remove the cappings. If heat is used then 'some' if only a little of the honey's benefits will be lost. Unless the cappings are cleanly removed some residual honey will remain in the combs. This is of little importance when, as is usual, the combs are given back to the bees to 'clean-up'. The 'wet' cappings retain a film of honey even after a day or so of straining. The washings are useful for mead making and the wax for an endless number of uses. The frame holding the honeycomb is held at an angle over a container and the knife used to slice the whole comb face off, the frame can then be reversed and the other side dealt with similarly..
See: uncapping fork, knife, plane and tray.

**uncapping fork** ***

U.S. 'honey scratcher'. A pronged device that is usually hand-held and with a bend of some 30 degrees about 3cm from the tips so that the fork can be held at a convenient angle to the comb as it is used to scratch open the cells.

**uncapping knife**

A knife with a specially designed hollow-ground, serried. cutting edge which allows the blade to pass with a saw-like motion, over the underlying honey without damaging the comb by suction. It requires some time to adapt to the manner of using them and manual, electric, hot, steam, vibrating and cold varieties are also available.

**uncapping plane**

An electrically heated, hand-operated slicer

for removing the thin layer of cappings surmounting a comb of sealed honey prior to extracting its contents. There are heavy-duty ones with variable heat control.

**uncapping tray**

A tray designed to hold and possibly to filter, honey from the cappings. They usually have an edge or small trap to hold a lug while the comb is being uncapped. The slice of honey-wet cappings is free to fall into the uncapping tray. There are various makes: Pratley, Granton etc.

**uncover**

To remove the covering and lay bare. When bees are uncovered (exposed) and unaccustomed light, open air and lower temperature strikes them, they give a characteristic, responsive buzz which may trigger off the release of 'alarm odour' and cause an increasing number of bees to investigate the cause of the disturbance. Smoke may be used to drive them back and cover cloths to minimize the adverse effects of cool air and light. It is always wise to wear a veil, have the smoker going and a cover-cloth ready before uncovering bees. A hive roof should be removed as carefully as possible and a crown board prized gently free before being 'twisted' off. The cap of a honeypot when taken off gives an immediate bouquet but should be replaced before any material in the air (dust or hairs) can settle on the contents.
See: opening hives, uncap and unsealed.

**undergrowth**

Tall grass, weeds, shrubs and small trees and various, usually uncultivated plants can spring up and make an herbaceous border around hives. Thistles, brambles and stinging nettles can interfere with manipulation, hide lost hive tools and are generally undesirable. So it is sensible to set up hives on stands or trestles to keep them at a useful working height, clear of the ground and allow a clear flight path for the bees. Due to lack of light in woods and forests hives would normally be placed at the edges or in clearings. Undergrowth can include useful flowering plants where daylight penetrates: rosebay, bramble and heather for example. One should always be mindful of the possibility of fire outbreaks when undergrowth becomes dry.

**under or over**

The question as to whether, when putting a new super on, it should simply be placed on top of the existing box or inserted between the super(s) already on and the brood chamber should be answered thus: persuasion for putting it above the brood chamber arise from whether or not the bees could use more room and whether there is foundation to be drawn. Putting it directly on top provokes the least disturbance but (especially if it is foundation) it could be wiser to tempt the bees into it by providing a 'bridge' i.e. swap central combs, two from upper with two from the lower to encourage the bees to occupy it straight away.

**undersized**

This could apply to hosts of miss-fitting objects or creatures. In the case of queens we always look for the normal. Drones (especially if reared from laying worker's eggs) can be dwarfed through being reared in worker sized cells. Bees of some races are slightly smaller (and their queens might well be able to squeeze through normal queen excluders) but it is suggested that if reared in cells already lined with several generations of 'last moults' they can be slightly smaller than usual. If objects are undersized it behoves you to consider what would be over-sized and what would be normal.

**underwing**

The second pair of wings which, on the honeybee, are kept folded underneath the forewings and lie on the back of the abdomen, unless they are extended for flight for cleaning purposes or because of disease. One hesitates to use an expression like 'akimbo' but workers can be seen wandering about in that condition. Their rear wings bear around 22 hooks on their leading edge and these are swept into engagement with ridges on the forewings when aligned for flight. Ability to fold the wings neatly on their 'back' makes entering their small cells possible. Some butterflies use the 'flash and scare' technique by suddenly exposing the brilliant underwing by way of putting off possible predators.

**undesirable**

Not wanted. A thing that does not fulfil the requirements. So we describe a site as undesirable as far as setting up an apiary is concerned or a race of bee as having 'undesirable' characteristics or honey having an undesirable aroma etc.

**undigested**

Food that the body cannot digest. The bee is really quite limited in the substances that its alimentary canal can cope with: water, nectar, honey, honeydew, pollen and closely related substitutes. Neither the wax nor the coating of pollen grains can be digested by humans. The former's benign passage through the body is doubtless cleansing and beneficial while the husks of pollen grains enable us to discover the diets or early humans.
See: pollen grain

**undress**

To disrobe, and when this applies to protective bee clothing you should not be in too much of a hurry to remove the veil and anyone allergic to stings should have someone check that their back is clear of bees and to stand perhaps in the shade of a tree or such like so that any bee emerging from a fold can fly towards the light. Lack of care in this respect has led to many an unwanted sting.

**uneven**

Not level or flat but having an irregular surface. Hives should always be made firm and level whether or not on even ground. The wax cappings on sealed honey are more easily uncapped and look better than uneven cappings and combs are spaced parallel to one another and separators used between rows of sections to encourage bees to finish off the comb with flat, even cappings. However when stop-go conditions result in one intake of nectar being sealed and then a later intake bulges adjacent cappings out further, you have either to cut deeper into the comb or use an uncapping fork before extracting.
See: level and uncapping.

**unfecundated**

Not fecundated – in other words a virgin queen that has failed to mate would be described in this way. Seeing that mating can only take place high up in the air and therefore requires fair weather conditions, it is all too common in this country for a virgin to go through a period of uncertain weather and if unfecundated become a 'drone breeder' should she be left to head a colony.
See: fecundated.

**Unguentum aegyptiacum**

A plaster made by boiling honey, vinegar and wintergreen was highly praised by Charles Butler. Unguent – ointment.

**unguiculi (ungues)**

Nails or claws of which the bee possesses in the form of two on each of its six feet.

**unguitractor**

A plate on the ventral side of the bee's tarsus operated by muscles in the femur and tibia, thus the claws are pulled down.

**unicellular**

Organisms consisting of only one cell yet capable of feeding, reproduction, motility etc. E.g. Protozoa - an amoeba for instance and disease of the honeybee's gut.

**unifloral honey**

This term has been used to mean normally extracted honey from a single plant source. Only possible in very unusual circumstances.

**union of stocks**

Making two or more stocks of similar or varied strengths into one. This is an autumn job as natural or attempted increase often results in colonies so small that even with a good queen and plenty of food they are unlikely to survive the winter on their own. The queen intended to survive should be carefully chosen and the other culled and the pricked newspaper method of uniting used after moving the stocks fairly close together first so as to avoid losing flying bees.

**unisexual**

Plants having carpels and stamens (the male and female parts) on separate flowers. E.g. hazel. Bees are often the most useful pollen transporters in such cases.

**unite**

To join. In beekeeping this usually refers to putting two or more colonies together to form one stronger colony with a single queen. A time honoured method of uniting includes the initial separation of the two lots by a sheet of pricked newspaper. It is even referred to as 'papering' in NZ. Other methods such as the exposure of all combs and bees to the open air for ten to fifteen minutes before re-assembly and the use of odours that mask the bees' own colony identity odour, spraying with water, dusting with icing sugar and other techniques are used.
See: uniting.

**United States of America**

Their first honeybees arrived from Europe between 1622 and 1640. By 1648 beeswax and honey was abundant in the Eastern States. Dubbed by the natives as 'the white man's fly'.

**uniting**

Causing two or more different colonies to amalgamate into one. If uniting is done without due care many bees may be killed in subsequent fighting. Fluke successes can be obtained when the weather is ideal. It is best to move the two colonies together, stage-by-stage if necessary, to feed sugar syrup to both, remove the least desirable queen about 6 hours before uniting and if valuable, cage the other. Then set the queenright lot over two sheets of newspaper with a possible central slit between the two middle top bars over the queenless lot. Release the queen after a week and see that the bees do not molest her. If they do re-cage her and check again later. Temporary immersion in tepid water will make a queen so humble that she will accept caresses which she might otherwise have spurned.
See: uniting aids.

**uniting aids**

Smoke, flour, fresh open air, shaking, dusting with icing sugar, the temporary obstacle of a sheet of pricked newspaper. Essences like: thyme, lavender, aniseed, vanilla, a fine water spray, the temporary caging of the queen or totally removing her 'bossiness' by keeping her away from other bees for about ten minutes or gently submerging her in tepid water for a moment, these and several other tricks help us to overcome the bees natural animosity to one another.

**unmanageable**

Impossible to govern or control. Weather, interference by predators and other conditions may cause a colony of honeybees to become unmanageable. Sometimes the genetic make-up of the bees or a state of unbalance within a colony (such as queenlessness) will cause it to be difficult or impossible to manage. Such bees may need to be anaesthetized in order to re-queen them with a more docile strain. The beginner should guard against assuming that a colony is unmanageable if he or she is mishandling it.
See: anaesthetic, temper and subjugation.

**unmarked queen**

This applies to a queen that has not been marked by the beekeeper. Sometimes supersedure shows up in this way when a colony that was known to have a marked queen and an unmarked queen shows up.

**unmated queen**

A newly emerged queen will normally take a few days to mature and become capable of flight and mating. This would ordinarily occur during the period from 5 – 15 days after emergence – weather and colony conditions permitting. Should a queen fail to mate within 20 – 30 days she is likely to become a drone breeder. See: virgin queen and mating.

**unopened queen cell**

A queen cell which is intact and has a live virgin sealed inside. Sometimes queen cells appear perfect but upon investigation hold a dead worker or drone, a rotting larva or are just empty. They normally remain sealed for six days. Towards the end of that time the tip may be nibbled bare by the nurse bees. Often sounds of a moving virgin can be heard from within. See: vacated queen cell.

**unpolished**

Having no shine. A cake of beeswax that is exhibited in a show is still a valid entry whether or not it is polished. An unpolished cake has a beauty of its own but may be brought to a high degree of shine by rubbing with velvet, the palm of the thumb or other non-hairy and non-absorbent soft material. The cells vacated by brood are highly polished by nurse bees and tainted with 'lay-here' pheromone.

**unpopular flowers**

Flowers that bees or humans do not like. Bees maybe dealt a hard blow as they delve for the nectary, as for example some legumes (lucerne) or they may find the aroma or plant material offered repellent as with elderflower and chrysanthemum. Modern agriculture has led to a war on weeds as witness arguments regarding Echium in Australia, attempts to have rosebay declared a noxious weed in the U.K. and the differences between beekeepers and other members of the public regarding the desirability of permitting certain wild flowers to exist. Charlock, dandelion, rosebay, water balsam which are useful to the bees but regarded as a menace by some.

**unripe honey**

Until nectar has been thoroughly ripened and capped over by the bees it is unripe honey. This has a high water content and lacks viscosity being apt to drip from a comb that is not held vertically or shoot out of one that is shaken. Such honey would seriously dilute good honey and cause it to ferment. Where sealed honey is to be extracted and some of the open cells contain unripe honey, it is best to spin the unripe honey out before uncapping so that this can be dealt with separately, i.e. either used for immediate consumption or fed back to the bees. See: green honey, dehumidifier and honey dryer.

**unsaleable**

Articles which for various reasons are unfit for sale. Honey that is fermenting can be very tasty and palatable but cannot be sold and might well overflow its jar. Unsealed honey may be spun from a comb that is not completely sealed so that the ripe honey is not diluted when subsequently extracted. Such unsealed honey will have a water content above 22%. It is usually very pleasant to eat but somewhat runny. It may be used for drinks, fed back to the bees, used to make honey cakes or biscuits or given away as presents provided the recipient is warned that it is best eaten before fermentation sets in.

**unsatisfactory**

This might apply to queens, bees, sites, methods, hives etc. Things which do not achieve the standard required. No matter how good other aspects may be if one has unsatisfactory weather or an unsatisfactory site, then any other good points will be offset and unsatisfactory results will ensue.

**unscrew**

A bee that has plunged its weapon (sting) into your flesh, will if allowed, often twist continuously in one direction so turning the barbs of the sting and enabling it to unscrew out of the wound. Such a bee may survive but is not at all likely to try to re-use its sting immediately. Screws are often difficult to start when you wish to undo them. A good screw driver, the head of which well matches the slot in the screw head, helps as does the application of beeswax to the spiral thread before the screw was driven home.

**unsealed**

Prior to being sealed or capped over and applying either to honey or brood as opposed to uncapped which implies that sealing or capping had already taken place. Unsealed honey may be ripe (when it stays-put if the comb is momentarily tipped from the vertical) but is usually still in the state of being processed by the bees.
See: uncapped and sealed brood/honey.

**unsound**

Not in a strong or healthy condition. This might well apply to the hive or the colony. Because bees can fill cracks and gaps with propolis, a hive may appear sound until one sets about trying to move it. Many a gift hive has been donated to a willing recipient who then discovers that it was a liability instead of an asset.
See: maintenance.

**unstable**

Unsteady or when referring to chemical substances compounds that decompose into other compounds either through age, heating, exposure to air etc. Pheromones and enzymes when their effectiveness is short-lived.

**untested queen**

A queen that is sold after it has begun to lay in its mating nucleus. 99% of all commercially raised queens are sold 'untested'.

**unwanted ingredients**

Honey should be free, as far as practicable, from moulds, insect parts, insect debris, brood and other organic or inorganic substances foreign to the composition of honey. Tiny particles of wax may rise to the surface together with bubbles to form 'froth' which should be carefully removed but better still prevented from getting there by allowing honey to 'stand' for a day or so in a temperature around 70F (22C) in a deep container (honey tank) which should give air bubbles time to rise when clear honey can be run off from the bottom. Neither pollen, wax nor propolis do harm but count as unwanted ingredients when pure honey is required.

**upper entrance**

Because bees normally construct a nest with stores at the top (often in wider cells) and develop a brood nest below it is usual for the foragers to depart from below. (A wasp nest gives a clear illustration of this), It is the part where debris can be ejected and where bees can arrive and deliver their loads without having to waste time in the turmoil of the brood nest nor having to go right up into the food store. Therefore in nearly every case hives are designed with bottom entrances. However, man, unlike the bee, is for every wanting to try new ideas. So in using multi queen hives or in raising a new queen 'upstairs', top (or upper) entrances may be used. The question of ventilation and security by guarding also make upper entrances seem less attractive.

**upside-down**

The wrong way up or turned right over. A queen cell is built hanging vertically and will be destroyed by the bees if it is inverted.

Pettigrew (the great skeppist) advocated setting a domed skep upside-down on a bucket when swarm cells were found if one wanted to stay the issue of a swarm. Other brood is still reared, whether or not it is inverted, or even when the comb is laid flat. (combs in a wasp nest lie flat). Because chambers are made with 'runners' (on which lugs rest) at the top, to set these upside-down would lead to certain problems.

**upward ventilation**

Moist air is lighter than drier air at the same temperature. As moisture-laden waste gases also behave in this way they require an upward escape route, yet care should be taken to prevent moist air impinging on a cold, metalled roof and forming condensation. Some beekeepers crack the crown board free so that matchsticks can be put under each corner for winter. Roofs are usually fitted with ventilators to prevent the accumulation of moisture and possibly mould. Bees are past-masters at using a relatively small entrance through which ambient air can be driven internally and to have its temperature and humidity accurately controlled, especially in the brood nest. Varroa has led to the widespread use of bottom screens instead of floor boards and this should be taken into account when considering other forms of ventilation.

**urate cells**

Excretory cells in the 'fat-body' of insects lacking Malpighian tubules including the juvenile form of the honeybee (larva).

**urea $CO(NH_2)_2$ (carbamide)**

A white crystalline substance found in the excretory product of mammals. It can be made synthetically and is used for making plastics, adhesives and fertilisers. These items have come into prominent use in apiculture in recent times.

**urgent**

Something demanding immediate attention. Inevitably things arise in the beekeeping world that require immediate attention. Careful planning, anticipation and sensible working routines will minimize the frequency of such occurrences which could otherwise throw a beekeeping project into chaos.

**urticaria**

A rash (hives) like nettle rash associated with itching following an irritation such as a sting. Itching weals that form raised patches on the flesh. They could indicate an allergic, or partially allergic condition. However it is often possible to alleviate such trouble by the use of suitable anti-histamines.

**USA Beekeeping Control**

In the 80's there were 4 million colonies and 90% of the beekeepers were hobbyists yet 60% of the county's extracted honey came from the honey farmers. Since 1952 there has been a Government support programme which resulted in a higher price for honey put into store resulting in a surplus equivalent to half the world's stocks. Nevertheless as a nation they still eat more honey than they produce, the 1982 exports were 4000 metric tons against imports of 46000.
up-date?

**used equipment**

With plastics, glues, fillers and materials to perform hitherto impossible tasks, used equipment can often be brought back into first-class condition. Be cautious however as regards equipment that has been used by someone else's bees. Bee diseases are always round the corner.
See: second-hand equipment.

**useful dimensions/figures**

Dead bees 2·75 per ml.
Mandril for queen cups 9·5mm (·163 - ·167").
Queen excluder slots 4·14 to max of 4·2mm or 4.11mm between wires.
Workers squeeze through 2·8mm slots or 3·2mm square holes while circular holes need to be 3·6mm.

The mesh of a queen cage 3-4mm.
See: apertures, cell sizes, interesting facts, temperatures, weights and worker larva.

**useful hints**

Given under: beeswax, bee venom, cautions, flat iron, hints, honey, lubrication, propolis, uncap, warnings and wax.

**uses of beeswax**

Candles, car polish, carbon paper, cold cream, cosmetics, dentistry, electrical insulation, floor polish, foundation, furniture cream, lipstick, lens grinding, lubrication, material re-enforcements, moulding, pharmacy, polishes, sports (archery/ bowls) water-proofing, wood filling, wood polish, wood preservative, and hosts of others.
See: lost wax and applicable recipes.

**uses of bee venom**

When inexpensive methods of isolating useful fraction from the main ingredients of bee venom become available, it is certain that many useful applications will be found. Already the whole bee venom is used for desensitisation and the use of an anti-inflammatory ingredient for arthritis sufferers.

**uses for honey**

Food: Feeding bees – especially when raising queens. Honey has that remarkable quality of improving the taste of all manner of foodstuffs. To glaze pork, chicken ham etc. All kinds of drinks including dry wine. Fruits that are normally considered sweet by themselves, like the various berries, gain acceptance with honey as do apples (baked or fresh), bananas, grape fruit, fruit sundaes etc. In cakes, biscuits and bread; marmalade (in recipes calling for sugar - replace at the rate of 1 lb honey being the equivalent of ¾lb sugar and ¼ pint of water) jam and preserves. Mixed 50/50 with butter it makes a nice dressing. It can be used to make dry, sweet or sparkling mead or left to ferment on into strong, tasty vinegar. As an ointment it is hygienic and benign, good for the skin the hair (shampoo) and proved invaluable for burns or scalds if applied immediately. Hospitals have been slow to make use of it and seem to have encouraged people to use Manuka (expensive) whereas other honeys, especially heather are every bit as good. It can be boiled to make toffee apples, fudge or toffee itself.

**uses of honeybee colonies**

Earliest man learnt how to make lighted tapers, and make beer from honey. When Mediterranean trade flourished wax was so unique and useful that it bid fair to be accepted even more than honey itself. As an educative medium a colony cannot be surpassed. They have been used in warfare from castles and ships. Their 'homing' ability put them alongside pigeons for passing messages. Their sense of smell has been used for forensic and other purposes. But their greatest, almost hidden value, lay in their ability to become (even portable) pollinators, increasing and sustaining fruit and seed yields so much so, that especially in the realm of luxuries like almonds, cranberries and top fruit, the apparent slow demise of the bee has caused the utmost consternation.

**uses of pollen**

As a human food it is well-balanced as regards carbohydrate, fat and protein content and has many attributes such as benefits from minerals, amino acids, vitamins and valuable 'trace elements.
For bees it is their one and only source of protein, essential for body building and nutrition of queens by way of royal jelly. Fresh pollen is infinitely better than older pollen. Because of its relative rarity on the market sportsmen and others are only slowly realizing its value. Coming from so many floral sources some are even more valuable as food than others.

**uses of propolis**

Dissolved in spirit it was used as a lacquer for varnishing violins. Its long-lasting, balsamic aroma adds credence to any pot pouri. As opposed to beeswax it is anti-biotic after being collected from plants and textured by bees – it is never used as food by the bees but for helping to stick their combs to certain surfaces and to fill gaps to exclude predators and pathogens. It has similar characteristics to the wax exuded by worker bees but melts at a different temperature, is very sticky when warm, brittle when cold but with a hot putty knife can disguise or fill cracks in wood. It has to be saponified (made soft and less sticky) before its value as a medicant (treatment for sore throats etc.) can be appreciated. See: properties of propolis.

**uses of royal jelly**
Owing to its significant role in the development of queens enabling them to grow larger, faster and to a perfect female imago quicker than workers, humans have often concluded that it must convey certain benefits when consumed by humans. Proof that is does benefit them is lacking. It deteriorates rapidly when taken from the cells, tastes rather unpleasant - strongly acidic, and like cow's milk is designed to develop almost abnormally rapid growth. See: royal jelly.

**U.S. herb garden**
Alyssum, basil, bergamot, borage, bugloss, catnip, dill, garlic, lavender, assorted marigolds, milfoil, moss rose, mugwort, parsley, spearmint, sweet marjoram, sweet William, tansy, thyme and yarrow. See: herbs.

**usquebaugh** **An Irish honey wine.**

**Utah**
The great seal of Utah consists of a beehive surrounded by flowers and the single word 'industry' overhead. (Mormans). Deseret was the former name for Utah and it means 'honeybee'. See: Mormon.

**UV (U/V)**
Ultra violet light which consists of rays that are invisible to us and beyond violet in the spectrum. The colours which bees can see omit red but extend to the ultra violet enabling them to derive much information about flower patterns which we can only see with special apparatus.

# V

**vacated queen cell**
A cell from which a virgin queen has recently emerged. Usually the now hinged tip will hang like a small flap though this can be pushed back into position creating the impression that the queen cell is still sealed. It has a neatly nibbled opening, unlike a cell that has been torn down and workers will climb in to lick up any royal jelly remaining. As the cell is of no further use house bees will demolish it in due course though bees don't hurry to do so.

**vacuole**
A small cavity in the cytoplasm of a biological cell. It will contain either air, partly digested food, sap, water or other materials.

**vagina**
That part of a queen's genitalia which lies between the median oviduct and the bursa. It is a sheath-like tube the posterior orifice of which takes the form of a horizontal slit. Inside and just below the orifice, a muscular fold in the vaginal floor is known as the valve fold. This has to be eased out of the way when instrumental insemination is being attempted. Above the valve fold lies the entry of the spermathecal duct. See: spermathecal duct and valve fold.

**Valerian ***V*aleriana officinalis* Valerianaceae**
A native perennial that forms dense clusters of pale pink flowers from June to August and offers nectar to the bees.

**valley**
An elongated hollow, possibly following a river or stream, with an outlet to lower ground. Because of the likely effect on local winds which tend to blow along the valley or down into the valley by night, the siting of an apiary within a valley may call for wind protection. An east/west valley may produce a longer flow from local flora due to one side warming up before the other. Isolation when the valley sides are mountainous is useful for queen breeding. Moisture will often be present and floods may be a danger. Access by road may not be simple depending on where the valley goes and whether there are exit routes in other directions.

**valve**
In the bee this is a muscular constriction which controls the flow or passage of fluids such as blood in the case of the heart and honey and pollen in the case of the proventricular valve. It is also a mechanical device that controls the flow of a fluid. Taps designed for extractors, honey tanks etc. are sometimes referred to as 'valves' and the Porter bee escape is a type of 'non-return' valve and is used to persuade bees to make a one-way passage out of a super or other cavity. See: valve honey.

**valve fold**
A muscular tongue-like structure found in the floor of a queen's vagina. It protrudes upwards in the form of a flap and is thought to be brought into use when a passing egg

has to be fertilized. The flap is capable of holding an egg so that sperm from the spermathecal duct may reach it before it goes on its way down through the bursa. It can close the passage between the vagina and the median oviduct of the queen. See: spermathecal duct and vagina.

**valve-heart**

Each of the five pairs of openings (ostia) of the bee's heart have flaps or one-way valves, which allow blood to enter the heart when it is dilated and to confine it and allow it to be forced forward when the muscular walls of the heart contract.

**valve – honey**

A shut-off cock that can be quickly and decisively operated so as to cut the flow of honey instantly. Its end of travel is in the fully closed position so that the flow of honey is controlled as desired. See: honey tap.

**valve-like plate**

Small, bilaterally symmetrical plates through which the bursal orifice opens to the entrance below the anus. These are probably mainly vestigial as other male hymenoptera often have more complex structures forming external copulatory organs.

**valve – oesophagus**

In the larva we find a simple valve at the end of the oesophagus that prevents the return or back-flow of substances from the ventriculus. In the adult this valve becomes the proventriculus at the anterior end of the foregut where the oesophagus has expanded to form the crop or 'honey stomach'. Its functions are now to control the passage of contents from the crop to the ventriculus and filter pollen. See: honey stopper.

**valve – proventriculus**

Situated at the end of the fore-gut it has four triangular lips, fringed with fine closely-set hairs, directed backwards toward the stomach. Gulping movements allow food to enter the central lumen of this valve where the hairs retain the pollen while the remaining liquid can be returned to the crop. As pouches within the valve become charged with pollen the small masses are then allowed to move on into the stomach. See: honey stopper.

**valve - spermathecal**

See: spermathecal valve.

**valve - ventricular**

The pyloric sphincter or valve formed by the thickening of the walls of the intestine at the posterior end of the ventriculus immediately behind the opening of the Malpighian tubules into the intestine. It is lined with recurved setae and regulates the passage of material from the ventriculus into the intestine.

**vandalism**

The wanton or malicious destruction or damage to property. To avoid this (a) we should always try to present the best possible image of beekeeping and its use to society and to the public. (b) Keep our hives in such a manner as to cause a minimum of interference with other folk and keeping the hives camouflaged or out-of-sight whenever possible. To safeguard against such an eventuality we should run more than one apiary and take out appropriate insurance. See: rustling.

**Vapona**

Dichlorvos, DDVP and Vaponette are permanently toxic to bees when it gets into beeswax. A slow release insecticide formulation. It is highly toxic to bees and is easily absorbed by many substances – wax and polyurethane, with long residual effects. See: residual effects.

**vaporize**

To escape from the solid or liquid form into vapour. Substances which are volatile

become gaseous quite rapidly and need to be kept in air-tight containers. We rely on the vaporization of substances such as PDB, (now banned) formalin and acetic acid to sterilize combs or deter wax moth. Vaporization is more rapid at higher temperatures and substances like benzaldehyde must be used according to temperature.
See: fumigants.

**vapour pressure**
The pressure created by the tendency for molecules of water vapour to escape from the surface of a liquid. It depends on the water content of air and fluctuates with temperature changes. Atmospherically it is expressed as 'relative humidity' (RH). The lower the RH, the lower the vapour pressure at any given temperature. Each honey sample has a relative humidity (vapour pressure) known as the equilibrium RH, at which no gain or loss takes place. This is about 60 – 75% RH at 20C but varies with the honey's water content. The viscosity of honey confines the intake or output of moisture to the surface layer, except over long periods.
See: vapour pressure deficit and equilibration.

**variable entrance size**
Slides are fitted on WBC hives so that the entrance can be narrowed to a single bee-way if required by moving the slides across the entrance until the bevelled edges touch in the centre. Other hives have entrance blocks (or cleats) which can be turned through 90 degrees to give two entrance sizes or left out altogether to give a full width entrance. Bielby made a rotatable disc entrance that gives various entrance segments: a ventilation screen or queen excluder slots. On occasions – when travelling or closing for spraying precautions call for a complete shutting of the entrance, foam rubber may be used but always remember to give them adequate ventilation when this is done.

**variation – (genetics)**
A deviation, either of structure or character of an offspring as compared with its parents.

**variation of pollen colours**
Howes mentions that Phacelia pollen is deep blue or pale yellow, darker in the pollen baskets. Elsewhere it has been noticed that its navy-blue pollen became lighter on drying. There are many factors that influence the reporting of pollen colours; consequently the colour is no hard and fast guide to identification. Nevertheless direct observation of bees at work on specific plants can be usefully linked to their arrival back at the hives with pollen loads. It was possible to ascertain by this procedure that a colony with bees 'cheating' by nibbling through the corollas of field beans, were also doing a good job of normal pollination.
See: pollen colours.

**variety –sub species**
A group of organisms within a species that differs from other members or groups within the species on one or more minor characteristics but not enough to justify another epithet. A distinct group within a species or subspecies. A slightly differing group within the species though less marked than a subspecies –often found in the wild. Honeybees of Apis mellifera have been extensively bred in different parts of the world and the terms line, race, specie, strain, subspecies, type and variety are used by various breeders.

**varii**
A Roman drink.
See: bracket

**various waxes**
Just as there are honey substitutes so there are a variety of different waxes besides beeswax. Many of these however can be useful as for example when blended with beeswax. The specific gravity, melting point and other physical properties can be varied to suit specific needs. In the hive though,

beeswax is infinitely superior and we must constantly guard against the use of any present-day substitutes which could get mixed with the pure material. Certain man-made plastics are an exception.
See: eclectic and waxes –various.

**varnish**

Resinous material dissolved in oil or spirit and used to improve appearance or to weather-proof the surface of wood by covering it with this film. Propolis has been used in the varnishing of violins. Wooden feeders are sometimes rendered water-proof by the application of varnish. Generally it is not advisable to use varnish as a coating on beehives as it is important to allow the wood to 'breathe' or for the colony moisture to exude. The bees line their nest cavity and fill gaps with a propolis varnish that can look quite shiny yet conforms to their requirements for air movement into and out of the wood.

**Varostan – Bayer**

A Japanese fumigant, toxic to bees but used in the treatment of varroasis.
See: varroasis and Varroa jacobsonii.

**Varro Marcus Terentius 116 – 27 B.C.**

A Roman master beekeeper –
'De re rustica' published 37 B.C. eight years before Virgil's 4th Georgia. He gave advice for the setting-up of small beekeeping concerns – mellarius, apiarium and the many types of hives from cloomed wicker to bark, logs and ceramics. Many references to sources of pollen, propolis and honey were made and honey from figs (could have been the juice of the fruit which splits very easily) is described as insipid. He mentions 'mulsum' the name used in labelling an Asiatic mite –ectoparasite on Apis.

**Varroa in Maryland USA**

Nov 16, 1979 – 2 mites were reportedly collected from a single drone honeybee and identified as Varroa jacobsonii by Dr. Ed Baker of the Insect Identification & beneficial Insect Introduction Institute USDA. Formic acid 80% was mentioned as treatment.

***Varroa jacobsonii* Oudemans 1904**

Family Dermanyssidae, Order Parasitiformes, Subfamily Varroidae. Originally a peaceful parasite on Apis cerana and dorsata (scutellata) - tropical honeybees, found in Asia and called the Asiatic mite. It moved westwards and onto A. mellifera which proved more susceptible and by the transportation of queens reached as far as Germany in 1980. Its dorsal shell 1·2 x 1·6 mm covers the idiasoma and hides the gnathosoma completely. The base of each tarsus is a lobed sucker with stiff ventral hairs that tangle with the bee making it difficult to knock off as the strong segmented feet have complex vacuum suckers. Mites cling to a bee near wing roots and under the abdomen and over-winter squeezed into the overlapping abdominal sclerites. The adult sucks blood through the intersegmental membrane but cannot lay eggs until the female has fed on larval blood. Advisory Leaflet 834.It results in drones with deformed wings and it impoverishes larvae that it feeds upon. Cappings look wavy but not perforated. It causes unwanted winter activity causing more food to be used and possibly dysentery

***Varroa jacobsonii* (female)**

Highly sclerotized. The body is oval and dorsoventrally flattened. The legs are short and stout, with the pretarsus developed onto a strong sucker. Long stiff setae are observed on the legs, especially on the tarsal segments. The dorsal shield is covered with numerous branched setae and has a row of thick, short and curved setae at its lateral edges. The significance of these morphological features in relation to the phoretic behaviour was discussed in the article (ABJ June '82) and it was stated that the mite jumps onto its host and attaches itself firmly maintaining a low profile.
See: phoresy.

**_Varroa jacobsonii_ - life cycle**
Males mate in the cells and die without parasitizing adults. Females lay eggs in open brood cells which hatch into mites in 7/9 days. Egg to protonymph to deutonymph, females living 7 – 10 months. The pregnant female uses highly motile palps to fix itself to an adult bee, (frequently drones) though it is nurse bees that carry it to the unsealed brood where it lays a few eggs, rests and feeds and then lays more. A six–legged larva hatches in 2 to 28 hours and within 48 hours becomes an 8-legged protonymph, feeding on haemolymph it sheds hair in summer and becomes an imago on the 8/9th day, lives for 2 – 3 months in summer, 5 – 8 in winter and can survive 5 – 7 days away from bees.

**varroasis symptoms**
A colony is unlikely to suffer much or show any signs in the first two years but in the third year may be seriously weakened often with 80% dead. Such colonies would have been a constant source of infection. The development of the disease is less virulent in cold regions where there is a break in brood rearing. Mites cause deformity, diminished vitality, flight difficulty and death. Weakened colonies are susceptible to other diseases such as nosema and various viruses.

**Varroasis treatment**
Because of the plethora of different treatments in the year 2010, none of which are put forward as cures and which call for several different chemicals which one person or another claim are effective – no particular treatment is put forward at this time.

**vasculum**
(L. little vessel)

**vascula**
were little hives in the times of the Romans. In biology they are containing vessels for the transmission or circulation of plant or animal fluids.

**vas deferens pl. vasa deferentia**
The sperm vessel, an organ that stores the sperm in the mature male – the seminal vesicle; while in the queen the spermatheca also does the job of storing the drone's sperm and being able to keep it viable by glandular secretions for several years.

**Vaseline (petroleum jelly)**
A trade name for a translucent, off-white grease derived from petroleum. Certain hive parts can be smeared with Vaseline to keep propolis at bay and the withdrawal of frames easier. It can be warmed to liquefy and applied with a brush (as can paraffin wax in the case of sections) but care must be taken not to get it on the top or bottom edges of the chambers for it is there that a good propolis seal is needed to make the hive secure against the weather, predators and movement in transporting. The usual parts to receive a smear are top bars, metal ends or Hoffman shoulders but it can be used at the tip of an 'instrument' for picking up tiny objects (mites?).

**V.B.B.A.**
The original name of Village Bee Breeders Association founded by Beowulf A. Cooper and re-named B.I.B.B.A.
The British Isles Bee Breeders Association.

**vector**
Maths: A line drawn to show both magnitude and direction. Bees communicate this information to one another when they perform the von Frisch dance. Biol: An organism which can act as a transmitter of diseases, parasites and other pathogens (especially viruses). When varroa was first called the 'destructor' it was not immediately realized just how many viruses could 'hitch a lift' by that vehicle.

**vegetable wax**
A substance derived from various plants such as the wax-palm. Resinous juices (usually secreted in the flakes by the epidermal cells) that some trees exude when wounded. The

waxy nature of these materials is used, often in conjunction with mineral or insect waxes, to form compound waxes for specific uses. The 'eclectic' use of some waxes enables a product with a different melting point than either of the waxes used in the mix.
Waxes can act as vectors for certain diseases.
See: waxes –various

**vegetation**

A plant community or communities. The total plant cover of an area and that part of which interests bees and therefore beekeepers, is referred to as 'forage' so we talk of 'foraging areas'.

**vegetative growth**

Developing as or like a plant. Normally descriptive of plant life it can also be used to describe the growing or spreading of diseases, such as Nosema apis, once it enters the vegetative stage and becomes active in the bee's gut.

**veil** ***

A light, flexible material which is more or less transparent, sewn into a protective net. A fine, black mesh is the commonest for bee veils. White reflects the light and hinders visibility. Worn over the face and held away from it by a stiff-brimmed hat and or circular bands. With the advent of B.J. Sherriff came a huge step forward when bee suits and veils were combined to give overall protection even to the most timid beekeeper. A veil needs to give clear visibility, to be airy enough to breathe through, to be able even to keep flies out, though some will allow gnats through.

**vein**

The bees' blood vessel is its many chambered heart but otherwise its blood is propelled around its body by moving diaphragms, the only 'veins' as such, are in the wings. The main ones carry a little blood and a few nerve fibres and were developed from the tracheal system of the larva. Particularly strong veins form the leading edge ribs of the forewings because they produce the 'power' stroke with the rest of the wing following. The veins have virtually become a strengthening framework for the membranous wings and comparison with the wings of other flying insects is interestingly different and leads to their identification.
See: wing venation.

**Velcro**

An ingenious 'touch and close' device which has formed a useful alternative to the 'zip' fastener. The mass of small hooks (not unlike the bees' hamuli) on a backing strip, lock into corresponding hooks on a facing strip. A tug will pull it undone quite easily, yet it makes an effective seal against bees or weather.
See: zip.

**velocity of light and sound**

Both are based on 'time' which we fail to completely understand. That of light is a 'constant' which helps us to understand much about the Universe. Sound, however, is registered as 331·7 metres per second in air. It travels randomly in all directions (unless deliberately controlled) until it strikes something. Wind has a dulling effect so that sounds are heard more clearly downwind. Bees are more sensitive to substrate borne sounds (vibrations) as transmitted through honeycomb. One might think of this as a form of 'touch' but should compare it with the sense we feel on our tympanic ear membranes. See: sounds.

**velum (fibula)**

Referred to by Mace and Herrod Hempsall as part of the antenna cleaner.

**venation** ***

The arrangement of veins on a leaf or an insect's wing.
See: cubital index, taxonomy and wing venation.

**venom**

A poisonous fluid secreted by the acid glands of a female honeybee. Although it comes from what is called the acid gland it is quite alkaline. Young bees have less poison but by 18 days their venom pouches are full. Fresh venom contains about 38% solids. Gas chromatography shows at least thirteen peaks showing it to be a highly complex mixture of enzymes, peptides and smaller molecules. Components include 1. hyaluronidase, MCD-peptide, protease inhibitor and some small molecules. 2. phospholipase A.melittin, 3. apimine, MCD-peptide, melittin, phospholipase and some small molecules. 4. hyaluronidase, phospholipase, melittin.

1. allows the poison to spread through the victim's system (antibodies in a beekeeper's sera prevent this).
2. ruptures blood cells.
3. Poison to CNS releases histamine and has other toxic properties.
4. Antigen/antibody reactions in hypersensitive persons can result in fatal anaphylactic shock. It depresses the rate of blood coagulation. Melittin is a main constituent (while that of Vespidae is polypeptide kinins). The main allergens are thought to be: phopholipase A. hyaluronidase, acid phosphate, allergen C and melittin.

See: bee venom, foreign proteins, bee venom activity.

**venom gland**

This is a long, forked tubule associated with the sting; the distal ends are slightly enlarged while the single proximal end widens to become contiguous with the venom sac.
See: poison gland

**venom sac**

The poison reservoir of the sting. It is relatively large and clearly visible when a sting has been planted in one's flesh. It can also, under the influence of the sting muscles, be seen to pulsate as the poison is pumped along the narrow duct of the sac into the bulb of the sting. When stung the sac needs to be scratched out and not compressed, so that the inflow of poison is minimized.
See: poison sac.

**ventilation**

Although ventilation means the provision of fresh air, within the hive we speak of ventilation in the wider sense of controlling humidity and getting rid of surplus moisture and waste gases. We use the term ventilation
See: control of honey houses, bee colonies, during winter and summer, upward and entrance size.

**ventilation cone *****

A brass or plastic cone commonly used of the front and rear of WBC hives with gabled roofs. As it tends to stick out and commercial hives are better flush alternative screened ventilation is used on roofs of other designs. Cones made of gauze have been widely used as bee escapes allowing as they do the ingress of air without bees or wasps. A cone escape is marketed.
See: conical.

**ventilation – control of**

Bees are able to fan collectively in chains and to move currents of air into and out of their hives. Generally speaking bees tend to use one entrance, filling other gaps with propolis, though it is not uncommon to find that bees in the wild, or in badly maintained hives, have more than one entrance. It is surprising how a colony will put up with a hive having perhaps no more than a 15mm ( ") hole as its entrance and this may be in the form of a tunnel, However once a hive becomes crowded it is quite likely that the colony will either swarm or even abscond.

**ventilation during summer**

As bees can ventilate adequately for most conditions by fanning during the summer it is only necessary to provide an entrance which can be adequately guarded in the active season. Sensible siting of the hives

and the apiary can also help bees to ventilate their hives as required. Ventilation not only helps to control brood nest temperatures and humidity but in the ripening of nectar into honey.
See: variable entrance, v. during winter and ventilation of honeybee colonies.

**ventilation during winter**

Matchsticks under crown board corners together with a well-ventilated roof. A winter colony has characteristics more akin to a mammal than to a number of insects. They produce heat by the vibration of their thoracic muscles and can vary the size of the cluster. A continual consumption of food is required and the queen is always adequately protected and even in the coldest winter might well have a small central brood nest. Loss of 'top' heat can be countered by extra packing in spring and since the advent of the varroa mite many beekeepers are wintering with a screen in place of a floor board.
See: ventilation in summer and for honeybee colonies.

**ventilation for honeybee colonies**

It is fatal to permit circumstances that could allow the colony temperature to rise unduly. Combs can collapse and bees drown in their own honey. Shade from hot and sensible top ventilation (top screens when moving) can help to avoid such a catastrophe. The situation in winter is entirely different in that instead of bees fanning at the entrance they form a tight cluster to ward off the cold, using muscle vibration and the intake of food to ensure that they stay well above the comatose level. Like a colony of penguins it is supposed that the composition of the cluster (repositioning of bees) takes place. Disturbance during the cold weather will be greeted by bees on the outside of the cluster protruding their stings and hissing. An aggressive display enough to discourage certain some predators.
See: ventilation – control of and variable entrance size.

**ventilation of honey house**

Control of temperature is most important in the honey house as the honey is processed far more readily at around 80F (26C) yet bees must be excluded when working conditions and the temperature call for the introduction of fresh air. Screens can be used to cover doorways and windows designed to let bees out but not in.
ventilation screen
A screen used to restrict the passage of bees but to provide the freedom of movement of air. These may be let into the floorboard or crownboard, in fact any part of the hive though care must be taken to exclude rain. Today things have given us a wide range of materials but they must be such that the bees cannot nibble them away, stretch or distort, are not objectionable to the bees and permit an adequate flow of air.
See: travelling screen.

**ventilator**

A gas permeable opening, important in the roof, is especially necessary to keep the area dry but to enable any rising moist (spent) air to escape. Whenever travelling with bees, either a complete colony or perhaps a swarm, attention should always be given to their ventilation – to suffocate a colony through neglecting this necessity would cause an unnecessary, depressing sight.

**ventral**

Situated on the abdominal side of the body or the underside of plants. In animals that adopt a horizontal posture it is the underside opposite to the dorsal. In humans it is the front side. In the honeybee wax pockets are situated ventrally while the scent gland is dorsal.

**ventral diaphragm**

A musculated skin equivalent to the dorsal diaphragm that stretches across and along the length of the abdomen floor. Its moving undulations cause blood to flow posteriorly, thus helping to bathe the viscera and emunctories and so transporting food and waste.

**ventral nerve trunks**

A pair of longitudinal commissures that run from the brain throughout the body connecting the seven ganglia of the adult. The first in the prothorax innervates the first pair of legs, the second is at the junction of meso and meta thoraces and it innervates those portions of the propodeum and the second true abdominal segment. In the abdomen, lying in the ventral sinus, 5 ganglia innervate segments I to VII. The posterior ganglion innervates all the segments posterior to it. Muscle co-ordination is not dependent on the brain as each ganglion has independent control over the portion it innervates. A bee without a head can sting, walk and even try to fly.

**ventral plate  sternite or sternum**

These exist as the under surface of the thorax and the abdomen.
See: sternite.

**ventricle**

A hollow vessel, cavity or chamber in an animal body as for instance the ventricles or chambers of a bee's heart.
See: heart.

**ventricular glands**

The proliferating cells of the epithelium of the ventriculus. They become detached and mingle with the food mass releasing enzymes which digest food such as the protein in the pollen. It is this glandular epithelia which is the site of the invasion and vegetative growth by the parasitic protozoan – *Nosema apis*.

**ventricular valve**

This is the pylorus or pyloric sphincter.
See: pyloric valve.

**ventriculus  chyle stomach**

The true stomach of the bee. Here the digestion and absorption of food takes place. A long thick cylindrical sac, looped in an 'S' bend across the abdomen. Its inner wall (the peritrophic membrane) consists of a thick layer of cellular epithelium which secretes enzymes and digestive juices and the folds of which offer an enormous surface for digestion and or expansion.
See: ventricular glands.

**venue**

A place chosen for a particular event suitable for the required purpose, e.g. Field Day, Committee meeting, talk or illustrated lecture, bring and buy sale and so on. Certain considerations such as distance for folk to travel, car parking facilities, temperature control of the building, facilities for making light refreshments, services such as electricity, water etc. seating capacity should all be taken into account.

***Verbascum thapsus*** ***

Known as Aaron's rod, Mary's candle, Hag's taper etc. it is a hardy perennial grows to 5 ft. and provides nectar and pollen for the bees. It has pale yellow flowers on long spikes rising from a winter rosette of large furry leaves from June to September. It naturalizes in dry uncultivated places.
See: mullein.

***Verbena officinalis*** *** **Verbenaceae**

This is the wild verbena of the British Isles but many annual and perennial cultivars have been introduced. It flowers over a long period from June – Oct. and gives rise to a dark amber honey with a good heavy flavour – says Whitehead. Most of the 250 spp. are tropical. Famous in myth and medicine. The Druids, Romans and Christians have all attached various significance to this inconspicuous plant.

**verge**

The edge, rim or margin. Road verges are often quite significant for the bee forage they contain though in the interests of visibility for motorists and ease of maintenance they tend to be sprayed and cut at short notice and otherwise mutilated to the detriment of any forage. See: road side.

**Vernon Frank G.**

A genius, erudite and capable of excellent bee and insect photography. A First class lecturer and beekeeper. Hon Member of IBRA Past President of Hampshire BKA Amongst other works 'Hogs at the Honey Pot' and 'Teach Yourself Beekeeping'.

**Veronica spp. *** Scrophulariaceae**

These include the speedwells and many other wild and garden varieties. They like a sunny position, bloom over a long period and offer nectar and pollen to the bees, the pollen is of a light-greyish brown. Some of the well-known ones include Germander and 'wall' and 'ivy-leafed' toadflax.

**vertex (crown)**

The region between the compound eyes. Non-existent on the drone.

**vertical**

Upright – perpendicular to the horizontal. Honeybee combs are always built 'upright' and hives should be aligned likewise. In the pitch darkness of the hive bees are aware of one direction , 'down'. Their combs are built hanging down. This clever arrangement has enabled them to build 'back-to-back' cells. Looked at from our point of view the combs are always upright though when using such a datum for communicating the direction of forage etc. the bees may be working from the basic datum of 'down'. Down being equated to 'down sun' or in the direction of shadows and that may be their starting angle. On occasions when a nest of bees is thrown into confusion because their tree has been blown down and their combs are shifted, the bees set to work at once re-aligning them until they regain their tidy vertical nature with parallel spacing. So when communicating foraging directions to one another they probably use their sense of downness on the comb to indicate angles of departure from the hive directly linking them with directions relative to the sun (or shadows). We need to apply this thinking to our handling of frames when tipping brood about and especially keeping queen cells vertical and by always putting hives on a firm base where they remain upright. See: level.

**vertical mode *****

The hexagons embossed onto beeswax foundation normally have points uppermost as opposed to flat sides

**vesicle – antenna**

A small bladder-like cavity under the base of the antenna which pulsates sending blood along a small vessel that runs to the tip.

**vesicula seminalis**

See: seminal vesicle.

***Vespa crabro* *** Vespidae**

The largest of the wasps but of similar habit though predacious on other wasps and also called the giant European hornet. Other species of Vespula include: vulgaris, germanica, sylvestris (of the woods) norvegica, maculifrons, lewisii (Japan). Females have needle-like stings which are unbarbed. They are beneficial in spring feeding on harmful insects such as the house fly, cabbage white butterfly etc. From mid-summer they turn to sweet foods and become a nuisance to bees being capable of cutting a bee in half and requiring at least two bees to repel it. Vespa orientalis is a hornet-like Israeli bee pest causing serious problems in late summer.

**vestibule**

A hollow or cavity acting as an entrance or approach to another cavity as in the case of the drone, where, just inside its seminal opening (phallotreme) we find the vestibule from either side of which spring the horns (cornua), conspicuous when extended.

**vestigial organ**

Visible evidence of an organ which no longer exists though it survives as identifiable evidence. In the adult such organs are diminutive and may be non-functional while the same organ in the embryo may have important links with evolutionary ancestors. Hypopharyngeal gland in the queen ovaries in normal workers are examples.
See: atrophied.

**vetch Vicia spp. *** Leguminosae**

A climbing herb and cattle forage plant. Annual, tolerant of more acid soils, this is a widely distributed relative if the field bean. Its extra-floral nectaries attract bees a week or two before the flowers bloom in June/July, these having deep nectaries which can prevent the honeybee from gaining access to the nectar. However where the bee can actually reach the surface of the nectar with its tongue, capillary action and surface tension enable it to take up the bulk of the nectar. It has yellow pollen and water-white honey similar to clover.

**VHC**

Used at honey shows to indicate that an entry has the equivalent rating of fourth prize or 'very highly recommended'. It is followed by HC, highly commended and C, commended.

**VHS**

Video helical scan. In 1984 the commonest format for domestic video recorders on ½" tape included Betamax VR 2000 Grundig and Phillips came into being but time marches on.
See: video.

**viable**

Able to live effectively and perform its full function. This might apply to seeds, spores, pollen 'being able to germinate.

**vibrating knife**

A power-operated, vibrating knife fastened to a frame by spring-steel mounts – vertically, horizontally or in an inclined position. It is steam-heated and vibrates in the direction of its length. The operator may draw the face of the combs across the knife or the machine may be fully automatic with a chain feed.

**vibration**

To make quiver or to move rapidly back and forth. When sound waves travel through the air they cause vibration when they impinge upon a surface (ear drum). Bees are sensitive to vibration and it has been reported that colonies kept close to railway lines are inclined to use more brace comb. A queen's piping causes substrate vibration which profoundly influences the bees that sense it. Dancing bees use it to signal to recruits. Steady vibration can force bees to leave a hive and in the days of skeppists bees were regularly 'driven' from the skeps to secure the honey by drumming on the sides of the skep. Some bee farmers travel with open entrances relying on vibration to keep bees from leaving their hives.
Uneven loading of an extractor can cause wobbling and vibration.
See: appreciation of sound, drummed bees, piping and quack.

**Viburnum *****

An evergreen, garden shrub that is very hardy and easy to grow, many varieties blooming during the winter. They have fragrant white or pinkish flowers and their colourful fruits attract birds.
Stratify ripe seeds.
See: laurustinus.

**vicious colonies**

Colonies that appear to be spiteful especially those whose reactions to the normal, reasonable use of smoke are unacceptable. Such colonies will attack almost anything that moves near their hive (killer bees?). re-queening offers the possibility of a cure but although the fault may be genetic it could as easily be caused by the actual conditions that exist in and around that apiary (wind swept?) Examples include: overhead power cables, unpleasant smells and interference. See: viciousness.

**viciousness**

Having an ugly disposition. The character of viciousness is attributed to many flying insects by the uninitiated. Honeybees have an essential defence system which can prove painful. The characteristic of being able, ready and willing to sting is essential if a creature with stores, so beloved by other living things is to survive. However races vary in their temperament and care should be taken not to eliminate good bees because the cause of their viciousness has not been investigated.

**vines - grapes**

Climbing plants with woody stems which bear grapes. Common in warmer climates than that of G.B. Although honeybees are often accused of doing harm to grapes because they can be seen sucking the sweet juices from any puncture in the skin, the mandibles of a bee are in fact incapable of starting a tear in such flexible material as the skin of a grape. It is most likely that other creatures (beetles – wasps) start the trouble and the bees come in afterwards.

**vinegar**

A liquid with a sharp acid taste. The product of two stages of fermentation. Five parts of water to one of honey are allowed to ferment naturally, exposed to the air but covered with nylon cloth to exclude dirt and flies. Prior boiling to exclude micro-organisms and improve the colour needs to be followed by filtering and the addition of yeast once the solution is tepid. Container should be of glass or wood. A constant temperature (wine +) is desirable. Cypress or beechwood shavings can be used to enhance the aerobic exposure of the maximum surface area during the acetic fermentation stage. Use 1oz to a gallon of water when wax rendering in order to acidulate the water.
See: honey vinegar and showing vinegar

**vinyl chloride $CH_2CHl$**

This is a colourless gas, used in the manufacturing of PVC (plastic) (TVL 10 ppm). The burning of PVC releases poisonous gases.

**Vipers bugloss *** *Echium vulgare* Boraginaceae**

A useful bee plant which can grow to a metre and has long, stiff stalks speckled like a viper. Bees plunder the bright, blue flowers for nectar and deep blue pollen in June and July. Bugloss means ox-tongued, leaf shape. Y. Allen described its pollen as almost colourless 12mu pollen. A hardy annual liking sunny borders and the honey is good flavoured and yellowish. In Australia the 'escape' known as Paterson's curse, grew too vigorously (and was blamed for harming cattle) and then called Salvation Jane where in lean times it saved cattle from starvation. Legislation was passed in 1979 for biological control. In 1980 in NZ where it is known as 'blueweed' it was withdrawn from schedules of noxious plants. See: Echium leaf miner.

**viral diseases**

Apis iridescent , Bee virus Y, Egyptian, Sac, APBV, CBPV, black queen cell virus (melanosis), Kashmir A & B, Slow paralysis, cloudy wing and this list continues to grow. See: virus and virology.

**Virgil 70-19 BC**

Responsible for 4th Georgic and others. A Roman poet who had a good knowledge of beekeeping and referred to bees more frequently than other poets. – Hives with high peaked domes – drowsy summer afternoons amongst the flowers with the humming of bees as they delve for golden dew. He knew about honeydew, fragrant nectar from thyme, the poisonous yews of Corsica and gave advice on the handling of swarms. 'Will tempt you to rest while they hum gently.' He dedicated a chapter to Aristaeus commending his accomplishments in a vividly poetical and noble style.

**virgin honey**
Acoeton – the very first honey taken from new comb. Virgin honey, which is the purest, of a late swarm which never bred bees.

**virgin queen**
A female imago that has not yet mated and is therefore infertile. She may be confined to her cell by the bees until they are ready to let her nibble her way out. She is then very active as she scrambles out past the hinged and hanging cell cap. Although quite sizeable at first, she is groomed for her mating flight and becomes smaller as she develops into a nubile queen. Once mated and fed for laying her abdomen extends more and more so that a mated queen in full lay is almost twice as large abdominally as her unmated sister.
See: evasive, furtive, imprisoned, nervous, laying, newly merged, unmated and undersized.

**virgin wax**
Pure white wax as first made into comb by the bees and before it has become used or 'travel stained'.

**Virginia creeper *** Parthenocissus Vitaceae**
Vars. quinquefolia and tricispidata. A climber that adorns the walls of houses, its leaves turning gold, then bright red prior to leaf fall. The inconspicuous flowers attract bees for their nectar and the volume of their buzzing is reminiscent of a swarm. It is the same family as the grape. The pollen is light brown to green. It bears bluish-black berries when successfully pollinated.

**virile**
Masculine or manly. Exhibiting a marked degree of masculine strength, vigour or forcefulness. Indicating masculine viability as of pollen, sperm, drones etc.

**virology**
The study of diseases. Generic names end with 'virus' and are normally from the name of the disease they produce. Names of families end 'viridae'.

**virus**
Sometimes crystalline, sometimes molecular but filterable through bacteria sized meshes. A sub-microscopic pathogen (protein particle) capable of using living cells to reproduce itself and in so doing to cause diseases appropriate to the particular virus.
See: viral diseases.

**viscus pl. viscera**
Entrails – Although this refers especially to the soft internal organs of the abdomen and thorax, it also includes soft internal cavities and organs. Not to be confused with 'viscous'.

**viscid**
As this word is used in definitions of honey, it is explained here as being viscous or of glutinous consistency – being thick, sticky or adhesive.

**viscin $C_{10}H_{10}O4$**
A sticky substance found on plants occasionally becoming threads which bind pollen grains as in Ericaceae, from mistletoe berries and some evening primroses. Viscin threads are thin ropes of sporopollenin and are believed to function as a means for insects to transmit masses of pollen.
See: pollinium.

**viscosity**
Stickiness or power of resisting movement. Honey has this property though its degree will vary slightly in various honeys and becomes much less when the honey is heated or diluted. A sample of honey will have a viscosity governed by its specific gravity, water content and consequent density. A standard type of honey with a water content between 17 and 20% will have a s.g. in the range of 1·414. Honey judges always look for a density (and therefore 'viscosity') that is neither too high – indicating heating, or

too low suggesting that the honey is not ripe. Filtration is helped by using warmth to lower viscosity and a graph of the latter suggests that you should extract in conditions as warm as you can comfortably bear.(Even 90F). Viscosity can be measured by the rate of fall of a ball bearing in a vertical tube of honey and a viscosity curve plotted for varying temperatures. See: density.

**viscous**

Having a high resistance to flow, being viscid or sticky' (*L. viscum* – mistletoe'. See: viscin and viscid.

**vision**

Ability to see – a veil that allows good vision and spectacles are often necessary if honeybee eggs are to be seen clearly on the comb. Generally speaking it is as well to put hives out of people's vision or to camouflage them so no blame for stings nor vandalism will be likely to occur. The vision of the bee does not allow for focusing, takes in a very wide sweep of its surroundings and ultra-violet (sky patterns and floral markings) and colours other than red are clearly distinguishable by them.

**visual aids**

Not long ago we were limited to portable slide projectors, films and such like but now with the advent of computers, lap tops and remarkable software definition, clarity and expert information has made education and learning much easier,

**vitality**

Vigour, energy, exuberance, zest.... A newly hived swarm or colony with a queen that has just started laying will exhibit these characteristics, but as a genetic factor it is associated with out-breeding and the introduction of new or rare sex alleles into the breeding population.

**vitamin**

A complex organic substance only required in small amounts but vital for body processes such as metabolism. A food factor which is essential in small quantities to maintain life yet which in itself provides no energy. Many different vitamins have been identified. Some found in pollen include: thiamine, riboflavin, ascorbic acid, pantothenic acid, vitamins D and E. See: honey vitamins.

**vitelline membrane**

The membrane that surrounds the egg yolk (the vitellus) and inside which is the peripheral cortical layer of cytoplasm which develops into the embryo. In some creatures it is the layer that forms around a fertilized egg to prevent other sperms from entering. See: chorion.

**viva voce**

An examination conducted in this way would be called 'oral'. Some beekeeping exams are conducted by word of mouth and without written questions and answers.

**viviparous**

Creatures that bring forth young (rather than eggs). This can lead to a very rapid replication, especially when the young are born pregnant as in the case of aphids. In plants it means producing young without seeds – forming bulbils or small plants on the leaves or stems. See: oviparous.

**void faeces**

To defecate, evacuate, empty or discharge the contents of the rectum. Honeybees normally defecate on the wing thus keeping their hive clean yet spreading valuable manure over the area surrounding the apiary. The queen's faeces are attended to by the nurse bees that tend her. A larval gut does not become
continuous until its feeding is complete so that the faeces released do not contaminate its food.
See: defecation, dysentery and staining washing.

**volatile**

1. Substances which evaporate rapidly and can be detected at some distance from the source. Capable of being vaporized i.e. rapidly evaporating oil that leaves no stain. Bees responses to odours have a marked effect on their choice of forage. Work by Gordon D. Waller and Dr. Roper has identified Ocimene, Myrcene, Limonene, Linelool, Citral A. and Geraniol as the main components of the Nasonov pheromone.
2. A computer memory which loses information when the power is cut.

**Volcano Island**

The Ruttner Bros, carried out experiments that proved conclusively for the first time that queens conjugated with several drones and not just one as had been repeatedly stated in earlier books. The contents of the spermatheca linked with visual evidence indicated that 3 – 11 matings would occur to impregnate a queen. Now it is known that in various weather conditions and with different races the numbers can be considerably greater. Drone sperm count could be another consideration.

**Voltarol**

Jasmine Edlin writing in BeeCraft July '85 told how she went into full anaphylactic shock after four stings when she had used an arthritic pain killer. She also mentioned others who had adverse reactions from wasp stings while taking Brufen. It appears that any anti-inflammatory drug is likely to cause this allergy.

**volume**

Space or size measured in three dimensions – cubic units. When speaking of brood it is usually to refer to brood area as meaning the comb surface occupied by brood though the significant aspect of combs or hives is their volumetric capacity. Long lugs lead to a wasteful amount of timber – compare three Langstroth shallows being the equivalent of five WBC shallows.

A volume can also form a book as a in part of a series.

**votator**

A small but highly efficient unit for cooling viscous materials like honey.

**VPD**

Vapour pressure deficit. This is a value expressed in millibars or their equivalent which gives the drying power of air. It is the difference between the existing vapour pressure of water in the air and the vapour pressure of water in saturated air in the same conditions of temperature and atmospheric pressure etc. VPD can be calculated when the RH and temperature of the air are known.

**Vraski hive**

Ancient Cretan beekeepers developed a vertical clay hive with movable combs. Wooden 'bars' virtually made an inner cover under a canopy of thatch which was often held down by a heavy, double-handled clay pot which could be up-turned for use as a feeder with Asphodel leaves to prevent drowning and as a drinking place in summer. The hive itself also had two handles and a 10mm hole as an entrance.

**vulnerable**

Weak to certain aspects of attack. A colony of bees would naturally choose a site high in a tree and isolated from other colonies. This gives it a measure of protection from weather and predators that is not possible when hives are grouped at ground level. Because a hive may be vulnerable to floods, gales, heavy snow falls, vandalism and interference from predators, every care should be taken in siting to use shelter, camouflage, strong level stands and checks whenever necessary especially at out-apiaries.

**vulva**

The external female genitalia or pudendum.

# W

**wad**
Linings for the caps of screw top honey jars. Waxed, wood pulp wads were widely used to render the caps of standard squat jars air-tight. Plastic linings and alternative containers are beginning to replace them. Caps using wads should be checked to ensure there is no dirt under the wad and from this point of view the newer, sealed-in, plastic cap (lid) is preferable.

**Wadey Herbert James**
Born 1899, editor of Bee Craft until 1978, he was a splendid beekeeper and his many contributions to beekeeping literature have enriched the craft.

**waggle dance**
The work of von Frisch have brought to light a host of different dances that worker bees perform as a means of encouraging and informing other workers of the merits and position of worthwhile sources.
See: wag-tail dance.

**wag-tail dance**
The dance of a successful scout bee as she waltzes across the comb at the same angle to the vertical as the source of forage is to 'down-sun' (direction of shadows). The comb is gripped so that she moves forward in a straight line while her body is excitedly waggled from side to side presumably sending out substrate signals through the comb while adjacent, potential recruits examine the performer with their antennae. Pollen collectors are the most likely to do this dance.

**waist - petiole or peduncle**
The narrow link between the thorax and the abdomen of an insect. The alimentary canal and nerve cord pass through this, though when severed the abdominal tip can still sting and the wings and legs keep the remainder mobile for a short time. The alignment of the thorax and abdomen on the bee tend to make the waist inconspicuous and lend a stream-lined aspect to the bee in flight..
See: neck.

**Waldron T. N.**
started beekeeping 1920
Realizing that the relatively flimsy, slotted zinc queen excluders could be pulled out of shape and were tenaciously sticky to pull off, he developed a wooden framed queen excluder consisting of straight wires kept in the right parallel spacing by cross members. The grid thus formed is enclosed in a stout frame with the same periphery as the hive. When propolised down the insertion of a hive tool and slight twist removes the excluder quite easily. Further modification was necessary to ensure correct bee-space above and below the grid.

**Wales Cymru**
Principality of G. B. crossed by mountains, drained by four rivers. There is a wide range of beekeeping terrain and beekeepers.
See: Welsh BKA.

**walk – perambulate**
Bees use far more energy for distance covered when walking (crawling) than when

flying. As combs are vertical strong hooks are essential so that the ladder-like combs can be scaled. All movement within the hive is by walking or occasionally 'falling' because there is insufficient room between the combs for flying. The queen needs to walk much further than is generally supposed for although she finishes with concentric ovals of eggs she searches in a criss-cross pattern for suitable cells which she must straddle before she can lower her ovipositor into a cell.. Her special legs must be strong to enable her to walk more miles than the workers fly. Who has calculated just how far she walks in her lifetime?

**wall** ***

Bee-boles were special hollows left in walls for the specific purpose of sheltering straw hives. Shelter is often found for hives by walls though the turbulence (eddies) caused by the wind can pull roofs off. Space should also be ample for an operator to get to the back of hives. Where exposed to baking-hot sun the reflected heat can last long after sun-set and such positions have led to swarming. Many useful bee plants like ivy and Virginia creeper can grow on walls. Hive walls are often referred to and there are single and double-walled hives.
See: bee-bole.

**wallflower** *** ***Cheiranthus cheiri***
**Cruciferae**

A fragrant perennial, the single varieties in spring being well-worked by bees when conditions are suitable. The pollen is greenish to pale yellow, Harwood gives greenish yellow, grey-green or grey and it has an 18mu grain.

**walnut** ***Juglans nigra/regia*** ***
**Amentaceae**

A large tree growing to almost 20m, its greenish-white catkins appear from May to June and when pollen is otherwise scarce bees collect its greenish loads. The female flowers resemble Italian wine flasks in shape and bear two recurved stigmata.

**warble**

Bees of nursing age have been reported as making a 'warbling' note when loaded with royal jelly which they cannot place. This sound was singled out by Woods with his electronic apparatus.
See: apidictor

**Wardecker waterer**

A device in the nature of a Miller feeder which fits onto the top of a hive with a view to providing the bees with water. It was invented by an American, A. L. Wardecker and reported in the ABJ. The bees have access to sponge material (foam rubber) which is kept constantly moistened while there is water in an adjacent compartment. This might well be useful when it is thought necessary to provide the bees with water in spring or during a very dry spell or when bees are in hot, arid conditions or during confinement caused by spraying when adequate ventilation should be given.

**warmth**

A moderate degree of heat. Honeybees are almost poikilothermic
- that is their temperature accords to that of their environment, however they are able to generate heat by the consumption of food and microvibrations of their thoracic muscles. A bee can maintain flight through quite cold air and yet not become comatose when landing on a cold surface. Individual bees are easily chilled but once a group or cluster is formed they are capable of performing like a warm-blooded animal. Wax secretion and their incubation of brood including warm cells where the queen is intended to lay, are possible because of this phenomenon. The 'feel' of warmth over the quilt (crownboard) can impart useful information to the beekeeper in late winter and early spring (the queen is laying).
See: warmth of the broodnest.

**warmth of the broodnest**

Honeybee brood has to be kept at a constant temperature of 94F (35C) while it is being incubated. Humidity too must be kept within narrow limits. Chilled brood can result in certain weather conditions or interference by the beekeeper and chalk brood can be encouraged. As warmth rises the best place to get foundation drawn is over the brood nest. The nature and position of a food canopy helps bees to insulate the brood nest area and wax making occurs when bees of the right age gorge themselves with food and raise their temperature to bring their wax glands into play.

**warm way**

This refers to the alignment of frames so that they run across the hive and parallel to the back as opposed to the 'cold' way when the parallel combs are perpendicular to the entrance. The constant entry of bees can, in the case of warm way alignment, cut part of the first comb away in the lower region.

**warping**

Well-seasoned, properly prepared timber should not 'warp' but where unsuitable timber or weather causes warping this can destroy bee-space and open gaps that predators might exploit.
Plastic hives should be guaranteed as free from this defect.

**wash**

To cleanse with the aid of a suitable fluid. It is customary to use water as it is so plentiful and suitable. Faeces should be wiped from clothing as quickly as possible using a damp cloth. Honey can be washed away with cold water but not wax or propolis which are soluble in oils and greases but not water. Strongly alkaline or really hot water may be used on hive parts. Bees' minute tongues, ending in a bouton, can sweep up material but cleansing of their body parts is carried out by caressing or stropping. Spouses have become allergic to bee-protein through washing clothes worn by their beekeeping partners!

**washboard dance**

Dr Bailey reported that in small colonies the only unusual behaviour of workers injected with sac brood virus was their reluctance to rear many workers and the so-called 'washboard' dance regularly performed by many individuals.

**washer**

A ring, usually small, of metal, rubber or plastic used either for protection or spacing – for instance to take up 'slack'.
See: shim.

**washing**

Bees often get a bad name for soiling clean washing that has been hung out to dry, especially when they need water in the spring. The tiny residue of detergent makes the water less viscous and easy to collect and with a secure foothold, in the sun, who could blame a bee for trying. It is as likely as not to react to the filling of its crop by emptying its rectum. By supplying what has become called a water 'fountain' bees can be encouraged to use this if washing is hung out near the hives.
See: water fountain.

**wasp *** Vespula spp. Vespidae Aculeata**

Insects about the size of a bee with a marked waist and brightly banded in yellow and black. Their colonies almost invariably die out when the frost have set in but by spring fertile females that have hibernated, start papery nests with a single-layer platform of down-facing hexagonal cells. Occupants rarely exceed a few thousands though in warm seasons or warmer climates their numbers can rival honeybees.. The females have an unbarbed, needle-like sting which can be used repeatedly. A wasp can easily overcome a single bee. Drones mate several times and do not die after mating. They do not store honey yet will rob weak bee colonies of theirs especially towards the end of summer.

Varieties include: *vulgaris* (common wasp), *sylvestris* (tree wasp) reputed to be the most aggressive, *norvegica*, *rufa* (red wasp) parasitized by *austriaca* (cuckoo wasp) and *crabro* (hornet). For the destruction of nests (and it is unwise to permit any within 1 km of the hives), use carbon tetrachloride, chlordane, insecticidal dusts BHC smokes or if there is no risk of fire – petrol.
See: hornet Vespula and wasp behaviours

**wasp behaviours**

Though not necessarily aggressive, once their guards are alerted by an emergency, the majority of adults (2 – 5,000) will be mobilized to vigorously defend their colony though unlike some strains of bee they will not follow beyond a distance of 10m or so. As it takes at least two bees to repel a wasp intent on getting into a hive, weak colonies (nuclei) are especially vulnerable in the late summer when wasps are searching for sweet things. Beekeepers therefore take wasp interference seriously. What is more because we have protective garb and to the layman are assumed to know all about wasps, we are frequently called upon to deal with them. Their stings can penetrate some bee suits so protection entails an extra layer (plastic raincoat?). They are all 'home' by nightfall and their nests are soft and vulnerable.
Like bees they are blind to red light which enables us to 'do things' to them. According to position (like awkward corners of a loft) one can use a long bamboo cane, cleft and holding a wad saturated with one of the multitude of wasp killing substances sold by pharmacies (carbon tetrachloride included) can be thrust right into the papery nest. This should aim not only at killing adults but brood as well because their brood can go on hatching without adults and set up afresh. Scrape earth away if the nest is in the ground until nest material is encountered, then insert a funnel and pour petrol in. No need to light it but if you do watch your eyebrows. Wasp traps work on the principal that they will struggle through quite a small hole to reach ale or diluted jam in which bees show little interest.

**wassail**

A salutation or toast once used as an expression of good will at festivities where mead might well have improved spirits.

**waste**

That which is unused, unproductive or not properly utilized. Weeds and flowers on waste ground or exudations (waste) from aphids and other sucking insects are avidly worked by bees and only recently has it become acknowledged how large a proportion of honeydew is included in what we regard as floral honey. Empty honey containers left within reach of bees can so easily cause diseases to spread. Scraps of wax, like propolis scraped from the frames should be put into a jar and not just discarded.
Bees work hard, like no other creature, to gain their provender, sweet honey washings, cappings, old comb etc. should never be wasted but put to good use.
See: debris.

**wasteland**

Uncultivated land set aside for some future potential use, land that has been allowed to go to waste. Even in populous, cultivated countries there is always the occasion when an area is allowed to run wild and such plants as can naturally establish themselves flourish and bloom. Areas around rubbish dumps, dis-used quarries, mines, railways that have been axed, often provide useful forage for bees and even the occasional site for an apiary.

**watch-dog**

It has been said that side-liners with out-apiaries sometimes keep an especially bad-tempered lot near the gate. This term might apply, at least during the day time, to a conspicuous hive in the active season.

**watch glass**

A small, curved glass dish – useful for mixing small quantities of pollen to put on a microscope slide or to hold syrup or honey to attract bees for marking or observation. Once the attention of a bee or wasp is fully engaged in the course of uploading sweet material it is possible to stealthily mark their thoraces.

**white tape**
stuck on with super glue lets you follow their departure flight path. Cross referencing from several points enabled Americans to track down nests in the days when feral bees were plentiful. Wasp's nests can be located in this way.

**watch strap**
Whether made of metal, leather, plastic or other material, by virtue of its position and likelihood of having a human smell, this is a favourite spot (like neck and ankles) for the bees to plant their stings. Best take it off unless you wear gauntlets

**water $H_2O$**
That two gases can combine to form this precious, life-giving liquid is one of the miracles of the universe. Like us the larva of the bee is largely composed of water. Pollen, nectar and propolis are vital to bees but nothing more so than water! Honey has been concentrated to the point where bees cannot use it until it is diluted. Yet while syrup may be gladly accepted within the hive water is usually ignored. For them it is far more than a food ingredient it is needed to control their humidity. We hardly notice the viscosity of water but to a bee it can be dangerously sticky. How far can a water-carrying bee fly on a cold spring day? They gather moisture in a subtle way by exposing honey to the moist hive atmosphere. Other aspects of water are its ability to make plants turgid, it leaches ground and in the form of snow can entice bees to their deaths as can the reflective surface of still water when they are fooled into believing it's the sky. Water can give life and take life. It is an element that makes life possible on this planet.

**water balsam *** *Impatiens roylei* Baslaminaceae**
Himalayan balsam a hardy annual which can grow as tall as 2m, has white, mauve or pink flowers giving it the name of policeman's helmet. It likes wet soils and spreads rapidly. A useful source of nectar and pollen. Along the banks of rivers and streams it can yield quite profusely into late in summer. Bees become dusted over with pale, golden pollen and the seed capsules give an explosive 'pop' when touched helping to spread seeds 'far and wide'. Not over popular with some folk yet it does encourage beekeepers in some areas (Welsh Harp) to migrate to it with their hives.

**water butt**
Bees will get into these and drown as the water can be deep and cold and a slight ripple will 'pull' a water collector onto the water. Floats can be provided or alternative means like netting to help them climb out. Safe nearby supplies where water is marginally warmer and good footing provided can reduce risk of losing bees in that way.
See: water carriers and water content.

**water carrier**
Containers suitable for water can also be used for the transportation of syrup (say to out-apiaries) but need to be pourable without spilling. Experiments in Germany showed that bees collecting water averaged 50 trips a day spending 2½ minutes in the hive between round trips of about 10 minutes. 5 water carriers coped with the requirements for 100 larvae and took a small food supply from a house bee before setting off on a trip. Having handed over collected water a worker will characteristically clean her tongue and turn to solicit food from another bee for the next flight. They also do their own special recruiting dance. A colony might use as much as 30 litres in a season.
See: water.

**water content**
Wax is hygroscopic to 1%. Honey's water content (it is a saturated solution) varies from 17½% to as high (heather) as 22%. 2

lb of sugar to one pint doubles its volume. Reservoir bees exist but water is not stored as such. The water content of nectars varies considerably but averages around 30%. Bees ignore weak nectars and find difficulty in imbibing dense nectars.

**water course**

The bed or channel of a waterway. When active these natural or man-made systems that effect drainage and allow the movement of heavy items are called canals, rivers and lakes. The quest for speed has led to their disuse and misuse so pollution is widespread and his affects bee forage along banks and edges. In terrain (rare in GB) where dried out beds offer what at first sight looks like a reasonably good spot for hives - don't risk it for flash floods are possible so don't take the gamble. A barge on a canal might seem a very reasonable place for hives and where rape abounds alongside it could be a gentle way of moving apiaries.

**water cress** *** ***Nasturtium officinale*** **Cruciferae**

This hairless creeping plant has hollow stems and frequents the streams and ditches all over the country. Its small white flowers are visited for nectar from June to August. It is edible.

**water fountain**

A device such as an up-turned sweet jar, full of water that allows its contents to spread out onto a none-too-cold surface where bees may collect it. Though 'fountain' implies squirting or spraying, for bees a slow drip or seepage is required and in this respect a fountain is a misnomer. Water supplies arranged so that a slow, steady trickle is presented in a manner that makes it safe and easy (good footholds) for the bees to collect. This is best put not far from the hives and in a sun-lit sheltered position. Sometimes beekeepers 'bait it with syrup to get the bees started on it but once water gatherers become accustomed to drinking from it, it is a good way to get bees out on cleansing flights and to get that all important (vital) supply.

**water gatherer**

A bee that sets out with just enough food to enable it to reach water and then rely on blood sugar for the return journey as its honey sac will then contain water so the only fuel the bee can now use is in its blood stream.
See: water carrier.

**water glass**

A solution of sodium silicate $NaSiO_3$ once used for preserving eggs. It will take propolis off the hands but is not good for the skin.

**waterlogged**

Hives should never be allowed to become waterlogged. A good, sound roof, well-drained site and hives raised off the ground are ways of keeping hives as dry as possible. When wet-rot or dry-rot gets into wooden top bars the lugs break off as the frames are prised out.

**water melon** *** ***Citrullus lanatus/vulgaris***

These are grown in the USA and their pollen is useful. They have a hard, green rind and watery, or pale reddish fruit.

**water method**

Of queen introduction. Like uniting it can be so effective that the operator gives the impression of being an expert. Introducing a strange queen to a queenless colony of young bees involves submerging the queen briefly in tepid water (say in a match box though a less smelly box would be better) and then releasing her directly into the brood nest (not the entrance). With a colony which includes older bees, flying bees and possibly laying workers, it would be wise to take greater precautions.

**water of crystallization**

When a crystal forms as from the glucose portion of honey, some water of crystallization is liberated. This can dilute the fructose content sometimes to a point where it may begin to ferment or as is sometimes seen, to cause the sample to form two distinct layers the granular glucose at the bottom surmounted by clear fructose honey at the top. Glucose honey is much less sweet than fructose.

**water-proofing canvass**

Use 1 gal Raw Linseed oil, 13 oz crude beeswax, 1 lb white lead and 12oz rosin. Boil and apply warm. Wet canvass on underside before applying.

**water-proofing leather**

(a) 3 parts Vaseline and one part beeswax. Mix by melting and rub well in.
(b) Melt together 1 part beeswax with 1 part tallow (rendered beef fat) and 1 part neat's foot oil – warm until liquid then mix thoroughly. Will not only water-proof but restores scuffed, worn leather to life.

**water-proofing for shoes**

4 oz beeswax, 4oz resin, 1 pint linseed oil and ¼ pint turpentine.
1. Melt wax an resin in jar placed in a pan of water. 2. Stir in the oil. 3. Remove from heat and add turpentine. When to be used melt if necessary and rub well into the leather.

**water repellent**

Many materials like paper and cloth can be rendered water repellent by dipping in molten beeswax. The inside of wooden feeders can be treated with liquid beeswax provided syrup is not poured into them when scalding hot. Articles made by Origami etc. can benefit by being partially stiffened in this fashion.

**water requirement**

The daily requirement of a normal colony in spring has been estimated at one third of a pint (200ml). Whenever brood has to be reared their food must include more water than pollen and honey. Fresh pollen and nectar boost colonies in the spring.

**water soluble**

Able to dissolve in water. Honey and sugar. The solubility of the former is such that only cold water is needed to get honey out of empty containers whereas hot water would melt and spread any particles of beeswax that might be present.

**water sources**

When nectar is plentiful bees have little need to collect water. Although water is not stored in cells, certain bees may carry supplies in their honey crops and we call them tanker or reservoir bees while they do that. They are also able to use the hygroscopic nature of honey in open cells to absorb moisture from the hive's atmosphere. Sugar feeders can, if gradually filled with increasingly weak solutions, still attract bees. Ponds, streams, water 'fountains' and even drains and cow pats might offer water warm enough but dew and rain that has fallen onto warmer surfaces can be welcome as can wet mossy ground. The activity of going out for water also acts as a remedial exercise as such bees seem to improve colony morale, voiding their faeces and eyeing the territory as well as performing a weather check.
See: water fountain.

**water table**

The under soil level at which the ground or pores in the rock are saturated. It roughly conforms to the configuration of the ground and depends too on season, drainage, rainfall and tides. The average level of the water table will have a positive influence on the type of flora and the amount of water in springs and wells. Where it rises above the ground a body of standing water exists. Its level will also give a clue as to the likelihood of flooding.

**water vapour**

Gaseous water or water molecules diffused in air especially at temperatures below boiling point. The amount that air can hold depends on the temperature and the

actual amount is called the relative humidity (RH). The significance of this is that fog, cloud formation, 'nectar secretion' and the hygroscopicity of honey are all dependent on this.
See: vapour pressure.

**waterways**

Rivers, canals and other stretches of water such as those made for the use of pleasure craft or transport. Apart from risks due to possible flooding a waterway provides good strips of forage and a wide variety of plants may be allowed to flourish along the banks Water balsam is one example but clovers, purple loosestrife, blackberry and a host of others may develop. A quiet waterway also offers an opportunity for transporting hives. Early Egyptian use of the Nile for migratory beekeeping comes in for many a mention.

**water-white**

This grade of honey makes it especially valuable on the market. It comes from plants like: acacia, rape, raspberry, rosebay willowherb and some clovers. Combs can be held up to the light so that the lighter honey can be detected and this honey extracted separately or blended to make a dark honey into one of medium grade.

**watery**

This can apply to substances like honey which contain more water than usual. Watery honey has a lower viscosity and is prone to fermentation.

**watt**

A unit of electrical power defined as one joule per second and it is the power of one amp current flowing across a potential difference of one volt.
See: joule.

**wattle**

1. Rods, stakes or wicker interwoven with twigs, branches or flat planks. They can be used to form an effective wind break.
2. It is also the Australian name for mimosa which produces spikes of fluffy, yellow flowers with a distinctive odour. It can grow out-of-doors in the warmer SW and associate islands but it blooms early Jan – March when it is likely to produce pollen but no nectar except when honeydew comes along.
   See: mimosa.

**wattle and daub**

In addition to straw skeps wattle smeared with mud and dried clay were used for making beehives.
See: cloom.

**wax** ***

A fat-like, yellow substance produced by bees. A white translucent substance. There are many other waxes derived from insects, plants and minerals. They are solid, non-greasy, insoluble in water and have a low softening or melting point. Even more uses for beeswax can be obtained by blending it with certain other waxes. It is sold in bulk, in ingot form, slab, flakes and powder. For use on the skin, for ear plugs when swimming or in the realm of polishes and water-proofers these natural products still come out on top. It has a pleasant balsamic aroma. Beeswax can carry pathogens and readily absorbs smells.
See: beeswax, judging wax, showing wax and waxes –various.

**wax bleaching**

Although pure beeswax has a beauty of its own there is a market for whitened, natural wax. It can be sun-bleached but chemicals which are used to bleach it include: activated charcoal (BDH), perchloric and persuccinic acids and sodium hypochlorite. But filtering can remove pigments i.e. – pressure filter through charcoal and fuller's earth, diatomaceous earth, chloramines, ozonized air, hydrogen peroxide.
See: bleaching and lighten wax.

**wax bloom**

After exposure to air for several weeks, wax develops a very thin layer of 'bloom'. This is a layer of wax which has undergone a physical re-arrangement of its molecules and it has a whitish-blue colouration. Its melting point is slightly lower than ordinary wax and its properties revert to normal after melting. Show pieces can be repolished with the palm of the thumb.

**wax cappings**

To store their honey bees cap each cell over with a slightly porous layer of wax. It is often slightly domed to include a pocket of air underneath. It can be removed cell-by-cell for economic use in winter and the capping is eaten (re-cycled). The flimsy nature means that they can easily be brushed or scratched. Once the beekeeper separates the cappings from the comb with a view to extracting the honey below, he or she is left with a mess of honey sticky wax that can be drained, washed and rendered into lovely golden ingots. See: cappings and honey cappings.

**wax composition**

The analysis of wax shows it to comprise: 70 – 72% esters, myricil palmitate, laceril palmitate, myricil oleo palmitate and so on. 14 – 15% free ceric acids inc. cerotic, montanic, melissic etc. 12% hydrocarbons, mostly saturated, range 15 – c 31) inc. pentacozane, heptocosane nonacosane. 1% free alcohols, 1% water and minerals, 0·5% lactone, 0·3% dye and dihydroxyflavone. There are over 100 non-volatile and over 200 volatile components. There is an excellent book on the subject by Ron Brown.

**waxed paper cartons**

These are very useful containers for honey and sell well when there is no need for the honey to be visible. They can be printed with attractive designs and are light in weight and fairly strong without being fragile.

**wax solubility**

Beeswax is soluble in benzene, carbon disulphide, carbon tetrachloride, chloroform and turpentine also in hot alcohol. When dissolved in turpentine it can then be saponified with a small amount of borax in a water solution as for making cold cream. See: cold cream and polishes.

**waxes various**

Mineral waxes include: paraffin*, microcrystalline ozocerite, petroleum, ceresin, Utah, and Montan. Vegetable: Carnauba*, candelilla, Chinese flower, fat, sandy, Japan, Ucuhuba, bayberry, myrtle, ouricury, otoba, cocoa butter, fiber, fir-bark. cotton, flax, sugar cane, rice oil – wax, resin or lacquer. Animal: Spermaceti, tallow, lanolin. Insect: Beeswax*, propolis, white (cera alba) yellow (c.flava), shellac, Pe La from *Ocus chinensis*.
See further information about those marked *.

**wax extraction**

The means whereby clean beeswax is extracted from the comb built by the bees. As the natural comb is reinforced in various ways by the bees i.e. with fibre, propolis and the pupal skins of hatching larvae, it is necessary to melt and filter it to obtain the pure, golden wax. In a solar wax extractor the heat of the sun is employed to do this, but steam heat is also used. When liquid it looks remarkably like honey. See: rendering.

**wax formation**

Bees, of wax producing age (with developed wax glands), hang in festoons where their temperature rises slightly and scales of wax secreted by four pairs of glands, form on the adjacent wax mirrors which lie under the preceding sternite. See: wax producing bees.

**wax foundation** ***

This is the name given to the thin sheets

of beeswax that have been embossed with the hexagonal shape of the cells as found in the septum of the comb. Only by offsetting one cell on one side against three on the other, can the amazingly clever (bearing in mind bees have been doing this for millions of years) back-to-back cell arrangement be achieved. Thin sheets can be obtained by setting in moulds by pouring into plastic or metal bases or by milling. In large scale production a continuous roll of wax is fed between rollers which compress and form the sheets which can be super fine, medium or of brood thickness. Used fresh the bees readily accept it and so the beekeeper is able to persuade the bees to build comb in frames of his chosen size. It can be reinforced with stainless steel wire to keep the comb vertical and to hold it together when whirled round in a honey extractor. It is sold in airtight plastic bags holding ten sheets (8 to the lb BS standard deep).
See: cell-size, reinforced, wired, wax foundation mill and worker.

**wax foundation mill**

Once the early pioneers Krechmer, Mehring and Schank developed milling rollers to make foundation many devices for home use came into being. A hinged press making sheets in a shallow tray came from Germany and alternatives came into being here. However making foundation and especially wiring it, takes lots of time and calls for facilities not easy to find in the kitchen. So the knee must be bent to those worthy manufacturers who make time-saving foundation available and give us more time to enjoy our beekeeping.

**wax from flowers**

Early writers - Aristotle, Celsus, Pliny and Columella evidently thought that bees collected wax from flowers. It is true bees do collect propolis from plants but our ever inquisitive scientists have now made a thorough investigation of beeswax and we are now quite certain where this unique material comes from.
See: waxes –various

**wax glands**

There are four pairs of glands on the inner sides of the worker bee's abdominal sternites. Wax oozes from the glands onto the wax 'mirrors' in the front part of the sternite where it hardens into scales. The sternites overlap, covering the scales which are just visible and are drawn out by the pollen combs on the bee's basitarsi. The wax glands are surmounted by the 'fat body' and are usually most active in bees 10 – 15 days old. The products of these glands which can produce wax even on a sugar diet, is often allowed to go to waste as many beekeepers fail to realize that the summer wax production is a normal activity in certain conditions which only exist during the prime months of the summer season so the advice should be 'get your foundation into the right part of the hive – in good time.
See: wax production.

**wax modelling**

Ceroplastics. The first International Congress in Ceroplastics in Science and the Arts was held in 1975. In Britain we are familiar with Madame Tussauds waxworks but in Cairo magical beeswax figures from Ancient Egypt span over 3000 years. The colour, feel and accuracy of detail makes beeswax a very suitable substance though it is not durable and is temperature sensitive. Our word 'sincere' came from the Roman masters who insisted that marble sculptures should be 'sin cere' meaning without wax.
Wax moth *Galleria mellonella* and *A. grisella.* These are the two distinct greater and lesser wax moths. They have been dubbed: bee miller, bee moth, dusty miller, greater honeycomb moth, honey moth, larger wax worm, web worm etc. The greater is the more troublesome of the two especially in warmer climates. In the UK Achroia grisella is very common but rarely takes hold on strong healthy colonies though it will destroy unprotected combs when temperatures allow it to be active. Infected material should be treated or destroyed as when left lying about it can only lead to the

spread of this predator.
Acetic acid is useful for destroying adults and more importantly preventing the almost invisible eggs from hatching into grubs. When they run riot in stored comb thick masses of web completely destroy the comb and present a threat as a source of further infestations.

**wax moulding**

Liquid wax can be run into a variety of moulds to form almost any shape. Not only can precise effigies be made but they can be encased in plaster (set to separate in two halves) and when the wax is removed it can be replaced with a more durable material. Wax cools very slowly and any attempt to hasten will result in cracks. Releasing agents are necessary when there is any likelihood of the wax sticking to the mould.
See: candle, lost wax process and releasing agent.

**wax nippers**

The bees use the pollen combs of the basitarsi to take wax scales from the pockets where they are formed and transfer them to their mandibles for chewing and adding other secretions in the formation of comb. The name wax nipper and also wax pincers have been used, especially in older books but it was not then clearly understood just how the workers dug the scales of wax out of the wax pockets.

**wax particles**

When you chew honeycomb until all the sweetness has been enjoyed you are left with particles of wax in your mouth. As wax is good for you it is better to swallow it before this stage. Screens at the bottom of a hive can result in particles of wax falling out of reach of the bees who would otherwise recycle them. Any sign of scattered pieces larger than particles indicates that predators (other bees) wasps, mice or some other activity has occurred because bees themselves always keep a neat and tidy home.

**wax plasticity**

The wafer thin walls of honeycomb can carry an enormous weight but only within a certain range of temperatures. It is quite brittle when cold but at hive temperatures it is on the way to becoming plastic. Beekeepers often fail to appreciate this when handling combs (especially brood) but once 90 -95F (c.37C) is reached a further increase will lead to comb collapse and lack of care when moving hives has led to bees drowning in their own honey.
See: temperatures.

**wax pockets**

The eight pockets are formed where the abdominal sternites overlap, adjacent to the four pairs of wax glands. When they are charged with wax the scales which are of variable shapes, protrude so that the white edges of the scales are just visible. The scale can be taken out artificially though in the normal course of events the stiff rows of comb-like hairs on the basitarsi are used to hook the small plates of wax from the pockets.

**wax press**

An apparatus that applies pressure to heated combs in order to force out the wax. – Not to be confused with a foundation press.
See: pollen press.

**wax producing bees**

Normally neither very old nor very young bees have active wax glands. Young bees are busy developing and using their brood food glands, older bees may have passed the stage where their wax glands can be activated. So whereas bees from 2 to 52 days old have been found in comb building clusters, the age of maximum wax production ranges from 7 days old to early foraging days around 28 to 38 days. Such bees charge themselves with nectar, honey or syrup and step up their temperature to produce eight plates of wax which are dug out of their wax pockets by means of the basitarsi. This is at around 92 – 97F (36C) by bees between 16 and 25 days. Deans states that the warmest part of the hive is the best place for wax production and the bees usually hang in festoons or chains when comb is being built. See: festoons and wax formation, scales, secretion.

**wax properties**
Brittle when cold, becomes plastic at 30C (86F) and then soft at
35C (95F) but melts at c.61C (142F). Like honey it has a high latent heat which means it heats up and cools slowly. In moulds cracking is likely unless 'slow' cooling takes place. It floats on water, makes a water-roof coating.
See: wax.

**wax purity**
Using a piece of pure beeswax for comparison check density by placing in water and adding alcohol until it sinks. Thus the s.g. or 'floatability' is compared. Smell, texture and appearance give further clues as to its adulteration or purity as does its smell when burned.
See: tests for wax purity.

**wax reclamation**
The salvaging of beeswax from brood or honey comb. Beekeepers are urged not to discard wax in any shape or form. Scrapings of burr and brace comb, comb built exterior to the frames, cappings, old or damaged combs, any waxy material produced by the bees can be salvaged. By heating (solar heat most effective) and filtering, pure, clean beeswax can be obtained. The M.G., Wakeford and other wax smelters use ingenious methods to reclaim wax using hot water techniques. Advantage is taken of the fact that liquid wax can be readily filtered and that it floats on water. Avoid alkaline (hard) water as this tends to saponify the wax spoiling the texture. The present day beekeeper really has no excuse for throwing this useful but potentially disease-spreading material about the apiary.
See: wax rendering.

**wax rendering**
Natural combs, though described as made of wax, actually include plenty of other materials. The cleaner the original material the better the end result but all processes require the liquefaction of the wax, usually by the sun's heat or that of water at its melting point is only 145F (62C). The liquid wax is then allowed to run clear of embedded impurities and is solidified into sheets, slabs or cast into ingots. African natives render wild nests by submerging the comb in sand at the bottom of a drum filled with water and subsequently heated to float clean wax to the top. More sophisticated methods include centrifuging, pressure filtering and solar heating.
See: preparing show wax.

**wax scales**
In 1684 Martin John demonstrated that the tiny scales of wax could be teased from beneath the abdominal sterna by the use of a small needle. These scales or 'plates' are thin, flat and somewhat irregular in shape though generally sub-oval. They are more transparent than wax though they have the whiteness of new comb. They often fall onto the floor of a hive and may even be cast from the entrance if there is nowhere for the bees to use them. When swarms are confined to boxes combs may be started and many loose scales can collect in their box. They have been described as looking like mica and are said to be more brittle at first than when they have finally been worked into comb. The wax mirrors may produce 2 or 3 sets of scales which are c.O·1mm (1/250th") some 3·2mm across and weigh about 1mg each.

**wax secretion**
Only worker bees of middle age (that is after nursing but before the deterioration leads to the atrophy of their wax glands) can make wax and then at a temperature between 95 -97F (37C) and when in a cluster with other bees all well-fed on carbohydrate, syrup, nectar or honey. Contrary to former notions that each pound of wax required the consumption of huge amounts of stores,

wax is a normal product and the bees almost certainly benefit by using their glands in this way.
See: wax producing bees.

**wax shears**
Like pincers and nippers this expression could only truly refer to the cutting of wax during its manipulation and mastication by the workers mandibles. However, faithful interpretations of parts of the legs such as the pollen press have in the past been thought by early writers to be involved. Now it seems most likely that the scales are hooked out by the combs on the basitarsi.

**wax smelter**
An apparatus that melts waxy material so that it can be separated from 'dross' (slum gum). Although smelting usually applies to the refining of metal by melting it has been borrowed by beekeepers to cover the process of extracting clean beeswax from the impurities such as cocoons, pollen grains and other unwanted material found in the old comb or scrapings from the hive.

**wax solubility**
Beeswax is soluble in benzene. carbon disulphide, carbon tetrachloride, chloroform and turpentine and is partially soluble in hot alcohol

**wax tenacity**
Beeswax can be extremely tenacious and this property together with the use of propolis enables the bees to suspend their heavy combs of honey from various surfaces – even plastic and glass though a preference for vegetable products such as straw or wood is evident. Releasing agents or specially prepared surfaces must be used where foundation mills or moulding processes are to allow the release the wax after treatment.

**wax texture**
Beeswax has a smooth, characteristic feel. It is of a dry 'soapy' nature at room temperatures though it becomes plastic, sticky and finally liquid as temperature is increased. It is quite brittle at low temperatures and can be grated or shredded for easy measurement or handling.

**wax- transparency**
Light can pass through a thin layer of beeswax and so can be described as 'translucent'. Combs of honey can be held up to the light to check the colour of the honey or to see whether any pollen has been stored in the comb. Wax scales are more transparent but once manipulated by the bees they produce frosty-white comb as if air and other substances have been incorporated.

**wayfaring tree** *** ***Viburnum lantana*** **Caprifoliaceae**
A shrub that grows by the roadsides having hairy leaves and clusters of white flowers and berries that turn from red to black. The pollen is slate grey. It likes calcareous soil.

**W.B.C. hive** ***
A hive within a hive. The prototype of this hive was introduced many years ago by W. Broughton Carr. It is a splendid looking hive and has many advantages which unfortunately do not include simplicity of construction, cost or ease of operation. They are however warm and dry and the lightest of hives to lift.
The brood area is 2000 sq.inches. When taking on such equipment care must be taken to ensure that the WBC in question is absolutely standard as many awkward, non-fitting variations are to be found. The floorboard is part of a stand with splayed legs. Light, inner deep and shallow boxes are enclosed by 3 'lifts' (outer covers appearing telescopic) the lower one incorporating a porch and arrangement for entrance changing slides. British standard frames are covered by a crown board and a characteristic sloping roof completes the rather splendid affair.
See: hive WBC.

**WBE**

Whole bee extract used without too much effort to desensitize against stings.

**weak**

A colony? a flow? syrup? So many things that are not up to normal strength can be described in this way. In the case of a colony according to the time of year it is possible to make an approximation as to the number of bees and combs covered. Then you might refer to it as being strong or weak. A honey flow would imply that the bees were bringing in nectar in better than average amounts but it would hardly be worthy of the description 'flow' if it were nothing more than weak. Syrup (that is sugar dissolved in water), can be regulated so that when aimed at stimulating brood rearing it is applied at a weaker strength than when, as in autumn, it is fed in a more concentrated form to get as much food as possible into colonies. The application of the word 'weak' has endless links with beekeeping ranging from leadership to moral or monetary support. See: strength.

**weal**

A small burning edge or area of itching skin as from an insect bite. In the case of a bee sting the tiny puncture soon becomes a red mark surrounded by a pale and somewhat puffy ring. This tends to disappear into a localized swelling where mild allergy exists (as is often the case with new beekeepers) and may in more severe cases become more widespread.

**wean**

This describes the change of diet from mother's milk to solid food and it is also appropriate for the change in the worker larva's diet from royal jelly through to pollen and honey. It is achieved by the gradation from healthy young bees with their brood food glands truly aroused feeding the youngest larvae and then as they become less productive they would feed older larvae. We only see this as a result of studying the development of larvae but the honeybee colony, through its wonderful system of pheromones produces a perfect range of feeding for workers, drones or queens.

**weapon**

Only a worker bee has an effective sting that can be used in colony defence. Those most inclined to use it have firm enough stings and poison enough to make a purposeful penetration of flesh. This precludes younger bees that, economically, would be making too big a sacrifice if they were to lose their lives by stinging while potentially having so much to offer the colony. For the same reason a queen withholds her sting (although unbarbed), reserving it only for rivals and requiring its life-long use an ovipositing organ. Worker's mandibles can be used to feebly bite but are of course no match for the array of armament carried by the wasp. Bees would have been eliminated millions of years ago but for this tiny but powerful sting.

**wearing gloves**

To be 'hand-in-glove' with the bees makes a lot of sense. Just as conditions in beekeeping almost invariably call for the wearing of gum boots (mud, snakes, rough terrain and of course a cover for vulnerable ankles) so with regard to the hands – apart from wanting to avoid stings so that work will not be impeded – it is nice to keep the hands free of that sticky insoluble 'propolis'. However, in truth there comes a time when familiarity and understanding allows a beekeeper to feel so 'at-home' with bees (stings regarded as no more significant than nettles or brambles) that to find that bees run away from hands laid carefully over the brood combs, that a queen can be carefully handled, and that greater care and dexterity comes from working with free hands it behoves one to aim at reaching that level of competence. But discretion is the better part of valour and one should never, through lack of gloves when bees are 'shirty', halt an inspection prematurely or rush hurriedly to get a job done.

**weather**

The various influences of the atmosphere particularly at the surface. As forecasting has improved immeasurably it is now possible to be able to plan what is likely to be possible the following day. The 'goings-on' in the pitch dark of the hive are in contrast to what goes on in the external environment to which bees respond unerringly.
The experienced beekeeper is almost subconsciously aware of the state of his or her colonies, the flora (forage possibilities) and the present and forthcoming weather. Rain can be tolerable if it is warm. Thunderstorms do change colony behaviour. In some summers, if we only manipulated on the only really suitable days, then these can be rare in some summers and you might find you have disturbed the bees on the very few days when they could have been usefully working. The sheltering of apiaries is important. The use of cover cloths and constant awareness that bees generally speaking are not inclined to appreciate our 'pulling their home about' should affect our linking weather conditions with bee manipulations. In particular, even on warm day brood needs to be kept at over 90F.
See: anticyclone, cross wind, cyclone, pressure and prevailing wind

**weather-proof**

Able to withstand the effects of weather. As hives spend most of their time in the open, it is important that they are weather proof. They must contend with rain, hail, ice, snow, blazing sun, gales and all that nature can throw at them. Particular attention should be given to roofs. They must be leak proof. The legs or parts of the hive that touch the ground, are also vulnerable to rot, while timber hives need to be treated at regular intervals with a protective coat.

**web (spider's)**

Bees have the utmost difficulty in breaking out of a spider's web and many are caught in or near the hive. Spiders do little harm in the roof space and could account for a moth or two. Wax moth larvae can spin an interlocking mass of web and it is important to keep these pests out of stored comb and to destroy their mess to prevent the hatching of hidden eggs and spread of the pest.
See: web worm.

**web worm**

The wax moth has been a long time predator of the honeybee. It can be very serious in warmer countries and if not checked will wipe out one colony after another. Fortunately although it is always ready to invade, in a strong colony it is kept well under control. However it has learnt how to insinuate itself into a colony and one moth can lay a lot of eggs and these are not easy to see. Just as soon as combs are taken away from the bee's protection they are at risk. The moth prefers combs that have had brood in them but not only will they destroy the newest of built combs but even foundation that has never been in a hive.
To use formaldehyde against them is going unnecessarily far but ascetic acid is not only fatal to them but leaves little or no sign of its having been used once the combs are aired.

**wedding certificate**

A humorous reference to the drone's genitalia trailing behind a newly mated virgin as she returns to the hive. Presumably the virgin, having copulated, can either free herself of the unnecessary part of the drone's genitalia or a subsequent encounter with another drone will do so. Beekeepers have been surprised as old ideas were swept aside and it became clear that for a queen to have such a large amount of spermatozoa in her receptacle (spermatheca) it must have required the work of several drones to achieve this. The workers soon 'tidy' the queen and but for the occasional report from an observant beekeeper this phenomenon is rarely noticed.

**wedge**

A wedge is a piece of hard material (e.g. wood or metal), suitably shaped so that it can be driven between two surfaces to separate them. When propolised boxes are eased apart by the insertion of a strong, thin wedge such as a 'hive-tool', further wooden wedges can be put in the space procured. It can even be a means of giving extra ventilation during hot weather. Because it is manually difficult for some to hold both smoker and hive tool at the same time, in these circumstances a wedge would help. 'The thin edge of the wedge' is an expression that implies once a thin edge has been inserted, thicker material can follow. The arrival of a disease pathogen can, to use another expression 'be nipped in the bud' if checked at the outset.

**Wedmore E. B.**

A much-quoted beekeeping writer and expert. 'A Manual of Beekeeping 1932 and 'The ventilation of beehives" in 1947.

**weed**

A wild herb that can grow where it is not wanted but they are especially common in uncultivated ground. The status of a weed is relative. They can be useful in some circumstances but a nuisance in others. A beekeeper is in a special position regarding weeds. He may not look unfavourably on dandelions, rosebay willow herb or bramble. The invasion of weeds useful in that they increase honey crops include: Echium lycopsis, Lythrum salicaria, Prosopis glandulosa, Melaleuca leucadendron, Chamerion angustifolium, Taraxacum officinalis, Cirsium arvense etc. See: Weed foundation.

**Weed foundation**

Milled wax not only embossed with cell walls but having slightly raised walls. Although Mehring produced the earliest foundation in 1851 it was not until 1895 that E.B. Weed built a wax-sheeting machine that could make continuous sheets of foundation with uniform thickness.

**weedicide**

A chemical weed killer. While often quite harmless to honeybees it is as well to consider the effects of any contamination if carried back to the hive. Possibilities include the smell causing colony members to treat the returning forager as if it were alien, harm being caused to brood and perhaps contamination of the comb or of the honey. The decimation of useful bee-forage is also possible through the use of weedicides. See: weed.

**week-end beekeeping**

Managing honeybee colonies by carrying out such work as is necessary during the week-ends, that is from Friday evening until Sunday evening. Even during the busiest part of the active season bees can be kept under control and prevented from swarming by weekly inspections. This plan can be adopted by those who work the whole week. Feeding, supering, swarm control and prevention, harvesting and preparation for winter can all be done in this way. See: weekly inspection

**weekly inspection**

The checking of a colony during the active season to ensure that they are healthy, queenright and have sufficient space for brood and food and show no signs of swarming. The steps towards swarming start with having drones on the wing, From then on race, colony development and weather will need to be considered. When queen 'cups' appear be sure the bees are not wanting for room either in the brood chamber or the supers. If an egg is found in a cup, swarm 'control' has not worked and 'prevention might be called for. There are times when colonies seem determined to swarm and may even start with a three day old larva and sometimes swarm before the cell is sealed. Quick inspections can be developed. The experienced beekeeper knows just where to look and won't be put off because there are masses of bees. Vital last minute action could involve moving the stock well away and leaving a new box on the site to harbour the returning fliers but see 'swarm control'.

**weeping**
The formation of wet drops on the face of the honeycomb. Weeping sections. Wax capping are permeable and in moist conditions weeping can destroy the clean, dry appetizing appearance of honey comb. Keep stored honey comb in the freezer or in a warm, dry place. See: bruised and guttation.

**weevil Cucurlionidae Rhynchophora**
This large family of small beetles has over 500 species and while storage of grain and other products absorbs much of their attention they also include pollen eating members and action against these is often harmful to foraging bees and therefore the honeybee colony as well. See: oilseed pests and rape.

**weighing**
Weighing bees (including bees for packages) hives of bees, swarms and hefting are headings under which weighing comes.

**weighing bees**
Bees can be sold by the pound (kilo) complete with a queen. There are some 4000 bees to a pound. Gauze cages of bees (packages) with so many pounds or kilos have been available though the advent of Varroa has brought in severe restrictions. Swarms may weigh around five pounds, casts usually much less and occasional swarms weighing as much as 15 lb. have been reported though it is likely that over-large swarms contain more than one queen. At times two swarms will join up. Swarming bees will have 'tanked up' whereas hungry bees might be as many as 5000 to a pound. See: weights.

**weighing hives**
Keeping a check on hive weights and possibly keeping a graph of the results can provide some very useful information. For instance the consumption of winter stores, surplus gained and time of flows, or the loss of a swarm are all obvious enough when a continuous record of hive weights is kept. A useful estimation of weight can be had by checking the half-weight by gently lifting the rear of the hive clear of the ground. We call this 'hefting'. See: hive scale.

**weighing honey**
Honey is usually 'weighed' as equivalent to a given volume so pound jars when filled to the requisite level are labelled as containing one pound or 454g. In Canada they say milk is weighed by the pint in summer and pound in the winter but joking apart the equivalent volume of honey will vary slightly according to its density. Cut comb honey can be sold according to its weight. The density, i.e. s.g., of honey is 1·414 making it almost one and a half times as heavy as water.

**weights**
Bee egg 0·13mg, 1 lb. wax can hold 30 lb. honey, a larva at heaviest is 93x heavier than the egg, emerging bee 112mg,, foraging bee 80 – 85mg, a queen 150 to 200mg (heavier ones better accepted).

**weld**
The uniting or fusing of two pieces of metal by applying heat and pressure. As much metal equipment is used in beekeeping welding plays a significant part. Where soldering is unacceptable as where heat is to be applied (say a smoker chamber) then welded joints are best. Solder also contains more lead that is permitted when in contact with honey. Plastics have now come to our rescue in many ways.

**well-filled**
Combs of honey, especially sections and cut comb look good when every cell has been filled and capped, empty or unsealed cells spoiling the otherwise perfect appearance. Combs that have few, or better still no cells that have not been capped, would be referred to as well-filled or possible well-finished.

**well-found**

Having a good reserve of food. Well-furnished with usable stores. Colonies should be well-found as winter approaches. The well-being of colonies demands that the reserve of stores never falls below six kilos and in Britain a food supply of twenty kilos is required if a colony is to remain unfed until the new season's nectar comes in at the end of late spring. A colony in spring should be described as 'fat and thriving'.
See: reserve, strength and winter feeding.

**well-made**

Constructed and made so that it will fulfil the purpose for which it was intended. All items purchased from equipment suppliers should be well-made and we would look for durability, interchangeability and competitively priced as well. Unfortunately warped or ill-fitting frames, unsuitable veils that let bees in and feeders that don't work properly have found their way onto the market. Shoddy workmanship should not be encouraged, it is usually better to pay a little more for an item that is well made.
The recent (June 2011) mad rush for beekeeping equipment has resulted in many substandard frames being sold, ones that are impossible to assemble without further cutting and shaping!

**well-kept**

An apiary or hive that is well-kept will be in a tidy, flourishing condition showing every indication that the beekeeper has in no way neglected her charges.

**well-proportioned**

Although it is not wise to judge a queen by appearance alone – obviously the characteristics of the colony she produces (her progeny) are more important than that; she should nevertheless be well-proportioned. That is she should be complete in every detail, nicely rounded and of good body length without being narrow or pointedly tapered. Races vary and experience counts but audiences never fail to exclaim 'Ah, there she is, what a beauty' when with regal gait the sedate hive mother parades herself on the comb and nearby bees pay homage.

**West cell-protector** ***

A coiled protector formed from a spiral of wire so that the narrow end is open for the queen to emerge when ready while at the top a wider diameter accommodates the wider, upper part of the cell. This is fitted with a thin metal plate that can slide in to prevent access to any part of the cell. The spiral of wire is open laterally between rings so that the cell benefits from the hive atmosphere. At the top end the spiral extends to form a projection which enables the queen cell to be attached to the comb.

**western red cedar**

Heartwood of Thuja plicata which is found on the western side of the west coast of North America. Indians used it for war canoes and totem poles but early settlers believed it to be related to the cedars of Lebanon. Eastern white cedar Thuja occidentalis provides an early source of pollen in Connecticut. Used as a hedge shrub making a dense, evergreen barrier but mellowing to a silver grey.
See: cedar and timber.

**wet**

You can use this word in regard to combs, supers, extracting equipment etc. when they are 'wet' with honey. Honey sticks to all manner of surfaces and in particularly it sticks to honey comb 'cappings' when they are cut off to open cells for extraction. The honey clinging to them can be partially drained off and a little more squeezed out or centrifuged out but you are still left with honey wet wax. The bees welcome back the honey-wet combs but to remove honey from utensils it is normal to use water. You can make good use of the washings from wet honeycomb. If not too dilute they can be fed back to the stock they came from or turned into mead or honey vinegar. Never expose honey-wet supers for bees to clean up in the open-air as this encourages disease and robbing.

**wetting agent**
A substance that is added to a liquid to increase its penetrative, spreading or wetting properties. It is likely that the merest trace of detergent, as found on a line of clothes hung out to dry, materially assists bees in the easy up-take of water. See: soiling.

**wheast**
A pollen substitute, sometimes fed in the form of a moist patty. It is dried yeast grown in whey with milk albumin and contains 57% protein. Rather expensive but experiments using such substances as fish meal are proceeding. It is rarely called for in the normal conditions existing in Britain.

**wheat *** Triticum spp.**
This cereal grass grown for the production of flour for bread making, would not, on the face of it, be of any use to honeybees. However secretions from extra floral nectaries and from accompanying aphids, have been known to attract honeybees. Wheat straw with long unbroken stems (minus ears) had a firm place in skep making and many a farm hand (cottager) would sit weaving a skep as a therapeutic relaxation when there was time to spare. See: skep making.

**wheeze**
The sound made when inhalation is difficult or partially obstructed by fluid. This condition is not uncommon amongst beekeepers and can be made worse by a bout of stinging but also in some cases by the mere exposure to bee-protein – bees themselves, their faeces, comb, wax, propolis etc. The wax moth, like other insects, can bring on asthma in some patients. See: prostaglandins and sneeze.

**whin**
Another name for gorse or furze. See these titles.

**whirl**
Honey is expelled from the cells by being whirled from uncapped comb while being spun around in a tank or suitable receptacle. When a bee engages a wasp, or more likely another bee, they spin or whirl round in their attempts to master one another. In the vortexes of a turbulent wind items can be whisked away. Bees wisely stay home but the beekeeper may find roofs torn off or other damage done when lack of shelter allows the wind to take its toll. See: fighting and honey extractor.

**white ant *****
A social insect of the tropics. Also known as the termite. Their colonies can become enormous and spread out with interconnected nests over a large area. In some cases queens are entrapped in sandstone cavities while worker castes carry away many thousands of eggs every day.

**white beam ****Sorbus intermedia/aria* Rosaceae**
Swedish white beam is worked for nectar in June. Hayes reports that masses of white flowers are very freely worked for nectar. It is widely used as a street tree being compact and its bright red crown of berries contrasts with it non-shiny leaves.

**whitebrush *Lippia ligustrina***
Privet lippia found for example in the Chihuahuan Desert of the Big Bend in West Texas. The shrub has been described 'nectarwise' as finicky, fickle, unreliable and undependable. Strong colonies though can build up a big surplus during the short blooming period. Adequate rainfall is required to initiate heavy bloom but like other members of the acacia family the flowers cannot tolerate contact with water.

**white bryony** ***Bryonia dioica*** **Curcurbitaceae**

A hardy perennial climbing plant with lobed leaves that can yield pollen and nectar over a long period May-Sept though mainly in late summer. It is common in hedgerows and has greenish-white flowers and scarlet berries.

**white cappings**

This is especially noticeable when bees have left a small air space between the capping and the honey. Most bees tend to seal honey with little or no airspace under the waxen capping and even when the honey is light in colour such comb does not look as white as that which has a gap between honey and capping. The trait of making white cappings was attributed to the one-time indigenous British black bee. Possibly in our damp climate the air space acted as a buffer so that moisture by osmosis through the slightly porous capping was not allowed to get into the stores.

**white clover Dutch** ***Trifolium repens***

Blooms June – July has rounded white flower heads and has a heady perfume. Once a floret has been serviced it flops down and becomes brown. It carries a 'heady' perfume. The honey is water-white to pale amber. The pollen is pale greenish-yellow also described as brown, dark brown and brownish-grey with a grain of 25mu. Cultivars S100 and S184 have probably hybridized with our white clover because it is not as commonly worked as heretofore. Nectar is yielded most freely when day temperatures reach between 70 and 80F (21-27C) and warm nights and moisture helps. The honey is slow to granulate. It was once declared "the honey-plant par excellence".

**white deadnettle ***** ***Lamium album*** **Labiatae**

Common in the south and flowers March to December. It is considered to be theoretically capable of yielding 500kg of nectar per hectare. Pollen dark orange. See: red deadnettle.

**white-eyed brood**

Uncapped brood can be assessed by the colour of the eyes. Initially they are white becoming pink, yellow and brown before finally turning black. Deformed adults due to a genetic fault can sometimes have white eyes.

**White Gilbert 1720-1793**

He was an English clergyman in Selborne, Hants and wrote on Natural History showing wit, meticulous accuracy and a great love of the countryside. Natural History and Antiquities of Selborne 1789 quote 'It does not appear from experiment that bees are in anyway capable of being affected by sounds. Nutt improved on White's Collateral box. See: Charles Butler and 'stand-still'.

**white spirit**

A turpentine substitute used as a solvent for cleaning brushes of oil-bound paints. It is used to soften or dissolve beeswax in the formation of polishes. Unlike turpentine which is derived from trees, white spirit is a mixture of petroleum hydrocarbons in the boiling range of 150 – 200C. The two can be mixed as they are when making shoe or furniture polish. See: furniture/turpentine.

**White Rev. Stephen**
**Rector of Holton in Suffolk.**

He described his two side-by-side bee boxes aiming at avoiding killing the bees when taking the honey. Published '*Collateral boxes*'1756 and stated that from a set of collateral boxes 100 lb. of honey have not infrequently been obtained. He doubtless influenced Thomas Nutt (1832).

**whitethorn Crataegus *****

Hawthorn a small, tough tree, thorny and widespread. It has bright coloured fruits 'may' and when in full blossom looks like a carpet of snow (there are pink and red varieties). A fickle nectar yielder but it can cause serious competition to apples when bees are moved in for pollination purposes. See: hawthorn.

**white wax (cera alba)**

Beeswax carefully refined and bleached is sold for pharmaceutical purposes in the pure white form. It makes superb cold cream and is a medically neutral (hypoallergenic) substance, being benign, cleansing and soothing to the skin and body. It does not have the rich aroma of golden beeswax as all the material other than wax has been removed. Paraffin wax is also called` 'white wax'.
See: paraffin wax.

**White William 1771**

Wrote a complete Guide to the Mystery and Management of Bees.
He called queens from casts 'governors'.

**whole bee extract WBE**

For some misguided reason sting allergy was for a long time treated with this substance, somewhat ineffectively by most accounts. A great measure of success now comes from using pure venom.
See: allergy and desensitization.

**whortleberry ***Vaccinium myrtillus* Vacciniaceae**

Bilberry myrtillus and cranberry oxycoccus, a low growing, shrubby plant that sheds its leaves in winter. It can cope in the deep shade cast by the surrounding heather but in places has ousted the heather. In spring the nodding greenish-pink flowers provide a feast of nectar for long-tongued insects like the hive bees. A delicate blue grey bloom forms on the fruit. Bees kept near the heather moors appreciate its help in the spring.

**wick**

A braid of soft threads forming a cord, strip or tube that will draw up melted beeswax, oil etc. Cotton is widely used as it is naturally absorbent. For making beeswax candles Clara Furness recommended professionally made, well woven, smooth wick that has no loose ends and tells us that for perfect combustion wicks are pickled in a dilute solution of mineral salts for about 24 hours. Her pamphlet 'How to make beeswax candles' may still be obtainable.
See: beeswax candle.

**wicker hives (alveary)**

These were made from privet and willow and cloomed with harl. 'If you shake them to their wicker hutch'. V. Sackville West.

**wide open entrances**

Because bees must always guard their stores the size of an entrance must always take into account the strength of the colony. Even in warm climates there is a tendency to use smaller entrances than we are used to, the bees themselves being able to take care of ventilation. During winter a wide open entrance covered by a mouse guard is quite usual though the coming of screen-type floorboards may have changed this attitude. Some Bee Farmers in Australia were reported as moving bees while leaving their entrances 'wide-open' The vibration of travelling kept the bees where they belonged.

**wide spacing**

The centre-to-centre spacing of the parallel brood combs needs to be quite accurate (1") to encourage the development of 'worker' brood. However in the supers, where honey is intended to be stored, the spacing can be extended by using wide spacers, Manley frames or other means. The spacing of 2" gives a nice wide comb that can be uncapped easily without fouling the side or top bar. Given 'free-reign' bees can draw their combs out to an even wider width but this is rarely convenient for the beekeeper. The giant honeybee which makes a single huge comb high up on rocky cliff faces puts honey up top in extra wide cells and then the comb tapers downwards to accommodate normal worker brood.

**widespread**

Spread over a wide region or something that occurs over a wide area. This could be the type of hive or system of management and of course it has on more than one occasion meant disease – the Isle of Wight disease and Varroa are examples. Only rigid counter measures can cope with serious outbreaks which unchecked, can 'sweep' across the country.

**width of**

Each make of hive will be of a specific width to accommodate a set number of frames. Frames too will conform to certain measurements of width and depth. In the three dimensional lay-outs width (or breadth), length and depth are specified, though conversion to metric figures often calls for conversion tables.

**Wighton John 1842 G. B.**

He observed that bees building comb added propolis to the wax. This would imply that bees need propolis, yet the only places they can store or position it can be anathema to some beekeepers.

**wild**

Growing without the care of man – fruit, flowers, honey etc. It also means violet, frantic or madly excited. We often hear mention of 'wild' bees which are not savage but feral. Although we have been able to domesticate honeybees they must be free to fly wherever they want and when it suits them, to depart from our care.
See: wild –feral.

**wild - feral**

In connection with honeybees this usually refers to 'native' (undomesticated) rather than to their temper. Plants or animals that survive in a state of nature – undomesticated, untamed, feral, ferine or savage. This applies to 'solitary' bees. It is unwise to refer to bad-tempered honeybees as 'wild'.

**wild carrot *Daucus carota***

Also known as Queen Anne's Lace. In the U.S. substantial amounts of carrot honey are produced whenever temperatures exceed 80F (27C). The honey is light amber.

**wild cherry *** *Prunus avium***

The gean, an elegant tree that is conspicuous in the spring. Ancestor of the cultivated sweet cherry. It likes open woods and can grow to 75 ft. (22m). Its timber is used for furniture making and its honey is almond flavoured.

**wild colonies**

Colonies that have established themselves without intervention by a beekeeper. They can be a source of unwanted swarms, disease, worried neighbours and possible stings. The whereabouts of wild colonies should be reported to Fera.

**wilderness**

An unchecked area of growing plants which may be isolated or desolate, in which only wild animals are found. Where formerly cultivated lands or gardens are allowed to go wild, a large variety of bee plants may be found. Hives of bees left in such places may completely disappear under a canopy of brambles and nettles. That they can survive in such circumstances points to the advantages of forage, shelter and discouragement of robber bees and lack of interference that such places provide.
See: ruderal.

**wild flowers**

Plants not domesticated or cultivated. Unlike weeds which are not wanted, flowers are usually welcomed, especially by beekeepers and other nature lovers. Attempts to save certain wild flowers from becoming extinct are being made by enlightened persons and the bees by pollinating continue helping to do this as they have through the ages.

**Wildman Thos**

He wrote 'A treatise on the Management of Bees 1768 and described the use of top bars in a box hive. In 1772 at a Bee Exhibition in Jubilee Gardens, Islington he or his nephew Daniel performed all manner of feats including appearing before George III on horseback, firing a pistol and wearing a swarm on his chin. A complete guide for the Management of Bees was published 1773 and he had commented that 'if the bees work comb under the board of their stand, it was a sure sign that the hive will not swarm that season.

**wild mustard *Sinapsis* or *Brassica arvensis***

A wild flower considered as a weed worthy of expensive extermination when crops like corn are cultivated.
See: charlock.

**wild white clover *Trifolium repens***

A hardy perennial yielding nectar and pollen and having prostrate flowers from June through summer. It naturalizes freely and likes limey soils.
See: clover.

**wild yeast**

Yeast that is present when no steps have been taken to eliminate it.
Whereas many of its forms can make a drinkable brew, greater certainty of successful fermentation is assured when Campden tablets (or heat) are used to destroy wild yeasts and then when the must has cooled to blood heat, yeasts of a desired variety can be cultured in. An example is Young's super wine yeast compound but to use brewer's yeast (as for bread) does not give the best results any more than the natural yeasts that might be present in honey.

**willow (sallow) *** Salix spp. Salicaceae**

Deciduous trees and shrubs. The dioecious male and female plants both produce early nectar - so useful for rearing early brood. Varieties include : crack, white, bay, osier (withy), pussy, sally, golden, weeping (*S.babylonica*). Nectar can be had at fairly low temperatures, they are rapid growing. The well-known 'pussy' willow 'lights-up' when its catkins are ablaze in March and early April. It is easily rooted from cuttings. The pollen is pale yellow with a greenish tinge, mustard coloured. S. viminalis used for commercial cultivation has rods or wands that are light, tough and supple – they can be bent into intricate shapes such as lobster pots, bird cages or the basic frame for a cloomed beehive.

**willowherb *****
***Epilobium/Chamerion angustifolium***
**Onagraceae**

A tall, showy, gregarious plant with willow-like leaves and bright, pinkish-purple spikes of flowers which bloom from July until the frosts. Once rare, its aggressive, creeping roots enable it to grow in stands and now it is common and vain attempts have been made to class it as a noxious weed. The white fluffy seeds are easily wind spread though it is best propagated from roots. The honey is water-white, slow to granulate and of high grade. Pollen grains often adhere to the sticky viscin produced and they sometimes trail from returning bees.
Colour: saxe-blue and deep greenish blue, with a grain of 72mu. Varieties: montanum, obscurum, palustre, parviflorum and tetragonum.
See: 'fireweed'.

**wind**

The movement of the air over the earth's surface. Its force affects: mating, foraging, need for shelter, ground speed, nectar secretion, plant development, temper of bees, direction and shape of foraging area.
See: prevailing wind, speed of a bee and shield entrance.

**windbreaks**

A natural or man-made barrier utilised to break the force of the wind. Solid barriers are not as suitable as those that restrict the wind without causing turbulent eddies which might bother the bees or blow roofs off (especially WBC's). A thicket of quick-growing willow and blackberry which also provide forage, is good or beech or evergreen hedges Such shelter may also have the advantage of causing the flight paths to lift over people's heads and will give the beekeeper greater privacy when manipulating.

**window**
The letting into a hive wall of a glass window was a method once used to see when the bees needed more room. As a means of observing the bees such a structure was of limited use. At times bees accidently fly straight at a glass window. This is because they regard the reflection of the sky as the reality.

**wind pollinated**
Plants that are fertilised by airborne pollen. These are numerous and include: hazel, pine trees, maize, and others. Oddly though, bees can benefit from many of them. Bees will collect pollen from hazel catkins and maize and pine trees can yield honeydew. On the other hand many plants capable of a partial set through wind benefit even more when bees join in the work. Wind alone often results in waves of fertilized seed across the direction of the wind.

**wind speed**
When the rate of air movement approaches the flying speed of a honeybee (c.12 m.p.h.) foraging becomes increasingly difficult. Mating can only occur when winds are light. Adonsonii queens do not fly when w/v reaches 5m per sec (c. 11 mph). Lighting smokers becomes more difficult, flying sparks can cause trouble and exposure of brood is unwise. See: windy site.

**windy site**
This can lead to bees becoming bad-tempered and (due to confinement) the increased likelihood of certain diseases. If a colony has to be kept in such a place shield them with such natural breaks a you can and provide man-made ones too. Also be prepared to feed and to move to a calmer site if you can arrange it.

**wine**
A beverage made by fermenting juices of fruit suitably sweetened and possibly spiced. Honey makes an excellent sweetener and blends with all kinds of fruit. With apple it is called 'cyser' and with grapes – 'pyment (from Roman times). Four pounds of honey to a gallon, made sterile, then suitable yeast added and fermentation valve inserted when initial fierce fermentation has subsided. See. drinks and mead.

**wing beats**
These create the characteristic flying sound of a bee and the rate has been stated as 240 c.p.s for workers in flight.

**wingless (apterous)**
Having no wings or merely rudimentary appendages. Several serious mites that are ectoparasitic and endoparasitic or commensal on the honeybee are also wingless.

**wing movements**
In flight the bee's wings make paddle-like, figure-of-eight movements with a downward power stroke and 'feathered' return. Trembling wing movements can occur for a variety of reasons, usually unhealthy. When a bee is stationary but fanning, the wings are usually unhooked. Short bursts of power may be used in the nest to enable bees to 'sniff' by drawing air over their antennae. See: flight, sniff, and wing beats.

**wing power**
The venation of the wing and the power of the flight muscles in the thorax combine to give 'wing power'. The same muscles are used in workers for the production of warmth to incubate brood and keep the winter cluster stable. Ventilation

including the requirement for brood temperature control, wax production and the processing of nectar all depend on the vibration of those thoracic muscles. See: qualities of a good queen.

**wings*****

1. Thin, flat, lateral extensions from the thorax – the specialised organs of flight. They are veined, membranous, transparent and dotted with thousands of small hairs. Although flimsy, particularly at the trailing edges, they are chitinous and reinforced by chitinous veins. One strong, double vein runs along the leading edge of the fore wing and absorbs the driving force imposed on it by the thoracic movements (vibrations). The pair of forewings latches onto a smaller pair of hind wings as the bee swivels them from the folded position on its back to the extended position at right angles to the body. They are used for fanning, flying and keeping the rain off a swarm cluster.
2. The horizontal side petals of a leguminous flower that act as an alighting platform for insects.
   See: flight, fore wing, hind wing, hooks and leguminous.

**wing swarm (archaic)**

A swarm that comes out after August (at the heather).

**wing venation**

Veins of the wings originated as tracheae in the prepupa when as small pouches they developed from the thorax. They harden into stiff tubes which strengthen the thin membrane of the wing and provide a frame which can be operated by the muscular thoracic movements. Their characteristic lay-out is specific to the honeybee but minor variations allow different strains to be identified.
See: cubital index, taxonomy and venation.
(Diagrams of each wing needed)

**Winnie the Pooh**

Popular story the author of which was a beekeeper visiting the London Zoo during World War I, to let his son see a special young bear originating from Winnipeg.

**winter**

The season that includes the end and the beginning of the year. Cold usually keeps colonies in a near comatose condition known as the 'winter cluster'. Good beekeepers will have prepared their apiaries in the autumn so that little more than security checks are required until the spring. Snow may make it difficult to reach hives but colonies survive when completely covered, at least for several days. Uncovering them to allow freezing winds to enter has been known to bring about their demise. Entrances should be shaded so that bright light reflected from the snow will not entice bees onto suicide missions. Guard against damp which is more damaging than cold. This is the time when articles separated from the bees can be sterilized. Propolis and wax are brittle, tricky things like excluders can be rendered clean and serviceable. Mouse guards on the hives should not be allowed to clog up with dead bees - a bent piece of wire can claw them clear. Honey and pollen can be kept fresh in the freezer.

**winter bees**

Bees physiologically conditioned to carry the colony through the winter. Both hypopharyngeal glands and the fat body are enlarged compared with other bees. That would have been helped by the consumption of pollen without the yielding of brood food. To quote Dr John B. Free 'As a result of physiological resources and diminished activity, many bees that emerge in August, September and October, survive the winter.
See: fat body and hypopharyngeal glands.

**winter cluster**

Bees begin to form small groups when the temperature reaches 14 – 19C (57F). So many things make the honeybee 'unique'

and this closing up together tends to change them to social insects with a mammalian tendency. The 'cluster' will have a density of about 33 bees per square centimetre with external temperature causing variations so that the cluster contracts when it falls. Cluster temperature diminishes from 24·5C at the centre to a surface shell of 6- 8C (43-46F). Below 9C bees go into a chill coma and fall to the floor from the cluster surface. Small groups join together and by -10C there is a single mass clinging together on the combs with many of the bees with their heads in the cells. Single bees or small isolated groups become immobile and eventually die. The queen is always kept alive until the very last and in colonies that have succumbed, it is not unusual to find her dead but still with a small faithful group around her.
See: comatose, pose and winter defence.

**winter confinement**
The period during which bees are confined to their hive due to unsuitable weather for flights causes the retention of considerable amounts of faeces in the rectum. This can become urgent if the period extends to more than 6 weeks. In ordinary circumstances bees make cleansing flights whenever daylight and weather conditions permit. When confinement exceeds six weeks it becomes really serious and there can be wholesale dysenteric droppings within the hive.

**winter defence**
When a cluster is disturbed in the winter the bees on the outside of the cluster exposed to light, cold or sudden interference raise their 'tails' high appearing to stand on 'tip-toe' while protruding their stings and frequently displaying a droplet of venom. This 'prickly'- looking mass must appear very off-putting to a small invader because of the sound and colour. When it is really cold they seldom attack, though 'kami-kazi-wise', the odd one might fly out but with little chance of it making it back again through the cold air.

**winter dwindling**
One report on an average strength colony tells us that 21,000 bees in November dwindled to 12,000 in March, which is about 74 bees per day, 518 per week and over two thousand a month.

**winter feeding**
Food is rarely put into hives (except in dire emergency), during the winter. Feeding for winter would be a more appropriate title but as that is done in late summer and autumn it is listed under autumn feeding. Active colonies that are able to fly have little hesitation in taking syrup fed to them in a suitable feeder. However by October colonies that may need feeding and seem unwilling to take liquid from a feeder, must know that such food would ferment – solid food in the form of candy is an alternative but it is a poor substitute for sealed stores.

**wintergreen *Gaultheria procumbens***
A low-growing plant found in eastern N. America and used to produce 'oil-of-wintergreen' which had been advocated as a preventative for Acarine.
See: oil-of-wintergreen.

**wintering chamber**
A room or building used for over-wintering colonies in countries like Canada where winter's cold makes open-air wintering uneconomical. Physical requirements include: complete darkness, a high capacity air conditioner keeping temperature within the range of 2 – 9C and humidity between 50 and 75% (R.H.). A re-circulated air flow of 0·10 litres/sec per 1 kg of bees in winter but increased up to 1·6litres/sec in autumn and spring when ambient air temperatures are higher. Often positioned below ground level, these well-insulated chambers are best with concrete floor, flight exits closed late autumn when hives are put into the chamber and re-opened in late winter. Hives often placed in tiers against a south-facing wall.

**winter kill**

The loss of colony(ies) due to causes associated with the winter. e.g. disease worsened by damp or cold, starvation, queen failure at a time when natural replacement is impossible. severe or freak weather, interference by creatures (mice etc.) which during the active season would present no difficulties.

**winter losses**

The number of colonies that do not survive the winter. In nature, many colonies die, the beekeeper however, having invested a considerable amount of time, money and effort in his or her bees, decides how many colonies are fit to over-winter safely and become honey gathering units next summer. Winter losses do occur however. Unusually long or severe periods of cold, the effect of floods, gales and the like, are added to interference by animals, disease and starvation. At the end of exceptionally bad winters the percentage of dead colonies can sometimes be all too high.

**winter moves**

Should you wish to reposition hives within an apiary during the winter, this can be done during an icy-cold spell lasting for several days, so that bees are not lost due to flying and returning to a former site. Needless to say great care should be taken so as to avoid jolting or knocking the hives. To move over any greater distance should wait until mouse guards can come off and pollen is going in.

**winter packing**

It has been said and verified, that the best packing for bees is bees. Nevertheless as man endeavours to over-winter honeybee colonies in many different regions and climes, it is not surprising that heavy insulation, as if to imitate the insulative properties of a tree, have been tried. As far as the U.K. is concerned, the local climate, type of bee and exposure of an apiary, must be taken into account. Some remove even the inner covers so that only a secure roof protects the bees themselves from the elements, some use a specially ventilated crown board. Few now use layers of insulated packing though this is sometimes indulged in once brood rearing starts in the spring. Since the advent of Varroa many beekeepers use ventilated screens instead of floorboards. See: hive insulation and insulation.

**winter passages**

At one time it was advocated that 'square strips' were laid over the top bars across the frames under the 'quilt'. A normal crown board with correct bee-space performs this function as do hives with top bee-space. Bees are more inclined to cross from one seam to another by going over the frames so by one means or another allow them to do that.

**wipe**

To rub lightly with a wet or dry cloth: this would apply to our hands but bees use their antennae in wiping or caressing movements. Cleaning of the antennae is done by passing through notches in the fore legs and this is often called 'stropping'. The tongue too and that can be inflated with blood and swells to a larger size to make it clean. On workers the bouton (or spoon) at the end of the tongue is used to wipe up or deliver minute amounts of liquid.

**wipe-out**

To remove, eradicate, destroy or annihilate. The British black bee was wiped-out by a syndrome of diseases and conditions which became called the "Isle of Wight' disease. Our attempts to wipe out certain bee diseases has resulted in much harm being done. Predators and benefiters alike are often removed at a stroke. Ramifications and side effects roll on long after treatments have been given. Why do vet's bills for cats increase after the wiping out of nests of wasps? Poison enters the cat's food chain.

**wire**

1. A length of flexible metal of circular cross-section and varying from fine fuse wire to hefty fencing wire. It has been much used to strengthen wax foundation to keep it taut and to fasten it into the frame and therefore vertical, also to prevent the resulting comb from flying to pieces in the extractor. The careful embedding of the wire in the 'septum' of the foundation is essential and it is too late to discover that strips of empty cells often follow the lines of the wire on account of its not having been fully embedded. A good quality wire: bright galvanized or Monel metal is used.
2. Spiral wire queen cell cages are another example of its use.
3. In the form of wire netting it has been used to cover hives to protect them from the ravages of green woodpeckers.
4. The Waldron type of queen excluder also makes use of spaced parallel wires.

See: spur embedder, wire foundation and wire tamer.

**wired foundation**

A 'jig' is set up so that the wire can be laid over the wax foundation and then a brief electrical charge melts the wire into the wax. This is efficiently and easily done in the factory with little extra cost. However many hobbyist spend hours (that would be better spent in the apiary) with jigs, transformers and such-like doing it at home. It is important that no bare wire is left as the bees do not like it – it must certainly conduct heat and perhaps magnetic charges away. It is always necessary to specify 'wired' foundation when ordering as much unwired foundation is used.
Especially with deep combs as wiring prevents sagging, keeps the resulting comb vertical and holds the comb together when the stresses imposed by centrifugal force in the extractor might otherwise lead to sagging or breakage.

**wire gauze**

Or wire cloth – this useful material has become expensive but is widely used to cover ventilation holes in roofs and crown boards. To some extent plastic netting has taken over and of course is easier to shape and cut. Screens, queen and queen cell cages, cages for package bees and many other applications have proved the usefulness of this sort of material to beekeepers. There is some risk of plastic being holed by smoker heat etc. and the all important 'grid' at the bottom of the smoker's fire chamber uses a heat resisting layer of perforated sheet.
See: wire netting.

**wire netting**

Available in meshes of various sizes it has filled all sorts of beekeeping requirements. Where wood peckers, cane toads or other predators are troublesome wire netting is useful. Along with wire gauze as an alternative it can be used for pollen traps, mouse guards and as a barrier to large animals and because bees prefer to fly over rather than through it, it can be used to force the bees' flight paths up to a more acceptable level.

**wire queen excluders**

Initially slotted metal (or plastic) sheets were used to confine the queen to the brood chamber and to prevent her entering the supers. The introduction of framed excluders providing bee- space above or below them came in the form of the Hertzog and Waldron. Being firmer but usually more expensive, these well-made (strong, permanently spaced wires 0·163" across)) excluders have become universally popular. The older slotted zinc excluders are still widely used but suffer from distortion as they are 'peeled' off, which propolisation makes necessary, and the 'unsticking' does nothing to please the bees.

**wire tamer**

A wooden stand that will hold a reel of embedding wire and allow it to unwind at whatever speed the operator chooses without risk of snagging.
See: embedder and wiring board.

**wiring board**
A flat board that holds either deep or shallow frames, keeping them square while wire is inserted and tensioned and for holding the frame while the foundation has the wire embedded into it. Listed in Thorne's catalogue.
See: tension.

**wiring frames and foundation**
When the wiring is a do-it-yourself job it is usual to put eyelets in the side bars and to fed the wire continuously through, anchoring it on a frame nail on the outside and pulling (stretching tightly) so that the wire pings with a note you become familiar with, before anchoring as described at the other end. A template which just fits over the frame has the wax sheet set upon it and the frame is lowered onto the wax so that the wires lay flat on the supported wax . A hot screw driver or 'safe' electrical device will then be used to embed the wires into the wax.
See: crimper, embedder, tension and Woiblet.

**Wisteria** ****W. chinensis*** **Leguminoseae**
This woody climber with its rugged stems, supporting pendant racemes of purple flowers towards the end of May, sometimes has bees delving for nectar when conditions are warm enough.

**with - withy**
A flexible twig, especially willow or band of osier as used for basketry, for the making of skeps and binding of various items.
See: wicker.

**withstand**
To successfully resist or oppose some potentially harmful element such as weather. Hives should be built and maintained so as to withstand the dampness of the British climate. This is the main reason for using Western red cedar which is not however widely used in the regions where it is grown. Hives should have a clear space underneath so that having been soaked by seemingly endless rain, they can become dry once more. Constant dampness causes mould, hinders health and can destroy timber. Barriers can help hives withstand the wind and beekeepers have to withstand the wrath of their spouses when work with the bees has unduly detained them.

**woad ***** ***Isatis tinctoria*** **Cruciferae**
The ancient historical significance of this yellow flowered, hardy biennial plant from which blue dye was obtained, is well known though it was superseded by indigo.
It blooms May/July and grows to 1m tall - a useful bee plant which produces both nectar and pollen.

**Woiblet**
Beekeepers who like to wire their own foundation can use this device which when warmed, runs along the wire placed flat on a sheet of wax foundation and with the right pressure will cause the wire to become embedded thus forming a satisfactory reinforcement.
See: spur embedder.

**women**
Ladies do make very successful beekeepers and some beekeeping partnerships could not perform without them. Working alone requires sensible applications so as to avoid having to lift heavy weights and generally speaking being less hirsute they seem to be stung less than the average male. Because two heads are better than one and many beekeeping jobs benefit by two people sharing the work, married couples where both partners are enthusiasts very often do wonders and are worth their weight in gold to Beekeeping Associations.

**wood**
1. A large area of trees, usually less extensive than a forest. Plants and animals with a specific name of 'sylvatica' are usually found in a wood, e.g. *Vespula sylvestris* (wood wasp). Where such

woods are of flowering trees, for example acacia, chestnut, eucalypt, they can be important honey producing areas.

2. Xylem the tough, fibrous, cellular tissue that underlies the bark of trees and consisting largely of lignin. When freshly cut its water content is high and considerable storage or kiln drying is required to bring this to a level where subsequent use as timber will not lead to warping or splitting. It may still be considered the most satisfactory material for the making of modern beehives, though inevitably plastics are taking hold. Its porous nature and the ease with which bees can grip it with their claws and fasten comb to it, together with its excellent insulative properties, have led to hollow trees making favourite places for bees to nest in.

See: lumber, timber and western red cedar.

**wood anemone**

Common plant of the woodlands its pollen is white.

**wood ant *** *Formica rufa***

The hill or horse ant. Beekeepers might well come across these when in the south they put hives close to the heather and near pine trees. These active creatures build ant hills (nests) covering a sizeable area yet up to five feet high (1·5m) and this is described as only the 'summer house'. They can 'nip' and it is unwise to get your eyes near the nest as they squirt formic acid at potential enemies. You would not wish to stumble on a nest in the half-light. In European forests much honeydew is collected by these ants and beekeepers move their colonies to harvest it too. The nurturing of the sucking insects that convert the pine sap into honeydew is done by the wood ants, a single female fecundrix being able to start a whole colony.

**Woodbury Committee**

In 1882 this BBKA committee laid down standard sizes for a British deep (17 x 8½) and shallow frame (17 x 5½). This was 31 years after Langstroth had tried to patent his movable frame hive. It had been hoped that this would make the transfer from fixed comb hives (skeps) more straight forward than if a plethora of sizes existed. Despite the new standard several different hives were developed and although the eleven frame National is still one of the more popular there still seems to remain antipathy towards American sizes.

See: Woodbury.

**Woodbury Thomas White**

Of Exeter whose writings gave impetus to the introduction of movable frame hives in G.B. The BBKA adopted the standard frames in 1882. He co-operated with A. Neighbour to send 4 stocks of Italian bees to Melbourne, Australia by steamship "Alhambra" 25.9.1862. Wrote '*Bees and Beekeeping*'.

**wooden**

Although plastics have begun to replace many articles of beekeeping the following examples show how widespread the use of wood is: crown board, divider, feeder, frame, framed excluder, hive, honeycomb section, mandril, wedge, trailer, house apiary.

**wooden hive**

These were first regarded with some skepticism by those who were successful with and thoroughly accustomed to, their straw skeps. As wood can be machined to reasonably fine limits and can be effectively stapled, nailed or glued, it is likely that as long as supplies last at acceptable prices this will continue to be the commonest material for beehives. Bees seem to get on well with wood, it's porous and easy for them to grip and to build on.

See: hives

**wood honey**

Defined by Sturges as floral contributions from trees. Tree honey also has connotations with honeydew and lachnid secretions from spruce etc.

**woodpecker *****

These birds have a liking for insects and their nests and can peck their way through the weaker parts of a wooden hive. Flapping strips of plastic sheets and chicken wire (wire netting) have been used to deter them. See: green woodpecker.

**wood preservatives**

To keep up-to-date with these products would not be easy. However you should always be aware of the danger of using anything with fumes offensive to bees or which contain insecticides. There are three main groups: organic solvent preservatives, water-borne preservatives and tar oils. Fungicides and insecticides are often added to these. The immersion of wooden hive parts in liquid paraffin wax has proved useful in countries where the heating of huge vats is not difficult. Acrylic paint is also used.
See: hot dip.

**wood sage *** *Teucrium scorodinia* Labiatae**

Wood germander has a pleasant aromatic scent and is found in dry woods, stony banks and thickets. Bees work it freely and in some locations get a surplus from it. A hardy perennial growing to 1 ft. high, small yellow flowers and sage-like leaves from July to August and will naturalise in sandy woodlands or heathland.

**Woods E. F.**

In 1956 he invented the 'Apidictor'. He was a leading sound engineer with the BBC. A microphone was inserted into a hive and the particular note coming from nurse bees loaded with royal jelly could be read as a forerunner of swarming. The apparatus could also read the 'queenless moan' and he went as far as to arrange for small lights to show green amber and red according to the readings obtained. At a meeting where Mrs. Roxina Clark followed his prognostications by actual investigation she was able to confirm the accuracy of his findings.

**wood swallow *Artamus personatus***

An Australian bird that is reported to alight and break a bee in two discarding the abdomen and sting.

**wood work**

Carpentry – Evening classes are run by skilled craftsmen and all the better if one is a beekeeper. Much can be learnt about the skills required for making hives and other equipment. Standards are laid down by BSI and hive plans etc. are available from BBKA.
woodworm   Anobium striatum /punctatum
The larva of this beetle (borer) makes tunnels into woodwork and this can include beehives. The usual methods of treatment utilize insecticides which are taboo as far as bees are concerned so their intrusion must be prevented by the use of repellent timber such as Western red cedar. Heat also provides another method of dealing with them. The bees do not object to these 'commensals' and unless accompanied by wet or dry rot they rarely seriously weaken timber, at the same time their spread should be discouraged.

**word processor**

The coming of this masterpiece has revolutionized the typing, storage and transmission of information. Systems take a while to get used to but once you have acquired the necessary skill they offer a new world of possibilities and are infinitely useful for so many aspects of beekeeping. Usually just an aspect of a PC - (computer), only the old and weary discover how much they are missing.

**work**

Exertion aimed at accomplishing something. The part-time activity of bees or humans. We talk about bees working (drawing-out) foundation, a flow or a particular crop. Although bees sometimes appear to be idle, they do look stationary at times when their glands are developing. We refer to their taking 'play-flights' and when a swarm takes (seemingly gleefully) to the air, it does seem for all the world, as if they are playing. Their industrious nature has astonished all observers because from birth to death they are always working for the well-being of their colony. Even drones?

**workable**

Things that can be made functional. Foundation can be worked and beeswax though brittle when cold can be 'worked' when heated to plasticity or to liquid for making polishes, creams, candles and in many other ways.

**worker**

The most numerous caste in ant, bee and wasp colonies. Workers are infertile although derived from the same egg source as a queen. They are highly specialized to perform their many different tasks. Its life span varies from as little as six weeks in the summer to as much as six months in the winter. Unless there is a change of queen all workers in a colony come from the same mother though the sperm father may change from time to time as different layers of spermatozoa are utilized from the queen's spermatheca and this may bring about a change in the worker's colour and characteristics. Workers carry a barbed sting, special tongue and pollen collecting apparatus and takes 21 days to rear in a worker sized cell.
See: queen, drone and laying worker.

**worker base foundation**

The cell base of the embossed sheets of wax which we call 'foundation' can be varied in cell size so as to accommodate either drones or workers. The worker base is fairly standard though can be varied slightly – e.g. the Mediterranean bees tend to be smaller than ours. Bees can change patches of comb from worker to drone size, produce 'pop' holes, make queen cells or add burr comb once the foundation is in the hive.
See: cell size and worker cells.

**worker brood**

When a fertile queen is present the bees select cells in a part of the hive where the brood nest temperature close to 36C (95-97F) exists. These they clean and polish and in doing so impart a message to the queen in the form of a pheromone which virtually tells her that it is OK to lay in them. This she does, inserting her head into the cell first to inspect and then manoeuvring her legs so that the tip of her abdomen can reach down and stick a tiny egg to the base of each cell. A good queen covers an area completely so that a compact, mutually warming brood mass can then follow the nurse bees' subsequent attentions. The shape which her abdomen feels and pheromonic invitation reminds her to pump one (or perhaps 2) spermatozoa onto each egg which leads to fertilization. Left to incubate for three days, a tiny larva then splits the chorion (shell) and is immediately given the richest food by young nurse bees. Over the next six days they are fed by increasingly older nurses thus weaning the larvae with a changing diet. They lie, incumbent on a bed of food, curled into the base of the cell until, feeding complete, they stretch out and begin to spin a cocoon which covers the top of the cell which is also enriched by a capping which workers bestow upon in. Thus enclosed, metamorphosis takes another twelve days when as an imago they nibble their way out of the cell and emerge to join their sisters immediately in the work of the colony.
See: cell size and worker cells.

**worker cells**

These are approximately two to the centimetre and accommodate developing worker brood just allowing as they do, for the queen's abdomen to reach down to the cell base (laying workers find this difficult). These worker cells are not only smaller but less domed than drone brood cells. Their hexagonal shape setting them cheek-by-jowl and backed by similar cells on the other side of the comb, warmth is mutually shared.

The same remarkable cells (constructed by unique beeswax) are used for bees to easily 'climb' over, to hold and permit the processing of nectar, to make a permanent store of honey and to hold pollen. Collectedly they form vertical combs which are spaced in a parallel fashion to allow passage and control of the colony's micro environment. They also act as a 'sound-base' upon which foraging directions can be given and the queen can 'pipe'.
See: cell-size, honeycomb, wax makers and worker comb.

**worker comb**

This is built in vertical slabs with a reasonably constant cell size for any particular race (the eastern honeybee is slightly smaller).Sizes go from around 5·3 - 6·3mm and a square decimetre of comb has around 615 cells when both sides are counted. The base of each cell is a hollow formed by the rhombic bases of three adjacent cells on the other side of the septum. When the cells are capped over they are domed but noticeably flatter than drone cells.
See: comb, cell sizes, honeycomb and worker cells.

**worker egg**

Eggs laid by a queen are visually indistinguishable from one another but those impregnated by spermatozoa are fertilized and will become workers or queens whereas those without sperm (unfertilized) become drones. Eggs begin in the germinal tissue at the tip of one of the queen's ovarioles and move along, accompanied by nurse cells. As it is nourished the nurse cells are absorbed and the egg attains its full shape. It is slightly curved, sausage-shaped, weighs about 0·13mg, has an apical diameter of mm and just exceeds 1½mm in length. It is glued to the cell floor perpendicularly. When looking for them in the cells you should hold the comb at a suitable height with the light shining over your shoulder from the rear.
See: egg and micropyle.

**worker jelly**

J. Beetsma differentiates between food given to the queen and the larvae. Queen larvae get secretions from both mandibular and hypopharyngeal worker glands and it is richer in both pantothenic acid and biopterin than the food given to workers.

**worker larva**

After incubating for three days in the egg stage it hatches from the egg as a minute larva, it receives ten thousand visits from nurse bees during the six days of feeding and careful weaning and rationing which controls their enzymes and causes the remarkable transition into a specialized worker. It then metamorphoses to become an adult after 12 days when it is capped over. It has been calculated by one observer that there were 143 separate feeds taking four hours while 125mg of honey and almost as much pollen was consumed. Bees taking part in the rearing of one larva including the capping of the cell number nearly 3000 totalling some ten hours of work.

**worker's duties**

1 – 3rd day Cleans cells and helps to maintain brood nest temperature.
3 – 6th Feeds older larvae with dilute honey and pollen and may attend queen.
6 – 15th Feeds younger larvae with brood food.
10 – 20th Processes nectar, makes wax, guards hive, helps maintain temperature, humidity and to shape comb.
After 20th she forages water, nectar, pollen, propolis or robbing and recruitment of foragers.
See: bees' duties and duties.

**working day**

Exterior efforts by a colony only take place when the sun is above the horizon except when they are interfered with. Long hours of daylight in suitable weather conditions greatly enhance a colony's chances of getting a good crop. In polar regions where daylight is almost continuous the bees are said to have a diurnal break from foraging. The polarized light patterns in the sky (regardless of cloud) enable bees to find their way about.

**working distance**

Bees rarely work flowers close to their hive but depart with a small amount of 'fuel' in their blood and some in their honey sac. This will enable them to go just so far and to forage for a certain time if they are going to make it home safely. This restricts the distance to which they can go. Naturally according to their 'reading' of other foragers success they will be recruited and work the most productive sources whether far or near. It has been found that when moving hives, some bees will have familiarized themselves with an area radiating out for, at the most – three miles. Malta has islands just under that distance apart and observers know that in certain wind conditions bees can work the thyme on one of the adjacent islands. When a colony is moved a bee-line distance of 3 miles a few workers might find their way back and certainly quite a few drones will do so. The most effective (economic) foraging distance is between a quarter and half-a-mile.

**world honey production**

In '83 China, 100 US 93, Mexico 64, Canada 35, Argentina 22 , Turkey 24, Ethiopia 2, Brazil.... Australia ... Metric tonnes.

**worms**

Used sometimes to describe the wax moth larvae (caterpillar). Parasitic worms can affect wasps and similar creatures but rarely honeybees. Maggots or larvae of many insects are sometimes referred to as 'worms'. See: Annelida.

**Wormit hive**

A commercial hive manufactured by Steele & Brodie is similar to the National but rebated joints are provided between the boxes making the hive more water-proof. The supers lock onto one another preventing movement. It is designed to take eleven British standard frames at 1 " centre-to-centre spacing or ten with a dummy board at each end.

**wormwood Artemisia spp.**

Mugwort – a strong smelling plant repellent to bees though in 'The Book' Edward Thomas wrote, 'there was a scent like honeycomb from mugwort dull (*A.vulgaris*). It was used to stuff geese or as a moth repellent Few realize Wormwood Scrubs must have received its name on account of this plant...

**worn-out**

Queens and workers show signs of being worn-out by developing an irregular 'tatty' edge to their wings and a lack of body hair. The queen's brood pattern (and propensity of the bees to build queen cells) will be further indications of a queen's ending her period of usefulness. A worker usually gets lost or is rejected from the hive once her days are over. Veils, overalls, smokers last well when care is taken over them.

**worried**

When a colony is unduly alert with many moving guard bees in evidence we might say that it is worried. This may be due to interference by wasps or other bees. Vibrations, smells or other intrusions associated with 'stress' may cause the bees to be 'tetchy'. An experienced eye should spot evidence of interference by mice, badgers, shrews, woodpeckers – the bees have many natural enemies but if the interference is prolonged their development will suffer.

**wound dressings**

The faith of those who from the earliest times have used honey for ailments of the flesh spring from the following facts: It is the cheapest and most effective ointment, is relatively sterile and has bactericidal qualities, it is hygroscopic and draws body moisture into itself avoiding tendency for dressings to stick to the

wound. It is non-irritant and stimulates the growth of healthy granulation tissue.
See: inhibine.

**woundwort Stachys spp. Labiatae**

There are annual, biennial and perennial varieties which flower from June – September. S.alpine bears purple-red flowers, grandiflora large purple flowers, *S. recta* yellow and *S.pulustris* is the marsh woundwort. Honey is white and of good flavour and slow to granulate.

**wreathe**

A coil or band of straw as in a skep. An eke for example might have five wreathes.

**wrinkle**

A clever device or way of doing things that usually come from experience and ingenuity. An artful method or procedure. Beekeeping is full of these and the practical bee-man is as likely if not more so than the scientist to produce them.
See: gadget, gimmick and implement.

**wrist**

The carpus or region joining the hand to the forearm. Like the ankles this can be attractive to the bee when it comes to stinging. The reasons are: movement, hairiness, perspiration (especially under a watch strap) and the likelihood of exposure of this part. Gauntlets help safeguard the wrists but where bare arms are used great attention to cleanliness is wise. Gum boots should give no access to bees from the top.

**writers**

Barbara Cartland who once graced the National Honey Show with her presence was a prolific novelist whose vim, vigour and good looks reinforced all the things she has said to put honey in its right perspective with the public.
See: Greek and Roman author

# X

**Xenophon 400 B.C.**

Greek historian and writer. He mentions a Greek beverage made of wheat, barley and honey. In Anabasis he wrote an account 'The Persian Expedition' covering the retreat of ten thousand Roman soldiers who are said to have raided honeybees and been affected by debilitating sickness after eating their honey. It was at Trebizond in Asia Minor and the plant source was thought to be Rhododendron /Azalea ponticum. Some soldiers were senseless for 24 hours and took 3 - 4 days to regain their strength.

**xeromorphic**

Able to withstand dry conditions.
Cactus etc.

**xylene (xylol) $C_6H_4(CH_3)2$**

Used to wash out clove oil when mounting microscopical preparations, also as the solvent for Canada balsam which hardens as the xylene evaporates. It is an oily, colourless liquid that comes from coal tar.

# Y

**yarrow** *** **Achillea millefolium**
Whitish flowers, sometimes used in medicine and as a tonic and astringent. Also called 'sneezewort'

**Year Book**
How now and Beekeeper's Annual

**yearly queen colour**
Years ending 1 and 6 white, 2 and 7 yellow, 3 and 8 red, 4 and 9 green, 5 and 10 blue.

**yeast**
A substance consisting of the aggregated cells of minute unicellular fungi which are used to induce fermentation in a 'must'. It can be obtained from wine making sources. There are of course wild yeasts as well. There's dried baker's yeast, granular and liquid yeast the strains of which are suitable for the production of mead. It is a living substance and does its work in a temperature range around 70F.
See: mother of vinegar, Torulopsis, Saccharomyces.

**yeast nutriments**
These exist in honey but as we dilute them so much in preparing the 'must' it is just as well to include nutrients: 2g ammonium phosphate, 2g tartaric acid or fruit juice (grape, apple etc.)

**yellow bee pirate** ***Philanthus diadema***
The banded bee pirate seen to catch a bee on the alighting board, common in warmer countries.
See: bee –wolf.

**yellow jacket**
**(common name for American wasp)**
A North American wasp with a body marked with a bright yellow. They will attack hives though strong colonies with entrances small enough to defend will usually be able to repulse them. *Vespula acadica, albida, arenaria* (small aerial nests).

**yellow loosestrife** ***Lysimachia vulgaris*** **Primulaceae**
A perennial that grows on the banks of streams. It is worked for pollen and may reach up to a metre. When burnt it drove away troublesome flies and gnats. It is pollinated by a tiny bee *Macropis labiata*, and wasps that visit it.

**yellow scabious** *** ***Cephalaria tartarica***
A hardy perennial that grows over a metre tall and yields both nectar and pollen.

**yellow toadflax** ***Linaria vulgaris***
When bees are working this flower they come back to the hive as if their thoraces had been marked with yellow paint.

**yew** *** **Taxus spp. Coniferae Taxaceeae**
An evergreen, coniferous tree with a fine grained elastic wood, Native of the old world, much-used around grave yards in U.K. It flowers in March yielding much wind-borne pollen. Favours chalky soils. Bees sometimes attempt to collect its light yellow (some say gold) pollen though it does not seem to yield nectar and the pips of its delicious looking velvet-red berries are poisonous as they contain a toxic alkaloid – taxine. Its fine-grained timber has great

elasticity and strength and selected straight staves were used for the long bows of archers. There are trees which have survived for 1500 years or more.

**yield**

The honey crop from a hive for one year. Also called the 'take', 'harvest' and 'crop'.
yield of honey per acre
Wedmore gives: buckwheat 95 – 150 lb. per acre. sainfoin 125, white mustard 10 – 13, OSR and ling 44 – 220, clover 60, white dead nettle, garden sage and robinia 500.
See: nectar yield.

**yield of wax**

Colonies kept for the production of honey also average around one kilo of wax from cappings, old comb etc. annually. Bees go through a natural cycle when their bodies are capable of producing wax and quite frequently their wax is actually cast out of the hive in the form of small sales amongst the debris. It can be false economy not to get colonies to build some new combs.

**yolk**

That part of the egg which does not enter directly into the formation of the embryo but is the nutritive material.

**Yorkshire spacer**

A pronged metal clip that can be attached to the lugs or side bars to give centre-to-centre spacing between frames. Unlike the WBC metal end it is fixed and will not slide off. Also the choice of position enables you to fix them on the lugs and at the bottom of the frame so that rigid spacing is provided for transport in the same way as Hoffman frames.
See: metal end and spacer.

**young adult**

The newly emerged drone or worker has a soft, downy appearance. They will not have developed 'colony odour' and are acceptable to any colony. Workers begin cell cleaning and eating pollen and carbohydrate to develop their brood food glands. Their wings and stings gradually toughen up but are at first too soft to permit their use. A newly emerged queen is much more active from the moment she climbs out of her cell.

**young queen**

A queen in her first year of laying. She has everything in her favour being less likely to carry pathogens or parasites, being prolific and less swarmy than an older queen. Every honey producing colony should have its queen replaced in good time to build up to peak foraging in time for the flow, unless it already has a queen from the previous year that is known to be fecund and highly active.

**Yucatan**

Great honey producing country with excellent reputation for its quality.

**Yugoslavia**

1981 centenary of Pchela their beekeeping journal. Honeydew from silver birch and endless floral sources. There have been political change in this area but bees go on as before!

# Z

**zest**
You might speak of a honeybee colony as working with zest, as if they display keen enthusiasm. In fact when any one colony is working harder than the others it may be because they are robbing and have found an easy source of supply, it may be because they have a new queen mated and laying or because they have been stimulated by feeding. However some strains are better 'go-getters' than others and 'zest' can be genetically transmitted.

**zinc queen excluder**
Slotted sheets of zinc that permit the passage of workers but not queen or drones. There are long and short slotted varieties. Being flimsy they soon buckle if used with bottom spaced hives as they have to be torn off from the propolised top bars. They are best framed but avoid heating accidently with the smoker.
See: excluder and queen excluder.

**zip**
Much use is made of zips to exclude bees in protective clothing though Velcro is also coming into use. Zips that 'catch' can be improved by rubbing them with beeswax.

**zygote**
A fertilized ovum or cell formed by the fusion of two gametes (or reproductive cells). A queen or worker egg begins as a zygote.

**zymase**
A group of enzymes functional in the fermentation of sugar into alcohol and carbon dioxide and also the inversion process of converting sucrose into glucose and fructose.
See: invertase.

**zythus**
Pure honey-beer for the richer ancient Gauls.

# Alphabetical Index

bird
bird cherry
bird's foot trefoil
biting
B.K.A.
B.L.
black
black bear
black bee
blackberry
black comb
blackcurrant
black netting
black queen cell virus
black robber disease
blackthorn
blueberry
blastocoel
bleaching
bleaching (slides)
blended honey
blind gut
blind louse
bloodbblood corpuscle
bloom
blooming times
blossom honey
blow
blowing bees
blowlamp
Blow Thomas
blue
bluebell
blueberry
blue orchard bee
blue titmouse
blue weed
board
BOD
body
body of bee
bog
boiler suit
bokhara clover
boll weevil
Bolton hive
Bombidae
Bombus muscorum
Bombylidae
Bombilious
boost
borage
Boraginaceae
borax
botanical name
botany
botchet
bottle
bottle brush
bottle feeder
bottling
bottom bar
bottom bee space
bottom board
bottom box
bottom entrance
botulism
bouquetbouton
box (container)
box (plant)
box hives
brace comb
bracken
bracket
bract
braggot
bragwort
brain
brain's trust
bramble
branch
Brand's cappings melter
branding
brass cone
Brassica
Brassica arvensis/napus
Braula coeca
Brazilian bee
breed
break crop
breaking dance
breathing
breathing pores
breeder colony
breeder drone
breeding
breeder queens
brickearth
bright
brightness

extra-floral nectary
exude
eye colour
eye compound
eyelet
eye simple

Fabre
face lotion
face pack
facet
factors
facts
faeces
Fagaceae
failing queen
faithfulness
family
family tree
famous sayings
fanners
farina
Fasciata
fat
fat body
fatal
father
Father of Bkpg
fatherless drone
fauna
Feb
fecund
feed hole
feeder
feeding
feeding larvae
feign death
fellowship
female
femur
fence
fennel
feral
ferment
fermentation valve
fertile
fertile worker
fertilisation
festoon
Fibonacci
fibre glass
fibula
fickle plants
finding queen
finger prints
finishing colony
fir

fire
fire blight
fire lighter
Fire thorn
fireweed
first flight
first skeletal ring
fixation
fixed combs
flabellum
flagellum
flame test
flanged cork
flank combs
flash-heater
flashlight
flashman
flat
flat iron
flavone
flavour
flax
flexor
flicker
fliers
flight
flight board
flight path
float
floccule
flood
floor
floor board
floorpolish
flora
floral nectary
florea
florescence
floret
flow
flowed in
flower
flower constancy
flower seed
flowering currant
flowering plant
flowers per tree
flowers rarely visited
fluid
fluid ounce
fluorescence
fly
fly catcher
flying
flying speed
foam rubber
foc us
Folbex
folic
follical
follow sheep
follower
follower board
food requirement
forest honey
fondant
food
food attractants
food chamber
food sharing
food transmission
foot
foot bath
footing
footprint pheromone
forage
forager
foraging
foraging area
foraging radius
force
forceps
fore legs
fore wings
foregut
foregut
foreign honey
foreign matter
foreign protein
forest
forest beekeeping
forest honey
forewarning
Forficula auricularia
Forget me not
fork lift truck
Formaldehyde
formalin
formic acid
formulation

fossa
foulbrood
foundation
foundation press
fowls
foxglove
frame
frame gripper
frame holder
frame jig
frame lifter
frame nailer
frame sizes
France
frass
free for all
freedom
freedom from granulation
freeze
freezer
French bee
French honeysuckle
frequency
Frisch
frog
frons
front slides
frost
frosting
froth
Frow
fructose
fruit
fruit sugar
fuchsin
fudge
fuel
fuel consumption
Fuller's earth
Fumagillin
fume board
Fumidil B
fumigation
function
function of honeybee colony
fundatrix
funeral
fungi
fungistat
funnel
furca
furcular
furniture cream
furniture polish
furze
Fuschia Onagraceae

goose feather
gooseberry
gorse
grading filters
grafting
grafting tool
GSL
grain
gram
Grange
Granton
granulated
granulated sugar
granulation
grape
grape fruit
grape sugar
grass
grass verge
grassland
grate
gravid
gravity
grease
Great Britain
great woodpecker
greater wax moth
Greece ancient
Greek bar hives
Greek bee
Greek beekeeping
Greek writers
Green belt
green honey
green sulphur
green woodpecker
greenfly
greenhouse
greensand
Gregarine
gregarious
grip
grooming
groove
ground
groundspeed
group
growth
grub
Gruit
Gruzinian
guajillo
guard bee
Guide bird
guide lines
Guiness book records
gumboots
gums
Gusathion
gustatory
gut
guttation
gynandromorphy

hiss
histamine
histology
hitch hiking
hive
Hive & the Honeybee
hive assembly
hive -baited
hive bar
hive bee
hive box
hive carrier
hive ceramic
hive clay
hive clips
hive closing
hive closure
hive cloth
hive collateral
hive colour
hive Dadant
hive decoy
hive density
hive double
hive entrance
hive escape
hive fastening
hive feet
hive fillet
hive fixed comb
hive floor
hive furniture
hive insulation
hive large
hive leaf
hive legs
hive lift
hive log
hive long-idea
hive maintenance
hive manufacture
hive materials
hive mates
hive mind
hive model
hive movable frame
hive multi-storey
hive national
hive number
hive observatory
hive odour
hive paint
hive plans
hive plastic
hive preservative
hive products
hive protection
hive records
hive roof
hive runners
hive scale
hive scraper
hive site
hive skyscraper
hive Smith
hive spacing
hive stand
hive standard
hive Stewarton
hive strain
hive strapping
hive straw
hive to
hive tool
hive top
hive twin
hive ventilation
hive WBC
hive wide idea
hive wooden
hives per hectare
hives under trees
hiving
hiving a swarm
hiving package bees
hiving sough
HMF
hobbyist beekeeper:
Hodges Dorothy
Hoffman frame
Hogs at the Honeypot
hogweed
holding down cage
Holland
holly
hollyhock
holm oak
Holometabola
Holotype
home yard

industry
infection
infestion
inflorescence
infuriate
ingestion
ingot
ingurgitate
inheritance
inhibine
inhibit
inhibitor
inner bodies
inner cover
innervate
inoculum
inorganic
inquiline
inquisitive bee
insect
insecticidal formulations
insecticidal fumigants
insecticide
insecticide toxicity
insectivore
inseminate
inseminatiion
insoluble
inspection
instar
instinct
instrumental insemination
insulation
insurance
integument
interchanging
intercommunication
interesting facts
interference
interior
intersegmental membrane
interstitial cells
intestine
in-the-flat
intine
introductory cage
intruder
invade
invagination
invention
inventory
inversion
inversion of queen cells
invert sugar
invertase
invertebrate
ion
iron
iron pan
irradiate
irritability
irritation
iso
iso-amylacetate
isohyet
isomer
isomerse syrup
isopentyl acetate
isophene
isopleth
isoprenaline
Israel
Italian bee
Italy
ivy

Lab diagnosis H,bee
label
labelling
labellum
labial glands
labial palp
Labiatae
labium
labour
labrum
laburnum
Lachnids
lacinia
lacquer
lactic acid
ladder
laevulootatory
Laidlaw
lake
lamina
laminated
lancets
Langstroth frame
L/D value
Langstroth hive
Langstroth Rev.L.L.
language
Lanius
large intestine
large scale
larva
larval colour
larval faeces
larval food
larval growth
larval skin
late
lateral
laurel
lavender
lawn
lay flat
laying
laying queen
laying rate
laying worker
LD. value
leaching
leaf hive
leaf honey
leaf-cutting bee
leaflets
leaking feeders
learner
learning by the bees
leather
leather polish
lecture
lecturer
lee
Lee James
leg
leg honey
legality
legume
Leguminosae flower
lemon
lemon balm
length
length of day
length of frame
length of life
length of the active season
length of time to mate and lay
length of tongue
length of winter
lens
Lepidoptera
lesser celandine
lesser wax moth
let alone beekeeping
lethargic
lettering
leucocytes
level
levulorotatory
levulose
liaison bees
lias
library
Libya
lid
life cycle
lift
light
light compass
reaction
lighten beeswax
light honey
light smoker

rectal glands
rectum
recycle
red
red cedar
red chestnut
red clover
red currant
red deadnettle
red-hotpoker
red pollen
red squirrel
reducing sugar
reduction
reduction division
reed
referee
refernce
refining
reflex
refractive index
refugium
regicidal knot
registration beekeepers
regulation
regurgitation
re-infection
re-enforce
rejection
relative humidity (r.h.)
release
releaser
releasing agent
religion
reluctance
remedial effects
remedy-antidote
remunerator
rendering wax
rent
repairs
repellent
reprints
reproduction isolation
reproductive organs (drone)
reproductive organs (queen)
reproductive system
Reptilia vertebrata
requeen
requeened (rqd)
requeening (by queencell)
research
Reseda odorata
reserve
reservoir
reservoir bees
residual effect
resin
resistance
Resmethrin
resolution
resonance
respiration
respiratory movement
respiratory system
response
rest
rest harrow
resting cell
restive
restless
restriction of brood
reticulation
retina
retinue
retinula
retrogressive
return
reverse side
reversible extractor
re-worker
R,H. – relative humidity
rhabdom
rhabdomeric microvilli
Rhamus cathartica
R,H,B. Ron Brown
rheumatism
Rhizobium
Rhododendron
rhododendron honey
rhodomel
rhomb
rhymes
Rhynchota
R.I..
Ribbands
Ribes grossularia
Ribes nigrum
Ribes sanguineum
riboflavin

tegula
telescopic hive
television
Tellian bee
telling the bees
telophase
TEM
temper
temperature control
temperatures
tension
tenth segment
tentorium
terebra
tergite
tergum
terminal filament
terminology
termites
terpenes
terrain
terramycin
territory
terror
test
test comb
test for beeswax
test for honey
test for insecticide
test guage
tested queen
testicular tubule
testis
tests
tethered queens
tetraploid
texture of wax
thanetosis
Theophrastus
theory
thermal conductivity
thermal vaporizer
thermometer
thermostat
thesis
thiamine
thin layer chromatograohy
third party
thistle
thixotrophy
thoracic salivary
thorax
Thorley
Thornes
threshold
thrift
throw
throwing out
Thuja plicata
thunder
thyme
thymol
tibia
tibio-tarsal joint
tie
tiering
Tilia
till
'tilt'
timber
timber for hives
time
time lag
time table
tin
tin smoker barrel
tin snips
Tinsley
tips
tissue
tit
toad
toad flax
tobacco
tomato
tomentum
ton
tongue
Tonsley
tool
top spacing
top bar
top bee space
top crossing
top entrance
top feeder
top fruit
top heavy
top packing
top ventilation

U.K. beekeeping
U.S.herb garden
ubiquitous
Ulex europeus
Ulmus sp.
Ulster B.K.A.
Ultra coelostroma
ultraviolet honey test
ultraviolet lt.
umbel
umbrella valves
unassembled
unblock
uncap
uncapped
uncapped brood
uncapped honey
uncapped queencell
uncapping
uncapping fork
uncapping knife
uncapping plane
uncapping tray
uncover
under or over
undergrowth
undersized
underwing
undesirable
undigested
undress
uneven
unfecundated
Unguentum
unguiculi
unguitractor
unicellular
unifloral
union of stocks
unisexual
unite
United states America
uniting
uniting aids
unmanageable
unmarked queen
unmated queen
unopened queencell
unpolished
unpopular flowers
unripe honey
unsaleable
unsatisfactory
unscrew
unsealed
unsound
unstable
untested queen
unwanted ingredients
upper entrance
upside down
upward ventilation
urate cells
urea
urgent
urticaria
USA beekeeping control
used equipment
useful dimensions
useful hints
uses for honey
uses for royal jelly
uses of bee venom
uses of beeswax
uses of honeybee colonies
uses of pollen
uses of propolis
U.K. beekeeping
U.S. herb garden
usquebaugh
Utah
UV

virile
virology
virus
viscid
viscin
viscus
viscosity
viality
vitamin
vitelline membrane
vitellus
vitis
vitreous body
Vitula edmandsae
vivo voce
viviparous
void faeces
volatile
Volcano Island
voltarol 50
volume
votator
VPD
Vraski hive
vulnerable
vulva
wad
Wadey
waggle dance
wag-tail dance
waist
Waldron
Wales
walk
wall
wallflower
walnut
warble
Wardecker
warm way
warmth
warmth of broodnest
warping
wash
washboard dance
washer
washing
wasp
wasp behaviour
wassail
waste
wasteland
watch dog
watch glass
watch strap
water
water balsam
water butt
water carrier
water content
water course
water cress
water fountain
water gatherer
water glass
water melon
water method
water of crystallization
water proofing leather
water proofing shoes
water repellent
water requirements
water soluble
water sources
water table
water vapour

worms
wormwood
worn out
worried
Worth cage
wound dressing
woundwort
wreathe
wrinkle
wrist
writers

Xenophon
xeromorphic
xylene

yarrow
Year Book
yearly queen colour
yeast
yeast nutriments
yellow bee pirate
yellow jacket
yellow loosestrife
yellow scabious
yellow toadflax
yew
yield
yield of honey per acre
yield of wax
yolk
Yorkshire spacer
young adult
young queen
Yucatan
Yugoslavia
zest
zinc queen excluder
zip
zygote
zymase
zythus

www.ingramcontent.com/pod-product-compliance
Lightning Source LLC
Chambersburg PA
CBHW080750300426
44114CB00020B/2683
* 9 7 8 1 9 0 8 9 0 4 2 1 8 *